ORGANIC REACTION MECHANISMS · 1981

ORGANIC REACTION MECHANISMS · 1981

An annual survey covering the literature dated December 1980 through November 1981

Edited by

A. C. KNIPE and W. E. WATTS,
The New University of Ulster,
Northern Ireland

An Interscience® Publication

JOHN WILEY & SONS
Chichester · New York · Brisbane · Toronto · Singapore

British Library Cataloguing in Publication Data:

Library of Congress Catalog Card Number 66–23143

Organic reaction mechanisms.—1981
1. Chemistry, Organic—Periodicals
2. Chemical reactions—Periodicals
547.13′9 QD258

ISBN 0 471 10459 0

Phototypeset by Speedlith Photo Litho Ltd., Manchester.
Printed at the Pitman Press, Bath, Avon.

Contributors

D. J. COWLEY	School of Physical Sciences, New University of Ulster
M. R. CRAMPTON	Department of Chemistry, Durham University
I. R. DUNKIN	Department of Pure and Applied Chemistry, University of Strathclyde
G. W. J. FLEET	Dyson Perrins Laboratory, South Parks Road, Oxford University
I. GOSNEY	Department of Chemistry, University of Edinburgh
M. C. GROSSEL	Department of Chemistry, Bedford and Royal Holloway Colleges, Egham, Surrey
A. F. HEGARTY	Chemistry Department, University College, Dublin, Ireland
A. J. KIRBY	University Chemical Laboratory, Cambridge
R. B. MOODIE	Department of Chemistry, University of Exeter
C. J. MOODY	Department of Chemistry, Imperial College of Science and Technology
A. E. MURRAY	Department of Chemistry, University of Dundee
D. C. NONHEBEL	Department of Pure and Applied Chemistry, University of Strathclyde
M. I. PAGE	Department of Chemical Science, Huddersfield Polytechnic, Huddersfield
R. M. PATON	Department of Chemistry, University of Edinburgh
J. SHORTER	Department of Chemistry, University of Hull

Preface

The present volume, the seventeenth in the series, surveys research on organic reaction mechanisms described in the literature dated December 1980 to November 1981. In order to limit the size of the volume, we must necessarily exclude or restrict overlap with other publications which review related specialist areas (e.g. photochemical reactions, biosynthesis, electrochemistry, surface chemistry, organometallic chemistry, heterogeneous catalysis). In order to minimize duplication, while ensuring comprehensive coverage, the editors conduct a survey of all relevant literature and allocate references to appropriate chapters. While a particular reference may be allocated to more than one chapter, we do assume that readers will be aware of the alternative chapters to which a borderline topic of interest may have been preferentially assigned.

We regret that Brian Capon has found it necessary to discontinue as an author and we wish to acknowledge the major contribution that he made to both the establishment and continuation of *Organic Reaction Mechanisms.* Brian, in conjunction with his co-editors Charles Rees and John Perkins, conceived and launched the series in 1965; he served as an editor of the volumes dated 1965–1972 but continued thereafter to contribute the chapter entitled 'Reactions of Aldehydes and Ketones and their Derivatives'. We have no doubt that the international community of chemists who take an interest in the mechanisms of organic reactions will have benefitted considerably from the contribution that Brian Capon made to this series over a period of sixteen years. We hope to maintain the quality and viability of *Organic Reaction Mechanisms* and to thereby ensure that benefit will continue to accrue from the initiative of the founder editors.

Once again we wish to thank the publication and production staff of John Wiley and Sons and our team of contributors, which has been rejoined by Dr. M. Page. We are also indebted to Dr. N. Cully who compiled the subject index.

A.C.K.
W.E.W.

Contents

1. **Reactions of Aldehydes and Ketones and their Derivatives** by M. I. Page . . . 1
2. **Reactions of Acids and their Derivatives** by A. J. Kirby 23
3. **Radical Reactions** by D. J. Cowley and D. C. Nonhebel 77
4. **Oxidation and Reduction** by G. W. J. Fleet 183
5. **Carbenes and Nitrenes** by C. J. Moody 239
6. **Nucleophilic Aromatic Substitution** by M. R. Crampton 263
7. **Electrophilic Aromatic Substitution** by R. B. Moodie 283
8. **Carbocations** by M. C. Grossel 299
9. **Nucleophilic Aliphatic Substitution** by J. Shorter. 323
10. **Carbanions and Electrophilic Aliphatic Substitution** by I. Gosney . . 353
11. **Elimination Reactions** by A. F. Hegarty 385
12. **Addition Reactions: Polar Addition** by R. M. Paton 409
13. **Addition Reactions: Cycloaddition** by I. R. Dunkin 433
14. **Molecular Rearrangements** by A. W. Murray 459

Author Index, 1981 559

Subject Index, 1981 621

Organic Reaction Mechanisms 1981
Edited by A. C. Knipe and W. E. Watts

CHAPTER 1

Reactions of Aldehydes and Ketones and their Derivatives

M. I. PAGE

Department of Chemical Sciences, Huddersfield Polytechnic

Formation and Reactions of Acetals and Ketals . . . 1
Hydrolysis and Formation of Glycosides . . . 4
Non-enzymic Reactions . . . 4
Enzymic Reactions . . . 4
Reactions and Formation of Nitrogen Bases . . . 5
Schiff Bases and Related Species . . . 5
Oximes, Hydrazones, and Related Compounds . . . 8
Aldol and Related Reactions . . . 10
Other Addition Reactions . . . 12
Enolization and Related Reactions . . . 14
Hydrolysis of Enol Ethers and Related Compounds . . . 16
Other Reactions . . . 17
References . . . 17

Formation and Reactions of Acetals and Ketals

The hydrolysis of substituted benzaldehyde ethyl hemiacetals, generated from the corresponding ethyl salicyl acetal, shows catalysis by hydronium ion, water, and hydroxide ion. The ρ value for hydronium ion catalysis is -1.9, in contrast to the value of -3.25 for the hydrolysis of benzaldehyde diethyl acetal, and hemiacetal breakdown will only be rate-limiting for the latter reaction with substituents in the phenyl ring having σ values < -0.5. The ρ value for the pH-independent hydrolysis of the hemiacetals is $+0.35$ and the solvent isotope effect k_{H_2O}/k_{D_2O} is 2.0 which suggest that proton transfer is important and is consistent with acid-catalysed breakdown of the hemiacetal anion (**1**).[1]

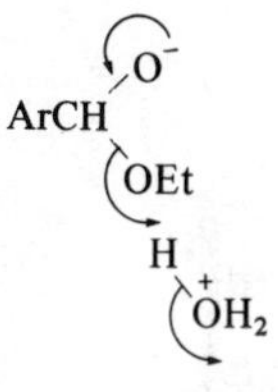

$$\mathrm{ArCH(O^-)(OEt)\cdots H\cdots \overset{+}{O}H_2}$$

(1)

The intramolecular addition of an alcohol to an aldehyde (**2**) to give a cyclic hemiacetal (**3**) is displaced towards cyclization in basic solution because of hemiacetal ionization. General base-catalysed cyclization occurs with a Brønsted β

value of 0.62. General acid catalysis is thought to occur by dual mechanisms, *viz.* general acid catalysis at carbonyl oxygen and specific-acid–general-base catalysis, which is used to account for the observation of different Brønsted α values for different classes of acids.[2]

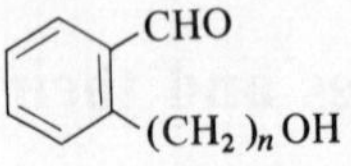

CHO (CH$_2$)$_n$OH

(2)

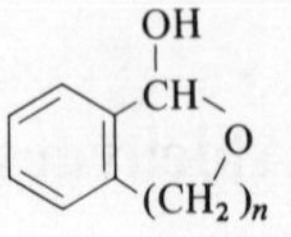

OH CH O (CH$_2$)$_n$

(3)

Glycosides with an equatorial leaving group normally hydrolyse faster than those with an axial group. In contrast, the axial anomer of the conformationally rigid acetal (**4**) hydrolyses 1.5 times faster than the equatorial anomer. Despite the small rate difference, this has been used to support different transition states for the two reactions.[3]

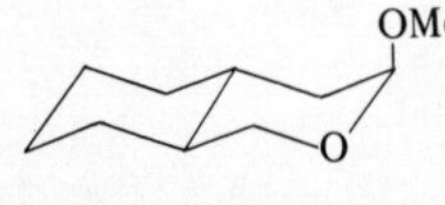

OMe O

(4)

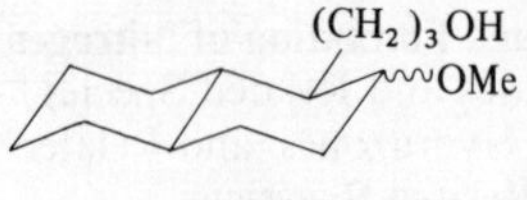

(CH$_2$)$_3$OH OMe

(5)

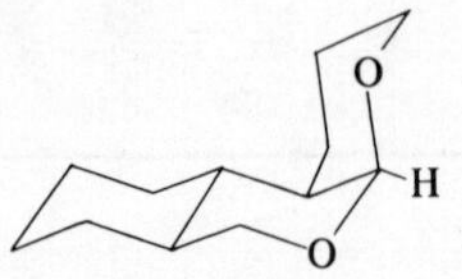

O H O

(6)

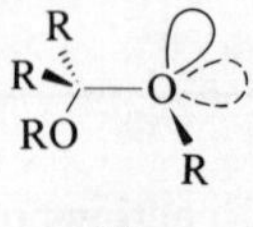

R R RO O R

(7)

The bicyclic hydroxypropyl methoxytetrahydropyran (**5**) gives only the *cis*-tricyclic acetal (**6**) under acid conditions. This is attributed to the development by the intermediate oxenium ion of a lone pair which becomes antiperiplanar to the newly formed C—O bond (**7**).[4] Acid-catalysed ring-opening of the tricyclic spiroketal (**8**) gives specifically the equatorial bicyclic ether-aldehyde (**9**) by stereoelectronically controlled hydride transfer to the intermediate oxenium ion (**10**) because the ether oxygen has a lone pair antiperiplanar to the newly formed C—H bond.[5]

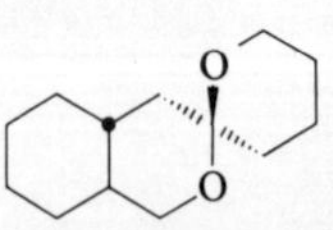

O O

(8)

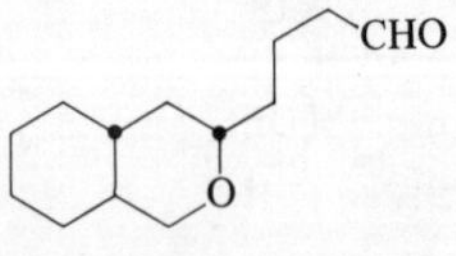

CHO O

(9)

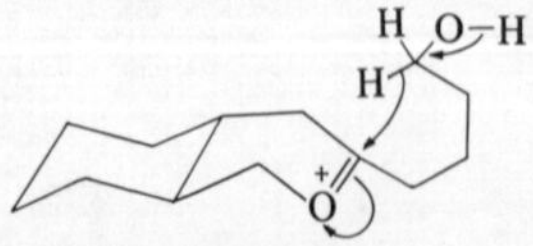

H O–H H O

(10)

Cyclization of the intermediate oxo-carbocation (**11**) in the formation of dioxolanes is a 5-*endo-trig* cyclization which is unfavourable by Baldwin's rules for ring-closure. It has therefore been suggested, with little supporting evidence, that ring-closure occurs *via* the protonated hemiacetal (**12**).[6] The spiro[5.5]undecane ketals exist in conformationally rigid systems (**13**) which minimize steric effects and maximize anomeric effects.[7]

Although the environment of a given chiral molecule (R) in the liquid phase will be diastereomeric with, and thus different from, that in the liquid of the racemic mixture (R, S) there is little evidence to suggest this corresponds to a measurable free-energy difference. For example, the conversions of (S)- and (R, S)-propane-1,2-diols into dioxolanes using the diol as solvent proceed at identical rates.[8]

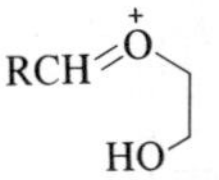

(11) 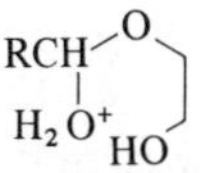 (12) 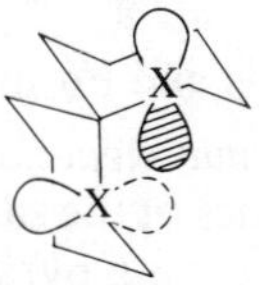 (13)

It has long been known that the rate or equilibrium differences among the members of a reaction series, which involves an interconversion of a trigonal carbon and a more crowded tetrahedral carbon, may be correlated with the relative size of the subsituent groups attached to the reaction centre. A calorimetric determination of the enthalpies of hydrolysis of a series of alkyl-substituted dimethyl acetals shows that the enthalpy of the acetal hydrolysis is remarkably insensitive to changes in the substituent alkyl group (approximate slope of acetal ΔH_r *vs.* E_s plot = 0.4), suggesting that polar interactions are negligible and that the larger effects seen with the ketals represent true steric effects. As the free-energy changes are well correlated with the E_s scale, entropy changes associated with solvation and internal contributions, particularly bond-angle bending, must make important contributions to steric effects.[9]

Neither the bicyclic ketal (**14**) nor its conjugate acid, nor the corresponding fully formed carbocation are severely strained but the geometry of (**14**) would be such as to inhibit resonance stabilization of the carbocation during its formation. Despite elegant rationalizations of why (**14**) should exhibit concerted proton transfer and C—O cleavage, it actually shows no buffer catalysis and a solvent hydrogen isotope effect (k_H/k_D) of 0.43. A conventional *A*1 mechanism is proposed.[10]

The rates of hydrolysis of 2-alkoxy-1,3-dioxolanes increase with increasing stability of the 1,3-dioxolenium cation (**15**) which is compatible with, but not proof

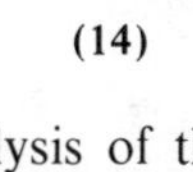

(14) 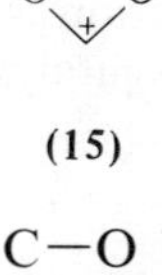 (15)

of, rate-limiting heterolysis of the exocyclic C—O bond.[11] The kinetics of the formation and hydrolysis of substituted dioxolanes have been reported.[12] 1,3-Dioxolanes give 1,3-dioxolan-2-ylium salts by hydride transfer on treatment with iodine monochloride.[13] The bromination of cyclic acetals is slowest for the six-membered ring, shows negative entropies of activation, and is thought to proceed by carbocation formation.[14] Halo-substituted boranes cause ring-opening of cyclic acetals by rate-limiting formation of an oxo-carbenium ion which can subsequently be reduced to a hydroxy-ether.[15] The rate-limiting step in the reaction of triethoxymethane, aniline, and dimedone is solvent-dependent.[16] The solvolysis of the thia analogue of methyl chloroformate, MeS_2CCl, proceeds by an S_N1 mechanism in methanol and at a rate faster than that of the oxygen derivative.[17] Sulphuryl chloride reacts with 1,3-oxathiolans and 1,3-dithiolans to give unstable *trans*-2,3-dichloro-1,4-oxathians and *trans*-2,3-dichloro-1,4-dithians, respectively.[18] Acetal formation with carboxamide dialkyl sulphate adducts has been described.[19]

Hydrolysis and Formation of Glycosides

Non-enzymic Reactions

The kinetics of the tautomerization of α-D- and β-D-galactopyranose in water show that the starting pyranose is simultaneously converted into the anomeric pyranose and into furanoses during the course of mutarotation. Furanose formation is *ca.* 10 times faster than pyranose formation, and rates of ring-opening of the furanoses are 10–35 times faster than those of the pyranoses.[20] Contrary to an earlier report,[21] the alkaline hydrolysis of phenyl β-D-glucopyranoside has been shown to proceed by the commonly accepted $S_{Ni}CB$ pathway.[22] The rate of mutarotation of *N*-acetyl-D-neuraminic acid (**16**) has a minimum at pD 5.4 whereas at pD 1.3 and 11.7 it is too fast to be measured by ^{1}H-NMR spectroscopy. With no suitable model studies, the rationalization of these observations is premature.[23]

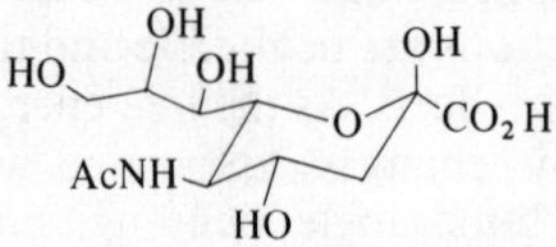

OH HO OH OH O CO_2H AcNH HO

(16)

In agreement with previous reports, removal of the 2′- or 3′-hydroxyl group enhances the rate of the acid-catalysed hydrolysis of adenosine nucleosides. Other substituent effects in the glycon may be interpreted by the *A*1 mechanism.[24] Anomeric hydroxy groups are protonated faster (< ten-fold) than non-anomeric hydroxy groups of monosaccharides in DMSO. Anomeric hydroxy groups of ketopyranoses and furanoses are protonated slightly faster than those of aldopyranoses, roughly in parallel with the rates of the acid-catalysed hydrolysis of the corresponding glycosides.[25] Glycosyl glycosides may be synthesized with α-stereoselectivity using trifluoromethanesulphonic anhydride as a catalyst.[26] 4-*O*-substituted uronic acids, treated with sodium methylsulphinylmethanide, undergo β-elimination in the ionization step.[27] The Hilbert–Johnson silyl nucleoside synthesis with Friedel–Crafts catalysts occurs by reaction of the sugar cation with the silylated base.[28]

Enzymic Reactions

Oxygen-18 leaving-group kinetic isotope effects are observed for the β-galactosidase-catalysed hydrolysis of 4-nitrophenyl β-D-galactoside but not for the 2,4-dinitrophenyl derivative. The latter is thought to hydrolyse by a S_N1 mechanism to give a galactosyl cation which partitions between a nucleophilic group on the enzyme and water. Substrates with poorer leaving groups hydrolyse by a S_N2

mechanism to give a covalent galactosyl enzyme.[29] The hydrolysis of aryl β-D-glucosidase from *Stachybotrys atra* proceeds through an intermediate glycosyl-enzyme which may be trapped with alcohols.[30] The kinetic α-deuterium isotope effect for the purine nucleoside phosphorylase-catalysed phosphorolysis of inosine is pH-dependent. This is interpreted in terms of a change in rate-limiting step, from one involving C—N cleavage at high pH to give an intermediate oxo-carbenium ion (**17**) to an earlier step near neutrality.[31]

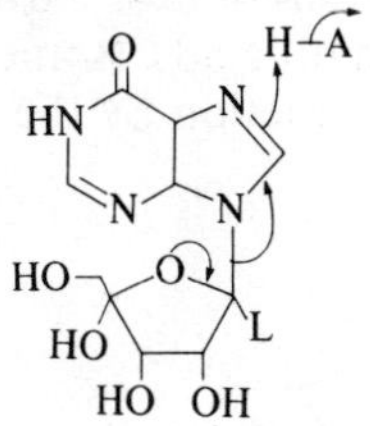

(17)

The initial rates of sucrose hydrolysis catalysed by invertase pass through a maximum and then decrease, through two plateaus, with increasing sucrose concentration. This is claimed to be due to increased intramolecular hydrogen bonding with increasing concentration.[32]

Reactions and Formation of Nitrogen Bases

Schiff Bases and Related Species

The hydrolysis of the oxazolidine (**18**) occurs in two separate reaction stages, *viz.* reversible ring-opening to the cationic Schiff base (**19**) followed by a slow formation of the hydrolysis products. The pK_a of the protonated oxazolidine is 6.2 and the ratio of this species to the cationic Schiff base is 1:19 and independent of acidity. Below pH 5, ring-opening is apparently pH-independent but represents acid-catalysed ring-opening of the neutral oxazolidine. Similarly, in this pH region, apparent general base catalysis in fact represents general acid-catalysed ring-opening of neutral (**18**) with a Brønsted α value of 0.70. Ring-opening also occurs by an uncatalysed mechanism above pH 5. The formation of hydrolysis products involves rate-limiting addition of water or hydroxide ion to the cationic Schiff base (**19**). This intermolecular reaction is much slower than the supposedly unfavourable intramolecular 5-*endo-trig* ring-closure.[33]

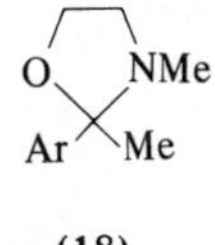

(18)

Ar = *p*-tolyl

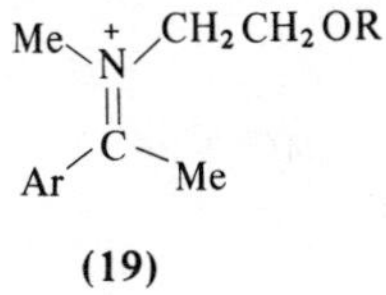

(19)

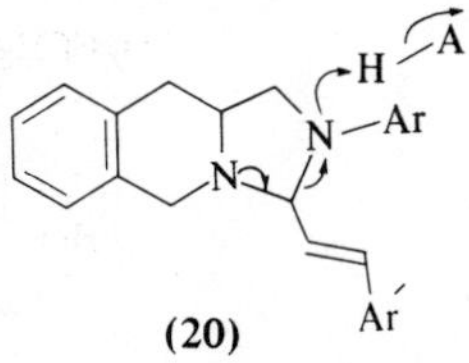

(20)

The ring-opening of imidazolidines to give a Schiff base is catalysed by general acids and shows a Brønsted α value of 0.7, indicative of a concerted reaction (**20**). The latter is attributed to carbocation stabilization and leaving-group effects rather than to the basicity of the ring nitrogens.[34] Mercury(II) chloride enhances the rate of hydrolysis of mono-protonated 9-(1-ethoxyethyl)purine and adenine but has no effect on the di-protonated species.[35] The enhanced rate of mono-protonated diamines reacting with ketones to give imines is attributed to intramolecular general acid-catalysed dehydration of the intermediate carbinolamine (**21**). Methylation of the carbon atoms between the amino groups generally decreases the reactivity which is rationalized by steric effects in the cyclic transition state.[36] Malondialdehyde, a biologically important breakdown product, reacts with the amino group of amino-acids to give enaminals.[37] Fluoren-9-one-1-carboxylic acid effects the transamination of primary amines and amino-acids. Isomerization of the intermediate imine may be facilitated by proton abstraction by the neighbouring carboxyl group.[38]

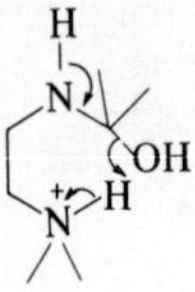

(**21**)

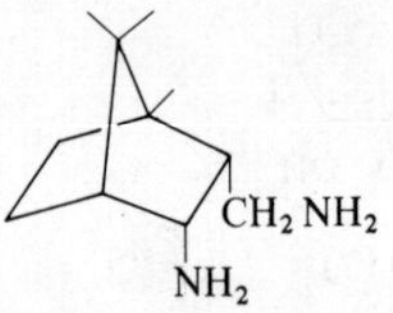

(**22**)

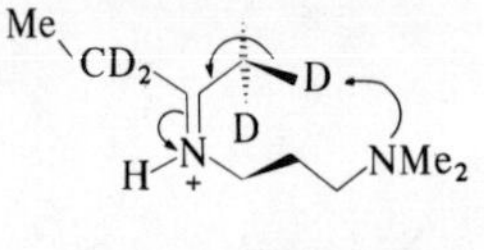

(**23**)

The chiral diamine (**22**) stereoselectively dedeuterates labelled acyclic ketones by forming an intermediate Schiff base in which the pro-*S* deuterons are removed preferentially by the intramolecular amino group acting as a general base catalyst. The transition state (**23**) is of lower energy than that involving removal of the pro-*R* deuterons because of less unfavourable steric effects.[39]

The kinetics of the catalysis of elimination of electronegative substituents from the β-position of α-amino-acids by pyridoxal have been investigated by NMR spectroscopy and the reaction is thought to occur through the di-, mono-, and non-protonated forms of the intermediate Schiff base. In the presence of metal ions, rate enhancements of up to 10-fold are observed.[40] The pyridoxal-catalysed β-elimination (*a*) and dealdolation (*b*) of β-hydroxyglutamic acid occur simultaneously, and in the pD range near the pK_a of the δ-carboxyl group decarboxylation follows elimination (**24**). Protonation of the azomethine and pyridine nitrogens retards expulsion of the hydroxide group but favours dealdolation.[41]

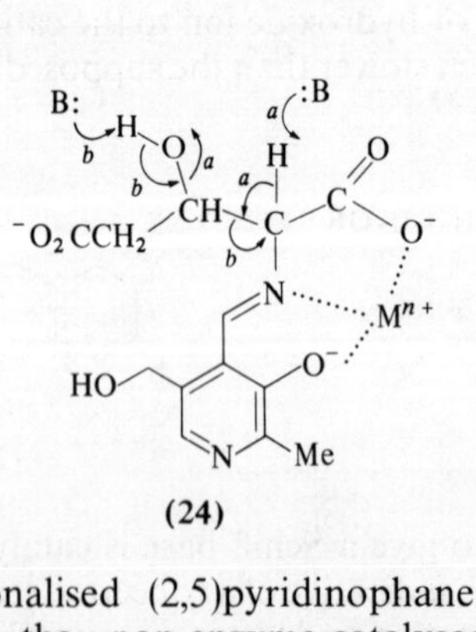

(**24**)

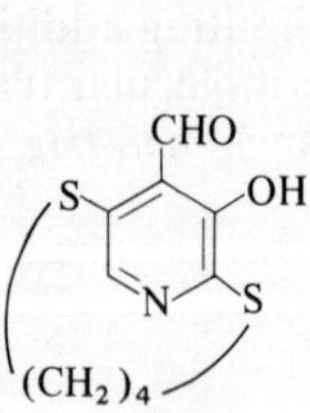

(**25**)

The functionalised (2,5)pyridinophane (**25**) is slightly more efficient than pyridoxal in the non-enzyme-catalysed racemization of monosodium L-glutamate.[42] The rate-limiting step for the liver alcohol dehydrogenase-catalysed reduction of aliphatic aldehydes is the dissociation of the enzyme–NAD^+ complex which in turn is controlled by a conformational change.[43] The Schiff-base mechanism enzyme, 2-keto-4-hydroxyglutarate aldolase, appears to have a non-specific anion binding site in addition to the anionic substrate analogue inhibition site.[44] The imidoyl azide structure (**26**), originally proposed, has in fact the enamine structure (**27**) which cyclizes to tetrazoles with a bell-shaped pH-rate profile. At low pH, *C*-protonation of the enimine is rate-limiting and is also general acid-catalysed with a Brønsted α value of 0.57. Increase in pH, or high buffer concentration, changes the rate-limiting step to isomerization of the protonated imine. The azide group acts as an efficient intramolecular trap for the imine in which the nitrogen lone pair is *cis* to the azide, to give the cyclic tetrazole.[45]

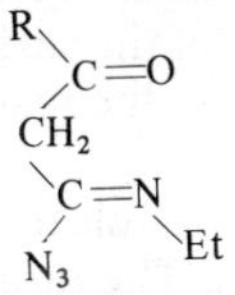

R–C(=O)–CH_2–C(=N–Et)–N_3

(26)

R–C(=O)–CH=C(–NHEt)–N_3

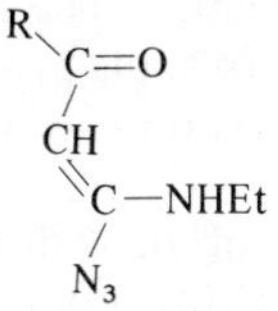

(27)

A positive ρ^+ value for the reaction of aniline with substituted aromatic aldehydes in acetonitrile has been used to suggest rate-limiting attack on the carbonyl group, although pyridine-4-aldehydes show deviations attributable to the influence of carbinolamine dehydration.[46] Acetic acid-catalysed Schiff-base formation from aromatic aldehydes and 2-amino-1,3,4-thiadiazoles involves rate-limiting dehydration of the carbinolamine and becomes independent of acid concentration at high concentrations of acetic acid.[47] The condensation of salicylaldehyde and primary diamines occurs within the coordination sphere of copper(II) and nickel(II).[48]

Schiff-base condensation reactions between 2,6-diacetylpyridine and 3,6-dioxa octane-1,8-diamine in the presence of Ba(II) and other metal ions can give complexes of the open-chain derivative formed from one molecule of dicarbonyl and two molecules of diamine.[49] These are the probable intermediates in the formation of the 30-membered [2 + 2] macrocycles.[50] The condensation of 3-methyl-1,2-diaminobenzene with aromatic aldehydes gives methylbenzimidazoles.[51] The reaction of Schiff bases with sodium then ethyl chloroformate gives 1,3,4,5-tetra arylimidazolidin-2-ones presumably *via* the dimeric dianion.[52] The stereoselectivity of the reaction of arylmethanephosphonate carbanions with Schiff bases varies with the cation and anion used to generate the carbanion.[53] A novel metal carbonyl-catalysed reaction of isocyanates with aldehydes to give imines is thought to occur by insertion of the aldehyde into a preformed isocyanate complex.[54] The rate of hydrolysis of substituted 4-(*N*,*N*-dimethylamino)benzylideneanilines (**28**) is acid-dependent below pH 3.5 but pH-independent from pH 3.5 to 6.5. The Hammett ρ value is positive under both conditions.[55]

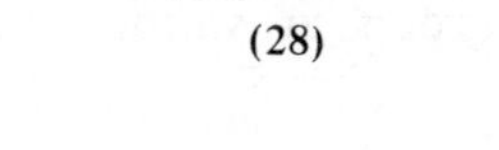

Me_2N–C_6H_4–CH=NAr

(28)

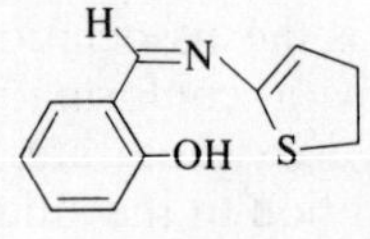

(29)

The hydrolysis of the imine (**29**) has been suggested to occur *via* intramolecular general base catalysis by the phenoxide ion but, because of kinetic ambiguity and lack of data for unsuitable model compounds, this requires more information. Copper(II) retards the reaction by complexation with the imine.[56] The rate of the acid-catalysed (but not the pH-independent) hydrolysis of imines formed from substituted 2-aminothiazoles and 2-hydroxybenzaldehydes is dependent upon the nature of the substituent.[57] The base-catalysed formation of Mannich bases from 3,5-dimethyl-4-nitroisoxazole proceeds by S_N2 attack of the carbanion of isoxazole on *N*-(hydroxymethyl)amine, whereas in acidic solution the reaction occurs *via* the imine cation.[58]

The methanolysis of benzylideneaniline derivatives of the type (**30**) causes cleavage of the C—MMe_3 bond, where M = Si or Sn. The rate of the uncatalysed cleavage is increased by electron-releasing substituents in both aromatic residues. A cyclic mechanism (**31**) is proposed. The rate of the base-catalysed cleavage is increased by electron-withdrawing groups in the aniline residue. Proton transfer to carbon is thought to lag behind C—$SnMe_3$ bond-breaking, but for C—$SiMe_3$ cleavage, proton transfer occurs to nitrogen.[59] Similar proposals have been made for the reactions of 1-methyl-2-trimethylsilylbenzimidazole.[60] Substituent, salt, and solvent effects on the stereochemistry of the addition of allylic organometallics to Schiff bases have been reported.[61]

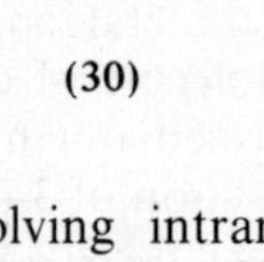

(30)

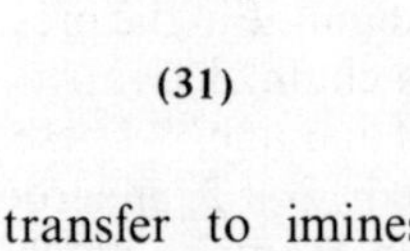

(31)

A reaction involving intramolecular hydride transfer to imines has been described.[62] The pH-rate profile for the hydrolysis of hexamethylenetetramine to formaldehyde is sigmoidal, corresponding to water- and acid-catalysed reactions of the conjugate acid.[63] Enamines with α- or β-hetero-atoms (with respect to the amine) have reduced conjugation between the nitrogen lone pair and the double bond. This is attributed to the molecules examined adopting conformations to minimize lone-pair–lone-pair repulsions.[64]

Oximes, Hydrazones, and Related Compounds

Variation in the secondary kinetic deuterium isotope effects in the addition of hydroxylamine to substituted cyclohexanones has been attributed to ring deformation and changes in the angle of approach of the nucleophile.[65] Large negative values for the entropies of activation for the formation of oximes from adamantyl ketones have been used to suggest rate-limiting dehydration.[66] The reactivity of the carbonyl group at various positions in steroids towards hydroxylamine has been examined.[67] The rate of formation of cyclohexanone oxime and its hydrolysis between pH 1 and 7 has been further studied.[68]

The cyclization of thiocyanatopropenal with hydroxylamine has been reported.[69] The effect of substituents on the barrier to isomerization of oximes,[70] and the pH dependence and substituent effects upon the rates of reactions of oximes have been reported.[71] Alkoxide substitution of (Z)-hydroximoyl chlorides (**32**) proceeds with >95% retention of configuration, and at half the rate of reaction of the corresponding bromide; the Hammett ρ value is 1.9. These observations are consistent with rate-limiting formation of a tetrahedral intermediate. It is suggested that stereoelectronically controlled loss of chloride ion from the tetrahedral intermediate (**33**) is faster than either loss of methoxide ion to regenerate starting material or stereomutation.[72]

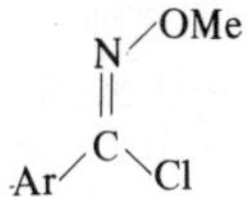

(**32**) 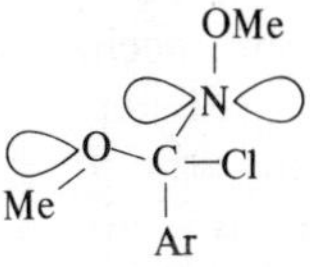 (**33**)

Benzil mono-oximes react with *N*-chlorosuccinimide, dimethyl sulphide, and triethylamine to afford a methylthiomethyl nitrone.[73] The conversion of benzaldoximes into benzaldehydes with manganese(III) acetate shows a Hammett ρ value of -0.57 and is thought to proceed by an electron-transfer mechanism to give an iminoxy radical.[74] The heterolytic cleavage of the N—O bond of 6-methyl-hept-5-en-2-one oxime produces Δ^1-pyrrolines by capture of the intermediate nitrilium ion by the alkene.[75] At low pHs where the formation of hydrazones usually occurs with rate-limiting formation of an intermediate carbinolamine and exhibits general acid catalysis, the reaction of phenylhydrazine and 2-hydroxybenzaldehyde shows no buffer catalysis. The rate of the pH-independent formation of 2-hydroxybenzaldehyde phenylhydrazone is 200 times faster than that of the 4-hydroxy derivative which is attributed to intramolecular hydrogen bonding (**34**). This also explains the absence of buffer catalysis.[76] It is interesting to note that a 2-hydroxy substituent in acetophenone decreases the rate of carbinolamine formation six-fold.

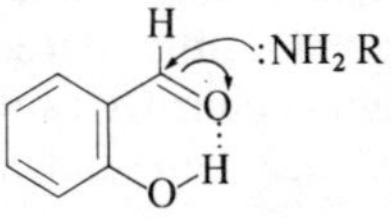

(**34**) 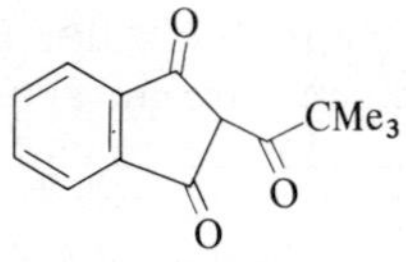 (**35**)

The mechanism of indole formation from 2-methoxyphenylhydrazones has been reviewed.[77] Thallium(III) acetate acts as an electrophilic catalyst by reversible coordination to the amino NH of substituted phenylhydrazones, facilitating nucleophilic attack at the methine carbon to give substituted hydrazines as products. By contrast, mercury(II) irreversibly binds to the imino nitrogen giving osazones as products.[78] Regioselectivity in ring-closure reactions of hydrazones with thionyl chloride may be rationalized by the greater acidity of methylene compared with methyl hydrogens.[79] 2-Pivaloylindane-1,3-dione (**35**) reacts with thiosemicarbazide in aqueous base, through two successive retro-Claisen reactions to give 2-acetylbenzoic acid which reacts with semicarbazide to give 1-hydroxy-4-methylphthalazine.[80]

Aldol and Related Reactions

The stereoselectivity of the aldol condensation has been reviewed.[81] In slightly basic media, ionization of the *gem*-diol group of β-bromopyruvate initiates bromide elimination to give β-hydroxypyruvate, which ionizes at pH >13 to give the carbanion enolate and undergoes an aldol condensation.[82] The Knorr–Paal cyclo-condensation shows a bell-shaped pH-rate profile, a large negative entropy of activation, a Brønsted α value of 0.25 for the substituted benzoic acid-catalysed condensation, and a Hammett ρ^+ value of -0.88, all of which are compatible with an addition–elimination mechanism.[83]

The sense and degree of stereoselectivity in the deprotonation of ketones can be critical in determining the overall stereoselectivity obtained in electrophilic substitution reactions employing enolate-like intermediates. For example, it usually assumed that the lithio intermediates, of stereochemistry (**36**) and (**37**), formed in deprotonations of active methylene-containing compounds are the kinetically formed products, and their formation can thus be rationalized from transition-state structures. However, some of these reactions may be reversible and hence thermodynamic factors need to be considered. Propionaldehyde dimethyl-hydrazone–lithium reagents, formed by deprotonation with lithium diisopropyl-amide, do not equilibrate under the reaction conditions and the ratio of the stereoisomers formed is far removed from equilibrium. In this case, distinct transition states for deprotonation must exist.[84]

H, R ; X⁻Li⁺, Y	R, H ; X⁻Li⁺, Y
(36)	**(37)**

Addition of an enolate to a chiral aldehyde can give two *erythro*-aldols resulting from attack at either of the diastereotopic faces of the aldehyde. A similar situation exists if the enolate is chiral. Several reagents are now available which give good diastereoselection in the aldol condensation. Double stereo-differentiation can occur if both reagents are chiral or if the solvent is chiral. Cram's rule predicting the selectivity in the addition to chiral aldehydes may be enhanced by double stereo-differentiation.[85]

Under kinetic control Z-enolates generally favour *erythro*-adducts and E-enolates the *threo*-isomers in aldol condensations. However, titanium enolates derived from ketones react with aldehydes to give preferentially the *erythro*-adduct which is attributed either to an acyclic transition state or a cyclic one in which chair-to-boat interconversion can occur.[86] Lithium enolates of hindred aryl esters condense with aldehydes to give predominantly *threo*-aldols.[87] Lithium enolates react with allylic acetates in a palladium-catalysed alkylation with retention of configuration.[88] Reactions of the chiral dianion (**38**) with aldehyde gives predominantly *erythro*-aldols. Variation of the *erythro*/*threo* ratios can be rationalized in terms of cyclic *vs.*

acyclic transition states. *Erythro*-selectivity may be rationalized by the minimization of steric interactions and maximal separation of negative charge provided by an acyclic *anti* transition state.[89]

(38) 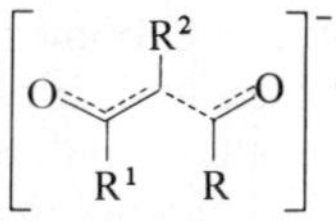 **(39)**

Boron enolates derived from chiral ethyl ketones show high stereoselectivity in the aldol reaction.[90] Boron enolates are the suggested intermediates in the aldol condensation of ketones in the presence of dialkylboryl triflates and tertiary amines. Variation of the boron ligands and solvent allow total stereo-control of the condensation.[91] High levels of aldol stereo-regulation may be obtained using zirconium enolates.[92] Chiral boron enolates are also effective in stereo-regulated aldol condensations,[93] and minimal non-bonded interactions in the pericyclic transition state for the aldol condensation with boron enolates has been proposed to account for chirality transfer.[94] Acyclic stereo-selection is also obtained from enolates with tris(diethylamino)sulphonium cation as the negligibly interacting counter-ion but, interestingly, is independent of enolate geometry. Rather than the normal pericyclic process *via* the metal-linked six-membered cyclic transition state, an extended geometry (**39**) is proposed to minimize electrostatic repulsion.[95]

Triphenyltin enolates undergo a rapid aldol condensation with aldehydes, without the need for the presence of Lewis acids, to give predominantly the *erythro*-product regardless of the geometry of the starting enolates. This cannot be explained by the conventional cyclic transition state but may be rationalized if the mechanism is step-wise.[96] Other stereospecific reactions of enolates have been reported.[97]

Despite the small differences in activation energies which may account for the relative proportion of isomeric alkenes formed in the Wittig reaction, changes in the steric course of the reaction with solvent have been used to support the equilibrium formation of betaine and oxaphosphetan intermediates.[98] The *cis*/*trans* ratio of the alkenes formed in the Wittig reaction of ylides from benzyltriaylphosphonium salts increases as the steric crowding at phosphorus increases.[99] An efficient silicon-mediated alkene synthesis from ketones produces almost exclusively (z)-alkenyl derivatives, which is attributed to the diastereoselective addition of the α-silyl carbanion to the carbonyl compound.[100] Reaction of α-silylated ester magnesium enolates with aldehydes gives only one diastereoisomer (of two possible β-hydroxysilanes) which yields E-unsaturated esters on acid work-up.[101]

The reaction of 3-chloroallyltrimethylsilane with aldehydes in the presence of Lewis acids gives methyl ethers of homoallyl alcohols.[102] Double stereo-differentiation using a chiral α-[(trimethylsilyl)oxy]ketones in an aldol reaction with chiral racemic aldehydes gives high diastereoface selectivity.[103] Silyl enol ethers, in the presence of $TiCl_4$, alkylate *exo*-2-norbornyl halides with retention of configuration.[104] The selectivity of the alkylation of cyclopropyl alkyl ketones depends on substituents attached to the carbonyl carbon and on the ring.[105] Solid hexamethylphosphotriamide can be used to control *C*- or *O*-alkylation with different metal ion enolates, presumably by interacting with the cations.[106] The influence of substituents, solvents, temperature, and concentration on the ratio of *O*- to *C*-methylation of enolate anions has been examined.[107]

Metallo enamines, as enolate equivalents, have introduced the possibility of a chiral environment and hence the control of the direction of electrophilic attack to give a novel asymmetric C—C bond-forming reaction. Thus, chiral imines may be prepared from cyclic ketones and chiral amines; formation of the lithioenamine and then alkylation gives, after hydrolysis, 2-alkylcycloalkanones in 87–100% enantiomeric purity.[108] Thioenolates generated from thioamides have the z-configuration and selectively give the *erythro*-isomer in aldol condensations.[109] The two-phase kinetics observed for the retro-aldol cleavage of pent-3-en-2-one to acetone and acetaldehyde are attributed to initial reversible hydration followed by C—C bond cleavage.[110] The formation of xanthyrones and glaucyrones from a "melt reaction" between pyrones, methyl methoxymethyleneacetoacetate, and dry sodium methoxide proceeds by Michael addition of the pyrone anion to the α,β-unsaturated ketone. The proportion of products results from competitive protonation and the leaving-group ability of the carbanion.[111] The product of the Perkin reaction of benzaldehyde and acetic anhydride in the presence of potassium acetate is, before hydrolysis, potassium cinnamate and K^+ (OAc, AcOH)$^-$ and not the commonly accepted mixed anhydride.[112]

Other Addition Reactions

In contrast to the cyclic transition-state structure proposed for the hydration of 1,3-dichloroacetone in water–dioxane, an acyclic structure is suggested for the reaction in micelles of bis(2-ethylhexyl)sodium sulphosuccinate (AOT) with the number of associated water molecules depending on the water–surfactant ratio. From labelling experiments, the linear dependence of the rate constant upon atom fraction of deuterium indicates that only one proton is moving in the transition state.[113] Solvent effects on the equilibrium constants for hydration of pyridine-4-carboxaldehyde have been reported.[114]

Theoretical models suggest that addition of a water molecule to formaldehyde in the gas phase gives a zwitterion of no finite life-time and the addition is therefore "enforced" to be concerted. Conversely, initial protonation of formaldehyde or addition of hydroxide ion are favourable processes.[115] The rates of the base-catalysed dehydration of oxalacetate hydrate are faster than those for enolization of oxalacetate; however, the converse holds when the reactions are catalysed by tertiary amines at high buffer concentration, but not at low concentrations. There is thus a rate cross-over with increasing concentration of tertiary amine which had led earlier workers to suggest the formation of a carbinolamine intermediate.[116] However, tertiary amines still show enhanced activity compared with other bases; the reason is not clear but electrostatic factors may be important.[117] The rates of addition of sulphite ion to aromatic aldehydes have been measured.[118] Analysis of structure–reactivity relationships for benzaldehyde cyanohydrin formation by the Taft and Yukawa–Tsuno dual-parameter equations indicate a product-like transition state.[119] Asymmetric addition of hydrogen cyanide to benzaldehyde catalysed by the cyclic dipeptide *cyclo*(L-phenylalanyl-L-histidine) gives an enantiomeric excess of 90% but the optical yield decreases with increasing reaction time.[120]

Ketols (**40**), in equilibrium with their hemiketals, undergo isomerization by base-catalysed intramolecular hydride transfer from the alcohol function to the carbonyl group.[121] The intramolecular Cannizzaro reaction of phthalaldehyde to 2-(hydroxymethyl)benzoate ion is a slow process because it proceeds *via* the cyclic hydrate dianion (**41**) which must undergo unfavourable ring-opening to (**42**) before intramolecular hydride transfer can occur.[122]

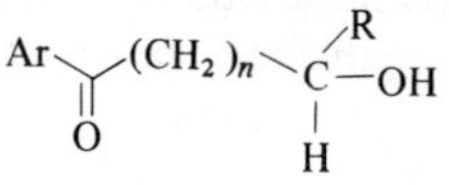

Ar–C(=O)–$(CH_2)_n$–C(R)(H)–OH

(40)

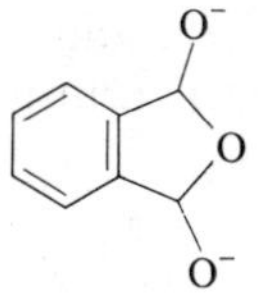

(41)

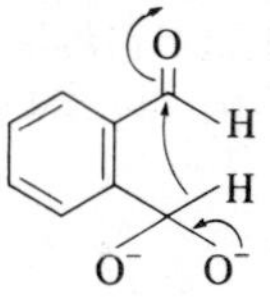

(42)

The rate of addition of anionic nucleophiles to carbonyl compounds may be affected by metal-ion complexation with either the nucleophile or the carbonyl oxygen. In the reaction of benzaldehyde with acetonitrile anion in THF, lithium ion facilitates the reaction by coordination to the carbonyl oxygen but potassium ion retards the rate by coordinating to the anion. Carbonyl complexation is favoured by electron-releasing substituents in the aldehyde but electron-withdrawing substituents encourage coordination to the nucleophile.[123] The unusual reactivity of constrained cage ketones has been reviewed.[124] It is generally agreed that nucleophilic addition to cyclohexanone, proceeding through an early reactant-like transition state, is directed into the equatorial position by steric hindrance but the reasons for the preference of some nucleophiles for a more hindered axial approach are uncertain. It has been suggested that transition-state stabilization, resulting from electron donation of the σ_{CH} bonds of cyclohexanone into a low-lying vacant σ^* orbital, associated with the σ-bond being formed in the reaction, favours the axial approach.[125] According to SCF perturbation theory, the directional nature of nucleophilic attack on carbonyl groups is dependent upon closed-shell repulsion, electrostatic interaction, and charge transfer. The frontier-orbital description of the charge-transfer term is therefore inadequate.[126] Attack of a reagent at an unsaturated site occurs such as to minimize anti-bonding secondary orbital interactions between the critical frontier molecular orbital of the reagent and those of the vicinal bonds.[127]

The Hammett ρ value for the reaction of methylmagnesium chloride with 1-aryl-2-phenylpropanones is 0.53.[128] Evidence supporting the single-electron-transfer mechanism in the reaction of Grignard reagents with ketones comes from the observation of an EPR spectrum during the reduction of aromatic ketones by β-hydrogen transfer from the alkyl group of the Grignard reagent.[129] The stereochemistry of the addition of phenyl- and methyl-magnesium bromide to chiral carbonyl compounds has been studied as a function of solvent polarity. In some, but not all, cases there is a correlation between stereoselectivity and the E_T solvent parameter.[130] The invariability of asymmetric induction as a function of substituent in the addition of CH_3MgBr to 1-aryl-2-phenylpropanones and the Hammett ρ value of 0.24 for the ratio of the rate constants have been used to substantiate a four-centre pericyclic concerted mechanism.[131] The formation of diols from reactions of benzylic Grignard reagents with aldehydes proceeds by nucleophilic addition of the aromatic *ortho*-carbon to the aldehydes.[132]

The reaction of *tert*-butyl isocyanide and trifluoroacetic acid with aldehydes and ketones to give an α-(acyloxy)carboxamide (Passerini reaction) is catalysed by pyridine acting as a general acid catalyst.[133] The reaction of benzaldehyde with Cl_3C^- in the presence of benzylquininium chloride as a phase-transfer catalyst gives 5.7% enantiomeric excess (R)-PhCH(OH)CCl_3.[134]

Volumes of activation have been used to suggest a concerted mechanism in the ene reaction of hex-1-ene with dimethyl mesoxalate.[135] A primary deuterium kinetic isotope effect ($k_H/k_D = 2.16$) for the addition of dimethyl mesoxalate to alkenes has been used to suggest a non-linear transition state for hydrogen transfer (**43**).[136]

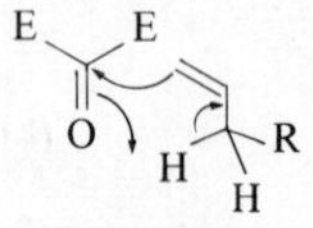

E = CO_2Me

(43)

Enolization and Related Reactions

Vinyl alcohol, the simplest enol, has been generated from several precursors and kept in solution below *ca.* −10° for several hours. Vinyl alcohol is converted into acetaldehyde *ca.* 100 times faster than is ethyl vinyl ether which, again, contradicts earlier assumptions.[137] The rates of iodination of β-piperidinopropiophenone and its *N*-methyl derivative are similar and up to 4000 times larger than those for suitable model compounds. This is attributed to intramolecular electrostatic stabilization of the developing negative charge on the carbonyl oxygen in the transition state (**44**).[138]

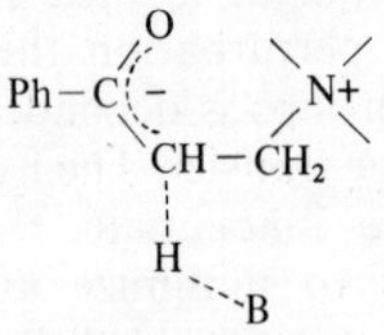

(44)

An unnecessarily complicated argument has been used to show that the pH-independent enolization of acetone represents water catalysis rather than the kinetically equivalent hydroxide-ion attack on the conjugate acid.[139] The equilibrium constants for tautomerism may easily be determined, because at low halogen concentration the rate of the acid-catalysed halogenation is first order in halogen and the rate-limiting addition of halogen to the enol is diffusion-controlled. Cyclohexanone contains 20 times more enol than does cyclopentanone. Contrary to previous suggestions, it has been shown that ketonization rate constants do not equal those for methyl enol ether hydrolysis and their ratio varies between 15 and 150, depending on enol structure. The transition state for acid-catalysed enolization has the proton less than half-transferred.[140] Variation in the rates of iodination of substituted acetophenones have been attributed to different degrees of resonance between the carbonyl group and the substituent.[141] A very small (<twice) rate enhancement is observed for the micellar-catalysed enolization of acetone.[142]

Glyoxalase I catalyses the conversion of glutathione and α-keto-aldehydes into the glutathione thiol ester of the corresponding α-hydroxy-acid. α-Keto-aldehydes react non-enzymically with glutathione to form hemithioacetals of which only one of the diastereoisomers binds to the enzyme.[143] The rearrangement of hemithioacetals (**45**) to α-hydroxythiol esters (**46**) is almost completely inhibited by the addition of flavin, a good trap for carbanions. This is indicative of an enediol intermediate (**47**) rather than the expected 1,2-hydride shift mechanism.[144]

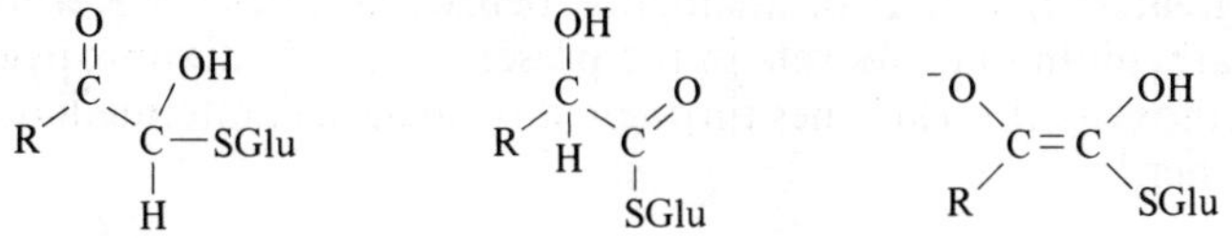

$$\underset{\textbf{(45)}}{RC(=O)\text{–}CH(OH)\text{–}SGlu} \qquad \underset{\textbf{(46)}}{RCH(OH)\text{–}C(=O)\text{–}SGlu} \qquad \underset{\textbf{(47)}}{(^{-}O)RC{=}C(OH)SGlu}$$

The kinetics of the enolization of cyclic ketones by amino-acids has been studied.[145] The kinetics and mechanisms of iodination reactions have been reviewed.[146] The rate of the acid-catalysed bromination of 2,4,6-trimethylacetophenone in 50% aqueous acetic acid is first order in bromine but is overall a rapid reaction. The reaction of the enol with bromine to give the bromonium ion is rate-limiting because of unfavourable non-bonded interactions.[147] Activation parameters and Brønsted β values for the iodination of cyclic ketones have been reported.[148] Cuprous iodide catalyses the acid-catalysed zero-order iodination of acetone.[149] The effect of the medium on the iodination of acetone has been discussed with reference to selective intermolecular forces.[150] The heterocyclic base-catalysed iodinations of acetophenone and acetonaphthones show Brønsted β values of 0.9–1.0.[151]

Halogens react rapidly with α,β-unsaturated ketones either by addition to the enol or by initial attack on the carbonyl oxygen, rather than by expected attack at the C=C bond.[152] The non-free-radical chlorination of ketones in methanol favours substitution at the least substituted carbon α to the carbonyl group. This is consistent with the formation and chlorination of the least hindered enol ether.[153] The gas-phase kinetic isotope effect, k_H/k_D, for the reaction of acetone with RO^- decreases in the order R = Me_3C > Br > Et > Me > H.[154]

The enolization reaction of alkyl mesityl ketones with alkylmagnesium bromide is first order in each reagent. Kinetic isotope effects ($k_H/k_D = 2.6$–3.1) with α-deuterio-substituted ketones have been used to support a step-wise mechanism, *viz.* pre-equilibrium coordination of the Grignard reagent to the carbonyl oxygen followed by rate-limiting hydrogen abstraction.[155] Triethylamine catalyses the enolization of ketones in the presence of Grignard reagents.[156]

Monomeric metaphosphate anion reacts with ketones to give enol phosphates and, in the presence of aniline, promotes Schiff-base formation. It is suggested that biosynthetic phosphorylations that require ATP may proceed through monomeric metaphosphate anion.[157] Silyl phosphites react with α-halocarbonyl compounds to give not only the enol phosphates (Perkow reaction) but also 1:1 carbonyl addition products and 2-oxo-phosphonates; the proportion of products depends upon substituents. The Perkow reaction probably proceeds *via* an initial attack of phosphite on the carbonyl carbon.[158] The powerful electrophile, 4,6-dinitrobenzofuroxan, reacts with monoketones and β-diketones, in the absence of added base, to give σ-adducts. Reaction presumably occurs *via* the enol of the ketone. With β-diketones both enolic and ketonic adducts are formed.[159] Allylation of acetals with allylsilanes is catalysed by iodotrimethylsilane to give the corresponding homoallyl ethers, with regiospecific transposition of the allyl group.[160]

The rates of tautomerization of di-*n*-alkyl β-diketones in mixed solvents compared with water have been analysed by consideration of the free energies of transfer of the transition states.[161] The degree of enolization of β-keto-esters of azines is greater than that of analogous benzoylacetates.[162] Ketones react with diethoxycarbenium fluoroborate in the presence of *N*,*N*-diisopropylethylamine to give α-(diethoxymethyl) ketones through, it is suggested, intermediate formation of the enol ether.[163]

Hydrolysis of Enol Ethers and Related Compounds

Hemiorthoesters, from hydration of ketene acetals, have been detected by ^{1}H- and ^{13}C-NMR spectroscopy. Their hydrolysis shows an acid-, base-, and water-catalysed reaction. The latter probably involves rate-limiting ionization of the hydroxyl group rapidly followed by breakdown of the mono-anion and hydronium ion within the first-formed encounter complex.[164] The mechanism and kinetics of reactions of ketenes have been reviewed.[165] The acid-catalysed hydrolysis of phenylketene acetals *O,O-* (**48**) and *O,S-* (**49**) acetals involves rate-limiting protonation of the double bond. A non-linear dependence of the rate upon formate buffer concentration is thought to be not attributable to a change in rate-limiting step as the rate is unaffected by the addition of mercaptoethanol.[166] However, as this thiol is not significantly ionized at low pH it may not be able to effectively compete with water attack on the intermediate carbocation.

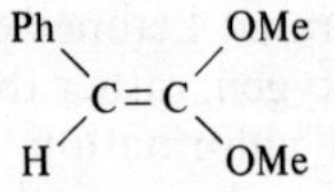

Ph(H)C=C(OMe)OMe **(48)**   Ph(H)C=C(SMe)OMe **(49)**

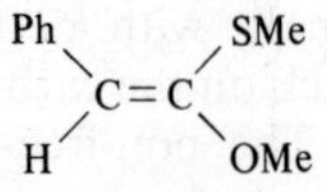

The rate of hydration of ketene to give acetic acid is pH-independent between pH 4 and 10 with a rate constant of $44\,s^{-1}$ and shows k_{H_2O}/k_{D_2O} of 1.9. This is compatible with rate-limiting uncatalysed addition of water to the ketene.[167] Dichloroketene reacts with dimethylketene dimethyl acetal to yield the cyclobutanone regiospecifically, but the reaction of methylchloroketene and ketene diethyl acetal yields an acyclic product, an acylketene acetal. This is consistent with a two-step mechanism involving a dipolar intermediate which can either lose a proton or cyclize.[168] The perfluoroacylketene (**50**), the first acylketene to be isolated, a related perfluorovinyl ketone (**51**), and the analogous sulphur derivatives have been prepared from a dimer of hexafluoropropene. The acylketene reacts as a diene to give adducts that are hydrolysis products of the vinyl ketone adducts.[169]

Kinetic solvent isotope effects ($k_{H_3O^+}/k_{D_3O^+} < 1$) and *ca.* 100% deuterium incorporation during the acidic hydrolysis of ketene selenoacetals indicate that a rapid and reversible protonation at the olefinic β-carbon occurs. The rate-limiting step could then be hydration of the carbocation or the breakdown of the

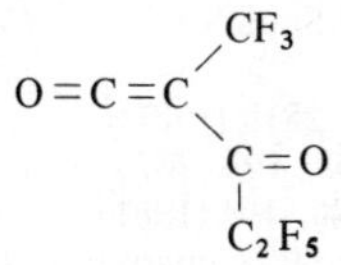

$O{=}C{=}C(CF_3)C(=O)C_2F_5$

(50)

$F_2C{=}C(CF_3)C(=O)C_2F_5$

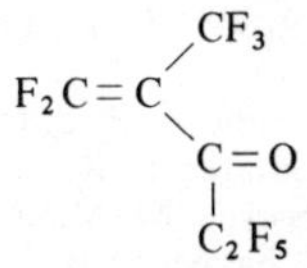

(51)

intermediate hemiorthoester.[170] The rate of the acid-catalysed hydrolysis of 9-methoxyoxacyclonon-2-ene (**52**) is 500 times faster than that of 2-methoxy-2,3-dihydropyran. The nine-membered ring compound exhibits non-linear buffer plots, compatible with reversible protonation and rate-limiting decomposition of the intermediate hemiacetal.[171] The effect of β-alkyl substitution in vinyl ether hydrolysis is not cumulative.[172] The acidic hydrolysis of dioxenes (**53**) occurs by *O*-protonation and rate-limiting ring-cleavage.[173]

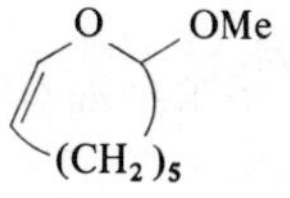

O OMe (CH$_2$)$_5$

(52)

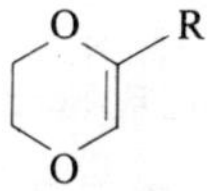

O R O

(53)

Other Reactions

The effect of solvent on the kinetics of the oxidation of cyclohexanone have been investigated.[174] The rate of the ruthenium(III)-catalysed oxidation of aromatic aldehydes by alkaline metaperiodate is independent of metaperiodate concentration and shows a Hammett ρ value of $+1.66$. This has been interpreted in terms of a change in mechanism from hydride loss in acid medium to proton loss in alkaline medium.[175] The oxidation of benzaldehyde to benzoic acid by acidic bromate occurs by slow formation of an intermediate bromate ester.[176] 3-Nitrobenzylidene dibromides and dichlorides decompose in base to give the aldehyde, but the difluorides give α-chloro-4,4′-dinitrostilbene oxide.[177]

The red colour produced in the Fujiwara reaction of *gem*-polyhalogen compounds with pyridine is due to the formation of a conjugated benzamidine.[178] The Hammett ρ value of -1.67 for the addition of substituted diaryldiazomethanes to chloranil indicates the development of a large positive charge on the diazo carbon in the transition state.[179] The reaction of two moles of an aromatic aldehyde with methyl thiocyanate in the presence of tributylphosphine gives both *S*-methylthiobenzoates and phenylacetonitriles probably by a novel disproportionation reaction.[180]

References

1 Przystas, T. J., and Fife, T. H., *J. Am. Chem. Soc.*, **103**, 4884 (1981).
2 Harron, J., McClelland, R. A., Thankachan, C., and Tidwell, T., *J. Org. Chem.*, **46**, 903 (1981).
3 van Eikeren, P., *J. Org. Chem.*, **45**, 4641 (1980).
4 Beaulieu, N., Dickinson, R. A., and Deslongchamps, P., *Can. J. Chem.*, **58**, 2531 (1980).
5 Deslongchamps, P., Rowan, D. D., and Pothier, N., *Can. J. Chem.*, **59**, 2787 (1981).
6 Reddy, C. P., Singh, S. M., and Rao, R. B., *Tetrahedron Lett.*, **22**, 973 (1981).

[7] Deslongchamps, P., Rowan, D. D., Pothier, N., and Saunders, J. K., *Can. J. Chem.*, **59,** 1105, 1122 (1981).
[8] Wynberg, H., and Lorand, J. P., *J. Org. Chem.*, **46,** 2538 (1981).
[9] Wiberg, K. B., and Squires, R. R., *J. Am. Chem. Soc.*, **103,** 4473 (1981).
[10] Wann, S. R., and Kreevoy, M. M., *J. Org. Chem.*, **46,** 419 (1981).
[11] Akhmatdinov, R. T., Chalova, O. B., Kantor, E. A., and Rakhmankulov, D. L., *Zh. Org. Chem.*, **16,** 962 (1980); *Chem. Abs.*, **93,** 185312 (1980).
[12] Cerveny, L., Marhoul, A., Krivska, M., and Ruzicka, V., *Chem. Prum.*, **30,** 526 (1980); *Chem. Abs.*, **94,** 155936 (1981).
[13] Goosen, A., and McCleland, C. W., *J. Chem. Soc., Perkin Trans. 1,* **1981,** 977.
[14] Kotlyar, S. A., Kamalov, G. L., Savranskaya, R. L., and Bogatskii, A. V., *Khim. Geterotsikl. Soedin,* **1981,** 176; *Chem. Abs.*, **95,** 23764 (1981).
[15] Bonner, T. G., Lewis, D., and Rutter, K., *J. Chem. Soc., Perkin Trans. 2,* **1981,** 1807.
[16] Uray, G., and Wolfbeis, O. S., *Monatsh. Chem.*, **112,** 627 (1981).
[17] La, S., Koh, K. S., and Lee, I., *Taehan Hwahakhoe Chi.*, **24,** 1, (1980); *Chem. Abs.*, **93,** 238144 (1980).
[18] Bulman-Page, P. C., Ley, S. V., Morton, J. A., and Williams, D. J., *J. Chem. Soc., Perkin Trans. 1,* **1981,** 457.
[19] Kantlehner, W., and Gutbrod, H. D., *Liebigs Ann. Chem.*, **1980,** 1677.
[20] Wertz, P. W., Garver, J. C., and Anderson, L., *J. Am. Chem. Soc.*, **103,** 3916 (1981).
[21] *Org. Reaction Mech.*, **1979,** 4.
[22] Moody, W., and Richards, G. N., *Carbohydr. Res.*, **93,** 83 (1981).
[23] Friebolin, H., Kunzelmann, P., Supp, M., Brossmer, R., Keilich, G., and Ziegler, D., *Tetrahedron Lett.*, **22,** 1383 (1981).
[24] York, J. L., *J. Org. Chem.*, **46,** 2171 (1981).
[25] Gillet, B., Nicole, D. J., and Delpuech, J. J., *J. Chem. Soc., Perkin Trans. 2,* **1981,** 1329.
[26] Pavia, A. A., and Ung-Chhun, S. N., *Can. J. Chem.*, **59,** 482 (1981).
[27] Shimizu, K., *Carbohydr. Res.*, **92,** 219 (1981).
[28] Vorbruggen, H., and Hofle, G., *Chem. Ber.*, **114,** 1256 (1981).
[29] Rosenberg, S., and Kirsch, J. F., *Biochemistry*, **20,** 3189, 3196 (1981).
[30] Aerts, G. M., and De Bruyne, C. K., *Biochim. Biophys. Acta,* **660,** 317 (1981).
[31] Stein, R. L., and Cordes, E. H., *J. Biol. Chem.*, **256,** 767 (1981).
[32] Combes, D., Monsan, P., and Mathlouthi, M., *Carbohydr. Res.*, **93,** 312 (1981).
[33] McClelland, R. A., and Somani, R., *J. Org. Chem.*, **46,** 4345 (1981).
[34] Fife, T. H., and Pellino, A. M., *J. Am. Chem. Soc.*, **103,** 1201 (1981).
[35] Lonnberg, H., Kappi, R., and Heikkinen, E., *Acta Chem. Scand.*, **35B,** 589 (1981).
[36] Hine, J., and Chou, Y., *J. Org. Chem.*, **46,** 649 (1981).
[37] Nair, V., Vietti, D. E., and Cooper, C. S., *J. Am. Chem. Soc.*, **103,** 3030 (1981).
[38] Panetta, C. A., and Dixit, A. S., *J. Org. Chem.*, **45,** 4503 (1980).
[39] Hine, J., and Zeigler, J. P., *J. Am. Chem. Soc.*, **102,** 7524 (1980).
[40] Tatsumoto, K., Martell, A. E., and Motekaitis, R. J., *J. Am. Chem. Soc.*, **103,** 6197 (1981).
[41] Tatsumoto, K., and Martell, A. E., *J. Am. Chem. Soc.*, **103,** 6203 (1981).
[42] Iwata, M., and Kuzuhara, H., *Chem. Lett.*, **1981,** 5.
[43] Hardman, M. J., *Biochem. J.*, **195,** 773, (1981).
[44] Grady, S. R., Wang, J. K., and Dekker, E. E., *Biochemistry*, **20,** 2497 (1981).
[45] Ahern, E. P., Dignam, K. J., and Hegarty, A. F., *J. Org. Chem.*, **45,** 4302 (1980).
[46] Maccorone, E., Mamo, A., and Perrini, G., *Gazz. Chim. Ital.*, **111,** 163, (1981); *Chem. Abs.*, **95,** 149506 (1981).
[47] Mahmoud, M. R., El-Shafei, A. K., and Adam, F. A., *Gazz. Chim. Ital.*, **110,** 221 (1980); *Chem. Abs.*, **94,** 29730 (1981).
[48] Rotondo, E., and Neri, G., *Congr. Naz. Chim. Inorg. (Atti), 13th,* **1980,** 303; *Chem. Abs.*, **95,** 23810 (1981).
[49] Nelson, S. M., Knox, C. V., McCann, M., and Drew, M. G. B., *J. Chem. Soc., Dalton Trans.*, **1981,** 1669.
[50] Drew, M. G. B., Nelson, J., and Nelson, S. M., *J. Chem. Soc., Dalton Trans.*, **1981,** 1678, 1685, 1691.
[51] Rao, C. V. C., Veeranagaiah, V., Reddy, K. K., and Rao, N. V. S., *Indian J. Chem.*, **17B,** 566 (1979); *Chem. Abs.*, **93,** 167168 (1980).
[52] Prasad, G., Singh, G., and Mehrotra, K. N., *Indian J. Chem.*, **19B,** 653 (1980); *Chem. Abs.*, **94,** 14770 (1981).
[53] Petrova, I., and Kirilov, M., in *Organometalliques Fonct. Ambidents, Recl. Commun., Colloq. Fr.-Bulg.*, **1980,** 272; *Chem. Abs.*, **95,** 149451 (1981).

[54] Drapier, J., Hoornaerts, M. T., Hubert, A. J., and Teyssie, P., *J. Mol. Catal.*, **11,** 53 (1981); *Chem. Abs.*, **95,** 79540, (1981).
[55] Saed, A. A. H., *Indian J. Chem.*, **17B,** 462 (1979); *Chem. Abs.*, **93,** 185530, (1980).
[56] Dash, A. C., Dash, B., and Praharaj, S., *J. Chem. Soc., Dalton Trans.*, **1981,** 2063.
[57] Dash, A. C., Dash, B., and Patra, M., *Indian J. Chem.*, **19B,** 492 (1980); *Chem. Abs.*, **94,** 14726 (1981).
[58] Jagannadham, V., Sethuram, B., and Rao, R. T., *Indian J. Chem.*, **17B**, 598 (1979); *Chem. Abs.*, **93,** 203542 (1980).
[59] Seconi, G., Pirazzini, G., Ricci, A., Fiorenza, M., and Eaborn, C., *J. Chem. Soc., Perkin Trans. 2*, **1980,** 1792.
[60] Seconi, G., and Eaborn, C., *J. Chem. Soc., Perkin Trans. 2*, **1981,** 1051.
[61] Arous-Chtara, R., Moreau, J. L., and Gaudemar, M., *J. Soc. Chim. Tunis*, **3,** 1 (1980); *Chem. Abs.*, **94,** 191527 (1981).
[62] Knollmuller, M., and Kosma, P., *Monatsh. Chem.*, **112,** 489 (1981).
[63] Strom, J. G., and Jun, H. W., *J. Pharm. Sci.*, **69,** 1261 (1980).
[64] Ahmed, M. G., Ahmed, S. A., and Hickmott, P. W., *J. Chem. Soc., Perkin Trans. 1*, **1980,** 2383.
[65] Boyer, B., Lamaty, G., and Moreau, C., *Rec. Trav. Chim. Pays-Bas*, **100,** 272 (1981).
[66] Ivanova, L. P., Polis, J., Grava, I., Raguel, B., Cherkasova, V. A., Hamad, H. J., and Kolobova, T. A., *Zh. Org. Chim.*, **17,** 325 (1981); *Chem. Abs.*, **95,** 6265 (1981).
[67] Bodor, A., Fey, L., and Breazu, D., *Rev. Roum. Chim.*, **25,** 1367 (1980).
[68] Egberink, H., and Van Heerden, C., *Chem. Anal. Chim. Acta*, **118,** 359 (1980); *Chem. Abs.*, **93,** 220106 (1980).
[69] Muhlstadt, M., Schulze, B., and Schubert, I., *Z. Chem.*, **21,** 326 (1981).
[70] Leroy, G., Nguyen, M.-T., Sana, M., and Villaveces, J.-L., *Bull. Soc. Chim. Belg.*, **89,** 1023 (1980).
[71] Kenley, R. A., Swidler, R., Manser, G. E., and Kirshen, N. A., *Gov. Rep. Announce Index* (*U.S.*), **80,** 2378 (1980); *Chem. Abs.*, **94,** 29734 (1981).
[72] Johnson, J. E., Nalley, E. A., Weidig, C., and Arfan, M., *J. Org. Chem.*, **46,** 3623 (1981).
[73] Ooi, N. S., and Wilson, D. A., *J. Chem. Res.* (*S*), **1980,** 366, 394; **1981,** 18.
[74] Rao, V. V., Sethuram, B., and Rao, T. N., *Indian J. Chem.*, **19A,** 1127 (1980); *Chem. Abs.*, **94,** 173900 (1981).
[75] Gawley, R. E., Termine, E. J., and Onan, K. D., *J. Chem. Soc., Chem. Commun.*, **1981,** 568.
[76] Bastos, M. P., Alves, K. B., Neto, G. O., and Amaral, L., *J. Org. Chem.*, **46,** 3342 (1981).
[77] Ishii, H., *Acc. Chem. Res.*, **14,** 275, (1981).
[78] Butler, R. N., Morris, G. J., and O'Donohue, A. M., *J. Chem. Soc., Perkin Trans. 2*, **1981,** 1243.
[79] Zimmer, O., and Meier, H., *Chem. Ber.*, **114,** 2938 (1981).
[80] Lemke, T. L., and Parker, G. R., *J. Heterocycl. Chem.*, **17,** 1519 (1980).
[81] Suzuki, A., and Hara, S., *Kagaku* (*Kyoto*), **35,** 406, (1980); *Chem. Abs.*, **94,** 14631 (1981). Mori, S., *Farumashia*, **17,** 825 (1981); *Chem. Abs.*, **95,** 167919 (1981).
Murata, S., and Noyori, R., *Kagaku Zokan* (*Kyoto*), **1981,** 117; *Chem. Abs.*, **95,** 96308 (1981). Maroni, P., in *Organometalliques Fonct. Ambidents, Recl. Commun. Colloq. Fr.-Bulg.*, **1980,** 75; *Chem. Abs.*, **95,** 60732 (1981).
[82] Fleury, D., and Fleury, M. B., *Tetrahedron*, **37,** 3403 (1981).
[83] Zeng, G. Z., Yan, J. S., and Shen, D. Z., *Hud Hsueh Hsueh Pao*, **39,** 215 (1981); *Chem. Abs.*, **95,** 149476 (1981).
[84] Davenport, K. G., Newcomb, M., and Bergbreiter, D. E., *J. Org. Chem.*, **46,** 3143 (1981).
[85] Heathcock, C. H., White, C. T., Morrison, J. J., and Van Derveer, D., *J. Org. Chem.*, **46,** 1296 (1981).
[86] Reetz, M. T., and Peter, R., *Tetrahedron Lett.*, **22,** 4691 (1981).
[87] Heathcock, C. H., Pirrung, M. C., Montgomery, S. H., and Lampe, J., *Tetrahedron*, **37,** 4087 (1981).
[88] Fiaud, J. C., and Malleron, J. C., *J. Chem. Soc., Chem. Commun.*, **1981,** 1159.
[89] Shieh, H. M., and Prestwich, G. D., *J. Org. Chem.*, **46,** 4319 (1981).
[90] Masamune, S., Choy, W., Kerdesky, F. A. J., and Imperiali, B., *J. Am. Chem. Soc.*, **103,** 1566 (1981).
[91] Evans, D. A., Nelson, J. V., Vogel, E., and Taber, T. R., *J. Am. Chem. Soc.*, **103,** 3099 (1981). Evans, D. A., Bartroli, J., and Shih, T. L., *J. Am. Chem. Soc.*, **103,** 2127 (1981).
[92] Evans, D. A., and McGee, L. R., *J. Am. Chem. Soc.*, **103,** 2876 (1981).
[93] Evans, D. A., and Taber, T. R., *Tetrahedron Lett.*, **21,** 4675 (1980).
[94] Evans, D. A., and Taber, T. R., *Tetrahedron Lett.*, **21,** 4675 (1980).
[95] Noyori, R., Nishida, I., and Sakata, J., *J. Am. Chem. Soc.*, **103,** 2106 (1981).
[96] Yamamoto, Y., Yatagai, H., and Maruyama, K., *J. Chem. Soc., Chem. Commun.*, **1981,** 162.
[97] Sato, F., Iijima, S., and Sato, M., *Tetrahedron Lett.*, **22,** 243 (1981). Hiyama, T., Kimura, K., and Nozaki, H., *Tetrahedron Lett.*, **22,** 1037 (1981). Yamamoto, Y., and Maruyama, K., *Tetrahedron Lett.*, **22,** 2895 (1981). Buss, A. D., and Warren, S., *J. Chem. Soc., Chem. Commun.*, **1981,** 100.

[98] Allen, D. W., *J. Chem. Res. (S)*, **1980,** 384.
[99] Allen, D. W., *Z. Naturforsch.*, **35B,** 1455 (1980); *Chem. Abs.*, **94,** 191529 (1981).
[100] Yamakado, Y., and Ishiguro, M., *J. Am. Chem. Soc.*, **103,** 5568 (1981).
[101] Larchevêque, M., and Debal, A., *J. Chem. Soc., Chem. Commun.*, **1981,** 877.
[102] Ochiai, M., and Fujita, E., *J. Chem. Soc., Chem. Commun.*, **1980,** 1118.
[103] Heathcock, C. H., Pirrung, M. C., Lampe, J., Buse, C. T., and Young, S. D., *J. Org. Chem.*, **46,** 2290 (1981).
[104] Reetz, M. T., Sauerwald, M., and Walz, P., *Tetrahedron Lett.*, **22,** 1101 (1981).
[105] Audo, D., Vincens, M., Dumont, C., and Vidal, M., *Can. J. Chem.*, **59,** 2199 (1981).
[106] Nee, G., Leroux, Y., and Seyden-Penne, J., *Tetrahedron*, **37,** 1541 (1981).
[107] Gompper, R., and Vogt, H.-H., *Chem. Ber.*, **114,** 2866 (1981).
[108] Meyers, A. I., Williams, D. R., Erickson, G. W., White, S., and Druelinger, M., *J. Am. Chem. Soc.*, **103,** 3081 (1981). Meyers, A. I., Williams, D. R., White, S., and Erickson, G. W., *J. Am. Chem. Soc.*, **103,** 3088 (1981).
[109] Tamaru, Y., Harada, T., Nishi, S., Mizutani, M., Hocki, T., and Yoshida, Z., *J. Am. Chem. Soc.*, **102,** 7806 (1980).
[110] Guthrie, J. P., *Can. J. Chem.*, **59,** 45 (1981).
[111] Baker, S. R., and Crombie, L., *J. Chem. Soc., Perkin Trans. 1*, **1981,** 172, 178.
[112] Poonia, N. S., Sen, S., Porwal, P. K., and Jayakumar, A., *Bull. Chem. Soc. Jpn.*, **53,** 3338 (1980).
[113] El Seoud, O. A., Vieira, R. A., and Farah, J. P. S., *J. Org. Chem.*, **46,** 1231 (1981).
[114] Abe, K., Endo, H., and Hirota, M., *Bull. Chem. Soc. Jpn.*, **53,** 466 (1981).
[115] Williams, I. H., Maggiora, G. M., and Schowen, R. L., *J. Am. Chem. Soc.*, **102,** 7831 (1980).
[116] *Org. Reaction Mech.*, **1978,** 16.
[117] Emly, M., and Leussing, D. L., *J. Am. Chem. Soc.*, **103,** 628 (1981).
[118] Basu, S., Schuster, P., and Wolsehann, P., *Monatsh. Chem.*, **112,** 421 (1981).
[119] Young, P. R., and McMahon, P. E., *J. Org. Chem.*, **45,** 5290 (1980).
[120] Oku, J. I., and Inoue, S., *J. Chem. Soc., Chem. Commun.*, **1981,** 229.
[121] Warnhoff, E. W., Wong, M. Y. H., and Raman, P. S., *Can. J. Chem.*, **59,** 688 (1981).
[122] McDonald, R. S., and Sibley, C. E., *Can. J. Chem.*, **59,** 1061 (1981).
[123] Loupy, A., Roux-Schmitt, M. C., and Seyden-Penne, J., *Tetrahedron Lett.*, **22,** 1685 (1981).
[124] Hirao, K., *Yakugaku Zasshi*, **100,** 473, (1980); *Chem. Abs.*, **93,** 203420 (1980).
[125] Cieplak, A. S., *J. Am. Chem. Soc.*, **103,** 4540 (1981).
[126] Stone, A. J., and Erskine, R. W., *J. Am. Chem. Soc.*, **102,** 7185 (1980).
[127] Caramella, P., Rondan, N. G., Paddon-Row, M. N., and Houk, K. N., *J. Am. Chem. Soc.*, **103,** 2438 (1981).
[128] Lasperas, M., Perez-Rubalcaba, A., Perez-Ossorio, R., and Quiroga, M. L., *An. Quim.*, **76C,** 191 (1980); *Chem. Abs.*, **95,** 42015 (1981).
[129] Ashby, E. C., and Goel, A. B., *J. Am. Chem. Soc.*, **103,** 4983 (1981).
[130] Arjona, O., Perez-Ossorio, R., Perez-Rubalcaba, A., and Quiroga, M. L., *J. Chem. Soc., Perkin Trans. 2*, **1981,** 597.
[131] Lasperas, M., Perez-Rubalcaba, A., and Quiroga-Feijoo, M. L., *Tetrahedron*, **36,** 3403 (1981).
[132] Bernardon, C., and Deberly, A., *J. Chem. Soc., Perkin Trans. 1*, **1980,** 2631.
[133] Lumma, W. C., *J. Org. Chem.*, **46,** 3668 (1981).
[134] Julia, S., and Ginebreda, A., *Afinidad*, **37,** 194 (1980); *Chem. Abs.*, **94,** 14740 (1981).
[135] Papadopoulos, M., and Jenner, G., *Tetrahedron Lett.*, **22,** 2773 (1981).
[136] Achmatowicz, O., and Szymoniak, J., *J. Org. Chem.*, **45,** 4774 (1980).
[137] Capon, B., Rycroft, D. S., Watson, T. W., and Zucco, C., *J. Am. Chem. Soc.*, **103,** 1761 (1981).
[138] Cox, B. G., De Maria, P., Fini, A., and Hassan, A. F., *J. Chem. Soc., Perkin Trans. 2*, **1981,** 1351.
[139] Stewart, R., and Srinivasan, R., *Can. J. Chem.*, **59,** 957 (1981).
[140] Dubois, J.-E., El-Alaoui, M., and Toullec, J., *J. Am. Chem. Soc.*, **103,** 5393 (1981).
[141] Ganapathy, K., and Ramanujam, M., *J. Indian Chem. Soc.*, **58,** 701 (1981); *Chem. Abs.*, **95,** 149686 (1981).
[142] Maruthamuthu, M., and Lakshmikanthan, N., *Int. J. Chem. Kinet.*, **13,** 695 (1981).
[143] Brown, C., Douglas, K. T., and Ghobt-Sharif, J., *J. Chem. Soc., Chem. Commun.*, **1981,** 944.
[144] Shinkai, S., Yamashita, T., Kusano, Y., and Manabe, O., *J. Am. Chem. Soc.*, **103,** 2070 (1981).
[145] Singh, K. J., and Bhadoria, G. P. S., *J. Indian Chem. Soc.*, **58,** 417 (1981); *Chem. Abs.*, **95,** 79756 (1981).
[146] Argentini, M., *EIR-Ber.*, **1979,** 393; *Chem. Abs.*, **93,** 237966 (1980).
[147] Pinkus, A. G., and Gopalan, R., *J. Chem. Soc., Chem. Commun.*, **1981,** 1016.
[148] Satyanarayana, N., and Sundaram, E. V., *J. Indian Chem. Soc.*, **57,** 1076 (1980); *Chem. Abs.*, **94,** 102386 (1981).

[149] Ghoneim, F. B., and Selim, E. A., *Egypt. J. Chem.*, **21,** 47 (1978); *Chem. Abs.*, **93,** 203547 (1980).
[150] Jin, S.-S., Zheng, X.-M., and Zhang, C.-J., *Hang-chou Ta Hsueh Hsueh Pao, Tzu Jan K'o Hsueh Pan*, **8,** 81 (1981); *Chem. Abs.*, **95,** 96523 (1981).
[151] Ananthakrishnanadar, P., and Gnanasekaran, C., *Indian J. Chem.*, **19A,** 646 (1980); *Chem. Abs.*, **94,** 29965 (1981).
[152] Heasley, V. L., Shellhamer, D. F., Carter, T. L., Gipe, D. E., Gipe, R. K., Green, R. C., Nordeen, J., Rempel, T. D., and Spaite, D. W., *Tetrahedron Lett.*, **22,** 2467 (1981).
[153] Gallucci, R. R., and Going, R., *J. Org. Chem.*, **46,** 2532 (1981).
[154] Noest, A. J., and Nibbering, N. M. M., *Int. J. Mass Spectrom. Ion. Phys.*, **34,** 383 (1980); *Chem. Abs.*, **94,** 14877 (1981).
[155] Pinkus, A. G., and Sabesan, A., *J. Chem. Soc., Perkin Trans. 2*, **1981,** 473.
[156] Tuulmets, A., and Kalbus, M., *Org. React. (Tartu)*, **17,** 112 (1981).
[157] Satterthwait, A. C., and Westheimer, F. H., *J. Am. Chem. Soc.*, **103,** 1177 (1981).
[158] Sekine, M., Okimoto, K., Yamada, K., and Hata, T., *J. Org. Chem.*, **46,** 2097 (1981).
[159] Terrier, F., Simonnin, M. P., and Pouet, M. J., *J. Org. Chem.*, **46,** 3537 (1981).
[160] Sakurai, H., Sasaki, K., and Hosomi, A., *Tetrahedron Lett.*, **22,** 745 (1981).
[161] Watarai, H., *Bull. Chem. Soc. Jpn.*, **53,** 3019 (1980).
[162] Stekhova, S. A., Zagulyaeva, O. A., Lapachev, V. V., and Mamaev, V. P., *Khim. Getesotsikl Soedin.*, **1980.** 822; *Chem. Abs.*, **93,** 203745 (1980).
[163] Mock, W. L., and Tsou, H. R., *J. Org. Chem.*, **46,** 2557 (1981).
[164] Capon, B., and Ghosh, A. K., *J. Am. Chem. Soc.*, **103,** 1765 (1981).
[165] Blake, P., in *The Chemistry of Ketenes, Allenes and Related Compounds*, (ed. Patai, S.), Wiley, Chichester–New York, 1980, p. 309.
[166] Okuyama, T., Kawao, S., and Fueno, T., *J. Org. Chem.*, **46,** 4372 (1981).
[167] Bothe, E., Dessouki, A. M., and Schulte-Frohlinde, D., *J. Phys. Chem.*, **84,** 3270 (1981).
[168] Brady, W. T., and Watts, R. D., *J. Org. Chem.*, **46,** 4047 (1981).
[169] England, D. C., *J. Org. Chem.*, **46,** 147, 153 (1981).
[170] Wautier, H., Desauvage, S., and Hevesi, L., *J. Chem. Soc., Chem. Commun.*, **1981,** 738.
[171] Burt, R. A., Chiang, V., and Kresge, A. J., *Can. J. Chem.*, **58,** 2199 (1981).
[172] Chiang, Y., Kresge, A. J., Tidwell, T. T., and Walsh, P. A., *Can. J. Chem.*, **58,** 2203 (1980).
[173] Tsivunin, V. S., Zaripova, V. G., and Zaripov, I. N., *Zh. Obshch. Khim.*, **51,** 318 (1981); *Chem. Abs.*, **95,** 96435 (1981).
[174] Kudesia, V. P., *Acta Cienc. Indica, (Ser.) Chem.*, **6,** 119 (1980); *Chem. Abs.*, **94,** 191448 (1981).
[175] Radhakrishnamurti, P. S., and Misra, P. C., *Indian J. Chem.*, **19A,** 427 (1980); *Chem. Abs.*, **93,** 185429 (1980).
[176] Anandan, S., and Gopalan, R., *Indian J. Chem.*, **17A,** 629 (1979); *Chem. Abs.*, **93,** 167263 (1980).
[177] Riad, Y., Abdallah, A. A., and Kassem, A. A., *Egypt. J. Chem.*, **22,** 413 (1979); *Chem. Abs.*, **95,** 149467 (1981).
[178] Uno, T., Okumura, K., and Kuroda, Y., *J. Org. Chem.*, **46,** 3175 (1981).
[179] Oshima, T., and Nagai, T., *Bull. Chem. Soc. Jpn.*, **53,** 3284 (1980).
[180] Kurauchi, M., Imamoto, T., and Yokoyama, M., *Tetrahedron Lett.*, **22,** 4985 (1981).

Organic Reaction Mechanisms 1981
Edited by A. C. Knipe and W. E. Watts

CHAPTER 2

Reactions of Acids and their Derivatives

A. J. KIRBY

University Chemical Laboratory, Cambridge

CARBOXYLIC ACIDS . . . 23
Tetrahedral Intermediates . . . 23
Intermolecular Catalysis and Reactions . . . 26
Reactions in Hydroxylic Solvents . . . 26
Reactions in Aprotic Solvents . . . 36
Intramolecular Catalysis and Neighbouring-group Participation . . . 39
Association-prefaced Catalysis . . . 42
Metal-ion Catalysis . . . 47
Decarboxylation . . . 48
Enzymic Catalysis . . . 50
Serine Proteinases . . . 50
Thiol Proteinases . . . 51
Acid Proteinases . . . 52
Metallo-proteinases . . . 52
Other Enzymes . . . 53
NON-CARBOXYLIC ACIDS . . . 54
Phosphorus-containing Acids . . . 54
Non-enzymic Reactions . . . 54
Enzymic Reactions . . . 61
Sulphur-containing Acids . . . 63
Other Acids . . . 66
References . . . 66

CARBOXYLIC ACIDS

Tetrahedral Intermediates

The tetrahedral intermediates involved in the hydrolysis and formation of esters are hemiorthoesters (**1**). Hemiorthoesters (**2**) are also formed in the hydrolysis of orthoesters (**3**), but generally these compounds exist only as transient intermediates, because their acid- and base-catalysed breakdown is much faster than their formation. However, recent work by Capon's group[1] has shown that simple hemiorthoesters (**2**) can be generated from suitable high-energy precursors in sufficient concentration to be detected by NMR techniques. An account of this work has appeared.[2] The most recent example is the generation of hemiorthoesters (**4**;

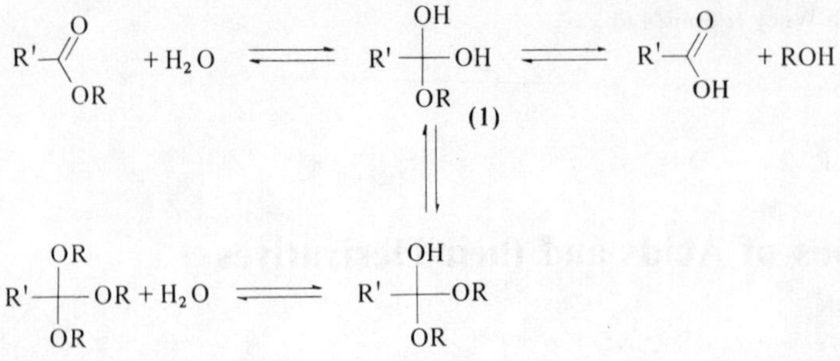

R = H, Me) by the hydration of the corresponding ketene acetals at low temperatures.[3] The breakdown of the pinacol derivative (**4**; R = Me) is rate-determining below pH 6, and can be followed at 205 nm to show acid- and base-catalysed and spontaneous reactions. The acid-catalysed reaction is thought to involve the usual mechanism for orthoester hydrolysis, with protonation of oxygen probably concerted with C—O cleavage, and is not a great deal faster for the hemiorthoester. The spontaneous reaction is not observed for simple orthoesters, and it has been suggested that the encounter ion-pair (**5**) generated by the ionization of (**4**) breaks down to form pinacol monoacetate (**6**) faster than the hydronium ion can diffuse away.[3]

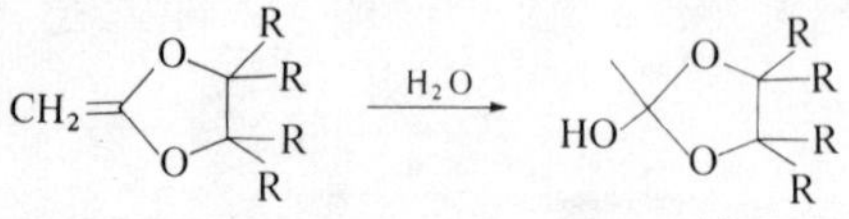

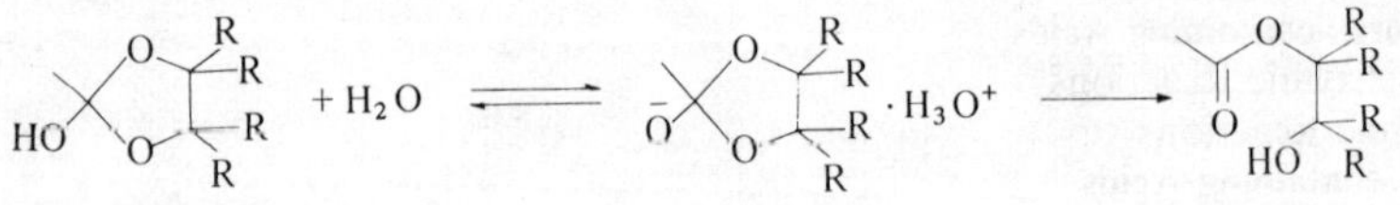

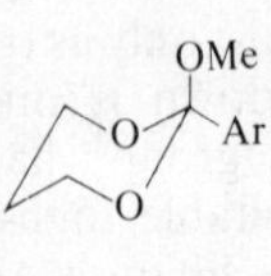

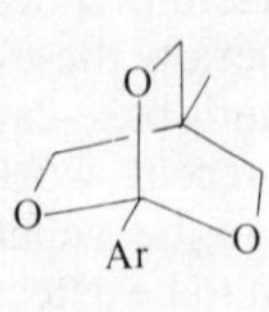

The breakdown of the hemiorthoester intermediate is also rate-determining in the hydrolysis of some cyclic orthoesters at low pH;[4,5] new work involves the six-membered ring compounds (**7**) and (**8**), and various methylated derivatives.[5] The first step, formation of a cyclic dioxocarbenium ion, is more favourable for the dioxanes than for the corresponding derivatives with five-membered rings, but the reactions are qualitatively similar. This initial C—O cleavage does not appear to be readily reversible for (**8**), since the same intermediate (**9**) can be generated rapidly from (**10**), under conditions where (**8**) is effectively stable, and shown to lead exclusively to the products of hydrolysis.[5] The same conclusion can be drawn from the effects of substituents on the rates of hydrolysis of these systems (**7**, **8** and the corresponding orthoacetates and orthoformates),[6] and contrasts with the cleavage

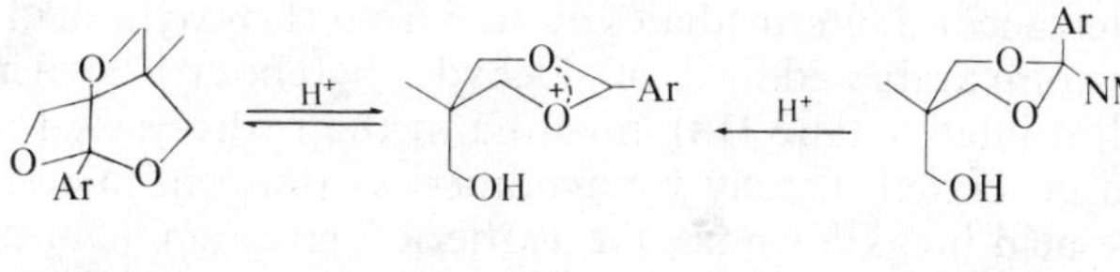

(**8**) (**9**) (**10**)

of related trioxaadamantane derivatives (**11**), where hydration of the dioxacarbenium ion is rate-determining.[7] C—O cleavage is also rate-determining for several cyclic orthoesters (**12**).[8]

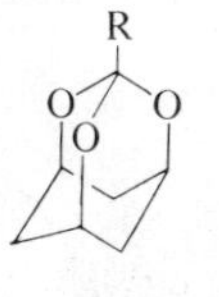

(**11**) 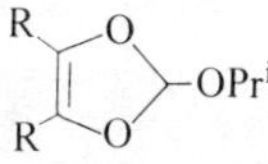 (**12**)

The tetrahedral intermediate (**15**) involved in the lactonization of ethyl 2-hydroxymethyl benzoate has been observed in the hydrolysis of orthoester (**13**), and has been generated independently from the cation (**14**).[9] The half-life of (**15**) reaches a maximum of 0.1 s at pH 3.5, and its partitioning between phthalide and hydroxy-ester gives 24 % of lactone below this pH, and 7.5 % in the base-catalysed region. This predominance of open-chain products is normal for the breakdown of a cyclic hemiorthoester, and, since the comparatively long life-time of the intermediate (**15**) at pHs near 3.5 allows ample time for conformational equilibrium to be attained, is unlikely to reflect a primary stereoelectronic effect.[10] In this system the breakdown of the intermediate (**15**) to products does not become the rate-determining step in

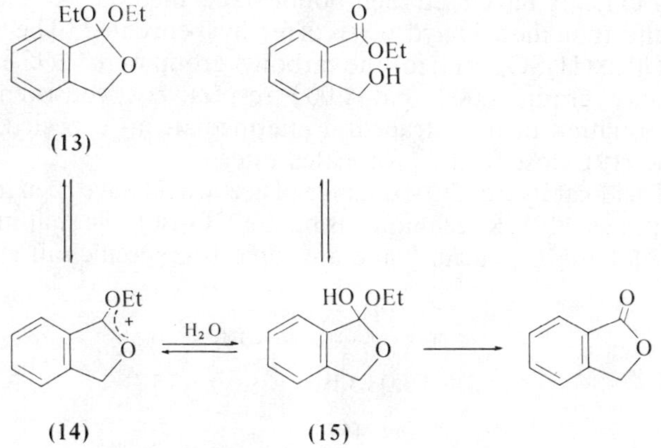

(**13**)

(**14**) (**15**)

the hydrolysis of the orthoester even at low pH; though it is rate-determining in the lactonization of the hydroxy-ester.[9]

By analysis of this system the equilibrium constant for the formation of the tetrahedral intermediate (**15**) from the hydroxy-ester has been estimated to be about 10^{-6}. This is almost 10^7 times less favourable than the equilibrium constant for the formation of the hemiacetal from 2-hydroxymethyl benzaldehyde; thus, it can be estimated that tetrahedral intermediates derived from carboxylic acid derivates are less stable than comparable adducts of aldehydes by about 10 kcal mol^{-1}.

The tetrahedral intermediate (**18**), involved in the hydrolysis of amide acetals (**16**)[11] has been generated directly by hydration of the benzimidatonium cation (**17**).[12] In dilute acid breakdown *via* the zwitterion gives amine and ester almost exclusively, but in moderately concentrated H_2SO_4 significant (though still small) amounts of amide are produced, particularly from the *p*-nitrobenzimidate derivative.[12]

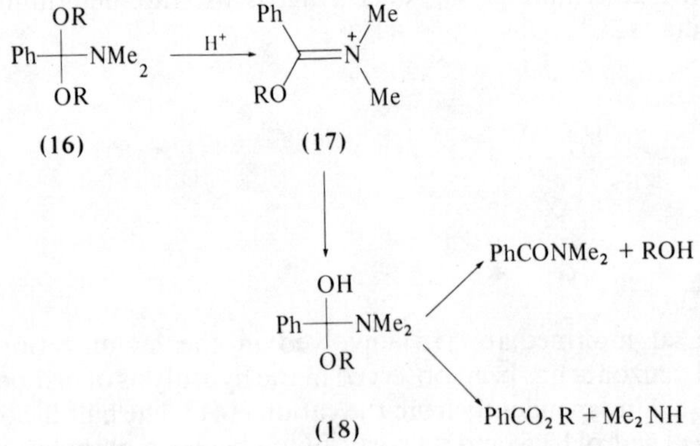

Ph–C(OR)₂–NMe₂ (**16**) $\xrightarrow{H^+}$ Ph(RO)C=N⁺Me₂ (**17**) → Ph–C(OH)(OR)–NMe₂ (**18**) → $PhCONMe_2$ + ROH, or $PhCO_2R$ + Me_2NH

(16) **(17)** **(18)**

Intermolecular Catalysis and Reactions

Reactions in Hydroxylic Solvents

The only review of any relevance this year, on acid–base catalysis, is short and in Czech.[13]

Marlier and O'Leary have used their double-label method[14] to measure heavy-atom isotope effects on the acid hydrolysis of methyl benzoate.[15] The ^{18}O- and ^{13}C-isotope effects (0.2M H_2SO_4, 91°) for the carbonyl group were 0.995 and 1.026, and for the methoxy group 1.002 and 1.000 respectively, consistent with rate-determining formation of the tetrahedral intermediate, as expected, *via* an early transition state (**19**), close to the protonated ester.

The rates of acid-catalysed ^{18}O-exchange of acetic acid have been measured by a ("nearly continuous") NMR technique using the ^{18}O-isotopic shift induced on the ^{13}C-signal of [1-^{13}C,$^{18}O_2$]-acid,[16] and are subject to specific salt effects.

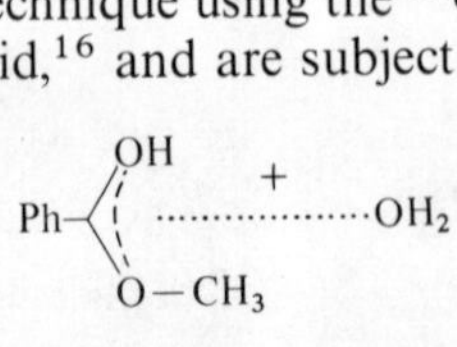

Ph–C(OH)(O–CH₃)$^{+}$ ·········· OH_2

(19)

In their classic paper on ester hydrolysis in strong sulphuric acid Yates and McClelland[17] suggested that a change-over from $A_{AC}2$ to an $A_{AL}1$ mechanism occurred in the hydrolysis of phenyl acetate at very high acid concentrations. Their results were complicated by sulphonation of the substrate at >12M acid, but their conclusions are consistent with new results in perchloric acid, where side-reactions are not a problem and a clean change of mechanism is indicated by various kinetic parameters.[18] The entropy of activation, for example, rises from −18.5 to +7.9 e.u. as the $HClO_4$ concentration is increased from 8 to 11.5M, as expected for a change from a bimolecular to a unimolecular mechanism. The acylium ion mechanism appears to be particularly favourable in perchloric acid, like other unimolecular processes in the presence of perchlorate ions.

The acid-catalysed hydrolysis of ethyl formate in aqueous dimethyl sulphoxide has been followed,[19] and a proton-inventory study of the hydrolysis of ethyl acetate catalysed by an acidic ion-exchange resin[20] (a reaction with activation parameters not significantly different from homogeneous acid-catalysed hydrolysis[21]) gave results consistent with, though not requiring, the usual mechanism with three vibrationally active protons in the transition state.

Activation parameters have been measured for the acid-catalysed hydrolysis of several ethyl α-hydroxy-esters,[22] and the influence of substituents,[23] including 13 amido substituents,[24] on the hydrolysis of γ-lactones (including two thia analogues) has been analysed in terms of polar and steric contributions.

In another re-examination of the Taft equation DeTar concludes,[25] with Charton,[26] that σ^* values do not represent polar effects at all, but the incomplete cancellation of steric effects. With the demise of the polarity term the Taft equation is simplified to

$$\log k = a + \sigma_s E_s$$

and a new set of E_s values is developed. (Previous modified steric substituent constants, such as Hancock's E^c_s and Charton's more radical use of van der Waals' radii, v, are considered to be "without special merit", and not worth retaining.) The simplified equation is applicable to all aspects of acyl-transfer reactions, including substitution in the nucleophile and leaving groups as well as at the acyl centre.

The most recent work on *ortho*-substituent constants, E^o_s, however, challenges Charton's contention[26] that only minor steric contributions are involved, and it is concluded that Taft's original opinion, that the *ortho*-effect involves an important steric contribution, is correct.[27,28]

The danger that the number of statistical analyses of this sort will soon exceed the number of data sets available is more apparent than real: while the same 50-year-old sets of data appear regularly in papers on the theory of linear free energy relationships, new studies of substituent effects are still very much in evidence. Yukawa and Tsuno and their co-workers, for example, have measured the rates of hydrolysis of 2-, 3- and 4-pyridyl and 2-, 3-, 4-, 5-, 6-, 7- and 8-quinolinyl benzoates[29] to establish substituent constants for all positions of the heteroaromatic rings. Linear free energy relationships have been established between the saponification rates of ethyl 3-substituted-phenylpropionates and the pK_as of the corresponding acids in several solvents;[30] and a single multi-parameter regression equation describes the effects of varying structure, solvent, and temperature on the rates of alkaline hydrolysis of aryl benzoates in aqueous dioxan mixtures.[31] The effects of acyl and leaving-group variation are additive,[32] and solvent effects on the saponification of aryl acetates in aqueous dioxan can be reversed by an appropriate choice of substituents.[33] Substituent effects have also been measured on the base-catalysed hydrolysis of benzhydryl benzoates,[34] as have solvent[35] and salt[36] effects on the saponification of half-esters of several dicarboxylic acids. The increased reactivity of methoxymethyl (MOM) and β-methoxyethoxymethyl (MEM) esters is due to the inductive effects in the acetal leaving group as well as possible metal cation co-ordination, as shown by their six-fold greater reactivity towards hydrolysis by tetra-*n*-butylammonium hydroxide in aqueous tetrahydrofuran, compared with the corresponding methyl and *n*-propyl esters.[37]

The hydrolysis of substituted-phenyl 4-nitrophenylacetates (**20**) in 80% dimethylsulphoxide–water involves the *E*1*cb* mechanism, as shown by the small solvent deuterium isotope effect and large Hammett $\rho(\sigma^-)$ of 4.4 for a range of leaving groups with electron-withdrawing substituents.[38] Conversion to the anion (**21**) is almost complete under the conditions [1.7×10^{-3}M $Ba(OH)_2$] and several compounds show evidence of a common intermediate with λ_{max}495 nm, which disappears with a rate constant of $0.55\,s^{-1}$, tentatively identified as *p*-nitrophenyl ketene (**22**). With a very good leaving group (Ar = *p*-nitrophenyl) the anion (**21**) does not accumulate. For leaving groups with electron-donating substituents the

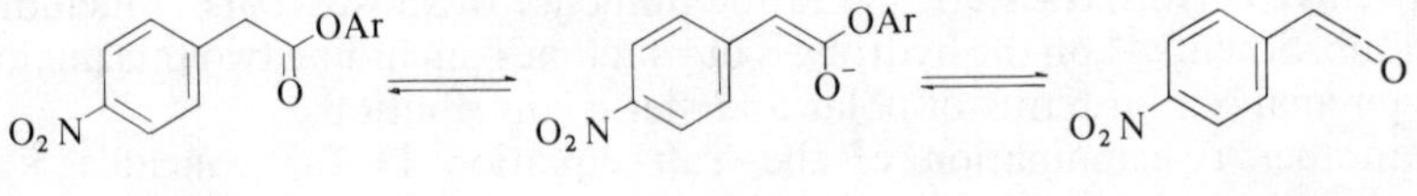

(20) (21) (22)

data for the disappearance of (**21**) are correlated by a smaller Hammett ρ (1.6), indicating a mechanism change; and a radical decomposition is likely.[38]

The question of relative leaving-group capabilities in *E*1*cB* reactions has been probed using isobasic plots for the breakdown of the anions derived from acetoacetate and fluorene-9-carboxylate esters.[38a] For a given pK_a it turns out that an oxygen leaving group (RO^-) departs 5000 and 79 times, respectively, more rapidly than RS^- from the two systems.

The alkaline hydrolysis of several *p*-nitrophenyl esters is accelerated by about 15% by a particular intensity of ultrasound.[39]

Solvent deuterium isotope effects on the acid- and base-catalysed hydrolysis of *n*-propyl and isopropyl acetates have been compared at several temperatures.[40]

A proton inventory study of the spontaneous (water-catalysed) hydrolysis of *p*-nitrophenyl dichloroacetate in water and in 10% mole fraction *t*-butanol–water has given results consistent with multiple involvement of water in the transition state.[41] An involvement of two to four molecules is consistent with the data, and cyclic transition states (*e.g.* **23**) have been suggested. The mechanism of hydrolysis of the mixed acetal–lactone (**24**) changes with pH from *A*1 at low pH to $B_{AC}2$ in base, with

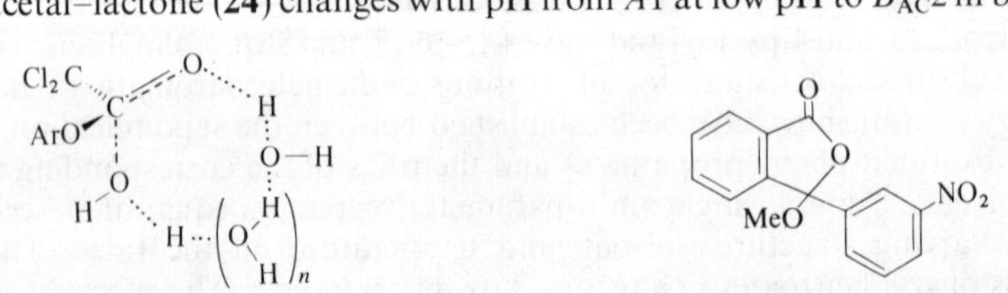

(23) (24)

a well-characterized ($\Delta S^{\neq} = -39.0$ *e.u.*, $k_{H_2O}/k_{D_2O} = 2.65$) spontaneous reaction near pH 5.[42]

The acid-catalysed esterification of substituted phenoxyacetic acids in methanol is faster than that of either the corresponding benzoic[43] or substituted phenylthio-acids,[44] and the rates are correlated by the Hammett equation ($\rho = 0.47$). The esterification of cinnamic acid in methanol[45] and of pentaerythritol by acetic acid[46] have also been studied, as have titanium(IV) fluoride[47] and alkoxide[48] salts as catalysts for the transesterification of dimethyl terephthalate with diols.

A large set of data for the esterification of 32 *ortho*-substituted benzoic acids with diphenyldiazomethane in 11 alcohol solvents have been collected and subjected to two forms of up-to-date correlation analysis.[49] The *ortho*-substituent has little effect on the way the solvating properties of the alcohol affect reactivity, and only *o*-OH shows a large H-bonding effect; though H-bond acceptors show important effects on the reaction in aprotic solvents.[50] The buttressing effect is shown clearly by an exaltation of reactivity of 2,3-dimethylbenzoic acid[51] (the effects of the two methyl groups of the 2,5-dimethyl isomer are additive), and the 2,6-dimethyl compound also shows a large positive deviation. The same reaction with diphenyldiazomethane has also been studied for naphthalene and biphenylcarboxylic acids in alcohol solvents,[52] for substituted thiophene-2-carboxylic acids in methanol,[53] and for several imidazole, pyrrole and pyrazole mono- and di-carboxylic acids in ethanol and dimethyl formamide.[54]

Alkyl-oxygen cleavage is also involved in the hydrolysis of substituted carbazole derivatives (**25**)[55] and in the formation of esters (**26**) of trifluoroacetic acid from the alcohol[56] (large negative ρ in both cases). Benzyl trifluoroacetates are formed more slowly, and the mechanism appears to change from reverse $A_{AL}1$ (electron-donating substituents) to reverse $A_{AC}2$ (electron-withdrawing substituents), as evidenced by a biphasic Hammett plot.[56]

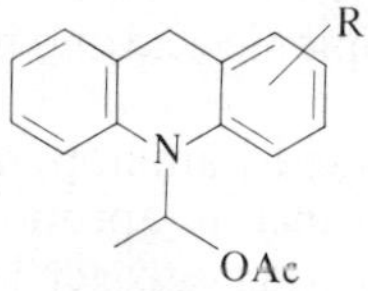

(25)

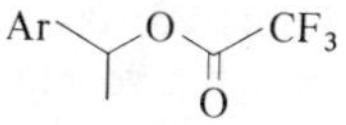

(26)

Two detailed investigations of the transition state for acyl transfer have been reported, using alternatively the classical method of substituent effects on rates and equilibria[57] and the secondary deuterium isotope effect technique[58] currently being developed by Schowen and his group.[59]

For the rates of transfer of a series of acyl groups from *p*-nitrophenoxide to hydroxide (*i.e.* the rates of alkaline hydrolysis of a set of *p*-nitrophenyl esters) ρ^* is 2.9. For transfer to the thiolate anion of *N*-acetyl-L-cysteine ρ^* is slightly larger (3.4), and for a series of phosphonate dianions $\beta = 0.3$ and $\rho^* = 2.4$ (for $ClCH_2PO_3^{2-}$).[57] These results have been interpreted in terms of an earlier transition state for the phosphonate reactions, and transition states close to the tetrahedral intermediate for the thiolate and hydroxide reactions. Whatever σ^* does measure (see above, page 5) it is a valid transition-state probe if used in a strictly comparative sense, and some care was taken to use acyl groups with a wide range of polarity. But the conclusion that the transition state for the hydroxide reaction is close to the tetrahedral intermediate (ρ^*

is close to ρ^* for the equilibrium addition of the nucleophile to RCHO) is not easy to accept, in view of isotope effect measurements,[58,60] and the low sensitivity of the rate to the basicity of the nucleophile for very basic oxy-anions. Steric effects are much smaller for the addition of nucleophiles to aldehydes compared with carboxylic acid derivatives (see above, page 4, for an apposite comparison) and the coincidence of ρ^* values may reflect the mongrel character of σ^* values.

The secondary deuterium isotope effects for the reactions of phenyl, p-nitrophenyl, and 2,4-dinitrophenyl acetates and CD_3-acetates with oxygen nucleophiles (acetate, phenoxide, and hydroxide) fall in the range (k_{CH_3}/k_{CD_3}) 0.950–0.980, consistent with early transition states in all cases, including reactions with weakly basic nucleophiles.[58,60] (Where comparable the results are consistent with those of Cordes and his group[61] based on α-deuterium isotope effects on the reactions of formate esters.)

For the reaction of a series of substituted phenolate anions with *p*-nitrophenyl acetate in ethanol $\beta_{nuc} = 0.57$ has been measured;[62] this is similar to that for attack by ArS^-, though the authors doubt that formation of the tetrahedral intermediate is rate-determining in the latter case.

The full paper[63] reporting Cox and Jencks' work on the methoxyaminolysis of phenyl acetate[64] has appeared. Bifunctional acid–base catalysts (cacodylate, bicarbonate, phosphonates, *etc.*) are up to 1000 times more effective than monofunctional general acids and bases, and "true" bifunctional catalysis (**27**), with no detectable intermediate stage between the two proton-transfer steps, is indicated.[65] The proton transfer is apparently so efficient that the attack of the amine on the ester is rate-determining. General acid–base catalysis is enforced by the instability of the zwitterionic tetrahedral intermediate $T^{\pm}$ (**27**) and the rate-determining step is thought to be the attack of the amine on the hydrogen-bonded complex of ester and general acid. A pre-association mechanism has also been suggested for general base catalysis of this reaction.

The formation of the hydroxamic acid from propionic acid and hydroxylamine at 90.5° is first order in each reactant.[66]

The reaction of *p*-nitrophenyl acetate with 2-, 3- and 4-aminopyridines in aqueous media involves hydrolysis (though amides are formed in aprotic solvents).[67] The pyridine acts as a nucleophilic catalyst through the ring nitrogen in almost all cases (even with 2-methyl substituents). Only in the case of (**28**) does the mechanism change to general base catalysis.

Rate constants for the reaction of acetic anhydride with substituted pyridines in water give a curved (concave upward) Brønsted plot, explained in terms of a change in rate-determining step from formation to breakdown of the tetrahedral intermediate as the basicity of the pyridine is decreased.[68] From the centre of curvature acetate appears to be equivalent in leaving-group capability to a pyridine of pK_a 6.1 (and is thus a poorer leaving group than a pyridine with the same pK_a).

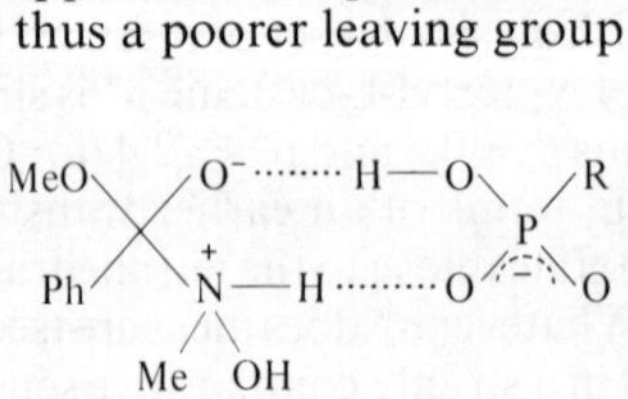

MeO, Ph, O^- ······· H—O, R, P, $\overset{+}{N}$—H ········ O, O, Me, OH

$T^{\pm}$

(**27**)

N

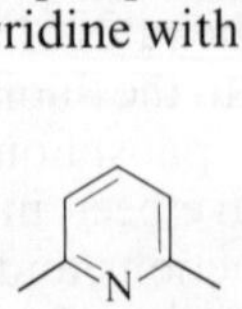

(**28**)

The rates of solvolysis of cinnamic anhydride catalysed by *N*-methylimidazole, pyridine, and 4-dimethylaminopyridine decrease with increasing proportions of acetonitrile in the mixed aqueous or alcoholic solvent.[69] Activation parameters have been measured for the reactions of *n*-propanol[70] and *n*-butanol[71] with maleic anhydride. Formation of the di-ester by reaction of alkyl alcohol with 3-methyl tetrahydrophthalic anhydride goes in two stages, of which only the second is catalysed by *p*-toluenesulphonic acid.[72]

Calculations on the reduction of unsymmetrically substituted succinic, maleic, and phthalic anhydrides show that the two carbonyl groups have differing reactivities, and that the size of the LUMO coefficient at carbonyl carbon is a reliable guide to the regioselectivity of the reaction.[73] This factor, and the possibility of cation-complexing in some systems, has to be considered in addition to steric approach control explanations of observed regioselectivity in these systems.

The rate of hydrolysis of ketene is pH-independent in the range 4.4–9.85, and the kinetic parameters ($k_{H_2O}/k_{D_2O} = 1.9$, $\Delta S^{\neq} = -16$ e.u.) are consistent with nucleophilic attack by water as the rate-determining step.[74] This is in contrast to the hydrolysis of di-*tert*-butylketene,[75] which is acid-catalysed, and shows a much slower spontaneous reaction. This difference, of some 10^6, in the rates of nucleophilic attack on the two compounds, is presumed to be due to a large steric effect of the two *t*-butyl groups.

Dipole moment measurements suggest that diaryl carbonates exist in solution in the Z,Z-conformation.[76]

The alkaline hydrolysis of a series of *S*-alkyl and *S*-aryl thiolacetates, including *S*-acetylcoenzyme A, have been studied as part of a wider programme of research into the *E*1*cb* reaction.[77] The low sensitivity to the basicity of the leaving group ($\beta_{LG} = -0.33$), the decreased reactivity relative to oxygen esters, and the lack of solvent deuterium incorporation into product acetate are all consistent with a normal, $B_{AC}2$ mechanism for the reaction.

The hydrolyses of hydroxythiolesters (**29**) and (**30**) are catalysed by boric acid (up to 80-fold by 0.04M borate at pH 9).[78] It is likely that a mechanism such as (**31**), involving an intermediate borate ester, is involved.

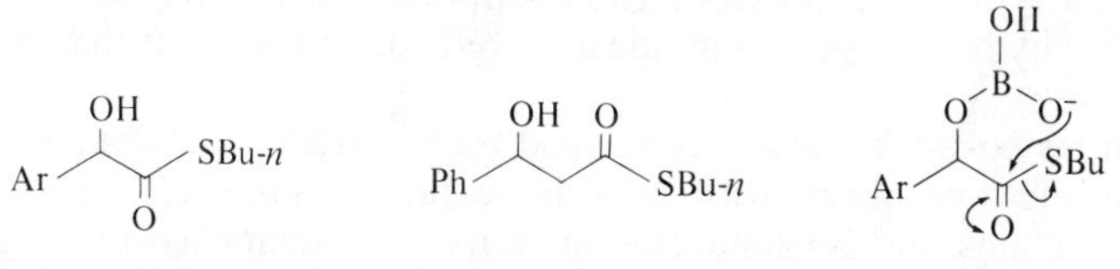

OH, Ar, SBu-*n*, O (29) OH O, Ph, SBu-*n* (30) OH, B, O, O⁻, Ar, SBu, O (31)

The protonation behaviour and hydrolysis of thioacetamide and thiobenzamide in sulphuric acid,[79] and the kinetics and mechanism of the hydrolysis of thioacetamide in the presence of various acids[80] have been studied.

An *ab initio* study of diacetamide gives the E,Z form (**32**) as the lowest energy conformation, largely because of the repulsion between the *syn*-oxygens in the Z,Z-conformer.[81] An intermolecular perturbation study of the acid-catalysed hydrolysis of diformamide has also been reported.[82]

The effects of substituents on nitrogen on the rate of rotation about the C—N bond of tertiary formamides and acetamides are predominantly steric.[83] The base-catalysed exchange of the H_E proton is faster than that of the H_Z proton for primary amides (**33**), as expected on several grounds. For secondary amides the H_Z proton (of the E conformer) exchanges faster, perhaps because of steric hindrance to solvation of the Z-imidate (**34**).[84] For primary amides in acid the intramolecular exchange occurs, in many cases, at the same rate as intermolecular exchange; this is evidence

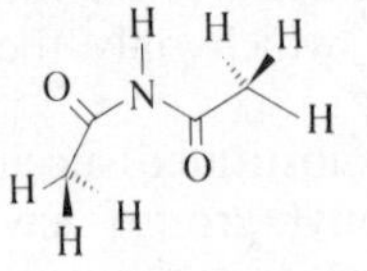

(**32**) 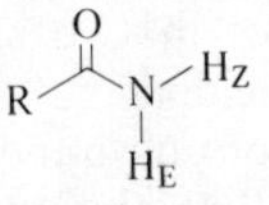 (**33**) 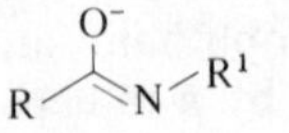 (**34**)

for a mechanism involving *N*-protonation,[85] but the three protons of the NH_3^+ group clearly do not have time to equilibrate during the life-time of the cation (**35**) because H_E is still lost more rapidly than H_Z. The *O*-protonation mechanism is still the best explanation of the exchange results with amides, like trichloracetamide, bearing strongly electron-withdrawing substituents.

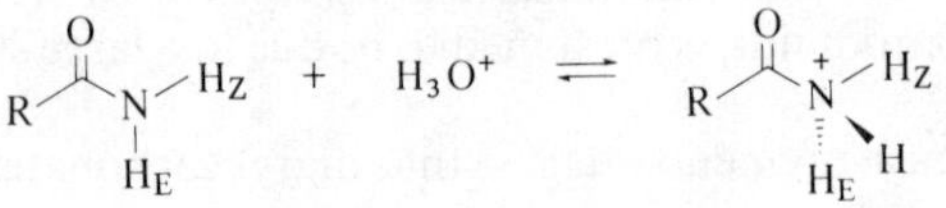

$$RCONH_EH_Z + H_3O^+ \rightleftharpoons RCON^+H_EH_ZH$$

(**35**)

The hydrolysis of formamide has been studied at 80° in the range pH 1–9:[86] the pH–rate profile shows a minimum near pH 6, and is dominated by the H^+- and OH^--catalysed reactions; but a spontaneous reaction appears to contribute up to 50% of the observed rate of hydrolysis near the rate minimum, and weak general acid catalysis is observed. General acid catalysis of the hydrolysis of simple amides has been observed only once previously (in the hydrolysis of *N*-*n*-butylacetamide at 220°), but the data for formamide hydrolysis are convincing, and the observation of a probable water reaction provides good supporting evidence.

The alkaline hydrolysis of formamide in mixed aqueous solvents has been studied independently.[87]

Yates and his co-workers have continued their systematic examination of acid-catalysed amide hydrolysis with two important studies of the reactions of benzamides, lactams, and benzimidatonium ions, using their *r*-hydration parameter, transition-state activity coefficient, and excess acidity methods.[88,89] All except special cases are hydrolysed by the $A_O{}^T2$ mechanism (**36**) (bimolecular, acid-catalysed, reaction of the *O*-protonated substrate *via* a tetrahedral intermediate), and rates show a strong dependence on water activity. The reactivity of lactams shows a strong dependence on ring size, with the δ-lactam particularly basic, and reactive. The β-lactam is hydrolysed by a different mechanism, involving rate-determining acylium ion formation from the unfavourable *N*-protonated form (**37**).[88] The *r*-hydration parameter is clearly negative for this reaction, consistent with rate-determining C—N cleavage. Apparently attack on the acylium ion by the β-amino group is slower than external attack by water, presumably because of the strain involved in four-membered ring-formation. The other atypical substrate found was (**38**), hydrolysed by an S_N2 reaction of water on the *O*-methyl group.[89]

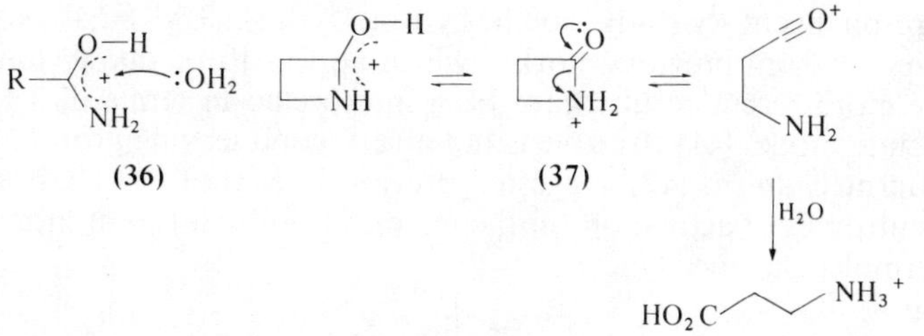

(36) (37)

Both the *r*-hydration parameter and the excess acidity treatments suggest that the details of the $A_O{}^T2$ mechanism change with increasing acidity. At low acidities the transition state appears to contain three molecules of water, but in concentrated solutions a single water molecule is involved as the nucleophile. (In the most concentrated solutions a reaction with bisulphate is also observed.) A comprehensive scheme for amide hydrolysis has been proposed,[89] part of which is shown in Scheme 1. The $3\,H_2O$ mechanism (**39**) appears to be a short cut to the cationic intermediate (**40**), which is an essential intermediate for hydrolysis but can be reached only by the step-wise $1\,H_2O$ pathway under conditions of low water activity. Accurate acidity constants for the substrates used in these investigations were obtained by the excess acidity method,[90,91] which has been extended to aqueous HCl and HBr.[92]

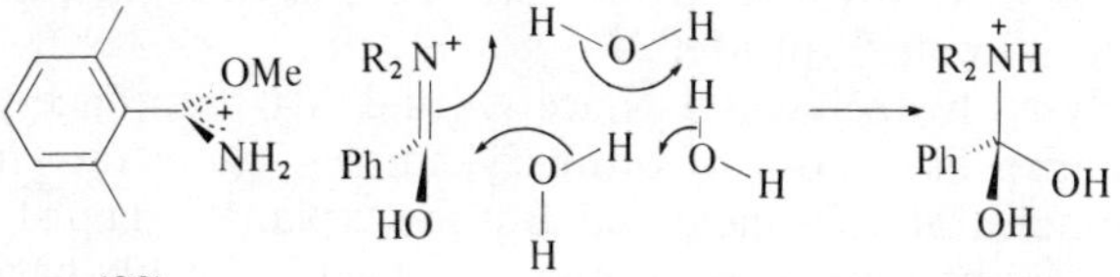

(38) (39) (40)

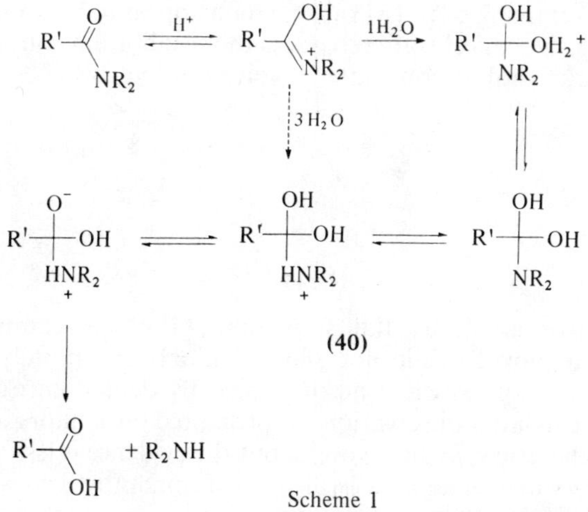

(40)

Scheme 1

The hydrolysis of methacrylamide has been studied in sulphuric acid,[93] and the acid-catalysed solvolyses of acetamide and propionamide in aqueous propanol[94] and dioxan.[95]

A new proton inventory study[96] of the hydrolysis of 1-acetyl-1,2,4-triazole reveals some differences from previous work[97] which appear to be due to ionic strength effects. The more recent results have been interpreted in terms of a four-proton transition-state model (**41**) (triazole is not a very good leaving group), though the three-proton mechanism (**42**) suggested previously cannot be ruled out, and the measured entropies of activation (in the region of −40 e.u.) seem more consistent with the termolecular mechanism.

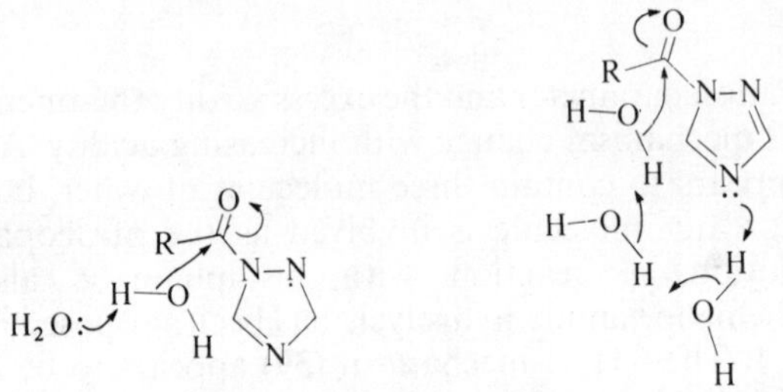

(**42**) (**41**)

This mechanism (**42**) is also thought to account for the hydrolysis of the 1-acetylimidazolium cation in water,[98] and results of a proton inventory study in water–acetonitrile mixtures[99] indicate that there is no qualitative change in mechanism up to 90 mole % of MeCN.

The acid-catalysed hydrolyses of *N*-acetyl- and *N*-trifluoroacetyl-pyrrole (and related compounds), like those of ordinary amides, show rate maxima as the sulphuric acid concentration is increased. But the explanation must be different for the acyl heterocycles because these compounds are too weakly basic to be largely protonated under the conditions.[100] Some at least of the substrates are hydrated under normal conditions,[101] and the acid-catalysed breakdown of the hydrate (*e.g.* **43**) is the rate-determining step. The rate maxima appear to be a result of electrolyte effects: increasing salt concentration reduces the equilibrium concentration of (**43**), which is the "tetrahedral intermediate" involved in hydrolysis.

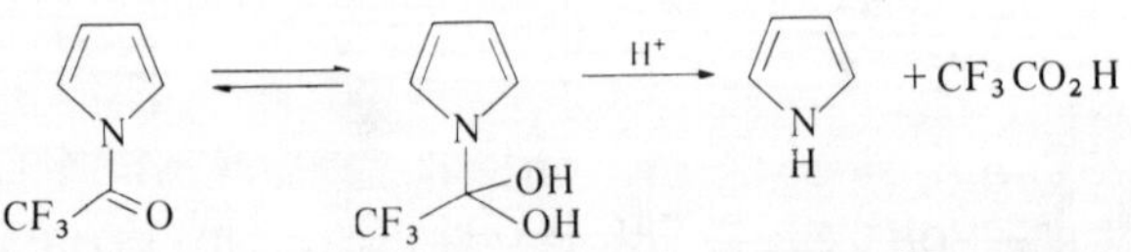

$$\xrightarrow{H^+} \quad + CF_3CO_2H$$

(**43**)

Another reactive amide is the sulphone (**44**). This compound, and the corresponding sulphoxide (but not the sulphide), are rapidly deacylated by methoxide in a reaction which is accompanied by deuterium exchange into the acetyl group.[102] This latter observation has prompted the authors to suggest that an *E*1*cb* type of mechanism may be involved, but the evidence is less than compelling.

The Hammett ρ value for alkaline hydrolysis of cinnamoyl azides is similar to that for the same reaction of the *p*-nitrophenyl esters.[103]

Rates of solvolysis of benzoyl chloride in aqueous alcohols are slower at lower water concentrations.[104] That of (**45**) is some one hundred times slower than that of normal phthaloyl chloride.[105] The acylation of butanol by butanoyl halides is

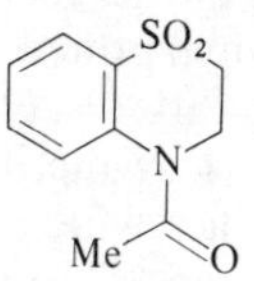

(44) 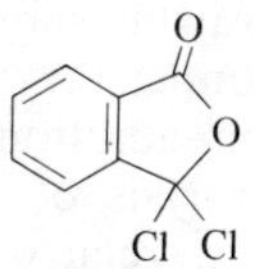 (45)

catalysed by diphenylphosphoric acid,[106] and the methanolyses of anisoyl chloride[107] and MeSCSCl[108] are S_N1-like. Both methanolysis[108] and hydrolysis[109] reactions of carbonic acid derivatives show more unimolecular character as oxygen is replaced by sulphur, though the trend may be reversed in solvent mixtures rich in acetonitrile and acetone.[109]

Acetonitrile is selectively hydrolysed to acetamide in the presence of a palladium complex,[110] and hydracrylonitrile is similarly hydrolysed to the amide at sulphuric acid concentrations below 6M.[111]

The addition of methanol to aromatic nitrile oxides to give methyl benzohydroximates (**46**) involves both acid-catalysed and spontaneous reactions, and both are characterized by negative Hammett ρ-values (-1.6 and -0.3, respectively).[112] Presumably the transition state is dominated by protonation (little C—O bond formation), and the small negative ρ for the spontaneous reaction indicates that proton transfer from the solvent is involved in the mechanism.

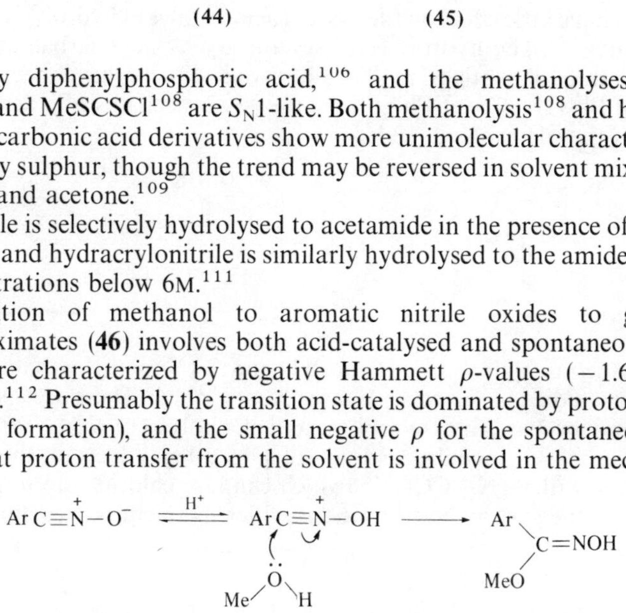

$$\mathrm{Ar\,C{\equiv}\overset{+}{N}{-}O^-} \xrightleftharpoons{H^+} \mathrm{Ar\,C{\equiv}\overset{+}{N}{-}OH} \longrightarrow \mathrm{Ar(MeO)C{=}NOH}$$

(46)

Mechanism studies of the reactions of *o*-phenylenediamine with aryl cyanates in aqueous dioxan to give benzimidazoles (**47**),[113] and of ethylenediamine with ureas to form imidazolidinones (*e.g.* **48**)[114] have been reported. The second step, and perhaps the first also, in this latter reaction is thought to involve an isocyanate intermediate. Ureas are also produced on heating salts (**49**), though in aqueous solution they lose COS to regenerate the amine,[115] and the hydrolysis of several substituted-urea herbicides has been studied.[116]

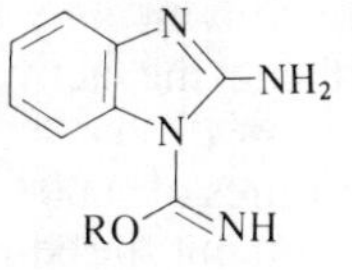

(47) 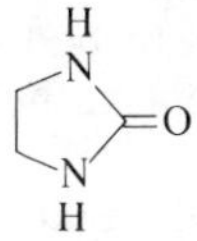 (48) 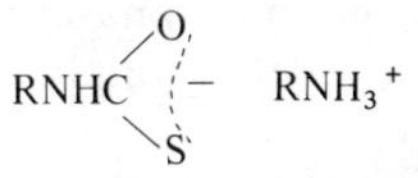 (49)

Although amides are protonated essentially exclusively on oxygen, the difference in basicity between O and N centres is much less in carbamates. In fact NMR evidence shows that compound (**50**), though protonated on oxygen at low acidities, gives the *N*-protonated cation in FSO_3H.[117] The Hammett plot derived from data for the alkaline hydrolysis of methyl carbanilates, $ArNHCO_2Me$, gives $\rho = 1.06$ for a series of electron-withdrawing substituents, but levels out to a much lower slope for electron-donating groups; this can be interpreted in terms of a change of mechanism to elimination–addition for the latter set.[118] Measurements of the kinetics of alkaline hydrolysis of 1-naphthyl *N*-methyl carbamate,[119] and of carbonates and carbamates related to (**51**), for which elimination to form (**52**) competes with hydrolysis, have also been reported.[120] Amine and pyridine groups in the side-chain have little effect on the rates of acid-catalysed hydrolysis of several *N*-aryl carbamates.[121] The hydrolyses of several aryl *N*-acyl carbamates and their mono and dithio derivatives in 20% aqueous dioxan have been studied.[122] Hydrolysis of PhNH—CS—SPh goes by the $A_{AC}2$ mechanism in acid, but by the *E*1*cb* route in the pH range 2–12.[123]

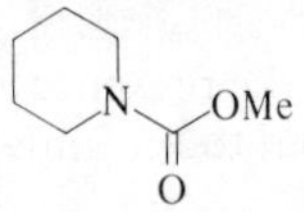

N OMe O OCONHR SO_2 SO_2

(**50**)  (**51**) (**52**)

The spontaneous hydrolysis of 3-acetyl- and 3-(*N*-methylcarbamoyl)-triazene (**53**) have been studied in mixed aqueous solvents.[124,125] There are also reports on the kinetics and mechanism of hydrolysis and alcoholysis of *N*,*N*,*N'*-tri-substituted chloroamidines,[126,127] including a study of substituent effects on the reactions of primary amines with ArN=CCl_2[128] (which can lose chloride ion to give rise to a nitrilium ion of the sort involved in the Bischler–Napieralski reaction).[129]

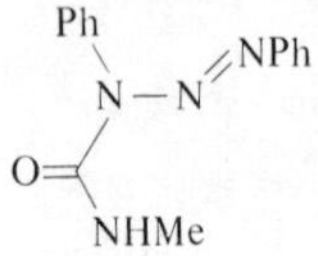

Ph NPh N—N O= NHMe

(**53**)

Reactions in Aprotic Solvents

A study of some positive ion-molecule reactions of simple acyl compounds by ion cyclotron resonance techniques suggests that the reactions of protonated carboxylic acids and thio-acids with nucleophiles (including ethers) go by acylium ion mechanisms, rather than by way of tetrahedral intermediates.[130] The appropriate link with condensed-phase chemistry is to be found more convincingly in related processes in very strongly acidic media rather than at enzyme active sites as suggested.

The main gas-phase reaction of methyl acetate at temperatures of 1000–1400 K, as revealed by ^{18}O-labelling experiments, is not ketene formation but the equilibration of the methyl group between the two oxygens.[131] The thermolyses of methylene phthalide (**54**) and the isomeric methylene coumaranone (**56**) under similar conditions give the rearrangement products (**55**) and (**57**), respectively.[132] An ester group (X = CO_2Et)[133] has a much larger effect on the rate of pyrolysis of ethyl acetate (**58**) than does an isopropenyl group[134] in the same position, and chlorine atoms in the acyl group have an even smaller activating influence;[135] further work on the steric effects of substituents in the alkoxy group has appeared.[136,137]

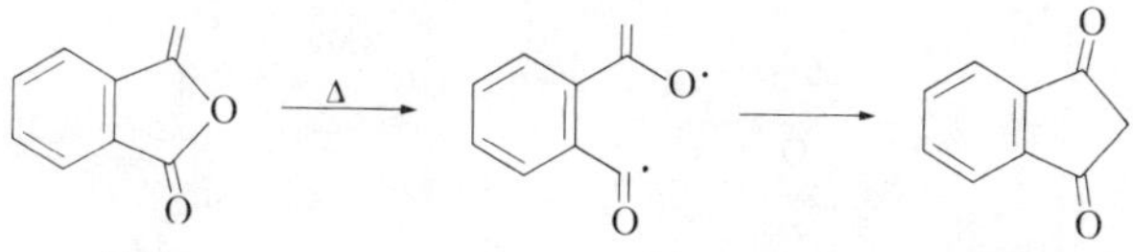

(54) (55)

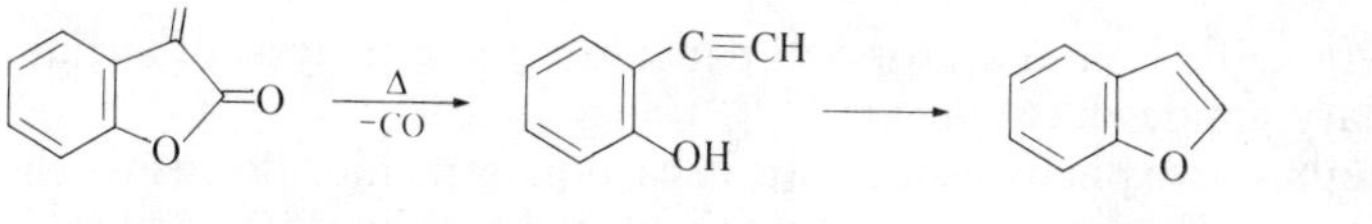

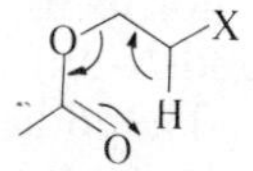

(56) (57) (58)

The thermolysis of some dialkyl dicarboxylates in quinoline gives some alkane as well as alkene products, and a hydride-transfer mechanism has been suggested.[138]

The morpholinolysis of aspirin in acetonitrile exhibits saturation kinetics,[139] much as previously observed for the reaction with *n*-butylamine,[140] and the effect of neighbouring quinoline nitrogen on the reaction of piperidine with aryl acetates has been studied in acetonitrile and chlorobenzene.[141] Proton transfers are always likely to be involved in reactions of this sort in aprotic solvents,[142] and catalysis by "naked" carboxylate of the *n*-butylaminolysis of *p*-nitrophenyl acetate has been observed in chlorobenzene.[143] In the absence of the amine simple exchange equilibria are set up, which in the case of *p*-nitrophenyl *o*-toluate and naked acetate in dipolar aprotic media give rise to a good yield of mixed anhydride.[144] The reaction is inhibited by the addition of protic solvents, and strongly accelerated by neighbouring quaternary ammonium centres: the reaction of (**59**) for example, being over 3000 times faster than that of the parent toluate,[144] presumably as a result of electrostatic catalysis. Both effects, rate enhancement by desolvation of the nucleophile and by electrostatic stabilization of transition states, are probable factors in enzyme catalysis.

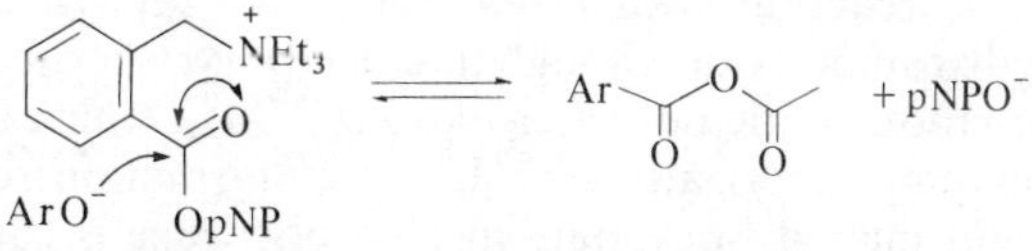

(59)

The benzylaminolysis of a series of substituted 4-benzylidene-2-phenyl oxazolinones (azlactones) (**60**) in acetonitrile has been studied.[145] *ortho*-Substituents in the benzylidene group cause some steric acceleration, but the corresponding substitution in Ar′ inhibits the reaction.

Bicarbonate is a highly efficient catalyst for the cleavage of *p*-nitro-*N*-methyltrifluoracetanilide in acetonitrile and dimethylformamide in the presence of crown ether. The naked bicarbonate anion is thought to act simultaneously as nucleophile and general acid, perhaps as shown in (**61**); the high reactivity compared with monobasic anions disappears in protic solvents, nor is it observed for the cleavage of *p*-nitrophenyl acetate.[146]

Rate and equilibrium constants have been measured for the transamidation of several aliphatic amides with aniline, diethylamine, and di-*n*-butylamine in

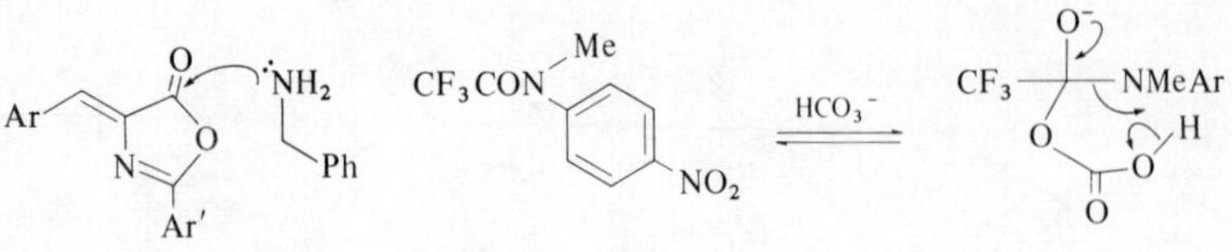

(60) **(61)**

chlorobenzene at 190–250°,[147] and a group of papers on the nitrolysis of tertiary and *N*-nitro-secondary amides has appeared.[148–152]

The addition of methanol to phenyl isocyanate in carbon tetrachloride, to give the methyl carbamate, is general-base-catalysed by substituted pyridines ($\beta = 0.49$).[153] The reaction with 2-pyridylmethanol (**62**) is over 10 times faster than that with the 4-isomer, consistent with intramolecular catalysis in this case. [The uncatalysed reaction with methanol appears to involve only one molecule of nucleophile, though the hydration of *p*-chlorophenyl isocyanate in ether depends on the third power of the water concentration.[154] A cyclic transition state (**63**) containing three molecules of water has been suggested for this and for the distinct second stage in which CO_2 is lost from the carbamic acid intermediate to give aniline. Consistent with this interpretation, both steps involve very low enthalpies and very large negative entropies of activation.[154]]

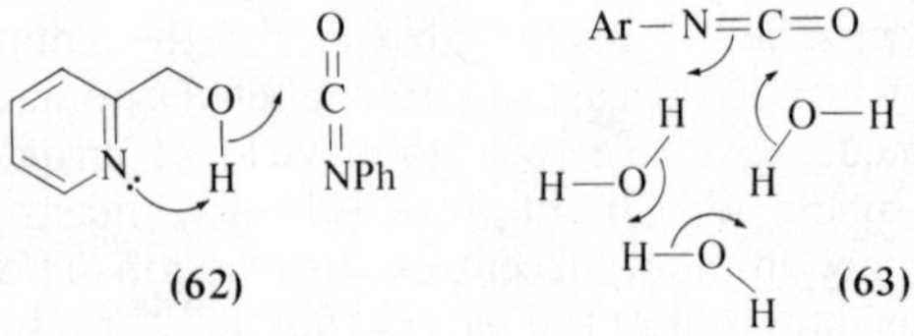

(62) **(63)**

A similar reaction may be involved between phenyl isocyanate and alcohols catalysed by alkanolamines in benzene, which also depends on the basicity of the basic group.[155] The reaction with *n*-butanol[155] is "catalysed" by dimethylformamide and dimethyl sulphoxide (though that of *n*-butyl isocyanate is inhibited),[156] and by various organometallic derivates like $Zr(OBu)_4$, which selectively enhance butanolysis (rather than trimerization).[157] Urethane formation from poly- and oligo-ethylene glycols and phenyl isocyanate in chlorobenzene is catalysed by sodium acetate (possibly with assistance from complexation of Na^+ by the polyether).[158]

The reaction of aniline with phenyl isocyanate, to give the urea, shows uncatalysed, aniline-catalysed and urea-catalysed terms;[159] reaction of butylamine with phosgene, to give *n*-butyl isocyanate, has been studied in the gas phase.[160]

Last year's report[161] that an *O*-acyl isourea is formed when *N*-carbobenzyloxyvaline reacts with *N,N'*-di-isopropylcarbodi-imide in $CDCl_3$ was wrong.[162] The autocatalytic reaction of aniline with phthalic anhydride has been studied in a number of aprotic solvents;[163] so too has the hydrolysis of benzoyl chloride.[164] The methanolysis of α-naphthoyl chloride in acetonitrile is faster than that of the β-isomer.[165] Benzoylation of benzyl alcohol in benzene[166] and other solvents[167] is subject to powerful nucleophilic catalysis by pyridines, and the butanolysis of butanoyl chloride in toluene is catalysed by various acids, including bifunctional diphenyl phosphoric acid, by several pathways.[168]

The Hammett ρ value for reaction of substituted benzoyl chlorides with azide ion, to give the benzoyl azides, is nearly five times greater than that for the acylation of *p*-toluidine under the same conditions (2.53 compared with 0.55 at 20°).[169]

Acylations (by benzoyl chlorides) of *N*-methylanilines,[170] 2,4-dinitroaniline,[171] and *N,N*-dimethylacetamide[172] (to give **64**) have been studied. The formation of benzoyl chloride from benzoic acid and phosgene, catalysed by caprolactam, appears to involve a similar intermediate (**65**).[173]

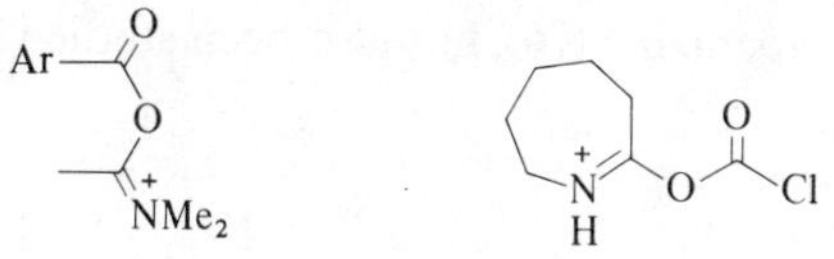

(**64**) (**65**)

Intramolecular Catalysis and Neighbouring-group Participation

Statistical and thermodynamic approaches to the estimation of effective molarities (*EM*) in intramolecular reactions have been discussed in a review.[174, 175] Rather low *EM*s are found for intramolecular nucleophilic catalysis by the pyridine and amine nitrogens, respectively, of the hydrolysis of carbonate esters (**66**) and (**67**).[176] Catalysis is much more efficient when five-membered cyclic transition states are involved, and apparently more efficient also for weakly basic amino groups, as in (**68**). These results for carbonate esters are in contrast to those for carboxylates, suggesting possible transition-state differences between the two systems.

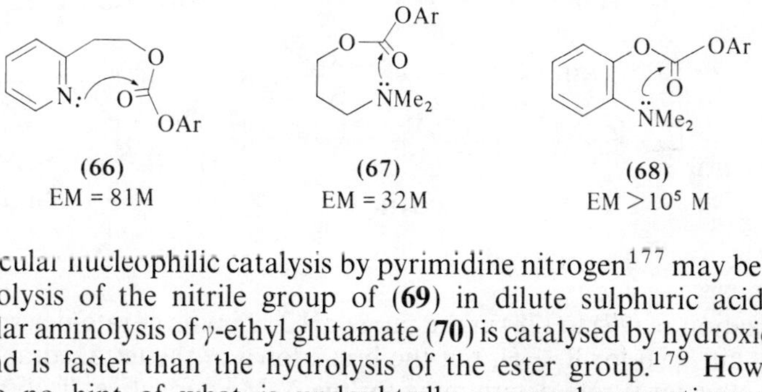

(**66**) EM = 81M (**67**) EM = 32M (**68**) EM $> 10^5$ M

Intramolecular nucleophilic catalysis by pyrimidine nitrogen[177] may be involved in the hydrolysis of the nitrile group of (**69**) in dilute sulphuric acid.[178] The intramolecular aminolysis of γ-ethyl glutamate (**70**) is catalysed by hydroxide ion, as expected, and is faster than the hydrolysis of the ester group.[179] However, the kinetics give no hint of what is undoubtedly a complex reaction; the rate-determining step does not change with pH and clear evidence of buffer catalysis could not be obtained. The intramolecular O → N acetyl-transfer reaction of *O*-acetylserine (**70a**), on the other hand, makes up for this deficiency: at pH > 7.6 the rate-determining steps are a pH-dependent proton switch ($T^0 \rightleftharpoons T^{\pm}$) and the

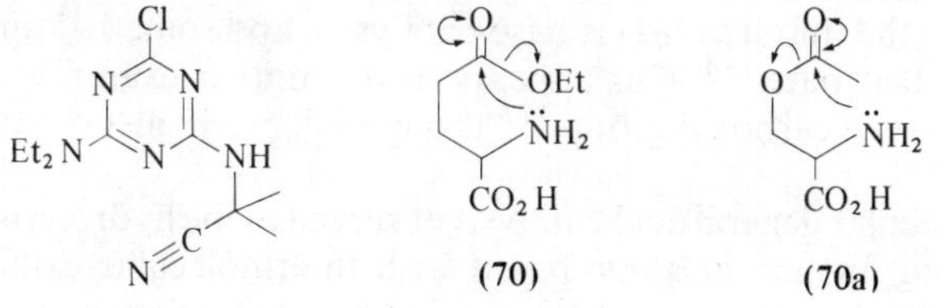

(**69**) (**70**) (**70a**)

reaction of hydroxide with $T^{\pm}$; below pH 7.6 the breakdown of T^{-} is rate-determining.[179]

Intramolecular O → N acyl transfers have also been observed in systems (**71**–**74**) in acetonitrile and dimethyl sulphoxide,[180] as part of the development of the "amine capture" strategy for peptide synthesis. Even though these reactions go by way of 9- and 12-membered ring intermediates, acyl transfer can be very rapid [half-lives as little as a few seconds for (**71**), and a few minutes for the others].

Intramolecular aminolysis of the β-lactam ring of the antibiotic cefadroxil (**75**), which gives the diketopiperazine (**76**), has also been studied as a function of pH.[181]

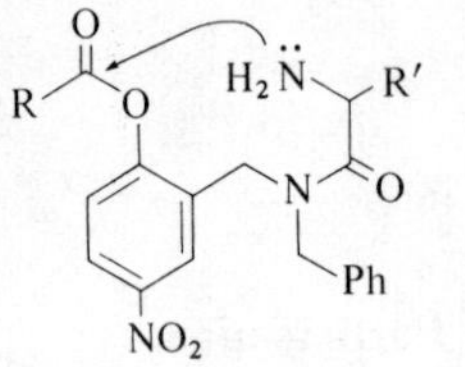

(71) 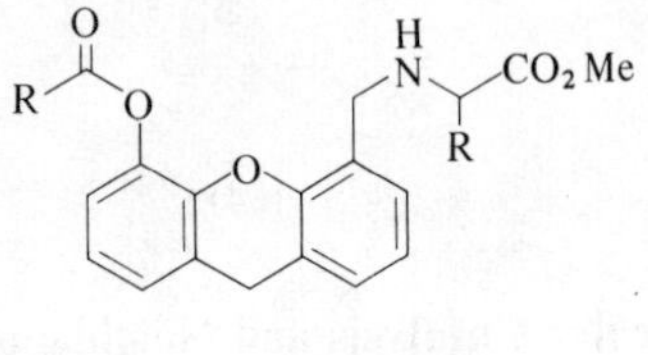 (72)

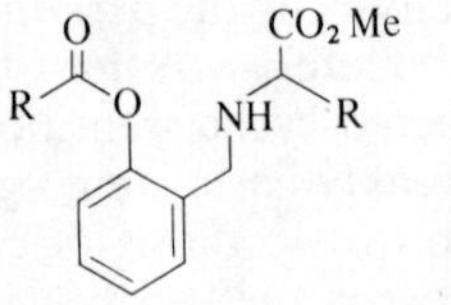

(73) 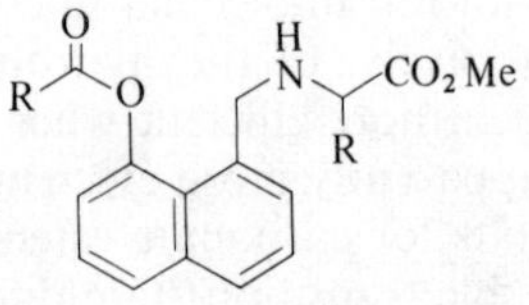 (74)

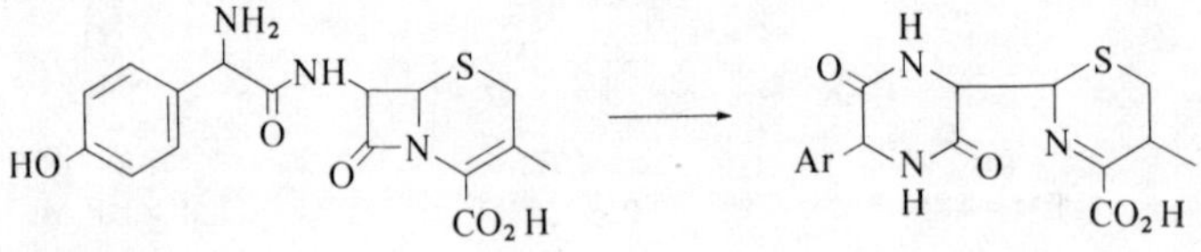

(75) (76)

The cyclization of (**77**) to (**78**) in the range pH 2–9 involves rate-determining initial attack by nitrogen for R = H, but the breakdown of the tetrahedral intermediate with loss of water is rate-determining for the *N*-Me derivative[182] above pH 5. The attempted transesterification of ester (**79**) with sodium butoxide in *n*-butanol led instead to the acid, by way of an intriguing sequence of events in which the desired product is formed, but rapidly dealkylated by the neighbouring (10-membered ring!) tertiary amino group (**80**). The final step is a Hoffmann elimination to regenerate the dimethylaminopropyloxy chain.[183]

Cyclization of the thiourea (**81**) is base-catalysed, and some 10^3 times faster than that of 2-ureidobenzoate.[184] This appears to be another example of nucleophilic attack on the ionized carboxyl group,[185] though full mechanistic details are not yet clear.

The intramolecular general acid catalysis observed in the hydrolysis of (**82**)[186] has been investigated further, and compared with intermolecular catalysis by acetic acid.[187] The authors express reasonable doubts about the mechanism (**82**) originally postulated, and prefer stabilization of the *N*-protonated substrate by the neighbouring carboxylate group, by H-bonding or electrostatic interaction (**83**); clear-cut mechanistic distinctions between the various alternatives do not appear to be possible at present. The complex pH-rate profiles for the hydrolysis of the sulphone (**84**), and the corresponding sulphoxide and sulphide (thioacylal) can be explained by the varying relative rates of hydrolysis of the two ester groups. For the SO compounds the acetate ester is cleaved first at pH <4, giving salicylate esters (which are hydrolysed at high pH with intramolecular general base catalysis by phenolate oxygen); above pH 4 the salicylate ester is hydrolysed preferentially, giving aspirin.[188]

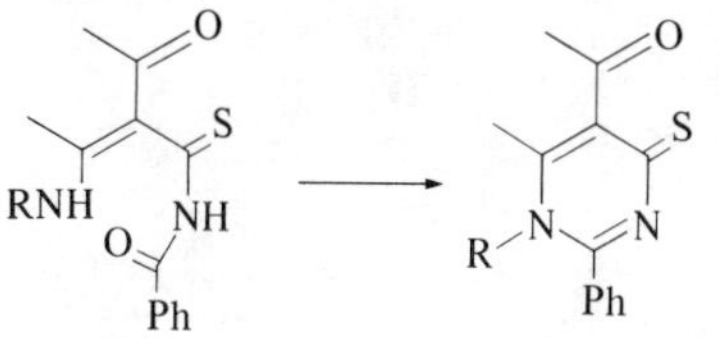

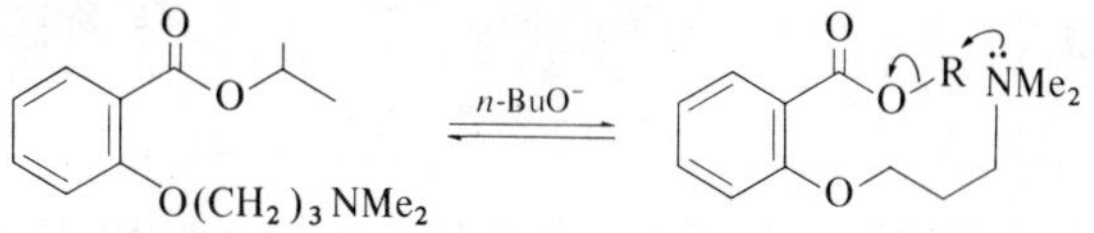

(77) (78) (79) (80)

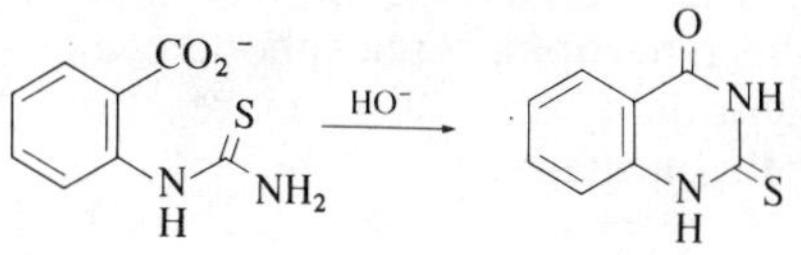

(81)

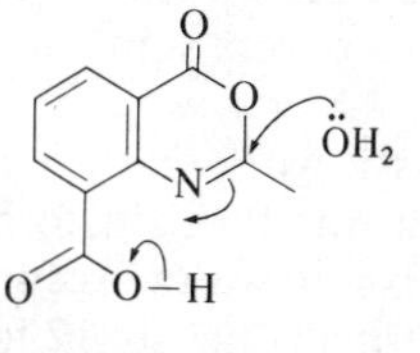

(82) 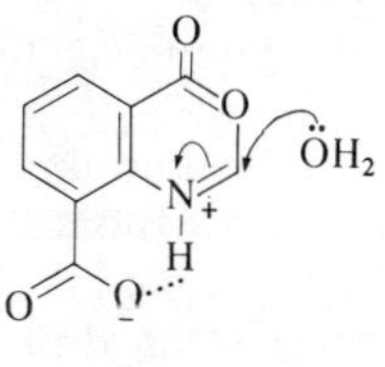 (83) 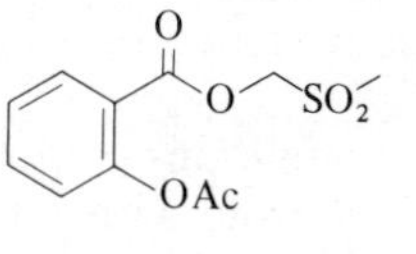 (84)

The dehydration of phthalamic acids (**85**) by vacuum thermolysis at 200° is favoured by bulky *N*-alkyl substituents,[189] and the use of ^{18}O-labelled material confirms that the amide group is acting as the nucleophile in the reaction.[190] *o*-Acetamidobenzamide (**86**) undergoes a similar transformation in the solid.[191] (For a case of intramolecular nucleophilic catalysis by the carboxylate group of a monoalkyl phthalate see below, page 26.)

Studies on the structure and reactivity of over 100 acid chlorides with neighbouring keto, acid, or ester groups rule out the existence of tautomeric equilibria between acid chloride and α-chloroacylal isomers[192] (see also Reference 105, above). Participation by the neighbouring ester group of (**87**) does, however, occur in the presence of strong Lewis acids,[193] and a symmetrical oxonium salt structure (**88**), deduced from the resulting NMR spectra, is formed.

The hydrolysis of *p*-nitrophenyl 3-phosphonopropionate (**89**) involves intramolecular nucleophilic catalysis by the phosphonate dianion ($EM = 7000$).[194]

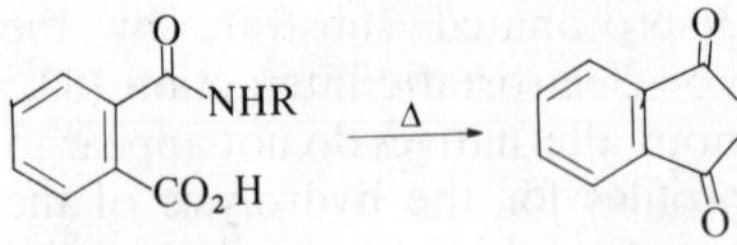

(85) 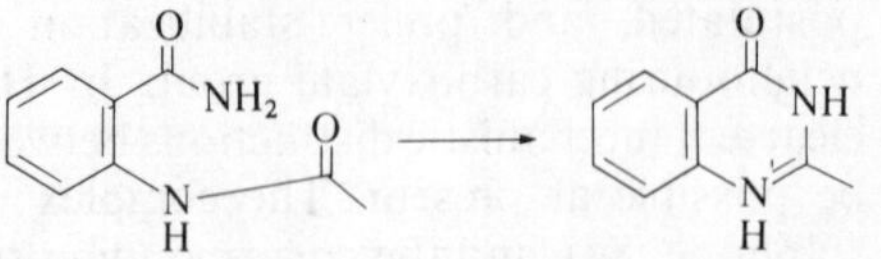 (86)

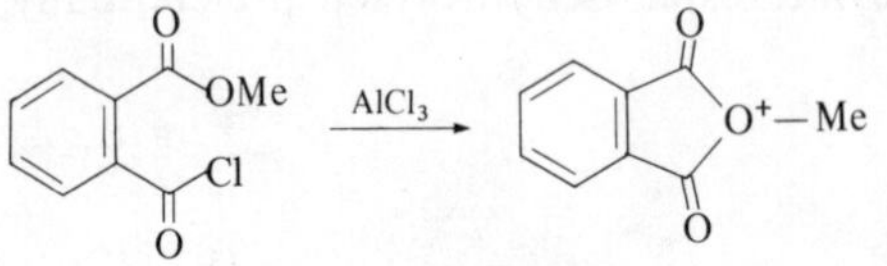

(87) (88) 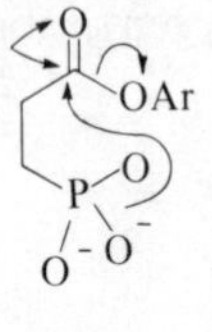 (89)

Unexpectedly, the source of the acceleration lies in the enthalpy term: $\Delta S^{\neq}$ is actually *less* favourable for the intramolecular reaction. Possible reasons are different rate-determining steps for the inter- and intra-molecular reactions [rate-determining breakdown of the tetrahedral intermediate for **(89)**] and an unusually important role for desolvation of the nucleophilic phosphonate dianion in the intermolecular reaction. ($\Delta S^{\neq}$ is unusually high for the reaction of methyl phosphonate dianion with *p*-nitrophenyl acetate.)[194]

Association-prefaced Catalysis

A useful review of enzyme-like synthetic catalysts, discussing precisely the topics covered in this section, has appeared.[195] A shorter review in Japanese of regioselective reactions of oriented asymmetric molecules covers some of the same ground.[196]

A reinvestigation is reported of the by now almost classic work on the reaction of *N*-decylimidazole with *p*-nitrophenyl decanoate, which was thought to be much accelerated by the formation of hydrophobic 1:1 complexes in water. Accurate data are difficult to obtain, but it seems clear that only small, up to 12-fold, rate accelerations are likely to be due to hydrophobic complexing in this and similar systems.[196a]

Work on micellar catalysis continues apace. Studies on some less conventional systems include a demonstration of quantitative differences in the activity of α-chymotrypsin in reverse micelles formed in isooctane by bis(2-ethylhexyl) sodium sulphosuccinate with 0.6–2.5% of water.[197] Oil-in-water emulsions stabilized by CTAB behave much like cationic micelles as a reaction medium for decarboxylation of 6-nitrobenzisoxazole-3-carboxylate ion and in the hydrolysis of bis-*p*-nitrophenyl carbonate.[198] The alkaline hydrolysis of ethyl benzoate is catalysed by inverse micelles absorbed on a platinum electrode.[199] The reaction of imidazole with *p*-nitrophenyl acetate in carbon tetrachloride is catalysed by non-ionic surfactants,[200] and the catalytic activity of imidazole in reverse micelles is also enhanced in this system. The activation parameters for aminolysis of ethyl *p*-nitrophenyl carbonate by dodecylamine in reverse micelles in chloroform stabilized by dodecylammonium diethylarsinate depend on the concentration of solubilized water.[201] The full paper on formation of bilayer assemblies from single-chain amphiphiles has appeared.[202]

Decarboxylation of (**90**), which is very sensitive to the hydrophobicity of the medium,[203] is catalysed by just about everything which reduces solvation by water; particularly effective are cationic surfactants and aggregates, including quaternized poly(4-vinylpyridines),[204] trimethyloctylammonium aggregates (which are more effective than dialkylammonium bilayer membranes), as well as CTAB.[205]

The effects of pressure on the hydrolysis of long-chain esters catalysed by CTAB[206] and on the hydrolysis and aminolysis of similar *p*-nitrophenyl esters by long-chain amines[207] have been studied. The alkaline hydrolysis of (**91**) is inhibited by cationic micelles like CTAB, becomes the cationic substrate remains in the aqueous phase while the hydroxide ion is concentrated in the micellar pseudo-phase. Added chloride ion displaces hydroxide from the micelles and relieves the inhibition.[208]

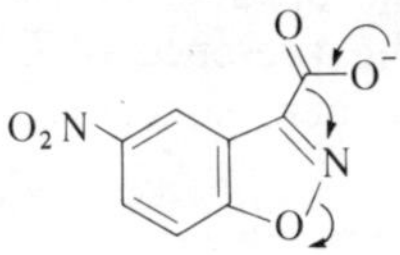

$Me_3\overset{+}{N}CH_2CO_2Et$

(**90**) (**91**)

Cationic micelles accelerate the *E*1*cb* reaction of *o*-nitrophenyl cyanoacetate by shifting the equilibrium for deprotonation in favour of the carbanion, which is stabilized on the surface of the micelle. But the breakdown of the carbanion to products is inhibited by this stabilization, so the reaction is slowed by the addition of micelles at high pH.[209]

The rate of alkaline hydrolysis of *p*-nitrophenyl acetate in hexanol in the presence of small amounts of water is strongly affected by partially lauroylated polyethyleneimine;[210] and retarded in water by increasing concentrations of KCl in the presence of CTAB.[211]

The stereoselective cleavage of diastereoisomeric dipeptide esters (**92**)[212] catalysed by CTACl involves mainly intramolecular cyclization to form the diketopiperazine (**93**) rather than hydrolysis.[213] This makes it easier to understand the contrasting selectivities of the "hydrolysis" reaction, which is faster for the D,L-isomer, and cleavage catalysed by the functionalized surfactant $C_{16}H_{33}\overset{+}{N}Me_2CH_2CH_2SH$, which is selective for the L,L-isomer.[212]

The acid-catalysed hydrolysis of *N*-trifluoroacetylindole (see above, page 12) is slightly accelerated by micelles of sodium lauryl sulphate.[214]

The basic hydrolysis of *N*-aryl-*N*-phenylbenzamides appears to involve a different rate-determining step in the presence of micelles of CTAB, as judged by a change in the ρ value, which falls from 2.63 to 1.99 in the micelle-catalysed reaction;[215] the medium change involved would otherwise have been expected to

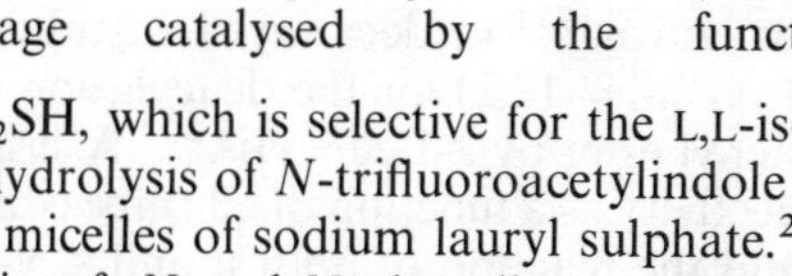

(**92**) (**93**)

cause the ρ values to be larger for reactions in the micellar phase. For substitution in the benzamide ring ρ also falls, from 1.79 to 1.62. It is thought that C—N cleavage is rate-determining in the water reaction. In the positive micellar environment loss of the amide anion may be more favourable, so that hydroxide attack becomes the rate-determining step of the reaction.[215]

Hydroxide attack is rate-determining even in water for the hydrolysis of amide (**94**), which has a stabilized amide anion leaving group, so there is no change of mechanism in the presence of CTAB micelles. Nevertheless the magnitude of catalysis is a complex function of pH, buffer, electrolyte, and surfactant concentration.[216]

Catalysis by micelles of CTAB of the reaction of anions of benzaldehyde oximes with *p*-nitrophenyl acetate is more efficient for less basic α-nucleophiles.[217] When the nucleophile has an *o*-hydroxy group the product (**95**) undergoes competing hydrolysis and cyclization reactions. In the presence of CTAB the hydrolysis reaction is inhibited and the final product is benzisoxazole.[218]

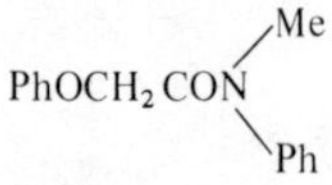

$PhOCH_2CON(Me)Ph$

(94)

N—OAc, O⁻ → N, O

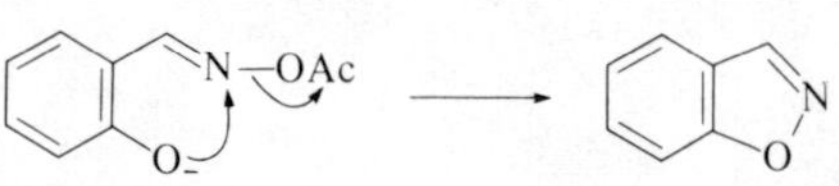

(95)

Moss and Bizzigotti[219] have prepared functionalized vesicles from $(n\text{-}C_{16}H_{33})\overset{+}{N}Me_2CH_2CH_2SH$ which show distinct "inside" and "outside" reactions with neutral esters (*p*-nitrophenyl acetate and hexanoate), which can pass through the vesicular membrane.

Mixed micelles of hydroxamic acids and cationic surfactants catalyse the hydrolysis of *p*-nitrophenyl alkanoates more or less efficiently depending on the lengths of the alkyl chains of all three components.[220] High reactivity is correlated with long, and preferably matching, chain-lengths.

When the hydroxamic acid is derived from an optically active acid, (N^{α}, N^{ε}-bis-benzyloxycarbonyl)-L-lysine, the mixed micelle with CTAB shows high enantioselectivity in the cleavage of *N*-lauroyl-Phe-pNP ($k_D/k_L = 5.68$ at 10°).[221] Selectivity is reversed for lauroyl[221] or decanoyl-L-histidine as the nucleophile[222] and is very high indeed ($k_L/k_D = 12.2$) for the deacylation of *N*-methoxycarbonyl phenylalanine *p*-nitrophenyl ester by Z-Leu—His.[223] A systematic examination of structure and enantioselectivity as a function of all three components in this system has been reported,[224] and the reaction of both R and S-*N*-acetylphenylalanine *p*-nitrophenyl esters catalysed by (R,S)- and (S,S)-(**96**) has been studied.[225] Bilayer systems composed of palmitoyl-L-histidine and double-chain surfactants $R_2NMe_2{}^+$ give higher enantioselectivity in the deacylation of *N*-long-chain-acyl phenylalanine *p*-nitrophenyl esters than comparable mixed micelles.[226] A functionalized vesicular membrane prepared from (**97**) also shows good enantioselectivity as well as catalytic efficiency for the cleavage of Z-Phe-pNP.[227]

Other catalysts tried for the hydrolysis of simple nitrophenyl esters include poly-iminomethylenes) derived from histidine and histamine [228] and imidazole-substituted polystyrene-based polyelectrolytes.[229]

Several bifunctional micellar surfactants based on the charge-relay system of the serine proteinases have been prepared. Compound (**98**) for example, contains both an aspartic acid and a histidine residue, and in a mixed micellar system with the alanine derivative (**99**) (itself not reactive) is an efficient catalyst for the hydrolysis of *p*-nitrophenyl hexanoate.[230] The imidazole group is evidently the effective catalytic function, because an aspartic acid derivative lacking this residue was also relatively ineffective. Moreover, the mechanism is predominantly general base catalysis, and only some 35% of the more usual nucleophilic pathway is observed, as indicated by solvent deuterium isotope effects, k_H/k_D, as large as 2.6; since (**98**) is 2.3 times more effective than the system lacking the carboxylate group it is at least not impossible that this assists catalysis by imidazole.[230]

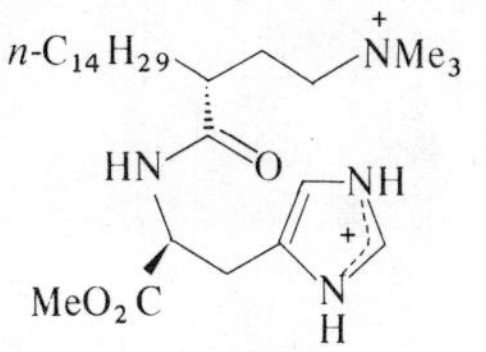

(96)

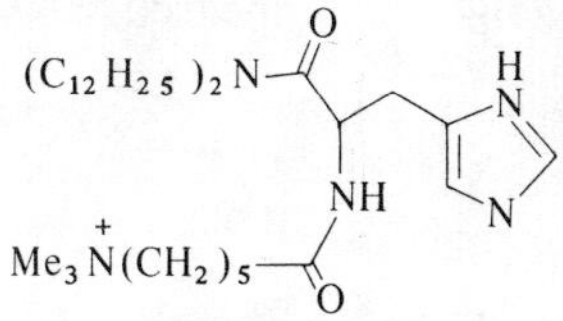

(97)

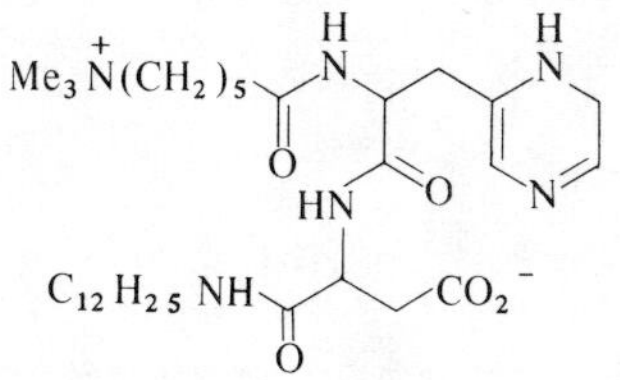

(98)

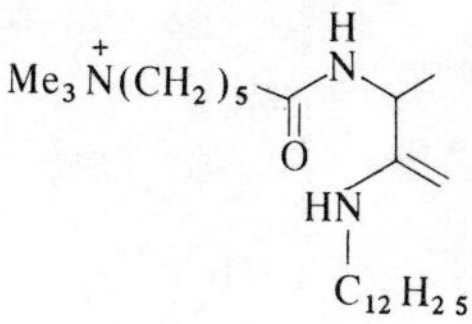

(99)

The usual nucleophilic mechanism, apparently suppresed by the tight structure of the micellar aggregate in the case of (**98**) and (**99**), accounts for catalysis of the hydrolysis of *p*-nitrophenyl hexanoate by functionalized micelles of (**100**) and (**101**).[231,232] The latter compound is more effective than chymotrypsin (towards this non-substrate) at pH 7.95, but the reactive form is the anion, and this is not significantly more reactive than a number of related derivatives;[231] see also O'Connor and Porter.[233]

A simple model for a metallo-enzyme can be asembled from two molecules of (**102**) and a zinc cation.[234] This is a highly effective catalyst for the hydrolysis of *p*-nitrophenyl pyridine-2-carboxylate in a mixed micellar system with CTAB, and there is evidence for rapid acylation of (**102**), followed by slow deacylation, in a 2:1 Zn^{2+} complex.

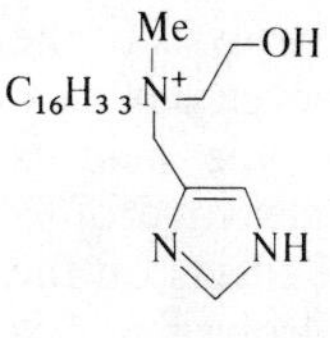

(100)

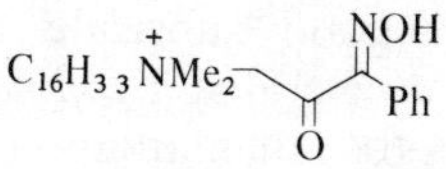

(101)

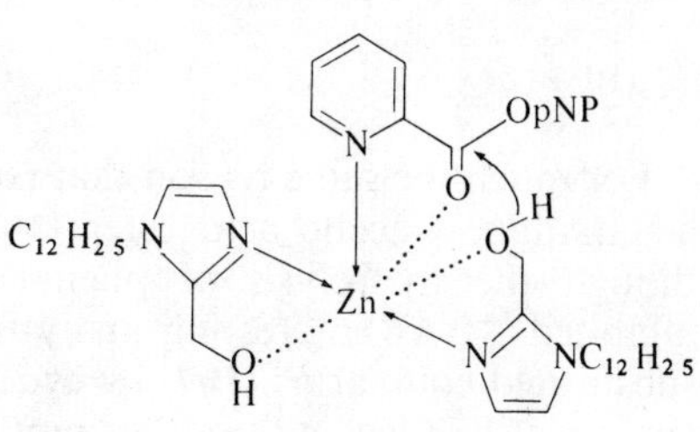

(102) **(102)**

The best-defined synthetic catalysts involving a well-defined binding step are molecules like (**103**) and (**104**) which can accommodate guests within their largely hydrophobic cavities. Paracyclophanes like (**103**) are very effective nucleophilic catalysts for the deacylation of active esters, though their subsequent deacylation is generally slow.[235] A new report concerns the reactivity of (**103**) in the presence of an imidazole surfactant similar to (**100**) (with two NMe groups and no ethanol side-chain) with *p*-nitrophenyl hexadecanoate. The active nucleophile in the ternary complex is the imidazole anion, but the acyl imidazole thus formed is rapidly deacylated by (**103**).[236]

The full paper describing catalysis (and inhibition) of the hydrolysis of activated esters by (**104**) has appeared.[237]

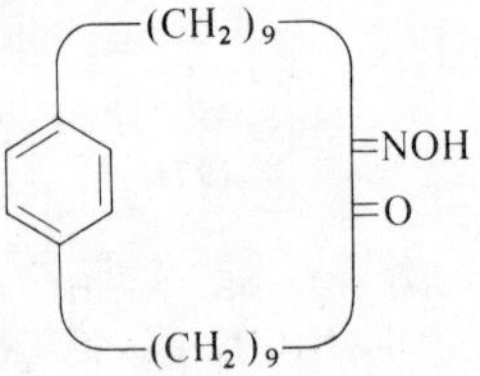

(**103**) 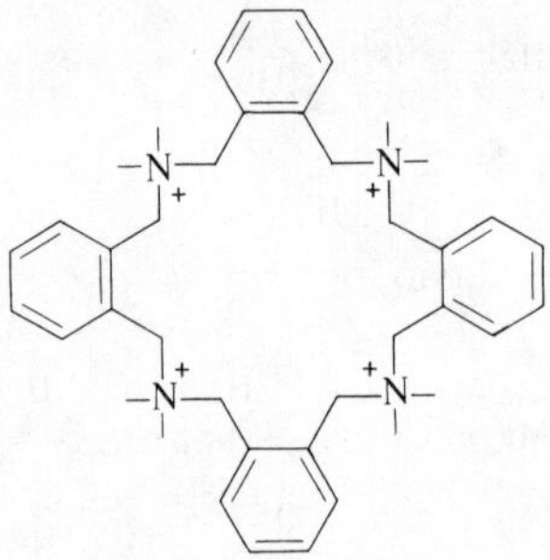 (**104**)

Trainor and Breslow have reported a further refinement of the ferrocenylacrylate system (**105**), which is the most effective substrate for catalysis by β-cyclodextrin.[238] By fixing the conformation of the side-chain in a ring system it has been shown that (**106**), rather than (**107**), models the optimum conformation, and is the most effective substrate yet devised.[239] Rate accelerations of the order of 10^7 can now be attained, and a 20-fold enantioselectivity is observed. By determining the absolute configuration of the more reactive enantiomer of (**106**) a geometric basis for the observed selectivity has been worked out.

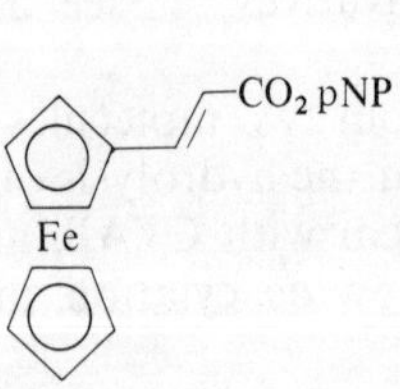

(**105**) 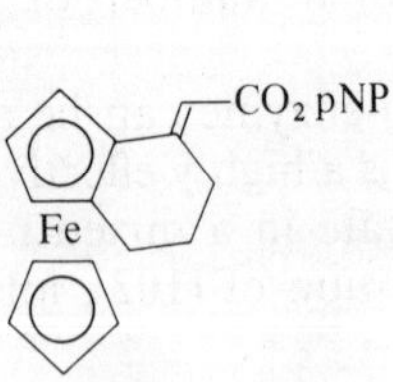 (**106**) 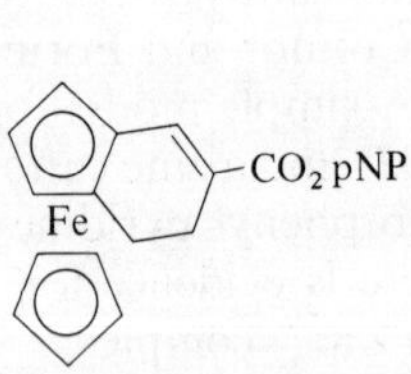 (**107**)

For much the same reason that (**105**) is a good "substrate" the hydrolysis of the adamantane-1-acetic acid ester (**108**) is accelerated by β- and γ-cyclodextrins, though the hydrolysis of phenol esters of adamantane-1-carboxylic acid is inhibited.[240] An interesting analysis of substituent effects on the cleavage of 7-substituted coumarins (**109**) by cyclodextrins involves separation of effects on the acyl (ρ_{para}) and leaving (ρ_{meta}) groups.[241] Compared with alkaline hydrolysis ρ_{meta} is increased on reaction with ($\alpha > \beta > \gamma$)-cyclodextrin (as expected for reaction in a

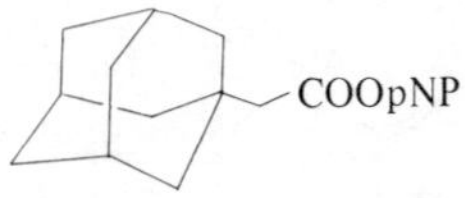

COOpNP

(108)

R O O

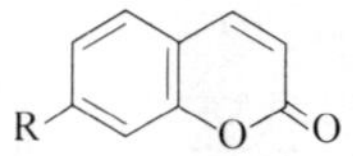

(109)

less good solvating medium, and as observed also in mixed aqueous–organic solvents), but ρ_{para} is *decreased* slightly.

The hydrolysis of 2,2,2-trifluoroethyl *p*-nitrobenzoate is modestly catalysed by α-cyclodextrin in a reaction that is notable for being apparently mostly general-base-catalysed by the alkoxide groups of the host.[242] (Most of the substrate is converted directly to *p*-nitrobenzoate, rather than acyl cyclodextrin.) If there were no restraints on the functional groups involved a nucleophilic mechanism would be expected, and it is likely that this is inhibited by the binding of the benzoate ring within the cavity (**110**). *β*-cyclodextrin equipped with an *N*-methylhydroxamic acid side-chain (**111**) is some 100 times more effective a catalyst for the cleavage of various mono- and di-nitrophenyl acetates than the parent cyclodextrin.[243] The selectivity amongst the small group of esters studied is different for the functionalized host.

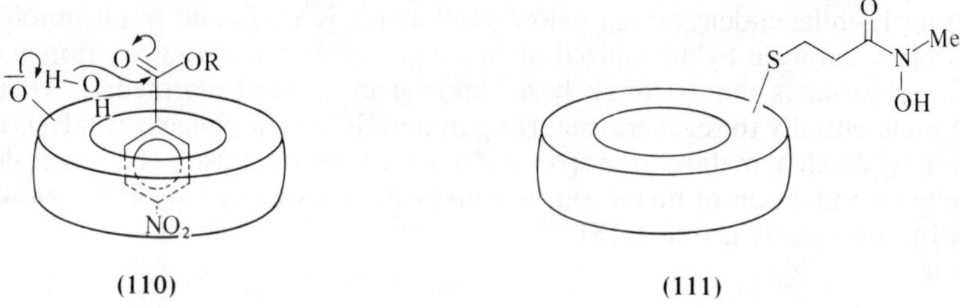

(110) **(111)**

Metal-ion Catalysis

The cyclization of (**112**) (and the COO^- form) to (**113**) is rapid in dilute acid and involves nucleophilic attack by co-ordinated water on the carboxyl group of bound glycine.[244] This "lactonization" reaction is much faster (10^7–10^{12}-fold) than the ^{18}O-exchange reaction of glycine, and buffer (general acid) catalysis is observed, characterized by a biphasic Brønsted plot with slopes, α, of 0 and 1.0 below and above p*K* 4.8. Evidently a diffusion-controlled proton-transfer step involving the tetrahedral addition intermediate formed from (**112**) is rate-determining.

$$\mathrm{en_2Co^{3+}(OH_2)(NH_2CH_2COOH)} \longrightarrow \mathrm{en_2Co(OC(O)CH_2NH_2)}$$

(112) **(113)**

A rather similar reaction involving a *four*-membered cyclic transition state (**114**) may be involved in the lactonization reaction catalysed by organotin oxides, which promises to be useful in the synthesis of macrocyclic lactones.[245] Compound (**115**) is hydrolysed in base, but not in acid, in two distinct stages to *N*,*N'*-dimethylethylenediamine and the oxamide complex (**116**).[246]

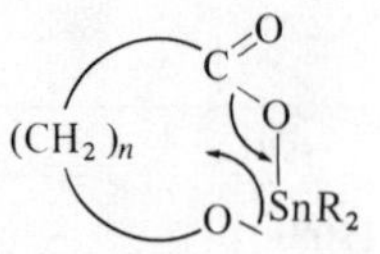

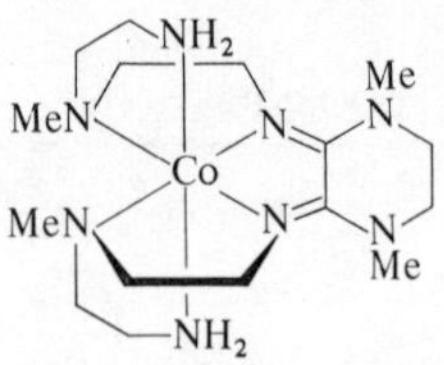

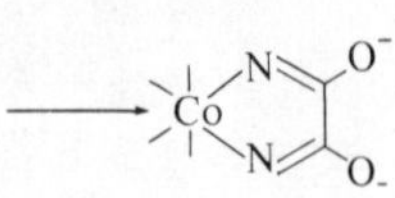

(114) (115) (116)

Nitriles are good ligands, and co-ordination to a metal cation sharply increases the electrophilic character of the group. The pK_a of the methylene protons of co-ordinated malononitrile in $(H_3N)_5CoN{\equiv}CCH_2CN^{3+}$, for example, is 5.7, compared with 11.3 for the unco-ordinated nitrile, and the rate of alkaline hydrolysis of acetonitrile is some 10^5 times faster when co-ordinated in this way. Unexpectedly, this acceleration is due exclusively to the more favourable entropy of activation, which is +4 e.u. in the complex, compared with −23 e.u. for the reaction of free acetonitrile. The hydrolysis of acrylonitrile co-ordinated to penta-ammineCo(III) is subject to nucleophilic catalysis by carbonate ion.[247]

The hydrolysis of the 2-pyridylmethyl half-ester of phthalic acid (**117**) at 90° shows a pH-independent region below pH 9 which is attributed to intramolecular nucleophilic catalysis by the ionized carboxyl group. This is a slow reaction because the leaving group is poor (strongly basic) and the tetrahedral intermediate will break down preferentially to regenerate starting materials. So the efficient catalysis of the reaction by divalent metal cations (Ni > Zn > Co), which is also pH-independent at a given concentration of metal ion, is convincingly explained by chelation of the metal by the leaving group (**118**).[248]

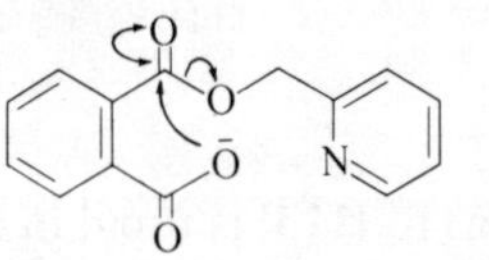

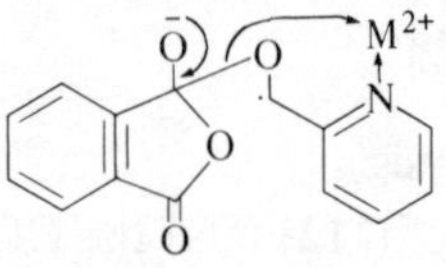

(117) (118)

Chelation of divalent metal cations by the substrate is also the likely explanation of the catalysis of the attack of nucleophiles (HO^-, RS^-) on *p*-nitrophenyl phosphonoacetate (**119**); Mg^{++} and Ca^{++} give rate accelerations of the order of 10^2 at saturating metal concentrations.[194] Catalysis of the hydrolysis of *S*-ethyl thiobenzoate esters by tetrachlorogold(III) in dilute acid is much more efficient than the acid-catalysed reaction. The absence of a solvent deuterium isotope effect, and the higher reactivity of the *p*-methoxy-ester compared with the *p*-nitro compound, suggest that the rate-determining step is co-ordination of sulphur to the metal (**120**), to generate an activated species which is rapidly hydrolysed by water.[249]

Decarboxylation

The hydration of CO_2 is catalysed by the Zn(II) complexes of two polydentate ligands (with **121** and **122**), which are offered as models for the active site of the enzyme carbonic anhydrase[250] (which has three histidine imidazoles and water bound to Zn^{++}).

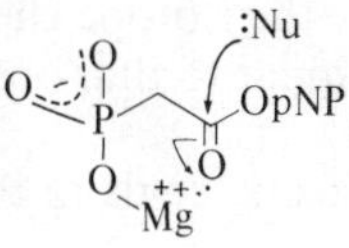

(119)

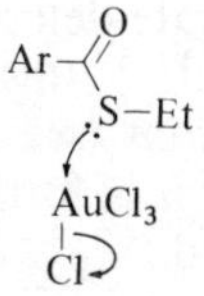

(120)

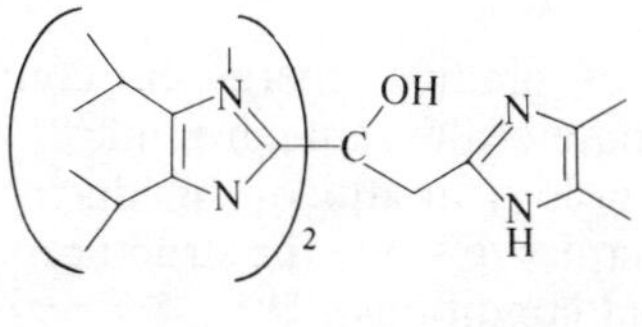

(121)

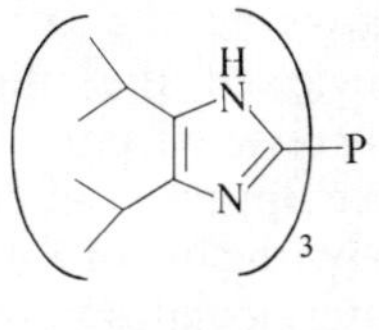

(122)

The synthesis is reported of α-lactylthiamine (**123**), the proposed intermediate in the thiamine-catalysed decarboxylation of pyruvate.[251] (**123**) is the reactive form at pH 7, where the rate of decarboxylation is pH-independent. At lower pH the rate goes through a maximum: the (pyrimidine) ring-protonated form is slightly more reactive than (**123**), but forms with the carboxyl group protonated are unreactive.

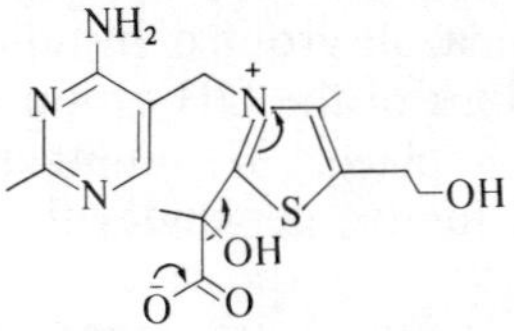

(123)

The effects of glycol solvents on the activation parameters for the decarboxylation of three alkyl-substituted malonic acids,[252] and the effect of the cation on the thermal stability of metal benzoates[253] have been studied.

The chlorodecarboxylation of salicylic acid effected by chloroamine-T is catalysed by chloride ion, and the rate levels off as the chloride concentration approaches that of the chloramine;[254] complexation has been suggested, but a simpler explanation might be the generation of Cl_2.

A review of oxidative decarboxylation has appeared[255] and the Ce(IV)-catalysed reaction of acetic and phenylacetic acids promoted by Ru(III) and Mn(II) has been studied.[256,257]

^{13}C-isotope effects have been measured for two enzyme-catalysed decarboxylations. k^{12}/k^{13} is 1.061 for the reaction of homoarginine catalysed by arginine decarboxylase,[258] and 1.018 for the glutamate decarboxylase reaction.[259] These are both pyridoxal phosphate reactions, and it has been suggested that the decarboxylation step is rate-determining in the former case, but that Schiff base formation and decarboxylation are both partially rate-determining for the glutamate enzyme (where there is a significant solvent deuterium isotope effect, $k_{H_2O}/k_{D_2O} \simeq 2$, on the reaction). Organic solvents appear to accelerate the decarboxylation step (though not the overall reaction) for arginine to the point where it is no longer rate-determining, because the isotope effect almost disappears in 24% ethylene glycol;[259] this behaviour is similar to that observed in comparable *in vitro* decarboxylations (see above, page 21).

The decarbonylation of formic and phenylacetic acids in the presence of strong acids has been studied.[260–2]

Enzymic Catalysis

Serine Proteinases

Jencks has reviewed the utilization of binding energy in coupled vectorial processes.[263] A review lecture covers some of the same ground.[264] Also, valuable summaries have appeared of cryoenzymology in aqueous media[265] and of X-ray cryoenzymology (the use of X-ray techniques to study the structures of what would be transient intermediates under normal conditions).[266]

There is a clear trend towards theoretical calculations on enzyme reactions. These generally start from the known three-dimensional structure of an active site, available from X-ray diffraction results, and set up models for the binding or catalytic steps in various ways. Some recent results have been summarized by Pincus and Scheraga,[267] and Warshel has described his method for calculation of activation energies for enzymic reactions.[268] A favourite objective of much of this work is to put the chemistry of the charge-relay system of the serine proteinases on a sound theoretical basis. One recent calculation, using the trypsin-inhibitor complex as a model, reveals that the imidazole group of histidine-57 can lower the activation energy for nucleophilic reactions of the OH group of serine-195 without proton transfer actually taking place;[269] another,[270] using a minimal formate–imidazole–methanol model, concludes that proton transfer to aspartic acid-102 is unlikely to occur.

Meanwhile structural investigations are approaching the same problem with increasing precision. The charge-relay system of γ-chymotrypsin at 1.9 Å resolution has no hydrogen bond between histidine-57 and serine-195, in common with trypsin and protease A[271] and with subtilisin, though the "artificial" enzyme thiosubtilisin does appear to have a hydrogen bond to the new SH group, according to NMR measurements.[272]

The technique most suited to locating individual protons in a structure is neutron diffraction; this has now been applied to probe the monoisopropyl phosphate (on serine-195) of β-trypsin, which is a model for the tetrahedral intermediate in the catalytic reaction of this and other serine proteinases.[273] The proton between aspartic acid-102 and histidine-57 has been unambiguously located on the imidazole group, in support of much other work suggesting that the aspartate does not act as a fully fledged general base in the reaction, but helps to orientate and stabilize the imidazolium cation electrostatically (which term can include hydrogen bonding). The structure also allows a detailed discussion of the deacylation step, in which the imidazole must act as a general base for the attack of a water molecule.

DeTar has developed a method for the computation of relative energies of transition states for the formation of acyl chymotrypsins, using tetrahedral intermediates as models, by applying the methods of molecular mechanics.[274,275]

The apparent pK_a of the active site histidine, as revealed by the pH dependence of the deacylation of a series of actyl chymotrypsins, varies over a whole pK_a unit.[276] It has been suggested that this results from an interaction of varying strength between the imidazolium group and the carboxylate group of aspartate-102. The hydrolysis and transesterification of furoyl-α-chymotrypsin give curved Arrhenius plots, possibly resulting from an equilibrium between two forms of the acyl enzyme.[277]

The aging of chymotrypsin (*i.e.* its loss of activity with time) is second order in enzyme (and hence "cannibalistic") near pH 7, but first order in enzyme and first order in hydroxide at high pH.[278] Cleavage of peptide bonds is clearly involved, and the pH-7 reaction is inhibited in the immobilized enzyme, or by conventional inhibitors of chymotrypsin action.[279]

The effects of varying temperature on the chymotrypsin-catalysed hydrolysis of *N*-benzoyl-L-tyrosine have been studied.[280]

Binding studies with a series of synthetic peptides show that the binding site of an α-lytic protease extends over a least six sub-sites, as k_{cat}/k_M increases by over 10^6 times on going from an *N*-acylamino-acid to a hexapeptide.[281]

The boronic acid (**124**) is a potent inhibitor of α-chymotrypsin (and subtilisin). The compound is an analogue of *N*-acetylphenylalanine amide, but binds more than 14,000 times more tightly to the enzyme and 20 times more tightly than its (S)-enantiomer.[282]

New "suicide" inhibitors for chymotrypsin are the *N*-nitrosoamide (**125**)[283] and haloenol-lactones (*e.g.* **126**).[284] The nitrosoamide is expected to be hydrolysed by the enzyme to phenyldiazomethane, which under the conditions would generate the benzyl carbenium ion and thus alkylate the closest available nucleophilic centre. Unexpectedly the enzyme is inhibited (with the incorporation of one benzyl group) by the unnatural enantiomer (**125**), perhaps because the *N*-benzyl group, rather than the benzyl group of the phenylalanine residue, binds in the hydrophobic pocket.[283]

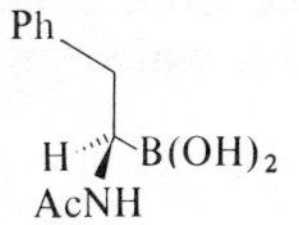

(**124**) 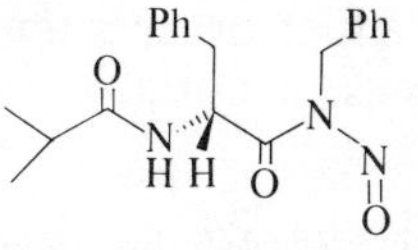 (**125**) 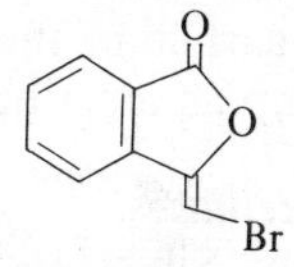 (**126**)

Chymotrypsin enriched in ^{13}C at the S-methyl group of methionine-192 has been prepared by demethylation of the corresponding dimethylsulphonium cation with the active-site directed reagent (**127**).[285] (Fortunately the labelled methyl group is not removed stereospecifically.)

Thiol Proteinases

Appropriate peptide diazomethyl ketones are specific inhibitors for various thiol proteinases, and unreactive towards simple thiols and other classes of proteinases.[286]

The hemithioacetal formed by the reaction of *N*-acetyl-L-phenylalanylglycinal (**128**) with the active-site SH group of papain[287] dissociates at least 10^8 times less rapidly than expected for an anionic tetrahedral intermediate free in solution, and is thus presumably in the neutral form.[288] The proton apparently comes originally from histidine-159, the pK_a of which is abnormally low ($\leqslant 3.0$) in the adduct.

A value of 3.45 for the pK_a of the same group has been obtained for the 'inactive' papain derivative with the active-site cysteine thiomethylated (thus,

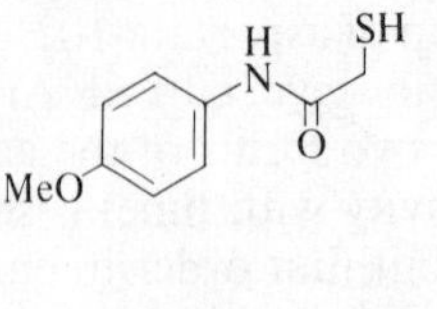

(127) 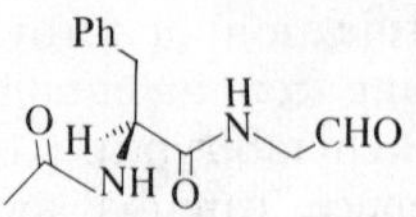 (128)

CysCH$_2$SSMe).[289] At pH 4.17 this group is the fully protonated end of an imidazolium–thiolate ion-pair when the SH group is free, as shown by the positions of proton resonances in the NMR spectrum.[289,290] Not only does the pK_a fall by about 4 pK units when the SH group is thiomethylated, but its heat of ionization also falls by about 8 kcal mol^{-1}.[291]

A proton inventory study of the hydrolysis by papain of several *N*-acylamino-acid *p*-nitrophenyl esters (reactions where deacylation is expected to be rate-determining) is consistent with a transition state involving a single mobile proton.[292] For this step, of course, histidine-159 can act as a general base because its pK_a will be low with the SH group acylated. The increased activity of papain consequent on hydroxynitrobenzylation of tryptophan-177[293] is greater towards specific substrates than for non-substrates, so that the binding step as well as the catalytic process must be affected.[294]

Thiolsubtilisin, a serine proteinase which has been converted chemically to a thiol-enzyme by substitution of sulphur in the side-chain of serine-221, is not active towards specific peptide substrates even though it shows high reactivity in reactions with non-substrates like 2,2′-dipyridyl disulphide.[295] It is suggested that the softer, weakly basic nucleophile (Cys—CH$_2$S$^-$) requires assistance from general acid catalysis earlier in the acyl-transfer process than serine—O$^-$, which an enzyme brought up as a serine proteinase is not able to provide.[295]

Acid Proteinases

A study by Silver and James[296,297] of the acceleration by small peptides of the cleavage of small peptide substrates by pepsin may have solved a long-standing problem in the chemistry of this enzyme. In some cases, particularly with dipeptide substrates, products apparently derived from amino-enzyme intermediates as well as acyl-enzyme intermediates may be formed.[298] Pepsin has an extended active site and is not an efficient catalyst for the cleavage of small peptides. It is now suggested that *two* small peptides may with advantage occupy the binding site simultaneously, and in some cases be converted into a larger peptide by a condensation which is the reverse of the usual proteolytic reaction. Rapid cleavage of the larger, better substrate at *other* peptide bonds can then produce the observed condensation products, which may have new amino-acid residues at either the acyl or the amino end, depending on where the new peptide is cleaved.[297]

Metallo-proteinases

A further refinement of the crystal structure of carboxypeptidase A at 1.75 Å resolution has improved the definition around the active-site zinc ion. In the native enzyme it appears to have a co-ordination number of five (two imidazole N, H$_2$O, and *both* oxygen atoms of the carboxylate group of glutamate-72).[299] Further to the demonstration that this enzyme catalyses H-D exchange at the enolizable position of a ketone analogue of a peptide substrate,[300] it has now been shown that the elimination of *p*-nitrothiophenolate from one enantiomer of (**129**) is also catalysed (though both enantiomers bind).[301]

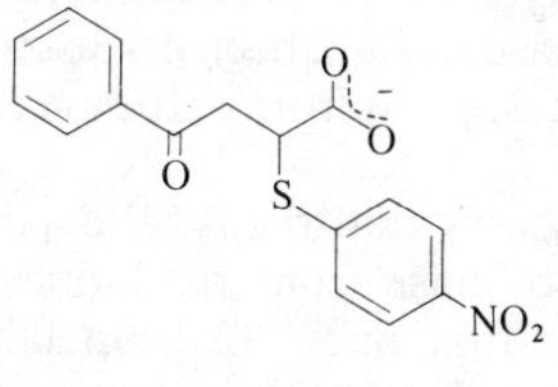

(**129**)

It has been suggested that the condensation of Z-L-Asp with L-Phe—OMe to give Z-Asp—Phe—OMe, catalysed by the enzyme thermolysin (which may resemble carboxypeptidase mechanistically), involves direct attack of the amino terminus on the carboxyl group of the acyl fragment.[302]

Other Enzymes

A number of inhibitors of β-lactamases (for a brief review see Reference 303) have been studied, including clavulanic acid (**130**)[304, 305] and its 9-deoxy derivative,[305] penicillanic acid sulphone (**131**)[306, 307] and related compounds,[308] and two derivatives (**132** and the parent alcohol) of olivanic acid.[309] Quinacillin sulphone (**133**) reacts specifically to acylate serine-70 of the *E. coli* enzyme[308] (*cf.* the enzyme from *B. cereus*[310]), which is likely to be the nucleophilic centre in the normal catalytic reaction of the enzyme. Attack on penicillanic sulphone appears to lead to cleavage of both rings of the penam (**134**),[306] and the reaction with clavulanic acid may be similar.[305]

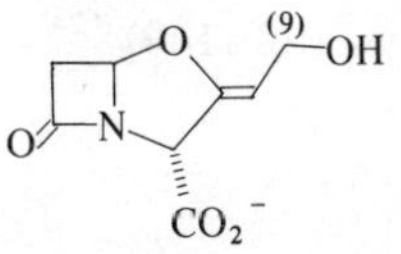

(**130**) 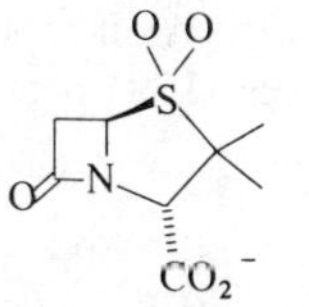 (**131**) 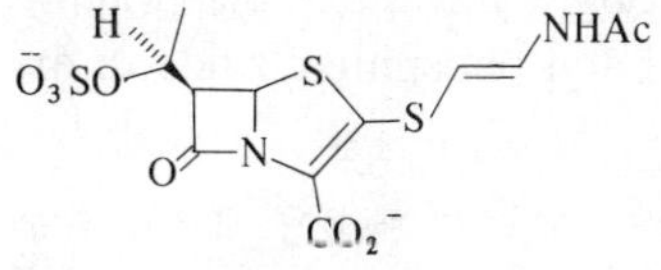 (**132**)

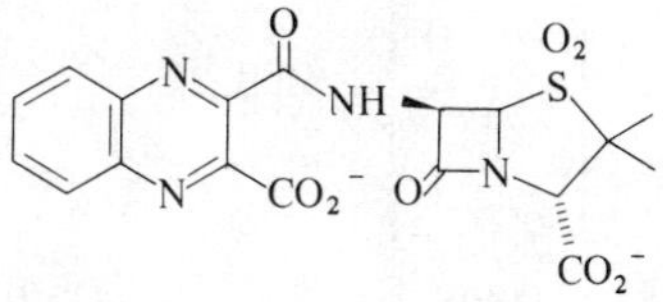

(**133**)

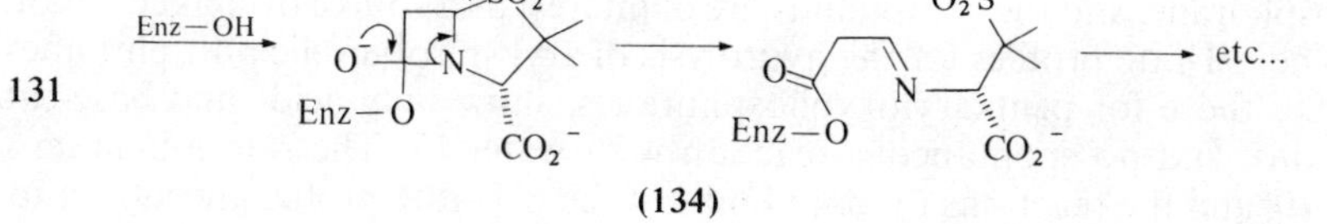

131 —Enz—OH→ ... → ... → etc...

(**134**)

The contribution of the positive charge on choline[311] increases the binding of the substrate to acetylcholinesterase by a factor of 3–10;[312] cleavage, by the enzyme, of several cyclic analogues of choline of known absolute configuration has been compared with their alkaline hydrolysis.[313] The structure and mechanism of action of bovine pancreatic phospholipase A_2 have been discussed in more detail,[314–6] and the structure of a similar enzyme isolated from rattlesnake venom has been refined to 2.5 Å resolution.[317]

^{18}O-labelling experiments show that the ATP-dependent hydrolysis of (**135**) to glutamate by the 5-oxo-L-prolinase from rat kidney retains all three substrate oxygens in the product, and that one ^{18}O from labelled water appears in the inorganic phosphate produced. A possible pathway consistent with these results involves a phosphorylated intermediate (**136**) and hydrolysis of γ-glutamyl phosphate (**137**) by exclusive P—O cleavage.[318]

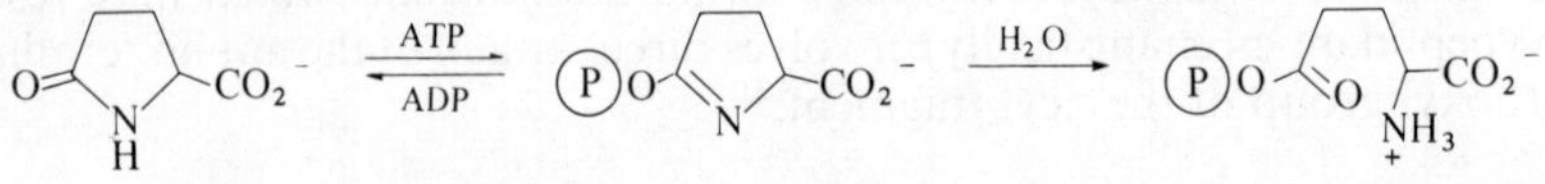

(**135**) (**136**) (**137**)

Non-carboxylic Acids

Phosphorus-containing Acids

Non-enzymic Reactions

The polymerization of spiro-acyloxyphosphoranes[319] and the current status of monomeric metaphosphate[320] have been reviewed. The full paper describing the generation of the monomeric metaphosphate anion by the Conant–Swan fragmentation[321] has appeared;[322] the phospho-Cope rearrangement of the sodium salt of allylvinylphosphonate (**138**) has also been reported.[323] Photolysis of di-*tert*-butyl phosphinic azide in methanol gives (**139**) presumably by way of (**140**).[324]

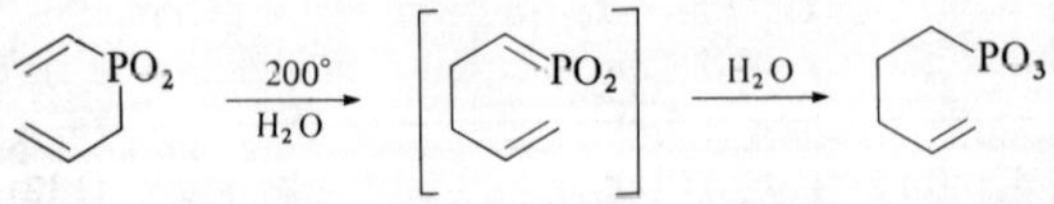

(**138**)

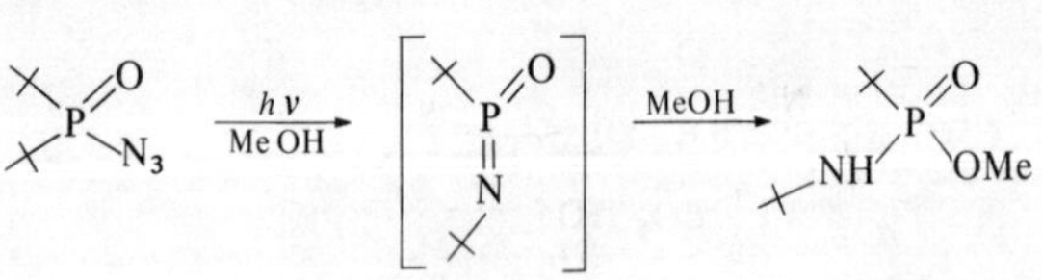

(**140**) (**139**)

Two *ionic* phosphonium fluorides (**141**; n = 2, 3) have been prepared; steric and electronic factors evidently combine to make (**141**) more stable than the isomeric phosphorane, and the compounds are of interest as a source of "naked" fluoride.[325]

The pH-rate profiles for the hydrolysis of aryl spirobicyclic phosphoranes (**142**), unlike those for pentaaryloxyphosphoranes, show only acid- and base-catalysed regions, and no spontaneous reaction with water.[326] The rate minimum is near pH 10, and the reactions on both limbs of the pH–rate profile are subject to buffer catalysis; general acid catalysis ($\alpha = 0.74$) appears to be analogous to orthoester hydrolysis, involving formation of a phosphonium intermediate (**143**) by the dissociative route shown. The Hammett ρ value (-1.0 for the H_3O^+ reaction) is even smaller than for the hydrolysis of an analogous orthoester, but this is taken to

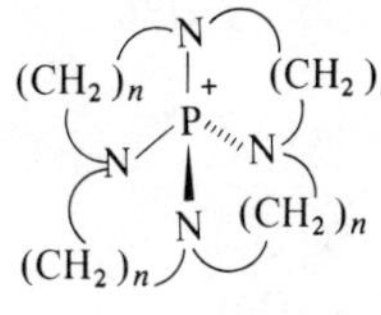

(141) 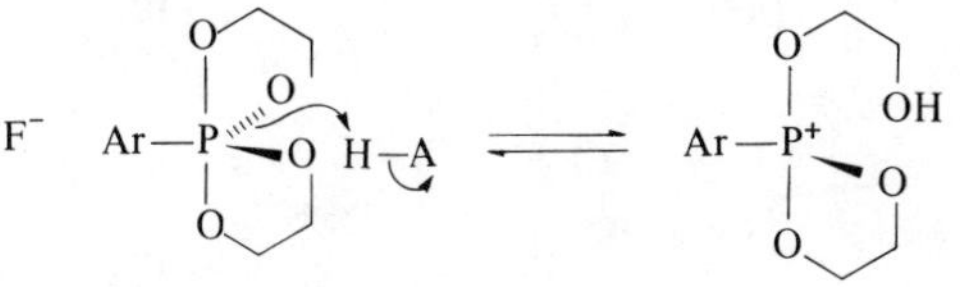 (142) (143)

be evidence for an early transition state. Alkaline hydrolysis, on the other hand, appears to involve an associative mechanism and thus a hexaco-ordinate intermediate (**144**) $\rho = 2.5, \Delta S^{\neq} = -19$ e.u.); buffer catalysis is explained in terms of a general base mechanism (**143a**).[326] The alcoholyses of several phosphoranes (*e.g.* **145**–**147**) have been studied by NMR methods in polyhalogenoalkanes.[327–329] The phosphonium and enol phosphonium salts from the protonation and ring-opening of (**146**; $n = 3, 4$) and (**147**) and several related compounds have been observed directly[328–329] in the presence of FSO_3H, but the suggestion[329] that a protonated oxyphosphorane is a full intermediate in reactions of this sort must be treated with reserve in view of the results of McClelland and co-workers discussed above.[326] Buck has reviewed some of the results of his group in this area.[330]

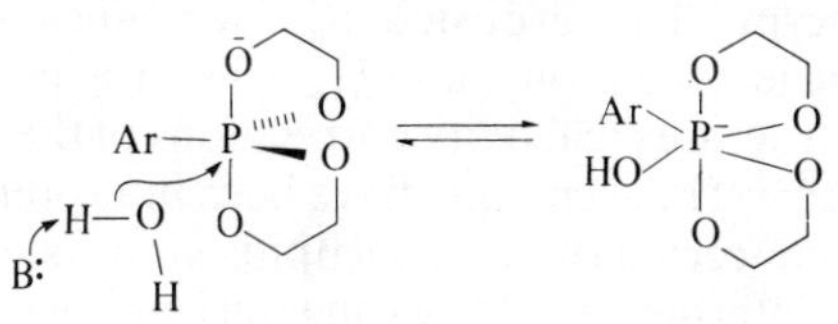

(143a) (144) 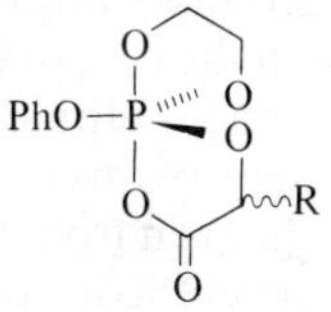 (145)

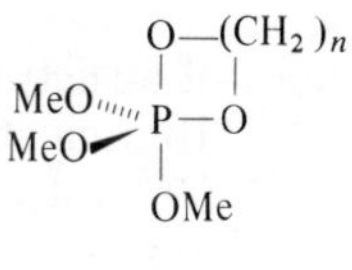

(146) 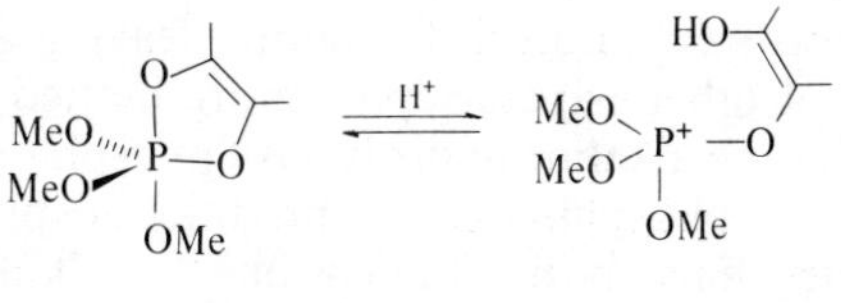 (147)

In the presence of trifluoroethoxide ion phosphoranes (*e.g.* **148**) with two or more alkoxy groups are in (favourable) equilibrium with hexaco-ordinate anions (**149**).[331] In this particular case isomeric products are possible, and the kinetic product is the *cis*-isomer (**149**), which is gradually converted to (**150**) under the conditions. The kinetic product from the addition of p-fluorophenoxide to (**151**) on the other hand, is the *trans*-isomer (**152**).[332] The factors which govern the stereochemistry of such addition reactions are not yet fully understood.

The full paper has appeared describing nucleophilic substitutions at the phosphorus centre of esters (**153**) and their equatorial epimers (**154**;

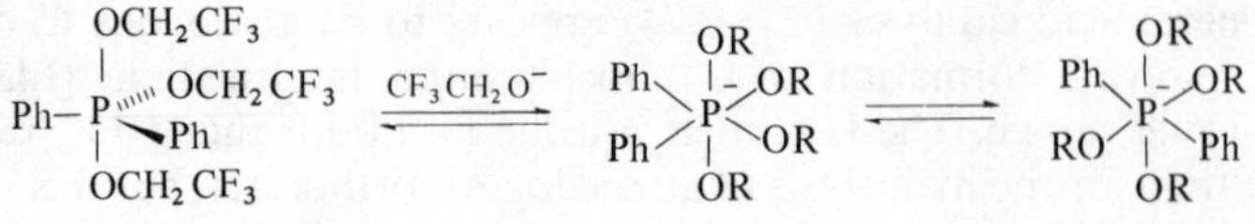

(148) (149) (150)

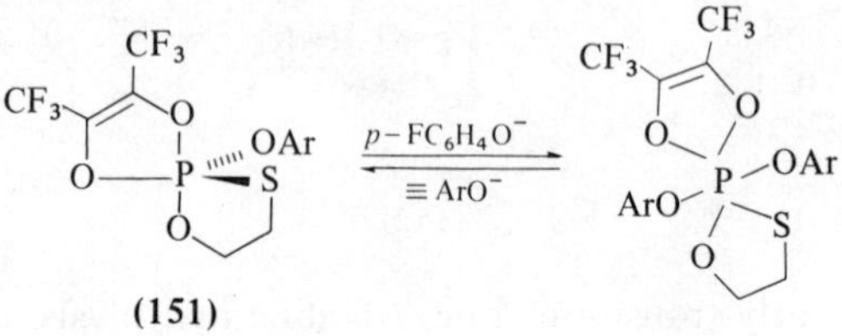

(151) (152)

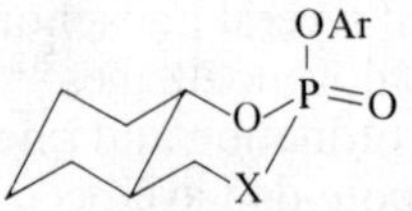

(153) 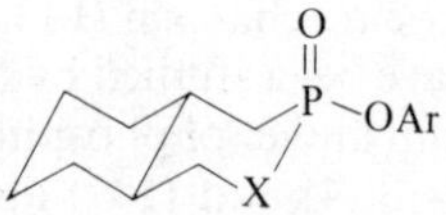 (154)

Ar = substituted-phenyl, X = O and in one case NH).[333,334] As observed previously, differences in reactivity between epimers are small, and largely reflect differences in ground-state energy. The Brønsted β_{nuc} for (nucleophilic) buffer catalysis falls with the pK_a of the leaving group, and β_{LG} falls with increasing pK_a of the buffer, leading to multiple structure–reactivity correlations of the type observed previously for esters of this sort.[335] These changes have been interpreted in terms of a change in mechanism from a concerted (weak nucleophiles) to a step-wise $S_N2(P)$ mechanism, *via* a pentacovalent intermediate, for strong nucleophiles. The step-wise mechanism, with an intermediate capable of pseudorotation, is also consistent with the retention of configuration observed (at least partially) for the reactions of all substrates with the most basic nucleophiles. A further complication is a trend towards general base catalysis for poorer leaving groups.

A number of other papers report more limited studies of structure–reactivity correlations for the alkaline hydrolysis of phosphate,[336–338] thiophosphate,[337,338] phosphonate,[339] phosphinate,[340] and thiophosphinate[341–344] esters. Effects of CTAB micelles have been studied on the alkaline hydrolysis of isobutyl *p*-nitrophenyl methylphosphonate[345] and on the nucleophilic reactions of the benzimidazole and naphth-2,3-imidazole anions with *p*-nitrophenyl diphenyl phosphate.[346] This latter reaction is also catalysed by the phase-transfer agents tri-*n*-octylethylammonium bromide and mesylate,[347] in reactions which appear to occur in small aggregates.[348] Dialkylammonium bilayer membranes catalyse the release of two equivalents of *p*-nitrophenoxide from ethyl di(*p*-nitrophenyl)phosphate,[349] and the reactions of three *p*-nitrophenyl phosphate tri-esters with hydroxide and fluoride ions in oil-in-water microemulsions have been studied.[350]

Benzoate is a better nucleophile than imidazole towards nitrophenyl diphenylphosphinates in acetonitrile containing small amounts of water, but no important term indicating general base catalysis by benzoate of attack by imidazole is indicated.[351] The alkaline hydrolysis of methyl di-isopropylphosphinate has been found to involve 25% methyl—O cleavage and 75% P—O cleavage;[352] amine catalysis of the ethanolysis of a series of *p*-nitrophenyl esters of phosphorus oxy- and thio-acids has been studied.[353]

Corriu and his co-workers have extended their investigations[354] of the stereochemistry of substitution at the phosphorus centre of phosphoro-chloridates and -fluoridates (**155**, and some P=S derivatives) by aryloxides.[355] One effect more

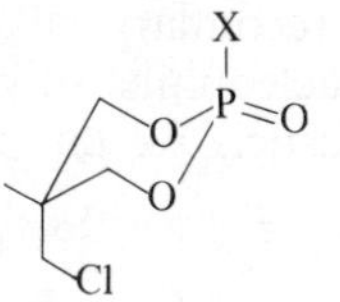

(*cis*-155) 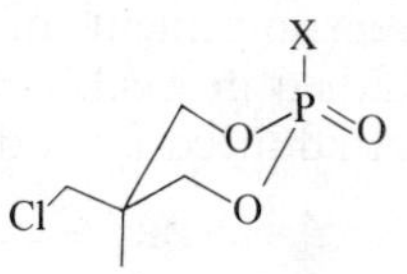 (*trans*-155)

clearly recognized now is the influence of the counter-ion of the nucleophile: Li^+ in particular appears to bias the result in favour of retention[356] whereas Me_4N^+ has the opposite effect. Paralleling the order X = F < Cl < Br favouring inversion, it is found that inversion is also favoured by more weakly basic nucleophiles. This order is that expected for decreasing stability of a pentacovalent intermediate, and is consistent with the results of Gorenstein (discussed above) for the $S_N2(P)$ reactions of epimeric esters: with a weakly basic nucleophile and a poor leaving group the pentacovalent species e.g. **156**) has only poor leaving groups (RO^-, X^-, Y^-) and thus lives longer, and has more time to pseudorotate. This is not necessary for inversion, but is for retention if both nucleophile and leaving group are successively to occupy apical positions. Corriu and his co-workers have stressed that for retention the geometry of attack should be different (**157**), as previously proposed for $S_N2(Si)$ reactions.[354, 355] (See further discussion of this question below.)

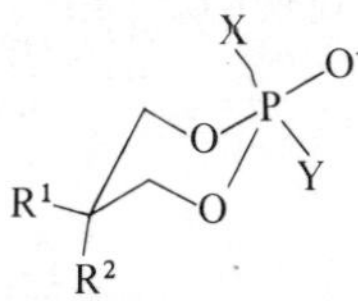

(**156**) 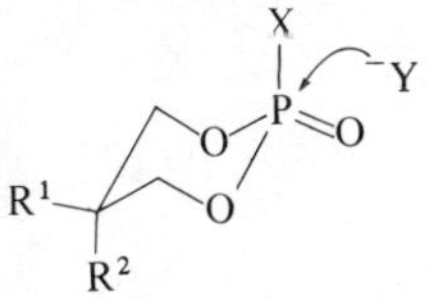 (**157**)

The same ratio of kinetic products is observed when fluoride (**158**) (and related cyclic phosphorofluoridates) react with phenols in the presence of CsF *whatever the isomeric composition of the reactant.*[357] Fluoride is a strong base, as well as a strong nucleophile for phosphorus under the conditions, and the pentacovalent intermediates (*e.g.* **159**) could have time to reach pseudorotation equilibrium. The authors suggest a more intriguing explanation whereby competing directions of attack on the *trans*-difluorophosphorane oxide (**160**) account for the stereochemistry of the products.[357]

The methanolysis of aryl esters (**155**; X = OAr) catalysed by $ZnCl_2$ gives some 60% of inversion, like the reaction in strong acid.[358]

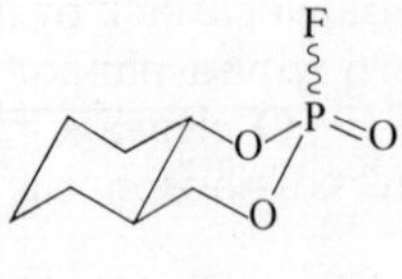

(158) 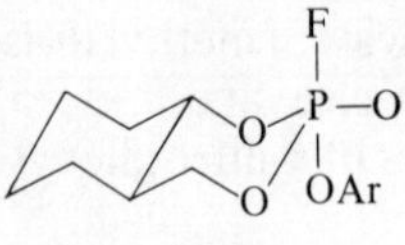 (159) 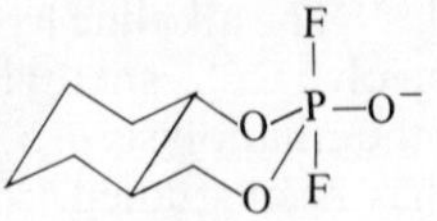 (160)

The thiono–thiolo rearrangement of secondary alkyl phosphorothionates (**161** ⇌ **162**) has been studied in weakly nucleophilic solvents like trifluoroacetic acid, and further support adduced for a dissociative ion-pair mechanism.[359,360]

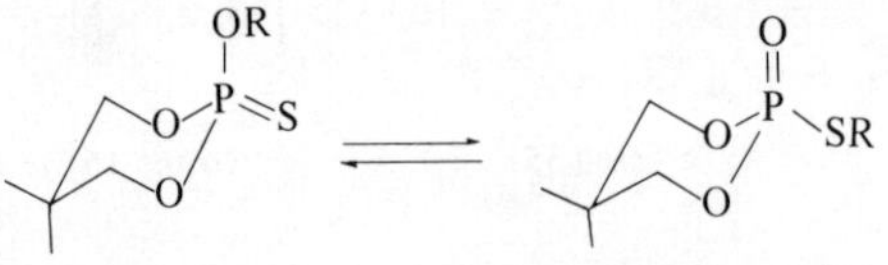

(161) (162)

The predominant retention of the ring in the hydrolysis and methanolysis of (**163**), reported last year[361] does not involve reversible ring-opening but alkyl-oxygen cleavage.[362] Reversible ring-opening has been suggested to account for the exchange of OCH_3 for OCD_3 groups of cyclic phosphate (**164**) in methanol-d_4.[328]

The first applications of chiral ^{16}O-, ^{17}O-, and ^{18}O-phosphates to the stereochemistry of non-enzymic reactions have been reported. The cyclizations of the diphenyl pyrophosphates of glucose-6-phosphate (**165**) and AMP to the 4,6- and 3′,5′-cyclic phosphates, respectively, involve inversion of configuration at phosphorus, as expected.[363,364]

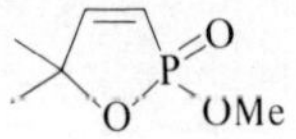

(163) 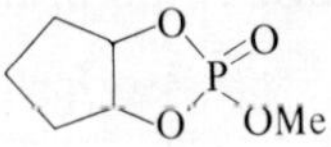 (164) 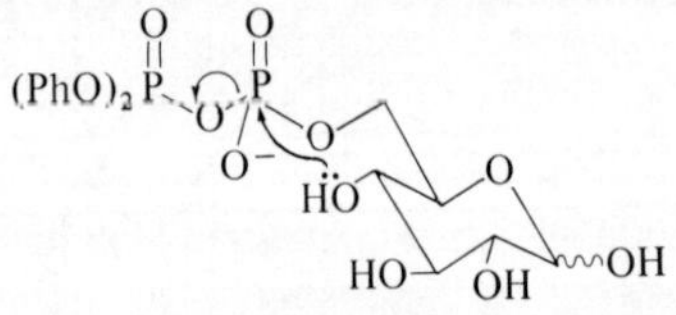 (165)

The acid-catalysed hydrolysis of ATP in $H_2{}^{18}O$–$HClO_4$ gives some pyrophosphate as well as inorganic phosphate and ADP. Reaction involves water attack on protonated triphosphate, probably at all three phosphorus centres, in the order $\gamma > \beta > \alpha$.[365] At higher pH in 50% aqueous acetonitrile the reaction is slower, and metaphosphate-intermediate mechanisms are indicated for the reactions of the tri-anion and tetra-anion (and possibly, in part, for the di-anion). The reaction of the tri-anion is faster than that of the tetra-anion (poorer leaving group), but the hydrolysis of the tetra-anion is selectively accelerated by Mg^{2+} and Ca^{2+} cations. Reactions of less highly ionized species involve $S_N2(P)$ mechanisms.[366] The rates are unaffected by Mg^{++} and Ca^{++} cations in acid, though the acid-catalysed hydrolyses of inorganic triphosphate and pyrophosphate are slower in the presence of trivalent cations in particular.[367]

The rate of release of *p*-nitrophenoxide from the penta-amminecobalt(III) complex of *p*-nitrophenyl phosphate (**166**), like that from similar complexes reported previously,[368] is at least 10^8 times faster than that from the unco-ordinated phosphate in the presence of 1 M ammonia and OH^-.[369] The cleavage reaction involves intramolecular nucleophilic attack on phosphorus by a deprotonated ammonia ligand (**167**) [*E*1*cb* elimination of the aryl phosphate from (**167**) is a competing reaction] to form a bidentate complex of phosphoramidate.

$(H_3N)_5\,CoO_3POpNP^+$

(**166**)

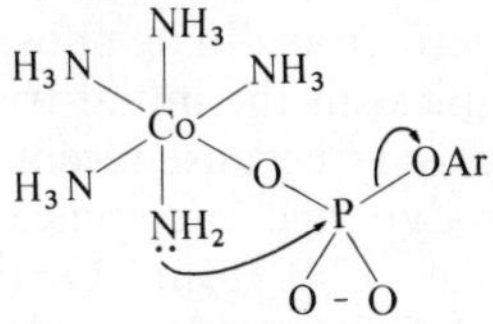

(**167**)

The hydrolysis of the corresponding complex of acetyl phosphate, $(H_3N)_5CoO.PO_2.OAc^+$, on the other hand, involves external attack by hydroxide on the carbonyl group, and a much more modest acceleration of the reaction.[370]

The hydrolysis of phosphorocreatine (**168**) in 2N HCl gives creatinine (**169**) as the predominant initial product, and is faster than either the cyclization of creatine or the hydrolysis of phosphorocreatinine.[371] It appears that cyclization and hydrolysis are concerted, in which case no satisfactory explanation is yet available.

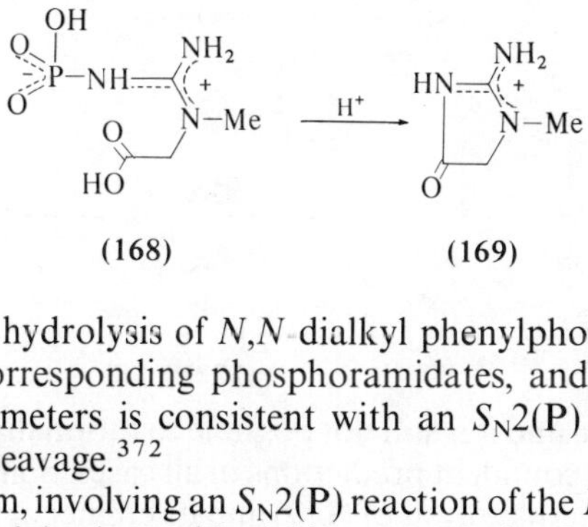

(**168**) $\xrightarrow{H^+}$ (**169**)

The acid-catalysed hydrolysis of *N*,*N*-dialkyl phenylphosphonamidates (**170**) is faster than that of corresponding phosphoramidates, and evidence from a large range of kinetic parameters is consistent with an $S_N2(P)$ mechanism (**171**) with considerable C—N cleavage.[372]

A similar mechanism, involving an $S_N2(P)$ reaction of the *N*-protonated form, has been postulated to explain the rapid trichloroacetolysis of cyclic phosphoramidate (**172**), which is several thousand times faster than that of acyclic or exocyclic nitrogen analogues.[373]

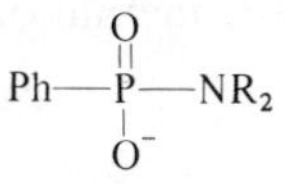

(**170**) 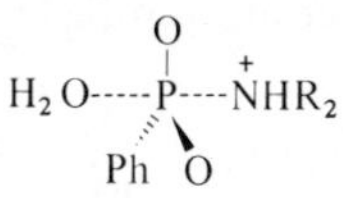 (**171**) 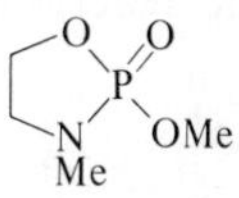 (**172**)

The alkaline hydrolysis of cyclic esters (**173**), to give (**174**), involves cleavage of the endocyclic C—O bond, and complete retention of configuration at phosphorus,[374] whereas good oxy-anion leaving groups are displaced from acyclic analogues with inversion. A new attempt to rationalize these and other related results defines a new concept, *apical potentiality*, (corresponding to what might be termed kinetic apicophilicity). If the stereochemical outcome of an $S_N2(P)$ reaction is determined by the geometry of attack of the nucleophile, as several authors now believe, then the balance between retention and inversion is determined not by relative rates of bond-cleavage and pseudorotation in pentacovalent intermediates, but by the factors controlling the direction from which the nucleophile adds. These factors are summarized by the new term, which refers to the relative likelihoods of *the original ligands* on tetraco-ordinate phosphorus being in-line with the incoming nucleophile, and thus apical in the initial trigonal bipyramidal intermediate formed. The new term is necessary because recent results make clear that this preference differs from the thermodynamic preference expressed as apicophilicity. For example, the formation of (**174**) from (**173**) (and similar observations with **175**)[375] can be rationalized if nitrogen has a *higher* apical potentiality than oxygen (or sulphur) in the cyclic cases, but *lower* in open-chain situations. Addition of hydroxide to (**173**) then generates an intermediate (**176**) with apical nitrogen, which can only lose oxygen after a pseudorotation. Attack of alkoxide on (**175**) produces a related species (**177**), which presumably breaks down directly with P—N cleavage, because this system gives (**178**) with inversion.[375]

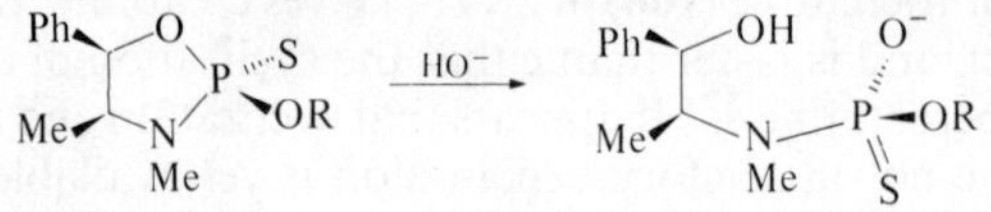

(**173**) (**174**) 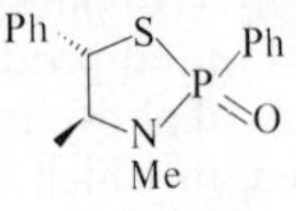 (**175**)

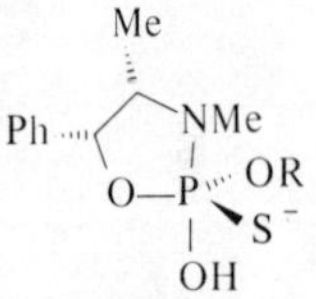

(**176**) 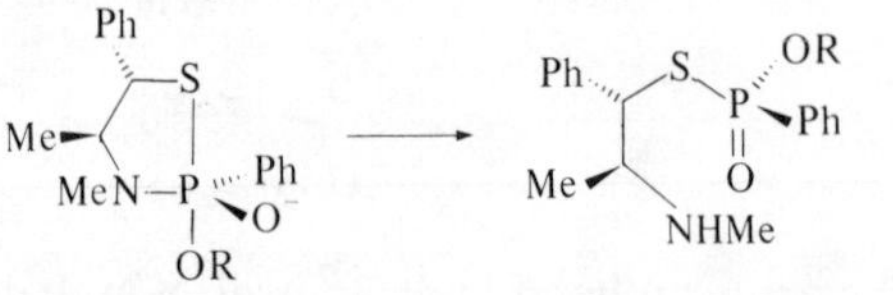 (**177**) (**178**)

This is a complex area, and it is still not possible to rationalize all the evidence now available, still less make confident predictions in all cases. Some new relevant results are the base-catalysed methanolysis of (**179**) and its epimer at phosphorus, which go with complete inversion,[376] and the (apparent) rapid hydrolysis of the amino group of the uridyluridine phosphoramidate (**180**).[377] There are no doubt innumerable related reactions as yet unexplored in P(III) chemistry: in one of the few cases observed so far the reactions of compound (**181**), with methanol, phenol, and

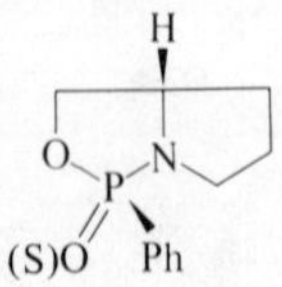

(**179**) 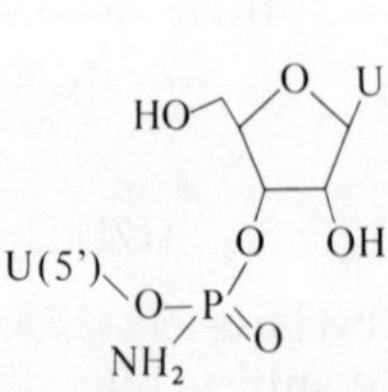 (**180**) 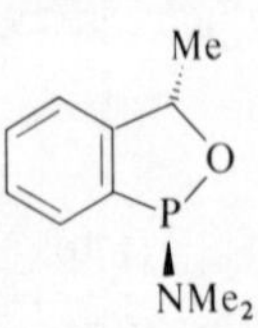 (**181**)

diethylamine, have been shown to involve exocyclic P—N cleavage with predominant inversion of configuration.[378]

The aminolysis reactions of phosphonamidic chlorides $RP(O)(NHBu^t)Cl$ with isopropylamine and *tert*-butylamine in methylene chloride appear to involve an elimination–addition mechanism, by way of a metaphosphonimidate intermediate.[379] The hydrolysis of cyclohexylphosphonodichloridate involves pyrophosphoryl intermediates[380] and the radiochloride exchange reactions of diaryl phosphorochloridates in acetonitrile show the same sensitivity to substitution as those of the corresponding chloridothioates.[381] Exchange reactions between $PhOPCl_2$ and carboxylic acid chlorides,[382] and between this and other P(III) chlorides, and chloridothioates, have been followed at high temperatures;[383] the reaction between $(MeO)_2P(O)SCl$ and tetra-*n*-propyl stannane has also been reported.[384]

Enzymic Reactions

The mechanism and regulation of ATP synthesis by the enzymes of the membrane-bound coupling factor F_1, involved in oxidative and photo-phosphorylation, have been reviewed[385] and phosphonate analogues of biological phosphates have been discussed in a review lecture.[386]

By using chiral [^{16}O, ^{17}O, ^{18}O]-esters the reaction catalysed by phosphoglucomutase, which interconverts glucose-1- and -6-phosphates by way of glucose-1,6-diphosphate and a phosphoryl-enzyme intermediate, has been shown to involve overall retention at phosphorus.[387] Phosphofructokinases from both bacterial and mammalian sources catalyse the transfer of phosphate from fructose-1,6-diphosphate to ADP with inversion.[388] The transfer of phosphate from ATP to the 3′-OH group of AMP catalysed by *E. coli* polynucleotide phosphorylase also involves inversion of configuration at phosphorus;[389] the same result has been observed for the hexokinase reaction[390] (confirming an earlier result using ATPγS)[391] and for the formation of phosphorocreatine catalysed by creatine kinase.[392] ATPγS has been used in a study of the stereochemistry of phosphate transfer to the 5′-OH groups of ApA catalysed by (bacteriophage) T4-induced polynucleotide kinase, which also proceeds with inversion.[393]

The formation of 3′,5′-cyclic AMP catalysed by adenylate cyclase (as deduced from the formation of the thiophosphate from ATPγS)[394,395] and its hydrolysis catalysed by a cyclic AMP phosphodiesterase[394,396] both (as one would hope!) go with inversion of configuration at phosphorus; as does the hydrolysis of the 2′-deoxy compound catalysed by the same enzyme.[397]

Nucleotide phosphate transfer (from dATP$\gamma S^{18}O_2$) catalysed by DNA polymerase I from *E. coli* has also been shown to involve inversion of configuration at phosphorus.[398] The exchange reaction between ADP and inorganic phosphate catalysed by polynucleotide polymerase from *M. luteus*, on the other hand, involves retention, suggesting a double-displacement mechanism for this reaction.[399] The final step of the T4 RNA ligase reaction involves inversion,[400] and the transfer of phosphate from chiral phenyl phosphate to (S)-propane-1,2-diol catalysed by liver acid phosphatase goes with retention of configuration,[401] consistent with a double-displacement mechanism involving a phosphoryl enzyme.

The phosphorylation of an enzyme from a PEP-dependent phosphotransferase system isolated from *E. coli* is accompanied by hydrogen-isotope transfer to C(3) of PEP, and a deuterium isotope effect of *ca.* 2, which shows that the proton transfer is partially rate-determining. Concerted mechanism (**182**) is one possibility.[402] NMR studies on creatine kinase have identified proton signals from three active site histidines,[403] two of which are involved in catalysis, and have shown that the equilibrium constant for bound substrates and products is close to 1.[404]

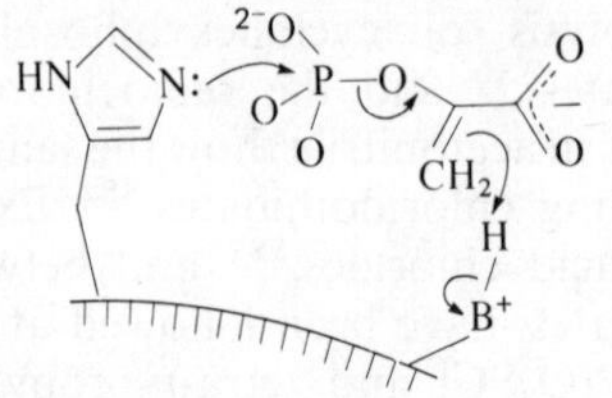

(182)

The rate of phosphate transfer to dTDP from nucleotide-5′-diphosphate kinase phosphoryl-enzyme varies by over an order of magnitude with the source of the phosphoryl group. Apparently the nucleoside triphosphate involved must leave some sort of "imprint" on the enzyme which affects its subsequent reactions even when the "old" nucleoside diphosphate has diffused away.[405]

NMR studies of a cyclic AMP-dependent protein kinase from bovine heart suggest an active enzyme–ATP–M^{n+} complex, with the metal bound to the $\beta\gamma$-pyrophosphate group. The distance between the γ-phosphate and the serine-OH group which is to be phosphorylated is large enough to suggest that a dissociative mechanism of phosphate transfer may be involved.[406]

Ethoxyformic anhydride, a reagent specific for histidine, modifies three residues of alkaline phosphatase per sub-unit, and inactivates the enzyme. Zinc ions inhibit both processes, as expected if all three histidines are involved in binding the metal at the active site.[407] X-ray structural data are beginning to appear for this enzyme, and various metal sites can be resolved.[408]

Further evidence for the scrambling of the β- and $\beta\gamma$-bridge protons of ATP, catalysed by carbamoyl phosphate synthetase in the presence of bicarbonate has appeared;[409,410] addition of ammonia stops the exchange.

A neutron-diffraction refinement of the structure of RNase A has modified some details of the conformations of three histidine residues.[411]

The apparently simple reaction catalysed by inorganic pyrophosphatase is gradually being unravelled, and—as is often the case with reactions of small substrates—is turning out to be enzymologically very complicated. In the presence of the essential Mg^{++} ions, substrate, and product, 20 distinct enzyme species must be taken into account. Cooperman and his co-workers[412] have developed an assay for pyrophosphate in the presence of an up to 10^4 molar excess of inorganic phosphate, and obtained results consistent with a model requiring three Mg^{2+} ions in the active site for enzyme–substrate complex formation. The equilibrium constant for (enzyme-bound) $PP_i \rightleftharpoons 2\,P_i$ is estimated as 4.8, and both hydrolysis and product-release steps are rate-determining. Mg^{++} is clearly involved in substrate binding, and $(H_3N)_4CoPP_i$ is a competitive inhibitor.[413] The active form of the substrate appears to be the Mg^{++} complex of fully ionized substrate, and pH studies of active site interactions have led to a fairly detailed proposal (**183**) for the mechanism of cleavage of pyrophosphate.[413]

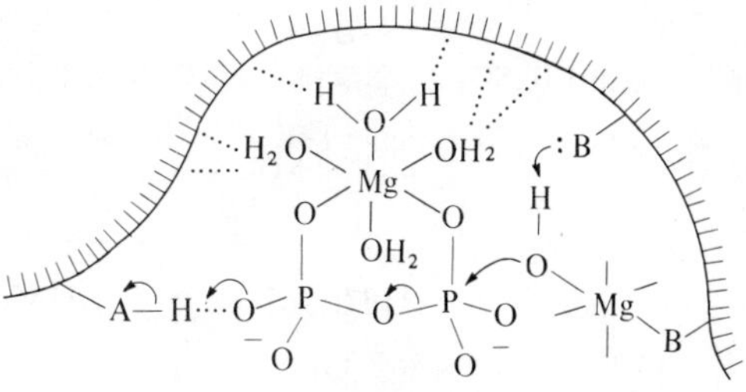

(183)

Sulphur-containing Acids

The persulphurane (**184**), the first analogue of SF_6 to be prepared with only C and O ligands, is a relatively stable crystalline solid, which rapidly breaks down in solution to the di-olefin (**185**).[414] The S—O bonds to the carboxyl groups are much (0.24 Å) longer than the ester S—O bonds, suggesting that the RO—S—OCO system is

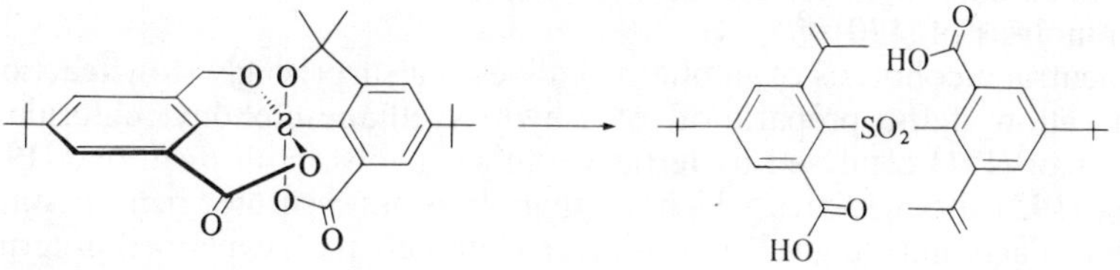

(185)

strongly polarized in the ground state, in the direction of an *E*1 transition state. An even greater difference (0.59)Å, in the same sense) is observed in the crystal structure of the sulphurane (**186**).[415]

Studies on the *E*1*cB* reactions of arylphenylmethane sulphonates[416] have been extended further with a comprehensive investigation of 2,4-dinitrophenyl esters (**187**).[417–419] A change of rate-determining step, from C—H cleavage to breakdown of the sulphene–dinitrophenoxide-ion encounter complex[416] is seen with increasing buffer concentration.[417] The charge distribution in the transition state has been investigated by varying the Ar group (**187**), and the reaction catalysed by pyridine shows a considerably greater sensitivity than the hydroxide reaction.[418] The corresponding reactions of $MeSO_2CH_2SO_2OAr$ at high pH involve both mono- and di-anions of the substrate, the latter presumably breaking down to a sulphene anion $MeSO_2\bar{C}{=}SO_2$.[420]

Several studies of S_N1 *vs.* S_N2 reactions in the solvolysis of alkyl tosylates[421–423] and alkyl fluorosulphates[424] have been reported.

The *n*-butylaminolysis of (**188**) in acetonitrile and toluene is second and third order in amine, respectively.[425] In toluene the reaction is catalysed by general bases and efficiency depends on hydrogen-bonding capability; when both nucleophile and hydrogen bond acceptor are present in the same molecule, $H_2NCH_2(CH_2OCH_2)_nCH_2NH_2$, reactivity is higher than for the corresponding long-chain α,ω-diamines. The poly-ether functions no doubt solvate and thus stabilise NH_3^+ groups during the reaction.

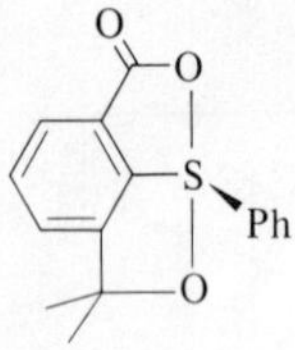

(186) 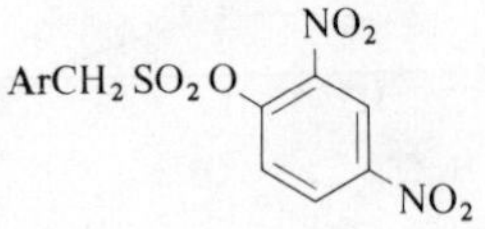 (187) 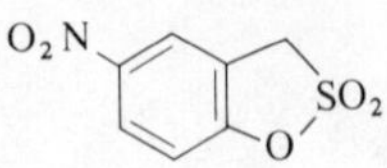 (188)

Ab initio calculations[426] on pentacovalent sulphur species are consistent with a mechanism involving such intermediates which was proposed to account for the hydrolysis of sulphonamides catalysed by a neighbouring carboxyl group:[427] a clear preference emerges for the trigonal bipyramidal configuration (**189**), of the formic acid–methanesulphonamide adduct, with the formate and ammonium groups axial; a study of salt effects on the reaction has been reported.[428]

The hydrolysis of *N*-substituted-benzyl phenylsulphonamides in strong sulphuric acid has been found to involve N—C cleavage,[429] and deuterium isotope effects and the Brønsted coefficient for general base catalysis have been measured for the *n*-butylaminolysis of (**190**).[430]

The neutral alcoholysis of sulphonyl halides is a surprisingly slow reaction, and this has allowed the preparation of 2-hydroxyethanesulphonyl chloride (**191**). Reactions of (**191**) catalysed by tertiary amines involve both β-sultone (**192**) and sulphene (**193**) intermediates.[431] The methanolysis of substituted benzenesulphonyl chlorides in acetonitrile gives a non-linear Hammett plot, explained in terms of a borderline S_N1/S_N2(S) mechanism.[432] Similar results are found for furan and thiophenesulphonyl chlorides.[433] The methanolysis of α- and β-naphthalenesulphonyl chlorides has also been studied in acetonitrile,[434] as have solvent and substituent effects on the solvolysis of arylsulphonyl chlorides in ethanol–trifluoroethanol mixtures.[435] The mesylation of *p*-chlorophenol catalysed by tertiary amines in benzene involves separate sulphene and general base catalysis routes,[436] and the same reaction with *p*-bromobenzenesulphonyl bromide is catalysed by substituted pyridines by the nucleophilic mechanism.[437] Activation parameters have been measured for the methanolysis of benzenesulphonyl chloride,[438] and for the reaction with substituted benzoate ions ($\rho = -0.35$) in methanol.[439]

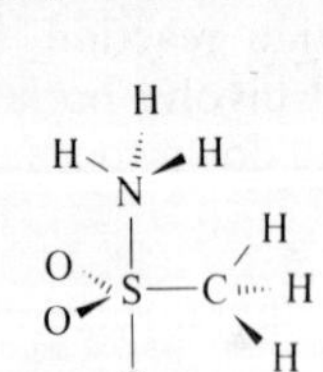

(189) 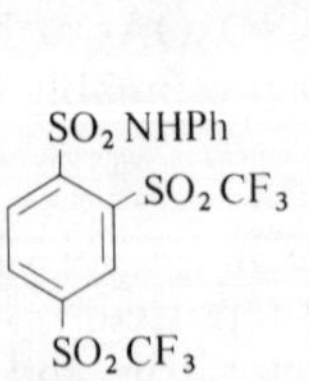 (190) 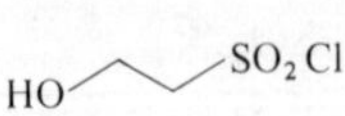 (191) 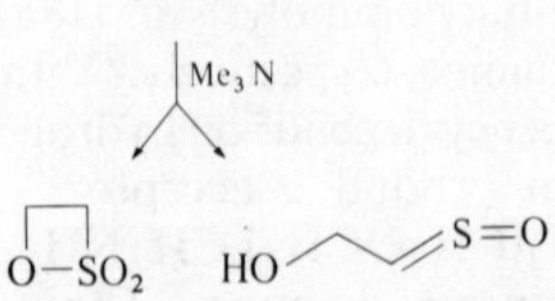 (192) (193)

Activation parameters are also reported for the solvolysis of Me_2NSO_2Cl in several mixed–aqueous solvents, and are consistent with a 'dissociative S_N2' mechanism.[440] *N*-Substituted imidazoles[441] and substituted pyridines react with arylsulphonyl halides by the nucleophilic mechanism to catalyse their reactions with *m*-chloroaniline, and the rate constants fall on the same Brønsted plot.[442] Acylation of an aniline by the *N*-sulphonylpyridine intermediate has been studied in the case of 4-dimethylaminopyridine.[443]

A study of the reactions of several nucleophiles (aniline, pyridine, and imidazole as well as acetate, nitrite, azide, and hydroxide) with 4- and 5-substituted thiophen-2-sulphonyl chlorides and fluorides in water has revealed several cases of U-shaped Hammett plots, which have been explained in terms of a shift in mechanism from S_N1- to S_N2(S)-type with more electron-withdrawing substituents.[444]

The *n*-butylaminolysis of 2,4-dinitrobenzenesulphonyl chloride leads eventually to the aniline, apparently by an S_NAr reaction with the sulphonamide initially produced.[445]

The reactions of benzenesulphonic anhydrides with anilines in nitrobenzene and its mixtures with cyclohexane are catalysed by pyridine-*N*-oxide, which acts as a nucleophile.[446]

Benzene-*o*-disulphonic ahydride (**194**) (the first reported example, originally thought to be the di-acid) is hydrolysed 200 times faster than $PhSO.SO_2Ph$, and 4000 times faster than the sulphinyl sulphone (**195**) formed spontaneously by the dehydration of another disulphinic acid (**196**). The reaction of (**194**) is characterized by a very low enthalpy of activation ($4.3\,kcal\,mol^{-1}$), compensated by a very negative $\Delta S^{\neq}$ of -49.7 e.u.[447] These data, and the solvent deuterium isotope effect, are consistent with a transition state for hydrolysis involving more than one molecule of water.

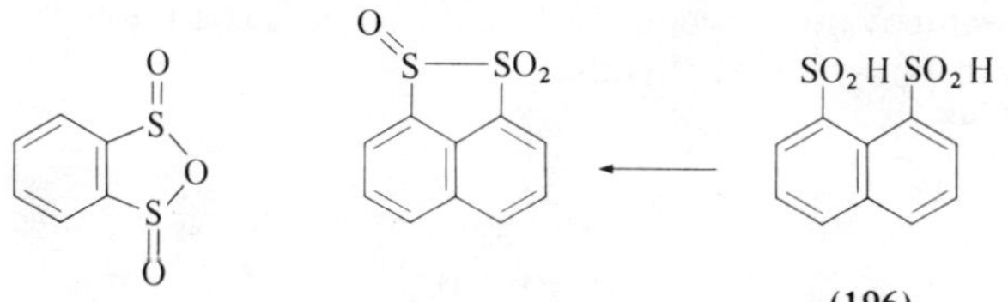

(**194**) (**195**) (**196**)

Another compound of this sort, di-*tert*-butyl sulphinic anhydride, is formed in the low-temperature oxidation of the thiosulphinate (**197**) by *m*-chloroperbenzoic acid in $CDCl_3$;[448] NMR evidence for the intermediacy of the disulphoxide (**198**) has been found.

Nucleophilic substitution reactions on alkyl α-disulphones generally involve elimination–addition mechanisms, with slow formation of sulphene intermediates. Only with weakly basic strong nucleophiles (like azide), and not too strongly acidic sulphones, is the direct substitution pathway favoured;[449] on the other hand direct substitution is favoured compared with the reactions of the corresponding sulphonyl halides. Estimates of the rate of the proton-removal step suggest that the mechanism is irreversible *E*1*cB*, or *E*2 so unsymmetrical as to be close to $E1cB_1$.

A kinetic and mechanistic study of nucleophilic substitution at the fluorine centre of perchloryl fluoride, $FClO_3$, by aryl sulphinate anions, $ArSO_2^-$, has been reported.[450]

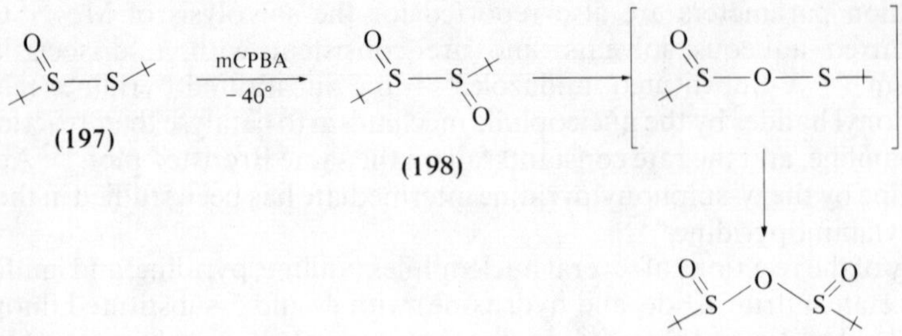

Other Acids

Nucleophilicity towards the nitrogen centre of alkyl nitrites depends on both the basicity and the polarizability of the nucleophile, as shown by the order of reactivity $N_3^- > AcO^- > NO_2^- > SCN^- > ArNH_2$ halide towards *n*-butyl nitrite in 70% dioxan–water.[451]

The acid-catalysed oxygen-exchange reaction of nitrous acid has been studied, using the isotopic shift caused by ^{18}O in the position of the ^{15}N-signal.[452] The reaction involves the conjugate acid, but the data do not allow a choice between *E–A* and *A–E* (or S_N2) mechanisms.

The temperature dependence of the deuterium isotope effect on the elimination reaction of benzyl nitrate to form benzaldehyde suggests that the proton-transfer step involves tunnelling.[453] This is considered inconsistent with the previously accepted *E*2-type mechanism.

Fluoride ion adds to bis(catecholato)silicon(IV) to give the square-pyramidal pentacovalent adduct (**199**). The corresponding Ph^- adduct is close to trigonal–bipyramidal, so a series of structures between these two extremes are to be expected for pentacovalent silicon compounds. There is an obvious similarity here to pentacovalent phosphorus chemistry.[454]

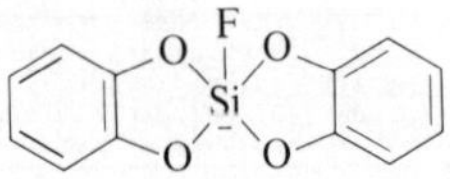

(199)

A quantitative interpretation of the auto-inhibition observed during the hydrolysis of triphenylsilane in aqueous acetonitrile is reported.[455]

References

1 See *Org. Reaction Mech.*, **1980**, 34; **1977**, 22.
2 Capon, B., Ghosh, A. K., and Grieve, D. McL. A., *Acc. Chem. Res.*, **14**, 306 (1981).
3 Capon, B., and Ghosh, A. K., *J. Am. Chem. Soc.*, **103**, 1765 (1981).
4 See *Org. Reaction Mech.*, **1979**, 28.
5 McClelland, R. A., Gedge, S., and Bohenek, J., *J. Org. Chem.*, **46**, 886 (1981).
6 Bouab, O., Lamaty, G., Moreau, C., and Pomares, O., *Nouveau J. Chim.*, **5**, 175 (1981).
7 See *Org. Reaction Mech.*, **1980**, 35.
8 Akhmatdinov, R. T., Chalova, O. B., Kantov, E. A., and Rakhmankulov, D. L., *Zh. Org. Khim.*, **16**, 962 (1980); *Chem. Abs.*, **93**, 185312 (1980).
9 McClelland, R. A., and Alibhai, M., *Can. J. Chem.*, **59**, 1169 (1981).
10 See *Org. Reaction Mech.*, **1980**, 34.

[11] See *Org. Reaction Mech.*, **1978,** 35.
[12] McClelland, R. A., and Potter, J. P., *Can. J. Chem.*, **58,** 2318 (1980).
[13] Sterba, V., *Sb. Ved. Pr., Vys. Sk. Chemickotechnol. Pardubice*, **42,** 43 (1980); *Chem. Abs.*, **94,** 191067 (1981).
[14] See *Org. Reaction Mech.*, **1979,** 27.
[15] Marlier, J. F., and O'Leary, M. H., *J. Org. Chem.*, **46,** 2175 (1981).
[16] Risley, J. M., and Van Etten, R. L., *J. Am. Chem. Soc.*, **103,** 4389 (1981).
[17] Yates, K., and McClelland, R. A., *J. Am. Chem. Soc.*, **89,** 2686 (1967).
[18] Said, Z., and Tillett, J. G., *J. Org. Chem.*, **46,** 2586 (1981).
[19] Singh, L., Singh, R. T., and Sha, R. C., *J. Indian Chem. Soc.*, **57,** 1089 (1980); *Chem. Abs.*, **94,** 83252 (1981).
[20] Mata-Segreda, J. F., *J. Phys. Chem.*, **85,** 1958 (1981).
[21] Zaki, A. B., and El-Sheikh, Y. M., *Z. Phys. Chem.* (*Leipzig*), **262,** 189 (1981).
[22] Mishra, J. P., and Bansal, N. L., *Z. Phys. Chem.* (*Leipzig*), **262,** 683 (1981).
[23] Hasan, M., Khan, N., and Latif, F., *Islamabad J. Sci.*, **5,** 35 (1978); *Chem. Abs.*, **94,** 64919 (1981).
[24] Rackham, D. M., Chakrabarti, J., and Davies, G. L. O., *Talanta*, **28,** 329 (1981); *Chem. Abs.*, **95,** 96456 (1981).
[25] DeTar, D. F., *J. Org. Chem.*, **45,** 5166 (1980).
[26] See *Org. Reaction Mech.*, **1975,** 33; **1977,** 28; **1978,** 36; **1979,** 30.
[27] Mager, H., *Tetrahedron*, **37,** 509 (1981).
[28] Mager, H., *Tetrahedron*, **37,** 523 (1981).
[29] Sawada, M., Ichihara, M. Ando, T., Yukawa, Y., and Tsuno, Y., *Tetrahedron Lett.*, **1981,** 4733.
[30] Fadnavis, N. W., and Bagavant, G., *Indian J. Chem.*, **17B,** 518 (1979); *Chem. Abs.*, **93,** 185531 (180).
[31] Sukhorukov, Yu. I., Plonov, V. M., Finkel'shtein, B. L., and Istomin, B. I., *Zh. Org. Khim.*, **17,** 1149 (1981); *Chem. Abs.*, **95,** 131911 (1981).
[32] Istomin, B. I., Finkel'shtein, B. L., and Eliseeva, G. D., *Zh. Org. Khim.*, **16,** 2268 (1980); *Chem. Abs.*, **94,** 46617 (1981).
[33] Eliseeva, G. D., Bazhenov, B. N., Istomin, B. I., and Finkel'shtein, B. L., *Izv. Vyssh. Uchebn. Zaved., Khim. Khim. Tekhnol.*, **23,** 790 (1980); *Chem. Abs.*, **94,** 14943.
[34] Kalfus, K., and Večeřa, M., *Commun.-Czech.-Pol. Colloq. Chem. Thermodyn. Phys. Org. Chem. 2nd*, **1980,** 215; *Chem. Abs.*, **95,** 131910 (1981).
[35] Holba, V., and Benko, J., *Collect. Czech. Chem. Commun.*, **45,** 2873 (1980).
[36] Holba, V., Benko, J., and Komadel, P., *Z. Phys. Chem.* (*Leipzig*), **262,** 445 (1981).
[37] Young, S. D., Coblens, K. E., and Ganem, B., *Tetrahedron Lett.*, **1981,** 4887.
[38] Broxton, T. J., and Duddy, N. W., *J. Org. Chem.*, **46,** 1186 (1981).
[38a] Douglas, K. T., and Alborz, M., *J. Chem. Soc., Chem. Commun.*, **1981,** 551.
[39] Kristol, D. S., Klotz, H., and Parker, R. C., *Tetrahedron Lett.*, **1981,** 967
[40] Mitzner, R., and Lemke, F., *Z. Chem.*, **21,** 189 (1981).
[41] Huskey, W. P., Warren, C. T., and Hogg, J. L., *J. Org. Chem.*, **46,** 59 (1981).
[42] Weeks, D. P., and Whitney, D. B., *J. Am. Chem. Soc.*, **103,** 3555 (1981).
[43] Baliah, V., and Gurumurthy, R., *Indian J. Chem.*, **20B,** 629 (1981); *Chem. Abs.*, **95,** 145719 (181).
[44] Baliah, V., and Gurumurthy, R., *Indian. J. Chem.*, **20B,** 725 (181); *Chem. Abs.*, **95,** 168066 (1981).
[45] Zalewski, R. I., *Zanco*, **6A,** 33 (1980); *Chem. Abs.*, **94,** 208053 (1981).
[46] Ostrizhko, F. N., Yuminov, V. S., and Panshin, Yu. A., *Sintez. Pentaplasta L*, **1979,** 5; *Chem. Abs.*, **94,** 64792 (1981).
[47] Moisa, C., *Mater. Plast.* (*Bucharest*), **17,** 85 (1980) *Chem. Abs.*, **93,** 185355 (1980).
[48] Tucek, E., and Dinse, H. D., *Acta Polym.*, **31,** 429 (1980); *Chem. Abs.*, **94,** 46377 (1981).
[49] Aslam, M. H., Burden, A. G., Chapman, N. B., Shorter, J., and Charton, M., *J. Chem. Soc., Perkin Trans.* **2**, **1981,** 500.
[50] Aslam, M., Chapman, N. B., Shorter, J., and Charton, M., *J. Chem. Soc., Perkin Trans. 2*, **1981,** 720.
[51] Aslam, M., Chapman, N. B., Charton, M., and Shorter, J., *J. Chem. Res.*, **1981,** 188(S), 2301(M).
[52] Ananthakrishnanadar, P., Kannan, N., and Tharunaraj, G. V., *Indian J. Chem.*, **19B,** 621 (1980); *Chem. Abs.*, **94,** 29752 (1981).
[53] Noto, R., Buscemi, S., Consiglio, G., and Spinelli, D., *J. Heterocycl. Chem.*, **18,** 735 (1981).
[54] Radojkovic-Velickovic, M., Misic-Vukovic, M., and Dimitrijevic, D., *Glas. Hem. Drus. Beograd*, **45,** 261 (1980); *Chem. Abs.*, **94,** 191178 (1981).
[55] Sutyagin, V. M., Lopatinskii, V. P., Tuzovskaya, S. A., and Ovchinnikova, L. V., *Izv. Vyssh. Uchebn. Zaved, Khim. Khim. Tekhnol.*, **23,** 1213 (1980); *Chem. Abs.*, **94,** 64787 (1981).
[56] Gillen, C. J., Knipe, A. C., and Watts, W. E., *Tetrahedron Lett.*, **1981,** 597.
[57] Shames, S. L., and Byers, L. D., *J. Am. Chem. Soc.*, **103,** 6170 (1981).

[58] Kovach, I. M., Elrod, J. P., and Schowen, R. L., *J. Am. Chem. Soc.*, **102,** 7530 (1980).
[59] See *Org. Reaction Mech.*, **1980,** 37.
[60] See *Org. Reaction Mech.*, **1978,** 38; **1979,** 27.
[61] See *Org. Reaction Mech.*, **1979,** 32.
[62] Guanti, G., Cevasco, G., Thea, S., Dell'Erba, C., and Petrillo, G., *J. Chem. Soc., Perkin Trans. 2,* **1981,** 327.
[63] Cox, M. M., and Jencks, W. P., *J. Am. Chem. Soc.*, **103,** 572 (1981).
[64] See *Org. Reaction Mech.*, **1978,** 32.
[65] Cox, M. M., and Jencks, W. P., *J. Am. Chem. Soc.*, **103,** 580 (1981).
[66] Jagdale, M. H., Salunkhe, M. M., Kadam, S. D., and Khodaskar, S. N., *Acta Cienc. Indica (Ser.) Chem.*, **6,** 9 (1980); *Chem. Abs.*, **94,** 3407 (1981).
[67] Deady, L. W., and Finlayson, W. L., *Aust. J. Chem.*, **33,** 2441 (1980).
[68] Castro, C., and Castro, E. A., *J. Org. Chem.*, **46,** 2939 (1981).
[69] Lin, S.-F., and Connors, K. A., *J. Pharm. Sci.*, **70,** 235 (1981).
[70] Tribunescu, P., Poraicu, M., Merca, E., and Facsko, O., *Bul. Stiint. Teh. Inst. Politeh. "Traian Vuia" Timosoara, Ser. Chim.*, **23,** 147 (1978); *Chem. Abs.*, **93,** 203607 (1980).
[71] Merca, E., Poraicu, M., and Tribunescu, P., *Bul. Stiint. Teh. Inst. Politeh. "Traian Vuia" Timosoara, Ser. Chim.*, **23,** 160 (1978); *Chem. Abs.*, **93,** 238179 (1980).
[72] Lychkin, I. P., Kontsova, L. V., and Berdutin, A. Ya., *Zh. Prikl. Khim.*, (*Leningrad*), **53,** 2461 (1980); *Chem. Abs.*, **94,** 120467 (1981).
[73] Kayser, M. M., and Eisenstein, O., *Can. J. Chem.*, **59,** 2457 (1981).
[74] Bothe, E., Dessouki, A. M., and Schulte-Frohlinde, D., *J. Phys. Chem.*, **84,** 3270 (1980).
[75] See *Org. Reaction Mech.*, **1979,** 35.
[76] Exner, O., and Jehlicka, V., *Collect. Czech. Chem. Commun.*, **46,** 856 (1981).
[77] Douglas, K. T., Naggi, N. F., and Mervis, C. M., *J. Chem. Soc., Perkin Trans. 2,* **1981,** 171.
[78] Okuyama, T., Nagamatsu, H., and Fueno, T., *J. Org. Chem.*, **46,** 1336 (1981).
[70] Lemetais, P., and Carpentier, J.-M., *J. Chem. Res.*, **1981,** 282(S), 3369(M).
[80] Mayanna, S. M. and Prakash, B. S. J., *J. Sci. Res.* (*Bhopal, India*), **2,** 145 (1980); *Chem. Abs.*, **94,** 83253 (1981).
[81] Radom, L., and Riggs, N. V., *Aust. J. Chem.*, **33,** 2337 (1980).
[82] Lee, I., and Yang, K., *Bull. Korean Chem. Soc.*, **1,** 109 (1980); *Chem. Abs.*, **94,** 83473 (1981).
[83] Yoder, C. Y., and Gardner, R. D., *J. Org. Chem.*, **46,** 64 (1981).
[84] Perrin, C. L., Johnston, E. R., Lollo, C. P., and Kobrin, P. A., *J. Am. Chem. Soc.*, **103,** 4691 (1981).
[85] Perrin, C. L., and Johnston, E. R., *J. Am. Chem. Soc.*, **103,** 4697 (1981).
[86] Hine, J., King, R. S. L. M., Midden, W. R., and Sinha, A., *J. Org. Chem.*, **46,** 3186 (1981).
[87] Sidahmed, I. M., Salem, S. M., and Abdel Halim, F. M., *Rev. Roum. Chim.*, **26,** 37 (1981).
[88] Wan, P., Modro, T. A., and Yates, P., *Can. J. Chem.*, **58,** 2423 (1980).
[89] Cox, R. A., and Yates, K., *Can. J. Chem.*, **59,** 2853 (1981).
[90] Cox, R. A., and Yates, K., *Can. J. Chem.*, **59,** 1560 (1981).
[91] Cox, R. A., Druet, L. M., Klausner, A. E., Modro, T. A., Wan, P., and Yates, K., *Can. J. Chem.*, **59,** 1568 (1981).
[92] Cox, R. A., and Yates, K., *Can. J. Chem.*, **59,** 2116 (1981).
[93] Chubarov, G. A., Obmelyukhina, T. N., Danov, S. M., and Pastukhova, G. L., *Zh. Prikl. Khim., Leningrad*, **53,** 1808 (1980); *Chem. Abs.*, **94,** 14738 (1981).
[94] Elsemongy, M. M., Amila, M. F., and Ahmed, A. M., *Indian J. Chem.*, **19A,** 658 (1980); *Chem. Abs.*, **94,** 3403 (1981).
[95] Elsemongy, M. M., and Amila, M. F., *J. Indian Chem. Soc.*, **57,** 506 (1980); *Chem. Abs.*, **93,** 238537 (1980).
[96] Patterson, J. F., Huskey, W. P., and Hogg, J. L., *J. Org. Chem.*, **45,** 4675 (1980).
[97] See *Org. Reaction Mech.*, **1978,** 41.
[98] See *Org. Reaction Mech.*, **1977,** 31.
[99] Huskey, W. P., and Hogg, J. L., *J. Org. Chem.*, **46,** 53 (1981).
[100] Cipiciani, A., Linda, P., Savelli, G., and Bunton, C. A., *J. Am. Chem. Soc.*, **103,** 4875 (1981).
[101] See *Org. Reaction Mech.*, **1978,** 41.
[102] Florio, S., Leng, J. L., and Stirling, C. J. M., *J. Heterocycl. Chem.*, **18,** 857 (1981).
[103] Suh, J., and Lee, B. H., *Taehan Hwahakhoe Chi*, **24,** 469 (1980); *Chem. Abs.*, **94,** 191440 (1981).
[104] Issa, I. M., Diefallah, E., Mahmoud, M., and Mousa, M. A., *Egypt J. Chem.*, **21,** 319 (1980); *Chem. Abs.*, **94,** 2974 (1981).
[105] Bhatt, M. V., El Ashry, S. H., and Somayaji, V., *Proc. Indian Acad. Sci.*, **89,** 7 (1980); *Chem. Abs.*, **93,** 167129 (1980).

[106] Litvinenko, L. M., Semenyuk, G. V., and Zhil'tsov, N. P., *Zh. Org. Khim.*, **17**, 961 (1981); *Chem. Abs.*, **95**, 96516 (1981).
[107] Jorge, J. A. L., Kiyan, N. Z., Miyata, Y., and Miller, J., *J. Chem. Soc., Perkin Trans. 2*, **1981**, 100.
[108] La, S., and Lee, I., *Taehan Hwahakhoe Chi*, **24**, 288 (1980); *Chem. Abs.*, **95**, 60881 (1981).
[109] La, S., Koh, K. S., and Lee, I., *Taehan Hwahakhoe Chi*, **24**, 8 (1980); *Chem. Abs.*, **93**, 203581 (1980).
[110] Villain, G., Constant, G., Gaset, E., and Kalck, P., *J. Mol. Catal.*, **7**, 355 (1980); *Chem. Abs.*, **94**, 208015 (1981).
[111] Chubarov, G. A., Danov, S. M., Logutov, V. I., Brovkina, G. V., and Kolesnikov, V. A., *Zh. Prikl. Khim. (Leningrad)*, **53**, 1376 (1980); *Chem. Abs.*, **93**, 203557 (1980).
[112] Beltrame, P., *Nouveau J. Chim.*, **5**, 453 (1981).
[113] Glatt, H. H., Bacaloglu, I., Truong, T. K., Boeriu, C., Bacaloglu, R., Martin, D., and Graubaum, H., *J. Prakt. Chem.*, **322**, 1053 (1980); *Chem. Abs.*, **94**, 174056 (1981).
[114] Butler, A. R., and Hussain, I., *J. Chem. Soc., Perkin Trans. 2*, **1981**, 317.
[115] Müller-Litz, W., and Mortag, M., *Z. Chem.*, **20**, 344 (1980); Müller-Litz, W., *Z. Chem.*, **20**, 344 (1980).
[116] Taran, P. N., and Rak, L. V., *Ukr. Khim. Zh.* '(*Russ. Ed.*), **46**, 866 (1980); *Chem. Abs.*, **93**, 238185 (1980).
[117] Battye, P. J., Cassidy, J. F., and Moodie, R. B., *J. Chem. Soc., Chem. Commun.*, **1981**, 68.
[118] Bergon, M., and Calmon, J.-P., *Tetrahedron Lett.*, **1981**, 937.
[119] Poncin, G., Plusquellec, D., and Martin, G., *J. Fr. Hydrol.*, **11**, 101 (1980); *Chem. Abs.*, **95**, 168009 (1981).
[120] Parkhomenko, P. I., Rybakova, M. V., Bezmenova, T. E., and Zaika, T. D., *Khim. Geterosikl. Soedin.*, **1980**, 1482; *Chem. Abs.*, **94**, 120472 (1981).
[121] Berner, J., and Heutzenroeder, K., *Wirkungsmech. Herbiz. Synth. Wachtumsregul.* (*Ber. Symp. Wiss. Koordinierungskonf.*) *12th*, **1979**, 20; *Chem. Abs.*, **93**, 238468 (1980).
[122] Mindl., J., Balcárek, P., Šilar, L., and Večeřa, M., *Collect. Czech. Chem. Commun.*, **45**, 3130 (1980).
[123] Mindl, J., Sulzar, J., and Večeřa, M., *Collect. Czech. Chem. Commun.*, **46**, 1970 (1981).
[124] See *Org. Reaction Mech.*, **1980**, 44.
[125] Pytela, O., Svoboda, P., and Večeřa, M., *Collect. Czech. Chem. Commun.*, **46**, 2091 (1981).
[126] Chicu, A., and Bacaloglu, R., *Rev. Roum. Chim.*, **26**, 435 (1981).
[127] Bacaloglu, R., Chicu, A., Bacaloglu, I., and Cotarcă, L., *Rev. Roum. Chim.*, **26**, 611 (1981).
[128] Kordi, T., and Bacaloglu, R., *Bul. Stiint. Teh. Inst. Politeh, "Traian Vuia" Timosoara, Ser. Chim.*, **24**, 84 (1979); *Chem. Abs.*, **95**, 131894 (1981).
[129] Nagubandi, S., and Fodor, G., *J. Heterocycl. Chem.*, **17**, 1457 (1980).
[130] Kim, J. K., and Caserio, M. C., *J. Am. Chem. Soc.*, **103**, 2124 (1981).
[131] Carlsen, R., Egsgaard, H., and Pagsberg, P., *J. Chem. Soc., Perkin Trans. 2*, **1981**, 1256.
[132] Bloch, R., and Orvane, P., *Tetrahedron Lett.*, **1981**, 3597.
[133] Chuchani, G., Martin, I., and Alonso, M. E., *J. Phys. Chem.*, **85**, 1241 (1981).
[134] Taylor, R., *J. Chem. Res.(S)*, **1981**, 250.
[135] Chuchani, G., Triana, J. L., Rotinov., A., and Caraballo, D. F., *J. Phys. Chem.*, **85**, 1243 (1981).
[136] Chuchani, G., and Dominguez, R. M., *Int. J. Chem. Kinet.*, **13**, 577 (1981).
[137] See *Org. Reaction Mech.*, **1980**, 46; **1979**, 35.
[138] Fedyanin, N. P., *Zh. Prikl. Khim. (Leningrad)*, **53**, 2620 (1980); *Chem. Abs.*, **94**, 102560 (1981).
[139] Dutka, F., Kömives, T., and Marton, A., *Magy. Kem. Foly.*, **86**, 558 (1980); *Chem. Abs.*, **95**, 6006 (1981).
[140] See *Org. Reaction Mech.*, **1978**, 50.
[141] Dutka, F., Kömives, T., and Marton, A., *Magy. Kem. Foly.*, **86**, 552 (1980); *Chem. Abs.*, **95**, 23737 (1981).
[142] Oleinik, N. M., Litvinenko, L. M., Sadovskii, Yu. S., Piskunova, Zh. P., and Popov, A. F., *Zh. Org. Khim.*, **16**, 1469 (1980); *Chem. Abs.*, **93**, 203585 (1980).
[143] Kömives, T., Marton, A. F., Dutka, F., Low, M., and Kisfaludy, L., *React. Kinet. Catal. Lett.*, **13**, 357 (1980); *Chem. Abs.*, **94**, 3548 (1981).
[144] Hadju, J., and Smith, G. M., *J. Am. Chem. Soc.*, **103**, 6192 (1981).
[145] Phelps, D. J., Godreau, P. V., and Nicholas, E. S., *J. Chem. Soc., Perkin Trans. 2*, **1981**, 140.
[146] Shinkai, S., Nakashima, N., and Kunitake, T., *Bull. Chem. Soc. Jpn*, **54**, 840 (1981).
[147] Spasskaya, R. I., Khitrin, S. V., and Zil'berman, E. N., *Izv. Vyssh. Uchebn. Zaved, Khim. Khim. Tekhnol.*, **23**, 1346 (1980); *Chem. Abs.*, **94**, 46429 (1981).
[148] Andreev, S. A., Lebedev, B. A., Tselinskii, I. V., and Shhobhor, I. N., *Zh. Org. Khim.*, **16**, 1353 (1980); *Chem. Abs.*, **93**, 238148 (1980).
[149] Andreev, S. A., Lebedev, B. A., and Tselinski, I. V., *Zh. Org. Khim.*, **16**, 1360 (1980); *Chem. Abs.*, **93**, 238149 (1980).

[150] Andreev, S. A., Lebedev, B. A., and Tselinski, I. V., *Zh. Org. Khim.*, **16,** 1370 (1980); *Chem. Abs.*, **93,** 238150 (1980).
[151] Andreev, S. A., Lebedev, B. A., and Tselinski, I. V., *Zh. Org. Khim.*, **16,** 1374 (1980); *Chem. Abs.*, **93,** 238151 (1980).
[152] Andreev, S. A., Lebedev, B. A., and Tselinski, I. V., *Zh. Org. Khim.*, **16,** 1365 (1980); *Chem. Abs.*, **94,** 3382 (1981).
[153] Moodie, R. B., and Sansom, P. J., *J. Chem. Soc., Perkin Trans. 2,* **1981,** 664.
[154] Stachell, R. S., and Nyman, R., *J. Chem. Soc., Perkin Trans. 2,* **1981,** 901.
[155] Locsei, V., Lucaciu, N., Prosteanu, N., and Pape, R. F., *Rev. Chim. (Bucharest)*, **31,** 969 (1980); *Chem. Abs.*, **94,** 174072 (1981).
[156] Dabi, S., and Zilkha, A., *Eur. Polymer J.*, **16,** 475 (1980); *Chem. Abs.*, **93,** 203559 (1980).
[157] Dabi, S., and Zilkha, A., *Eur. Polymer J.*, **16,** 471 (1980); *Chem. Abs.*, **93,** 167191 (1980).
[158] Bakalo, L. A., Chirkova, L. I., and Lipatova, T. E., *Ukr. Khim. Zh. (Russ. Ed.)*, **46,** 645 (1980); *Chem. Abs.*, **93,** 185329 (1980).
[159] Dragolov, V. V., Chimishkyan, A. L., and Sukhareva, I. P., *Izv. Vyssh. Uchebn. Zaved., Khim. Khim. Tekhnol.*, **23,** 1557 (1980); *Chem. Abs.*, **94,** 174102 (1981).
[160] Kulagin, N. I., Mogilyanskii, A. I., Brailovskii, S. M., Kukalenko, S. S., and Volodkovich, S. D., *Khim. Primen. Pestits. Prep.*, **1977,** 31; *Chem. Abs.*, **93,** 167177 (1980).
[161] See *Org. Reaction Mech.*, **1980,** 46.
[162] Benoiton, N. L., and Chen, F. M. F., *J. Chem. Soc., Chem. Commun.*, **1981,** 543.
[163] Sadovnikov, A. I., and Kuritsyn, L. U., *Izv. Vyssh. Uchebn. Zaved., Khim. Khim. Tekhnol.*, **24,** 316 (1981); *Chem. Abs.*, **95,** 23809 (1981).
[164] Ivanova, N. S., Vorob'ev, N. K., and Kulikova, T. V., *Vopr. Kinet. i Kataliza, Ivanovo,* **1979,** 33; *Chem. Abs.*, **95,** 60918 (1981).
[165] Yoon, S. K., Uhm, T. S., and Sung, D. D., *Taehan Hwahakhoe Chi,* **24,** 347 (1980); *Chem. Abs.*, **94,** 120494 (1981).
[166] Litvinenko, L. M., Kirichenko, A. I., and Bondarenko, L. I., *Zh. Org. Khim.*, **16,** 2503 (1980); *Chem. Abs.*, **94,** 102410 (1981).
[167] Kirichenko, A. I., Litvinenko, L. M., and Bondarenko, L. I., *Dokl. Akad. Nauk SSSR*, **254,** 151 (1980); *Chem. Abs.*, **94,** 46407 (1981).
[168] Zhil'tsor, N. P., and Semenyuk, G. V., *Ukr. Khim. Zh.(Russ. Ed.)*, **46,** 1087 (1980); *Chem. Abs.*, **94,** 46406 (1981).
[169] Kosolapov, V. T., Zlobin, V. A., Ioganov, K. M., and Tarasov, A. K., *Kinet. Katal.*, **22,** 318 (1981); *Chem. Abs.*, **95,** 96542 (1981).
[170] Shpan'ko, I. V., Litvinenko, L. M., Goncharov, A. N., and Korzhilova, O. I., *Zh. Org. Khim.*, **17,** 965 (1981); *Chem. Abs.*, **95,** 96517 (1981).
[171] Norokhatka, D. A., Kheifets, V. I., Morozova, V. Ya., Brandina, L. I., Lyubimova, T. B., Shmelev, V. A., and Shumskii, F. Z., *Monomery dlya Polikondensatsii, Tula,* **1979,** 17; *Chem. Abs.*, **94,** 3427 (1981).
[172] Sokolenko, V. N., and Drupp, P. V., *Vopr. Khim. Khim. Tekhnol.*, **59,** 92 (1980); *Chem. Abs.*, **95,** 23756 (1981).
[173] Nemirova, L. I., Kondratenko, V. I., and Mantulo, A. P., *Katal. Katal.*, **18,** 56 (1980); *Chem. Abs.*, **95,** 41818 (1981).
[174] Berezin, I. V., and Yatsimirskii, A. K., *Zh. Fiz. Khim.*, **54,** 1633 (1980); *Chem. Abs.*, **93,** 203570 (1980).
[175] See *Org. Reaction Mech.*, **1980,** 47.
[176] Fife, T. H., and Hutchins, J. E. C., *J. Am. Chem. Soc.*, **103,** 4194 (1981).
[177] See *Org. Reaction Mech.*, **1978,** 52.
[178] Grayson, B. T., *Pestic. Sci.*, **11,** 493 (1980); *Chem. Abs.*, **94,** 173869 (1981).
[179] Caswell, M., Chaturvedi, R. K., Lane, S. M., Zvilichovsky, B., and Schmir, G. L., *J. Org. Chem.*, **46,** 1585 (1981).
[180] Kemp, D. S., Kerkman, D. J., Leung, S.-L., and Hanson, G., *J. Org. Chem.*, **46,** 490 (1981).
[181] Tsuji, A., Nakashima, E., Deguchi, Y., Nishida, K., Shimizu, T., Horiuchi, S., Ishikawa, R., and Yamana, T., *J. Pharm. Sci.*, **70,** 1120 (1981).
[182] Macháček, S., El-Bahaie, and Šteřba, V., *Collect. Czech. Chem. Commun.*, **46,** 256 (1981).
[183] Su, T. T., and Chang, C. T., *J. Org. Chem.*, **46,** 2812 (1981).
[184] Kavález, J., Kotyk, M., El-Bahaie, and Šteřba, V., *Collect. Czech. Chem. Commun.*, **46,** 246 (1981).
[185] See *Org. Reaction Mech.*, **1978,** 51.
[186] See *Org. Reaction Mech.*, **1978,** 48.
[187] Rozario, A. P., Craig, D. J., Hudson, R. F., and Williams, A., *J. Chem. Soc., Perkin Trans. 2,* **1981,** 590.
[188] Loftsson, T., and Bodor, N., *J. Pharm. Sci.*, **70,** 750 (1981).
[189] Verbicky, J. W., and Williams, L., *J. Org. Chem.*, **46,** 175 (1981).

[190] Dellacoletta, B. A., Ligon, W. V., Verbicky, J. W., and Williams, L., *J. Org. Chem.*, **46,** 3923 (1981).
[191] Errede, L. A., Etter, M. C., Williams, R. C., and Darnauer, S. M., *J. Chem. Soc., Perkin Trans. 2*, **1981,** 233.
[192] Bhatt, M. V., El Ashry, S. H., and Somayaji, V., *Indian J. Chem.*, **19B,** 473 (1980); *Chem. Abs.*, **93,** 238410 (1980).
[193] Rao, M. V., Shaller, H. E. A., and Bhatt, M. V., *Tetrahedron Lett.*, **1981,** 145.
[194] Shames, S. L., and Byers, L. D., *J. Am. Chem. Soc.*, **103,** 6177 (1981).
[195] Boyer, G. P., *Adv. Catal.*, **29,** 197 (1980).
[196] Sugimoto, T., and Tanimoto, S., *Yuki Gosei Kagaku Kyokai-Shi*, **38,** 747 (1980); *Chem. Abs.*, **94,** 3316 (1981).
[196a] Guthrie, J. P., and O'Leary, S., *Can. J. Chem.*, **59,** 2358 (1981). See *Org. Reaction Mech.*, **1975,** 43.
[197] Barbaric, S., and Luisi, P., *J. Am. Chem. Soc.*, **103,** 4329 (1981).
[198] Bunton, C. A., and de Buzzacarini, F., *J. Phys. Chem.*, **85,** 3139 (1981).
[199] Franklin, T. C., and Iwunre, M., *J. Am. Chem. Soc.*, **103,** 5937 (1981).
[200] Kon-No, I., Inoue, T., Hanada, T., Nakamura, K., and Kitahara, A., *Yukagaku*, **29,** 670 (1980); *Chem. Abs.*, **94,** 83200 (1981).
[201] El Seoud, M. I., and El Seoud, O. A., *J. Org. Chem.*, **46,** 2686 (1981).
[202] Kunitake, T., Okahata, Y., Shimomura, M., Yasunami, S., and Takarabe, K., *J. Am. Chem. Soc.*, **103,** 5401 (1981).
[203] See *Org. Reaction Mech.*, **1978,** 66.
[204] Shinkai, S., Hirakawa, S., Shimomura, M., and Kunitake, T., *J. Org. Chem.*, **46,** 868 (1981).
[205] Kunitake, T., Okahata, Y., Ando, R., Shinkai, S., and Hirakawa, S., *J. Am. Chem. Soc.*, **102,** 7877 (1980).
[206] Taniguchi, Y., Makimoto, S., and Suzuki, K., *J. Phys. Chem.*, **85,** 2218 (1981).
[207] Morild, E., and Aksnes, G., *Acta Chem. Scand.*, **A35,** 169 (1981).
[208] Al-Lohedan, H., Bunton, C. A., and Romsted, L. S., *J. Phys. Chem.*, **85,** 2123 (1981).
[209] Al-Lohedan, H., and Bunton, C. A., *J. Org. Chem.*, **46,** 3929 (1981).
[210] Ishiwatari, T., Okubo, T., and Ise, N., *Polym. Prepr., Am. C.S. Div. Polym. Chem.*, **20,** 634 (1979); *Chem. Abs.*, **94,** 155945 (1981).
[211] Kudryavtseva, L. A., Il'ina, O. M., Bel-skii, V. E., Fedorov, S. B., and Ivanov, B. E., *Izv. Akad. Nauk SSSR, Ser. Khim.*, **1980,** 1915; *Chem. Abs.*, **93,** 167199 (1980).
[212] See *Org. Reaction Mech.*, **1980,** 52.
[213] Moss, R. A., and Lee, Y.-S., *Tetrahedron Lett.*, **1981,** 2353.
[214] Cipiciani, A., Linda, P., Savelli, G., and Bunton, C. A., *J. Org. Chem.*, **46,** 911 (1981).
[215] Broxton, T. J., Fernando, D. R., and Rowe, J. E., *J. Org. Chem.*, **46,** 3522 (1981).
[216] Broxton, T. J., Fernando, D. R., and Rowe, J. E., *Aust. J. Chem.*, **34,** 1615 (1981).
[217] Meyer, G., and Viout, P., *Tetrahedron*, **37,** 2269 (1981).
[218] Soto, R., Meyer, G., and Viout, P., *Tetrahedron*, **37,** 2977 (1981).
[219] Moss, R. A., and Bizzigotti, G. O., *J. Am. Chem. Soc.*, **103,** 6512 (1981).
[220] Ueoka, R., and Matsumoto, Y., *J. Chem. Res.*, **1981,** 242(S), 2660(M).
[221] Ono, S., Shosenji, H., and Yamada, K., *Tetrahedron Lett.*, **1981,** 2391.
[222] Ihara, Y., Hosako, R., Nango, M., and Kuroki, N., *J. Chem. Soc., Chem. Commun.*, **1981,** 393.
[223] Ihara, Y., Kunikiyo, N., Kunimasa, T., Nango, M., and Kuroki, N., *Chem. Lett.(Tokyo)*, **1981,** 67.
[224] Ohkubo, K., Sugahara, K., Ohta, H., Tokuda, K. and Ueoka, R., *Bull. Chem. Soc. Japan*, **54**, 576 (1981).
[225] Brown, J. M., Elliott, R. C., Griggs, C. G., Helmchen, G., and Nill, G., *Angew. Chem. Int. Ed. Engl.*, **20,** 890 (1981).
[226] Ueoka, R., Matsumoto, Y., Ninomiya, Y., Nakagawa, Y., Inouye, Y., and Ohkubo, K., *Chem. Lett. (Tokyo)*, **1981,** 785.
[227] Murakami, Y., Nakano, A., Yoshimatsu, A., and Fukuya, K., *J. Am. Chem. Soc.*, **103,** 728 (1981).
[228] van der Eijk, J. M., Nolte, R. J. M., Richteeeers, V. E. M., and Drenth, W., *Rec. Trav. Chim.*, **100,** 222 (1981).
[229] Kawabe, H., Tanabe, A. and Suda, H., *Polym. Prepr., Am. Chem. Soc. Div. Polym. Chem.*, **20,** 641 (1979); *Chem. Abs.*, **94,** 173848 (1981).
[230] Murakami, Y., Nakano, A., Yoshimatsu, A., and Matsumoto, K., *J. Am. Chem. Soc.*, **103,** 2750 (1981).
[231] Anoardi, L., Fornasier, R., and Tonellato, U., *J. Chem. Soc., Perkin Trans. 2*, **1981,** 260.
[232] See *Org. Reaction Mech.*, **1977,** 50.
[233] O'Connor, C. J., and Porter, A. J., *Aust. J. Chem.*, **34,** 1603 (1981).
[234] Eiki, T., Mori, M., Kawada, S., Matsushima, K., and Tagaki, W., *Chem. Lett. (Tokyo)*, **1980,** 1431.
[235] See *Org. Reaction Mech.*, **1977,** 51; **1978,** 57.

[236] Sunamoto, J., Kondo, H., Okamoto, H., and Terada, O., *Nagasaki Daigaku Kogakubu Kenkyu Hokoku*, **14,** 103 (1980); *Chem. Abs.*, **93,** 185305 (1980).
[237] Tabushi, I., Kimura, Y., and Yamamura, K., *J. Am. Chem. Soc.*, **103,** 6486 (1981).
[238] See *Org. Reaction Mech.*, **1980,** 54.
[239] Trainor, G. L., and Breslow, R., *J. Am. Chem. Soc.*, **103,** 154 (1981).
[240] Komiyama, M., and Inoue, S., *Bull. Chem. Soc. Japan*, **53,** 3266 (1980).
[241] Uekama, K., Lin, C.-L., Hirayama, F., Otagiri, M., Takadate, A., and Goya, S., *Chem. Lett.* (*Tokyo*), **1981,** 563.
[242] Komiyama, M., and Inoue, S., *Bull. Chem. Soc. Japan*, **53,** 3334 (1980).
[243] Tabushi, I., Kuroda, Y., and Sakata, Y., *Heterocycles*, **15,** 815 (1981); *Chem. Abs.*, **95,** 41755 (1981).
[244] Boreham, C. J., Buckingham, D. A., Francis, D. J., Sargeson, A. M. and Warner, L. G., *J. Am. Chem. Soc.*, **103,** 1975 (1981).
[245] Steliou, K., Szczygielska-Nowosielska, A., Favre, A., Poupart, M. A., and Hanessian, S., *J. Am. Chem. Soc.*, **102,** 7578 (1980).
[246] Searle, G. H., *Aust. J. Chem.*, **33,** 2159 (1980).
[247] Creaser, I. I., Harrowfield, J. M., Keene, F. R., and Sargeson, A. M., *J. Am. Chem. Soc.*, **103,** 3559 (1981).
[248] Fife, T. H., and Przystas, T. J., *J. Am. Chem. Soc.*, **102,** 7297 (1980).
[249] Patel, G., Satchell, R. S., and Satchell, D. P. N., *J. Chem. Soc., Perkin Trans. 2*, **1981,** 1406.
[250] Huguet, J., and Brown, R. S., *J. Am. Chem. Soc.*, **102,** 7571 (1980).
[251] Kluger, R., Chin, J., and Smyth, T., *J. Am. Chem. Soc.*, **103,** 884 (1981).
[252] Biswas, M. K., and Majumdar, D. K., *Bull. Chem. Soc. Japan*, **54,** 2213 (1981).
[253] Shimanouchi, N., and Kawasaki, M., *Kochi Kogyo Koto Semmon Gako Gakujutsu Kiyo*, **16,** 71 (1980); *Chem. Abs.*, **94,** 174016 (1981).
[254] Vivekananda, S., Venkatarao, K., Santappa, M., and Shanmuganathan, S., *Indian J. Chem.*, **20A,** 86 (1981); *Chem. Abs.*, **94,** 191242 (1981).
[255] Serguchev, Yu. A., and Beletskaya, I. P., *Usp. Khim.*, **49,** 2257 (1980); *Chem. Abs.*, **94,** 120309 (1981).
[256] Rao, Y. R., and Saiprakash, P. K., *Proc. Natl. Symp. Catal.*, *4th*, **1978,** 341; *Chem. Abs.*, **94,** 46397 (1981).
[257] Sarma, Y. R., Rajanna, K. C., and Saiprakash, P. K., *Indian J. Chem.*, **19A,** 872 (1980); *Chem. Abs.*, **94,** 46428 (1981).
[258] O'Leary, M. H., Yamada, H., and Yapp, C. J., *Biochemistry*, **20,** 1476 (1981).
[259] O'Leary, M. H., and Piazza, G. J., *Biochemistry*, **20,** 2743 (1981).
[260] Vinnik, M. I., Kozlov, V. A., and Popkova, I. A., *Kinet. Katal.*, **22,** 903 (1981); *Chem. Abs.*, **95,** 168237 (1981).
[261] Ciuhandu, G., and Dumitreanu, A., *J. Prakt. Chem.*, **323,** 595 (1981); *Chem. Abs.*, **95,** 132038 (1981).
[262] Dumitreanu, A., Ciuhandu, G., Chiriac, A., and Simon, Z., *An. Univ. Timosoara, Ser. Stiinke Fiz.-Chem.*, **17,** 41 (1979); *Chem. Abs.*, **94,** 173891 (1981).
[263] Jencks, W. P., *Adv. Enz.*, **51,** 75 (1980).
[264] Page, M. I., *Chem. Ind.* (*London*), **1981,** 144.
[265] Douzou, P., *Adv. Enz.*, **51,** 1 (1980).
[266] Fink, A. L., and Petsko, G. A., *Adv. Enz.*, **52,** 177 (1981).
[267] Pincus, M., and Scheraga, H. A., *Acc. Chem. Res.*, **14,** 299 (1981).
[268] Warshel, A., *Biochemistry*, **20,** 3167 (1981).
[269] Umeyama, H., Nakagawa, S., and Kudo, T., *J. Mol. Biol.*, **150,** 409 (1981).
[270] Kollman, P. A., and Hayes, D. M., *J. Am. Chem. Soc.*, **103,** 2955 (1981).
[271] Cohen, G. H., Silverton, E. W., and Davies, D. R., *J. Mol. Biol.*, **148,** 449 (1981).
[272] Jordan, F., and Polgár, L., *Biochemistry*, **20,** 6366 (1981).
[273] Kossiakoff, A. A., and Spencer, S. A., *Biochemistry*, **20,** 6462 (1981).
[274] DeTar, D. F., *Biochemistry*, **20,** 1730 (1981).
[275] DeTar, D. F., *J. Am. Chem. Soc.*, **103,** 107 (1981).
[276] Hamilton, S. E., and Zerner, B., *J. Am. Chem. Soc.*, **103,** 1827 (1981).
[277] Wang, C.-L. A., Calvo, K. C., and Klapper, M. H., *Biochemistry*, **20,** 1401 (1981).
[278] Wu, H.-L., Wastell, A., and Bender, M. L., *Proc. Natl. Acad. Sci. U.S.A.*, **78,** 4116 (1981).
[279] Wu, H.-L., Lace, D. A., and Bender, M. L., *Proc. Natl. Acad. Sci. U.S.A.*, **78,** 4118 (1981).
[280] Seaone, F., *Rev. Roum. Chim.*, **26,** 539 (1981).
[281] Bauer, C.-A., Brayer, G. D., Sielecki, A. R., and James, M. N. G., *Eur. J. Biochem.*, **120,** 289 (1981).
[282] Matteson, D. S., Sadhu, K. M., and Lienhard, G. E., *J. Am. Chem. Soc.*, **103,** 5421 (1981).
[283] White, E. H., Jelinski, L. W., Politzer, I. R., Branchini, B. R., and Roswell, D. F., *J. Am. Chem. Soc.*, **103,** 4231 (1981).

[284] Kraft, G. A., and Katzenellenbogen, J. A., *J. Am. Chem. Soc.*, **103,** 5459 (1981).
[285] Matta, M. S., Henderson, P. A., and Patrick, T. B., *J. Biol. Chem.*, **256,** 4172 (1981).
[286] Green, G. D. J., and Shaw, E., *J. Biol. Chem.*, **256,** 1923 (1981).
[287] See *Org. Reaction Mech.*, **1977,** 55.
[288] Frankfater, A., and Kuppy, T., *Biochemistry*, **20,** 5517 (1981).
[289] Johnson, F. A., Lewis, S. D., and Shafer, J. A., *Biochemistry*, **20,** 44 (1981).
[290] Lewis, S. D., Johnson, F. A., and Shafer, J. A., *Biochemistry*, **20,** 48 (1981).
[291] Johnson, F. A., Lewis, S. D., and Shafer, J. A., *Biochemistry*, **20,** 52 (1981).
[292] Sjawelski, R. J., and Wharton, C. W., *Biochem. J.*, **199,** 681 (1981).
[293] See *Org. Reaction Mech.*, **1979,** 49.
[294] Evans, B. L. B., Knopp, J. A., and Horton, H. R., *Arch. Biochem. Biophys.*, **206,** 362 (1981).
[295] Brocklehurst, K., and Malthouse, J. P. G., *Biochem. J.*, **193,** 819 (1981).
[296] Silver, M. S., and James, S. L. T., *Biochemistry*, **20,** 3177 (1981).
[297] Silver, M. S., and James, S. L. T., *Biochemistry*, **20,** 3183 (1981).
[298] See *Org. Reaction Mech.*, **1977,** 56.
[299] Rees, D. C., Lewis, M., Honzatko, R. B., Lipscomb, W. N., and Hardman, K. D., *Proc. Natl. Acad. Sci. U.S.A.*, **78,** 3408 (1981).
[300] See *Org. Reaction Mech.*, **1980,** 62.
[301] Nashed, N. T., and Kaiser, E. T., *J. Am. Chem. Soc.*, **103,** 3611 (1981).
[302] Oyama, K., Kihara, K., and Nonaka, Y., *J. Chem. Soc., Perkin Trans. 2*, **1981,** 356.
[303] Waley, S. G., *Chem. Ind. (London)*, **1981,** 131.
[304] Reading, C., and Farmer, T., *Biochem. J.*, **199,** 779 (1981).
[305] Charnas, R. L., and Knowles, J. R., *Biochemistry*, **20,** 3214 (1981).
[306] Brenner, D. G., and Knowles, J. R., *Biochemistry*, **20,** 3680 (1981).
[307] Kemal, C., and Knowles, J. R., *Biochemistry*, **20,** 3688 (1981).
[308] Fisher, J., Charnas, R. L., Bradley, S. M., and Knowles, J. R., *Biochemistry*, **20,** 2276 (1981).
[309] Charnas, R. L., and Knowles, J. R., *Biochemistry*, **20,** 2733 (1981).
[310] See *Org. Reaction Mech.*, **1980,** 61.
[311] See *Org. Reaction Mech.*, **1980,** 62.
[312] Hasan, F. B., Elkind, J. L., Cohen, S. G., and Cohen, J. B., *J. Biol. Chem.*, **256,** 7781 (1981).
[313] Lambrecht, G., *Arch. Pharm. (Weinheim)*, **313,** 368 (1980); *Chem. Abs.*, **93,** 185299 (1980).
[314] Dijkstra, B. W., Drenth, J., and Kalk, K. H., *Nature (London)*, **289,** 604 (1981).
[315] Dijkstra, B. W., Kalk, K. H., Hol, W. G. H., and Drenth, J., *J. Mol. Biol.*, **147,** 97 (1981).
[316] See *Org. Reaction Mech.*, **1980,** 63.
[317] Keith, C., Feldman, D. S., Deganello, S., Glick, J., Ward, K. B., Jones, E. O., and Sigler, P. B., *J. Biol. Chem.*, **256,** 8602 (1981).
[318] Griffith, D. W., and Meister, A., *J. Biol. Chem.*, **256,** 9981 (1981).
[319] Kobayashi, S., and Saegusa, T., *Pure Appl. Chem.*, **53,** 1663 (1981).
[320] Westheimer, F. H., *Chem. Rev.*, **81,** 313 (1981).
[321] See *Org. Reaction Mech.*, **1980,** 64.
[322] Satterthwait, A. C., and Westheimer, F. H., *J. Am. Chem. Soc.*, **103,** 1177 (1981).
[323] Loewus, D. I., *J. Am. Chem. Soc.*, **103,** 2292 (1981).
[324] Harger, M. J. P., and Stephen, M. A., *J. Chem. Soc., Perkin Trans. 2*, **1981,** 736.
[325] Richman, J. E., and Flay, R. B., *J. Am. Chem. Soc.*, **103,** 5265 (1981).
[326] McClelland, R. A., Patel, G., and Cirinna, C., *J. Am. Chem. Soc.*, **103,** 6432 (1981).
[327] Kobayashi, S., Narukawa, Y., Hashimoto, T., and Saegusa, T., *Chem. Lett. (Tokyo)*, **1980,** 1599.
[328] van Aken, D., McPaulissen, L., and Buck, H. M., *J. Org. Chem.*, **46,** 3189 (1981).
[329] Castellijns, M. M. C. F., Schipper, P., van Aken, D., and Buck, H. M., *J. Org. Chem.*, **46,** 47 (1981).
[330] Buck, H. M., *Rec. Trav. Chim.*, **100,** 217 (1981).
[331] Denney, D. B., Denney, D. Z., Hammond, P. J., and Wang, Y.-P., *J. Am. Chem. Soc.*, **103,** 1785 (1981).
[332] Font Freide, J. J. H. M., and Trippett, S., *J. Chem. Res.(S)*, **1981,** 218.
[333] Rowell, R., and Gorenstein, D. G., *J. Am. Chem. Soc.*, **103,** 5894 (1981).
[334] See *Org. Reaction Mech.*, **1980,** 66.
[335] See *Org. Reaction Mech.*, **1970,** 492.
[336] Waeschke, H., and Mitzner, R., *Z. Chem.*, **20,** 381 (1980).
[337] Ponomarchuk, N. P., Kasukhim, L. F., Budilova, I. Yu., and Gololobov, Yu. G., *Zh. Obshch. Khim.*, **50,** 1937 (1980); *Chem. Abs.*, **94,** 46388 (1981).
[338] Wolfe, N. L., *Chemosphere*, **9,** 571 (1980); *Chem. Abs.*, **95,** 23963 (1981).
[339] Bel'skii, V. E., Kudryavtseva, L. A., Kurguzova, A. M., and Ivanov, B. E., *Izv. Akad. Nauk SSSR, Ser. Khim.*, **1980,** 2811; *Chem. Abs.*, **94,** 155972 (1981).

[340] Sukhorukhova, N. A., Istomin, B. I., and Kalabina, A. V., *Zh. Obshch. Khim.*, **50,** 2141 (198); *Chem. Abs.*, **94,** 29766 (1981).
[341] Istomin, B. I., Eliseeva, G. D., and Kalabina, A. V., *Zh. Obshch. Khim.*, **50,** 1186 (1980); *Chem. Abs.*, **93,** 167154 (1980).
[342] Eliseeva, G. D., Istomin, B. I., and Kalabina, A. V., *Zh. Obshch. Khim.*, **50,** 1901 (1980); *Chem. Abs.*, **94,** 3388 (1981).
[343] Zhdankovich, E. L., Istomin, B. I., and Voronkov, M. G., *Izv. Akad. Nauk SSSR, Ser. Khim.*, **1981,** 1264; *Chem. Abs.*, **95,** 131876 (1981).
[344] Istomin, B. I., Voronkov, M. G., Zhdankovich, E. L., and Bazhenov, B. N., *Dokl. Akad. Nauk SSSR*, **258,** 659 (1981); *Chem. Abs.*, **95,** 149493 (1981).
[345] Begunov, A. V., and Rutkovskii, G. V., *Zh. Org. Khim.*, **16,** 1607 (1980); *Chem. Abs.*, **94,** 14748 (1981).
[346] Bunton, C. A., Hong, Y. S., Romsted, L. S., and Quan, C., *J. Am. Chem. Soc.*, **103,** 5784 (1981).
[347] Bunton, C. A., Hong, Y. S., Romsted, L. S., and Quan, C., *J. Am. Chem. Soc.*, **103,** 5788 (1981).
[348] See *Org. Reaction Mech.*, **1977,** 49.
[349] Okahata, Y., Ihara, H., and Kunitake, T., *Bull. Chem. Soc. Japan*, **54,** 2072 (1981).
[350] Mackay, R. A., and Hernansky, C., *J. Phys. Chem.*, **85,** 739 (1981).
[351] Wallerberg, G., and Haake, P., *J. Org. Chem.*, **46,** 43 (1981).
[352] Rahil, J., and Haake, P., *J. Org. Chem.*, **46,** 3048 (1981).
[353] Bel'skii, V. E., Kudryavtseva, L. A., Derstunagova, K. A., Fedorov, S. B., and Ivanov, B. E., *Zh. Obshch. Khim.*, **50,** 1997 (1980); *Chem. Abs.*, **94,** 46390 (1981).
[354] See *Org. Reaction Mech.*, **1980,** 70.
[355] Corriu, R. J. P., Dutheil, J. P., and Lanneau, G. F., *Tetrahedron*, **37,** 3681 (1981).
[356] Hall, C. R., Inch, T. D., and Pottage, C., *Phosphorus, Sulfur*, **10,** 229 (1981); *Chem. Abs.*, **95,** 149817 (1981).
[357] Corriu, R. J. P., Dutheil, J. P., and Lanneau, G. F., *J. Chem. Soc., Chem. Commun.*, **1981,** 101.
[358] Wadsworth, W. S., *J. Org. Chem.*, **46,** 4080 (1981).
[359] Bruzik, K., and Stec, W., *J. Org. Chem.*, **46,** 1618, 1625 (1981).
[360] See *Org. Reaction Mech.*, **1980,** 68.
[361] See *Org. Reaction Mech.*, **1980,** 67.
[362] Macomber, R. S., and Krudy, G. A., *J. Org. Chem.*, **46,** 4038 (1981).
[363] Jarvest, R. L., Lowe, G., and Potter, B. V. L., *J. Chem. Soc., Chem. Commun.*, **1980,** 1142.
[364] Cullis, P. M., Jarvest, R. L., Lowe, G., and Potter, B. V. L., *J. Chem. Soc., Chem. Commun.*, **1981,** 245.
[365] Hutchings, G. J., and Banks, B. E. C., *Biochemistry*, **20,** 5809 (1981).
[366] Ramirez, F., Marecek, J. F., and Szamosi, J., *J. Org. Chem.*, **45,** 4748 (1980).
[367] Watanabe, M., Matsuura, M., and Yamada, T., *Bull. Chem. Soc. Japan*, **54,** 738 (1981).
[368] See *Org. Reaction Mech.*, **1977,** 66.
[369] Harrowfield, J. M., Jones, D. R., Lindoy, L. F., and Sargeson, A. M., *J. Am. Chem. Soc.*, **102,** 7733 (1980).
[370] Buckingham, D. A., and Clark, C. R., *Aust. J. Chem.*, **34,** 1769 (1981).
[371] Lillocci, C., and Vernon, C. A., *Gazz. Chim. Ital.*, **111,** 23 (1981).
[372] Rahil, J., and Haake, P., *J. Am. Chem. Soc.*, **103,** 1723 (1981).
[373] Modro, T. A., and Graham, D. H., *J. Org. Chem.*, **46,** 1923 (1981).
[374] Hall; C. R., and Inch, T. D., *J. Chem. Soc., Perkin Trans. 1*, **1981,** 2368.
[375] Hall, C. R., and Williams, N. E., *J. Chem. Soc., Perkin Trans. 1*, **1981,** 2746.
[376] Koizumi, T., Yanada, R., Takagi, H., Hirai, H., and Yoshii, E., *Tetrahedron Lett.*, **1981,** 477.
[377] Tomasz, J., and Simoncsits, A., *Tetrahedron Lett.*, **1981,** 3905.
[378] Dahl, O., *Tetrahedron Lett.*, **1981,** 3281.
[379] Harger, M. J. P., *Tetrahedron Lett.*, **1981,** 4741.
[380] Andreev, N. A., and Grishina, O. N., *Zh. Obshch. Khim.*, **51,** 566 (1981); *Chem. Abs.*, **95,** 41801 (1981).
[381] Reimschüssel, W., Mikolajczyk, M., Ślebocka-Tilk, H., and Gajl, M., *Int. J. Chem. Kinet.*, **12,** 979 (1980).
[382] Gazizov, T. Kh., Belyalov, R. U., and Pudovik, A. N., *Zh. Obshch. Khim.*, **50,** 1673 (1980); *Chem. Abs.*, **93,** 220092 (1980).
[383] Khokhlov, P. S., and Sokolova, G. D., *Zh. Obshch. Khim.*, **50,** 1214 (1980); *Chem. Abs.*, **93,** 185336 (1980).
[384] Kutyrev, G. A., Vinokurov, A. I., Cherkasov, R. A., and Pudovik, N. A., *Dokl. Akad. Nauk SSSR*, **252,** 1138 (1980); *Chem. Abs.*, **94,** 3393 (1981).
[385] Cross, R. L., *Ann. Rev. Biochem.*, **50,** 681 (1981).
[386] Blackburn, G. M., *Chem. Ind.* (*London*), **1981,** 134.
[387] Lowe, G., and Potter, B. V. L., *Biochem. J.*, **199,** 693 (1981).

[388] Jarvest, R. L., Lowe, G., and Potter, B. V. L., *Biochem. J.*, **199,** 427 (1981).
[389] Jarvest, R. L., and Lowe, G., *Biochem. J.*, **199,** 273 (1981).
[390] Lowe, G., and Potter, B. V. L., *Biochem. J.*, **199,** 227 (1981).
[391] See *Org. Reaction Mech.*, **1979,** 58.
[392] Hansen, D. E., and Knowles, J. R., *J. Biol. Chem.*, **256,** 5967 (1981).
[393] Bryant, F. R., Benkovic, S. J., Sammons, D., and Frey, P. A., *J. Biol. Chem.*, **256,** 5965 (1981).
[394] Eckstein, F., Ronaniuk, P. J., Heideman, W., and Storm, D. R., *J. Biol. Chem.*, **256,** 9118 (1981).
[395] See *Org. Reaction Mech.*, **1980,** 71.
[396] Jarvest, R. L., and Lowe, G., *J. Chem. Soc., Chem. Commun.*, **1980,** 1145.
[397] Coderre, J. A., Mehdi, S., Demou, P. C., Weber, R., Traficante, D. D., and Gerlt, J. A., *J. Am. Chem. Soc.*, **103,** 1870 (1981).
[398] Brody, R. S., and Frey, P. A., *Biochemistry*, **20,** 1245 (1981).
[399] Marlier, J. F., Bryant, F. R., and Benkovic, S. J., *Biochemistry*, **20,** 2213 (1981).
[400] Bryant, F. R., and Benkovic, S. J., *J. Am. Chem. Soc.*, **103,** 696 (1981).
[401] Saini, M. S., Buchwald, S. L., Van Etten, R. L., and Knowles, J. R., *J. Biol. Chem.*,**256,** 10453 (1981).
[402] Hoving, H., Lolkema, J. S., and Robillard, G. T., *Biochemistry*, **20,** 87 (1981).
[403] Rosevear, P. R., Desmeules, P., Kenyon, G. L., and Mildvan, A. S., *Biochemistry*, **20,** 6155 (1981).
[404] Rao, B. D. N., and Cohn, M., *J. Biol. Chem.*, **256,** 1716 (1981).
[405] Katz, M., and Westley, J., *Arch. Biochem. Biophys.*, **204,** 464 (1980).
[406] Granoth, J., Mildvan, A. S., and Kaiser, E. T., *Arch. Biochem. Biophys.*, **205,** 1 (1980).
[407] McCracken, S. and Meighey, E. A., *J. Biol. Chem.*, **256,** 3945 (1981).
[408] Sowadski, J. M., Foster, B. A., and Wyckoff, H. W., *J. Mol. Biol.*, **150,** 245 (1981).
[409] Rubio, V., Britton, H. G., Grisola, S., Sproat, B. S., and Lowe, G., *Biochemistry*, **20,** 1969 (1981).
[410] See *Org. Reaction Mech.*, **1980,** 73; **1979,** 62.
[411] Wlodawer, A., and Sjölin, L., *Proc. Natl. Acad. Sci. U.S.A.*, **78,** 2853 (1981).
[412] Springs, B., Welsh, K. M., and Cooperman, B. S., *Biochemistry*, **20,** 6384 (1981).
[413] Knight, W. B., Fitts, S. W., and Dunaway-Mariano, D., *Biochemistry*, **20,** 4079 (1981).
[414] Lam, W. Y., and Martin, J. C., *J. Am. Chem. Soc.*, **103,** 120 (1981).
[415] Lam, W. Y., Duesler, E. N., and Martin, J. C., *J. Am. Chem. Soc.*, **103,** 127 (1981).
[416] See *Org. Reaction Mech.*, **1980,** 72; **1979,** 63.
[417] Thea, S., Kashefi-Naini, N., and Williams, A., *J. Chem. Soc., Perkin Trans. 2*, **1981,** 65.
[418] Thea, S., and Williams, A., *J. Chem. Soc., Perkin Trans. 2*, **1981,** 72.
[419] Thea, S., Harun, M. G., Kashefi-Naini, N., and Williams, A., *J. Chem. Soc., Perkin Trans. 2*, **1981,** 78.
[420] Thea, S., Guanti, G., and Williams, A., *J. Chem. Soc., Chem. Commun.*, **1981,** 535.
[421] Sendega, R. V., Gorbatenko, N. G., and Vizgert, R. V., *Org. React. (Tartu)*, **17,** 138 (1980).
[422] Sendega, R. V., Gorbatenko, N. G., and Vizgert, R. V., *Org. React. (Tartu)*, **17,** 247 (1980).
[423] Sendega, R. V., Morozov, A. I., and Sytilin, M. S., *Zh. Org. Khim.*, **17,** 50 (1981); *Chem. Abs.*, **95,** 23984 (1981).
[424] Cafferata, L. F. R., Desvard, O. E., and Sicre, J. E., *J. Chem. Soc., Perkin Trans. 2*, **1981,** 940.
[425] Ciuffarin, E., Isola, M., and Leoni, P., *J. Org. Chem.*, **46,** 3064 (1981).
[426] Graafland, T., Nieuwpoort, W. C., and Engberts, J. B. F. N., *J. Am. Chem. Soc.*, **103,** 4490 (1981).
[427] See *Org. Reaction Mech.*, **1980,** 74.
[428] Graafland, T., Kirby, A. J., and Engberts, J. B. F. N., *J. Org. Chem.*, **46,** 215 (1981).
[429] Gawali, B. B., Neelakantan, P., Rao, S. N., Iyangar, D. S., and Bhaleran, U. T., *Indian J. Chem.*, **20B,** 616 (1981); *Chem. Abs.*, **95,** 149489 (1981).
[430] Vizgert, R. V., and Sheiko, S. G., *Katal. Stiint. Org. Soedin. Sery*, **1979,** 201; *Chem. Abs.*, **93,** 238188 (1980).
[431] Kim, J. F., and Hillhouse, J. H., *J. Chem. Soc., Chem. Commun.*, **1981,** 295.
[432] Lee, I., and Koo, I. S., *Taehan Hwahakhoe Chi*, **25,** 7 (1981); *Chem. Abs.*, **95,** 41797 (1981).
[433] Ballistreri, F. P., Cantone, A., Maccarone, E., Tomaselli, G. A., and Tripolone, M., *J. Chem. Soc., Perkin Trans. 2*, **1981,** 438.
[434] Kang, H. K., Lee, I., Song, H. B., and Lee, J. K., *Bull. Korean Chem. Soc.*, **2,** 33 (1981); *Chem. Abs.*, **95,** 131853 (1981).
[435] Lee, I., Koo, I. S., and Kang, H. K., *Bull. Korean Chem. Soc.*, **2,** 41 (1981); *Chem. Abs.*, **95,** 169317 (1981).
[436] Skrypnik, Yu., and Bezrodnyi, V., *Org. React. (Tartu)*, **17,** 168 (1980).
[437] Vizgert, R. V., Maksimenko, N. N., and Panov, E. P., *Zh. Org. Khim.*, **17,** 605 (1981); *Chem. Abs.*, **95,** 6067 (1981).
[438] Lee, H. W., Song, K. B., Song, H. B., and Lee, I., *Kich'o Kwahak Yonguso Nonmunjip 'Inha Taehakkyo)*, **2,** 27 (1981); *Chem. Abs.*, **95,** 131852 (1981).

[439] Banjoko, O., and Okwuiwe, R., *J. Org. Chem.*, **45,** 4966 (1980).
[440] Lee, B. C., and Lee, I., *Taehan Hwahakhoe Chi*, **24,** 342 (1980); *Chem. Abs.*, **94,** 102400 (1981).
[441] Makarova, R. A., Rybachenko, V. I., and Titov, E. V., *Katal. Sint. Org. Soedin. Sery*, **1979,** 205; *Chem. Abs.*, **93,** 238561 (1980).
[442] Zaslavskii, V. G., Savelova, V. A., Litvinenko, L. M., and Salomoichenko, T., *Zh. Org. Khim.*, **16,** 1247 (1980); *Chem. Abs.*, **93,** 185324 (1980).
[443] Litvinenko, L. M., Savelova, V. A., Dadali, P. A., Belousova, I. A., Zubareva, T. M., Simanenko, Yu. S., and Lapshin, S. A., *Dokl. Akad. Nauk SSSR*, **255,** 878 (1980); *Chem. Abs.*, **94,** 208028 (1981).
[444] Arcoria, A., Ballistreri, F. P., Musumarra, G., and Tomaselli, G. A., *J. Chem. Soc., Perkin Trans. 2*, **1981,** 221.
[445] Popov, A. F., Sheiko, S. G., and Mitchenko, E. S., *Dopov. Akad. Nauk Ukr. RSR., Ser. B, Geol., Khim., Biol. Nauki*, **1980,** 59; *Chem. Abs.*, **94,** 46414 (1981).
[446] Maleeva, N., and Savelova, V. A., *Ukr. Khim. Zh. (Russ. Ed.)*, **46,** 960 (1980); *Chem. Abs.*, **94,** 14760 (1981).
[447] Kice, J. L., and Liao, S.-T., *J. Org. Chem.*, **46,** 2691 (1981).
[448] Freeman, F., and Angeletakis, C. N., *J. Am. Chem. Soc.*, **103,** 6232 (1981).
[449] Farng, L. O., and Kice, J. L., *J. Am. Chem. Soc.*, **103,** 1137 (1981).
[450] Vigalok, I. V., Pterova, G. G., Lukashina, G. S., Vigolok, A. A., and Levin, Ya. A., *Zh. Org. Khim.*, **16,** 1442 (1980); *Chem. Abs.*, **94,** 3384 (1981).
[451] Bhattacharjee, G., *Indian J. Chem.*, **20B,** 526 (1981); *Chem. Abs.*, **95,** 79602 (1981).
[452] Van Etten, R. L., and Risley, J. M., *J. Am. Chem. Soc.*, **103,** 5633 (1981).
[453] Kwart, H., George, T. J., Horgan, A. G., and Lin, Y. T., *J. Org. Chem.*, **46,** 1970 (1981).
[454] Harland, J. J., Day, R. O., Vollano, J. F., Sau, A. C., and Holmes, R. R., *J. Am. Chem. Soc.*, **103,** 5269 (1981).
[455] Tondeur, J.-J., Borghese, A., and Vandendungen, G., *J. Chem. Res.*, **1981,** 134(S), 1628(M).

Organic Reaction Mechanisms 1981
Edited by A. C. Knipe and W. E. Watts

CHAPTER 3

Radical Reactions

D. J. Cowley

School of Physical Sciences, New University of Ulster

and D. C. Nonhebel

Department of Pure and Applied Chemistry, University of Strathclyde

Introduction . . . 78
Structure, Stereochemistry, and Stability . . . 78
Carbon Radicals . . . 78
Nitrogen Radicals . . . 80
Oxygen Radicals . . . 81
Miscellaneous Radicals . . . 82
Peroxides . . . 82
Autoxidations . . . 87
Azo Compounds . . . 90
Diazonium Salts . . . 91
Combination and Disproportionation Reactions . . . 92
Atom-abstraction Reactions . . . 93
Carbon-centred Radicals . . . 94
Halogenation . . . 95
Hydrogen Abstraction by Atoms . . . 96
Hydroxyl, Alkoxyl, and Peroxyl Radicals . . . 96
Other Hetero Radicals . . . 98
Halogen Atom Abstraction . . . 98
Addition Reactions . . . 99
Carbon-centred Radicals . . . 99
Atoms . . . 101
Hetero Radicals . . . 101
Intramolecular Addition . . . 103
Fragmentations . . . 105
Rearrangements . . . 107
Homolytic Aromatic Substitution . . . 111
Biradicals . . . 113
Theory and Structure . . . 113
Carbon–Carbon Centred Biradicals . . . 114
Hetero-centred Biradicals . . . 116
Carbonyl-derived Biradicals . . . 116
Nitroxyls and Spin-trapping . . . 116
Structural and Synthetic Aspects . . . 116
Reactions of Nitroxyls . . . 118
Spin-trapping . . . 119

S_H2 Reactions 120
Phosphoranyl and Sulphuranyl Radicals 120
Homolytic Oxidation and Reduction 122
Electron-transfer Reactions 131
Radical Ions 142
Photolysis 149
Fragmentation 149
Hydrogen-atom Transfers 150
Carbonyl Compounds 151
Miscellaneous Compounds 152
Pyrolysis 153
Hydrocarbons 153
Other Compounds 153
Radiolysis 154
Compounds of Biological Interest 154
Other Compounds 155
CIDNP and CIDEP Methods 156
References 157

Introduction

A collection of papers presented at a symposium have been published[1] and books on metal-catalysed oxidations of organic compounds, and oxygen and oxy-radicals in chemistry have appeared.[2,3] The relationship between energy profiles and structure–reactivity relationships has been reviewed.[4] The relative rates of decomposition of substituted dibenzylmercury compounds have been used to derive a new scale of substituent constants, $\sigma\cdot$, applicable to radical reactions; $\sigma\cdot$ values are positive for both electron-donating and electron-withdrawing substituents.[5] Other topics which have been reviewed include a theoretical treatment of the structure and properties of radicals,[6] conformations of biradicals,[7] cyclopropyl radicals (including a discussion of the preferred stereochemistry and reactivity of cyclopropyl radicals),[8] the stereochemistry of homolytic C—C bond fission in small rings,[9] hetero-aromatic radicals with Group VI and Groups V and VI hetero-atoms,[10] organometallic radicals,[11] homolytic organometallic reactions,[12] organogermyl radicals,[13] radical photo-substitution of hetero-aromatic compounds,[14] radicals derived from acylnitrosoamines,[15] regio- and stereo-selectivity of radical reactions (with special reference to the cyclization of hex-5-enyl and related radicals, the ring-opening of cycloalkylcarbinyl radicals, and S_H reactions),[16] polar effects in radical reactions,[17] magnetic and spin effects in radical reactions.[18] Electron-transfer chain catalysis of substitution reactions has been discussed:[19] the best known examples are the familiar $S_{RN}1$ reactions. Other categories include $S_{ON}2$ reactions [bimolecular nucleophilic substitutions involving one-electron oxidation — see equations (1)–(4), *cf.* p. 37), $S_{OE}1$ reactions (unimolecular electrophilic substitutions involving one-electron oxidation), and $S_{RE}2$ reactions (bimolecular electrophilic substitutions involving one-electron reduction).

Structure, Stereochemistry, and Stability

Carbon Radicals

The bond dissociation energies of the C_2H_5—H and Bu^t—H bonds are 100.3 ± 1.0 and $93.9 \pm 1.0\,kcal\,mol^{-1}$, respectively, relative to that for the CH_3—H bond of

$$RX \xrightarrow{-e} RX^{+\cdot} \qquad \text{Initiation} \quad (1)$$

$$RX^{+\cdot} + Nu^- \longrightarrow [RX.Nu]\cdot \qquad (2)$$

$$[RX.Nu]\cdot \longrightarrow RNu^{+\cdot} + X^- \qquad \text{Propagation} \quad (3)$$

$$RNu^{+\cdot} + RX \longrightarrow RNu + RX^{+\cdot} \qquad (4)$$

104.4 kcal mol^{-1}.[20] An IR matrix study of Bu$^t\cdot$ is consistent with pyramidal geometry with C_{3v} symmetry.[21] *Ab initio* MO studies also favour this geometry.[22,23] The origin of the pyramidalization has been attributed to simultaneous maximizing of hyperconjugative stabilization and minimizing of torsional interaction between the methyl groups and the radical centre.[24]

The preferred conformation of the cyclobutylmethyl radical (**1**) contrasts with that for $Me_2CH\dot{C}H_2$ in which there is eclipsing between the β-C—H bond and the SOMO.[25] In acyclic radicals C—H hyperconjugation is more important than C—C

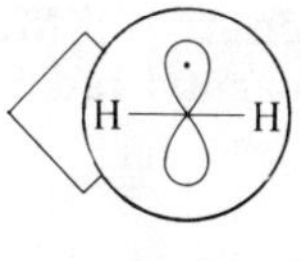

(**1**)

hyperconjugation whereas the reverse is the case with small-ring cycloalkylmethyls: the weaker C—C bond with greater π-character in small-ring cycloalkanes favours C—C hyperconjugation while the stronger C—H bonds disfavour C—H hyperconjugation. SCF calculations indicate that the β-C—H bonds in the cubyl radical are stronger than in acyclic alkyl radicals.[26] The 2-methylnorbornyl radical is pyramidal at the radical centre with the methyl group bent in the *endo*-direction.[27] In the radical derived from 2-ethylidenenorbornane, the allylic fragment is planar but there is some distortion of the bicyclo[2.2.1]heptane skeleton.

Ab initio MO calculations predict that the radical centre in $\dot{C}H_2CH_2Cl$ is slightly non-planar.[28] There is no evidence for chlorine bridging: hyperconjugation determines the preferred eclipsed conformation of the radical. The conformations of the radicals derived from addition of $Et_3Si\cdot$ to methylenecycloalkanes have been determined by ESR spectroscopy.[29]

The barriers to internal rotation ($\sim$2 kcal mol^{-1}) about the C_β—C_γ bond in 1-substituted butyl radicals are consistent with an eclipsed conformation for the radical.[30] The α-oxyalkyl radicals derived from oxacyclanes are very much more bent at the radical centre than the corresponding cycloalkyl radicals.[31] 1-Hydroxycyclohexyl radicals are non-planar with an inversion barrier of 2 kcal mol^{-1}.[32,33] Long-range coupling of the γ- and δ-protons in $R_2\dot{C}OCHO$[34] and $R_2\dot{C}CR^1_2OCHO$[35] indicate that the radicals exist in the conformations (**2**) and (**3**) respectively.

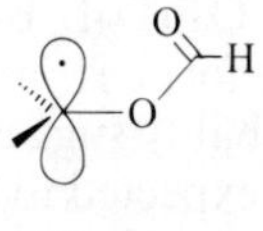

(**2**) 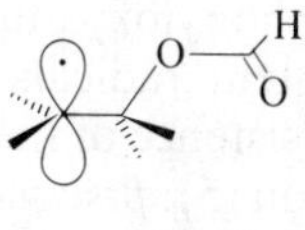 (**3**)

The α-chloro-α-fluoroalkyl radicals, $\dot{C}Cl_2F$, $\dot{C}FClCF_2Cl$, and $\dot{C}FCl_2$ are all non-planar but less so than $\dot{C}F_3$.[36,37] The stabilization energies of the α-aminoalkyl radicals, $\dot{C}H_2NH_2$, $\dot{C}H_2NHMe$, and $\dot{C}H_2NMe_2$ are 10, 17, and 20 kcal mol^{-1}, respectively.[38] The delocalization which accounts for this stability is described by a three-electron bond which requires that two electrons are in a bonding orbital and the third electron is in a low-lying anti-bonding orbital. The stability of this class of radical presumably accounts for the driving force in the Stevens rearrangement.

ESR studies indicate that the unpaired spin in the capto-dative radicals $Ar_2N\dot{C}(CN)Ar$ and $Ar_2N\dot{C}(CN)_2$ is much more extensively delocalized than in radicals with two electron-donating or electron-withdrawing substituents.[39] The ease of interconversion of *meso*- and (±)-ArC(CN)(SEt)C(CN)(SEt)Ar has been attributed to the stability of the capto-dative stabilized $Ar\dot{C}(CN)SEt$.[40] Capto-dative stabilization is also responsible for the enhanced rates of thermal isomerization of ArAr′C=CArAr′ (Ar = *p*-$MeOC_6H_4$, Ar′ = *p*-NCC_6H_4) relative to those of ArPhC=CPhAr (Ar = *p*-$MeOC_6H_4$ or *p*-NCC_6H_4),[41] and for the failure of the half-wave oxidation potentials of ArAr′C=CArAr′ to correlate with σ^+ or σ^- substituent constants when Ar = *p*-$MeOC_6H_4$ and Ar′ = *p*-NCC_6H_4.[42] The persistence of the 1,4-benzothiazinyl radical (**4**) has also been attributed to capto-dative stabilization.[43]

(4)

Addition of $R_3M\cdot$(M = C, Si, Ge) to thioketones occurs more readily than to ketones as the unpaired spin is delocalized more efficiently by the sulphur than by the oxygen[44] of the resultant spin adducts in which the S—MR_3 bond is eclipsed with the SOMO.[45] Persistent radicals are obtained by addition of $R_3M\cdot$(R = Si, Ge, Sn) to fluorenone and xanthone.[46]

The rotational barrier in allyl radicals is 15.7 ± 1.0 kcal mol^{-1}: from this value, the stabilization energy has been estimated to be 14.0–14.5 kcal mol^{-1}.[47] The barrier to rotation in pentadienyl radicals has been estimated to be 104 kJ mol^{-1} (24.9 kcal mol^{-1}).[48] The stabilization energies of $HC{\equiv}\dot{C}CHCH_3$ and $CH_3C{\equiv}CCH_2\cdot$ have been estimated to be 10.0 ± 2.2 and 7.8 ± 2.2 kcal mol^{-1} respectively.[49] The 1-naphthylmethyl and the 9-anthrylmethyl radicals have been estimated using VLPP to have stabilization energies 3.7 and 8.2 kcal mol^{-1} greater than that of benzyl radicals.[50] The stabilities of the 2-, 3-, and 4-picolinyl radicals are very similar to each other[51] and 10-aryltelluroxanthyl radicals have been shown to be persistent.[52] The conformations of the alkyl groups in a series of 1-alkyl-4-methoxycarbonyl radicals have been deduced.[53]

Nitrogen Radicals

N-*tert*-Butyl-*N*-phosphonyloxyaminyls [$Bu^t\dot{N}$—$OP(O)L_2Bu^t$ ^-N—$\overset{+\cdot}{O}P(O)L_2$] are a new class of persistent radicals: the dipolar form is less important than for $Bu^t\dot{N}OBu^t$.[54] The persistence of the radicals probably suggests that they are in the staggered conformation (**5**): β-scission would be expected to occur readily from the eclipsed conformation. The unpaired spin in 1,2,3-dithiazolyl radicals (**6**) is mainly on the nitrogen, C(1), and C(3): the nitrogen coupling constant is increased by electron-donating substituents at C(6) and reduced by electron-withdrawing substituents.[55–58] The two aryl groups in the persistent hydrazyl radical (**7**) are non-equivalent.[59] A persistent radical is obtained from the reaction of Et_2Zn with $Bu^tN{=}CHCH{=}NBu^t$.[60]

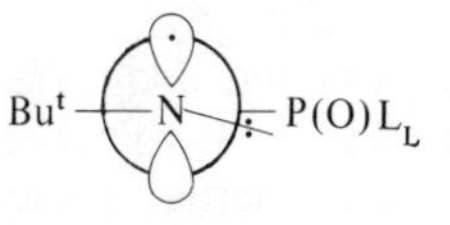

(**5**) 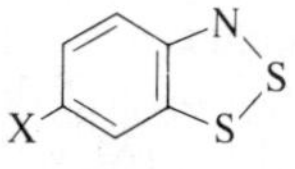 (**6**) 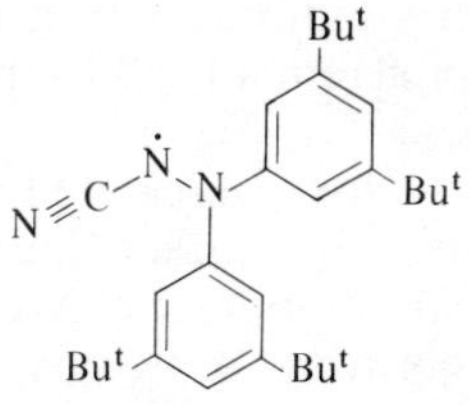 (**7**)

ESR studies of aryldiazenyl radicals indicate that free rotation occurs about the C(1)—N(1) bond except when the aryl radical possesses *ortho*-substituents, *cf*. (**8**) in which coupling occurs to only one *meta*-proton.[61] ESR studies indicate that triazenyl radicals $RN{-}\dot{N}{-}NR'$ are σ- and not π-radicals as had been proposed.[62,63] Twisting occurs in the hydrazinyl radical (**9**): the barrier to rotation is low due to compensation of the loss of π–π delocalization by the gain of σ–π delocalization[64] (radical cations of hydrazones also exist in a twisted conformation).[65]

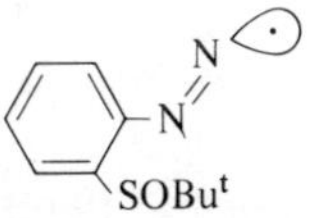

(**8**) 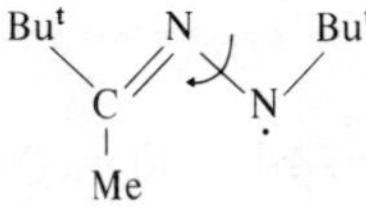 (**9**)

The ESR parameters of a number of acyclic *N*-alkylcarboxamidyls, $R\dot{N}COR'$, can be interpreted in terms of a π-electronic configuration in which the unpaired electron lies mainly in an orbital perpendicular to the RNC plane;[66] *ab initio* MO calculations predict that the ground state of these radicals is a mixture of π- and Σ-configurations, but that sulphonamidyls have the π_N-configuration in the ground state.[67] The ESR spectra of cyclic oxyamidyls have been reported.[68]

NMR studies show that there is no significant interaction of the unpaired electron in the 3-*tert*-butyl-1([2.2]paracyclophan-4-yl)-5-phenylverdazyl radical[69] and ESR studies of *N*-1, *N'*-1-linked biverdazyls with a [2.2]paracyclophenylene bridge indicate that the verdazyl rings are twisted out of conjugation with the [2.2]paracyclophane bridge.[70] The ESR spectra of complexes of a verdazyl derivative of benzo-15-crown-5 with sodium thiocyanate provide evidence for a 2:1 complex.[71]

Oxygen Radicals

The distinct existence of σ- and π-acetoxy radicals has been proposed: σ-acetoxy radicals, in which the ground state correlates with the ground state of carbon dioxide, decarboxylate much more readily than π-acetoxy radicals.[72] MO calculations on the Σ- and π-states of formyloxyl and formylaminyl radicals have been reported.[73] Sterically hindered galvinoxyls behave as phenoxyl radicals with equilibrating quinonoid–benzenoid rings and not as a delocalized system.[74] The 4,4′-(biphenyl-3,3′-diyl)bis(3,5-di-*tert*-butylphenoxyl) biradical exists in the *cis*-conformation.[75] A novel galvinoxyl-related silicon tetraradical has been obtained.[76]

ESR studies indicate that in the iminoxyl radicals, ArC(X)=N—O·, the aryl groups are preferentially *cis* to the oxygen if the aryl group does not possess *ortho*-substituents.[77]

A range of oxygen-centred radicals has been obtained from reaction of phenanthraquinone, other quinones, or 1,2-diketones with metal carbonyls, aluminium trichloride, or $MCl_{3-n}Ph_n$ (M = As, Sb): metal species rapidly equilibrate between the two oxygens.[78–81]

Miscellaneous Radicals

Persistent phosphinyls and arsinyls $\dot{M}[CH(SiMe_3)_2]$ and $\dot{M}[N(SiMe_3)_2]$ (M = P, As) have been reported.[82] There is considerable delocalization of spin onto phosphorus in λ^3-phosphorin radicals (**10**)[83] but not in λ^5-phosphorin radicals (**11**), which resemble cyclohexadienyl radicals.[84]

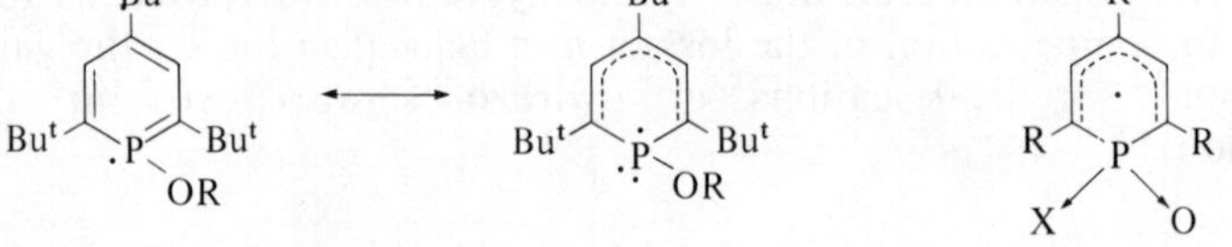

Bu^t Bu^t Bu^t P OR ⟷ Bu^t Bu^t Bu^t P OR   R R R P X O

(10)   **(11)**

The ESR spectra of benzyldimethylgermyl radicals, $PhCH_2\dot{G}eMe_2$, indicate partial phenyl bridging to germanium.[85] Triarylgermyl radicals, $Ar_3Ge\cdot$ (Ar = 2, 6-$Me_2C_6H_3$ or 2,4,6-$Me_3C_6H_2$) are longer-lived than the analogous silicon- and tin-centred radicals, $Ar_3Si\cdot$ and $Ar_3Sn\cdot$: the lesser stability of $Ar_3Si\cdot$ is due to their tendency to abstract hydrogen whereas for $Ar_3Sn\cdot$ it is a consequence of the larger size of Sn which permits Sn–Sn dimerization.[86,87] The *g*-values for the radicals $R_3Ph_{3-n}Sn\cdot$ fall as the number of phenyl groups increases: the radicals become progressively more pyramidal.[88] $R_3Sn\cdot$ radicals with bulky alkyl groups are also significantly non-planar.[89]

Peroxides

The chemistry of peroxides has been reviewed.[90]

Di-*tert*-butyl peroxide undergoes induced decomposition in benzene and perfluoromethylcyclohexane.[91] The photo-sensitized decomposition of this peroxide with aromatic hydrocarbons results in efficient triplet energy transfer to the peroxide with the formation of two *tert*-butoxyl radicals.[92] The decomposition of organosilicon peroxides has been studied.[93] Butyllithium catalyses the decomposition of $(Bu^tO)_2$ and organosilicon peroxides in an electron-transfer process.[94] The triazinyl peroxides, $XYC_3N_3OOBu^t$ and $XYC_3N_3OOC_3XY$, have been synthesized: they decompose to give $XYC_3N_3O\cdot$ radicals, which cannot be directly detected, but have been trapped by ethylene and trimethyl phosphite to give

$XYC_3N_3OCH_2CH_2\cdot$ and $XYC_3N_3O\dot{P}(OMe)_3$ respectively.[95] The stabilities of these peroxides are more akin to those of dialkyl peroxides: this implies that there is little delocalization of unpaired spin in $XYC_3N_3O\cdot$ and that these radicals are σ-species. Alkyl radicals are generated in the reaction of trialkylboranes with dimethyl(methylperoxy)borane.[96]

$$R_3B + MeOOBMe_2 \rightarrow R_3\dot{B}OMe + Me_2BO\cdot$$
$$R_3\dot{B}OMe \rightarrow R_2BOMe + R\cdot$$

The alkyl radicals generated in the reaction of trialkylboranes with $Me_3SiOOBu^t$ have been spin-trapped with Bu^tNO.[97] $CF_3O\cdot$ radicals are generated in pyrolyses of CF_3OOCF_3 and CF_3OOOCF_3.[98]

Thermolysis of 1-butoxy-1-*tert*-butylperoxyethane gives both butyl formate and butyl acetate, the latter resulting from induced decomposition of the peroxide (Scheme 1).[99] A series of 2-*tert*-butylperoxyoxacyclanes behave similarly.[100] The peroxytetrahydrofuran (**12**) breaks down as shown in Scheme 2.[101]

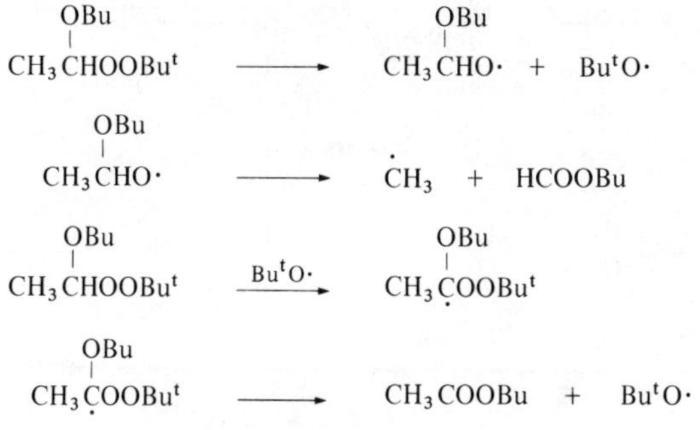

$$CH_3CH(OBu)OOBu^t \longrightarrow CH_3CH(OBu)O\cdot + Bu^tO\cdot$$
$$CH_3CH(OBu)O\cdot \longrightarrow \dot{C}H_3 + HCOOBu$$
$$CH_3CH(OBu)OOBu^t \xrightarrow{Bu^tO\cdot} CH_3\dot{C}(OBu)OOBu^t$$
$$CH_3\dot{C}(OBu)OOBu^t \longrightarrow CH_3COOBu + Bu^tO\cdot$$

SCHEME 1

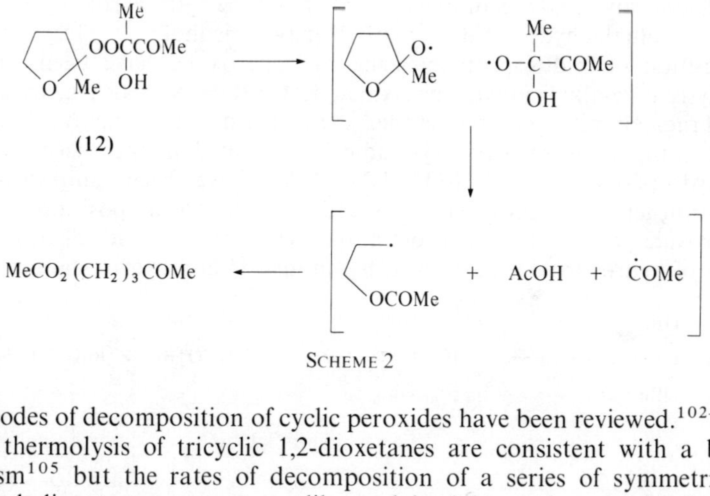

SCHEME 2

The modes of decomposition of cyclic peroxides have been reviewed.[102–104] The rates of thermolysis of tricyclic 1,2-dioxetanes are consistent with a biradical mechanism[105] but the rates of decomposition of a series of symmetric alkyl-substituted dioxetanes are not readily explained by such a mechanism.[106,107]

Further examples of electron-exchange chemiluminescence from sterically hindered cyclobutadiene-1,2-dioxetanes have been reported.[108] The yield of triplet acetone from thermolysis of dimethyldioxetanone is much lower than from tetramethyldioxetane as intersystem crossing in the biradical to the triplet manifold is less efficient with dimethyldioxetanone due to rapid loss of carbon dioxide.[109]

The activation parameters for the thermolysis of 3,3,5,5-tetramethyl-1,2-dioxolane are consistent with a step-wise process.[110] Estensive dehydrogenation occurs in the vacuum pyrolysis of 3,6-dimethyl-1,2-dioxacyclohexane and 2,3-dioxabicyclo[2.2.2]octane (**13**) (*cf.* Scheme 3).[111] In contrast 2,3-dioxabicyclo-[2.2.1]heptane (**14**) gives 90% of the epoxyketone (**15**): steric constraints in the intermediate biradical prevent the hydrogens becoming sufficiently close for dehydrogenation to occur.[112] A similar reaction occurs in the photolysis of 1,5-dimethyl-6,7-dioxabicyclo[2.2.1]heptane.[113] 1,2-Dioxolanes have been prepared from reaction of hydroperoxides, *e.g.* $PhCHR^1CH_2C(OOH)R^2R^3$, with lead(IV) acetate.[113]

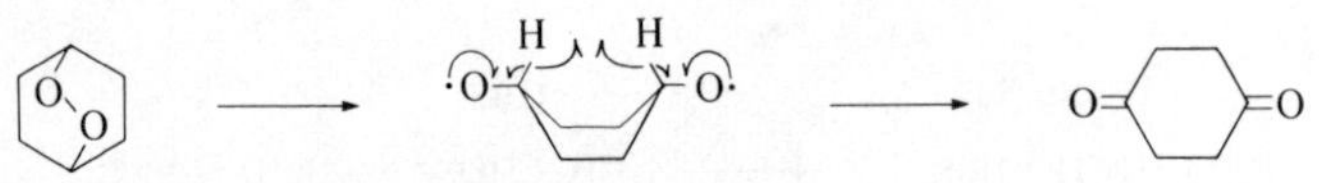

(**13**)

SCHEME 3

(**14**) (**15**)

Antimony pentachloride catalyses both the homolytic and heterolytic cleavage of Bu^tOOH.[114] 1,1-Dibutoxyethane hydroperoxide decomposes as shown in Scheme 4.[115] Decomposition of 1-phenylcyclohexyl hydroperoxide gives a mixture of 1-phenylcyclohexanol and hexanophenone.[116] *p*-Xylene hydroperoxide breaks down to give *p*-tolualdehyde and *p*-methylbenzyl alcohol.[117] The kinetics of decomposition of *m*-diisopropylbenzene hydroperoxides have been studied.[118] Thermolysis of α-alkylazohydroperoxides, $R_2C(OOH)N{=}NR'$, gives alkyl and hydroxyl radicals, nitrogen and ketone.[119] Further details of the hydroxylation of aromatic compounds by hydroxyl radicals generated in the thermolysis of α-arylazohydroperoxides, $PhCH(OOH)N{=}NAr$, have been published.[120] An electron-transfer mechanism is involved in the decomposition of cumyl hydroperoxide with the hemeprotein-metmyoglobin,[121] and also in decompositions of hydroperoxides catalysed by amines[122] and hydroxylamines.[123]

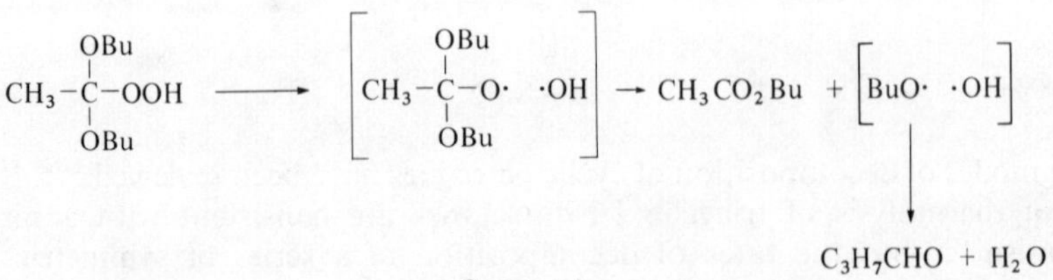

$$CH_3C(OBu)_2OOH \longrightarrow [CH_3C(OBu)_2O\cdot \;\; \cdot OH] \rightarrow CH_3CO_2Bu + [BuO\cdot \;\; \cdot OH] \rightarrow C_3H_7CHO + H_2O$$

SCHEME 4

A series of Hammett substituent constants $\sigma^{\cdot H}$ suitable for use in free-radical reactions, *e.g.* the NBS bromination of substituted toluenes, have been evaluated from the rates of decomposition of *tert*-butyl peroxyarylacetates, $ArCH_2CO_3Bu^t$.[124] The rates of decomposition of these peresters are retarded by electron-withdrawing substituents.[125] The rates of decomposition of *tert*-butyl bridgehead peroxycarboxylates are determined mainly by polar effects in the transition state.[126] Thermolysis of $Pr_3GeCH_2CH_2CO_3Bu^t$ gives rise to $Pr_3Ge\cdot$ and $Pr_3GeCH_2CH_2\cdot$ radicals: migration of the Pr_3Ge group from carbon to oxygen occurs under certain conditions[127] (*cf.* the migration of the R_3Si group in the thermolysis of $R_3SiCH_2CH_2CO_3Bu^t$).[128] The rates of thermolyses of 1-phenylethyl peroxyacetate and substituted peroxybenzoates together with the values of kinetic isotope effects indicate that O—O bond homolysis occurs: the observation of chemiluminescence indicates that $PhCOCH_3$ is formed in an excited state.[129]

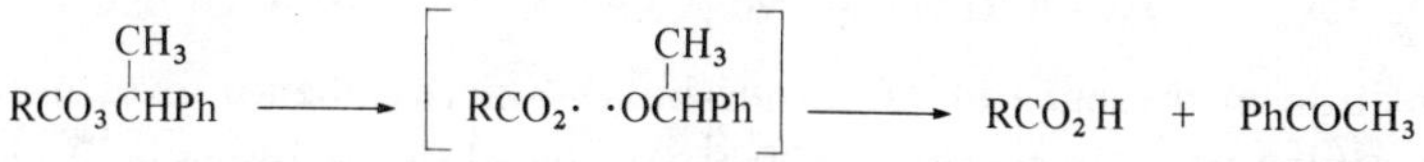

$$RCO_3\overset{\overset{\large CH_3}{|}}{C}HPh \longrightarrow \left[RCO_2\cdot \ \cdot O\overset{\overset{\large CH_3}{|}}{C}HPh \right] \longrightarrow RCO_2H + PhCOCH_3$$

Decompositions of these peresters is accelerated by electron-rich aromatic compounds as a result of an electron-transfer process (Scheme 5). $DPPH_2$ catalyses the decomposition of peresters.[130] Thermolysis of di-*tert*-butyl dipermuconate results in O—O bond homolysis followed by fragmentation of the radical fragments.[131]

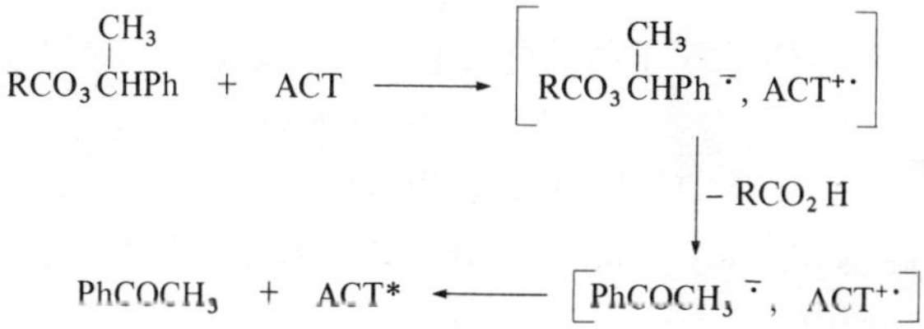

$$RCO_3\overset{\overset{\large CH_3}{|}}{C}HPh + ACT \longrightarrow \left[RCO_3\overset{\overset{\large CH_3}{|}}{C}HPh^{\overline{\cdot}},\ ACT^{+\cdot} \right]$$

$$\downarrow -RCO_2H$$

$$PhCOCH_3 + ACT^* \longleftarrow \left[PhCOCH_3^{\overline{\cdot}},\ ACT^{+\cdot} \right]$$

SCHEME 5

The decomposition of lauroyl peroxide in benzene at 50° occurs by a first-order process and by an induced mechanism.[132] 2,2,6,6-Tetramethylpiperidin-1-oxyl induces the decomposition of diacyl peroxides.[133] Photolysis of unsymmetrical diacyl peroxides at −78° gives 50–70 % yields of mixed dimers.[134] Increasing chain-length and chain-branching in acetyl alkanoyl peroxides decreases the importance of a radical mechanism.[135] The rates of decomposition of perfluoroalkanoyl peroxides have been reported.[136] CIDNP and product studies on the photolysis of acyl benzoyl peroxides indicates that the rate of decarboxylation of acyloxy radicals is very high.[137] The thermolysis of dibenzoyl peroxide in presence of alkyl iodides, RI, gives PhR and PhI.[138] Triethylamine catalyses the decomposition of dibenzoyl peroxide: CIDNP studies show that an electron-transfer process is involved.[139a] Amines induce the decomposition of *p*-nitrobenzoyl peroxide.[139b] Electron-transfer processes are also involved in the reactions of dibenzoyl peroxide with *N,N,N′,N′*-tetramethyl-*p*-phenylenediamine[140] and with anispinacolone.[141]

$$An_3CCOAn + (PhCO_2)_2 \rightarrow [An_3CCOAn]^{+\cdot} + PhCO_2\cdot + PhCO_2^-$$
$$[An_3CCOAn]^{+\cdot} \rightarrow An_3C^+ + An\dot{C}O$$
$$An\dot{C}O + (PhCO_2)_2 \rightarrow AnCOOCOPh + PhCO_2\cdot$$
$$An_3C^+ + PhCO_2^- \rightarrow PhCO_2CAn_3$$

In the decomposition of the trifunctional peroxide $(Bu^tO_3CCH_2CH_2CO_2)_2$ O—O bond homolysis is accompanied by C—O and C—C bond homolysis.[142]

Benzoyloxy radicals are generated in the thermolysis of dibenzoyl monoperoxycarbonate.[143] A CIDNP study of the thermolysis of diisopropylperoxydicarbonate suggests the importance of induced decomposition.[144]

$$MeCHOCO_2^{\cdot} + (Me_2CHOCO_2)_2 \rightarrow Me_2CHOCO_2H + \dot{C}Me_2OCO_2.O_2COCHMe_2$$
$$\dot{C}Me_2OCO_2.O_2COCHMe_2 \rightarrow Me_2CO + CO_2 + Me_2CHOCO_2\cdot$$

Decomposition in presence of 2,6-di-*tert*-butyl-4-methylphenol gives no acetone: the $Me_2CHOCO_2\cdot$ radicals are trapped by the aryloxyl radicals.[145] Extensive induced decomposition with the formation of high yields of acetone also occurs in the thermolysis of $Me_2CHOCO_3Bu^t$.[146]

Thermal decarboxylation of peroxycholanic acid gives the epimeric 16-hydroxynorcholanes: the peracid breaks down to give a primary alkyl radical which undergoes intramolecular 1,5-hydrogen transfer to give a secondary alkyl radical which undergoes an S_H2 reaction with the peracid to give the product.[147] In the decomposition of peroxydecanoic acid in CCl_4, the electrophilic $\dot{C}Cl_3$ radicals abstract hydrogen from the peracid to give acylperoxyl radicals.[148]

$$\dot{C}Cl_3 + RCO_3H \rightarrow RCO_3\cdot + CHCl_3$$
$$RCO_3\cdot \rightarrow R\dot{C}O + O_2$$
$$R\dot{C}O + CCl_4 \rightarrow RCOCl + \dot{C}Cl_3$$

The anion of peroxycarboxyimidic acid RC(=NH)OOH induces the decomposition of hydrogen peroxide.[149]

The relative rates, $k_1:k_2:k_3$, of the three self-reactions of $CH_3O_2\cdot$ are 0.32:0.68:0.08.[150]

$$2\,CH_3O_2\cdot \xrightarrow{k_1} 2\,CH_3O\cdot + O_2$$
$$2\,CH_3O_2\cdot \xrightarrow{k_2} CH_3OH + CH_2O + O_2$$
$$2\,CH_3O_2\cdot \xrightarrow{k_3} CH_3OOCH_3 + O_2$$

The overall rates of the self-reaction of $MeO_2\cdot$ have been reported.[151,152] The self-reaction of $MeO_2\cdot$ is predominantly terminating at room temperature but becomes less so as the temperature rises.[153] The principal reaction of $MeO_2\cdot$ in the oxidation of methane is second-order propagation.[154] The reaction $2\,Bu^tO_2\cdot \rightarrow (Bu^tO)_2 + O_2$, which had been regarded as a cage reaction, occurs in the gas phase.[153] The rate of the self-reaction of $PhCMe_2OO\cdot$ in solution at room temperature is $1.5 \times 10^4 M^{-1} s^{-1}$.[155] The rates of combination of $Me_2C(CN)OO\cdot$,[156] p-$MeC_6H_4CH_2OO\cdot$,[157] and the peroxyl radical intermediates in the oxidations of 1-methyl-3-phenylindan and related compounds have been measured.[158]

The kinetics of formation of peroxyl radicals from the reactions of $\dot{C}Cl_3$ and $\dot{C}F_3$ with O_2 have been studied.[159] Ethyl radicals react with oxygen to give ethylene and $HO_2\cdot$ in a disproportionation step. Ethylene oxide and hydroxyl radicals are formed in a competitive reaction: the initial peroxyl radical undergoes a 1,5-hydrogen shift (Scheme 6).[160] The equilibrium between allyl radicals and oxygen and allylperoxyl radicals has been studied: the bond energy of the allyl—OO bond has been estimated[161] to be $17.2 \pm 1.0\,\mathrm{kcal\,mol^{-1}}$. The kinetics of the reactions of benzyl radicals with oxygen and NO have been investigated.[162,163] Butoxyl radicals and oxygen undergo a disproportionation reaction to give butanal and $HO_2\cdot$.[164] The rates of the reactions of $Bu^tO_2\cdot$ and PhNHMe,[165] $CH_3O_2\cdot$ and SO_2,[166] $PhCMe_2OO\cdot$ and $C(CH_2OCOPr)_4$,[167] and $Me_2C(CN)OO\cdot$ and 2,6-di-*tert*-butyl-4-methylphenoxyl[168] have been studied. The rate constants for hydrogen abstraction by $Bu^tO_2\cdot$ from $Me_3CCH_2CH_3$ at 333 K are 1.5×10^{-3} (CH_2), 4×10^{-5} (Me_3C) and 6×10^{-5} (CH_3) $\mathrm{M^{-1}\,s^{-1}}$;[169] the relative reactivities of 1°, 2°, and 3° hydrogens in acyclic alkanes towards abstraction by Bu^tO_2 are 1:35–70:325–3000.

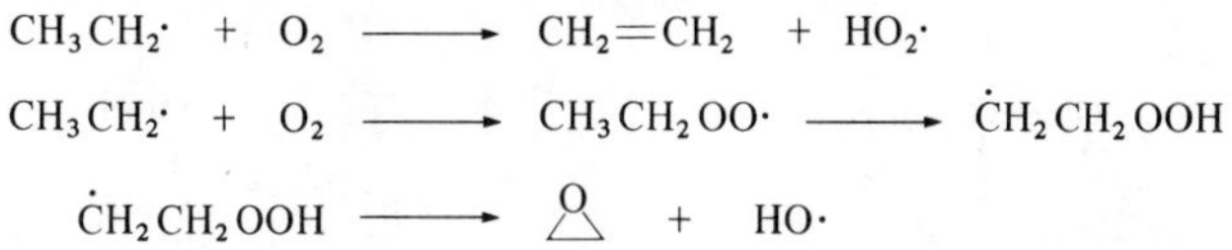

$$CH_3CH_2\cdot + O_2 \longrightarrow CH_2{=}CH_2 + HO_2\cdot$$

$$CH_3CH_2\cdot + O_2 \longrightarrow CH_3CH_2OO\cdot \longrightarrow \dot{C}H_2CH_2OOH$$

$$\dot{C}H_2CH_2OOH \longrightarrow \text{(ethylene oxide)} + HO\cdot$$

SCHEME 6

Further studies on the cyclization of alkenylperoxyl radicals have been reported:[170] the hydroperoxide (**16**) forms radical (**17**) which cyclizes stereospecifically to give (**18**) and thence the hydroperoxy epidioxide (**19**); the radical (**20**) generated in the same reaction cyclizes to give the cyclopropylcarbinyl radical (**21**) which is trapped to give the hydroperoxide (**22**).[171] A similar sequence (Scheme 7) is involved in the formation of bicyclo endoperoxides, related to prostaglandins, from methyl linolenate; cyclization again occurs stereospecifically to give the *cis*-radical.[172] The peroxyl radical (**23**) cyclizes to give the six-membered peroxide.[173]

Autoxidations

Cyclohexylperoxyl radicals are intermediates in the photochemical oxidation of cyclohexane.[174] Autoxidation of 1,3-and 1,4-dimethylcyclohexanes proceeds *via* tertiary hydroperoxides.[175] Egg lecithin is more resistant to autoxidation than would be expected on the basis of the chemical composition of the lecithin bilayers.[176a] This has been attributed to the polarity of peroxyl radicals [$R{-}\ddot{O}{-}O\cdot \leftrightarrow R{-}\overset{+\cdot}{\ddot{O}}{-}O^-$]: as a consequence of the high polarity of the peroxyl portion the radical is able to diffuse out of the non-polar autoxidizable hydrocarbon environment in which it was formed. The rôle of nucleotide–iron complexes in lipid peroxidation has been discussed.[176b] Peroxyl radicals react very rapidly with β-carotene and other polyenes.[177] Autoxidations of cycloalkenes have been reported[178] and the importance of polar effects in autoxidations of substituted toluenes have been assessed.[179] Aryloxyl radicals are generated in the heat treatment of polycyclic aromatics in presence of oxygen.[180]

β-Hydroperoxysulphides, which can be readily reduced to the corresponding β-hydroxysulphides, are obtained by thiol-oxygen co-oxidation of alkenes. The

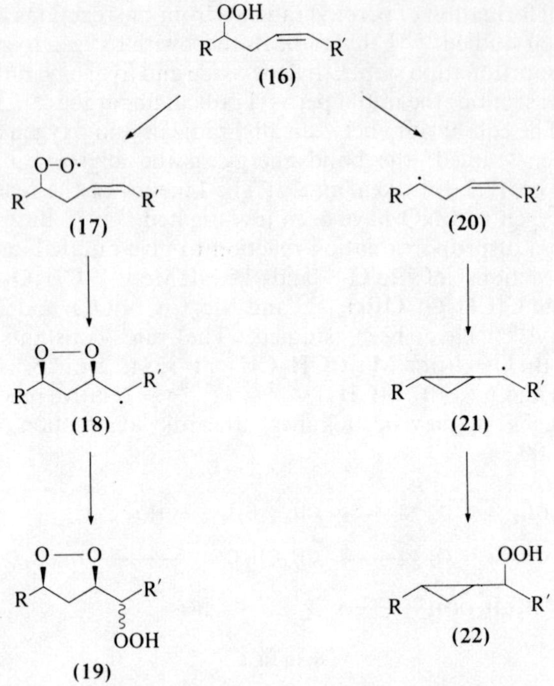

(16)

(17)

(20)

(18)

(21)

(19)

(22)

R = $[CH_2]_5$ Me, R′ = $[CH_2]_7$ CO_2 Me

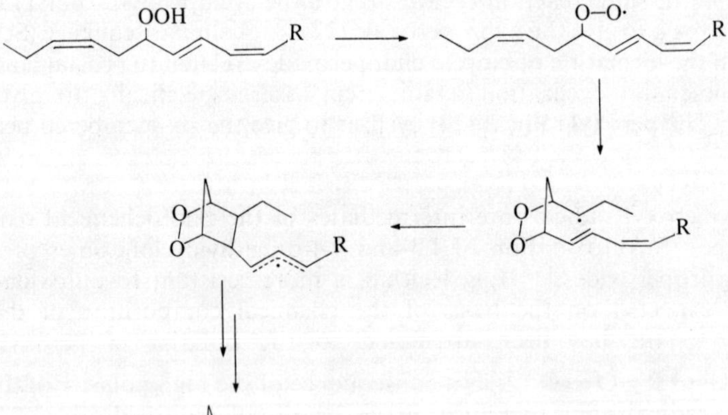

(23)

R = $(CH_2)_7$ CO_2 Me

SCHEME 7

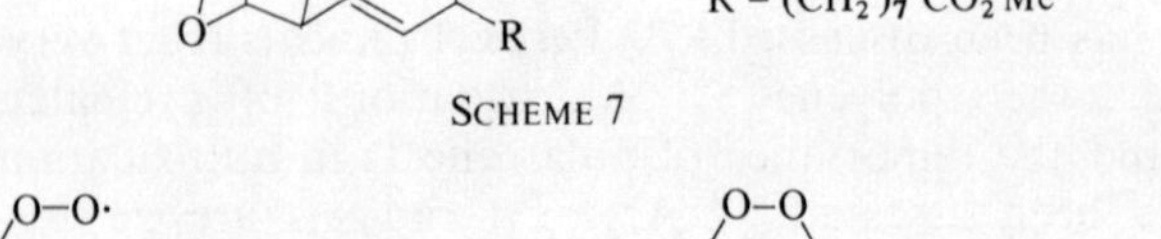

stereochemistry of the products indicate that the intermediate β-thioalkyl radical couples with oxygen mainly *anti* to the RS group, but that rotation about the $C_\alpha-C_\beta$ bond competes with trapping of this radical with oxygen (Scheme 8).[181]

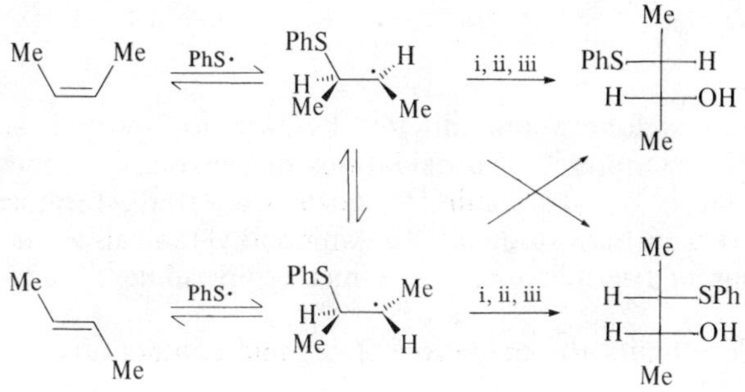

i: O_2, ii: PhSH, iii: Ph_3P

SCHEME 8

Autoxidation of *cis*- and *trans*-β-methylstyrenes gives mainly epoxides and benzaldehyde: very little allylic oxidation occurs (Scheme 9).[182] Epoxides are also formed in the autoxidation of *cis*-β-alkoxystyrene.[183] In the photo-oxidation of styrene and benzoin, PhĊO radicals and thence $PhCO_3\cdot$ are generated; the $PhCO_3\cdot$ radicals add to the styrene and the resultant radicals give the epoxide in an S_Hi reaction.[184]

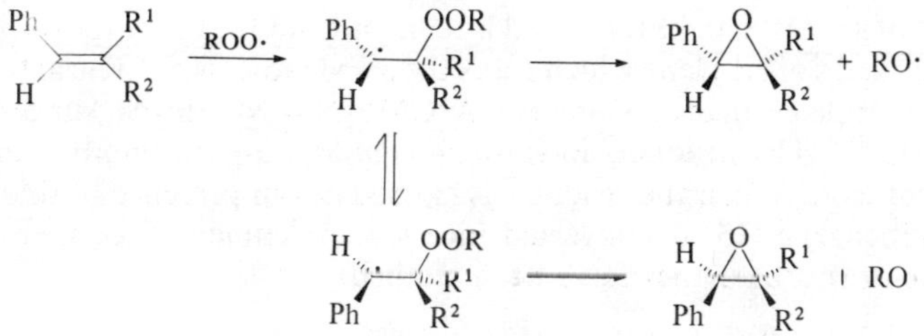

SCHEME 9

$CHCl_2OO\cdot$ and $CH_2ClOO\cdot$ are intermediates in the chlorine atom-initiated oxidations of CH_2Cl_2 and CH_3Cl.[185] Phenol, benzaldehyde, and phenyl benzoate are formed in the benzophenone photo-induced oxidation of benzyl phenyl ether which proceeds *via* PhCH(OPh)OOH.[186] The induced decomposition of 1,1-dibutoxyethane[187] and the reaction of ozone[188] with 1,3-dioxolanes proceeds *via* α,α-dialkoxyalkyl radicals.

α-Tocopherol (**24**) is about 10 times as efficient as an anti-oxidant as 4-methyl-2,3,5,6-tetramethylphenol ($k_{inhib} = 23.5 \times 10^5$ and $2.1 \times 10^5\ M^{-1}\,s^{-1}$ respectively): the high efficiency is due to the fact that the fused ring structure holds the *p*-type lone pair of the chroman oxygen more or less perpendicular to the plane of the aromatic system thus stabilizing the aryloxyl radical by about 3 kcal mol^{-1} more than if the oxygen lone pair were in the plane of the aromatic ring as occurs in 4-methoxy-2,3,5,6-tetramethylphenol.[189,190]

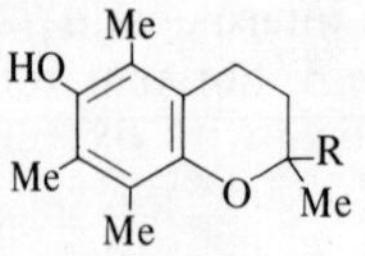

HO, Me, Me, Me (chroman ring) R, Me, O

$R = CH_2(CH_2CH_2\overset{CH_3}{C}HCH_2)_3H$

(24)

Autoxidation of cyclohexanone initiated by 2-cyano-2-propyl radicals gives 2-hydroperoxycyclohexanone.[191] Autoxidations of benzoin,[192] 2-methyl-5-phenylfuran-3-carboxylates,[193] L-adrenalin,[194] nucleoside trialkyl phosphite,[195] and alkylzirconocenes have been studied.[196] α-Aminoalkyl radicals are intermediates in the autoxidations of triethylamine,[197] *N*-methylpiperidine,[198] and *N*,*N*-diethylacetamide.[199]

Stannic iodide inhibits the oxidation of organic compounds by reacting with peroxyl radicals.[200]

Azo Compounds

The decomposition of azo compounds in which the azo group is incorporated into the skeleton of a bicycloalkane, *e.g.* 2,3-diazabicyclo[2.2.2]oct-2-ene, have been reviewed.[201]

The rates of decomposition of azoethane and azoisopropane have been determined.[202–204] The decomposition of azomethane has been shown to proceed in part by a chain mechanism.[205]

$$CH_3\cdot + CH_3N{=}NCH_3 \rightarrow CH_4 + CH_3N{=}NCH_2\cdot$$

This type of process can be repressed by carrying out the decompositions in presence of a large excess of ethylene which traps the alkyl radicals.[204] The activation volume for the decomposition of azocumenes, $ArCMe_2N{=}NCMe_2Ar$, suggests a two-bond mechanism.[206] The products include α-*ortho* and α-*para*-semibenzenes formed in the solvent cage. When the reaction is carried out in presence of thiophenol, the α-*para*-semibenzene (**25**) is converted into *p*-cumylcumene. The thermolyses of *cis*- and *trans*-azonorbornanes have been studied.[207]

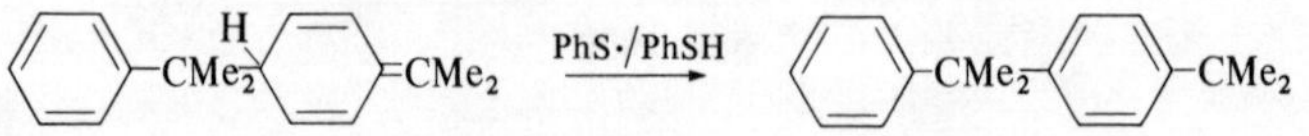

$$C_6H_5\text{–}CMe_2\text{–}C_6H_5(H){=}CMe_2 \xrightarrow{PhS\cdot/PhSH} C_6H_5\text{–}CMe_2\text{–}C_6H_4\text{–}CMe_2$$

(25)

α-Hydroxyazo compounds have been used to effect the hydroalkylation of alkenes and azobenzene (Scheme 10).[208] *β*-Cyanoalkyl radicals are generated in the decomposition of (**26**) in CCl_4.[209]

$$Ph_2C(OH){-}N{=}N{-}R \longrightarrow Ph_2\dot{C}OH + N_2 + R\cdot$$

$$R\cdot + CH_2{=}CHY \longrightarrow RCH_2\dot{C}HY$$

$$RCH_2\dot{C}HY + Ph_2C(OH){-}N{=}N{-}R \longrightarrow RCH_2CH_2Y + Ph_2CO + N_2 + R\cdot$$

SCHEME 10

$$Me_2\underset{\substack{|\\ OH}}{C}-N{=}N-CH_2CHRCN \;+\; \dot{C}Cl_3$$

(26)

$$\longrightarrow CHCl_3 + Me_2CO + N_2 + \dot{C}H_2CHRCN$$

The activation parameters for the decomposition of 3-methyl-1-*p*-tolyltriazene are similar to those for azoarylalkanes suggesting a similar mode of decomposition: the products obtained suggest that both tautomers, ArN=NNHMe and ArNHN=NMe are involved.[210] The photo-decomposition of ArNMeN=NNMeAr gives ArNMeNMeAr: this is formed both from in-cage and out-of-cage processes.[211] The formation of (**28**) from the photolysis of (**27**) indicates that one-bond scission of the tetrazene occurs.[212]

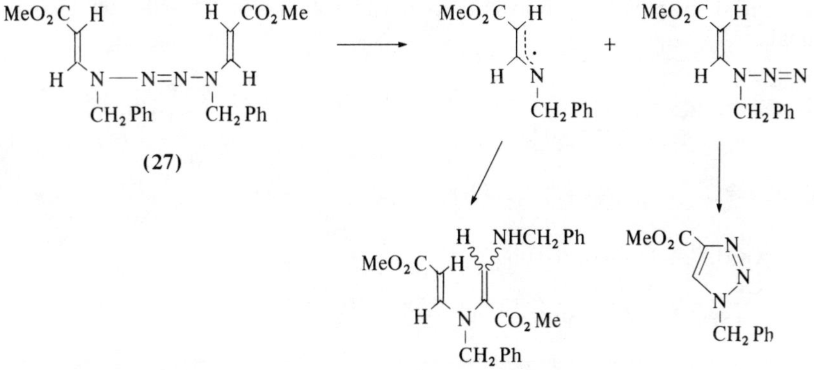

MeO₂C–CH=CH–N(CH₂Ph)–N=N–N(CH₂Ph)–CH=CH–CO₂Me **(27)** ⟶ MeO₂C–CH=CH–N•(CH₂Ph) + MeO₂C–CH=CH–N(CH₂Ph)–N=N

MeO₂C–CH=CH–N(CH₂Ph)–CH=C(NHCH₂Ph)CO₂Me + Other products

1-benzyl-4-(methoxycarbonyl)-1,2,3-triazole **(28)**

Diazonium Salts

CIDNP effects, kinetic studies, and product analysis suggest that aryl radicals are generated from diazonium salts as depicted in Scheme 11.[213–215] The dediazoniation of diazonium salts in weakly alkaline solution in presence of oxygen proceeds by a radical-chain mechanism involving phenoxide anions.[216]

$$ArN_2^+ + 2\,HO^- \longrightarrow ArN_2O^- + H_2O$$

$$ArN_2O^- + ArN_2^+ \rightleftharpoons ArN_2{-}O{-}N_2Ar$$

$$ArN_2{-}O{-}N_2Ar \longrightarrow [ArN_2\cdot \;\; \cdot ON_2Ar]$$

$$[ArN_2\cdot \;\; \cdot ON_2Ar] \xrightarrow{-N_2} [Ar\cdot + \cdot ON_2Ar]$$

$$Ar\cdot + ArH \longrightarrow \text{products}$$

SCHEME 11

$$Ar\cdot + {}^{-}O\text{-}Ar'\text{-}H \rightarrow Ar\text{-}\dot{A}r'(H)\text{-}O^{-}$$

$$Ar\text{-}\dot{A}r(H)\text{-}O^{-} + ArN_2^{+} \rightarrow Ar\text{-}Ar'O^{-} + ArN_2\cdot + H^{+}$$

Aryl radicals are generated in reductions of diazonium salts by Sn^{2+}, Fe^{2+}, ferrocene $Fe(CN)_6^{4-}$, ascorbic acid,[217] semiquinone radical anions,[218,219] $\dot{C}H_2OH$, $\dot{C}Me_2OH$,[219] tetrathiofulvenes,[220] and copper.[221] Pulse radiolysis of diazonium fluoroborates yields aryl radicals[222] as does photolysis of *p*-*N*,*N*-dialkylaminobenzene diazonium fluoroborates.[223] A new method of generating aryl radicals involves treatment of diazonium salts with Bu_3SnH in ether.[224] This has been applied to *o*-alkenyloxyarene diazonium salts (**29**): the *o*-alkenyloxyaryl radicals (**30**) cyclize to give dihydrobenzofurans (**31**). The same radicals can also be generated by reductions with sodium iodide in acetone, sodium thiophenoxide in DMSO,[224] and by nitroxides.[225] Further details of the reaction of diazonium salts with mesityl oxide in presence of titanium(II) salts have been published.[226] Adenine reacts with benzene diazonium salts to give triazenes which break down to give aryl radicals.[227]

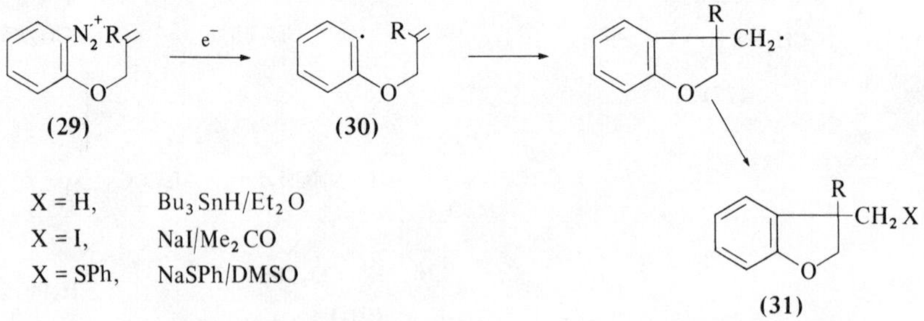

(29) (30) (31)

X = H, Bu_3SnH/Et_2O
X = I, NaI/Me_2CO
X = SPh, NaSPh/DMSO

Combination and Disproportionation Reactions

The reversible recombination of radicals has been reviewed[228] and a theoretical treatment of the recombination kinetics of contact radical pairs and separated radicals has been presented.[229] A simple analysis for the investigation of substituent effects and favourable reaction paths, involving pair-wise orbital transformations, has been applied to the combination of methyl radicals and the disproportionation of ethyl radicals.[230] There is now good agreement between theory and experiment for the $C_2H_6 \rightleftharpoons 2\,Me\cdot$ system in the light of the revised heat of formation for the Me· radical of 35.1 ± 0.15 kcal mol^{-1} (298 K).[231,232] The high-pressure rate constant for the reaction $Me\cdot + H\cdot \rightarrow CH_4$ has been measured and related to theoretical models.[233] An absolute rate for the combination of methylidine radical with atomic oxygen has been determined.[234] Theoretical appraisal of the combination of methylidine radicals reveals a sudden change in spin coupling at $R_{CC} \sim 6$ a.u. associated with a 0.33 eV barrier on the potential energy surface.[235]

A photolysis technique utilizing harmonically modulated excitation has been developed and applied to the diffusion-controlled termination reactions of $Bu^t\cdot$, $Pr^i\cdot$, and $PhCH_2\cdot$ radicals.[236] Combination and disproportionation rate constants for $Pr^n\cdot$ and $Pr^i\cdot$ radicals have been measured on flash photolyses of azopropanes.[237]

The rate of combination of *n*-pentyl radicals in toluene has been determined.[238] Pentamethylcyclopentadienyl radicals combine reversibly to form a dehydro dimer but disproportionate irreversibly to yield tetramethylfulvene.[239]

Rates of self-reaction of $\cdot CF_2Cl$ radicals and their reaction with Cl· atoms have been measured.[240] The recombination of $\cdot CH_2OH$ radicals in MeOH at elevated temperatures is diffusion-controlled.[241] Rates of reaction of $CH_3O\cdot$ and $\cdot CH_2OH$ radicals with H· atoms are in a 2:3 ratio, and hydrogen abstraction accounts for 75% of each reaction.[242] The rates of the combination reactions $H\dot{C}O + H\dot{C}O$ and $H\dot{C}O$ and Me· have been determined using intracavity laser spectroscopy.[243] While on energetic grounds the self-termination of benzoyl radicals could yield electronically excited benzil products, such an occurrence is forbidden by state symmetry control.[244] The self-reaction of $Bu^tO\cdot$ radicals has a rate constant of $1.3 \pm 0.5 \times 10^9\ M^{-1}\ s^{-1}$ at 293 K in the gas phase.[245]

The termination of *N*-hydro-4-acetylpyridinyl radicals in Pr^iOH yielding the respective pinacol is not diffusion-controlled.[246] A versatile apparatus for the study of thin films by optical spectroscopy over a wide temperature range has been described[247] and has been used in the (re)-investigation of monomer–dimer equilibria involving 1-alkyl-2-(carbomethoxy)pyridinyl radicals;[248] 1-alkyl-4-(carboalkoxy) and 1-alkyl-4-carbamidopyridinyl radicals[249] in solution and in thin films. The equilibrium between the fairly persistent 1-methyl-4-phenylpyridinyl radical and its dimer has also been investigated.[250] It has been demonstrated that the UV absorption spectrum attributed to the 1,1′-trimethylenebis-4-(carbomethoxy)pyridinyl diradical contains contributions from intramolecularly cyclized forms.[251] Such "cyclomers" yield diradical metal complexes in the presence of metal ions such as Li^+ and Mg^{2+}.

Solvation effects appear to influence the diffusion-controlled dimerization of 2,6-diphenyl-4-methoxyphenoxyl radicals.[252] For the *o*-organosilyl-substituted phenoxyl radicals (**32**) log *k* (dimerization) is linearly related to the sum of steric constants of the alkyl groups on the silicon. Biphenyls are formed after migration of the silicon onto the phenoxyl oxygen.[253] Attack of radical (**33**) on the phenoxyl radical (**34**) occurs at the sterically more hindered *ortho* position of (**34**).[254] Reaction of thioamides, $R^1\dot{C}SNHR^2R^3$, with *N*-nitrosoacetanilide or phenylazotriphenylmethane, both of which are sources of phenyl radicals, give α-thioalkyl radicals $R^1\dot{C}(SPh)NR^2R^3$ which undergo hydrogen abstraction with $PhN_2O\cdot$ or $Ph_3C\cdot$ in a disproportionation reaction.[255] The product depends on the nature of R^1, R^2, and R^3; $Ph\dot{C}(SPh)NHPh$ gives $PhC(SPh){=}NPh$, $Ph_2CH\dot{C}(SPh)NHPh$ gives a mixture of $Ph_2CHC(SPh){=}NPh$ and $Ph_2C{=}C(SPh)NHPh$, while $Ph_2CH\dot{C}(SPh)NEt_2$ gives $Ph_2C{=}C(SPh)NEt_2$.

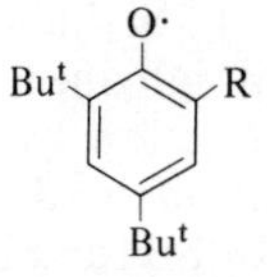

R = Me_3Si, Me_2SiEt, $MeSiEt_2$ and Et_3Si (**34**) (**33**)

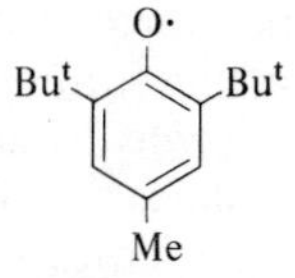

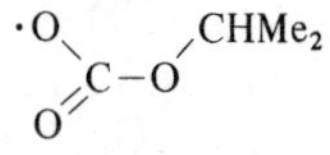

(**32**)

Atom-abstraction Reactions

INDO-UHF calculations have confirmed that relative reactivities in hydrogen abstraction from hydrocarbons (RH) by radicals (X·) are explicable[256] *via* the interaction between the highest occupied σ-orbital of RH and the singly occupied *p*-

orbital of X·. BEBO calculations have been performed for hydrogen abstractions by allyl radical.[257] The use of two-parameter (polar and resonance substituent constants) correlations which permit inferences on the structure of transition states has been applied *inter alia* to hydrogen abstraction from toluenes.[258] Correlations of $O(^3P)$ and ·OH radical addition and abstraction rates with ionization potentials and C—H bond dissociation energies reveals differences between simple alkenes and aromatics/halogenated compounds which advises caution in interpretation.[259,260]

Carbon-centred Radicals

The kinetics of reactions of methylidine radicals in hydrocarbon combustion systems at room temperature have been reported.[261] Relative rates of reaction of CH_3· radicals with HCl and DCl[262] are given by the expression

$$k_{HCl}/k_{DCl} = (1.23 \pm 0.06)\exp[(22.3 \pm 15)K/T].$$

Hydrogen abstraction from acetylene by CH_3· radicals is twenty times slower than from ethane[263] in gas-phase shock-tube systems but methyl radicals are over a hundred times more reactive than peroxyl radicals towards hydrocarbons at 85° in the liquid phase.[264] Arrhenius parameters have been determined for abstraction reactions involving methyl radicals with nitromethane,[265] azomethane,[266] and ozone.[267]

The deuterium kinetic isotope effects for abstraction from both the acetyl and methoxy groups of methyl acetate by CH_3· ($k_H/k_D = 11$ and 8, respectively) and CF_3· ($k_H/k_D = 24$ and 4, respectively) have been determined.[268]

Hydrogen transfer from ethyl radicals to molecular oxygen is of minor importance relative to the combination reaction under tropospheric conditions.[269] Abstraction rates by Et· radical from propan-1-ol follow the expected pattern.[270]

In the reaction of Bu^t· radicals with $CHCl_3$ the hydrogen to chlorine abstraction rates are in the ratio 1.4:1 and independent of temperature indicating the importance of polar effects in such transfer reactions of Bu^t· radicals.[271] Alkyl radicals abstract hydrogen from CH_2Cl_2 over eight times faster than from $ClCH_2CH_2Cl$ whereas the chlorine abstraction rates are nearly equal.[272] Relative hydrogen abstraction rates from 1,1-dialkoxyalkanes by phenyl radicals[273] and undecyl radicals[274] have been determined.

Towards heptane (at 50°), CF_3· radicals are 3.6 times more reactive than are C_2F_5· radicals.[275] The frequency factor ratio A_H/A_D of 4.6 for H(D) abstraction from $CF_3CO_2H(D)$ by CF_3· radicals is abnormally high due to A_D being unusually low.[276] Activation parameters for hydrogen abstraction by CF_3· radicals from Cl_3SiH and Cl_3GeH in the gas phase[277] can be rationalized on the basis of polar effects, as can the reactions of CCl_3· radicals with Me_3SiH and Et_3SiH,[278] which are much more reactive than Cl_3SiH. The strong retarding influence of chlorine substituents in silanes is also manifest in the abstraction reactions of chloromethylidine (CCl) radical with methylhalosilanes.[279]

Activation energies for hydrogen atom abstraction by CCl_3· radicals from cyclanes in the liquid phase[280] follow the order expected on the basis of ring strain namely, cyclohexane > cyclododecane > cyclooctane. Abstraction rates from cyclohexene by CCl_3· radicals yield an estimate of 85 ± 1 kcal mol^{-1} for the allylic C—H bond strength.[281] The activation energy for hydrogen abstraction by CCl_3·

radicals from secondary alcohols is 1.8 kcal mol^{-1} lower than that for the secondary hydrogens in alkanes.[282]

Treatment of $Cl_3CCH_2CHCl(CH_2)_5CH_3$ with $Fe(CO)_5$ and $(Me_2N)_2PO$ leads to $Cl_2CHCH_2CHCl(CH_2)_nCHCl(CH_2)_{4-n}Me$ products, where $n = 1, 2, 3$, and 4, indicating the existence of long-distance migrations of the radical centre by intramolecular hydrogen abstractions (dependent on the concentration of $(Me_2N)_2PO$).[283]

Halogenation

Halogenation of cycloalkane derivatives[284] and mechanisms of halogenation[285] have been surveyed. The temperature-dependent balance between substitutive chlorination of $ClCH_2CH_2Cl$ and additive chlorination of C_2Cl_4 is accounted for by a competition between deactivation of excited $CCl_3CCl_2\cdot$ radicals by collision and their loss of chlorine atom.[286] Selectivity in the yield of CH_3CCl_3 from chlorination of CH_3CHCl_2 is influenced by aromatic solvents which can π-complex chlorine atoms.[287,288] In the chlorination of the *meta*-xylenes, *m*-MeC_6H_4R ($R = CH_3$, CH_2Cl, $CHCl_2$) and *m*-$ClCH_2C_6H_4CH_2Cl$ the C—H bond of a methyl group is four times more reactive than that of a CH_2Cl group.[289]

Correlations of chlorination rates of alkylaromatic hydrocarbons by $PhICl_2$ with σ^+ substituent constants reveal large negative ρ values; the kinetic deuterium isotope effect decreases in proportion to increasing substrate reactivity but this is not true for ρ.[290]

Chlorination of the amidines (**35**) by CCl_4[291] to give (**36**) appears to be initiated by electron transfer, the resulting $CCl_3\cdot$ radical then abstracting an α-hydrogen *etc.* The preponderance of 2-chlorohexane over 3-chlorohexane in the photochlorination of hexane by *N*-chloroammonium perchlorate has been attributed to restrictions on the accessibility of the bulky *tert*-aminium radicals.[292] Halogenation of ethyl 1,2-dimethylcyclopropenecarboxylate by Bu^{t}OCl in the dark involves an intermediate radical of structure (**37**) rather than (**38**).[293] Selectivity in chlorination, with sulphuryl chloride, of rotationally frozen 9-*tert*-butyl-1,2,3,4-tetrachlorotriptycenes[294] is ascribed to neighbouring-group participation of the *peri*-chloro substituent.[295] Halogenation of 3,5-dehydronoriceane (**39**) proceeds by front-side attack (route a) with halonium ion but by corner-side attack (route b) with halogen radicals.[296]

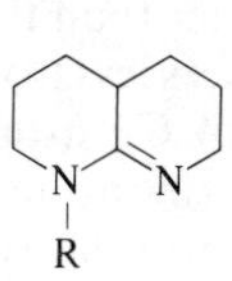

N N R (**35**) R = H, Me

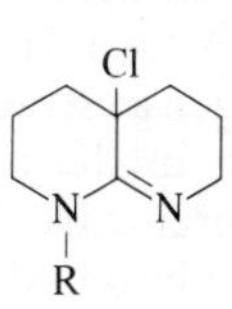

Cl N N R (**36**)

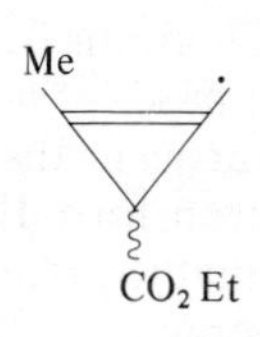

Me CO_2Et (**38**)

Me CO_2Et (**37**)

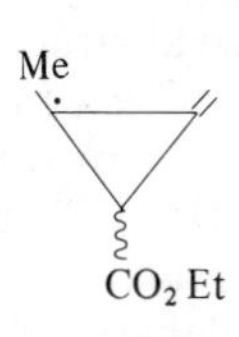

(b) H (a)

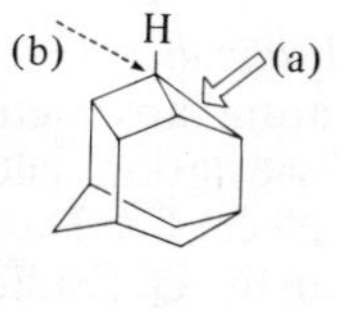

(**39**)

The same pattern of selectivities towards butanes in chlorination by *N*-chloro-, and in bromination by *N*-bromo-, 4,4-dimethyl-2-oxazolidinones (NXDMO), 2-oxazolidinones (NXO), and succinimide (NXS) reveals the intermediacy of hydrogen-abstracting nitrogen-centred chain-carrying radicals: at 80° the relative reactivity towards tertiary C—H as compared with primary C—H depends on the nitrogen radical structure, being 200, 70, and 11 for NXDMO, NXO, and NXS respectively.[297]

In the bromination of cyclohexane and cyclopentane[298] competitive cage reversal, "internal return":

$$Br\cdot + RH \rightleftharpoons (R\cdot, HBr)_{cage} \rightarrow etc.$$

approaches 30% of the rate of bromination but is fortuitously cancelled by the "external return" route involving

$$\begin{aligned} R\cdot &+ HBr \rightarrow RH + Br\cdot \\ R\cdot &+ Br_2 \rightarrow RBr + Br\cdot \\ R\cdot &+ HBr_3 \rightarrow RH + Br_3\cdot \\ HBr &+ Br_2 \rightleftharpoons HBr_3 \end{aligned}$$

despite differences in the structure of radical R·. Serious errors in competitive bromination studies can arise from these phenomena. Formation of $Ph_2C{=}CPh_2$ in the photo-bromination of $Ph_2CHCHPh_2$ is thought to arise from reaction of $Ph_2\dot{C}{-}CHPh_2$ with Br_2 to give $Ph_2CBrCHPh_2$ followed by elimination of HBr.[299] Contrary to an earlier report the photo-induced bromination of 10-substituted 9-methylanthracenes by $CBrCl_3$ is insensitive to the 10-substituent *if HBr is scavenged*,[300] thus preventing the reversible hydrogen transfer of radicals with HBr, a reaction previously reported but often ignored.

The iodination of toluenes at 80° yields a ρ value of -1.6; electron donors weaken the benzylic C—H bond dissociation energy by ~3 kcal mol^{-1}/unit change in σ.[301]

Hydrogen Abstraction by Atoms

Absolute rate constants for hydrogen abstraction by H· and D· from methanol at 500–600 K have been determined.[302] The temperature coefficient of the rate of the stratospherically important reaction of Cl· atom with CH_4 has been measured.[303] Hydrogen abstraction by Cl· atoms from CH_3Cl is 1.31 times faster than that from CH_2Cl_2 at 298 K in the gas phase.[304] β-Hydrogen abstraction suffers less from steric hindrance than does α-hydrogen abstraction in the reaction of Cl· atoms with $CH_2ClCHCl_2$.[305] Fourier transform IR measurements of the chlorine atom–formyl chloride reaction[306] show that the reaction is a hundred times slower than chlorine atom–aldehyde/alkane abstractions. Revised rate constants for the reaction of Br· and Cl· atoms with $H_2C{=}O$ have been reported.[307,308]

Hydroxyl, Alkoxyl, and Peroxyl Radicals

Rate coefficients for hydrogen atom abstractions by ·OH radicals have been correlated with C—H bond dissociation enthalpies with due allowance for temperature and reduced mass effects.[309] The reactivity of fluorochloroalkanes towards hydrogen atom abstraction by ·OH radical goes through a maximum as the number of fluorine atoms relative to the number of chlorine atoms increases.[310]

Hydrogen atom abstraction in the reaction of ·OH radicals with allene forms only a minor (<12%) pathway at low gas pressures.[311]

The kinetics of hydroxyl radical reaction with H_2S, MeSH, Me_2S, and MeSSMe, which include hydrogen-abstraction modes, have been studied.[312] Since only a small fraction (~23%) of ·OH radicals (and H· atoms) captured by $Me_2S{=}O$ produce methane in aqueous acidic media, methane production does not serve as a reliable test for the identification of methyl radicals in biological systems.[313]

Abstraction reactions of ·OH radicals with $H_2C{=}O$,[314] butanals and pentanals in flowing $H_2O_2/NO_2/CO$ mixtures,[315] and acrolein[316] have been investigated.

Selectivity for abstraction by ·OH radicals at the 2-position of *N*-substituted-1,3-oxazacyclopentanes is not altered appreciably by the substituent.[317] Abstractions by hydroxyl radical from α-D-glucose[318] and *myo*-inositol[319] are unselective but the resulting α,β-dioxygen-substituted radicals undergo acid-catalysed fragmentations; those with axial β-hydroxy groups decompose faster.

Methoxyl radicals exhibit a reactivity pattern similar to *tert*-butyl radicals with regard to hydrogen abstraction and addition to double bonds.[320] In aqueous solution the Bu^{t}O· radical is electrophilic like the ·OH radical, but unlike the latter it reacts with prop-2-en-1-ol by abstraction of hydrogen rather than addition[321]. Activation energies for hydrogen atom abstractions of hydrogen rather than addition[321]. Activation energies for hydrogen atom abstractions from CH_3CHO and $(CH_3)_2CO$ by alkoxyl radicals have been determined and compared with semi-empirical estimates.[322]

High rates of abstraction by Bu^{t}O· radicals occur for ethene, acetals, and orthoformates[323] in situations where the C—H bond makes a small dihedral angle (≤30°) with the adjacent oxygen *p*-orbital: at large dihedral angles, *e.g.* 90°, the abstraction rate is much retarded (×5–×10); molecules with high conformational mobility, *e.g.* THF, show high reactivity. Such effects have also been noticed in abstractions from substituted 1,3-dioxanes by Bu^{t}O·.[324] Hydrogen atom abstraction rates from oxygen- and sulphur-containing 1,3-diheterocycloalkanes by Bu^{t}O· radicals,[325] follow the sequences *O,O* < *O,S* < *S,S* and 5- > 7- > 6-membered ring. Di-*tert*-alkoxymethanes are converted to *tert*-alkyl formates and alkanes by $(Bu^tO)_2$.[326] Ring-opening of the 2-methyltetrahydrofuranoxyl radical provides a convenient reference for the measurement of relative hydrogen atom abstraction rates by an alkoxyl radical.[327]

While polar structures contribute to the transition state in the reaction of Bu^{t}O· with amines the contribution is less than that for excited triplet carbonyl compounds.[328] Such stereoelectronic effects have been explored in a comparative study of the reactions of Bu^{t}O·, Bu^{t}O$_2$·, and $^3(Ph_2C{=}O)$ with amines.[329]

Absolute rate constants of 3.3×10^8 and $1.6 \times 10^9\ \text{M}^{-1}\,\text{s}^{-1}$ have been measured for the first time for the reaction of Bu^{t}O· radicals with phenol and *p*-methoxyphenol, respectively.[330] Large kinetic isotope effects are found and pyridine retards the reactions *via* strong hydrogen bonding to the phenols.

π-Acetoxyl radicals show low selectivity in hydrogen abstractions.[331] Cumyloxyl radicals react fifteen times faster with tetrakis(allyloxymethyl)methane than with 2,2,4-trimethylpentane[332] and <15% of the radicals are consumed by addition to the allyl groups. C—H Bond energies of the *ortho*- and *meta*-methyl groups of 2-chloro-*p*-xylene and the methyl groups of 2,5-dichloro-*p*-xylene have been estimated as 90.6, 82.0, and 89.0 kcal mol^{-1}, respectively, from evaluation of the reactivities of the xylenes to oxidation by cumylperoxyl radicals.[333] A C(2) methyl group decreases the rate of reaction of 1,3-dioxolane, -dioxane, and -dioxepane with cumylperoxyl radicals but dimethyl substitution at C(4) or C(5) has little effect.[334]

The anti-oxidant activity of vitamin E and related compounds has been pursued (see p. 13).

Other Hetero Radicals

The kinetics of hydrogen transfer between dihydropyridine and mono-, bis-, and tris-hydrazyls of the tricyanobenzene series have been studied.[335] An inverse isotope effect of 0.75 in the DPPH-inhibited thermal polymerization of 2,6-dideuterio-styrene is consistent with hydrogen transfer to DPPH from a Diels–Alder dimer of styrene.[336] The gas-phase radical nitration of low-molecular-weight aliphatic hydrocarbons by HNO_3 has been studied and modelled.[337] Quantum-mechanical tunnelling is suspected in the reaction of $(CF_3)_2NO\cdot$ with toluene and toluene-d_8 in fluorochlorocarbon solvents since the activation energy difference, $E_D - E_H$, is large (1.6 kcal mol^{-1}) and both frequency factors are approximately equal and low.[338] Unlike $Bu^tO\cdot$ radicals, $MeS\cdot$ radicals selectively abstract allylic hydrogen at position 4 from 2-methyloxazoline and 2-methylthiazoline.[339] Aliphatic thiols catalyse the photo-reduction of benzophenone by primary and secondary amines *via* a sequence of hydrogen atom abstractions which convert a disproportionating nitrogen-centred radical into a reducing carbon-centred radical; aromatic thiols cause retardation.[340]

Arrhenius parameters for the forward and reverse processes in the reaction

$$(CH_3)_3Si\cdot + D_2 \rightleftharpoons (CH_3)_3SiD + D\cdot$$

have been determined relative to the combination reaction of $(CH_3)_3Si\cdot$ radicals.[341]

Phosphite radicals abstract hydrogen from ethane thiol and penicillamine at the same rate ($k = 3 \times 10^8\ \text{M}^{-1}\,\text{s}^{-1}$) but the equilibrium constants for

$$PO_3{}^{2-} + RSH \rightleftharpoons RS\cdot + HPO_3{}^{2-}$$

are 800 and 1500, respectively.[342]

Halogen Atom Abstraction

The gas-phase kinetics of the iodine abstraction, $Cl\cdot + C_2F_5I \rightarrow ICl + C_2F_5\cdot$,[343] and the bromine abstraction, $C_2F_5\cdot + Br_2 \rightarrow C_2F_5Br + Br\cdot$,[344] have been measured.

Dibenzo-18-crown-6 ether complexes the undecyl radical and increases its selectivity in chlorine abstraction reactions in solution.[345]

Transfer rates of chlorine from $CFCl_3$ to cyclohexyl radicals in liquid cyclohexane are low due to an increased C—Cl bond energy, estimated as 74.4 kcal mol^{-1}, caused by the presence of fluorine.[346] Energetic chain branching is found in the gas-phase reactions of CF_3OF with chloro- and fluoro-methanes.[347]

Cyclohexyl radicals react 10^4 times faster than $\cdot CH_2CN$ radicals in chlorine abstraction from CCl_4 and CH_3CCl_3 but are more selective.[348]

In reactions with alkyl halides $SiF_3\cdot$ radicals are less reactive and more selective than $SiCl_3\cdot$ or $Me_3Si\cdot$ radicals; however, the rates of halogen abstraction do not correlate with the strength of the bond being broken.[349] Using thermal radical sources to investigate the reaction

$$Me_3Si\cdot + Me_2SiCl_2 \rightarrow Me_3SiCl + Me_2\dot{S}iCl$$

proved to be unsuccessful; a photochemical approach gave unconventional Arrhenius parameters (low *A* and *E*), which were attributed to the distinctive ability of silicon to form a cyclic dichlorine-bridged intermediate of low energy.[350] The activation energies for radical-chain dechlorination of chloroethanes were 2–3 kcal mol^{-1} lower in liquid Et_3SiH than in cyclohexane, arising from the chlorine-transfer reactions involving $Et_3Si\cdot$ radicals.[351] Such radicals abstract halogen from benzyl bromide and chloride with rate constants of 1.4×10^9 and 1.4×10^7 M^{-1} s^{-1}, respectively, and are thus more reactive than trialkyltin radicals.[352]

Appreciable build-up of negative charge at the exocyclic carbon atom, and the importance of delocalization in the transition state, is inferred in the abstraction of chlorine from heteroarylmethyl chlorides by $Ph_3Sn\cdot$ radicals.[353]

Experimental attention has been focused for the first time on the effect of β-substituents on the configurational stability of pyramidal cyclopropyl radicals. In the reduction of 1-substituted 7-chloro-7-fluoronorcaranes (**40**, **a**–**f**) with Bu_3SnH the stereospecificity in the product (**42**) decreases in the order **e** $\sim$ **d** > **f** > **c** > **b** > **a** indicating that in (**d**) and (**e**) the pyramidal configuration of the intermediate radical (**41**) is stabilized relative to (**f**) whereas in (**a**), (**b**), and (**c**) destabilization occurs.[354]

The balance between fragmentation and defluorination of perfluorodienes and cyclobutenes is critically dependent on the nature of the metal catalyst (Fe, Pt), its surface characteristics and the temperature, but can be rationalized.[355]

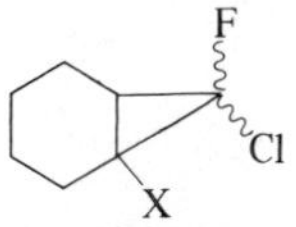

(**40**) 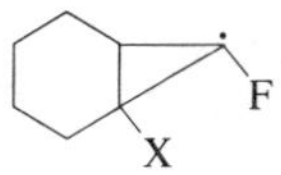 (**41**) 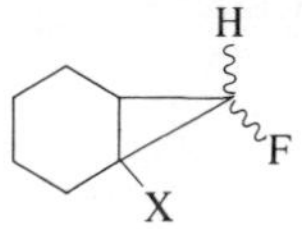 (**42**)

	(a)	(b)	(c)	(d)	(e)	(f)
X =	F	MeO	OAc	Me_3Si	Me	H

Addition Reactions

Carbon-centred Radicals

Activation energies for homolytic addition reactions in the telomerization of olefins with alkyl bromides correlate linearly with localization energies.[356] MO analysis predicts that ethylene and acetylene should exhibit the same differences in reactivity towards radicals as they do towards other electrophilic/nucleophilic reagents.[357] In contrast to olefins, acetylenes yield a high ratio of cyclic to acyclic adducts in reaction with allyl radicals.[358,359] Theoretical studies of the additions of vinyl radicals to ethylene, and of phenyl radical to benzene, reveal that in each case the transition state is "early" on a reaction coordinate in which charge transfer from the substrate occurs in advance of back-donation.[360] Additions of cyclohexyl radicals to styrenes, $PhCHX{=}CH_2$, acrylic esters, $CH_2{=}CXCO_2R$, and acrylonitriles, $CH_2{=}CXCN$, exhibit Hammett ρ values of 3.1, 3.2, and 3.8, respectively; substituents X at the non-attacked vinyl carbon atom exert mainly a polar effect.[361] MINDO/3-UHF calculations reveal that a tighter transition state occurs for addition of methyl radical at a non-terminal carbon than at a terminal carbon of olefins and carbonyl systems.[362] Rate constants have been obtained for the addition of Me· to CO[363] and MeCOEt,[364] and for addition of $Pr^n\cdot$ radicals to HCHO.[365]

No evidence has been obtained for cycloaddition of pendant nitrile groups in polyacrylonitrile and related model compounds initiated by alkyl/aryl free radicals.[366] TCNE, tetracyanopyrazine, and especially 7,7,8,8-tetracycanoquinodimethane readily trap carbon-centred radicals in the gas phase, as determined by chemical ionization mass spectrometry.[367,368]

A strongly curved Arrhenius plot allows dissection of rate constants for addition and chlorine abstraction in the reaction of $CF_3\cdot$ radicals with cyanogen chloride.[369] Polar effects on the reactivity and orientation of addition of fluoromethyl radicals to fluoroethylenes have been related theoretically to electron delocalization.[370] The reactivity of olefins towards perfluoroalkyl radicals decreases, in general, with the number of fluorine substituents in the olefin. The $C_2F_5\cdot$ radical is more reactive than $CF_3\cdot$ or $(CF_3)_2CF\cdot$ radicals in such additions.[371]

Both activation energies and frequency factors for additions of $CF_3\cdot$ radicals to acetylenes are larger than those for additions to similarly substituted olefins, but the resultant rates are thereby similar. Addition of $CF_3\cdot$ to acetylene gives predominantly E-vinyl radicals, whereas addition to HC≡CMe, HC≡CCF_3, and MeC≡CMe yields mainly Z-vinyl radicals.[372] The tendency for formation of *trans*-1,4-adducts in the photo-initiated addition of CF_3I to CHR=CH−CH=CH_2 (R = H, Me) and CH_2=CH−C≡CH has been attributed to steric factors and the propensity of the dienes to adopt the *s-trans* conformation.[373]

Unlike the addition of phenoxides, condensation of CF_2BrX (X = Br, Cl, or CF_2Br) with enamines and ynamines proceeds *via* a radical chain as shown in Scheme 12.[374]

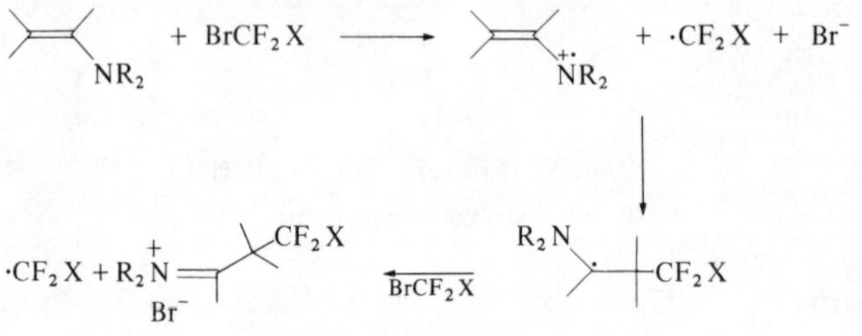

NR_2 + $BrCF_2X$ ⟶ $\dot{N}R_2^{+}$ + $\cdot CF_2X$ + Br^-

$\cdot CF_2X$ + $R_2\overset{+}{N}$= ... CF_2X Br^- ⟵($BrCF_2X$) R_2N ... CF_2X

SCHEME 12

Telomerization of vinyl chloride with ICF_2SO_2F yields $FSO_2CF_2(CH_2CHCl)_nI$, $n = 1$–3.[375] Oxidation of $CCl_3\cdot$ radicals occurs by initial addition of O_2 to give $CCl_3O_2\cdot$ radicals which then yield $CCl_3O\cdot$ radicals; the latter lose Cl· to form $COCl_2$.[376]

Strong mesomeric effects, as well as inductive effects, are found in the addition of $CCl_3\cdot$ radicals to the chloroethylenes, *cis*- and *trans*-$C_2Cl_2H_2$ and C_2Cl_3H in liquid C_6H_{12}–CCl_4 mixtures.[377] 1,5-Hydrogen migrations are observed in the adduct radicals derived from CCl_2=$CH(CH_2)_3Cl$ and electrophilic (*e.g.* RS·, $CCl_3\cdot$) or nucleophilic (*e.g.* $PhCH_2\cdot$) radicals using mainly $Fe(CO)_5$ in DMF as initiator.[378–380] In the homolytic addition of CCl_4 and $BrCCl_3$ to β-halostyrenes, the β-halo radicals are key intermediates which undergo telomerization, halogen transfer or β-cleavage; β-elimination of Br· generates molecular bromine which scavenges the β-bromo radical intermediates.[381]

New products have been identified in the reaction of hydroxy-substituted alkyl radicals with *trans*-2-butene in the presence of O_2 and NO_2.[382]

Only terminal attack occurs upon addition of $\cdot CH(CN)_2$ radicals to allene whereas for 1,1-dimethylallene only the central carbon is attacked.[383] Additions of Me_2O to fluorocycloalkenes,[384] and of alkyl propanoates and *iso*butyrates to dialkyl maleates[385,386] have been reported. Activation energies for the formation of 2-hexenylsuccinic anhydride and 1,2-bis(succinic anhydridyl)hex-3-ene from 1-hexene and maleic anhydride are 10.4 and 0 kcal mol^{-1}, respectively.[387]

The kinetics of the free-radical teleomerization of acetone with propylene[388,389] and isobutene,[390] of isopropanol with 3,3,3-trifluoro-1-propene,[391] and of isobutyric acid with hexafluoropropene[392] have been investigated. Acetonyl radical addition to $HC{\equiv}CCH_2OH$ yields $CH_3COCH_2CH{=}CHCH_2OH$ as the only product, while addition to $HC{\equiv}CCH_2OAc$ yields only $CH_3COCH{=}CHCH_2CH_2OAc$ by contrast.[393]

Atoms

Theoretical calculations have been performed to determine the kinetic isotope effect of chemically activated vinyl radicals formed on addition of hydrogen atoms to acetylene.[394] Contrary to prediction no isotope effects due to differences of deuterium substitution in ethylene are observed for the addition of H· or D· atoms.[395] Activation energies at the low and high pressure limits for addition of H· atoms to allene are 2.1 and 3.5 kcal mol^{-1}, respectively.[396] Studies of the addition of H· atoms to isobutylene[397] have led to an estimate of 10.6 ± 0.5 kcal mol^{-1} for heat of formation for the $Bu^t\cdot$ radical;[398] this value is in agreement with a modified structural model for $Bu^t\cdot$ but incompatible with a result based on iodination data.

Rate constants for reactions of hydrogen and oxygen atoms with fluoroethylenes correlate well with an INDO-derived reactivity index which includes allowance for the attacking atom,[399] but the success is probably fortuitous!

The quenching of excited HF radiation (by the reaction $F\cdot + H_2 \rightarrow HF^*$) has been used to obtain rate constants for the addition of F· atoms to ethylene and benzene,[400] propylene and isobutylene,[401] and also to acetone and acetonitrile after allowance for hydrogen atom abstraction.[402] The terminal/central addition ratios for reaction of thermal $^{18}F\cdot$ atoms with propyne and 3,3,3-trifluoropropyne are 2.6 and 3.7 respectively.[403] An activated radical intermediate is formed in the addition of Cl· atoms to C_2BrF_3 at 30.5°.[404]

In the low-temperature (40–90 K) hydrobromination of ethylene the rate-determining step is the formation of $\cdot C_2H_4Br$ radical rather than its subsequent transformation by hydrogen atom transfers.[405]

A comprehensive study of the brominolysis of *trans*-1,2-diarylcyclopropanes,[406,407] in which the rate constants range over five orders of magnitude, has shown quantitatively that the magnitude of the substituent influence (*i.e.* ρ value) at either involved carbon centre is a function of the substituent at the other centre. Thus a continuum of transition-state structures exists. Homolytic halogenation of 3,3-dichloro-1,2-bis(trifluoromethyl)- and 1,3-dichloro-2,3-bis(trifluoromethyl)-cyclopropenes have been reported.[408]

In the absence of silver ions terminal alkynes appear to react with iodine in methanol by a free-radical mechanism to yield 1,2-diiodoalkenes.[409]

Hetero Radicals

Fragmentation of $Bu^tO\cdot$ in aqueous solution is rapid but addition reactions[321] to vinyl ethers and furans are detectable ($k > 10^6$ M^{-1} s^{-1}). Retinyl acetate and β-

carotene intercept peroxyl radicals such as $Me_2C(CN)O_2\cdot$ and $PhCHMeO_2\cdot$ as well as radicals such as $\cdot CMe_2CN$.[410] Reaction of NH· radicals with liquid ethylene and propylene leads *inter alia* to addition.[411] The gas-phase reaction of C_3H_6 and $NO_3\cdot$ radicals is initiated by addition.[412] The factors influencing the competition between 1,2-addition and allylic hydrogen atom abstraction in the photo-induced reaction of *N*-haloamides to olefins have been investigated. The addition is favoured by (*i*) decrease in temperature, (*ii*) increase of electronegativity of Z in the *N*-haloamide ZCONRX, (*iii*) use of chloro-(X = Cl) rather than bromo-(X = Br)amide, (*iv*) increase in the size of R.[413]

Two-step aminobromination of phenylethylenes and α-olefins with diethyl *N*,*N*-dibromophosphoramidate proceeds *via* a regiospecific anti-Markownikoff radical-chain addition.[414]

DPPH radical intercepts the Diels–Alder dimer of styrene to yield the 1,2,3,4-tetrahydro-4-phenyl-1-naphthyl radical; styrene trimers, but not dimers such as 1,2-diphenylcyclobutanes, are also eliminated by DPPH.[415] Verdazyl radicals add to the ligand in copper(II) diketonates in the rate-determining step of a process leading to "radical-nucleophilic" fission of the ligand to yield copper alkanoates.[416]

The electronic character of phosphinyl radicals, $R_2P\cdot$ (R = H, Me, F), in their addition to ethylene and fluoroethylenes has been explored using a theoretical solvent polarity parameter (ΔΔ) criterion.[417]

Chain-transfer constants have been determined for the γ-ray-initiated addition of $EtSiHCl_2$ to 1-hexene.[418] 2-Adamantyl oxytrimethylsilyl radicals are formed by addition of $Et_3Si\cdot$ radicals to 2-adamantanone.[419] In the formation of adduct radicals from 4-(alkylthio)- and 4-(alkylsulphonyl)-pyridines with $Ph_3Si\cdot$ and $Ph_3Ge\cdot$ radicals, the latter radicals act as electron acceptors of the heterocyclic nitrogen electron.[420] Addition of $Bu^n{}_3Sn\cdot$ and $Ph_3Sn\cdot$ radicals to tri-substituted ethylenes, *e.g.* $MeCH{=}CPhCO_2Me$ or $PhCH{=}CMeCN$, has been shown to be stereoselective but reversible.[421]

That the reactivity of $Bu_{3-n}Cp_nSn\cdot$ radicals towards ethylene, biacetyl, and other unsaturated compounds increases as *n* increases is due possibly to carbon–metal hyperconjugation [as in η^1-tin(IV) compounds] but this is not certain.[422]

Buten-2-yl and thiyl free radicals act as chain carriers in the addition–elimination processes leading to thermal isomerization of *cis*-but-2-ene in the presence of H_2S.[423]

Radicals (**43**) and (**44**) have been detected in the photo-induced addition of $(MeS)_2$ to cyclopentene: (**44**) is not formed by direct hydrogen atom abstraction.[424] Steric effects and reversibility in the chain addition of benzenethiol to monoalkyl-

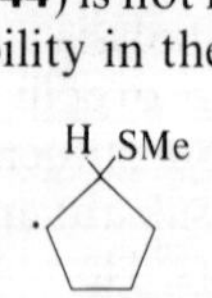

H SMe   H SCH₂·

(**43**) 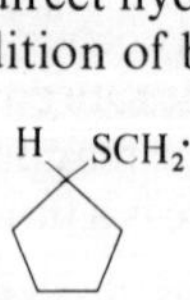   (**44**)

substituted allenes, $RCH{=}C{=}CH_2$, have been examined.[425] For R = Et, Pr^i, Bu^n, and Bu^i the ratio of products arising from attack at C(2) to those from attack at C(3) is 83:17. Isotopic studies have established that addition is not reversible and that the kinetically favoured products are formed.[426] Fluorine substituents appear to destabilize a radical centre on the same carbon since both PhSH and Pr^nCHO add to n-$C_{10}H_{21}CH{=}CF_2$ *via* initial addition at the terminal carbon.[427] Additions of thiophenol to ethyl (β-phenylthio)crotonate[428] and to ketones[429] have been investigated; in the latter reaction both possible radical adducts have been spin-trapped. Addition of thiyl radicals RS· to $CCl_2{=}CHCl$ is followed by subsequent loss of Cl· atom to yield $CCl_2{=}CHSR$ product.[430]

Analysis of kinetic data for the addition of the thiyl radicals, p-XC_6H_4S·, to vinyl monomers, $CH_2{=}CHY$ (Y = OBu^i, OAc, CN), suggests that the polar effects on reactivity are related to the effects on the thermodynamic stability of the p-XC_6H_4S· radical and on polar resonance structures in the transition state, whose contribution varies with the electron donor/acceptor ability of the group Y in the vinyl monomer.[431]

Reversibility in the addition of benzothiazole-2-thiyl radicals to vinyl monomers is important in determining the decay characteristics of this radical.[432]

A Markovnikov-type radical addition, involving L-cysteine and protohaemin (iron protoporphyrin IX) in micellar solution in the presence of oxygen, forms the basis of a novel synthesis of haemin c.[433] The thermal radical addition of phenylareneselenosulphonates to olefins occurs in an anti-Markovnikov fashion and non-stereospecifically, but complements the Markovnikov addition achieved by the use of $BF_3 \cdot Et_2O$.[434]

The photo-stimulated substitution of vinyl mercurial halides in the overall process $R^1CH{=}CHHgX + R^2YYR^2 \rightarrow R^1CH{=}CHYR^2 + XHgYR^2$ where R^1 = alkyl, Y = S, Se or Te, and R^2 = alkyl, phenyl *etc*., does not occur *via* S_N attack but rather involves a β-mercurio radical $R^1\dot{C}HCH(HgX)YR^2$ whose radical centre interacts with the metal centre.[435]

Intramolecular Addition

The regioselectivity of intramolecular double-bond additions is little influenced by interaction between the radical MO and the HOMO of the intervening σ-bond centres.[436] Cyclization of (**45**), equivalent to a 1,2-disubstituted hex-5-enyl radical, is regiospecific, yielding only five-membered ring products (**46**), (**47**), and (**48**), and stereoselective in favouring (**47**). The behaviour of (**45**) is thus close to that of a 1-substituted radical.[437,438] The related radical (**49**) closes[439] predominantly to give the *cis* configuration product (**50**) in 56% yield with little (under 4%) of the expected major product (**51**); however, this can be reconciled with the frontier orbital theory guideline that requires overlap of the unpaired electron orbital with the π^*-orbital of the double bond. Thus radical (**52**) has a high relative *exo* cyclization rate due to configurational control by the five-membered ring, but closure in (**53**) is slow—in strict conformity with the guidelines. The rate constant for the cyclization of hex-5-enyl radicals has been revised.[238]

Free-radical annelation of (**54**; R^1 = H, R^2 = H, X = Cl, SPh, and SePh) is regiospecific in forming (**55**) by *endo* mode closure of the intermediate radical (**54**; X = unpaired e).[440] Some *exo* mode closure to give (**56**) is observed with (**54**; X = Cl, $R^1 = CO_2Me$, $R^2 = CO_2Bu^t$).[441]

While the γ-silicon in $CH_2{=}CHSiMe_2CH_2CH_2CH_2$· radicals seems not to influence greatly the rate and regioselectivity of intramolecular addition as compared to hex-5-enyl radicals, the α-silicon in the radical $CH_2{=}CHCH_2CH_2SiMe_2CH_2$· produces a reversed selectivity by diminishing the *exo* cyclization mode.[442] The o-allyltetrafluorophenylthio radical, o-$CH_2{=}CHCH_2C_6F_4S$·, cyclizes to give a mixture of both five- and six-membered products.[445]

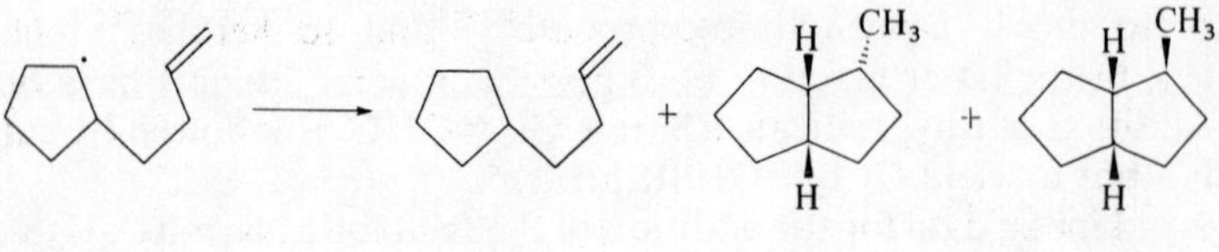

(45) → **(46)** 39.5% + **(47)** 54% + **(48)** 6.5%

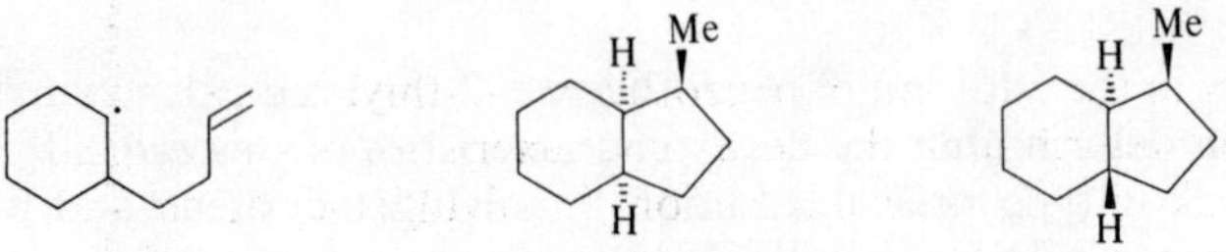

(49) **(50)** **(51)**

(52) **(53)**

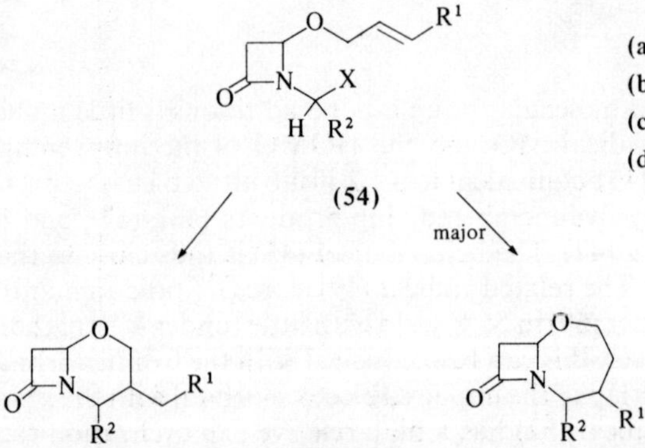

(54)

	R^1	R^2
(a)	H	H
(b)	H	CO_2Bu^t
(c)	CO_2Me	CO_2Bu^t
(d)	Ph	CO_2Bu^t

major

(56) **(55)**

Phenyl *o*-styrylphenyl iminyl radicals cyclize to give a mixture of isoquinoline and 1-*H*-isoindole derivatives in a 4:3 ratio.[443] *N*-chloro-3-aryl-5-hexenylamines can be cyclized to 2-chloromethyl-4-phenylpiperidines.[444] Good, or even total, stereospecificity can be achieved in free-radical cyclization of *cis*- and *trans*-1-methyl-4-hexenyl-*N*-chloroamines by complexation of the intermediate radical with metal ions, especially Fe(II) and Co(II).[446]

Radical cyclizations of unsaturated aldehydes have been reviewed.[447] Intramolecular addition of oxyl radicals generated from *A*-homo-4*A*-cholesten-3-ol hypoiodites is a faster process then β-cleavage and leads to the formation of bridged oxabicyclic products, *e.g.* 3α,5-epoxy-4$\alpha\beta$-iodo-*A*-homo-5β-cholestane.[448]

Intramolecular addition to an azido group has been observed[449] (Scheme 13).

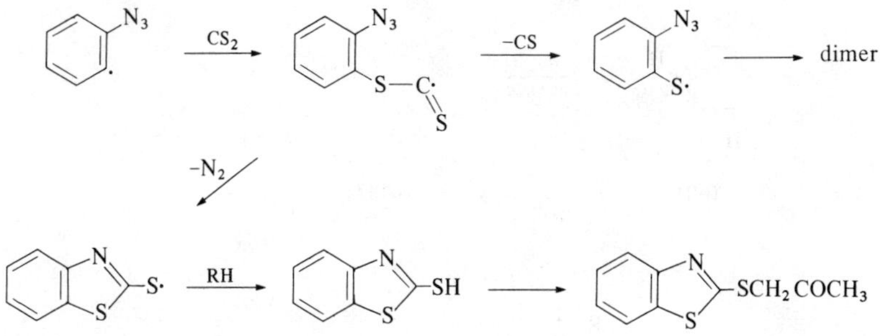

SCHEME 13

Fragmentations

The rate of fragmentation of $Me_3CO\cdot$ radicals has been determined.[450,451] In the oxidation of ethylene glycol by Ag(II), $CH_2OHCH_2O\cdot$ radicals fragment to give $\dot{C}H_2OH + CH_2O$.[452] The course of β-scission of 9-decalinoxyl and related radicals is very sensitive to changes in the molecular geometry of the system:[453] this is possibly a reflection of the rôle of stereoelectronic factors in determining the direction of scission. The importance of stereoelectronic considerations has been demonstrated in the behaviour of the epimeric alcohols (**57**) and (**58**) on treatment with HgO/Br_2:[454] the former gives the corresponding ketone and a ring-opened bromide (**59**) whilst the latter gives no ring-opened products. It has been suggested that depending on whether the summation of the lone-pair interactions in the parent alcohol is *trans*-antiperiplanar with a C—C or C—H bond, then ring-opening or oxidation occurs.

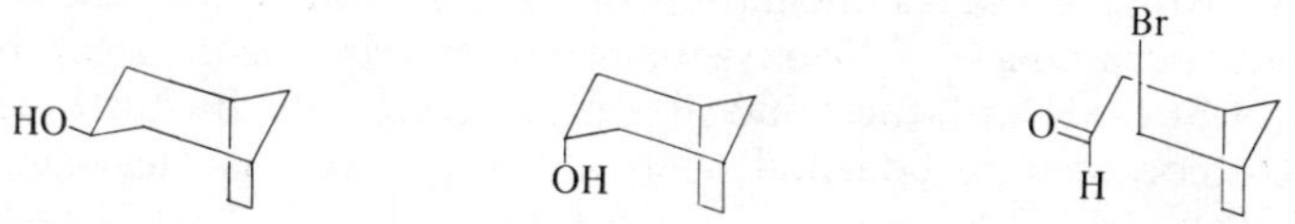

(**57**) (**58**) (**59**)

The radicals derived from dialkoxymethanes fragment to give alkyl radicals and alkyl formates:[455–457] there is relatively little discrimination in the preferred direction of scission of radical from unsymmetrical dialkoxymethanes.[458] Similar fragmentation of radicals derived from 1,3-dioxolanes and 1,3-dioxanes occurs;[459] there is no evidence for concerted two-bond scission in radicals from 1,3-dioxanes as encountered for radicals from 1,3-dioxolanes.[460,461]

The 2-methyltetrahydrofuryl-2-oxyl radical fragments to give $CH_3COCH_2CH_2CH_2O\cdot$.[462,463] The radical (**61**) produced in the acetophenone photo-induced hydrogen abstraction from (**60**) undergoes irreversible endocyclic ring-opening to give (**62**) and exocyclic ring-opening with loss of Me to give (**63**).[464] The extent of ring-opening is not dependent on whether (**62**) is primary or tertiary as the yield of ring-opened products is 20% when R = Me, R′ = H, but 29% when R=H, R′ = Me. The radical (**64**) undergoes fragmentation of the weaker C—C bond rather than the C—O bond to give (**65**).[465] The radicals derived from 2-alkyl-1,3-oxa-azacyclopentanes undergoes C—N bond scission to give $RCONR'CH_2CH_2\cdot$.[466,467]

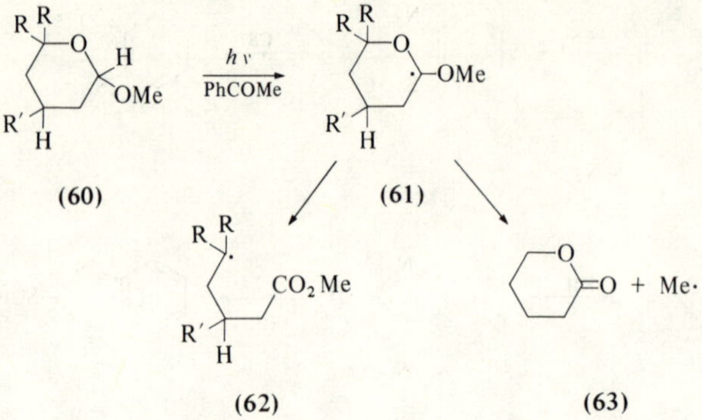

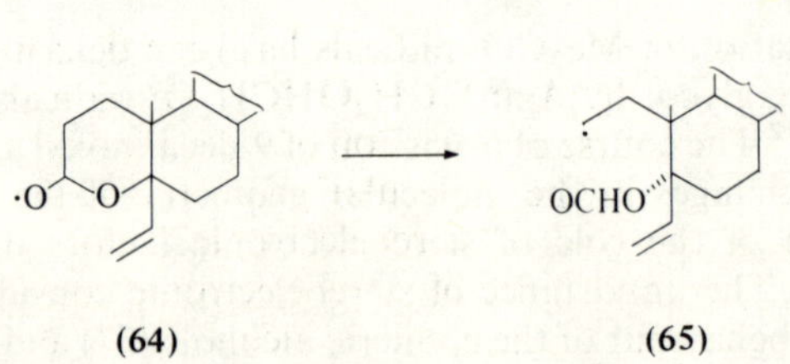

(60) (61) (62) (63)

(64) (65)

The adduct radical obtained from reaction of PhS· with $PhCO_2Ph$ fragments to give PhO· and PhCOSPh.[468]

The positive value for the volume of activation for the loss of Bu^t· from $Bu^t_2C{=}N$· indicates significant C—C bond stretching in the transition state.[469]

Reactions involving the fragmentation of radicals have found extensive use in natural-product chemistry.[470] Deoxygenation of primary alcohols can be effected by reaction of Bu_3SnH with their thiocarbonyl esters;[471] the Bu_3SnH reduction of α,β-epoxy-*O*-thiocarbonylimidazolide derivatives of alcohols yields unsaturated alcohols (Scheme 14).[472] Several procedures for deamination have been developed. Thus, isonitriles can be reduced to alkanes by Bu_3SnH; the intermediate imidoyl radicals undergo β-scission to give alkyl radicals.[473,474]

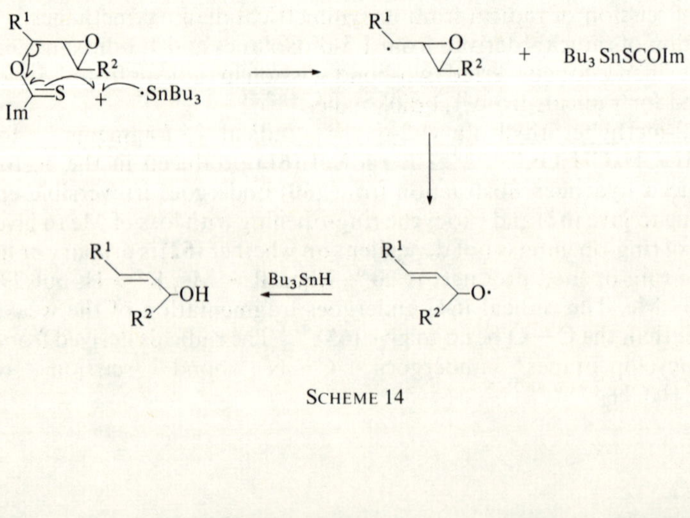

SCHEME 14

$$R{-}N{\equiv}C + Bu_3Sn\cdot \rightarrow R{-}N{=}\dot{C}{-}SnBu_3$$
$$R{-}N{=}\dot{C}{-}SnBu_3 \rightarrow R\cdot + Bu_3SnCN$$
$$R\cdot + Bu_3SnH \rightarrow RH + Bu_3Sn\cdot$$

Isothiocyanates and 1-isocyano-2-xanthates can similarly be reduced to alkanes with Bu_3SnH.[474] A similar mode of fragmentation is encountered in imidoyl radicals obtained by addition of Ph· and $Me_3Si\cdot$ to isonitriles.[475]

The radicals derived from α- and β-D-glucose and from myoinositol with HO· undergo acid-catalysed loss of water: elimination of an axial β-OH is favoured over loss of an equatorial β-OH because of the favourable overlap of the SOMO and the bond undergoing scission, *e.g.* (**66**) gives (**67**).[476,477]

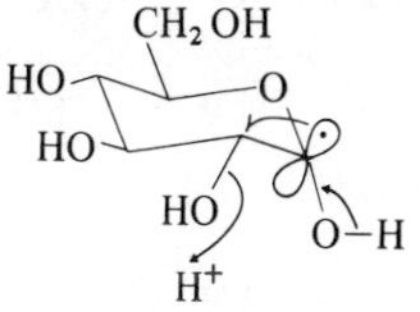

(**66**) 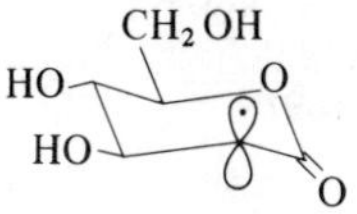 (**67**)

Rearrangements

Radical rearrangements have been reviewed.[478]

The kinetics of isomerization of *cis*- and *trans*-pent-2-en-4-ynyl radicals have been investigated.[479]

An *ab initio* study of free-radical migrations indicates that whereas 1,5- and 1,6- but not 1,2-hydrogen migrations should occur readily, the reverse is true for chlorine atom migrations.[480] 1,5-, 1,6-, 1,7-, and 1,8-Hydrogen migration is encountered in reactions of $CH_3(CH_2)_5CHClCH_2CCl_3$.[481] The rates of chlorine migrations in $RCH_2\dot{C}MeCCl_3$, $RCH_2\dot{C}HCCl_2F$, and $CCl_3CH_2\cdot$ are 1×10^4, 1×10^5, and $1 \times 10^6\,s^{-1}$, respectively.[482]

Allylic hexafluoroisopropyl sulphides rearrange under both thermal and photochemical conditions to give the internal sulphide by an elimination–addition mechanism.[483]

$$CH_2{=}CMeCMe_2SR_F \rightarrow [CH_2{=}CMe\dot{C}Me_2\,\cdot SR_F] \rightarrow Me_2C{=}CMeCH_2SR_F$$

The ring-opening of cyclopropylcarbinyl and related radicals continues to attract attention.[484] Evidence in support of the dipolar transition state (**68**) proposed by Beckwith comes from the direction of ring-opening of the radicals derived from substituted glycidols (**69**),[485] and of 2,2-difluorocyclopropylmethyl radicals (72).[486] The ring-opening of (**72**) to give (**73**) was attributed to weakening of the bond opposite the CF_2; this would suggest that ring-opening of (**72**) occurs faster than the unsubstituted radical, whereas Beckwith's postulate would suggest that the ring-opening of (**72**) would be slower. In the ring-opening of (**69**; R = $COCH_3$), the rearranged radical (**70**) undergoes a 1,5-acetyl shift to give (**71**): the analogous reaction is not observed for (**70**; R = COEt or COPr). The rate of ring-opening of *c*-$C_3H_5\dot{C}HCH_2OMe$ was estimated to be $1 \times 10^5\,s^{-1}$; this seems anomalously slow.[487] There is evidence that monoradicals in biradicals undergo rearrangements with similar rate constants to those of the analogous monoradicals: thus the radical

c-$C_3H_5\dot{C}HCH_2CH_2\dot{C}(OH)Ph$ rearranges at much the same rate as the cyclopropylmethyl radical.[488]

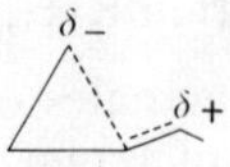

(68)

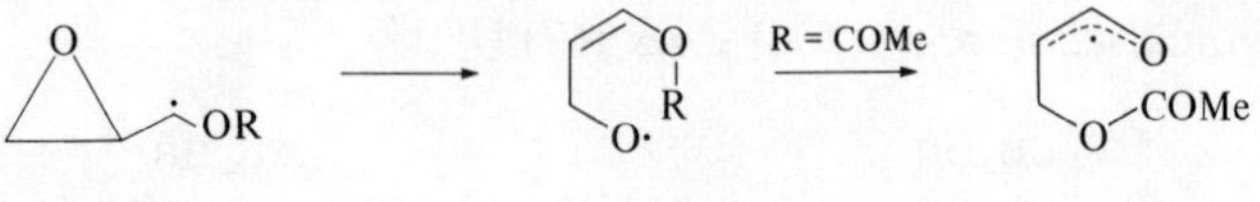

(69) (70) (71)

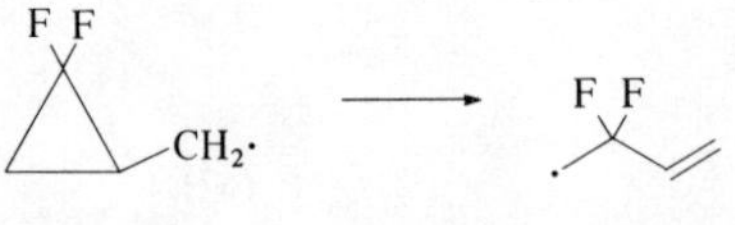

(72) (73)

The rate of rearrangement of norbornenyl radicals (**74**; X = H) to nortricycyl radicals (**75**; X = H) is $6 \times 10^2\ s^{-1}$; the activation parameters for this rearrangement suggests that less ring strain is developed in ring-closure to (**75**) than in the but-3-enyl–cyclopropylmethyl rearrangement.[489] Only substituted nortricycyl radicals (**75**) can be detected from the addition of $Bu^tS\cdot$, $Bu^tO\cdot$, and $CF_3O\cdot$ to norbornadiene between -60 and $-130°$. The addition of $Bu^tO\cdot$ at $-150°$ similarly yields only (**75**; X = Bu^tO) but at $-70°$ this rearranges further to (**76**). The same allylic radical is obtained from reaction of $Bu^tO\cdot$ with hexamethyl(Dewar)benzene and hexamethylprismane as a result of a series of cyclopropylcarbinyl–homoallyl radical rearrangements.[490]

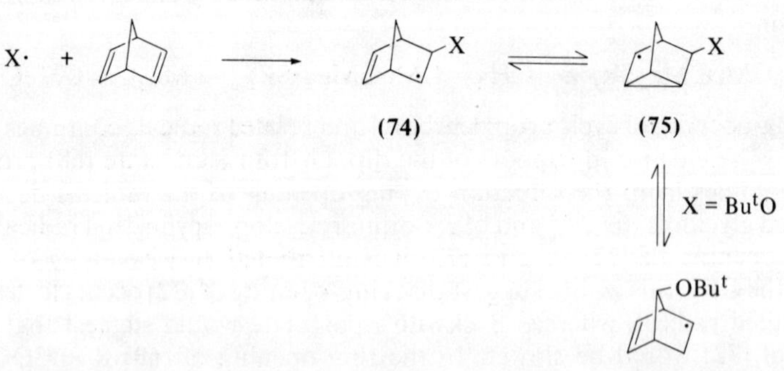

(74) (75)

X = Bu^tO

(76)

The rate constants for the ring-opening of cyclobutylmethyl, 1-cyclobutyl-1-methylethyl, and cyclobut-2-enylmethyl radicals have been determined by kinetic ESR spectroscopy: the values of $\log(k/s^{-1})$ are $(13.1 \pm 1.4) - (49.8 \pm 7.5$ kJ mol^{-1})/2.3RT, (13.6 ± 1.6) − (58.1 ± 9.0 kJ mol^{-1})/2.3RT and (12.2 ± 1.4) − (4.2 ± 8.0 kJ mol^{-1})/2.3RT, respectively.[491] A number of unprecedented radical rearrangements have been postulated for the pyrolysis of the diallyldisilane $CH_2{=}CHCH_2SiMe_2SiMe_2CH_2CH{=}CH_2$ which initially proceeds *via* the silicon-substituted cyclobutylmethyl radical (**77**).[492] β-Scission and 1,2-aryl-migration reactions of radicals derived from methylindans and tetralin have been reported.[493]

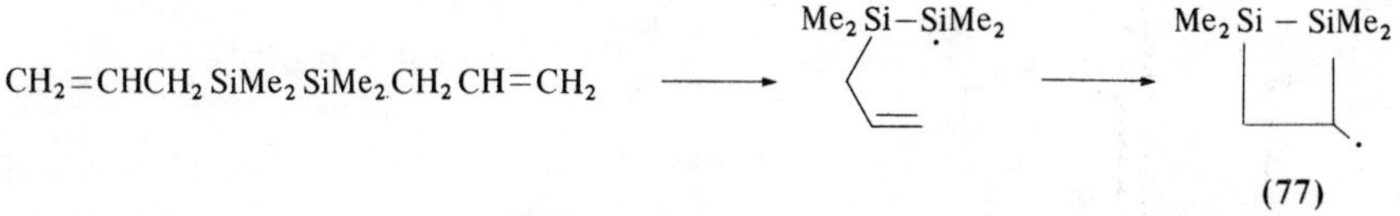

$CH_2{=}CHCH_2SiMe_2SiMe_2CH_2CH{=}CH_2$ ⟶ $Me_2Si{-}\dot{S}iMe_2$ ⟶ $Me_2Si - SiMe_2$

(**77**)

The β-cyanoalkyl radical $\dot{C}H_2CHMeCN$ rearranges to $CH_3\dot{C}HCH_2CN$ at a rate that is very much faster than that of analogous homopropargyl radicals.[494] The acid-catalysed rearrangement of β-hydroxylalkyl radicals proceeds *via* a radical cation intermediate.[495]

$$\cdot CH_2CMe_2OH + H^+ \rightarrow [CH_2CMe_2]^{+\cdot} + H_2O \rightarrow Me_2\dot{C}CH_2OH + H^+$$

The Bu_3SnH reduction of methyl 6β-isothiocyanatopenicillinate (Scheme 15) is accompanied by intramolecular radical capture and cleavage of the C(2)—S bond to give the thiazoline (**78**).[496] A possible mechanism for penicillin biosynthesis involves rearrangement of a 5-isothiazolidinoyl radical to a β-lactam; chemical studies have shown that model radicals do not lead to β-lactams.[497]

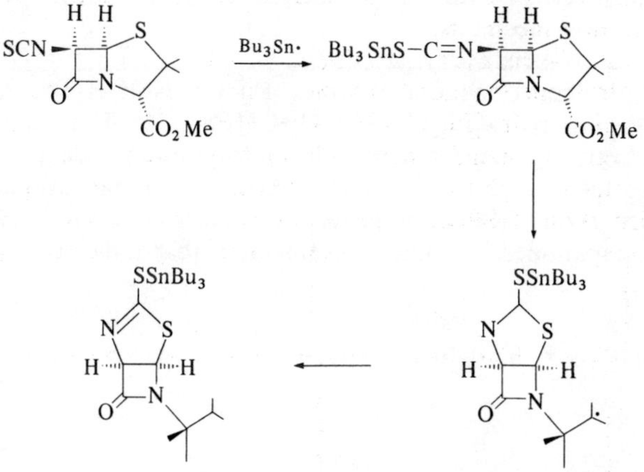

SCN, H H, S, O, N, CO_2Me —$Bu_3Sn\cdot$→ $Bu_3SnS{-}C{=}N$, H H, S, O, N, CO_2Me

$SSnBu_3$, N, S, H, H, O, N (**78**) ← $SSnBu_3$, N, S, H, H, O, N

(**78**)

Scheme 15

Further evidence for radical intermediates in reactions of cobaloximes is provided by the formation of enones resulting from a 1,2-acyl shift in the intermediate radicals produced in the photolysis of 2-acyl-2-methoxycarbonylpropylcobaloximes (Scheme 16).[498] 1,2-Acyl shifts in radicals are also observed in the reaction of (**79**) with Bu_3SnH.[498] Migrations of groups (H, D, acyl, Me_2SnCl) between oxygen atoms of catechol derivatives have been studied.[499]

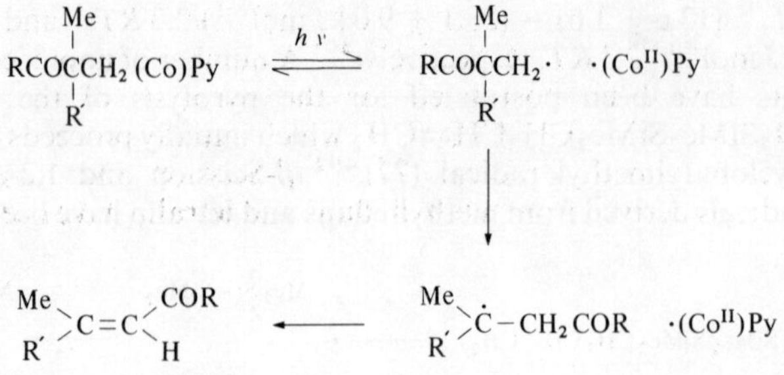

$$\mathrm{RCOC(Me)(R')CH_2(Co)Py \underset{}{\overset{h\nu}{\rightleftharpoons}} RCOC(Me)(R')CH_2\cdot \;\; \cdot(Co^{II})Py}$$

$$\downarrow$$

$$\mathrm{Me(R')C{=}CH(COR) \longleftarrow Me(R')\dot{C}{-}CH_2COR \;\; \cdot(Co^{II})Py}$$

SCHEME 16

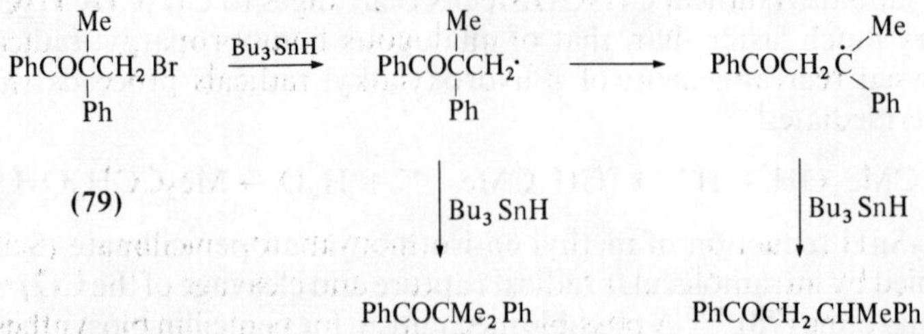

$$\mathrm{PhCOC(Me)(Ph)CH_2Br \xrightarrow{Bu_3SnH} PhCOC(Me)(Ph)CH_2\cdot \longrightarrow PhCOCH_2\dot{C}(Me)Ph}$$

(79)

$$\mathrm{PhCOC(Me)(Ph)CH_2\cdot \xrightarrow{Bu_3SnH} PhCOMe_2Ph \qquad PhCOCH_2\dot{C}(Me)Ph \xrightarrow{Bu_3SnH} PhCOCH_2CHMePh}$$

1-Benzyl-3-pyrazolin-5-ones undergo photo-induced benzyl migration to give 5-benzyloxypyrazoles.[500] The thermally induced benzyl migration in the rearrangement of 2-benzyloxytropones to 3- and 5-benzyltropolones proceeds by an addition–elimination mechanism.[501]

Thermolysis of *N*-benzhydryl-α,α-diphenyl nitrone, $Ph_2C{=}N^+(O^-)CHPh_2$, gives *o*-benzhydrylbenzophenone oxime, $Ph_2C{=}NOCH_2Ph$, by a process involving the radical pair $[Ph_2C{=}N{-}O\cdot\ \dot{C}HPh_2]$.[502] A similar mechanism is involved in the rearrangement of benzhydryl *p*-tolyl sulphoxide, $Ph_2CHS(O)Ar$, to benzhydryl *p*-toluenesulphonate, $Ph_2CHSOAr$.[503] A radical-pair mechanism occurs in the formation of (**80**) from the reaction of *tert*-butyl sulphinyl chloride with *N*-hydroxysulphonamides.[504] The Stevens-type rearrangement of open-chain

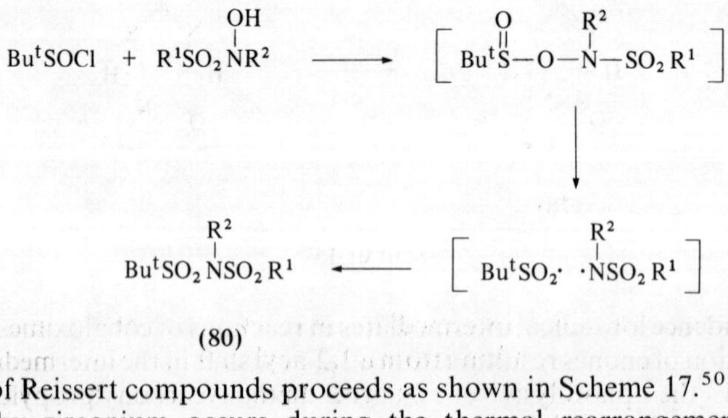

$$\mathrm{Bu^tSOCl + R^1SO_2N(OH)R^2 \longrightarrow [Bu^tS(O){-}O{-}N(R^2){-}SO_2R^1]}$$

$$\downarrow$$

$$\mathrm{Bu^tSO_2N(R^2)SO_2R^1 \longleftarrow [Bu^tSO_2\cdot \;\; \cdot N(R^2)SO_2R^1]}$$

(80)

analogues of Reissert compounds proceeds as shown in Scheme 17.[505] Anchimeric assistance by zirconium occurs during the thermal rearrangement of methyl zirconocenyldiphenylmethyl ether, $Ph_2C(ZrCp_2Cl)OMe$, to $Ph_2O(Me)OZrCp_2Cl$.[506] The product from the reaction of 2-benzylpyridine-1-

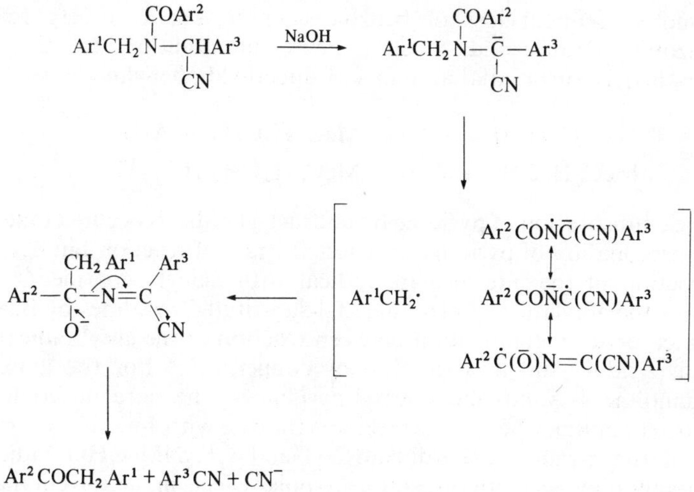

$$\mathrm{Ar^1CH_2N(COAr^2)-CH(CN)Ar^3} \xrightarrow{\text{NaOH}} \mathrm{Ar^1CH_2N(COAr^2)-\bar{C}(CN)-Ar^3}$$

$$\downarrow$$

$$\left[\mathrm{Ar^1CH_2\cdot} \quad \begin{matrix} \mathrm{Ar^2CO\dot{N}\bar{C}(CN)Ar^3} \\ \updownarrow \\ \mathrm{Ar^2CO\bar{N}\dot{C}(CN)Ar^3} \\ \updownarrow \\ \mathrm{Ar^2\dot{C}(\bar{O})N{=}C(CN)Ar^3} \end{matrix} \right] \longrightarrow \mathrm{Ar^2{-}C(CH_2Ar^1)(O^-){-}N{=}C(Ar^3)CN}$$

$$\downarrow$$

$$\mathrm{Ar^2COCH_2Ar^1 + Ar^3CN + CN^-}$$

SCHEME 17

oxide and *N*-phenylbenzimidoyl chloride rearranges by a radical-pair process.[507] 2-Oxidoanilinium ylides rearrange in part by a radical-pair mechanism.[508,509]

3-*p*-Tolyl-1,2,3,-ferrocenylcyclopropene rearranges *via* the biradical $\mathrm{Fe\dot{C}{=}CFc{-}\dot{C}FcTol}$.[510]

Homolytic Aromatic Substitution

Free-radical aromatic substitution has been reviewed.[511] A review of the formation of aryl–aryl bonds including the use of radical processes has been published.[512]

A theoretical study of the formation of phenylcyclohexadienyl radicals from the reaction of phenyl radicals with benzene indicates that the transition state is early on in the reaction path: the exothermicity of the reaction has been calculated to be 22.5 kcal mol^{-1}.[513] Resonance energies of π-hydrocarbon radicals give a reasonable correlation for the rates of phenylation of polycylic aromatics.[514] The isomer distribution of methylbiphenyls obtained in the photochemical phenylation of benzene is governed by the frontier electron densities of the ground state: this indicates that the transition state is reactant-like.[515]

The kinetic isotope effect for the *p*-tolylation of pyridine is greater than unity but the composition of the reaction mixture is unchanged: it is deduced that the isotope effect results from the disproportionation step and not from reversibility of the initial step.[516] Additives, *e.g.* *o*-chloranil, ArNO, $ArNO_2$, $Cu(OCOPh)_2$, $Fe(OCOPh)_3$, result in the formation of nearly theoretical yields of ArCOOH and biaryl in the thermolysis of diaroyl peroxides in arenes: partial rate factors obtained under these conditions correlate well with calculated values and are free from the uncertainties caused by dimerization.[517] Aryl cyanides have also been shown to be capable of oxidizing cyclohexadienyl radicals.[518]

Photolysis of halopyridines in benzene gives phenylpyridines.[519] Bis(dinitrogen)-bis[1,2-bis(diphenylphosphino)ethene]molybdenum, $[Mo(N_2)(dppe)_2]$ breaks down in chlorobenzene to give phenyl radicals and thence chlorobiphenyls.[520] The photo-decomposition of dibenzoyl peroxide in toluene gives 2,2′-, 2,3′-, and 3,3′-dimethylbiphenyls, formed *via* dimerization of (benzoyloxy)methylcyclohexadienyl radicals and subsequent loss of benzoic acid.[521] The Gomberg reaction of toluenediazonium salts and xylenes gives, in addition to the expected trimethylbiphenyls, substantial amounts of dimethyldiphenylmethanes.[522]

$$C_6H_4Me_2 + Ar\cdot \rightarrow MeC_6H_4CH_2\cdot + ArH$$
$$MeC_6H_4CH_2\cdot + ArH \rightarrow MeC_6H_4CH_2Ar + [H\cdot]$$

Homolytic substitution of pyridine by undec-1-yl radicals occurs at the 2- and 4-positions: protonation of pyridine increases the rate of reaction but decreases the rate of reaction of the intermediate radical with diacyl peroxide.[523] The rate constants for the alkylation of protonated 4-substituted pyridines by Bu· and $Bu^t\cdot$ has been measured by competition between reaction of the alkyl radical with the protonated pyridine and its oxidation by copper(II).[524] For the more reactive bases—quinoline, 4-cyano- and 4-acetyl-pyridines—the more nucleophilic $Bu^t\cdot$ is more reactive than Bu·, whereas the reverse is the case with 4-methoxypyridine. The reactivity of the π-radicals, 1,4-dioxan-2-yl and $\dot{C}H_2NMeCHO$ radicals, with substituted quinolines is more susceptible to polar effects in the solvent than are the reactions of $Ph\dot{C}O$ and $\dot{C}ONMe_2$, which are both σ-radicals: reaction of alkyl radicals is least effected by solvent polarity.[525]

Substituted cyclohexadienyl radicals from the reactions of trialkylsilyl radicals with benzene have been detected by ESR.[526]

Hydrogenation of 9.10-dimethylanthracene to 9,10-dimethyl-9,10-dihydro-anthracene with hydridopentacarbonylmanganese(I) proceeds by a two-stage mechanism *via* the 9,10-dimethyl-9,10-dihydroanthryl radical.[527] A cyclo-hexadienyl radical is also formed in the reaction of hydrogen atoms with benzimidazolium cations.[528] Chlorination of *o*-dichlorobenzene proceeds *via* initial formation of a π-complex between Cl· and *o*-$C_6H_4Cl_2$ followed by formation of a σ-cyclohexadienyl radical.[529] Cyanation of arenes has been accomplished using cyanogen and by use of a gaseous plasma; extensive *ipso*-substitution occurs in these reactions.[530,531]

ipso-Homolytic aromatic substitution has been reviewed.[532] The extent of *ipso*-substitution is determined largely by polar factors.[533] 2,5-Diacetylpyridine, but not 2,4-diacetylpyridine, undergoes *ipso*-substitution with 1,4-dioxolanyl radicals at both the 2- and 5-positions in neutral conditions, whereas in acidic conditions attack occurs at the 2- and 4-positions which are those with the highest positive charge.[534] The large amount of *ortho*-substitution in reactions of phenyl and cyclohexyl radicals with mono-substituted benzenes has been attributed to *ipso*-attack followed by a 1,2-shift.[535] 3-Substituted indoles react at the 3-position with dibenzoyl peroxide.[536] Biphenyl-2-carboxyl radicals cyclize intramolecularly to give mainly δ-lactones; the same products are formed from the corresponding 2-alkoxybiphenyl-2′-carboxyl radicals as a result of *ipso*-substitution. It was suggested that these radicals are π-radicals, and that acyloxy radicals which decarboxylate readily are in a Σ-ground state (cf. p. 000).[537]

The reactions of hydroxyl radicals with benzene,[538,539] toluene,[538,539] xylenes,[539–541] and phenol[542] have been studied; the extent of *ipso*-attack leading to dealkylated products is said to be extensive with *o*- and *p*-xylenes but very little with toluene. The extent of side-chain attack with methylbenzenes increases with increasing temperature and is the dominant reaction above 500°.[538,539] The rates of reaction of HO· with anthracene, pyrene, and benz[*a*]pyrene are 1.09×10^{10},

2.67×10^{10} and $6.9 \times 10^{10}\,\text{M}^{-1}\,\text{s}^{-1}$, respectively.[543] The reaction of peroxydisulphate with a mixture of benzene and nitrobenzene gives nitrophenols; these are not formed in absence of PhH suggesting that the hydroxylating agent is the hydroxycyclohexadienyl radical.[544]

$$PhH + SO_4^{\cdot -} \rightarrow PhH^{\cdot +} + SO_4^{2-}$$
$$PhH^{+\cdot} + H_2O \rightarrow [HOPhH]\cdot \rightarrow HO\cdot + PhH$$
$$HO\cdot + PhNO_2 \rightarrow [HOPhNO_2]\cdot \rightarrow HOC_6H_4NO_2$$

Hydroxylation of arenes has also been accomplished by α-hydroxyazo compounds,[208] and by carbonyl oxides.[545] Hydroxylation of thiophene and substituted thiophenes with hydroxyl radicals proceeds *via* the initial formation of the thiophene radical cation followed by addition of HO· at the 2- or 5-positions: a similar mechanism occurs in reactions of $Cl_2^{-\cdot}$ and $H_3^{+\cdot}$.[546]

One of the first authenticated examples of a reaction, proceeding by an $S_{ON}2$ mechanisms (cf. p. 2) involves the substitution of fluoride by acetate in the anodic, Co(III) or Ag(II) oxidation of *p*-fluoroanisole in the presence of acetate.[547]

$$ArF - e \rightarrow ArF^{+\cdot}$$
$$ArF^{+\cdot} + AcO^{-} \rightarrow [AcO.ArF]\cdot$$
$$[AcO.ArF]\cdot \rightarrow ArOAc^{+\cdot} + F^{-}$$
$$ArOAc^{+\cdot} + ArF \rightarrow ArOAc + ArF^{+\cdot}$$

Reaction of *p*-fluoroanisole with dibenzoyl peroxide in presence of acetate proceeds by a similar pathway to give a mixture of *p*-methoxyphenyl acetate and benzoate.[548]

The rate of oxidation of hydroxycyclohexadienyl radicals by $Fe(CN)_6^{4-}$ is $1.8 \times 10^{7}\,\text{M}^{-1}\,\text{s}^{-1}$.[549]

Biradicals

Theory and Structure

Theoretical considerations indicate that the singlet biradical state of tetracoordinate *planar* carbon is 25–30 kcal mol^{-1} lower in energy than the closed shell ($^1A_1$) state, and is likely to be best stabilized with delocalized π-substrates and molecular charge.[550] Diradicals containing four π-electrons[551] and hypothetical strain-free oligo-radicals,[551] with ways of increasing stabilization, have been surveyed. Progress is being made in the harmonization of theory and experimental data for 2-methylenecyclopentane-1,3-diyl (MCP) and trimethylenemethane (TMM). The theoretical energy difference of the $^3A_2$ and $^1B_1$ states of TMM is ~ 14 kcal mol^{-1}, and the $^1B_2$ state lies 2–3 kcal mol^{-1} above the $^1B_1$ state. In MCP the experimental singlet–triplet separation is only 1–4 kcal mol^{-1}. However, calculations reveal that the $^1A_1$ state of MCP is significantly lower in energy than the $^1B_2$ state, with other states similar to those in TMM.[553] Kinetic evidence has been presented to prove that it is the singlet biradical form of 2-isopropylidenecyclopentane-1,3-diyl which is active in cycloadditions with olefins.[554] Despite the presence of a hetero-atom to life degeneracy, the ground state of *m*-naphthoquinomethane is a triplet.[555] The rôle of σ- assistance, and its modulation of intramolecular reactivity, has been examined in *inter alia* biradicals.[556] Non-concerted ring-openings have been analysed in an excellent review.[557]

Carbon–Carbon Centred Biradicals

The *cis*-conformer of perfluoroacetylmethylene, $CF_3\ddot{C}C(=O)CF_3$, is kinetically 5 kcal mol^{-1} more stable than the *trans*-conformer. The first excited singlet state of alkynes is analogous to a *trans*-vinyl biradical.[558] Thus photolysis of cyclononyne leads to bicyclo[4.3.0]non-1-ene *via* an intramolecular hydrogen atom abstraction followed by cyclization.[559]

Secondary deuterium isotope effects on the rate- and product-determining steps in the thermolysis of 4-methylene-1-pyrazoline can be rationalized only by a rate-determining C—N bond rupture to give a diazenyl diradical which then closes to methylenecyclopropane.[560] Transfer of optical activity in the decomposition of (+)-*trans*-3,5-diphenyl-1-pyrazoline is dominated by pathway (a) (Scheme 18), but pathway (b) accounts for 30% of the overall reaction.[561]

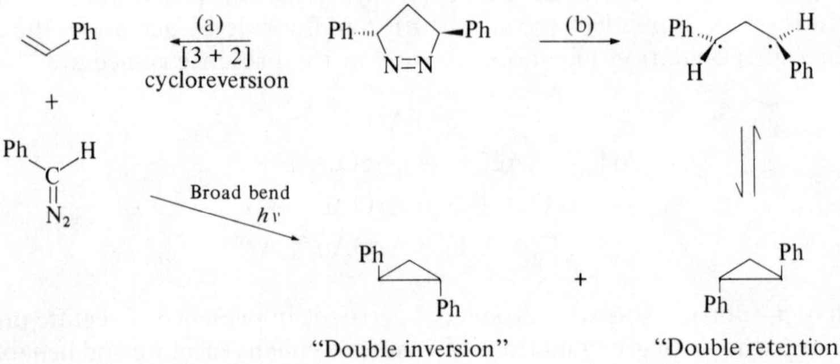

(a) [3 + 2] cycloreversion; (b); Broad bend $h\nu$

"Double inversion" + "Double retention"

SCHEME 18

The intermediacy of an effectively orthogonal biradical (**81**) in the thermal rearrangement of (**82**) and various derivatives, is indicated by the 2:1 statistical ratio of 1,3-shift (**83**) and 3,3-shift (**84**) products.[562] Less than 5% of the singlet biradical (**85**) undergoes retrograde di-π-methane rearrangement to the cyclopropyl dicarbinyl diradical (**86**) and thence to (**87**), and this is reduced by stabilizing benzhydryl or spirocyclopropyl groups (R_2) α to the radical site.[563]

Thermally equilibrated singlet and triplet (**88**) cyclize exclusively to (**89**) but vibrationally hot (**88**) yields some (**90**).[564]

Use has been made of an intramolecular diyl trapping in the total synthesis of the mold metabolite, (±)-hirsutene.[565] 2-Chloronaphthalene-1,3-diyl has been generated by reaction of 1-bromo-3,4-benzo-6,6-dichlorobicyclo[3.1.0]hexane with $KOBu^t$ in THF-d_8.[566]

Kinetic evidence has been obtained for the formation of discrete 1,4-dehydrobenzene intermediates in the solution pyrolysis of HC≡CC(R)=C(R)C≡CH (R = Et, Pr^n). 1,5-Hydrogen transfer in the intermediate biradicals involves a 5 kcal mol^{-1} barrier whereas the activation energy for ring-opening is ~10 kcal mol^{-1}.[567]

The adduct of singlet 1,8-dehydronaphthalene with CS_2 rapidly eliminates CS rather than internally cyclizing; the resultant biradical can close, or add a further molecule of CS_2 and then close.[568]

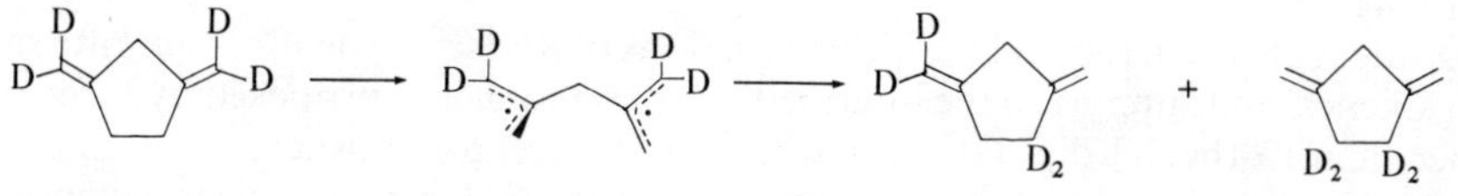

(82) (81) (83) (84)

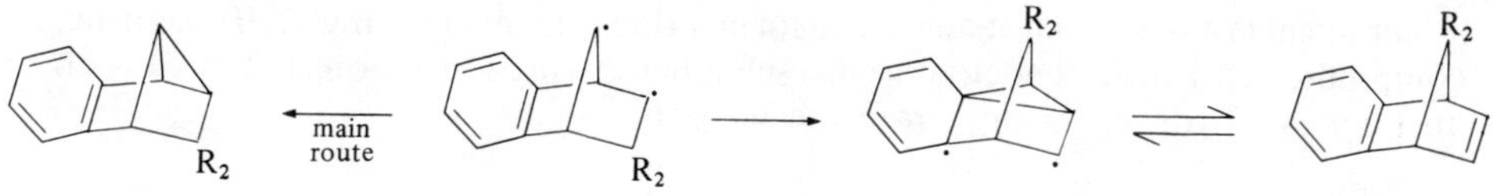

main route

(85) (86) (87)

(a) R = H (b) R + R = $-CH_2-CH_2-$ (c) R + R = $=C(Ph)_2$

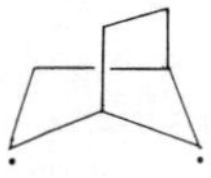

(88) 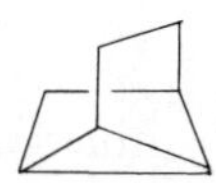 (89) 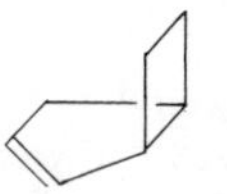 (90)

Diradical (**91**) is a common intermediate in the thermolysis of *cis* and *trans*-1,3-divinylcyclobutane.[569] The formation of alternating copolymers, and the suppression of cyclo-adducts, in the reaction of *p*-methoxystyrene with vinyl monomers in the presence of dimethylcyanofumarate or trimethylenetricarboxylate indicates the participation of tetramethylene biradical intermediates.[570] Diradical (**92**), the common precursor of products in the photo-sensitized addition of citraconic anhydride to 3,3-dimethyl-1-butyne, undergoes intramolecular disproportionation.[571] Sigmatropic [2,3]rearrangement products are found from a biradical intermediate following sulphide extrusion from methyl sulphonium salts of the sulphur-bridged sesquiterpenes, mint sulphide and *iso*-mint sulphide, by methyllithium.[572] 1,2,3-Benzothiadiazole traps triplet Ph_2C: to yield a biradical, after N_2 extrusion.[573]

Diradical (**93**) is deduced to be a common intermediate in the decompositions of *anti*-tricyclo[4.2.1.1^{2,5}]deca-3,7-diene, *endo*-[2 + 4]dicyclopentadiene, and *anti-cis*-[2 + 2]dicyclopentadiene.[574] Intense red chemiluminescent emission is produced on recombination of biradicals derived from lepidopterene, *sym*-dibenzylidenedimethyltriasterane, and 2,6-diphenylbicyclo[3.3.0]octa-2,6-6-diene.[575]

(91) 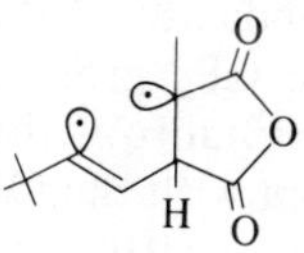 (92) 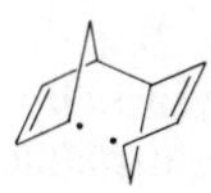 (93)

Hetero-centred Biradicals

Substituent effects in the cycloaddition reactions of Reissert hydrofluoroborate salts with alkenes lend support to the intermediacy of biradicals as proposed by Firestone rather than to the 1,3-dipolar non-radical mechanism of Huisgen.[576]

cis-3-Alkyl-2-phenyloxetanes fragment to alkene and benzaldehyde in a step-wise diradical fashion; the *trans*-epimers yield mainly alkenylbenzenes and formaldehyde by routes as yet uncertain.[577]

Diradical (**94**) shows no tendency to form a three-membered ring (1-*H*-diazirine) compound, and little tendency to five-membered ring-formation, but prefers to abstract two hydrogen atoms from the solvent.[578]

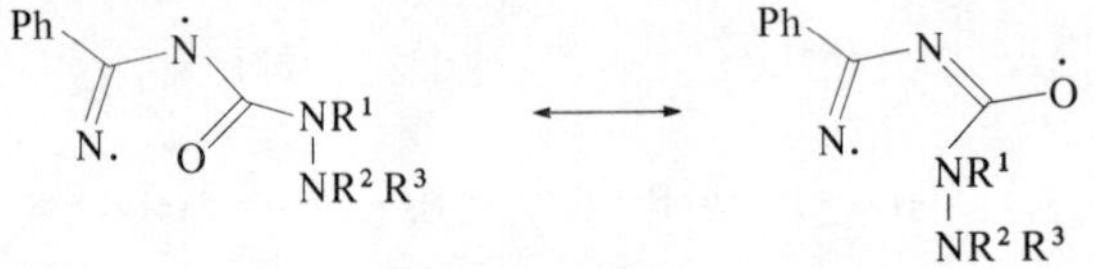

(94)

Carbonyl-derived Biradicals

Photo-cyclization yields of *para*-substituted 2,4,6-triisopropylbenzophenones do not correlate with the *para*-substituent constants σ^+ whereas the triplet life-times do. The Hammett ρ value of -0.35 for the latter contrasts markedly with that of $+0.60$ obtained for unhindered benzophenones.[579,580]

A new species, absorbing at 535 nm, has been detected in the Paterno–Buchi reaction of $Ph_2C{=}O$ and 1,4-dioxene,[581] and is believed to be a 1,4-biradical triplet; the species undergoes heterolytic cleavage to yield $(Ph_2C{=}O)^{\overline{\cdot}}$ with a rate of $6.3 \times 10^8\ s^{-1}$.

Evidence has been found that the mercury photo-sensitized fission of cyclohexanone gives a randomized vibrationally excited triplet acyl–alkyl biradical (by α-C—C bond cleavage) which can transform *inter alia* into a penta-1,5-diyl, giving ultimately 1-pentene and cyclopentane products.[582]

No co-operative interaction between the two radical sites is found in the β-fission of $Ph\dot{C}(OH)CH_2CH_2\dot{C}HCH_2X$ (X = Cl, Br, I) biradicals. However, very efficient loss is observed in the biradicals derived from 4-halo-1,4-dimethyl-1-benzoylcyclohexanones where the axial halogen is held *anti* to the unpaired electron on the adjacent atom.[583] Studies of the Norrish II photo-reactions of α-allylbutyrophenone confirm that biradicals rearrange at rates comparable to, and characteristic of, monoradical equivalents.[584]

Nitroxyls and Spin-trapping

Structural and Synthetic Aspects

The oxygen-17 h.f.s. of 23.6 G in enriched $(CF_3)_2NO\cdot$ radicals confirms that the spin density on oxygen is higher than in dialkyl nitroxyls.[585] IR and Raman studies on $(CF_3)_2NOH$ and $(CF_3)_2NO\cdot$ indicate[586] that the latter should be considered as planar and that the barriers to CF_3 group rotation are $\geq 6\ kJ\ mol^{-1}$. However, alkoxylalkyl and hydroxyalkyl trifluoromethyl nitroxyls $F_3CN(O\cdot)CR^1R^2OR^3$, are probably non-planar at nitrogen with the OR^3 group in an eclipsed position with respect to the nitrogen 2*p*-orbital due to conformational preference and an anomeric effect.[587] ESR studies on ^{17}O-labelled $Bu^tN(O\cdot)R$ and $MeN(O\cdot)COMe$ indicate that the radicals are non-planar about the nitrogen.[588] The β-hydrogens in $(Me_2N)_2P(O)NMeCH_2N(O\cdot)R$ radicals become non-equivalent on cooling below $-20°$ when R = Bu^t, and remain non-equivalent up to 80° when R = 2,3,5,6-Me_4C_6H.[589] Diastereoisomeric nitroxyls possessing two asymmetric carbon centres, formed from phenyl *N-tert*-butyl nitrone and ether/amine radicals, have been characterized and distinguished by ESR.[590]

The proportion of ene-thiol tautomer (**95**) correlates inversely with the N(1) h.f.s. in a series of fourteen solvents.[591] The ESR h.f.s. of the hydroxyl radical adduct of 5,5-dimethylpyrolline-*N*-oxide depend on basicity. A p*K* of 12.96 has been determined for the acid dissociation of the hydroxyl proton.[592] Association with solvents of various polarity of piperidinyl nitroxyls,[593] $Bu^tN(O\cdot)OBu^t$,[594] and $Bu^t_2NO\cdot$[595] has been investigated.

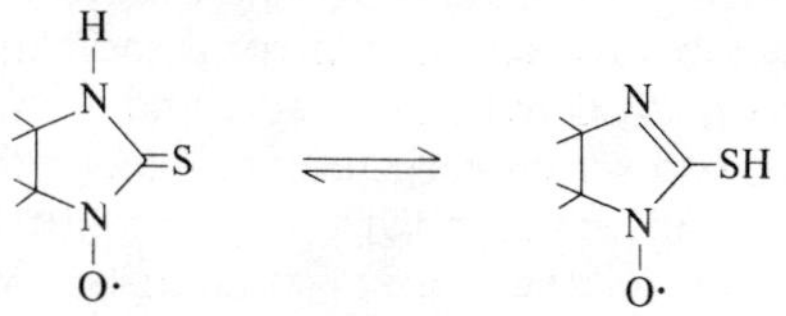

(**95**)

The applications of nitroxyl radicals in probing structure continue unabated. Nitroxyl derivatives of acetylacetone have been used as spin-labelled ligands for metal ions such as Co(III), Pd(II), Pt(II), and Rh(III).[596] While exerting little effect on chain conformation at room temperature, *gem*-substituents R in (**96**) decrease the distance between radical sites at low temperatures.[597] Analogues of dissacharides with the interglycosidic oxygen-bridge replaced by an hydroxyimino group, readily oxidizable to nitroxyl, have been synthesized as potential spin markers in biological studies.[598] Radical (**97**) closely mimics cholesterol in its properties and has been used as a probe of cholesterol–protein interactions in human high-density lipoprotein particles.[599] Strong spin exchange is found in solutions of the hetero biradical (**98**).[600]

Dimerization of (**99**) and $Bu^tN(O\cdot)CMe_2CMe{=}NOH$ radicals occurs *via* hydrogen bonding of oxime groups.[601]

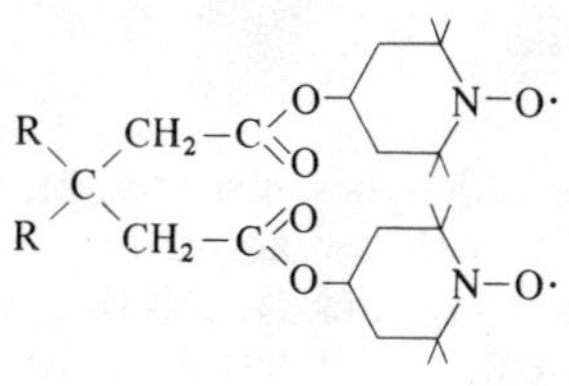

(**96**)

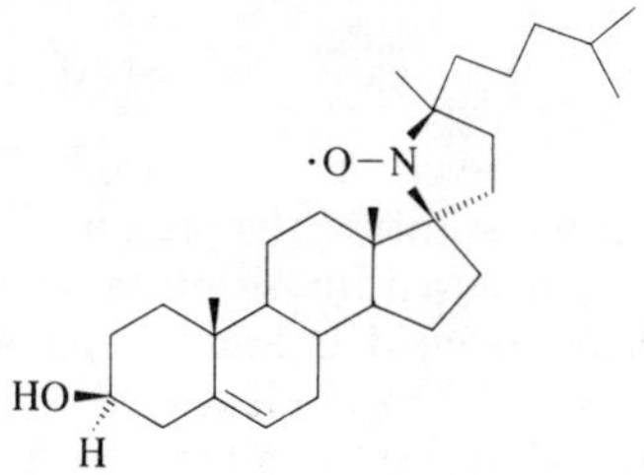

(**97**)

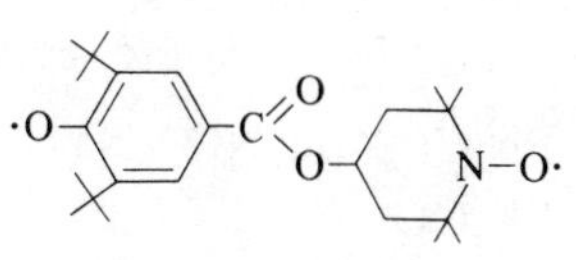

(**98**)

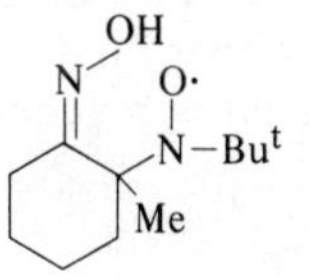

(**99**)

A clear and timely warning has been sounded that the "spontaneous formation" of nitroxyl species from nitroso compounds, ascribed to molecule-assisted homolysis, may arise in many cases from inadvertent photolysis of the nitroso compound. Normal visible light (present in most laboratories) may be sufficient to lead to accumulation of radicals by spin-trapping.[601] The ESR spectrum ascribed[603] to $RSN(N{=}O)O\cdot$ radicals in the reaction of thiophenols with $NaNO_2$ in methanol has been reassigned[604] to an iron-centred dinitrosyl species! Phenyl radicals appear to be genuine intermediates in the reaction of phenylhydrazones with nitrosobenzene and are formed from $PhN_2\cdot$ radicals expelled in a nitrone "reverse-trapping" reaction.[605] Mechanistic and ESR studies have been reported on the formation of nitroxyl radicals by red-light photolysis of humulene nitrosite $C_{15}H_{24}N_2O_3$.[606] Acid-catalysed transposition of groups from C(2) to C(3) in diastereoisomeric indolines has been confirmed by ESR of nitroxyl radicals formed by subsequent oxidation.[607] 2,2-Diphenylbenzoquinoline nitroxyl radicals have been obtained by the action of PhMgBr on 2-phenyl-, 2-cyano-, and 4-cyano-benzoquinoline-*N*-oxides.[608] (**100**), obtained by cyclodimerization of 3-methylbut-3-en-2-one oxime at 80°, yields a stable nitroxyl by reaction with poly(*N*-methylacrylamide) radical.[609] 1,3-Radical additions of silanes and stannanes $MHR^1_2R^2$ to nitrones $PhCH{=}N(O)R^3$ yield[610] *C*- and *O*-metallohydroxylamines and the nitroxyl radicals $R_2{}^1R^2MCH_3NR^3O\cdot$. Oxidative coupling of amines and hydroxylamines with nitrones[611,612] yields amidinyl-*N*-oxides and -*N,N'*-dioxides as secondary radicals, *e.g.* (**101**) whose conformations have been elucidated by ESR; when R^2 = Ar, radical (**101**) eliminates ArH to give (**102**).[612] Solvation effects on the formation, *inter alia*, of $PhN(O\cdot)Me$ in the reaction of PhNO with $KOBu^t$ and other bases have been discussed.[613]

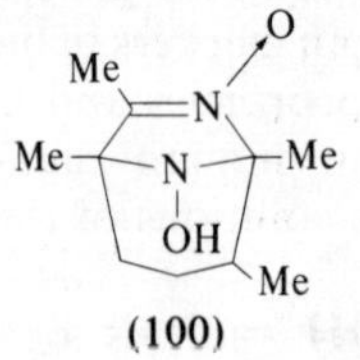

(100) 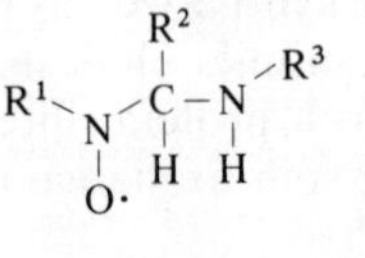 (101) 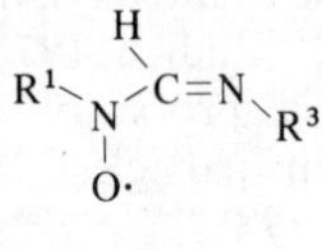 (102)

R^1 = Me, Bu^t, Ph, $PhCH_2$

R^3 = Me, Pr^n, Bu^n, Bu^t

Reactions of Nitroxyls

While $Bu^t_2NO\cdot$ is stable in the presence of styrene, aldehydes, nitrones, phenol, and water, it is pyrolysed in solution above 90°C, with an activation energy of 33 kcal mol^{-1}, to yield a low conversion to Bu^tNO and $Bu^t_2NOBu^t$ via rate-determining C—N bond cleavage.[614] In an attempt to investigate the possible nucleophilic behaviour of nitroxyls in S_N2 reactions, by reacting $Bu^t_2NO\cdot$ with $CH_3OS(O_2)CF_3$, the unexpected formation of *N,N*-di-*tert*-butylhydroxylamine radical cation was observed in triflic acid (CF_3SO_3H) as solvent.[615] Reaction of $Bu^t_2NO\cdot$ with either Ph_3C^- or Ph_3C^+ generates $Ph_3C\cdot$ radicals,[616] but in the former case the nitroxyl is regenerated, in the presence of O_2, *via* the sequence

$$Ph_3C^-Na^+ + Bu^t_2NO\cdot \rightarrow Ph_3C\cdot + Bu^t_2NO^-Na^+$$

$$Ph_3C\cdot + O_2 \rightarrow Ph_3CO_2\cdot$$

$$Ph_3CO_2\cdot + Bu^t_2NO^-Na^+ \rightarrow Ph_3CO_2{}^-Na^+ + Bu^t_2NO\cdot$$

The photochemical reactivity of $Bu^t_2NO\cdot$ in pentane and in CCl_4 solution has been investigated.[617]

$(CF_3)_2NO\cdot$ radicals ($Y\cdot$) react with the ethers MeOX (X = Me or Ph) to form YCH_2OX at room temperature with $(YCH_2)_2O$ as a minor by-product; with Et_2O, however, multiple hydrogen atom abstractions occur.[618] Reaction of benzyl cyanide (similarly with the chloride) with $(CF_3)_2NO\cdot(Y\cdot)$ in equimolar proportions gives PhCHYCN; this reacts with further nitroxyl to give $Ph\dot{C}(CN)ON(CF_3)$ which in turn decomposes to $(CF_3)_2N\cdot$ radical and benzoyl cyanide, or reacts with $Y\cdot$ to give the combination product $PhCY_2CN$.[619] Radical processes are involved in fluoroalkylation of aromatic compounds[620] with $(CF_3)_2NO$ and in its reactions with diphenyl-ketene, -acetic acid, -acetyl chloride, and -acetamide.[621]

A thorough exploratory study of the formation and reactions of *tert*-butyl-*p*-formylphenyl nitroxyl and related nitroxyls has been published.[622]

Direct acyloxylations and methoxylation of 1,2-dihydro-2,2-disubstituted-3-oxo-3*H*-indole-1-oxyl radicals *via* oxoammonium salt intermediates have been achieved.[623] Reversible absorption of O_2 by $Ph_2CHN(O\cdot)C_6H_4R$ radicals has been studied.[624] Rate constants for reaction of hydroxyl radicals with the stable nitroxyls TANO and TEMPO have been measured by pulse radiolysis.[625]

The novel rôle of nitroxyls in biosynthetic processes has been proposed. A flavin-based nitroxyl has been suggested as the hydroxylating agent in the flavin mono-oxygenase *ortho*-hydroxylation of phenolic substrates.[626] Step-wise oxidation of some isoquinoline alkaloids by Fremy's salt to aminium radical, followed by proton loss from α- or δ-positions, leads to aromatization of ring B.[627]

Spin-trapping

Spin-trapping has been reviewed.[628,629] Spin adducts with 3,3,5,5-tetramethyl-pyrroline-*N*-oxide are more persistent than with DMPO; this new trap is especially useful for alkyl, aryl, and benzoyloxyl radicals.[630] Sodium 3,5-dibromo-4-nitrosobenzene sulphonate is a new water-soluble trap with improved ESR spectral resolution due to the absence of *meta*-splittings.[631]

An ESR spin-trap method has been applied to the determination of the rate constant for α-hydrogen atom abstraction from PhEt by triplet excited anthraquinone.[632] Addition rate constants of $CCl_3\cdot$ and $ClCH_2CH_2CCl_2\cdot$ radicals to Bu^tNO, nitrosodurene, and $PhCH{=}N(O)Bu^t$ have been measured.[633]

The spin adduct of diphenyliminyl radicals with nitrosodurene breaks down to give duryl radicals.[634]

$$Ph_2C{=}N\cdot + DurNO \rightarrow Ph_2C{=}N{-}N(O\cdot)Dur$$

$$Ph_2C{=}N{-}N(O\cdot)Dur \rightarrow Ph_2CO + DurN{=}N\cdot$$

$$DurN{=}N\cdot \rightarrow Dur\cdot + N_2$$

A complexity exists in many spin-trap systems which points to the need for caution in interpretation. Thus, with a nitrone trap α-hydroxyalkyl radicals generated photolytically form normal adduct radicals ($RA\cdot$) which are, however, photochemically labile in yielding a hydroxylamine (B). The nitrone trap itself forms an oxaziridine (A) on irradiation. Photo-reaction of (B) with O_2 gives $RA\cdot$ radicals as does the reaction of (A) and (B) in the dark.[635] The use of traps also requires a proper consideration of the redox potentials of radicals and traps.[636] Disconcertingly, many nitrosoaromatics seem to react with hydrocarbons[637] as well as with benzene and chlorobenzene[638] to yield nitroxyls (but note Reference 601). PBN reacts efficiently with singlet oxygen in aqueous solution ($k = 1.4 \times 10^8\ \text{M}^{-1}\ \text{s}^{-1}$), but many common nitrone traps are much less efficient. Hence, the formation of hydroxyl radical adducts in low concentration could reflect singlet oxygen rather than hydroxyl radical generation in some systems.[639]

Azolyl radicals, formed from azolyl anions and azoles on reaction with Bu^{t}NO, can be trapped.[640] A spin-trapping study of the Bz_2O_2 photo-induced reactions of amino acids and peptides[641] shows decarboxylation to be the major process, except in the case of valine which undergoes hydrogen atom abstraction.[642] Photo-fission of the C—S bond in Bu^{t}SSBu^t and $(PhCH_2S)_2$ has been demonstrated by spin-trapping methods.[643] Photolyses of sulphanilamide, and 4-aminobenzoic acid in aqueous solution at pH 7, yield $\cdot NH_2$ radicals and H· atoms respectively, as judged by spin-trapping with DMPO.[644] The first evidence for a selenylnitroxyl RSeN(O·)R′ has been obtained; the spin density is low on the selenium atom.[645]

S_H2 Reactions

The metathetical reaction of H· atoms with $(EtS)_2$ has a high $\Delta S^{\neq}$, which suggests the breakage of partial hydrogen bonding in $(EtS)_2$ in the transition state.[646] Rate constants for reactions of thermal ^{18}F atoms with methyl and trifluoromethyl halides,[647,648] and of thermal ^{38}Cl atoms with $PbMe_4$ and $PbEt_4$[649] in the gas phase have been measured. Allyl radicals react with NO_2 at 300 K to give C_3H_5O and NO.[650] Cyclohexyl radicals participate in an S_H2 reaction with $(C_6F_5)_2S_2$ as well as reactiing with pentafluorobenzenesulphenyl chloride.[651]

The general reaction R· + PhSeSePh → PhSeR + PhSe has provided a route to bridgehead selenides, *e.g.* 1-adamantyl phenyl selenide, which can be oxidized and then pyrolysed to give bridgehead alcohols or olefins.[652]

The α-carbon of the benzyl ligand in benzylbis(dimethylglyoximato)pyridinecobalt(III) is attacked directly by $CCl_3\cdot$ radical to yield $PhCH_2CCl_3$.[653] In substituted benzyl systems the rate and yield of this S_H2 product is determined by the substituent in the order: polymethyl ≥ Me > H > 4-Cl > 4-NO_2. Homolytic displacement of cobaloxime(II) from 3-methylbut-3-enylbis(dimethylglyoximato)-pyridinecobalt(III) by regiospecific attack of *N,N*-dimethylsulphamoyl radicals on the γ-carbon of the allyl ligand produces 1,1,*N,N*-tetramethylallylsulphonamide.[654] Reaction of $Cl_3Si\cdot$ radicals with cyclohexyl octyl ether leads to cyclohexane and *n*-nonane in 61% and 7% yield, respectively.[655]

S_H2 reactions on silicon–carbon bonds in benzosilacyclobutenes by triplet ketone, *e.g.* $Ph_2C{=}O$, lead to ring-opened diradicals, which subsequently cyclize to six-membered rings.[656] $R_3Sn\cdot$ radical-induced sulphur–C(2) bond cleavage shown in Scheme 15 (see p. 33) is the reverse of the proposed mechanism for thiazolidine ring formation in penicillin biosynthesis.[657] Triphenylverdazyl radical causes homolysis of the C—Br bond in Ph_3CBr by an S_H2 process,[658] but no general mechanism is applicable to its reaction with $Ph_2C(NO_2)Br$, PhCH(CN)Br, $CH(CN)_2Br$, $(AcO)_2C(NO_2)Br$, and $EtCO_2CH(CN)Br$.[659]

Bromine atoms attack 1,3-diarylcyclopropanes in an S_H2 reaction to give 1,3-diaryl-1,3-dibromopropanes.[660]

Phosphoranyl and Sulphuranyl Radicals

The chemistry of phosphoranyl radicals has been reviewed.[661,662] The $Me_2P(S)Br^{\overline{\cdot}}$ radical anion is tetrahedral, with the unpaired electron in the σ^*-orbital of the

P—Br bond.[663] Single-crystal ESR studies of two orientated spirophosphoranyl radicals with nitrogen substituents, having trigonal bipyramidal geometry and the unpaired electron in an equatorial position, have indicated the advisability of deducing the geometry of such radicals in solution from *both* phosphorus and nitrogen ESR h.f.s.[664] Radical (**103**) is formed from radical (**104**) precursor on X-ray irradiation of the parent compound.[665] Evidence has been obtained for ligand exchange in phosphoranyl radicals *via* a Berry pseudo-rotation mechanism, *i.e.* pair-wise interconversion of ligands around the unpaired electron orbital as a pivot.[666]

The rate of pseudo-rotation in the spin-labelled phosphorane (**105**), generated by cyclization of triazarylphosphoranyl radical, has been determined from the exchange rate (for which $\log_{10} k = 12.7 \pm 0.3 - 21.8 \pm 1.5/2.303\,RT\,\text{kJ mol}^{-1}$).[667]

Radicals (**106**) with a *m*- or *p*-carboranyl group have a trigonal pyramidal structure with the carborane ring in an apical position; in (**107**) with an *o*-carboranyl group the structure is tetrahedral with the unpaired electron localized on the carborane ring.[668] Apical–equatorial ring-proton exchange rates in the trigonal–bipyramidal (TBP) phosphoranyl radicals (**107**) have been estimated from ESR line-shape effects. The rate increases as the energy difference, between the TBP C_{2v} structure and a σ^* (P–S) intermediate of C_{3v} symmetry, decreases. The apicophilicity of RS is greater than for R′O ligand, but it has been suggested that the apicophilicity in TBP and σ^*-isomers is related to the ease of heterolytic dissociation of the phosphorus–ligand bond rather than to ligand electronegativity alone.[669]

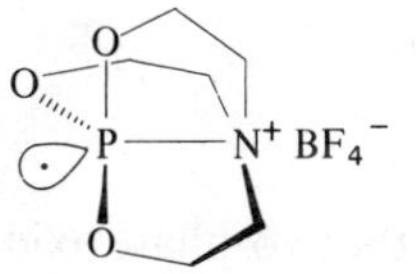

O P N$^+$ BF$_4^-$ O O

(103)

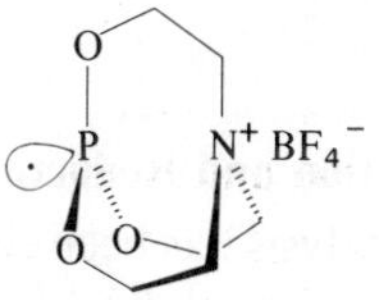

O P N$^+$ BF$_4^-$ O O

(104)

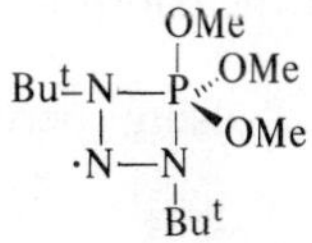

OMe
Bu^t–N—P–OMe
OMe
·N—N
Bu^t

(105)

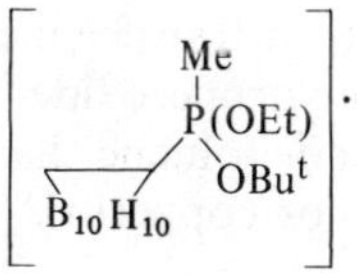

[Me P(OEt) OBu^t $B_{10}H_{10}$]·

(106)

[CH_2CH_2 O O P R^{2}S OR^1]·

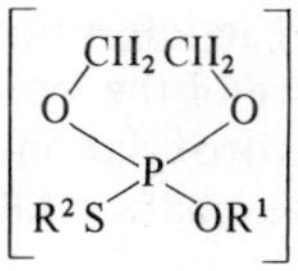

(107)

Kinetic data on the reaction of Me· radicals with Et_3PO_4 indicate a phosphoranyl structure as the transition state.[670] β-Scission of Bu^{t}OP(OEt)$_3$· has a very low activation volume as expected for an early transition state of a highly exothermic reaction.[671] Recombination and α-cleavage of (**108**), and also pseudo-rotation in the radical, have been studied by ESR.[672] Solvent polarity effects on the reaction of Bu^{t}O· with phosphoryl compounds have been investigated.[673] A laser flash photolysis study of the formation, fragmentation, and rearrangement of *tert*-butoxyphosphoranyl radicals has provided much useful data.[674] The β-scission activation energy obtained for the Bu^{t}O$\dot{P}$(OEt)$_3$ species agrees well with that obtained earlier by ESR; the hypervalent Bu^{t}O(EtO)$_2\dot{P}$NCO radical rearranges to the ligand π-radical, $Bu^tO(EtO)_2P{=}N{-}\dot{C}{=}O$; reaction of $Bu^tO\cdot$ with $PhP(OEt)_2$ yields a ligand π-type of radical; $Bu^tO\dot{P}Bu^n{}_3$ radical undergoes α-scission liberating $Bu^n\cdot$ radicals. Radical $X\cdot$ (X = RO, R, RCO) addition to $(Me_3Si)_2NP{=}NSiMe_3$ gives persistent radicals having a pyramidal geometry at phosphorus.[675] Rate constants for addition of alkyl radicals to diethyl vinylphosphonate follow the order Me < primary R < secondary R < Bu^t and are in interpreted in terms of the overriding importance of polar effects in determining the barrier to addition.[676]

c-$C_6H_{11}SO_2\cdot$ and *c*-$C_6H_{11}\dot{S}(O)$ radicals are intermediates in the chain decomposition of *c*-$C_6H_{11}SO_2Cl$ initiated at 9–60° by *c*-$C_6H_{11}O_2COOCO_2C_6H_{11}$-*c*.[677]

Low-temperature photolysis of perester (**109**) yields the bridged radical (**110**) having nine electrons around a three-coordinate sulphur atom.[678a]

At 130° benzylhydryl methyl sulphoxide decomposes simply to yield $CH_3SO\cdot$ radicals; the decomposition of benzhydryl *p*-tolylsulphoxide to *p*-toluenesulphinyl radicals is complicated by an equilibrium of the reactant with benzhydryl *p*-toluenesulphenate.[678b]

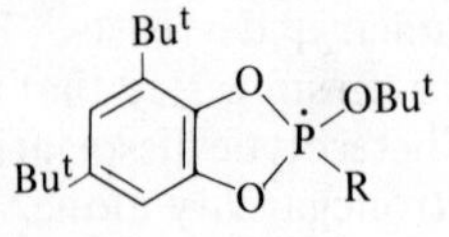

(**108**) 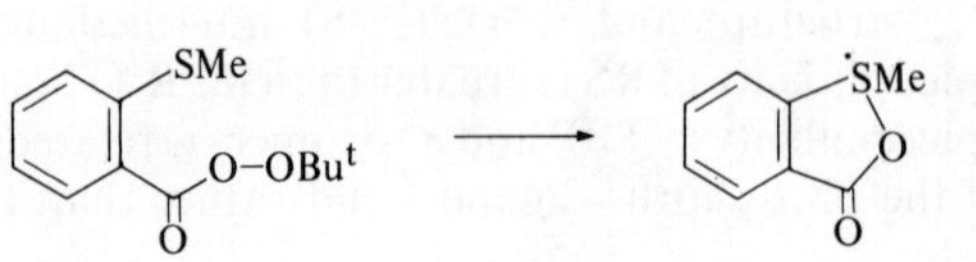 (**109**) (**110**)

Homolytic Oxidation and Reduction

Phase-transfer catalysis has been employed in the two-phase oxidation of toluene by CAN using the lipophilic 4,4′-dialkoxy-2,2′-bipyridine-1,1′-dioxide as the catalyst.[679] 18-Crown-6 with potassium peroxydisulphate has been used to initiate the polymerization of vinyl monomers in hydrocarbon solvents.[680]

Photo-oxidation of alkenes in presence of iron(III) chloride gives α-chloroketones: the reaction proceeds *via* an electron-transfer mechanism within the coordination sphere of the iron *via* a β-chlorohydroperoxide.[681]

Hydroxylation of benzene and toluene has been achieved using peroxydiphosphate catalysed by iron(II) or copper(II).[682]

$$H_2P_2O_8{}^{2-} + Fe^{2+} \rightarrow HPO_4{}^{\dot{-}} + HPO_4{}^{2-} + Fe^{3+}$$
$$HPO_4{}^{\dot{-}} + H^+ \rightarrow H_2PO_4\cdot$$
$$H_2PO_4\cdot + PhH \rightarrow H_2PO_4{}^- + PhH^{+\cdot}$$
$$PhH^{+\cdot} + H_2O \rightarrow [HO.PhH]\cdot + H^+$$
$$[HO.PhH]\cdot + H_2PO_4\cdot \rightarrow PhOH + H_3PO_4$$

The rate-determining step in the oxidation of methylbenzenes by CAN depends on the nature of the substrate: $k_2 \gg k_{-1}[Ce(III)]$ for 5-*tert*-butyl-1,2,3-trimethylbenzene but $k_2 \lesssim k_{-1}[Ce(III)]$ for the more readily oxidizable *p*-methoxytoluene and hexamethylbenzene.[683] Deprotonation of alkylbenzene radical cations is not only a simple base-catalysed proton abstraction; the kinetics of the electrochemical oxidation of hexamethylbenzene show that step 3 is also important.[684,685] The liquid-phase oxidations of toluenes with

$$ArCH_3 + Ce^{4+} \underset{k_{-1}}{\overset{k_1}{\rightleftharpoons}} ArCH_3^{+\cdot} + Ce^{3+}$$

$$ArCH_3^{+\cdot} + B \xrightarrow{k_2} ArCH_2^{\cdot} + BH^+$$

$$ArCH_3^{+\cdot} + ArCH_2^{-} \underset{k_{-3}}{\overset{k_3}{\rightleftharpoons}} ArCH_3 + ArCH_2$$

$Co(OAc)_2$–$Cu(OAc)_2$–NaBr[686] and with manganese(III)[687] to give mixtures of benzyl acetates and benzaldehyde, and with $CuBr_2$/Bu^{t}OOH/AcOH[688] to give mixtures of benzyl bromides and benzoic acids, proceed by initial electron transfer followed by proton loss to give benzyl radicals. Oxidations of *p*-toluic acid by $Co(OAc)_3$ and other metal-ion oxidants in the presence of alkali metal bromides give side-chain and nuclear-substituted bromides.[689,690] Benzyl radicals are also intermediates in the oxidation of electron-rich alkylbenzenes to carbonyl compounds with $S_2O_8^{2-}/Cu^{2+}$.[691] Oxidations of cumene and *p*-cymene with the $S_2O_8^{2-}/Fe^{2+}/AcO^-/AcOH$ system affords excellent yields of the lactone (**111**; R = H or Me): The reaction of *p*-cymene also gives *p*-isopropylbenzyl acetate and 4,4′-diisopropyl-1,2-diphenylethane resulting from *p*-isopropylbenzyl radicals: when the reaction is carried out in presence of quinoxaline both *p*-isopropylbenzyl and *p*-methylcumyl radical adducts are formed.[692] The major product from anodic or $Fe(CN)_6^{3-}$ oxidation of 1,2,5-trimethylpyrrole in presence of cyanide is 2,5-dicyano-1,2,5-trimethylpyrroline: the pyrrole radical cation is formed initially and this undergoes nucleophilic capture by cyanide.[693]

$$ArCHMe_2 + SO_4^{-\cdot} \longrightarrow ArCHMe_2^{+\cdot} \longrightarrow Ar\dot{C}Me_2 + H^+$$

$$CH_3COO^- + SO_4^{-\cdot} \longrightarrow CH_3^{\cdot} + CO_2 + SO_4^{2-}$$

$$CH_3^{\cdot} + CH_3COOH \longrightarrow CH_4 + \dot{C}H_2COOH$$

$$Ar\dot{C}Me_2 \xrightarrow{-e^-,\,-H^+} ArC(=CH_2)Me$$

$$ArC(=CH_2)Me + \dot{C}H_2COOH \longrightarrow Ar\dot{C}(Me)CH_2CH_2COOH$$

$$Ar\dot{C}(Me)CH_2CH_2COOH \xrightarrow{-e^-,\,-H^+} \text{(111)}$$

(111)

Azulenes react with copper(II) salts, *e.g.* $Cu(OSO_2Ph)_2$, to give high yields of substituted azulenes: the reaction is postulated rather improbably to proceed *via* the azulene radical anion.[694,695]

The anodic dimerization of 4,4′-dimethoxystilbene involves coupling of the radical cation with a second molecule of the stilbene.[696] Dimeric products are obtained by oxidation of phenolic ethers with iron(III) chloride on silica *via* the corresponding radical cation.[697]

In suitably substituted reticulines radicals derived from oxidation of both phenolic rings have been detected by ESR providing *prima facie* evidence for intramolecular coupling of these biphenols: thus (**112**; X = Y=R′=H,R=Me) gives both (**112**; X = ·, Y = R′=H, R = Me) and (**112**; X = R′=H,Y=·, R = Me) but (**112**; X = Y=H, R = R′=Me) gives only (**112**; X = H, Y = ·, R = R′=Me).[698] Cation radicals are generated in the PbO_2 oxidation of sulphonated phenols in FSO_3H.[699] The oxidative coupling of 2-bromo-4,6-di-*tert*-butylphenol with ferricyanide gives 1,4-dihydro-4-bromo-2,4,6,8-tetra-*tert*-butyl-dibenzofuran: the initial step of the reaction involves *ortho–ortho* coupling of the phenoxyl radicals.[700] The oxidative coupling of (s)-(+)-2-hydroxy-3,4,8-trimethyl-5,6,7,8-tetrahydronaphthalene yields completely stereospecifically the optically active (s,s)-(+)-*trans*-dinaphthol.[701] Intramolecular C—O oxidative coupling of 2,4′-dihydroxydiphenyl ethers affords *p*-benzoquinone acetals.[702,703] C—O coupled products are also formed in the lead(IV) acetate oxidation of perfluoronaphthols.[704]

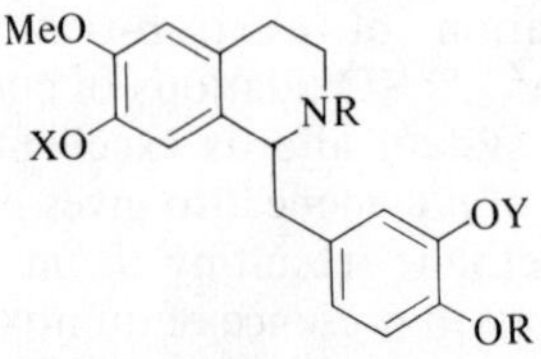

(**112**)

Propan-2-ol induces a chain reaction between peroxydiphosphate and iron(II) and is itself oxidized to acetone.[705,706]

$$P_2O_8^{4-} + Fe^{2+} + 5H^+ \rightarrow H_2PO_4\cdot + H_3PO_4 + Fe^{3+}$$
$$Me_2CHOH + H_2PO_4\cdot \rightarrow Me_2\dot{C}OH + H_3PO_4$$
$$Me_2\dot{C}OH + P_2O_8^{4-} + 4H^+ \rightarrow Me_2CO + H_2PO_4\cdot + H_3PO_4$$
$$H_2PO_4\cdot + Fe^{2+} + H^+ \rightarrow H_3PO_4 + Fe^{3+}$$

Alkoxy radicals and subsequently α-hydroxyalkyl radicals are generated in the iron(III) photo-oxidation of alcohols.[707]

$$RCH_2OH + Fe^{3+} \xrightarrow{h\nu} RCH_2O\cdot + Fe^{2+} + H^+$$
$$RCH_2O\cdot + RCH_2OH \longrightarrow RCH_2OH + R\dot{C}HOH$$
$$R\dot{C}HOH + Fe^{3+} \longrightarrow RCHO + Fe^{2+} + H^+$$

The oxidative coupling of ketone enolates has been brought about with iron(III) chloride.[708]

$$Bu^tCOCH_2^- + Fe^{3+} \rightarrow Bu^tCOCH_2\cdot + Fe^{2+}$$
$$2\,Bu^tCOCH_2\cdot \rightarrow Bu^tCOCH_2CH_2COBu^t$$

Oxidation of (**113**) with MnO_2 gives the delocalized radical (**114**) which couples to give the 1′,3′- and 3,3′-dimers.[709] Oxidation of alkanone cyanohydrins (**115**) with $S_2O_8^{2-}/Ag^+$ leads to δ-ketonitriles (**117**) *via* the α-cyanoalkoxyl radicals (**116**); these radicals, in addition to undergoing 1,5-hydrogen transfers, also fragment.[710]

Oxidative cleavage of 1,2-diols with *N*-iodosuccinimide probably gives initially a hypoiodite intermediate which undergoes O—I homolysis.[711]

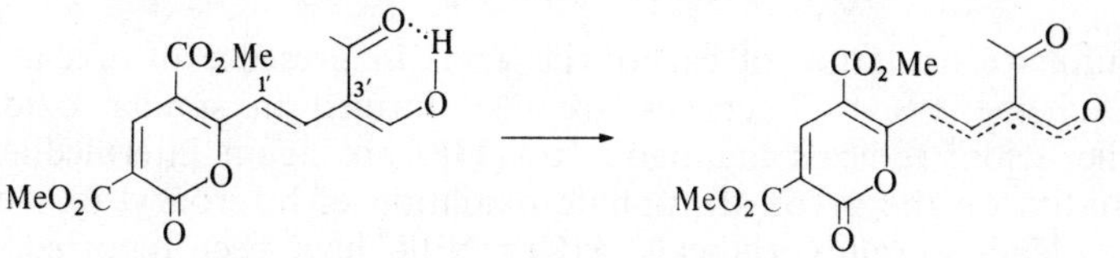

(113) (114)

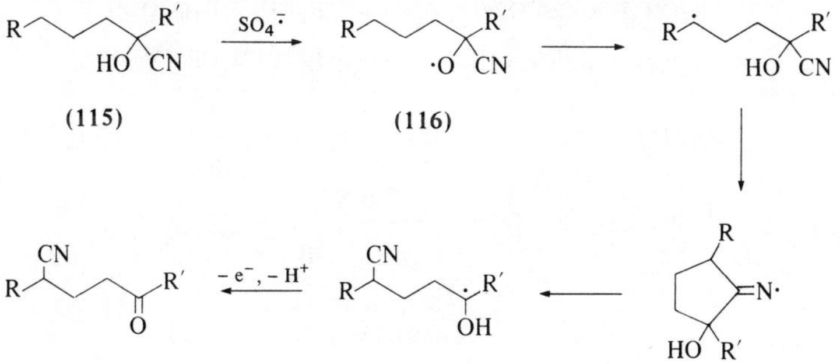

(115) (116)

(117)

The radical coupling of thiols occurs on oxidation of organometallic thiol complexes, *e.g.* $[Fe(C_5H_5)(CO)(L)SPh]^+$.[712] Photo-excited tris(2,2′-bipyridine)-ruthenium(II) is quenched by thiophenolate ions in single-electron transfer process.[713]

$$Ru(bpy)_3^{2+*} + PhS^- \rightarrow Ru(bpy)_2^{+} + PhS\cdot$$

Enzymatic oxidation of alkyl sulphides by hydroxyl radicals and by the enzyme system Cytochrome P-450 gives both *S*-oxygenated and *S*-dealkylated products: the reaction proceeds *via* initial electron transfer followed by proton loss.[714]

$$ArSCH_2 - e^- \rightarrow Ar\overset{+\cdot}{S}CH_2X \rightarrow ArS\dot{C}HX + H^+$$

Iron(III) chloride brings about the oxidative coupling of ketone enolates to give diketones.[715] α-Chlorination of acetophenones occurs on reaction with manganese(III) acetate in presence of chloride.[716]

Oxidative decarboxylation of carboxylic acids has been reviewed.[717]

Oxidation of carboxylic acids with $TiCl_3/H_2O_2$ gives more or less random attack by HO· supplemented by preferential abstraction of β-C—H through a complexed oxidizing species in which the titanium is coordinated both to the carboxyl group and the attacking radical.[718] Oxidation of arylacetic acids by peroxydisulphate in presence of copper(II) in acetic acid leads to benzylic acetates: the intermediate phenylacetoxy radicals are generated by electron transfer from the carboxylate anion and to a lesser extent from the aromatic nucleus.[719]

$$ArCH_2COO^- + SO_4^{\cdot -} \rightarrow ArCH_2COO\cdot + SO_4^{2-}$$
$$ArCH_2COO\cdot \rightarrow ArCH_2\cdot + CO_2$$
$$ArCH_2COOH + SO_4^{\cdot -} \rightarrow ArCH_2COOH + SO_4^{2-}$$
$$ArCH_2COOH \rightarrow ArCH_2\cdot + CO_2 + H^+$$

Peroxydisulphate oxidation of carboxylic acids in presence of alkenes leads to lactones (Scheme 19).[720] Lactones are also formed in similar oxidations of dicarboxylic acids: radicals analogous to (**118**) are again intermediates.[721,722] Further studies on the peroxydisulphate oxidation of heteroaryl oxime *O*-acetic acids, which leads to iminyl radicals, ArC(=N·)R, have been reported.[723,724]

$$R_2CHCOO^- + SO_4^{\cdot -} \longrightarrow R_2\dot{C}H + CO_2 + SO_4^{2}$$
$$R_2\dot{C}H + R_2CHCOOH \longrightarrow R_2CH_2 + R_2\dot{C}COOH$$
$$R_2\dot{C}COOH + R'_2{}^1C{=}CH_2 \longrightarrow R'_2\dot{C}CH_2CR_2COOH \quad \textbf{(118)}$$
$$R_2\dot{C}CH_2CR_2COOH \xrightarrow{-e^-,\,-H^+} \text{(lactone: R, R / R, R)}$$

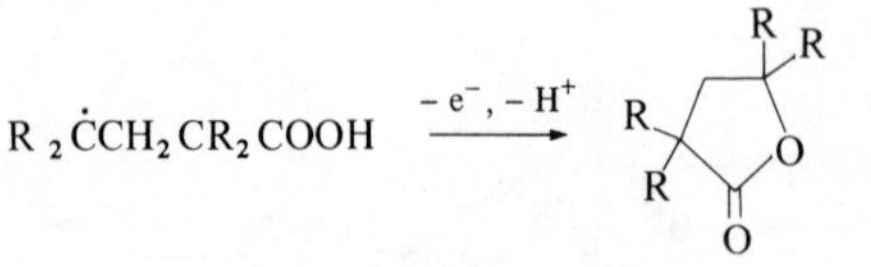

SCHEME 19

Tetraphenylbenzidine is obtained by oxidation of triphenylamine as a result of coupling of $Ph_3\overset{+\cdot}{N}$.[725,726] 1,4-Diaminobenzene radical cations are generated under anodic conditions and on reaction with bromine: reaction with a second mole of the diamine gives the radical cation dimer.[727,728] α-Aminoalkyl radicals are generated by horseradish peroxidase-catalysed oxidation of *N*,*N*-dialkylanilines by hydrogen peroxide.[729] The *N*-methylacridanyl radical has been spin-trapped by Bu^{t}NO in the oxidation of *N*-methylacridan by 2,3-dicyano-1,4-benzoquinone thus providing direct evidence for an electron, proton, electron-transfer mechanism.[730]

$$DH + A \rightleftharpoons DH^{+\cdot}A^{\cdot -} \rightarrow D + HA$$

The oxidation of methylviologen radical cation to the dication is brought about by metal-ion oxidants, benzoquinone, 2,6-dichloroindophenolate and 2,3,5-triphenyltetrazolium chloride *via* a SET mechanism.[731] The oxidation with hydrogen peroxide also generates hydroxyl radicals which are immobilized by attachment to the dication.[732]

$$MV^+ + H_2O_2 \rightarrow MV^{2+} + HO\cdot + HO^-$$

Oxidation of dimethylcobalt(III) macrocycles either by chemical oxidants or anodically leads to transient dimethylcobalt(IV) which undergoes spontaneous homolysis to give methyl radicals.[733]

Benzyl radicals, generated in the photolysis of dibenzyl ketones, are efficiently scavenged by copper(II) chloride to give benzyl chloride.[734] Galvinoxyl radicals can be oxidized by $RuCl_3$ and $RhCl_3$ but are reduced by iron(II) and cobalt(II) acetylacetonates.[735]

Reductive cleavage of 9,9-diarylfluorenes proceeds by formation of the radical anion followed by disproportionation to give the dianion, which undergoes cleavage to give the 9-arylfluorene and aryl anions.[736] Reduction of 9,10-bis(phenylethynyl)-9,10-dihydroanthracene-9,10-diol by $SnCl_2$ gives the 9,10-bis(phenylethynyl)-10-hydroxy-9,10-dihydroanthracene-9-yl radical, which can cyclize onto the 10-ethynyl group.[737] Radicals are generated by reduction of tropylium salts with chromium(II).[738]

Cathodic reduction of bridgehead iodides gives the corresponding radicals.[739,740] The correlation between half-wave potentials for the polarographic reduction of substituted benzyl chlorides and the σ-substituent constants is consistent with formation of a radical anion intermediate.[741] Reduction of trichloroacetate anions by vanadium(II) gives $\dot{C}Cl_2COO^-$.[742] The addition of carbon tetrachloride to styrene and oct-1-ene is promoted by the Cu_2O or CuO/Et_2NH system which acts as a redox catalyst.[743,744]

$$Cu^+L_n + CCl_4 \rightleftharpoons [Cu^{2+}L_nCl^-CCl_3] \rightarrow Cu^{2+}L_nCl^- + \dot{C}Cl_3$$
$$RCH{=}CH_2 + \dot{C}Cl_3 \rightarrow R\dot{C}HCH_2CCl_3$$
$$R\dot{C}HCH_2CCl_3 + Cu^{2+}L_nCl^- \rightarrow RCHClCH_2CCl_3 + Cu^{2+}L_n$$

The same reaction is catalysed by $[Mo_2(CO)_6(\eta\text{-}Cp)_2]$.[745] Cyclopropylidene radical anions are proposed as intermediates in the reduction of 1,1-dihalogenocyclopropanes.[746] Acyl radicals are generated in the reaction of acyl halides with iron pentacarbonyl.[747]

The leaving-group abilities in the cathodic reduction of benzyl, allyl, and cinnamyl derivatives, RX, are in the order: $X = Et_3\overset{+}{N} > Ph_3P > Np^{\alpha}CO_2^-$.[748]

Allyl and benzyl methyl ethers react with sodium and chlorotrimethylsilane to yield allyl and benzyl trimethylsilanes.[749]

$$PhCH_2OMe + Na \rightarrow PhCH_2OMe^{\overline{\cdot}} + Na^+$$
$$PhCH_2OMe^{\overline{\cdot}} \rightarrow PhCH_2\cdot + MeO^-$$
$$PhCH_2\cdot + Na \rightarrow PhCH_2^- + Na^+$$
$$PhCH_2^- + Me_3SiCl \rightarrow PhCH_2SiMe_3 + Cl^-$$

The observed half-wave reduction potentials of unsymmetrical benzophenones possessing an electron-donating substituent in one ring and an electron-withdrawing substituent in the second ring indicate that the benzophenone ketyl radical anions are more stable than expected as a result of capto-dative stabilization[750] (*cf.* p. 4). Reduction of mesityl oxide gives the pinacol dimer through coupling at the α-position with Ti or V or at the γ-position with Cr or Fe, depending on the site of greatest spin density in the complexed radical anions.[751] Scheme 20 depicts the detailed mechanism for the reduction of camphor by alkali metals in liquid ammonia to give borneol, isoborneol, and two isomeric pinacols, which result from formation of the ketyl radical anion followed by disproportionation, hydrogen abstraction from the medium, and dimerization.[752] In the presence of ammonium ions, the ketyl radical anion is protonated and the resulting α-hydroxyalkyl radical reduced to the carbanion and protonated; very little disproportionation occurs. Reductive coupling of pyridine-4-carboxaldehyde gives a mixture of *meso*- and $(\pm)$-pinacols; the $(\pm)$:*meso* ratio increases from 0.57 at pH 6.1 to 1.92 at pH 13.3.[753] The variation has been attributed to differences in the

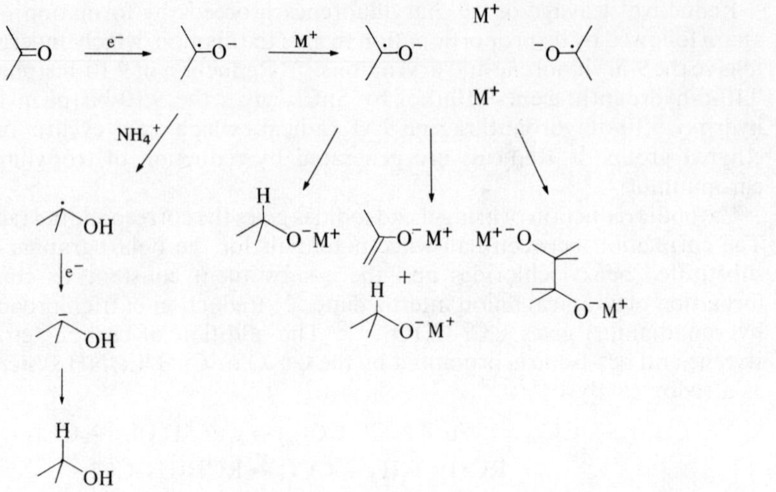

SCHEME 20

surface orientation of radical cations, radicals, and radical anions at different pH values. Complexation of ethyl cinnamate, acetophenone, benzophenone, and benzaldehyde with β-cyclodextrin results in very efficient protonation of the radical anions. The acetophenone complex gives a 1:1 mixture of the products (**119**) and (**120**), which are formed as depicted in Scheme 21.[754] The cleavage of aryl-substituted cyclopropyl ketones with zinc and zinc chloride to form acyclic ketones

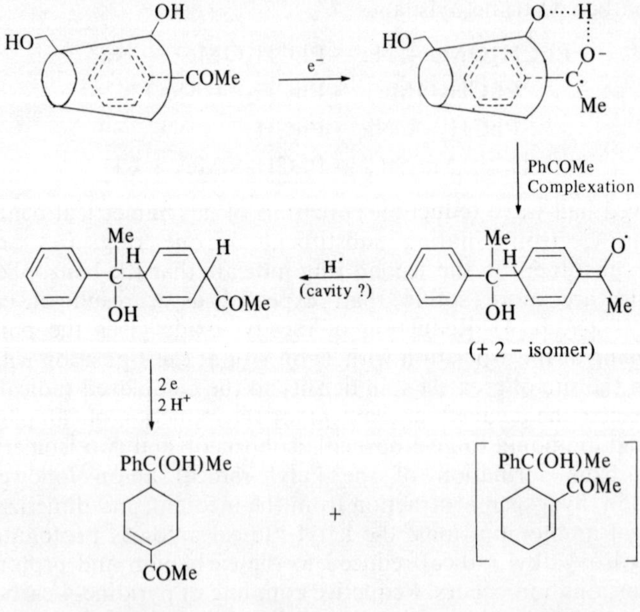

SCHEME 21

has been attributed to the initial formation of the ketyl radical anion followed by ring-opening to the acyclic radical anion;[755] it is at least feasible that the ring-opening is induced by zinc chloride acting as a Lewis acid.[756] The ketyl radical anion (**122**) obtained from reduction of the ketone (**121**) cyclizes to give the vinyl radical anion (**123**), which is either further reduced to (**124**) or to a lesser extent inverts to give (**125**) and thence (**126**).[757]

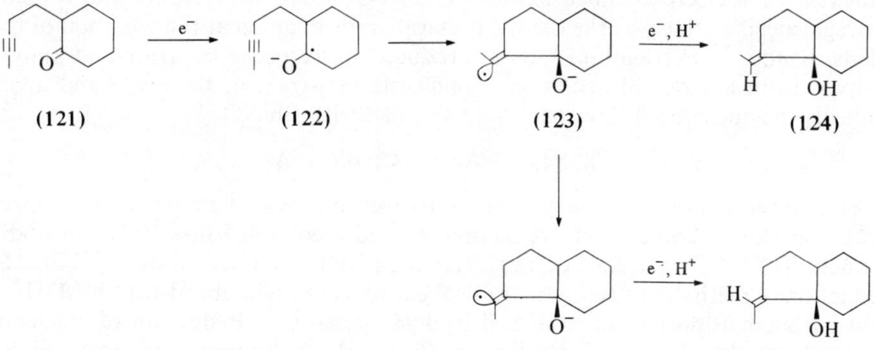

(**121**) (**122**) (**123**) (**124**) (**125**) (**126**)

Reduction of benzoyl cyanide with titanium(III) chloride gives the pinacol dimer.[758] Electro-reduction of carbon dioxide generates $CO_2^{\cdot -}$, which in presence of butadiene gives pent-3-enoic acid and hex-3-enedioic acid (Scheme 22).[759] Oxalates undergo fragmentation on cathodic reduction to give alkenes (Scheme 23).[760]

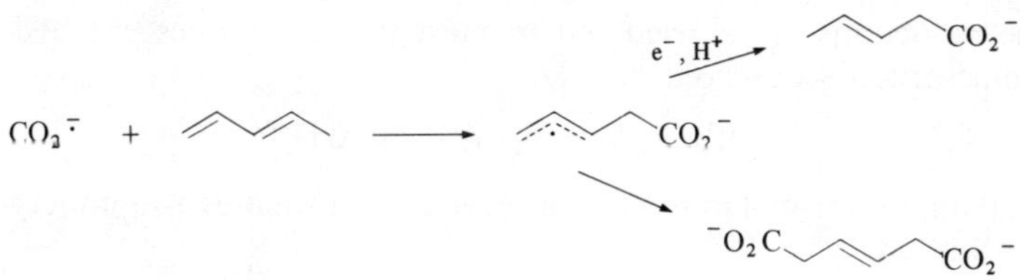

SCHEME 22

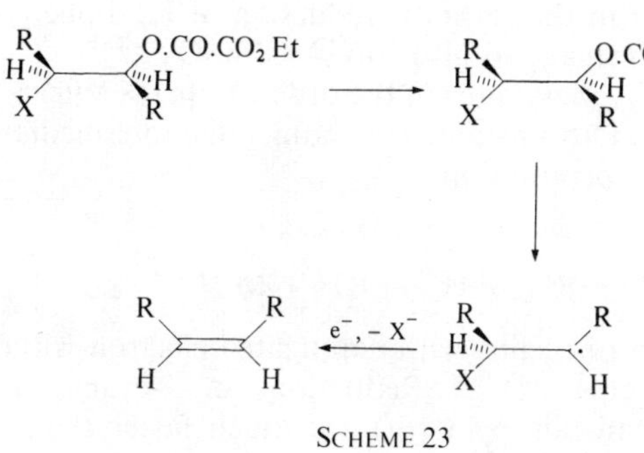

SCHEME 23

Cathodic reduction of 4,5-bis(alkylthio)-1-3-dithiol-2-ones in aprotic media followed by alkylation gives tetrathioethylenes quantitatively.[761] Electro-hydrodimerization of activated olefins, *e.g.* methyl cinnamate, involves coupling of the intermediate radical anion with the substrate and not dimerization of the radical anion when the concentration of proton donors is low;[762–764] at higher proton donor concentrations, reduction rather than electrohydrodimerization occurs.[765] Electro-hydrodimerization of cyclohex-2-enones leads to α,α-, α,γ-, and γ,γ-dimers.[766] The electrochemical addition of allyl groups to diethyl fumarate proceeds by regiospecific attack by the diethyl fumarate radical anion at the α-carbon of the allylic halide.[767] Ferricinium salts are reduced to ferrocene by azide and anions derived from a series of azoles, *e.g.* imidazole or pyrazole; the azide and azole radicals thus generated have been spin-trapped with Bu^tNO.[768–770]

$$(Cp_2Fe)^+ + Az^- \rightarrow Cp_2Fe + Az.$$

Electro-reduction of carbazoles leads to bicarbazoles.[771] Bis(tetrahydrodiazepinyls) are formed on cathodic reduction of 1,4-diazepinium salts.[772] The cathodic reduction of NAD to NADH occurs by consecutive single-electron steps.[773] Photo-reduction of methylviologen to its radical cation is brought about with NADH[774] and by haematoporphyrin catalysed by hydrogenase.[775] Reduction of lucigenin proceeds by consecutive single-electron transfers.[776] Radicals are obtained on reduction of actinomycin D and phenoxazone analogues using sodium borohydride.[777] Oxoamides are formed by reduction of bicyclic oxaziridines with iron(II) sulphate.[778] Reductions of indoxyls give the corresponding radical anions.[779]

Reduction of diaryl sulphides gives the corresponding disulphide radical anions which fragment:[780,781]

$$ArSSAr + e^- \rightarrow ArSSAr^{\cdot -} \rightarrow ArS\cdot + ArS^-$$

When the reductions are carried out in presence of a fluorescent hydrocarbon, chemiluminescence is observed.[782]

$$RS\cdot + ArH^{\cdot -} \rightarrow RS^- + ArH^*$$

The radical anions formed in reductions of trithiocarbonates also undergo scission of the S—S bond:[783,784]

$$ArSCSSR^{\cdot -} \rightarrow ArS^- + RS\dot{C}{=}S$$

The cathodic reduction of aromatic dithio-esters gives 1,2-diarylethynes.[785] The radical anions generated in the cathodic reduction of 1,2-diphenylthiiren dioxide undergoes C—S bond cleavage to give $Ph\dot{C}{=}CPhSO_2^-$.[786]

Radicals are reduced by metal ions in the order: Cr(II) > V(II) > Ti(III) > Fe(II). The reactions proceed *via* an unstable organometallic intermediate which breaks down on reaction with a proton donor.[787]

$$R\cdot + M^{n+} \rightarrow M{-}R^{n+}$$
$$M{-}R^{n+} + H^+ \rightarrow RH + M^{(n+1)+}$$

The reductive alkylation of olefins with conjugated electron-withdrawing groups has been studied (Scheme 24).[788] Reductions of α-cyano, α-acetyl, and α-methoxycarbonylalkyl radicals by Cr(II) are much faster than of simple alkyl

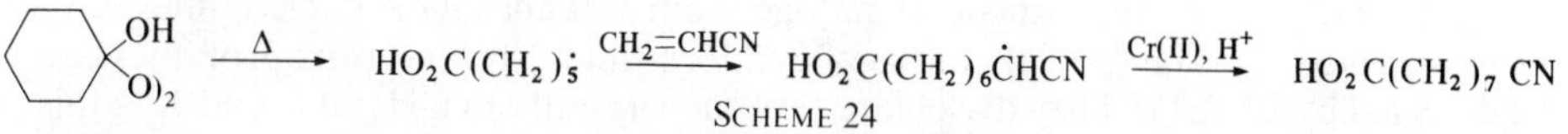

$$\text{(cyclohexyl-OH, O)}_2 \xrightarrow{\Delta} HO_2C(CH_2)_5^{\cdot} \xrightarrow{CH_2{=}CHCN} HO_2C(CH_2)_6\dot{C}HCN \xrightarrow{Cr(II),\ H^+} HO_2C(CH_2)_7CN$$

SCHEME 24

radicals. The reactions of methyl radicals with iron(II) and iron(III) deuterio-porphyrins apparently occur at the metal since they are much faster than with the iron-free porphyrin.[789]

Electron-transfer Reactions

It has been suggested that nitration of arenes may proceed by an electron-transfer mechanism *via* the ion pair $[ArH^{+\cdot}\cdot NO_2]$.[790] It has now been shown that $PhH^{+\cdot}$ reacts with NO_2 to give a σ-bonded Wheland intermediate.[791] The reaction of $NpH^{+\cdot}$ with NO_2 gives 1- and 2-nitronaphthalenes in the ratio 40:1;[792] this is different from the ratio observed in the normal nitration of naphthalene and hence it was concluded that the electron-transfer route was not operative (see also Reference 793). However, CIDNP studies provide evidence for this mechanism in the nitration of *N,N*-dimethylaniline.[794,795] The electron-transfer route has also been invoked in nitrations of phenothiazines[796] and phenoxazines.[797]

S_{RN} reactions have been reviewed.[798] The rates of reaction of *p*-nitrophenyl radicals with Cl^-, Br^-, and I^- are 2.8×10^7, 5.5×10^8, and $5 \times 10^9\ M^{-1}\ s^{-1}$, respectively; the relative rates can be rationalized in terms of initial interaction between the LUMO of the radical and the HOMO of the halide.[799] The S_{RN} reaction of halomesitylenes with $K^+NH_2^-$ and $CH_2{=}C(CH_3)O^-K^+$ leads to both $ArCH_2COCH_3$ and the reduced product ArH; the ratio of these products is independent of the nature of the leaving group, indicating that the halide has departed before the aryl radical reacts further with the enolate anion or with solvated electrons.[800] The enolate anion (**127**; R = H), but not (**127**; R = Ph), is reactive towards phenyl radicals, as in the latter case the extra delocalization of negative charge makes it a poorer electron donor.[801] The first example of a phase-transfer catalysed photo-stimulated S_{RN} carbonylation of an aryl halide using cobalt carbonyl to give a benzoic acid has been reported.[802] The carbonylation of aryl halides with $NaCo(CO)_4$ has also been reported.[803]

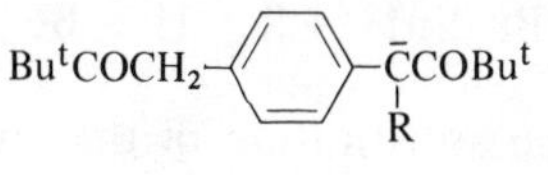

$$Bu^tCOCH_2\text{-}C_6H_4\text{-}\bar{C}(R)COBu^t$$

(127)

$$Ar\cdot + Co(CO)_4^- \rightarrow ArCo(CO)_4^{\cdot -}$$

$S_{RN}1$ reactions of haloarenes with diphenyl arsenide[804] and phenyl selenide[805] have been reported. 4-Azaindoles have been synthesized by the $S_{RN}1$ reaction of 3-amino-2-chloropyridine with ketone enolates.[806] $S_{RN}1$ reactions of other chloro heterocycles have also been studied.[807] 5-Nitrothienyl radicals, generated from reactions of 2-halo-5-nitrothiophenes with base followed by loss of halide from the radical anion, dimerize to give 2,2′-dinitro-5,5′-bithienyl.[808,809]

An electron-transfer mechanism competes with the nucleophilic substitution mechanism in reactions of 4-nitrobenzylidene chlorides with bases:[810]

$$ArCHCl_2 + HO^- \rightarrow Ar\bar{C}Cl_2 + H_2O$$
$$Ar\bar{C}Cl_2 + ArCHCl_2 \rightarrow Ar\dot{C}Cl_2 + ArCHCl_2^{\cdot -}$$

The electron-transfer reactions of 1-halogenonitroalkanes with nucleophiles have been reviewed.[811] The radical anions of 2-chloro- and 2-bromo-nitropropanes can be detected by ESR.[812] They dissociate extremely rapidly to $\dot{C}H_2NO_2$ and X^-; this is because in the staggered conformation the SOMO of the radical anion mixes with the unoccupied anti-bonding orbital associated with the C—X bond in an in-phase manner.[813] An *ab initio* study of nucleophilic substitutions by $CH_2NO_2^-$ has been carried out.[813] Electron-transfer reactions of Me_2CXNO_2 with thiolate anions,[814] and enolate anions of ketones[815], β-diketones and β-keto-esters[816] have been reported. Ion-pairing phenomena have been shown to influence markedly the relative reactivity of different anions towards $Me_2\dot{C}NO_2$.[817] The reaction of sodium enolates of β-dicarbonyl compounds with carbon tetrachloride to give α-chloro-β-dicarbonyl compounds proceed by an electron-transfer mechanism.[818] Other examples of nucleophilic substitution reactions that proceed by an electron-transfer mechanism, which have been reported, include the reactions of methyl *p*-nitrobenzenesulphonate with lithium isobutyrophenone,[819] tri-*p*-tolylsulphonium bromide with sodium isopropoxide[820] and $PhSO_2CH_2Br$ with 9-arylfluorene anions.[821]

The reduction of 2-bromo-*N*,*N*-dimethyl-2,2-diphenylacetamide with sodium methoxide proceeds by a radical anion radical-chain mechanism.[822]

$$Ph_2C(Br)CONMe_2 + MeO^- \rightarrow Ph_2C(Br)CONMe_2^{\cdot -} + CH_3O\cdot$$
$$Ph_2C(Br)CONMe_2^{\cdot -} \rightarrow Ph_2\dot{C}CONMe_2 + Br^-$$
$$Ph_2\dot{C}CONMe_2 + CH_3O^- \rightarrow Ph_2CHCONMe_2 + CH_2O^{\cdot -}$$
$$Ph_2C(Br)CONMe_2 + CH_2O^{\cdot -} \rightarrow Ph_2C(Br)CONMe_2^{\cdot -} + CH_2O$$

Dimers of $Ph_2\dot{C}CONMe_2$ are also produced in this reaction. A similar mechanism is operative in the reaction of 5-nitrofurfuryl halides with bases.[823]

Nitro groups are replaced by hydrogen in reactions of nitro compounds with Bu_3SnH.[824]

$$R_3CNO_2 + Bu_3Sn\cdot \rightarrow R_3CNO_2^{\cdot -} + Bu_3Sn^+$$
$$R_3CNO_2^{\cdot -} \rightarrow R_3C\cdot + NO_2^-$$
$$R_3C\cdot + Bu_3SnH \rightarrow R_3CH + Bu_3Sn\cdot$$

tert-Nitro compounds undergo substitution of the nitro group on reaction with

$CH_2NO_2^-$ or $CH_2SOCH_3^-$.[825]

$$RCMe_2NO_2 + CH_2NO_2^- \rightarrow RCMe_2NO_2^{\cdot -} + \dot{C}H_2NO_2$$
$$RCMe_2NO_2^{\cdot -} \rightarrow R\dot{C}Me_2 + NO_2^-$$
$$R\dot{C}Me_2 + CH_2NO_2^- \rightarrow RCMe_2CH_2NO_2^{\cdot -}$$
$$RCMe_2CH_2NO_2^{\cdot -} + RCMe_2NO_2 \rightarrow RCMe_2CH_2NO_2 + RCMe_2NO_2^{\cdot -}$$

The nitro compound (**128**) reacts with benzene thiolate with retention of configuration at high thiolate concentration but with competing inversion at lower concentrations; at high thiolate concentrations the derived radical (**129**) is trapped faster than it yields a planar radical.[826] Electron-transfer radical-chain substitution reactions of amidentate anions, *e.g.* $Me_2CNO_2^-$ or $PhSO_2^-$, give only one type of product.[827]

(128) → (129)

$$ArCMe_2NO_2 + PhSO_2Cl \rightarrow ArCMeSO_2Ph + Cl^-$$
$$90\%$$

For the above reaction, there was no evidence for the formation of the *O*-alkylated product which has been shown to rearrange to the *S*-alkylated compound under the reaction conditions.

$$ArCMe_2OSOPh^{\bar{\cdot}} \rightarrow Ar\dot{C}Me_2 + PhSO_2^- \rightarrow ArCMe_2SO_2Ph^{\bar{\cdot}}$$

The rearrangement is intermolecular and is inhibited by Bu^t_2NO and *m*-dinitrobenzene. Some *O*-alkylation occurs in the reaction of $ArCHBu^tNO_2$ with $Me_2CNO_2^-$, as evidenced by the formation of $ArCOBu^t$.

The radical anions of *vic*-dinitro compounds (**130**; $X = NO_2$)[828] and *β*-nitrosulphones (**130**; $X = SO_2R$),[828,829] which are generated from the neutral compounds by electron-transfer reactions, undergo fragmentation to yield alkenes. A very similar sequence is involved in the the cathodic reduction of acylated *vic*-nitro-alcohols (**130**; $X = OCOR$ or OSO_2R) except that the radical anion initially loses NO_2^-; the resultant radical is further reduced to the carbanion which gives the alkene as a result of loss of X^-.[830] *N*-Benzyl-1,4-dihydronicotinamide reacts with α-nitrosulphones by a one-electron transfer mechanism with resultant loss of the sulphonyl group.[831]

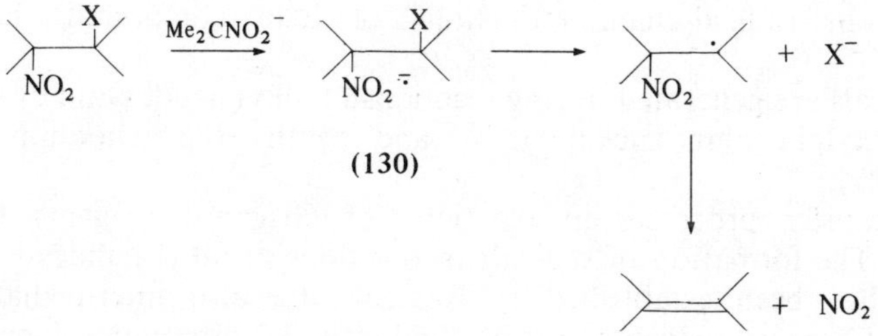

(130)

Further details of the reaction of trimethyltinsodium with alkyl halides have been published: free radicals generated in the reaction can be efficiently trapped with dicyclohexylphosphine.[832,833] Secondary and tertiary bromides react by the SET route.

$$RBr + Me_3Sn^- \rightarrow [R\cdot\ Br^-\cdot SnMe_3] \rightarrow R\cdot + Me_3Sn\cdot + Br^-$$
$$R\cdot + Me_3Sn\cdot \rightarrow Me_3SnR$$
$$R\cdot + R'_2PH \rightarrow RH + R'_2P\cdot$$

Additional evidence for the radical route comes from formation of both norbornenyl and nortricyclyl trimethyltin compounds in reactions of either 5-halonorborn-2-enes or 3-halonortricyclanes.[834] Radicals are generated upon reaction of secondary alkyl bromides with trimethylgermylsodium, as seen by the formation of cyclic products from 6-bromohept-1-ene:[835]

$$RBr + Me_3Ge^- \rightarrow RBr^{\overline{\cdot}} + Me_3Ge\cdot$$
$$RBr^{\overline{\cdot}} \rightarrow R\cdot + Br^-$$

Radicals are also produced by reaction of alkyl halides with lithium diisopropylamide and a variety of metal halides. Triarylmethyl radicals formed by reaction of Ph_3CX with Bu^tO^-,[836] metal hydrides (*e.g.* $LiAlH_4$, AlH_3, MgH_2, CpMgH),[837–841] and trialkylaluminiums can be detected by ESR.[842]

$$Ph_3CX + MH \rightarrow Ph_3CX^{\overline{\cdot}} + MH^+$$
$$Ph_3CX^{\overline{\cdot}} \rightarrow Ph_3C\cdot + X^-$$

The formation of 1,1,3-trimethylcyclopentane from reactions of 1-halo-2,2-dimethylhex-5-enes with metal hydrides[841] and $LiNPr_2$ [836] is indicative of a free-radical process yielding 2,2-dimethylhex-5-enyl radicals which cyclize to give 3,3-dimethylcyclopentylmethyl radicals. Hydrocarbon radical anions are also generated in reactions of polycyclic aromatic hydrocarbons with metal hydrides,[838,840,843,844] $LiNPr^i_2$,[836] or R_3Al;[842] these same reductants and also lithium thioalkoxides give ketyl radical anions on reaction with dimesityl ketone.[845–847]

$$ArH + MH \rightarrow ArH^{\overline{\cdot}} + MH^+$$
$$Ar_2CO + MH \rightarrow Ar_2CO^{\overline{\cdot}} + MH^+$$

ESR evidence supports a SET mechanism for the LAH reduction of secondary and tertiary alcohols to hydrocarbons.[848] The $BH_3^{\overline{\cdot}}$ radical anion has been detected in reactions of borohydride with *tert*-butoxyl radicals and is implicated as an intermediate in radical-chain reductions of halo compounds by sodium borohydride.[849]

Alkyl radicals are generated during reactions of alkyl halides and cyanides with bis(tricyclohexylphosphine)nickel(o),[850] and with the rhodium complex $Rh(dmgH)_2PPh_3$.[851]

Electron-transfer processes in reactions of Grignard reagents have been reviewed.[852] The formation of radicals in reactions of alkyl halides with atomic magnesium has been established.[853] Radicals are also intermediates in the formation of Grignard reagents from *cis*- and *trans*-1,2-dibromocyclopropanes.[854] Further details of the evidence for radical intermediates in the formation of 1,2-addition products from the reactions of 1,1- and 2,2-dimethylhex-5-enylmagnesium chlorides with benzophenone have been published.[855,856] ESR studies give direct evidence for SET in the reduction of dimesityl ketone with Grignard reagents derived from primary, secondary, and tertiary halides.[857] The ketyl radical anion of benzyl 2,4,6-trimethylphenyl ketone is formed upon reaction of the ketone with PhMgBr.[858] The reaction of aromatic carboxylic acids with Grignard reagents gives, initially, anion radicals of the carboxylate anions and subsequently ketyl radical anions.[859] The reaction of Bu^tMgCl with $ArCH{=}CHCO_2Et$ gives *inter alia* 1,3-addition products; these are formed from attack by the nucleophilic *tert*-butyl radical at the 3-position to give, initially, $Ar\dot{C}HCHBu^tCO_2Et$.[860] The carbonation of 1-tryptycyllithium with CO_2 proceeds by an electron-transfer combination mechanism.[861a]

$$\text{Trip}^-\text{Li}^+ + CO_2 \rightarrow [\text{Trip}\cdot\cdot\text{LiCO}_2] \rightarrow \text{TripCO}_2\text{Li}$$

The reaction of lithium with 1-fluoro-1-methyl-2,2-diphenylcyclopropane involves single-electron transfer from the metal into the phenyl substituent to give the cyclopropane radical anion; this undergoes ring-opening or loss of fluoride to give a cyclopropyl radical.[861b] Some organocuprates are effective electron-transfer reagents in reactions of α,β-unsaturated ketones.[862] Radical intermediates have been invoked for the formation of cyclopropanes from the copper-catalysed reaction of diethyl dibromomalonate with alkenes.[863]

Cyclopropyl radicals are generated in the Hunsdiecker reaction of silver cyclopropanecarboxylates.[864] The thallium(I) carboxylate modification of the Hunsdiecker reaction also involves radical intermediates.[865] Hexaalkylditin compounds react with *p*-benzoquinone in an electron-transfer process to give 4-trialkylstannoxyphenoxyls.[866]

$$R_6Sn_2 + C_6H_4O_2 \rightarrow [R_6Sn_2^{+\cdot}\ C_6H_4O_2^{-\cdot}] \rightarrow p\text{-}R_3SnOC_6H_4O + R_3Sn$$

3,6-Di-*tert*-butyl-*o*-benzoquinone reacts with sodium methoxide in an electron-transfer process to give, initially, the radical pair (**131**); the methoxyl radical then substitutes the *o*-semiquinone radical anion to give, eventually, (**132**).[867a] Radicals are generated in an electron-transfer reaction between triorganotin hydroxides and catechols.[867b] Homolysis of the cobalt–carbon bond occurs in the vitamin B_{12}–1,2-bis(*p*-chlorophenyl)chloroethane complex.[868] The thermally or photochemically induced reaction of ArXXAr (X = S or Se) with alkylbis(dimethylglyoximato)-pyridine cobalt(III) complexes yields RXAr.[869a]

$$\text{RCo}^{\text{III}}(\text{dmgH})_2\text{py} \rightleftharpoons \text{R}\cdot + \text{Co}^{\text{II}}(\text{dmgH})_2\text{py}$$

$$\text{R}\cdot + \text{ArXXAr} \rightarrow \text{RXAr} + \text{ArX}\cdot$$

Co—C homolysis of organocobalamins has been reported.[869b]

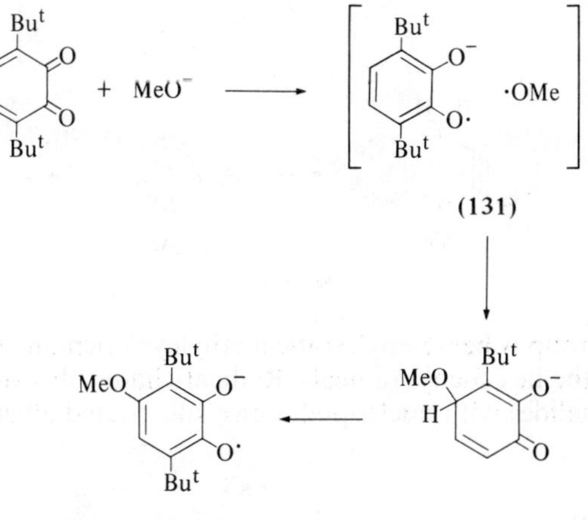

(131)

(132)

The mechanism of the hydrogen-atom transfer between the 3,3,5-trimethyl-2-morpholinon-3-yl radical (**133**) and benzils involves electron transfer followed by proton transfer; the kinetic data show a linear correlation with σ^+ for the

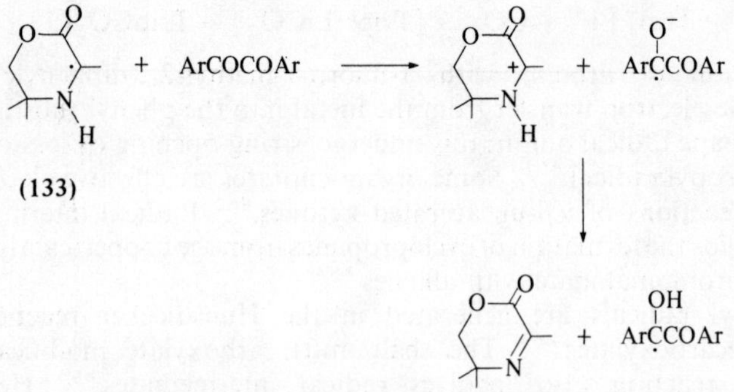

(133) + ArCOCOAr → + $ArC(O^-)COAr$ → + $ArC(OH)COAr$

substituents on the benzil.[870] Other electron-transfer reactions of this radical have also been observed. The correlation between Hammett σ constants and rates of reduction of *N*-methylacridinium ion by 1-aryl-1,4-dihydronicotinamides is consistent with an electron, proton, electron-transfer mechanism (Scheme 25).[871] An electron-transfer radical-chain mechanism is involved in the hydromercuration of alkylmercury compounds with *N*-benzyl-1,4-dihydronicotinamide:[872]

$$R\cdot + BNAH \rightarrow RH + BNA\cdot$$
$$BNA\cdot + RHgOAc \rightarrow BNA^+ + RHg\cdot + AcO^-$$
$$RHg\cdot \rightarrow R\cdot + Hg$$

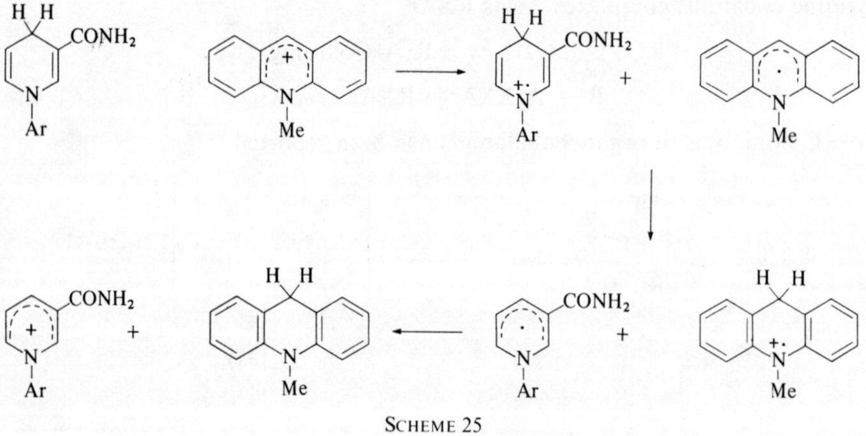

SCHEME 25

When the alkyl group is hex-5-enyl, some methylcyclopentane is formed as a result of cyclization of the hex-5-enyl radicals. Radical-chain substitution reactions of 1-alkenylmercury halides with nucleophiles give substituted alkenes (Scheme 26).[873]

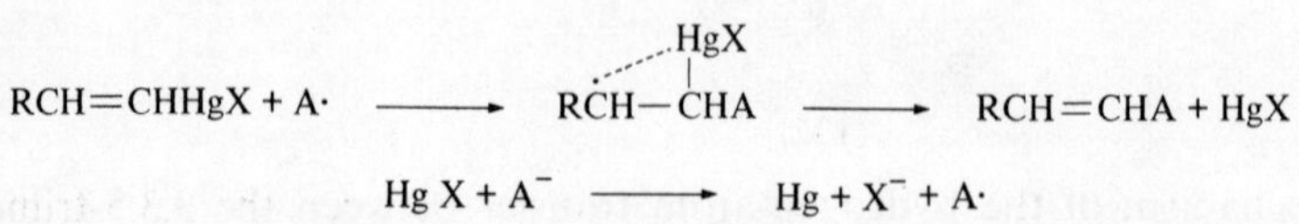

$$RCH{=}CHHgX + A\cdot \longrightarrow R\dot{C}H{-}CH(HgX)A \longrightarrow RCH{=}CHA + HgX$$

$$HgX + A^- \longrightarrow Hg + X^- + A\cdot$$

SCHEME 26

Radicals are generated in the reaction of complexed dialkylmercurys with carbon tetrachloride,[874a] and in the decomposition of 2-thiophenemercury(II) complexes.[874b]

Arylation of carbanions by iodonium salts proceeds *via* ion-pair formation, electron transfer and radical coupling.[875]

$$Ar_2I^+ + R^- \rightarrow [Ar_2I^+,R^-] \rightarrow [Ar_2I\cdot \cdot R]$$
$$[Ar_2I\cdot \cdot R] \rightarrow [Ar\cdot \cdot R] + ArI$$
$$[Ar\cdot \cdot R] \rightarrow ArR$$

Substitution of 6-phenyl-1,4-dithiafulvenes at the 6-position by bromine proceeds by an electron-transfer mechanism with radical cation intermediates.[876]

The radical cation (**134**) generated in the electron-transfer reaction of methionine with hydroxyl radicals undergoes intramolecular electron transfer to give (**135**),

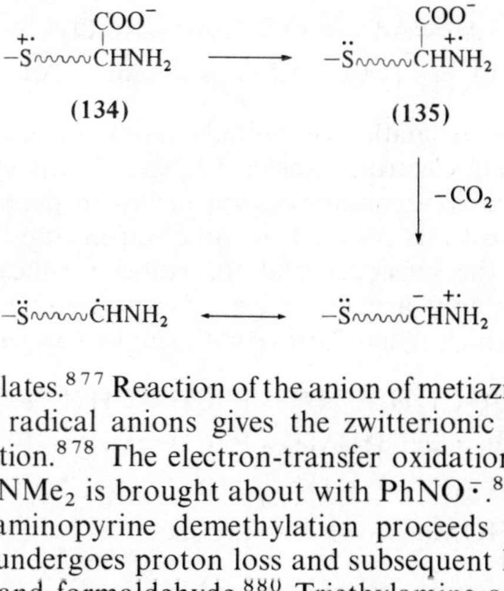

$$-\overset{+\cdot}{S}\sim\sim\sim\underset{}{\overset{COO^-}{\overset{|}{C}}}HNH_2 \longrightarrow -\ddot{S}\sim\sim\sim\overset{COO^-}{\overset{|\ +\cdot}{C}}HNH_2$$

(**134**) (**135**)

$\downarrow -CO_2$

$$-\ddot{S}\sim\sim\sim\dot{C}HNH_2 \longleftrightarrow -\ddot{S}\sim\sim\sim\overset{-}{C}H\overset{+\cdot}{N}H_2$$

which then decarboxylates.[877] Reaction of the anion of metiazinic acid, MZ^-, with a variety of radical or radical anions gives the zwitterionic radical $MZ^{\pm\cdot}$ by an electron-transfer reaction.[878] The electron-transfer oxidation of the carbanion of $PhCH_2CH(NH_2)CONMe_2$ is brought about with $PhNO^{\cdot-}$.[879] The prostaglandin synthase-dependent aminopyrine demethylation proceeds *via* the aminopyrine radical cation which undergoes proton loss and subsequent hydrolysis to give the demethylated amine and formaldehyde.[880] Triethylamine and dialkylamines are oxidized in an electron-transfer process to their respective radical cations by hexachloroacetone[881] and methyl *N,N*-dinitroamine.[882] The decomposition of diazoacetophenone is induced in an electron-transfer reaction with an aminium salt to give benzoylcarbene, which can be trapped by cyclohexene or MeOH.[883]

$$PhCOCHN_2 + Ar_3N^{+\cdot} \rightarrow [PhCO\dot{C}HN_2{}^+ Ar_3N] \rightarrow PhCOCH: + ArN + N_2$$
$$PhCOCH: + MeOH \rightarrow PhCOCH_2OMe + PhCH_2CO_2Me$$

The rates of electron transfer between $Ph_2\dot{C}OH$ and electron acceptors correlate well with various measures of the electron-withdrawing ability of these compounds.[884] A linear correlation has been obtained between Hamett σ^+ constants and the rates of reaction of galvinoxyl with aromatic amines.[885a] The rates of electron transfer between *N*-vinylcarbazole radical cations and Et_3N, Ph_2NH, and $PhNMe_2$ are 1.3×10^9, 2.0×10^9, and $3.9 \times 10^9\ M^{-1}\ s^{-1}$ respectively.[885b] Phenothiazines oxidize $Me_2\dot{C}CN$ radicals to the corresponding carbanions.[886]

Singlet oxygen reacts with electron-rich substrates to give superoxide; the yield decreases from 80% with *N,N*-dimethyl-*p*-anisidine to 0% with *N,N*-dimethylaniline.[887] Oxygen acts as an electron acceptor in the photodimerization of *p*-methoxy- and *p*-methyl-styrenes to give 1,2-diarylcyclobutanes (**136**).[888] Oxidation of adamantylideneadamantane with aminium salts in presence of oxygen gives the

$$ArCH{=}CH_2 + O_2 \xrightarrow{h\nu} ArCH{=}CH_2{}^{+\cdot} + {}^1O_2$$

$$ArCH{=}CH_2 \quad ArCH{=}CH_2{}^{+\cdot} \xrightarrow{e} \text{1,2-diarylcyclobutane (Ar, Ar)}$$

(136)

dioxetane derived from adamantylideneadamantane.[889a] The dioxetane is also obtained if the adamantylideneadamantane radical cation is generated electrochemically.[889b]

$$Ad = Ad + Ar_3N^{+\cdot} \rightarrow Ad{=}Ad^{+\cdot} + Ar_3N$$
$$Ad{=}Ad^{+\cdot} + O_2 \rightarrow [Ad{=}Ad.O_2]^{+\cdot}$$
$$[Ad{=}Ad.O_2]^{+\cdot} + Ad{=}Ad \rightarrow \text{dioxetane} + Ad{=}Ad^{+\cdot}$$

Photo-oxygenation of aromatic compounds sensitized with electron acceptors proceeds by two distinct electron-transfer mechanisms involving $O_2^{\bar{\cdot}}$ and O_2.[890]

The photo-sensitized oxygenation of aryl olefins in presence of stilbene using cyanoaromatics as sensitizers proceeds by an electron-transfer mechanism to give the radical anion of the sensitizer and the stilbene radical cation: subsequent electron transfer to a second aryl olefin, *e.g.* 1,1-diphenyl-2-methoxyethylene, leads to its radical cation which reacts further with singlet oxygen.[891]

$$PhCH{=}CH_2 + 9{,}10\text{-}C_{14}H_8(CN)_2^* \rightarrow PhCH{=}CH_2{}^{+\cdot} + 9{,}10\text{-}C_{14}H_8(CN)_2{}^{\bar{\cdot}}$$
$$PhCH{=}CH_2{}^{+\cdot} + Ph_2C{=}CHOMe \rightarrow PhCH{=}CH_2 + Ph_2C{=}CHOMe^{+\cdot}$$
$$9{,}10\text{-}C_{14}H_8(CN)_2{}^{\bar{\cdot}} + O_2 \rightarrow 9{,}10\text{-}C_{14}H_8(CN)_2 + O_2{}^{\bar{\cdot}}$$
$$Ph_2C{=}CHOMe^{+\cdot} + O_2{}^{\bar{\cdot}} \rightarrow \text{products}$$

The possible involvement of radical intermediates in the transfer of dioxygen from the 4*a*-hydroperoxyflavin anion to the 3-position of substituted indoles has been examined.[892] The tetraphenylporphyrin-sensitized photo-oxidation of thiophenolates involves attack of $^1O_2$ on ArS^- to give ArS· and $O_2{}^{\bar{\cdot}}$.[893]

The use of electron-transfer sensitization in synthetic processes has been reviewed.[894]

Photo-induced cyanomethylation of arenes has been studied using $ClCH_2CN$ as the source of $\dot{C}H_2CN$.[895] The photocyanation of arenes with cyanide occurs in the presence of electron acceptors, *e.g.* *p*-dicyanobenzene or peroxydisulphate (Scheme 27).[896,897] The arene radical cation intermediate reacts with cyanide to give a cyanocyclohexadienyl radical. Some *ipso*-substitution also occurs in the reaction of 1-methoxynaphthalene. The photo-Birch reduction of arenes with sodium borohydride in presence of *p*-dicyanobenzene as sensitizer proceeds similarly.[898]

$$ArH^{+\cdot} + BH_4{}^- \rightarrow [ArH.H]\cdot$$
$$[ArH.H]\cdot + DCNB^{\bar{\cdot}} \rightarrow [ArH.H]^- + DCNB$$
$$[ArH.H]^- + H^+ \rightarrow H.ArH.H$$

$$ArH + DCNB \xrightarrow{h\nu} ArH^{+\cdot} + DCNB^{-\cdot}$$

$$ArH^{+\cdot} + CN^- \longrightarrow [ArH.CN]\cdot$$

$$[ArH.CN]\cdot \xrightarrow{-e^-,\ -H^+} ArCN$$

SCHEME 27

Photochemical benzylation of 1,4-dicyanonaphthalene with toluene proceeds by initial electron transfer followed by *ipso*-attack of the benzyl radicals ($PhCH_3^{+\cdot} \rightarrow PhCH_2\cdot + H^+$) on the 1,4-dicyanonaphthalene radical anion.[899] Irradiation of *p*-dicyanobenzene in propan-2-ol gives isopropyl *p*-cyanobenzimidate (**137**).[900] The photo-substitution of 2-methylfuran by 1,1-diphenylethylene to give 5-(2,2-diphenylethyl)-2-methylfuran is sensitized by 1-cyanonaphthalene and proceeds by attack of $Ph_2C{=}CH_2^{+\cdot}$ on the 2-methylfuran.[901] The photo-induced cross-coupling of 5-bromo-1,3-dimethyluracil and 1,4-dimethylnaphthalene involves attack of the uracil radical on the 1,4-dimethoxynaphthalene radical cation.[902]

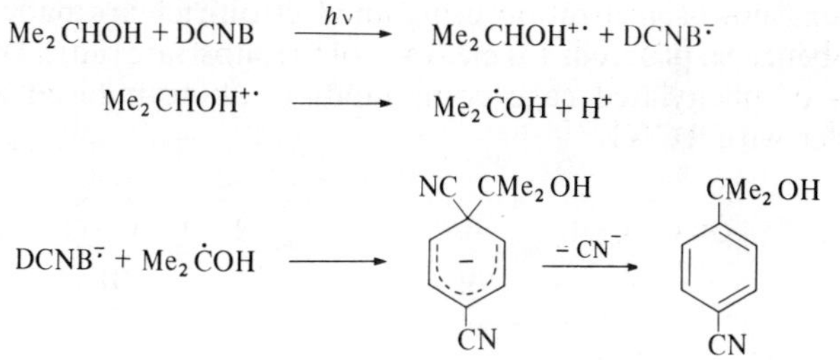

$$Me_2CHOH + DCNB \xrightarrow{h\nu} Me_2CHOH^{+\cdot} + DCNB^{-\cdot}$$

$$Me_2CHOH^{+\cdot} \longrightarrow Me_2\dot{C}OH + H^+$$

$$DCNB^{-\cdot} + Me_2\dot{C}OH \longrightarrow \text{[1-NC-1-}CMe_2OH\text{-4-CN cyclohexadienyl anion]} \xrightarrow{-CN^-} p\text{-}NC\text{-}C_6H_4\text{-}CMe_2OH$$

(**137**)

Photo-additions to alkenes sensitized by cyanoaromatics have been reviewed.[903] The photochemical addition of $MeNPr^i_2$ in acetonitrile (Scheme 28) gives the adduct derived from addition of the less substituted $Pr^i_2N\dot{C}H_2$ radical: proton loss from the intermediate radical cation is subject to stereoelectronic control.[904,905] The reaction of Me_2NCH_2R, where R is a group capable of delocalizing unpaired spin at the adjacent carbon ($R = CO_2Et$, $CH{=}CH_2$ or $C{\equiv}CH$), with stilbene proceeds in the same way in acetonitrile but in hexane the major product is that resulting from addition of $Me_2N\dot{C}HR$:[906] the extent of stilbene–amine charge transfer is much smaller in hexane than in acetonitrile and consequently there is a change from electron transfer (followed by proton transfer) to hydrogen-atom transfer in less polar solvents. The triplet state of flavins oxidizes the methylene carbon of *N*,*N*-

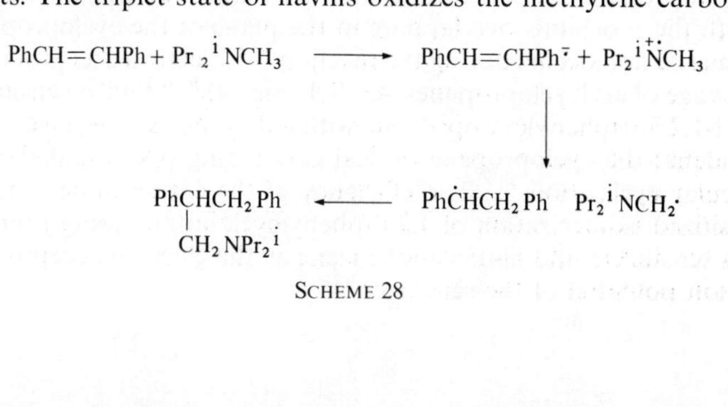

$$PhCH{=}CHPh + Pr_2^iNCH_3 \longrightarrow PhCH{=}CHPh^{-\cdot} + Pr_2^i\overset{+\cdot}{N}CH_3$$

$$\downarrow$$

$$PhCH(CH_2NPr_2^i)CH_2Ph \longleftarrow Ph\dot{C}HCH_2Ph \quad Pr_2^iNCH_2\cdot$$

SCHEME 28

dimethylglycine, whereas the flavoenzymes monoamine oxidase oxidizes the methyl carbon, possibly reflecting a change from the atom-transfer to electron-transfer mechanisms. The photochemical addition of iminium salts to alkenes gives the amino-ether (**138**) and the aminoalkene (**139**).[907] Photo-cyclization of *N*,allyliminium salts yields pyrollidones by the same mechanism.[908] Irradiation of *N*-methylphthalimides in presence of alkenes in methanol produces a radical-anion–radical-cation pair which couples and adds methanol to give the product.[909] In the aromatic hydrocarbon-photo-sensitized reactions of furans, $Ph_2C{=}CH_2$, and indene with *p*-dicyanobenzene, the donor radical cation ($D^{+\bullet}$) is captured either by the nucleophilic solvent, to give $\cdot D{-}OR$, or by a second molecule of substrate, in which case the dimeric radical cation $D_2{}^{+\bullet}$ is formed.[910] The photo-initiated dimerization of 2-quinolinecarbonitrile in acidified propan-2-ol proceeds *via* an electron-transfer process.[911] The photo-sensitized cycloaddition between alkyl vinyl ethers and aryl vinyl ethers, using *p*-dicyanobenzene as sensitizer, gives mainly the head-to-head dimers (*i.e.* 1-alkoxy-2-aryloxycyclobutanes), resulting from coupling of the two radical cations.[912] The stereospecific ring-cleavage of 1-phenoxy-1,2,2*a*,3,4,8*b*-hexahydrocyclobuta[*a*]naphthalene-8*b*-carbonitrile, by redox photo-sensitization using an electron-rich aromatic hydrocarbon and *p*-dicyanobenzene, proceeds *via* cleavage of the substrate radical cation.[913] The cycloreversion of phenylated cage compounds is photo-induced as a result of electron transfer with TCNE.[914]

$$>\overset{+}{N}{=}C< \;+\; >C{=}C{-}C{-}H \xrightarrow{h\nu} >N{-}\dot{C}< \quad >C^{+\bullet}{-}C{-}C{-}H$$

$$>N{-}\dot{C}< \quad >C^{+\bullet}{-}C{-}C{-}H \xrightarrow{MeOH,\ -H^+} >N{-}\dot{C}< \quad {-}C(OMe){-}\dot{C}{-}C{-}H \longrightarrow >N{-}C{-}C(CH){-}C{-}OMe \quad (\mathbf{138})$$

$$>N{-}\dot{C}< \quad >C{=}C{-}\dot{C}< \longrightarrow >N{-}C{-}C{-}C{=}C< \quad (\mathbf{139})$$

CIDNP and product studies on the photo-reaction of *cis*- and *trans*-diphenylcyclopropanes with chloranil indicate that a non-planar radical cation is formed with the π-orbitals overlapping in the plane of the cyclopropane ring; no photo-isomerization occurs during the reaction.[915] Electron acceptors promote the photo-cleavage of arylcyclopropanes (see Scheme 29).[916] Photo-sensitized reaction of 3-methyl-1,2,3-triphenylcyclopropene with *p*-dicyanobenzene gives 1-methyl-2,3-diphenylindene; the cyclopropene radical cation ring-opens and then undergoes intramolecular cyclization.[917] The efficiency of the cation-radical transfer in the photo-sensitized isomerization of 1,2-diphenylcyclobutane using aromatic hydrocarbons as sensitizers and *m*-dicyanobenzene as the electron acceptor depends on the oxidation potential of the sensitizer.[918]

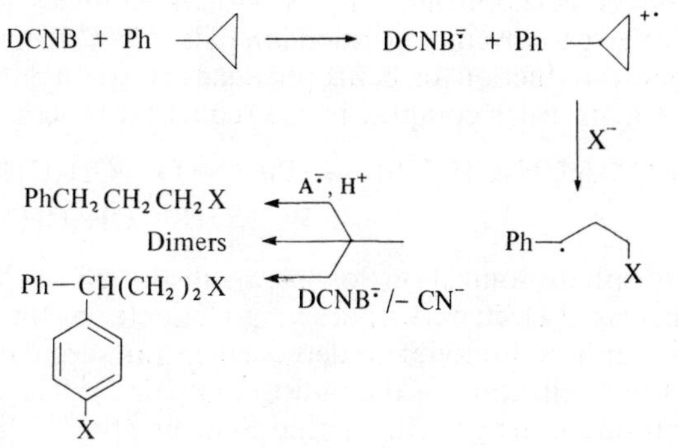

DCNB + Ph–cyclopropyl ⟶ DCNB$^{\cdot-}$ + Ph–cyclopropyl$^{+\cdot}$

X^-

$PhCH_2CH_2CH_2X$ ← $A^{\cdot}, H^+$

Dimers

Ph—CH$(CH_2)_2X$ (p-X-C$_6$H$_4$) ← DCNB$^{\cdot}$/– CN^-

Ph–$\dot{C}$HCH_2CH_2X

SCHEME 29

The radical pair generated in the photolysis of 1,2,2-tri(*p*-methoxyphenyl)vinyl bromide undergoes electron transfer to give the $An_2C{=}\overset{+}{C}An\ Br^-$ ion pair.[919] The photo-reduction of 5-bromouracils in the triplet manifold proceeds *via* an electron-transfer mechanism (Scheme 30).[920] Acetone acts as a photo-sensitizer in the coupling of 5-bromouridine to tryptophan derivatives.[921]

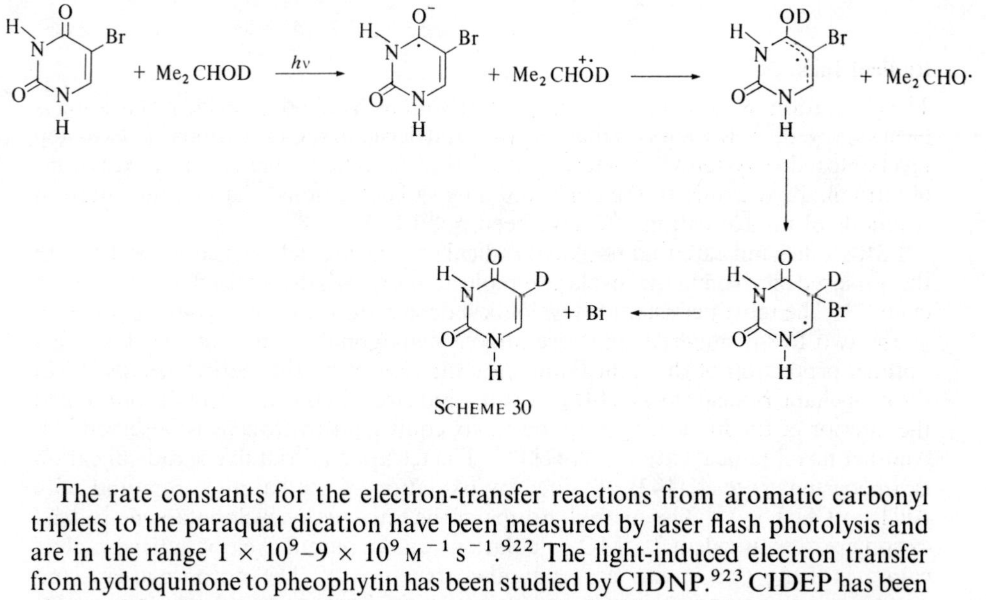

5-bromouracil + Me_2CHOD —$h\nu$→ radical anion + $Me_2CH\overset{+\cdot}{O}D$ ⟶ (OD) radical + $Me_2CHO\cdot$

↓

5-deuterouracil + Br ← (D, Br) radical

SCHEME 30

The rate constants for the electron-transfer reactions from aromatic carbonyl triplets to the paraquat dication have been measured by laser flash photolysis and are in the range 1×10^9–$9 \times 10^9\ \mathrm{M}^{-1}\ \mathrm{s}^{-1}$.[922] The light-induced electron transfer from hydroquinone to pheophytin has been studied by CIDNP.[923] CIDEP has been observed for the reaction of the chloropyll *a* radical cation and benzosemiquinone radical anion.[924] There is evidence that bacteriochloropyll *a* and bacteriopheophytin *a* can act as electron acceptors.[925] Electron-transfer reactions of quinones, hydroquinones, and methylviologen are photo-sensitized by tris(2,2′-bipyridine)-ruthenium(II).[926] Electron-transfer reactions between methylviologen and porphyrins, phthalocyanines, and their metal derivatives is effected by anaerobic irradiation of the system.[927]

Photo-reduction of benzophenone by *N,N*-dialkylanilines proceeds by rapid electron transfer to give a solvent-separated ion pair.[928,929] Picosecond absorption studies on the photo-reduction of benzophenone by triethylamine establish the existence of a charge-transfer complex in the reduction process.[930]

$$^{3}Ph_2CO^{*} + CH_3CH_2NEt_2 \rightarrow [Ph_2\dot{C}{-}O^{-}\ CH_3CH_2\overset{+\cdot}{N}Et_2]$$
$$\rightarrow Ph_2\dot{C}OH + CH_3\dot{C}HNEt_2$$

The efficiency of photo-ionization of *para*-substituted *N,N*-dimethylanilines decreases with increased electron-withdrawing character of the substituent.[931] 2-Dialkylamino-1,4-naphthoquinones undergo an intramolecular electron-transfer reaction followed by cyclization of the radical ion pair.[932] The dehalogenation of haloarenes is photo-induced by triethylamine (Scheme 31).[933] The photo-oxidation of carbazole in ethanol in presence of carbon tetrachloride yields 1- and 3-carboethoxycarbazoles; the carbon tetrachloride induces the formation of the carbazole radical cation.[934]

$$Et_3N + ArCl \xrightarrow{h\nu} Et_3N^{+\cdot} + ArCl^{\cdot}$$
$$ArCl^{\cdot} \longrightarrow Ar\cdot + Cl^{-}$$
$$Ar\cdot + Et_3N \longrightarrow ArH + CH_3\dot{C}HNEt_2$$

SCHEME 31

Radical Ions

The generation of radical cations and radical anions by use of ionizing radiation has been surveyed.[935] Reviews on the structure and decay of radical cations,[936] two-step reversible redox systems,[937] heterocyclic radical cations,[938] one-electron oxidations of tetraalkylhydrazines to the corresponding radical cations[939] and fragmentation reactions of radical cations[940] have been published.

ESR studies indicate that *n*-alkane radical cations are delocalized σ-species with the unpaired electron in the in-plane σ molecular orbitals delocalized over the entire chain.[941] The most persistent biscycloalkylidene radical cation is (**140**; $m = n = 2$) as the two five-membered rings are almost orthogonal to each other: this offers optimal protection of the radical and cationic centres by the methyl groups.[942] In the propellane radical cation (**141**) the unpaired electron is in a σ-orbital confined to the interior of the molecule; coupling to six equivalent hydrogens is observed.[943a] Another novel radical cation is (**142**).[943b] The tetraphenyltriafulvene radical cation is a persistent species (**143**) obtained by oxidation of the parent compound with $AlCl_3$ in CH_2Cl_2.[944] The methyl groups of the tetramethylcyclobutadiene radical cation are all equivalent.[945] MO calculations on the parent cyclobutadiene radical cation indicate that it has a rectangular structure.[946] The norbornadiene and quadricyclane radical cations are distinct species.[947] Arene radical cation salts, $ArH^{+\cdot}PF_6^{-}$, can be obtained by anodic oxidation of a solution of the arene in CH_2Cl_2 in presence of NH_4PF_6.[948] The main contributor to the resonance hybrid of p-$Me_2NC_6H_4CN^{+\cdot}$ is (**144**).[949] The benzidine radical cation has been reported.[950] Trialkylaluminium compounds react with aza-aromatics (az), *e.g.* pyrazine, to give persistent radical cations of the type $(Me_2Al)_2az^{+\cdot}$.[951,952] The ESR spectra of radical cations of *peri*-bridged naphthalenes indicate that the nature of the

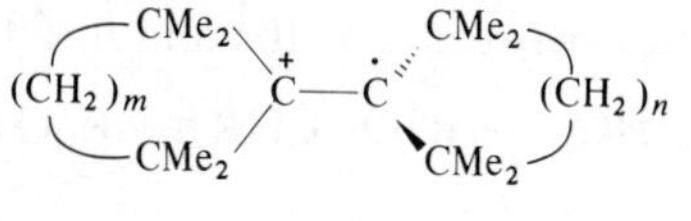

(140)

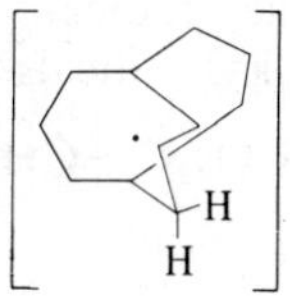

(141)

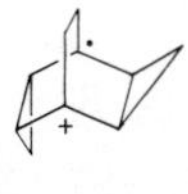

(142)

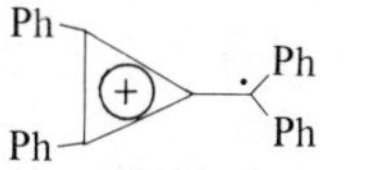

(143)

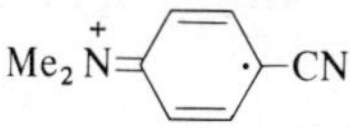

(144)

bridge dominates the spin and charge distribution in the radical cation.[953] The OH group lies in the plane of the ring in the 2,6-di-*tert*-butylphenol radical cation despite the steric interaction with the *o*-*tert*-butyl groups.[954] The ESR spectra of $Si_2Me_6^{+\cdot}$ and $Ge_2Me_6^{+\cdot}$ indicate that the unpaired electron is in a σ-bonding orbital between the metal atoms.[955] The higher value of the proton coupling constant in $Me_2S^{+\cdot}$ and $Me_2Se^{+\cdot}$ compared with $Me_2SSMe_2^{+\cdot}$ and $Me_2SeSeMe_2^{+\cdot}$ has been attributed to the anti-bonding character of the σ^* SOMO.[956] The N—S—S—N portion of the radical cation derived from Me_2N—S—S—NMe_2 is almost planar, in marked contrast to the neutral molecule.[957]

Radical cations of hydrazones, $R^1R^2\overset{+\cdot}{N}$—N=$CR^3R^4$, can possess σ, π, or mixed σ–π electronic structures depending on the nature of the substituents: the mixed σ–π electronic configuration (**145**) exists in a twisted conformation.[65] Protonation of 1,4-dihydro-1,2,4-benzotriazinyl radicals leads to the corresponding radical cations.[958] Photo-reduction of 1- and 4-azaphenanthrene in presence of phenol or DMF gives the azaphenanthrene radical cation.[959] Phenothiazine radical cations are good probes of inductive effects: the magnitude of a_N varies with the length of the polymethylene chain and the polarity of the terminal substituent as measured by σ_I.[960] Phenothiazine and phenoxazine radical cations are planar species.[961] The spin density in alloxazine and isoalloxazine radical cations is greatest at N(5).[962] The spin density distribution in the naphtho[1,8-*c*,*d*][1,2,6]thiadiazine radical cation is similar to that in the isoelectronic phenalenyl radical.[963]

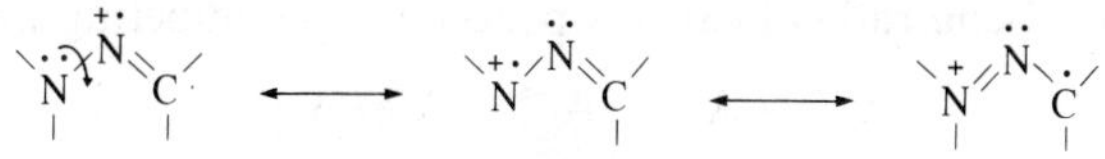

(145)

π-Cation radicals of iron(II) combined with free base isobacteriochlorines are formed as a result of electron loss from the macrocycle rather than the metal.[964] Chloroiron(III) tetraphenylporphyrin has been oxidized to its radical cation.[965]

Radical cations from 2,3-di-*tert*-butyl-2,3-diazabicyclo[2.2.1]heptane and its 7,7-spirocyclopropyl analogue decompose thermally with loss of Bu^t; this relieves the strain caused by adjacent *tert*-butyl groups.[966]

Dialkoxylalkene radical cations react with hydroxide to give mainly alkoxycarbonylalkyl radicals at rates between 2×10^8 and $6 \times 10^9\ \text{M}^{-1}\,\text{s}^{-1}$.

$$(R'O)_2\overset{+}{C}-\dot{C}HR^2 + HO^- \rightarrow R^1O\underset{\underset{O}{\|}}{C}\dot{C}HR^2 + R^1OH$$

Reaction of $HPO_4{}^{2-}$ and water give $(R^1O)_2(PO_4{}^{2-})C\dot{C}HR^2$ and $(R^1O)_2C(OH)$-$\dot{C}HR^2$ respectively.[967] Cyclic radical cations (**146**) are formed in the reaction of HO with vinyl sulphides: these radical cations then ring-open on reaction with a second molecule of the vinyl sulphide.[968]

$$HO\cdot + RSCH{=}CH_2 \longrightarrow RS\dot{C}HCH_2OH \xrightarrow{H^+} \textbf{(146)}$$

(146)

$$\textbf{(146)} \xrightarrow[H_2O]{RSCH=CH_2} RS\dot{C}HCH_2\underset{}{C}H(SR)CH_2OH$$

The three *cis/trans* isomers of hexa-2,4-diene radical cations are distinguishable, indicating that rotation about the C=C bonds is not facile. The 1,4-hexadiene radical cation isomerizes to a 2,4-hexadiene radical cation.[969] The allylbenzene radical cation isomerizes to a mixture of the propenylbenzene radical cation and the indane radical cation.[970] Low-energy molecular ions from primary and secondary aliphatic amines undergo NH/CH exchange of hydrogen atoms prior to fragmentation.[971]

$$R\overset{+\cdot}{N}HCH_2CH_2CH_2CH_2R' \rightleftharpoons R\overset{+}{N}H_2CH_2CH_2CH_2\dot{C}HR'$$

Naphthalene radical cations react with nucleophiles to give substituted naphthalenes.[972]

$$NpH^{+\cdot} + Nu^- \rightarrow [NpH.Nu]\cdot$$
$$[NpH.Nu]\cdot - e^- \rightarrow [NpH.Nu]^+ \rightarrow NpNu + H^+$$

The reaction of perylene radical cation with fluoride proceeds similarly.[973]

9,10-Diphenylanthracene and 10-phenylthiazine radical cations disproportionate.[974]

$$2\,A^{+\cdot} \rightarrow A + A^{2+}$$

Dimerization of perylene radical cations proceeds by competing processes.[975]

$$2\,A^{+\cdot} \rightarrow A_2{}^{2+};\ A^{+\cdot} + A \rightarrow A_2{}^{+\cdot}$$

The disproportionation of pyrazine radical cations has been studied.[976]

The Diels–Alder reaction of cyclohexa-1,3-diene is catalysed by $Ar_3N^{+\cdot}$; it is probable that the process involves reaction between the dienophile radical cation ($DP^{+\cdot}$) and the neutral diene (D) to give the [4 + 2]adduct radical cation ($A^{+\cdot}$).[977]

$$DP + Ar_3N^{+\cdot} \rightarrow DP^{+\cdot} + Ar_3N$$
$$DP^{+\cdot} + D \rightarrow A^{+\cdot}$$
$$A^{+\cdot} + DP \rightarrow A + DP^{+\cdot}$$

Thermal decomposition of diazodiphenylmethane, Ph_2CN_2, in presence of oxygen and sulphides gives $Ph_2CN_2^{+\cdot}$ which effects the chain decomposition of Ph_2CN_2.[978]

$$Ph_2CN_2^{+\cdot} + Ph_2CN_2 \rightarrow Ph_2\dot{C}{-}\overset{+}{C}Ph_2 + 2\,N_2$$

$$Ph_2\dot{C}{-}\overset{+}{C}Ph_2 + Ph_2CN_2 \rightarrow Ph_2C{=}CPh_2 + Ph_2CN_2^{+\cdot}$$

Phenazine, phenothiazine, and tetraphenylbenzidine radical cations reduce triaryl aminium radical cations to give Ar_3N and the donor dication.[979]

Ab initio MO calculations suggest that $CH_3F^{\overline{\cdot}}$ and $CH_3Cl^{\overline{\cdot}}$ are loose C_{3v} complexes between $CH_3\cdot$ and X^-, and that $SiH_3F^{\overline{\cdot}}$ and $SiH_3Cl^{\overline{\cdot}}$ are T-shaped complexes with one unique hydrogen atom.[980] The radical anions $SiH_3Br^{\overline{\cdot}}$, $MeSiH_2Br^{\overline{\cdot}}$, and $Me_2SiHBr^{\overline{\cdot}}$ have trigonal bipyramidal structures with bromine and hydrogen in the axial positions: the corresponding iodo species have C_{3v} symmetry.[981] $EtSiH_2Br^{\overline{\cdot}}$ and $CH_2{=}CHSiH_2Br^{\overline{\cdot}}$ also have trigonal bipyramidal structures with equatorial ethyl and vinyl groups respectively: the vinyl group is equatorial as its π-system serves as an electron donor.[982] ESR studies indicate that $CH_2CF_2^{\overline{\cdot}}$ has perpendicular geometry with the unpaired electron in an orbital which can be designated as σ^* for the CF_2 group and π for the CH_2 group.[983] Both carbons of $C_2F_4^{\overline{\cdot}}$ have pyramidal geometry.[984] The radical anion of tetraneopentyldiborane is remarkably persistent ($t_{1/2}$ = 15 min at 140°): the *tert*-butyl groups shield the borons from above and below.[985] An ESR study on the fulvene radical anion has been published.[986] An ELDOR study indicates that $Ph{-}Ph^{\overline{\cdot}}$ is planar.[987] The hyperfine splitting constants of highly twisted phenyl-substituted radical anions, *e.g.* 9,10-di-*o*-tolylanthracene, indicates that there is significant σ–π mixing, *i.e.* the phenyl group can exert a hyperconjugative effect.[988] ESR studies of 5,5′- and 6,6′-bisazulenyl radical anions indicate that in the latter but not the former there is a strong interaction between the two azulene π-systems.[989] The unpaired electron in [2](1,3)-azuleno[2]paracyclophane is mainly in the azulene ring.[990] The radical anion of $[2_4]$(1,2,4,5)cyclophane is present as the free ion in DMF and as a tight ion pair in THF.[991] ESR studies of thiaheterohelicene radical anions indicate that conjugation determines the spin distribution.[992] The spin and charge densities in the pentafluoropyridine radical anions are mainly at the 1- and 4-positions: coupling occurs at the 4-position.[993] The spin density in pyridine-*N*-oxide radical anions is greatest at the 2- and 4-positions.[994] The spin density distribution in the 3,3′-bipyridine[995] and benzo[*c*]cinnoline[996] radical anions has been determined.

The σ^* orbital containing the unpaired electron in $C_6F_5X^{\overline{\cdot}}$ (X = Cl, Br, I) is largely confined to the C—X bond, though for the *ortho*-disubstituted species there is some delocalization into the neighbouring C—X bond.[997] π-Conjugative interactions are the dominant factor in determining the preferred configuration of $PhOH^{\overline{\cdot}}$ and $PhCN^{\overline{\cdot}}$, but for $PhMe^{\overline{\cdot}}$ inductive effects are more important.[998] The spin density at the *para*-position of $ArSiR_{3-n}(OR)_n^{\overline{\cdot}}$ increases with n.[999] The spin density distributions in trimethylsilyl-substituted naphthalenes, anthracenes, and phenanthrenes,[1000] $PhNO_2^{\overline{\cdot}}$,[1001] and dimethylphosphine-substituted benzenes[1002] have been obtained. Reduction of ArCN and $ArNO_2$ with alkali metals in presence of $NaBPh_4$ gives radical ion pairs, which are in equilibrium with triple ions $Na^+ A^{\overline{\cdot}} Na^+$.[1003]

The radical anions $CH_3\dot{C}(O^-)OH$ and $CH_3\dot{C}(O^-)NH_2$ are non-planar even though the parent molecules are planar: this non-planarity results in $a(\beta\text{-}CH_3)$

decreasing from 23.0 G, for a planar species, to 13.0 G.[1004] There is a linear correlation between the rate of disproportionation of $Ar_2\dot{C}O^-$ and the spin density on the carbonyl carbon.[1005] There is free rotation about the *C*-phenyl bond in the radical anions of α-,β-, and γ-benzoylpyridines but only about the *C*-pyridine bond for γ-benzoylpyridine.[1006] There is spin delocalization throughout the whole system in the radical anions of benzophenone–chromium tricarbonyl complexes.[1007] The influence of substituent, solvent, and steric effects on the extent of intramolecular hydrogen bonding in semiquinone and semidione radical anions has been investigated.[1008] Semidione radicals R—C(O·)=C(O⁻)R have been obtained for R = 2-pyridyl, 2-quinolinyl, and 2-benzthiazolyl.[1009] INDO calculations indicate that the radical anions derived from ascorbic acid and α-hydroxytetronic acid have the structure (**147**); R = $CH(OH)CH_2OH$ and H respectively.[1010] The semi-

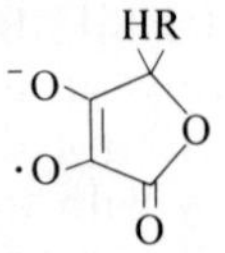

(147)

quinone radical anions derived from 3,4-dihydroxyphenylalanine form metal chelates with divalent metal ions.[1011] There is a simple correlation between the hyperfine splittings in semiquinones derived from naturally occurring methoxybenzoquinones and related alkyl aryl ether radical cations.[1012] Semiquinone radical anions are formed by UV photolysis of melanin precursors.[1013] ESR spectra of hydroxyanthrasemiquinones have been obtained.[1014] Triple ions are formed on reduction of *o*-diaroylbenzenes with group(III) metals[1015] and 1,2,4,5-tetramesitoylbenzene with sodium.[1016] ESR evidence suggests that the energy barrier for electron migration between quinone rings in tryptycene bis-quinone monoanion radicals is greater than in the trisanion radical.[1017] The relative spin density at carbon and sulphur in thioketyl radical anions $[R_2\dot{C}—S^- \leftrightarrow R_2\bar{C}—S\cdot]$ is 3:1; the spin density for selenoketyl radical anions is distributed similarly.[1018]

The influence of crown ethers on Li^+ fluorene$^{\cdot-}$,[1019,1020] Na^+ dihydrotetracene$^{\cdot-}$,[1021] and Na^+ and K^+ 9-fluorenone$^{\cdot-}$[1022] ion pairs have been studied: the crown ether affects the rate of migration of the cations and also the rate of decay of these systems. Addition of 18-crown-6 to a solution of Et_3GeK and benzophenone or an aromatic hydrocarbon gives $Ph_2CO^{\cdot-}$ or $ArH^{\cdot-}$, respectively.[1023]

The stability of alkyl halide radical anions $RX^{\cdot-}$ has been discussed:[1024] these radical anions can be formed even when the configuration and hybridization of the completely dissociated radical differs from that in $RX^{\cdot-}$.[1025,1026] The rates of dehalogenation of radical anions of cyanobenzyl halides and halobenzonitriles have been measured. The rates of halide loss from halobenzonitriles range from 10^4 to $>10^7\ s^{-1}$: the rates decrease in the order: Br > Cl > F and *ortho* > *para* ≫ *meta*. The absolute rates are about 10^5 faster than for the corresponding nitro compounds.[1027] Halogen-substituted acetophenone radical anions undergo intramolecular transfer and halide loss.[1028]

$$XC_6H_4\dot{C}(O^-)CH_3 \rightarrow \cdot C_6H_4COCH_3 + X^-$$

The rates of bromide loss from *m*- and *p*-$BrCH_2COC_6H_4NO_2^{\cdot -}$ are 1.5×10^2 and $4.1 \times 10^4\,s^{-1}$, respectively; this is much faster than the rate of decay of *p*-$BrCH_2CH_2C_6H_4NO_2^{\cdot -}$ ($k \sim 0.2\,s^{-1}$). The nitroaromatic radical anions allow sufficient electron density to reach the halide through a CH_2 or CH_2CO group to bring about dehalogenation: the rate depends on the nature of the halogen and its position relative to the nitro group.[1029] The radical anions (**148**; $n = 3$ or 4) undergo

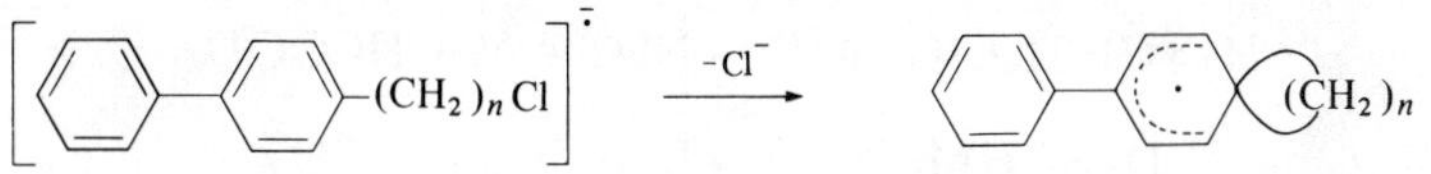

(148)

intramolecular carbanionic displacement by the biphenyl radical anion on the chloride to give the corresponding cyclohexadienyl radical.[1030] The radical anion $BrCH_2CO_2H^{\cdot -}$ loses bromide to give $\dot{C}H_2COOH$.[1031] Isopropylthiobenzenes react with sodium to give their radical anions which fragment to give arene thilates.[1032]

$$ArSCHMe_2^{\cdot -} \rightarrow ArS^- + Me_2\dot{C}H$$

On treatment with alkali metals aryl-substituted cyclopropanes rearrange *via* trimethylene radical anion intermediates.[1033] 6*b*,7*a*-Dihydro-7*H*-cycloprop[*a*]-acenaphthylene radical anion (**149**) undergoes disrotatory ring-opening to give (**150**);[1034] (**151**) which cannot undergo disrotatory ring-opening is longer-lived. 4,4′-Disubstituted diphenylsulphone radical anions cyclize to give 3,6-disubstituted dibenzothiophene sulphone radical anions.[1035] Anthracene radical anions react with phenol by initial protonation and then electron transfer to give the anion followed by a second protonation.[1036] Anthracene radical anions with electron-withdrawing substituents dimerize reversibly to give dianion dimers.[1037] CIDNP studies indicate that the aromatic hydrocarbon-sensitized photo-isomerization of stilbenes proceeds by a radical anion mechanism.[1038,1039] *Cis–trans* isomerization of ferrocenylcyanoethylenes is induced by electron transfer with cyclooctatetraene dianion.[1040] A radical anion intermediate is involved in the *cis–trans* isomerization of 3-(5-nitro-2-furyl)-2-(2-furyl)acrylamide.[1041]

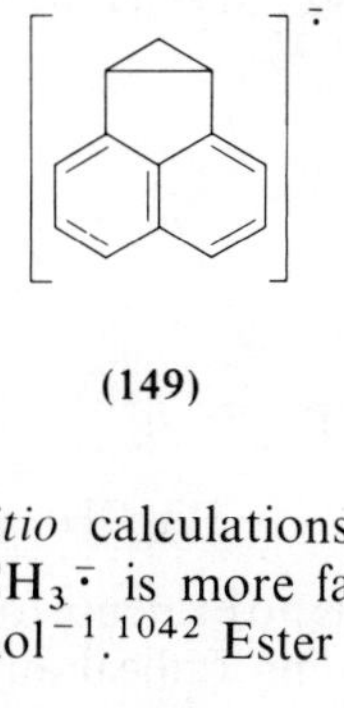

(149) 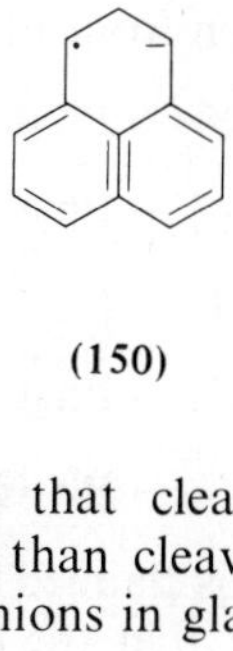 (150) 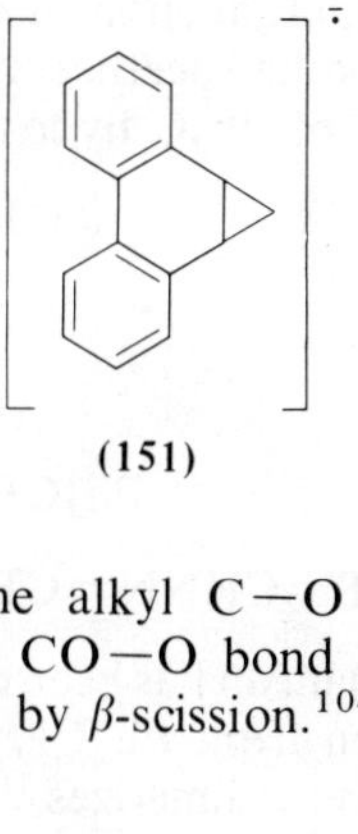 (151)

Ab initio calculations indicate that cleavage of the alkyl C—O bond in $HCOOCH_3^{\cdot -}$ is more favourable than cleavage of the CO—O bond by about $40\,kcal\,mol^{-1}$.[1042] Ester radical anions in glasses decay by β-scission.[1043]

$$R\dot{C}O(O^-)R' \rightarrow RCO_2^- + R'\cdot$$

Dialkyl carbonate radical anions also undergo β-scission.[1044]

$$(RO)_2\dot{C}-O^- \rightarrow R\cdot + ROC(=O)O^-$$

The α-^{13}C coupling constant indicates that the dialkyl carbonate radical anion is pyramidal at carbon and that there is appreciable spin delocalization onto the oxygen. Amino acid radical anions deaminate, whereas the radical cations decarboxylate.[1045]

$$H_3\overset{+}{N}CHMe\dot{C}(O^-)OD \rightarrow NH_3 + Me\dot{C}HCOOD$$

$$D_2\overset{+\cdot}{N}CHMeCO_2^- \rightarrow D_2N\dot{C}HMe + CO_2$$

There is no evidence that deuterated radical anions of carboxylic acids decay to give acyl radicals;[1046] the results suggest that abstraction of hydrogen from the parent acid occurs:

$$RCH_2\dot{C}(OD)_2 + RCH_2COOD \rightarrow RCH_2CH(OD)_2\ (\rightarrow RCH_2CHO) + R\dot{C}HCOOD$$

The half-lives of bicyclo[3.1.0]hex-3-en-2-one radical anions are very short except when the compound has a 4-phenyl group.[1047] The disproportionation of ascorbate radical anions proceeds *via* initial dimerization to the dimeric dianion.[1048]

$$2A^{\cdot-} \rightarrow A_2^{2-}$$

$$H^+ + A_2^{2-} \rightarrow HA^- + A$$

6-Halogeno-2*H*,3*H*-benzo[*b*]thiophene-2,3-semidiones undergo C(6)—O coupling followed by loss of halide: the spin density in the semidione is highest at C(6).[1049] The corresponding 5-halogeno semidiones do not dimerize. Maleimide radical anions dimerize to give 1,2,3,4-butanetetracarboxylic-1,2:3,4-diimide..[1050]

The phenylnitrene radical anion adds to carbonyl compounds to give radical anions which then fragment: the preferred route leads to the more stable radical (Scheme 32).[1051] The phenylnitrene radical anion shows both radical and anionic character in its reaction with trifluoroacetone.[1052]

$$PhN^{\cdot-} + CH_3COCF_3 \rightarrow Ph\dot{N}H + CF_3COCH_2^-$$

$$PhN^{\cdot-} + CH_3COCF_3 \rightarrow Ph\bar{N}COCF_3 + CH_3\cdot$$

Cathodic reduction of diazodiphenylmethane gives benzophenone azine as the major product together with diphenylmethane: the proposed mechanism involves formation of Ph_2C by loss of nitrogen from $Ph_2CN_2^-$.[1053]

$$Ph_2CN_2 + e^- \rightarrow Ph_2CN_2^{\cdot-}$$

$$Ph_2CN_2^{\cdot-} \rightarrow Ph_2C^{\cdot-} + N_2$$

$$Ph_2C^{\cdot-} + SH \rightarrow Ph_2CH^- + S\cdot$$

$$Ph_2CH^{\cdot-} + PhCN_2 \rightarrow Ph_2CH\bar{N}N{=}CPh_2$$

$$Ph_2CH\bar{N}N{=}CPh_2 + Ph_2C^{\cdot-} \rightarrow Ph_2CN{=}NCPh_2 + Ph_2CH\cdot$$

This mechanism has been disputed: it is claimed that there is no evidence for the loss of nitrogen from $Ph_2CN_2^{\cdot-}$ [1054] and that the corresponding radical anion from diazofluorene dimerizes.[1055]

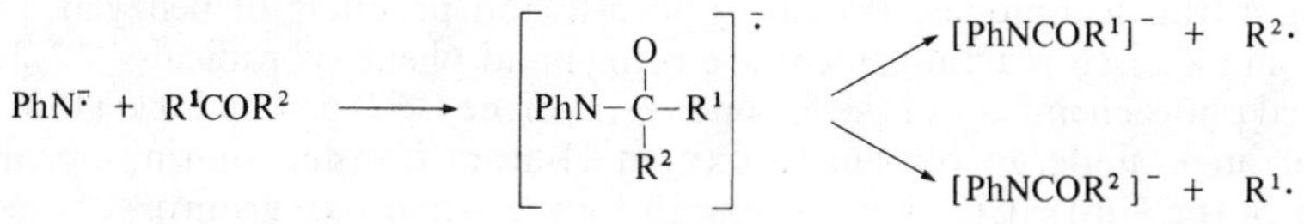

$$\mathrm{PhN^{\bar{\cdot}} + R^1COR^2 \longrightarrow \left[PhN{-}\underset{R^2}{\overset{O}{C}}{-}R^1 \right]^{\bar{\cdot}} \longrightarrow [PhNCOR^1]^- + R^2\cdot \text{ or } [PhNCOR^2]^- + R^1\cdot}$$

SCHEME 32

The radical anion intermediates formed in the autoxidation of catecholamines form complexes with triorganotin or diorganothallium cations thus enabling their detection by ESR spectroscopy.[1056] Radical anions of pentacarbonylmetal complexes are formed on reaction of pyrazine radical anions with $Mo(CO)_6$ and $W(CO)_6$.[1057] Electron transfer occurs in reactions of cyclooctatetraene radical anions with substituted cyclooctatetraenes and [16]annulene.[1058]

Oxidation of hydrophenazines, reduced flavins, and related compounds with superoxide proceeds by a hydrogen-atom transfer mechanism and not by electron transfer: the reaction with *N*-methyl-*N*′-hydrophenazine does not give the radical cation, $CH_3PhenH^{+\cdot}$.[1059]

$$FlH_2 + O_2^{\bar{\cdot}} \rightarrow Fl + HO\cdot + HO^-$$
$$CH_3PhenH^{+\cdot} + O_2^{\bar{\cdot}} \rightarrow CH_3Phen\cdot + HO_2^-$$

Alkyl peroxide anions are formed in reactions of $O_2^{\bar{\cdot}}$ with methylviologen[1060] and with N^5-ethyl-3-methyllumiflavin.[1061] The reaction of aromatic halides containing electron-withdrawing substituents with $O_2^{\bar{\cdot}}$ involves nucleophilic attack by $O_2^{\bar{\cdot}}$ and does not proceed by an electron-transfer mechanism.[1062] The reaction of *N*-benzylamides with $O_2^{\bar{\cdot}}$ in presence of crown ethers gives *ortho*- and *para*-hydroxylated products (*cf*. Scheme 33).[1063]

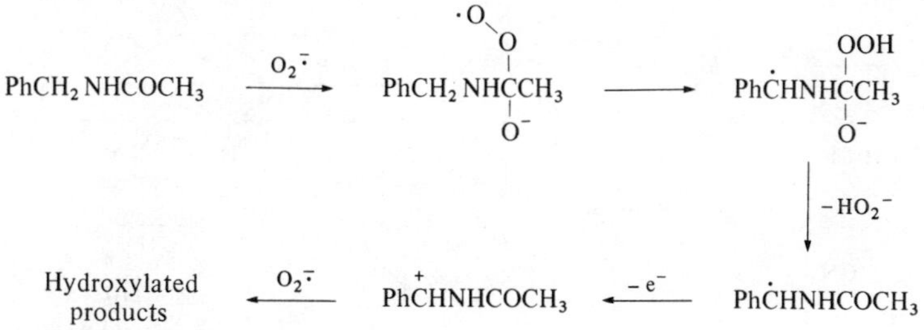

$$\mathrm{PhCH_2NHCOCH_3 \xrightarrow{O_2^{\bar{\cdot}}} PhCH_2NH\underset{O^-}{\overset{O-O\cdot}{C}}CH_3 \longrightarrow Ph\dot{C}HNH\underset{O^-}{\overset{OOH}{C}}CH_3 \xrightarrow{-HO_2^-} Ph\dot{C}HNHCOCH_3 \xrightarrow{-e^-} Ph\overset{+}{C}HNHCOCH_3 \xrightarrow{O_2^{\bar{\cdot}}} \text{Hydroxylated products}}$$

SCHEME 33

Photolysis

Fragmentations

The photochemistry and photo-cyclization of aryl halides has been reviewed.[1064] Vibrationally excited $Pr^n\cdot$ radicals, produced on photolysis of iodopropane, decompose mainly to Me· radical and ethylene.[1065] Photolysis of aliphatic dicarboxylic acids in solution results mainly in α-cleavage and photo-reduction by solvent. Emission ESR spectra found for radicals formed by C—C(=O) bond cleavage of α-dimethyl-substituted acids indicate a triplet precursor.[1066] First results of a systematic study of the photolysis of di-,[1067] tri-, and poly-peptides and aromatic amino acids[1068] have covered trends in decarboxylation and deamination using spin-trap techniques. Primary photo-fission products of benzylaryl ethers undergoing Claisen rearrangement are benzyl and phenoxyl radicals.[1069] Triplet-sensitized photochemistry of the 5,6-epoxy-1,3-diene (**152**) proceeds *via* (**153**) which can rearrange, undergo oxygen to oxygen H-atom transfer, or ring-opening by scission of the C(6)—C(7) bond (induced by the 7-hydroxy group).[1070]

The availability of secondary hydrogens is more important than the presence of silicon in the vacuum UV photolysis of $EtMe_3Si$.[1071] Cleavage of triplet Ph_3P appears to involve a hitherto unknown type of disproportionation: cleavage of *ortho*-hydrogen from phenyl radical to form dehydrobenzene.[1072] Rates of reaction of alkyl radicals scavenged by metal ions during photolysis of alkylcobaloximes have been determined.[1073] The imidyl radical (**154**) has been detected on flash photolysis of *N*-bromo-3,3-dimethylglutarimide in CCl_4.[1074] C—S bond cleavage occurs on photolysis of penicillamine disulphide and related compounds in solution, especially when the carbon-centred radical is stabilized by alkyl or aryl substituents.[1075]

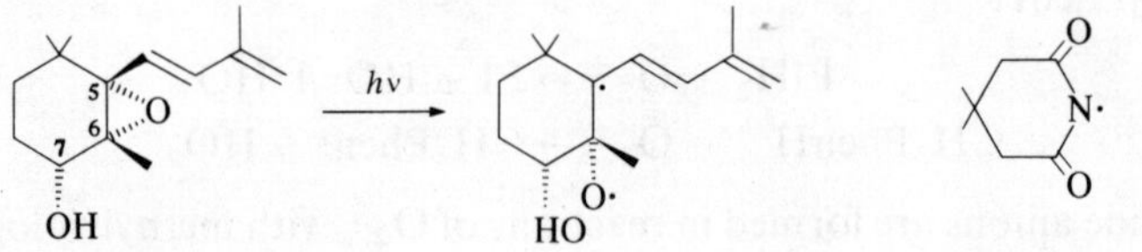

(152) (153) (154)

Benzisothiazoles react photochemically with dimethyl acetylenedicarboxylate as shown in Scheme 34.[1076]

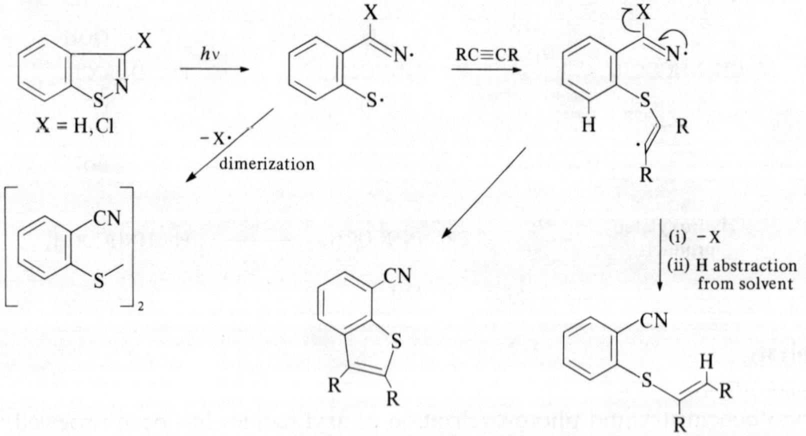

SCHEME 34

Hydrogen-atom Transfers

Photo-oxidation of the carboxylic acids RCO_2H (R = H, Me, Et) with aqueous H_2O_2 is initiated by hydroxyl radical abstraction of hydrogen.[1077] Et_2NOH is more efficient than $C_6H_{13}SH$ as a quencher of $n\pi^*$ excited carbonyl compounds but the latter traps 1,4-biradicals more efficiently.[1078] The contribution of allylic hydrogen atom abstraction to the total quenching of triplet benzophenone by olefins and dienes ranges from 0 for conjugated dienes to 94% for 1,4-cyclohexadiene. The hydrogen abstraction process has a higher activation energy than quenching by the double bond, hence its importance rises with increase in temperature.[1079] Acetone-sensitized triplet π-excitation of *syn*-sesquinorbornene (**155**) pushes the four ethylene bridges back in the opposite sense to that in the ground state; this leads to hydrogenations on the "hindered" *endo*-face by hydrogen atom abstractions from acetone by excited (**155**).[1080]

Photo-reduction of benzil by Et_3N does not involve a long-lived triplet exciplex but rather hydrogen abstraction followed by a very rapid carbonyl group reduction ($k = 1.8 \times 10^9\ \text{M}^{-1}\,\text{s}^{-1}$, 22°) by α-aminoalkyl radicals.[1081] Differences in rates of abstraction (with subsequent cyclization to cyclopropyl thiols) at the β-position of the λ-aralkyl thione (**156**) have been measured for two excited states, $S_2(\pi,\pi^*)$ and $S_1(n,\pi^*)$: hydrogen atom abstraction of S_1 shows a large ($k_H/k_D > 17$) kinetic isotope effect while a small inverse k.i.e. for S_2 implicates a biradical intermediate.[1082] Intramolecular hydrogen atom transfer in a β,γ-unsaturated amide, *via* a seven-membered transition state, forms biradical (**157**), which fails to undergo reverse hydrogen transfer but cyclizes exclusively to yield a pyrrolidinone.[1083]

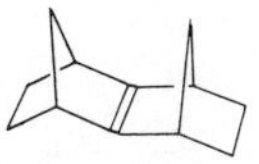

(155) 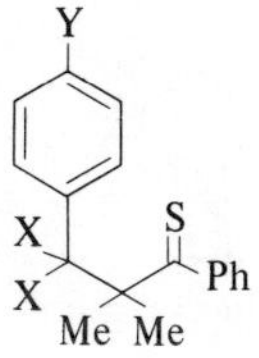 **(156)** **(157)**

X=H, D

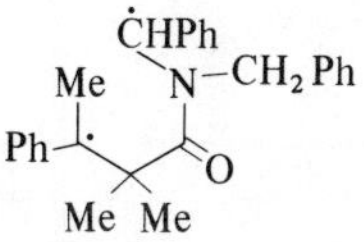

Carbonyl Compounds

Experimental support exists for the Wagner proposal that radical cleavage of the carbon–halogen bond in α-halocyclohexanones occurs from the triplet state.[1084] A radical is highly unlikely to be the intermediate in the *o*-nitrobenzaldehyde photo-rearrangement.[1085] The mechanisms of the photo-decarboxylation of α-keto-acids and of their photo-epoxidation of $PhMeC{=}CH_2$ have been elucidated.[1086] The rates of recombination of (**158**) radicals, obtained by laser flash photolysis, decrease under high pressure being limited by translational diffusion.[1087] Secondary *O*-benzoylbenzamides (**159**; $R^1 = H$) are photochemically inert *via* reversible cyclization to (**160**). Tertiary compounds, *e.g.* (**159**; $R^1 = R^2 = Et, Bu^n$), are deactivated, not by intramolecular hydrogen atom abstraction but rather by cleavage of the amide C—N bond. The radical thus formed cyclizes to yield a 3-phenylphthalid-3-yl radical which either captures hydrogen or couples.[1088] Photo-cyclizations of *N*-phenacyl and *N*-acetonyl-lactams afford 1-azabicyclo-(*n*.2.0.)alkanes in good yields.[1089] Photolysis of benzophenone hydrazones in CCl_4 in the presence of O_2 gives the corresponding azine, by N—N bond cleavage, as the major product together with the hydrazone hydrochloride. The *N*-acetyl derivatives yield a photo-labile hydroperoxide which yields benzophenone.[1090]

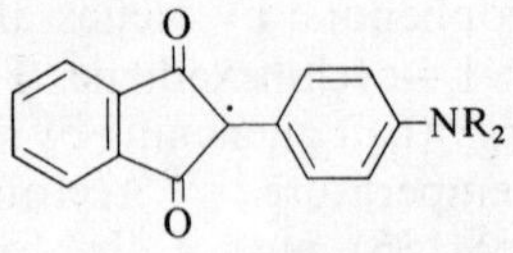

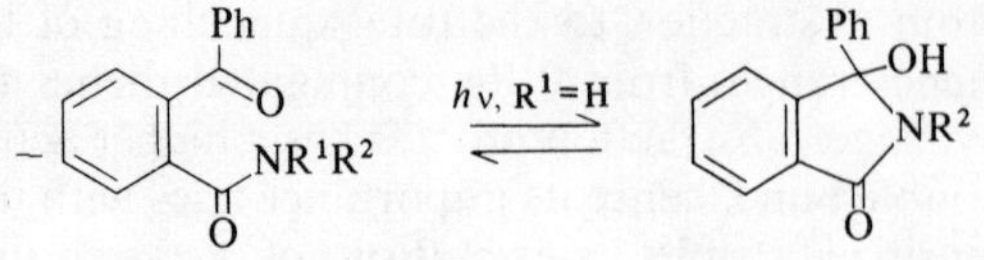

(158)

R = Me, Ph

(159)

(160)

R^2 = Et, Pr^i, PhCN

Internal radical addition to an ester group provides a new route to chromones from furanones (**161**).[1091] Triplet excited 2-arenesulphonamido-2-cyclohexenones desulphonate with high efficiency with concurrent intramolecular migration of the aryl substituent.[1092]

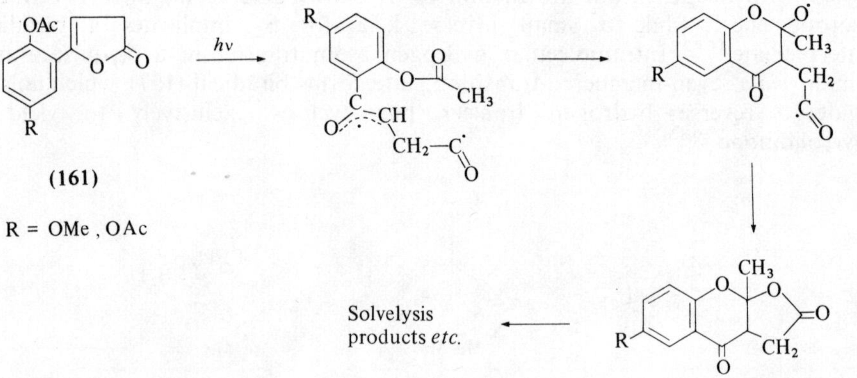

(161)

R = OMe, OAc

Solvelysis products *etc.*

The oxetan (**162**), obtained by photo-addition of $MeCOSiMe_3$ to *trans*-$MeO_2CCH{=}CHCO_2Me$, opens photolytically to the 1,4-biradical (**163**) which rearranges to the 1,3-biradical (**164**) by migration of the silicon atom before closure.[1093]

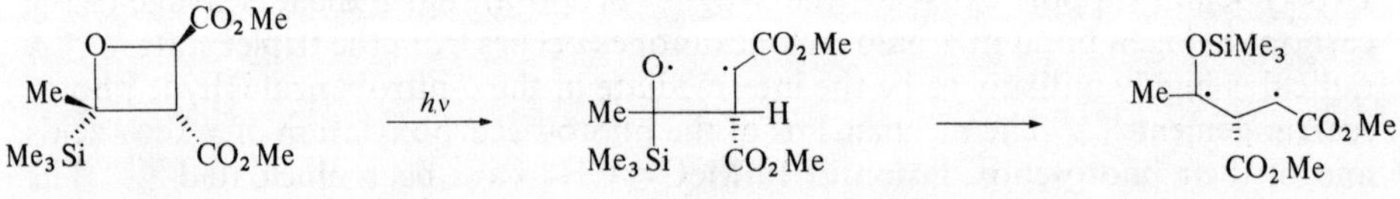

(162)

(163)

(164)

Miscellaneous Compounds

From the orientations in photo-substitution products of Ph· and Me· radicals with toluene it has been concluded, and confirmed by MO calculations, that the methyl radicals react with photo-excited toluene.[1094] Photo-reduction and photo-substitution/addition of PhCl and chloroanisoles with methanol,[1095] and of bromoanthracenes with Me_2NPh,[1096] have been studied. Photo-solvolysis of iodomethylnaphthalenes in methanol does not involve the direct formation of a carbonium ion but proceeds *via* homolytic C—I bond fission.[1097] *N*-Benzyl-4-(carboalkoxy)pyridinyl radicals lose benzyl radicals on photolysis with light of 300 and 395, but not 630 nm wavelength. The $PhCH_2\cdot$ radicals abstract hydrogen from pyridinyl radicals, thus giving dihydropyridines.[1098] 3-Substituted benzo-[6]thiophen-1,1-dioxides give only a head-to-head photo-dimer, *anti* with respect to the cyclobutane ring, *via* a 1,4-biradical; 2-substituted derivatives give mixtures of photo-dimers.[1099]

Pyrolysis

Hydrocarbons

A set of self-consistent Arrhenius parameters for over 500 reactions encountered in the thermal degradation of *n*-alkanes has been compiled for use in modelling studies.[1100] Rate constants for the decomposition of Me· radical in the temperature range 2150–2850 K have been determined.[1101] Induction-period studies of the pyrolysis of ethane at 903 K have enabled a determination of several rate constants for Et· radical reactions.[1102] A detailed kinetic study of the pyrolysis of propane, including the temperature dependence of the product distribution, leads to a readjustment of the rate constants for reactions of Me· radicals with propane and of hydrogen atoms with ethylene.[1103] Reaction between ethylene and cyclopentene is not concerted, but does not involve a radical-chain mechanism.[1104] VLPP of hex-1-yne, [1105] 4-methylpent-2-yne, and 4,4-dimethylpent-2-yne,[1106] and 3-methylbut-1-yne and 3-methylbuta-1,2-diene[1107] have been investigated. Fission of the C—C bond adjacent to the acetylenic bond occurs and estimates of the resonance stabilization energies of methyl-substituted propargyl radicals can be made.

Allene lowers the activation energy of the thermal decomposition of 1-hexene by 81 kJ mol^{-1} by an initiating disproportionation reaction.[1108]

Activation energies for pyrolysis of cyclohexane[1109] and 1,2-, 1,3- and 1,4-dimethylcyclohexanes[1110] are 39.1, 55.1, 56.0, and 70.9 kcal mol^{-1}, respectively.

High-pressure rate constants have been obtained for the unimolecular decomposition by *β*-C—C bond homolysis of alkylbenzenes PhR, where R = Et, Pr^i, Bu^t,[1111] Pr^n, Bu^i, and neopentyl.[1112] *trans*-1,2-Diphenylethene, formed by direct and indirect coupling of 1,2-diphenylethyl radicals, toluene, and 1,1-diphenylethane, are the major products on pyrolysis of 1,2-diphenylethane.[1113]

Electronic and steric effects on the ground state and on the transition state leading to cyclopenta-1,3-diyls have been assessed from the kinetics of the skeletal inversions of *cis*-*exo*-2,3-dideuteriobicyclo[2.1.0]pentane, its 1-methyl derivative, and of *endo*- and *exo*-5-methylbicyclo[2.1.0]pentane, its 1-methyl derivative, and of *endo*- and *exo*-5-methylbicyclo[2.1.0]pentanes.[1114]

While the intermediate 1,4-biradicals have roughly the same energy the ground-state strain energy is higher in [2.2.0]hexane and [2.2.2]propellane than in [4.2.2]propellane, thus producing differences in activation energies for decomposition of the hydrocarbons.[1115]

Other Compounds

Thermal transformations of spiropentane systems have received considerable attention.[1116–1120] Pyrolyses of optically active *trans*-1,2-dimethylspiropentane and of *syn*-4,4-dideuterio-*cis*-1,2-dimethylspiropentane proceed with rate constants of 1.36×10^{-5} and $8.76 \times 10^{-5}\ s^{-1}$, respectively, at 290° and occur *via* cleavage of the C(1)—C(2) bond in a disrotatory fashion with double inversion, contrary to some theoretical expectations. In contrast pyrolysis studies of *trans, medial*- and *trans, proximal*-1,2,4-trimethylspiropentanes indicate that interconversions occur predominantly by C(4)—C(5) bond cleavage.[1116] (**166**) is the only product formed on pyrolysis of (**165**) when X = H or F, Y = Z = H, the bond opposite to the CF_2 group being weakened.[1118] The second CF_2 in (**165**, X = Z = F, Y = H) has a dramatic effect in weakening the C(3)—C(5) bond and leads to (**167**; X = H) as the sole product *via* the biradical (**168**).[1119] Similarly, (**165**; X = Y = Z = F) leads to (**167**; X = F).[1120]

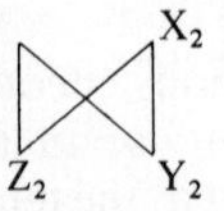

X_2 Z_2 Y_2 (**165**) X_2 (**166**) F_2 F_2 CX_2 (**167**) $CF_2\cdot$ F_2 $CH_2\cdot$ (**168**)

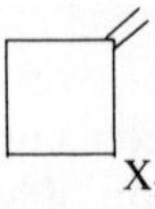

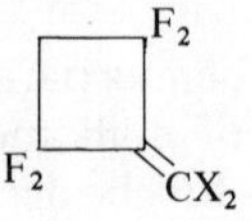

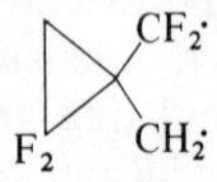

On activation energy arguments the thermal rearrangement of 2,2-difluorovinylcyclopropane is thought to be concerted rather than to involve a diradical.[1121] The high-efficiency conversion of *N,N*-dihalo derivatives of 1-adamantyl- and neopentyl-amine to the corresponding halides may involve an intramolecular homolytic cleavage.[1122]

Relative rates of hydrogen atom abstraction from toluene by isomeric picolinyl radicals do not correlate with the dissociation energies necessary to form the radicals by pyrolysis of ethylpyridines.[1123] Kinetic and mechanistic studies have been made of the thermal decompositions of 2-picoline-*N*-oxide,[1124] nitramines,[1125] furazans and furoxans,[1126] 4-methyl-2,3-pentanedione,[1127] ethanol influenced by butadiene,[1128] and benzaldehyde diethyl acetal in the liquid phase.[1129] Flash thermolysis of methylenephthalide does not result in decarbonylation but solely in ring-closure following bond rotation in the biradical formed by C—O bond fission.[1130]

The chain decomposition of cyclohexanesulphonyl chloride in the liquid phase yields a value of 59.3 kJ mol^{-1} for the C—S bond energy;[1131] oxygen effects have been investigated.[1132]

CIDNP studies have established that the α-azidosulphoxide PhS(=O)C(N_3)HPh decomposes at 70° to the radical pair [PhSO·, PhCHN_3·] but decomposition of the α-azidosulphide PhSCHPhN_3 is concerted with neighbouring-group participation.[1133]

A self-consistent set of ΔH_f values for the silyl radicals $SiMe_xCl_yH_{3-(x+y)}$ have been calculated.[1134] Allyltrimethylsilane undergoes Si–allyl bond rupture initially and forms vinyltrimethylsilane in a rate-determining bimolecular step.[1135]

Improvements of the σ· scale, which is based on the thermal homolysis of benzyl mercurials in solution, have been made[1136] and the scale discussed in relation to other measures of radical stabilization.[1137]

Radiolysis

The rôle of radical processes in radiation chemistry has been reviewed.[1138]

Compounds of Biological Interest

Absolute rate constants for the oxidation of the anion of metiazinic acid, an effective anti-inflammatory agent, by common radicals, *e.g.* ·OH, $Br_2^{\cdot-}$, $(MeSMe)_2^{+\cdot}$ etc., are high (10^9–10^{10} $M^{-1} s^{-1}$) except in the case of ·CH_2CHO radical

(2×10^8 M^{-1} s^{-1}).[1139] Hydroxyl radical-induced oxidation of methionine in aqueous solution leads eventually to a disulphur-centred radical cation of the $(R_2S)_2{}^{+\cdot}$ type.[1140] Nearly 90% of H· atoms react with thionine cation TH^+ to form the semiquinone $\cdot TH_3{}^{2+}$, which dismutates to TH^+ and the leucodye ($2k = 2.4 \times 10^9$ M^{-1} s^{-1}); hydrogen adducts are formed by a second attack mode (10%) on the aromatic ring carbons.[1141] Half-life determination for semiquinones derived from anti-tumour quinones, adriamycin, mitomycin C and 2,5-diaziridinyl-3,6-bis(carboethoxy)amine-1,4-benzoquinone (50, 100, and 200 μs, respectively) has indicated that these semiquinone radicals must be produced close to the site of their biological effect in tumour tissue.[1142]

Nickel(II)-coordinated glycine radicals are produced in the reaction of ·OH radicals with nickel(II) glycine complexes, but not with $Br_2{}^{\cdot-}$ radicals which oxidize the metal centre.[1143] Radiolysis of nucleic acid constituents continues to attract interest.[1144–1147] Pyrimidines undergo only hydrogen atom addition to the 5,6-double bond; deoxyribose undergoes abstraction of C—H hydrogen from two initial sites.[1144] Free radicals from various halo derivatives of uracil, uridine, and deoxyuridine, by reaction with e^-, ·OH, and light, have been spin-trapped.[1145] Radicals formed by ·OH radical addition at C(6) in uracil and methyl- and carboxyl-substituted uracils oxidize TMPD by one-electron transfer; C(5) hydroxyl radical adducts reduce $C(NO_2)_4$ but, in the case of orotic acid derivatives, undergo base-catalysed dehydration to yield radicals capable of oxidizing TMPD.[1146] DNA base cation radicals have been generated in frozen aqueous $NaClO_4$ and LiCl glasses.[1147]

Other Compounds

Methyl radicals with half-lives reckoned in hours, even in the presence of O_2, have been produced on basic and neutral alumina at 300 K.[1148] The ESR spectral parameters of alkyl-radical–halide-ion adducts from electron capture by alkyl bromides and iodides do not support a σ^* structure but rather charge-transfer from the halide into the radical *p*-orbital.[1149]

Protonation of acetone ketyl radical has been observed directly by conductometric pulse radiolysis.[1150] Rates of radical recombination in γ-irradiated CCl_4–MeOH mixtures are a sensitive function of temperature, related to the phase transitions of the glassy matrices.[1151]

The main process occurring in the γ-irradiation of chloroformyl- and fluoroformyl-perfluoropolyethers is electron capture to yield acyl radicals; acyl fluorides undergo direct splitting of the C(O)F group.[1152a] Radiolyses of unsaturated acids, RCH=CHCH=CHCOOH, give the conjugated allylic radicals $RCH_2\dot{C}HCH{=}CHCOOH$.[1152b]

The nitrogen atom appears to influence the β-position reactivity towards hydrogen atom abstraction from R_4N^+ cations (R = Me → Bu) by ·OH, $SO_4{}^{\cdot-}$, $HPO_4{}^{\cdot-}$, and $Cl_2{}^{\cdot-}$.[153] Electrons bring about deamination of $PhCH_2\overset{+}{N}R_3$ cations to yield benzyl radicals; hydrogen atom abstraction processes increase in importance as the chain length of the alkyl group R increases.[1154]

Reaction of $SO_4{}^{\cdot-}$ radicals with 3-phenylpropanoic acid leads to $PhCH_2CH_2\cdot$ radicals having optical absorption at λ_{max} 309 nm, $\varepsilon = 5000$ M^{-1} cm^{-1}. This negates earlier claims that these radicals were generated by electron attachment to 2-haloethylbenzenes.[1155] The high yield of *meta*-product from the reaction of ·OH radicals with pyridine exemplifies the electrophilic nature of the radicals. The low yield of *para*-product is caused by inefficient oxidation of the intermediate *para*-radical adduct.[1156]

CIDNP and CIDEP Methods

Magnetid field and magnetic isotope effects in photochemical reactions have been reviewed,[1157] as have cage reactions of radical pairs and biradicals in homogeneous and micellar solutions, applied especially to dibenzyl ketone photolyses.[1158] Enrichments of ^{13}C in this system have been fully investigated.[1159] A time-resolved laser flash study of triplet spin correlated geminate radical pairs from dibenzyl ketones has probed two intersystem crossing routes—a fast ($k \sim 2 \times 10^6\,s^{-1}$) hyperfine induced ISC and a slower ($k \sim 2 \times 10^4\,s^{-1}$) spin lattice relaxation mode.[1160] Cage disproportionation of radical pairs on photolysis of deoxybenzoins in micellar solution, giving benzaldehydes and α-methylstyrenes is enhanced by a factor of 10 over that occurring in homogeneous solution and involves a competition between escape of the triplet radical pair from the micelle and hyperfine induced ISC.[1161] CIDNP effects in hydrogen atom abstraction from phenols by excited benzophenones has been investigated; in C_6D_6 complexation of reactants results in a predominantly singlet reaction. With tetrafluorophenol, multiplet effects are superimposed on the CIDNP spectra due to interaction of strongly coupled fluorine nuclei with the more weakly coupled proton.[1162] The rate of decarbonylation of the phenylacetyl radical has been estimated by CIDNP measurements.[1163] Pair substitution effects due to cation radical deprotonation have been used to probe the surface structure of proteins *via* nuclear spin polarization of adenosine-5′-monophosphate moieties.[1164] Incorporation of deuterium into the enolizable positions of the cyclic ketones (**169**) and (**170**) alters the magnetic field dependence of the photo-CIDNP effects by changing the ISC probabilities rather than by modifying conformations.[1165] A balance between proton hyperfine coupling induced ISC and spin-orbit, product-selective, coupling is a pre-requisite for CIDNP generation in biradical products, as ascertained from studies of the 1,4-biradical from the Norrish II photo-reaction of valerophenone.[1166]

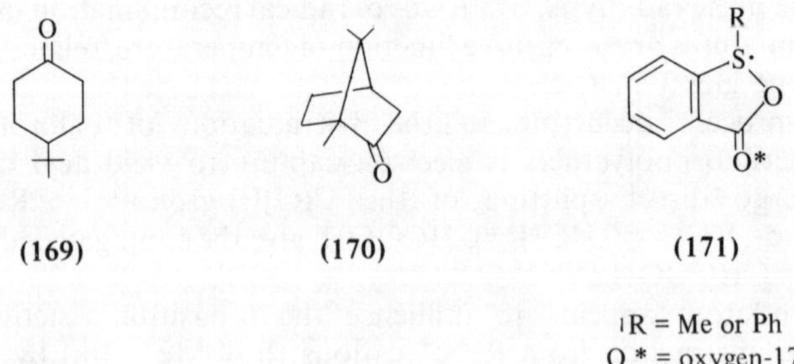

R = Me or Ph
O * = oxygen-17

Cage recombination and escape processes after photo-sensitized C—S bond cleavage in *ortho*-substituted phenyl methyl sulphoxides are significantly perturbed by the nature of the substituent.[1167] One-bond, N—S, fission is deduced from ^1H- and ^{13}C-CIDNP studies on the thermal decomposition of arylazoaryl sulphones in C_2Cl_4 at 100°C.[1168] Oxygen-17 NMR studies of products from decomposition of *tert*-butyl-*o*-methylthio- and -*o*-phenylthio-peroxybenzoates demonstrate the intermediacy of radicals (**171**) having a bridged sulphuranyl structure.[1169]

The *cis–trans* isomerization of maleic acid, catalysed by 4-hydroxy-2,2,6,6-tetramethylpiperidinoxyl, is accelerated in an external magnetic field.[1170] Cage effects and translational motion of radical pairs, both intra- and/or inter-particulate, on a silica gel surface have been demonstrated.[1171]

Failure to observe CIDNP effects, arising from the annihilation of geminate-pair-induced polarizations by free-ion polarizations on a long time-scale, can be circumvented by the use of sub-microsecond time-resolved CIDNP which probes only the former.[1172] Laser flash photolysis with NMR detection and with greater time resolution (0.1 μs) has allowed the separation of geminate and random-phase pair processes. With this technique it is possible to study the dynamics of precursor triplet states as well as those of singlet and triplet biradicals from ketone photolyses.[1173]

On pair-wise production of radicals, *e.g.* in photolyses, geminate pairs dominate the polarization processes.[1174]

The ESR spectra of radicals formed by attack of Ti(III)/H_2O_2 generated hydroxyl radicals on the anions of acetylenedicarboxylic acid and benzene-1,3,5-tricarboxylic acid, above pH 8, appear wholly in emission due to electron transfer from Ti(III) to the radicals in an asymmetric complex.[1175] Time-resolved CIDEP of the photoreduction of furaldehyde, furyl methyl ketone, and acetylthiophene by phenols shows them to be essentially carbonyl-like in behaviour.[1176] The photo-oxidation of vitamin C by pyruvic acid, duroquinone and vitamin K, has been investigated by time-resolved CIDEP.[1177]

References

1. Pryor, W. A., *Frontiers of Free Radical Chemistry*, Academic Press, New York, 1980.
2. Sheldon, R. A., and Kochi, J. K., *Metal-Catalyzed Oxidations of Organic Compounds*, Academic Press, New York, 1981.
3. Rodgers, M. A. J., and Powers, E. L. (Eds.), *Oxygen and Oxy-Radicals in Chemistry and Biology*. Academic Press, New York, 1981.
4. Agmon, N., *Int. J. Chem. Kinet.*, **13**, 332 (1981).
5. Dinçtürk, S., Jackson, R. A., Townson, M., Ağirbaş, H. B., Billingham, N. C., and March, G., *J. Chem. Soc., Perkin Trans. 2*, **1981**, 1121; Dinçtürk, S., and Jackson, R. A., *J. Chem. Soc., Perkin Trans. 2*, **1981**, 1127.
6. Leroy, G., *Nato Adv. Study Inst. Ser., Ser. C.*, **1981**, 67; *Chem. Abs.*, **95**, 131754 (1981).
7. Firestone, R. A., *Lect. Heterocycl. Chem.*, **5**, S89 (1980); *Chem. Abs.*, **93**, 203417 (1980).
8. Walborsky, H. M., *Tetrahedron*, **37**, 1625 (1981).
9. Gajewski, J. J., *Israel J. Chem.*, **21**, 169 (1981).
10. Hanson, P., *Adv. Heterocycl. Chem.*, **27**, 31, (1981).
11. Milaev, A. G., and Okhlobystin, O., Yu., *Usp. Khim.*, **49**, 1879 (1980); *Chem. Abs.*, **94**, 3325 (1981).
12. Davies, A. G., *J. Organomet. Chem.*, **200**, 87 (1980).
13. Mochida, K., *Yuki Gosei Kagaku Kyokaishi*, **39**, 32 (1981); *Chem. Abs.*, **95**, 79361 (1981).
14. Párkányi, C., *Bull. Soc. Chim. Belg.*, **90**, 599 (1981).
15. Cadogan, J. I. G., *Adv. Free-Radical Chem.*, **6**, 155 (1980).
16. Beckwith, A. L. J., *Tetrahedron*, **37**, 3073 (1981).
17. Minisci, F., and Citterio, A., *Adv. Free-Radical Chem.*, **6**, 65 (1980).
18. Molin, Yu. N., Sagdaev, R. Z., Leshina, T. V., Podopelov, A. V., Dushkin, A. V., Grishin, Yu., A., and Veinev, L. M., *Magn. Reson. Relat. Phenom. Proc. Congr. AMPERE, 20th*, **1978**, 49; *Chem. Abs.*, **93**, 238240 (1980).
19. Alder, R. W., *J. Chem. Soc., Chem. Commun.*, **1980**, 1184.
20. Castelhano, A. L., Marriott, P. R., and Griller, D., *J. Am. Chem. Soc.*, **103**, 4262 (1981).
21. Pacansky, J., and Chang, J. S., *J. Chem. Phys.*, **74**, 5539 (1981).
22. Overill, R. E., and Guest, M. F., *Mol. Phys.*, **41**, 119 (1980).
23. Yoshimine, M., and Pacansky, J., *J. Chem. Phys.*, **74**, 5168 (1981).

[24] Paddon-Row, M. N., and Houk, K. N., *J. Am. Chem. Soc.*, **103,** 5046 (1981).
[25] Ingold, K. U., Maillard, B., and Walton, J. C., *J. Chem. Soc., Perkin Trans. 2,* **1981,** 970.
[26] Schubert, W., Yoshimine, M., and Pacansky, J., *J. Phys. Chem.*, **85,** 1340 (1981).
[27] Faucitano, A., Buttafava, A., Faucitano Martinotti, F., and Cesca, S., *J. Phys. Chem.*, **85,** 367 (1981).
[28] Hopkinson, A. C., Lien, M. H., and Csizmadia, I. G., *Chem. Phys. Lett.*, **71,** 557 (1980).
[29] Lunazzi, L., Placucci, G., and Grossi, L., *J. Chem. Soc., Perkin Trans. 2,* **1980,** 1761.
[30] Lloyd, R. V., *J. Phys. Chem.*, **85,** 1440 (1981).
[31] Naimushin, A. I., Chuvylin, N. D., Lebedev, V. L., Zlotskii, S. S., and Rakhmankulov, D. L., *Dopov. Akad. Nauk Ukr. RSR, Ser. B: Geol. Khim. Biol. Nauki*, **1981,** 42: *Chem. Abs.*, **95,** 114391 (1981).
[32] Lloyd, R. V., and Causey, J. G., *J. Chem. Soc., Perkin Trans. 2,* **1981,** 1143.
[33] Micheau, J. C., Despax, B., Paillous, N., Lattes, A., Castellano, A., Catteau, J. P., Lablache-Combier, A., *Nouveau J. Chim.*, **5,** 257 (1981).
[34] Smith, P., and Karukstis, K. K., *J. Magn. Reson.*, **42,** 208 (1981).
[35] Smith, P., and Karukstis, *J. Magn. Reson.*, **43,** 122 (1981).
[36] Wang, H.-C., Pace, M. D., and Kispert, L. D., *J. Chem. Phys.*, **74,** 246 (1981).
[37] Faucitano, A., Buttafava, A., Faucitano Martinotti, F., Caporiccio, G., and Corti, C., *J. Chem. Soc., Perkin Trans. 2,* **1981,** 425.
[38] Griller, D., and Lossing, F. P., *J. Am. Chem. Soc.*, **103,** 1586 (1981).
[39] Aurich, G., and Deuschle, E., *Liebigs Ann. Chem.*, **1981,** 719.
[40] Stella, L., Pochat, F., and Merenyi, R., *Nouveau J. Chim.* **5,** 55 (1981).
[41] Leigh, W. J., and Arnold, D. R., *Can. J. Chem.*, **59,** 609 (1981).
[42] Leigh, W. J., and Arnold, D. R., *Can. J. Chem.*, **59,** 3061 (1981).
[43] Ciminale, F., Liso, G., and Trapani, G., *Tetrahedron Lett.*, **22,** 1455 (1981).
[44] Adeleke, B. B., Chen, K. S., and Wan, J. K. S., *J. Organomet. Chem.*, **208,** 317 (1981).
[45] Alberti, A., Colonna, F. P., Guerra, M., Bonini, B. F., Mazzanti, G., Dinya, Z., and Pedulli, G. F., *J. Organomet. Chem.*, **221,** 47 (1981).
[46] Hillgärtner, H., Neumann, W. P., Schulten, W., and Zarkadis, A., *J. Organomet. Chem.*, **201,** 197 (1980).
[47] Korth, H.-G., Trill, H., and Sustmann, R., *J. Am. Chem. Soc.*, **103,** 4483 (1981).
[48] Davies, A. G., Griller, D., Ingold, K. U., Lindsay, D., and Walton, J. C., *J. Chem. Soc., Perkin Trans. 2,* **1981,** 633.
[49] Nguyen, T. T., and King, K. D., *J. Phys. Chem.*, **85,** 3130 (1981).
[50] McMillen, D. F., Trevor, P. L., and Golden, D. M., *J. Am. Chem. Soc.*, **102,** 7400 (1980).
[51] Barton, B. D., and Stein, S. E., *J. Chem. Soc., Faraday Trans. 1,* **77,** 1755 (1981).
[52] Panov, V. B., Sadekov, I. D., Ladatko, A. A., Okhlobystin, O. Yu., and Minkin, V. I., *Khim. Geterotsikl. Soedin*, **1980,** 1420: *Chem. Abs.*, **94,** 64699 (1981).
[53] Akiyama, K., Kubota, S., and Ikegami, Y., *J. Phys. Chem.*, **85,** 120 (1981).
[54] Negareche, M., Boyer, M., and Tordo, P., *Tetrahedron Lett.*, **22,** 2879 (1981).
[55] Mayer, R., Bleisch, S., and Domschke, G., *Z. Chem.*, **21,** 146 (1981).
[56] Mayer, R., Domschke, G., Bleisch, S., Bartl, A., and Staško, A., *Z. Chem.*, **21,** 264 (1981).
[57] Gey, E., Fabian, J., Bleisch, S., Domschke, G., and Mayer, R., *Z. Chem.*, **21,** 265 (1981).
[58] Mayer, R., Domschke, G., Bleisch, S., and Bartl, A., *Z. Chem.*, **21,** 324 (1981).
[59] Balaban, A. T., Caproiu, M. T., and Negoiţă, N., *Rev. Roum. Chim.*, **25,** 1327 (1980).
[60] Jastrzebski, J. T. B. H., Klerks, J. M., Van Koten, G., and Vrieze, K., *J. Organomet. Chem.*, **210,** 49 (1981).
[61] Suehiro, T., Tashiro, T., and Nakausa, R., *Chem. Lett.*, **1980,** 1339.
[62] Brand, J. C., and Roberts, B. P., *J. Chem. Soc., Chem. Commun.*, **1981,** 748.
[63] *Org. Reaction Mech.*, **1978,** 95.
[64] Berndt, A., Schnaut, R., and Ahrens, W., *Tetrahedron Lett.*, **22,** 4043 (1981).
[65] Berndt, A., Bolze, R., Schnaut, R., and Woynar, H., *Angew. Chem. Int. Ed.*, **20,** 390 (1981).
[66] Sutcliffe, R., Griller, D., Lessard, J., and Ingold, K. U., *J. Am. Chem. Soc.*, **103,** 624 (1981).
[67] Teeinga, H., Nieuwpoort, W. C., and Engberts, J. B. F. N., *Z. Naturforsch.*, **36B,** 279 (1981).
[68] Forrester, A. R., and Irikawa, H., *J. Chem. Soc., Chem. Commun.*, **1981,** 253.
[69] Neugebauer, F. A., Fischer, H., and Brunner, H., *Tetrahedron*, **37,** 1391 (1981).
[70] Neugebauer, F. A., and Fischer, H., *J. Chem. Soc., Perkin Trans. 2,* **1981,** 896.
[71] Mukai, K., Yano, T., and Ishizu, K., *Tetrahedron Lett.*, **22,** 4661 (1981).
[72] Skell, P. S., and May, D. D., *J. Am. Chem. Soc.*, **103,** 967 (1981).
[73] Kikuchi, O., *Bull. Chem. Soc. Jpn.*, **53,** 3149 (1980).
[74] Kirste, B., Harrer, W., Kurreck, H., Schubert, K., Bauer, H., and Gierke, W., *J. Am. Chem. Soc.*, **103,** 6280 (1981).

[75] Mukai, K., and Inagaki, N., *Bull. Chem. Soc. Jpn.*, **53,** 2695 (1980).
[76] Kirste, B., Harrer, W., and Kurreck, H., *Angew. Chem. Int. Ed.*, **20,** 873 (1981).
[77] Alberti, A., Barbaro, G., Battaglia, A., Guerra, M., Bernardi, F., Dondoni, A., and Pedulli, G. F., *J. Org. Chem.*, **46,** 742 (1981).
[78] Weir, D., and Wan, J. K. S., *J. Organomet. Chem.*, **220,** 323 (1981).
[79] Creber, K. A. M., and Wan, J. K. S., *J. Am. Chem. Soc.*, **103,** 2101 (1981).
[80] Barker, P. E., Hudson, A., and Jackson, R. A., *J. Organomet. Chem.*, **208,** C1 (1981).
[81] Alberti, A., Hudson, A., Maccioni, A., Podda, G., and Pedulli, G. F., *J. Chem. Soc., Perkin Trans. 2*, **1981,** 1274.
[82] Gyane, M. J. S., Hudson, A., Lappert, M. F., Power, P. P., and Goldwhite, H., *J. Chem. Soc., Dalton Trans.*, **1980,** 2428.
[83] Dimroth, K., and Heide, W., *Chem. Ber.*, **114,** 3019 (1981).
[84] Dimroth, K., and Heide, W., *Chem. Ber.*, **114,** 3004 (1981).
[85] Mochida, K., Kira, M., and Sakurai, H., *Chem. Lett.*, **1981,** 645.
[86] Gynane, M. J. S., Lappert, M. F., Riley, P. I., Rivière, P., and Rivière-Baudet, M., *J. Organomet. Chem.*, **202,** 5 (1980).
[87] Buschhaus, H.-U., Neumann, W. P., and Apoussidis, T., *Liebigs Ann. Chem.*, **1981,** 1190.
[88] Lehnig, M., and Dören, K., *J. Organomet. Chem.*, **210,** 331 (1981).
[89] Lehnig, M., Buschhaus, H.-U., Neumann, W. P., and Apoussidis, Th., *Bull. Soc. Chim. Belg.*, **89,** 907 (1980).
[90] Hiatt, R. R., *Front. Free Radical Chem.*, [*Pap. Symp.*], **1979,** 225; *Chem. Abs.*, **94,** 102261 (1981).
[91] Quintans, M. T., and Cafferata, L. F. R., *An. Asoc. Quim. Argent.*, **68,** 129 (1980); *Chem. Abs.*, **95,** 6199 (1981).
[92] Scaiano, J. C., and Wubbels, G. G., *J. Am. Chem. Soc.*, **103,** 640 (1981).
[93] Gorbatov, V. V., Yablokova, N. V., Kheidorov, V. P., and Aleksandrov, Yu., *Zh. Obshch. Khim.*, **50,** 2482 (1980); *Chem. Abs.*, **94,** 120618 (1981).
[94] Kurskii, Yu. A., Baryshnikov, Yu. N., Vesnovskaya, G. I., Kaloshina, N. N., and Aleksandrov, Yu. A., *Dokl. Akad. Nauk SSSR*, **258,** 936 (1981); *Chem. Abs.*, **95,** 114414 (1981).
Kurskii, Yu. B., Baryshnikov, Yu. N., Kaloshina, N. N., Vesnovskaya, G. I., *Zh. Obshch. Khim.*, **51,** 362 (1981); *Chem. Abs.*, **94,** 191284 (1981).
[95] Davies, A. G., and Sutcliffe, R., *J. Chem. Soc., Perkin Trans. 2*, **1981,** 1512.
[96] Huschens, R., Rensch, R., and Friebolin, H., *Chem. Ber.*, **114,** 3581 (1981).
[97] Razuvaev, G. A., Dodonov, V. A., Grishin, D. F., and Cherkasov, V. K., *Dokl. Chem.*, **253,** 321 (1981).
[98] Francisco, J. S., Findeis, M. A., and Steinfeld, J. I., *Int. J. Chem. Kinet.*, **13,** 627 (1981); Czarnowski, J., and Schumacher, H. J., *Int. J. Chem. Kinet.*, **13,** 639 (1981).
[99] Maillard, B., Filliatre, C., Manigand, C., and Villenave, J. J., *Bull. Soc. Chim. Belg.*, **90,** 915 (1981).
[100] Maillard, B., Manigand, C., Villenave, J.-J., and Filliatre, C., *Bull. Soc. Chim. Fr. II*, **1981,** 255; Villenave, J.-J., Filliatre, C., Maillard, B., and Tarassova, N. P., *Bull. Soc. Chim. Fr. II*, **1981,** 268; Tarassova, N. P., Villenave, J.-J., Filliatre, C., and Maillard, B., *Bull. Soc. Chim. Fr. II*, **1981,** 261.
[101] Glukhovtsev, V. G., Il'in, Yu. V., Ignatenko, A. V., Slavinskii, N. V., and Nikishin, G. I., *Izv. Akad. Nauk SSSR, Ser. Khim.*, **1980,** 2156; *Chem. Abs.*, **94,** 29810 (1981).
[102] Adam, W., *Pure Appl. Chem.*, **52,** 2591 (1980).
[103] Adam, W., and Bloodworth, A. J., *Top. Curr. Chem.*, **97,** 121 (1981).
[104] Balci, M., *Chem. Rev.*, **81,** 91 (1981).
[105] Kopecky, K. R., Lockwood, P. A., Gomez, R. R., and Ding, J.-Y., *Can. J. Chem.*, **59,** 851 (1981).
[106] Bechara, E. J. H., and Wilson, T., *J. Org. Chem.*, **45,** 5261 (1980).
[107] Baumstark, A. L., and Wilson, C. E., *Tetrahedron Lett.*, **22,** 4363 (1981).
[108] Adam, W., Zinner, K., Krebs, A., and Schmalsteig, H., *Tetrahedron Lett.*, **22,** 4567 (1981).
[109] Schmidt, S. P., Vincent, M. A., Dykstra, C. E., and Schuster, G. B., *J. Am. Chem. Soc.*, **103,** 1292 (1981).
[110] Richardson, W. H., McGinness, R., and O'Neal, H. E., *J. Org. Chem.*, **46,** 1887 (1981).
[111] Bloodworth, A. J., and Baker, D. S., *J. Chem. Soc., Chem. Commun.*, **1981,** 547.
[112] Marshall, R. M., and Rekers, J. W., *J. Am. Chem. Soc.*, **103,** 207 (1981).
[113] Kropf, H., and von Wallis, H., *Synthesis*, **1981,** 237.
[114] Petrov, L. V., and Solyanikov, V. M., *Bull. Acad. Sci. USSR*, **29,** 1081 (1980).
[115] Estrina, G. Ya., Kuramshin, E. M., Imashev. U. B., Zlotskii, S. S., and Rakhmankulov, D. L., *J. Org. Chem. USSR*, **16,** 1355 (1980).
[116] Velyutin, L. P., and Potekhin, V. M., *Izv. Vyssh. Uchebn. Zaved., Khim. Khim. Tehknol.*, **23,** 952 (1980); *Chem. Abs.*, **94,** 3534 (1981).

[117] Ariko, N. G., Putyrskaya, G. V., and Mitskevich, N. I., *Oxid. Commun.*, **1,** 275 (1980); *Chem. Abs.*, **95,** 42005 (1981).

[118] Tovstokhat'ko, F. I., Ovchinnikov, V. I., and Potekhin, V. M., *Zh. Prikl. Khim.* (*Leningrad*), **53,** 2286 (1980); *Chem. Abs.*, **94,** 64894 (1981).

[119] Schulz, M., and Missol, U., *J. Prakt. Chem.*, **322,** 417 (1980).

[120] Tezuka, T., Narita, N., Ando, W., and Oae, S., *J. Am. Chem. Soc.*, **103,** 3045 (1981); *Org. Reaction Mech.*, **1980,** 94.

[121] Griffin, B. W., and Ramirez, D., *Bioorg. Chem.*, **10,** 177 (1981).

[122] Cholvad, V., and Tkáč, A., *Collect. Czech. Chem. Commun.*, **46,** 1071 (1981).

[123] Cholvad, V., Staško, A., Tkáč, A., Buchachenko, A. L., and Malik, L., *Collect. Czech. Chem. Commun.*, **46,** 823 (1981).

[124] Funahashi, T., *J. Sci. Hiroshima Univ. Ser. A. Phys. Chem.*, **44,** 331 (1981); *Chem. Abs.*, **95,** 6212 (1981).

[125] Maillard, B., Villenave, J. J., and Filliatre, C., *Thermochim. Acta*, **39,** 205 (1980); *Chem. Abs.*, **93,** 220214 (1980).

[126] Rüchardt, C., Golzke, V., and Range, G., *Chem. Ber.*, **114,** 2769 (1981).

[127] *Org. Reaction Mech.*, **1980,** 95.

[128] Razuvaev, G. A., Brevnova, T. N., and Chesnokova, T. A., *Zh. Obschch. Khim.*, **51,** 813 (1981); *Chem. Abs.*, **95,** 114514 (1981).

[129] Dixon, B. G., and Schuster, G. B., *J. Am. Chem. Soc.*, **103,** 3068 (1981).

[130] Kucher, R. V., Luk'yanenko, L. V., and Turovskii, A. A., *J. Org. Chem. USSR*, **17,** 734 (1981).

[131] Fedorova, V. A., Darmograi, M. I., and Panchenko, Yu. V., *Khim. Tekhnol.* (*Kiev*), **1980,** 29; *Chem. Abs.*, **94,** 29918 (1981).

[132] Liu, Y.-C., Liu, Y.-C., Lei, X.-G., Lu, H.-L., *Hua Hsueh Hsueh Pao*, **38,** 213 (1981); *Chem. Abs.*, **94,** 3527 (1981).

[133] Moad, G., Rizzardo, E., and Solomon, D. H., *Tetrahedron Lett.*, **22,** 1165 (1981).

[134] Feldhues, M., and Schäfer, H. J., *Tetrahedron Lett.*, **22,** 433 (1981).

[135] Stankevich, A. I., Zyat'kov, I. P., Lazerera, A. M., and Ei'nitskii, A. P., *J. Org. Chem. USSR*, **16,** 1543 (1980).

[136] Serov, S. I., Zhuravlev, M. V., Sass, V. P., and Sokolov, S. V., *J. Org. Chem. USSR*, **16,** 1360 (1980).

[137] Nedelec, J. Y., and Lefort, D., *Tetrahedron*, **36,** 3199 (1980).

[138] Hegedic, D., *Naucno-Teh. Pregl.*, **30,** 32 (1980); *Chem. Abs.*, **95,** 42009 (1981).

[139a] Markaryan, Sh. A., and Beileryan, N. M., *Teor. Eksp. Khim.*, **17,** 424 (1981); *Chem. Abs.*, **95,** 114427 (1981).

[139b] Morsi, S. E., Zaki, A. B., and Habib, A., *Z. Phys. Chem.* (*Leipzig*), **262,** 196 (1981).

[140] Efremova, E. P., Chikhacheva, I. P., Motov, S. A., Stavrova, S. D., and Pravednikov, A. N., *Dokl. Akad. Nauk SSSR*, **258,** 354 (1981); *Chem. Abs.*, **95,** 96562 (1981).

[141] Okamoto, K., Takeuchi, K., and Murai, O., *Tetrahedron Lett.*, **22,** 2785 (1981).

[142] Artym, I. I., Gorbachevskaya, K. R., Kovbuz, M. A., Konovalenko, V. V., and Ivanchev, S. S., *Zh. Obshch. Khim.*, **51,** 940 (1981); *Chem. Abs.*, **95,** 96668 (1981).

[143] Schuster, G. B., and Hurst, J. R., *Report*, **1979,** TR-22; *Chem. Abs.*, **93,** 167274 (1980).

[144] Duynstee, E. F. J., Esser, M. L., and Schellekens, R., *Eur. Polym. J.*, **16,** 1127 (1980).

[145] Schellekens, R., Cosemans, J., Linnartz, T., Beulen, J., and Duynstee, E. F. J., *Chem. Ind.* (*London*), **1981,** 31.

[146] Maillard, B., Bourgeois, J., Campagnole, M., Filliatre, C., and Villenave, J. J., *Therm. Anal.*, [*Proc. Int. Conf. Therm. Anal.*], *6th* **2,** 501 (1980); *Chem. Abs.*, **94,** 64893 (1981).

[147] Bégué, J.-P., Lefort, D., and Thac, T. D., *J. Chem. Soc., Chem. Commun.*, **1981,** 1086.

[148] Sorba, J., Fossey, J., Lefort, D., and Nedelec, J. Y., *Tetrahedron*, **37,** 69 (1981).

[149] Sawaki, Y., and Ogata, Y., *Bull. Chem. Soc. Jpn.*, **54,** 793 (1981).

[150] Niki, H., Maker, P. D., Savage, C. M., and Breitenbach, L. P., *J. Phys. Chem.*, **85,** 877 (1981).

[151] Sander, S. P., and Watson, R. T., *J. Phys. Chem.*, **85,** 2960 (1981).

[152] Adachi, H., Basco, N., and James, D. G. L., *Int. J. Chem. Kinet.*, **12,** 949 (1980).

[153] Kirsch, L. J., and Parkes, D. A., *J. Chem. Soc., Faraday Trans. 1*, **77,** 293 (1981).

[154] Mantashyan, A. A., Khachatryan, L. A., Niazyan, O. M., and Arsent'ev, S. D., *Kinet. Katal.*, **22,** 500 (1981); *Chem. Abs.*, **95,** 114460 (1981).

[155] Howard, J. A., Bennett, J. E., and Brunton, G., *Can. J. Chem.*, **59,** 2253 (1981).

[156] Goldberg, V. M., Lukacs, J., Vasvari, G., and Gal, D., *Oxid. Commun.*, **1,** 189 (1980); *Chem. Abs.*, **94,** 64805 (1981).

[157] Kenigsberg, T. P., Ariko, N. G., and Mitskevich, N. I., *Dokl. Akad. Nauk BSSR*, **24,** 817 (1980); *Chem. Abs.*, **94,** 29842 (1981).

158 Rafikova, V. S., Maizus, Z. K., Skibida, I. P., and Volkov, R. N., *Int. J. Chem. Kinet.*, **13,** 111 (1981).

159 Cooper, R., Cumming, J. B., Gordon, S., and Mulac, W. A., *Radiat. Phys. Chem.*, **16,** 169 (1980); *Chem. Abs.*, **93,** 185397 (1980).

160 Baldwin, R. R., Pickering, I. A., and Walker, R. W., *J. Chem. Soc., Faraday Trans. 1*, **76,** 2374 (1980).

161 Ruiz, R. P., Bayles, K. D., Macpherson, M. T., and Pilling, M. J., *J. Phys. Chem.*, **85,** 1622 (1981).

162 Ebata, T., Obi, K., and Tanaka, I., *Chem. Phys. Lett.*, **77,** 480 (1981).

163 Ebata, T., Kanayama, M., Obi, K., and Tanaka, I., *Kosoku Hanno Toronkai Koen Yokoshu, 14th,* **1979,** 64; *Chem. Abs.*, **94,** 14779 (1981).

164 Niki, H., Maker, P. D., Savage, C. M., and Breitenbach, L. P., *J. Phys. Chem.*, **85,** 2698 (1981).

165 Tavadyan, L. A., Mardoyan, V. A., and Nalbandyan, A. B., *Dokl. Akad. Nauk SSSR*, **259,** 1143 (1981); *Chem. Abs.*, **95,** 168120 (1981).

166 Kan, C. S., Calvert, J. G., and Shaw, J. H., *J. Phys. Chem.*, **85,** 1126 (1981).

167 Pozdeeva, N. N., Denisov, E. T., and Martemyanov, V. S., *Kinet. Katal.*, **22,** 591 (1981); *Chem. Abs.*, **95,** 114461 (1981).

168 Rubtsov, V. L., Roginskii, V. A., Miller, V. B., and Zaikov, G. E., *Kinet. Katal*, **21,** 612 (1980); *Chem. Abs.*, **94,** 29793 (1981).

169 Howard, J. A., and Chenier, J. H. B., *Can. J. Chem.*, **58,** 2808 (1980).

170 *Org. Reaction Mech.*, **1977,** 91: **1980,** 100.

171 Frankel, E. N., Weisleder, D., and Neff, W. E., *J. Chem. Soc., Chem. Commun.*, **1981**, 766.

172 O'Connor, D. E., Mihelich, E. D., and Coleman, M. C., *J. Am. Chem. Soc.*, **103,** 222 (1981).

173 Porter, N. A., Roe, A. N., and McPhail, A. T., *J. Am. Chem. Soc.*, **102,** 7574 (1980).

174 Galimova, L. G., Maslennikov, S. I., and Nikolaev, A. I., *Bull. Acad. Sci. USSR*, **29,** 1731 (1980).

175 Mushenko, V. D., Selivanov, N. T., and Proskuryakov, V. A., *Zh. Prikl. Khim.* (*Leningrad*), **53,** 2393 (1980); *Chem. Abs.*, **94,** 64895 (1981).

176a Barclay, L. R. C., and Ingold, K. U., *J. Am. Chem. Soc.*, **102,** 7792 (1980).

176b Hochstein, P., *Isr. J. Chem.*, **21,** 52 (1981).

177 Kasaikina, O. T., Kartasheva, Z. S., and Gagarina, A. B., *Bull. Acad. Sci. USSR*, **30,** 381 (1981).

178 Blau, K., Müller, U., Pritzhow, W., Schmidt-Renner, W., and Sedshaw, W., *J. Prakt. Chem.*, **322,** 915 (1980).

179 Patnaik, L. N., Mallick, N., Rout, M. K., and Rout, S. P., *Can. J. Chem.*, **58,** 2754 (1980).

180 Lewis, I. C., and Singer, L. S., *J. Phys. Chem.*, **85,** 354 (1981).

181 Beckwith, A. L. J., and Wagner, R. D., *J. Org. Chem.*, **46,** 3638 (1981).

182 Fuchs, K. P., and Pritzkow, W., *J. Prakt. Chem.*, **322,** 353 (1980).

183 Kanno, T., Hisaoka, M., Sakuragi, H., and Tokumaru, K., *Bull. Chem. Soc. Jpn.*, **54,** 2330 (1981).

184 Sawaki, Y., and Ogata, Y., *J. Am. Chem. Soc.*, **103,** 2049 (1981).

185 Niki, H., Maker, P. D., Savage, C. M., and Breitenbach, L. P., *Int. J. Chem. Kinet*, **12,** 1001 (1980).

186 Baumann, H., and Timpe, H.-J., *J. Prakt. Chem.*, **322,** 865 (1980).

187 Kuramshin, E. M., Gumerova, V. K., Imashev, U. B., Zlotskii, S. S., and Rakhmankulov, D. L., *React. Kinet. Catal. Lett.*, **16,** 213 (1981); *Chem. Abs.*, **95,** 149651 (1981).

188 Brudnik, B. M., Kuramshin, É. M., Imashev, U. B., Zlotskii, S. S., and Rakhmankulov, D. L., *J. Org. Chem. USSR*, **17,** 608 (1981).

189 Burton, G. W., Le Page, Y., Gabe, E. J., and Ingold, K. U., *J. Am. Chem. Soc.*, **102,** 7791 (1980).

190 Burton, G. W., and Ingold, K. U., *J. Am. Chem. Soc.*, **103,** 6472 (1981).

191 Kucher, R. V., Opeida, I. A., Nechitailo, L. G., and Simonov, M. A., *J. Org. Chem. USSR*, **17,** 93 (1981).

192 Franklin, T. C., and Ogiya, N., *Int. J. Chem. Kinet.*, **12,** 1045 (1980).

193 Graziano, M. L., and Scarpati, R., *J. Chem. Soc., Perkin Trans. 1*, **1981,** 1811.

194 Stegmann, H. B., Stolze, K., Bergler, H. U., and Scheffler, K., *Tetrahedron Lett.*, **22,** 4057 (1981).

195 Gaida, T. M., Sopchik, A. E., and Bentrude, W. G., *Tetrahedron Lett.*, **22,** 4167 (1981).

196 Brindley, P. B., and Scotton, M. J., *J. Chem. Soc., Perkin Trans., 2*, **1981,** 419.

197 Aleksandrov, A. I., *Bull. Acad. Sci. USSR*, **29,** 1740 (1980).

198 Aleksandrov, A. L., and Krisanova, L. D., *Bull. Acad. Sci. USSR*, **29,** 1735 (1980).

199 Aleksandrov, A. L., *Bull. Acad. Sci. USSR*, **29,** 1896 (1980).

200 Kovtun, G. A., Lysenko, D. L., Berenblyum, A. S., and Moiseev, I. I., *Izv. Akad. Nauk SSSR, Ser. Khim.*, **1980,** 2425; *Chem. Abs.*, **94,** 14817 (1981).

201 Engel, P. S., and Nalepa, C. J., *Pure Appl. Chem.*, **52,** 2621 (1980).

202 Ács, G., Péter, A., and Huhn, P., *Magy. Kem. Foly.*, **86,** 529 (1980); *Chem. Abs.*, **94,** 191385 (1981).

203 Szirovicza, L., *Acta Phys. Chem.*, **25,** 147 (1979); *Chem. Abs.*, **93,** 203760 (1980).

204 Ács, G., Péter, A., and Huhn, P., *Int. J. Chem. Kinet.*, **12,** 993 (1980).

205 Durban, P. C., and Marshall, R. M., *Int. J. Chem. Kinet.*, **12,** 1001 (1980).

206 Neumann, R. C., and Amrich, M. J., *J. Org. Chem.*, **45,** 4629 (1980).

207 Schmittel, M., Schulz, A., Rüchardt, C., and Hädicke, E., *Chem. Ber.*, **114,** 3533 (1981).

208 Yeung, D. W. K., and Warkentin, J., *Can. J. Chem.*, **58,** 2386 (1980).

209 Nazran, A. S., and Warkentin, J., *J. Am. Chem. Soc.*, **103,** 236 (1981).

210 Vaughan, K., and Liu, M. T. H., *Can. J. Chem.*, **59,** 923 (1981).

211 Bae, D.-H., and Shine, H. J., *J. Org. Chem.*, **45,** 4448 (1980).

212 Lübbe, F., Grosz, K.-P., Hillebrand, W., and Sucrow, W., *Tetrahedron Lett.*, **22,** 227 (1981).

213 Dreher, E.-L., Niederer, P., Rieker, A., Schwarz, W., and Zollinger, H., *Helv. Chim. Acta*, **64,** 488 (1981).

214 Besse, J., Schwarz, U., and Zollinger, H., *Helv. Chim. Acta*, **64,** 504 (1981).

215 Schwarz, W., and Zollinger, H., *Helv. Chim. Acta*, **64,** 513 (1981).

216 Besse, J., and Zollinger, H., *Helv. Chim. Acta*, **64,** 529 (1981).

217 Galli, C., *J. Chem. Soc., Perkin Trans. 2*, **1981,** 1459.

218 Jirkovský, J., Fojtík, A., and Becker, H. G. O., *Collect. Czech. Chem. Commun.*, **46,** 1560 (1981).

219 Packer, J. E., Mönig, J., and Dobson, B. C., *Aust. J. Chem.*, **34,** 1433 (1981).

220 Kampars, V., Bumbure, G., Kokars, V., and Neilands, O., *Zh. Obshch. Khim.*, **50,** 2057 (1980); *Chem. Abs.*, **94,** 29866 (1981).

221 Sikkar, R., and Martinson, P., *Acta Chem. Scand.*, **34B,** 551 (1980).

222 Brede, O., Mehnert, R., Naumann, W., and Becker, H. G. O., *Ber. Bunsenges. Phys. Chem.*, **84,** 666 (1980).

223 Rehorek, D., and Marx, J., *J. Prakt. Chem.*, **322,** 872 (1980).

224 Beckwith, A. L. J., and Meijs, G. F., *J. Chem. Soc., Chem. Commun.*, **1981,** 136.

225 Beckwith, A. L. J., and Meijs, G. F., *J. Chem. Soc., Chem. Commun.*, **1981,** 595.

226 Citterio, A., Ramperti, M., and Vismara, E., *J. Heterocycl. Chem.*, **18,** 763 (1981); *Org. Reaction Mech.*, **1980,** 102.

227 Chin, A., Hung, M.-H., and Stock, L. M., *J. Org. Chem.*, **46,** 2203 (1981).

228 Khudyakov, I. V., Levin, P. P., and Kuz'min, V. A., *Usp. Khim.*, **49,** 1990 (1980); *Chem. Abs.*, **94,** 14634 (1981).

229 Bagdasar'yan, Kh. S., *Dokl. Akad. Nauk SSSR*, **258,** 1133 (1981); *Chem. Abs.*, **95,** 168122 (1981).

230 Fujimoto, H., Koga, N., Endo, M., and Fukui, K., *Tetrahedron Lett.*, **22,** 3427 (1981).

231 Skinner, G. B., Rogers, D., and Patal, K. B., *Int. J. Chem. Kinet.*, **13,** 481 (1981).

232 Baghal-Vayjooee, M. H., Colussi, A. J., and Benson, S. W., *Int. J. Chem. Kinet.*, **11,** 147 (1979).

233 Patrick, R., Pilling, M. J., and Rogers, G. F., *Chem. Phys.*, **53,** 279 (1980); *Chem. Abs.*, **94,** 102404 (1981).

234 Messing, I., Carrington, T., Filseth, S. V., and Sadowski, C. M., *Chem. Phys. Lett.*, **74,** 56 (1980); *Chem. Abs.*, **93,** 238351 (1980).

235 Raimondi, M., Simonetta, M., and Gerratt, J., *Chem. Phys. Lett.*, **77,** 12 (1981); *Chem. Abs.*, **94,** 174071 (1981); Krohn, H., Leuschner, R., and Dohrmann, J. K., *Ber. Bunsenges. Phys. Chem.*, **85,** 139 (1981).

236 Huggenberger, C., and Fischer, H., *Helv. Chim. Acta*, **64,** 338 (1981).

237 Adachi, H., and Basco, N., *Int. J. Chem. Kinet.*, **13,** 367 (1981).

238 Ingold, K. U., Maillard, B., and Walton, J. C., *J. Chem. Soc., Perkin Trans. 2*, **1981,** 970.

239 Davies, A. G., and Lusztyk, J., *J. Chem. Soc., Perkin Trans., 2*, **1981,** 692.

240 Komerov, V. S., Kozliner, M. Z., and Chelibanov, V. P., *Kinet. Katal.*, **21,** 1587 (1980); *Chem. Abs.*, **94,** 102459 (1981).

241 Zimina, G. M., *Khim. Vys. Energ.*, **15,** 86 (1981); *Chem. Abs.*, **94,** 173932 (1981).

242 Hoyermann, K., Loftfield, N. S., Sievert, R., and Wagner, H. G., *Symp. (Int.) Combust.*, [*Proc.*] *1980, 18th*, **1981,** 831; *Chem. Abs.*, **95,** 61021 (1981).

243 Mulenko, S. A., *Zh. Prikl. Spektrosk.*, **33,** 35 (1980); *Chem. Abs.*, **94,** 14793 (1981).

244 Huggenberger, C., Lipscher, J., and Fischer, H., *J. Phys. Chem.*, **84,** 3467 (1980).

245 Wong, S. K., *Int. J. Chem. Kinet.*, **13,** 433 (1981).

246 Krohn, H., Leuschner, R., and Dohrmann, J. K., *Ber. Bunsenges. Phys. Chem.*, **85,** 139 (1981); *Chem. Abs.*, **95,** 23833 (1981).

247 Hermolin, J., Levin, M., and Kosower, E. M., *J. Am. Chem. Soc.*, **103,** 4801 (1981).

248 Hermolin, J., Levin, M., Ikegami, Y., Sawayanagi, M., and Kosower, E. M., *J. Am. Chem. Soc.*, **103,** 4795 (1981).

249 Hermolin, J., Levin, M., and Kosower, E. M., *J. Am. Chem. Soc.*, **103,** 4808 (1981).

250 Akiyama, K., Kubota, S., and Ikegami, Y., *Chem. Lett.*, **1981,** 469.

251 Hermolin, J., and Kosower, E. M., *J. Am. Chem. Soc.*, **103,** 4813 (1981).

[252] Khudyakov, I. V., Levin, P. P., Kuz'min, V. A., Hageman, H. J., and de Jonge, C.R.H.I., *J. Chem. Soc., Perkin Trans. 2*, **1981,** 1234.

[253] Muslin, D. V., Lyapina, N. Sh., Klimov, E. S., Kirilicheva, V. G., and Razuvaev, G. A., *Izv. Akad. Nauk SSSR, Ser. Khim.*, **1980,** 1385; *Chem. Abs.*, **93,** 185385 (1980).

[254] Schellekens, R., Cosemans, J., Linnartz, T., Bevlen, J., and Duynstee, E. F. J., *Chem. Ind.* (*London*), **1981,** 31.

[255] Kandror, I. I., Brazina, I. O., and Freidlina, R. Kh., *Izv. Akad. Nauk SSSR, Ser. Khim.*, **1981,** 1167; *Chem. Abs.*, **95,** 96556 (1981).

[256] Ohkubo, K., Nakashima, Y., Masumoto, K., and Gal, D., *Oxid. Commun.*, **1,** 197 (1980); *Chem. Abs.*, **94,** 64806 (1981).

[257] Löser, U., Scherzer, K., and Rohde, R., *Z. Chem.*, **21,** 266 (1981).

[258] Afanas'ev, I. B., *Int. J. Chem. Kinet.*, **13,** 173 (1981).

[259] Atkinson, R., *Int. J. Chem. Kinet.*, **12,** 761 (1980).

[260] Gaffney, J. S., and Levine, S. Z., *Int. J. Chem. Kinet.*, **12,** 767 (1980).

[261] Butler, J. E., Fleming, J. W., Goss, L. P., and Lin, M. C., *Chem. Phys.*, **56,** 355 (1981); *Chem. Abs.*, **95,** 96529 (1981).

[262] Pohjonen, M. L., *Acta Chem. Scand.*, **A34,** 597 (1980); *Chem. Abs.*, **94,** 156001 (1981).

[263] Rogers, D., and Skinner, G. B., *Int. J. Chem. Kinet.*, **13,** 741 (1981).

[264] Selivanov, N. T., *Zh. Prikl. Khion.* (*Leningrad*), **53,** 2274 (1980); *Chem. Abs.*, **94,** 102432 (1981).

[265] Ballod, A. P., Fedorova, T. V., Chikvaidze, N., Titarchuk, T. A., and Kumova, A. A., *Kinet. Katal.*, **21,** 1095 (1980); *Chem. Abs.*, **94,** 29825 (1981).

[266] Durban, P. C., and Marshall, R. M., *Int. J. Chem. Kinet.*, **12,** 1031 (1980).

[267] Ogryzlo, E. A., Paltenghi, R., and Bayes, K. D., *Int. J. Chem. Kinet.*, **13,** 667 (1981).

[268] Arthur, N. L., and Newitt, P. J., *Aust. J. Chem.*, **34,** 727 (1981).

[269] Plumb, I. C., and Ryan, K. R., *Int. J. Chem. Kinet.*, **13,** 1011 (1981).

[270] Peter, A., Acs, G., Horvath, I., and Huhn, P., *Magy. Kem. Foly.*, **87,** 131 (1981); *Chem. Abs.*, **95,** 114388 (1981).

[271] Dütsch, H.-R., and Fischer, H., *Int. J. Chem. Kinet.*, **13,** 527 (1981).

[272] Dzheiranishvili, M. S., Chkhubianishvili, N. G., and Afanas'ev, I. B., *Izv. Akad. Nauk. Gruz. SSR, Ser. Khim.*, **7,** 23 (1981); *Chem. Abs.*, **95,** 114418 (1981).

[273] Batyrbaev, N. A., Zorin, V. V., Imashev, U. B., Zlotskii, S. S., and Rakhmankulov, D. L., *Zh. Org. Khim.*, **16,** 1594 (1980); *Chem. Abs.*, **94,** 14801 (1981).

[274] Batyrbaev, N. A., Zorin, V. V., Imashev, U. B., Zlotskii, S. S., and Rakhmankulov, D. L., *Zh. Prikl. Khim.* (*Leningrad*), **53,** 1338 (1980); *Chem. Abs.*, **93,** 203648 (1980).

[275] Serov, S. I., Sass, V. P., and Sokolov, S. V., *Zh. Org. Khim.*, **16,** 2478 (1980); *Chem. Abs.*, **94,** 102465 (1981).

[276] Arthur, N. L., and David, L. F., *Aust. J. Chem.*, **34,** 1535 (1981).

[277] Ratajczak, E., and Pieniazrk, M., *Bull. Acad. Pol. Sci., Ser. Sci. Chima*, **28,** 341 (1980); *Chem. Abs.*, **95,** 149508 (1981)

[278] Baruch, G., and Horowitz, A., *J. Phys. Chem.*, **84,** 2535 (1980).

[279] James, F. C., Choi, H. K. J., Strausz, O. P., and Bell, T. N., *Chem. Phys. Lett.*, **73,** 522 (1980).

[280] Alfassi, Z. B., and Feldman, L., *Int. J. Chem. Kinet.*, **13,** 517 (1981).

[281] Alfassi, Z. B., and Feldman, L., *Int. J. Chem. Kinet.*, **13,** 771 (1981).

[282] Feldman, L., and Alfassi, Z. B., *J. Phys. Chem.*, **85,** 3060 (1981).

[283] Freidlina, K. Kh., Rybakova, N. A., Dostovalova, V. I., and Kiseleva, L. N., *Dokl. Akad. Nauk SSSR*, **255,** 887 (1980); *Chem. Abs.*, **94,** 208140 (1981).

[284] Tedder, J. M., and Walton, J. C., *Adv. Free-Radical Chem.*, **6,** 155 (1980); *Chem. Abs.*, **94,** 14635 (1981).

[285] Traynham, J. G., *Front Free Radical Chem.*, [*Pap. Symp.*] *1979*, (Ed. Pryor, W. A.), Academic Press, New York, 1980, pp. 283–296; *Chem. Abs.*, **94,** 102263 (1981).

[286] Aver'yanov, V. A., Staroverova, N. V., and Erokhina, N. G., *Deposited Doc.*, **1980,** Viniti, 1013–80; *Chem. Abs.*, **95,** 79632 (1981).

[287] Aver'yanov, V. A., Zarytovskii, V. M., Shvets, V. F., Treger, Yu. A., and Emel'yanov, V. I., *Zh. Org. Khim.*, **17,** 36 (1981); *Chem. Abs.*, **95,** 23983 (1981).

[288] Aver'yanov, V. A., Zarytovskii, V. M., and Shvets, V. F., *Zh. Org. Khim.*, **17,** 45 (1981); *Chem. Abs.*, **95,** 23834 (1981).

[289] Kosheleva, L. M., Rozenberg, V. R., and Motsarev, G. V., *Zh. Org. Khim.*, **16,** 1890 (1980); *Chem. Abs.*, **94,** 14815 (1981).

[290] Dneprovskii, A. S., and Kasatochkin, A. N., *Zh. Org. Khim.*, **17,** 793 (1981); *Chem. Abs.*, **95,** 79638 (1981).

[291] Löfas, S., and Ahlberg, P., *J. Chem. Soc., Chem. Commun.*, **1981,** 998.
[292] Fuller, S. E., Smith, J. R. L., Norman, R. O. C., and Higgins, R., *J. Chem. Soc., Perkin Trans. 2*, **1981,** 545.
[293] Vincens, M., Choubani, S., and Vidal, M., *Tetrahedron Lett.*, **22,** 4695 (1981).
[294] Seki, S., Moringa, T., Kikuchi, H., Mitsuhashi, T., Yamamoto, G., and Oki, M., *Bull. Chem. Soc. Jpn.*, **54,** 1465 (1981).
[295] *Org. Reaction Mech.*, **1980,** 107.
[296] Katsushima, T., Yamaguchi, R., Lemura, S., and Kawanisi, M., *Bull. Chem. Soc. Jpn.*, **53,** 3324 (1980).
[297] Migita, T., Nakayama, M., Watanuki, T., Suzuki, M., and Mumamoto, T., *Bull. Chem. Soc. Jpn.*, **54,** 822 (1981).
[298] Tanner, D. D., Ruo, T. C.-S., Takiguchi, H., and Guillaume, A., *Can. J. Chem.*, **59,** 1368 (1981).
[299] Smith, W. B., *J. Org. Chem.*, **46,** 187 (1981).
[300] Tanner, D. D., Blackburn, I. V., Reed, D. W., and Satiloane, B. P., *J. Org. Chem.*, **45,** 5183 (1980).
[301] Pryor, W. A., Church, D. F., Tang, F. Y., and Tang, R. H., *Front. Free Radical Chem.*, [*Pap. Symp.*] *1979*, (Ed. Pryor, W. A.), Academic Press, New York, 1980, p. 355; *Chem. Abs.*, **94,** 174042 (1981).
[302] Hoyermann, K., Sievert, R., and Wagner, H. G., *Ber. Bunsenges. Phys. Chem.*, **85,** 149 (1981); *Chem. Abs.*, **94,** 155976 (1981).
[303] Heneghan, S. P., Knoot, P. A., and Benson, S. W., *Int. J. Chem. Kinet.*, **13,** 677 (1981).
[304] Niki, H., Maker, P. D., Savage, C. M., and Breitenbach, L. P., *Int. J. Chem. Kinet.*, **12,** 1001 (1980).
[305] Aver'yanov, V. A., Somov, G. V., and Dryndin, V. M., *Deposited Doc.*, **1980,** Viniti, 582; *Chem. Abs.*, **94,** 120750 (1981).
[306] Niki, H., Maker, P. D., Savage, C. M., and Breitenbach, L. P., *Int. J. Chem. Kinet.*, **12,** 915 (1980).
[307] Foulet, G., Laverdet, G., and Le Bras, G., *J. Phys. Chem.*, **85,** 1892 (1981).
[308] Nava, D. F., Michael, J. V., and Stief, L. J., *J. Phys. Chem.*, **85,** 1896 (1981).
[309] Heicklen, J., *Int. J. Chem. Kinet.*, **13,** 651 (1981).
[310] Paraskevopoulos, G., Singleton, D. L., and Irwin, R. S., *J. Phys. Chem.*, **85,** 561 (1981).
[311] Hoyermann, K., Sievert, R., and Wagner, H. Gg., *Oxid. Commun.*, **1,** 145 (1980); *Chem. Abs.*, **94,** 29815 (1981).
[312] Wine, P. H., Kreutter, N. M., Gump, C. A., and Ravishankara, A. R., *J. Phys. Chem.*, **85,** 2660 (1981).
[313] Koulkes-Pujo, A. M., Moreau, M., and Sutton, J., *FEBS Lett.*, **129,** 52 (1981); *Chem. Abs.*, **95,** 168090 (1981).
[314] Stief, L. J., Nava, D. F., Payne, W. A., and Michael, J. V., *J. Chem. Phys.*, **73,** 2254 (1980); *Chem. Abs.*, **94,** 46379 (1981).
[315] Audley, G. J., Baulch, D. C., and Campbell, I. M., *J. Chem. Soc., Faraday Trans., 1*, **77,** 2541 (1981).
[316] Maldotti, A., Chiorboli, C., Bignozzi, C. A., Bartocci, C., and Carassiti, V., *Int. J. Chem. Kinet.*, **12,** 905 (1980).
[317] Lapshova, A. A., Zorin, V. V., Shuvalov, V. F., Moravskii, A. P., Zlotskii, S. S., Karakhanov, R. A., and Rakhmankulov, D. L., *Izv. Akad. Nauk SSSR, Ser. Khim.*, **1980,** 1197; *Chem. Abs.*, **93,** 167221 (1980).
[318] Gilbert, B. C., King, D. M., and Thomas, C. B., *J. Chem. Soc., Perkin Trans. 2*, **1981,** 1186.
[319] Gilbert, B. C., King, D. M., and Thomas, C. B., *J. Chem. Soc., Perkin Trans. 2*, **1980,** 1821.
[320] Abuin, E., Mujica, C., and Lissi, E., *Rev. Latinoam. Quim.*, **11,** 78 (1980); *Chem. Abs.*, **94,** 64799 (1981).
[321] Gilbert, B. C., Marshall, P. D. R., Norman, R. O. C., Pineda, N., and Williams, P. S., *J. Chem. Soc., Perkin Trans. 2*, **1981,** 1392.
[322] Al Akeel, N. Y., Selby, K., and Waddington, D. J., *J. Chem. Soc., Perkin Trans., 2*, **1981,** 1036.
[323] Malatesta, V., and Ingold, K. U., *J. Am. Chem. Soc.*, **103,** 609 (1981).
[324] Beckwith, A. L. J., and Easton, C. J., *J. Am. Chem. Soc.*, **103,** 615 (1981).
[325] Batyrbaev, N. A., Zorin, V. V., Zlotskii, S. S., Gren, A. I., and Rakhmankulov, D. L., *Zh. Prikl, Khim. (Leningrad)*, **53,** 1411 (1980); *Chem. Abs.*, **93,** 203649 (1980).
[326] Rol'nik, L. Z., Kalashnikov, S. M., Pastushenko, E. V., Zlotskii, S. S., and Rakhmankulov, D. L., *Zh. Org. Khim.*, **17,** 706 (1981); *Chem. Abs.*, **95,** 61121 (1981).
[327] Zorin, V. V., Zlotskii, S. S., Glukhovtsev, V. G., Il'in, Yu. V. Nikishin, G. I., and Rakhmankulov, D. L., *Zh. Org. Khim.*, **17,** 281 (1981); *Chem. Abs.*, **95,** 41847 (1981).
[328] Encina, M. V., Diaz, S., and Lissi, E., *Int. J. Chem. Kinet.*, **13,** 119 (1981).
[329] Griller, D., Howard, J. A., Marriott, P. R., and Scaiano, J. C., *J. Am. Chem. Soc.*, **103,** 619 (1981).
[330] Das, P. K., Encinas, M. V., Steenken, S., and Scaiano, J. C., *J. Am. Chem. Soc.*, **103,** 4162 (1981).
[331] Skell, P. S., and May, D. D., *J. Am. Chem. Soc.*, **103,** 967 (1981).
[332] Chodak, I., Bakos, D., and Zimanyova, E., *Collect. Czech. Chem. Commun.*, **46,** 484 (1981).

[333] Khokhryakova, N. A., Ariko, N. G., and Mitskevich, N. I., *Dokl. Akad. Nauk BSSR*, **24,** 715 (1980); *Chem. Abs.*, **93,** 220133 (1980).
[334] Estrina, G. Ya, Agisheva, S. A., Kuramshin, E. M., Imashev, U. B., Zlotskii, S. S., and Rakhmankulov, D. L., *Neftekhimiya*, **20,** 264 (1980); *Chem. Abs.*, **93,** 185361 (1980).
[335] Abou-Elenien, G., Rieser, J., Ismail, N., and Wallenfels, K., *Z. Naturforsch.*, **36B,** 391 (1981).
[336] Kopecky, K. R., and Hall, M. C., *Can. J. Chem.*, **59,** 3090 (1981).
[337] Toromanova-Petrova, P., *Khim. Ind. (Sofia)*, **1981,** 152; *Chem. Abs.*, **95,** 114372 (1981).
[338] Malatesta, V., and Ingold, K. U., *J. Am. Chem. Soc.*, **103,** 3094 (1981).
[339] Grossi, L., Lunazzi, L., and Placucci, G., *Tetrahedron Lett.*, **22,** 251 (1981).
[340] Stone, P. G., and Cohen, S. G., *J. Phys. Chem.*, **85,** 1719 (1981).
[341] Ellul, R., Potzinger, P., Reimann, B., and Camilleri, P., *Ber. Bunsenges. Phys. Chem.*, **85,** 407 (1981); *Chem. Abs.*, **95,** 79631 (1981).
[342] Schafer, K., and Asmus, K.-D., *J. Phys. Chem.*, **85,** 852 (1981).
[343] Ahonkhai, S. I., and Whittle, E., *Int. J. Chem. Kinet.*, **13,** 427 (1981).
[344] Evans, B. S., and Whittle, E., *Int. J. Chem. Kinet.*, **13,** 59 (1981).
[345] Zorin, V. V., Batyrbaev, N. A., Zlotskii, S. S., Rakhmankulov, D. L., and Karakhanov, R. A., *Dokl. Akad. Nauk SSSR*, **255,** 626 (1980); *Chem. Abs.*, **94,** 120543 (1981).
[346] Baruch, G., Rajbenbach, L. A., and Horowitz, A., *Int. J. Chem. Kinet.*, **13,** 473 (1981).
[347] Teitel'boim, M. A., Shoikhet, A. A., Kaplunov, M. G., and Vedeneev, V. I., *Kinet. Katal.*, **22,** 298 (1981); *Chem. Abs.*, **95,** 96532 (1981).
[348] Gonen, Y., Horowitz, A., and Rajbenbach, L. A., *Int. J. Chem. Kinet.*, **13,** 219 (1981).
[349] Cadman, P., and Owen, H. L., *J. Chem. Soc., Faraday Trans. 1*, **77,** 1913 (1981).
[350] Davidson, I. M. T., and Matthews, J. I., *J. Chem. Soc., Faraday Trans. 1*, **77,** 2277 (1981).
[351] Aloni, R., Rajbenbach, L. A., and Horowitz, A., *Int. J. Chem. Kinet.*, **13,** 23 (1981).
[352] Chatgilialoglu, C., Ingold, K. U., Scaiano, J. C., and Woynar, H., *J. Am. Chem. Soc.*, **103,** 3231 (1981).
[353] Soppe-Mbang, H., and Gleicher, G. J., *J. Am. Chem. Soc.*, **103,** 4100 (1981).
[354] Ando, T., Ishihara, T., Ohtani, E., and Sawada, H., *J. Org. Chem.*, **46,** 4446 (1981).
[355] Chambers, R. D., Lindley, A. A., Fielding, H. C., Moillet, J. S., and Whittaker, G., *J. Chem. Soc., Perkin Trans. 1*, **1981,** 1064.
[356] Myshkin, V. E., Shostenko, A. G., and Malkov, A. V., *Tr.-Mosk. Khim. Tekhnol. Inst. im. D. I. Mendeleeva*, **107,** 142 (1979); *Chem. Abs.*, **95,** 41865 (1981).
[357] Volovik, S. V., Dyadyusha, G. G., and Staninets, V. I., *Ukr. Khim. Zh. (Russ. Ed.)*, **47,** 720 (1981); *Chem. Abs.*, **95,** 114419 (1981).
[358] Nohara, D., and Sakai, T., *Ind. Eng. Chem. Fundam.*, **19,** 340 (1980); *Chem. Abs.*, **93,** 167184 (1980).
[359] Nohara, D., and Sakai, T., *Sekiyu Gakkaishi*, **24,** 122 (1981); *Chem. Abs.*, **95,** 96443 (1981).
[360] Arnaud, R., Douady, J., and Subra, R., *Nouveau J. Chim.*, **5,** 181 (1981).
[361] Giese, B., and Meixner, J., *Chem. Ber.*, **114,** 2138 (1981).
[362] Koehler, H. J., and Knoll, H., *J. Prakt. Chem.*, **323,** 166 (1981); *Chem. Abs.*, **95,** 131929 (1981)
[363] Parkes, D. A., *Chem. Phys. Lett.*, **77,** 527 (1981); *Chem. Abs.*, **94,** 191130 (1981).
[364] Knoll, H., *React. Kinet. Catal. Lett.*, **15,** 431 (1980); *Chem. Abs.*, **95,** 79629 (1981).
[365] Knoll, H., Nacsa, A., Foergeteg, S., and Berces, T., *React. Kinet. Catal. Lett.*, **15,** 481 (1980); *Chem. Abs.*, **95,** 61005 (1981).
[366] Forrester, A. R., Irikawa, H., Thomson, R. H., Woo, S. O., and King, T. J., *J. Chem. Soc., Perkin Trans. 1*, **1981,** 1712.
[367] McEwen, C. N., and Rudat, M. A., *J. Am. Chem. Soc.*, **103,** 4343 (1981).
[368] Rudat, M. A., and McEwen, C. N., *J. Am. Chem. Soc.*, **103,** 4345 (1981).
[369] Cosa, F., Oexler, E. V., and Staricco, E. H., *J. Chem. Soc., Faraday Trans. 1*, **77,** 253 (1981).
[370] Volovik, S. V., Dyadyusha, G. G., Pen'kovskii, V. V., and Staninets, V. I., *Dopov. Akad. Nauk Ukr. RSR, Ser. B: Geol., Khim. Biol. Nauki*, **1981,** 43; *Chem. Abs.*, **95,** 96541 (1981).
[371] Serov, S. I., Zhuravlev, M. V., Sass, V. P., and Sokolov, S. V., *Zh. Org. Khim.*, **17,** 53 (1981); *Chem. Abs.*, **94,** 208073 (1981).
[372] El Soueni, A., Tedder, J. M., and Walton, J. C., *J. Chem. Soc., Faraday Trans. 1*, **77,** 89 (1981).
[373] El Soueni, A., Tedder, J. M., and Walton, J. C., *J. Fluorine Chem.*, **17,** 51 (1981); *Chem. Abs.*, **95,** 6333 (1981).
[374] Rico, I., Cantacuzene, D., and Wakselman, C., *Tetrahedron Lett.*, **22,** 3405 (1981).
[375] Grigor'ev, N. A., Volkov, N. D., Ger'man, L. S., Nazaretyan, V. P., Yagupol'skii, L. M., and Freidlina, R. Kh., *Izv. Akad. Nauk SSSR, Ser. Khim.*, **1981,** 1082; *Chem. Abs.*, **95,** 114401 (1981).
[376] Ohta, T., and Mizoguchi, I., *Int. J. Chem. Kinet.*, **12,** 717 (1980).
[377] Horowitz, A., and Baruch, G., *Int. J. Chem. Kinet.*, **12,** 883 (1980).

[378] Kruglova, N. V., Dostovalova, V. I., and Freidlina, R. Kh., *Bull. Acad. Sci. USSR*, **29,** 628 (1980).
[379] Kruglova, N. V., and Freidlina, R. Kh., *Bull. Acad. Sci. USSR*, **29,** 1332 (1981).
[380] Kruglova, N. V., and Freidlina, R. Kh., *Izv. Akad. Nauk SSSR, Ser. Khim.*, **1980,** 1857; *Chem. Abs.*, **93,** 238269 (1980).
[381] Kim, S. S., *Bull. Korean Chem. Soc.*, **1,** 45 (1980); *Chem. Abs.*, **94,** 102451 (1981).
[382] Martinez, R. I., Huie, R. E., and Herron, J. T., *Chem. Phys. Lett.*, **72,** 443 (1980); *Chem. Abs.*, **93,** 203653 (1980).
[383] Bartels, H. M., and Boldt, P., *Liebigs Ann. Chem.*, **1981,** 40.
[384] Chambers, R. D., Kelly, N., Musgrave, W. K. R., Jones, W. G. M., and Rendell, R. W., *J. Fluorine Chem.*, **16,** 351 (1980); *Chem. Abs.*, **94,** 29806 (1981).
[385] Ogibin, Yu. N., Elinson, M. N., and Nikishin, G. I., *Bull. Acad. Sci. USSR*, **29,** 747 (1980).
[386] Ogibin, Yu, N., Elinson, M. N., Nikishin, G. I., Kadentsev, V. I., and Chizhov, O. S., *Bull. Acad. Sci. USSR*, **29,** 1320 (1980).
[387] Sirazhitdinova, D. S., Aminov, S. N., and Shostenko, A. G., *Dokl. Akad. Nauk Uzb. SSR*, **1978,** 38; *Chem. Abs.*, **93,** 167151 (1980).
[388] Lugovoi, Yu. M., Tarasova, N. P., and Shostenko, A. G., *Deposited Doc.*, **1979,** VINITI 3784, 481–4; *Chem. Abs.*, **94,** 138897 (1981).
[389] Tarasova, N. P., Shostenko, A. G., and Lugovoi, Yu. M., *Zh. Org. Khim.*, **16,** 1793 (1980); *Chem. Abs.*, **94,** 29817 (1981).
[390] Tarasova, N. P., Shostenko, A. G., Smetannikov, Yu. V., and Lugovoi, Yu. M., *React. Kinet. Catal. Lett.*, **15,** 221 (1980); *Chem. Abs.*, **94,** 173937 (1981).
[391] Zamyslov, R. A., Shostenko, A. G., Dobrov, I. V., and Myshkin, V. E., *Zh. Org. Khim.*, **16,** 897 (1980); *Chem. Abs.*, **93,** 167229 (1980).
[392] Myshkin, V. E., Shostenko, A. G., and Malkov, A. V., *Deposited Doc.*, **1979,** VINITI 3784, 479–80; *Chem. Abs.*, **94,** 138896 (1981).
[393] Gardrat, C., Montaudon, E., and Laland, E., *Bull. Soc. Chim. Belg.*, **89,** 1039 (1980).
[394] Kowari, K., Sugawara, K., Sato, S., and Nagase, S., *Bull. Chem. Soc. Jpn.*, **54,** 1222 (1981).
[395] Sugawara, K., Okazaki, K., and Sato, S., *Chem. Phys. Lett.*, **78,** 259 (1981); *Chem. Abs.*, **94,** 208047 (1981).
[396] Aleksandrov, E. N., Arutyunov, V. S., Dubrovina, I. V., and Kozlov, S. N., *Kinet. Katal.*, **21,** 1323 (1980); *Chem. Abs.*, **94,** 29826 (1981).
[397] Canosa, C. E., Marshall, R. M., and Sheppard, A., *Int. J. Chem. Kinet.*, **13,** 295 (1981).
[398] Canosa, C. E., and Marshall, R. M., *Int. J. Chem. Kinet.*, **13,** 303 (1981).
[399] Sugawara, K., Okazaki, K., and Sato, S., *Bull. Chem. Soc. Jpn.*, **54,** 358 (1981).
[400] Pukhal'skaya, G. V., Chebotarev, N. F., Kolovskii, V. B., and Pshezhetskii, S. Ya., *Kinet. Katal.*, **21,** 1063 (1980); *Chem. Abs.*, **93,** 238276 (1980).
[401] Pukhal'skaya, G. V., and Pshezhetskii, S. Ya., *Kinet. Katal.*, **21,** 1066 (1980); *Chem. Abs.*, **93,** 238277 (1980).
[402] Pukhal'skaya, G. V., and Pshezhetskii, S. Ya., *Kinet. Katal.*, **21,** 1589 (1980); *Chem. Abs.*, **94,** 102460 (1981).
[403] Concannon, C., and Rowland, F. S., *J. Phys. Chem.*, **85,** 89 (1981).
[404] Brarda, M. T., and Staricco, E. H., *An. Asos. Quim. Argent.*, **69,** 147 (1981); *Chem. Abs.*, **95,** 96528 (1981).
[405] Barkalov, I. M., Gol'danskii, V. I., Kiryukhin, D. P., and Zanin, A. M., *Chem. Phys. Lett.*, **73,** 273 (1980); *Chem. Abs.*, **93,** 203572 (1980).
[406] Applequist, D. E., and Gdanski, R. D., *J. Org. Chem.*, **46,** 2502 (1981).
[407] *Cf. Org. Reaction Mech.*, **1976,** 144.
[408] Birchall, J. M., Burger, K., Haszeldine, R. N., and Nona, S. N., *J. Chem. Soc., Perkin Trans. 1*, **1981,** 2080.
[409] Heasley, V. L., Shellhamer, D. F., Heasley, L. E., Yaeger, D. B., and Heasley, G. E., *J. Org. Chem.*, **45,** 4649 (1980).
[410] Kasashkina, O. T., Kartasheva, Z. S., and Gagarina, A. B., *Izv. Akad. Nauk SSSR, Ser. Khim.*, **1981,** 536; *Chem. Abs.*, **95,** 41866 (1981).
[411] Kitamura, T., Tsunashima, S., and Sato, S., *Bull. Chem. Soc. Jpn.*, **54,** 55 (1981).
[412] Bandow, H., Okuda, M., and Akimoto, H., *J. Phys. Chem.*, **84,** 3604 (1980).
[413] Lessard, J., Mondon, M., and Touchard, D., *Can. J. Chem.*, **59,** 431 (1981).
[414] Zawadzki, S., and Zwierzak, A., *Tetrahedron*, **37,** 2675 (1981).
[415] Kopecky, K. R., and Hall, M. C., *Can. J. Chem.*, **59,** 3095 (1981).
[416] Maletin, Yu. A., Strizhakova, N. G., and Sheka, I. A., *Zh. Obshch. Khim.*, **51,** 1119 (1981); *Chem. Abs.*, **95,** 96560 (1981).

417 Pen'kovskii, V. V., *Teor. Eksp. Khim.*, **17,** 531 (1981); *Chem. Abs.*, **95,** 168125 (1981).
418 Lugovoi, Yu. M., Myshkin, V. E., and Shostenko, A. G., *Deposited Doc.*, **1979,** VINITI 3784, 475–8; *Chem. Abs.*, **94,** 138895 (1981).
419 Icli, S., *Chim. Acta Turc.*, **7,** 261 (1979); *Chem. Abs.*, **94,** 3437 (1981).
420 Alberti, A., Guerra, M., Cabiddu, S., and Pedulli, G. F., *Gazz. Chim. Ital.*, **109,** 647 (1979); *Chem. Abs.*, **93,** 203632 (1980).
421 Podesta, J. C., Chopa, A. B., and Ayala, A. D., *J. Organomet. Chem.*, **212,** 163 (1981).
422 Davies, A. G., and Hawari, J. A. A., *J. Organomet. Chem.*, **201,** 221 (1980).
423 Richard, C., Boiveaut, A., and Martin, R., *Int. J. Chem. Kinet.*, **12,** 921 (1980).
424 Lunazzi, L., Placucci, G., and Grossi, L., *J. Chem. Soc. Perkin Trans. 2*, **1981,** 703.
425 Pasto, D. J., Warren, S. E., and Morrison, M. A., *J. Org. Chem.*, **46,** 2837 (1981).
426 Pasto, D. J., and Warren, S. E., *J. Org. Chem.*, **46,** 2842 (1981).
427 Suda, M., *Tetrahedron Lett.*, **22,** 2395 (1981).
428 Churkina, T. D., Petrova, R. G., and Dostovalova, V. I., *Izv. Akad. Nauk SSSR, Ser. Khim.*, **1981,** 156; *Chem. Abs.*, **94,** 173888 (1981).
429 Petrova, R. G., Gasanov, R. G., and Churkina, T. D., *Izv. Akad. Nauk SSSR, Ser. Khim.*, **1981,** 379; *Chem. Abs.*, **95,** 6095 (1981).
430 Mirskova, A. N., Martynov, A. V., and Voronkov, M. G., *J. Org. Chem. USSR*, **16,** 1766 (1981).
431 Ito, O., and Matsuda, M., *J. Am. Chem. Soc.*, **103,** 5871 (1981).
432 Ito, O., Nogami, K., and Matsuda, M., *J. Phys. Chem.*, **85,** 1365 (1981).
433 Kojo, S., and Sano, S., *J. Chem. Soc., Perkin Trans. 1*, **1981,** 2864.
434 Back, T. G., and Collins, S., *J. Org. Chem.*, **46,** 3249 (1981).
435 Russell, G. A., and Hershberger, J., *J. Am. Chem. Soc.*, **102,** 7603 (1980).
436 Volovik, S. V., Dyadyusha, G. G., and Staninets, V. I., *Zh. Org. Khim.*, **17,** 915 (1981); *Chem. Abs.*, **95,** 96875 (1981).
437 See *Org. Reaction Mech.*, **1980,** 116.
438 Wolff, S., and Agosta, W. C., *J. Chem. Res. (S)*, **1981,** 78.
439 Beckwith, A. L. J., Phillipou, G., and Serelis, A. K., *Tetrahedron Lett.*, **22,** 2811 (1981).
440 Bachi, M. D., and Hoornaert, C., *Tetrahedron Lett.*, **22,** 2693 (1981).
441 Bachi, M. D., and Hoornaert, C., *Tetrahedron Lett.*, **22,** 2689 (1981).
442 Wilt, J. W., *J. Am. Chem. Soc.*, **103,** 5251 (1981).
443 Atmaram, S., Forrester, A. R., Gill, M., and Thomson, R. H., *J. Chem. Soc., Perkin Trans. 1*, **1981,** 1721.
444 Stella, L., Raynier, B., and Surzur, J.-M., *Tetrahedron*, **37,** 2843 (1981).
445 Brooke, G. M., and Wallis, D. I., *J. Chem. Soc., Perkin Trans. 1*, **1981,** 1659.
446 Bougrois, J.-L., Stella, L., and Surzur, J.-M., *Tetrahedron Lett.*, **22,** 61 (1981).
447 Chatzopoulos, M., and Montheard, J. P., *Rev. Roum. Chim.*, **26,** 275 (1981).
448 Suginome, H., Isayama, S., Maeda, N., Furusaki, A., and Katayama, C., *J. Chem. Soc., Perkin Trans. 1*, **1981,** 2963.
449 Benati, L., and Montevecchi, P. C., *J. Org. Chem.*, **46,** 4570 (1981).
450 Choo, K. Y., and Benson, S. W., *Int. J. Chem. Kinet.*, **13,** 833 (1981).
451 Gilbert, B. C., Marshall, P. D. R., Norman, R. O. C., Pineda, N., and Williams, P. S., *J. Chem. Soc., Perkin Trans. 2*, **1981,** 1392.
452 Kumar, A., *J. Am. Chem. Soc.*, **103,** 5179 (1981).
453 Macdonald, T. L., and O'Dell, D. E., *J. Org. Chem.*, **46,** 1501 (1981).
454 Bensadoun, N., Brun, P., Casanova, J., and Waegell, B., *J. Chem. Res. (S)*, **1981,** 236.
455 Rol'nik, L. Z., Kalashnikov, S. M., Pastushenko, E. V., Zlotskii, S. S., and Rakhmankulov, D. L., *J. Org. Chem. USSR*, **17,** 614 (1981).
456 Rol'nik, L. Z., Kalashnikov, S. M., Pastushenko, E. V., Zlotskii, S. S., Imashev, U. B., and Rakhmankulov, D. L., *Zh. Obshch. Khim.*, **51,** 778 (1981); *Chem. Abs.*, **95,** 79648 (1981).
457 Rol'nik, L. Z., Kalashnikov, S. M., Pastushenko, E. V., Zlotskii, S. S., Imashev, U. B., and Rakhmankulov, D. L., *Zh. Org. Khim.*, **17,** 539 (1981); *Chem. Abs.*, **95,** 41863 (1981).
458 Kalashnikov, S. M., Zlotskii, S. S., Imashev, U. B., and Rakhmankulov, D. L., *J. Org. Chem. USSR*, **16,** 1379 (1980).
459 Ramazanov, O. M., Kalashnikov, S. M., Skurko, M. R., Pastushenko, E. V., and Karakhanov, R. A., *Izv. Akad. Nauk SSSR, Ser. Khim.*, **1980,** 2635; *Chem. Abs.*, **94,** 120527 (1981).
460 Petryaev, E. P., Kosobutskii, V. S., and Shadyro, O. I., *J. Org. Chem. USSR*, **17,** 770 (1981).
461 Petryaev, E. P., Shadyro, O. I., and Vasil'ev, G. N., *Zh. Org. Khim.*, **16,** 227 (1980); *Chem. Abs.*, **92,** 214559 (1980).
462 Glukhovtsev, V. G., Zaitseva, G. I., and Il'in, Yu. V., *J. Org. Chem. USSR*, **17,** 438 (1981).

463 Zorin, V. V., Zlotskii, S. S., Glukhovtsev, V. G., Il'in, Yu. V., Nikishin, G. I., and Rakhmankulov, D. L., *J. Org. Chem. USSR*, **17,** 230 (1981).
464 Babcock, B. W., Dimmel, D. R., Graves, D. P., and McKelvey, R. D., *J. Org. Chem.*, **46,** 736 (1981).
465 Suginome, H., Isayama, S., Maeda, N., Furusaki, A., and Katayama, C., *J. Chem. Soc., Perkin Trans. 1*, **1981,** 2963.
466 Lapshova, A. A., Stashina, G. A., Zorin, V. V., Zlotskii, S. S., Zhulin, V. M., and Rakhmankulov, D. L., *Zh. Org. Khim.*, **16,** 1251 (1980); *Chem. Abs.*, **93,** 203741 (1980).
467 Lapshova, A. A., Zorin, V. V., Zlotskii, S. S., Karakhanov, R. A., and Rakhmankulov, D. L., *J. Org. Chem. USSR*, **16,** 2163 (1980).
468 Petrova, R. G., Gasanov, R. G., and Churkina, T. D., *Bull. Acad. Sci. USSR*, **30,** 298 (1981).
469 Marriott, P. R., and Griller, D., *J. Am. Chem. Soc.*, **103,** 1521 (1981).
470 Barton, D. H. R., and Motherwell, W. B., *Pure Appl. Chem.*, **53,** 15, 1081 (1981).
471 Barton, D. H. R., Motherwell, W. B., and Stange, A., *Synthesis*, **1981,** 743.
472 Barton, D. H. R., Motherwell, R. S. H., and Motherwell, W. B., *J. Chem. Soc., Perkin Trans. 1*, **1981,** 2363.
473 Barton, D. H. R., Bringmann, G., and Motherwell, W. B., *J. Chem. Soc., Perkin Trans. 1*, **1980,** 2665.
474 Barton, D. H. R., Bringmann, G., Lamotte, G., Motherwell, W. B., Motherwell, R. S. H., and Porter, A. E. A., *J. Chem. Soc., Perkin Trans. 1*, **1980,** 2657.
475 Kim, S. S., *Taehan Hwahakhoe Chi.*, **24,** 250 (1980); *Chem. Abs.*, **94,** 102416 (1981).
476 Gilbert, B. C., King, D. M., and Thomas, C. B., *J. Chem. Soc., Perkin Trans. 2*, **1980,** 1821.
477 Gilbert, B. C., King, D. M., and Thomas, C. B., *J. Chem. Soc., Perkin Trans. 2*, **1981,** 1186.
478 Freidlina, R. Kh., and Terent'ev, A. B., *Adv. Free-Radical Chem.*, **6,** 1 (1980).
479 Roberts, C., and Walton, J. C., *J. Chem. Soc., Perkin Trans. 2*, **1981,** 553.
480 Fossey, J., and Nedelec, J.-Y., *Tetrahedron*, **37,** 2967 (1981).
481 Freidlina, R. Kh., Rybakova, N. A., Dostovalova, V. I., and Kiseleva, L. N., *Dokl. Chem.*, **255,** 566 (1980).
482 Gasanov, R. G., *Bull. Acad. Sci. USSR*, **29,** 1583 (1980).
483 Burton, D. J., and Inouye, Y., *J. Fluorine Chem.*, **18,** 459 (1981).
484 *Org. Reaction Mech.*, **1980,** 118–120; Beckwith, A. L. J., and Moad, G., *J. Chem. Soc., Perkin Trans. 2*, **1980,** 1473.
485 Davies, A. G., Hawari, J. A.-A., Muggleton, B., and Tse, M. W., *J. Chem. Soc., Perkin Trans. 2*, **1981,** 1132.
486 Dolbier, W. R., Al-Sader, B. H., Sellers, S. F., and Koroniak, H., *J. Am. Chem. Soc.*, **102,** 2138 (1981).
487 Stavinoha, J. L., and Mariano, P. S., *J. Am. Chem. Soc.*, **103,** 3136 (1981).
488 Wagner, P. J., Liu, K. C., and Noguchi, Y., *J. Am. Chem. Soc.*, **103,** 3837 (1981).
489 Wong, P. C., and Griller, D., *J. Org. Chem.*, **46,** 2327 (1981).
490 Effio, A. A., and Ingold, K. U., *J. Org. Chem.*, **46,** 96, (1981).
491 Ingold, K. U., Maillard, B., and Walton, J. C., *J. Chem. Soc., Perkin Trans. 2*, **1981,** 970.
492 Barton, T. J., and Jacobi, S. A., *J. Am. Chem. Soc.*, **102,** 7979 (1980).
493 Franz, J. A., and Camaioni, D. M., *J. Org. Chem.*, **45,** 5247 (1980).
494 Nazran, A. S., and Warkentin, J., *J. Am. Chem. Soc.*, **103,** 236 (1981).
495 Gilbert, B. C., Norman, R. O. C., and Williams, P. S., *J. Chem. Soc., Perkin Trans. 2*, **1981,** 1401.
496 John, D. I., Tyrrell, N. D., and Thomas, E. J., *J. Chem. Soc., Chem. Commun.*, **1981,** 901.
497 Baldwin, J. E., Beckwith, A. L. J., Davis, A. P., Procter, G., and Singleton, K. A., *Tetrahedron*, **37,** 218 (1981).
498 Okabe, M., Osawa, T., and Tada, M., *Tetrahedron Lett.*, **22,** 1899 (1981).
499 Solodovnikov, S. P., Prokof'ev, A. I., Bubnov, N. N., Malisheva, N. A., Belostotskaya, I. S., Ershov, V. V., and Kabachnik, M. I., *Magn. Reson. Relat. Phenom. Proc. Cong. AMPERE, 20th*, **1978,** 174; *Chem. Abs.*, **93,** 238395 (1980).
500 Singh, G., Singh, D., and Ram, R. N., *Tetrahedron Lett.*, **22,** 2213 (1981).
501 Takeshita, H., Mametsuka, H., Chisaka, A., and Matsuo, N., *Chem. Lett.*, **1981,** 73.
502 Villareal, J. A., and Grubbs, E. J., *J. Org. Chem.*, **46,** 260 (1981).
503 Mizuno, H., Matsudo, M., and Iino, M., *J. Org. Chem.*, **46,** 520 (1981).
504 Bleeker, I. P., and Engberts, J. B. F. N., *J. Org. Chem.*, **46,** 1012 (1981).
505 Stamegna, A. P., and McEwen, W. E., *J. Org. Chem.*, **46,** 1653 (1981).
506 Erker, G., and Rosenfeldt, F., *Tetrahedron Lett.*, **22,** 1379 (1981).
507 Abramovitch, R. A., Abramovitch, D. A., and Tomasik, P., *J. Chem. Soc., Chem. Commun.*, **1981,** 561.
508 Ollis, W. D., Somanathan, R., and Sutherland, I. O., *J. Chem. Soc., Chem. Commun.*, **1981,** 573.
509 Ollis, W. D., Sutherland, I. O., and Thebtaranonth, Y., *J. Chem. Soc., Perkin Trans. 1*, **1981,** 1981.

510 Fry, A. J., Jain, P. S., and Krieger, R. L., *J. Organomet. Chem.*, **214,** 381 (1981).
511 Minisci, F., and Porta, O., *Chim. Ind. (Milan)*, **62,** 769 (1980); *Chem. Abs.*, **94,** 64664 (1980).
512 Sainsbury, M., *Tetrahedron*, **36,** 3327 (1980).
513 Arnaud, R., Douady, J., and Subra, R., *Pure Appl. Chem.*, **5,** 181 (1981).
514 Herndon, W. C., *J. Org. Chem.*, **46,** 2119 (1981).
515 Ogata, Y., Tomizawa, K., Furuta, K., and Kato, H., *J. Chem. Soc., Perkin Trans. 2*, **1981,** 110.
516 Nakabayashi, T., Horii, T., Kawamura, S., and Abe, S., *Bull. Chem. Soc. Jpn.*, **54,** 2535 (1981).
517 Bolton, R., Dailly, B. N., Hirakubo, K., Lee, K. H., and Williams, G. H., *J. Chem. Soc., Perkin Trans. 2*, **1981,** 1109.
518 Neta, P., and Behar, D., *J. Am. Chem. Soc.*, **103,** 103 (1981).
519 Terashima, M., Seki, K., Yoshida, C., and Kanaoka, V., *Heterocycles*, **15,** 1075 (1981).
520 Nardelli, M., Pelizzi, C., Predieri, G., and Chiusoli, G. P., *J. Organomet. Chem.*, **204,** 75 (1981).
521 Sakaguchi, Y., Hayashi, H., and Nagakura, S., *Bull. Chem. Soc. Jpn.*, **53,** 3059 (1980).
522 Novroćik, J., Movrocíková, M., and Titz, M., *Collect. Czech. Chem. Commun.*, **45,** 3140 (1980).
523 Sebedio, J. L., Sorba, J., Fossey, J., and Lefort, D., *Tetrahedron*, **37,** 2829 (1981).
524 Citterio, H., Minisci, F., and Franchi, V., *J. Org. Chem.*, **45,** 4752 (1980).
525 Palla, G., *Tetrahedron*, **37,** 2917 (1981).
526 Kira, M., and Sakurai, H., *Chem. Lett.*, **1981,** 927.
527 Sweany, R., Butler, S. C., and Halpern, J., *J. Organomet. Chem.*, **213,** 487 (1981).
528 Berberova, N. T., Ivakhnenko, E. P., Bubnov, N. N., and Okhlobystin, O. Yu., *Khim. Geterosikl. Soedin*, **1980,** 1568; *Chem. Abs.*, **94,** 138880 (1981).
529 Aver'yanov, V. A., Zarytovskii, V. M., and Shvets, V. F., *J. Org. Chem. USSR*, **17,** 40 (1981).
530 So, Y.-H., and Miller, L. L., *J. Am. Chem. Soc.*, **102,** 7119 (1980).
531 So, Y.-H., and Miller, L. L., *J. Am. Chem. Soc.*, **103,** 4204 (1981).
532 Tiecco, M., *Pure Appl. Chem.*, **53,** 239 (1981).
533 Volovik, S. V., Dyadyusha, G. G., and Staninets, V. I., *J. Org. Chem. USSR*, **17,** 481 (1981).
534 Maiolo, F., Testaferri, L., Tiecco, M., and Tingoli, M., *Tetrahedron Lett.*, **22,** 2023 (1981).
535 Volovik, S. V., *Dopov. Akad. Nauk Ukr., RSR, Ser. B: Geol., Khim. Biol. Nauki*, **1980,** 38; *Chem. Abs.*, **94,** 173909 (1981).
536 Colonna, M., Greci, L., and Poloni, M., *Tetrahedron Lett.*, **22,** 1143 (1981).
537 Glover, S. A., Golding, S. L., Goosen, A., and McCleland, C. W., *J. Chem. Soc., Perkin Trans. 1*, **1981,** 842.
538 Tully, F. P., Ravishankara, A. R., Thompson, R. L., Nicovich, J. M., Shah, R. C., Kreutter, N. M., and Wine, P. H., *J. Phys. Chem.*, **85,** 2262 (1981).
539 Kenley, R. A., Davenport, J. E., and Hendry, D. G., *J. Phys. Chem.*, **85,** 2740 (1981).
540 Nicovich, J. M., Thompson, R. L., and Ravishankara, A. R., *J. Phys. Chem.*, **85,** 2913 (1981).
541 Apatu, J. O., Chapman, D. C., and Heaney, H., *J. Chem. Soc., Chem. Commun.*, **1981,** 1079.
542 Bottura, G., Bubani, B., and Barker, G. C., *J. Electroanal. Chem. Interfacial Electrochem.*, **119,** 331 (1981); *Chem. Abs.*, **95,** 6115 (1981).
543 Chekulaev, V. P., Shevchuk, I. M., *Eesti NSV Tead, Akad. Toim, Keem* **30,** 138 (1981); *Chem. Abs.*, **95,** 41889 (1981).
544 Eberhardt, E., *J. Am. Chem. Soc.*, **103,** 3876 (1981).
545 Sawaki, Y., Kato, H., and Ogata, Y., *J. Am. Chem. Soc.*, **103,** 3832 (1981).
546 Gilbert, B. C., Norman, R. O. C., and Williams, P. S., *J. Chem. Soc., Perkin Trans. 2*, **1981,** 207.
547 Eberson, L., and Jönsson, L., *J. Chem. Soc., Chem. Commun.*, **1980,** 1187.
548 Eberson, L., and Jönsson, L., *J. Chem. Soc., Chem. Commun.*, **1981,** 133.
549 Madhavan, V., and Schuler, R. H., *Radiat. Phys. Chem.*, **16,** 139 (1980).
550 Crans, D. C., and Snyder, J. P., *J. Am. Chem. Soc.*, **102,** 7152 (1980).
551 Borden, W. T., and Davidson, E. R., *Acc. Chem. Res.*, **14,** 69 (1981).
552 Hoffmann, R., Eisenstein, O., and Balaban, A. T., *Proc. Natl. Acad. Sci. U.S.A.*, **77,** 5588 (1980).
553 Dixon, D. A., Dunning, T. H., Eades, R. A., and Kleier, D. A., *J. Am. Chem. Soc.*, **103,** 2878 (1981).
554 Mazur, M. R., and Berson, J. A., *J. Am. Chem. Soc.*, **103,** 684 (1981).
555 Seeger, D. E., Hilinski, E. F., and Berson, J. A., *J. Am. Chem. Soc.*, **103,** 720 (1981).
556 Verhoeven, J. W., *Recl. Trav. Chim. Pays-Bas*, **99,** 369 (1980).
557 Gajewski, J. J., *Isr. J. Chem.*, **21,** 169 (1981).
558 Murai, H., Ribo, J., Torres, M., and Strausz, O. P., *J. Am. Chem. Soc.*, **103,** 6422 (1981).
559 Inoue, Y., Ueda, Y., and Hakushi, T., *J. Am. Chem. Soc.*, **103,** 1806 (1981).
560 Chang, M. H., and Crawford, R. J., *Can. J. Chem.*, **59,** 2556 (1981).
561 Schneider, M. P., and Bippi, H., *J. Am. Chem. Soc.*, **102,** 7363 (1980).
562 Gajewski, J. J., and Salazar, J. Del C., *J. Am. Chem. Soc.*, **103,** 4145 (1981).

563 Adam, W., and De Lucchi, O., *J. Org. Chem.*, **46,** 4133 (1981).
564 Adam, W., Carballeira, N., and De Lucchi, O., *J. Am. Chem. Soc.*, **103,** 6406 (1981).
565 Little, R. D., and Muller, G. W., *J. Am. Chem. Soc.*, **103,** 2744 (1981).
566 Billups, W. E., Buynak, J. D., and Butler, D., *J. Org. Chem.*, **45,** 4636 (1980).
567 Lockhart, T. P., Comita, P. B., and Bergman, R. G., *J. Am. Chem. Soc.*, **103,** 4082 (1981).
568 Nakayama, J., Dan, S., and Hoshino, M., *J. Chem. Soc., Perkin Trans. 1*, **1981,** 413.
569 Schwarz, W., Trautman, W., and Musso, H., *Chem. Ber.*, **114,** 990 (1981).
570 Hall, H. K., and Abdelkader, M., *J. Org. Chem.*, **46,** 2948 (1981).
571 Mayer, W., Wendisch, D., Born, L., and Hartmann, W., *Chem. Ber.*, **114,** 1287 (1981).
572 Uyehara, T., Ohnuma, T., Saito, T., Kato, T., Yoshida, T., and Takahashi, K., *J. Chem. Soc., Chem. Commun.*, **1981,** 127.
573 Benati, L., Montevecchi, P. C., Spagnolo, P., and Tundo, A., *J. Chem. Soc., Perkin Trans. 1*, **1981,** 1544.
574 Grimme, W., Schumachers, L., Roth, W. R., and Breuckmann, R., *Chem. Ber.*, **114,** 3197 (1981).
575 Kaupp, G., and Schmitt, D., *Chem. Ber.*, **113,** 3932 (1980).
576 McEwen, W. E., Hernandez, M. A., Ling, C.-F., Marmugi, E., Padronaggio, R. M., Zepp, C. M., and Lubinkowski, J. J., *J. Org. Chem.*, **46,** 1656 (1981).
577 Imai, T., and Nishida, S., *Can. J. Chem.*, **59,** 2503 (1981).
578 Adembri, G., Camparini, A., Ponticelli, F., and Tedeschi, P., *J. Chem. Soc., Perkin Trans. 1*, **1981,** 1703.
579 Ito, Y., Umehara, Y., Yamada, Y., and Matsuura, T., *J. Chem. Soc., Chem. Commun.*, **1980,** 1160.
580 Ito, Y., Nishimura, H., Matsuura, T., and Hayashi, H., *J. Chem. Soc., Chem. Commun.*, **1981,** 1187.
581 Freilich, S. C., and Peters, K. S., *J. Am. Chem. Soc.*, **103,** 6255 (1981).
582 Baulch, D. L., Colburn, A., Lenney, P. W., and Montague, D. C., *J. Chem. Soc., Faraday Trans. 1*, **77,** 1803 (1981).
583 Wagner, P. J., Lindstrom, M. J., Sedon, J. H., and Ward, D. R., *J. Am. Chem. Soc.*, **103,** 3842 (1981).
584 Wagner, P. J., Liu, K.-C., and Noguchi, Y., *J. Am. Chem. Soc.*, **103,** 3837 (1981).
585 Chatgilialoglu, C., Malatesta, V., and Ingold, K. U., *J. Phys. Chem.*, **84,** 3597 (1980).
586 Compton, D. A. C., Chatgilialoglu, C., Mantsch, H. H., and Ingold, K. U., *J. Phys. Chem.*, **85,** 3093 (1981).
587 Chatgilialoglu, C., and Ingold, K. U., *Can. J. Chem.*, **59,** 1745 (1981).
588 Aurich, H. G., and Czepluch, H., *Tetrahedron*, **36,** 3543 (1980).
589 Gasanov, R. G., *Izv. Akad. Nauk SSSR, Ser. Khim.*, **1981,** 1134; *Chem. Abs.*, **95,** 96854 (1981).
590 Kotake, Y., and Kuwata, K., *Bull. Chem. Soc. Jpn.*, **54,** 394 (1981).
591 Darcy, R., *J. Chem. Soc., Perkin Trans. 2*, **1981,** 1089.
592 Kirino, Y., Ohkuma, T., and Kwan, T., *Chem. Pharm. Bull.*, **29,** 29 (1981); *Chem. Abs.*, **94,** 173934 (1981).
593 Liu, Y.-C., Jiang, Z.-Q., and Wu, S.-P., *Kao Teng Hsueh Hsiao Hua Hsueh Hsueh Pao*, **1,** 67 (1980); *Chem. Abs.*, **95,** 6106 (1981).
594 Janzen, E. G., and Coulter, G. A., *Tetrahedron Lett.*, **22,** 615 (1981).
595 Draney, D., and Kingsbury, C. A., *J. Am. Chem. Soc.*, **103,** 1041 (1981).
596 Plancherel, D., and Eaton, D. R., *Can. J. Chem.*, **59,** 156 (1981).
597 Barbarin, F., Fabre, C., Germain, J. P., Michon, J., Rassat, A., and Tchapla, A., *Nouveau J. Chim.*, **4,** 437 (1980); *Chem. Abs.*, **93,** 238669 (1980).
598 Tronchet, J. M. J., Winter-Mihaly, E., Habashi, F., Schwarzenbach, D., Likic, U., and Geoffrey, M., *Helv. Chim. Acta*, **64,** 610 (1981).
599 Keana, J. F. W., Tamura, T., McMillen, D. A., and Jost, P. C., *J. Am. Chem. Soc.*, **103,** 4904 (1981).
600 Mukai, K., Yano, H., and Ishizu, K., *Tetrahedron Lett.*, **22,** 1093 (1981).
601 Tkacheva, O. P., Martin, V. V., Volodarskii, L. B., and Buchachenko, A. L., *Izv. Akad. Nauk SSSR, Ser. Khim.*, **1981,** 1153; *Chem. Abs.*, **95,** 96401 (1981).
602 Chatgilialoglu, C., and Ingold, K. U., *J. Am. Chem. Soc.*, **103,** 4833 (1981).
603 Yang, G. C., and Joshi, A., *J. Phys. Chem.*, **84,** 228 (1980).
604 Boyer, M. P., Morton, J. R., and Preston, K. F., *J. Phys. Chem.*, **84,** 2989 (1980).
605 Angert, J. L., Hurst, A. F., Lowry, S. A., and Landolt, R. G., *J. Org. Chem.*, **46,** 168 (1981).
606 MacAlpine, D. K., Porte, A. L., and Sim, G. A., *J. Chem. Soc., Perkin Trans. 1*, **1981,** 2533.
607 Berti, C., Greci, L., and Poloni, M., *J. Chem. Soc., Perkin Trans. 1*, **1981,** 1610.
608 Colonna, M., Greci, L., and Poloni, M., *J. Heterocycl. Chem.*, **17,** 1473 (1980).
609 Ota, T., Masuda, S., and Tanka, H., *Chem. Lett.*, **1981,** 411.
610 Rivière, P., Richelme, S., Rivière-Baudet, M., Satgé, J., Riley, P. I., Lappert, M. F., Dunoguès, J., and Calas, R., *J. Chem. Res.*, **1981,** (S) 130, (M) 1663.

[611] Aurich, H. G., Duggal, S. K., Höhlein, P., and Klingelhöfer, H.-G., *Chem. Ber.*, **114,** 2431 (1981).
[612] Aurich, H. G., Duggal, S. K., Höhlein, P., Klingelhöfer, H.-G., and Weiss, W., *Chem. Ber.*, **114,** 2440 (1981).
[613] Aurich, H. G., Dersch, W., and Heinrich, J.-M., *Liebigs Ann. Chem.*, **1981,** 1271.
[614] Lissi, E. A., Rubio, M. A., Araya, D., and Zanocco, G., *Int. J. Chem. Kinet.*, **12,** 871 (1980).
[615] Smith, R. J., and Pagni, R. M., *J. Org. Chem.*, **46,** 4307 (1981).
[616] Singh, H., and Tedder, J. M., *J. Chem. Soc., Chem. Commun.*, **1981,** 70.
[617] Keute, J. S., Anderson, D. R., and Koch, T. H., *J. Am. Chem. Soc.*, **103,** 5434 (1981).
[618] Banks, R. E., Brown, A. K., Haszeldine, R. N., and Jefferson, J., *J. Chem. Soc., Perkin Trans. 1*, **1981,** 1068.
[619] Banks, R. E., Birchall, J. M., Haszeldine, R. N., Hughes, R. A., Nona, S. N., and Stephens, C. W., *J. Chem. Soc., Perkin Trans. 1*, **1981,** 455.
[620] Cowell, A. B., and Tamborski, C., *J. Fluorine Chem.*, **17,** 345 (1981).
[621] Banks, R. E., Haszeldine, R. N., and Murray, M. B., *J. Fluorine Chem.*, **17,** 561 (1981).
[622] Forrester, A. R., Henderson, J., and Hepburn, S. P., *J. Chem. Soc., Perkin Trans. 1*, **1981,** 1165.
[623] Berti, C., and Greci, L., *J. Org. Chem.*, **46,** 3060 (1981).
[624] Kamimori, M., Umeda, M., Sakuragi, H., Tokumaru, K., and Yoshida, M., *Nippon Kagaku Kaishi*, **1981,** 469; *Chem. Abs.*, **95,** 60973 (1981).
[625] Anastassopoulou, J. D., Chandrinos, J. D., and Rakintzis, N. Th., *Radiat. Phys. Chem.*, **17,** 55 (1981); *Chem. Abs.*, **94,** 173926 (1981).
[626] Frost, J. W., and Rastetter, W. H., *J. Am. Chem. Soc.*, **103,** 5242 (1981).
[627] Castedo, L., Puga, A., Saa, J. M., and Suau, R., *Tetrahedron Lett.*, **22,** 2233 (1981).
[628] Perkins, M. J., *Adv. Phys. Org. Chem.*, **17,** 1 (1980); *Chem. Abs.*, **94,** 173741 (1981).
[629] Rehorek, D., *Z. Chem.*, **20,** 325 (1980).
[630] Janzen, E. G., Shetty, R. V., and Kunanec, S. M., *Can. J. Chem.*, **59,** 756 (1981).
[631] Kaur, H., Leung, K. H. W., and Perkins, M. J., *J. Chem. Soc., Chem. Commun.*, **1981,** 142.
[632] Moger, G., Simon, P., and Rockenbauer, A., *React. Kinet. Catal. Lett.*, **14,** 301 (1980); *Chem. Abs.*, **94,** 191263 (1981).
[633] Gasanov, R. G., Ivanova, L. V., and Freidlina, R. Kh., *Dokl. Akad. Nauk SSSR*, **255,** 1156 (1980); *Chem. Abs.*, **94,** 173935 (1981).
[634] Suehiro, T., Masuda, S., Motoyama, N., and Sasaki, M., *Chem. Lett.*, **1980,** 1329.
[635] Coxon, J. M., Gilbert, B. C., and Norman, R. O. C., *J. Chem. Soc., Perkin Trans. 2*, **1981,** 379.
[636] Belevskii, V. N., and Yarkov, S. P., *Dokl. Akad. Nauk SSSR*, **254,** 1417 (1980); *Chem. Abs.*, **94,** 64832 (1981).
[637] Simon, P., Sumegi, L., Rockenbauer, A., Nemes, I., and Tudos, F., *React. Kinet. Catal. Lett.*, **15,** 493 (1980); *Chem. Abs.*, **95,** 96533 (1981).
[638] Zubarev, V. E., *Vestn. Mosk., Univ., Ser. 2: Khim*, **21,** 535 (1980); *Chem. Abs.*, **94,** 120541 (1981).
[639] Harbour, J. R., Issler, S. L., and Hair, M. L., *J. Am. Chem. Soc.*, **102,** 7778 (1980).
[640] Babin, V. N., Gumenyuk, V. V., Solodovnikov, S. P., and Belousov, Yu. A., *Z. Naturforsch.*, **36B,** 400 (1981); *Chem. Abs.*, **95,** 41860 (1981).
[641] See *Org. Reaction Mech.*, **1980,** 162.
[642] Rosenthal, I., Mossoba, M. M., and Riesz, P., *J. Phys. Chem.*, **85,** 2398 (1981).
[643] Joshi, A., and Yang, G. C., *J. Org. Chem.*, **46,** 3736 (1981).
[644] Chignell, C. F., Kalyanaraman, B., Sik, R. H., and Mason, R. P., *Photochem. Photobiol.*, **34,** 147 (1981).
[645] Franzi, R., and Geoffrey, M., *J. Organomet. Chem.*, **218,** 321 (1981).
[646] Ekwenchi, M. M., Safarik, I., and Strausz, O. P., *Int. J. Chem. Kinet.*, **13,** 799 (1981).
[647] Iyer, R. S., and Rowland, F. S., *J. Phys. Chem.*, **85,** 2488 (1981).
[648] Iyer, R. S., and Rowland, F. S., *J. Phys. Chem.*, **85,** 2493 (1981).
[649] Kikuchi, M., Lee, F. S. C., and Rowland, F. S., *J. Phys. Chem.*, **85,** 84 (1981).
[650] Slagle, I. R., Yamada, F., and Gutman, D., *J. Am. Chem. Soc.*, **103,** 149 (1981),
[651] Harris, J. F., *J. Org. Chem.*, **46,** 268 (1981).
[652] Perkins, M. J., and Turner, E. S., *J. Chem. Soc., Chem. Commun.*, **1981,** 139.
[653] Bougeard, P., Dass Gupta, B., and Johnson, M. D., *J. Organomet. Chem.*, **206,** 211 (1981).
[654] Bougeard, P., and Johnson, M. D., *J. Organomet. Chem.*, **206,** 221 (1981).
[655] Jackson, R. A., Malek, F., and Ozaslan, N., *J. Chem. Soc., Chem. Comm.*, **1981,** 956.
[656] Okazaki, R., Kang, K.-T., and Inamoto, N., *Tetrahedron Lett.*, **22,** 235 (1981).
[657] John, D. I., Tyrrell, N. D., and Thomas, E. J., *J. Chem. Soc., Chem. Commun.*, **1981,** 901.
[658] Tomilenko, E. I., Staninets, V. I., and Beletskaya, I. P., *Zh. Org. Khim.*, **16,** 1091 (1980); *Chem. Abs.*, **93,** 167232 (1980).

659 Tomilenko, E. I., Staninets, V. I., Kukhar, V. P., and Beletskaya, I. P., *Dokl. Akad. Nauk SSSR*, **252,** 129 (1980); *Chem. Abs.*, **93,** 220124 (1980).
660 Applequist, D. E., and Gdanski, R. D., *J. Org. Chem.*, **46,** 2502 (1981).
661 Solodovnikov, S. P., and Bubnov, N. N., *Russ. Chem. Rev.*, **49,** 1 (1980).
662 Rogerts, B. P., *Adv. Free-Radical Chem.*, **6,** 225 (1980); *Chem. Abs.*, **94,** 29615 (1981).
663 Evans, J. C., and Mishra, S. P., *J. Inorg. Nucl. Chem.*, **43,** 481 (1981); *Chem. Abs.*, **95,** 79913 (1981).
664 Hamerlinck, J. H. H., Schipper, P., and Buck, H. M., *J. Chem. Soc., Chem. Commun.*, **1981,** 104.
665 Hamerlinck, J. H. H., Schipper, P., and Buck, H. M., *J. Chem. Soc., Chem. Commun.*, **1981,** 1148.
666 Hamerlinck, J. H. H., Hermkens, P. H. H., Schipper, P., and Buck, H. M., *J. Chem. Soc., Chem. Commun.*, **1981,** 358.
667 Brand, J. C., and Roberts, B. P., *J. Chem. Soc., Chem. Commun.*, **1981,** 1107.
668 Rumanskii, B. L., Degtyarev, A. N., Bubnov, N. N., Solodovnikov, S. P., Bregadze, V. I., Godovikov, N. N., and Kabachnik, M. I., *Izv. Akad. Nauk SSSR, Ser. Khim.*, **1980,** 2627; *Chem. Abs.*, **94,** 102438 (1981).
669 Giles, J. R. M., and Roberts, B. P., *J. Chem. Soc., Perkin Trans. 2*, **1981,** 1211.
670 Levin, Ya. A., Abul'khanov, A. G., and Nefedov, V. P., *Zh. Obshch. Khim.*, **51,** 348 (1981); *Chem. Abs.*, **94,** 191283 (1981).
671 Marriott, P. R., Griller, D., *J. Am. Chem. Soc.*, **103,** 1521 (1981).
672 Tumanskii, B. L., Khodak, A. A., Solodornikov, S. P., Bubnov, N. N., Gilyarov, V. A., and Kabachnik, M. I., *Izv. Akad. Nauk SSSR, Ser. Khim.*, **1981,** 1014; *Chem. Abs.*, **95**, 114400 (1981).
673 Abul'khanov, A. G., Levin, Ya, A., and Ryzhkina, I. S. *Zh. Obshch. Khim*, **51,** 571 (1981); *Chem. Abs.*, **95,** 24062 (1981).
674 Roberts, B. P., and Scaiano, J. C., *J. Chem. Soc., Perkin Trans. 2*, **1981,** 905.
675 Roberts, B. P., and Singh, K., *J. Chem. Soc., Perkin Trans. 2*, **1981,** 866.
676 Baban, J. A., and Roberts, B. P., *J. Chem. Soc., Perkin Trans. 2*, **1981,** 161.
677 Komissarov, V. D., and Safiullin, R. L., *React. Kinet. Catal. Lett.*, **14,** 67 (1980); *Chem. Abs.*, **94,** 14794 (1981).
678a Perkins, C. A., Martin, J. C., Arduengo, A. J., Lau, W., Alegria, A., and Kochi, J. K., *J. Am. Chem. Soc.*, **102,** 7753 (1980).
678b Mizuno, H., Matsuda, M., and Ino, M., *J. Org. Chem.*, **46,** 520 (1981).
679 Skarzewski, J., *J. Chem. Res. (S)*, **1980,** 410.
680 Rasmussen, J. K., and Smith, H. K., *J. Am. Chem. Soc.*, **103,** 730 (1981).
681 Kohda, A., Ueda, K., and Sato, T., *J. Org. Chem.*, **46,** 509 (1981).
682 Tomizawa, K., and Ogata, Y., *J. Org. Chem.*, **46,** 2107 (1981).
683 Baciocchi, E., Rol, E., and Mandolini, L., *J. Am. Chem. Soc.*, **102,** 7598 (1980).
684 Baumberger, R. S., and Parker, V. D., *Acta Chem. Scand.*, **B34,** 537 (1980).
685 Parker, V. D., *Acta Chem. Scand.*, **B35,** 123 (1981).
686 Okada, T., and Kamiya, Y., *Bull. Chem. Soc. Jpn.*, **54,** 2724 (1981).
687 Chaintreau, A., Adrian, G., and Couturier, D., *J. Org. Chem.*, **46,** 4562 (1981).
688 Chaintreau, A., Adrian, G., and Couturier, D., *Synth. Commun.*, **11,** 669 (1981).
689 Beletskaya, I. P., Dem'yanov, P. I., Makhon'kov, D. I., and Zeldis, I. M., *J. Org. Chem., USSR*, **17,** 591 (1981).
690 Beletskaya, I. P., Dem'yanov, P. I., Makhon'kov, D. I., Zeldis, I. M., and Burenko, S. N., *J. Org. Chem. USSR*, **17,** 583 (1981).
691 Bhatt, M. V., and Perumal, P. T., *Tetrahedron Lett.*, **22,** 2605 (1981).
692 Giordano, C., Belli, A., Citterio, A., and Minisci, F., *Tetrahedron*, **36,** 3559 (1980).
693 Eberson, L., *Acta Chem. Scand.*, **B34,** 747 (1980).
694 Nefedov, V. A., Tarygina, L. K., Kryuchkova, L. V., and Ryabokobylko, Yu. S., *J. Org. Chem. USSR*, **17,** 487 (1981).
695 Cf. *Org. Reaction Mech.*, **1980,** 138.
696 Aalstad, B., Ronlán, A., and Parker, V. D., *Acta Chem. Scand.*, **B35,** 247 (1981).
697 Jempty, T. C., Gogins, K. A. Z., Mazur, Y., and Miller, L. L., *J. Org. Chem.*, **46,** 4545 (1981).
698 Hewgill, F. R., and Pass, M. C., *Tetrahedron Lett.*, **22,** 2125 (1981).
699 Rudenko, A. P., Zarubin, M. Ya., and Aver'yanov, S. F., *Zh. Org. Khim.*, **16,** 1106 (1980); *Chem. Abs.*, **93,** 167285 (1980).
700 Tashiro, M., Yoshiya, H., and Fukata, G., *J. Org. Chem.*, **46,** 3784 (1981).
701 Feringa, B., and Wynberg, H., *J. Org. Chem.*, **46,** 2547 (1981).
702 Coutts, I. G., Hamblin, M. R., and Welsby, S. E., *J. Chem. Soc., Perkin Trans. 1*, **1981,** 493.
703 Lewis, J. R.. and Paul, J. G., *J. Chem. Soc., Perkin Trans. 1*, **1981,** 770.
704 Kovtonyuk, V. N., Kobrina, L. S., and Yakobson, G. G., *J. Fluorine Chem.*, **18,** 578 (1981).

705 Maruthamuthu, P., and Subramanian, N., *J. Chem. Soc., Perkin Trans. 2*, **1981,** 1280.
706 Levey, G., Bida, G., and Edwards, J. O., *Int. J. Chem. Kinet.*, **13,** 497 (1981).
707 Walling, C., and Humphreys, R. W. R., *J. Org. Chem.*, **46,** 1260 (1981).
708 Frazier, R. H., and Harlow, R. L., *J. Org. Chem.*, **45,** 5408 (1980).
709 Baker, S. R., Begley, M. J., and Crombie, L. J., *J. Chem. Soc., Perkin Trans. 1*, **1981,** 190.
710 Ogibin, Yu. N., Velibekova, D. S., Katsin, M. I., Troyanskii, E. I., and Nikishin, G. I., *Bull. Acad. Sci. USSR*, **30,** 127 (1981).
711 Beebe, T. R., Hii, P., and Reinking, P., *J. Org. Chem.*, **46,** 1927 (1981).
712 Treichel, P. M., and Rosenheim, L. D., *J. Am. Chem. Soc.*, **103,** 691 (1981).
713 Miyashita, S., and Matsuda, M., *Bull. Chem. Soc. Jpn.*, **54,** 1740 (1981).
714 Watanabe, Y., Numata, T., Iyanagi, T., and Oae, S., *Bull. Chem. Soc. Jpn.*, **54,** 1163 (1981).
715 Frazier, R. H., and Harlow, R. L., *J. Org. Chem.*, **45,** 5408 (1980).
716 Kurosawa, K., and Yamaguchi, K., *Bull. Chem. Soc. Jpn.*, **54,** 1757 (1981).
717 Serguchev, Yu. A., and Beletskaya, I. P., *Russian Chem. Rev.*, **49,** 1119 (1980).
718 Hewgill, F. R., and Proudfoot, G. M., *Aust. J. Chem.*, **34,** 335 (1981).
719 Giordano, C., Belli, A., Citterio, A., and Minisci, F., *J. Chem. Soc., Perkin Trans. 1*, **1981,** 1574.
720 Giordano, C., Belli, A., Casagrande, F., Guglielmetti, G., and Citterio, A., *J. Org. Chem.*, **46,** 3149 (1981).
721 Ogibin, Yu. N., Troyanskii, E. I., Kadentsev, V. I., Lubuzh, E. D., Chizhov, O. S., and Nikishin, G. I., *Izv. Akad. Nauk SSSR, Ser. Khim.*, **1981,** 1070; *Chem. Abs.*, **95,** 96555 (1981).
722 Ogibin, Yu. N., Troyanskii, E. I., Kadentsev, V. I., Lubuzh, E. D., Chizhov, O. S., and Nikishin, G. I., *Bull. Acad. Sci. USSR*, **30,** 850 (1981).
723 Forrester, A. R., Napier, R. J., and Thomson, R. H., *J. Chem. Soc., Perkin Trans. 1*, **1981,** 984.
724 *Org. Reaction Mech.*, **1979,** 121.
725 Atamanyuk, V. Yu., Koshecko, V. G., and Pokhodenko, V. D., *J. Org. Chem. USSR*, **16,** 1619 (1980).
726 Kemp, T. J., Moore, P., and Quick, G. R., *J. Chem. Res. (S)*, **1981,** 33.
727 Albery, W. J., Compton, R. G., and Kerr, I. S., *J. Chem. Soc., Perkin Trans. 2*, **1981,** 825.
728 Ernstbrunner, E. E., Girling, R. B., Grossman, W. E. L., Majer, E., Williams, K. P. J., and Hester, R. E., *J. Raman. Spectrosc.*, **10,** 161 (1981); *Chem. Abs.*, **95,** 96705 (1981).
729 Galliani, G., and Rindone, B., *Bioorg. Chem.*, **10,** 283 (1981).
730 Lai, C. C., and Colter, A. K., *J. Chem. Soc., Chem. Commun.*, **1980,** 1115.
731 Pavlova, E. K., Uzienko, A. B., and Yasnikov, A. A., *Dopov. Akad. Nauk RSR Ser. B, Geol. Khim. Biol. Nauki*, **1981,** 69; *Chem. Abs.*, **95,** 114474 (1981).
732 Levey, G., Rieger, A. L., and Edwards, J. O., *J. Org. Chem.*, **46,** 1255 (1981).
733 Tamblyn, W. H., Klingler, R. J., Hwang, W. S., and Kochi, J. K., *J. Am. Chem. Soc.*, **103,** 3161 (1981).
734 Turro, N. J., *Pure Appl. Chem.*, **53,** 259 (1981).
735 Voevodskaya, M. V., and Khudyakov, I. V., *Fiz.-Khim. Protzessy Gazov. Kondens. Fazakh*, **1979,** 34; *Chem. Abs.*, **93,** 203678 (1980).
736 Walsh, T. D., and Megremis, T. L., *J. Am. Chem. Soc.*, **103,** 3897 (1981).
737 Hanhela, P. J., and Paul, D. B., *Aust. J. Chem.*, **34,** 1669 (1981).
738 Takeuchi, K., Kurosaki, T., Yokomichi, Y., Kimura, Y., Kubota, Y., Fujimoto, H., and Okamoto, K., *J. Chem. Soc., Perkin Trans. 2*, **1981,** 670.
739 Abeywickrema, R. S., and Della, W. S., *J. Org. Chem.*, **46,** 2352 (1981).
740 Hess, U., and Huhn, D., *J. Prakt. Chem.*, **1981,** 323, 381.
741 Tanner, D. D., Plambeck, J. A., Reed, D. W., and Mojelsky, T. W., *J. Org. Chem.*, **45,** 5177 (1980).
742 Adamčikova, L., *Collect. Czech. Chem. Commun.*, **45,** 3287 (1981).
743 Hájek, M., Šilhavý, P., and Málek, J., *Collect. Czech. Chem. Commun.*, **45,** 3488 (1980).
744 Hájek, M., Šilhavý, P., and Málek, J., *Collect. Czech. Chem. Commun.*, **45,** 3502 (1980).
745 Davis, R., and Groves, I. F., *J. Organomet. Chem.*, **215,** c23 (1981).
746 Oku, A., Tsuji, H., Yoshida, M., and Yoshiura, N., *J. Am. Chem. Soc.*, **103,** 1244 (1981).
747 Luh, T.-Y., Lee, K. S., and Tam, S. W., *J. Organomet. Chem.*, **219,** 345 (1981).
748 Pardini, V. L., Roullier, L., Utley, J. H. P., and Webber, A., *J. Chem. Soc., Perkin Trans. 2*, **1981,** 1520.
749 Tzeng, D., and Weber, W. P., *J. Org. Chem.*, **46,** 265 (1981).
750 Leigh, W. J., Arnold, D. R., Humphreys, R. W. R., and Wong, P. C., *Can. J. Chem.*, **58,** 2537 (1980).
751 Pons, J.-M., Zahra, J.-P., and Santelli, M., *Tetrahedron Lett.*, **22,** 3965 (1981).
752 Rantenstrauch, V., Willhalm, B., Thommen, W., and Burger, U., *Helv. Chim. Acta*, **64,** 2109 (1981).
753 Rusling, J. R., and Zuman, P., *J. Org. Chem.*, **46,** 1906 (1981).
754 Smith, C. Z., and Utley, J. H. P., *J. Chem. Soc., Chem. Commun.*, **1981,** 492.
755 Murphy, W. S., and Wattanasin, S., *Tetrahedron Lett.*, **22,** 695 (1981).

756 Cf. McCormick, J. P., and Barton, D. L., *J. Chem. Soc., Chem. Commun.*, **1975,** 303.
757 Pradhan, S. K., Kadam, S. R., Kolhe, J. N., Radhakrishnan, T. V., Sohani, S. V., and Thaker, V. B., *J. Org. Chem.*, **46,** 2622 (1981).
758 Clerici, A., Porta, O., and Riva, M., *Tetrahedron Lett.*, **22,** 1043 (1981).
759 van Tilborg, W. J. M., and Smit, C. J., *Recl. Trav. Chim. Pays-Bas*, **100,** 437 (1981).
760 Sopher, D. W., and Utley, J. H. P., *J. Chem. Soc., Chem. Commun.*, **1981,** 134.
761 Falsig, M., and Lund, H., *Acta Chem. Scand.*, **B34,** 591 (1980).
762 Parker, V. D., *Acta Chem. Scand.*, **B35,** 279 (1981).
763 Parker, V. D., *Acta Chem. Scand.*, **B35,** 147 (1981).
764 Parker, V. D., *Acta Chem. Scand.*, **B35,** 149 (1981).
765 Parker, V. D., *Acta Chem. Scand.*, **B35,** 295 (1981).
766 Tissot, P., Surbeck, J.-P., Gülaçar, F. O., and Magaretha, P., *Helv. Chim. Acta*, **64,** 1570 (1981).
767 Satoh, S., Suginome, H., and Tokuda, M., *Tetrahedron Lett.*, **22,** 1895 (1981).
768 Babin, V. N., Gumenyuk, V. V., Solodovnikov, S. P., and Belousov, Yu. A., *Z. Naturforsch.*, **36B,** 400 (1981).
769 Babin, V. N., Belousov, Yu. A., Gumenyuk, V. V., Materikova, R. B., Salimov, R. M., and Kochetkova, N. S., *J. Organomet. Chem.*, **214,** C13 (1981).
770 Babin, V. N., Belousov, Yu. A., Lyatifov, I. R., Materikova, R. B., and Gumenyuk, V. V., *J. Organomet. Chem.*, **214,** C11 (1981).
771 Kubáček, P., *Collect. Czech. Chem. Commun.*, **46,** 40 (1981).
772 Lloyd, D., Vincent, C. A., and Walton, D. J., *J. Chem. Soc., Perkin Trans. 2*, **1981,** 801.
773 Hermolin, J., Kirowa-Eisner, E., and Kosower, E., *J. Am. Chem. Soc.*, **103,** 1591 (1981).
774 Martens, F. M., and Verhoeven, J. W., *Recl. Trav. Chim. Pays-Bas*, **100,** 228 (1981).
775 Okura, I., and Thuan, N. K., *J. Chem. Soc., Faraday Trans. 1*, **76,** 2209 (1980).
776 Ahlberg, A., Hammerich, O., and Parker, V. D., *J. Am. Chem. Soc.*, **103,** 844 (1981).
777 Nakazawa, H., Chou, F. E., Andrews, P. E., and Bachur, N. R., *J. Org. Chem.*, **46,** 1493 (1981).
778 Black, D. St. C., and Johnstone, L., *Angew. Chem. Int. Ed.*, **20,** 669 (1981).
779 Berti, C., Greci, L., Marchetti, L., Andruzzi, R., and Trazza, A., *J. Chem. Res. (S)*, **1981,** 340.
780 Simonet, J., Carriou, M., and Lund, L., *Liebigs Ann. Chem.*, **1981,** 1665.
781 Tagaya, H., Aruga, T., Ito, O., and Matsuda, M., *J. Am. Chem. Soc.*, **103,** 5484 (1981).
782 Pragst, F., *J. Electroanal. Chem., Interfacial Electrochem.*, **119,** 315 (1981).
783 Falsig, M., and Lund, H., *Acta Chem. Scand.*, **B34,** 545 (1980).
784 Falsig, M., Lund, H., Nadjo, L., and Savéant, J. M., *Acta Chem. Scand.*, **B34,** 685 (1980).
785 Falsig, M., and Lund, H., *Acta Chem. Scand.*, **B34,** 585 (1980).
786 Fry, A. J., Anker, K., and Handa, V. K., *J. Chem. Soc., Chem. Commun.*, **1981,** 120.
787 Gold, V., Pemberton, S. M., and Wood, D. L., *J. Chem. Soc., Perkin Trans. 2*, **1981,** 1230.
788 Citterio, A., Minisci, F., and Serravalle, M., *J. Chem. Res. (S)*, **1981,** 198.
789 Brault, D., and Neta, P., *J. Am. Chem. Soc.*, **103,** 2705 (1981).
790 *Org. React. Mech.*, **1977,** 132; **1978,** 147; **1979,** 127.
791 Schmitt, R. J., and Ross, D. S., *J. Am. Chem. Soc.*, **103,** 5265 (1981).
792 Eberson, L., and Radner, F., *Acta Chem. Scand.*, **B34,** 739 (1980).
793 Olah, G. A., Narang, S. C., and Olah, J. A., *Proc. Natl. Acad. Sci. USA*, **78,** 3298 (1981).
794 Helsby, P., Ridd, J. H., and Sandall, J. P. B., *J. Chem. Soc., Chem. Commun.*, **1981,** 825.
795 Ridd, J. H., and Sandall, J. P. B., *J. Chem. Soc., Chem. Commun.*, **1981,** 402.
796 Morkovnik, A. S., Dobaeva, N. M., Okhlobystin, O. Yu., *Khim. Geterotsikl. Soedin*, **1980,** 1135; *Chem. Abs.*, **94,** 14739 (1981).
797 Morkovnik, A. S., Okhlobystin, O. Yu., *Khim. Geterotsikl. Soedin*, **1981,** 123; *Chem. Abs.*, **94,** 191274 (1981).
798 Simig, G., *Magy. Kem. Lapja*, **35,** 347 (1980); *Chem. Abs.*, **94,** 64649 (1981).
799 Parker, V. D., *Acta Chem. Scand.*, **B35,** 533 (1981).
800 Tremelling, M. J., and Bunnett, J. F., *J. Am. Chem. Soc.*, **102,** 7377 (1980).
801 Alonso, R. A., and Rossi, R. A., *J. Org. Chem.*, **45,** 4760 (1980).
802 Brunet, J.-J., Sidot, C., and Caubere, P., *Tetrahedron Lett.*, **22,** 1013 (1981).
803 Brunet, J.-J., Sidot, C., and Caubere, P., *J. Organomet. Chem.*, **204,** 229 (1981).
804 Rossi, R. A., Alonso, R. A., and Palacios, S. M., *J. Org. Chem.*, **46,** 2498 (1981).
805 Rossi, R. A., and Peñéñory, A. B. B., *J. Org. Chem.*, **46,** 4580 (1981).
806 Fontan, F., Galvez, C., and Viladoms, P., *Heterocycles*, **16,** 1473 (1981).
807 Carver, D. R., Komin, A. P., Hubbard, J. S., and Wolfe, J. F., *J. Org. Chem.*, **46,** 294 (1981).
808 Sosonkin, I. M., Strogov, G. N., and Fedyainov, N. V., *Dokl. Chem.*, **253,** 349 (1980).

[809] Sosonkin, I. M., Strogov, G. N., and Fedyainov, N. V., *Dokl. Akad. Nauk SSSR*, **253**, 365 (1980); *Chem. Abs.*, **94**, 46472 (1981).
[810] Goh, S. H., and Kam, T. S., *J. Chem. Soc., Perkin Trans. 1*, **1981**, 423.
[811] Fridman, A. L., Surkov, V. D., and Novikov, S. S., *Russ. Chem. Rev.*, **49**, 1068 (1980).
[812] Symons, M. C. R., and Bowman, W. R., *Tetrahedron Lett.*, **22**, 4549 (1981).
[813] Bigot, B., Roux, D., and Salem, L., *J. Am. Chem. Soc.*, **103**, 5271 (1981).
[814] Bowman, W. R., and Richardson, G. D., *Tetrahedron Lett.*, **22**, 1551 (1981).
[815] Russell, G. A., Mudryk, B., and Jawdosiuk, M., *J. Am. Chem. Soc.*, **103**, 4610 (1981).
[816] Russell, G. A., Mudryk, B., and Jawdosiuk, M., *Synthesis*, **1981**, 62.
[817] Russell, G. A., Ros, F., and Mudryk, B., *J. Am. Chem. Soc.*, **102**, 7601 (1980).
[818] Terpigorev, A. N., Shcherbinin, M. B., Bazanov, A. G., and Tselinskii, I. V., *J. Org. Chem. USSR*, **16**, 2175 (1980).
[819] Jackman, L. M., and Lange, B. C., *J. Am. Chem. Soc.*, **103**, 4494 (1981).
[820] Chung, S. K., and Sasamoto, K., *J. Chem. Soc., Chem. Commun.*, **1981**, 346.
[821] Bordwell, F. G., and Clemens, A. H., *J. Org. Chem.*, **46**, 1035 (1981).
[822] Simig, Gy., and Lempert, K., *Acta Chim. Acad. Sci. Hung.*, **107**, 375 (1981).
[823] Prousek, J., *Collect. Czech. Chem. Commun.*, **45**, 3347 (1980).
[824] Tanner, D. D., Blackburn, E. V., and Diaz, E. D., *J. Am. Chem. Soc.*, **103**, 1557 (1981).
[825] Kornblum, N., and Erickson, A. S., *J. Org. Chem.*, **40**, 1037 (1981).
[826] Norris, R. K., and Smyth-King, R. J., *J. Chem. Soc., Chem. Commun.*, **1981**, 79.
[827] Kornblum, N., Ackermann, P., and Swiger, R. T., *J. Org. Chem.*, **45**, 5294 (1980).
[828] Ono, N., Miyake, H., Tamura, R., Hamamoto, I., and Kaji, A., *Chem. Lett.*, **1981**, 1139.
[829] Ono, N., Tamura, R., Nakatsuka, T., Hayami, J., and Kaji, A., *Bull. Chem. Soc. Jpn.*, **53**, 3295 (1980).
[830] Petsom, A., and Lund, H., *Acta Chem. Scand.*, **B34**, 614 (1980).
[831] Ono, N., Tamura, R., Tanikaga, R., and Kagi, A., *J. Chem. Soc., Chem. Commun.*, **1981**, 71.
[832] *Org. Reaction Mech.*, **1980**, 146, 148.
[833] Smith, G. F., Kuivila, H. G., Simon, R., and Sultan, L., *J. Am. Chem. Soc.*, **103**, 833 (1981).
[834] Alnajjar, M. S., and Kuivila, H. G., *J. Org. Chem.*, **46**, 1053 (1981).
[835] Kitching, W., Olszowy, H., and Harvey, K., *J. Org. Chem.* **46**, 2423 (1981).
[836] Ashby, E. C., Goel, A. B., and DePriest, R. N., *J. Org. Chem.*, **46**, 2429 (1981).
[837] Singh, P. R., Khurana, J. M., and Nigam, A., *Tetrahedron Lett.*, **22**, 2901 (1981).
[838] Goel, A. B., and Ashby, E. C., *J. Organomet. Chem.*, **214**, C1 (1981).
[839] Ashby, E. C., and Goel, A. B., *J. Org. Chem.*, **46**, 3934 (1981).
[840] Ashby, E. C., Goel, A. B., and DePriest, R. N., *Tetrahedron Lett.*, **22**, 3729 (1981).
[841] Ashby, E. C., DePriest, R. N., and Goel, A. B., *Tetrahedron Lett.*, **22**, 1763 (1981).
[842] Ashby, E. C., and Goel, A. B., *J. Organomet. Chem.*, **221**, C15 (1981).
[843] Ashby, E. C., Goel, A. B., DePriest, R. N., and Prasad, H. S., *J. Am. Chem. Soc.*, **103**, 973 (1981).
[844] Ashby, E. C., and Goel, A. B., *Tetrahedron Lett.*, **22**, 4783 (1981).
[845] Ashby, E. C., Goel, A. B., and DePriest, R. N., *Tetrahedron Lett.*, **22**, 4355 (1981).
[846] Ashby, E. C., Goel, A. N., and DePriest, R. N., *J. Am. Chem. Soc.*, **102**, 7779 (1980).
[847] Ashby, E. C., Goel, A. B., and Park, W. S., *Tetrahedron Lett.*, **22**, 4209 (1981).
[848] Ashby, E. C., and Goel, A. B., *Tetrahedron Lett.*, **22**, 1879 (1981).
[849] Giles, J. R. M., and Roberts, B. P., *J. Chem. Soc., Chem. Commun.*, **1981**, 360.
[850] Morvillo, A., and Turco, A., *J. Organomet. Chem.*, **208**, 103 (1981).
[851] Espenson, J. H., and Tinner, U., *J. Organomet. Chem.*, **212**, C43 (1981).
[852] Okubo, M., and Maruyama, K., *Kagaku (Kyoto)*, **35**, 338, 467 (1980); *Chem. Abs.*, **94**, 14632, 14633 (1981).
[853] Sergeev, G. B., Smirnov, V. V., and Zagorsky, V. V., *J. Organometal. Chem.*, **201**, 9 (1980).
[854] Seetz, J. W. F. L., Akkerman, O. S., and Bickelhaupt, F., *Tetrahedron Lett.*, **22**, 4857 (1981).
[855] Ashby, E. C., and Bowers, J. C., *J. Am. Chem. Soc.*, **103**, 2242 (1981).
[856] *Org. React. Mech.*, **1980**, 146.
[857] Ashby, E. C., and Goel, A. B., *J. Am. Chem. Soc.*, **103**, 4983 (1981).
[858] Okubo, M., Morigami, Y., and Suenaga, R., *Bull. Chem. Soc. Jpn.*, **53**, 3029 (1980).
[859] Stăsko, A., Malik, L., Mat'ašova, E., and Tkáč, A., *Org. Magn. Reson.*, **17**, 74 (1981).
[860] Jalander, L., *Acta Chem. Scand.*, **B35**, 419 (1981).
[861a] Kawada, Y., and Iwamura, H., *J. Org. Chem.*, **46**, 3357 (1981).
[861b] Walborsky, H. M., and Powers, E. J., *Israel J. Chem.*, **21**, 210 (1981).
[862] Hannah, D. J., Smith, R. A. J., Teok, I., and Weavers, R. T., *Aust. J. Chem.*, **34**, 181 (1981).
[863] Kawabata, N., and Tanimoto, M., *Tetrahedron*, **36**, 3517 (1980).

864 Seetz, J. W. F. L., Akkerman, O. S., and Bickelhaupt, F., *Tetrahedron Lett.*, **22,** 4855 (1981).
865 Cambie, R. C., Hayward, R. C., Jurlina, J. L., Rutledge, P. S., and Woodgate, P. D., *J. Chem. Soc., Perkin Trans. 1*, **1981,** 2608.
866 Kozima, S., Fujita, H., Hitomi, T., Kobayashi, K., and Kawanisi, M., *Bull. Chem. Soc. Jpn.*, **53,** 2953 (1980).
867a Prokof'ev, A. I., Vol'eva, V. B., Novikova, I. A., Belostotskaya, I. S., Ershov, V. V., and Kabachnik, M. I., *Bull. Acad. Sci. USSR*, **29,** 1883 (1980).
867b Stegmann, H. B., Schrade, R., Saur, H., Schuler, P., and Scheffler, K., *J. Organomet. Chem.*, **214,** 197 (1981).
868 Nome, F., and Zanette, D., *Can. J. Chem.*, **58,** 2402 (1981).
869a Deniau, J., Duong, K. N. V., Gaudemer, A., Bougeard, P., and Johnson, M., *J. Chem. Soc., Perkin Trans. 2*, **1981,** 393.
869b Schrauzer, G. N., and Grate, J. H., *J. Am. Chem. Soc.*, **103,** 541 (1981).
870 Burns, J. M., Wharry, D. L., and Koch, T. H., *J. Am. Chem. Soc.*, **103,** 849 (1981).
871 Ohno, A., Shio, T., Yamamoto, H., and Oka, S., *J. Am. Chem. Soc.*, **103,** 2045 (1981).
872 Kurosawa, H., Okada, H., and Hattori, T., *Tetrahedron Lett.*, **22,** 4495 (1981).
873 Russell, G. A., and Hershberger, J., *J. Am. Chem. Soc.*, **102,** 7604 (1980).
874a Razuvaev, G. A., Shabanov, A. V., Egorochkin, A. N., Kuznetsov, V. A., and Zhiltsov, S. F., *J. Organomet. Chem.*, **202,** 363 (1980).
874b Obafemi, C. A., *J. Organomet. Chem.*, **219,** 1 (1981).
875 Sheradsky, T., and Nov, E., *J. Chem. Soc., Perkin Trans. 1*, **1980,** 2781.
876 Lakshmikantham, M. V., and Cava, M. P., *J. Org. Chem.*, **46,** 3246 (1981).
877 Hiller, K.-O., Masloch, B., Göbl, M., and Asmus, K. D., *J. Am. Chem. Soc.*, **103,** 2734 (1981).
878 Bahnemann, D., Asmus, K. D., and Willson, R. L., *J. Chem. Soc., Perkin Trans. 2*, **1981,** 890.
879 Guthrie, R. D., Hrovat, D. A., Prahl, F. G., and Swan, J., *J. Org. Chem.*, **46,** 498 (1981).
880 Lasker, J. M., Sivarajah, K., Mason, R. P., Kalyanaraman, B., Abou-Donia, M. B., and Eling, T. E., *J. Biol. Chem.*, **256,** 7764 (1981).
881 Talley, J. J., *Tetrahedron Lett.*, **22,** 823 (1981).
882 Shcherbinin, M. B., Bedin, M. P., Bazanov, A. G., and Tselinskii, I. V., *Zh. Obshch. Khim.*, **51,** 709 (1981); *Chem. Abs.*, **95,** 23910 (1981).
883 Jones, C. R., *J. Org. Chem.*, **46,** 3370 (1981).
884 Kemp, T. J., and Martins, L. J. A., *J. Chem. Soc., Perkin Trans. 2*, **1980,** 1708.
885a Koshkin, L. V., *Neftekhimiya*, **21,** 613 (1981); *Chem. Abs.*, **95,** 168126 (1981).
885b Washio, M., Tagawa, S., and Tabata, Y., *J. Phys. Chem.*, **84,** 2876 (1980).
886 Fujihara, H., Fuke, S., Yoshihara, M., and Maeshima, T., *Chem. Lett.*, **1981,** 1271.
887 Saito, I., Matsuura, T., and Inoue, K., *J. Am. Chem. Soc.*, **103,** 188 (1981).
888 Kojima, M., Sakuragi, H., and Tokumaru, K., *Tetrahedron Lett.*, **22,** 2889 (1981).
889a Nelsen, S. F., and Akaba, R., *J. Am. Chem. Soc.*, **103,** 2096 (1981).
889b Clennan, E. L., Simmons, W., and Almgren, C. W., *J. Am. Chem. Soc.*, **103,** 2098 (1981).
890 Santamaria, J., *Tetrahedron Lett.*, **22,** 4511 (1981).
891 Steichen, D. S., and Foote, C. S., *J. Am. Chem. Soc.*, **103,** 1855 (1981).
892 Muto, S., and Bruice, T. C., *J. Am. Chem. Soc.*, **102,** 7559 (1980).
893 Jongsma, S. J., and Cornelisse, J., *Tetrahedron Lett.*, **22,** 2919 (1981).
894 Albini, A., *Synthesis*, **1981,** 249.
895 Lapin, S., and Kurz, M. E., *J. Chem. Soc., Chem. Commun.*, **1981,** 817.
896 Bunce, N. J., Bergsma, J. P., and Schmidt, J. L., *J. Chem. Soc., Perkin Trans. 2*, **1981,** 713.
897 Yasuda, M., Pac, C., and Sakurai, H., *J. Chem. Soc., Perkin Trans. 1*, **1981,** 746.
898 Yasuda, M., Pac, C., and Sakurai, H., *J. Org. Chem.*, **46,** 788 (1981).
899 Albini, A., Fasani, E., and Oberti, R., *J. Chem. Soc., Chem. Commun.*, **1981,** 50.
900 Sugimori, A., Nishijimi, M., and Yashima, T., *Chem. Lett.*, **1981,** 303.
901 Mizuno, K., Ishii, M., and Otsuji, Y., *J. Am. Chem. Soc.*, **103,** 5570 (1981).
902 Ito, S., Saito, I., and Matsuura, T., *Tetrahedron*, **37,** 45 (1981).
903 Arnold, D. R., Wong, P. C., Maroulis, A. J., and Cameron, T. S., *Pure Appl. Chem.*, **52,** 2609 (1980).
904 Lewis, F. D., Ho, T.-I., and Simpson, J. T., *J. Org. Chem.*, **46,** 1077 (1981).
905 *Org. Reaction Mech.*, **1980,** 149.
906 Lewis, F. D., and Simpson, J. T., *J. Am. Chem. Soc.*, **102,** 7593 (1980).
907 Stavinoha, J. L., and Mariano, P. S., *J. Am. Chem. Soc.*, **103,** 3136 (1981).
908 Stavinoha, J. L., Mariano, P. S., Leone-Bay, A., Swanson, R., and Bracken, C., *J. Am. Chem. Soc.*, **103,** 3148 (1981).
909 Mazzocchi, P. H., and Khachik, F., *Tetrahedron Lett.*, **22,** 4189 (1981).

910 Majima, T., Pac, C., Nakasone, A., and Sakurai, H., *J. Am. Chem. Soc.*, **103,** 4499 (1981).
911 Caronna, T., Morrocchi, S., and Vittimberga, B. M., *J. Org. Chem.*, **46,** 34 (1981).
912 Mizuno, K., Ueda, H., and Otsuji, Y., *Chem. Lett.*, **1981,** 1237.
913 Majima, T., Pac, C., and Sakurai, H., *J. Chem. Soc., Perkin Trans. 1,* **1980,** 2705.
914 Mukai, T., Sato, K., and Yamashita, Y., *J. Am. Chem. Soc.*, **103,** 670 (1981).
915 Roth, H. D., and Schilling, M. L. M., *J. Am. Chem. Soc.*, **102,** 7956 (1980).
916 Mizuno, K., Ogawa, J., and Otsuji, Y., *Chem. Lett.*, **1981,** 741.
917 Padwa, A., Chou, C. S., and Riecker, W. F., *J. Org. Chem.*, **45,** 4555 (1980).
918 Gotoh, T., Kato, M., Yamamoto, M., and Nishijima, Y., *J. Chem. Soc., Chem. Commun.*, **1981,** 90.
919 Schnabel, W., Naito, I., Kitamura, T., Kobayashi, S., and Taniguchi, H., *Tetrahedron*, **36,** 3229 (1980).
920 Swanson, B. J., Kutzer, J. C., and Koch, T. H., *J. Am. Chem. Soc.*, **103,** 1274 (1981).
921 Ito, S., Saito, I., and Matsuura, T., *J. Am. Chem. Soc.*, **102,** 7535 (1980).
922 Das, P. K., *Tetrahedron Lett.*, **22,** 1307 (1981).
923 Maruyama, K., Furuta, H., and Otsuki, J., *Chem. Lett.*, **1981,** 709, 1025.
924 Warden, J. T., *Israel J. Chem.*, **21,** 338 (1981).
925 Forman, A., Davis, M. S., Fujita, I., Hanson, L. K., Smith, K. M., and Fajer, J., *Israel J. Chem.*, **21,** 265 (1981).
926 Darwent, J. R., and Kalyanasundaram, K., *J. Chem. Soc., Faraday Trans. 2*, **77,** 373 (1981).
927 Maillard, P., Gaspard, S., Krausz, P., and Giannotti, C., *J. Organomet. Chem.*, **212,** 185 (1981).
928 Simon, J. D., and Peters, K. S., *J. Am. Chem. Soc.*, **103,** 6403 (1981).
929 Inbar, S., Linschitz, H., and Cohen, S. G., *J. Am. Chem. Soc.*, **103,** 1048 (1981).
930 Schaefer, C. G., and Peters, K. S., *J. Am. Chem. Soc.*, **102,** 7566 (1980).
931 Forster, M., and Hester, R. E., *J. Chem. Soc., Faraday Trans. 2*, **77,** 1521 (1981).
932 Gritsan, N. P., and Bazhin, N. M., *Bull. Acad. Sci. USSR*, **30,** 210 (1981).
933 Davidson, R. S., and Goodin, J. W., *Tetrahedron Lett.*, **22,** 163 (1981).
934 Zelent, B., and Durochar, G., *J. Org. Chem.*, **46,** 1496 (1981).
935 Symons, M. C. R., *Pure Appl. Chem.*, **53,** 223 (1981).
936 Maier, J. P., *Angew. Chem. Int. Ed.*, **20,** 638 (1981).
937 Hünig, S. M., and Berneth, H., *Top. Curr. Chem.*, **92,** 1 (1980).
938 Morkovnik, A. S., and Okhlobystin, O. Yu., *Khim. Geterotsikl. Soedin*, **1980,** 1011: *Chem. Abs.*, **93,** 237984 (1980).
939 Nelsen, S. F., *Acc. Chem. Res.*, **14,** 131 (1981).
940 Schwarz, H., *Top. Curr. Chem.*, **97,** 1 (1981).
941 Toriyama, K., Nunome, K., and Iwasaki, M., *J. Phys. Chem.*, **85,** 2149 (1981).
942 Gerson, F., Lopez, J., Krebs, A., and Rüger, W., *Angew. Chem. Int. Ed.*, **20,** 95 (1981).
943a Alder, R. W., Sessions, R. B., and Symons, M. C. R., *J. Chem. Res. (S)*, **1081,** 82.
943b Gölitz, P., and de Meijere, A., *Angew. Chem. Int. Ed.*, **20,** 298 (1981).
944 Komatsu, K., Moriyama, T., Nishiyama, T., and Okamoto, K., *Tetrahedron*, **37,** 721 (1981).
945 Broxterman, Q. B., Hogeveen, H., and Kok, D. M., *Tetrahedron Lett.*, **22,** 173 (1981).
946 Borden, W. T., Davidson, E. R., and Feller, D., *J. Am. Chem. Soc.*, **103,** 5725 (1981).
947 Roth, H. D., and Schilling, M. L. M., *J. Am. Chem. Soc.*, **103,** 1246 (1981).
948 Kröhnke, C., Enkelmann, V., and Wegner, G., *Angew. Chem. Int. Ed.*, **19,** 912 (1981).
949 Forster, M., and Hester, R. E., *J. Chem. Soc., Faraday Trans. 2*, **77,** 1535 (1981).
950 Hester, R. E., and Williams, R. P. J., *J. Chem. Soc., Faraday Trans. 2*, **77,** 541 (1981).
951 Kaim, W., *J. Organomet. Chem.*, **201,** C5 (1980).
952 Kaim, W., *J. Organomet. Chem.*, **214,** 325, 337 (1981).
953 Bock, H., Brähler, G., Dauplaise, D., and Meinwald, J., *Chem. Ber.*, **114,** 2622 (1981).
954 Nemoto, F., Tsuzuki, N., Mukai, K., and Ishizu, K., *J. Phys. Chem.*, **85,** 2450 (1981).
955 Wang, J. T., and Williams, F., *J. Chem. Soc., Chem. Commun.*, **1981,** 666.
956 Wang, J. T., and Williams, F., *J. Chem. Soc., Chem. Commun.*, **1981,** 1184.
957 Bock, H., Schulz, W., and Stein, U., *Chem. Ber.*, **114,** 2632 (1981).
958 Neugebauer, F. A., and Umminger, I., *Chem. Ber.*, **114,** 2423 (1981).
959 Ivanov, V. L., Al'-Ainen, S. A., and Kuz'min, M. G., *J. Org. Chem. USSR*, **17,** 720 (1981).
960 Hanson, P., Isham, W. J., Lewis, R. J., and Stockburn, W. A., *J. Chem. Soc., Perkin Trans. 2*, **1981,** 1492.
961 Hester, R. E., and Williams, K. P. J., *J. Chem. Soc., Perkin Trans. 2*, **1981,** 852.
962 Müller, F., Grande, H. J., Harding, L. J., Dunham, W. R., Visser, A. J. W. G., Reinders, J. H., Hemmerich, P., and Ehrenberg, A., *Eur. J. Biochem.*, **116,** 17 (1981).
963 Gerson, F., Plattner, G., Bartetzko, R., and Gleiter, R., *Helv. Chim. Acta*, **63,** 2144 (1980).

964 Chang, C. K., Hanson, L. K., Richardson, P. F., Young, R., and Fajer, J., *Proc. Natl. Acad. Sci. U.S.A.*, **78,** 2652 (1981).

965 Gans, P., Marchon, J. C., Reed, C. A., and Regnard, J. R., *Nouveau J. Chim.*, **5,** 201 (1981).

966 Nelsen, S. F., and Parmelee, W. P., *J. Org. Chem.*, **46,** 3453 (1981).

967 Behrens, G., Bothe, E., Koltzenburg, G., and Schulte-Frohlinde, D., *J. Chem. Soc., Perkin Trans. 2,* **1981,** 143.

968 Gilbert, B. C., Norman, R. O. C., and Williams, P. S., *J. Chem. Soc., Perkin Trans. 2,* **1981,** 1066.

969 Benz, R. C., Dunbar, R. C., and Claspy, P. C., *J. Am. Chem. Soc.*, **103,** 1799 (1981).

970 Andrews, L., Harvey, J. A., Kelsall, B. J., and Duffey, D. C., *J. Am. Chem. Soc.*, **103,** 6415 (1981).

971 Hammerum, S., *Tetrahedron Lett.*, **22,** 157 (1981).

972 Fritz, H. P., and Ecker, P., *Chem. Ber.*, **114,** 3463 (1981).

973 Stephenson, M. T., and Shine, H. J., *J. Org. Chem.*, **46,** 3139 (1981).

974 Bancroft, E. E., Pemberton, J. E., and Blount, H. N., *J. Phys. Chem.*, **84,** 2557 (1980).

975 Kotowski, S., Grabner, E. W., and Brauer, H. D., *Ber. Bunsengers. Phys. Chem.*, **84,** 1140 (1980).

976 Roullier, L., and Laviron, E., *Electrochim. Acta*, **25,** 795 (1980).

977 Belville, D. J., Wirth, D. D., and Bauld, N. L., *J. Am. Chem. Soc.*, **103,** 718 (1981).

978 Benati, L., Montevecchi, P. C., and Spagnolo, P., *J. Chem. Soc., Perkin Trans. 2,* **1981,** 1437.

979 Koshechko, V. G., Atamanyuk, V. Yu., and Pokhodenko, V. D., *J. Org. Chem. USSR*, **16,** 1852 (1980).

980 Clark, T., *J. Chem. Soc., Chem. Commun.*, **1981,** 515.

981 Hasegawa, A., Uchimura, S., and Hayashi, M., *J. Chem. Soc., Perkin Trans. 2,* **1980,** 1690.

982 Hasegawa, A., Nagayama, A., and Hayashi, M., *Bull. Chem. Soc. Jpn.*, **54,** 2620 (1981).

983 Wang, J. T., and Williams, F., *J. Am. Chem. Soc.*, **103,** 2902 (1981).

984 Symons, M. C. R., *J. Chem. Res. (S)*, **1981,** 286.

985 Klusik, H., and Berndt, A., *Angew. Chem. Int. Ed.*, **20,** 870 (1981).

986 Davies, A. G., Giles, J. R. M., and Lusztyk, J., *J. Chem. Soc., Perkin Trans. 2,* **1981,** 747.

987 van der Drift, E., Dammers, E., and Smidt, J., *J. Phys. Chem.*, **85,** 829 (1981).

988 Grein, K., Kirste, B., and Kurreck, H., *Chem. Ber.*, **114,** 254 (1981).

989 Gerson, F., Lopez, J., Metzger, A., and Jutz, C., *Helv. Chim. Acta*, **63,** 2135 (1980).

990 Iwaizumi, M., Fukazawa, Y., Kato, N., and Ito, S., *Bull. Chem. Soc. Jpn.*, **54,** 1299 (1981).

991 Gerson, F., Lopez, J., and Boekelheide, V., *J. Chem. Soc., Perkin Trans. 2,* **1981,** 1298.

992 Tanaka, H., Ogashiwa, S., and Kawazura, H., *Chem. Lett.*, **1981,** 585.

993 Chambers, R. D., Musgrave, W. K. R., Sargent, C. R., and Drakesmith, F. G., *Tetrahedron*, **37,** 591 (1981).

994 Sharma, K. K., and Boyd, R. J., *Theor. Chim. Acta*, **57,** 163 (1980).

995 Fujita, H., Deguchi, Y., Ohya-Nishiguchi, H., and Ishizu, K., *Org. Magn. Reson.*, **16,** 245 (1981).

996 Fujita, H., and Deguchi, Y., *Bull. Chem. Soc. Jpn.*, **54,** 1565 (1981).

997 Symons, M. C. R., *J. Chem. Soc., Faraday Trans. 1*, **77,** 783 (1981).

998 Guerra, M., Bernardi, F., and Pedulli, G. F., *Chem. Phys. Lett.*, **81,** 289 (1981).

999 Makarov, I. G., Kazakova, V. M., Chechulina, I. N., Kopylov, V. M., Prikhod'ko, P. L., and Komapenkova, N. G., *Zh. Obshch. Khim.*, **51,** 1328 (1981); *Chem. Abs.*, **95,** 149425 (1981).

1000 Reffy, J., and Karger-Kocsis, J., *Z. Phys. Chem. (Leipzig)*, **261,** 1081 (1980).

1001 Lendzian, F., Plato, M., and Möbius, K., *J. Magn. Reson.*, **44,** 20 (1981).

1002 Kaim, W., and Bock, H., *Chem. Ber.*, **114,** 1576 (1981).

1003 Barzaghi, M., Oliva, C., Gamba, A., and Saba, A., *J. Chem. Soc., Perkin Trans. 2,* **1980,** 1617.

1004 Suryanarayana, D., and Sevilla, M. D., *J. Phys. Chem.*, **84,** 3045 (1980).

1005 Egashira, N., Takita, Y., and Hori, F., *Denki Kagaku Oyobi Kogyo Butsuri Kagaku*, **49,** 217 (1981); *Chem. Abs.*, **95,** 41831 (1981).

1006 Sevenster, A. J. L., and Tabner, B. L., *J. Chem. Soc., Perkin Trans. 2,* **1981,** 1148.

1007 Gogan, N. J., Dickinson, I. L., Doull, J., and Patterson, J. R., *J. Organomet. Chem.*, **212,** 71 (1981).

1008 Loth, K., and Graf, F., *Helv. Chim. Acta*, **64,** 1910 (1981).

1009 Pünkt, J., Ulrici, B., Rehorek, D., and Weissenfels, M., *Z. Chem.*, **21,** 281 (1981).

1010 Thomson, C., *J. Mol. Struct.*, **67,** 133 (1980).

1011 Felix, C. C., and Sealy, R. C., *J. Am. Chem. Soc.*, **103,** 2831 (1981).

1012 Holton, D. M., and Murphy, D., *J. Chem. Soc., Perkin Trans. 2,* **1980,** 1757.

1013 Felix, C. C., and Sealy, R. C., *Photochem. Photobiol.*, **34,** 423 (1981).

1014 Pedersen, J. A., and Thomson, R. H., *J. Magn. Reson.*, **43,** 373 (1981).

1015 Loiá, M. C. B., Herold, B. J., and Atherton, N. M., *J. Chem. Soc., Perkin Trans. 2,* **1980,** 1788.

1016 Celina, M., Lazana, R. L. R., Herold, B. J., and Atherton, N. M., *Tetrahedron*, **37,** 605 (1981).

1017 Russell, G. A., Suleman, N. K., Iwamura, H., and Webster, O. W., *J. Am. Chem. Soc.*, **103,** 1560 (1981).
1018 Helling, D., Klages, C.-P., and Voss, J., *J. Phys. Chem.*, **84,** 3638 (1980).
1019 Tabner, B. J., and Walker, T., *J. Chem. Soc., Perkin Trans. 2*, **1981,** 1295.
1020 Rothwell, E. J., and Tabner, B. J., *J. Chem. Soc., Perkin Trans. 2*, **1981,** 1384.
1021 Tabner, B. J., and Walker, T., *J. Chem. Soc., Perkin Trans. 2*, **1981,** 1508.
1022 Nakamura, K., *Bull. Chem. Soc. Jpn.*, **53,** 2792 (1980).
1023 Bravo-Zhivotovskii, D. A., Kruglaya, O. A., and Vyazankin, N. S., *Zh. Obshch. Khim.*, **50,** 2144 (1980); *Chem. Abs.*, **94,** 29867 (1981).
1024 Symons, M. C. R., *Chem. Phys. Lett.*, **72,** 559 (1980).
1025 Wang, J. T., and Williams, F., *Chem. Phys. Lett.*, **72,** 557 (1980).
1026 Cf. *Org. Reaction. Mech.*, **1980,** 155.
1027 Neta, P., and Behar, D., *J. Am. Chem. Soc.*, **103,** 103 (1981).
1028 Behar, D., and Neta, P., *J. Am. Chem. Soc.*, **103,** 2280 (1981).
1029 Behar, D., and Neta, P., *J. Phys. Chem.*, **85,** 690 (1981).
1030 Kigawa, H., Takamuku, S., Toki, S., Kimura, N., Takeda, S., Tsumori, K., and Sakurai, H., *J. Am. Chem. Soc.*, **103,** 5176 (1981).
1031 Lund, A., Samskog, P. O., and Nilsson, G., *Chem. Phys.*, **57,** 399 (1981).
1032 Maialo, F., Testaferri, L., Tiecco, M., and Tingoli, M., *J. Org. Chem.*, **46,** 3070 (1981).
1033 Boche, G., and Wintermayr, H., *Angew. Chem. Int. Ed.*, **20,** 874 (1981).
1034 Dodd, J. R., Pagni, R. M., and Watson, C. R., *J. Org. Chem.*, **46,** 1688 (1981).
1035 Staško, A., Malik, L., Tkač, A., and Pelikan, P., *Z. Phys. Chem.* (*Leipzig*), **261,** 1205 (1980).
1036 Parker, V. D., *Acta Chem. Scand.*, **B35,** 349 (1981).
1037 Hammerich, O., and Parker, V. D., *Acta Chem. Scand.*, **B35,** 341 (1981).
1038 Leshina, T. V., Belyaeva, S. G., Mar'yasova, V. I., Sagdeev, R. Z., and Molin, Yu. N., *Dokl. Akad. Nauk. SSSR*, **255,** 141 (1980); *Chem. Abs.*, **94,** 102547 (1981).
1039 Leshina, T. V., Belyaeva, S. G., Maryasova, V. I., Sagdeev, R. Z., and Molin, Yu. N., *Chem. Phys. Lett.*, **75,** 438 (1980).
1040 Todres, Z. V., Chernyshova, T. M., Koridze, A. A., Denisovich, L. I., and Kursanov, D. N., *Dokl. Chem.*, **254,** 463 (1980).
1041 Kalyanaraman, B., Mason, R. P., Rowlett, R., and Kispert, L. D., *Biochim. Biophys. Acta*, **660,** 102 (1981).
1042 Cremaschi, P., Morosi, G., and Simonetta, M., *Angew. Chem. Int. Ed.*, **20,** 673 (1981).
1043 Sevilla, M. D., Morehouse, K. M., and Swartz, S., *J. Phys. Chem.*, **85,** 923 (1981).
1044 Hudson, R. L., and Williams, F., *J. Phys. Chem.*, **85,** 510 (1981).
1045 Samskog, P.-O., Nilsson, G., Lund, A., and Gilbro, T., *J. Phys. Chem.*, **84,** 2819 (1980).
1046 Sevilla, M. D., Swartz, S., Bearden, R., Morehouse, K. M., and Vartanian, T., *J. Phys. Chem.*, **85,** 918 (1981).
1047 Jullien, R., Stahl-Lariviere, H., Zann, D., and Nadjo, L., *Tetrahedron*, **37,** 3159 (1981).
1048 Bielski, B. H. J., Allen, A. O., and Schwarz, H. A., *J. Am. Chem. Soc.*, **103,** 3516 (1981).
1049 Alberti, A., Ciminale, F., Pedulli, G. F., Testaferri, L., and Tiecco, M., *J. Org. Chem.*, **46,** 751 (1981).
1050 Barradas, R. G., Gust, S. L., and Porter, J. D., *Tetrahedron Lett.*, **22,** 4579 (1981).
1051 McDonald, R. N., and Chowdury, A. K., *J. Am. Chem. Soc.*, **103,** 674 (1981).
1052 Pellerite, M. J., and Brauman, J. I., *J. Am. Chem. Soc.*, **103,** 676 (1981).
1053 McDonald, R. N., Triebe, F. M., January, J. R., Borhani, K. J., and Hawley, M. D., *J. Am. Chem. Soc.*, **102,** 7867 (1980).
1054 Parker, V. D., and Bethell, D., *Acta Chem. Scand.*, **B35,** 72 (1981).
1055 Parker, V. D., and Bethell, D., *Acta Chem. Scand.*, **B34,** 617 (1980).
1056 Stegmann, H. B., Bergler, H. U., and Scheffler, K., *Angew. Chem. Int. Ed.*, **20,** 389 (1981).
1057 Kaim, W., *Inorg. Chim. Acta*, **53,** L151 (1981).
1058 Stevenson, G. R., and Forch, B. E., *J. Phys. Chem.*, **85,** 378 (1981).
1059 Nanni, E. J., and Sawyer, D. T., *J. Am. Chem. Soc.*, **102,** 7591 (1980).
1060 Nanni, E. J., Angelis, C. T., Dickson, J., and Sawyer, D. T., *J. Am. Chem. Soc.*, **103,** 4268 (1981).
1061 Nanni, E. J., Sawyer, D. T., Ball, S. S., and Bruice, T. C., *J. Am. Chem. Soc.*, **103,** 2797 (1981).
1062 Gareil, M., Pinson, J., and Savéant, J. M., *Nouveau J. Chim.*, **5,** 311 (1981).
1063 Galliani, G., and Rindone, B., *Tetrahedron*, **37,** 2313 (1981).
1064 Grimshaw, J., and de Silva, A. P., *Chem. Soc. Rev.*, **10,** 181 (1981).
1065 Carr, R. W., and Topor, M. G., *J. Photochem.*, **16,** 51 (1981); *Chem. Abs.*, **95,** 79649 (1981).
1066 Wymann, L., Kaiser, T., Paul, H., and Fischer, H., *Helv. Chim. Acta*, **64,** 1739 (1981).
1067 Lion, Y. F., Kuwabara, M., and Riesz, P., *J. Phys. Chem.*, **84,** 3378 (1980).

1068 Lion, Y., Kuwabara, M., and Riesz, P., *Photochem. Photobiol.*, **34,** 297 (1981).
1069 Bögel, H., and Haucke, G., *Collect. Czech. Chem. Commun.*, **46,** 219 (1981).
1070 Frei, B., Jeger, O., Nakamura, N., and Wolf, H. R., *Tetrahedron*, **37,** 3339 (1981).
1071 Doyle, D. J., and Koob, R. D., *J. Phys. Chem.*, **85,** 2278 (1981).
1072 Levin, Ya. A., Gol'dfarb, E. I., and Vorkunova, E. I., *Zh. Obshch. Khim.*, **50,** 1981 (1980); *Chem. Abs.*, **94,** 29812 (1981).
1073 Golding, B. T., Kemp, T. J., and Sheena, H. H., *J. Chem. Res.*, **1981,** (S) 34, (M) 334.
1074 Yip, R. W., Chow, Y. L., and Beddard, C., *J. Chem. Soc., Chem. Commun.*, **1981,** 955.
1075 Morine, G. H., and Kuntz, R. R., *Photochem. Photobiol.*, **33,** 1 (1981); *Chem. Abs.*, **95,** 6077 (1981).
1076 Sindler-Kulyk, M., and Neckers, D. C., *Tetrahedron Lett.*, **22,** 525 (1981).
1077 Ogata, Y., Tomizawa, K., and Takagi, K., *Can. J. Chem.*, **59,** 14 (1981).
1078 Encina, M. V., Lissi, E. A., and Soto, H., *J. Photochem.*, **16,** 43 (1981); *Chem. Abs.*, **95,** 96549 (1981).
1079 Encinas, M. V., and Scaiano, J. C., *J. Am. Chem. Soc.*, **103,** 6393 (1981).
1080 Bartlett, P. D., Roof, A. A. M., and Winter, W. J., *J. Am. Chem. Soc.*, **103,** 6520 (1981).
1081 Scaiano, J. C., *J. Phys. Chem.*, **85,** 2851 (1981).
1082 Couture, A., Gomez, J., and de Mayo, P., *J. Org. Chem.*, **46,** 2010 (1981).
1083 Aoyama, H., Inoue, Y., and Omote, Y., *J. Org. Chem.*, **46,** 1967 (1981).
1084 Purohit, P. C., and Sonawane, H. R., *Tetrahedron*, **37,** 873 (1981).
1085 Filby, W. G., and Günther, K., *Z. Phys. Chem. (Wiesbaden)*, **125,** 21 (1981).
1086 Sawaki, Y., and Ogata, Y., *J. Am. Chem. Soc.*, **103,** 6455 (1981).
1087 Yasmenko, A. I., Khudyakov, I. V., Darmanjan, A. P., Kuzmin, V. A., and Clasesson, S., *Chem. Scr.*, **18,** 49 (1981).
1088 Gramain, J.-C., and Lhomme, M.-F., *Bull. Soc. Chim. Fr. II*, **1981,** 141.
1089 Gramain, J.-.C., Quazzani-Chahdi, L., and Troin, Y., *Tetrahedron Lett.*, **22,** 3185 (1981).
1090 Suginome, H., and Uchida, T., *Bull. Chem. Soc. Jpn.*, **53,** 3225 (1980).
1091 Martinez-Utrilla, R., and Miranda, M. A., *Tetrahedron*, **37,** 2111 (1981).
1092 Cossy, J., and Pete, J.-P., *Tetrahedron*, **37,** 2287 (1981).
1093 Dalton, J. C., and Bourque, R. A., *J. Am. Chem. Soc.*, **103,** 699 (1981).
1094 Ogata, Y., Tomizawa, K., Furuta, K., and Kato, H., *J. Chem. Soc., Perkin Trans. 2*, **1981,** 110.
1095 Soumillion, J. Ph., and DeWolf, B., *J. Chem. Soc., Chem. Commun.*, **1981,** 436.
1096 Fulara, J., and Latowski, T., *Z. Naturforsch.*, **36B,** 846 (1981).
1097 Slocum, G. H., Kaufmann, K., and Schuster, G. B., *J. Am. Chem. Soc.*, **103,** 4625 (1981).
1098 Takagi, K., and Ogata, Y., *J. Org. Chem.*, **46,** 989 (1981).
1099 el Faghi el Amoudi, M. S., Geneste, P., and Olive, J. L., *Nouveau J. Chim.* **5,** 251 (1981).
1100 Allara, D. L., and Shaw, R., *J. Phys. Chem. Ref. Data*, **9,** 523 (1980); *Chem. Abs.*, **94,** 14900 (1981).
1101 Roth, P., Barner, U., and Loehr, R., *Shock Tubes Waves, Proc. Int. Symp. 12th 1979*, **1980,** 621; *Chem. Abs.*, **94,** 29908 (1981).
1102 Pacey, P. D., and Wimalasena, J. H., *Chem. Phys. Lett.*, **76,** 433 (1980); *Chem. Abs.*, **94,** 102566 (1981).
1103 Hautman, D. J., Santoro, R. J., Dryer, F. L., and Glassman, I., *Int. J. Chem. Kinet.*, **13,** 149 (1981).
1104 Benson, S. W., *Int. J. Chem. Kinet.*, **12,** 755 (1980).
1105 King, K. D., *Int. J. Chem. Kinet.*, **13,** 273 (1981).
1106 King, K. D., and Nguyen, T. T., *Int. J. Chem. Kinet.*, **13,** 255 (1981).
1107 Nguyen, T. T., and King, K. D., *J. Phys. Chem.*, **85,** 3130 (1981).
1108 Korzun, N. V., Magaril, R. Z., and Trushkova, L. V., *Neftekhimiya*, **21,** 239 (1981); *Chem. Abs.*, **95,** 96661 (1981).
1109 Aribike, D. S., Susu, A. A., and Oguyne, A. F., *Thermochim. Acta*, **47,** 1 (1981); *Chem. Abs.*, **95,** 79784 (1981).
1110 Satanova, R. B., Kalinenko, R. A., and Nametkin, N. S., *Kinet. Katal.*, **21,** 1390 (1980); *Chem. Abs.*, **94,** 102565 (1981).
1111 Robaugh, D. A., and Stein, S. E., *Int. J. Chem. Kinet.*, **13,** 445 (1981).
1112 Robaugh, D. A., Barton, B. D., and Stein, S. E., *J. Phys. Chem.*, **85,** 2378 (1981).
1113 Miller, R. E., and Stein, S. E., *J. Phys. Chem.*, **85,** 580 (1981).
1114 Baldwin, J. E., and Ollerenshaw, J., *J. Org. Chem.*, **46,** 2116 (1981).
1115 Wiberg, K. B., and Matturro, M. G., *Tetrahedron Lett.*, **22,** 3481 (1981).
1116 Gajewski, J. J., and Chang, M. J., *J. Am. Chem. Soc.*, **102,** 7542 (1980).
1117 Dolbier, W. R., Sellers, S. F., Al-Sader, B. H., Fielder, T. H., Elsheimer, S., and Smart, B. E., *Isr. J. Chem.*, **21,** 176 (1981).
1118 Dolbier, W. R., Sellers, S. F., Al-Sader, B. H., and Elsheimer, S., *J. Am. Chem. Soc.*, **103,** 715 (1981).

1119 Dolbier, W. R., Sellers, S. F., Al-Sader, B. H., and Fielder, T. H. Jr., *J. Am. Chem. Soc.*, **103,** 717 (1981).
1120 Dolbier, W. R., and Sellers, S. F., *Tetrahedron Lett.*, **22,** 2953 (1981).
1121 Dolbier, W. R., Al-Sader, B. H., Sellers, S. F., and Koroniak, H., *J. Am. Chem. Soc.*, **103,** 2138 (1981).
1122 Roberts, J. T., Rittberg, B. R., Kovacic, P., Scalzi, F. V., and Seely, M. J., *J. Org. Chem.*, **45,** 5239 (1980).
1123 Barton, B. D., and Stein, S. E., *J. Chem. Soc., Faraday Trans. 1*, **77,** 1755 (1981).
1124 Ohsawa, A., Kawaguchi, T., and Igeta, H., *Chem. Pharm. Bull.*, **29,** 1481 (1981); *Chem. Abs.*, **95,** 96755 (1981).
1125 Lur'e, B. A., and Ivakhov, V. N., *Tr.-Mosk. Khim.-Tekhnol. Inst. im. D. I. Mendelaeva*, **104,** 12 (1979); *Chem. Abs.*, **93,** 238438 (1980).
1126 Prokudin, V. G., Nazin, G. M., and Manelis, G. B., *Dokl. Akad. Nauk SSSR*, **255,** 917 (1980); *Chem. Abs.*, **95,** 23952 (1981).
1127 Knoll, H., Barry, I., and Scherzer, K., *J. Prakt. Chem.*, **323,** 393 (1981); *Chem. Abs.*, **95,** 132029 (1981).
1128 Richard, C., and Martin, R., *Int. J. Chem. Kinet.*, **12,** 1021 (1980).
1129 Kalashnikov, S. M., Imashev, U. B., Zlotskii, S. S., Glukhova, S. S., and Rakhmankulov, D. L., *Zh. Org. Khim.*, **16,** 990 (1980); *Chem. Abs.*, **93,** 185370 (1980).
1130 Bloch, R., and Orvane, P., *Tetrahedron Lett.*, **22,** 3597 (1981).
1131 Komissarov, V. D., and Safiullin, R. L., *Kinet. Katal.*, **21,** 594 (1980); *Chem. Abs.*, **93,** 203767 (1980).
1132 Komissarov, V. D., Safiullin, R. L., and Denisov, E. T., *Dokl. Akad. Nauk SSSR*, **252,** 1177 (1980); *Chem. Abs.*, **93,** 238437 (1980).
1133 Jarvis, B. B., Nicholas, P. E., and Midiwo, J. O., *J. Am. Chem. Soc.*, **103,** 3878 (1981).
1134 Bell, T. N., Perkins, K. A., and Perkins, P. G., *J. Chem. Soc., Faraday Trans. 1*, **77,** 1779 (1981).
1135 Davidson, I. M. T., and Wood, I. T., *J. Organomet. Chem.*, **202,** C65 (1980).
1136 Dinctürk, S., Jackson, R. A., Townson, M., Agirbas, H., Billingham, N. C., and March, G., *J. Chem. Soc., Perkin Trans. 2*, **1981,** 1121.
1137 Dinctürk, S., and Jackson, R. A., *J. Chem. Soc., Perkin Trans. 2*, **1981,** 1127.
1138 M. Z. Hoffmann, *et al.*, *J. Chem. Educ.*, **58,** 84 (1981) and subsequent papers.
1139 Bahnemann, D., Asmus, K.-D., and Willson, R. L., *J. Chem. Soc., Perkin Trans. 2*, **1981,** 890.
1140 Hiller, K.-O., Masloch, B., Göbl, M., and Asmus, K.-D., *J. Am. Chem. Soc.*, **103,** 2734 (1981).
1141 Solar, S., Getoff, N., Solar, W., and Mark, F., *Can. J. Chem.*, **59,** 2719 (1981).
1142 Svingen, B. A., and Powis, G., *Arch. Biochem. Biophys.*, **209,** 119 (1981).
1143 Bhattacharyya, S. N., and Neta, P., *J. Phys. Chem.*, **85,** 1527 (1981).
1144 Riederer, H., Hüttermann, J., and Symons, M. C. R., *J. Phys. Chem.*, **85,** 2789 (1981).
1145 Kuwabara, M., Lion, Y., and Riesz, P., *Int. J. Radiat. Biol. Relat. Stud. Phys., Chem. Med.*, **39,** 491 (1981); *Chem. Abs.*, **95,** 149352 (1981).
1146 Fujita, S., and Steenken, S., *J. Am. Chem. Soc.*, **103,** 2540 (1981).
1147 Sevilla, M. D., Suryanarayana, D., and Morehouse, K. M., *J. Phys. Chem.*, **85,** 1027 (1981).
1148 Oduwole, D. A., and Wiseall, B., *Bull. Chem. Soc. Jpn.*, **54,** 2551 (1981).
1149 Symons, M. C. R., and Smith, I. G., *J. Chem. Soc., Perkin Trans. 2*, **1981,** 1180.
1150 Janata, E., and Schuler, R. H., *J. Phys. Chem.*, **84,** 3351 (1980).
1151 Kovalev, M. D., and Koritskii, A. T., *Khim. Vys. Energ.*, **15,** 318 (1981); *Chem. Abs.*, **95,** 114420 (1981).
1152a Faucitano, A., Buttafava, A., Faucitano Martinotti, F. F., Caporiccio, G., and Corti, C., *J. Chem. Soc., Perkin Trans. 2*, **1981,** 425.
1152b Muszkat, L., *J. Phys. Chem.*, **85,** 2916 (1981).
1153 Bobrowski, K., *J. Phys. Chem.*, **84,** 3524 (1980).
1154 Bobrowski, K., *J. Phys. Chem.*, **85,** 382 (1981).
1155 McAskill, N. A., and Sangster, D. F., *Aust. J. Chem.*, **34,** 721 (1981).
1156 Selvarajan, N., and Raghavan, N. V., *J. Phys. Chem.*, **84,** 2548 (1980).
1157 Turro, N. J., and Kraeutler, B., *Acc. Chem. Res.*, **13,** 369 (1980).
1158 Turro, N. J., *Pure Appl. Chem.*, **53,** 259 (1981).
1159 Turro, N. J., Chow, M.-F., Chung, C.-J., and Kraeutler, B., *J. Am. Chem. Soc.*, **103,** 3886 (1981).
1160 Turro, N. J., Chow, M.-F., Chung, C.-J., Tanimoto, Y., and Weed, G. C., *J. Am. Chem. Soc.*, **103,** 4574 (1981).
1161 Turro, N. J., and Mattay, J., *J. Am. Chem. Soc.*, **103,** 4220 (1981).
1162 Schilling, M. L. M., *J. Am. Chem. Soc.*, **103,** 3077 (1981).
1163 Lehr, G. F., and Turro, N. J., *Tetrahedron*, **37,** 3411 (1981).

1164 Scheek, R. M., Stob, S., Schleich, T., Alma, N. C. M., Hilbers, C. W., and Kaptein, R., *J. Am. Chem. Soc.*, **103,** 5930 (1981).

1165 Doubleday, C., *Chem. Phys. Lett.*, **81,** 164 (1981); *Chem. Abs.*, **95,** 149525 (1981).

1166 Kaptein, R., de Kanter, F. J. J., and Rist, G. H., *J. Chem. Soc., Chem. Commun.*, **1981,** 499.

1167 Khait, I., Lüdersdorf, R., Muszkat, K. A., and Praefcke, K., *J. Chem. Soc., Perkin Trans. 2*, **1981,** 1417.

1168 Yoshida, M., Furuta, N., and Kobayashi, M., *Bull. Chem. Soc. Jpn.*, **54,** 2356 (1981).

1169 Nakanishi, W., Jo, T., Miura, K., Ikeda, Y., Sugawara, T., Kawada, Y., and Iwamura, H., *Chem. Lett.*, **1981,** 387.

1170 Ponomarev, V. S., and Stepukhovich, A. D., *Zh. Fiz. Khim.*, **54,** 1769 (1980); *Chem. Abs.*, **93,** 203747 (1980).

1171 Aunir, D., Johnston, L. J., De Mayo, P., and Wong, S. K., *J. Chem. Soc., Chem. Commun.*, **1981,** 958.

1172 Closs, G. L., and Sitzmann, E. V., *J. Am. Chem. Soc.*, **103,** 3217 (1981).

1173 Closs, G. L., and Miller, R. J., *J. Am. Chem. Soc.*, **103,** 3586 (1981).

1174 Wong, S. K., Chiu, T.-K., and Bolton, J. R., *J. Phys. Chem.*, **85,** 12 (1981).

1175 Symons, M. C. R., Foxall, J., Gilbert, B. C., Norman, R. O. C., and Winters, J. N., *J. Chem. Soc., Chem. Commun.*, **1980,** 1207.

1176 Chen, K. S., and Wan, J. K. S., *Can. J. Chem.*, **59,** 116 (1981).

1177 Depew, M. C., Adeleke, B. B., and Wan, J. K. S., *Can. J. Chem.*, **59,** 2708 (1981).

Organic Reaction Mechanisms 1981
Edited by A. C. Knipe and W. E. Watts

CHAPTER 4

Oxidation and Reduction

G. W. J. Fleet

Dyson Perrins Laboratory, South Parks Road, Oxford

Oxidation by Metal Ions . . . 183
Chromium . . . 183
Manganese . . . 185
Copper, Silver, Gold, Mercury, and Thallium . . . 185
Cerium and Lead . . . 189
Bismuth, Uranium, Vanadium, and Molybdenum . . . 190
Group VIII Metals . . . 191
Oxidation by Compounds of Non-metallic Elements . . . 193
Nitrogen . . . 193
Sulphur . . . 193
Selenium and Tellurium . . . 195
Halogens, including Periodate and Bromate . . . 196
Miscellaneous Oxidations . . . 199
Ozonolysis and Ozonation . . . 200
Peracids, Peroxides, and Superoxide . . . 202
Atomic Oxygen and Singlet Oxygen . . . 206
Autoxidation and Other Reactions of Triplet Oxygen . . . 208
Reduction by Complex Hydrides . . . 211
Aluminium Hydrides . . . 211
Borohydrides . . . 212
Other Hydride Reductions . . . 214
Reduction by Metals, Metal Complexes, and Ions . . . 217
Miscellaneous Reductions . . . 218
Hydrogenation . . . 219
Reductions and Oxidations of Biological Interest . . . 221
References . . . 221

Oxidation by Metal Ions

Chromium

In the three-electron co-oxidation of 2-hydroxy-2-methylbutanoic acid and propan-2-ol, ^{13}C and deuterium isotope studies have provided evidence for a step-wise breakdown of the chromium(VI) intermediate (**1**); formation of a chromium(IV) species (**2**), which undergoes exchange with labelled hydroxymethylbutanoic acid, is followed by fragmentation to chromium(III) (Scheme 1). It appears that chromium(VI) is remarkably selective for oxidation of the C—H bond of the alcohol whereas chromium(IV) is equally selective for oxidation of the C—C bond of the hydroxy-

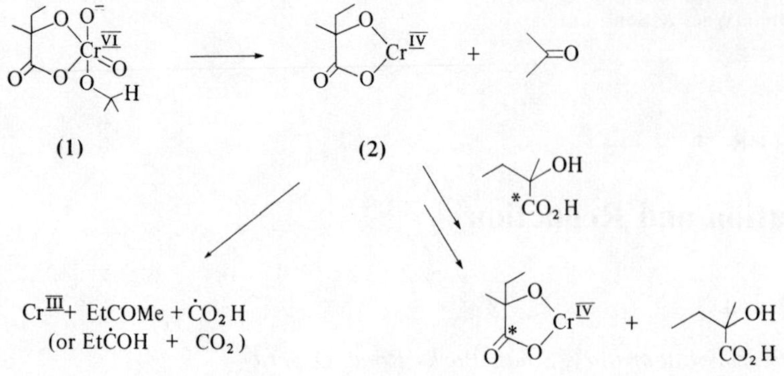

(1) (2)

Cr^{III} + EtCOMe + $\dot{C}O_2H$
(or Et$\dot{C}$OH + CO_2)

SCHEME 1

acid.[1] The oxidation of alcohols by pyridinium chlorochromate is catalysed by haloacetic acids.[2] The equilibrium constants for cycloalkanols and the corresponding ketones have been used to determine their relative energies and have been correlated with rates of oxidation of the alcohols by chromium trioxide.[3] The relative reactivities of the rotamers of 9-(2-formyl-1-naphthyl)fluorenol and related alcohols have been determined and discussed.[4] Pyridinium dichromate (PDC) has been used in the oxidation of α-ynol–iodine complexes to α,β-unsaturated α-iodo-aldehydes (Scheme 2); only one diastereoisomeric aldehyde is formed.[5] Poly(vinylpyridinium) dichromate has been shown to be a useful polymer-bound chromium reagent for the oxidation of alcohols.[6] The chromium(VI) oxidations of ferrocenyl,[7] benzyl,[8] tertiary,[9] and other[10] alcohols have been the subjects of kinetic investigations.

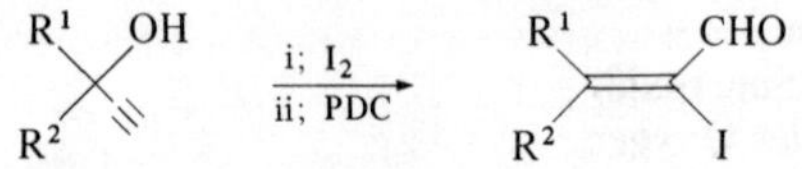

i; I_2
ii; PDC

SCHEME 2

The rate of oxidation of methyl phenyl sulphoxides is considerably affected by steric effects.[11,12] Triarylamines are oxidized to the corresponding cation radicals by chromium(VI) by one-electron transfer; the chromium(V) so formed also oxidizes the tertiary amines to cation radicals.[13] The relative rates of formation of quinones from hydroquinone silyl ethers, on treatment with pyridinium chlorochromate, qualitatively parallel the electrochemical oxidation potentials.[14] A cyclic intermediate formed from chromic acid and picolinic acid is responsible for the oxidation of cyclohexanone.[15] The chromium(VI) oxidation of ethyl 3-phenylacrylate involves rate-determining formation of a chromium(IV) ester.[16] The reactions of a series of alkanes with chromium trioxide–sulphuric acid solutions were characterized by large kinetic H/D isotope effects, relatively rapid tertiary C—H bond cleavage, and a dependence of the rate on the Hammett acidity function.[17] Iridium(IV) chloride complexes have been shown to catalyse the oxidation of alkanes by chromium(VI).[18,19] Kinetic studies on the chromium(VI) oxidations of ascorbic acid,[20,21] malonic acid,[22] benzimidazole,[23] and polyarylpyrroles[24] have been reported. The oxidation of organic compounds by chromium(VI) and permanganate has been reviewed.[25]

Manganese

The factors affecting formation of α-ketols and -diones in the oxidation of alkenes by permanganate have been discussed.[26] Both glycols and ketols arise from common intermediate manganese(V) esters (**3**) which undergo hydrolysis to diols or oxidative decomposition to cyclic manganate esters (**4**), followed by hydrolysis of the resulting manganese(IV) esters.[27] A similar scheme has been proposed for the mechanism of the permanganate oxidation of uracils,[28] involving initial [3 + 2]-cycloaddition across the 5,6-double bond of uracil, followed by formation of a soluble manganese(IV) species. A soluble manganese(IV) species has been detected in the permanganate oxidation of *trans*-cinnamic acid.[29] A manganese intermediate with a coordination number greater than four has been observed in the oxidation of hexa-1,5-diene.[30] No significant stereoselectivity was observed in the oxidation of olefins by potassium permanganate in the presence of chiral phase-transfer catalysts.[31]

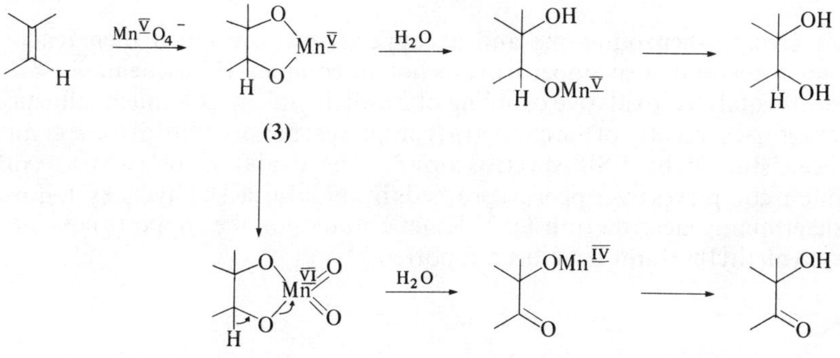

(3)

(4)

The mechanism of phenol oxidation by alkaline permanganate has been studied.[32] Initially, phenol and chlorophenols give a manganese(VI) ion[33] which also oxidizes phenols by a radical-chain mechanism.[34] The lack of effect of pressure on the rate of oxidation of benzhydrol by alkaline permanganate supports a mechanism of rate-determining hydrogen transfer.[35] Unsaturated secondary alcohols are oxidized to ketones by potassium permanganate adsorbed on solid support without any accompanying oxidation of the double bond;[36] solid permanganate under heterogeneous conditions oxidizes allylic alcohols more slowly than saturated alcohols.[37] Mechanistic studies on the permanganate oxidations of tartaric[38] and propionic[39] acids, phenylalanine,[40] *N,N*-disubstituted adenosines,[41] piperidones,[42] and cyclohexanone[43] have been described. The inhibiting effect of manganese(II) on the oxidation of *meta*-substituted toluenes (Scheme 3) can be removed by performing the oxidation with potassium permanganate in the presence of an aliphatic acid, its anhydride, and a halide salt such as potassium bromide.[44] Kinetic studies have been reported on the manganese(III) oxidations of alcohols and acids,[45] aliphatic esters,[46] and cyclohexanone.[47] Manganese(IV) oxide has been shown to be an efficient oxidant for the conversion of hydroquinones into 1,4-benzoquinones.[48]

Copper, Silver, Gold, Mercury, and Thallium

The oxidation of catechols by copper(II)–pyridine–methanol[49] systems involves a dicopper(II) catecholate intermediate followed by electron transfer to two copper(II)

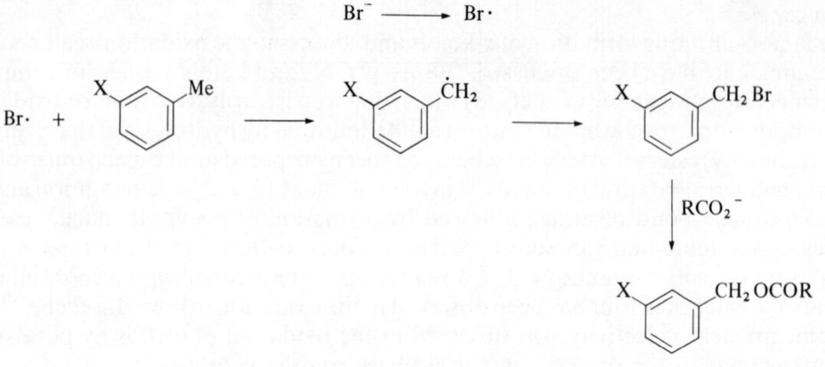

$Br^- \longrightarrow Br\cdot$

$Br\cdot$ + (3-X-toluene) → (3-X-benzyl radical, $\dot{C}H_2$) → (3-X-benzyl bromide, CH_2Br) $\xrightarrow{RCO_2^-}$ (3-X-benzyl ester, CH_2OCOR)

SCHEME 3

centres giving *o*-benzoquinone and a copper(I) reagent which then reacts with catechol regenerating dicopper(II) catecholate complex (**5**) (Scheme 4).[50,51] The copper(II)-catalysed oxidative coupling of 2,6-dialkylphenols is much influenced by the tetragonal geometry of the copper(II) complexes; the structure of these complexes had been studied by ESR spectroscopy.[52] The oxidation of (α-hydroxyalkyl)-chromium complexes by copper(II) proceeds by initial attack at hydroxyl followed by rate-determining electron transfer.[53] Kinetic studies on the copper(II) oxidation of (hydroxymethyl)furfural have been reported.[54]

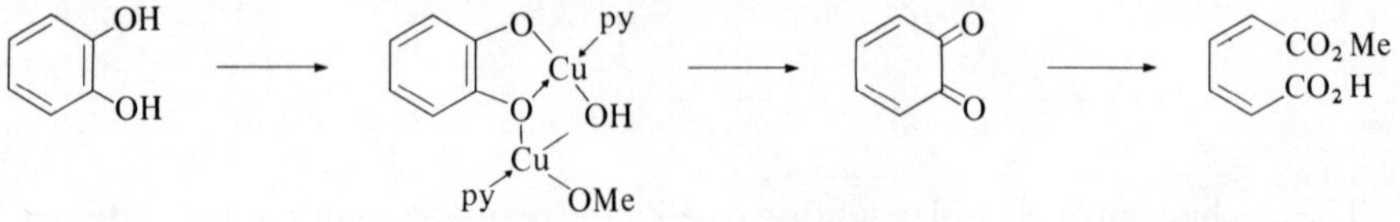

Catechol (OH, OH) → complex (**5**) [catecholate O,O–Cu(py)(OH); O–Cu(py)(OMe)] → *o*-benzoquinone (O, O) → CO_2Me / CO_2H diene

(**5**)

SCHEME 4

The oxidation of furfural by silver(I) and mercury(I) ions involves a two-electron mechanism forming mainly furan-2-carboxylic acid which is decarboxylated in the presence of mercury(I) ions.[55] The oxidative carbon–carbon bond-cleavage of 1,2-diols by silver(II) proceeds by initial complexation followed by electron transfer from hydroxyl to silver(II) giving an alkoxyl radical (**6**) which undergoes *β*-scission and further one-electron oxidation (Scheme 5).[56] Free-radical intermediates occur in the silver(II) oxidation of formamides.[57,58] Competing reactions involving Ag^{2+} and $(AgOH)^+$ have been identified in the silver(II) perchlorate oxidation of alcohols.[59] A two-electron-transfer process from alcohol to silver (III) is the rate-determining step in the oxidation of alcohols by di(periodato)argentate(III).[60] The gold(III) oxidation of peptide and disulphides bonds has been discussed in terms of a mechanism for gold toxicity.[61]

Concurrent electrophilic and oxidative pathways have been identified in the reactions of α-hydroxy- and α-alkoxy-alkyl complexes of chromium(III) with mercury(II) ions.[62] Mercury(II)–EDTA dehydrogenates the pyrrolidine (**7**) to (**8**) in which neighbouring-group participation by the *N*-oxide is followed by a second two-

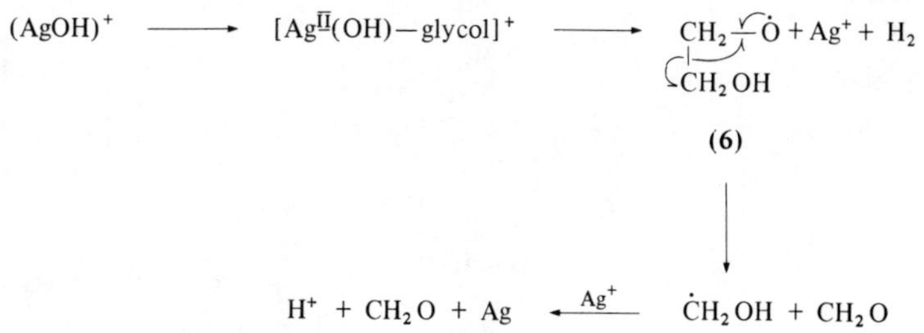

$(AgOH)^+ \longrightarrow [Ag^{II}(OH)\text{—glycol}]^+ \longrightarrow CH_2\text{—}\dot{O} + Ag^+ + H_2$ / CH_2OH **(6)**

$\downarrow$

$H^+ + CH_2O + Ag \xleftarrow{Ag^+} \dot{C}H_2OH + CH_2O$

SCHEME 5

electron oxidation and hydrolysis to give (**9**) (Scheme 6).[63] An allylic mercurial is not an intermediate in the oxidation of allylbenzene to *trans*-cinnamyl ethers by mercury(II) oxide in tetrafluoroboric acid.[64] Initial aminomercuriation of 1-phenylprop-2-ynol (**10**) is followed by allylic oxidation by mercury(II) to the imine (**11**).[65]

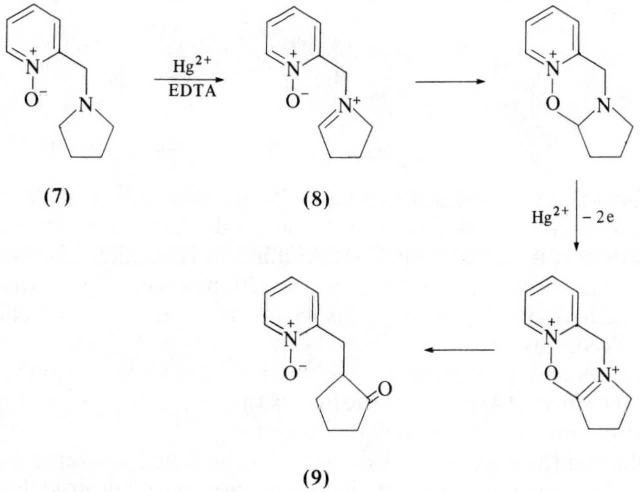

(7) $\xrightarrow[\text{EDTA}]{Hg^{2+}}$ (8) $\longrightarrow$ ... $\xrightarrow{Hg^{2+}, -2e}$ ... $\longrightarrow$ (9)

SCHEME 6

(10) $\xrightarrow[Hg^{II}]{ArNH_2}$ ... $\longrightarrow$ ... $\longrightarrow$ (11)

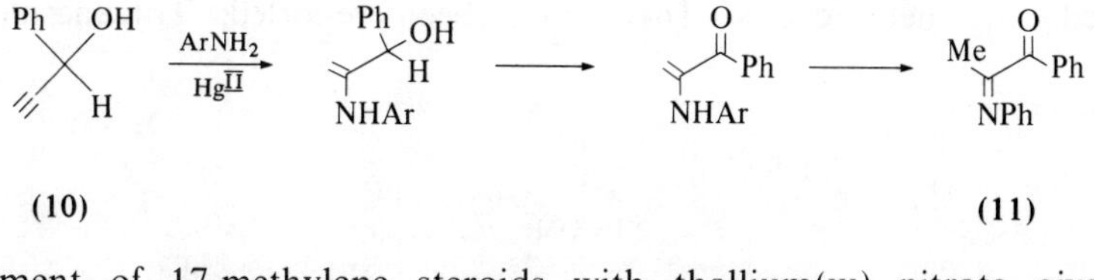

Treatment of 17-methylene steroids with thallium(III) nitrate gives ring-enlargement to a ketone (**12**) which undergoes enolization, followed by oxythallation and methanolysis of the carbon–thallium bond to give (**13**).[66] Olefinic cyclizations of *o*-prenylphenols by thallium(III) has been studied (Scheme 7).[67] In the reaction of thallium(III) with crotonaldehyde, 2,3-dihydroxybutanal is formed in two

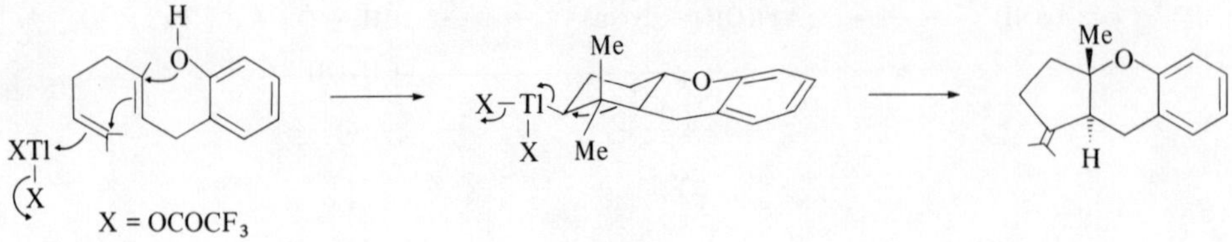

$X = OCOCF_3$

SCHEME 7

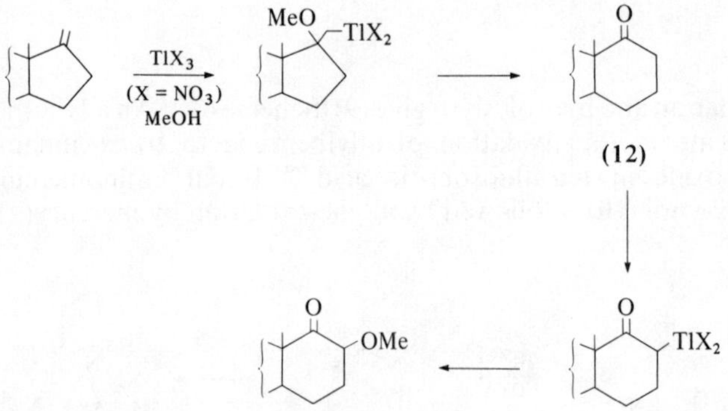

TlX_3 ($X = NO_3$) MeOH

(12)

(13)

consecutive one-electron-transfer processes;[68] the mechanism of the thallium(III) oxidative dimerization of 4-alkoxycinnamic acids has been studied.[69] In the diacetoxylation of conjugated dienes with thallium(III) acetate, 1,2-diacetoxylation predominates over alternative 1,4-attack; the reaction proceeds by acetoxythallation, followed by dethallation by acetate displacement.[70] Kinetic salt effects on the oxidation of alkenes by thallic salts have been investigated.[71] The oxidation of the monomethyl ether of naphthalene-1,5-diol by thallium(III) nitrate in the presence of ethylene glycol to form (**14**) provides the first example of a direct oxidation of a *para*-unsubstituted phenol to a monoacetal of a quinone.[72]

In the palladium(II)-catalysed oxidation of substituted benzenes to biaryls by thallium(III) trifluoroacetate, initial thallation of the arene is followed by substitution of thallium by palladium to form an unstable palladium(II) species which oxidatively dimerizes.[73] The thallium(III) oxidations of methyl 3,5-dibromo-4-hydroxyphenyl-pyruvate oxime to spiro-oxazoles[74] and of aromatic bis-amides (**15**) to generate five-membered ring heterocycles (**16**)[75] have been reported. Toluenesulphonyl-

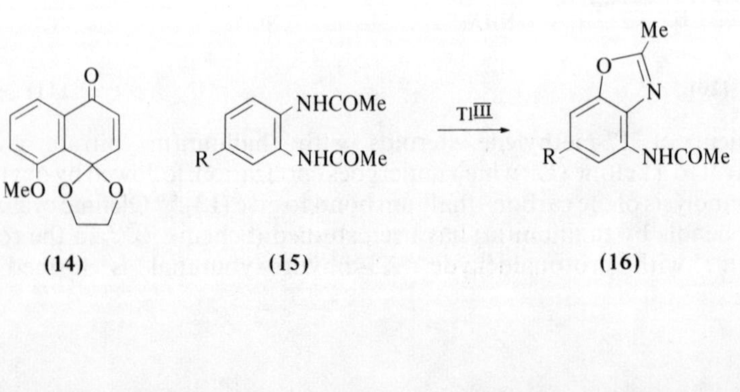

Tl^{III}

(14) **(15)** **(16)**

hydrazones are cleanly oxidized to the corresponding carbonyl compounds by thallium(III), in contrast to the mixtures of products obtained with mercury(II) and lead(IV).[76] The thallium oxidation of phenylhydrazines proceeds *via* initial thallation at nitrogen, followed by a two-electron transfer.[77] A change from an ionic to a radical mechanism occurs in the presence of ruthenium(III) in the oxidation of aniline by thallium(III).[78]

Cerium and Lead

Among other studies on the Belousov–Zhabotinskii oscillating reaction [bromate–cerium(IV)–malonate],[79–83] the effect of secondary alcohols on the reaction has been examined; the inhibiting or initiating activity of the alcohol depends on the reactant ratio.[84] The oxidation of ethyl acetoacetate by cerium(IV) is an inner-sphere reaction which proceeds through an intermediate complex; the modified Belousov–Zhabotinskii reaction system containing ethyl acetoacetate as substrate is noteworthy in that regular oscillations of cerium(IV) ion concentration arise in an undisturbed system without any induction period.[85]

The rate-determining step in the cerium(IV) oxidation of several α,β-unsaturated aldehydes is the acid-catalysed hydration of the aldehyde;[86] in contrast, rate-determining complexation of cerium with a saturated aldehyde has been proposed as the slow step in the oxidation of propanal.[87] A change-over from rate-determining electron transfer to rate-determining proton transfer has been observed in the oxidation of alkyl side-chains of substituted aromatic compounds by cerium(IV).[88] The initial step in oxidative cyclization and fragmentation of steroidal alcohols induced by ceric ammonium nitrate is a one-electron oxidation followed by subsequent abstraction of a γ-hydrogen.[89] Naphthoquinones are formed by the cerium(IV) oxidative cyclization of (**17**).[90] Radicals have been suggested as intermediates in the cerium(IV) oxidation of catechol,[91] carboxylic acids,[92–94] deoxybenzoin,[95] glycol,[96] aliphatic amines,[97] and phenacyl bromides.[98]

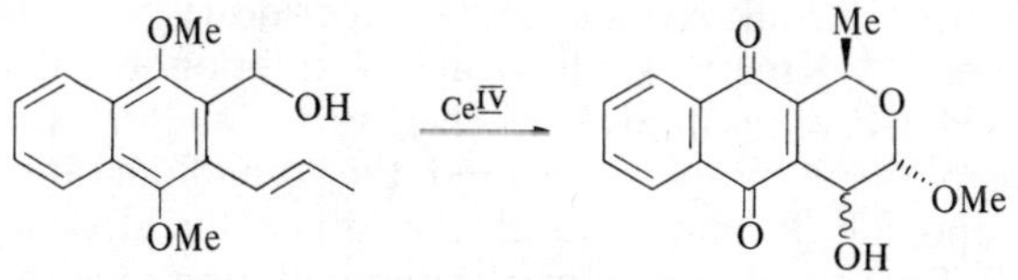

OMe OH OMe Ce^{IV} O Me O OMe O OH

(**17**)

The oxidation of organic compounds by lead tetraacetate has been reviewed.[99] The stereochemistry of fragmentations of alcohols and carboxylic acids by lead tetraacetate is determined by steric interactions and torsional strain factors in the process of acetoxyl transfer.[100,101] The differences in reactions of lead tetraacetate with ArXR (X = S, NR′,O) depend on the combined effect of the first ionization potential of the hetero-atom X, the availability of the hetero-atom lone pair for reaction electrophiles, and its ability to expand its electron octet.[102] The oxidation of 3-amino-2-(arylalkyl)quinazolones with lead(IV) gives *N*-amino nitrenes which react intramolecularly with electron-rich aromatic aryl groups *via* a seven-membered transition state.[103] α,β-Epoxy-alcohols are oxidatively cleaved by lead tetra-acetate[104] (Scheme 8). The lead(IV) oxidations of benzoic acids[105] and of some methoxyacetophenones[106] have been studied.

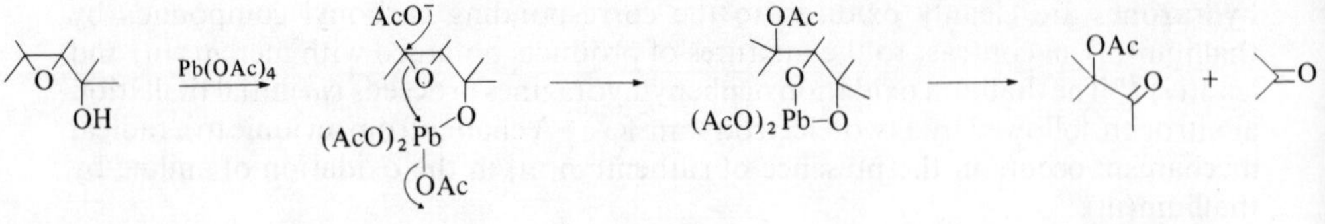

SCHEME 8

Bismuth, Uranium, Vanadium, and Molybdenum

A catalytic method has been reported (Scheme 9)[107] for the cleavage of α-glycols by a catalytic amount of triphenylbismuth in the presence of potassium carbonate, using *N*-bromosuccinimide as the oxidant for regenerating bismuth(V). The kinetics of the reaction of uranium(III) with halopropanoic acids have been studied.[108]

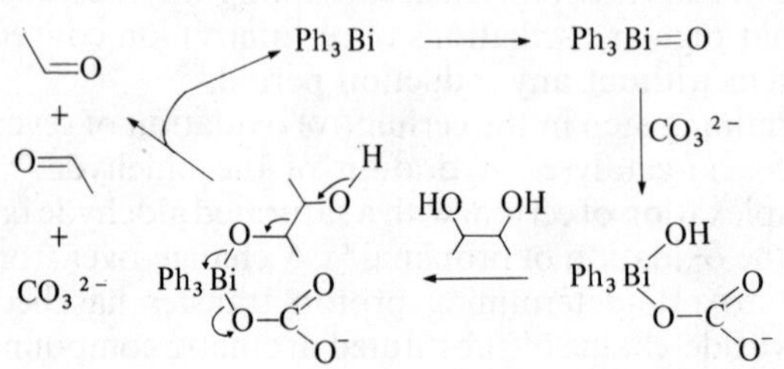

SCHEME 9

Further examples of the preferred transition-state geometries in the oxidative cyclization of indole and isoquinoline derivatives by vanadium(V) show that chemical cyclization reactions tend to form six- rather than five-membered transition states.[109] One-electron-transfer processes are involved in the vanadium(V) oxidations of dihydroisoquinolines,[110] 2-alkoxyethanols,[111] piperidones,[112] butanone,[113] and 4-methylpentan-2-one.[114] Kinetic studies on epoxidation of alkenes by hydroperoxides catalysed by vanadium(V) acetylacetonate suggest that two different complexes between the hydroperoxide and the vanadium are formed; one is responsible for epoxide formation and the other for decomposition of the hydroperoxide.[115] The rôle of vanadium in the epoxidation consists of activating the hydroperoxide for the electrophilic reaction.[116] Although non-stereoselective vanadium- catalysed epoxidation of some 2-benzylidenecyclohexanols has been reported,[117] considerable discrimination was found by vanadium(V)–hydroperoxide epoxidation of pseudomonic acid *C* to pseudomonic acid *A*.[118] The mechanism for oxo–peroxo oxygen exchange in peroxovanadium(V) and peroxomolybdenum(VI) with ^{18}O-labelled hydrogen peroxide has been elucidated.[119] Though selective epoxidation of olefins by molybdenum(VI)-catalysed peroxy-bond heterolysis is much used, the nature of the oxidizing species is not clear; in particular, the requirement for olefin activation through coordination to the metal centre prior to oxygen transfer is still a matter of controversy. With molybdenum porphyrins, steric hindrance of the macrocyclic ligand would render simultaneous complexing of olefin and hydroperoxide unlikely. The catalytic activity found supports the mechanism proposed by Sharpless and Sheldon in which direct attack of olefin on activated hydroperoxide occurs without coordination of the olefin to the metal.[120] A molybdenum(V) oxalate complex has been used as a homogeneous or an anion-exchange supported catalyst for the epoxidation of olefins by hydroperoxides; in both the homogeneous and heterogeneous systems, molybdenum(V) is oxidised to molybdenum(VI) which undergoes further conversion to a molybdenum peroxooxalate complex.[121] Molybdenum(VI)-catalysed oxidation of silyl enol ethers with *tert*-butyl hydroperoxide cleaves the olefin to carbonyl compounds (Scheme 10).[122] The mechanism of molybdenum-catalysed oxidation of sulphides and alkenes by hydrogen peroxide have been studied.[123]

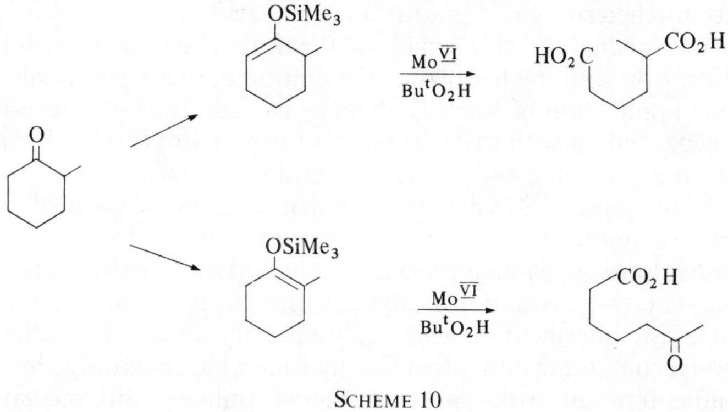

OSiMe$_3$ — Mo^{VI}, Bu^tO_2H → HO_2C ... CO_2H

OSiMe$_3$ — Mo^{VI}, Bu^tO_2H → CO_2H ... O

SCHEME 10

Group VIII Metals

Some selective reactions of transition-metal complexes have been reviewed.[124] A mechanism has been proposed for the oxidation of cyclohexanone by a nickel(III)–(macrocyclic ligand) complex.[125] Cobalt(III) acetate oxidation of olefins gives exclusively allylic acetates with no 1,2-addition or products arising from skeletal rearrangements; cobalt(III) attack is sterically hindered since cobalt(III) is a bulky dimeric species.[126] The reaction of oxygen with aryl-substituted olefins in the presence of bis[dimethylglyoximato]chloro(pyridine)cobalt(III) and sodium borohydride gives regiospecific conversion to alcohols (Scheme 11).[127] The influence of water on the cobalt(III) oxidation of toluene has been investigated.[128] Iron(III) nitrate in methanol has been used for oxygenation of ketenimines.[129] Ketone enolates are oxidatively coupled by dry anhydrous ferric chloride to 1,4-diketones.[130] Iron(III) chloride as a reagent supported on silica is a suitable oxidant for coupling of phenols

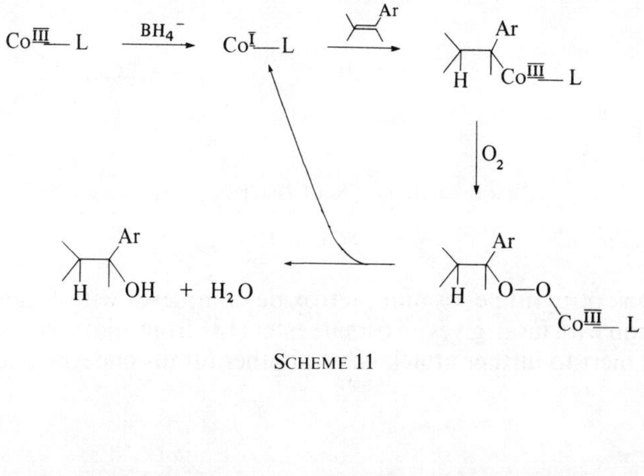

Co^{III}—L $\xrightarrow{BH_4^-}$ Co^{I}—L → (Ar olefin) → H, Ar, Co^{III}—L

O_2

H, Ar, OH + H_2O ← H, Ar, O—O—Co^{III}—L

SCHEME 11

to biaryls.[131] Photolysis of iron(III) perchlorate in primary alcohols yields a complex mixture, although initial production of alkoxy radicals can account for the distribution of products; the photolysis rates parallel the importance of a charge-transfer band in the near UV, implying it is involved in the primary photochemical process.[132] Photolysis of mono- and di-substituted olefins in pyridine in the presence of iron(III) chloride gives α-chloroketones while tri- and tetra-substituted olefins yield dichloroketones with accompanying carbon–carbon bond cleavage; the results can be interpreted in terms of an electron-transfer mechanism within the coordination sphere of the iron ion.[133]

The oxidation of heterocyclic[134] and aromatic[135] acid hydrazides by iron(III) in the presence of 1,10-phenanthroline involves the formation of a complex of the phenanthroline, iron, and the hydrazine. Estimation of redox potentials of organic radicals by the application of Marcus' theory has been used in discussions of the oxidation of benzenediols with tris(1,10-phenanthroline)iron(III).[136] The oxidations of the following systems by hexacyanoferrate(III) have been investigated: phenols,[137,138] mercapto-,[139] hydroxy-,[140] and other carboxylic acids,[141] *N*-benzyl-1,4-dihydronicotinamide,[142] benzoins,[143] and disaccharides.[144]

Several studies have appeared on ruthenium(III)-catalysed oxidations of alcohols. The ruthenium-catalysed oxidation of allylic alcohols by molecular oxygen involves a ruthenium alkoxide intermediate which subsequently undergoes β-elimination to give the carbonyl compound and ruthenium hydride which is oxidized by oxygen.[145] Although ruthenium alkoxides are considered unlikely intermediates in the ruthenium-catalysed oxidation of alcohols and aldehydes by iodosyl benzene,[146] a ruthenium alkoxide is proposed in the catalytic oxidation of primary hydroxyl groups using $(Ph_3P)_3RuCl_2$ as a catalyst; this catalyst is highly effective for selective oxidation of a primary hydroxyl group in the presence of a secondary alcohol; *e.g.* 10-hydroxyundecanal is formed in excellent yield from undecane-1,10-diol.[147]

Selective monoalkylation of arylacetonitriles by alcohols with a ruthenium catalyst has been achieved (Scheme 12).[148] The mechanisms of ruthenium(III)-catalysed oxidations of aromatic aldehydes[149] and unsaturated acids[150] and of ruthenium(IV) oxidation of alkanes[151] have been investigated. Addition of acetonitrile to the traditional carbon tetrachloride–water solvent system for ruthenium tetraoxide oxidation of olefins greatly increases the efficiency of the oxidation.[152]

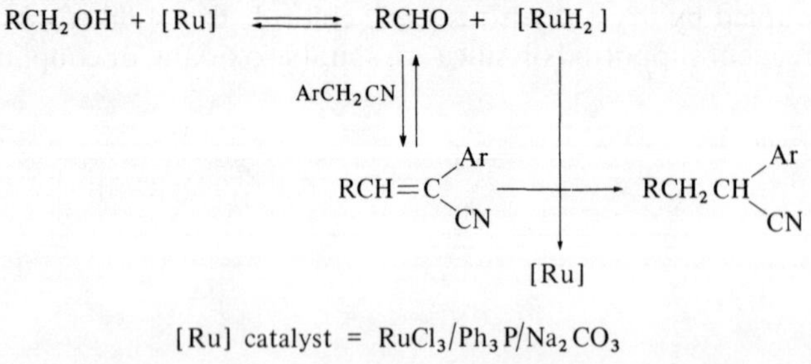

$$RCH_2OH + [Ru] \rightleftharpoons RCHO + [RuH_2]$$

$$RCHO \xrightarrow{ArCH_2CN} RCH{=}C(Ar)CN \xrightarrow{[RuH_2]\ \rightarrow\ [Ru]} RCH_2CH(Ar)CN$$

$$[Ru]\ \text{catalyst} = RuCl_3/Ph_3P/Na_2CO_3$$

SCHEME 12

The reactions of pyridine–osmium-tetroxide complexes with furans have been studied; reaction with furan gives an osmate ester (**18**), from addition across C(2) and C(5), which is inert to further attack–several other furans undergo addition across

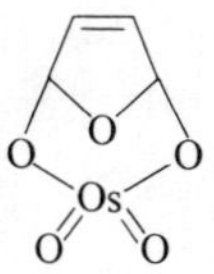

(18)

C(2) and C(3) giving products which are then subject to further oxidation.[153] The reaction of silyl enol ethers with osmium tetroxide produces α-ketols in high yield.[154] The kinetics of the osmium(VIII) oxidations of cyclic ketones,[155] cyclic alcohols,[156] halocarboxylic acids,[157,158] and glycol[159] have been investigated. Osmate(VI) esters chelated with optically active *trans-N,N',N'*-tetramethylcyclohexane-1,2-diamine are useful derivatives for determination of the stereochemistry of sub-milligram quantities of glycols by ^{1}H-NMR spectroscopy.[160]

The mechanisms of palladium(II)-catalysed oxidations of ethylene,[161] styrenes,[162] benzene,[163] and toluene,[164] and of iridium(IV)-catalysed oxidations of pyruvate[165] and oxalic acid[166] have been studied.

Oxidation by Compounds of Non-metallic Elements

Nitrogen

Oxidation reactions with diethyl azodicarboxylate have been reviewed.[167] Cation radicals of triphenylamines have been used for the one-electron oxidations of triphenyl derivatives of nitrogen sub-group elements; a mechanism has been proposed for the conversion of triphenylamine radical cation to the tetraphenylbenzidine radical cation.[168] The oxidation of formic acid by nitrous acid involves formation of the cation radical of the acid (Scheme 13).[169] The formation of quinones

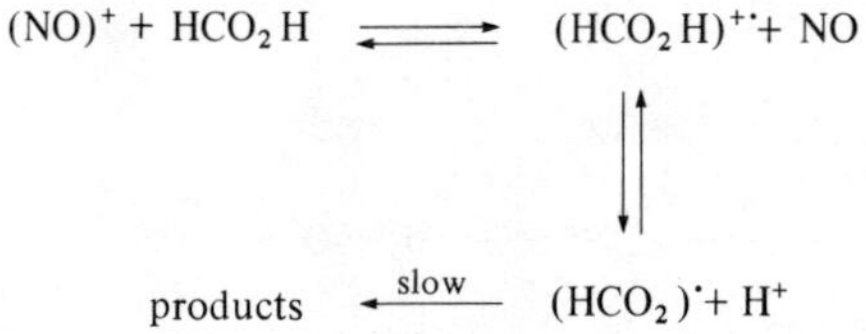

$$(NO)^+ + HCO_2H \rightleftharpoons (HCO_2H)^{+\cdot} + NO$$

$$(HCO_2H)^{+\cdot} \rightleftharpoons (HCO_2)^{\cdot} + H^+$$

$$(HCO_2)^{\cdot} + H^+ \xrightarrow{\text{slow}} \text{products}$$

SCHEME 13

from phenols by nitric acid[170,171] (in which the rate-determining step is the formation of dinitrogen tetroxide) and by inorganic nitrate salts in trifluoroacetic anhydride[172] has been reported. The transition-metal nitro–nitroso redox couple has been exploited in the catalytic oxidation of olefins to ketones by acetonitrile(nitro)-palladium(II) complexes; at least two intermediates with carbon–palladium bonds have been observed.[173] Ethyl nitroacetate oxidizes alcohols to aldehydes under neutral conditions in the presence of triphenylphosphine and diethyl azodicarboxylate; the carbonyl group is formed by decomposition of an intermediate *aci*-nitroester (**19**).[174]

Sulphur

The rate-determining step in the conversion of benzoin into benzil by thionyl chloride is the formation of (**20**) which then extrudes sulphur monoxide to give benzil.[175] Aryl

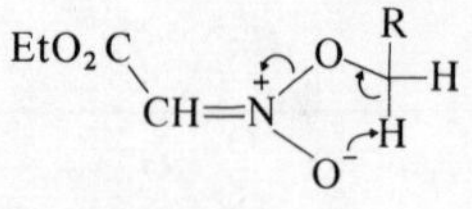

(19)

(20)

sulphinamides have been used (Scheme 14) in an approach to the dehydrogenation of amines to imines.[176] The mechanism of the ring-expansion of benzothiazolines to 1,4-benzothiazines by dimethyl sulphoxide has been investigated.[177] In a study of metal oxide catalysis of dimethyl sulphoxide oxidation of alcohols to carbonyl compounds, molybdenum oxides have been found to be the only useful catalytic reagents (Scheme 15).[178]

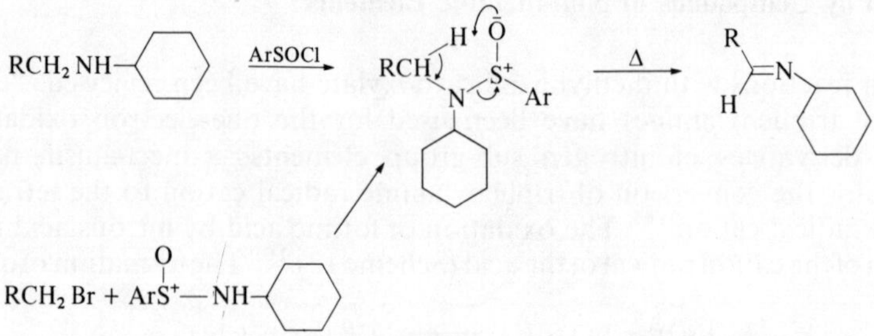

SCHEME 14

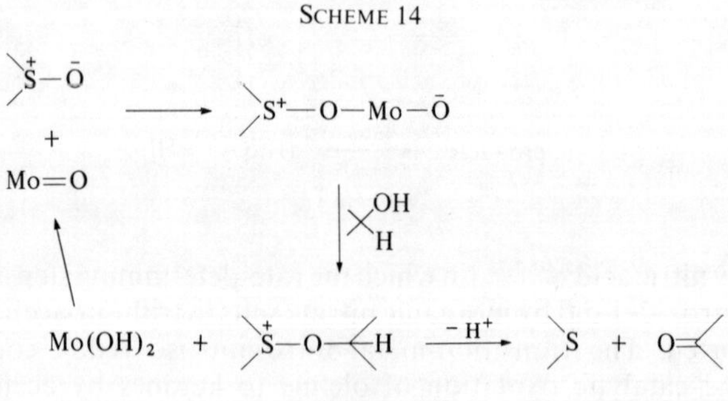

SCHEME 15

Thioureas are oxidized to stable dication disulphides, $(R_2N)_2\overset{+}{C}SS\overset{+}{C}(NR_2)_2$, by trifluoromethanesulphonic anhydride.[179] Aminium radicals, $R_3N^{+\cdot}$, are possible intermediates in the oxidation by Fremy salt of some isoquinoline alkaloids.[180] Potassium peroxomonosulphate has been used in the chemoselective oxidation of sulphides to sulphones.[181] The oxidation of pyridine to pyridine-*N*-oxide by peroxomonosulphate is catalysed by ketones; the active species for the oxidation is probably a dioxirane (**21**).[182]

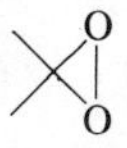

(21)

Two different pathways have been observed in the peroxydisulphate oxidation of arylacetic acids: one pathway proceeds by intermediate formation of an aromatic radical cation and the alternative by direct decarboxylation of arylacetate ion.[183] Mechanisms involving radicals and a sulphate radical ion have been proposed for the peroxydisulphate oxidation of dimethylamino-alcohols,[184] aromatic alcohols,[185] and aromatic amines.[186–188] The rate of oxidation of benzylamine by persulphate is decreased in the presence of both anionic and cationic micelles; the intermediate oxidation product, benzamide, becomes trapped in the hydrophobic region of the micelles and is not oxidized further to 2-hydroxybenzamide.[189] The crown-ether–potassium peroxydisulphate initiator system is very efficient; the activation energy in organic systems is clearly very different from that in aqueous solution.[190]

Copper(II) catalysis of peroxydisulphate oxidation of electron-rich benzylic hydrocarbons gives carbonyl compounds *via* one-electron-transfer processes.[191] Oxidation of lactic acid involves reaction of a copper–lactate complex.[192] The difference in the rôles of silver(I) and copper(II) in catalysis of peroxydisulphate oxidation has been discussed; silver(I) initiates a chain-reaction by direct interaction with the oxidant whereas copper(II) does not participate in chain-initiation.[193] Mechanisms involving radical intermediates have been proposed for the silver(I)-catalysed peroxydisulphate oxidation of alcohols,[194,195] alkanone cyanohydrins,[196] thiourea,[197] atrolactic acid,[198] and acid amides;[199,200] a twelve-step mechanism has been proposed on the basis of kinetic evidence for the oxidation of *N*-methylacetamide.[201]

Selenium and Tellurium

Oxidation of 1,1-disubstituted hydrazines with benzeneseleninic acid to tetrazenes does not involve intermediate nitrenes;[202] other hydrazines are converted into the corresponding diazenes, while acid hydrazides form seleno-esters (**22**) and diacylhydrazides (**23**).[203] Azasteroid lactams undergo dehydrogenation with benzeneseleninic anhydride[204] whereas phenols form quinones by a mechanism involving a [2,3]-sigmatropic shift (Scheme 16).[205] The regio- and stereo-specific allylic oxidation of germacrane-type sesquiterpene lactones with selenium dioxide and *tert*-butyl hydroperoxide may be rationalized by sequential ene and [2,3]-sigmatropic shift reactions.[206] Alcohols are oxidized to carbonyl compounds in the presence of phenylthio and phenylseleno groups by dimesityl diselenide and *tert*-butyl hydroperoxide.[207] Selenoxides may be used as oxidizing agents as alternatives to amine oxides in osmium-tetroxide-catalysed oxidations of olefins to diols.[208]

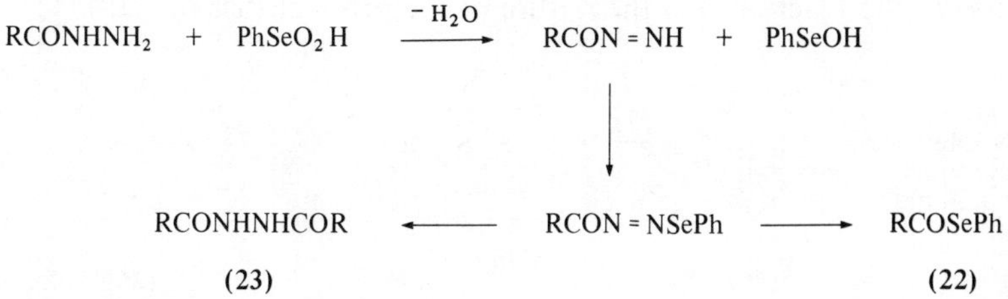

$$RCONHNH_2 + PhSeO_2H \xrightarrow{-H_2O} RCON{=}NH + PhSeOH$$

$$\downarrow$$

$$\underset{(23)}{RCONHNHCOR} \longleftarrow RCON{=}NSePh \longrightarrow \underset{(22)}{RCOSePh}$$

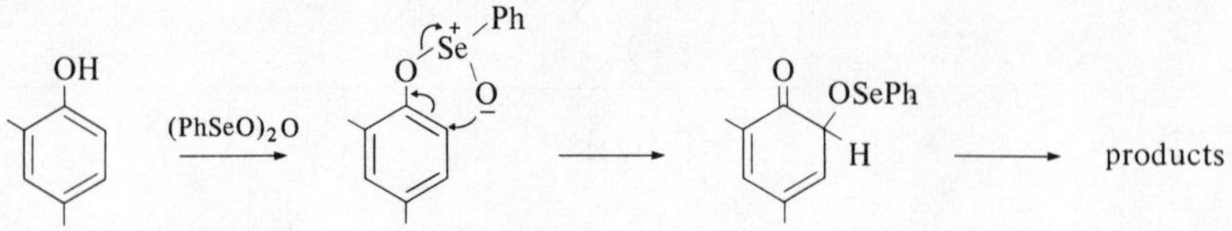
OH
$(PhSeO)_2O$
Ph
Se
O
O
O
OSePh
H
products

SCHEME 16

Diaryl telluroxides have been used for several oxidations such as phosphines to phosphine oxides, thiols to disulphides, and thiocarbonyl compounds to ketones.[209] Although normal oxidation of olefins by tellurium dioxide produces alkane-1,2-diol mono- and di-acetates together with elemental tellurium, a suitably positioned hydroxy group leads to production of a stable tellurium(IV) species; naturally, 5-*exo*-ring-closure is favoured over the competing 6-*endo*-process (Scheme 17).[210]

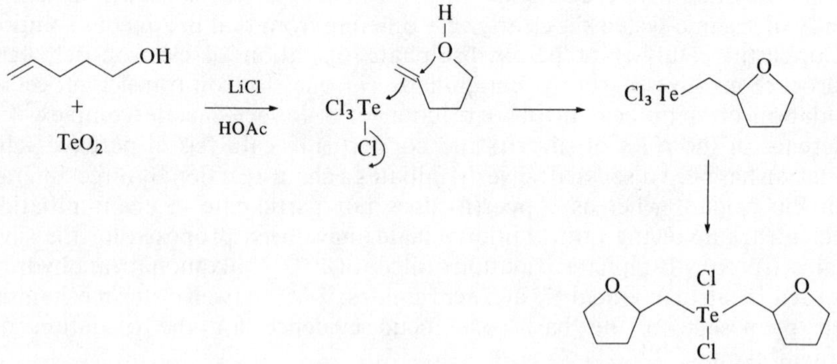
OH
+
TeO_2
LiCl
HOAc
H
O
Cl_3Te
Cl
O
Cl_3Te
O
Cl
Te
Cl
O

SCHEME 17

Halogens, including Periodate and Bromate
Oxidation of diphenylacetylene to benzil by iodine(VII) and iodine(V) requires the presence of iodine for initial attack by I^+; attack by electrophilic oxygen is not involved in the reaction.[211] A million-fold rate increase relative to aliphatic sulphoxides is observed in the oxidation of 5-methyl-1-thia-5-azacyclooctane (**24**) by iodine;[212] a similar rate enhancement is observed in the oxidation of 1,5-dithiacyclooctane (**25**);[213] the acceleration and insensitivity to changes in pH have been ascribed to intramolecular catalysis by the transannular hetero-atoms. The bicyclic alkoxysulphonium salt (**26**), formed by transannular hydroxyl participation in the iodine oxidation of 5-hydroxy-1-thiacyclooctane, has been isolated and the crystal structure determined.[214] Neighbouring-group participation by hydroxyl in the oxidation of hydroxysulphides by hexabutyldistannoxane–bromine determines the course of the reaction and the nature of the product (Scheme 18).[215]

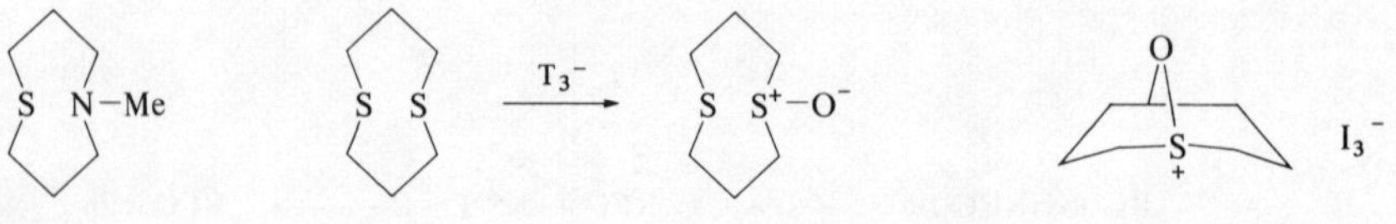
S
N−Me
S
S
I_3^-
S
$S^+−O^-$
O
S
+
I_3^-

(24) **(25)** **(26)**

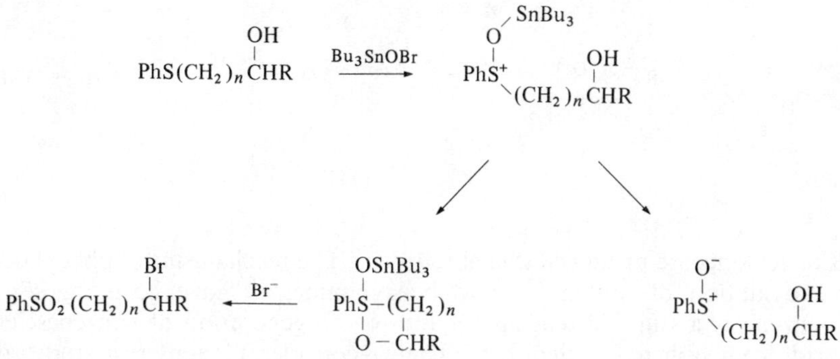

$$PhS(CH_2)_nCH(OH)R \xrightarrow{Bu_3SnOBr} PhS^+(OSnBu_3)(CH_2)_nCH(OH)R$$

$$PhSO_2(CH_2)_nCH(Br)R \xleftarrow{Br^-} PhS(OSnBu_3)\text{–}(CH_2)_n\text{–}O\text{–}CHR \qquad PhS^+(O^-)(CH_2)_nCH(OH)R$$

SCHEME 18

Sodium hypochlorite is an efficient oxidant for the conversion of α,β-unsaturated aldehydes to the corresponding acids[216] and of *N*-(arenesulphonyl)sulphilimines to *N*-(arenesulphonyl)sulphoximines.[217] Photo-promoted hypochlorite oxidation of amino-acids (the Strecker degradation) proceeds *via* rapid formation of *N*-chloramino-acids (**27**) which slowly decompose.[218] The kinetics of oxidations by aqueous halogens of glycine,[219] hydroxy-acids,[220] hydroxy-esters,[221] and alcohols[222] have been determined. The oxidation of amines by chlorine dioxide involves initial hydrogen abstraction either from the α-carbon, or, if these are not available, from the nitrogen atom.[223]

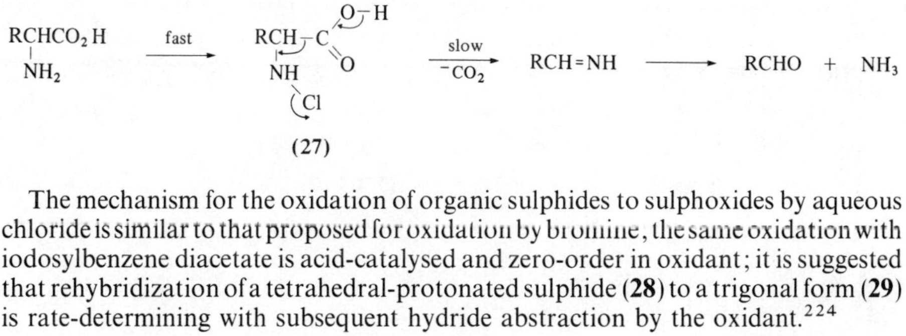

$$RCH(NH_2)CO_2H \xrightarrow{\text{fast}} RCH(NHCl)\text{–}C(=O)\text{–}OH \xrightarrow[-CO_2]{\text{slow}} RCH{=}NH \longrightarrow RCHO + NH_3$$

(**27**)

The mechanism for the oxidation of organic sulphides to sulphoxides by aqueous chloride is similar to that proposed for oxidation by bromine; the same oxidation with iodosylbenzene diacetate is acid-catalysed and zero-order in oxidant; it is suggested that rehybridization of a tetrahedral-protonated sulphide (**28**) to a trigonal form (**29**) is rate-determining with subsequent hydride abstraction by the oxidant.[224]

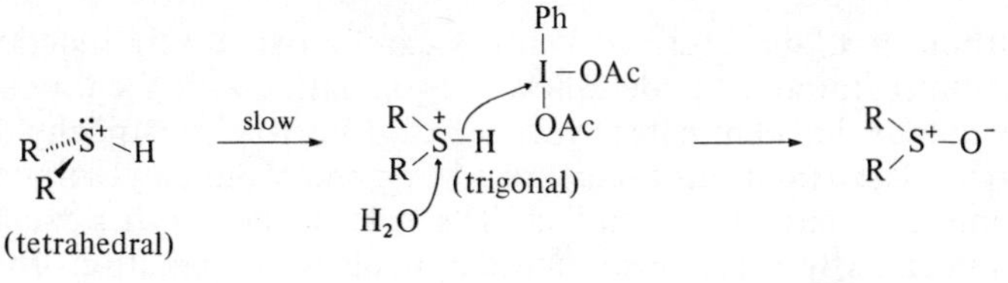

$$R_2\ddot{S}^+\text{–}H \xrightarrow{\text{slow}} R_2\overset{+}{S}\text{–}H \;(+\;PhI(OAc)_2,\ H_2O) \longrightarrow R_2S^+\text{–}O^-$$

(tetrahedral) (trigonal)

(**28**) (**29**)

Oxygen-atom transfer from iodosobenzene to ketenes produces α-lactones (**30**) and causes decarboxylation of α-keto-acids *via* mixed anhydrides (**31**); initial attack in these reactions is nucleophilic attack by oxygen.[225] Ketones are converted into acyloins by iodosobenzene in methanol; in the absence of acid hydrolysis, dimethyl

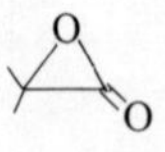

(30) 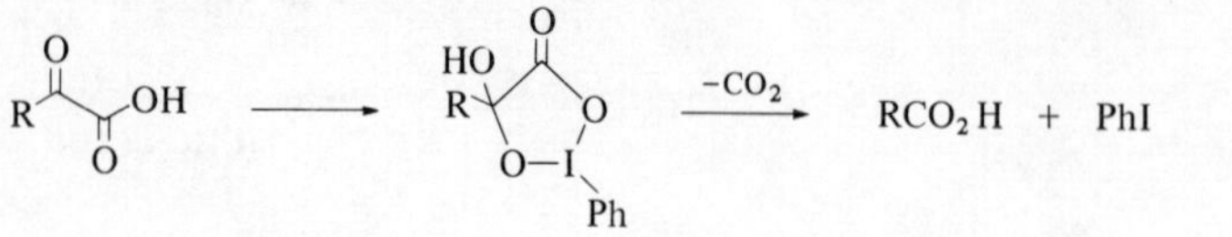 (31)

$RCO_2H + PhI$

ketals of acyloins are produced (Scheme 19).[226] The mechanisms of phenyliodosoacetate oxidation of toluene[227] and benzylamines[228] have been investigated. Iodylbenzene is a suitable reagent for the *in situ* generation of benzeneseleninic anhydride from diphenyl diselenide.[229] There is considerable activity in studies of the oxidation of organic compounds with *N*-halogen systems. *N*-Iodosuccinimide oxidatively cleaves 1,2-diols by fragmentation of an intermediate hypoiodite (**32**).[230]

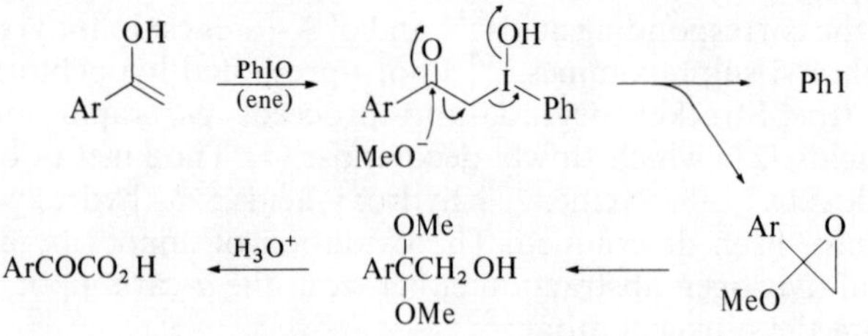

PhIO (ene) PhI H_3O^+ $ArCOCO_2H$ $ArC(OMe)_2CH_2OH$

SCHEME 19

(32)

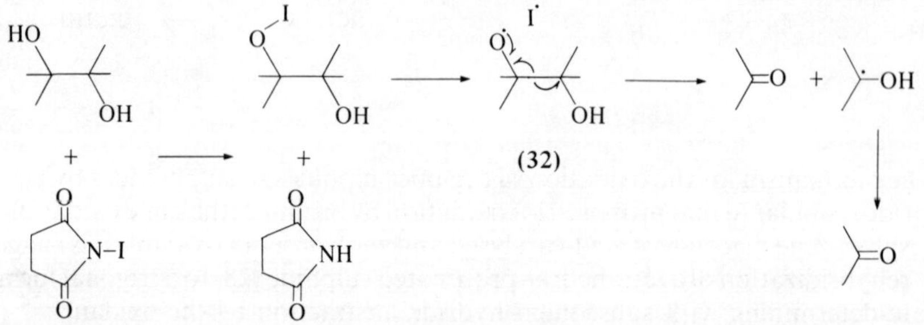

The *β*-carbon of *α*,*β*-unsaturated ketones can be oxidatively functionalized by Michael addition of thiophenoxide, followed by oxidation with *N*-chlorosuccinimide to a chlorosulphide that eliminates hydrogen chloride to give 3-(phenylthio)enones (Scheme 20).[231] The oxidations of alcohols[232,233] and of dimethyl sulphoxide by *N*-bromoacetamide[234] have been studied. The rôle of mercury(II) catalysis in the oxidation of alcohols by *N*-bromosuccinimide involves complexation of mercury and halogen and the suppression of increase in proton concentration.[235] Unsaturated lactones are converted into chlorolactones on treatment with chloramine-*T* in the presence of methanesulphonic acid.[236] Kinetic studies on the oxidation of the following systems by chloramine-*T* and other *N*-halosulphonamides have been reported; alcohols,[237,238] sulphoxides,[239,240] aldehydes,[241] and aroyl hydrazines.[242] Bromate esters are common intermediates in the acid-catalysed sodium

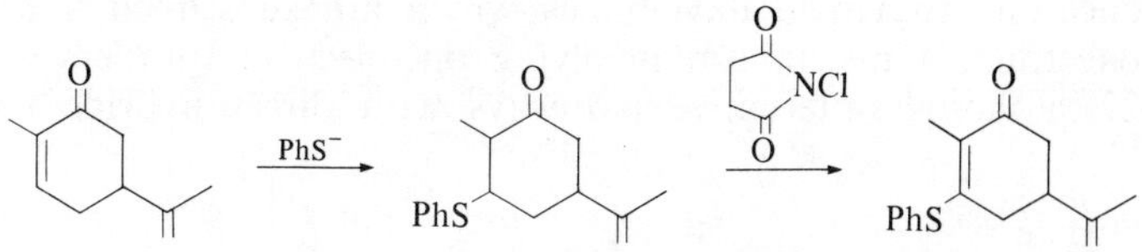

SCHEME 20

bromate oxidation of phenols,[243] naphthols,[244] alcohols,[245] aromatic aldehydes,[246,247] and amino-acids.[248] Ruthenium(III) catalysis in the vanadium(IV)-catalysed bromate oxidation of phenol is attack of $(BrO_2)^+$ at the *ortho*-position.[251] Manganese(II) catalysis in bromate oxidation involving oscillating reactions has been investigated.[252–255] The oxidizing action of iodic acid on organic compounds, including aromatic amines,[256] has been reviewed.[257] Sodium periodate selectively oxidizes unsymmetrical thiosulphinic *S*-esters to thiosulphonic *S*-esters:[258]

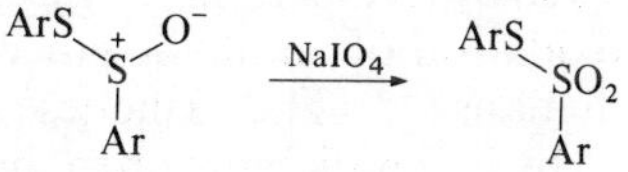

$$\mathrm{ArS{-}\overset{+}{S}(O^-){-}Ar} \xrightarrow{NaIO_4} \mathrm{ArS{-}SO_2{-}Ar}$$

Kinetic studies on the periodate oxidations of arenes,[259,260] aminocarboxylic[261,262] and sulphonic[263] acids, salicylic acid,[264] and *N*-(*p*-bromobenzoyl)palytoxin[265] have been reported.

Miscellaneous Oxidations

A biomimetic oxidative deamination of primary amines to aldehydes by imine transposition has been developed using pyridine-2-aldehyde (Scheme 21).[266]

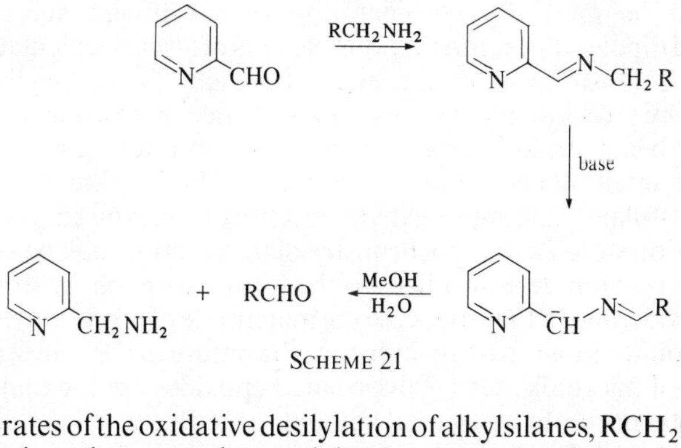

SCHEME 21

The relative rates of the oxidative desilylation of alkylsilanes, $RCH_2CH_2SiR_3$, with triphenylcarbenium ion may be explained on purely electronic grounds; the transition state for the hydride transfer requires anti-coplanar arrangements of the silicon moiety and the β-hydrogen being extruded.[267] Reaction of trityl cation with *threo*-3-deuterio-2-(trimethylstannyl)butane to produce but-2-ene proceeds with 99% *anti*-stereoselectivity; the mechanism is interpreted as involving abstraction of hydride to give a σ–π-conjugated carbocation which subsequently loses the trimethylstannyl group; the difference between the energies of open and bridged carbocation intermediates is small, and there is no evidence for stabilization by bridging in these types of reaction.[268,269] Tertiary amines are oxidized by hexachloroacetone to vinyldialkylamines which are subsequently acylated by hexachloroacetone; a mechanism involving one-electron transfers is proposed (Scheme 22) although alternative pathways with direct hydride transfer are possible.[270]

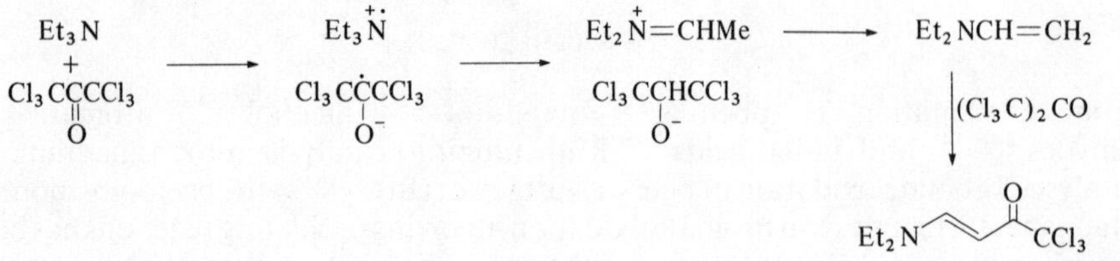

$Et_3N + Cl_3CC(=O)CCl_3 \longrightarrow Et_3\overset{+\cdot}{N}\ \ Cl_3C\dot{C}(O^-)CCl_3 \longrightarrow Et_2\overset{+}{N}{=}CHMe\ \ Cl_3CCH(O^-)CCl_3 \longrightarrow Et_2NCH{=}CH_2 \xrightarrow{(Cl_3C)_2CO} Et_2NCH{=}CHC(=O)CCl_3$

SCHEME 22

Oxidation of phenylhydrazines by quinones proceeds by reversible formation of a charge-transfer complex followed by rate-determining hydrogen transfer.[271,272] The first example of a quinone oxidation of an allylic ether has been observed in the oxidation of aryltetralin lignans.[273] The kinetics of the oxidation of 2-mercaptopropanoic acid by 2,6-dichlorophenolindophenol have been investigated.[274]

Ozonolysis and Ozonation

The stereochemistry of the ozonolysis of alkenes can be controlled by factors associated with either the primary ozonide or the carbonyl oxide.[275] Restricted Hartree–Fock calculations have been used to determine the electronic properties of alkyl-substituted primary ozonides (more stable in envelope conformations) and final ozonides (more stable in twist forms). Predictions of stereochemistry of ozonolysis can be made by consideration of substituent–substituent, substituent–ring, and dipole–dipole interactions. Results of *ab initio* calculations show that the nature of the transition state changes as a function of crowding in the alkene.[276] The thermochemical character of primary ozonide decomposition to carbonyl oxide changes from being endothermic for ethylene and terminal alkenes, weakly exothermic for small alkenes, and exothermic for bulky alkenes — implying by Hammond's postulate a late, intermediate, and early transition state, respectively, for the decomposition step. The stereochemistry of the reaction may be either carbonyl-oxide- or primary-ozonide-controlled; in the former case the larger stability of the *syn*-carbonyl oxide, and in the latter case the magnitude of alkyl–alkyl interactions in the primary ozonide, most strongly influence the nature of the transition state.[277] In the ozonolysis of fluoroalkenes,[278] fluorinated epoxides are the major products of ozonolysis of alkenes with fluorine substituents at both ends of the double bond; in contrast, alkenes with fluorine substituents on only one of the carbon atoms form ozonides as major products — these results are consistent with electrophilic attack of ozone as a 1,3-dipolar reagent.[279]

The products of the gas-phase reaction between ozone and olefins have been identified with a photo-ionization mass spectrometer coupled to a stirred-flow reactor; products observed are characteristic of a primary Criegee split to a carbonyl compound and a carbonyl oxide, followed by reactions of the Criegee intermediates such as unimolecular decomposition and ozonide formation.[280] The reaction of the vinyl sulphide (**33**) gives an oxirane and provides support for the suggestion that oxiranes are intermediates in reactions of ozone with vinyl derivatives where the final products have unmodified carbon chains.[281] The major products formed in the ozonation of hexamethylbenzene may all be rationalized as being derived from the carbonyl oxide (**34**).[282] The anomalous products from ozonolysis of α,β-unsaturated carbonyl compounds may all be explained by rearrangements of intermediate carbonyl oxides.[283] The effect of strain on the reactions of olefins with ozone has been discussed,[284] and kinetic investigations of the reaction of ozone with alkenes in water[285] and under atmospheric conditions[286] have been reported. The reactions of carbonyl oxides with various organic substrates have been studied; olefins give small amounts of epoxides but mainly carbon–carbon bond-cleavage or products from allylic hydrogen abstraction. The reactions of carbonyl oxides with a range of sulphur compounds indicate that nucleophilic oxygen transfer is a characteristic reaction, as evidenced by a positive ρ value for reaction with substituted diphenyl sulphoxides.[287] The nature of intermediates in the acidolysis of ozonides has been elucidated.[288]

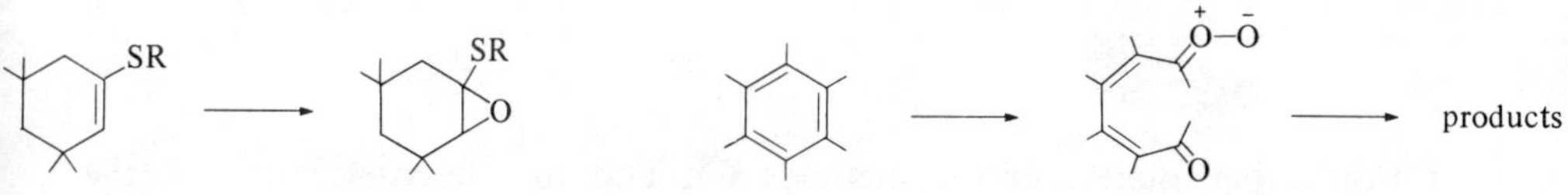

SR → SR O

(33)

products

(34)

There have been relatively few investigations of the reaction of ozone with alkynes. Ozonation of di-*tert*-butylacetylene gives products which could arise from an intermediate carbonyl oxide and are consistent with a modification of the Criegee mechanism (Scheme 23); the *tert*-butyl groups do not confer any great stability on the intermediates.[289]

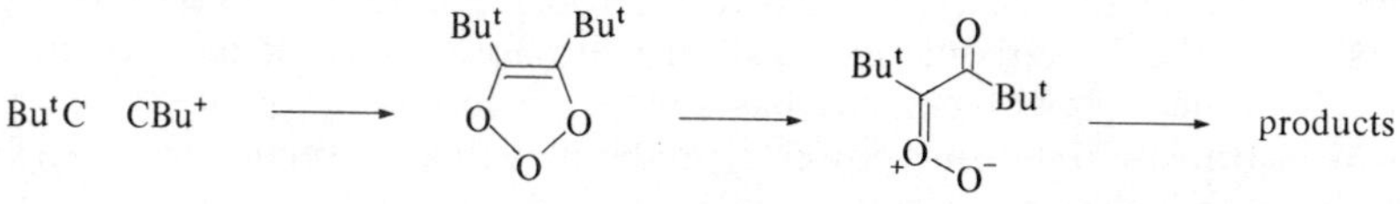

$Bu^tC \equiv CBu^t$ → products

SCHEME 23

The reaction of epoxides with ozone has been interpreted in terms of a perepoxide intermediate (**35**).[290] Treatment of di-*tert*-butylthioketen with ozone gives the sulphoxide (**36**) in quantitative yield.[291] The reactions of ozone with pyridine[292] and with sodium benzenesulphonate[293] have been investigated. Several studies on the reactions of ozone with carbon–hydrogen bonds have appeared. A 1:1 complex between cumene and ozone has been characterized.[294] Isotopic studies on the ozonation of ethanol indicate that breaking of the α-C—H bond is rate-determining.[295,296] Radical mechanisms have been proposed for the reactions of ozone with cyclohexanol[297] and 1,1-diethoxyethanehydroperoxide.[298] The

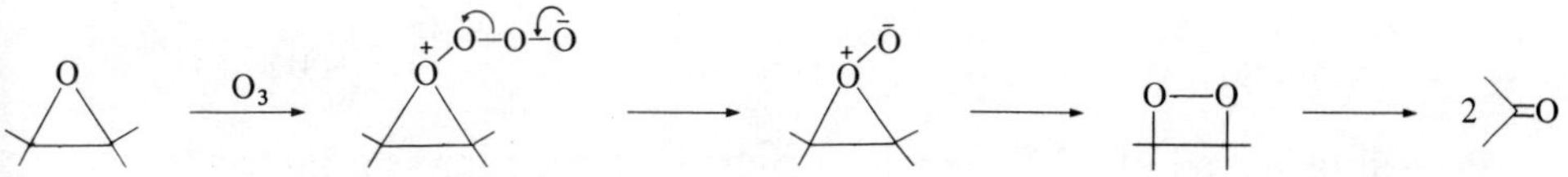

$\xrightarrow{O_3}$ → → → 2 >=O

(35)

reactions of ozone with bi- and tri-cyclic alkanes have been discussed in terms of stereoelectronic effects.[299]

Peracids, Peroxides, and Superoxide

Baeyer–Villiger and related reactions of bridged bicyclic ketones have been reviewed.[300] The Baeyer–Villiger oxidation of chiral cyclohexanone substituted at C(2) with deuterium, catalysed by cyclohexanone oxygenase, occurs with complete retention of configuration at the migrating carbon.[301] The unusual cleavage of ring *D* in keto-steroid (**37**) to give (**38**) on treatment with alkaline hydrogen peroxide proceeds *via* an initial Baeyer–Villiger reaction.[302] Use of fluoride ion enhances the reactivity of *m*-chloroperbenzoic acid in Baeyer–Villiger reactions of aromatic aldehydes and in the epoxidation of olefins.[303]

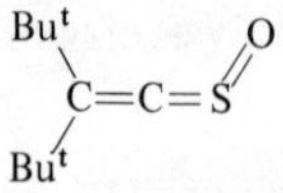

$Bu^t_2C{=}C{=}S{=}O$

(36) 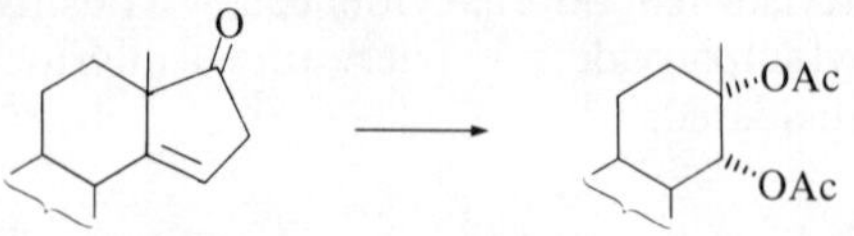 **(37)** **(38)**

Considerable interest in the methods for, and the stereoselectivity of, the epoxidation of olefins continues; a review on the synthetic and mechanistic aspects of metal-catalysed epoxidation with hydroperoxides has appeared.[304] Epoxidations with ethylbenzenehydroperoxide,[305] 2-(benzenesulphonyl)-3-aryloxaziridines (**39**),[306] and the pyrazoline (**40**)[307] have been reported. The effective oxidant in the epoxidation of olefins by hydrogen peroxide in the presence of tetrachloroacetone is thought to be the hydroperoxide (**41**);[308] dioxiranes, rather than carbonyl oxides, are the active species in epoxidation reactions by potassium peroxomonosulphate with ketones.[309] Nucleophilic oxygen abstraction from oxaziridines has been studied.[310] An analysis of the diastereomeric transition states for stereoselective epoxidation of acyclic allylic alcohols with peracids has shown that consideration of $A^{[1,2]}$ and $A^{[1,3]}$ strain in the transition state has powerful predictive value and relates observed kinetic selectivity to the relative energies of the diastereomeric transition states.[311] Among other studies on stereoselective epoxidations of olefins,[312–314] epoxidation of bay-region diols of benzophenanthrenes[315] and benzanthracenes[316] give stereoselectively *syn*-diol epoxides rather than the expected isomeric *anti*-diol epoxides, indicating a dominant *cis*-directing effect of the benzylic rather than the allylic axial hydroxyl groups. Kinetic studies on the epoxidation of olefins have been reported[317–319] and epoxidation of α,β-unsaturated carbonyl compounds by hydrogen peroxide has been investigated.[320–323]

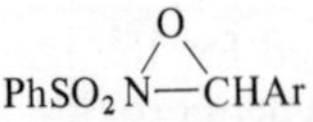

$PhSO_2N$—CHAr (oxaziridine) $Cl_2CHC(OH)(OOH)CHCl_2$

(39) 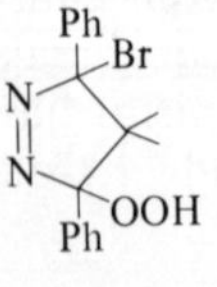 **(40)** 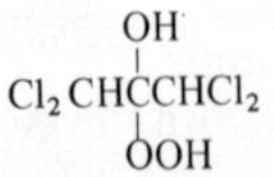 **(41)**

Peracid oxidation of 1,2,3-trimethoxy-5-methylbenzene gives 2,3-dimethoxy-5-methyl-1,4-benzoquinone in which the oxygen at C(1) is derived from the oxidant.[324]

The products from the hydrogen peroxide oxidation of 3,5-di-*tert*-butyl-*o*-benzoquinone are formed from the two intermediates (**42**) and (**43**);[325] other hydrogen peroxide oxidations of quinones have been described.[326] Anthracene is oxidized to anthraquinone by *tert*-butyl hydroperoxide.[327] Alkylbenzenes may be

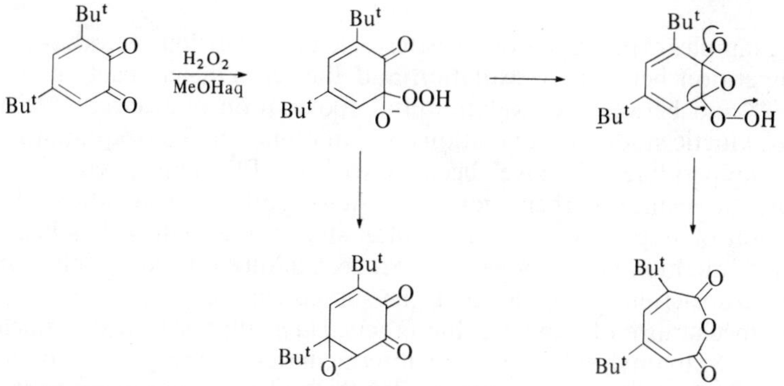

H_2O_2, MeOHaq

(42) **(43)**

hydroxylated with hydrogen peroxide in hydrogen fluoride/boron trifluoride (Scheme 24); polyhydroxylation is suppressed since the product phenols are protonated under the reaction conditions and so deactivated to further electrophilic attack.[328] Hydroxylation of xylenes by bis(trimethylsilyl) peroxide and aluminium

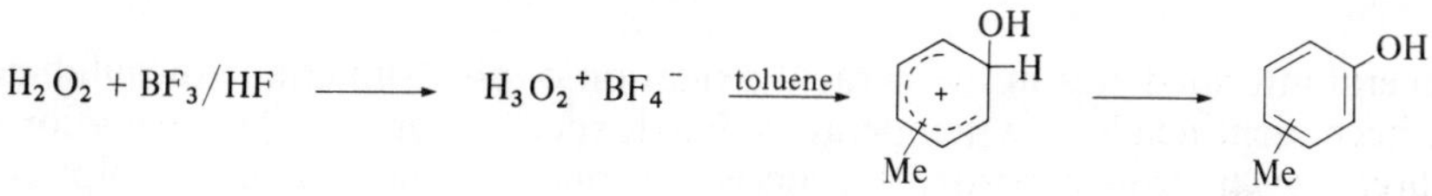

$$H_2O_2 + BF_3/HF \longrightarrow H_3O_2{}^+\,BF_4{}^- \xrightarrow{\text{toluene}}$$

SCHEME 24

chloride proceeds by *ipso*-attack (**44**) followed by rearrangement to (**45**).[329] The rate of hydroxylation of aromatic compounds with peroxymonophosphoric acid is comparable to the rate with peroxytrifluoroacetic acid; the hydroxylating species are both protonated and unprotonated peroxymonophosphoric acid.[330] In contrast, a radical mechanism is involved in aromatic hydroxylation by potassium peroxydiphosphate.[331] The hydroxylations of phenol,[332] phenylalanine,[333] and aniline[334] have been studied. A novel lactone synthesis is derived from peracid oxidation of furan derivatives (Scheme 25)[335] and polyene-dione-functionalized macrocycles are obtained from *m*-chloroperbenzoic acid treatment of cyclic furan-acetone oligomers.[336] The reactions of paraquat,[337] indole derivatives,[338] and 3-aminopyridine[339] with peroxide reagents have been investigated.

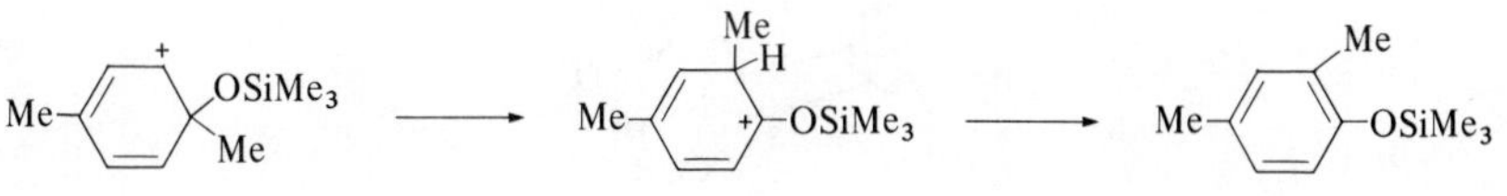

(44) **(45)**

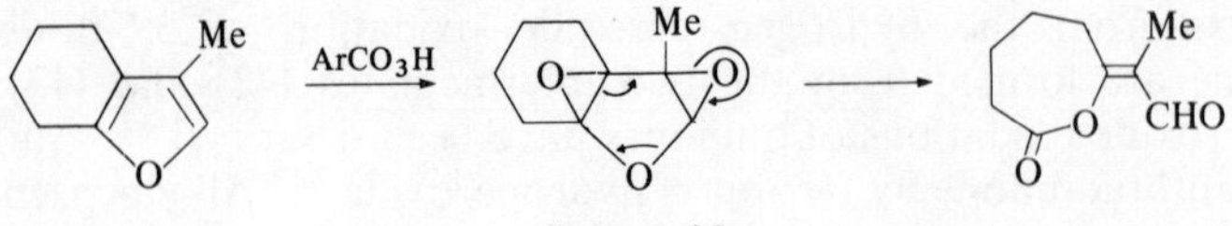

SCHEME 25

Many examples of peroxidations of hetero-elements with lone pairs have appeared. The competition between *N*-oxidation and Baeyer–Villiger reactions of pyridyl ketones,[340] and between *N*-oxidation and epoxidation of azaarenes[341] has been discussed; kinetic studies on oxidations of quinolines and benzoquinolines[342] and tetramethylpiperidines[343] have been described. Phosphites are oxidized to phosphates by *m*-chloroperbenzoic acid.[344] Among other mechanistic studies on the peroxidation of sulphides,[345,346] the initial step in the hydroxyl-radical-induced oxidation of methionine is competition between addition to the sulphur atom and hydrogen abstraction; in strong acid, the subsequent steps are identical with the oxidation mechanism of aliphatic thio-ethers, *via* a sulphur-centred radical cation complex.[347] Sulphinyl sulphones are intermediates in the peracid oxidation of thiosulphonic *S*-esters to α-disulphones.[348] Peracid oxidation of neopentanethiolsulphinate (**46**) gives (E)- and (Z)-*S*-oxides (**47**).[349]

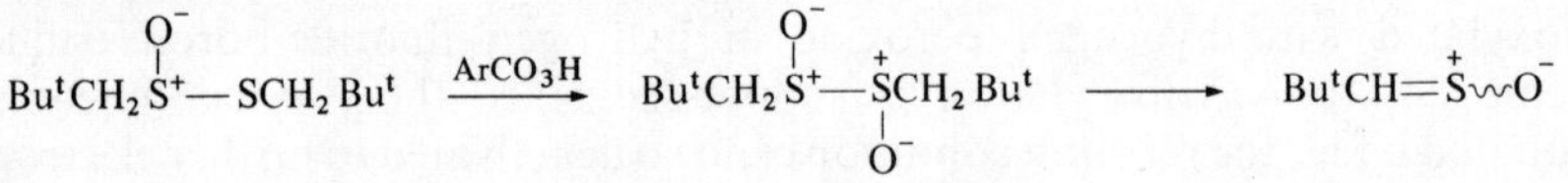

$$\mathrm{Bu^tCH_2\overset{O^-}{S^+}{-}SCH_2Bu^t} \xrightarrow{\mathrm{ArCO_3H}} \mathrm{Bu^tCH_2\overset{O^-}{S^+}{-}\underset{O^-}{\overset{+}{S}}CH_2Bu^t} \longrightarrow \mathrm{Bu^tCH{=}\overset{+}{S}{\sim}O^-}$$

(46) **(47)**

Several intermediates in the peracid oxidation of phenylmethanethiosulphinate have been identified by low-temperature NMR spectroscopy.[350] The oxidations of thioureas by hydrogen peroxide proceed by nucleophilic attack by sulphur at oxygen.[351,352] Peracid oxidation of β-aminoalkyl selenides yields allylic amides by selenoxide fragmentation; the anomalous thermal stability of selenoxide (**48**) is ascribed to intramolecular hydrogen bonding.[353] The reaction of equimolar quantities of aryl diselenides and *tert*-butyl hydroperoxide leads to a mixture of seleninic anhydride, ArSe(O)OSe(O)Ar, and unreacted diselenide and *not* to a substantial amount of the corresponding selenenic anhydride, ArSeOSeAr, as has been previously reported.[354] The oxidation of alkyl iodides by peracids proceeds through intermediate alkyliodoso compounds; this reaction is related to cytochrome *P*-450 oxidation of organic halides.[355] Peracid oxidations of di- and tri-silanes,[356] diazirines,[357] aldehydes,[358] nitrobenzofuroxans,[359] and dimethyl sulphoxide,[360] and hydrogen peroxide oxidations of keto-steroids,[361] steroidal imines,[362] terminal olefins,[363] hydrocarbons,[364] alkylpyridines,[365] tartaric acid,[366] and other carboxylic acids[367] have been studied from a mechanistic standpoint.

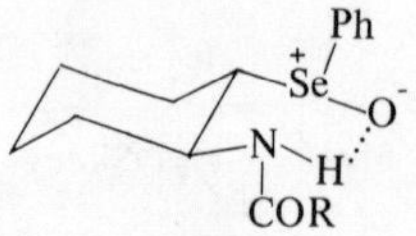

(48)

Superoxide has been shown to be involved in the photo-oxygenation of 1,4-dimethylnaphthalene, formed in an initial one-electron-transfer step (Scheme 26).[368]

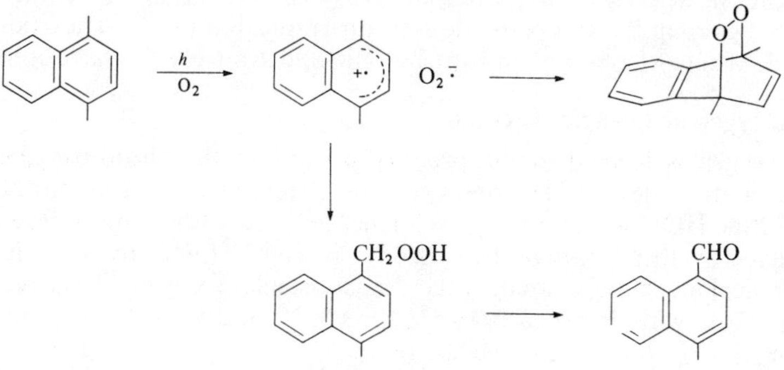

$h\nu$, O_2 ; $O_2^{\cdot -}$; CH_2OOH ; CHO

SCHEME 26

N-Benzylamides (**49**) give *ortho*- and *para*-hydroxylated products on treatment with potassium superoxide in benzene in the presence of 18-crown-6-as a phase-transfer catalyst; the proposed mechanism involves nucleophilic attack by superoxide ion on the amide carbonyl, followed by hydrogen abstraction from the benzylic position in the substrate–superoxide adduct (**50**).[369] Nucleophilic attack by superoxide and electron-transfer processes are suggested as the mode of oxidation of a range of sulphur and other compounds; several peroxysulphur intermediates have been identified in the course of the reactions.[370,371] A revised structure (**51**) has been suggested for the product of potassium superoxide oxidation of 2,2,5,7,8-pentamethylchroman-6-ol, an α-tocopherol model compound.[372]

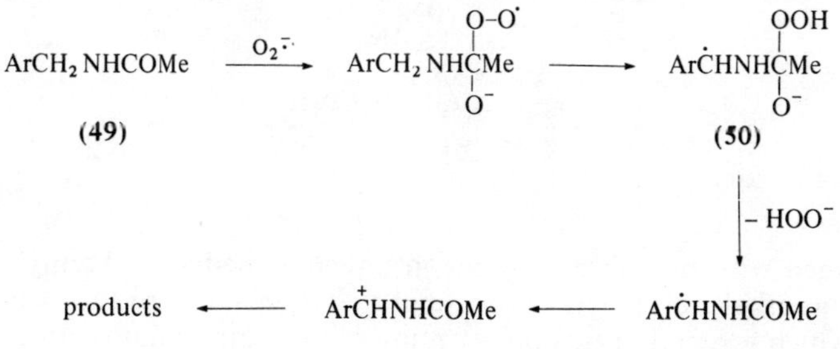

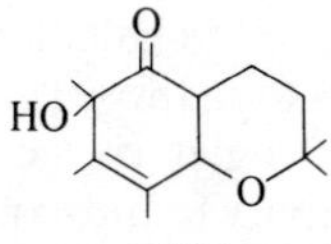

$ArCH_2NHCOMe$ (**49**) $\xrightarrow{O_2^{\cdot -}}$ $ArCH_2NHC(O{-}O^{\cdot})(O^-)Me$ $\longrightarrow$ $Ar\dot{C}HNHC(OOH)(O^-)Me$ (**50**)

(**50**) $\xrightarrow{-\,HOO^-}$ $Ar\dot{C}HNHCOMe$ $\longrightarrow$ $Ar\overset{+}{C}HNHCOMe$ $\longrightarrow$ products

HO, O

(**51**)

Organic halides are rapidly oxidized by superoxide ion in dipolar aprotic media; the reaction is multi-step, and the rates of degradation of the organic halides increase with increasing halogen substitution.[373] Among studies of the oxidation of flavin-related compounds by superoxide,[374,375] it has been shown that superoxide is a selective oxidant for substrate susceptible to oxidation by hydrogen-atom transfer.[376] There is strong evidence against the formation of singlet oxygen in non-catalysed disproportionation of superoxide.[377] The reaction of superoxide ion and a radical cation derived from paraquat provides one of the first examples of stoichiometric coupling of superoxide ion with a radical cation.[378] The oxidations by superoxide of keto-steroids[379] and arachidonic acid in micelles[380] have been studied.

Atomic Oxygen and Singlet Oxygen

Atomic oxygen is formed in the primary process in the photo-oxygenation by aromatic amine oxides.[381] The primary reaction between atomic oxygen and ethanol produces MeCHOH which undergoes further reaction with atomic oxygen to form acetaldehyde by abstraction of the hydroxyl hydrogen; $O(^3P)$ reacts with the first-formed radical much more rapidly than does molecular oxygen.[382] The reactions of atomic oxygen with hydrocarbons,[383] allene,[384] acetylene,[385,386] and sulphur compounds[387,388] have been investigated.

Two general reviews on singlet oxygen[389,390] have been accompanied by accounts of the photo-oxidation of sulphur compounds[391] and of the ene reaction of singlet oxygen and olefins.[392] The competitive rubrene method has been used for determining the rate constants for the quenching of $^1O_2$ by strained molecules.[393] Nitrone spin-traps react chemically with singlet oxygen at a significant rate to lead to predominantly diamagnetic products, and a technique of oxygen uptake offers a method of continuous monitoring of chemical reactions of singlet oxygen in aqueous solution.[394] Unexpected solvent deuterium isotope effects on the life-time of singlet oxygen have been observed.[395] Singlet oxygen is generated in the gas phase at atmospheric pressure using heterogeneous photo-sensitization when oxygen flows through an irradiated tube coated with Rose Bengal.[396] The water-soluble endoperoxide (**52**) is a convenient and mechanistically simple source for reactions of

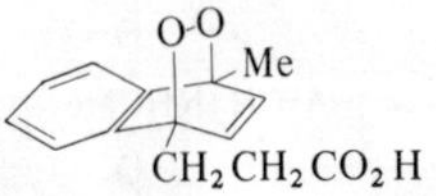

(52)

singlet oxygen with biological systems in aqueous media.[397] Methyl-substituted poly(vinylnaphthalene) reacts with singlet oxygen to produce endoperoxide polymers which regenerate $^1O_2$ on warming.[398] The air oxidation of 1-benzyl-3,4-dihydroisoquinoline has been shown to be a self-sensitized photo-oxidation involving singlet oxygen.[399,400] *Ab initio* calculations indicate that concerted ene reactions are lower energy pathways than those *via* either 1,4-diradical or perepoxide intermediates for the reaction of singlet oxygen with simple alkenes; the high stereospecificity and low regioselectivity found experimentally are strong arguments against diradical intermediates.[401,402] *syn*-Regioselectivity has been observed in the $^1O_2$ ene reactions of 1-alkylcycloalkenes[403] and a series of conjugated dienic propellanes.[404] The products of singlet oxygen ene reactions with cycloalkenes,[405] 1,1-dimethyl-3-sila(or germa)cyclopentenes,[406] and limonene[407] have been identified. Photo-sensitized oxygenations of several other olefins have been studied.[408–410]

The stereochemistry of singlet oxygen capture by 1,4-dimethoxynaphthalenes laterally fused to bridged bicyclic systems depends on contributions made by the σ-electrons of the bicyclic moieties to the aromatic π-orbitals; the tilting caused by such interactions is thought to be the source of the experimental *exo/endo*-ratios.[411,412] The photo-sensitized oxygenation of some phenylpyruvate derivatives has been used as a model for *p*-hydroxyphenylpyruvate dioxygenase systems; (**53**) in dipolar aprotic solvents forms endoperoxides whereas hydroxylated products are obtained in methanol (Scheme 27).[413] Intramolecular silyl migration in the singlet oxygen

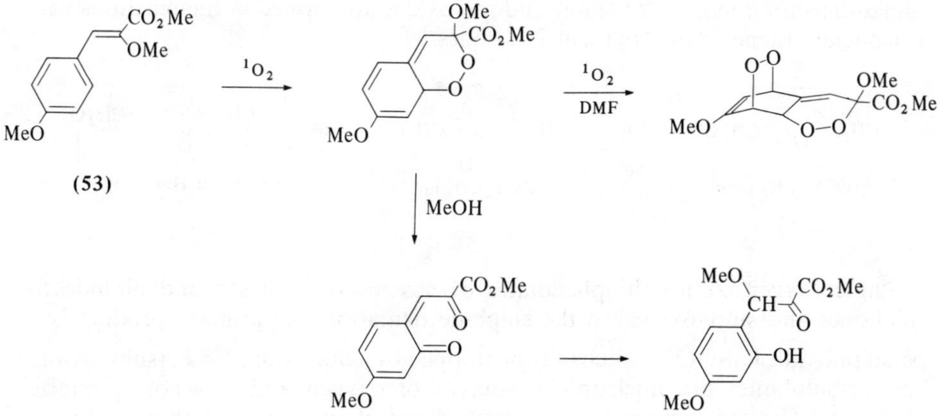

SCHEME 27

oxygenation of 2-methyl-5-trimethylsilylfuran (**54**) gives trimethylsilyl 4-oxopent-2-enoate (**56**) by rearrangement of the intermediate dioxirane (**55**).[414] The existence of intermediate carbonyl oxides in oxygen transfer by furan endoperoxides has been

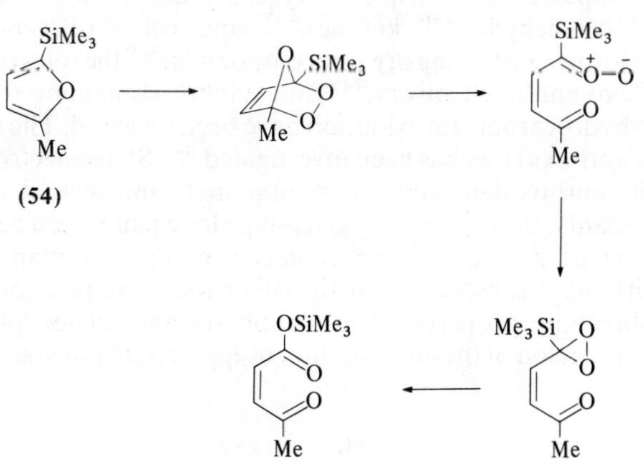

investigated.[415] Anomalous ozonolysis products have been isolated from the singlet oxygen addition to 2-methoxymethylfuran (Scheme 28);[416] these types of reactions have been used in the synthesis of a macrocyclic lactone derived from 2,5-furo-18-

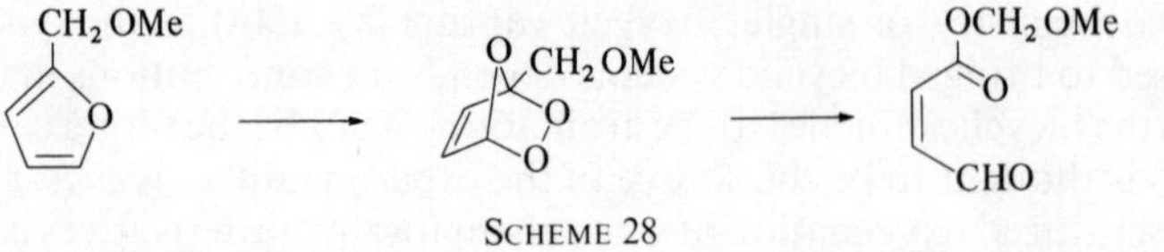

SCHEME 28

crown-6 (**57**).[417] Initial [4 + 2]-concerted reactions of singlet oxygen to give endoperoxides have been identified as the primary steps in the photo-sensitized oxygenations of oxazoles,[418] pyrazolobenzotriazole,[419] polyalkoxybenzenes,[420] and α-diazoquinone;[421,422] stable endoperoxides are formed in the reactions with cyclooctatetraene[423] and pyrazin-2(*H*)-ones.[424]

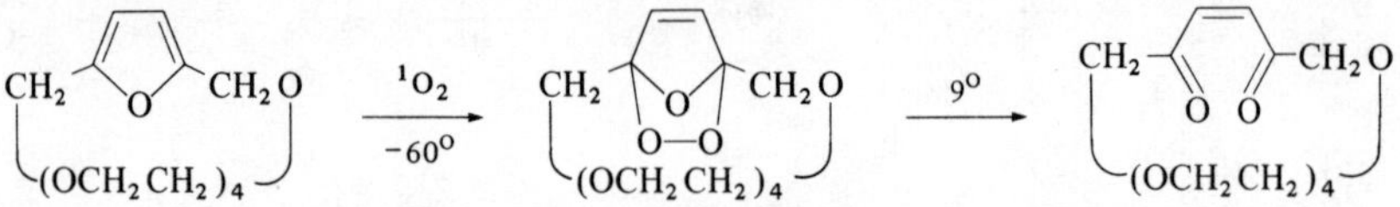

(**57**)

Singlet oxygen oxidizes thiophenolates to benzenesulphonates[425] and sulphides to sulphones and sulphoxides; in the sulphide oxidation, the primary product is a persulphoxide, $R_2\overset{+}{S}OO^-$, which may be trapped by sulphoxides.[426] Persulphoxides and persulphones are nucleophilic sources of oxygen and will not epoxidize olefins.[427] The mechanism for the photochemical oxidation of thioketones to carbonyl compounds has been investigated.[428] Singlet oxygenations of menaquinones,[429] quinolylindan-1,3-diones,[430] and of formaldehyde and polyphenol[431] have been reported.

Autoxidation and Other Reactions of Triplet Oxygen

Reviews have appeared on triplet oxygenations of hydrocarbons,[432–434] alkylaromatics,[435] aldehydes,[436] ketones,[437] and other oxygen-,[438] sulphur-, nitrogen-, and chlorine-containing organic compounds;[439] the rôles of organometallic complexes,[440] quantum chemistry,[441] and sulphur-containing transition-metal complexes[442] in hydrocarbon autoxidation have been reviewed. The mode of action of arylamines as anti-oxidants has been investigated.[443] Stereoelectronic factors are important in the anti-oxidant activity of vitamin E and related chain-breaking phenolic anti-oxidants; the ethereal oxygen *p*-type lone pair is held perpendicular to the aromatic ring as a result of the geometry of the chroman ring—this in combination with alkyl substitution at the other four ring-positions explains the superior chain-breaking properties of α-tocopherol and related phenols such as (**58**).[444,445] A comparison of the autoxidation of egg lecithin phosphatidylcholine in

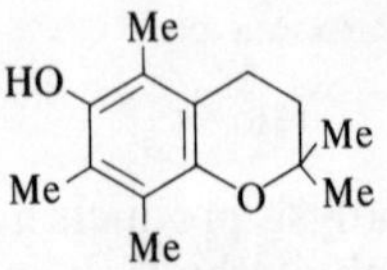

(**58**)

water and in chlorobenzene has been made as a model study of membrane autoxidation — the physical structure of lecithin bilayers renders them more resistant to oxidation than would be expected on the basis of their chemical composition. Thus, although the bilayers follow normal kinetics for autoxidation, initiation is rather inefficient because of the high micro-viscosity and the oxidizability appears to be reduced because of removal of peroxy radicals from the autoxidizable region of the bilayer.[446,447]

Air oxidation of the isoindene (**59**) gives rise to the isobenzofuran (**62**); an intermediate endoperoxide (**60**) rearranges to (**61**) which loses acetone in retro-Diels–Alder reaction.[448] Enol ethers react with triplet oxygen in the dark at elevated

(**59**) (**60**) (**61**) (**62**)

temperatures to produce ketones; dioxetanes are chemiluminescent intermediates in the reaction — the oxidations of enol ethers with singlet and triplet oxygen are very similar, proceeding through the same dioxetane intermediates.[449] Cyclohexenone is the major product of sulphur-dioxide-induced allylic oxidation of cyclohexene with molecular oxygen in the presence of acetate ion as the base catalyst; the reaction proceeds *via* a radical anion produced from sulphur-dioxide-stabilized peroxide ions.[450] Photochemical and thermal epoxidation of propene in the presence of sulphur dioxide to give propene oxide occurs in high yield at room temperature *via* radical cations and superoxide ion (Scheme 29).[451]

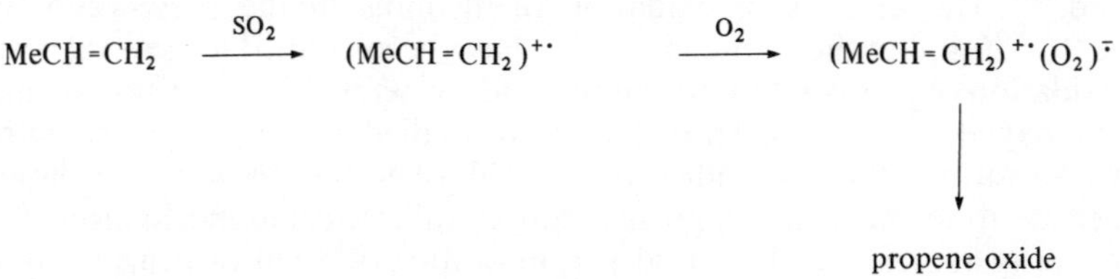

$$\mathrm{MeCH{=}CH_2} \xrightarrow{\mathrm{SO_2}} (\mathrm{MeCH{=}CH_2})^{+\cdot} \xrightarrow{\mathrm{O_2}} (\mathrm{MeCH{=}CH_2})^{+\cdot}(\mathrm{O_2})^{\overline{\cdot}} \longrightarrow \text{propene oxide}$$

SCHEME 29

Photo-oxidation of benzoin in the presence of olefins leads to mainly *trans*-epoxides; an acylperoxy radical is a probable intermediate.[452] Epoxides are also formed in the oxidation of alkoxystyrenes, although competing reactions lead to carbonyl products.[453] Transition-metal–Schiff-base complexes catalyse the oxidation of (**63**) to carpanone (**64**) by molecular oxygen.[454] The inhibition of autoxidation of styrene by copper(II)–Schiff-base complexes has been investigated.[455] The reactions of α-methylstyrene with triplet oxygen in the presence of metal catalysts have been studied.[456,457] The rôle of oxygen as an activator in olefin metathesis by metal catalysis is due to the formation of epoxides and the corresponding metallaoxacyclobutanes (**65**); initiating metallacarbenes are produced from fission of (**65**).[458] Irradiation of CT bands of some styrenes with oxygen-induced dimerization of the olefin through the styrene radical cations,

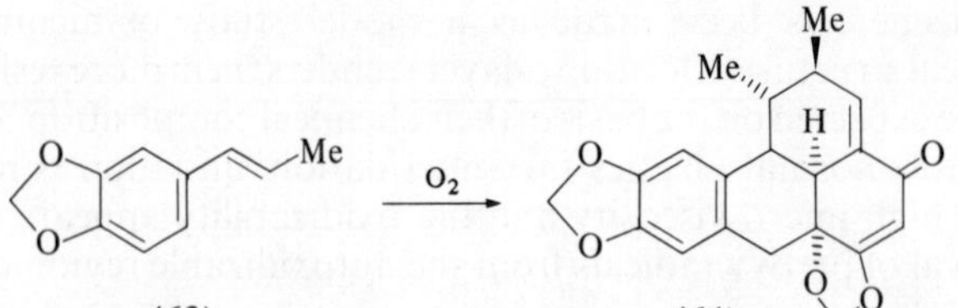

(63) (64) 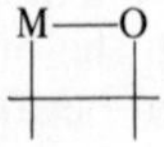 (65)

generated by electron transfer to oxygen on CT excitation.[459] Other studies of the reactions of olefins with molecular oxygen include unsaturated alicyclic compounds,[460] conversion into ketones in the presence of manganese porphyrin,[461] the autoxidation of acetoxycyclohexene,[462] and propene in acetic acid in the presence of copper(II).[463] The oxygenation reactions of the adamantylideneadamantane radical cation with triplet oxygen have been reported.[464]

Continuing interest, mostly Russian, is shown in the oxidation of alkyl-substituted aromatic hydrocarbons and related compounds. Thus, reports on the kinetics of the catalysed and uncatalysed oxidations of alkylbenzenes,[465–484] fused alkylaromatic compounds,[485,486] and heterocyclic systems[487–489] have appeared.

It has been shown that propene is sole precursor of propene oxide in the oxygenation of propane.[490,491] The initial product in the liquid-phase oxidation of 1-methylcyclohexane is the 1-hydroperoxide;[492] however, cyclohexyloxy and cyclohexylperoxy radicals are not intermediates in the oxidation of cyclohexane to cyclohexene in the presence of nitrous oxide.[493] The results of charge-distribution calculations on dimethylcyclohexanes have been used to rationalize the rates and product distributions of their oxidation.[494] The effects of tin(II)[495] and chromium salts[496] on the course of autoxidation of cyclohexane have been investigated, as have the effects of elevated temperatures on the autoxidation of butane.[497] The mechanism of formation of cleavage products in *n*-hexadecane autoxidation has been elucidated.[498] The kinetics of oxidation of alcohols in the presence of metal catalysts[499–502] and in their absence[503–507] have been studied for several systems.

The oxidation of propanal to propanoic acid by peracid is much slower than by molecular oxygen;[508] the reaction has been studied by mass spectrometry.[509] Fourier-transform infra-red studies of the chlorine-atom-initiated oxidation of formaldehyde to formic acid supports a peroxy radical-chain mechanism.[510] The main source of radicals in the initial period of autoxidation of benzaldehyde by molecular oxygen is the degenerate branching of chains with participation of perbenzoic acid; the branching capability of the peracid is masked in the later stages by its high reactivity in the heterolytic oxidation of benzaldehyde, and by the autocatalysis of this ionic oxidation by benzoic acid.[511]

Cyclohexane-1,2-dione and 2-hydroxycyclohexanone are the primary products in the oxygenation of cyclohexanone;[512] the primary product of oxidation of cyclohex-2-enone is **(66)** which subsequently decomposes in competing unimolecular and bimolecular (with starting material) processes.[513] The effects of temperature on chain-terminals in the oxidation of β-ionone,[514] and of manganese(II) in the initiation

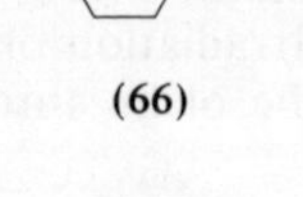

(66)

of autoxidation of pentadecan-8-one[515] have been studied. The sites of initial hydrogen abstraction in the oxidation of several esters[516–520] and acids[521] have been identified. The oxidations of *N*-methylpiperidine,[522] and of triethylamine and *N*,*N*-dimethylcyclohexylamine[523] proceed by radical-chain pathways. Free-radical oxidation provides a method for the stereospecific oxidation of nucleoside trialkyl phosphite to the corresponding phosphates and has allowed a novel preparation of the individual diastereoisomeric ^{18}O-labelled thymidine 3′,5′-cyclic monophosphates.[524] Among other studies on the base-catalysed autoxidations of phenols,[525–527] the products of oxygenation of 4-aryl-2,6-di-*tert*-butylphenols arise from a first-formed π-complex of the phenolate ion and molecular oxygen which is then attacked by the oxygen predominately at position 4.[528]

Reduction by Complex Hydrides

Aluminium Hydrides

Much evidence has been presented for a single-electron-transfer mechanism being the predominant pathway in the reduction of several organic substrates with lithium aluminium hydride and several other main-group hydrides. Direct observation by EPR spectroscopy of radical intermediates has been reported in the $LiAlH_4$ reductions of alkyl halides[529] and secondary and tertiary alcohols[530] to hydrocarbons. Reduction of 6-halohex-1-enes and of 6-halo-5,5-dimethylhex-1-enes produces both straight-chain and cyclized reduction products; the formation of cyclized products clearly indicates the presence of radical intermediates in these reductions.[531] Radical-anion–radical-cation pairs are intermediates in the reduction of polynuclear hydrocarbons with aluminium hydride and other hydrides.[532] The reduction of ketones, alkyl halides, and other organic substrates by lithium tetrakis(*N*-dihydropyridyl)aluminate takes place *via* single-electron-transfer processes.[533]

Reduction of maleimide with $LiAlD_4$ yields pyrrolidine with five (rather than the expected six) carbon–deuterium bonds, and *N*-deuteriated maleimide with $LiAlH_4$ gives (**67**); these results require migration of hydrogen from nitrogen to carbon, favouring reduction over initial abstraction of an acidic hydrogen.[534] Lithium-aluminium-hydride–aluminium-hydride reduction of sultones yields mercapto alcohols from sulphur–oxygen bond-cleavage for secondary and tertiary sultones; in contrast, primary sultones gives sulphonic acids arising from breaking of the carbon–oxygen bond.[535]

D N H

(**67**)

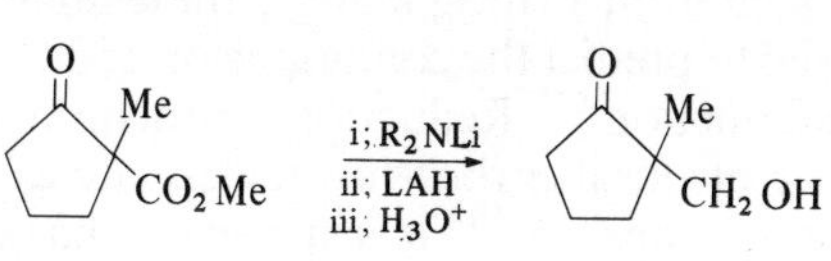

O Me CO_2Me i; R_2NLi ii; LAH iii; H_3O^+ O Me CH_2OH

SCHEME 30

Lithium tris[(3-ethyl-3-pentyl)oxy]aluminium hydride is selective for the reduction of aldehydes in the presence of ketones;[536] selective reduction of esters in β-keto-esters can be achieved by protection of the ketone moiety as an enolate (Scheme 30).[537] A mechanistic rationale has been provided for the influence of electronegative substituents at the *peri*-positions on the lithium aluminium hydride reductions of

9,10-anthraquinones.[538] Hydrogenolysis of 3,5-*O*-benzylidene acetals with $LiAlH_4$–$AlCl_3$ reagents in methyl D-xylofuranosides gives mainly 5-benzyl ethers as products.[539,540] Selective reductions of functional groups in steroidal dicarbonyl compounds[541] and α-oximinoketones[542] by aluminium hydrides have been reported.

The influence of intramolecular amino groups on the reductions of ketones can be correlated with the formation of *N*-coordinated cyclic transition states.[543] Asymmetric reductions of ketones with lithium aluminium hydride modified by chiral amino-1,2-diols[544] and chiral 2,2′-dihydroxy-1,1′-binaphthyl[545] have been described; the reactions of α,β-unsaturated ketones[546] and of isopropyl phenyl ketone[547a] with chiral β-branched trialkylaluminium compounds lead to chiral alcohol products by transfer of the β-hydrogen of the alkyl group in a six-membered transition state with the carbonyl group complexed to aluminium. Lithium mono-*tert*-butoxyaluminium hydride has a radically different stereoselectivity from that of lithium aluminium hydride and of lithium tri-*tert*-butoxyaluminium hydride in the reduction of 3,3,5-trimethylcyclohexanone.[547b]

Stereoselective reductions of isoxazolines[548] and of steroidal enones[549] by lithium aluminium hydride have been used for preparation of aminopolyols and of side-chain cholesterol alcohols, respectively.

Stereoselectivity in the lithium aluminium hydride reduction of monoimines of benzil to the corresponding amino-alcohols can be rationalized by preferred hydride attack on the least hindered side of the imine, after making various assumptions about the relative energies of the different conformers on the imine.[550] The stereochemical course of the reduction of optically active styrene-2,2-d_2 oxide by aluminium hydride has been elucidated.[551]

Borohydrides

Although several reports of kinetic studies have appeared,[552–554] the main interest in the sodium borohydride reduction of carbonyl compounds has been in stereochemical features of the reaction. Twist-angle approach to stereoselectivity in the reduction of ketones by sodium borohydride indicates that the stereochemical results can be rationalized by preferred hydride approach to the carbonyl group from the side of the obtuse angle with respect to the C—O bond; the twist incorporates the electronic nature of the substituent and helps to explain its steric effect. Molecular-orbital calculations show that the π^*-orbital of the carbonyl group of asymmetric ketones deviates from orthogonality; these small deviations, measured by twist angles, are used to predict the configuration of diastereomeric alcohols produced on borohydride reduction.[555] Reduction of prochiral ketones with sodium borohydride in the presence of chiral carbohydrate derivatives and carboxylic acids can lead to asymmetric induction;[556–558] it is suggested that sodium borohydride is converted into a lipophilic acyloxy borohydride which then binds covalently to the hydroxyl group of the sugar, producing a chiral anion.[556] The asymmetric reduction of ketones has also been achieved with chiral aminoboranes.[559,560] Although the stereoselective reduction of imines has not been well studied, very high stereoselectivity has been observed in imine reaction with substitute alkali metal borohydrides;[561] a novel asymmetric reduction of imines with chiral sodium triacyloxyborohydrides, derived from reaction of sodium borohydride and *N*-acyl-L-prolines, has been reported.[562]

The reduction of unsymmetrically substituted cyclic anhydrides, such as (**68**), by sodium borohydride occurs as shown at the most hindered carbonyl group; this cannot be rationalized by "steric hindrance along preferred reaction path" since both carbonyl groups are equally accessible by non-perpendicular approach.[563] The stereoselective reduction of tertiary alkyl, benzyl, and allyl halides to hydrocarbons with lithium 9,9-di-*n*-butyl-9-borabicyclo[3.3.1]nonanate proceeds *via* an intermediate ion pair *e.g.* (**69**).[564] The mechanism of the photo-reduction of aromatic compounds by sodium borohydride in the presence of dicyanobenzene involves electron transfer from the excited state of the arene to dicyanobenzene followed by nucleophilic attack of the borohydride on the resulting radical cation.[565]

Mechanistic features of the reductions of triazines,[566] cyclopropenium ions,[567] and aromatic halides[568] have been investigated. Esters are reduced to the corresponding alcohols by sodium borohydride in polyethylene glycol; the reducing ability of sodium borohydride is enhanced by the formation of polymeric dialkoxyborohydrides.[569] Carboxylic acid amides are reduced to amines by sodium borohydride in DMSO in the presence of methanesulphonic acid; a protonated amide is implicated as an intermediate.[570] Reaction of sodium borohydride with acid chlorides in DMF at $-78°$ produces (**70**) which may be quenched by propanoic acid/dilute hydrochloric acid/ethyl vinyl ether to give good yields of aldehydes.[571] Sodium borohydride normally reduces enones predominantly to allylic alcohols, whereas in the presence of complexing amines saturated alcohols are the major products;[572] reduction of cyclohex-2-enones under conditions of phase transfer also leads mainly to saturated alcohols.[573] Under phase-transfer conditions, sodium borohydride reduces halides and sulphonate esters to alkanes.[574]

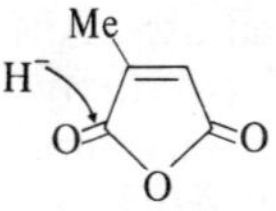

Me H O O O

(**68**)

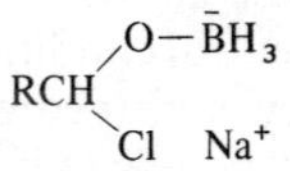

$RCH(O-\bar{B}H_3)(Cl)$ Na^+

(**70**)

Bu^t Me Cl → [Bu^t + Me Bu^n Bu^n $\bar{B}$] → Bu^t H Me

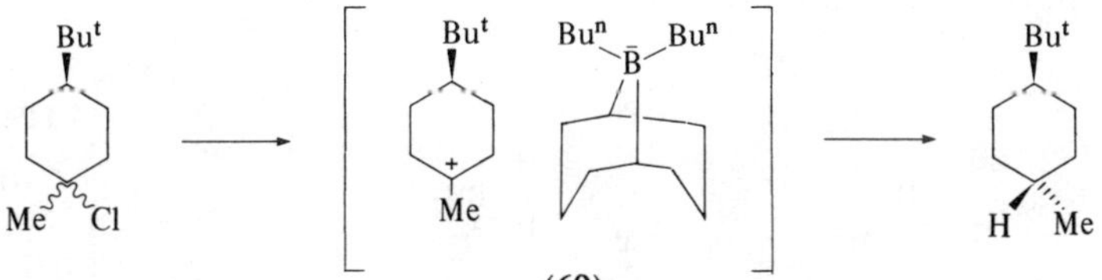

(**69**)

Electrophilic hydrides give products of conjugate reduction with α-oxoketene dithioacetals (Scheme 31).[575] Mild selective reductions have been described with bis(trifluoroacetoxy)borane[576] and bis(benzoyloxy)borane;[577] asymmetric synthesis with chiral boranes has been reviewed.[578] Although triphenylphosphine borane, Ph_3PBH_3, is inert under normal conditions, reaction with methyl iodide forms triphenylmethylphosphonium iodide and releases diborane, which may then be used for hydroboration.[579]

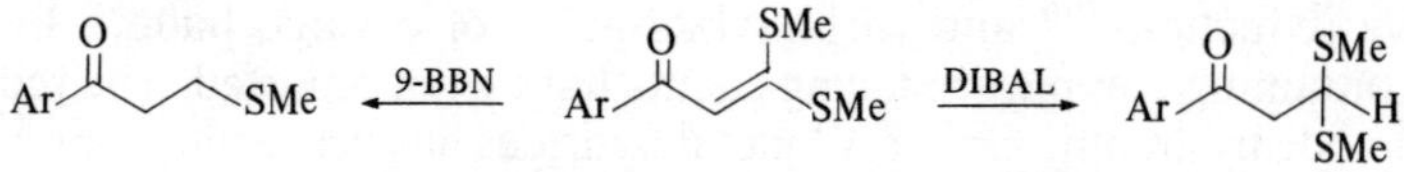

ArCOCH$_2$CH$_2$SMe ←(9-BBN)— ArCOCH=C(SMe)$_2$ —(DIBAL)→ ArCOCH$_2$CH(SMe)$_2$

SCHEME 31

Mechanistic rationalizations and comments have appeared on the sodium cyanoborohydride reductions of isopropylidene acylmalonates, 5-acylbarbituric acids, and 3-acyl-4-hydroxycoumarins to their corresponding alkyl derivatives,[580] and on the reductive amination of ketones by cyanoborohydride and ammonium acetate.[581] Conversion of amides into thioamides and alkylation with methyl iodide to give thioimonium salts, followed by treatment with cyanoborohydride, provides a convenient route for the conversion of amides into amines.[582] *O*-acyl oximes are reduced by cyanoborohydride to hydroxylamine derivatives without cleavage of the N—O bond.[583] The use of tetraalkylammonium cyanoborohydrides for selective reductive aminations in non-polar media has been reported.[584] Earlier misinterpretations regarding the selectivity of tetraalkylammonium borohydrides have been corrected.[585] *erythro*-α,β-Epoxy-alcohols have been prepared in high stereoselectivity by zinc borohydride reduction of the corresponding α,β-epoxyketones;[586] comparison of the rates of reduction of various functional groups shows that zinc borohydride is a stronger reducing agent in the mixture of THF and ether than in THF alone.[587] The major rôle of lanthanide(III) ions in the regioselective reduction of α-enones to allylic alcohols by sodium borohydride is catalysis of decomposition of borohydride ion by hydroxylic solvents to alkoxyborohydrides, which are responsible for the observed regioselectivity; the stereoselectivity of the process is also modified by lanthanide ions since axial attack on cyclohexanones is enhanced.[588]

Among other work on the use of bis(triphenylphosphine)copper(I) tetrahydroborate (**71**),[589] aromatic azides may be reduced to arylamines by the reagent;[590] iminophosphoranes were shown not to be intermediates. Although (**71**) will not reduce ketones under neutral conditions, addition of acid catalysts leads to reduction to alcohols with a very high degree of stereoselectivity; under these conditions aldehydes are chemoselectively reduced in the presence of ketones.[591] The reducing capabilities of (**71**) have been compared with those of the "diphos" (**72**) and *o*-phenanthroline (**73**) complexes of copper(I) borohydride; there is not a great deal of difference in their reducing properties.[592]

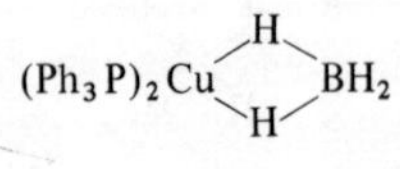

(**71**) 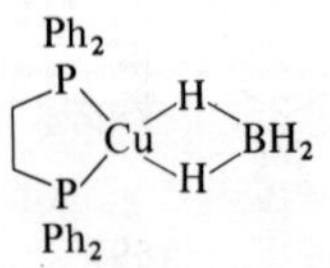 (**72**) 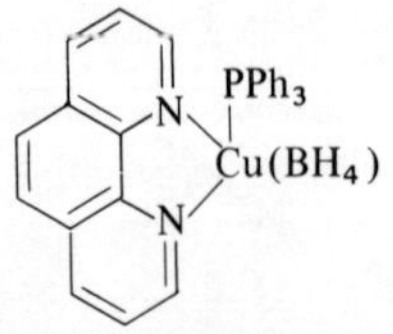 (**73**)

Other Hydride Reductions

General reviews have appeared on hydride reducing agents[593] and on asymmetric reduction of ketones.[594] Hydromagnesiation of prop-2-ynylic alcohols with *tert*-butylmagnesium chloride proceeds with complete regio- and stereo-selectivity in the presence of dicyclopentadienyltitanium(IV) dichloride to give alkenylmagnesium chlorides (Scheme 32).[595]

Selective reductions[596] and carbonylations[597] of organic halides by sodium-hydride-containing complex reducing agents have been reported. The reduction of carbon dioxide by anionic Group VI metal hydrides has been studied by ^{13}C-NMR spectroscopy.[598]

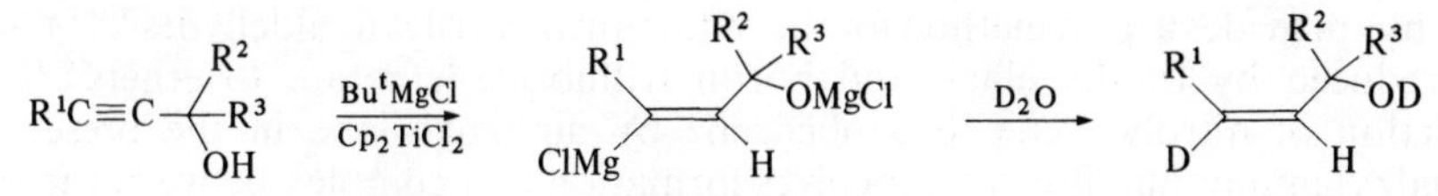

$$R^1C{\equiv}C{-}C(R^2)(R^3)OH \xrightarrow[Cp_2TiCl_2]{Bu^tMgCl} (ClMg)R^1C{=}CH{-}C(R^2)(R^3)OMgCl \xrightarrow{D_2O} (D)R^1C{=}CH{-}C(R^2)(R^3)OD$$

SCHEME 32

Tri(organo)tin hydride reduction of methyl 6β-isothiocyanatopenicillanate is accompanied by intramolecular radical capture and cleavage of the sulphur–carbon bond to give thiazolines (Scheme 33).[599] The reduction of 1-substituted 7-chloro-7-

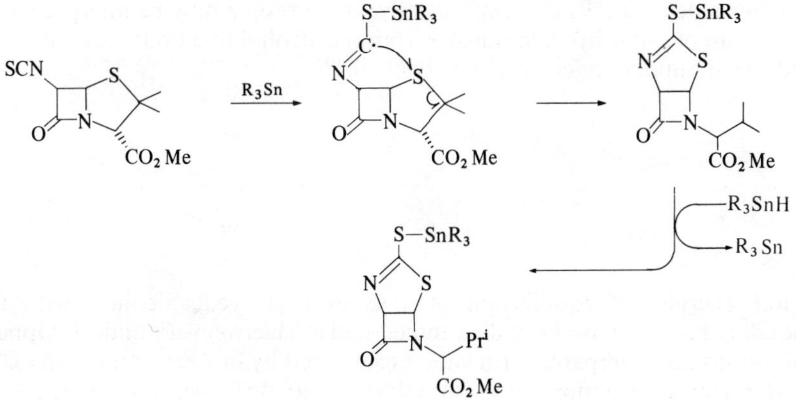

SCHEME 33

fluoronorcaranes (**74**) with tributyltin hydride varies in stereospecificity as a function of the nature of R, suggesting that the configurational stability of the 7-fluoro radical (**75**) is affected by the substituent at the position β to the radical centre; thus β-trimethylsilyl and β-methyl stabilize the pyramidal configuration of the cyclopropyl radical, whereas the pyramidal arrangement is destabilized by β-fluoro, β-methoxy, and β-carbomethoxy groups.[600] Radical mechanisms are involved in the tributyltin hydride reductions of peroxyorganomercurials (**76**) to peroxides (**77**),[601] of 3-*O*-hexofuranosyl *S*-methyldithiocarbonates,[602] and of acid chlorides to aldehydes in the presence of palladium catalysts.[603] Triethylsilane reduces nitrilium ions to imines

(74) (75)

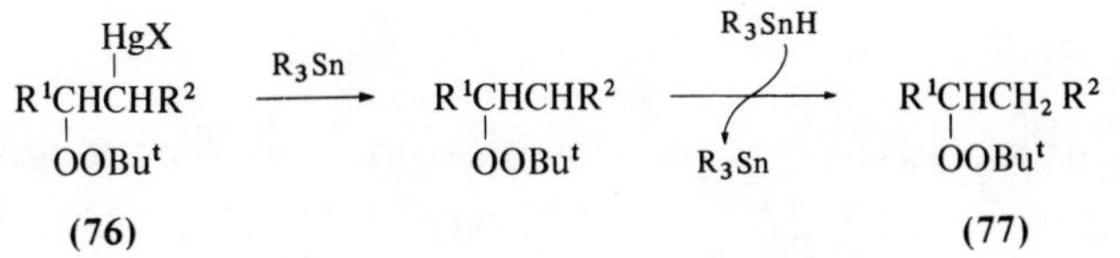

$$R^1CH(HgX)CHR^2(OOBu^t)\ (76) \xrightarrow{R_3Sn} R^1CH\dot{C}HR^2(OOBu^t) \xrightarrow{R_3SnH \to R_3Sn} R^1CHCH_2R^2(OOBu^t)\ (77)$$

and thus provides a mild method for the reduction of nitriles to aldehydes.[604] Lactols are reduced by triethylsilane and boron trifluoride–etherate to ethers.[605] The reduction of nitrobenzene to azobenzene by diphenylsilane in the presence of tetraalkylammonium fluorides involves formation of a complex between the silane and the ammonium salt followed by electron transfer to nitrobenzene to form a radical anion.[606] Alkoxysilanes may be activated by potassium fluoride to provide an effective reagent for reduction of carbonyl groups in the presence of many other functionalities.[607]

Evidence for stereoelectronic control in hydride transfer to cyclic oxenium ions is provided in the oxido-reduction of (**78**) under acid catalysis to give the equatorial bicyclic ether-aldehyde (**79**) stereospecifically; these results may be interpreted by invoking intramolecular hydride transfer from the alcohol to a cyclic oxenium ion, delivered axially under stereoelectronic control.[608]

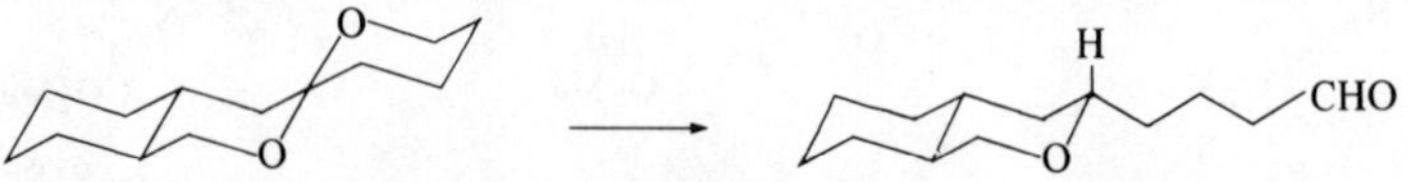

(**78**) → (**79**)

The free energies of equilibrium of a series of bicyclic alcohols with the corresponding ketones have been determined under Meerwein–Ponndorf–Oppenauer conditions and compared with values calculated by molecular mechanics.[609] Hydride-ion-transfer reactions from formaldehyde to aldehydes and ketones have been studied in the presence of metal salt catalysts.[610] A hydroxamic acid derivative (**80**) has been shown to be an intermediate in the reductive formylation of *N,N*-dimethyl-*p*-nitrosoaniline by glyoxylic acid.[611] Formate ion reductions of aldehydes and ketones in dipolar aprotic solvents,[612] 1,1′-benzylidenedipiperidine,[613] and of dinitroaromatic compounds to mononitroarenes in the presence of palladium(0)[614] have been studied. Selective hydrogen transfer from formic acid to α,β-unsaturated ketones by polymer-bound $IrCl(CO)(PPh_3)_2$ gives saturated carbonyl compounds: although formic acid forms hydrogen and carbon dioxide in the presence of the iridium catalyst, the reduction proceeds by direct transfer from formate.[615] A novel reducing agent (**81**), derived from formic acid and two equivalents of a Grignard reagent, is chemoselective for the reduction of aldehydes; ketones are reduced only slowly by (**81**).[616] Isopropylmagnesium bromide in the presence of $(Ph_3P)_2NiCl_2$ reduces vinyl sulphides to olefins, with no saturation of the carbon–carbon double bond.[617]

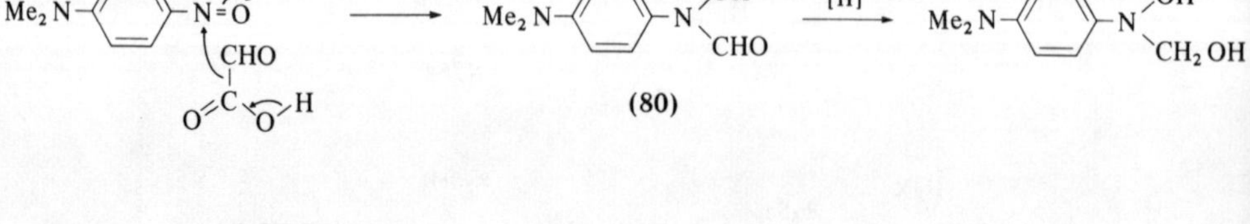

$Me_2N-C_6H_4-N{=}O$ + OHC–CO$_2$H → $Me_2N-C_6H_4-N(OH)CHO$ (**80**) $\xrightarrow{[H]}$ $Me_2N-C_6H_4-N(OH)CH_2OH$

$$HCO_2H \; + \; 2\,RMgX \longrightarrow RCH(OMgX)_2 \xrightarrow{R'CHO} R'CH_2OMgX$$

(**81**)

Reduction by Metals, Metal Complexes, and Ions

Ab initio molecular-orbital calculations on cyclohexa-1,3- and -1,4-diene show that, for substituted cyclohexadienes, the 1,3-isomer is always favoured; kinetic protonation of cyclohexadienyl anions frequently does not produce the thermodynamic product.[618] However, protonation of a dianion always leads to the most stable monoanion.[619] Thiopyrylium cations are reduced by alkali metals to produce thiabenzene radicals.[620] Reductions of dimethoxynaphthyl ketones[621] and cannabine *N*-oxide[622] by sodium in liquid ammonia have been investigated. The regiospecific reduction of α,β-unsaturated carbonyl compounds with neighbouring hydroxyl groups can be explained by intra- *versus* inter-molecular protonation.[623] Frontier-molecular-orbital analysis has been used to rationalize reductive cyclizations involving attack of radical anions from carbonyl compounds on isolated carbon–carbon multiple bonds.[624,625] The stereochemistry of the reduction of steroidal ketones by alkali metals in ethanol can be interpreted in terms of the frontier-molecular-orbital picture of the ketyl; reduction of enolizable ketones by potassium in liquid ammonia is controlled by steric factors connected with hydrogen abstraction by a dianion from the position α to a ketone.[626] The mechanism for reduction of ketones proposed by House may be dominant only when ammonium chloride is the proton source.[627] α,β-Unsaturated amides and lactams are reduced to saturated amides by magnesium in methanol.[628] 4-Phenyl-1,3-dioxans are reduced by sodium–potassium alloy to phenylbutanediols (**82**) *via* benzylic carbon–oxygen bond-cleavage followed by recombination with the ketyl from formaldehyde.[629] The dark-blue solution of sodium–potassium alloy, solubilized by 18-crown-6-ether, has been used for the deoxygenation of both hindered and unhindered esters,[630] the hydrogenolysis of alkyl fluorides,[631] and the deoxygenation of *N*,*N*-dialkylamino-(thiocarbonyloxy)alkanes.[632] The reductions of alkyl halides[633] and benzophenone azine[634] by sodium–naphthalene have been studied.

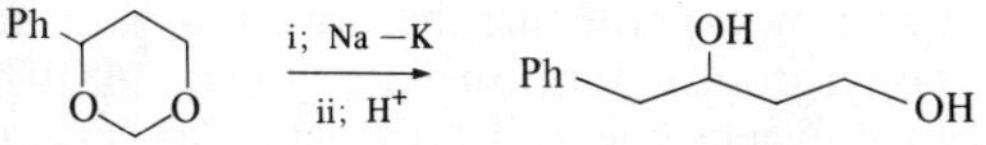

Ph, O, O; i; Na – K; ii; H^+; OH; Ph; OH

(82)

Allyl methyl ethers are reduced with sodium and trimethylchlorosilane to give allyltrimethylsilane; the reaction can be rationalized in terms of a mechanism of electron transfer to yield allylic radical and anionic intermediates (Scheme 34).[635] Sulphides may be deoxygenated to sulphides by zinc and trimethylchlorosilane.[636] *cis*-Stilbene is the major product of the Clemmensen reduction of benzoin in anhydrous ethereal hydrogen chloride.[637] Aryl-substituted cyclopropyl rings (**83**) are cleaved to acyclic ketones (**84**) by zinc in alcohol, whereas alkyl-substituted

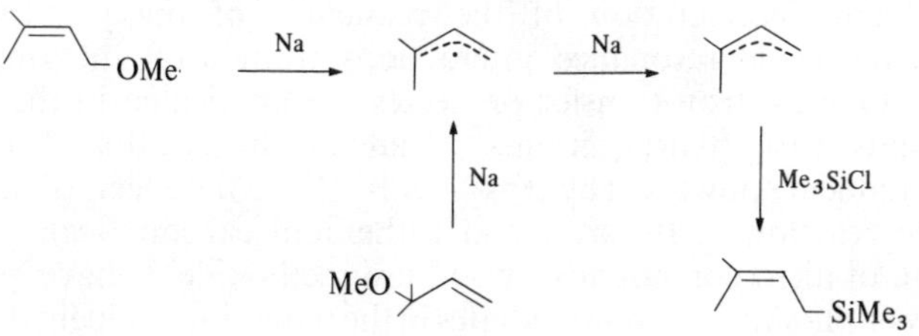

OMe; Na; Na; Na; Me_3SiCl; MeO; $SiMe_3$

SCHEME 34

cyclopropyl ketones are unaffected.[638] The reduction of diaroylbenzenes by Group III metals has been investigated by ESR spectroscopy.[639] The deoxygenation of arene 1,4-endoxides to arenes by low-valent metal complexes[640] has been reported. Titanium-induced reductive couplings of carbonyl compounds[641–644] proceed *via* radical-anion intermediates. The reductions of diazines and triazines by titanium(III)[645] and of (phenylthiomethyl)carbinyl benzoate (**85**) to alkenes with titanium metal[646] have been described. More than 90% of carbon monoxide reduction to methanol by vanadium(II)–catechol complexes in alkaline methanol proceeds without intermediate formation of free formaldehyde.[647] Reductive acylation of α-keto-imines (**86**) by acetylcobalttetracarbonyl, generated *in situ*, occurs only at the carbon–nitrogen double bond to give β-keto-amides (**87**).[648] Acid chlorides are reduced to α-diketones by diiodosamarium by a mechanism involving one-electron-transfer processes.[649]

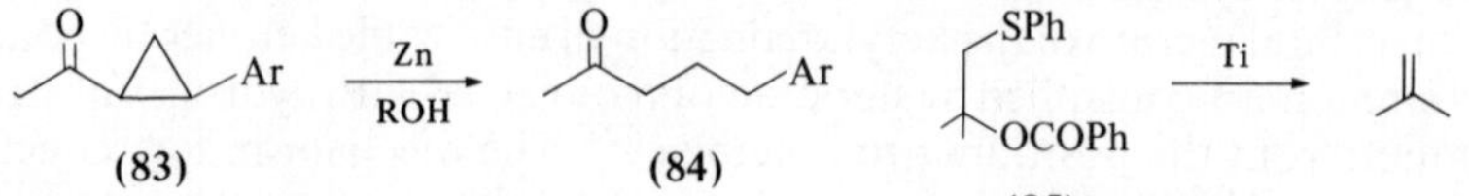

(83) —Zn, ROH→ (84) (85) —Ti→

(86) —MeCO—Co(CO)$_4$→ (87)

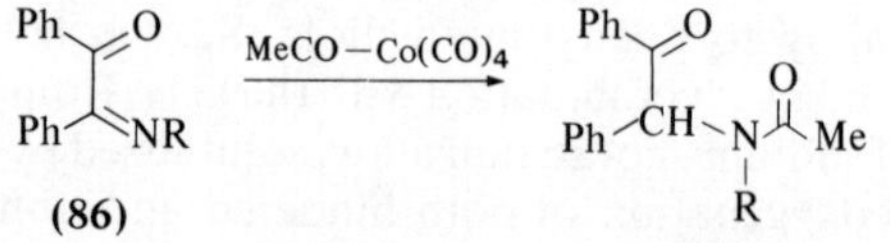

Miscellaneous Reductions

The deoxygenation of butanal and butanone by atomic carbon has been studied by MNDO calculations and by experiment; the preferred path involves addition of carbon across the carbonyl group to form an oxiranylidene (**88**) which rearranges to a ketene (**89**) which has sufficient energy to dissociate to carbon monoxide and a

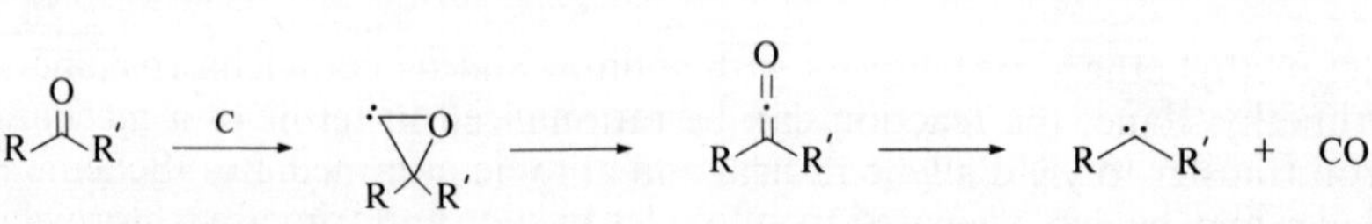

RCOR′ —C→ (88) → (89) → RC̈R′ + CO

carbene. Experimentally, an intermediate ketene was trapped by addition of water, following reaction of arc-generated carbon atoms with frozen butanal.[650] Steric interactions cause acceleration of the reduction of bulky sulphoxides with dichlorocarbene, while resonance interactions from aromatic rings retard the reduction.[651] One-electron-transfer processes are implicated in the reductions of diazonium salts with hydroquinones[652] and with tetrathiafulvalenes.[653] The comparative reducing powers of hydroxy-acids[654] and the effects of the metal on the course of the reaction in the iron and ruthenium carbonyl-catalysed reductive carbonylation of nitro compounds by sodium methoxide[655] have been discussed.

α-Hydroxysulphinates are intermediates in the reduction of aldehydes and ketones by sodium dithionite.[656] The reduction of cyclohexanones by sodium dithionite gives mainly equatorial alcohols whereas bicyclic ketones form mainly *endo*-alcohols; the reaction proceeds by one-electron-transfer processes.[657] Iodomethyl ketones react with thiols and selenols by nucleophilic attack on iodine to give reduction to methyl ketones; in contrast, bromomethyl ketones give mainly substitution products.[658] α-Haloketones are reduced to ketones by PI_3 and by P_2I_4; an enolate is formed as an intermediate by nucleophilic attack by phosphorus either directly on the halogen or by initial attack on the carbonyl group.[659] The fixation and deoxygenation of carbon dioxide to form furan derivatives by phosphorus(III) reagents[660] and the reduction of *gem*-dibromides such as (**90**) and (**91**) to the corresponding monobromides by diethylphosphite[661] have been reported.

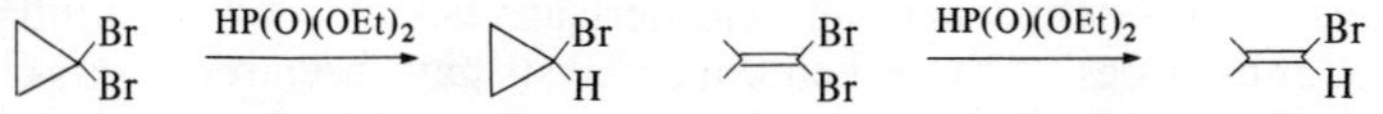

Br Br —HP(O)(OEt)$_2$→ Br H (**90**)   Br Br —HP(O)(OEt)$_2$→ Br H (**91**)

The rapid desulphurization of trisulphides to disulphides by tris(dialkylamino)-phosphines involves rate-determining formation of a phosphonium salt; the central sulphur atom is lost in the case of diaryl trisulphides whereas dialkyl trisulphides lose a terminal sulphur (Scheme 35).[662]

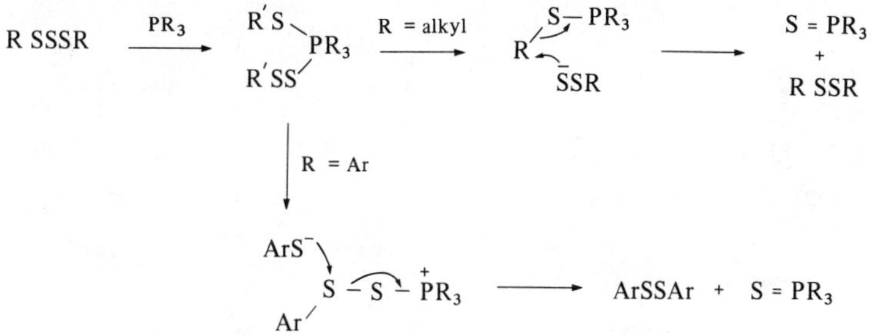

R SSSR —PR_3→ R′S, R′SS PR_3 —R = alkyl→ R S—PR_3, $\bar{S}$SR ⟶ S = PR_3 + R SSR

R = Ar ↓

ArS^- Ar S—S—$\overset{+}{P}R_3$ ⟶ ArSSAr + S = PR_3

SCHEME

The reductive cyclizations of keto-acids to polycyclic hydrocarbons by hydriodic acid and red phosphorus[663] and the reductive deimination of sulphoximides and sulphilimides with toluenesulphonyl nitrite[664] have been investigated.

Hydrogenation

Mechanistic aspects of homogeneous catalytic hydrogenation and related processes have been reviewed.[665]

There have been first attempts[666] to characterize intermediates in homogeneous asymmetric catalysis by chiral phosphine–rhodium complexes by measurements of extended X-ray fine structure. Among further ^{31}P-NMR studies,[667] the mode of bidentate binding of prochiral olefin substrates by chiral pyrrolidinodiphosphine-rhodium complexes has been found to be very regioselective; "induced fit" phenomena have been observed with the chiral rhodium complexes.[668] It has been found that the temperature of the reaction, the pressure of hydrogen, and the relative basicity of added amines all have considerable effect on the enantioselectivity of the reduction of amino-acid precursors by chiral phosphine–rhodium complexes.[669]

Rhodium catalysts with chiral chelating six-membered ring diphosphines, "skewphos" (**92**) and "chairphos" (**93**), are efficient for the hydrogenation of amino-acid precursors; "skewphos" adopts a chiral conformation and leads to high optical yields, whereas "chairphos" adopts an achiral conformation and is ineffective as a chiral hydrogenation catalyst.[670] The rhodium catalyst based on (**94**) is considerably more enantioselective in the hydrogenation of prochiral precursors to amino-acids than the catalyst based on (**95**), suggesting that the amino groups provide additional binding sites for the substrates.[671] The uses of a very bulky cyclohexylphosphine ligand giving a stereochemically rigid chelating group[672] and of polymer catalysts containing optically active pendant alcohols[673] have been investigated in asymmetric rhodium-catalysed hydrogenation reactions. Asymmetric hydrogenations of enol phosphinates[674] and dehydroamino-acid derivatives[675,676] by rhodium catalysts, and of cyclic anhydrides[677] by ruthenium catalysts have been reported. The optical yields of asymmetric transfer hydrogenation to α-methylcrotonic acid and its esters by alcohols with chiral diphosphine complexes of ruthenium(II) is much affected by the alcohol used; the mechanism involves complexation of the alcohol hydride donor with the ruthenium complex.[678] Reductions of pyruvamides containing (S)-chiral centres give (S)-lactamides in high optical yields with diastereoisomer ratios up to 98:2; it is proposed that the asymmetric induction arises from adsorption of the phenyl group of the substrate on to the surface of the palladium catalyst (**96**) before reduction.[679]

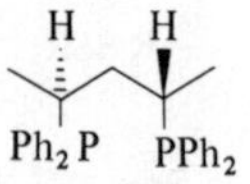

(**92**) 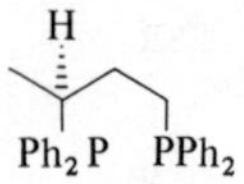 (**93**) 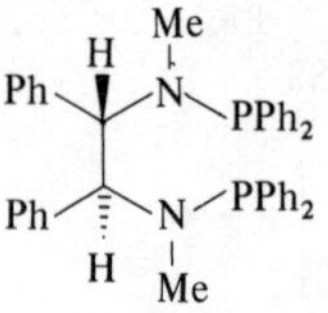 (**94**) 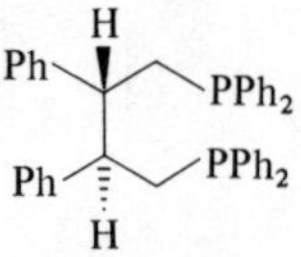 (**95**)

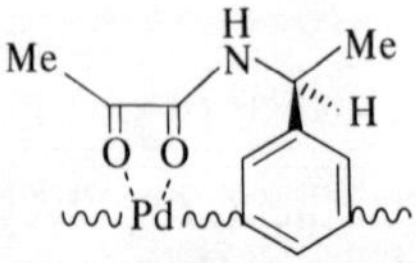

(**96**)

The products from the homogeneous hydrogenation of carbon monoxide are all formed from formaldehyde as a key intermediate; although the formation of formaldehyde from hydrogen and carbon monoxide is thermodynamically unfavourable, the concentration of formaldehyde as a transient intermediate is sufficiently high to account for the products.[680] The hydrogenation of hindered steroidal olefins with a complex iridium catalyst proceeds from the α-face of the olefin with no reduction of carbonyl groups, halogens, or cyclopropane rings present.[681] Investigations of hydrogenations with molybdenum hydrides, formed in the reaction of molybdenum atoms with tetrahydrofuran, have been reported.[682] The mechanisms of the homogeneous hydrogenations of cyclohexene by ruthenium[683] and rhodium[684] complexes, of unsaturated carboxylic acids,[685] unsaturated ketones,[686] Dewar benzenes,[687] aromatic nitro compounds,[688,689] and styrene oxide[690] have been studied. Polymer-bound rhodium hydride catalysts reduce aromatic compounds to cycloalkanes,[691] and hydrogenation of alkynes with a palladium-anchored polystyrene catalyst gives predominantly *cis*-alkenes, though less selectively than with the Lindlar catalyst.[692] Possible mechanisms have been proposed for the deoxygenation of ketones[693] and amides[694] to alkanes, for the hydrogenolysis of cyclohexene epoxides by palladium,[695] and for the reduction of phloroglucinol trioxime to 1,3,5-triaminobenzene.[696] Nickel boride has been shown to be more susceptible to poisoning than Raney nickel in hydrogenation reactions.[697]

Reductions and Oxidations of Biological Interest

Asymmetric reduction of α-keto-esters by Hantsch esters is catalysed by Grignard reagents;[698] bis(1,4-dihydropyridine)-zinc and -magnesium complexes are both stereo- and regio-selective in the reduction of aromatic nitrogen heterocycles.[699] α-Nitrosulphones are reduced to nitro compounds by *N*-benzyl-1,4-dihydronicotinamide by a non-chain free-radical process involving one-electron transfer;[700] in contrast, the reductions of 5-nitroisoquinolinium cations with 1,4-dihydronicotinamides are probably hydric in nature.[701] Strong evidence has been provided that reductions of *N*-methylacridinium ions[702,703] and trifluoroacetophenones[704] involve one-electron-transfer processes. A ruthenium-complex-mediated photoreduction of olefins with 1-benzyl-1,4-dihydronicotinamide has been used as a probe for electron-transfer processes with NAD(P)H-mode compounds.[705] The reduction of hexachloroacetone in chloroform is much enhanced in the presence of 1,6-bis(1-benzyldihydronicotinamide)hexane due to intramolecular electronic interactions of a charge-transfer character;[706] high enantioselectivity has been achieved in asymmetric reductions with 1,4-dihydropyridines contained in chiral macrocycles[707] and with chiral bis(NADH) model compounds.[708]

Among other model studies on NADH compounds,[709–714] isotope studies indicate that hydrogen transfer is rate-determining in the reductions of 2-acylpyridines by 1-benzyl-1,4-dihydronicotinamide.[715] The first reactive polymer-bound NADH model has been synthesized.[716] Many model studies on the mechanisms of action of flavins have been reported.[717–729] There have been several examples of the use of microbial stereo-differentiating reductions of carbonyl compounds.[730–735] Many mechanistic studies on oxidation–reduction reactions in enzymic systems have been reported.[736–772]

References

1 Ramesh, S., Mahapatro, S. N., Liu, J. H., and Rocek, J., *J. Am. Chem. Soc.*, **103,** 5172 (1981).

2 Panigrahi, G. P., and Mahapatro, D. D., *Indian J. Chem.*, **19A,** 579 (1980); *Chem. Abs.*, **93,** 203718 (1980).

3 Mueller, P., and Blanc, J., *Helv. Chim. Acta*, **63,** 1759 (1980).

4 Oki, M., and Saito, R., *Chem. Lett.*, **1981,** 649.

5 Antonioletti, R. D'Auria, M., Piancatelli, G., and Scettri, A., *Tetrahedron Lett.*, **22,** 1041 (1981).

6 Frechet, J. M. J., Darling, P., and Farrall, M. J., *J. Org. Chem.*, **46,** 1728 (1981).

7 Handlir, K., Holecek, J., and Nadvornik, M., *Proc. Conf. Coord. Chem. 8th*, **1980,** 115; *Chem. Abs.*, **95,** 79665 (1981).

8 Saran, N. K., Dash, M. N., and Acharya, R. C., *Acta Cienc. Indica,* [*Ser.*] *Chem.*, **6,** 56 (1980); *Chem. Abs.*, **94,** 3589 (1981).

9 Gupta, K. D., and Guha, D., *J. Inst. Chem.* (*India*), **53,** 17 (1981); *Chem. Abs.*, **95,** 23913 (1981).

10 Vasil'eva, V. N., Vasil'ev, V. P., Dmitrieva, N. G., Raskova, O. G., Poletaeva, N. K., and Ryabikova, V. N., *Vopr. Kinet. i Kataliza, Ivanovo*, **1979,** 36; *Chem. Abs.*, **95,** 79698 (1981).

11 Baliah, V., and Satyanarayana, P. V. V., *Indian J. Chem.*, **19B,** 620 (1980); *Chem. Abs.*, **94,** 29964 (1981).

[12] Baliah, V., and Satyanarayana, P. V. V., *Indian J. Chem.*, **19B,** 619 (1980); *Chem. Abs*, **94,** 14847 (1981).
[13] Sherstyuk, V. P., Koshechko, V. G., and Atamanyuk, V. Yu., *Zh. Obshch. Khim.*, **50,** 2153 (1980); *Chem. Abs.*, **94,** 46497 (1981).
[14] Willis, J. P., Gogins, K. A. Z., and Miller, L. L., *J. Org. Chem.*, **46,** 3215 (1981).
[15] Lin, T.-Y., *J. Chin. Chem. Soc. (Taipei)*, **28,** 149 (1980); *Chem. Abs.*, **95,** 131993 (1981).
[16] Mehta, S. B., and Vynas, D. N., *J. Inst. Chem. (India)*, **52,** 179 (1980); *Chem. Abs.*, **94,** 64842 (1981).
[17] Rudakov, E. S., Tishchenko, N. A., and Lutsyk, A. I., *Dokl. Akad. Nauk SSSR*, **252,** 893 (1980); *Chem. Abs.*, **93,** 203707 (1980).
[18] Arzamaskova, L. N., Romanenko, A. V., and Ermakov, Yu. I., *React. Kinet. Catal. Lett.*, **13,** 395 (1980); *Chem. Abs.*, **93,** 238312 (1980).
[19] Arzamaskova, L. N., Romanenko, A. V., and Ermakov, Yu. I., *React. Kinet. Catal. Lett.*, **13,** 391 (1980); *Chem. Abs.*, **93,** 220164 (1980).
[20] Baldea, I., and Munteanu, L., *Stud. Univ. Babes-Bolyai, [Ser.] Chem.*, **25,** 24 (1980); *Chem. Abs.*, **94,** 173990 (1981).
[21] Banas, B., *Inorg. Chim. Acta*, **53,** L13 (1981).
[22] Ansari, A. H., *J. Sci. Res. (Bhopal, India)*, **2,** 43 (1980); *Chem. Abs.*, **94,** 14866 (1981).
[23] Chattopadhyay, S., *Indian J. Chem.*, **20(B),** 242 (1981); *Chem. Abs.*, **95,** 41929 (1981).
[24] Petrusi, S., Lamartina, L., Migliara, O., and Spiro, V., *J. Chem. Soc., Perkins Trans. 1*, **1981,** 2642.
[25] Lee, D. G., (*Open Court Publ. Co.: La Salle, Ill.*), 1980; *Chem. Abs.*, **94,** 83123 (1981).
[26] Lee, D. G., and Srinivasan, N. S., *Can. J. Chem.*, **59,** 2146 (1981).
[27] Wolve, S., Ingold, C. F., and Lemieux, R. U., *J. Am. Chem. Soc.*, **103,** 938 (1981).
[28] Freeman, F., Fuselier, C. O., Armstead, C. R., Dalton, C. E., Davidson, P. A., Karchefski, E. M., Krochman, D. E., Johnson, M. N., and Jones, N. K., *J. Am. Chem. Soc.*, **103,** 1154 (1981).
[29] Simandi, L., and Jaky, M., *Magy. Kem. Foly.*, **86,** 263 (1980); *Chem. Abs.*, **93,** 185441 (1980).
[30] Wolfe, S., and Ingold, C. F., *J. Am. Chem. Soc.*, **103,** 940 (1981).
[31] Inoue, K., Noguchi, H., Hidai, M., and Uchida, Y., *Yukagaku*, **29,** 397 (1980); *Chem. Abs.*, **94,** 102469 (1981).
[32] Bobkov, V. N., *Khim. i Biokhim. Okislenie Sistem, Soderzhashch. D-elementy, Chelyabinsk*, **1979,** 31; *Chem. Abs.*, **94,** 46501 (1981).
[33] Lee, D. G., and Sebastian, C. F., *Can. J. Chem.*, **59,** 2776 (1981).
[34] Lee, D. G., and Sebastian, C. F., *Can. J. Chem.*, **59,** 2780 (1981).
[35] Isaacs, N. S., and Heremans, K. A. H., *Tetrahedron Lett.*, **22,** 4759 (1981).
[36] Noureldin, N. A., and Lee, D. G., *Tetrahedron Lett.*, **22,** 4889 (1981).
[37] Menger, F. M., and Lee, C., *Tetrahedron Lett.*, **22,** 1655 (1981).
[38] Purohit, R., Ameta, S. C., and Chowdry, H. C., *Z. Phys. Chem. (Leipzig)*, **262,** 737 (1981).
[39] Zielinski, M., *Radiochem. Radioanal. Lett.*, **47,** 193 (1981); *Chem. Abs.*, **95,** 96593 (1981).
[40] Ameta, S. C., Pande, P. N., Gupta, H. L., and Chowdry, H. C., *Z. Phys. Chem. (Leipzig)*, **261,** 1222 (1980).
[41] Kato, T., Ogawa, S., and Ito, I., *Tetrahedron Lett.*, **22,** 3205 (1981).
[42] Pillay, M. K., and Thirunavukkarasu, A., *Indian J. Chem.*, **20B,** 583 (1981); *Chem. Abs.*, **95,** 149588 (1981).
[43] Kudesia, V., *Acta Cienc. Indica, [Ser.] Chem.*, **6,** 119 (1980); *Chem. Abs.*, **94,** 191448 (1981).
[44] Chaintreau, A., Adrian, G., and Couturier, D., *J. Org. Chem.*, **46,** 4562 (1981).
[45] Ahmad, F., Baswani, V. S., and Kumar, S., *J. Indian Chem. Soc.*, **58,** 162 (1981); *Chem. Abs.*, **95,** 6151 (1981).
[46] Rao, V. V., Sethuram, B., and Rao, T. N., *Z. Phys. Chem. (Leipzig)*, **261,** 1171 (1980).
[47] Nagori, R. R., Mehta, M., and Mehrotra, R. N., *J. Chem. Soc., Dalton Trans.*, **1981,** 581.
[48] Bruce, J. M., Fitzjohn, S., and Pardasani, R. T., *J. Chem. Res. (S)*, **1981,** 252.
[49] Speier, G., and Tyeklar, Z., *J. Mol. Catal.*, **9,** 233 (1980); *Chem. Abs.*, **94,** 120569 (1981).
[50] Demmin, T. R., Swerdloff, M. D., and Rogic, M. M., *J. Am. Chem. Soc.*, **103,** 5795 (1981).
[51] Demmin, T. R., and Rogic, M. M., *J. Org. Chem.*, **45,** 4210 (1980).
[52] Tsuruya, S., Kuse, T., Masai, M., and Imamura, S., *J. Mol. Catal.*, **10,** 285 (1981); *Chem. Abs.*, **95,** 6048 (1981).
[53] Bakac, A., and Espenson, J. H., *J. Am. Chem. Soc.*, **103,** 2721 (1981).
[54] Krupenskii, V. I., *Izv. Vyssh. Uchebn, Zaved., Khim. Khim. Tekhnol.*, **24,** 553 (1981); *Chem. Abs.*, **95,** 114468 (1981).
[55] Krupenskii, V. I., *Izv. Vyssh. Uchebn. Zaved., Khim. Khim Tekhnol.* **23,** 676 (1980); *Chem. Abs.*, **93,** 238349 (1980).
[56] Kumar, A., *J. Am. Chem. Soc.*, **103,** 5179 (1981).
[57] Ahmad, F., and Baswani, V. S., *Indian J. Chem.*, **19A,** 423 (1980); *Chem. Abs.*, **93,** 185428 (1980).

58 Ahmad, F., and Baswani, V. S., *Int. J. Chem. Kinet.*, **13,** 565 (1981).
59 Baiocchi, C., and Mentasti, E., *Transition Met. Chem. (Weinheim, Ger.)*, **5,** 259 (1980); *Chem. Abs.*, **94,** 46480 (1981).
60 Rao, P. J. P., Sethuram, B., and Rao, T. N., *Indian J. Chem.*, **20A,** 733 (1981); *Chem. Abs.*, **95,** 131995 (1981).
61 Witkiewicz, P. L., and Shaw, C. F., *J. Chem. Soc., Chem. Commun.*, **1981,** 1111.
62 Espenson, J. H., and Bakac, A., *J. Am. Chem. Soc.*, **103,** 2728 (1981).
63 Möhrie, H., Tröster, G., Linden, M., Mootz, D., and Wunderlich, H., *Tetrahedron*, **37,** 2881 (1981).
64 Barluenga, J., Alonso-Cires, L., and Asensio, G., *Tetrahedron Lett.*, **22,** 2239 (1981).
65 Barluenga, J., Aznar, F., and Liz, R., *J. Chem. Soc., Chem. Commun.*, **1981,** 1181.
66 Forcellese, M. L., Camerini, E., Ruffini, B., and Mincione, E., *J. Org. Chem.*, **46,** 3326 (1981).
67 Yamada, Y., Nakamura, S., Iguchi, K., and Hosaka, K., *Tetrahedron Lett.*, **22,** 1355 (1981).
68 Melichercik, M., and Treindl, L., *Chem. Zvesti*, **34,** 310 (1980); *Chem. Abs.*, **93,** 185439 (1980).
69 Taylor, E. C., Andrade, J. G., Rall, G. J. H., Steliou, K., Jagdmann, G. E., and McKillop, A., *J. Org. Chem.*, **46,** 3078 (1981).
70 Uemura, S., Miyoshi, H., Tabata, A., and Okano, M., *Tetrahedron*, **37,** 291 (1981).
71 Strasak, M., and Majer, J., *Collect. Czech. Chem. Commun.*, **46,** 693 (1981).
72 Crouse, D. J., Wheeler, M. M., Goemann, M., Tobin, P. S., Basu, S. K., and Wheeler, D. M. S., *J. Org. Chem.*, **46,** 1814 (1981).
73 Ryabov, A. D., Deiko, S. A., Yatsimirsky, A. K., and Berezin, I. V., *Tetrahedron Lett.*, **22,** 3793 (1981).
74 Noda, H., Niwa, M., and Yamamura, S., *Tetrahedron Lett.*, **22,** 3247 (1981).
75 Lau, K. S. Y., and Basiulis, D. I., *Tetrahedron Lett.*, **22,** 1175 (1981).
76 Butler, R. N., Morris, G. J., and O'Donohue, A. M., *J. Chem. Res. (S)*, **1981,** 61.
77 Banthiya, U. S., Joshi, B. C., and Gupta, Y. K., *Indian J. Chem.*, **20A,** 43 (1981); *Chem. Abs.*, **95,** 41907 (1981).
78 Radhakrishnamurti, P. S., and Pati, S. N., *Indian J. Chem.*, **19A,** 980 (1980); *Chem. Abs.*, **94,** 102528 (1981).
79 Farage, V. J., and Janjic, D., *React. Kinet. Catal. Lett.*, **15,** 487 (1980); *Chem. Abs.*, **95,** 60931 (1981).
80 Edelson, D., *Int. J. Chem. Kinet.*, **13,** 1175 (1981).
81 D'Alba, F., and Serravalle, G., *J. Chim. Phys. Phys.-Chim. Biol.*, **78,** 131 (1981); *Chem. Abs.*, **95,** 6139 (1981).
82 Kovalenko, A. S., Tikhonova, L. P., and Yatsimirskii, K. B., *Teor. Eksp. Khim.*, **17,** 493 (1981); *Chem. Abs.*, **95,** 149615 (1981).
83 Farage, V. J., and Janjic, D., *Chimia*, **34,** 342 (1980); *Chem. Abs.*, **93,** 203727 (1980).
84 Zueva, T. S., and Sipershtein, I. N., *Teor. Eksp. Khim.*, **17,** 277 (1981); *Chem. Abs.*, **95,** 23922 (1981).
85 Treindl, L., and Kaplan, P., *Chem. Zvesti*, **35,** 145 (1981); *Chem. Abs.*, **95,** 23924 (1981).
86 Melichercik, M., and Treindl, L., *Chem. Zvesti*, **35,** 153 (1981); *Chem. Abs.*, **95,** 41962 (1981).
87 Suprun, V. Ya., and Mokryi, E. N., *Visn. L'viv. Politekh. Inst.*, **139,** 142 (1980); *Chem. Abs.*, **94,** 138943 (1981).
88 Baciocchi, E., Rol, C., and Mandolini, L., *J. Am. Chem. Soc.*, **102,** 7597 (1980).
89 Balasubrramanian, V., and Robinson, C. H., *Tetrahedron Lett.*, **22,** 501 (1981).
90 Chorn, T. A., Giles, R. G. F., Mitchell, P. R. K., and Green, I. R., *J. Chem. Soc., Chem. Commun.*, **1981,** 534.
91 Yadav, R. L., and Verma, R. G., *J. Indian Chem. Soc.*, **58,** 768 (1981); *Chem. Abs.*, **95,** 132001 (1981).
92 Rao, G. V., Rao, Ch. N., and Saiprakash, P. K., *Kinet. Katal.*, **22,** 633 (1981); *Chem. Abs.*, **95,** 96675 (1981).
93 Calvaruso, G., Cavasino, F. P., and Sbriziolo, C., *Int. J. Chem. Kinet.*, **13,** 1029 (1981).
94 Calvaruso, G., Cavasino, F. P., and Sbriziolo, C., *Int. J. Chem. Kinet.*, **13,** 135 (1981).
95 Sinha, B. K., and Behera, G. B., *Indian J. Chem.*, **19A,** 583 (1980); *Chem. Abs.*, **93,** 20370 (1980).
96 Rao, B. M., *Z. Phys. Chem. (Leipzig)*, **261,** 1113 (1980).
97 Rao, M. A., Sethuram, B., and Rao, T. N., *Indian J. Chem.*, **19A,** 747 (1980); *Chem. Abs.*, **94,** 29874 (1981).
98 Kahan, B. H., and Sundaram, E. V., *J. Indian Chem. Soc.*, **57,** 1979 (1980); *Chem. Abs.*, **94,** 102502 (1981).
99 Deberitz, J., *Lect.-Hydride Symp., 3rd*, **1979,** 199; *Chem. Abs.*, **93,** 203395 (1980).
100 Mihailovic, M. L., Andrejevic, V., and Teodorovic, A. V., *Glas. Hem. Drus. Beograd*, **45,** 315 (1980); *Chem. Abs.*, **94,** 208250 (1981).
101 Mihailovic, M. L., Andrejevic, V., and Teodorovic, A. V., *Glas. Hem. Drus. Beograd*, **45,** 327 (1980); *Chem. Abs.*, **94,** 208251 (1981).
102 Galliani, G., and Rindone, B., *J. Chem. Res. (S)*, **1981,** 74.

[103] Atkinson, R. S., Malpass, J. R., and Woodthorpe, K. L., *J. Chem. Soc., Chem. Commun.*, **1981,** 160.
[104] Brocksom, T. J., Tercio, J., Ferreira, B., and Braga, A. L., *J. Chem. Res. (S)*., **1981,** 334.
[105] Rudenko, A. P., Zarubin, M. Y., and Barsheva, N. S., *Zh. Obshch. Khim.*, **50,** 1663 (1980); *Chem. Abs.*, **94,** 46303 (1981).
[106] Singh, V., Sharma, T. C., and Bokadia, M. M., *Indian J. Chem.*, **17B,** 644 (1979); *Chem. Abs.*, **93,** 167286 (1980).
[107] Barton, D. H. R., Motherwell, W. B., and Stobie, A., *J. Chem. Soc., Chem. Commun.*, **1981,** 1232.
[108] Adamcikova, L., Lucivjansky, P., and Treindl, L., *React. Kinet. Catal. Lett.*, **15,** 451 (1980); *Chem. Abs.*, **95,** 41830 (1981).
[109] Powell, M., and Sainsbury, M., *Tetrahedron Lett.*, **22,** 4751 (1981).
[110] Tsitini-Tsamis, M., Chaigneau, M., Likforman, J., and Hamon, M., ANALUSIS, **8,** 428 (1980); *Chem. Abs.*, **94,** 173953 (1981).
[111] Nath, N., and Dubey, R. C., *Colloid Polym. Sci.*, **258,** 944 (1980); *Chem. Abs.*, **94,** 173945 (1981).
[112] Selvaraj, K., Senthilnathan, V. P., and Ramalingam, K., *Indian J. Chem.*, **17A,** 589 (1979); *Chem. Abs.*, **93,** 185411 (1980).
[113] Pillai, G. C., Rajaram, J., and Kuriacose, J. C., *Rev. Roum. Chim.*, **26,** 21 (1981).
[114] Pillai, G. C., Rajaram, J., and Kuriacose, J., *Indian J. Chem.*, **19A,** 585 (1980); *Chem. Abs.*, **93,** 203721 (1980).
[115] Sapunov, V. N., Sharykin, V. G., Rubtsova, I. I., Litvintsev, I. Yu., and Lebedev, N. N., *Deposited Doc.*, **1980,** VINITI 814; *Chem. Abs.*, **94,** 120586 (1981).
[116] Sapunov, V. N., Sarykin, V. G., Vylyudnova, S. M., Litvintsev, I. Yu., and Lebedev, N. N., *Deposited Doc.*, **1980,** VINITI 815; *Chem. Abs.*, **94,** 120585 (1981).
[117] Sanghvi, Y. S., and Rao, A. S., *Indian J. Chem.*, **19B,** 608 (1980); *Chem. Abs.*, **94,** 46645 (1981).
[118] Kozikowski, A. P., Schmiesing, R. J., and Sorgi, K. L., *Tetrahedron Lett.*, 22, 2059 (1981).
[119] Bortolini, O., Di Furia, F., and Modena, G., *J. Am. Chem. Soc.*, **103,** 3924 (1981).
[120] Ledon, H. J., Durbut, P., and Varescon, F., *J. Am. Chem. Soc.*, **103,** 3601 (1981).
[121] Konishevskaya, G. A., Filippov, A. P., Sobchak, Yu., Zyulkovskii, Yu. Yu., Yatsimirskii, K. B., and Belousov, V. M., *Kinet. Katal.*, **21,** 933 (1980); *Chem. Abs.*, **94,** 46470 (1981).
[122] Kaneda, K., Kii, N., Jitsukawa, K., and Teranishi, S., *Tetrahedron Lett.*, **22,** 2595 (1981).
[123] Bortolini, O., Di Furia, F., Modena, G., Scardellato, C., and Scrimin, P., *J. Mol. Catal.*, **11,** 107 (1981); *Chem. Abs.*, **95,** 41948 (1981).
[124] Muetterties, E. L., *Inorg. Chim. Acta*, **50,** 1 (1981).
[125] Welsh, W. A., Otutakowski, J., and Henry, P. M., *Can. J. Chem.*, **59,** 697 (1981).
[126] Hirano, M., Nakamura, K., and Morimoto, T., *J. Chem. Soc., Perkin Trans. 2*, **1981,** 817.
[127] Okamoto, T., and Oka, S., *Tetrahedron Lett.*, **22,** 2191 (1981).
[128] Czytko, M. P., and Bub, G. K., *Ind. Eng. Chem. Prod. Res. Dev.*, **20,** 481 (1981); *Chem. Abs.*, **95,** 79700 (1981).
[129] Perumal, S. I., and Barker, M. W., *Indian J. Chem.*, **19B,** 508 (1980); *Chem. Abs.*, **93,** 220170 (1980).
[130] Frazier, R. H., and Harlow, R. L., *J. Org. Chem.*, **45,** 5408 (1980).
[131] Jempty, T. C., Gogins, K. A. Z., Mazur, Y., and Miller, L. L. *J. Org. Chem.*, **46,** 4545 (1981).
[132] Walling, C., and Humphreys, R. W. R., *J. Org. Chem.*, **46,** 1260 (1981).
[133] Kohda, A., Ueda, K., and Sato, T., *J. Org. Chem.*, **46,** 509 (1981).
[134] Frank, M. S., and Rao, P. V. K., *Indian J. Chem.*, **17A,** 632 (1979); *Chem. Abs.*, **93,** 167264 (1980).
[135] Frank, M. S., and Rao, P. V. K., *Indian J. Chem.*, **19A,** 538 (1980); *Chem. Abs.*, **93,** 220175 (1980).
[136] Kimura, M., Yamabe, S., and Minato, T., *Bull. Chem. Soc. Jpn.*, **54,** 1699 (1981).
[137] Feringa, B., and Wynberg, H., *J. Org. Chem.*, **46,** 2547 (1981).
[138] Tashiro, M., Yoshiya, H., and Fukata, G., *J. Org. Chem.*, **46,** 3784 (1981).
[139] Kapoor, R. C., Potter, P. C., Kachhwaha, O. P., and Sinha, B. P., *Indian J. Chem.*, **20A,** 89 (1981); *Chem. Abs.*, **95,** 41908 (1981).
[140] Singh, R. N., and Tikoo, P. K., *Indian J. Chem.*, **19A,** 1210 (1981); *Chem. Abs.*, **94,** 173991 (1981).
[141] Singh, B., Singh, B. B., and Singh, S., *J. Indian Chem. Soc.*, **57,** 662 (1980); *Chem. Abs.*, **93,** 238333 (1980).
[142] Boiko, T. S., Grishin, O. M., and Yasnikov, A. A., *Ukr. Khim. Zh. (Russ. Ed.)*, **47,** 848 (1981); *Chem. Abs.*, **95,** 149606 (1981).
[143] Jarrar, A. A., El-Zaru, R., and Mbarak, M. S., *Dirasat, [Ser.]: Nat. Sci. (Univ. Jordon)*, **6,** 7 (1979); *Chem. Abs.*, **95,** 23889 (1981).
[144] Gupta, K. C., Sharma, A., and Misra, V. D., *Tetrahedron*, **37,** 2887 (1981).
[145] Matsumoto, M., and Ito, S., *J. Chem. Soc., Chem. Commun.*, **1981,** 907.
[146] Muller, P., and Godoy, J., *Tetrahedron Lett.*, **22,** 2361 (1981).
[147] Tomioka, J., Takai, K., Oshima, K., and Nozaki, H., *Tetrahedron Lett.*, **22,** 1605 (1981).
[148] Grigg, R., Mitchell, T. R. B., Sutthivaiyakit, S., and Tongpenyai, N., *Tetrahedron Lett.*, **22,** 4107 (1981).

[149] Radhakrishnamurti, P. S., and Misra, P. C., *Indian J. Chem.*, **19A,** 427 (1980); *Chem. Abs.*, **93,** 185429 (1980).
[150] Radhakrishnamurti, P. S., and Panda, H. P., *Gazz. Chim. Ital.*, **109,** 637 (1979); *Chem. Abs.*, **93,** 185425 (1980).
[151] Rudakov, E. S., Tret'yakov, V. P., Min'ko, L. A., and Tishchenko, N. A., *React. Kinet. Catal. Lett.*, **16,** 77 (1981); *Chem. Abs.*, **95,** 61080 (1981).
[152] Carlsen, P. H. J., Katsuki, T., Martin, V. S., and Sharpless, K. B., *J. Org. Chem.*, **46,** 3936 (1981).
[153] Hartman, R. F., and Rose, S. D., *J. Org. Chem.*, **46,** 4340 (1981).
[154] McCormick, J. P., Tomasik, W., and Johnson, M. W., *Tetrahedron Lett.*, **22,** 607 (1981).
[155] Rao, M. P., Sethuram, B., and Rao, T. N., *Indian J. Chem.*, **19A,** 582 (1980); *Chem. Abs.*, **93,** 203719 (1980).
[156] Rao, M. P., Sethuram, B., and Rao, T. N., *Proc. Natl. Symp. Catal., 4th*, **1978,** 466; *Chem. Abs.*, **94,** 46487 (1981).
[157] Singh, B., Singh, B. B., and Singh, R. P., *J. Inorg. Nucl. Chem.*, **43,** 1283 (1981); *Chem. Abs.*, **95,** 114448 (1981).
[158] Singh, B., Rai, J. P., Shukla, R. K., and Krishna, B., *Proc. Natl. Symp. Catal., 4th*, **1978,** 313; *Chem. Abs.*, **94,** 46485 (1981).
[159] Anand, B. H., and Menghani, G. D., *J. Indian Chem. Soc.*, **58,** 374 (1981); *Chem. Abs.*, **95,** 79706 (1981).
[160] Resch, J. F., and Meinwald, J., *Tetrahedron Lett.*, **22,** 3159 (1981).
[161] Kuznetsova, N. I., Likholobov, V. A., and Ermakov, Yu. I., *Kinet. Katal.*, **22,** 157 (1981); *Chem. Abs.*, **94,** 208121 (1981).
[162] Yatsimirskii, A. K., Ryabov, A. D., Zagorodnikov, V. P., Sakodinskaya, I. K., Kavetskaya, O. I., and Berezin, I. V., *Inorg. Chim. Acta*, **48,** 163 (1981); *Chem. Abs.*, **95,** 41930 (1981).
[163] Yatsimirskii, A. K., and Berezin, I. V., *Dokl. Akad. Nauk SSSR*, **255,** 1193 (1980); *Chem. Abs.*, **94,** 173982 (1981).
[164] Starchevskii, M. K., Vargaftik, M. N., and Moiseev, I. I., *Kinet. Katal.*, **21,** 1451 (1980); *Chem. Abs.*, **94,** 120576 (1981).
[165] Sengupta, K. K., Chatterjee, U., Maiti, S., Sen, P. K., and Bhattacharya, B., *J. Inorg. Nucl. Chem.*, **42,** 1177 (1980); *Chem. Abs.*, **93,** 238374 (1980).
[166] Chatterjee, U., Maiti, S., and Sengupta, K. K., *Z. Phys. Chem. (Leipzig)*, **261,** 1231 (1980).
[167] Yoneda, F., *Yuki Gosei Kagaku Kyokaishi*, **38,** 679 (1980); *Chem. Abs.*, **93,** 203387 (1980).
[168] Atamanyuk, V. Yu., Koshechko, V. G., and Pokhodenko, V. D., *Zh. Org. Khim.*, **16,** 1901 (1980); *Chem. Abs.*, **94,** 29870 (1981).
[169] Phelan, K. G., and Stedman, G., *J. Chem. Res. (S)*, **1981,** 224.
[170] Bazanova, G. V., and Stotskii, A. A., *Zh. Org. Khim.*, **16,** 1679 (1980); *Chem. Abs.*, **94,** 29851 (1981).
[171] Bazanova, G. V., and Stotskii, A. A., *Zh. Org. Khim.*, **16,** 1674 (1980); *Chem. Abs.*, **94,** 3478 (1981).
[172] Crivello, J. V., *J. Org. Chem.*, **46,** 3056 (1981).
[173] Andrews, M. A., and Kelly, K. P., *J. Am. Chem. Soc.*, **103,** 2894 (1981).
[174] Mitsunobu, O., and Yoshida, N., *Tetrahedron Lett.*, **22,** 2295 (1981).
[175] Okumura, Y., Hase, N., Chikyu, H., Mori, N., and Szuki, T., *Rep. Fac. Sci., Shizuoka Univ.*, **14,** 27 (1980); *Chem. Abs.*, **93,** 220059 (1980).
[176] Trost, B. M., and Liu, G., *J. Org. Chem.*, **46,** 4617 (1981).
[177] Liso, G., Trapani, G., Latrofa, A., and Marchini, P., *J. Heterocycl. Chem.*, **18,** 279 (1981).
[178] Masuyama, Y., Tsuhako, A., and Kurusu, Y., *Tetrahedron Lett.*, **22,** 3973 (1981).
[179] Maas, G., and Stang, P. J., *J. Org. Chem.*, **46,** 1606 (1981).
[180] Castedo, L., Puga, A., Saa, J. M., and Suau, R., *Tetrahedron Lett.*, **22,** 2233 (1981).
[181] Trost, B. M., and Curran, D. P., *Tetrahedron Lett.*, **22,** 1287 (1981).
[182] Gallopo, A. R., and Edwards, J. O., *J. Org. Chem.*, **46,** 1684 (1981).
[183] Giordano, C., Belli, A., Citterio, A., and Minisci, F., *J. Chem. Soc., Perkin Trans. 1*, **1981,** 1574.
[184] Mkhitaryan, R. P., Gukasyan, T. T., and Beileryan, N. M., *Arm. Khim. Zh.*, **33,** 269 (1980); *Chem. Abs.*, **93,** 203699 (1980).
[185] Srivastava, S. P., Sharma, R. G., Singhal, S. K., Singh, K. P., and Sandilya, R. K., *Acta Cienc. Indica, [Ser.] Chem.*, **6,** 111 (1980); *Chem. Abs.*, **94,** 3479 (1981).
[186] Srivastava, S. P., and Gupta, V. K., *Natl. Acad. Sci. Lett. (India)*, **3,** 25 (1980); *Chem. Abs.*, **93,** 238332 (1980).
[187] Srivastava, S. P., and Gupta, V. K., *Oxid. Commun.*, **1,** 251 (1980); *Chem. Abs.*, **95,** 23904 (1981).
[188] Gupta, V. K., and Srivastava, S. P., *Indian Chem. Manuf.*, **18,** 1 (1980); *Chem. Abs.*, **94,** 173951 (1981).
[189] Gevorkyan, M. G., Israelyan, S. Zh., Kishoyan, V. S., and Beileryan, N. M., *Arm. Khim. Zh.*, **33,** 447 (1980); *Chem. Abs.*, **93,** 238320 (1980).
[190] Rasmussen, J. K., and Smith, H. K., *J. Am. Chem. Soc.*, **103,** 730 (1981).

191 Bhatt, M. V., and Thirumalai Perumal, P., *Tetrahedron Lett.*, **22,** 2605 (1981).
192 Agarwal, S. C., and Saxena, L. K., *J. Inorg. Nucl. Chem.*, **42,** 932 (1980); *Chem. Abs.*, **93,** 167314 (1980).
193 Agarwal, S. C., Singh, M., and Agrawal, V. B., *Chem. Era*, **15,** 13 (1979); *Chem. Abs.*, **93,** 238316 (1980).
194 Srivastava, S. P., Gupta, V. K., Gupta, J. C., and Maheshwari, M. K., *J. Indian Chem. Soc.*, **57,** 797 (1980); *Chem. Abs.*, **93,** 238365 (1980).
195 Srivastava, S. P., Kumar, A., Mittal, A. K., and Gupta, V. K., *Oxid. Commun.*, **1,** 265 (1980); *Chem. Abs.*, **95,** 41919 (1981).
196 Ogibin, Yu. N., Velibekova, D. S., Katsin, M. I., Troyanskii, E. I., and Nikishin, G. I., *Izv. Akad. Nauk SSSR, Ser. Khim.*, **1981,** 149; *Chem. Abs.*, **94,** 173976 (1981).
197 Qaiser, A., and Nand, K. C., *J. Sci. Res.* (*Bhopal, India*), **2,** 121 (1980); *Chem. Abs.*, **94,** 83339 (1981).
198 Singh, H., Prasad, M., Saksena, S. C., and Kansak, B. C., *J. Chin. Chem. Soc.* (*Taipei*), **27,** 119 (1980); *Chem. Abs.*, **94,** 14839 (1981).
199 Jagdale, M. H., and Sankpal, S. G., *J. Shivaji Univ., Sci.*, **17,** 73 (1977); *Chem. Abs.*, **93,** 185449 (1980).
200 Jagdale, M. H., Sankpal, S. G., and Patil, N., *J. Shivaji Univ., Sci.*, **17,** 67 (1977); *Chem. Abs.*, **94,** 3467 (19-1).
201 Singh, H., Chauhan, K. S., and Rathi, U. K., *J. Indian Chem. Soc.*, **57,** 809 (1980); *Chem. Abs.*, **93,** 203729 (1980).
202 Back, T. G., *J. Chem. Soc., Chem. Commun.*, **1981,** 530.
203 Back, T. G., Collins, S., and Kerr, R. G., *J. Org. Chem.*, **46,** 1564 (1981).
204 Back, T. G., *J. Org. Chem.*, **46,** 1442 (1981).
205 Barton, D. H., Brewster, A. G., Ley, S. V., Read, C. M., and Rosenfeld, M. N., *J. Chem. Soc., Perkin Trans. 1*, **1981,** 1473.
206 Haruna, M., and Ito, K., *J. Chem. Soc., Chem. Commun.*, **1981,** 483.
207 Shimizu, M., Urabe, H., and Kuwajima, I., *Tetrahedron Lett.*, **22,** 2183 (1981).
208 Abatjoglou, A. G., and Bryant, D. R., *Tetrahedron Lett.*, **22,** 2051 (1981).
209 Ley, S. V., Meerholz, C. A., and Barton, D. H. R., *Tetrahedron Lett.*, **37** (Suppl. 9), 213 (1981).
210 Bergman, J., and Engman, L., *J. Am. Chem. Soc.*, **103,** 5196 (1981).
211 Gebeyehu, G., and McNelis, E., *J. Org. Chem.*, **45,** 4280 (1980).
212 Doi, J. T., Musker, W. K., de Leeuw, D. L., and Hirschon, A. S., *J. Org. Chem.*, **46,** 1239 (1981).
213 Doi, J. T., and Musker, W. K., *J. Am. Chem. Soc.*, **103,** 1159 (1981).
214 Hirschon, A. S., Beller, J. D., Olmstead, M. M., Doi, J. T., and Musker, W. K., *Tetrahedron Lett.*, **22,** 1195 (1981).
215 Ueno, Y., Miyano, T., and Okawara, M., *Bull. Chem. Soc. Jpn.*, **53,** 3615 (1980).
216 Bal, B. S., Childers, W. E., and Pinnick, J. W., *Tetrahedron*, **37,** 2091 (1981).
217 Furukawa, N., Akutagawa, K., Yoshimura, T., and Oae, S., *Tetrahedron Lett.*, **22,** 3989 (1981).
218 Ogata, Y., Kimura, M., and Kondo, Y., *Bull. Chem. Soc. Jpn.*, **54,** 2057 (1981).
219 Kudesia, V. P., and Sharma, N. K., *React. Kinet. Catal. Lett.*, **16,** 49 (1981); *Chem. Abs.*, **95,** 79711.
220 Tapodi, H. P., Shanker, R., and Bakore, G. V., *Indian J. Chem.*, **19A,** 330 (1980); *Chem. Abs.*, **93,** 185426 (1980).
221 Kaushik, D., Malkani, R. K., and Bakore, G. V., *J. Indian Chem. Soc.*, **57,** 1074 (1980); *Chem. Abs.*, **94,** 83334 (1981).
222 Manglik, A., Sharma, S. K., and Kudesia, V. P., *React. Kinet. Catal. Lett.*, **15,** 467 (1980); *Chem. Abs.*, **95,** 61066 (1981).
223 Jander, J., and Reich, K. P., *Z. Anorg. Allgem. Chem.*, **465,** 41 (1980); *Chem. Abs.*, **94,** 29732 (1981).
224 Humffray, A. A., and Imberger, H. E., *J. Chem. Soc., Perkin Trans. 2*, **1981,** 382.
225 Moriarty, R. M., Gupta, S. C., Hu, H., Berenschot, D. R., and White, K. B., *J. Am. Chem. Soc.*, **103,** 686 (1981).
226 Moriarty, R. M., Hu, H., and Gupta, S. C., *Tetrahedron Lett.*, **22,** 1283 (1981).
227 Pati, S. C., and Dev, B. R., *Z. Phys. Chem.* (*Leipzig*), **262,** 176 (1981).
228 Pati, S. C., and Mahapatro, R. C., *Indian J. Chem.*, **19A,** 1126 (1980); *Chem. Abs.*, **94,** 173988 (1981).
229 Barton, D. H. R., Morzycki, J. W., Motherwell, W. B., and Ley, S. V., *J. Chem. Soc., Chem. Commun.*, **1981,** 1044.
230 Beebe, T. R., Hii, P., and Reinking, P., *J. Org. Chem.*, **46,** 1927 (1981).
231 Bakuzis, P., and Bakuzis, M. L. F., *J. Org. Chem.*, **46,** 235 (1981).
232 Mukherjee, J., and Banerji, K. K., *J. Org. Chem.*, **46,** 2323 (1981).
233 Negi, S. C., Bhatia, I., and Banerji, K. K., *J. Chem. Res.* (*S*), **1981,** 360.
234 Radhakrishnamurti, P. S., and Sahu, N. C., *Indian J. Chem.*, **20A,** 269 (1981); *Chem. Abs.*, **95,** 41927 (1981).
235 Gopalakrishnan, G., Pai, B. R., and Venkatasubramanian, N., *Indian J. Chem.*, **19B,** 293 (1980); *Chem. Abs.*, **93,** 238380 (1980).

[236] Damin, B., Forestiere, A., Garapon, J., and Sillion, B., *J. Org. Chem.*, **46,** 3552 (1981).
[237] Uma, K. V., and Mayanna, S. M., *Int. J. Chem. Kinet.*, **12,** 861 (1980); *Chem. Abs.*, **94,** 14871 (1981).
[238] Swamy, R., Yathirajan, H. S., and Mahadevappa, D. S., *Rev. Roum. Chim.*, **26,** 565 (1981).
[239] Mahadevappa, D. S., Jadhav, M. B., and Naidu, H. M. K., *React. Kinet. Catal. Lett.*, **14,** 15 (1980); *Chem. Abs.*, **94,** 14842 (1981).
[240] Jadhav, M. B., Naidu, H. M. K., and Mahadevappa, D. S., *J. Indian Chem. Soc.*, **57,** 693 (1980); *Chem. Abs.*, **93,** 238343 (1980).
[241] Mahadevappa, D. S., Jadhav, M. B., and Naidu, H. M. K., *J. Indian Chem. Soc.*, **58,** 454 (1981); *Chem. Abs.*, **95,** 41959 (1981).
[242] Ramaiah, A. K., and Rao, P. V. K., *Indian J. Chem.*, **19A,** 1120 (1980); *Chem. Abs.*, **94,** 173987 (1981).
[243] Lakshmi, V., and Sundaram, E. V., *Indian J. Chem.*, **20A,** 300 (1981); *Chem. Abs.*, **95,** 6159 (1981).
[244] Manikyamba, P., Vijayalakshmi, and Sundaram, E. V., *J. Indian Chem. Soc.*, **58,** 574 (1981); *Chem. Abs.*, **95,** 79722 (1981).
[245] Vijayalakshmi, and Sundaram, E. V., *J. Indian Chem. Soc.*, **57,** 800 (1980); *Chem. Abs.*, **93,** 203728 (1980).
[246] Murthy, N. K., Rao, P. R., and Sundaram, E. V., *Indian J. Chem.*, **17A,** 630 (1979); *Chem. Abs.*, **93,** 185412 (1980).
[247] Anandan, S., and Gopalan, R., *Indian J. Chem.*, **17A,** 629 (1979); *Chem. Abs.*, **93,** 167263 (1980).
[248] Ramalingam, V., Srinivasan, S., and Subramanian, P. S., *Indian J. Chem.*, **19A,** 1012 (1980); *Chem. Abs.*, **94,** 138925 (1981).
[249] Radhakrishnamurti, P. S., and Sarangi, L. D., *Indian J. Chem.*, **20A,** 301 (1981); *Chem. Abs.*, **95,** 6160 (1981).
[250] Radhakrishnamurti, P. S., and Sarangi, L. D., *Indian J. Chem.*, **19A,** 1124 (1980); *Chem. Abs.*, **94,** 174114 (1981).
[251] Parimala, N., Lakshmi, V., and Sundaram, E. V., *J. Indian Chem. Soc.*, **58,** 667 (1981); *Chem. Abs.*, **95,** 131981 (1981).
[252] Habon, I., and Koros, E., *Ann. Univ. Sci. Budap. Rolando Eotvos Nominatae, Sect. Chim.*, **15,** 23 (1979); *Chem. Abs.*, **94,** 83341 (1981).
[253] Maselko, J., *Chem. Phys.*, **51,** 473 (1980); *Chem. Abs.*, **94,** 46492 (1981).
[254] Reddy, C. S., Lakshmi, V., and Sundaram, E. V., *Indian J. Chem.*, **19A,** 741 (1980); *Chem. Abs.*, **94,** 29872 (1981).
[255] Reddy, C. S., Lakshmi, V., and Sundaram, E. V., *Indian J. Chem.*, **19A,** 544(1980); *Chem. Abs.*, **93,** 203716 (1980).
[256] Rao, M. D. P., and Padmanabha, J., *Indian J. Chem.*, **19A,** 984 (1980); *Chem. Abs.*, **94,** 102529 (1981).
[257] Avico, U., Gagliardi, L., Ciranni-Signoretti, E., Amato, A., and Chiavarelli, S., *Cron. Chim.*, **59,** 15 (1979); *Chem. Abs.*, **95,** 23602 (1981).
[258] Takata, T., Kim, Y. H., and Oae, S., *Bull. Chem. Soc. Jpn.*, **54,** 1443 (1981).
[259] Bhatt, M. V., and Gopinathan, M. B., *Indian J. Chem.*, **20B,** 72 (1981); *Chem. Abs.*, **95,** 23886 (1981).
[260] Radhakrishnamurti, P. S., and Panda, H. P., *React. Kinet. Catal. Lett.*, **14,** 193 (1980); *Chem. Abs.*, **93,** 238368 (1980).
[261] Srivastava, S. P., Bhattacharjee, G., Gupta, V. K., and Pal, S., *Indian J. Chem.*, **19A,** 744 (1980); *Chem. Abs.*, **94,** 29873 (1981).
[262] Srivastava, S. P., Bhattacharjee, G., Pal, S., and Gupta, V. K., *React. Kinet. Catal. Lett.*, **14,** 219 (1980); *Chem. Abs.*, **94,** 29846 (1981).
[263] Srivastava, S. P., Bhattacharjee, G., Gupta, V. K., and Pal, S., *React. Kinet. Catal. Lett.*, **13,** 231 (1980); *Chem. Abs.*, **93,** 167268 (1980).
[264] Srivastava, S. P., Bhattacharjee, G., Kumar, A., and Pal, S., *Indian J. Chem.*, **19A,** 578 (1980); *Chem. Abs.*, **93,** 203717 (1980).
[265] Moore, R. E., Woolard, F. X., and Bartolini, G., *J. Am. Chem. Soc.*, **102,** 7370 (1980).
[266] Babler, J. H., and Invergo, B. J., *J. Org. Chem.*, **46,** 1937 (1981).
[267] Washburne, S. S., and Szendroi, R., *J. Org. Chem.*, **46,** 691 (1981).
[268] Hannon, S. J., and Traylor, T. G., *J. Org. Chem.*, **46,** 3645 (1981).
[269] Traylor, T. G., and Koermer, G. S., *J. Org. Chem.*, **46,** 3651 (1981).
[270] Talley, J. J., *Tetrahedron Lett.*, **22,** 823 (1981).
[271] Rykova, L. A., Kiprianova, L. A., and Gragerov, I. P., *Teor. Eksp. Khim.*, **17,** 545 (1981); *Chem. Abs.*, **95,** 149617 (1981).
[272] Rykova, L. A., Kiprianova, L. A., and Gragerov, I. P., *Teor. Eksp. Khim.*, **16,** 825 (1980); *Chem. Abs.*, **94,** 102514 (1981).
[273] Ward, R. S., Satyanarayana, P., and Gopala Rao, B. V., *Tetrahedron Lett.*, **22,** 3021 (1981).

274 Rastogi, M., Mishra, K. K., and Sinha, B. P., *Indian J. Chem.*, **20B,** 726 (1981); *Chem. Abs.*, **95,** 168177 (1981).
275 Cremer, D., *Angew. Chem. Int. Ed.*, **20,** 888 (1981).
276 Cremer, D., *J. Am. Chem. Soc.*, **103,** 3619 (1981).
277 Cremer, D., *J. Am. Chem. Soc.*, **103,** 3627 (1981).
278 Cremer, D., *J. Am. Chem. Soc.*, **103,** 3633 (1981).
279 Agropovich, J. W., and Gillies, C. W., *J. Am. Chem. Soc.*, **102,** 7572 (1980).
280 Martinez, R. I., Herron, J. T., and Huie, R. E., *J. Am. Chem. Soc.*, **103,** 3807 (1981).
281 Morin, L., Barillier, D., Strobel, M.-P., and Paquer, D., *Tetrahedron Lett.*, **22,** 2267 (1981).
282 Saito, K., Niki, E., and Kamiya, Y., *Bull. Chem. Soc. Jpn.*, **54,** 253 (1981).
283 Yamamoto, Y., Niki, E., and Kamiya, Y., *J. Org. Chem.*, **46,** 250 (1981).
284 Paquette, L. A., and Carr, R. V. C., *J. Am. Chem. Soc.*, **102,** 7553 (1980).
285 Miyata, T., and Arai, M., *Mizu Shori Gijutsu*, **22,** 145 (1981); *Chem. Abs.*, **95,** 23894 (1981).
286 Adeniji, S. A., Kerr, J. A., and Williams, M. R., *Int. J. Chem. Kinet.*, **13,** 209 (1981).
287 Sawaki, Y., Kato, H., and Ogata, Y., *J. Am. Chem. Soc.*, **103,** 3832 (1981).
288 Miura, M., Nojima, M., Kusabayashi, S., and Nagase, S., *J. Am. Chem. Soc.*, **103,** 1789 (1981).
289 Jenkins, J. A., and Mendenhall, G. D., *J. Org. Chem.*, **46,** 3997 (1981).
290 Ito, Y., and Matsuura, T., *J. Chem. Soc., Perkin Trans. 1*, **1981,** 1871.
291 Jayathirtha Rao, V., and Ramamurthy, V., *J. Chem. Soc., Chem. Commun.*, **1981,** 638.
292 Tyupalo, N. F., and Bernashevskii, N. V., *Dokl. Akad. Nauk SSSR*, **253,** 896 (1980); *Chem. Abs.*, **94,** 3477 (1981).
293 Yakobi, V. A., *Khimiya Elementoorgan. Soedin.*, (*Gor'kii*), **1979,** 104; *Chem. Abs.*, **93,** 185440 (1980).
294 Shereshovets, V. V., Komissarov, V. D., and Galimova, L. G., *Izv. Akad. Nauk SSSR, Ser. Khim.*, **1980,** 2632; *Chem. Abs.*, **94,** 102623 (1981).
295 Shereshovets, V. V., Shafikov, N. Ya., and Komissarov, V. D., *Kinet. Katal.*, **21,** 1596 (1980); *Chem. Abs.*, **94,** 102518 (1981).
296 Shereshovets, V. V., Shafikov, N. Ya., and Komissarov, V. D., *Zh. Fiz. Khim.*, **54,** 1288 (1980); *Chem. Abs.*, **93,** 167277 (1980).
297 Korotkova, N. P., Syroezhko, A. M., and Proskuryakov, V. A., *Zh. Prikl. Khim.* (*Leningrad*), **54,** 660 (1981); *Chem. Abs.*, **95,** 23907 (1981).
298 Brudnik, B. M., Komissarova, I. N., Kuramshin, E. M., Imashev, U. B., Zlotskii, S. S., and Rakhmankulov, D. L., *React. Kinet. Catal. Lett.*, **13,** 97 (1980); *Chem. Abs.*, **93,** 167269 (1980).
299 Popov, A. A., Blinov, N. N., Vorob'eva, N. S., Zaikov, G. E., and Karpova, S. G., *Kinet. Katal.*, **22,** 139 (1981); *Chem. Abs.*, **95,** 41906 (1981).
300 Krow, G. R., *Tetrahedron*, **37,** 2697 (1981).
301 Schwab, J. M., *J. Am. Chem. Soc.*, **103,** 1876 (1981).
302 Bialer, M., *Tetrahedron Lett.*, **22,** 2683 (1981).
303 Camps, F., Coll, J., Messeguer, A., and Pericas, M. A., *Tetrahedron Lett.*, **22,** 3895 (1981).
304 Sheldon, R. A., *J. Mol. Catal.*, **7,** 107 (1980); *Chem. Abs.*, **94,** 173732 (1981).
305 Lu, N.-C., Li, G.-F., Lu, Y.-Z., and Zhao, X.-J., *Hua Kung Hsueh Pao*, **1980,** 365; *Chem. Abs.*, **94,** 156037 (1981).
306 Davis, F. A., Abdul-Malik, N. F., Awad, S. B., and Harakal, M. E., *Tetrahedron Lett.*, **22,** 917 (1981).
307 Baumstark, A. L., Chrisope, D. R., and Landis, M. E., *J. Org. Chem.*, **46,** 1964 (1981).
308 Stark, C. J., *Tetrahedron Lett.*, **22,** 2089 (1981).
309 Curci, R., Fiorentino, M., Troisi, L., Edwards, J. O., and Pater, R. H., *J. Org. Chem.*, **45,** 4758 (1980).
310 Hata, Y., and Watanabe, M., *J. Org. Chem.*, **46,** 610 (1981).
311 Narula, A. S., *Tetrahedron Lett.*, **22,** 2017 (1981).
312 Anastasia, M., Allevi, P., Fiecchi, A., and Scala, A., *J. Org. Chem.*, **46,** 3265 (1981).
313 Tan, R. Y., Russell, R. A., and Warrener, R. N., *Aust. J. Chem.*, **34,** 421 (1981).
314 Rossiter, B. E., Katsuki, T., and Sharpless, K. B., *J. Am. Chem. Soc.*, **103,** 464 (1981).
315 Sayer, J. M., Yagi, H., Criosy-Delcey, M., and Jerina, D. M., *J. Am. Chem. Soc.*, **103,** 4970 (1981).
316 Lee, H., and Harvey, R. G., *Tetrahedron Lett.*, **22,** 1657 (1981).
317 Schneider, H.-J., Becker, N., and Philippi, K., *Chem. Ber.*, **114,** 1562 (1981).
318 Sapunov, V. N., Riko, E. A., and Litvintsev, I. Yu., *Kinet. Katal.*, **21,** 794 (1980); *Chem. Abs.*, **93,** 203691 (1980).
319 Stojanowa-Antoszczyn, M., Atanasow, P., and Zielinski, A. Z., *Chem. Stosow.*, **24,** 211 (1980); *Chem. Abs.*, **94,** 173949 (1981).
320 Badasyan, G. V., Gabrielyan, S. M., Kamalov, G. L., and Treger, Yu. A., *Arm. Khim. Zh.*, **33,** 789 (1980); *Chem. Abs.*, **94,** 83344 (1981).

[321] Tishchenko, I. G., and Revinski, I. F., *Vestsi Akad. Navuk BSSR, Ser. Khim. Navuk*, **1981,** 90; *Chem. Abs.*, **95,** 41897 (1981).
[322] Badasyan, G. V., Gabrielyan, S. M., Kamalov, G. L., and Treger, Yu. A., *Arm. Khim. Zh.*, **33,** 794 (1980); *Chem. Abs.*, **94,** 102526 (1981).
[323] Khante, R. N., and Chandalia, S. B., *Indian Chem. J.*, **15,** 33 (1981); *Chem. Abs.*, **95,** 79730 (1981).
[324] Morimoto, H., Hayashi, N., and Kawakaru, T., *Chem. Ber.*, **114,** 1963 (1981).
[325] Speier, G., and Tyeklar, Z., *J. Chem. Soc., Perkin Trans. 2*, **1981,** 1176.
[326] Gellerstedt, G., Hardell, H. L., and Lindfors, E. L., *Acta Chem. Scand.*, **34B,** 669 (1980).
[327] Muller, P., and Bobillier, C., *Tetrahedron Lett.*, **22,** 5157 (1981).
[328] Olah, G. A., Fung, A. P., and Keumi, T., *J. Org. Chem.*, **46,** 4305 (1981).
[329] Apatu, J. O., Chapman, D. C., and Heaney, H., *J. Chem. Soc., Chem. Commun.*, **1981,** 1079.
[330] Ogata, Y., Sawaki, Y., Tomizawa, K., and Ohno, T., *Tetrahedron*, **37,** 1485 (1981).
[331] Tomizawa, K., and Ogata, Y., *J. Org. Chem.*, **46,** 2107 (1981).
[332] Khante, R. N., and Chandalia, S. B., *Indian J. Technol.*, **18,** 407 (1980); *Chem. Abs.*, **94,** 191302 (1981).
[333] Shimidzu, T., Iyoda, T., and Kanda, N., *J. Chem. Soc., Chem. Commun.*, **1981,** 1206.
[334] Panda, A. K., Mahapatro, S. N., and Panigrahi, G. P., *J. Org. Chem.*, **46,** 4000 (1981).
[335] Gingerich, S. B., Campbell, W. H., Bricca, C. E., Jennings, P. W., and Campana, C. F., *J. Org. Chem.*, **46,** 2589 (1981).
[336] Williams, P. D., and LeGoff, E., *J. Org. Chem.*, **46,** 4143 (1981).
[337] Levey, G., Rieger, A. L., and Edwards, J. O., *J. Org. Chem.*, **46,** 1255 (1981).
[338] Ouannes, C., and Thal, C., *Tetrahedron Lett.*, **22,** 951 (1981).
[339] Mahapatro, S. N., Panda, A. K., and Panigrahi, G. P., *Bull. Chem. Soc. Jpn.*, **54,** 2507 (1981).
[340] Crooks, P. A., Damani, L. A., and Cowan, D. A., *Chem. Ind. (London)*, **1981,** 335.
[341] Engelhardt, U., and Schaefer-Ridder, M., *Tetrahedron Lett.*, **22,** 4687 (1981).
[342] Lokhov, R. E., *Khim. Geterotsikl, Soedin.*, **1981,** 89; *Chem. Abs.*, **94,** 191320 (1981).
[343] Wang, C.-H., and Yang, S.-Y., *Hua Hsueh Tung Pao*, **1981,** 211; *Chem. Abs.*, **95,** 96600 (1981).
[344] Ogilvie, K. K., and Nemer, M. J., *Tetrahedron Lett.*, **22,** 2531 (1981).
[345] Tutubalina, V. P., *Izv. Vyssh Uchebn. Zaved., Khim. Khim. Tekhnol.*, **23,** 1371 (1980); *Chem. Abs.*, **94,** 120575 (1981).
[346] Panigrahi, G. P., and Nayak, R. N., *Curr. Sci.*, **49,** 740 (1980); *Chem. Abs.*, **94,** 29868 (1981).
[347] Hiller, K.-O., Masloch, B., Gobl, M., and Asmus, K.-D., *J. Am. Chem. Soc.*, **103,** 2734 (1981).
[348] Oae, S., and Takata, T., *Chem. Lett.*, **1981,** 845.
[349] Freeman, F., Angeletakis, C. N., and Maricich, T. J., *Tetrahedron Lett.*, **22,** 1867 (1981).
[350] Freeman, F., and Angeletakis, C. N., *J. Org. Chem.*, **46,** 3991 (1981).
[351] Suszka, A., *Pol. J. Chem.*, **54,** 2289 (1980); *Chem. Abs.*, **95,** 96594 (1981).
[352] Indukumari, P. V., and Joshua, C. P., *Indian J. Chem.*, **19B,** 667 (1980); *Chem. Abs.*, **94,** 29869 (1981).
[353] Toshimitsu, A., Owada, H., Aoai, T., Uemura, S., and Okano, M., *J. Chem. Soc., Chem. Commun.*, **1981,** 546.
[354] Gancarz, R. A., and Kice, J. L., *Tetrahedron Lett.*, **22,** 1661 (1981).
[355] Macdonald, T. L., Narasimhan, N., and Burka, L. T., *J. Am. Chem. Soc.*, **102,** 7760 (1980).
[356] Razuvaev, G. A., Brevnova, T. N., and Semenov, V. V., *Zh. Obshch. Khim.*, **50,** 1806 (1980); *Chem. Abs.*, **94,** 3469 (1981).
[357] Liu, M. T. H., Palmer, G. E., and Chishti, N. H., *J. Chem. Soc., Perkin Trans. 2*, **1981,** 53.
[358] Pikh, Z. G., Fedevich, M. D., and Yatchishin, I. K., *Dopov. Akad. Nauk Ukr. RSR, Ser. B: Geol., Khim. Biol. Nauki*, **1980,** 48; *Chem. Abs.*, **94,** 29844 (1981).
[359] Boyer, J. H., and Huang, C., *J. Chem. Soc., Chem. Commun.*, **1981,** 365.
[360] Ogata, Y., Sawaki, Y., and Tsukamoto, Y., *Bull. Chem. Soc. Jpn.*, **54,** 2061 (1981).
[361] Holland, H. L., Daum, U., and Riemland, E., *Tetrahedron Lett.*, **22,** 5127 (1981).
[362] Dadoun, H., Alazard, J.-P., and Luzinchi, X. *Tetrahedron*, **37,** 1525 (1981).
[363] Roussel, M., and Mimoun, H., *J. Org. Chem.*, **45,** 5387 (1980).
[364] Nagiev, T. M., Bairamov, F. G., Nagieva, Z. M., and Agaeva, S. I., *Azerb. Khim. Zh.*, **1980,** 32; *Chem. Abs.*, **94,** 156032 (1981).
[365] Nagiev, M., Mamed'yarov, G. M., and Ali-Zade, N. I., *Azerb. Khim. Zh.*, **1979,** 10; *Chem. Abs.*, **93,** 203697 (1980).
[366] Tamaki, K., Hiraga, S., and Nishino, J., *Yuki Gosei Kagaku Kyokaishi*, **38,** 306 (1980); *Chem. Abs.*, **94,** 102557 (1981).
[367] Ogata, Y., Tomizawa, K., and Takagi, K., *Can. J. Chem.*, **59,** 14 (1981).
[368] Santamaria, J., *Tetrahedron Lett.*, **22,** 4511 (1981).
[369] Galliani, G., and Rindone, B., *Tetrahedron*, **37,** 2313 (1981).

370 Oae, S., Takata, T., and Kim, Y. H., *Bull. Chem. Soc. Jpn.*, **54,** 2712 (1981).
371 Oae, S., Takata, T., and Kim, Y. H., *Tetrahedron,* **37,** 37 (1981).
372 Matsumoto, S., Matsuo, M., and Iitaka, Y., *Tetrahedron Lett.*, **22,** 3649 (1981).
373 Roberts, J. L., and Sawyer, D. T., *J. Am. Chem. Soc.*, **103,** 712 (1981).
374 Nanni, E. J., Sawyer, D. T., Ball, S. S., and Bruice, T. C., *J. Am. Chem. Soc.*, **103,** 2797 (1981).
375 El-Sukkary, M. M. A., and Speier, G., *J. Chem. Soc., Chem. Commun.*, **1981,** 745.
376 Nanni, E. J., and Sawyer, D. T., *J. Am. Chem. Soc.*, **102,** 7591 (1980).
377 Aubry, J. M., Rigaudy, J., Ferradini, C., and Pucheault, J., *J. Am. Chem. Soc.*, **103,** 4965 (1981).
378 Nanni, E. J., Angelis, C. T., Dickson, J., and Sawyer, D. T., *J. Am. Chem. Soc.*, **103,** 4268 (1981).
379 Alvarez, E., Betancor, C., Freire, R., Martin, A., and Suarez, E., *Tetrahedron Lett.*, **22,** 4335 (1981).
380 Fridovich, S. E., and Porter, N. A., *J. Biol. Chem.*, **256,** 260 (1981).
381 Ogawa, Y., Iwasaki, S., and Okuda, S., *Tetrahedron Lett.*, **22,** 2277 (1981).
382 Washida, N., *J. Chem. Phys.*, **75,** 2715 (1981).; *Chem. Abs.*, **95,** 168168 (1981).
383 Andresen, P., and Luntz, A. C., *J. Chem. Phys.*, **72,** 5842 (1980); *Chem. Abs.*, **94,** 3459 (1981).
384 Aleksandrov, E. N., Arutyunov, V. S., and Kozlov, S. N., *Kinet. Katal.*, **21,** 1327 (1980); *Chem. Abs.*, **94,** 29882 (1981).
385 Aleksandrov, E. N., Arutyunov, V. S., and Kozlov, S. N., *Kinet. Katal.*, **22,** 513 (1981); *Chem. Abs.*, **95,** 79694 (1981).
386 Loehr, R., and Roth, P., *Ber. Bunsenges. Phys. Chem.*, **82,** 153 (1981); *Chem. Abs.*, **94,** 173938 (1981).
387 Nip, W. S., Singleton, D. L., and Cvetanovic, R. J., *J. Am. Chem. Soc.*, **103,** 3526 (1981).
388 Cvetanovic, R. J., Singleton, D. L., and Irwin, R. S., *J. Am. Chem. Soc.*, **103,** 3530 (1981).
389 Gorman, A. A., and Rodgers, M. A. J., *Chem. Soc. Rev.*, **10,** 205 (1981).
390 Wasserman, H. H., and Ives, J. L., *Tetrahedron,* **37,** 1825 (1981).
391 Ando, W., *Sulfur Rep. 1981*, **1,** 147 (1981); *Chem. Abs.*, **95,** 79385 (1981).
392 Stephenson, L. M., Grdina, M. J., and Orfanopoulos, M., *Acc. Chem. Res.*, **13,** 419 (1980).
393 Turro, N. J., Chow, M.-F., Kanfer, S., and Jacobs, M., *Tetrahedron Lett.*, **22,** 3 (1981).
394 Harbour, J. R., Issler, S. L., and Hair, M. L., *J. Am. Chem. Soc.*, **102,** 7778 (1980).
395 Ogilby, P. R., and Foote, C. S., *J. Am. Chem. Soc.*, **103,** 1219 (1981).
396 Eisenberg, W. C., Snelson, A., Butler, R., Veltman, J., and Murray, R. W., *Tetrahedron Lett.*, **22,** 377 (1981).
397 Saito, I., Matsuura, T., and Inoue, K., *J. Am. Chem. Soc.*, **103,** 188 (1981).
398 Saito, I., Nagata, R., and Matsuura, T., *Tetrahedron Lett.*, **22,** 4231 (1981).
399 Martin, N. H., and Jefford, C. W., *Helv. Chim. Acta*, **64,** 2189 (1981).
400 Martin, N. H., and Jefford, C. W., *Tetrahedron Lett.*, **22,** 3949 (1981).
401 Yamaguchi, K., Yabushita, S., Fueno, T., and Houk, K. N., *J. Am. Chem. Soc.*, **103,** 5043 (1981).
402 Yamaguchi, K., Fueno, T., Saito, I., Matsuura, T., and Houk, K. N., *Tetrahedron Lett.*, **22,** 749 (1981).
403 Jefford, C. W., and Rimbault, C. G., *Tetrahedron Lett.*, **22,** 91 (1981).
404 Landheer, I., and Ginsburg, D., *Tetrahedron,* **37,** 143 (1981).
405 Lopez, S., and Juan, A., *Afinidad*, **38,** 153 (1981); *Chem. Abs.*, **95,** 114450 (1981).
406 Laporterie, A., Dubac, J., and Mazerolles, P., *J. Organomet. Chem.*, **202,** C89 (1980).
407 Clark, B. C., Jones, B. B., and Iacobucci, G. A., *Tetrahedron*, **37** (Suppl. 9), 405 (1981).
408 Ceccherelli, P., Curini, M., and Pellicciari, R., *J. Chem. Res.* (*S*), **1981,** 77.
409 Mohanraj, S., and Herz, W., *J. Org. Chem.*, **46,** 1362 (1981).
410 Nowicki, A. W., and Turner, A. B., *J. Chem. Res.* (*S*), **1981,** 110.
411 Paquette, L. A., Carr, R. V. C., Arnold, E., and Clardy, J., *J. Org. Chem.*, **45,** 4907 (1980).
412 Paquette, L. A., Bellamy, F., Bohm, M. C., and Gleiter, R., *J. Org. Chem.*, **45,** 4913 (1980).
413 Kotsuki, H., Saito, I., and Matsuura, T., *Tetrahedron Lett.*, **22,** 469 (1981).
414 Adam, W., and Rodriguez, A., *Tetrahedron Lett.*, **22,** 3505 (1981).
415 Adam, W., and Rodriguez, A., *Tetrahedron Lett.*, **22,** 3509 (1981).
416 Feringa, B. L., and Butselaar, R. J., *Tetrahedron Lett.*, **22,** 1447 (1981).
417 Feringa, B. L., *Tetrahedron Lett.*, **22,** 1443 (1981).
418 Wasserman, H. H., Pickett, J. E., and Vinnick, F. S., *Heterocycles*, **15,** 1069 (1981); *Chem. Abs.*, **95,** 6132 (1981).
419 Albini, A., Bettinetti, G. F., Minoli, G., and Vasconi, G., *J. Chem. Soc., Chem. Commun.*, **1981,** 1089.
420 Matsumoto, M., Kuroda, K., and Suzuki, Y., *Tetrahedron Lett.*, **22,** 3253 (1981).
421 Ando, W., Miyazaki, H., Ueno, K., Nakanishi, H., Sakurai, T., and Kobayashi, K., *J. Am. Chem. Soc.*, **103,** 4949 (1981).
422 Ryang, H.-S., and Foote, C. S., *J. Am. Chem. Soc.*, **103,** 4951 (1981).
423 Adam, W., Cueto, O., and De Lucchi, O., *J. Org. Chem.*, **45,** 5220 (1980).
424 Nishio, T., Nakajima, N., and Omote, Y., *Tetrahedron Lett.*, **22,** 753 (1981).

[425] Jongsma, S. J., and Cornelisse, J., *Tetrahedron Lett.*, **22,** 2919 (1981).
[426] Gu, C., Foote, C. S., and Kacher, M. L., *J. Am. Chem. Soc.*, **103,** 5949 (1981).
[427] Sawaki, Y., and Ogata, Y., *J. Am. Chem. Soc.*, **103,** 5947 (1981).
[428] Ramnath, N., Ramesh, V., and Ramamurthy, V., *J. Chem. Soc., Chem. Commun.*, **1981,** 112.
[429] Creed, D., Werbin, H., and Daniel, T., *Tetrahedron Lett.*, **22,** 2039 (1981).
[430] Kuramoto, N., and Kitao, T., *J. Chem. Soc., Perkin Trans. 2*, **1980,** 1569.
[431] Lichszteld, K., and Kruk, I., *Z. Phys. Chem. (Leipzig)*, **262,** 673 (1981).
[432] Emanuel, N. M., *Oxid. Commun.*, **1,** 125 (1980); *Chem. Abs.*, **94,** 29602 (1981).
[433] Rudakov, E. S., *Izv. Sib. Otd. Akad. Nauk SSSR, Ser. Khim. Nauk*, **1980,** 161; *Chem. Abs.*, **93,** 237980 (1980).
[434] Mill, T., and Hendry, D. G., *Compr. Chem. Kinet.*, **16,** 1 (1980); *Chem. Abs.*, **94,** 29612 (1981).
[435] Beletskaya, I. P., and Makhon'kov, D. I., *Usp. Khim.*, **50,** 1007 (1981); *Chem. Abs.*, **95,** 114224 (1981).
[436] Sajus, L., and Seree de Roch, I., *Compr. Chem. Kinet.*, **16,** 89 (1980); *Chem. Abs.*, **93,** 237992 (1980).
[437] Kulicki, Z., and Stec, Z., *Zesz. Nauk. Politech. Slask., Chem.*, **633,** 65 (1980); *Chem. Abs.*, **94,** 120324 (1981).
[438] Denisov, E. T., *Compr. Chem. Kinet.*, **16,** 125 (1980); *Chem. Abs.*, **94,** 29613 (1981).
[439] Trimm, D. L., *Compr. Chem. Kinet.*, **16,** 205 (1980); *Chem. Abs.*, **94,** 29614 (1981).
[440] Ohkatsu, Y., *Kagaku Kogyo*, **32,** 908 (1981); *Chem. Abs.*, **95,** 167914 (1981).
[441] Haber, J., and Witko, M., *Acc. Chem. Res.*, **14,** 1 (1981).
[442] Howard, J. A., *Front. Free Radical Chem.*, [*Pap. Symp.*], **1979,** 237; *Chem. Abs.*, **94,** 102262 (1981).
[443] Rotschova, J., and Pospiril, J., *Chem. Ind. (London)*, **1980,** 393.
[444] Burton, G. W., Le Page, Y., Gabe, E. J., and Ingold, K. U., *J. Am. Chem. Soc.*, **102,** 7791 (1980).
[445] Burton, G. W., and Ingold, K. U., *J. Am. Chem. Soc.*, **103,** 6472 (1981).
[446] Barclay, L. R. C., and Ingold, K. U., *J. Am. Chem. Soc.*, **102,** 7792 (1980).
[447] Barclay, L. R. C., and Ingold, K. U., *J. Am. Chem. Soc.*, **103,** 6478 (1981).
[448] Johansson, E., and Skramstad, J., *J. Org. Chem.*, **46,** 3752 (1981).
[449] Meijer, E. W., and Wynberg, H., *Tetrahedron Lett.*, **22,** 785 (1981).
[450] Tempesti, E., Montoneri, E., Giuffré, L., and Airoldi, G., *J. Org. Chem.*, **45,** 4278 (1980).
[451] Sasaki, T., *J. Am. Chem. Soc.*, **103,** 3882 (1981).
[452] Sawaki, Y., and Ogata, Y., *J. Am. Chem. Soc.*, **103,** 2049 (1981).
[453] Kanno, T., Hisaoka, M., Sakuragi, H., and Tokamaru, K., *Bull. Chem. Soc. Jpn.*, **54,** 2330 (1981).
[454] Matsumoto, M., and Kuroda, K., *Tetrahedron Lett.*, **22,** 4437 (1981).
[455] Kovtun, G. A., Lysenko, D. L., Kaverinskii, V. A., Berenblyum, A. S., and Moiseev, I. I., *Izv. Akad. Nauk SSSR, Ser. Khim.*, **1980,** 1266; *Chem. Abs.*, **93,** 220155 (1980).
[456] Ivanov, S., Kateva, I., Dimitrov, V., Dahlmann, J., and Hoeft, E., *J. Prakt. Chem.*, **323,** 230 (1981); *Chem. Abs.*, **95,** 79680 (1981).
[457] Grasselli, R. K., Burrington, J. D., Suresh, D. D., Friedrich, M. S., and Hazle, M. A. S., *J. Catal.*, **68,** 109 (1981); *Chem. Abs.*, **95,** 96580 (1981).
[458] Ivin, K. J., Reddy, B. S. R., and Rooney, J. J., *J. Chem. Soc., Chem. Commun.*, **1981,** 1062.
[459] Kojima, M., Sakuragi, H., and Tokumaru, K., *Tetrahedron Lett.*, **22,** 2889 (1981).
[460] Batog, A. E., Opeida, I. A., and Nosyreva, O. V., *Dopov. Akad. Nauk Ukr. RSR, Ser. B: Geol., Khim. Biol. Nauki*, **1981,** 36; *Chem. Abs.*, **95,** 114435 (1981).
[461] Perrée-Fauvet, M., and Gaudemer, A., *J. Chem. Soc., Chem. Commun.*, **1981,** 874.
[462] Gudimenko, Yu. I., Agabekov, V. E., and Mitskevich, N. I., *Oxid. Commun.*, **1,** 285 (1980); *Chem. Abs.*, **95,** 23906 (1981).
[463] Gusevskaya, E. V., Likholobov, V. A., and Ermakov, Yu. I., *Kinet. Katal.* **21,** 639 (1980); *Chem. Abs.*, **93,** 203690 (1980).
[464] Nelsen, S. F., and Akaba, R., *J. Am. Chem. Soc.*, **103,** 2096 (1981).
[465] Subramanian, T. V., and Jagannadhaswamy, B., *Indian J. Technol.*, **18,** 4 (1980); *Chem. Abs.*, **94,** 3500 (1981).
[466] Dimitrov, D., *Hung. J. Ind. Chem.*, **8,** 417 (1980); *Chem. Abs.*, **94,** 138934 (1981).
[467] Digurov, N. G., Kashirskii, V. F., Shevyreva, E. V., and Lebedev, N. N., *Kinet. Katal.*, **2,** 364 (1981); *Chem. Abs.*, **95,** 61060 (1981).
[468] Agaev, F. M., Abdullaev, F. Z., Gashimov, A. G., and Mekhtiev, D. S., *Dokl. Akad. Nauk Az. SSSR*, **36,** 40 (1980); *Chem. Abs.*, **94,** 83338 (1981).
[469] Ivanov, S., Khinkova, M., Karshalukov, K., and Boneva, M., *Neftekhimiya*, **20,** 702 (1980); *Chem. Abs.*, **94,** 14874 (1981).
[470] Ovchinnikov, V. I., Tovstokhat'ko, F. I., and Potekhin, V. M., *Zh. Prikl. Khim. (Leningrad)*, **54,** 1118 (1981); *Chem. Abs.*, **95,** 114459 (1981).

471 Brilev, V. V., Ariko, N. G., and Mitskevich, N. I., *Vestsi Akad. Navuk BSSR, Ser. Khim. Navuk*, **1980,** 33; *Chem. Abs.*, **94,** 14870 (1981).

472 Samodumov, S. A., and Ostanina, V. A., *Izv. Vyssh. Uchebn. Zaved., Khim. Tekhnol.*, **23,** 1079 (1980); *Chem. Abs.*, **94,** 14864 (1981).

473 Nitteberg, E. B., and Rokstad Odd, A., *Acta Chem. Scand.*, **34A,** 199 (1980); *Chem. Abs.*, **93,** 238334 (1980).

474 Tmenov, D. N., Shcherbina, F. F., and Belous, N. P., *Dopov. Akad. Nauk Ukr. RSR, Ser. B: Beol., Khim. Biol. Nauki*, **1980,** 66; *Chem. Abs.*, **95,** 6121 (1981).

475 Navazio, G., and Carbini, M., *Atti Ist. Veneto Sci., Lett. Arti, Cl. Sci. Mat. Nat.*, **137,** 169 (1979); *Chem. Abs.*, **95,** 6124 (1981).

476 Velyutin, L. P., Potekhin, V. M., and Ovchinnikov, V. I., *Zh. Org. Khim.*, **16,** 984 (1980); *Chem. Abs.*, **93,** 167283 (1980).

477 Pavlov, G. P., Sinovich, I. D., and Bykova, N. V., *Zh. Org. Khim.*, **16,** 1253 (1980); *Chem. Abs.*, **93,** 167290 (1980).

478 Opeida, I. A., Shendrik, A. N., and Galat, V. F., *Ukr. Khim. Zh. (Russ. Ed.)*, **46,** 1197 (1980); *Chem. Abs.*, **94,** 64857 (1981).

479 Hanotier, J., and Hanotier-Bridoux, M., *J. Mol. Catal.*, **12,** 133 (1981); *Chem. Abs.*, **95,** 149339 (1981).

480 Bawa, J. S., Rawat, D. S., Dabral, R. P., and Bhattacharyya, K. K., *Proc. Natl. Symp. Catal., 4th*, **1978,** 319; *Chem. Abs.*, **94,** 156011 (1981).

481 Beyrich, J., Regenass, W., and Richarz, W., *Chimia*, **34,** 244 (1980); *Chem. Abs.*, **93,** 203677 (1980).

482 Khodzhaev, O. M., Shik, G. L., Chernikov, V. V., and Shakhtakhtinskii, T. N., *Azerb. Khim. Zh.*, **1980,** 22; *Chem. Abs.*, **94,** 120579 (1981).

483 Kulicki, Z., and Baj, S., *Zesz. Nauk. Politech. Slask., Chem.*, **633,** 73 (1980); *Chem. Abs.*, **94,** 138908 (1981).

484 Starchevskii, M. K., Vargaftik, M. N., and Moiseev, I. I., *Kinet. Katal.*, **22,** 622 (1981); *Chem. Abs.*, **95,** 114463 (1981).

485 Gaevskii, V. F., and Evmenenko, N. P., *Neftekhimiya*, **20,** 846 (1980); *Chem. Abs.*, **94,** 120573 (1981).

486 Tkacheva, G. D., Dorfman, Ya. A., and Suvorov, B. V., *Izv. Akad. Nauk Kaz. SSR, Ser. Khim.*, **1981,** 50; *Chem. Abs.*, **94,** 173995 (1981).

487 Kuz'min, A. K., Chistyakov, A. N., and Proskuryakov, V. A., *Zh. Prikl. Khim. (Leningrad)*, **54,** 120 (1981); *Chem. Abs.*, **95,** 23871 (1981).

488 Antonova, V. V., Bespalova, A. M., Smirnova, T. I., and Ustavschchikov, B. F., *Okislenie Org. Soedin. Zhidk. Faze*, **1978,** 49; *Chem. Abs.*, **93,** 203725 (1980).

489 Leichenko, A. A., Shchedrinskaya, T. V., and Volkov, M. N., *Khim. Geterotsikl. Soedin.*, **1980,** 924; *Chem. Abs.*, **93,** 238331 (1980).

490 Moshkina, R. I., Polyak, S. S., Romanovich, L. B., and Nalbandyan, A. B., *Kinet. Katal.*, **21,** 866 (1980); *Chem. Abs.*, **93,** 238356 (1980).

491 Moshkina, R. I., Polyak, S. S., Romanovich, L. B., and Nalbandyan, A. B., *Kinet. Katal.*, **21,** 1379 (1980); *Chem. Abs.*, **94,** 102515 (1981).

492 Musaev, M. R., Alimardanov, Kh. M., Medzhidov, A. A., and Safarov, A. S., *Azerb. Khim. Zh.*, **1980,** 40; *Chem. Abs.*, **94,** 156033 (1981).

493 Nagiev, T. M., and Gasanova, L. M., *Azerb. Khim. Zh.*, **1980,** 13; *Chem. Abs.*, **95,** 41947 (1981).

494 Mushenko, V. D., Mushenko, D. V., and Fedorov, V. A., *Zh. Prikl. Khim. (Leningrad)*, **53,** 1588 (1980); *Chem. Abs.*, **93,** 238338 (1980).

495 Geletii, Yu. V., Karasevich, E. I., Moravskii, A. P., and Shteinman, A. A., *Kinet. Katal.*, **22,** 349 (1981); *Chem. Abs.*, **95,** 7960 (1981).

496 Chichagov, V. N., Solyanikov, V. M., Bondarenko, T. G., Tkacheva, G. A., Zhavoronkov, A. P., and Denisov, E. T., *Neftekhimiya*, **20,** 559 (1980); *Chem. Abs.*, **94,** 3475 (1981).

497 Jensen, R. K., Korcek, S., Mahoney, L. R., and Zinbo, M., *J. Am. Chem. Soc.*, **103,** 1742 (1981).

498 Poladyan, E. A., and Mantashyan, A. A., *Arm. Khim. Zh.*, **33,** 529 (1980); *Chem. Abs.*, **94,** 29847 (1981).

499 Kozhevnikov, I. V., Taraban'ko, V. E., and Matveev, K. I., *Kinet. Katal.*, **21,** 947 (1980); *Chem. Abs.*, **93,** 238358 (1980).

500 Kozhevnikov, I. V., Taraban'ko, V. E., and Matveev, K. I., *Kinet. Katal.*, **21,** 940 (1980); *Chem. Abs.*, **93,** 238357 (1980).

501 Ariko, N. G., Ivanyutina, Z. M., and Mitskevich, N. I., *Vestsi Akad. Navuk BSSR, Ser. Khim. Navuk*, **1980,** 40; *Chem. Abs.*, **93,** 167298 (1980).

502 Fiege, H., and Wedemeyer, K., *Angew. Chem. Int. Ed.*, **20,** 783 (1981).

503 Bothe, E., and Schulte-Frohlinde, D., *Z. Naturforsch.*, **35b,** 1035 (1980); *Chem. Abs.*, **94,** 3474 (1981).

504 Wan, P., and Yates, K., *J. Chem. Soc., Chem. Commun.*, **1981,** 1023.
505 Sogomonyan, B. M., and Beileryan, N. M., *Arm. Khim. Zh.*, **33,** 885 (1980); *Chem. Abs.*, **94,** 83480 (1981).
506 Bao Iglesias, M., Delgado Diaz, S., Diaz Rodriguez, F., Fernandez Gonzalez, J., and Alvarez Diaz, M., *Ing. Quim. (Madrid)*, **1981,** 13; *Chem. Abs.*, **95,** 79727 (1981).
507 Sogomonyan, B. M., and Beileryan, N. M., *Arm. Khim. Zh.*, **33,** 880 (1980); *Chem. Abs.*, **94,** 83479 (1981).
508 Horvath, I., and Yatchishin, I. I., *Neftekhimiya*, **20,** 573 (1980); *Chem. Abs.*, **93,** 238353 (1980).
509 Kaiser, E. W., *Chem. Phys. Processes Combust.*, **1979,** Paper No. 18, 4; *Chem. Abs.*, **94,** 208134 (1981).
510 Niki, H., Maker, P. D., Savage, C. M., and Breitenbach, L. P., *Chem. Phys. Lett.*, **72,** 71 (1980); *Chem. Abs.*, **93,** 185427 (1980).
511 Ivanov, A. M., Mikhailovskaya, T. N., and Chervinskii, K. A., *Neftekhimiya*, **20,** 683 (1980); *Chem. Abs.*, **94,** 14872 (1981).
512 Kucher, R. V., Opeida, I. A., Nechitailo, L. G., and Simonov, M. A., *Kinet. Katal.*, **22,** 332 (1981); *Chem. Abs.*, **95,** 61059 (1981).
513 Kosmacheva, T. G., Agabekov, V. E., and Mitskevich, N. I., *Vesti Akad. Navuk BSSR, Ser. Khim. Navuk*, **1981,** 28; *Chem. Abs.*, **94,** 191325 (1981).
514 Evteeva, N. M., and Gagarina, A. B., *Izv. Akad. Nauk SSSR, Ser. Khim.*, **1980,** 2715; *Chem. Abs.*, **94,** 120588 (1981).
515 Dmitrieva, O. P., Agabekov, V. E., and Mitskevich, N. I., *Dokl. Akad. Nauk BSSR*, **25,** 329 (1981); *Chem. Abs.*, **95,** 23908 (1981).
516 Martem'yanov, V. S., Borisov, I. M., Degtyareva, T. G., and Kivganova, V. I., *Neftekhimiya*, **21,** 444 (1981); *Chem. Abs.*, **95,** 114470 (1981).
517 Martem'yanov, V. S., Borisov, I. M., Denisov, E. T., and Nabiullina, Z. V., *Izv. Akad. Nauk SSSR, Ser. Khim.*, **1981,** 91; *Chem. Abs.*, **94,** 138940 (1981).
518 Klyuchkinskii, A. I., Krip, I. M., and Donets, G. I., *Vestn. L'vov, Politekhn. In-ta*, **1980,** 58; *Chem. Abs.*, **94,** 83345 (1981).
519 Butovskaya, G. V., Agabekov, V. E., and Mitskevich, N. I., *Dokl. Akad. Nauk BSSR*, **25,** 722 (1981); *Chem. Abs.*, **95,** 149603 (1981).
520 Martem'yanov, V. S., Abdullin, M. I., Orlova, T. E., and Minsker, K. S., *Neftekhimiya*, **21,** 123 (1981); *Chem. Abs.*, **95,** 41904 (1981).
521 Azarko, V. A., Agabekov, V. E., and Mitskevich, N. I., *Vestsi Akad. Navuk BSSR, Ser. Khim. Navuk*, **1981,** 21; *Chem. Abs.*, **95,** 149604 (1981).
522 Aleksandrov, A. L., and Krisanova, L. D., *Izv. Akad. Nauk SSSR, Ser. Khim.*, **1980,** 2469; *Chem. Abs.*, **94,** 120560 (1981).
523 Aleksandrov, A. L., *Izv. Akad. Nauk SSSR, Ser. Khim.*, **1980,** 2474; *Chem. Abs.*, **94,** 120561 (1981).
524 Gajda, T. M., Sopchik, A. E., and Bentrude, W. G., *Tetrahedron Lett.*, **22,** 4167 (1981).
525 Beltrame, P., Beltrame, P. L., Bussola, M., and Carniti, P., *Gazz. Chim. Ital.*, **110,** 141 (1980); *Chem. Abs.*, **94,** 3460 (1981).
526 San Clemente, M. R., Sarkanen, K. V., and Sundin, S. E., *Sven. Papperstidn.*, **1981,** 84, R1; *Chem. Abs.*, **94,** 191326 (1981).
527 Vorotyntsev, V. M., Kurnetsova, E. P., Pyatnitskii, Yu. I., and Golodets, G. I., *React. Kinet. Catal. Lett.*, **13,** 373 (1980); *Chem. Abs.*, **93,** 220163 (1980).
528 Nishinaga, A., Shimizu, T., Fujii, T., and Matsuura, T., *J. Org. Chem.*, **45,** 4997 (1980).
529 Ashby, E. C., Goel, A. B., and DePriest, R. N., *Tetrahedron Lett.*, **22,** 3729 (1981).
530 Ashby, E. C., and Goel, A. B., *Tetrahedron Lett.*, **22,** 1879 (1981).
531 Ashby, E. C., DePriest, R. N., and Goel, A. B., *Tetrahedron Lett.*, **22,** 1763 (1981).
532 Ashby, E. C., and Goel, A. B., *J. Org. Chem.*, **46,** 3934 (1981).
533 Ashby, E. C., Goel, A. B., DePriest, R. N., and Prasad, H. S., *J. Am. Chem. Soc.*, **103,** 973 (1981).
534 Nayer, B. M. S., and Callery, P. S., *J. Org. Chem.*, **46,** 1044 (1981).
535 Smith, M. B., and Wolinsky, J., *J. Org. Chem.*, **46,** 101 (1981).
536 Krishnamurthy, S., *J. Org. Chem.*, **46,** 4628 (1981).
537 Kraus, G. A., and Frazier, K., *J. Org. Chem.*, **45,** 4262 (1980).
538 Shyamasundar, N., and Caluwe, P., *J. Org. Chem.*, **46,** 1552 (1981).
539 Lipták, A., Neszmélyi, A., Kovác, P., and Hirsch, J., *Tetrahedron*, **37,** 2379 (1981).
540 Fugedi, P., and Liptak, A., *J. Chem. Soc., Chem. Commun.*, **1980,** 1234.
541 Palialunga, Paradisi, M., and Pagani Zecchini, G., *Tetrahedron*, **37,** 971 (1981).
542 Lee, Y. T., and Song, B. S., *Yongnam Taehakkyo Nonmunjip, Chayon Kwahak Pyon*, **13,** 69 (1979); *Chem. Abs.*, **94,** 173944 (1981).

543 Angiolini, L., Costa Bizzarri, P., Scapini, G., and Tramontini, M., *Tetrahedron*, **37,** 2137 (1981).
544 Morrison, J. D., Grandbois, E. R., Howard, S. I., and Weisman, G. R., *Tetrahedron Lett.*, **22,** 2619 (1981).
545 Nishizawa, M., and Noyori, R., *Kagaku, Zokan* (*Kyoto*), **1981,** 181; *Chem. Abs.*, **95,** 96306 (1981).
546 Giacomelli, G., Caporusso, A. M., and Lardicci, L., *Tetrahedron Lett.*, **22,** 3663 (1981).
547 (a) Giacomelli, G., and Lardicci, L., *J. Org. Chem.*, **46,** 3116 (1981).
(b) McMahon, R. J., Wiegers, K. E., and Smith, S. G., *J. Org. Chem.*, **46,** 99 (1981).
548 Jager, V., Schwab, W., and Buss, V., *Angew. Chem. Int. Ed.*, **20,** 601 (1981).
549 Ishiguro, M., Koizumi, N., Yasuda, M., and Ikekawa, N., *J. Chem. Soc., Chem. Commun.*, **1981,** 115.
550 Alcaide, B., Fernandez de la Pradilla, R., Lopez-Mardomingo, C., Perez-Ossorio, R., and Plumet, J., *J. Org. Chem.*, **46,** 3234 (1981).
551 Elsenbaumer, R. L., Mosher, H. S., Morrison, J. D., and Tomaszewski, J. E., *J. Org. Chem.*, **46,** 4034 (1981).
552 Stewart, R., and Teo, K. C., *Can. J. Chem.*, **58,** 2491 (1980).
553 Stewart, R., Teo, K. C., and Ng, L. K., *Can. J. Chem.*, **58,** 2497 (1980).
554 Ananthakrishnanadar, P., and Kannan, N., *Indian J. Chem.*, **19B,** 923 (1980); *Chem. Abs.*, **94,** 102495 (1981).
555 Giddings, M. R., and Hudec, J., *Can. J. Chem.*, **59,** 459 (1981).
556 Morrison, J. D., Grandbois, E. R., and Howard, S. I., *J. Org. Chem.*, **45,** 4229 (1980).
557 Mirao, A., Nakahama, S., Mochizuki, H., Itsuno, S., and Yamazaki, N., *J. Org. Chem.*, **45,** 4231 (1980).
558 Kirao, A., Ohwa, M., Itsuno, S., Mochizuki, H., Nakahara, S., and Yamazaki, N., *Bull. Chem. Soc. Jpn.*, **54,** 1424 (1981).
559 Grundon, M. F., McCleery, D. G., and Wilson, J. W., *J. Chem. Soc., Perkin Trans. 1*, **1981,** 231.
560 Hirao, A., Itsuno, S., Nakahama, S., and Yamazaki, N., *J. Chem. Soc., Chem. Commun.*, **1981,** 315.
561 Yamada, K., Takeda, M., and Iwakuma, T., *Tetrahedron Lett.*, **22,** 3869 (1981).
562 Wrobel, J. E., and Gauem, B., *Tetrahedron Lett.*, **22,** 3447 (1981).
563 Kayser, M. M., and Morand, P., *Can. J. Chem.*, **58,** 2484 (1980).
654 Toi, H., Yamamoto, Y., Sonoda, A., and Murahashi, S.-I., *Tetrahedron*, **37,** 2261 (1981).
565 Yasuda, M., Pac, C., and Sakurai, H., *J. Org. Chem.*, **46,** 788 (1981).
566 Sasaki, T., Minamoto, K., and Harada, K., *J. Org. Chem.*, **45,** 4594 (1980).
567 Mata-Segreda, J. F., and Schowen, R. L., *J. Org. Chem.*, **46,** 644 (1981).
568 Meunier, B., *J. Organomet. Chem.*, **204,** 345 (1981).
569 Santaniello, E., Ferraboschi, P., and Sozzani, P., *J. Org. Chem.*, **46,** 4584 (1981).
570 Wann, S. R., Thorsen, P. T., and Kreevoy, M. M., *J. Org. Chem.*, **46,** 2579 (1981).
571 Babler, J. H., and Invergo, B. J., *Tetrahedron Lett.*, **22,** 11 (1981).
572 D'Incan, E., Loupy, A., and Maia, A., *Tetrahedron Lett.*, **22,** 941 (1981).
573 D'Incan, E., and Loupy, A., *Tetrahedron*, **37,** 1171 (1981).
574 Rolla, F., *J. Org. Chem.*, **46,** 3909 (1981).
575 Gammill, R. B., Sobieray, D. M., and Gold, P. M., *J. Org. Chem.*, **46,** 3555 (1981).
576 Maryanoff, B. E., McComsey, D. F., and Nortey, S. O., *J. Org. Chem.*, **46,** 355 (1981).
577 Kabalka, G. W., and Summers, S. T., *J. Org. Chem.*, **46,** 1217 (1981).
578 Brown, H. C., Jadhav, P. K., and Mandal, A. K., *Tetrahedron*, **37,** 3547.
579 Pelter, A., Rosser, R., and Mills, S., *J. Chem. Soc., Chem. Commun.*, **1981,** 1014.
580 Nutaitis, C. F., Schultz, R. A., Obaza, J., and Smith, F. X., *J. Org. Chem.*, **45,** 4606 (1980).
581 Tadanier, J., Hallas, R., Martin, J. R., and Stanaszek, R. S., *Tetrahedron*, **37,** 1309 (1981).
582 Sundberg, R. J., Walters, C. P., and Bloom, J. D., *J. Org. Chem.*, **46,** 3730 (1981).
583 Sternbach, D. D., and Jamison, W. C. L., *Tetrahedron Lett.*, **22,** 3331 (1981).
584 Hutchins, R. O., and Markowitz, M., *J. Org. Chem.*, **46,** 3571 (1981).
585 Raber, D. J., Guida, W. C., and Shoenberger, D. C., *Tetrahedron Lett.*, **22,** 5107 (1981).
586 Nakata, T., Tanaka, T., and Oishi, T., *Tetrahedron Lett.*, **22,** 4723 (1981).
587 Kim, J. D., and Kwak, S. T., *Yongnam Taehakkyo Nonmunjip, Chayon Kwahak Pyon*, **13,** 49 (1979); *Chem. Abs.*, **94,** 138899 (1981).
588 Gemal, A. L., and Luche, J.-L., *J. Am. Chem. Soc.*, **103,** 5454 (1981).
589 Sorrell, T. M., and Pearlman, P. A., *J. Org. Chem.*, **46,** 2603 (1981).
590 Clarke, S. J., Fleet, G. W. J., and Irving, E. M., *J. Chem. Res.* (*S*), **1981,** 17.
591 Fleet, G. W. J., and Harding, P. J. C., *Tetrahedron Lett.*, **22,** 675 (1981).
592 Davey, P. N., Fleet, G. W. J., and Harding, P. J. C., *J. Chem. Res.* (*S*), **1981,** 336.
593 Kim, S. O., *Hwahak Kwa Kongop Ui Chinbo*, **20,** 293 (1980); *Chem. Abs.*, **94,** 102222 (1981).
594 Mukaiyama, T., and Asami, M., *Kagaku, Zokan* (*Kyoto*), **1981,** 151. *Chem. Abs.*, **95,** 96307 (1981).

[595] Sato, F., Ishikawa, H., Watanabe, H., Miyake, T., and Sato, M., *J. Chem. Soc., Chem. Commun.*, **1981,** 718.
[596] Vanderesse, R., Brunet, J.-J., and Caubere, P., *J. Org. Chem.*, **46,** 1270 (1981).
[597] Brunet, J.-J., Sidot, C., and Caubere, P., *J. Org. Chem.*, **46,** 3147 (1981).
[598] Darensbourg, D. J., Rokicki, A., and Darensbourg, M. Y., *J. Am. Chem. Soc.*, **103,** 3223 (1981).
[599] John, D. I., Tyrrell, N. D., and Thomas, E. J., *J. Chem. Soc., Chem. Commun.*, **1981,** 901.
[600] Ando, T., Ishihara, T., Ohtani, E., and Sawada, H., *J. Org. Chem.*, **46,** 4446 (1981).
[601] Bloodworth, A. J., and Courtneidge, J. L., *J. Chem. Soc., Chem. Commun.*, **1981,** 1117.
[602] Fuller, T. S. S., and Stick, R. V., *Aust. J. Chem.*, **33,** 2509 (1980).
[603] Four, P., and Guibe, F., *J. Org. Chem.*, **46,** 4439 (1981).
[604] Fry, J. L., and Ott, R. A., *J. Org. Chem.*, **46,** 602 (1981).
[605] Kraus, G. A., Frazier, K. A., Roth, B. D., Taschner, M. J., and Neuenschwander, K., *J. Org. Chem.*, **46,** 2417 (1981).
[606] Tartakovskaya, L. M., Makarov, I. G., Kazakova, V. M., Kipylov, V. M., and Zhdanov, A. A., *Izv. Akad. Nauk SSSR, Ser. Khim.*, **1980,** 2818; *Chem. Abs.*, **94,** 138932 (1981).
[607] Boyer, J., Corriu, R. J. P., Perz, R., and Réyé, C., *J. Chem. Soc., Chem. Commun.*, **1981,** 121.
[608] Deslongchamps, P., Rowan, D. D., and Pothier, N., *Heterocycles*, **15,** 1093 (1981); *Chem. Abs.*, **95,** 6347 (1981).
[609] Muller, P., and Blanc, J., *Tetrahedron Lett.*, **22,** 715 (1981).
[610] Cook, J., and Maitlis, P. M., *J. Chem. Soc., Chem. Commun.*, **1981,** 924.
[611] Corbett, M. D., and Corbett, B. R., *J. Org. Chem.*, **46,** 466 (1981).
[612] Babler, J. H., and Sarussi, S. J., *J. Org. Chem.*, **46,** 3367 (1981).
[613] Fukawa, H., Suzuki, K., and Sekiya, M., *Chem. Pharm. Bull.*, **28,** 2863 (1980); *Chem. Abs.*, **94,** 83433 (1981).
[614] Terpko, M. O., and Heck, R. F., *J. Org. Chem.*, **45,** 4992 (1980).
[615] Azran, J., Buchman, O., and Blum, J., *Tetrahedron Lett.*, **22,** 1925 (1981).
[616] Babler, J. H., and Invergo, B. J., *Tetrahedron Lett.*, **22,** 621 (1981).
[617] Trost, B. M., and Ornstein, P. L., *Tetrahedron Lett.*, **22,** 3463 (1981).
[618] Birch, A. J., Hinde, A. L., and Radom, L., *J. Am. Chem. Soc.*, **103,** 284 (1981).
[619] Rabideau, P. W., Peters, N. K., and Huser, D. L., *J. Org. Chem.*, **46,** 1593 (1981).
[620] Joo, W.-C., and Kim, C.-K., *Bull. Korean Chem. Soc.*, **1,** 75 (1980); *Chem. Abs.*, **94,** 102504 (1981).
[621] Chatterjee, A., Raychaudhuri, S. R., and Chatterjee, S. K., *Tetrahedron*, **37,** 3653 (1981).
[622] Iwasa, K., Chinnasamy, P., and Shamma, M., *J. Org. Chem.*, **46,** 1378 (1981).
[623] Iwata, C., Miyashita, K., Tomita, K., and Yamada, M., *J. Chem. Soc., Chem. Commun.*, **1981,** 461.
[624] Pradhan, S. K., Kadam, S. R., Kolhe, J. N., Radhakrishnan, T. V., Sohani, S. V., and Thaker, V. B., *J. Org. Chem.*, **46,** 2622 (1981).
[625] Pradhan, S. K., Kadam, S. R., and Kolhe, J. N., *J. Org. Chem.*, **46,** 2633 (1981).
[626] Pradhan, S. K., and Sohani, S. V., *Tetrahedron Lett.*, **22,** 4133 (1981).
[627] Rautenstrauch, V., Willhalm, B., and Thommen, W., *Helv. Chim. Acta*, **64,** 2109 (1981).
[628] Brettle, R., and Shibib, S. M., *J. Chem. Soc., Perkin Trans. 1*, **1981,** 2912.
[629] Bailey, W. F., and Cioffi, E. A., *J. Chem. Soc., Chem. Commun.*, **1981,** 155.
[630] Barrett, A. G., Godfrey, C. R., Hollinshead, D. M., Prokopiou, P. A., Barton, D. H. R., Boar, R. B., Joukhadar, L., McGhie, J. F., and Misra, S. C., *J. Chem. Soc., Perkin Trans. 1*, **1981,** 1501.
[631] Ohsawa, T., Takagaki, T., Haneda, A., and Oishi, T., *Tetrahedron Lett.*, **22,** 2583 (1981).
[632] Barrett, A. G., Prokopiou, P. A., and Barton, D. H. R., *J. Chem. Soc., Perkin Trans. 1*, **1981,** 1510.
[633] Walsh, T. D., and Dabestani, R., *J. Org. Chem.*, **46,** 1222 (1981).
[634] Handoo, K. L., and Handoo, S. K., *Indian J. Chem.*, **19B,** 684 (1980); *Chem. Abs.*, **94,** 14855 (1981).
[635] Tzeng, D., and Weber, W. P., *J. Org. Chem.*, **46,** 265 (1981).
[636] Schmidt, A. H., and Russ, M., *Chem. Ber.*, **114,** 822 (1981).
[637] Gurudutt, K. N., and Ravindranath, B., *Indian J. Chem.*, **20B,** 152 (1981); *Chem. Abs.*, **95,** 6152 (1981).
[638] Murphy, W. S., and Wattanasin, S., *Tetrahedron Lett.*, **22,** 695 (1981).
[639] Loia, M. C., Herold, B. J., and Atherton, N. M., *J. Chem. Soc., Perkin Trans. 2*, **1980,** 1788.
[640] Hart, H., and Nwokogu, G., *J. Org. Chem.*, **46,** 1251 (1981).
[641] Castedo, L., Saá, J. M., Suau, R., and Tojo, G., *J. Org. Chem.*, **46,** 4292 (1981).
[642] Clerici, A., Porta, O., and Riva, M., *Tetrahedron Lett.*, **22,** 1043 (1981).
[643] Dams, R., Malinowski, M., Westdorp, I., and Geise, H. J., *J. Org. Chem.*, **46,** 2407 (1981).
[644] Pons, J.-M., Zahra, J.-P., and Santelli, M., *Tetrahedron Lett.*, **22,** 3965 (1981).
[645] Armand, J., Chekir, K., and Pinson, J., *J. Heterocycl. Chem.*, **17,** 1237 (1980).
[646] Welch, S. C., and Loh, J.-P., *J. Org. Chem.*, **46,** 4072 (1981).
[647] Isaeva, S. A., Nikonava, L. A., and Shilov, A. E., *Nouveau J. Chim.*, **5,** 21 (1981).

648 Alper, H., and Amaratunga, S., *Tetrahedron Lett.*, **22,** 3811 (1981).
649 Girard, P., Couffignal, R., and Kagan, H. B., *Tetrahedron Lett.*, **22,** 3959 (1981).
650 Dewar, M. J. S., Nelson, D. J., Shevlin, P. B., and Biesiada, K. A., *J. Am. Chem. Soc.*, **103,** 2802 (1981).
651 Dyer, J. C., and Evans, S. A., *J. Org. Chem.*, **45,** 5350 (1980).
652 Rykova, L. A., Kiprianova, L. A., and Gragerov, I. P., *Teor. Eksp. Khim.*, **17,** 542 (1981); *Chem. Abs.*, **95,** 149616 (1981).
653 Kampars, V., Bumbure, G., Kokars, V., and Neilands, O., *Zh. Obshch. Khim.*, **50,** 2057 (1980); *Chem. Abs.*, **94,** 29866 (1981).
654 Mathur, P. C., Kumar, S., and Srivastava, S. S., *Chem. Era*, **15,** 18 (1979); *Chem. Abs.*, **93,** 220265 (1980).
655 Alper, H., and Hashem, K. E., *J. Am. Chem. Soc.*, **103,** 6514 (1981).
656 de Vries, J. G., and Kellogg, R. M., *J. Org. Chem.*, **45,** 4126 (1980).
657 Krapcho, A. P., and Seidman, D. A., *Tetrahedron Lett.*, **22,** 179 (1981).
658 Seshadri, R., Pegg, W. J., and Israel, M., *J. Org. Chem.*, **46,** 2596 (1981).
659 Denis, J. N., and Krief, A., *Tetrahedron Lett.*, **22,** 1431 (1981).
660 Griffiths, D. V., and Tebby, J. C., *J. Chem. Soc., Chem. Commun.*, **1981,** 607.
661 Hirao, T., Masunaga, T., Ohshiro, Y., and Agawa, T., *J. Org. Chem.*, **46,** 3745 (1981).
662 Harpp, D. N., Ash, D. K., and Smith, R. A., *J. Org. Chem.*, **45,** 5155 (1980).
663 Platt, K. L., and Oesch, F., *J. Org. Chem.*, **46,** 2601 (1981).
664 Oae, S., Iida, K., and Takata, T., *Tetrahedron Lett.*, **22,** 573 (1981).
665 Halpern, J., *Inorg. Chim. Acta*, **50,** 11 (1981).
666 Stults, B. R., Friedman, R. M., Koenig, K., Knowles, W., Gregor, R. B., and Lytle, F. W., *J. Am. Chem. Soc.*, **103,** 3235 (1981).
667 Descotes, G., Lafont, D., Sinou, D., Brown, J. M., Chaloner, P. A., and Parkey, D., *Nouveau J. Chim.*, **5,** 167 (1981).
668 Ojima, I., Kogure, T., and Yoda, N., *J. Org. Chem.*, **45,** 4728 (1980).
669 Sinou, D., *Tetrahedron Lett.*, **22,** 2987 (1981).
670 MacNeil, P. A., Roberts, N. K., and Bosnich, B., *J. Am. Chem. Soc.*, **103,** 2273 (1981).
671 Fiorini, M., and Giongo, G. M., *J. Mol. Catal.*, **7,** 411 (1980); *Chem. Abs.*, **94,** 120569 (1981).
672 Riley, D. P., and Shumate, R. E., *J. Org. Chem.*, **45,** 5187 (1980).
673 Baker, G. L., Fritschel, S. J., and Stille, J. K., *J. Org. Chem.*, **46,** 2960 (1981).
674 Hayashi, T., Kanehira, K., and Mukai, T., *Tetrahedron Lett.*, **22,** 4417 (1981).
675 Baker, G. L., Fritschel, S. J., Stille, J. R., and Stille, J. K., *J. Org. Chem.*, **46,** 2954 (1981).
676 Meyer, D., Poulin, J.-C., Kagan, H. B., Levine-Pinto, H., Morgat, J.-L., and Fromageot, P., *J. Org. Chem.*, **45,** 4680 (1980).
677 Osakada, K., Obana, M., Ikariya, T., Saburi, M., and Yoshikawa, S., *Tetrahedron Lett.*, **22,** 4297 (1981).
678 Ohkubo, K., Terada, I., Sugahara, K., and Yoshinaga, K., *J. Mol. Catal.*, **7,** 421 (1980); *Chem. Abs.*, **94,** 120570 (1981).
679 Harada, K., Munegumi, T., and Nomoto, S., *Tetrahedron Lett.*, **22,** 111 (1981).
680 Fahey, D. R., *J. Am. Chem. Soc.*, **103,** 136 (1981).
681 Suggs, J. W., Cox, S. D., Crabtree, R. H., and Quirk, J. M., *Tetrahedron Lett.*, **22,** 303 (1981).
682 Reid, A. H., Shevlin, P. B., Yun, S. S., and Webb, T. R., *J. Am. Chem. Soc.*, **103,** 709 (1981).
683 Khan, M. M. T., and Mohiuddin, R., *Proc. Natl. Symp. Catal., 4th*, **1978,** 374; *Chem. Abs.*, **94,** 64845 (1981).
684 Zuber, M., Banas, B., and Pruchnik, F., *J. Mol. Catal.*, **10,** 143 (1981); *Chem. Abs.*, **94,** 156026 (1981).
685 Shanthalakshmy, P., Vancheesan, S., Rajaram, J., and Kuriacose, J. C., *Indian J. Chem.*, **19A,** 901 (1980); *Chem. Abs.*, **94,** 83328 (1981).
686 Veselova, M. E., and Sul'man, E. M., *Svoistva Veshchestv i Stroenie Molekul, Kalinin*, **1980,** 140; *Chem. Abs.*, **94,** 64848 (1981).
687 Van Rantwijk, F., and Van Bekkum, H., *Delft Prog. Rep.*, **5,** 14 (1980); *Chem. Abs.*, **94,** 46636 (1981).
688 Banerjee, T. K., and Saha, C. R., *Indian J. Chem.*, **19A,** 964 (1980); *Chem. Abs.*, **94,** 120578 (1981).
689 Chepaikin, E. G., Khidekel, M. L., Ivanova, V., Zakhariev, A., and Shopov, D., *J. Mol. Catal.*, **10,** 115 (1980); *Chem. Abs.*, **94,** 120571 (1981).
690 Fujitsu, H., Shirahama, S., Matsumura, E., Takeshita, K., and Mochida, I., *J. Org. Chem.*, **46,** 2287 (1981).
691 Ward, M. D., and Schwartz, J., *J. Am. Chem. Soc.*, **103,** 5253 (1981).
692 Holy, N. L., and Shelton, S. R., *Tetrahedron*, **37,** 25 (1981).
693 Maier, W. F., Bergmann, K., Bleicher, W., and Schleyer, P. von R., *Tetrahedron Lett.*, **22,** 4227 (1981).
694 Van Tamelen, E. E., and Webber, B. D., *Proc. Natl. Acad. Sci. USA*, **78,** 1321 (1981).

[695] Accrombessi, G. C., Geneste, P., Olive, J.-L., and Pavia, A. A., *J. Org. Chem.*, **45,** 4139 (1980).
[697] Schreifels, J. A., Maybury, P. C., and Swartz, W. E., *J. Org. Chem.*, **46,** 1263 (1981).
[698] Fushimi, M., Baba, N., Oda, J., and Inouye, Y., *Bull. Inst. Chem. Res., Kyoto Univ.*, **58,** 357 (1981); *Chem. Abs.*, **94,** 46478 (1981).
[699] De Konig, A. J., Budzelaar, P. H., Boersma, J., and Van der Kerk, G. J., *J. Organomet. Chem.*, **199,** 153 (1980).
[700] Ono, N., Tamura, R., Ranikaga, R., and Kaji, A., *J. Chem. Soc., Chem. Commun.*, **1981,** 71.
[701] Bunting, J. W., and Sindhuatmadja, S., *J. Org. Chem.*, **46,** 4211 (1981).
[702] Ohno, A., Shio, T., Yamamoto, H., and Oka, S., *J. Am. Chem. Soc.*, **103,** 2045 (1981).
[703] Shinkai, S., Tsuno, T., and Manabe, O., *Chem. Lett.*, **1981,** 1203.
[704] Ohno, A., Yamamoto, H., and Oka, S., *J. Am. Chem. Soc.*, **103,** 2041 (1981).
[705] Pac, C., Ihama, M., Yasuda, M., Miyauchi, Y., and Sakurai, H., *J. Am. Chem. Soc.*, **103,** 6495 (1981).
[706] Murakami, Y., Aoyama, Y., and Kikuchi, J., *J. Chem. Soc., Chem. Commun.*, **1981,** 444.
[707] Jouin, P., Troostwijk, C. B., and Kellogg, R. M., *J. Am. Chem. Soc.*, **103,** 2091 (1981).
[708] Seki, M., Baba, N., Oda, J., and Inouye, Y., *J. Am. Chem. Soc.*, **103,** 4613 (1981).
[709] Gase, R. A., and Pandit, U. K., *Recl. Trav. Chim. Pays-Bas*, **99,** 334 (1980).
[710] Yoneda, F., Yamato, H., and Ono, M., *J. Am. Chem. Soc.*, **103,** 5943 (1981).
[711] DiCosimo, R., Wong, C.-H., Daniels, L., and Whitesides, G. M., *J. Org. Chem.*, **46,** 4622 (1981).
[712] Shinkai, S., Hamada, H., Kuroda, H., and Manabe, O., *J. Org. Chem.*, **46,** 2333 (1981).
[713] Ono, A., Yasui, S., and Oka, S., *Bull. Chem. Soc. Jpn.*, **53,** 3244 (1980).
[714] Bunting, J. W., and Sindhuatmadja, S., *J. Org. Chem.*, **45,** 5411 (1980).
[715] Ohno, A., Yasui, S., Gase, R. A., Oka, S., and Pandit, U. K., *Bioorg. Chem.*, **9,** 199 (1980); *Chem. Abs.*, **93,** 167289 (1980).
[716] Mathis, R., Dupas, G., Decoreille, A., and Queguiner, G., *Tetrahedron Lett.*, **22,** 59 (1981).
[717] Loechler, E. L., and Hollocher, T. C., *J. Am. Chem. Soc.*, **102,** 7328 (1980).
[718] Loechler, E. L., and Hollocher, T. C., *J. Am. Chem. Soc.*, **102,** 7322 (1980).
[719] Yoneda, F., Kuroda, K., and Kamishimoto, M., *J. Chem. Soc., Chem. Commun.*, **1981,** 1160.
[720] Loechler, E. L., and Hollocher, T. C., *J. Am. Chem. Soc.*, **102,** 7312 (1980).
[721] Shepherd, P. T., and Bruice, T. C., *J. Am. Chem. Soc.*, **102,** 7774 (1980).
[722] Fried, H. E., and Kaiser, E. T., *J. Am. Chem. Soc.*, **103,** 182 (1981).
[723] Muto, S., and Bruice, T. C., *J. Am. Chem. Soc.*, **102,** 7559 (1980).
[724] Shannon, P., and Bruice, T. C., *J. Am. Chem. Soc.*, **103,** 4580 (1981).
[725] Frost, J. W., and Rastetter, W. H., *J. Am. Chem. Soc.*, **103,** 5242 (1981).
[726] Ball, S., and Bruice, T. C., *J. Am. Chem. Soc.*, **103,** 5494 (1981).
[727] Baumstark, A. L., and Chrisope, D. R., *Tetrahedron Lett.*, **22,** 4591 (1981).
[728] Addink, R., and Berends, W., *Tetrahedron*, **37,** 833 (1981).
[729] Shinkai, S., Yamashita, T., Kusano, Y., and Manabe, O., *J. Org. Chem.*, **45,** 4947 (1980).
[730] Desrut, M., Kergomard, A., Renard, M. F., and Veschambre, H., *Tetrahedron*, **37,** 3825 (1981).
[731] Mori, K., *Tetrahedron*, **37,** 1341 (1981).
[732] Fauve, A., and Kergomard, A., *Tetrahedron*, **37,** 899 (1981).
[733] Nakazaki, M., Chikamatsu, H., and Asao, M., *J. Org. Chem.*, **46,** 1147 (1981).
[734] Nakazaki, M., Chikamatsu, H., Fujii, T., and Nakatsuji, T., *J. Org. Chem.*, **46,** 585 (1981).
[735] Nakazaki, M., Chikamatsu, H., Nishino, M., and Murakami, H., *J. Org. Chem.*, **46,** 1151 (1981).
[736] Hill, R. K., Rhee, S.-W., Leete, E., and McGaw, B. A., *J. Am. Chem. Soc.*, **102,** 7344 (1980).
[737] de Smet, M.-J., Witholt, B., and Wynberg, H., *J. Org. Chem.*, **46,** 3128 (1981).
[738] Caspi, E., Shapiro, S., and Piper, J. U., *Tetrahedron*, **37,** 3535 (1981).
[739] Caspi, E., Piper, J., and Shapiro, S., *J. Chem. Soc., Chem. Commun.*, **1981,** 76.
[740] Quigley, F. R., and Floss, H. G., *J. Org. Chem.*, **46,** 464 (1981).
[741] Dalton, H., Golding, B. T., Waters, B. W., Higgins, R., and Taylor, J. A., *J. Chem. Soc., Chem. Commun.*, **1981,** 482.
[742] Akhtar, M., Calder, M. R., Corina, D. L., and Wright, J. N., *J. Chem. Soc., Chem. Commun.*, **1981,** 129.
[743] Hanson, L. K., Chang, C. K., Davis, M. S., and Fajer, F., *J. Am. Chem. Soc.*, **103,** 663 (1981).
[744] Silverman, R. B., Hoffman, S. J., and Catus, W. B., *J. Am. Chem. Soc.*, **102,** 7126 (1980).
[745] Ortiz de Montellano, P. R., and Kunze, K. L., *J. Am. Chem. Soc.*, **102,** 7373 (1980).
[746] Burka, L. T., Thorsen, A., and Guengerich, F. R., *J. Am. Chem. Soc.*, **102,** 7615 (1980).
[747] Gunter, M. J., Mander, L. N., and Murray, K. S., *J. Chem. Soc., Chem. Commun.*, **1981,** 799.
[748] Chang, C. K., and Ebina, F., *J. Chem. Soc., Chem. Commun.*, **1981,** 778.
[749] Chevrier, B., Weiss, R., Lange, M., Chottard, J.-C., and Mansuy, D., *J. Am. Chem. Soc.*, **103,** 2899 (1981).
[750] Lauffer, R. B., Heistand, R. H., and Que, L., *J. Am. Chem. Soc.*, **103,** 3947 (1981).

[751] Silverman, R. B., *J. Am. Chem. Soc.*, **103,** 5939 (1981).
[752] Rajananda, V., and Brown, S. B., *Tetrahedron Lett.*, **22,** 4331 (1981).
[753] Klinman, J. P., Humphries, H., and Voet, J. G., *J. Biol. Chem.*, **255,** 1648 (1980).
[754] White, R. E., Silgar, S. G., and Coon, M. J., *J. Biol. Chem.*, **255,** 1108 (1980).
[755] Grimshaw, C. E., Cook, P. F., and Cleland, W. W., *Biochemistry*, **20,** 5665 (1981).
[756] Damgaard, S. E., *Biochemistry*, **20,** 5662 (1981).
[757] Vallari, R. C., and Pietruszko, R., *Arch. Biochem. Biophys.*, **212,** 9 (1981).
[758] Gomes, B., Fendrich, G., and Abeles, R. H., *Biochemistry*, **20,** 1481 (1981).
[759] Damgaard, S. E., *Biochem. J.*, **191,** 613 (1980).
[760] Akiyama, S. K., and Hammes, G. G., *Biochemistry*, **20,** 1491 (1981).
[761] Robertson, P. J., Fridovich, S. E., Misra, H. P., and Fridovich, I., *Arch. Biochem. Biophys.*, **207,** 282 (1981).
[762] Cook, P. F., Oppenheimer, N. J., and Cleland, W. W., *Biochemistry*, **20,** 1817 (1981).
[763] Kishore, G. M., and Snell, E. E., *J. Biol. Chem.*, **256,** 4228 (1981).
[764] Nagoaka, S., Yu. L., and King, T. E., *Arch. Biochem. Biophys.*, **208,** 334 (1981).
[765] May, S. W., Phillips, R. S., Mueller, P. W., and Herman, H. H., *J. Biol. Chem.*, **256,** 8470 (1981).
[766] Cook, P. F., and Cleland, W. W., *Biochemistry*, **20,** 1805 (1981).
[767] Davies, H. C., Smith, L., and Nava, M. E., *Arch. Biochem. Biophys.*, **210,** 49 (1981).
[768] Kalhorn, T., Becker, A. R., and Sternson, L. A., *Bioorg. Chem.*, **10,** 144 (1981).
[769] Winterbourn, C. C., *Arch. Biochem. Biophys.*, **209,** 159 (1981).
[770] Ortiz de Montellano, P. R., and Kunze, K. L., *Arch. Biochem. Biophys.*, **209,** 710 (1981).
[771] Hardman, M. J., *Biochem. J.*, **195,** 773 (1981).
[772] Iyanagi, T., Makino, R., and Anan, F. K., *Biochemistry*, **20,** 1722 (1981).

Organic Reaction Mechanisms 1981
Edited by A. C. Knipe and W. E. Watts

CHAPTER 5

Carbenes and Nitrenes

C. J. MOODY

Department of Chemistry, Imperial College of Science and Technology, London, SW7

Structure and Reactivity	239
Generation	240
Carbenes	240
Nitrenes	242
Addition	243
Intermolecular	243
Intramolecular	245
Insertion and Abstraction	246
Intermolecular	246
Intramolecular	247
Rearrangement	248
Carbenes	248
Nitrenes	251
Aromatics	252
Carbenes	252
Nitrenes	252
Nucleophiles and Electrophiles	255
Silylenes	256
Transition-metal Complexes	257
References	258

Structure and Reactivity

MINDO/3 calculations on carbenes have been reviewed.[1]

Single–triplet gaps for 30 carbenes have been computed by STO-3G methods, and show a linear correlation with theoretical and empirical measures of the π-donation by substituents; increasing π-donation stabilizes the singlet. Electronegativity arguments are thought to be unnecessary.[2]

ESR studies on *o*-carboranylcarbene at 5 K indicate little or no delocalization of π-spin density into the carboranyl moiety.[3] The absolute decay rates of diphenylcarbene and fluorenylidene have been measured by ESR in a matrix, and are pseudo first order from reaction of the carbene with the host.[4] Singlet fluorenylidene exhibits radical-like abstraction reactions,[5] and also gives cyclopropanes non-stereospecifically with *cis*-olefins.[6] The reaction with *trans*-olefins is stereospecific.

Conformational barriers in 1- and 2-naphthylcarbenes have been determined by ESR in a matrix at 77 K, and show, for example, that interconversion of (**1**) and (**2**) is much slower than the reaction with the host. The barrier is estimated at >19–$26\ \mathrm{kJ\,mol^{-1}}$.[7] Similar studies on quinolylcarbenes have been reported. The 3-, 4-, and 8-quinolylcarbenes all show two similar, but non-identical, triplet signals in the ESR spectrum, assigned to the rotamers with different orientation of the carbene hydrogen. In contrast, the number of species observed for the 1-isomer (**3**) is matrix-dependent. The zero-field parameters for all the carbenes are quite similar to those of naphthylcarbene, implying, rather surprisingly, that the close proximity of a nitrogen atom to a divalent carbon does not have a substantial effect on π-spin distribution.[8]

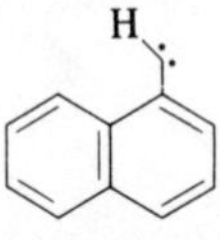

(**1**) 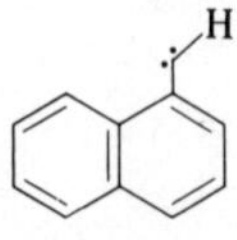 (**2**) 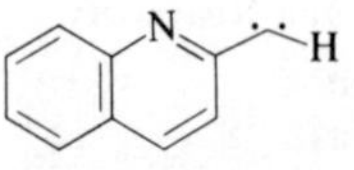 (**3**)

Ab initio calculations on the lowest singlet of cyclopentadienylidene show two energy minima corresponding to localized and delocalized π-bonds having 4π- and 6π-electrons respectively.[9] The related, but previously unknown, 2*H*-imidazolylidene (**4**) has been generated, and shown to be a very reactive, highly electrophilic carbene with strong diradical character.[10] A theoretical study of the fragmentation of carbene (**5**) shows that a step-wise process is of lower energy than concerted fragmentation, but that free carbenes are unlikely to be involved in olefin-forming reactions.[11] MNDO calculations suggest that cycloheptatetraene (**6**) is more stable than the planar isomeric cycloheptatrienylidene by $96\ \mathrm{kJ\,mol^{-1}}$.[12]

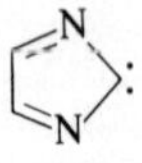

(**4**) 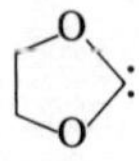 (**5**)  (**6**) 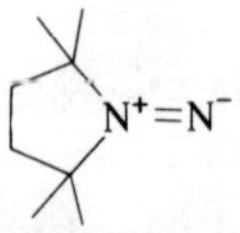 (**7**)

The aminonitrene (**7**), stable in solution at -78°C, decomposes on photolysis to a mixture of the corresponding tetrazene and four hydrocarbons, formed by loss of nitrogen.[13]

Generation

Carbenes

The acid-catalysed decomposition of α-diazoketones has been reviewed.[14] *Ab initio* calculations on the decomposition of diazomethane give potential energy characteristics and rate constants that agree with experimental data, and indicate that $^1A_1$ methylene is the only species generated on pyrolysis.[15] Photochemical generation of methylene from ketene gives acetylene and molecular hydrogen as a result of triplet methylene recombination.[16] In the matrix photochemical generation of cyclopentadienylidene from diazocyclopentadiene, the loss of nitrogen is irreversible, and formation of fulvalene is shown to be a true dimerization of the carbene. Photolysis with plane-polarized light gives a matrix which exhibits linear dichroism.[17]

(8) SCHEME 1 **(9)**

Attempted reduction of the nitro compound (**8**) to the amine gives unexpected products derived from the carbene (**9**) formed as shown in Scheme 1.[18] The carbene (**10**) has been generated and shown to be highly nucleophilic.[19] Cycloheptatrienylidene is formed on desilylation of the tropylium salt (**11**) with fluoride ion.[20] The tris-fused cycloheptatrienylidene (**12**) has been generated from the corresponding tosylhydrazone. Its properties are consistent with a facile singlet–triplet equilibrium.[21] The generation of cyclononatetraenylidene (**13**) has been reported; it dimerizes *via* the triplet.[22]

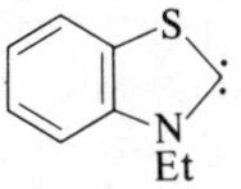

(10) 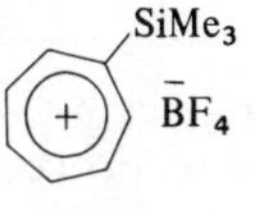 **(11)** 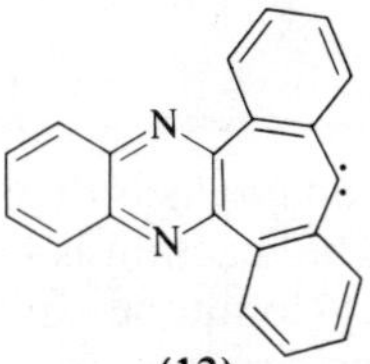 **(12)** 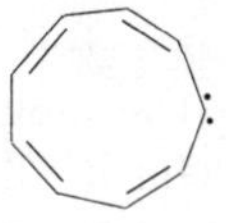 **(13)**

Treatment of 7,7-dichlorobicyclo[4.1.0]heptane with KOBu^t generates carbene (**14**) *via* rearrangement of (**15**).[23] Similarly, treatment of (**16**) with KOBu^t proceeds through the bicyclo[4.1.0]heptatriene (**17**), thus constituting an independent generation of an intermediate involved in arylcarbene rearrangement. The products derived from (**16**) possibly involve rearrangement of (**17**) to 2-tolylcarbene.[24]

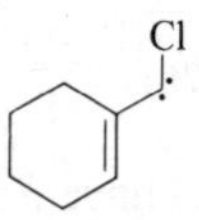

(14) 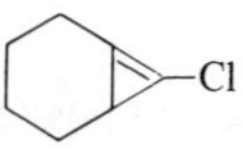 **(15)** 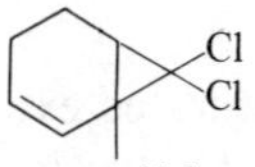 **(16)** 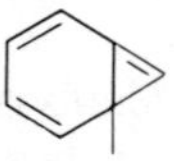 **(17)**

Treatment of 2-nitrobenzyl chloride with aqueous alkali gives 2,2′-dinitrostilbene. The kinetics are in accord with a rate-determining unimolecular generation of 2-nitrophenylcarbene.[25] The aryldinitromethane salt, $Ar\bar{C}(NO_2)_2\overset{+}{A}g$, undergoes α-elimination to give the aryl(nitro)carbene. The products are derived from the resonance form of the carbene (**18**).[26] A new method of arylcarbene generation involves elimination of trimethylsilanol from the precursors (**19**). The method is also applicable to silylcarbenes using 1,1-bis(trimethylsilyl)-substituted alcohols.[27]

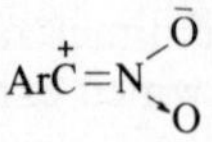

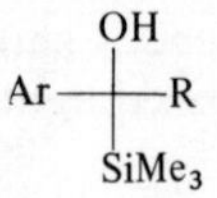

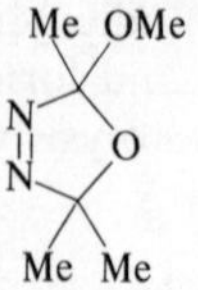

(18) **(19)** **(20)**

Thermal decomposition of the oxadiazoline (**20**) generates dimethylcarbene and methoxy(methyl)carbene *via* fragmentation of the intermediate carbonylylid.[28] Spirocyclopropenes (**21**) ring-open on heating at 140°C to give vinylcarbenes which are trapped by reaction with CS_2.[29] γ,δ-Epoxyenones (**22**) undergo photochemical ring-opening to give products derived from the carbene (**23**).[30] Photolysis of propellanes incorporating cyclobutanones gives oxacarbenes *via* ring-expansion.[31]

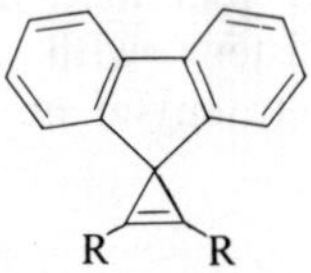

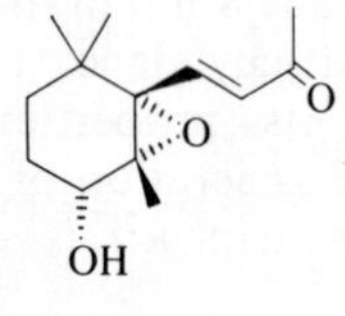

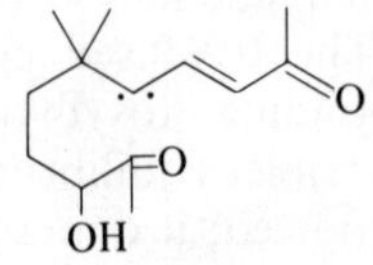

(21) **(22)** **(23)**

Thermolysis of tricycloheptylaziridines (**24**) gives 3-cyanocyclohexene and the olefin $R^1CH{=}CR^2R^3$. The reaction is believed to proceed through retro-carbene ring-opening of the bicyclobutane unit to generate carbene (**25**) which then fragments to the observed products.[32]

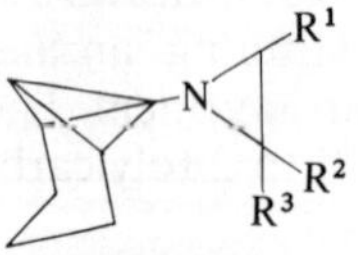

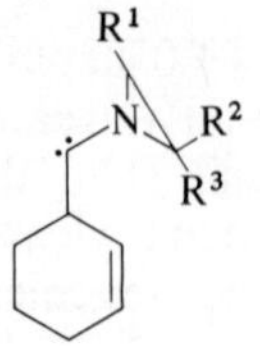

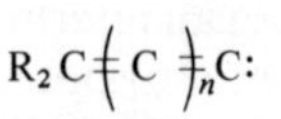

(24) **(25)** **(26)**

Full details of highly unsaturated extended carbenes (**26**; $n = 4$) have appeared. The carbenes are unencumbered electrophilic singlets.[33] Less is known about the carbenes (**26**) where n is an odd number; however, the carbene (**26**; $n = 3$, R = Me) has now been generated.[34] The synthesis of small and strained ring compounds *via* these unsaturated carbenes has been reviewed.[35]

Nitrenes

The decomposition of phenyl azide in acetic acid has been shown to proceed through the singlet nitrene rather than an azide conjugate acid.[36] Phthalimido-diazene-1-oxides (**27**), formed by trapping of oxidatively generated phthalimidonitrene with nitroso compounds, regenerate the nitrene on photolysis, and hence are useful alternative photo-precursors of phthalimidonitrene.[37,38] Cyanonitrene can be generated from sodium cyanamide by treatment with Bu^{t}OCl in methanol at 0–10°C; at lower temperatures the α-elimination of NaCl does not occur. Excellent evidence for the intermediacy of the nitrene is presented, including the observation of an ESR signal for the triplet state.[39] A variety of nitrenes has been generated by thermal α-deoxysilylation of bis-trimethylsilylhydroxylamines (**28**).[40]

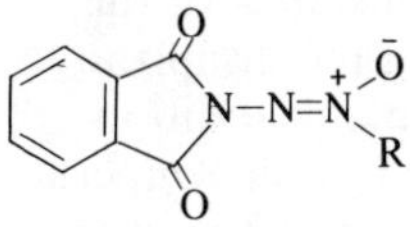

(27) 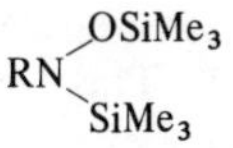 **(28)** 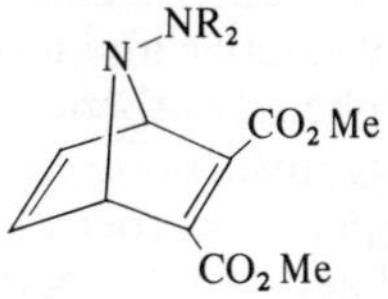 **(29)**

Photolysis of 1-benzyloxy-1,2,3-benzotriazole leads to products derived from phenylnitrene. The nitrene is formed by fragmentation of the initial diradical intermediate with loss of benzaldehyde.[41] Treatment of *N*-aminopyrroles with dimethyl acetylenedicarboxylate at 60°C generates aminonitrenes *via* fragmentation of the Diels–Alder adducts (**29**). In certain cases, a previously unreported reaction of dialkylaminonitrenes was observed. For example, reaction of *N*-dimethylamino-pyrrole with the acetylene gave products derived from addition of the azomethinimine tautomer of the nitrene to the triple bond.[42] Thiazyl fluoride, FS≡N, is a tautomer of fluorosulphenylnitrene. Cycloaddition products derived from the nitrene are formed on photolysis of FS≡N in fluoro-olefins.[43] Nitrous acid deamination of 1,1-disubstituted arylhydrazines generates triplet nitrenium ions.[44]

Addition

Intermolecular

It has been shown that contrary to common understanding the selectivity of singlet carbenes in 1,2-addition to olefins can increase with increasing carbene reactivity. The FMO model shows that whilst halocarbenes obey the classical reactivity–selectivity relationship, other carbenes, for example dimethoxycarbene, follow the inverse relationship.[45] The temperature dependence of the linear free energy relationship of carbene cycloadditions has been determined. The selectivity of carbenes with substituents of weak electron-withdrawing ability increases with increasing temperature.[46] Selective additions to 6,6-dimethylfulvene constitute a useful probe for carbene philicity. Electrophilic carbenes attack exclusively at the endocyclic double bonds, whereas the exocyclic double bond is attacked by nucleophilic carbenes.[47]

Weakly nucleophilic acceptors such as enol ethers can facilitate the addition reactions of monohalocarbenoids. The high yields and stereoselection towards the more highly hindered *syn*-adducts are explained by complexation of the oxygen atom of the substrate with the carbenoid metal atom.[48] The cyclopropanation of olefins by irradiation of the olefin with diiodomethane is synthetically useful, and is significantly less subject to steric effects than the Simmons–Smith reaction. The iodomethyl cation is thought to be the active methylene-transfer species.[49] Cyclopropanation reactions of dimethylcarbene are much less common. Although reasonable yields are obtained with 2,3-dimethylbut-2-ene, other olefins give poor yields of cyclopropanes, indicating that this carbene puts severe demand on the nucleophilicity of the olefin substrate.[50] The solvent effects on the reaction of tetracyanoethylene with diphenyldiazomethane have been measured, and the second-order rate constants decrease with increasing solvent basicity. The reaction has been proposed as a new parameter for solvent basicity.[51]

Chloro(trifluoromethyl)carbene adds to olefins stereospecifically, despite the fact that the electron-withdrawing CF_3 group destabilizes the carbene, and hence makes it highly reactive and unselective.[52] The absolute rate constants for the addition of chloro(phenyl)carbene have been measured by following the disappearance of the carbene absorption at 320 nm.[53] The intermolecular addition of cyclopropyl(phenyl)carbene to 2-methylpropene is favoured at lower temperatures. The ratio of addition to rearrangement to 1-phenylcyclobutene increases from 3:63 at 25°C to 42:10 at −128°C. If both products arise from competitive reactions of the singlet, the results suggest that the activation energy for rearrangement must be greater than that for addition by 5.4 kJ mol^{-1}. Although not impossible, this was thought unlikely. Some evidence for triplet involvement was found, although no firm conclusions could be drawn.[54] Adamantylidene (**30**) adds stereospecifically to olefins to give spirocyclopropanes, and also shows more selectivity than expected for a dialkylcarbene in its addition reactions. Selectivity is also shown in C—H insertion reactions where steric hindrance disfavours insertion into tertiary C—H bonds.[55]

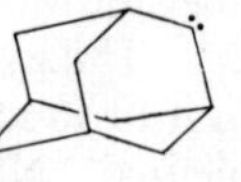

CO_2Me CO_2Me Cl Cl

(30) 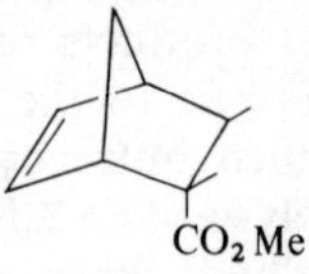 **(31)** 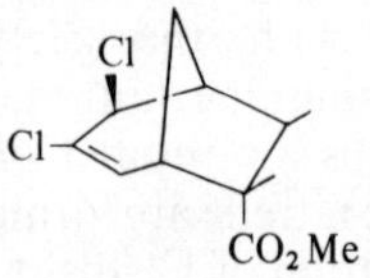 **(32)**

Dichlorocarbene, generated from $PhHgCBrCl_2$, adds to (**31**) to give exclusively (**32**) after rearrangement. If the carbene is generated from chloroform by the phase-transfer method, mixtures of products result.[56] The addition of phase-transfer-generated dihalocarbenes to allyl halides proceeds to give mixtures of dihalocyclopropanes and allyl trihalomethanes *via* nucleophilic attack of trihalomethyl anion.[57] 1,1-Dichlorocyclopropane, formed by the addition of methylene generated photochemically from ketene to 1,1-dichloroethylene, is formed in a chemically activated state, and rapidly rearranges to 1,1-dichloro- and 2,3-dichloropropene. The rate constants have been measured.[58]

The ratio of *exo*-1,2- to homo-1,4-addition of halocarbenes to norbornadienes (**33**) has been taken as a measure of carbene nucleophilic character. However, it was not clear whether the ratio was affected by an increase in the homo-1,4-nucleophilic addition or a decrease in the 1,2-electrophilic addition. This has now been resolved by the measurement of partial rate constants for the addition reactions of dichlorocarbene which show that it is the *exo*-1,2-addition mode that is affected more by changes in the norbornadiene substituent, R.[59] The addition of carbenes to 1,2,2-trimethylbicyclo[1.1.0]butane (**34**) gives the diene (**35**) as the major product. A concerted two-bond pluck mechanism in which the central and one side bond are simultaneously cleaved as shown is preferred. This mechanism has analogy in the fragmentation of cyclopropylidene to acetylene and ethylene.[60]

MNDO calculations on the reaction of atomic carbon with aldehydes indicate that the preferred path should involve addition across the C=O bond to give an

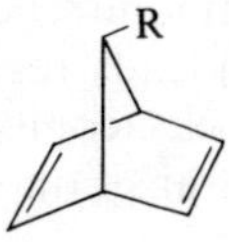

(33) 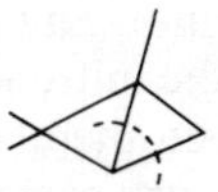 (34) 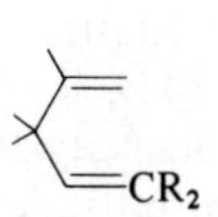 (35)

oxiranylidene. The initial "carbene" addition is followed by rearrangement to a ketene, which is formed with enough excess energy to undergo dissociation to CO and another carbene. Experimental observations support the calculations.[61]

Several studies on the addition reactions of ethoxycarbonylnitrene have been reported. The stereochemistry of the addition to norbornene is found to be predominantly *exo*,[62] and in the addition to cyclic vinyl ethers the preferred attack leads to the aziridine (**36**), particularly if the substituent, R, is axial. The aziridines are readily *trans*-ring-opened, and the reaction has been applied to the synthesis of amino-sugars.[63] The addition of the nitrene to azoisopropane gives an azimine, photolysis of which leads to the first example of a triaziridine (**37**).[64] Addition to mesoionic 4-thiazolones (**38**) proceeds *via* addition of the nitrene to the 1,3-dipole system. This is the first reported example of a [1 + 3]cycloaddition of a nitrene.[65]

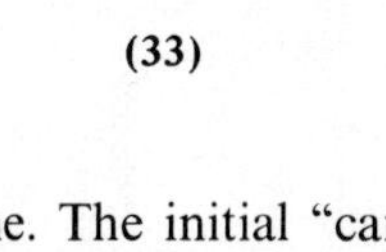

(36) 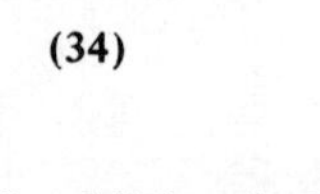 (37) 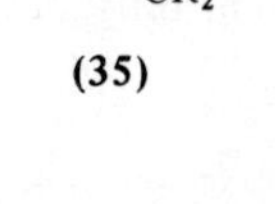 (38)

2,4-Dinitrophenylsulphenylnitrene adds readily to electron-rich olefins to give aziridines. There is no reaction with electron-deficient olefins, and the addition to *cis*-β-methylstyrene is non-stereospecific.[66]

Intramolecular

Generation of the carbene (**39**) leads to 1-methylbenzvalene *via* an intramolecular 1,4-addition reaction. MINDO/3 calculations predict a bisected conformation of the singlet ground state.[67] The intramolecular 1,1-cycloaddition reaction of diazoalkanes can be viewed as a formal concerted cycloaddition of the singlet nitrene tautomer (**40**). The addition is stereospecific with exclusive formation of the *exo*-isomer (**41**). A step-wise reaction followed by very rapid collapse of the dipole cannot be ruled out.[68]

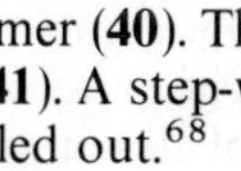

(39) 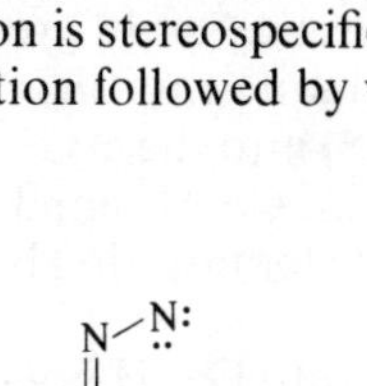 (40) 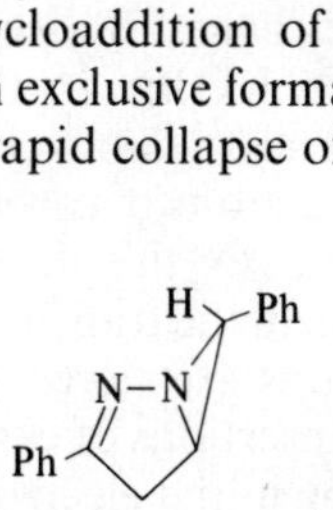 (41)

The competitive intramolecular additions of aminonitrenes have been investigated. Oxidation of the *N*-aminoquinazolone (**42**) leads regio- and stereospecifically to (**43**). The attack of the nitrene is stereoelectronically controlled proceeding exclusively through a seven-membered ring transition state in a stepwise fashion *via* fully developed dipolar species.[69]

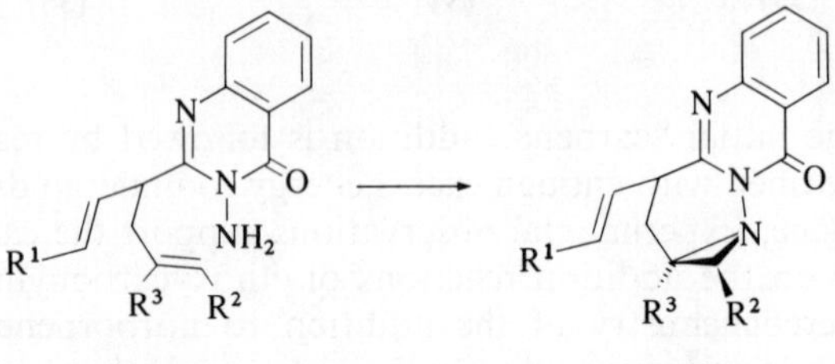

(**42**) (**43**)

Insertion and Abstraction

Intermolecular

Concurrent generation of difluorocarbene and hydrogen atoms by photolysis of CF_2N_2 and HI or H_2S in a matrix at 14 K leads to a prominent IR absorption of $\cdot CHF_2$, demonstrating that this is the primary product.[70] Clear evidence for insertion reactions of halogens with bromo(fluoro)carbene, generated by reaction of atomic oxygen with $F_2C{=}CFBr$, has been obtained. The triplet carbene was monitored in a discharge flow system.[71]

Good yields of C—H insertion products are obtained in the rhodium(II)-catalysed decomposition of ethyl diazoacetate in hydrocarbons at 22°C. Rhodium(II) derivatives of strong organic acids ($pK < 1$) are particularly effective catalysts, and the reaction probably proceeds through a rhodium carbenoid, since different C—H selectivity is observed from the photochemically generated free carbene. The C—H insertion selectivity is also catalyst-dependent, illustrating the importance of carbenoid–lipophilic interactions.[72] The Wolff rearrangement of benzoyl(phenyl)carbene is suppressed in the presence of transition metals. Bis(acetylacetonato)copper(II)-catalysed decomposition of azibenzil in the presence of primary amines leads to products derived from insertion of the carbene into the N—H bond.[73]

Thermolysis of tetrachlorocyclopropene generates tetrachlorovinylcarbene (**44**), which gives good yields of C—H insertion products with a variety of hydrocarbon and ether solvents.[74] Insertion into similar solvents has also been observed for aryl(halo)carbenes. These free carbenes, generated by the reaction of arylidene halides with KOBu^t in the presence of 18-crown-6, show similar selectivity to dihalocarbenes.[75] A novel oxy-anion substituent effect has been noted in the C—H insertion reactions of carbene (**45**) into the α-C—H bond of alkoxides. The effect is possibly due to the decrease in C—H bond energy in the alkoxide, or the stabilization of the transition state for insertion by the electron-donating oxy-anion. The insertions are stereospecific.[76]

Formal insertions of carbenes into O—H bonds can proceed through a one-step insertion, by initial electrophilic attack on oxygen, or by initial protonation of the carbene. These possibilities have been investigated for cyclopentadienylidene and cycloheptatrienylidene using CH_3OD and C_2H_5OD. Not surprisingly the

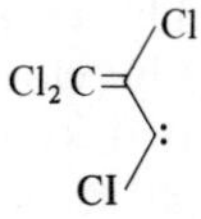

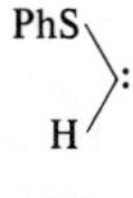

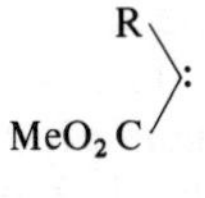

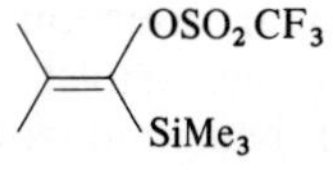

(44) (45) (46) (47)

electrophilic cyclopentadienylidene is found to react by electrophilic attack on oxygen, whereas initial protonation is the favoured mechanism in the case of cycloheptatrienylidene.[77] Although O—H insertion is the major reaction of methoxycarbonyl(phenyl)carbene (**46**; R = Ph) with alcohols at room temperature, in a rigid matrix of the alcohol a dramatic increase in C—H insertion occurs, probably through the ground-state triplet carbene. No such increase in C—H insertion occurs for methoxycarbonylcarbene (**46**; R = H), suggesting that this carbene is a ground-state singlet.[78]

Treatment of (**47**) with fluoride ion generates isopropylidenecarbene which reacts with thioketones *via* an insertion into the S—H bond of the enethiol form to give divinyl sulphides in good yields.[79] Extended unsaturated carbenes (**26**; $n = 2$) react with silicon, germanium, and tin hydrides through insertion into the M—H bond to give novel cumulenes.[80]

Intramolecular

Arylcarbenes, generated by thermal elimination of trimethylsilanol from α-silylbenzyl alcohols, undergo intramolecular insertion into methyl groups. For example, pyrolysis of (**48**) gives 2,3-dihydrobenzofuran, and 1,1-dimethylindane is formed from (**49**) by initial rearrangement of the *para*-arylcarbene to the *ortho*-isomer followed by insertion.[27]

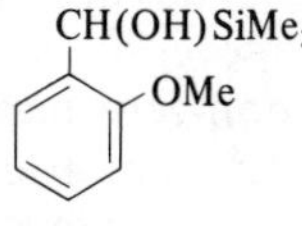

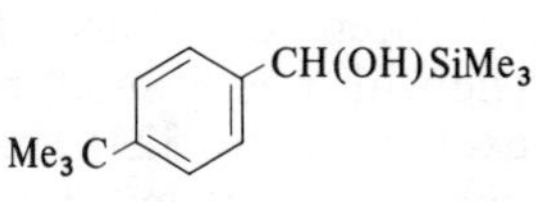

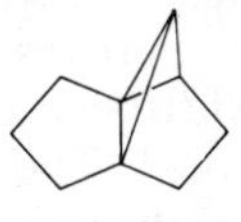

(48) (49) (50) (51)

Generation of the tricyclic cyclopropylidene (**50**), as its lithium carbenoid, leads to (**51**), the product of intramolecular insertion. Small amounts of the olefinic dimer of (**50**) are also formed.[81] This olefin can be considered a derivative of the unknown tetra-*tert*-butylethylene. A re-examination of the pyrolysis of diphenylmethyl propiolate shows that the primary product (**52**) is the one derived from insertion of the methylene carbene (**53**), an isomer of the starting acetylene.[82]

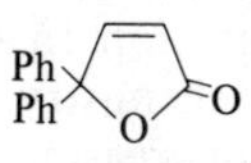

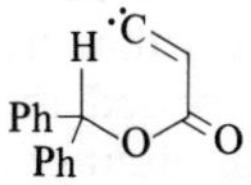

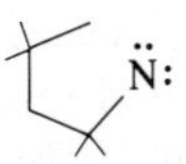

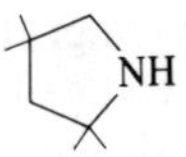

(52) (53) (54) (55)

Intramolecular nitrene insertion has been observed in the pyrolysis of di-*tert*-octylthiadiaziridine-1,1-dioxide. The initially formed diradical is thought to fragment into *tert*-octylnitrene (**54**). The products are rationalized by formation and further reaction of (**55**) by insertion.[83]

Intramolecular insertion of nitrene (**56**) occurs most readily under conditions which favour the triplet nitrene. The singlet preferentially closes onto the unsubstituted benzimidazole nitrogen atom.[84] In the related system (**57**), no intramolecular insertion into the 5-methyl group was observed even under triplet conditions. The major product (**58**) was formed by ring-closure onto N(2). This unexpected behaviour is possibly due to the effect of the phenazine ring which limits the radical insertion reactivity of the triplet, so that in the presence of the neighbouring nucleophilic nitrogen, triplet reactions are indistinguishable from singlet, (**58**) being formed from both spin states of the nitrene. The alternative explanation involving a rapid singlet–triplet equilibrium with the triplet being unreactive was rejected.[85]

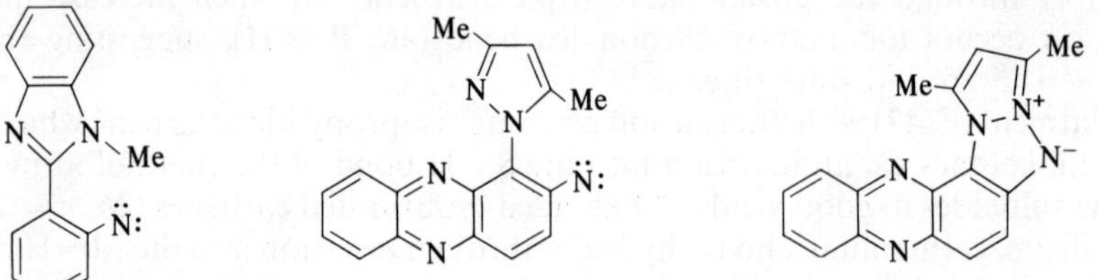

(**56**) (**57**) (**58**)

Rearrangement

Carbenes

An extensive review on the rearrangement of carbenes and nitrenes has appeared.[86]

The barriers to 1,2-shifts have been calculated for several carbenes. Methylcarbene rearranges to ethylene with a very small but non-zero barrier of $9\,\mathrm{kJ\,mol^{-1}}$. The rearrangement is exothermic by $323\,\mathrm{kJ\,mol^{-1}}$.[87] The transition state for the rearrangement of dihydroxycarbene to formic acid has been located by *ab initio* methods, and is shown to be close in energy to $\cdot H + \cdot CO_2H$.[88] Calculations suggest that there is no barrier to 1,2-hydrogen shift in the rearrangement of fluorovinylidene to fluoroacetylene. In contrast, the corresponding 1,2-fluorine shift in difluorovinylidene has a barrier of $152\,\mathrm{kJ\,mol^{-1}}$. Therefore this carbene should have a significant life-time, in agreement with experiment.[89] Experimental studies on the competing 1,2-hydrogen and phenyl shifts in carbene (**59**) have produced some surprising results. It has always been assumed that there is little or no barrier to 1,2-hydrogen shift, however, as the temperature is decreased from ambient to −110°C, the proportion of phenyl migration increases, suggesting that in this case phenyl migration has a lower activation energy than hydrogen migration. In a solid matrix at −196°C phenyl migration is suppressed, and the stereochemistry of the 1,2-hydrogen migration is also altered, the amount of the Z-olefin increasing in the solid phase.[90]

Both 1,2-hydrogen and 1,2-aryl shifts occur in benzocyclobuten-1-ylcarbene (**60**). The ring-expansion predominates, and indene is the major product. Deuterium labelling has established that, as expected, it is the aryl bond and not the alkyl bond which shifts. No evidence for the involvement of pseudoindene (**61**) has been found.[91]

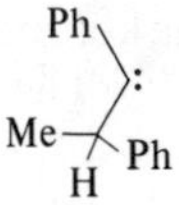

(59) 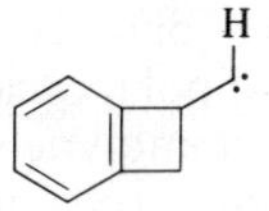 (60) 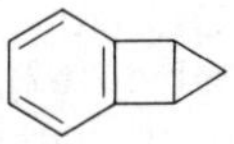 (61)

Direct evidence for the interconversion of ketocarbenes has been obtained in the photolysis of isomeric α-diazoketones (**62**) and (**63**). For example photolysis of both (**62**; R = Me) and (**63**; R = Me) gives products derived from the Wolff rearrangement of acetyl(phenyl)carbene. The carbene can also be trapped intermolecularly with *cis*-but-2-ene providing the first evidence for free singlet carbenes in the Wolff rearrangement. Addition of triplet sensitizers reduces the yield of Wolff rearrangement products, suggesting that the photochemical rearrangement proceeds through the excited singlet. No evidence for oxirene involvement has been found.[92] An alternative intermediate to an oxirene has been proposed to explain the scrambling of the carbon label in the photolysis of azibenzil. It is suggested that intermediate (**64**) could be formed by reversible intramolecular 1,3-dipolar cycloaddition of the diazo 1,3-dipole to the C=O bond. However, it has been concluded that any actual or time-averaged intermediate, either (**64**) or the oxirene, that is formed in the photolysis, does not revert to starting material at a rate competitive with its rearrangement to the ketene. The oxirene remains the most

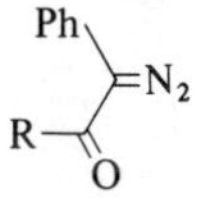

(62) 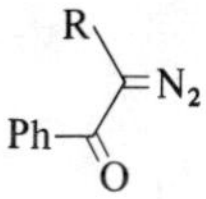 (63) 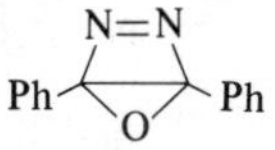 (64)

likely intermediate.[93] Further studies on the photochemical Wolff rearrangement reveal the importance of conformation. The rearrangement is thought to take place directly from the singlet excited state of the Z-conformer of the α-diazoketone, whereas the E-conformer gives singlet ketocarbene, which in turn may give the ketene or other carbene products. Thus the predominantly E-diazoketone (**65**) gives carbene products, whereas the locked Z-conformer (**66**) reacts exclusively by Wolff rearrangement.[94] The Wolff rearrangement of benzoyl(phenyl)carbene is completely suppressed in a matrix at −196°C, whereas the rearrangement of benzoylcarbene is hardly suppressed at all. These results reflect the effect of substituents on the ground-state multiplicity of the carbene, the latter carbene being a ground-state singlet.[78]

Azibenzil undergoes the normal Wolff rearrangement when heated in the presence of diarylmethanimines, $Ar_2C{=}NH$. The products, $Ph_2CHCON{=}CAr_2$, are simply derived by addition of the imine to diphenylketene. However, no Wolff

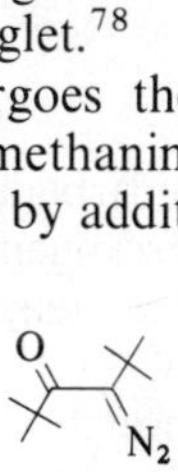

(65) 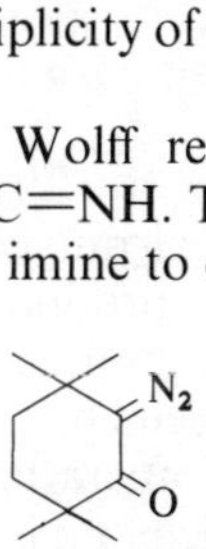 (66) 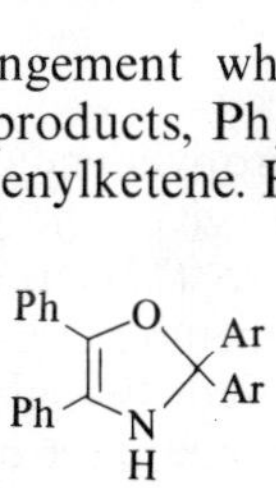 (67)

rearrangement is observed on photolysis, and the final product, $Ph_2C(CN)OCHAr_2$, is derived by 1,3-dipolar cycloaddition of the ketocarbene to the C=N to give (**67**), followed by homolytic ring-opening.[95] The related α-cyanoiminocarbenes have a more pronounced tendency to react as a 1,3-dipole than α-ketocarbenes. Thus, the carbene (**68**) does not rearrange by a 1,2-methyl shift in the presence of dipolarophiles, but adds to give, for example with acetonitrile, the imidazole (**69**). The carbene even adds as a 1,3-dipole to benzene to give (**70**).[96]

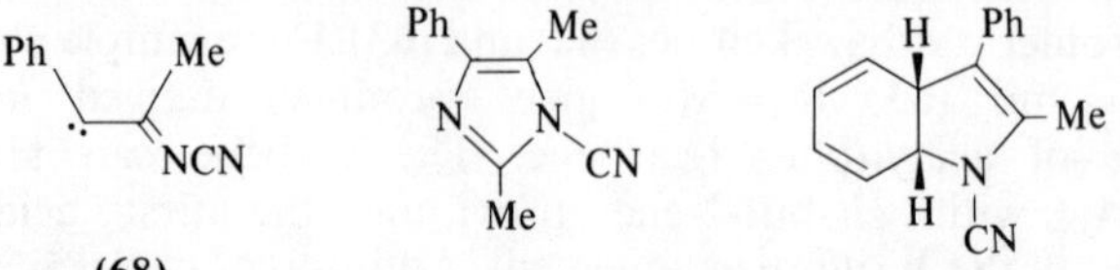

(**68**) (**69**) (**70**)

1,2-Shifts in silylcarbenes lead to silene intermediates. An IR study in an argon matrix indicates that the pyrolysis of Me_3SiCHN_2 gives singlet trimethylsilylcarbene which rearranges to $Me_2Si{=}CHMe$; the silene subsequently dimerizes.[97] The 6-silafulvene (**71**) has been generated by a 1,2-rearrangement of a carbene.[98] The silene (**72**), formed by a 1,2-shift of the trimethylsilyl group in carbene (**73**), rearranges further by migration of the OEt group to silicon to give a ketene.[99]

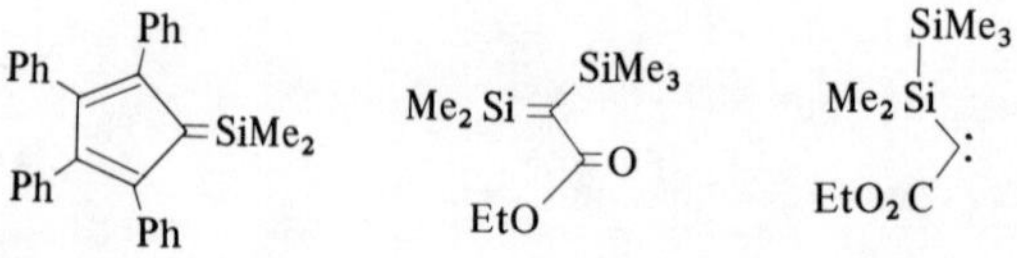

(**71**) (**72**) (**73**)

Phosphenes are also formed by 1,2-rearrangement of carbenes. Diphenylphosphonyl(phenyl)carbene (**74**; R = Ph), formed by photolysis of the corresponding diazo compound, rearranges by a 1,2-phenyl shift to give the phosphene.[100] Bis(diphenylphosphonyl)carbene (**74**; R = Ph_2PO) behaves similarly.[101] The phosphenes are trapped by the addition of HX or by cycloaddition to C=O compounds.

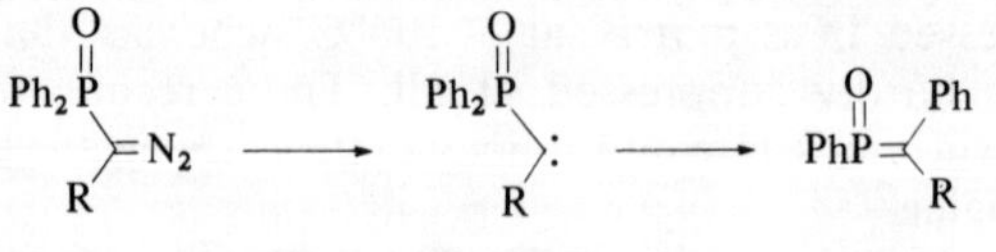

(**74**)

The vinylcyclopropylidene (**75**) rearranges readily to carbene (**76**). Independent generation of (**76**) gives the same products. The rearrangement of vinylcyclopropylidene itself to (**77**) proceeds more easily at lower temperatures than the competing ring-opening rearrangement to $H_2C{=}C{=}CHCH{=}CH_2$. Allene formation is favoured by higher temperatures.[102] Although vinylcyclobutylidene (**78**) does not rearrange, if the double bond is incorporated into a strained system as in bicyclo[4.1.1]oct-2-en-7-ylidene (**79**), then rearrangement occurs readily to give products derived from carbene (**80**).[103]

(75) (76)

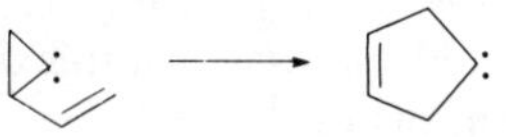

(77)

(78)

(79) (80)

Nitrenes

Calculations on the rearrangement of aminonitrene to *trans*-diimide suggest that the barrier to the 1,2-hydrogen shift is 394 kJ mol^{-1}. The barrier to a bimolecular hydrogen exchange *via* the cyclic dimer (**81**), however, is only 17 kJ mol^{-1}.[104] The thermolysis of azidomethane and 1,2-rearrangement of methylnitrene have been analysed by photo-electron spectroscopy.[105] Thermolysis of the azidocyclopropane (**82**; R = Me) gives the expected rearrangement product, the 1-azacyclobutene. In the corresponding *S*-allyl compound (**82**; R = allyl) rearrangement is suppressed, and intramolecular nitrene addition to the double bond supervenes.[106]

The photolysis of phenyl azide in the presence of "naked" acetate ion gives 3*H*-azepinone by rearrangement of phenylnitrene to benzazirine followed by addition of acetate.[107] The ring-contraction rearrangement of phenylnitrene to cyanocyclopentadiene generally requires the nitrene to be generated at very high temperatures. Decomposition of the complexed phenyl azide, η^6-azidobenzene-η^5-cyclopentadienyliron hexafluorophosphate gives the ring-contraction product, cyanoferrocene, at 120°C. The complexation of the nitrene to the electron-withdrawing $CpFe^+$ unit, and the nature of the product facilitate the rearrangement.[108]

Stereoelectronic control is important in the thermal fragmentation of 2-acyl-5-azidofurans (**83**) to RCOCOCH=CHCN, and fragmentation appears to be favoured when the carbonyl group is out-of-plane with the furan ring. Thus the *tert*-butyl derivative (**83**; R = Bu^t) fragments with $t_{1/2} = 11$ h at 20°C, whereas the corresponding aldehyde (**83**; R = H) has $t_{1/2} > 900$ h.[109]

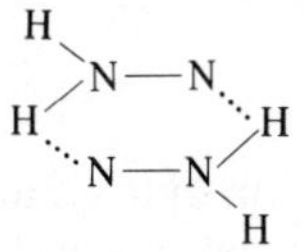

H N—N H H N—N H

SR N$_3$

N$_3$ O R O

(81)  (82) 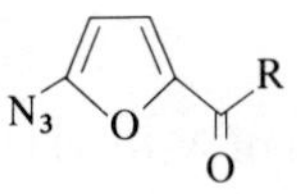 (83)

Aromatics

Carbenes

The reaction of the aromatic dianion (**84**) with excess chlorocarbene leads to *s*-benzvalenobenzobenzvalene (**85**).[110] The attack of dichlorocarbene on phenolate ion normally occurs mainly in the *ortho*-position. However, in the presence of α-cyclodextrin *para*-attack predominates. Complexation of the carbene to cyclodextrin is thought to be important.[111]

The formation of cycloheptatrienes from diazoacetates and aromatic hydrocarbons is catalysed by rhodium(II) carboxylates. The observed regioselectivity is in accord with the attack of a highly electrophilic carbenoid species.[112] Cycloheptatrienes are formed in the reaction of the carbene (**86**),[113] and (**87**)[114] with benzene. The intermediacy of the spiro-cycloheptatriene (**88**) from carbene (**87**) has been proved by independent synthesis. The carbene (**89**) also gives cycloheptatrienes with aromatics. In all cases a single adduct is formed, steric influences being more important than the electronic directing effect of the aromatic substituents. In the case of mesitylene, the equilibrium between the cycloheptatriene and the initial bicyclo[4.1.0]heptadiene can be observed by ^{13}C-NMR spectroscopy.[115]

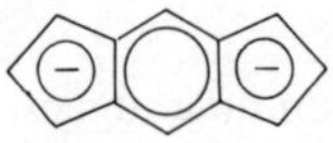

(**84**) 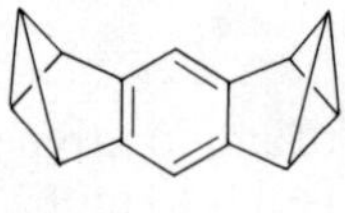 (**85**) 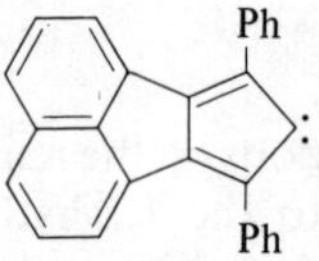 (**86**)

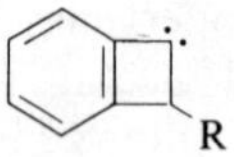

(**87**) 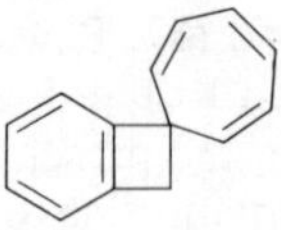 (**88**) 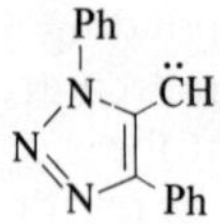 (**89**)

Intramolecular carbene attack on the aromatic ring of carbene (**90**) can proceed in two ways. In the case of X=CH_2 or NH the major product is (**91**). Studies with substituted compounds have established that (**91**; X=NH) is formed by σ-insertion rather than *via* a spiro-diene intermediate. The cycloheptatriene (**92**) formed by π-insertion followed by ring-expansion is the major product from carbene (**90**; X = O or S).[116] The formation of (**93**) in the flash vacuum pyrolysis of phenyl tetrolate is explained by equilibration of the acetylene with the methylene carbene followed by intramolecular attack on the aromatic ring.[117]

Nitrenes

The formation of *N*-tosylazepine from the decomposition of tosyl azide in benzene at 160°C is dramatically increased if the reaction is carried out under a pressure of nitrogen. The other product, *N*-tosylaniline, is also converted into the azepine by

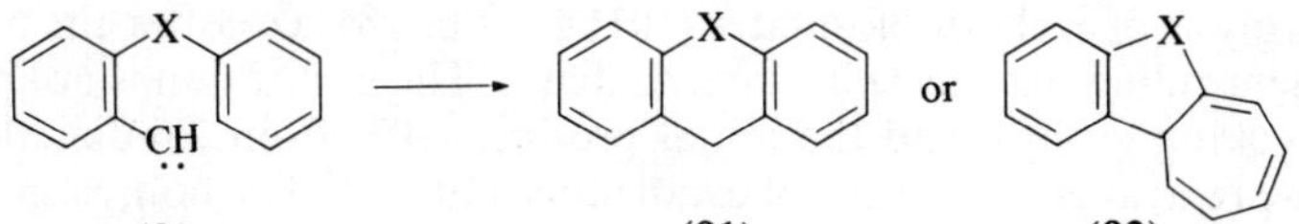

(90) (91) (92)

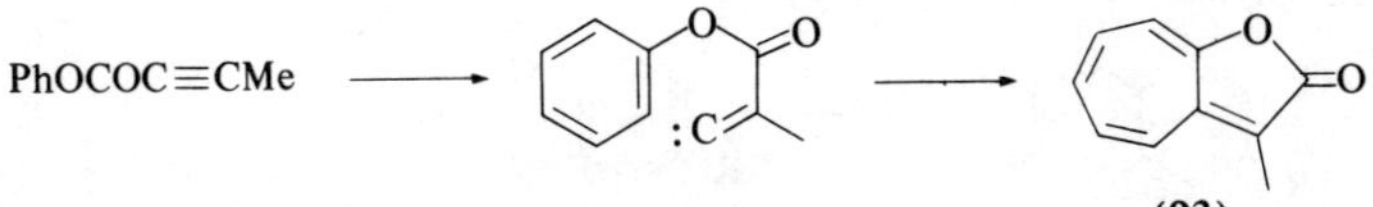

(93)

heating in benzene under nitrogen pressure.[118] Photolysis of the azide (**94**; X = N_3) in aromatic solvents gives the anilides (**94**; X = NHAr) *via* the triplet nitrene. No rearrangement to the isocyanate has been observed.[119]

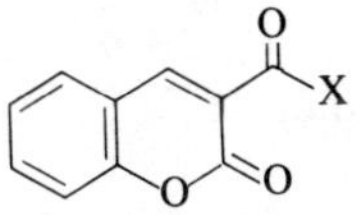

(94) 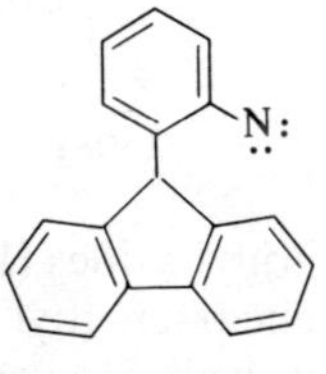 (95) 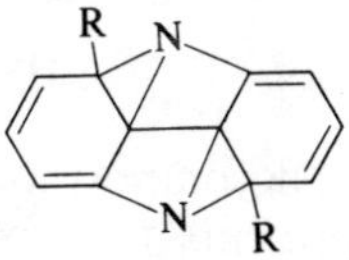 (96)

Intramolecular attack by arylnitrenes on aromatic rings has been observed in several cases. Photolysis of 1-azatriptycene generates 2-(9-fluorenyl)phenylnitrene (**95**) which gives several products derived by intramolecular attack on the aromatic ring.[120] Small amounts of phenazines are unexpectedly formed on photolysis of 2,2′-diazidobiphenyls in a matrix at 77 K, in addition to the expected benzo[*c*]cinnolines. One possible explanation involves a double intramolecular attack by the bis-nitrene to give the intermediate (**96**).[121] Thermolysis of the azide (**97**) gives the azepinoindole (**98**) and smaller amounts of the acridine (**99**).[122] Carbazoles are formed *via* spiro intermediates on pyrolysis of aryl 2-azidobenzoates (**100**).[123]

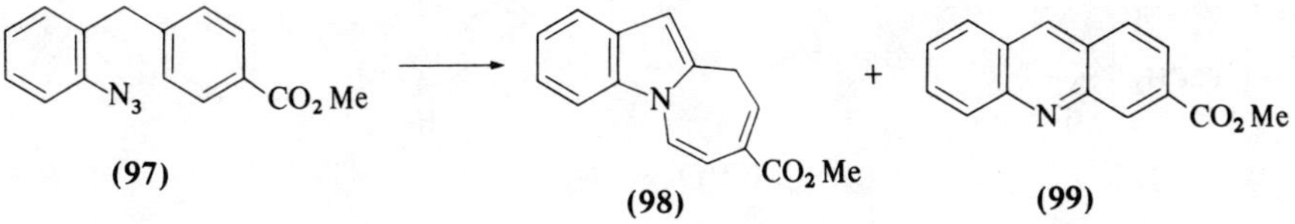

(97) (98) (99)

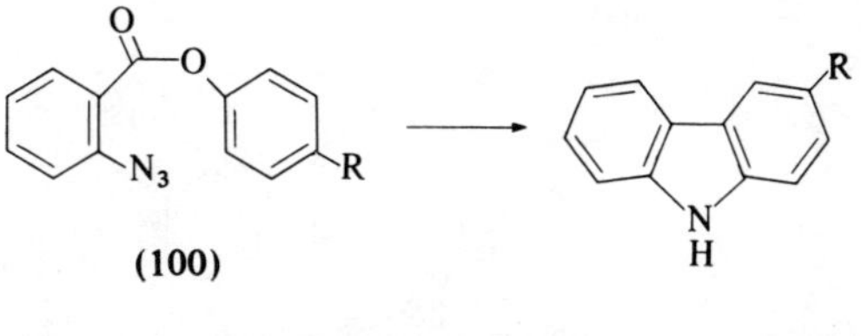

(100)

Optically active 3-methyl-2-phenyl-2*H*-azirine gives 2-methylindole on thermolysis at a rate 2040 times slower than it racemizes. This is best rationalized by the intermediacy of a vinylnitrene which exhibits a preference for reclosure to the azirine rather than intramolecular attack on the aryl ring to give the indole.[124]

Spray pyrolysis of aryl azidoformates (**101**) gives benzoxazolones, the position of the substituent ruling out any spiro intermediates. The cyclization is not prevented by *ortho*-blocking groups, and the initial product (**102**) from 2,6-dimethylphenyl azidoformate rearranges to the cyclohexadienone (**103**).[125] The homologous benzyl azidoformates also cyclize on spray pyrolysis. The isolated products are dimers of the azepines (**104**).[126]

(**101**)

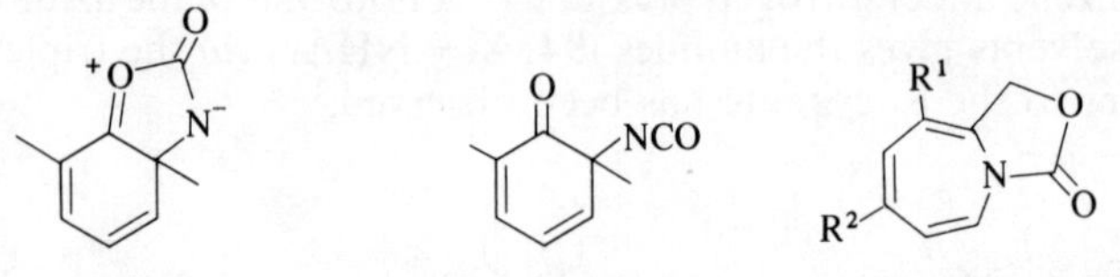

(**102**) (**103**) (**104**)

The decomposition of the sulphonyl azides (**105**; $n = 2$–5) has been investigated, and presents a complex picture. The phenethyl derivative (**105**; $n = 2$) gives a poor yield of the sultam on solution pyrolysis. However, on flash vacuum pyrolysis good yields of the cyclopentapyridine are obtained. This remarkable rearrangement product is thought to be formed by attack of the nitrene on the aromatic C=C, followed by loss of SO_2 and subsequent ring-opening *via* diradical intermediates.[127] The homologue (**105**; $n = 3$) gives the analogous rearrangement product, 5,6,7,8-tetrahydroquinoline, albeit in poor yield, the major products being the sultam and 1,2,3,4-tetrahydroquinoline formed at the expense of the sultam at higher temperatures. The next higher homologue (**105**; $n = 4$) gives the eight-membered sultam as the major product, but some side-chain C—H insertion occurs to give the six-membered ring sultam. Side-chain insertion is the only reaction observed in the solution pyrolysis of (**105**; $n = 5$).[128] The reactions are summarized in Scheme 2.

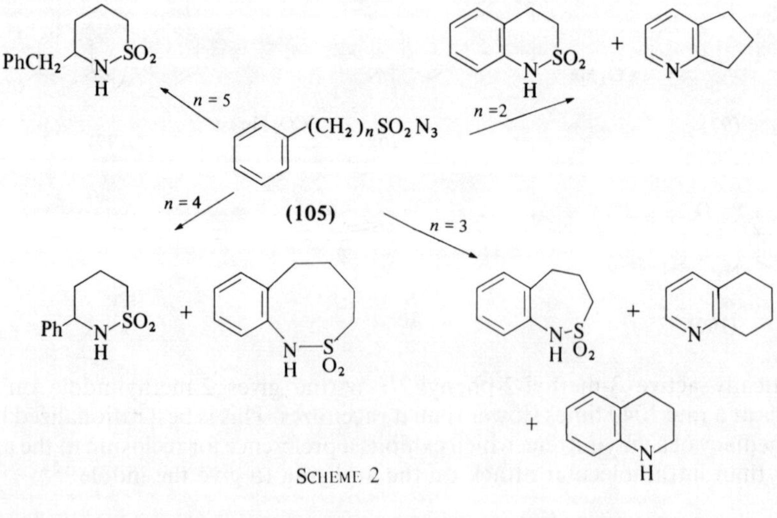

(**105**)

SCHEME 2

The preference for a seven-membered ring transition state in the intramolecular addition of aminonitrenes to alkenes[69] is paralleled in the corresponding intramolecular attack on aromatic rings. Thus oxidation of the *N*-amino-quinazolone (**106**) gives a mixture of (**107**; R^1 = H, R^2 = OMe) and the isomeric (**107**; R^1 = OMe, R^2 = H). This preferential attack of the nitrene through a seven-membered transition state avoids unfavourable eclipsing in the six-membered transition state derived by *ipso*-attack.[129]

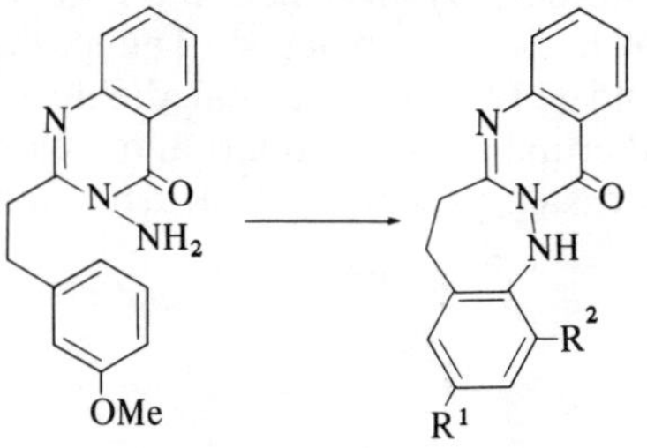

(**106**) (**107**)

Nucleophiles and Electrophiles

The attack of carbenes on divalent sulphur has often been assumed to proceed *via* the singlet. The first definitive example of attack *via* the triplet is provided in the reaction of diphenylcarbene with benzothiadiazole.[130] The presence of oxygen causes a marked change in the product pattern. A chain reaction with the radical cation of the diazo precursor as chain carrier has been suggested.[131]

Sulphoxides undergo facile deoxygenation with phase-transfer-generated dichlorocarbene. Steric acceleration occurs with bulky sulphoxides.[132] Dichlorocarbene is much more effective than peracids in the deamination of (**108**) to substituted naphthalenes.[133] Ureas, R_2NCONH_2, are dehydrated by dichlorocarbene to give cyanamides, R_2NCN.[134]

Isopropylidene, generated from (**47**), reacts with isocyanates by electrophilic attack on oxygen to give the virtually unknown enol derivatives, vinyl carbamates, $RNHCOOCH{=}CMe_2$, on hydrolytic work-up.[135]

One possible explanation for the isotope scrambling in the reaction of tosyl azide with ^{15}N-labelled azide ion involves the addition of tosylnitrene to tosyl azide to give a tetrazene.[136]

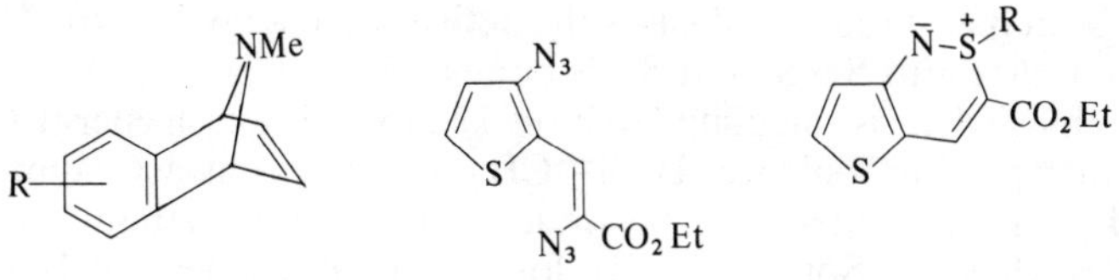

(**108**) (**109**) (**110**)

Intramolecular nucleophilic attack on a vinylnitrene by a thiophen sulphur atom is thought to account for the unprecedented ring-cleavage of the 3-azidothiophen (**109**) with extrusion of acetylene. The corresponding furan does not undergo ring-cleavage.[137] The nitrenes derived from 3-azidothiophens can also be intercepted intramolecularly by a suitable sulphur nucleophile, and cyclic sulphur–nitrogen ylids (**110**) are formed in high yield.[138]

Silylenes

Calculations on the geometry of the lowest singlet and triplet states of methylsilylene show that the singlet is considerably more bent than the triplet. The lowest triplet is 80 kJ mol^{-1} above the ground-state singlet.[139] The mass spectra of several silicon compounds, *e.g.* (**111**) and (**112**), indicate that dichlorosilylene is formed under electron impact.[140] The thermal cyclo-elimination reaction of 7-silanorbornadienes (**113**) is shown to be a two-step process by ESR detection of, and trapping of, the biradical intermediates.[141]

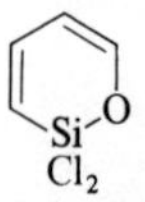

(**111**) 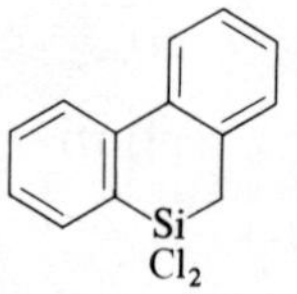 (**112**) 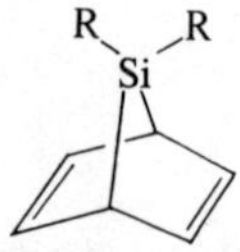 (**113**)

Difluorosilylene reacts with propylene in the gas phase through a silirane intermediate.[142] The reaction with fluoro-olefins, however, involves insertion of the silylene into the C—F bond.[143] Methyl(phenyl)silylene, generated photochemically from $Me_3SiSiMe(Ph)SiMe_3$, adds to 2,3-dimethylbuta-1,3-diene to give the expected 1-silacyclopent-3-ene *via* a silirane intermediate.[144] Addition of phenyl(trimethylsilyl)silylene to both *cis*- and *trans*-MeCH=CHCl proceeds stereospecifically to give *cis*- and *trans*-MeCH=CHSi(Cl)(Ph)$SiMe_3$, respectively. The addition of the silylene to the double bond is already known to be stereospecific, hence the ring-opening of the silirane is also stereospecific.[145]

Ab initio calculations on the insertion of silylene into molecular hydrogen have put the barrier at 36 kJ mol^{-1}, compared with the experimental estimate of 23 kJ mol^{-1}; in contrast, methylene inserts into hydrogen with zero energy barrier.[146] The kinetics of the insertion reactions of dimethylsilylene into Si—H bonds have been measured; in all cases the activation energy is zero.[147] The same silylene also inserts into S—Si and S—S bonds.[148]

Singlet methylsilylene is calculated to be 49 kJ mol^{-1} lower in energy than its 1,2-rearrangement product silene, $H_2Si{=}CH_2$.[139] In contrast, dimethylsilene, $Me_2Si{=}CH_2$, is calculated to be much more stable than the isomeric ethyl(methyl)silylene.[149] Some evidence for the rearrangement of $H_2C{=}SiHMe$ into dimethylsilylene in the gas phase has been obtained.[150]

Dimethylsilylene deoxygenates cyclooctene oxide to cyclooctene probably by electrophilic attack on oxygen followed by extrusion of dimethylsilanone.[151] In the corresponding reaction of dimesitylsilylene with epoxides, the extruded silanone was trapped by the epoxide to give the first example of a silanone–epoxide adduct (**114**), formed as shown in Scheme 3.[152]

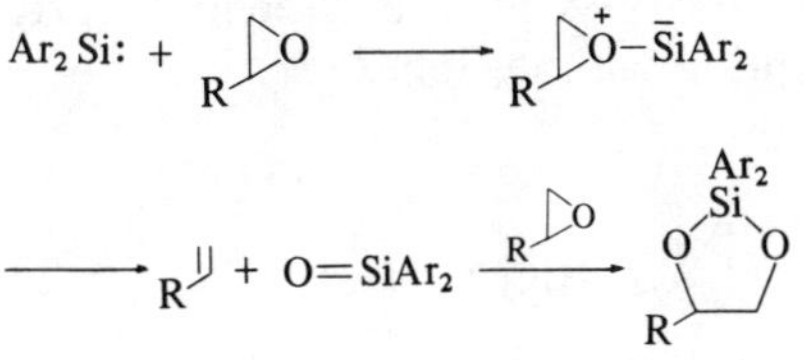

(114)

Ar = 2, 4, 6-$Me_3C_6H_2$–

SCHEME 3

The initial adduct (**115**) formed by electrophilic attack of dimethylsilylene on $Me_2C{=}CHCH_2OMe$ rearranges by a [2,3]-sigmatropic shift.[153]

(115)

Transition-metal Complexes

Reaction of tungsten carbene complexes (**116**; R = Ph) with $HAuCl_4$ gives new gold complexes, ClAu=CPh(OMe).[154] The tungsten complex (**116**; R = Me) is an effective catalyst for olefinic amine metathesis.[155] The reactivity of metallocarbenes such as (**116**) towards olefins and acetylenes depends on their stability. Stabilized metallocarbenes react much faster with triple bonds that with double bonds; the reverse is true for unstabilized metallocarbenes.[156]

Reaction of the bridged carbene complex (**117**) with diazomethane gives the triply bridged complex (**118**), the first compound with two different carbene bridges.[157]

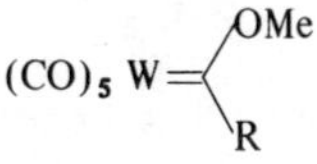

(116) 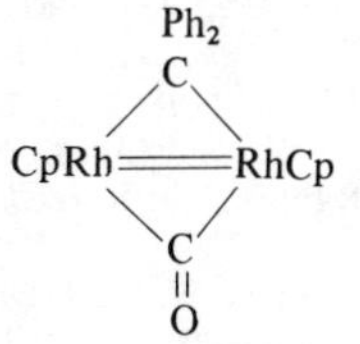 (117) 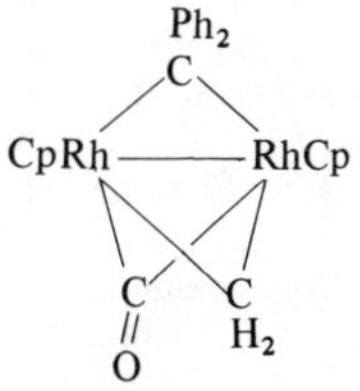 (118)

The rates of migration of R to the carbene centre in the zirconoxycarbene complexes of niobocene (**119**) have been determined, and found to follow the order H ≫ Me > CH_2Ar.[158] The vitamin-B_{12}-mediated formation of *trans*-4,4′-dichlorostilbene from (**120**) is thought to proceed in two steps by nucleophilic attack of Co(I) with displacement of chloride followed by α-elimination to generate a cobalt complex of the carbene $Ar_2CH-\ddot{C}H$. The carbenoid rearranges to the stilbene too fast to allow its interception with cyclohexene.[159]

$Cp_2Nb{=}C(H)OZr(H)Cp^*_2$ (with R on Nb) $(4\text{-}ClC_6H_4)_2CHCHCl_2$

$Cp = \eta^5-C_5H_5$, $Cp^* = \eta^5-C_5Me_5$

(119) **(120)**

References

1 Minkin, V. I., and Minyaev, R. M., *Izv. Sib. Otd. Akad. Nauk SSSR.* **1980,** 87; *Chem. Abs.*, **93,** 237974 (1980)
2 Mueller, P. H., Rondan, N. G., Houk, K. N., Harrison, J. F., Hooper, D., Willen, B. H., and Liebman, J. F., *J. Am. Chem. Soc.*, **103,** 5049 (1981).
3 Hutton, R. S., Roth, H. D., and Charl, S., *J. Phys. Chem.*, **85,** 753 (1981).
4 Senthilnathan, V. P., and Platz, M. S., *J. Am. Chem. Soc.*, **102,** 7637 (1980).
5 Wong, P. C., Griller, D., and Scaiano, J. C., *J. Am. Chem. Soc.*, **103,** 5934 (1981).
6 Zupancic, J. J., Grasse, P. B., and Schuster, G. B., *J. Am. Chem. Soc.*, **103,** 2423 (1981).
7 Senthilnathan, V. P., and Platz, M. S., *J. Am. Chem. Soc.*, **103,** 5503 (1981).
8 Hutton, R. S., Roth, H. D., Manion-Schilling, M. L., and Suggs, J. W., *J. Am. Chem. Soc.*, **103,** 5147 (1981).
9 Shepard, R., and Simon, J.,., *Int. J. Quantum Chem., Quantum Chem. Symp.*, **14,** 349 (1980); *Chem. Abs.*, **94,** 139053 (1981).
10 Bru, N., and Vilarrasa, J., *Chem. Lett.*, **1980,** 1489.
11 Feller, D., Davidson, E. R., and Borden, W. T., *J. Am. Chem. Soc.*, **103,** 2558 (1981).
12 Waali, E. E., *J. Am. Chem. Soc.*, **103,** 3604 (1981).
13 Schultz, P. G., and Dervan, P. B., *J. Am. Chem. Soc.*, **103,** 1563 (1981).
14 Smith, A. B., and Dieter, R. K., *Tetrahedron*, **37,** 2407 (1981).
15 Yano, K., and Itoh, R., *Waseda Daigaku Rikogaku Kenkyusho Hokoku*, **1981,** 48; *Chem. Abs.*, **95,** 132111 (1981).
16 Banyard, S. A., Canosa-Mas, C. E., Ellis, M. D., Frey, H. M., and Walsh, R., *J. Chem. Soc., Chem. Commun.*, **1980,** 1156.
17 Baird, M. S., Dunkin, I. R., Hacker, N., Poliakoff, M., and Turner, J. J., *J. Am. Chem. Soc.*, **103,** 5190 (1981).
18 Jones, G., Sliskovic, Ḋ. R., Foster, B., Rogers, J., Smith, A. K., Wong, M. Y., and Yarham, A. C., *J. Chem. Soc., Perkins Trans. 1*, **1981,** 78.
19 Balli, H., Grüner, H., Maul, R., and Schepp, H., *Helv. Chim. Acta*, **64,** 648 (1981).
20 Hoffmann, R. W., Lotze, M., Reiffen, M., and Steinbach, K., *Liebigs Ann. Chem.*, **1981,** 581.
21 Dürr, H., and Hackenberger, A., *J. Chem. Res.* (*S*), **1981,** 178.
22 Waali, E. E., and Wright, C. W., *J. Org. Chem.*, **46,** 2201 (1981).
23 Arct, J., and Migaj, B., *Tetrahedron*, **37,** 953 (1981).
24 Billups, W. E., Reed, L. E., Casserly, E. W., and Lin, L. P., *J. Org. Chem.*, **46,** 1326 (1981).
25 Abdallah, A. A., El-Nahas, H. M., and Raid, Y., *Egypt. J. Chem.*, **22,** 345 (1979); *Chem. Abs.*, **95,** 41758 (1981).
26 Coutouli-Argyropoulou, E., and Alexandrou, N. E., *J. Org. Chem.*, **45,** 4158 (1980).
27 Sekiguchi, A., and Ando, W., *J. Org. Chem.*, **45,** 5286 (1980).
28 Bekhazi, M., and Warkentin, J., *J. Am. Chem. Soc.*, **103,** 2473 (1981).
29 Ege, G., Gilbert, K., and Nader, F. W., *Chem. Ber.*, **114,** 1074 (1981).
30 Nakamura, N., Schweizer, W. B., Frei, B., Wolf, H. R., and Jeger, O., *Helv. Chim. Acta*, **63,** 2230 (1980).
31 Odaira, Y., Sakai, Y., Fukuda, Y., Negoro, T., Hirata, F., Tobe, Y., and Kimura, K., *J. Org. Chem.*, **46,** 2977 (1981).
32 Zoch, H.-G., Kinzel, E., and Szeimies, G., *Chem. Ber.*, **114,** 968 (1981).
33 Stang, P. J., and Ladika, M., *J. Am. Chem. Soc.*, **103,** 6437 (1981).
34 le Noble, W. J., Basak, S., and Srivastava, S., *J. Am. Chem. Soc.*, **103,** 4638 (1981).

[35] Stang, P. J., *Isr. J. Chem.*, **21,** 119 (1981).
[36] Takeuchi, H., and Koyama, K., *J. Chem. Soc., Chem. Commun.*, **1981,** 202.
[37] Hoesch, L., and Köppel, B., *Helv. Chim. Acta*, **64,** 864 (1981).
[38] Hoesch, L., *Helv. Chim. Acta*, **64,** 890 (1981).
[39] Hutchins, M. G. K., and Swern, D., *Tetrahedron Lett.*, **22,** 4599 (1981).
[40] Chang, Y. H., Chiu, F.-T., and Zon, G., *J. Org. Chem.*, **46,** 342 (1981).
[41] Feld, W. A., Paessun, R., and Servé, M. P., *J. Heterocycl. Chem.*, **17,** 1309 (1980).
[42] Schultz, A. G., Shen, M., and Ravichandran, R., *Tetrahedron Lett.*, **22,** 1767 (1981).
[43] Bludssus, W., and Mews, R., *Chem. Ber.*, **114,** 1539 (1981).
[44] De Rosa, M., and Haberfield, P., *J. Org. Chem.*, **46,** 2639 (1981).
[45] Schoeller, W., *Angew. Chem. Int. Ed.*, **20,** 698 (1981).
[46] Giese, B., and Lee, W.-B., *Chem. Ber.*, **114,** 3306 (1981).
[47] Moss, R. A., Young, C. M., Perez, L. A., and Krogh-Jespersen, K., *J. Am. Chem. Soc.*, **103,** 2413 (1981).
[48] Barlet, R., LeGoaller, R., and Gey, C., *Can. J. Chem.*, **59,** 621 (1981).
[49] Kropp, P. J., Pienta, N. J., Sawyer, J. A., and Polniaszek, R. P., *Tetrahedron*, **37,** 3229 (1981).
[50] Fischer, P., and Schaefer, G., *Angew. Chem. Int. Ed.*, **20,** 863 (1981).
[51] Oshima, T., Arikata, S., and Nagai, T., *J. Chem. Res. (S)*, **1981,** 204.
[52] Moss, R. A., Guo, W., Denney, D. Z., Houk, K. N., and Rondan, N. G., *J. Am. Chem. Soc.*, **103,** 6164 (1981).
[53] Turro, N. J., Butcher, J. A., Moss, R. A., Guo, W., Munjal, R. C., and Fedorynski, M., *J. Am. Chem. Soc.*, **102,** 7576 (1980).
[54] Moss, R. A., and Wetter, W. P., *Tetrahedron Lett.*, **22,** 997 (1981).
[55] Moss, R. A., and Chang, M. J., *Tetrahedron Lett.*, **22,** 3749 (1981).
[56] Schomburg, D., and Landry, D. W., *J. Org. Chem.*, **46,** 170 (1981).
[57] Mandel'shtam, T. V., Kharicheva, E. M., Labeish, N. N., and Kostikov, R. R., *Zh. Org. Khim.*, **16,** 2513 (1980); *Chem. Abs.*, **94,** 120501 (1981).
[58] Eichler, K., and Heydtmann, H., *Int. J. Chem. Kinet.*, **13,** 1107 (1981).
[59] Klumpp, G. W., and Kwantes, P. M., *Tetrahedron Lett.*, **22,** 831 (1981).
[60] Mock, G. B., and Jones, M., *Tetrahedron Lett.*, **22,** 3819 (1981).
[61] Dewar, M. J. S., Nelson, D. J., Shevlin, P. B., and Biesiada, K. A., *J. Am. Chem. Soc.*, **103,** 2802 (1981).
[62] Kozlowska-Gramsz, E., *Pol. J. Chem.*, **54,** 1607 (1980); *Chem. Abs.*, **95,** 6326 (1981).
[63] Kozlowska-Gramsz, E., and Descotes, G., *Tetrahedron Lett.*, **22,** 563 (1981).
[64] Leuenberger, C., Hoesch, L., and Dreiding, A. S., *J. Chem. Soc., Chem. Commun.*, **1980,** 1197.
[65] Sheradsky, T., and Zbaida, D., *Tetrahedron Lett.*, **22,** 1639 (1981).
[66] Atkinson, R. S., and Judkins, B. D., *J. Chem. Soc., Perkin Trans. 1*, **1981,** 2615.
[67] Burger, U., Gandillon, G., and Mareda, J., *Helv. Chim. Acta*, **64,** 844 (1981).
[68] Padwa, A., and Rodriguez, A., *Tetrahedron Lett.*, **22,** 187 (1981).
[69] Atkinson, R. S., Malpass, J. R., Skinner, K. L., and Woodthorpe, K. L., *J. Chem. Soc., Chem. Commun.*, **1981,** 549.
[70] Jacox, M. E., *J. Mol. Spectrosc.*, **81,** 349 (1980); *Chem. Abs.*, **93,** 238237 (1980).
[71] Purdy, J. R., and Thrush, B. A., *Int. J. Chem. Kinet.*, **13,** 873 (1981).
[72] Demonceau, A., Noels, A. F., Hubert, A. J., and Teyssié, P., *J. Chem. Soc., Chem. Commun.*, **1981,** 688.
[73] Singh, S. B., and Mehrotra, K. N., *Can. J. Chem.*, **59,** 2475 (1981).
[74] Dehmlow, E. V., and Naser-ud-Din, *J. Chem. Res. (S)*, **1981,** 144.
[75] Steinbeck, K., and Klein, J., *Angew. Chem. Int. Ed.*, **20,** 773 (1981).
[76] Harada, T., and Oku, A., *J. Am. Chem. Soc.*, **103,** 5965 (1981).
[77] Kirmse, W., Loosen, K., and Sluma, H.-D., *J. Am. Chem. Soc.*, **103,** 5935 (1981).
[78] Tomioka, H., Okuno, H., and Izawa, Y., *J. Chem. Soc., Perkin Trans. 2*, **1980,** 1636.
[79] Stang, P. J., and Christensen, S. B., *J. Org. Chem.*, **46,** 823 (1981).
[80] Stang, P. J., and White, M. R., *J. Am. Chem. Soc.*, **103,** 5429 (1981).
[81] Warner, P., Chang, S.-C., Powell, D. R., and Jacobson, R. A., *Tetrahedron Lett.*, **22,** 533 (1981).
[82] Brown, R. F. C., Eastwood, F. W., Chaichit, N., Gatehouse, B. M., Pfeiffer, J. M., and Woodroffe, D., *Aust. J. Chem.*, **34,** 1467 (1981).
[83] Timberlake, J. W., Alender, J., Garner, A. W., Hodges, M. L., Özmeral, C., and Jacobus, J. O., *J. Org. Chem.*, **46,** 2082 (1981).
[84] Hawkins, D., Lindley, J. M., McRobbie, I. M., and Meth-Cohn, O., *J. Chem. Soc., Perkin Trans. 1*, **1980,** 2387.
[85] Albini, A., Bettinetti, G. F., and Minoli, G., *J. Chem. Soc., Perkin Trans. 1*, **1981,** 4.
[86] Jones, W. M., *Org. Chem. (N.Y.)*, **42,** 95 (1980); *Chem. Abs.*, **94,** 3319 (1981).

[87] Nobes, R. H., Radom, L., and Rodwell, W. R., *Chem. Phys. Lett.*, **74,** 269 (1980); *Chem. Abs.*, **94,** 3583 (1981).
[88] Feller, D., Borden, W. T., and Davidson, E. R., *J. Comput. Chem.*, **1,** 158 (1980); *Chem. Abs.*, **94,** 29933 (1981).
[89] Frisch, M. J., Krishman, R., Pople, J. A., and Schleyer, P. von R., *Chem. Phys. Lett.*, **81,** 421 (1981); *Chem. Abs.*, **95,** 149713 (1981).
[90] Tomioka, H., Ueda, H., Kondo, S., and Izawa, Y., *J. Am. Chem. Soc.*, **102,** 7817 (1980).
[91] O'Leary, M. A., and Wege, D., *Tetrahedron*, **37,** 801 (1981).
[92] Tomioka, H., Okuno, H., Kondo, S., and Izawa, Y., *J. Am. Chem. Soc.*, **102,** 7123 (1980).
[93] Blaustein, M. A., and Berson, J. A., *Tetrahedron Lett.*, **22,** 1081 (1981).
[94] Tomioka, H., Okuno, H., and Izawa, Y., *J. Org. Chem.*, **45,** 5278 (1980).
[95] Mehrotra, K. N., and Prasad, G., *Bull. Chem. Soc. Jpn.*, **54,** 604 (1981).
[96] Danion, D., Arnold, B., and Regitz, M., *Angew. Chem. Int. Ed.*, **20,** 113 (1981).
[97] Mat'tsev, A. K., Korolev, V. A., Khabashesku, V. N., and Nefedov, O. M., *Dokl. Akad. Nauk SSSR*, **251,** 1166 (1980); *Chem. Abs.*, **93,** 220193 (1980).
[98] Sekiguchi, A., and Ando, W., *J. Am. Chem. Soc.*, **103,** 3579 (1981).
[99] Ando, W., Sekiguchi, A., and Sato, T., *J. Am. Chem. Soc.*, **103,** 5573 (1981).
[100] Regitz, M., and Eckes, H., *Tetrahedron*, **37,** 1039 (1981).
[101] Regitz, M., Bennyarto, F., and Heydt, H., *Liebigs Ann. Chem.*, **1981,** 1044.
[102] Brinker, U. H., and Ritzer, J., *J. Am. Chem. Soc.*, **103,** 2116 (1981).
[103] Brinker, U. H., and König, L., *J. Am. Chem. Soc.*, **103,** 212 (1981).
[104] Kemper, M. J. H., and Buck, H. M., *Can. J. Chem.*, **59,** 3044 (1981).
[105] Bock, H., Dammel, R., and Horner, L., *Chem. Ber.*, **114,** 220 (1981).
[106] Jorritsma, R., Steinberg, H., and de Boer, Th. J., *Recl. Trav. Chim. Pays-Bas*, **100,** 307 (1981).
[107] Colman, R., Scriven, E. F. V., Suschitzky, H., and Thomas, D. R., *Chem. Ind.* (*London*), **1981,** 249.
[108] Lee, C. C., Azogu, C. I., Chang, P. C., and Sutherland, R. G., *J. Organomet. Chem.*, **220,** 181 (1981).
[109] Newcombe, P. J., and Norris, R. K., *Tetrahedron Lett.*, **22,** 699 (1981).
[110] Gandillon, G., Bianco, B., and Burger, U., *Tetrahedron Lett.*, **22,** 51 (1981).
[111] Komiyama, M., and Hirai, H., *Bull. Chem. Soc. Jpn.*, **54,** 2053 (1981).
[112] Anciaux, A. J., Demonceau, A., Noels, A. F., Hubert, A. J., Warin, R., and Teyssié, P., *J. Org. Chem.*, **46,** 873 (1981).
[113] Balasubramanian, P., and Narasimhan, K., *Tetrahedron Lett.*, **22,** 685 (1981).
[114] O'Leary, M. A., Richardson, G. W., and Wege, D., *Tetrahedron*, **37,** 813 (1981).
[115] Bedford, C. D., Bruckmann, E. M., and Smith, P. A. S., *J. Org. Chem.*, **46,** 679 (1981).
[116] Crow, W. D., and McNab, H., *Aust. J. Chem.*, **34,** 1037 (1981).
[117] Brown, R. F. C., and Eastwood, F. W., *J. Org. Chem.*, **46,** 4588 (1981).
[118] Ayyangar, N. R., Bambal, R. B., and Lugade, A. G., *J. Chem. Soc., Chem. Commun.*, **1981,** 790.
[119] Elkasaby, M. A., and Noureldin, N. A., *Indian J. Chem.*, **19B,** 1080 (1980); *Chem. Abs.*, **95,** 23827 (1981).
[120] Sugawara, T., and Iwamura, H., *J. Am. Chem. Soc.*, **102,** 7134 (1980).
[121] Yabe, A., *Bull. Chem. Soc. Jpn.*, **53,** 2933 (1980).
[122] Carde, R. N., Hayes, P. C., Jones, G., and Cliff, C. J., *J. Chem. Soc., Perkins Trans. 1*, **1981,** 1132.
[123] Clancy, M. G., Hesabi, M. M., and Meth-Cohn, O., *J. Chem. Soc., Chem. Commun.*, **1980,** 1112.
[124] Isomura, K., Ayabe, G.-I., Hatano, S., and Taniguchi, H., *J. Chem. Soc., Chem. Commun.*, **1980,** 1252.
[125] Meth-Cohn, O., and Rhouati, S., *J. Chem. Soc., Chem. Commun.*, **1981,** 241.
[126] Meth-Cohn, O., and Rhouati, S., *J. Chem. Soc., Chem. Commun.*, **1980,** 1161.
[127] Abramovitch, R. A., Holcomb, W. D., and Wake, S., *J. Am. Chem. Soc.*, **103,** 1525 (1981).
[128] Abramovitch, R. A., Hendi, S. B., and Kress, A. O., *J. Chem. Soc., Chem. Commun.*, **1981,** 1087.
[129] Atkinson, R. S., Malpass, J. R., and Woodthorpe, K. L., *J. Chem. Soc., Chem. Commun.*, **1981,** 160.
[130] Benati, L., Montevecchi, P. C., Spagnolo, P., and Tundo, A., *J. Chem. Soc., Perkin Trans. 1*, **1981,** 1544.
[131] Benati, L., Montevecchi, P. C., and Spagnolo, P., *J. Chem. Soc., Perkin Trans. 2*, **1981,** 1437.
[132] Dyer, J. C., and Evans, S. A., *J. Org. Chem.*, **45,** 5350 (1980).
[133] Gribble, G. W., Allen, R. W., Le Houllier, C. S., Eaton, J. T., Easton, N. R., Slayton, R. I., and Sibi, M. P., *J. Org. Chem.*, **46,** 1025 (1981).
[134] Schroth, W., Kluge, H., Matthias, M., Krieg, R., and Schädler, H.-D., *Z. Chem.*, **21,** 25 (1981).
[135] Stang, P. J., and Anderson, G. H., *J. Org. Chem.*, **46,** 4585 (1981).
[136] Casewit, C., Wenninger, J., and Roberts, J. D., *J. Am. Chem. Soc.*, **103,** 6248 (1981).
[137] Moody, C. J., Rees, C. W., and Tsoi, S. C., *J. Chem. Soc., Chem. Commun.*, **1981,** 550.

[138] Moody, C. J., Rees, C. W., Tsoi, S. C., and Williams, D. J., *J. Chem. Soc., Chem. Commun.*, **1981**, 927.
[139] Goddard, J. D., Yoshioka, Y., and Schaefer, H. F., *J. Am. Chem. Soc.*, **102**, 7644 (1980).
[140] Bochkarev, V. N., Polivanov, A. N., Slyusarenko, T. F., Bernadskii, A. A., Silkina, N. N., and Klimentov, B. N., *Zh. Obshch. Khim.*, **51**, 824 (1981); *Chem. Abs.*, **95**, 96385 (1981).
[141] Mayer, B., and Neumann, W. P., *Tetrahedron Lett.*, **21**, 4887 (1980).
[142] Shiau, C. C., Hwang, T. L., and Liu, C. S., *J. Organomet. Chem.*, **214**, 31 (1981).
[143] Hwang, T. L., Pai, Y. M., and Liu, C. S., *J. Am. Chem. Soc.*, **102**, 7519 (1980).
[144] Ishikawa, M., Nakagawa, K., Enokida, R., and Kumada, M., *J. Organomet. Chem.*, **201**, 151 (1980).
[145] Ishikawa, M., Nakagawa, K., Katayama, S., and Kumada, M., *J. Am. Chem. Soc.*, **103**, 4170 (1981).
[146] Gordon, M. S., *J. Chem. Soc., Chem. Commun.*, **1981**, 890.
[147] Davidson, I. M. T., and Ostah, N. A., *J. Organomet. Chem.*, **206**, 149 (1981).
[148] Chihi, A., and Weber, W. P., *J. Organomet. Chem.*, **210**, 163 (1981).
[149] Hanamara, M., Nagase, S., and Morokuma, K., *Tetrahedron Lett.*, **22**, 1813 (1981).
[150] Conlin, R. T., and Wood, D. L., *J. Am. Chem. Soc.*, **103**, 1843 (1981).
[151] Goure, W. F., and Barton, T. J., *J. Organomet. Chem.*, **199**, 33 (1980).
[152] Ando, W., Ikeno, M., and Hamada, Y., *J. Chem. Soc., Chem. Commun.*, **1981**, 621.
[153] Tzeng, D., and Weber, W. P., *J. Org. Chem.*, **46**, 693 (1981).
[154] Aumann, R., and Fischer, E. O., *Chem. Ber.*, **114**, 1853 (1981).
[155] Edwige, C., Lattes, A., Laval, J. P., Mutin, R., Basset, J. M., and Nouguier, R., *J. Mol. Catal.*, **8**, 297 (1980); *Chem. Abs.*, **94**, 173853 (1981).
[156] Katz, T. J., Savage, E. B., Lee, S. J., and Nair, M., *J. Am. Chem. Soc.*, **102**, 7942 (1980).
[157] Bauer, C., and Herrmann, W. A., *J. Organomet. Chem.*, **209**, C13 (1981).
[158] Threlkel, R. S., and Bercaw, J. E., *J. Am. Chem. Soc.*, **103**, 2650 (1981).
[159] Nome, F., and Zanette, D., *Can. J. Chem.*, **58**, 2402 (1980).

Organic Reaction Mechanisms 1981
Edited by A. C. Knipe and W. E. Watts

CHAPTER 6

Nucleophilic Aromatic Substitution

MICHAEL R. CRAMPTON

Department of Chemistry, Durham University

General 263
The S_NAr Mechanism 266
Heterocyclic Systems 271
Meisenheimer and Related Adducts 274
Benzyne and Related Intermediates 277
References 278

General

Both ionic and free-radical pathways are available for reactions of arenediazonium ions in the presence of nucleophiles. Volumes of activation for the dediazoniation of substituted benzenediazonium ions in water vary little with the nature of the substituent.[1] The value of $+10.8\,cm^3$ is in accord with cleavage to give aryl cations. Oscillatory evolution of nitrogen has been observed during the decomposition in water of benzenediazonium chloride,[2] but the phenomenon is physical rather than chemical in origin. Kinetic parameters have been reported[3] for the thermal decomposition of diazonium salts of sulphanilic acid and sulphanilamide and for the photo-decomposition of the diazonium salt of *p*-dimethylaminoaniline. It is well known that arenediazonium ions may form complexes with crown ethers and that this increases both their solubility in non-polar solvents and also their thermal stability. Now rate constants for the dediazoniation of complexed diazonium ions and equilibrium constants for the association as a function of ring size of the crown ether have been reported[4] and a theoretical study (by CNDO/2) of the complexing process has been made.[5] A variety of co-solvents in addition to crown ethers may be used to increase solubility[6] and it has been suggested that yield enhancement in reactions in non-polar solutions may result from suppression of side-reactions rather than by anion activation or specific complexation.

There has been renewed interest in the mechanism of formation of fluorene derivatives by dediazoniation of 2-alkylbiphenyl-2′-yl diazonium salts. The isolation[7] of a benzyl alcohol derivative among the products is evidence for the intermediacy of a benzyl cation which is probably derived from the initially formed phenyl cation.

The diazonio group, as well as being an excellent leaving-group, is strongly activating towards nucleophilic substitution of the S_NAr type at *ortho*- and *para*-

positions. Thus 2-nitro-4-chlorobenzenediazonium ions hydrolyse in weakly acidic media by denitration while the isomeric 2,4-compound suffers denitration (70 %) or dechlorination (30 %); the kinetics indicate that general base-catalysed addition of a hydroxyl group is rate-limiting.[8] A further possible reaction is base addition to the diazonio function, and fast-reaction techniques have been applied[9] to the reactions of hydroxide ions with benzenediazonium ions carrying electronegative substituents (NO_2, Cl). Here the formation of the *syn*-diazotate, from the initially produced diazohydroxide, is followed by production of the thermodynamically more stable *anti*-diazotate. Various nucleophiles add reversibly at the 10-position of 10-phenyl-9-anthracenediazonium salts to yield neutral adducts[10] such as (**1**; R = H, Me). The derivative formed by dimethylamine addition isomerizes to the triazene (**2**). Triazenes are also formed by reaction of benzenediazonium ions with adenine and its derivatives;[11] they decompose in basic aqueous solutions, probably by a radical pathway, to produce 8-aryladenines. Also reported[12] is the use of the reaction of piperidinotriazenes, formed from diazonium ions and piperidine, with hydrogen fluoride to synthesize ^{18}F-labelled estrogens.

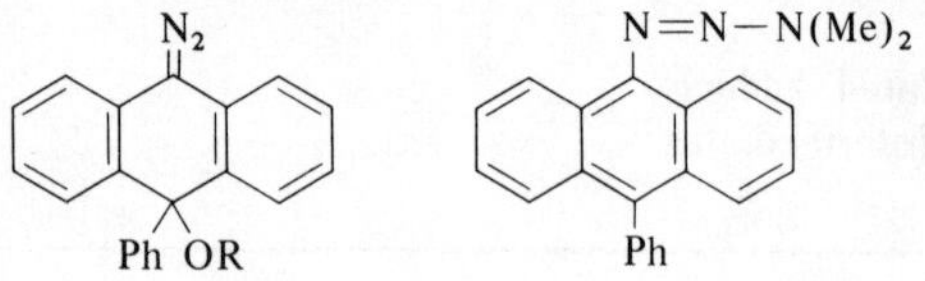

N_2 / Ph OR — **(1)** $N{=}N{-}N(Me)_2$ / Ph — **(2)**

It has been shown[13] that arenediazonium salts carrying alkenyloxy or alkenylamino substituents in *ortho*-positions will give ring-closed hydroxylamine derivatives *via* a free-radical mechanism. Further work[14] on the mechanism of the Sandmeyer reaction of hydrogenodediazoniation confirms the two-step nature of the reaction (equations 1 and 2). Several new reducing agents have been shown to be effective in the first step although the second step specifically requires a Cu(II) salt. The palladium(0)-catalysed arylation of olefins by arenediazonium salts probably involves the formation of arylpalladium species,[15] and palladium(0) also catalyses[16] the reaction of arenediazonium salts with carbon monoxide and sodium carboxylates, to give mixed acid anhydrides.

$$PhN_2^+ + Cu^+ \rightarrow Ph\cdot + N_2 + Cu^{2+} \quad (1)$$

$$Ph\cdot + CuX_2 \rightarrow PhX + CuX \quad (2)$$

Mechanisms of free-radical aromatic substitutions have been reviewed.[17] The radical-chain $S_{RN}1$ mechanism has proved to be a valuable method for achieving substitutions in unactivated aryl halides. Reaction is initiated by electron transfer to the substrate to give a radical anion $[ArX]^{\cdot-}$ which may expel a nucleofugic group to give an aryl radical which then reacts with nucleophiles. Competition between acetone enolate ions and amide ions, for reaction with mesityl radicals produced from four 2-halomesitylenes, shows that the reactivity ratio $k_{enolate}/k_{NH_2^-}$ is independent of the nature of the halogen.[18] This observation is in agreement with the $S_{RN}1$ mechanism since the leaving-group is not present when product selection occurs. Nevertheless the proportion of side-products such as the reduction product mesitylene does vary with the nucleofugal group. This is explained by the previously advanced hypothesis that reaction occurs during mixing of the reagents. Hence the concentration of solvated electrons present at the time of leaving-group departure depends on the rate of fragmentation of $[ArX]^{\cdot -}$. There have been reports of the photo-stimulated $S_{RN}1$ reactions in liquid ammonia of haloarenes with potassium diphenylarsenide,[19] of halogenated pyrimidines, pyridazines, and pyrazines with ketone enolates,[20] and of *p*-dihalobenzenes with ketone enolates.[21] The scope of the $S_{RN}1$ reaction with ketone enolates has been examined,[22] showing that the inefficient reaction of dialkyl-substituted ketone and ester enolates is due to hydrogen atom transfer between the carbon adjacent to the enolate anion and the transient phenyl radical. Intramolecular coupling has been shown[22] to lead to the formation of six-, seven-, eight-, and ten-membered rings. Another synthetically useful example of the $S_{RN}1$ mechanism is the condensation of aryl radicals carrying an *ortho*-amino substituent with enolates derived from ketones and aldehydes leading to the indole ring system.[23] The cyclization of *ortho*-halogenated *N*-acylbenzylamines provides a synthesis of (±)-cherylline and may involve $S_{RN}1$ and benzyne intermediates.[24] The $S_{RN}1$ mechanism has also been invoked in the reaction of aryl bromides with cobalt carbonyl which, in the presence of phase-transfer catalysts, gives good yields of arylcarboxylic acids.[25] However, despite ESR evidence for the presence of radicals, the reactions[26] of 2-bromo- or 2-iodo-5-nitrothiophene with base, giving coupled and reduced products, are unlikely to involve the $S_{RN}1$ mechanism.

The more general opportunities for electron-transfer chain catalysis (ETC) of substitution reactions have been discussed.[27] One possibility is the $S_{ON}2$ mechanism outlined in Scheme 1. This consists of an initial one-electron oxidation step (3)

$$ArX \longrightarrow [ArX]^{\cdot +} \quad (3)$$

$$[ArX]^{\cdot +} + Nu^- \longrightarrow [ArXNu]^{\cdot} \quad (4)$$

$$[ArXNu]^{\cdot} \longrightarrow [ArNu]^{\cdot +} + X^- \quad (5)$$

$$[ArNu]^{\cdot +} + ArX \longrightarrow ArNu + [ArX]^{\cdot +} \quad (6)$$

$$[ArX] + Nu^{\cdot} \longrightarrow [ArXNu]^{\cdot} \quad (7)$$

SCHEME 1

followed by *ipso*-attack of a nucleophile, step (4), and loss of X^- to give a new radical, step (5). The chain-transfer step is the oxidation of ArX by the radical cation, step (6). Two possible examples[28] of the operation of this mechanism are the anodic oxidation of 4-fluoroanisole in the presence of acetate ions to give 4-acetoxyanisole, and the Cu(III)-catalysed reaction of chloro- and fluoro-benzenes to give phenol. A further possible entry into the $S_{ON}2$ mechanism is *ipso*-attack by Nu· on the parent, step (7), instead of steps (3) and (4). Thus it has been shown[29] that 4-fluoroanisole can be converted into a mixture of 4-methoxyphenyl acetate and benzoate by decomposing benzoyl peroxide in acetic acid with potassium acetate.

Free-radical substitutions of protonated heteroaromatic bases by carbon radicals are synthetically important reactions: rate constants for the homolytic alkylation of protonated pyridines and quinolines by *n*-butyl and *tert*-butyl radicals have been measured[30] by competition methods and indicate the importance of polar effects relative to steric effects. Examples of nucleophilic substitutions involving the *ipso*-attack of carbon radicals have been summarized[31] and the synthetic utility of alkyldenitrations and alkyldeacylations has been noted. The reaction[32] of diphenylmethane with NaK to give a benzyl-substituted biphenyl and toluene may also involve *ipso*-substitution with a benzyl anion as leaving group. It is reported[33] that the reaction of mono-substituted benzenes with cyanogen in the plasma zone results in substitution of hydrogen and also *ipso*-substitution. ESR measurements have been made[34] on the radicals formed from halobenzene derivatives on exposure to ^{60}Co γ-rays; *p*-bromo and *p*-iodophenol give species containing

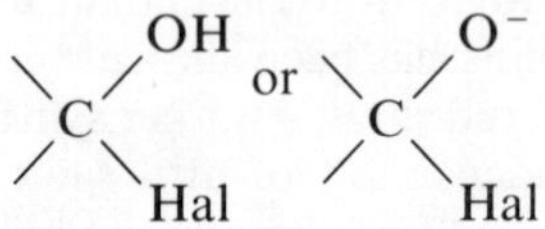

$$\gt\dot{C}(OH)(Hal) \quad \text{or} \quad \gt\dot{C}(O^-)(Hal)$$

groups and the occurrence of halogen atom migration is indicated.

Electro-generated superoxide ions $[O_2]^{\cdot-}$ in dimethyl sulphoxide are unreactive towards unactivated haloaromatics. However, when the aromatic halides are activated by electron-attracting groups they react readily to yield the corresponding phenols. Kinetic and electrochemical data for these reactions indicate that the primary step is not electron transfer from the superoxide ion to substrate but rather an S_NAr attack of $[O_2]^{\cdot-}$ at the halo-substituted ring position.[35] The hydroxylation of benzylamides by potassium superoxide in benzene containing crown ether has been reported.[36] Further work,[37] including rate accelerations in the presence of peroxides, has been reported on the reactions of 2,4-diactivated halobenzenes with tertiary amines.

The S_NAr Mechanism

There has been renewed interest[38] in the mechanism of base catalysis in substitutions involving amines (Scheme 2). Two interpretations of the occurrence of base catalysis currently advocated are (*i*) rate-limiting proton transfer from the intermediate (**3**) to a general base followed by rapid loss of X^-, or (*ii*) rapid equilibrium between (**3**) and its deprotonated form with rate-limiting general-acid-catalysed loss of X^-; the latter is known as the SB–GA (specific-base–general-acid) mechanism. The reactions of piperidine and pyrrolidine with 2,4-dinitrophenyl and 2,4-dinitro-6-methylphenyl ethers in 60% dioxane/40% water are all catalysed by hydroxide ions. However the ratio k_3/k_{-1} is an order of magnitude larger for the pyrrolidine than for the piperidine reactions. Values of k_{-1} will vary little and it has been argued[38] that variation of the amine would not be expected to greatly affect the rate of proton transfer from the intermediate (**3**) to base. However, rate-limiting detachment of the nucleofuge might well account for the slower reaction of

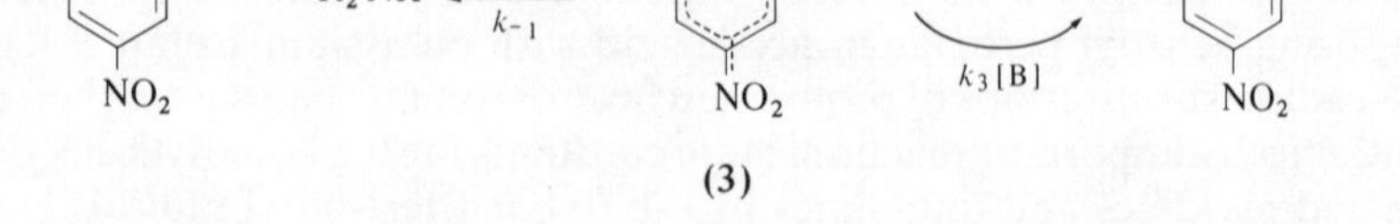

X, R′, NO_2, NO_2 + R_2NH $\underset{k_{-1}}{\overset{k_1}{\rightleftharpoons}}$ X $\overset{+}{N}HR_2$, R′, NO_2, NO_2 (**3**) $\xrightarrow[k_3[B]]{k_2}$ NR_2, R′, NO_2, NO_2

Scheme 2

piperidine than with pyrrolidine. Hence the SB–GA mechanism is favoured in this case. Relevant to this issue is a study[39] of the kinetics of reaction of 2,4-dinitro-1-naphthyl ethyl ether with the two amines in DMSO. Here reaction occurs in two distinct stages, the fast formation of the σ-adduct (**4**) followed by its slow decomposition (equation 8). It is significant that the value of k_4 in the pyrrolidine system is about 11,000 times greater than in the piperidine system. This huge difference between apparently similar systems is thought to arise from steric

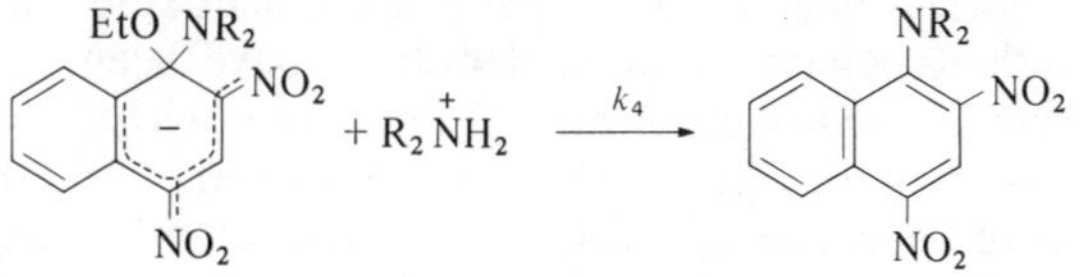

(4) (8)

interactions forced by differences in conformation between the amino moieties in the σ-adducts as they release the nucleofuge. Similar stereoelectronic factors are also thought to account for the very much slower rate of reaction of piperidine with its *N*-(2,4-dinitro-1-naphthyl) derivative to form the σ-adduct (**5**) than of the analogous reactions of pyrrolidine or butylmethylamine with their corresponding *N*-(2,4-dinitronaphthyl) derivatives.[40]

Previous work has shown that the *uncatalysed* decomposition of the zwitterionic intermediate (step k_2 in Scheme 2) is likely to involve intramolecular proton transfer rather than an SB–GA mechanism with solvent acting as the base: further support for this view has come from kinetic study[41] of the reaction of phenyl 2,4,6-trinitrophenyl ether with aniline in solvents of varying basicity. For this reaction in benzene[41,42] there is a third-order dependence on the aniline concentration and the cyclic transition state (**6**) has been suggested. The reactions of 1-fluoro- and 1-chloro-2,4-dinitrobenzenes with 2,2,2-trifluoroethylamine (TFE) in DMSO and acetonitrile are not base-catalysed showing that the initial base attack is rate-limiting.[43] This contrasts with the reaction in acetonitrile of the fluoro compound with aniline (an amide of similar basicity to TFE) where base catalysis was observed. It has been suggested[43] that this results from the intrinsically higher value of k_{-1} for aromatic than for aliphatic amines.

Kinetic studies of the hydrolysis of picrylimidazole in aqueous buffers between pH 0.47 and 10.6 show[44] that reaction can occur by attack on the parent or its conjugate acid. Buffer catalysis is most probably due to concerted general base catalysis of water addition as represented by (**7**).

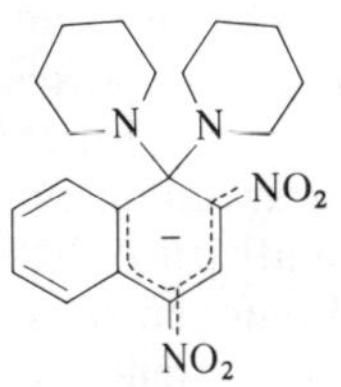

(**5**)

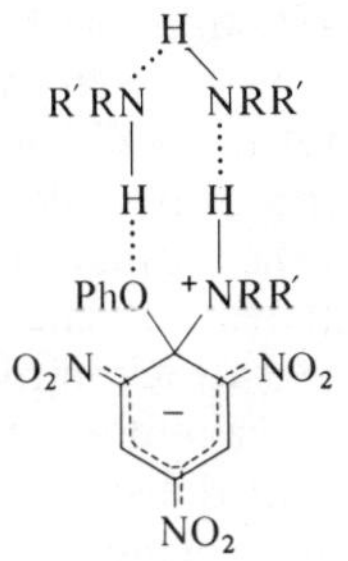

(**6**)

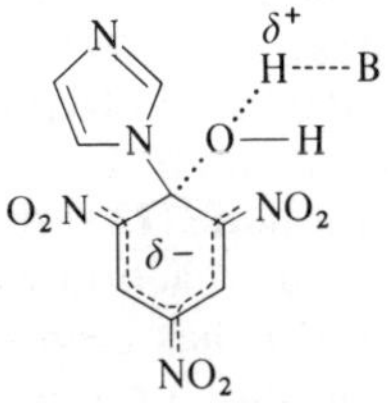

(**7**)

Mechanisms and reactivity in aromatic nucleophilic substitution reactions have been reviewed in Spanish.[45] The reactions of 2,4- and 2,6-dinitroanisole with piperidine and *N*-methylpiperidine in benzene give not only the expected aromatic substitution products but also the corresponding 2,4-dinitrophenols formed by S_N2 attack at the methyl group.[46]

Competition between S_NAr displacement of a nitro group and S_N2 displacement on the alkyl groups has been reported in the reactions of *p*-nitrobenzoate and *o*-nitrobenzoate esters with azide, alkoxide, and thiophenoxide ions.[47] Further examples of the ambident behaviour of nucleophiles have been reported. Thus, reaction of 2-aminothiazole (**8**) or its 4-methyl derivative with 2,4-dinitrofluorobenzene may involve the ring nitrogen or the amine nitrogen,[48] and the reaction of the enolate ion of ethyl acetoacetate with 2,4-dinitroactivated substrates may give *C*- or *O*-arylated products. In this latter reaction the effects of varying the nature of the leaving-group[49] and of the cation[50] have been investigated.

Nucleophilic substitution reactions have been used to examine substituent effects: kinetic data for the reaction of picryl chloride with 2-substituted and 2,5-disubstituted anilines indicate the presence of steric interactions,[51] while crystallographic measurements on 1-chloro-2,4-dinitro-3,6-dimethylbenzene show that the nitro groups are twisted from the ring plane so that reaction with 4-hydroxyaniline is inhibited.[52] Kinetic data for reaction of substituted 4-aminothianes with 1-chloro-2,4-dinitrobenzene show that compounds with an axial $C(4)-NH_2$ bond (**9**) react faster than their equatorial epimers (**10**) due to steric and solvation effects.[53] There have been reports of the effects of substituents at the 4′-position on the alkaline hydrolyses of 4-nitrodiphenylsulphones,[54,55] and of the transmission of substituent effects in aminocarbazoles.[56]

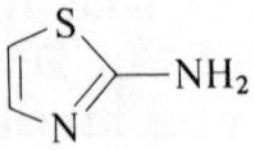

S N NH_2 (**8**) NH_2 S (**9**) NH_2 S (**10**)

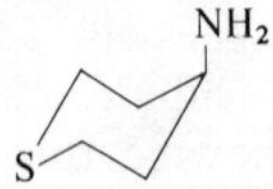

The nucleophilic substitution of picryl chloride with sodium carboxylates gives intermediates (**11**) which in the presence of alkoxide ions yield carboxylic esters in a synthetically useful reaction.[57] There have been kinetic studies of the reaction of picryl halides with nitrite ion in acetonitrile[58] and of the reaction[59] of picryl chloride with hydrazides of aromatic phosphoric acids [e.g. $Ph_2P(O)NHNH_2$].

1,3,5-Trifluorotrinitrobenzene is much more susceptible to nucleophilic attack than its trichloro analogue and its reactions with a number of nitrogen, oxygen, carbon, and halogen nucleophiles have been described;[60] reaction with the ambident dinitromethide ion may result in *C*- or *O*-arylation depending on the cation present. Study[61] of the reaction of perchloroindane (**12**) with nucleophiles shows that contrary to previous ideas the aromatic chlorines are more susceptible to substitution than the aliphatic chlorines. The chlorines in 2,3-dichloro-1,4-naphthoquinone (**13**) are susceptible to attack in pyridine by arylamines[62] or by 2-aminophenol, reaction with the latter reagent leading to a synthesis of heterocyclic quinones.[63] It has been reported[64] that the substitution of halogen by astatine in chloro-, bromo-, or iodo-benzene can be achieved in a solvent system consisting of 85% halobenzene/15% butylamine.

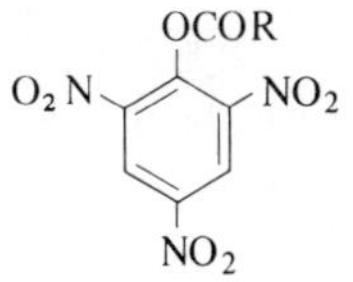

(11) 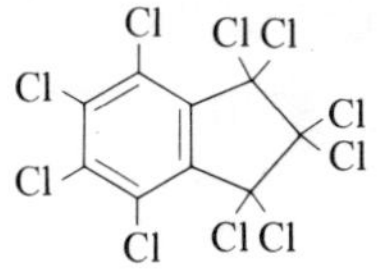 **(12)** 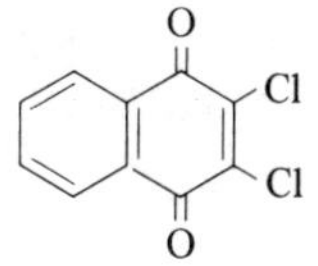 **(13)**

Nucleophilic substitutions are often carried out with advantage in dipolar aprotic solvents and in such solvents the nitro groups in 2- and 4-nitronaphthalic 1,8-anhydrides can be substituted by amines.[65] There has been further study[66] of the reaction of 1-chloro-2,4-dinitrobenzene with methoxide ions in DMSO where it is now well known that base attack initially occurs at an unsubstituted ring position. Nevertheless even on treatment with the very basic medium of potassium *tert*-butoxide in hexamethylphosphoric triamide the three isomeric trichlorobenzenes fail to participate in the base-catalysed halogen dance. However 1,2,3,5- and 1,2,4,5-tetrachlorobenzenes undergo disproportionation to penta- and tri-chlorobenzenes as well as interconversion, and pentachlorobenzene also disproportionates.[67] The mechanism involves the initial formation of a carbanion by acid–base reaction. The carbanion may then abstract a Cl^+ moiety from a second molecule to give a new chlorocarbon and a new carbanion; substitution reactions to form aryl *tert*-butyl ethers are observed as side-reactions.

In non-polar solvents anionic nucleophiles may be strongly ion-paired with cations thus reducing their reactivities; however the presence of crown ethers capable of complexing the cation has been found to enhance reactivity in these solvents. Thus, rate constants for reaction of 4-nitrobromobenzene with potassium phenoxide in dioxane in the presence of various crown ethers have been found to increase linearly with the stability constant of the crown K^+ complex.[68] Also reported is the reduction of halobenzenes by cryptand-activated potassium hydride in tetrahydrofuran.[69] However, polyethylene glycol as a co-solvent can effectively replace crown ether in the photochemical substitution reaction of anisole with potassium cyanide in dichloromethane.[70] A novel method for the phase-transfer-catalysed sulphodechlorination of 1-chloro-2,4-dinitrobenzene using protonated tertiary amines has been reported.[71] The reaction of *n*-hexylamine with 2,4-dinitrochlorobenzene has been found to be faster in micro-emulsions of *n*-octane, *n*-hexylamine, and cetyltrimethylammonium bromide (CTABr) than in water, probably due to an increased concentration of reactants in the droplets.[72] There have been further studies[73] of the catalysis by micelles of CTABr of reactions between anionic nucleophiles and substrates carrying anionic substituents and it appears that the ambident nitrite ion is particularly well solubilized.

Kinetic studies have been reported[74] of the reversible Smiles rearrangement of *S*-(2,4-dinitrophenyl)cysteine (**14**) to *N*-(2,4-dinitrophenyl)cysteine (**16**) *via* the intermediate σ-adduct (**15**). Further studies have been made[75,76] of the photo-induced $O \rightarrow N$-type Smiles rearrangements of 1-(*p*-nitrophenoxy)-ω-anilino-alkanes, and it is reported that in the presence of base (2-hydroxyethyl)- and (3-hydroxy-*n*-propyl)-methyl(*p*-nitrophenyl)sulphonium ions undergo $S \rightarrow O$-type rearrangements.[77] Intramolecular substitutions involving *ortho*-substituents are observed in the reactions of chloronitrobenzenes and chloropyridines with mono-substituted hydrazines leading to the 8-azapurine and benzopyrazole ring systems.[78] Benzo-fused ring systems may also be formed by use of suitably

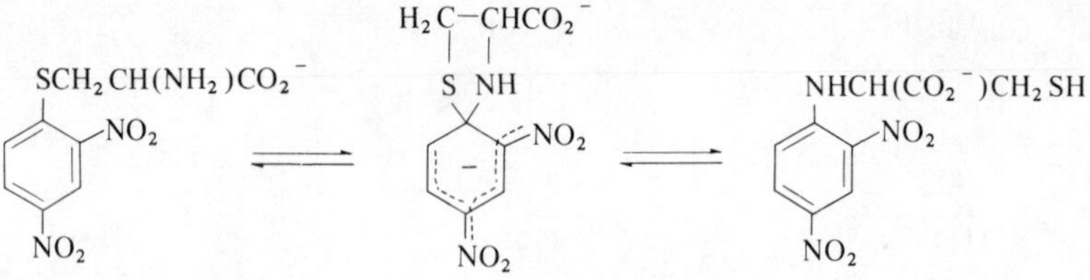

$H_2C-CHCO_2^-$ / S NH; $SCH_2CH(NH_2)CO_2^-$; NO_2; NO_2; $NHCH(CO_2^-)CH_2SH$; NO_2; NO_2; NO_2; NO_2

(14) ⇌ **(15)** ⇌ **(16)**

substituted aryloxazolines bearing an *o*-methoxyl group; the oxazoline ring facilitates the intramolecular substitution of the methoxyl group by its ability to co-ordinate a metal cation.[79] Intramolecular substitutions are also involved in the reactions of spiroepoxycyclohexa-2,4-dienones with the pentachlorothiophenoxide anion leading to phenoxanthiins.[80]

π-Complexing of aromatic molecules with transition-metal ligands, such as chromium tricarbonyl, greatly increases the ease of nucleophilic attack, and calculations[81] have shown that regioselectivity may depend not only on the substituent but also on the conformation of the $Cr(CO)_3$ unit. There has been a survey[82] of the addition of carbon nucleophiles to such complexes providing a method for the formation of new carbon–carbon bonds. Kinetic studies[83] of the methoxydehalogenation of π-complexed halogenobenzenes show that activation increases through the series $Cr(CO)_3 < Mo(CO)_3 \ll (\eta^5\text{-}C_5H_5)Fe^+ < Mn(CO)_3^+$ and there is evidence that the cationic complexes form ion-pairs with methoxide ions resulting in a reduction in their reactivity. Reaction of methoxide with the $Mn(CO)_3^+$-activated chlorobenzene occurs in two distinct stages *via* the intermediate (**17**). Treatment[84] of the η^5-cyclopentadienyl–η^6-benzenecobalt dication with methoxide ions in methanol results in the stereospecific (*exo*) addition of two methoxide ions to give (**18**).

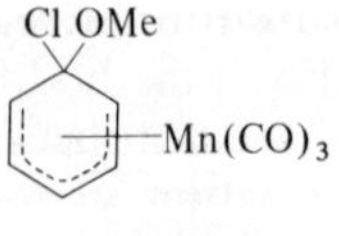

Cl OMe; $Mn(CO)_3$

(17)

OMe; OMe; Co

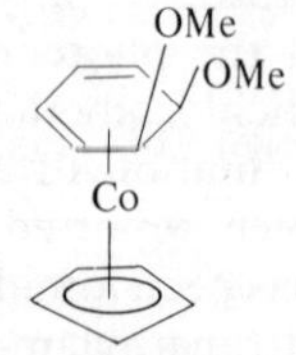

(18)

Copper-catalysed substitutions of aryl halides are known as Ullman condensations. Kinetic studies[85] of the reaction of 1-amino-4-bromoanthraquinone-2-sulphonic acid with aromatic amines in bicarbonate or phosphate buffers suggest the formation of bifunctional catalysts, such as $CuCO_3^-$, which aid the departure of a proton and a bromide ion as shown in (**19**). Some reactions of piperidine with bromonaphthoic acids[86] and 8-chloronaphthalene-1-sulphonic acids[87] are catalysed by copper salicylate, and catalysis by copper compounds of the reactions of 2-nitrochlorobenzene with ammonia[88] and of 4,4′-dinitrodiphenyl sulphone derivatives with sodium sulphite[89] has been reported. In hexamethylphosphoric acid triamide containing copper(I) iodide, non-activated aryl iodides react with arenethiolate ion to give aryl sulphides,[90] while in dimethylformamide containing sodium hydride and a cuprous halide intramolecular cyclization of *N*-(2-

bromo-4,5-dimethoxyphenethyl)acetamide gives an indole derivative;[91] the latter reaction may involve (**20**). Other studies have been reported of the catalysis by palladium(0) of the substitution reactions of halotropolones with olefins which probably involves the formation of tropolonylpalladium(II) σ-bonded complexes,[92] and of the catalysis by rhodium(III) or palladium(II) salts of the dehalogenation of aromatic halides in basic alcohol.[93]

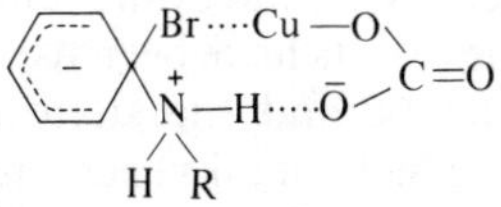

(19)

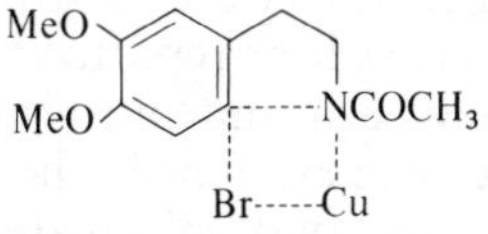

(20)

Heterocyclic Systems

Kinetic data for the reactions of a series of 2-bromo-3-nitro-5-*X*-thiophens with a series of substituted anilines in methanol have been reported;[94] a first-order dependence on both reactants indicates that here nucleophilic attack is rate-limiting. Surprisingly the Hammett ρ and Brønsted β values are insensitive to the reactivities of the starting materials and it has been suggested that this apparent violation of the reactivity–selectivity principle results from intramolecular hydrogen bonding between the incipient ammonium hydrogen and the 3-nitro group in the rate-determining transition state (**21**). The piperidino-demethoxylation reactions of (**22**; $X = CO_2Me, Ac$) in methanol are catalysed by both piperidine and methoxide.[95] As in the case of benzene derivatives base catalysis might signify rate-limiting proton transfer from the initially-formed intermediate, or general acid catalysis of leaving-group departure. The evidence here favours the latter, SB–GA, mechanism. Kinetic measurements on the morpholino-demethoxylation of (**23**; Z = NMe, O) show that nucleophilic attack on the pyrilium ring occurs 3×10^6 times faster than on the pyridinium ring. Both reactions are subject to general base catalysis and it appears that while catalysis with the pyrilium derivative is of the SB–GA type that of the pyridinium derivative involves rate-limiting proton transfer.[96] Base catalysis of the reactions between 2-halo-6-nitrobenzothiazoles and aliphatic amines in benzene is thought to be of the bifunctional type.[97] There have been studies[98] of base catalysis and oxygen *versus* sulphur reactivity in the reactions of 7-nitroquinoline derivatives with piperidine in benzonitrile.

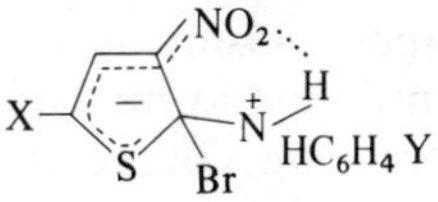

(21) 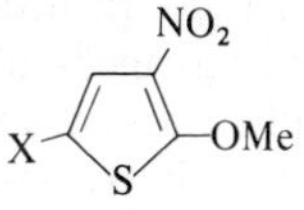 **(22)** 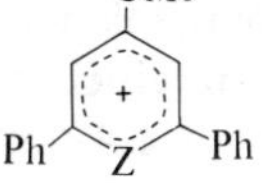 **(23)**

Kinetic measurements have been reported of the hydrolyses of a number of 2-fluoro nitrogen heterocycles in aqueous hydrochloric acid.[99] With the more reactive substrates, such as 2-fluoropyrimidine, the data indicate rate-determining nucleophilic attack by water on the fluorine-bearing carbon atom of the protonated substrates while with less reactive compounds, such as 2-fluoropyridine, it is likely that nucleophilic attack may be assisted by proton transfer to a second water molecule. There is evidence also that the hydrolysis of the chloro-1,3,5-triazine (**24**) in acid solutions involves mono- and di-protonated forms.[100] Triazinyl peroxides have been prepared[101] by reaction of chlorotriazines and peroxide nucleophiles. Experimental and theoretical studies[102,103] of the effects of intramolecular hydrogen bonding in haloazines show that with a *para*-orientation, as in (**25**), the halogen is activated to nucleophilic substitution.

Rate measurements of the exchange between $^{36}Cl^-$ and cyano- and nitro-activated 2-chloropyridines have been reported.[104] The results have been used to compare the activating effects of aza and nitro substituents, and the steric effect of methyl substituents have been examined. The polar effects of alkyl substituents have been investigated in the alkoxy-dechlorination reactions of a series of alkyl-substituted 2- and 6-chloropyrimidines.[105] The use of *N–N*-linked heterocycles in the regiospecific syntheses of 4-substituted pyridines has been reported;[106] thus, nucleophilic attack on the *N*-pyrrolylpyridinium ion (**26**) occurs preferentially at the 4-position due to steric shielding by the pyrrole methyl groups.[107] An alternative method[108] of producing 4-alkylpyridines also relies on steric effects to inhibit attack at the 2- and 6-positions; the key idea here is to introduce a phosphonyl group at the 4-position of 1-ethoxycarbonylpyridine to give an intermediate which is selectively alkylated.

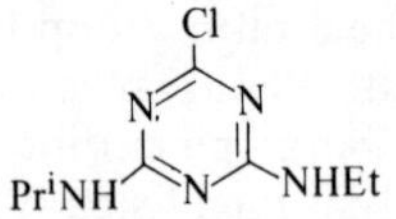

(**24**) 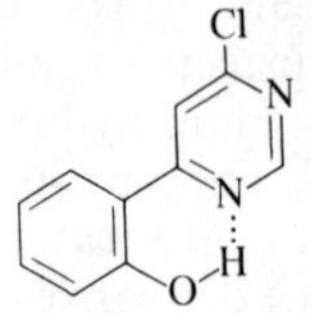 (**25**) 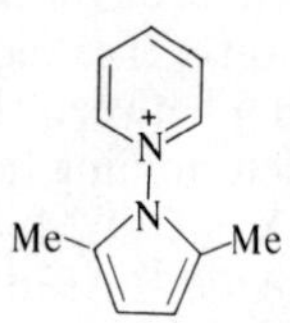 (**26**)

Diverse mechanistic studies have confirmed that the geminal amination of aminocyclotriphosphazenes involves formation of the conjugate base,[109] that nucleophilic substitution of thiamin by sulphite occurs *via* the bisulphite addition product,[110] and that the Chichibabin amination of heterocycles[111] involves σ-adduct intermediates. There have been reports of the synthesis and substitution reactions of furans containing CH:C(Y)CN substituents at the 2-position,[112] and of the synthesis of polyethereal macrocycles incorporating 1,8- or 1,5-naphthyridine rings.[113]

Nucleophilic attack at a ring carbon atom adjacent to that carrying a nucleofugic group may lead to *cine*-substituted products. Thus treatment of 1,4-dinitro-3-methylpyrazole with nucleophiles gives 5-substituted 3-methyl-4-nitropyrazoles in a reaction which has been used in a synthesis of the antibiotic formycin.[114] The reaction of 1,4-dinitropyrazole with pyrazole[115] involves two consecutive *cine*-substitutions giving the tripyrazole derivative (**27**). Although reaction of the 3-nitroimidazo[1,2-*a*]pyridine derivative (**28**) with thioglycolate anion results in normal substitution of the nitro group[116] it has been reported that reaction of the drug metronidazole (**29**) with 2-aminoethanethiol yields both normal and *cine*-substituted products.[117] Crystallographic data show that the product of reaction of 2,5-dimethyl-3,4-dinitrothiophen with morpholine[118] is the *trans*-adduct (**30**; NR_2 = morpholino), and that the formation of the phenylphenothiazine derivative (**31**) by intramolecular substitution of a nitro group involves a Smiles rearrangement.[119]

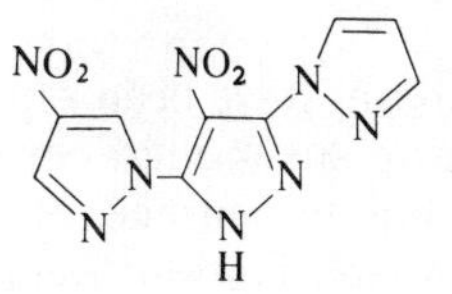

(27) 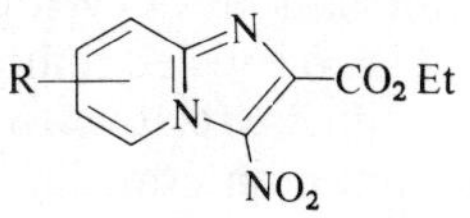 (28)

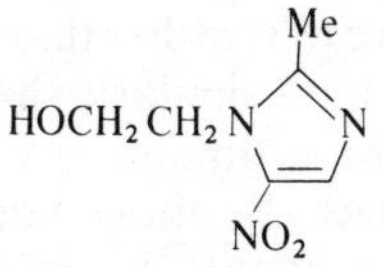

(29) 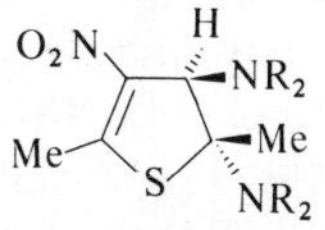 (30) 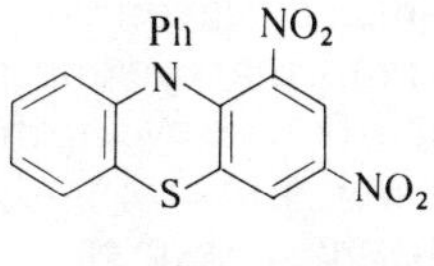 (31)

There has been a review of the cleavage of pyridine derivatives under the influence of nucleophiles and their recyclization to give new ring systems.[120] Quaternization to give pyridinium ions greatly enhances reactivity in these systems. Kinetic studies[121] show that the initial reaction of pyrilium cations with primary amines to give ring-opened intermediates is fast for strongly basic amines and is base-catalysed for weak amines; the ring-closure of the intermediates to give pyridinium derivatives is subject to steric and electronic hindrance and is acid-catalysed. The reactions of the strongly activated 1-substituted 3,5-dinitro-4-pyridones with diethyl sodio-3-oxopentanedioate give 1-substituted 3,5-bis(ethoxycarbonyl)4-pyridones with the elimination of sodio-1,3-dinitro-2-propane and it is likely that *meta*-bridged adducts (**32**) are intermediates.[122] 1-Substituted 3,5-dinitro-4-pyridones may also be used as

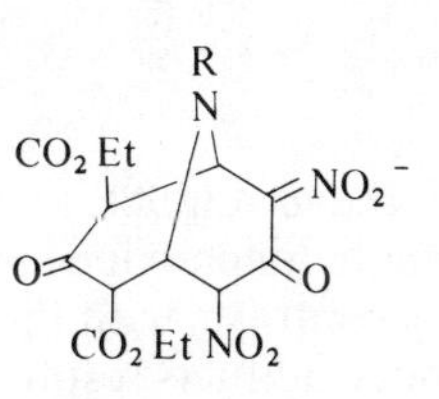

(32)

protecting groups for primary amines.[123] A novel pyrimidine-to-pyridine ring transformation resulting in the synthesis of pyrido[2,3-*d*]pyrimidines has been reported by reaction of 1,3-dimethyluracil derivatives with 1,3-ambident nucleophiles,[124] and reaction of 5-aminopyrimidine with *N*-alkyl(aryl)imidoyl chlorides in the presence of phosphorus oxychloride yields 4-formyl-1,2-dialkyl(aryl)imidazoles.[125] It has also been reported[126] that displacement of hydrazine hydrochloride from 1,4-dichlorophthalazines by bidentate nucleophiles, such as 1,2-diaminobenzenes, results in the formation of 2,2′(1,2-arylene)bis-benzimidazoles. Other studies have shown that ring-opened intermediates may be involved in the reactions of 2-bromo-4*H*-thiopyran-4-ones with amines which may yield aminopyridones,[127] and in the reactions of tetracyano-1,4-dithiin and tetracyanothiophene with nucleophiles.[128]

Meisenheimer and Related Adducts

Ab initio MO calculations of cyclohexadienyl anions have been reported.[129] They show that the rates of nucleophilic substitution of substituted benzenes correlate well with the stabilities, relative to the corresponding benzene, of the substituted cyclohexadienyl anion intermediates in the S_NAr mechanism. Related theoretical studies[130] concerned with the formation of cyclohexadienes by Birch reduction of substituted benzenes have shown that the kinetically favoured site of protonation of the cyclohexadienyl anion intermediates frequently differs from the thermodynamically favoured position. These studies[129] have confirmed previous work indicating that cyclohexadienyl anions have the planar structure (**33**) rather than the homoaromatic cyclopentadienyl structure (**34**). NMR data[131] have similarly shown that the 4-hydropyridyl anion is planar with no 1,5-homoaromatic overlap occurring. However 1,6-dihydro-3-aryl(alkyl)-1,2,4,5-tetrazines do show homoaromatic character[132] as do their conjugate acids and conjugate bases (**35**; R = alkyl, aryl). Low-temperature ^{1}H-NMR measurements on these systems show that the two hydrogens at the 6-position are markedly non-equivalent, the hydrogen oriented towards the ring showing substantial shielding consistent with the presence of an induced ring current. ^{13}C-NMR measurements[133] show that the homoaromatic adducts (**36**; R = alkyl, aryl) formed from 3-substituted 1,2,4,5-tetrazines in liquid ammonia are anionic, the driving force for the deprotonation being an increase in resonance stabilization in the homoaromatic ring.

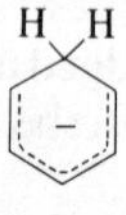

(33) 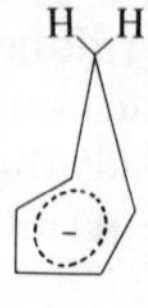 (34) 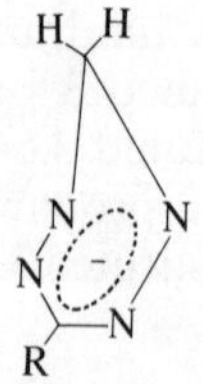 (35) 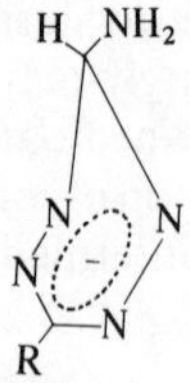 (36)

Spectroscopic studies[134] of the σ-adduct formed from 1,3,5-trinitrobenzene (TNB) and potassium 2,4,6-trimethylphenoxide in DMSO show that bonding is *via* the oxygen of the aryloxide; this contrasts with the behaviour of the phenoxide ion itself where carbon–carbon bonded adducts result. Kinetic and equilibrium data[135] for reaction of several nitro compounds with base in *tert*-butyl alcohol containing water indicate that the reactive nucleophile is the hydroxide ion rather than *t*-butoxide; σ-adducts are initially formed from TNB and picryl chloride by hydroxide addition at unsubstituted ring positions, although the latter substrate and 1-chloro-2,4-dinitrobenzene eventually yield the corresponding phenols. Kinetic studies of the reactions in DMSO between TNB and aliphatic amines (Scheme 3) show that the proton-transfer step may be kinetically significant; reduction of the rate constants for proton transfer below the values expected for diffusion control is attributed to a steric effect which is more pronounced in the reactions involving piperidine than in the reactions of primary amines.[136] It has been reported that the reaction of TNB with trifluoroethylamine is base-catalysed in acetonitrile, but, surprisingly, is uncatalysed in DMSO.[43]

It is now well known[137] that 1,1-dialkoxy adducts can strongly complex certain alkali-metal cations by a chelation effect; further evidence for this phenomenon has

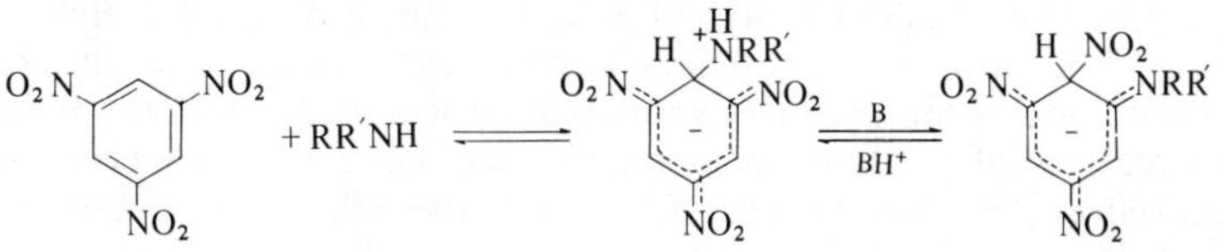

SCHEME 3

come from kinetic studies of the reaction of 1-methoxy-2-nitro-4-cyanonaphthalene with metal methoxides in methanol.[138] Salt effects on the rates of decomposition of 1,1-dimethoxy complexes in water have been reported,[139] and studies[140,141] of the effects of pressure on the kinetics of σ-complex formation in hydroxylic solvents have allowed the determination of volumes of activation.

There is spectroscopic evidence[142] for the formation of the 1,1-dipiperidyl adduct (**5**) by reaction of 2,4-dinitro(1-piperidyl)naphthalene with piperidine in DMSO, and for the formation of 1,1- and 1,3-adducts by the reaction of the analogous 2,4,5-trinitronaphthalene derivative. The stability of the 1,1-adducts is reduced by stereoelectronic factors.[40] There have also been reports of the formation of σ-adducts by hydroxide addition to 2,4,7-trinitrodiphenylenesulphone[143] and by *tert*-butyl peroxide addition to trinitroarenes.[144]

The trifluoromethylsulphonyl ring substituent is strongly activating towards σ-adduct formation and a comparison of the visible spectra of adducts containing NO_2 and/or SO_2CF_3 substituents has been reported.[145] The oxidation of σ-adducts of 1,3,5-tris(trifluoromethylsulphonyl)benzene yields 1-substituted 2,4,6-tris(trifluoromethylsulphonyl)benzenes and there is NMR evidence[146] that the 1-cyano compound so formed reacts with cyanide ions to give (**37**). Reactions of the troponoid adduct (**38**) to give the corresponding oxime or to give 1-nitro-2-methoxycarbonylnaphthalene have been reported.[147] σ-Adducts have been observed during reactions of methyl and phenyl picrates and nucleophiles.[148]

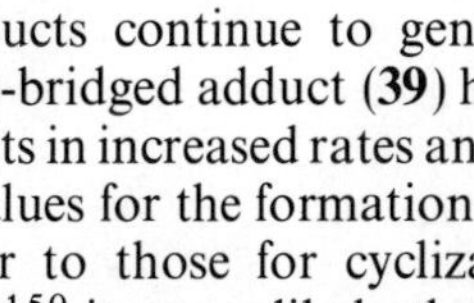

(37)

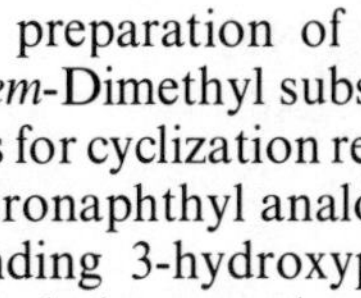

(38)

Spiro-adducts continue to generate interest and the preparation of the 1,8-naphthalene-bridged adduct (**39**) has been reported.[149] *gem*-Dimethyl substitution usually results in increased rates and equilibrium constants for cyclization reactions. However, values for the formation of (**40**) and its 2,4-dinitronaphthyl analogue are quite similar to those for cyclization of the corresponding 3-hydroxypropoxy compounds;[150] it seems likely that the presence of the methyl groups here causes increased steric crowding. NMR studies[151] have shown that in acid solutions the dioxolane ring of the spiro-picryl complex of adenosine opens stereoselectively to give 3′-*O*-(2,4,6-trinitrophenyl)adenosine. There have also been reports of a spectrophotometric study of the cyclization of 1,3-bis(β-oxyethoxy)-2,4,6-trinitrobenzene[152] and of the reaction of 2,4,6-trinitroanisole with 1,4-butylene glycolate.[153]

The preparation of adducts (**41**; M = Si, Ge, Sn) containing metals has been reported.[154] The nitration of propylphenazone yields at least 27 products many of which give σ-adducts on treatment with alkaline acetone.[155] Kinetic studies of the Janovsky reaction of 1,3-dinitrobenzene with acetonate ions have been concerned with salt effects,[156] solvent effects,[157] and the effects of electron-acceptor compounds.[158] There has also been a report of the mechanism of the decomposition of Janovsky adducts by proton-donor reagents.[159]

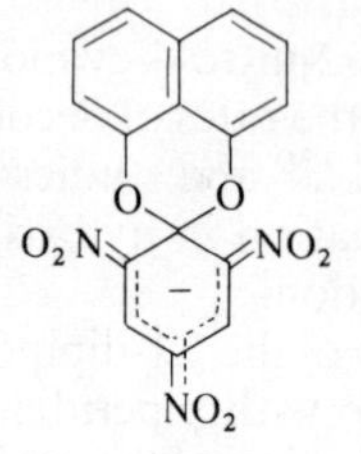

(39) 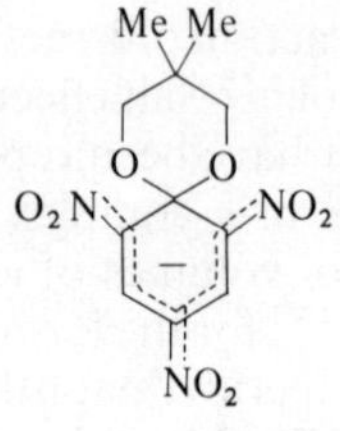 **(40)** 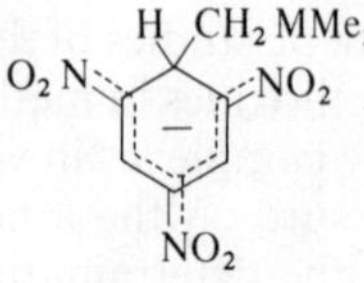 **(41)**

Reactions with base of ring-activated toluenes and anilines are complicated by the possibility that transfer of a side-chain proton may compete with base addition. Kinetic and spectroscopic studies[160] of the reaction of 2,4,6-trinitrobenzyl chloride (TNBC1) with alkoxides have allowed the identification of three processes resulting from 1:1 interaction; these are base addition at the 3-position to give (**42**) or, at the 1-position to give (**43**), or proton transfer to give (**44**). In contrast with the behaviour of 2,4,6-trinitrotoluene (TNT) the more stable σ-adduct results from addition at the 1-position and has a stability comparable to that of the conjugate base; steric and electronic factors are responsible. Reaction of TNT and TNBCl with sulphite, a good nucleophile but poor base, gives 1:1 and 1:2 adducts by sulphite addition at unsubstituted ring positions.[161] NMR measurements have shown[162] that the major products of interaction of 2,4-dinitroaniline or its *N*-alkylated derivatives with methoxide ions in methanol–DMSO are the conjugate bases formed by transfer of an amino proton. However, with 2,6-dinitroaniline base, addition at the 3-position effectively competes with proton loss. Measurements by visible spectroscopy[163] have shown that the 1:1 and 1:2 interactions of *N*-picrylethylenediamine hydrochloride with base in DMSO–methanol involve deprotonation at nitrogen centres and that in the presence of excess base the dianion, (**45**), is formed.

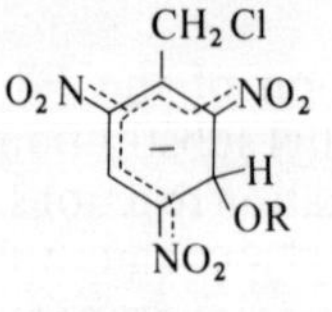

(42) 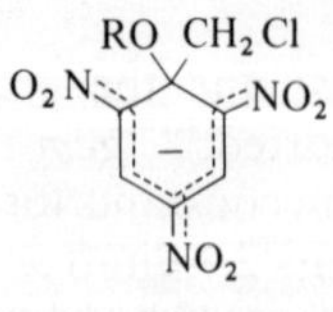 **(43)** 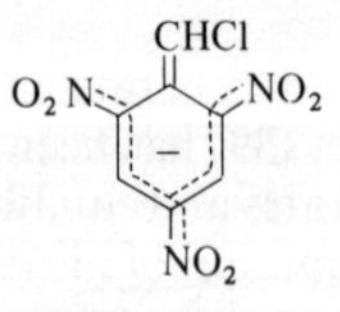 **(44)** 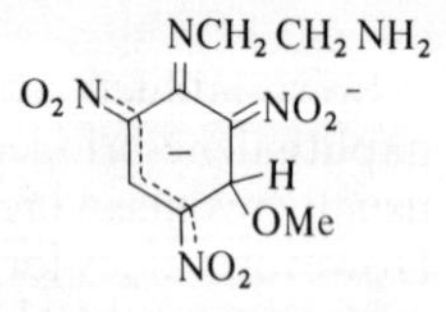 **(45)**

4,6-Dinitrobenzofuroxan (**46**), DNBF, is an exceedingly strong electrophile and reacts with ketones in DMSO even in the absence of base to give carbon-bonded σ-adducts. NMR measurements show[164] that addition occurs at the 7-position and that with β-diketones enolic and/or ketone adducts are formed. The presence of two chiral centres in the DNBF–cyclopentanone adduct allowed the characterization of diastereomeric complexes. NMR data have been reported also for the adducts of DNBF with ethanethiol and L-cysteine.[165] The initial site of methoxide attack on 4-nitrobenzofurazan (**47**; X = O) is the 5-position although conversion to the thermodynamically more stable 7-methoxy adduct follows; a comparison[166] with the sulphur and selenium analogues (**47**; X = S, Se) gives a reactivity sequence O ≫ Se > S. ^{1}H-NMR measurements[167] of methoxide addition to 2,4,6-triphenylpyrilium, 2,4,6-triphenylthiopyrilium, and 1,2,4,6-tetraphenylpyridinium ions show that the thermodynamically stable adducts usually result from attack at the 2-positions of the substrates. ^{1}H and ^{13}C-NMR studies[168] of amide additions to 1,*X*-naphthyridines have been reported. These show that at low temperatures rearrangement of the adducts, initially formed under kinetic control, to their thermodynamically more stable isomers, may be slow, thus allowing a rationalization of the variation with temperature of the position of Chichibabin amination of these compounds. Kinetic studies have been made[169] of the kinetics of the hydride-transfer reaction between 5-nitroisoquinolinium cations and 1,4-dihydronicotinamides.

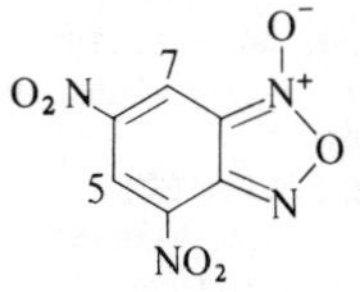

(46) 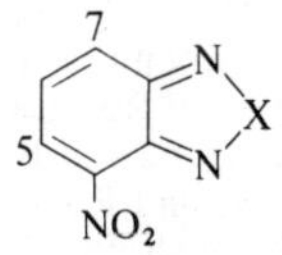 **(47)**

Benzyne and Related Intermediates

There have been reviews of the gas-phase chemistry of carbynes[170] and of the reactions of benzynes with heterocyclic compounds.[171] The 1,3-cycloaddition of benzyne to thiophene derivatives is reported to give low yields of benzo[*b*]-thiophens,[172] while reaction with 2-*tert*-butyl-3-methylbenzothiazoline yields a heterocyclic sulphur ylide.[173] It has also been reported[174] that substituted benzynes can be trapped intramolecularly by an attached furan moiety in a reaction which has been used in the synthesis of mansonone E, a naturally occurring *o*-naphthoquinone.

The reaction of *o*- and *m*-halobenzamides with potassium amide in liquid ammonia is reported[175] to yield *o*-hydroxyphenylamidines instead of the expected benzoxazole derivatives. There is evidence that the benzoxazoles (**48**) are initially formed *via* aryne intermediates but that they are readily aminated to give the observed products.

The 2-pyridyl cation (**49**) might be expected to have aryne character (**50**); however, there is no evidence[176] for participation in aryne-like cycloaddition reactions and the products obtained result from normal electrophilic substitution.

(48)  **(49)**  **(50)**

References

[1] Kuokkanen, T., *Finn. Chem. Lett.*, **1980**, 189.
[2] Bowers, P. G., and Dick, Y. M., *J. Phys. Chem.*, **84**, 2498 (1980).
[3] Isac, D., Mracec, M., Prosteanu, N., and Simon, Z., *Rev. Roum. Chim.*, **26**, 29 (1981).
[4] Nakazumi, H., Szele, I., and Zollinger, H., *Tetrahedron Lett.*, **22**, 3053 (1981).
[5] Bartsch, R. A., and Carsky, P., *J. Org. Chem.*, **45**, 4782 (1980).
[6] Korzeniowski, S. H., Leopold, A., Beadle, J. R., Ahern, M. F., Sheppard, W. A., Khanna, R. K., and Gokel, G. W., *J. Org. Chem.*, **46**, 2153 (1981).
[7] Stumpe, R. W., *Tetrahedron Lett.*, **1980**, 4891.
[8] Pibulik, I. I., Weber, R. V., and Zollinger, H., *Helv. Chim. Acta*, **64**, 1777 (1981).
[9] Goodman, P. D., Kemp, T. J., and Pinot de Moira, P., *J. Chem. Soc., Perkin Trans. 2*, **1981**, 1221.
[10] Rigaudy, J., Barcelo, J., and Valt, M.-H., *C.R. Hebd. Seances Acad. Sci., Ser. C*, **291**, 211 (1980).
[11] Chin, A., Hung, M.-H., and Stock, L. M., *J. Org. Chem.*, **46**, 2203 (1981).
[12] Ng, J. S., Katzenellenbogen, J. A., and Kilbourn, M. R., *J. Org. Chem.*, **46**, 2520 (1981).
[13] Beckwith, A. L. J., and Meijs, G. F., *J. Chem. Soc., Chem. Commun.*, **1981**, 595.
[14] Galli, C., *J. Chem. Soc., Perkin Trans. 2*, **1981**, 1459.
[15] Kikukawa, K., Nagira, K., Wada, F., and Matsuda, T., *Tetrahedron*, **37**, 31 (1981).
[16] Kikukawa, K., Kono, K., Nagira, K., Wada, F., and Matsuda, T., *J. Org. Chem.*, **46**, 4413 (1981).
[17] Minisci, F., and Porta, O., *Chim. Ind. (Milan)*, **62**, 769 (1980); *Chem. Abs.*, **94**, 64664 (1981).
[18] Tremelling, M. J., and Bunnett, J. F., *J. Am. Chem. Soc.*, **102**, 7375 (1980).
[19] Rossi, R. A., Alonso, R. A., and Palacios, S. M., *J. Org. Chem.*, **46**, 2498 (1981).
[20] Carver, D. R., Komin, A. P., Hubbard, J. S., and Wolfe, J. F., *J. Org. Chem.*, **46**, 294 (1981).
[21] Alonso, R. A., and Rossi, R. A., *J. Org. Chem.*, **45**, 4760 (1980).
[22] Semmelhack, M. F., and Bargar, T., *J. Am. Chem. Soc.*, **102**, 7765 (1980).
[23] Beugelmans, R., and Roussi, G., *Tetrahedron*, **37**, *Supplement No. 9*, 393 (1981).
[24] Kessar, S. V., Singh, P., Chawla, R., and Kumar, P., *J. Chem. Soc., Chem. Commun.*, **1981**, 1074.
[25] Brunet, J.-J., Sidot, C., and Caubere, P., *Tetrahedron Lett.*, **22**, 1013 (1981).
[26] Sosonkin, I. M., Strogov, G. N., and Fedyainov, N. V., *Dokl. Akad. Nauk SSSR*, **253**, 365 (1980); *Chem. Abs.*, **94**, 46472 (1981).
[27] Alder, R. W., *J. Chem. Soc., Chem. Commun.*, **1980**, 1184.
[28] Eberson, L., and Jönsson, L., *J. Chem. Soc., Chem. Commun.*, **1980**, 1187.
[29] Eberson, L., and Jönsson, L., *J. Chem. Soc., Chem. Commun.*, **1981**, 133.
[30] Citterio, A., Minisci, F., and Franchi, V., *J. Org. Chem.*, **45**, 4752 (1980).
[31] Tiecco, M., *Pure Appl. Chem.*, **53**, 239 (1981).
[32] Collins, C. J., Hombach, H.-P., Maxwell, B. E., Benjamin, B. M., and McKamey, D., *J. Am. Chem. Soc.*, **103**, 1213 (1981).
[33] So, Y.-H., and Miller, L. L., *J. Am. Chem. Soc.*, **102**, 7119 (1980).
[34] Mishra, S. P., and Symons, M. C. R., *J. Chem. Soc., Perkin Trans. 2*, **1981**, 185.
[35] Gareil, M., Pinson, J., and Saveant, J. M., *Nouveau J. Chim.*, **5**, 311 (1981).
[36] Galliani, G., and Rindone, B., *Tetrahedron*, **37**, 2313 (1981).
[37] Ivanova, T. M., and Shein, S. M., *Zh. Org. Khim.*, **16**, 1014, 1221 (1980); *Chem. Abs.*, **93**, 167160, 167175 (1980).
[38] Bunnett, J. F., and Cartano, A. V., *J. Am. Chem. Soc.*, **103**, 4861 (1981).
[39] Bunnett, J. F., Sekiguchi, S., and Smith, L. A., *J. Am. Chem. Soc.*, **103**, 4865 (1981).
[40] Sekiguchi, S., and Bunnett, J. F., *J. Am. Chem. Soc.*, **103**, 4871 (1981).
[41] Banjoko, O., and Otiono, P., *J. Chem. Soc., Perkin Trans. 2*, **1981**, 399.
[42] Banjoko, O., and Khalil-Ur-Rahman, *J. Chem. Soc., Perkin Trans. 2*, **1981**, 1105.
[43] Bamkole, T. O., Hirst, J., and Onyido, I., *J. Chem. Soc., Perkin Trans. 2*, **1981**, 1201.
[44] de Rossi, R. H., and de Vargas, E. B., *J. Am. Chem. Soc.*, **103**, 1533 (1981).
[45] Nudelman, N. S., *An. Acad. Nac. Cienc. Exactas, Fis. Nat. (Buenos Aires)*, **32**, 109 (1980); *Chem. Abs.*, **94**, 173737 (1981).
[46] Nudelman, N. S., and Palleros, D., *J. Chem. Soc., Perkin Trans. 2*, **1981**, 995.
[47] Logue, M. W., and Han, B. H., *J. Org. Chem.*, **46**, 1638 (1981).
[48] Forlani, L., De Maria, P., Foresti, E., and Pradella, G., *J. Org. Chem.*, **46**, 3178 (1981).
[49] Kurts, A. L., Kogan, G. O., and Bundel, Yu. G., *Vestn. Mosk. Univ. Ser. 2: Khim.*, **22**, 86 (1981); *Chem. Abs.*, **95**, 6042 (1981).
[50] Kurts, A. L., Kogan, G. O., and Bundel, Yu. G., *Vestn. Mosk. Univ. Ser. 2: Khim.*, **22**, 186 (1981); *Chem. Abs.*, **95**, 60925 (1981).
[51] Emokpae, T. A., Nwaedozie, J. M., and Hirst, J., *J. Chem. Soc., Perkin Trans. 2*, **1981**, 883.

[52] Henichart, J. P., Bernier, J. L., Vaccher, C., Houssin, R., Warin, V., and Baert, F., *Tetrahedron*, **36**, 3535 (1980).
[53] Subramanian, P. K., Ramalingam, K., Satyamurthy, N., and Berlin, K. D., *J. Org. Chem.*, **46**, 4384 (1981).
[54] Lerman, Z. A., Ivanov, A. V., and Gitis, S. S., *Kinet. Katal.*, **21**, 359 (1980); *Chem. Abs.*, **93**, 167359 (1980).
[55] Ivanov, A. V., Lerman, Z. A., Shakhel'dyan, I. V., Gitis, S. S., and Shakel'dyan, A. R., *Sint. Anal. Strukt. Org. Soedin* (*Tula*), **1979**, 52; *Chem. Abs.*, **93**, 238145 (1980).
[56] Litvinenko, L. M., Shved, E. N., Popova, R. S., Popov, A. F., and Pirgo, M. D., *Dokl. Akad. Nauk. SSSR*, **258**, 359 (1981); *Chem. Abs.* **95**, 114572 (1981).
[57] Takimoto, S., Inanaga, J., Katsuki, T., and Yamaguchi, M., *Bull. Chem. Soc. Jpn.*, **54**, 1470 (1981).
[58] Hayami, J., Asaki, M., Tamura, R., and Ono, N., *Bull. Inst. Chem. Res. Kyoto Univ.*, **58**, 222 (1980); *Chem. Abs.*, **94**, 46372 (1981).
[59] Yanchuk, N. I., *Kinet. Katal.*, **20**, 1404 (1979); *Chem. Abs.*, **93**, 203532 (1980).
[60] Koppes, W. M., Lawrence, G. W., Sitzman, M. E., and Adolph, H. G., *J. Chem. Soc., Perkin Trans. 1*, **1981**, 1815.
[61] Ballester, M., Riera, J., Julia, L., Castaner, J., and Ros, F., *J. Chem. Soc., Perkin Trans. 1*, **1981**, 1690.
[62] Agarwal, N. L., and Schäfer, W., *J. Org. Chem.*, **45**, 5139 (1980).
[63] Agarwal, N. L., and Schäfer, W., *J. Org. Chem.*, **45**, 5144 (1980).
[64] Vasaros, L., Norseev, Yu. V., Nhan, D. D., and Khal'kin, V. A., *Radiochem. Radioanal. Lett.*, **47**, 313, 403 (1981); *Chem. Abs.*, **95**, 79593, 131854 (1981).
[65] Alexiou, M., Tyman, J., and Wilson, I., *Tetrahedron Lett.*, **22**, 2303 (1981).
[66] Sharon, M., and Dhavale, D. D., *Natl. Acad. Sci. Lett.* (*India*), **3**, 78 (1980); *Chem. Abs.*, **94**, 120512 (1981).
[67] Mach, M. H., and Bunnett, J. F., *J. Org. Chem.*, **45**, 4660 (1980).
[68] Movsumzade, M. M., Shabanov, A. L., and Abdullabekov, I. M., *Dokl. Akad. Nauk Az. SSR*, **36**, 60 (1980); *Chem. Abs.*, **94**, 3386 (1981).
[69] Handel, H., Pasquini, M. A., and Pierre, J. L., *Tetrahedron*, **36**, 3205 (1980).
[70] Suzuki, N., Shimazu, K., Ito, T., and Izawa, Y., *J. Chem. Soc., Chem. Commun.*, **1980**, 1253.
[71] Gisler, M., and Zollinger, H., *Angew. Chem. Int. Ed.*, **20**, 203 (1981).
[72] Bunton, C. A., and de Buzzaccarini, F., *J. Phys. Chem.*, **85**, 3142 (1981).
[73] Broxton, T. J., *Aust. J. Chem.*, **34**, 969 (1981).
[74] Kondo, H., Moriuchi, F., and Sunamoto, J., *J. Org. Chem.*, **46**, 1333 (1981).
[75] Mutai, K., and Kobayashi, K., *Bull. Chem. Soc. Jpn.*, **54**, 462 (1981).
[76] Yokoyama, K., Nakagaki, R., Nakamura, J., Nagakura, S., and Mutai, K., *Koen. Yoshishu-Bunshi Kozo Sogo Toronkai*, **1979**, 236; *Chem. Abs.*, **93**, 220114 (1980).
[77] Irie, T., and Tanida, H., *J. Org. Chem.*, **45**, 4961 (1980).
[78] De Fusco, A. A., and Strauss, M. J., *J. Heterocycl. Chem.*, **18**, 351 (1981).
[79] Meyers, A. I., Reuman, M., and Gabel, R. A., *J. Org. Chem.*, **46**, 783 (1981).
[80] Baldwin, J. E., Cacioli, P., and Reiss, J. A., *Tetrahedron Lett.*, **1980**, 4971.
[81] Albright, T. A., and Carpenter, B. K., *Inorg. Chem.*, **19**, 3092 (1980).
[82] Semmelhack, M. F., Clark, G. R., Garcia, J. L., Harrison, J. J., Thebtaranonth, Y., Wulff, W., and Yamashita, A., *Tetrahedron*, **37**, 3957 (1981).
[83] Knipe, A. C., McGuinness, S. J., and Watts, W. E., *J. Chem. Soc., Perkin Trans. 2*, **1981**, 193.
[84] Lai, Y.-H., Tam, W., and Vollhardt, K. P. C., *J. Organomet. Chem.*, **216**, 97 (1981).
[85] Vrba, Z., *Collect. Czech. Chem. Commun.*, **46**, 92 (1981).
[86] Lisitsyn, V. N., and Orlova, G. V., *Izv. Vyssh. Uchebn. Zaved., Khim. Khim. Tekhnol.*, **24**, 33 (1981); *Chem. Abs.*, **94**, 191229 (1981).
[87] Lisitsyn, V. N., and Stankevich, G. S., *Izv. Vyssh. Uchebn. Zaved., Khim. Khim. Tekhnol.*, **23**, 400 (1980); *Chem. Abs.*, **93**, 185310 (1980).
[88] Kondratov, S. A., and Shein, S. M., *Kinet. Katal.*, **21**, 1232 (1980); *Chem. Abs.*, **94**, 64770 (1981).
[89] Gaca, J., Kozlowski, K., Dzierzkiewicz, H., and Kucybala, Z., *Pr. Wydz. Nauk Techn., Bydgoskic. Tow. Nauk, Ser. A*, **14**, 41 (1980); *Chem. Abs.*, **94**, 29753 (1981).
[90] Suzuki, H., Abe, H., and Osuka, A., *Chem. Lett.*, **1980**, 1363.
[91] Kametani, T., Ohsawa, T., and Ihara, M., *J. Chem. Soc., Perkin Trans. 1*, **1981**, 290.
[92] Horino, H., Inoue, N., and Asao, T., *Tetrahedron Lett.*, **22**, 741 (1981).
[93] Okamoto, T., and Oka, S., *Bull. Chem. Soc. Jpn.*, **54**, 1265 (1981).
[94] Consiglio, G., Arnone, C., Spinelli, D., Noto, R., and Frenna, V., *J. Chem. Soc., Perkin Trans. 2*, **1981**, 388.

95 Consiglio, G., Arnone, C., Spinelli, D., and Noto, R., *J. Chem. Soc., Perkin Trans. 2*, **1981**, 642.
96 Aveta, R., Doddi, G., Illuminati, G., and Stegel, F., *J. Am. Chem. Soc.*, **103**, 6148 (1981).
97 Forlani, L., and Todesco, P., *Gazz. Chim. Ital.*, **110**, 561 (1980); *Chem. Abs.*, **94**, 173893 (1981).
98 Cidda, C., and Sleiter, G., *Gazz. Chim. Ital.*, **110**, 155 (1980); *Chem. Abs.*, **94**, 64733 (1981).
99 Clark, H. R., Beth, L. D., Burton, R. M., Garrett, D. L., Miller, A. L., and Muscio, O. T., *J. Org. Chem.*, **46**, 4363 (1981).
100 Plust, S. J., Loehe, J. R., Feher, F. J., Benedict, J. H., and Herbrandson, H. F., *J. Org. Chem.*, **46**, 3661 (1981).
101 Davies, A. G., and Sutcliffe, R., *J. Chem. Soc., Perkin Trans. 2*, **1981**, 1512.
102 Petrenko, O. P., Krivopalov, V. P., Lapachev, V. V., and Mamaev, V. P., *Izv. Akad. Nauk SSSR, Ser. Khim.*, **1980**, 1611; *Chem. Abs.*, **93**, 220083 (1980).
103 Petrenko, O. P., Krivopalov, V. P., Lapachev, V. V., and Mamaev, V. P., *Izv. Akad. Nauk SSSR, Ser. Khim.*, **1981**, 227; *Chem. Abs.*, **94**, 191211 (1981).
104 Gore, P. H., Hundal, A. S., and Morris, D. F. C., *Tetrahedron*, **37**, 167 (1981).
105 Salickaite, L. P., Jasinskas, L. L., and Dienys, G. J., *Org. React. (Tartu)*, **17**, 286 (1980).
106 Katritzky, A. R., Ibrahim, M. H., Valnot, J. Y., and Sammes, M. P., *J. Chem. Res.*, **1981**, (S) 70, (M) 859.
107 Katritzky, A. R., Beltrami, H., and Sammes, M. P., *J. Chem. Res.*, **1981**, (S) 133, (M) 1684.
108 Akiba, K., Matsuoka, H., and Wada, M., *Tetrahedron Lett.*, **22**, 4093 (1981).
109 Gabay, Z., and Goldschmidt, J. M. E., *J. Chem. Soc., Dalton Trans.*, **1981**, 1456.
110 Zoltewicz, J. A., and Uray, G., *J. Org. Chem.*, **46**, 2398 (1981).
111 Inoyatova, D. A., and Otroschchenko, O. S., *Sb. Nauchn. Tr. Tashk. Gos. Univ. in V.I. Lenina*, **622**, 35 (1980); *Chem. Abs.*, **95**, 114377 (1981).
112 Knoppova, V., Kada, R., and Kovac, J., *Chem. Zvesti*, **34**, 556 (1980); *Chem. Abs.*, **94**, 83205 (1981).
113 Newkome, G. R., Garbis, S. J., Majestic, V. K., Fronczek, F. R., and Chiari, G., *J. Org. Chem.*, **46**, 833 (1981).
114 Buchanan, J. G., Stobie, A., and Wightman, R. H., *Can. J. Chem.*, **58**, 2624 (1980).
115 Berbee, R. M., and Habraken, C. L., *J. Heterocycl. Chem.*, **18**, 559 (1981).
116 Teulade, J.-C., Grassy, G., Escale, R., and Chapat, J.-P., *J. Org. Chem.*, **46**, 1026 (1981).
117 Goldman, P., and Wuest, J. D., *J. Am. Chem. Soc.*, **103**, 6224 (1981).
118 Mugnoli, A., Dell'Erba, C., Guanti, G., and Novi, M., *J. Chem. Soc., Perkin Trans. 2*, **1980**, 1764.
119 Aleksandrov, G. G., Struchkov, Yu. T., Knyazev, V. N., Mozhaeva, T. Ya., and Drozd, V. N., *Izv. Akad. Nauk SSSR, Ser. Khim.*, **1980**, 1802.
120 Kost, A. N., Gromov, S. P., and Sagitullin, R. S., *Tetrahedron*, **37**, 3423 (1981).
121 Katritzky, A. R., and Manzo, R. H., *J. Chem. Soc., Perkin Trans. 2*, **1981**, 571.
122 Matsumura, E., Ariga, M., and Tohda, Y., *Bull. Chem. Soc. Jpn.*, **53**, 2891 (1980).
123 Matsumura, E., Ariga, M., Tohda, Y., and Kawashima, T., *Tetrahedron Lett.*, **22**, 757 (1981).
124 Hirota, K., Kitade, Y., Senda, S., Halat, M. J., Watanabe, K. A., and Fox, J. J., *J. Org. Chem.*, **46**, 846 (1981).
125 Gonczi, C., Swistun, Z., and van der Plas, H. C., *J. Org. Chem.*, **46**, 608 (1981).
126 Kanoktanaporn, S., MacBride, J. A. H., and King, T. J., *J. Chem. Res.*, **1980**, (S) 406, (M) 4901.
127 Gaoni, Y., and Greenberg, F. H., *J. Org. Chem.*, **46**, 74 (1981).
128 Simmons, H. E., Vest, R. D., Vladuchick, S. A., and Webster, O. W., *J. Org. Chem.*, **45**, 5113 (1980).
129 Birch, A. J., Hinde, A. L., and Radom, L., *J. Am. Chem. Soc.*, **102**, 6430 (1980).
130 Birch, A. J., Hinde, A. L., and Radom, L., *J. Am. Chem. Soc.*, **103**, 284 (1981).
131 Olah, G. A., and Hunadi, R. J., *J. Org. Chem.*, **46**, 715 (1981).
132 Counotte-Potman, A., van der Plas, H. C., and van Veldhuizen, B., *J. Org. Chem.*, **46**, 2138 (1981).
133 Counotte-Potman, A., van der Plas, H. C., and van Veldhuizen, B., *J. Org. Chem.*, **46**, 3805 (1981).
134 Buncel, E., Moir, R. Y., Norris, A. R., and Chatrousse, A.-P., *Can. J. Chem.*, **59**, 2470 (1981).
135 AlAruri, A. D. A., and Crampton, M. R., *J. Chem. Soc., Perkin Trans. 2*, **1981**, 807.
136 Crampton, M. R., and Gibson, B., *J. Chem. Soc., Perkin Trans. 2*, **1981**, 533.
137 AlAruri, A. D. A., and Crampton, M. R., *J. Chem. Res.*, **1980**, (S) 140, (M) 2157.
138 Sekiguchi, S., Aizawa, T., and Aoki, M., *J. Org. Chem.*, **46**, 3657 (1981).
139 Golopolosova, T. V., Gitis, S. S., Glaz, A. I., Kaminskii, A. Ya., and Savinova, L. N., *Zh. Org. Khim.*, **16**, 1022 (1980); *Chem. Abs.*, **93**, 185533 (1980).
140 Sasaki, M., Takisawa, N., Amita, F., and Osugi, J., *J. Am. Chem. Soc.*, **102**, 7268 (1980).
141 Sasaki, M., Amita, F., Takisawa, N., and Osugi, J., *Kosoku Hanno Toronkai Koen Yokoshu 14th*, **1979**, 36; *Chem. Abs.*, **94**, 120625 (1981).
142 Sekiguchi, S., Hikage, R., Obana, K., Matsui, K., Ando, Y., and Tomoto, N., *Bull. Chem. Soc. Jpn.*, **53**, 2921 (1980).

[143] Alekhina, N. N., Gitis, S. S., Ivanov, A. V., Nikitin, N. R., and Agafonova, A. S., *Sint. Anal. Strukt. Org. Soedin* (*Tula*), **1979**, 3; *Chem. Abs.*, **93**, 186083 (1980).
[144] Kropf, H., Ball, M., Siegfriedt, K.-H., and Wagner, S., *J. Chem. Res.*, **1981**, (S) 339, (M) 4001.
[145] Il'chenko, A. Ya., Boiko, V. N., Ignat'ev, N. V., Shchupak, G. M., and Yagupol'skii, L. M., *Ukr. Khim. Zh.* (*Russ. Ed.*), **47**, 843 (1981); *Chem. Abs.*, **95**, 170966 (1981).
[146] Ignat'ev, N. V., Boiko, V. N., and Yagupol'skii, L. M., *Zh. Org. Khim.*, **16**, 1501 (1980); *Chem. Abs.*, **94**, 3827 (1981).
[147] Cavazza, M., Morganti, G., and Pietra, F., *Tetrahedron Lett.*, **22**, 1459 (1981).
[148] Artamkina, G. A., Egorov, M. P., Beletskaya, I. P., and Reutov, O. A., *Zh. Org. Khim.*, **17**, 29 (1981); *Chem. Abs.*, **95**, 24413 (1981).
[149] Knyazev, V. N., Drozd, V. N., and Mozhaeva, T. Ya., *Zh. Org. Khim.*, **16**, 2012 (1980); *Chem. Abs.*, **94**, 84029 (1981).
[150] Crampton, M. R., Routledge, P. J., and Wilson, P. M., *J. Chem. Res.*, **1981**, (S) 152, (M) 1972.
[151] Ah-Kow, G., Terrier, F., Pouet, M.-J., and Simonnin, M.-P., *J. Org. Chem.*, **45**, 4399 (1980).
[152] Nikitin, N. R., and Gitis, S. S., *Sint. Anal. Strukt. Org. Soedin* (*Tula*), **1979**, 38; *Chem. Abs.*, **94**, 3340 (1981).
[153] Mel'nikov, A. I., Gitis, S. S., Glaz, A. I., Shakhel'dyan, I. V., and Nikitin, N. R., *Sint. Anal. Strukt. Org. Soedin*, **9**, 56 (1979); *Chem. Abs.*, **95**, 25013 (1981).
[154] Artamkina, G. A., Tuzikov, A. B., Beletskaya, I. P., and Reutov, O. A., *Izv. Akad. Nauk SSSR, Ser. Khim.*, **1980**, 2430; *Chem. Abs.*, **94**, 83163 (1981).
[155] Kovar, K.-A., Rohlfes, W., and Auterhoff, H., *Arch. Pharm.* (*Weinheim Ger.*), **314**, 532 (1981).
[156] Golopolosova, T. V., Gitis, S. S., Savinova, L. N., Kaminskii, A., Ya., and Glaz, A., *Org. React.* (*Tartu*), **17**, 22 (1980).
[157] Golopolosova, T. V., Gitis, S. S., Savinova, L. N., Kaminskii, A. Ya., and Glaz, A., *Org. React.* (*Tartu*), **17**, 11 (1980).
[158] Savinova, L. N., Gitis, S. S., Kaminskii, A. Ya., and Golopolosova, T. V., *Sint. Anal. Strukt. Org. Soedin* (*Tula*), **1979**, 7; *Chem. Abs.*, **93**, 203583 (1980).
[159] Gitis, S. S., Kaminskii, A. Ya., Savinova, L. N., and Illarionova, L. V., *Sint. Anal. Strukt. Org. Soedin* (*Tula*), **1979**, 26; *Chem. Abs.*, **93**, 203584 (1980).
[160] Brooke, D. N., Crampton, M. R., Corfield, G. C., Golding, P., and Hayes, G. F., *J. Chem. Soc., Perkin Trans. 2*, **1981**, 526.
[161] Brooke, D. N., and Crampton, M. R., *J. Chem. Soc., Perkin Trans. 2*, **1980**, 1850.
[162] Crampton, M. R., and Wilson, P. M., *J. Chem. Soc., Perkin Trans. 2*, **1980**, 1854.
[163] Buncel, E., Hamaguchi, M., and Norris, A. R., *Can. J. Chem.*, **59**, 795 (1981).
[164] Terrier, F., Simonnin, M. P., Pouet, M. J., and Strauss, M. J., *J. Org. Chem.*, **46**, 3537 (1981).
[165] Strauss, M. J., De Fusco, A., and Terrier, F., *Tetrahedron Lett.*, **22**, 1945 (1981).
[166] Deicha, C., and Terrier, F., *J. Chem. Res.*, **1981**, (S) 312, (M) 3642.
[167] Aveta, R., Doddi, G., Insam, N., and Stegel, F., *J. Org. Chem.*, **45**, 5160 (1980).
[168] van der Haak, H. J. W., van der Plas, H. C., and van Veldhuizen, B., *J. Org. Chem.*, **46**, 2134 (1981).
[169] Bunting, J. W., and Sindhuatmadja, S., *J. Org. Chem.*, **46**, 4211 (1981).
[170] James, F. C., Choi, H. K. J., Ruzsicska, B., Strausz, O. P., and Bell, T. N., *Front. Free Radical Chem.* [*Pap. Symp.*], **1979**, 139; *Chem. Abs.*, **94**, 102260 (1981).
[171] Bryce, M. R., and Vernon, J. M., *Adv. Heterocyclic Chem.*, **28**, 183 (1981).
[172] Del Mazza, D., and Reinecke, M. G., *J. Chem. Soc., Chem. Commun.*, **1981**, 124.
[173] Hori, M., Kataoka, T., Shimizu, H., and Ueda, N., *Tetrahedron Lett.*, **22**, 3071 (1981).
[174] Best, W. M., and Wege, D., *Tetrahedron Lett.*, **22**, 4877 (1981).
[175] El-Sheikh, M. I., Marks, A., and Biehl, E. R., *J. Org. Chem.*, **46**, 3256 (1981).
[176] Bunnett, J. F., and Singh, P., *J. Org. Chem.*, **46**, 4567 (1981).

Organic Reaction Mechanisms 1981
Edited by A. C. Knipe and W. E. Watts

CHAPTER 7

Electrophilic Aromatic Substitution

R. B. MOODIE

University of Exeter

General 283
Halogenation 284
Hydrogen Exchange 285
Nitration and Nitrosation 286
Azo-coupling Reactions 289
Sulphonation 290
Friedel–Crafts and Related Reactions 290
Metallation 292
Miscellaneous Reactions 292
References 294

General

A review of substituent interactions in substituted benzenes, largely based on the results of *ab initio* calculations using the STO-3G basis set, has appeared.[1] Free electron superdelocalizability indices have been used to rationalize partial rate factors,[2] and relative reactivities of different aromatic systems.[3] Calculations based on the HMO model suggest that hyperconjugative interaction through the methylene group leads to a degree of aromaticity in protonated aromatics.[4] The use of point charges to simulate medium effects, when calculating localization energies for toluene, has been described.[5] Localization energies of pyrrole, furan, and thiophene have been calculated by the CNDO/2 method.[6] Charge density calculations by HMO and Del Re methods have been made on some reactive aromatic compounds.[7] A refined version of the so-called "static" model of aromatic reactivity is claimed to give a better description of the reactivity of polycyclic arenes than does the localization approach.[8]

Charge exchange is the predominant process when CH_3^+, NO^+, NO_2^+, and $O_2NCH_2^+$ react with mono-substituted benzenes in the gas phase; adducts of the Wheland intermediate type are not formed.[9] Nitration, bromination, and azo coupling of 3-aminophenalone (**1**) occurs at C(2).[10] The nitration and bromination of [1]benzothieno[3,2-*d*]pyrimidine,[11] and the electrophilic aromatic substitution reactions of arsabenzene,[12] have been investigated. The reactivities of selenolo[3,2-*b*]-selenophene and -thiophene and thieno[3,2-*b*]thiophene towards electrophilic substitution decreases in the order given.[13] The relative reactivities of (**2**; n = 1–3) and of veratrole in electrophilic substitutions have been compared.[14] A review of electrophilic substitution in the benzene ring of indole compounds has appeared.[15] Phenols and their ethers in micellar solutions of anionic and cationic detergents exhibit non-uniform upfield shifts of their different protons, suggesting preferred average orientations, which might lead to regioselectivity in aromatic substitutions.[16]

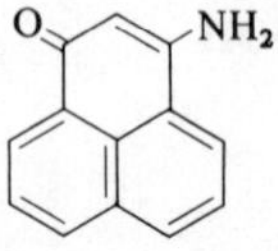

O NH₂ (CH₂)ₙ O O

(1) 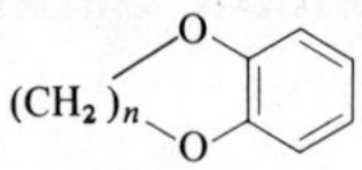 **(2)**

Halogenation

The use of xenon difluoride as a fluorinating agent has been further investigated.[17] With boron trifluoride etherate as catalyst, XeF_2 gives with 2-bromo-4,5-dimethylphenol the dienone (**3**), together with (**4**) and, unexpectedly, (**5**).[18] The reaction of the fluoroxysulphate ion, SO_4F^- with aromatics gives some fluorination products, but the mixture of products is generally a complex one; it is suggested that initial electrophilic attack can be followed by free-radical side-reactions.[19] Acetyl hypofluorite, prepared from F_2 and sodium acetate, may prove to be a useful electrophilic fluorinating agent.[20]

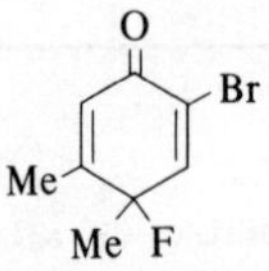

O Br Me Me F OH F Br Me Me OH Br F Me Me

(3) 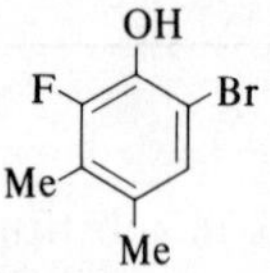 **(4)** 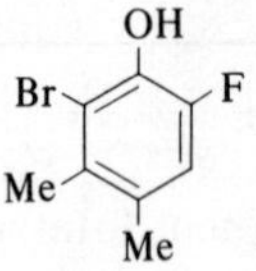 **(5)**

Products from the reaction of 2- and 3-methylbenzocyclopropenes with halogens and HCl are those expected from a consideration of the relative stabilities of the Wheland intermediates formed by electrophilic attack at the carbon atoms joining the two rings.[21] *Para*/*ortho* ratios in the chlorination of alkylbenzenes catalysed by chemically-modified silicas are much higher than is usual.[22] The kinetics of chlorination of anisole and *p*-nitroanisole by *N*-chlorosuccinimide,[23] and of the chlorination of phenols by *N*-chlorosaccharin,[24] have been investigated. 2-(2-Arylazophenoxy)-2-methylpropanoic acids react with thionyl chloride to form 4-(2-chloroanilino) benzoxazinones. Substituent effects show that the chlorination is electrophilic, and the transition state (**6**) has been suggested.[25] The bromination of anisole in aqueous acetic acid shows a primary kinetic isotope effect when the reagent is Br_3^-, but not when it is Br_2.[26] Quantitative studies have been made of the effects of bulky alkyl groups on the reactivity of anisoles towards bromination.[27] Ring-substituted bromophenols derived from *o*-, *m*-, and *p*-cresol have been converted by further bromination into 4-bromo-2,5-dienones; the rearrangements of the latter in sulphuric acid have been investigated.[28] Full details of a study of the bromination of substituted phenols and anisoles in SbF_5-HF have now appeared;

meta-substitution is held to arise through reaction of the *O*-protonated substrate.[29,30] The kinetics of the bromination of several reactive aromatic compounds with *N*-bromosuccinimide have been investigated,[31,32] and there has been a study of the kinetics of bromination of some substituted thiophenes in acetic acid.[33] Monobromination of 1-*R*-pyrroles (R = H, Me, benzyl, Ph) with *N*-bromosuccinimide gives the corresponding 2-bromopyrroles regiospecifically; if bromine is used, the thermodynamically more stable 3-bromopyrroles are formed because the hydrogen bromide formed catalyses isomerization.[34] Bromination of 2-methoxy- and 2-acetamido-3-bromobiphenylenes gives benzocyclooctatetraene derivatives, not biphenyls as previously reported.[35] Mono- and poly-bromination of (7) has been achieved[36] (attempts at chlorination, nitration, and sulphonation were unsuccessful). Bromination of deactivated aromatics can be effected with potassium

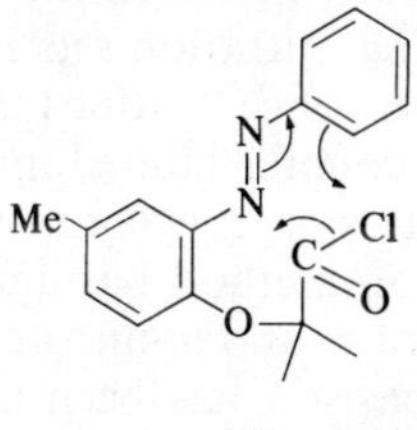

(6) 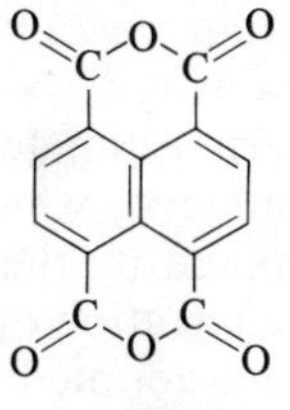 **(7)**

bromate in aqueous sulphuric acid.[37] There has been an extensive review of the kinetics and mechanisms of iodination reactions.[38] The kinetics of iodination of nitrophenols with hydrated monopositive iodine,[39] of substituted anilines[40] and of phenols[41] by iodine monochloride, and of phenol by *N*-iodosuccinimide,[42] have been investigated.

Hydrogen Exchange

Hypersurfaces for protonation of benzene by H^+ and H_3O^+ have been produced by the MINDO/3 method; desolvation of the proton occurs after the transition state has been passed.[43] Partial rate factors for protodetritiation from each position of chrysene (**8**),[44] 4*H*-cyclopenta[*def*]phenanthrene (**9**), and hexahelicene (**10**), are as shown; those of (**9**) are correctly predicted by Hückel localization energies,[45] and those of (**10**) show by comparison with other helicenes the effect of ground-state

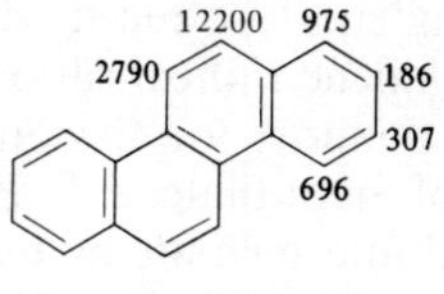

(8) 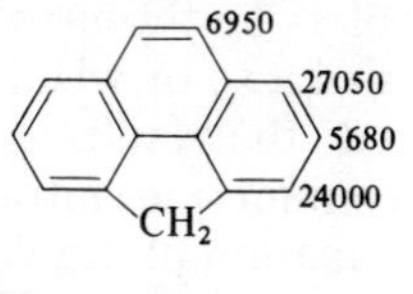 **(9)**

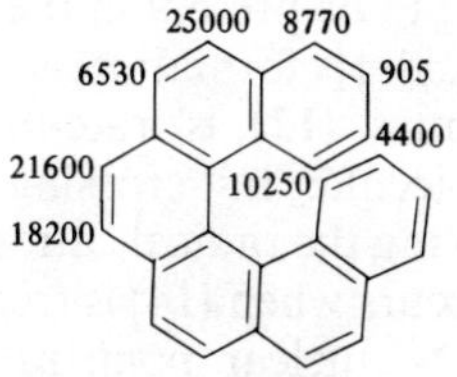

(10)

destabilization through ring distortion.[46] There have been studies of hydrogen deuterium exchange of some 9-*n*-alkylanthracenes in CF_3COOH and FSO_3H,[47] of some polymethylnaphthalenes and hexahydropyrene in $FSO_3H–SbF_5$,[48a] and of azulene or 1,3,5-trimethoxybenzene with acetic acid in various solvents.[48b] Rate constants have been measured for some thermoneutral isotope exchange reactions of protonated aromatic molecules with, for instance, D_2O in the gas phase.[49]

Nitration and Nitrosation

A book entitles "Aromatic Nitration", a comprehensive review of mechanistic aspects, has appeared.[50]

Conditions under which the nitrations of reactive aromatics in phosphoric acid are rate-limited by formation of nitronium ion, by encounter of nitronium ion and aromatic, or by a subsequent step, have been investigated; because of the high viscosity of the medium, encounter-controlled nitration sets in at a relatively low level of aromatic reactivity.[51] Nitronium ion concentrations in 63–100% aqueous nitric acid have been estimated, and encounter-limited nitrations have been recognized.[52] The influence which mass diffusion effects can have upon apparent relative rates determined by the competition method has again been illustrated; product ratios in the nitration of mixtures of *o*- and *m*-nitrotoluenes depend upon the method of mixing adopted.[53] A comparison has been made of the rates of nitration of methylnitrobenzenes in aqueous sulphuric acid with the corresponding rates derived, using a theoretical mixing-reaction model, from the results of mixing-disguised nitrations with nitronium hexafluorophosphate.[54]

Consideration has been given to the possibility of controlling isomer distributions in nitrations; nitronium ion carriers are not considered to provide a useful approach, but it is pointed out that Hg(II) and Pd(II) catalyse the nitration of toluene and provide a novel isomer distribution.[55] Nitrations with *N*-nitropyrazole show similar positional selectivity, but different substrate selectivity, to nitronium ion nitrations.[56] Ammonium nitrate and metal nitrates in trifluoroacetic anhydride have been used for nitrations, notably of polymers with aromatic groups.[57] Nitrations with silver nitrate in acetonitrile, with BF_3 catalyst, have also been investigated.[58]

Radical intermediates in nitrations have continued to excite attention. There has been an interesting report on nitrations of, for instance, benzene and toluene in liquid N_2O_4, reactions which proceed rapidly when NO and O_2 are bubbled into the solution.[59] Ratios of 1- and 2-nitronaphthalenes produced when naphthalene is nitrated by NO_2^+ or NO_2, or when naphthalene radical cation reacts with NO_2, have been compared; the results give no support for the one-electron transfer mechanism of nitronium ion nitration of naphthalene.[60] Evidence has been presented for the formation of the Wheland intermediate by reaction of NO_2 and benzene radical cation dimer in the high-pressure gas phase.[61] The *ipso*-Wheland intermediate (**11**) has been shown to be present in the nitration of *N*,*N*-dimethyl-*p*-toluidine in aqueous sulphuric acid. Its formation is catalysed by nitrous acid, and its rearrangement to the product (**12**) is rate-limited by proton loss. This rearrangement is largely intramolecular, but crossing experiments show there is an intermolecular component involving the radical cation (**13**; $R^1 = H$, $R^2 = Me$) and NO_2.[62] Nitro group exchange occurs when (**14**) is treated with $H^{15}NO_3$ in H_2SO_4, and the labelled (**14**) shows ^{15}N nuclear polarization; (**13**; $R^1 = R^2 = Me$) is therefore an intermediate in the exchange.[63] ^{15}N polarization of the *para* nitro product of nitrous-acid-catalysed nitration of *N*,*N*-dimethylaniline with $H^{15}NO_3$ in 85–90% H_2SO_4 suggests the intermediacy of (**13**; $R^1 = R^2 = H$).[64]

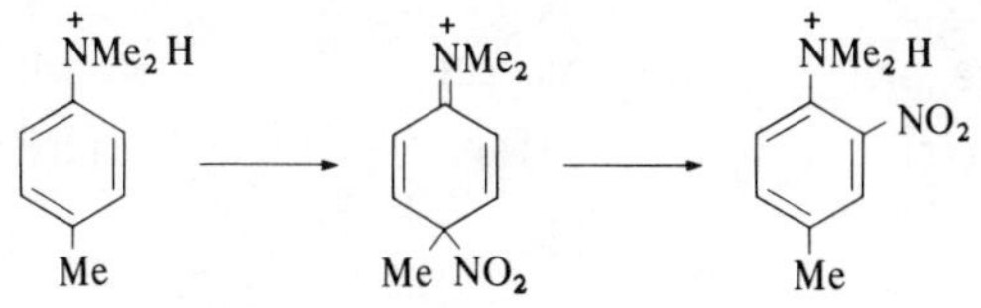

2,4,5-Trimethylneopentylbenzene reacts with nitric acid to give 3-(4,5-dimethyl-2-nitrooxymethylphenyl)-2-methyl-1-nitrobut-2-ene, whilst with ceric ammonium nitrate a mixture of benzyl nitrates is produced.[65]

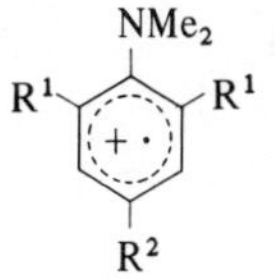

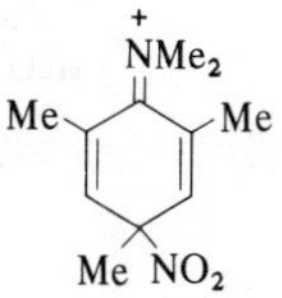

Acylpentamethylbenzenes and 1,3-diacyltetramethylbenzenes, treated with nitric acid in dichloromethane give the benzyl nitrates or nitromethanes resulting from modification of the methyl group at the most crowded site. *Ipso*-attack *meta* to the acyl group(s), followed by proton loss from the methyl group *para* to the site of attack, then transformation of the diene into the product *via* a benzyl cation/nitrite anion ion-pair, has been suggested.[66] Substantial yields of 4-nitro-*tert*-butylbenzene are obtained when 4,4′-di-*tert*-butyldiphenylmethane is nitrated under a variety of conditions; this and similar results highlight the importance of *ipso*-attack in the nitration of electron-rich substituted diphenylmethanes.[67] The configurations of each of the 1,4-nitronium acetate adducts (**15**) formed by nitrating *p*-xylene in acetic anhydride, and of the related dienol (**16**) and its methyl ether, have been established.[68] In the solvolysis of (**15**) in aqueous ethanol there is rate-limiting loss of nitrite to give (**17**; R = COMe) followed by migration of the acetoxyl group to give 2,5-dimethylphenyl acetate, but (**16**) gives mainly 2,4-dimethylphenol and migration of the hydroxy group of (**17**; R = H) is rate-limiting.[69] Rearomatization of adducts similar to (**15**) and (**16**) formed from 4-ethyltoluene has also been studied.[70] Nitration of (**18**; R = Me) or of (**18**; R = Et) by nitric acid in acetic anhydride containing ethanol or methanol respectively leads to a mixture of (**19**; R = Me) and (**19**; R = Et); the intermediacy of (**20**) is indicated.[71] Nitration of (**18**; R = Et or Ph) with dilute nitric acid in acetic anhydride gives (**19**; R = H) as the major product, but (**21**) gives only the normal nitro products.[72] Nitration of *o*-cresol in Ac_2O/HNO_3 at $-60°C$ gives the dienone (**22**), which rearranges regiospecifically on warming to $-20°C$ to (**23**); other alkylphenols behave similarly.[73] Nitration of (**24**) with fuming nitric acid gives rise to (**25**), the structure and stereochemistry of which have been established by X-ray crystal structure analysis.[74] Nitrations of 2,4,5-trimethylphenol, 3,4,6-trimethyl-2-nitrophenol, and of their acetates, all give rise to the dienone (**26**) in high yield.[75] Evidence for *ipso* attack in the nitration of 5-bromo- and 5-methyl-2-cyclopropylthiophenes has been presented.[76]

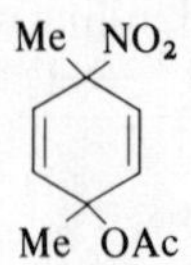

(15) 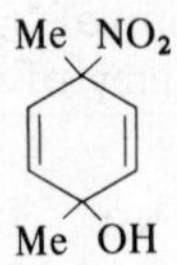 (16) 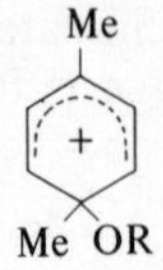 (17)

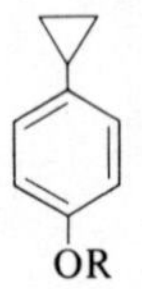

(18) 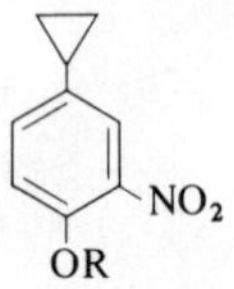 (19) 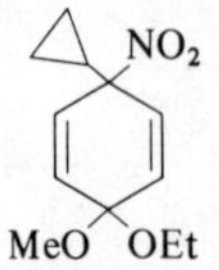 (20) 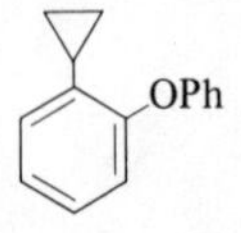 (21)

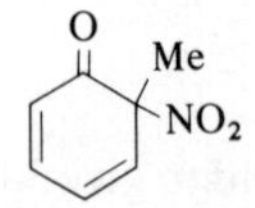

(22) 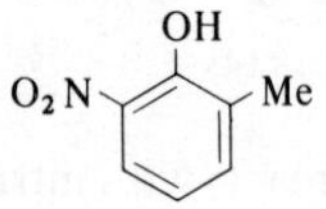 (23)

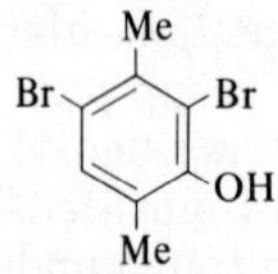

(24) 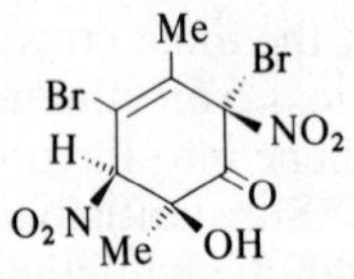 (25) 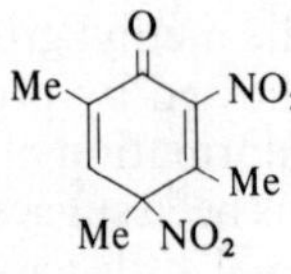 (26)

There have been some CNDO/2 calculations relating to possible hydrogen-bonded complexes involved in the nitration of nitrobenzene in mixed acid.[77] Partial rate factors for nitration of nitrobenzene and methyl phenyl sulphone, fit on a σ^+ plot for nitration by nitronium hexafluorophosphate in nitromethane, unlike the corresponding quantities for nitration by nitric acid in sulphuric acid; the rôle of hydrogen-bonded complexes in the latter medium has been discussed.[78] Ratios of mononitro products from the nitration of chlorobenzene are apparently slightly changed when the reaction vessel is subjected to a strong electric field.[79] Kinetics and products of nitration of cinnamic acids and related styrene derivatives have been examined; the further reactions of the first-formed ring nitro compounds are electrophilic additions to the exocyclic double bond, and may be followed by elimination or retro-aldol reaction.[80] 1-Methyl-2-phenyl-3-*X*-indoles (X = N_2Ph, CH_2OH, COMe, CHO) give, on nitration or nitrosation, 1-methyl-2-phenyl-3-nitroindole, as well as other products.[81] The products from the nitration of 2,5-dimethylthiophene and its 3,4-dibromo derivative have been examined; the former gives (**27**) and (**28**), the latter, (**29**).[82]

The rate of nitrous-acid-catalysed nitration of *p*-nitrophenol to 2,4-dinitrophenol shows a non-linear dependence on nitrous acid concentration; it has been suggested that there is substantial formation of a nitrite ester which undergoes acid-catalysed

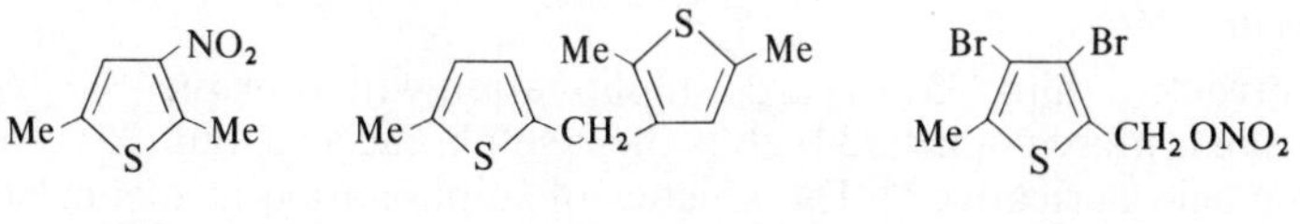

(27) (28) (29)

rearrangement, followed by oxidation.[83] Under conditions for nitrosation of phenol by nitrous acid, *o*-nitrosophenol is converted into the more stable *para*-isomer by an intermolecular process.[84] Concentrations of *p*-nitrosophenol and *p*-nitrophenol as functions of time, following dissolution of phenol in water containing sulphuric acid, nitrous acid, and nitric acid, have been calculated.[85] Nitrosations or diazotizations of aniline derivatives by propyl nitrite in acidified propan-1-ol show a non-linear dependence of rate constant upon the concentration of catalyst X^- (halide ion or thiourea); the rate-determining step is either *N*-nitrosation of free basic aniline by NOX present in equilibrium concentration, or subsequent proton loss.[86] Rate constants for *C*-nitrosation of Ph_2NR (R = Me, Et, Ph, Ac) have been determined, and their dependence on the concentration of acid catalyst examined.[87]

Azo-coupling Reactions

A negatively charged phase-transfer catalyst, tetrakis[3,5-bis(trifluoromethyl)phenyl] borate, has been used to aid azo coupling; the kinetics confirm phase-transfer catalysis.[88] An investigation has been made, using stopped- and quenched-flow kinetics and rapid-scanning UV spectrophotometry, of the rates and equilibria of the several steps involved in the reactions of nitro-substituted arenediazonium ions with hydroxide ion.[89] Rates of coupling of substituted benzenediazonium ions with 2,3-dihydroxynaphthalene-6-sulphonate ion have been investigated.[90] Base catalysis and isotope effects occur in the coupling of benzenediazonium ion with *N*,*N*-dialkylanilines in some non-aqueous solvents.[91] 1-Diazonio-2-naphthol-4-sulphonate reacts with 1-naphthol preferentially at the 2-position.[92] An investigation of side-products in the dediazotization of (**30**) suggests that (**31**) and (**32**) are intermediates in the formation of the fluorene derivative (**33**).[93]

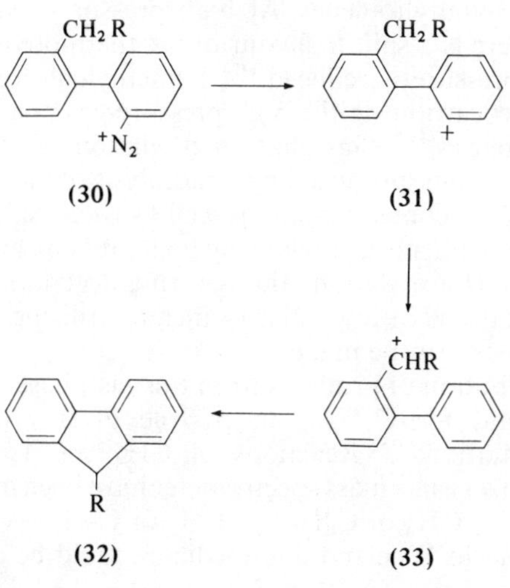

(30) (31)

(32) (33)

Sulphonation

(**34**) is less reactive than (**35**) towards sulphonation with dioxane–SO_3 because its annulene ring is more planar; (**34**) gives the 2-sulphonic acid, but (**35**) gives the 2,7-disulphonic acid derivative.[94] The kinetics of sulphonation in oleum of some *N*-alkyl- and *N,N*-dialkyl-anilines[95] and of benzene with H_2SO_4 in a heterogeneous system have been examined.[96] Mass-transfer effects are important in determining

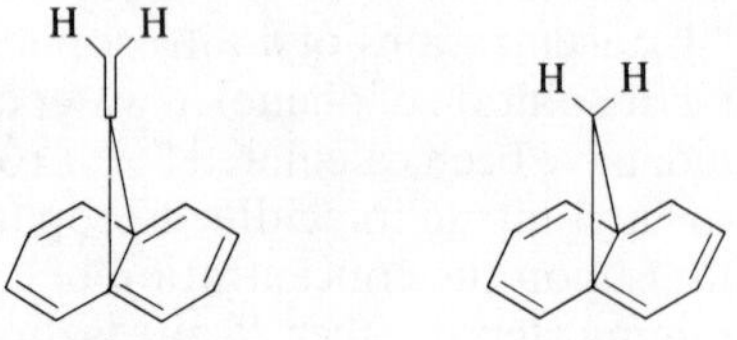

(**34**) (**35**)

the amount of by-product (Ph_2SO) formed in the sulphonation of benzene with SO_3.[97] Rate constants for sulphonation of pyrene and other arenes by sulphuric acid in nitrobenzene have been correlated with basicities.[98] Evidence has been presented[99] that the reaction of *p*-dichlorobenzene with $ClSO_3H$ gives $Cl_2C_6H_3SO_2Cl$ *via* $Cl_2C_6H_3SO_2OH$. The sulphonation of 2-*tert*-butylphenol with chlorosulphonic acid gives thirteen products which are sulphonated derivatives of phenol, 2- and 4-*tert*-butylphenols, and 2,4-di-*tert*-butylphenol; an analysis not only of the products but also of the kinetics of this complicated series of reactions has been achieved.[100]

Friedel–Crafts and Related Reactions

A review of Friedel–Crafts reactions has appeared,[101] and applications of isotopic labelling in these reactions have been examined.[102] There has been a review, in Russian, of mechanisms of alkylation of aromatic hydrocarbons.[103]

There have been several studies of alkylations and related reactions in the gas phase. Butyl cations obtained by radiolysis of butane, react with toluene in the gas phase to give only *sec*-butyltoluenes. At high pressures, *ortho-para* substitution predominates, but there is a shift in favour of the thermodynamically more stable *meta*-isomer as the pressure is reduced.[104] Dimethylchloronium ion reacts with several aromatic hydrocarbons in the high-pressure gas phase by Me^+ transfer to yield σ-bonded complexes.[105] Gas-phase methylation of phenol and anisole by dimethylhalogenonium ions, produced by γ-radiolysis of the corresponding methyl halide, leads to extensive *O*-methylation as well as ring-methylation.[106] Similarly, gas-phase acetylation of phenols, at relatively high pressure and in the presence of a base (NH_3), favours *O*-acetylation, though ring-acetylation is also observed; analogies with acetylations in low-polarity media, utilizing poorly co-ordinating Friedel–Crafts catalysts, can be made.[107]

The isomeric distribution of products from the gas-phase electrophilic attack of $^3He^3H^+$, Me_2F^+, and *tert*-$C_4H_9{}^+$ on pyrrole gives evidence, unusually, of predominant β-orientation.[108] Reactions of $CH_4{}^+$, $CH_3{}^+$, and other simple carbocations with benzene in a mass spectrometer have been investigated; $CH_3{}^+$ for instance gives $C_6H_5{}^+ + CH_4$ or $C_6H_6{}^+ + CH_3$ or $C_7H_7{}^+ + H_2$ in each case with hydrogen scrambling. No Wheland intermediates could be detected.[109]

The alkylation of toluene with 1-butene in the liquid phase gives initially *o*- and *p*-(2-butyl)toluene, which isomerize towards an equilibrium mixture containing mainly the *meta*-isomer to an extent which depends upon the catalyst and reaction conditions.[110] Nine alkylation products from the reaction of biphenyl with *tert*-butyl chloride and $AlCl_3$ have been identified.[111] Studies have been made of the kinetics of transalkylations of alkylbenzenes in the presence of $AlBr_3$,[112] and of the kinetics and mechanisms of the exchange reactions between diphenylalkanes and isotopically labelled benzene in the presence of $AlCl_3$.[113] The reaction of *o*-xylyl chloride with benzene in the presence of $AlCl_3$ gives some toluene and diphenylmethane, as well as *o*-benzyltoluene; the intermediacy o-tolylaluminium dichloride is suggested.[114] Ethyl cyclopropanecarboxylate with benzene and $AlCl_3$ gives 2-methyl-1-indanone.[115] Some aryl cyclopropyl ketones (**36**) cyclize to (**37**) in the presence of Lewis acid catalysts.[116] Deuterium labelling experiments have been used to show that the conversion of (**38**) to (**39**), effected by boron trifluoride–etherate, occurs both directly and by *ipso*-attack and rearrangement (Scheme 1).[117] Friedel–Crafts alkylations of benzene with alkyl nitro compounds

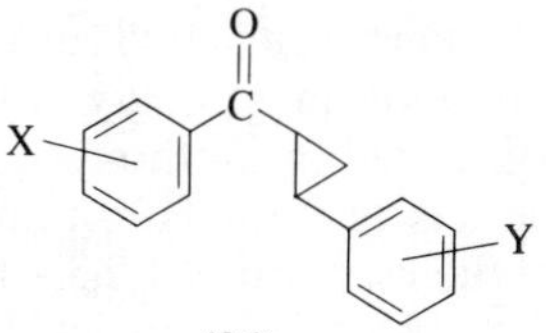

(**36**)

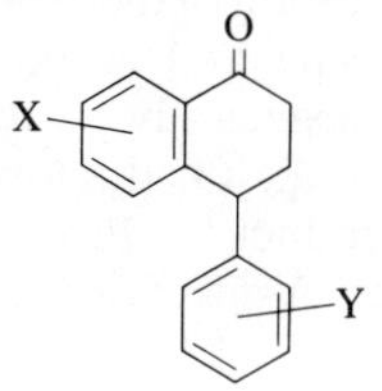

(**37**)

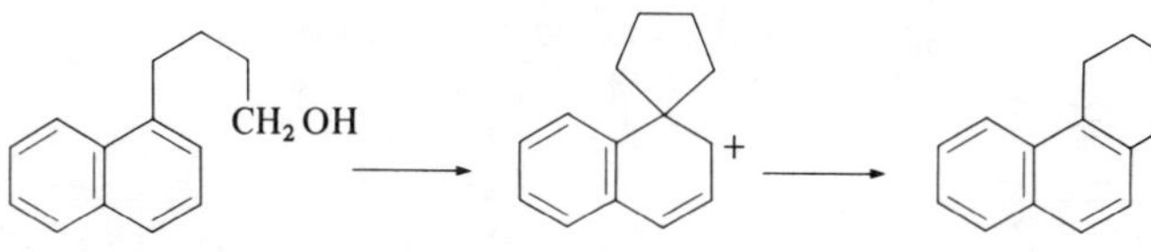

(**38**) CH_2OH (**39**)

Scheme 1

have been investigated,[118] as have the reactions of allylic chlorides with benzene and toluene, catalysed by WCl_6.[119] Phenol reacts with *n*-propanol in the presence of alumina at 250–350°C to give mono- to penta-*n*-propylphenols, together with the products of *O*-alkylation; only three per cent of the C_3H_7 groups in the product mixture are isopropyl.[120] With isopropanol under similar conditions, phenol gives mainly mono- and di-isopropylphenols.[121] The kinetics of alkylation of phenol by 1-decene, using a cation-exchange resin catalyst, have been examined.[122] The logarithm of the rate constant for isopropylation of benzene with isopropanol in aqueous sulphuric acid has a linear, unit slope correlation with $-H_0$.[123] Isopropylation of *tert*-butylbenzene in methanesulphonic acid and nitrobenzene has been investigated.[124]

Pronounced solvent and substituent effects upon the ratio of yields of 1-acyl- and 2-acylnaphthalene in Friedel–Craft acylations have been reported.[125] The ratio depends also on temperature and reaction time and it has been suggested that 2-

acetylation is first order, whilst 1-acetylation is second order, in acylating agent.[126] Acetylation of anthracene by acetyl chloride and $AlCl_3$ has been investigated.[127] Acetylmesitylene and acetyldurene undergo acetal exchange with ^{14}C-labelled acetyl chloride in nitromethane, in the presence of $AlCl_3$; deuterium labelling experiments have been used to show that exchange takes place *via* an *ipso*-complex, and not by acetyldeprotonation–protodeacetylation or by protodeacetylation–acetyldeprotonation.[128] The kinetics of acetylation of mesitylene with acetic anhydride co-catalysed by metal halides and HCl have been studied.[129] Friedel–Crafts β-acylation of pyrroles can be achieved by prior *N*-phenylsulphonylation.[130] Acylation of phenols and phenol ethers with nitriles and trifluoromethanesulphonic acid has been investigated.[131] Radical cations are intermediates in the reaction of 2-*X* heptafluoronaphthalenes (X = F, H, Me) with pentafluorobenzene in SbF_5 to give 3-*X* -undecafluoro-1-phenylnaphthalenes and 3-*X* -pentadecafluoro-1,5-diphenylnaphthalenes.[132]

The reaction in Scheme 2, with *p*-toluenesulphonic acid as catalyst, may be useful in synthesis.[133] Alkyl arene-thiocarboxylates and -dithiocarboxylates have been prepared by alkylating COS and CS_2 respectively with alkyl fluoride and SbF_5 in SO_2, followed by addition of the aromatic.[134] α-Ferrocenylalkyl carbenium ions are formed from ferrocene and carbonyl compounds in strongly acidic media.[135] Reactive aromatics are readily 1,3-dithiolan-2-ylated by 2-ethoxy-1,3-dithiolane in the presence of $BF_3 \cdot Et_2O$; the formal products of formylation are obtained by hydrolysis of the product.[136] Products from the reaction of 1,2,3,4-tetrahydrocarbazoles[137] and their *N*-methyl derivatives[138] with the Vilsmeier–Haack reagent have been examined.

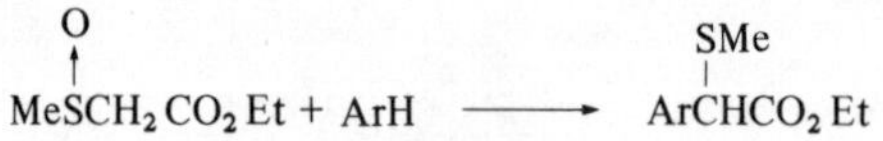

$$\mathrm{Me\overset{\overset{O}{\uparrow}}{S}CH_2CO_2Et + ArH \longrightarrow Ar\overset{\overset{SMe}{|}}{C}HCO_2Et}$$

Scheme 2

Metallation

Plumbylation of benzene and toluene in dichloroacetic acid and in trifluoroacetic acid is a typical electrophilic substitution which is more selective than thalliation or mercuration.[139] Thalliation is slower than mercuration, has a ρ^+ value of -7.4, and shows a primary kinetic hydrogen isotope effect.[140] Transthalliation of anisole and chlorobenzene by phenylthallium(III) bis(trifluoroacetate) has been studied.[141] Mercuration of ferrocene with mercuric acetate proceeds with rapid complex formation followed by rate-determining conversion of the complex to products.[142]

Miscellaneous Reactions

Four papers have appeared concerning electrophilic hydroxylation. It is involved in the reaction of various peroxides with *o*- and *p*-xylenes, and proceeds with *ipso*-attack followed by methyl migration.[143] Hydroxylations of mesitylene, phenol, and anisole with peroxymonophosphoric acid,[144] and of alkylbenzenes with H_2O_2 in HF/BF_3 at low temperature[145] have been investigated. Hydroxylation of *p*- and 2,6-di-alkylated phenols by H_2O_2 in SbF_5–HF occurs at the *meta* position, and it has been suggested that the electrophile attacks the *O*-protonated substrate.[146]

Relative rates of σ-complex formation between benzene derivatives and H_2PtCl_6 in CF_3CO_2H are reported to yield a ρ^+ of -1.5; these complexes are intermediates in the oxidation of aromatic hydrocarbons by H_2PtCl_6.[147] Aryloxenium ions (**40**) generated by thermolysis of *N*-(aryloxy)pyridinium tetrafluoroborates, react with solvent anisole to give the products of attack by both oxygen and carbon centres of (**40**); the former predominate when X is electron-withdrawing. When anisole is replaced by benzonitrile as the solvent, the main product is a benzoxazole.[148] The electrophilic substitution of aromatic amines with 2-methylthio-1,3-dithiolylium salts has been described.[149] Rate constants for protiodesilylation of substituted 2-trimethylsilylthiophens can be correlated with those for similarly substituted benzene derivatives.[150]

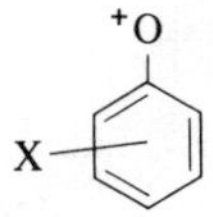

$^+$O

X

(**40**)

The acid-catalysed reactions of *N*-arylhydroxylamines with benzene and toluene give diphenylamines with high regioselectivity when the catalyst is trifluoroacetic acid, and aminobiphenyls with low regioselectivity when the catalyst is trifluoromethanesulphonic acid. There is evidence that the latter reaction involves dications (Scheme 3); in accord with this conclusion is the observation that

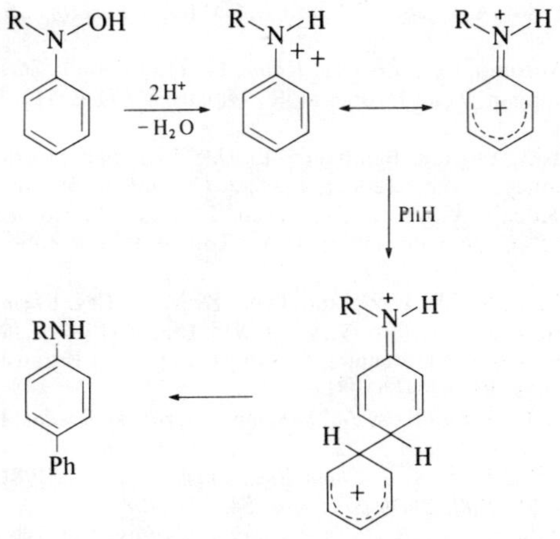

R–N–OH

$2H^+$

$-H_2O$

R–N–H ++

R–$\overset{+}{N}$–H

PhH

R–$\overset{+}{N}$–H

H

H

RNH

Ph

Scheme 3

nitrosobenzene and azoxybenzene react with benzene to give analogous products in the stronger acid.[151] 2-Aminoindophenol (**41**) is hydrolysed in acid, but in alkali cyclizes *via* its mono-anion to (**42**) which undergoes air oxidation to (**43**).[152] Selective deuteration at possible ring-closure sites, and product analyses, reveal a primary kinetic isotope effect in the cyclization in Scheme 4.[153]

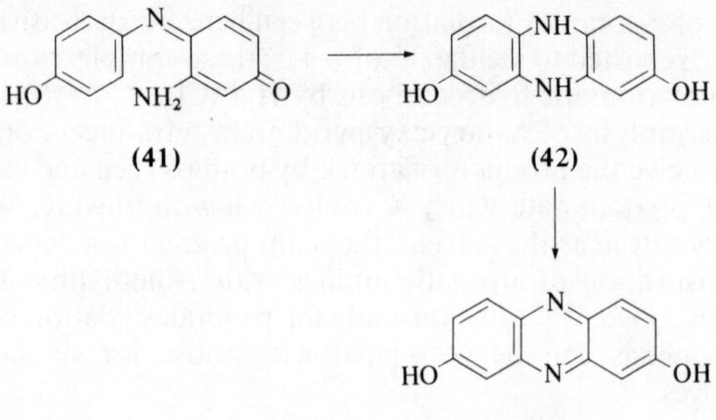

HO NH$_2$ N O
(41)

HO NH NH OH
(42)

HO N N OH
(43)

Ar +N =N −N R → Ar R H N=N

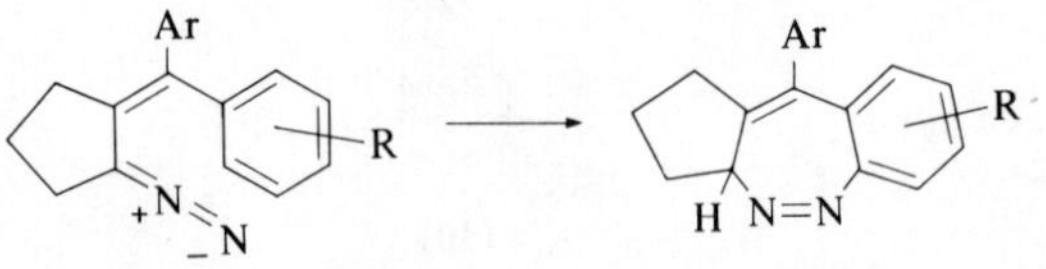

Scheme 4

References

[1] Pross, A., and Radom, L., *Prog. Phys. Org. Chem.*, **13,** 1 (1981); *Chem. Abs.*, **95,** 41612 (1981).
[2] Von Szentpaly, L., *Chem. Phys. Lett.*, **77,** 352 (1981); *Chem. Abs.*, **94,** 174084 (1981).
[3] Zhang, J.-C., Liu, X.-H., Wei, W., and Liu, R.-Z., *Kao. Teng. Hsueh Hsiao Hua Hsueh Hsueh Pao*, **2,** 122 (1981); *Chem. Abs.*, **94,** 208232 (1981).
[4] Aihara, J., *Bull. Chem. Soc. Jpn.*, **54,** 2268 (1981).
[5] Abronin, I. A., Cherkasov, A. V., and Zhidomirov, G. M., *Zh. Fiz Khim.*, **54,** 1888 (1980); *Chem. Abs.*, **93,** 185344 (1980).
[6] Belenkii, L. I., and Abronin, I. A., *Zh. Org. Khim.*, **17,** 1129 (1981); *Chem. Abs.*, **95,** 168302 (1981).
[7] Nanjan, M. J., Kannappan, V., and Ganesan, R., *Monatsch. Chem.*, **112,** 581 (1981); *Chem. Abs.*, **95,** 96729 (1981).
[8] Kachurin, O., Vysotsky, Yu., and Balabanov, E., *Org. Reactions* (*Tartu*), **17,** 267 (1980).
[9] Morrison, J. D., Stanney, K., and Tedder, J. M., *J. Chem. Soc., Perkin Trans. 2*, **1981,** 967.
[10] Solodar, S. L., and Kochin, V. A., *Zh. Org. Khim.*, **17,** 828 (1981); *Chem. Abs.*, **95,** 79576 (1981).
[11] El-Kashef, H., Rault, S., Cugnon de Sévricourt, M., Touzot, P., and Robba, M., *J. Heterocycl. Chem.*, **17,** 1399 (1980).
[12] Ashe, A. J., Chan, W.-T., Smith, T. W., and Taba, K. M., *J. Org. Chem.*, **46,** 881 (1981).
[13] Gronowitz, S., Konar, A., and Litvinov, V., *Chem. Scr.*, **15,** 206 (1980); *Chem. Abs.*, **94,** 120466 (1981).
[14] Dauksas, V., Purvaneckas, G., Udrenaite, E., Gineitite, V., and Borauskaite, A., *Heterocycles*, **15,** 1395 (1981); *Chem. Abs.*, **95,** 42041 (1981).
[15] Budylin, V. A., Yudin, L. G., and Kost, A. N., *Khim. Geterotsikl. Soedin.*, **1980,** 1189; *Chem. Abs.*, **94,** 14627 (1981).
[16] Suckling, C. J., and Wilson, A. A., *J. Chem. Soc., Chem. Commun.*, **1981,** 613.
[17] Sket, B., and Zupan, M., *Bull. Chem. Soc. Jpn.*, **54,** 279 (1981).
[18] Koudstaal, H., and Olieman, C., *Recl. Trav. Chim. Pays-Bas.*, **100,** 246 (1981).
[19] Ip, D. P., Arthur, C. D., Winans, R. E., and Appelman, E. H., *J. Am. Chem. Soc.*, **103,** 1964 (1981).
[20] Lerman, O., Tor, Y., and Rozen, S., *J. Org. Chem.*, **46,** 4629 (1981).
[21] Bee, L. K., Garratt, P. J., and Mansuri, M. M., *J. Am. Chem. Soc.*, **102,** 7076 (1980).
[22] Konishi, H., Yokota, K., Ichihashi, Y., Okano, T., and Kiji, J., *Chem. Lett.*, **1980,** 1423.
[23] Radhakrishnamurti, P. S., and Sasmal, B. M., *Indian J. Chem.*, **19A,** 880 (1980); *Chem. Abs.*, **94,** 64788 (1981).
[24] Anandasundaresan, P., Penchatsharum, V. S., Nagarajan, K., Balasubramanian, V., and Venkatasubramanian, N., *Indian J. Chem.*, **19A,** 576 (1980); *Chem. Abs.*, **93,** 203601 (1980).
[25] Byers, M., Forrester, A. R., John, I. L., and Thomson, R. H., *J. Chem. Soc., Perkin Trans. 1*, **1981,** 1092.

[26] Joseph, N., and Gnanpragasam, N. S., *Indian J. Chem.*, **19A**, 653 (1980); *Chem. Abs.*, **93**, 238209 (1980).
[27] Dubois, J.-E., and Balou, D., *Nouveau J. Chim.*, **5**, 33 (1981).
[28] Brittain, J. M., de la Mare, P. B. D., and Newman, P. A., *J. Chem. Soc., Perkin Trans. 2*, **1981**, 32.
[29] Jaquesey, J.-C., Jouannetaud, M.-P., and Makani, S., *Nouveau J. Chim.*, **4**, 747 (1980).
[30] Jacquesy, J. C., and Jouannetaud, M. P., *Tetrahedron*, **37**, 747 (1981).
[31] Sivakamasundari, S., and Ganesan, R., *Int. J. Chem. Kinet.*, **12**, 837 (1980).
[32] Radhakrishnamurti, P. S., and Rao, B. V. D., *Indian J. Chem.*, **20A**, 473 (1981); *Chem. Abs.*, **95**, 79598 (1981).
[33] Kannappan, V., Nanjan, M. J., and Ganesan, R., *Indian J. Chem.*, **19A**, 1183 (1981); *Chem. Abs.*, **94**, 155978 (1981).
[34] Gilow, H. M., and Burton, D. E., *J. Org. Chem.*, **46**, 2221 (1981).
[35] Kidokoro, H., Sato, M., and Ebine, S., *Chem. Lett.*, **1981**, 1269.
[36] Solomatin, G. G., Filippov, M. P., and Shigalevskii, V. A., *Zh. Org. Khim.*, **17**, 877 (1981); *Chem. Abs.*, **95**, 96464 (1981).
[37] Harrison, J. J., Pellegrini, J. P., and Selwitz, C. M., *J. Org. Chem.*, **46**, 2169 (1981).
[38] Argentini, M., *EIR-Ber.*, **1979**, 393; *Chem. Abs.*, **93**, 237966 (1980).
[39] Sircar, J. K., Yadav, J. R., Yadava, K. L., and Dey, A. K., *Z. Phys. Chem. (Leipzig)*, **261**, 964 (1980); *Chem. Abs.*, **93**, 29758 (1980).
[40] Rao, M. D. P., and Padmanabha, J., *Indian J. Chem.*, **19A**, 1179 (1981); *Chem. Abs.*, **94**, 155977 (1981).
[41] Rao, M. D. P., and Padmanabha, J., *Indian J. Chem.*, **20A**, 133 (1981); *Chem. Abs.*, **95**, 23782 (1981).
[42] Radhakrishnamurti, P. S., and Janardhana, C., *Indian J. Chem.*, **19A**, 550 (1980); *Chem. Abs.*, **93**, 203600 (1980).
[43] Sordo, T., Arumi, M., and Bertrán, J., *J. Chem. Soc., Perkin Trans. 2*, **1981**, 708.
[44] Archer, W. J., Taylor, R., Gore, P. H., and Kamounah, F. S., *J. Chem. Soc., Perkin Trans. 2*, **1980**, 1828.
[45] Archer, W. J., and Taylor, R., *J. Chem. Soc., Perkin Trans. 2*, **1981**, 1153.
[46] Archer, W. J., Shafig, Y. E., and Taylor, R., *J. Chem. Soc., Perkin Trans. 2*, **1981**, 675.
[47] Van de Griendt, F., and Cerfontain, H., *Tetrahedron*, **37**, 643 (1981).
[48] (a) Lammertsma, K., *J. Am. Chem. Soc.*, **103**, 2062 (1981).
[48] (b) Serebryanskaya, A. I., Kurenkova, V. M., and Shatenshtein, A. I., *Zh. Obshch. Khim.*, **50**, 2789 (1980); *Chem. Abs.*, **94**, 120682 (1981).
[49] Ausloos, P., and Lias, S. G., *J. Am. Chem. Soc.*, **103**, 3641 (1981).
[50] Schofield, K. *Aromatic Nitration*, Cambridge University Press, Cambridge, 1980.
[51] Gibbs, H. W., Main, L., Moodie, R. B., and Schofield, K., *J. Chem. Soc., Perkin Trans. 2*, **1981**, 848.
[52] Draper, M. R., and Ridd, J. H., *J. Chem. Soc., Perkin Trans. 2*, **1981**, 94.
[53] Fujiwara, K., Tamura, M., and Yoshida, T., *Helv. Chim. Acta*, **63**, 1897 (1980).
[54] Manglik, A. K., Moodie, R. B., Schofield, K., Dedeoglu, E., Dutly, A., and Rys, P., *J. Chem. Soc., Perkin Trans. 2*, **1981**, 1358.
[55] Stock, L. M., *Report* 1980, ARO-13957.4-CX, Order No. AD-A086286, avail. NTIS. From *Gov. Rep. Announce Index (U.S.)*, **80**, 4693 (1980); *Chem. Abs.*, **94**, 102709 (1981).
[56] Olah, G. A., Narang, S. C., and Fung, A. P., *J. Org. Chem.*, **46**, 2706 (1981).
[57] Crivello, J. V., *J. Org. Chem.*, **46**, 3056 (1981).
[58] Olah, G. A., Fung, A. P., Narang, S. C., and Olah, J. A., *J. Org. Chem.*, **46**, 3533 (1981).
[59] Ross, D. S., and Blucher, W., *Report* 1980 ARO-13831.3-CX, Order No. AD-A085324, avail. NTIS. From *Gov. Rep. Announce Index (U.S.)*, **80**, 4255 (1980); *Chem. Abs.*, **94**, 120451 (1981).
[60] Olah, G. A., Narang, S. C., and Olah, J. A., *Proc. Natl. Acad. Sci. U.S.A.*, **78**, 3298 (1981).
[61] Schmitt, R. J., Ross, D. S., and Buttrill, S. E., *J. Am. Chem. Soc.*, **103**, 5265 (1981).
[62] Al-Omran, F., Fujiwara, K., Giffney, J. C., Ridd, J. H., and Robinson, S. R., *J. Chem. Soc., Perkin Trans. 2*, **1981**, 518.
[63] Helsby, P., Ridd, J. H., and Sandall, J. P. B., *J. Chem. Soc., Chem. Commun.*, **1981**, 402.
[64] Ridd, J. H., and Sandall, J. P. B., *J. Chem. Soc., Chem. Commun.*, **1981**, 402.
[65] Suzuki, H., Nagae, K., Maeda, H., and Osuka, A., *J. Chem. Soc., Chem. Commun.*, **1980**, 1245.
[66] Suzuki, H., Hashihama, M., and Mishina, T., *Bull. Chem. Soc. Jpn.*, **54**, 1186 (1981).
[67] Tashiro, M., Yamato, T., Fukata, G., and Fukuda, Y., *J. Org. Chem.*, **46**, 2376 (1981).
[68] Fischer, A., Henderson, G. N., Smyth, T. A., Einstein, F. W. B., and Cobbledick, R. E., *Can. J. Chem.*, **59**, 584 (1981).
[69] Geppert, J. T., Johnson, M. W., Myhre, P. C., and Woods, S. P., *J. Am. Chem. Soc.*, **103**, 2057 (1981).
[70] Fischer, A., and Henderson, G. N., *Can. J. Chem.*, **59**, 2314 (1981).

[71] Mochalov, S. S., Karpova, V. V., and Shabarov, Yu. S., *Zh. Org. Khim.*, **16,** 2238 (1980); *Chem. Abs.*, **94,** 65011 (1981).
[72] Mochalov, S. S., Karpova, V. V., and Shabarov, Yu. S., *Zh. Org. Khim.*, **16,** 2092 (1980); *Chem. Abs.*, **94,** 156452 (1981).
[73] Fischer, A., and Henderson, G. N., *Tetrahedron Lett.*, **21,** 4661 (1980).
[74] Hartshorn, M. P., Ing, H. T., Richards, K. E., and Robinson, W. T., *J. Chem. Soc., Chem. Commun.*, **1981,** 225.
[75] Hartshorn, M. P., Richards, K. E., Thompson, R. S., and Vaughan, J., *Aust. J. Chem.*, **34,** 1345 (1981).
[76] Mochalov, S. S., Surikova, T. P., Abdel-razek, F. M., Zakharova, V. D., and Shabarov, Yu. S., *Khim. Geterotsikl. Soedin*, **1981,** 189; *Chem. Abs.*, **95,** 23765 (1981).
[77] Mitsuru, A., Takayuki, A., Fujiwara, K., Tamura, M., and Yoshida, T., *Kogyo Kayaku*, **41,** 71 (1980); *Chem. Abs.*, **94,** 102601 (1981).
[78] Fujiwara, K., Andoh, T., Arai, M., Tamura, M., and Yoshida, T., *J. Org. Chem.*, **45,** 4973 (1980).
[79] Sohrabi, M., and Kaghazchi, T., *Pakistan J. Sci. Ind. Res.*, **23,** 93 (1980); *Chem. Abs.*, **94,** 155959 (1981).
[80] Moodie, R. B., Schofield, K., Taylor, P. G. and Baillie, P. J., *J. Chem. Soc., Perkin Trans. 2*, **1981,** 842.
[81] Colonna, M., Greci, L., and Poloni, M., *J. Chem. Soc., Perkin Trans. 2*, **1981,** 628.
[82] Suzuki, H., Hidaka, I., Iwasa, A., Mishina, T., and Osuka, A., *Bull. Chem. Soc. Jpn.*, **54,** 771 (1981).
[83] Gosney, A. P., and Page, M. I., *J. Chem. Soc., Perkin Trans. 2*, **1980,** 1783.
[84] Bazanova, G. V. and Stotskii, A. A., *Zh. Org. Khim.*, **16,** 2416 (1980); *Chem. Abs.*, **94,** 64782 (1981).
[85] Bazanova, G. V. and Stotskii, A. A., *Zh. Org. Khim.*, **16,** 2421 (1980); *Chem. Abs.*, **94,** 64783 (1981).
[86] Aldred, S. E., and Williams, D. L. H., *J. Chem. Soc., Perkin Trans. 2*, **1981,** 1021.
[87] Strepikheev, Yu. A., Nikitenkova, L. P., Naumova, I. I., and Karakotov, S. D., *Deposited Doc.* 1979, VINITI 3784, 495; *Chem. Abs.*, **94,** 138868 (1981).
[88] Kobayashi, H., Sonoda, T., Iwamoto, H., and Yoshimura, M., *Chem. Lett.*, **1981,** 579; *Chem. Abs.*, **95,** 149441 (1981).
[89] Goodman, P. D., Kemp, T. J., and Pinot de Moira, P., *J. Chem. Soc., Perkin Trans. 2*, **1981,** 1221.
[90] Cox, A., Goodman, P. D., Kemp, T. J., and Pinot de Moira, P., *Int. J. Chem. Kinet.*, **13,** 97 (1981).
[91] Hashida, Y., Tanabe, F., and Matsui, K., *Nippon Kagaku Kaishi*, **1980,** 865; *Chem. Abs.*, **93,** 185321 (1980).
[92] Kishimoto, S., Kitahara, S., Manabe, O., and Hiyama, H., *Nippon Kagaku Kaishi*, **1981,** 428; *Chem. Abs.*, **95,** 60879 (1981).
[93] Stumpe, R. W., *Tetrahedron Lett.*, **21,** 4891 (1980).
[94] Lambrechts, H. J. A., and Cerfontain, H., *Recl. Trav. Chim. Pays-Bas*, **100,** 291 (1981).
[95] Khelevin, R. N., *Zh. Org. Khim.*, **17,** 1486 (1981); *Chem. Abs.*, **95,** 131913 (1981).
[96] Poppl, R. F., Quiroga, O. D., Capretto de Castillo, M. E., and Gonzo, E. E., *Rev. Latinoam. Ing. Quim. Apl.*, **11,** 25 (1981); *Chem. Abs.*, **95,** 114384 (1981).
[97] Beenackers, A. A. C. M., and Van Swaaj, W. P. M., *Phosphorus Sulphur*, **10,** 217 (1981); *Chem. Abs.*, **95,** 114383 (1981).
[98] Velichko, L. I., Zaraiskii, A. P., Fedorchuk, E. S., and Kachurin, O. I., *Ukr. Khim. Zh.* (*Russ. Ed.*), **46,** 1193 (1980); *Chem. Abs.*, **94,** 83237 (1981).
[99] Koshelev, V. I., and Borodaev, S. V., *Zh. Prikl. Khim.* (*Leningrad*), **53,** 2298 (1980); *Chem. Abs.*, **94,** 64784 (1981).
[100] Brandström, A., Stradlund, G., and Lagerström, P., *Acta Chem. Scand.*, **B34,** 467 (1980).
[101] Olah, G. A., and Meidar, D., *Kirk-Othmer Encycl. Chem. Technol., 3rd Ed.*, **11,** 269 (1980); *Chem. Abs.*, **94,** 3294 (1981)).
[102] Roberts, R. M., and Gibson, T. L., *Isot. Org. Chem.*, **5,** 103 (1980); *Chem. Abs.*, **94,** 46280 (1981).
[103] Lipovich, V. G., *Sint. Prevrasch. Alkilaromat. Soedin*, **1979,** 5; *Chem. Abs.*, **94,** 191068 (1981).
[104] Cacace, F., Ciranni, G., and Giacomello, P., *J. Am. Chem. Soc.*, **103,** 1513 (1981).
[105] Stone, J. A., Lin, M. S., and Varah, J., *Can. J. Chem.*, **59,** 2412 (1981).
[106] Pepe, N., and Speranza, M., *J. Chem. Soc., Perkin Trans. 2*, **1981,** 1430.
[107] Speranza, M., and Sparapani, C., *Radiochim. Acta*, 28, **87** (1981); *Chem. Abs.*, **95,** 23816 (1981).
[108] Speranza, M., *J. Chem. Soc., Chem. Commun.*, **1981,** 1177.
[109] Morrison, J. D., Stanney, K., and Tedder, J. M., *J. Chem. Soc., Perkin Trans. 2*, **1981,** 838.
[110] Kulsrestha, G. N., Shanker, U., Pathania, B. S., Negi, J., and Bhattacharyya, K. K., *Indian J. Chem.*, **20B,** 298 (1981); *Chem. Abs.*, **95,** 79572 (1981).
[111] Häfelinger, G., and Beyer, M., *Liebigs Ann. Chem.*, **1980,** 2012.
[112] Chenets, V. V., Istomin, B. I., Sergeeva, O. R., and Lipovich, V. G., *Zh. Org. Khim.*, **16,** 1447 (1980); *Chem. Abs.*, **93,** 238152 (1980).

[113] Bakalova, E. P., Chenets, V. V., and Lipovich, V. G., *Kinet. Katal.*, **21,** 1446; *Chem. Abs.*, **94,** 102406 (1981).

[114] Apte, D. V., and Deodhar, K. D., *Indian J. Chem.*, **19B,** 758 (1980); *Chem. Abs.*, **94,** 83240 (1981).

[115] Pinnick, H. W., Brown, S. P., McLean, E. A., and Zoller, L. W., *J. Org. Chem.*, **46,** 3758 (1981).

[116] Murphy, W. S., and Wattanasin, S., *J. Chem. Soc., Perkin Trans. 1*, **1981,** 2920.

[117] Jackson, A. H., Shannon, P. V. R., and Taylor, P. W., *J. Chem. Soc., Perkin Trans. 2*, **1981,** 286.

[118] Bonvino, V., Casini, G., Ferappi, M., Cingolani, G. M., and Pietroni, B. R., *Tetrahedron*, **37,** 615 (1981).

[119] Mochida, I., Takashita, K., Ohgai, M., and Seiyama, T., *Ind. Eng. Chem. Prod. Res. Dev.*, **19,** 392 (1980); *Chem. Abs.*, **93,** 167134 (1980).

[120] Klemm, L. H., and Taylor, D. R., *J. Org. Chem.*, **45,** 4320 (1980).

[121] Klemm, L. H., and Taylor, D. R., *J. Org. Chem.*, **45,** 4326 (1980).

[122] Belov, P. S., Grigoreva, E. N., and Tsvetkov, O. N., *Khim. Tekhnol. Topl. Masel*, **1981,** 11; *Chem. Abs.*, **95,** 114385 (1981).

[123] Zerkalenkov, A. A., and Kachurin, O. I., *Kinet. Katal.*, **21,** 1442 (1980); *Chem. Abs.*, **94,** 102405 (1981).

[124] Kachurin, O. I., and Dereza, N. A., *Ukr. Khim. Zh. (Russ. Ed.)*, **47,** 732 (1981); *Chem. Abs.*, **95,** 114382 (1981).

[125] Dumpis, M., Kudryashova, N. I., and Khromov-Borisov, N. V., *Zh. Org. Khim.*, **16,** 2113 (1980); *Chem. Abs.*, **94,** 65010 (1981).

[126] Andreou, A. D., Bulbulian, R. V., Gore, P. H., Kamounah, F. S., Miri, A. Y., and Waters, D. N., *J. Chem. Soc., Perkin Trans. 2*, **1981,** 376.

[127] Ahmed, A. M., Elsemongy, M. M., and Amira, M. F., *Indian J. Chem.*, **19B,** 593 (1980); *Chem. Abs.*, **94,** 29751 (1981).

[128] Andreou, A. D., Bulbulian, R. V., Gore, P. H., Morris, D. F. C., and Short, E. L., *J. Chem. Soc., Perkin Trans. 2*, **1981,** 830.

[129] Yamanaka, T., and Imai, T., *Bull. Chem. Soc. Jpn.*, **54,** 1288 (1981).

[130] Rokach, J., Hamel, P., Kakushima, M., and Smith, G. M., *Tetrahedron Lett.*, **22,** 4901 (1981).

[131] Booth, B. L., and Noori, G. F. M., *J. Chem. Soc., Perkin Trans. 1*, **1980,** 2894.

[132] Selivanov, B. A., Pozdnyakovitch, Yu. V., Chuikova, T. V., Osina, O. I., and Shteingarts, V. D., *Zh. Org. Khim.*, **16,** 1910 (1980); *Chem. Abs.*, **94,** 46401 (1981).

[133] Tamura, Y., Choi, H.-D., Shindo, H., Uenishi, J., and Ishibashi, H., *Tetrahedron Lett.*, **22,** 81 (1981).

[134] Olah, G. A., Bruce, M. R., and Clouet, F. L., *J. Org. Chem.*, **46,** 438 (1981).

[135] Herrmann, R., and Ugi, I., *Tetrahedron*, **37,** 1001 (1981).

[136] Jo, S., Tanimoto, S., Sugimoto, T., and Okano, M., *Bull. Chem. Soc. Jpn.*, **54,** 2120 (1981).

[137] Murakami, Y., and Ishii, H., *Chem. Pharm. Bull.*, **29,** 699 (1981); *Chem. Abs.*, **95,** 23785 (1981).

[138] Murakami, Y., and Ishii, H., *Chem. Pharm. Bull.*, **29,** 711 (1981); *Chem. Abs.*, **95,** 23786 (1981).

[139] Stock, L. M., and Wright, T. L., *J. Org. Chem.*, **45,** 4645 (1980).

[140] Roberts, R. M. G., *Tetrahedron*, **36,** 3281 (1980).

[141] Kooyman, E. C., Huygens, A. V., de Lepper, J. P. J., de Vos, D., and Wolters, J., *Recl. Trav. Chim. Pays-Bas.*, **100,** 24 (1981).

[142] Fung, C. W., and Roberts, R. M. G., *Tetrahedron*, **36,** 3289 (1980).

[143] Apatu, J. O., Chapman, D. C., and Heaney, H., *J. Chem. Soc., Chem. Commun.*, **1981,** 1079.

[144] Ogata, Y., Sawaki, Y., Tomizawa, K., and Ohno, T., *Tetrahedron*, **37,** 1485 (1981).

[145] Olah, G. A., Fung, A. P., and Keumi, T., *J. Org. Chem.*, **46,** 4305 (1981).

[146] Gesson, J.-P., Jacquesy, J.-C., and Souannetund, M.-P., *J. Chem. Soc., Chem. Commun.*, **1980,** 1128.

[147] Shul'pin, G. B., and Kitaigorodskii, A. N., *Zh. Fiz. Khim.*, **55,** 266 (1981); *Chem. Abs.*, **94,** 139047 (1981).

[148] Abramovitch, R. A., Alvernhe, G., Bartnik, R., Dassanayake, N. L., Inbasekaran, M. N., and Kato, S., *J. Am. Chem. Soc.*, **103,** 4558 (1981).

[149] Nakayama, J., Matsumaru, N., and Hoshino, M., *J. Chem. Soc., Chem. Commun.*, **1981,** 565.

[150] Seconi, G., and Eaborn, C., *J. Chem. Soc., Perkin Trans. 2*, **1981,** 931.

[151] Shudo, K., Ohta, T., and Okamoto, T., *J. Am. Chem. Soc.*, **103,** 645 (1981).

[152] Brown, K. C., and Corbett, J. F., *J. Chem. Soc., Perkin Trans. 2*, **1981,** 886.

[153] Miller, J. K., Sharp, J. T., Thomas, G. J., and Thompson, I., *Tetrahedron Lett.*, **22,** 1537 (1981).

Organic Reaction Mechanisms 1981
Edited by A. C. Knipe and W. E. Watts

CHAPTER 8

Carbocations

M. C. GROSSEL

Department of Chemistry, Bedford and Royal Holloway Colleges, Egham, Surrey

Bicyclic Systems . . . 299
Polycyclic Systems . . . 301
Participation by Double and Triple Bonds . . . 304
Organometallic Systems . . . 304
Stable Carbocations: their Structures and Reactions . . . 308
Other Reactions . . . 314
Theoretical Calculations . . . 316
References . . . 317

Bicyclic Systems

Evidence for the structure of the norborn-2-yl cation has been discussed.[1,2] From a comparison of the relative rates of hydrolysis of 6-substituted norbornyl tosylates (**1**) and an analysis of the products formed, it is suggested that there is significant bridging between C(6) and C(2) during ionization when the C(6)-substituent (X) is a σ-electron donor, but that this effect is suppressed when X is electron-withdrawing.[3] ^{13}C-NMR chemical shifts and ^{13}C–^{1}H coupling constants calculated for classical 2-methylnorborn-2-yl and norborn-2-yl cations are in very close agreement with the experimental results which have been reported for the former. In contrast, values predicted for a bridged-ion structure are much closer to the experimental data for the norborn-2-yl cation itself.[4] Empirical force-field calculations on the energetics for the rearrangement of secondary norbornyl cations into tertiary isomers suggest that the secondary ion is *ca.* 7.6 kcal mol^{-1} more stable than acyclic models.[5] This work has been challenged, principally because of some of the thermochemical data chosen; MINDO/3 calculations on such secondary-tertiary cation equilibrations suggest that the norborn-2-yl cation is more stable than "normal" secondary cations by *ca.* 6 $\pm$ 1 kcal mol^{-1} in superacid media.[6] β-Deuterium isotope effects on the ^{13}C chemical shifts of carbocations appear to provide direct insight into the extent of σ-bridging. For example, the isotope shift of -2.2 ppm observed for the 2-(methyl-$^{2}H_3$)norborn-2-yl cation is attributed to non-classical behaviour.[7]

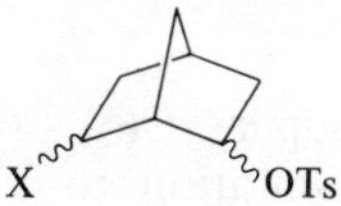

X OTs

(**1**)
X = CO_2Na, F

A detailed study of several methylated 2-arylnorborn-2-yl cations has appeared. Whilst the 1-methyl-2-phenylnorborn-2-yl cation has yet to be observed directly by spectroscopy, the 1-methyl-2-*p*-anisylnorborn-2-yl cation (**2**) has been prepared at low temperatures; interference between the methyl group and the aromatic ring significantly reduces the degree of benzylic stabilization (as demonstrated by the low energy barrier for aryl group rotation), and the cation readily isomerizes to the 6-*endo*-methyl-2-*p*-anisyl isomer (**3**) above $-10°$. The 3-*exo*- and 3-*endo*-methyl-2-phenylnorborn-2-yl cations are in equilibrium with each other, though this equilibrium lies much in favour of the *exo*-isomer. Attempts to prepare the 1,3-dimethyl-2-phenyl cation at low temperatures lead to a complex mixture of ions all of which are unexpectedly converted at $-50°$ into the 3-*exo*,6-endo-dimethyl-2-phenylnorborn-2-yl cation. Both *syn*- and *anti*-1,7-dimethyl-2-phenylnorborn-2-yl cations readily rearrange to the 5-*exo*,6-*endo*-dimethyl-2-phenylnorborn-2-yl cation.[8]

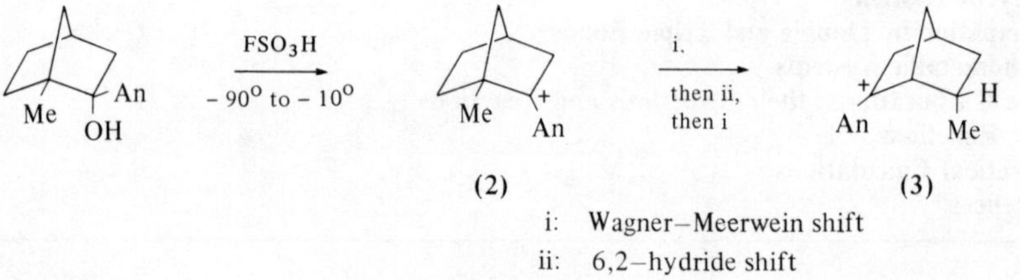

Me An OH $\xrightarrow[-90° \text{ to } -10°]{FSO_3H}$ Me + An (2) $\xrightarrow[\text{then ii, then i}]{\text{i.}}$ + An H Me (3)

i: Wagner–Meerwein shift
ii: 6,2–hydride shift

A study of the product distribution from acetolysis of a series of 2-aryl-3-oxonorborn-2-yl methanesulphonates (**4**) reveals that significantly more rearranged acetate and tricyclic ketone is formed as the electronegativity of the aryl substituent increases. A kinetic study in ethanol shows a marked discontinuity in behaviour between electron-donating and electron-withdrawing substituents. These results have been ascribed to π-electron stabilization of the intermediate cation by the carbonyl substituent (**5**).[9]

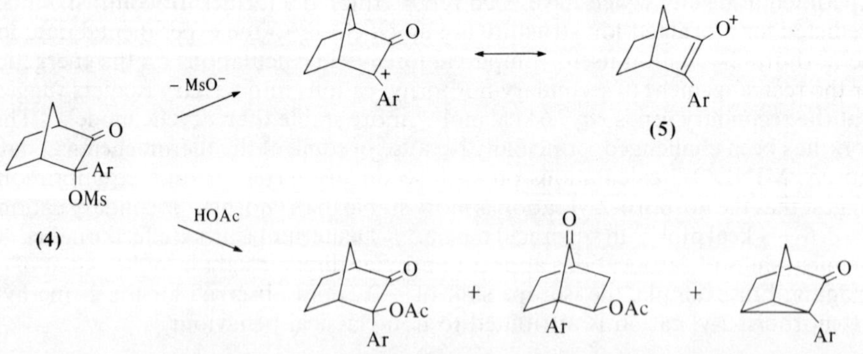

O Ar OMs (4) $\xrightarrow{-\,MsO^-}$ O + Ar $\longleftrightarrow$ O^+ Ar (5)

(4) $\xrightarrow{HOAc}$ O OAc Ar + O OAc Ar + O Ar

Introduction of a methyl group at C(6) of the norborn-5-en-2-*endo*-ols (**6**) encourages acid-catalysed fragmentation to (**7**) in place of the predominant hydration observed for the parent compound. However, for both reaction pathways, protonation of the 5,6-double bond remains the rate-determining step.[10]

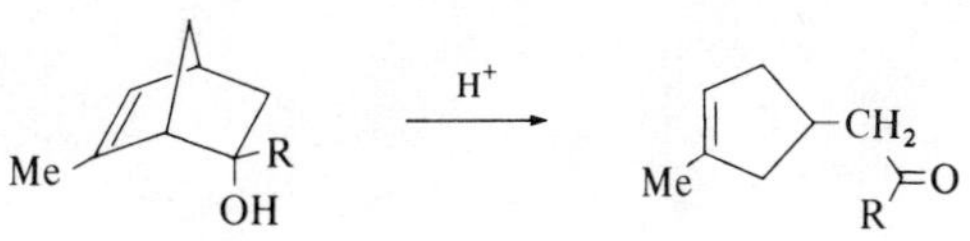

R = H, Me (6) (7)

The activation volume for degenerate equilibration of the 7-methylnorbornadien-7-yl cation is *ca.* zero as predicted for a process in which non-classical species are interconverted by a classical transition state.[11]

The influence of differential bridging strain on the *exo*/*endo* solvolysis rate ratios in the formation of bridged ions from norborn-2-yl, bicyclo[3.2.1]oct-2- and -8-yl, bicyclo[3.3.1]non-2-yl, and bicyclo[4.2.1]non-2-yl esters has been discussed.[12]

Polycyclic Systems

In liquid SO_2, 9-chloro-9-methoxy-*endo*-tricyclo[$4.2.1.0^{2,5}$]nona-3,7-diene (**8**) is in equilibrium with the cation (**9**); whilst the skeleton of the former is thermodynamically the more stable, the opposite is true for the corresponding cations and quenching of the reaction medium with nucleophiles more reactive than chloride ion leads to derivatives (**10**).[13]

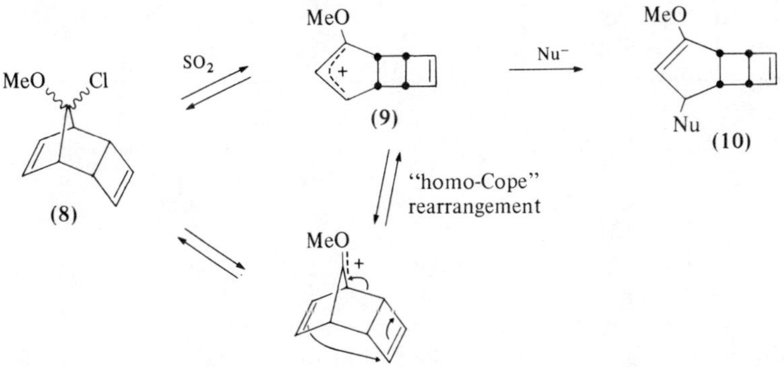

(8) (9) (10)

"homo-Cope" rearrangement

The semibullvalen-1-ylmethyl cation is implicated as an intermediate in the hydrolysis of the methanesulphonate (**11**) but only ring-expanded products are isolated.[14] NMR spectroscopic studies of the isotope-labelled 9-barbaralyl cation provide evidence for a structure (**12**) which undergoes degenerate equilibration *via* a series of divinylcyclopropylcarbinyl–divinylcyclopropylcarbinyl cationic rearrangements.[15] The 9-methylbarbaralyl cation is formed when (**13**) is protonated and *not* the 2-methylbicyclo[3.2.2]nonatrienyl cation as previously suggested.[16]

Bicyclohomoaromaticity has been implicated to explain the significantly higher reactivity of *anti*-bicyclo[4.3.2]undecatetraen-7-ol (**14**; X = OH) and its derivatives in comparison with their *syn*-isomers. Products from reactions of the former show complete retention of configuration whereas in the latter case both (**14**) and armilenyl derivatives are obtained.[17] The [3.5.3]armilenium cation (**15**) is formed when bicyclo[4.4.1]undeca-2,4,7,9-tetraen-11-ol is dissolved in fluorosulphonic acid at low temperatures.[18]

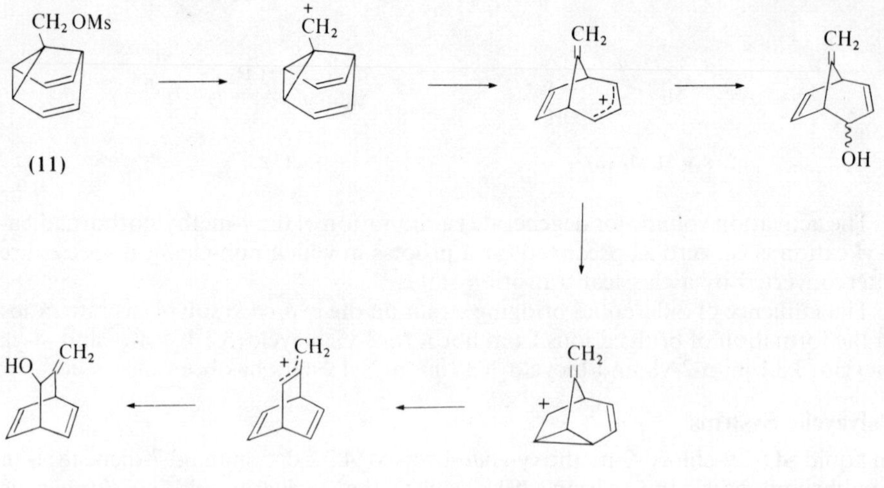

(11)

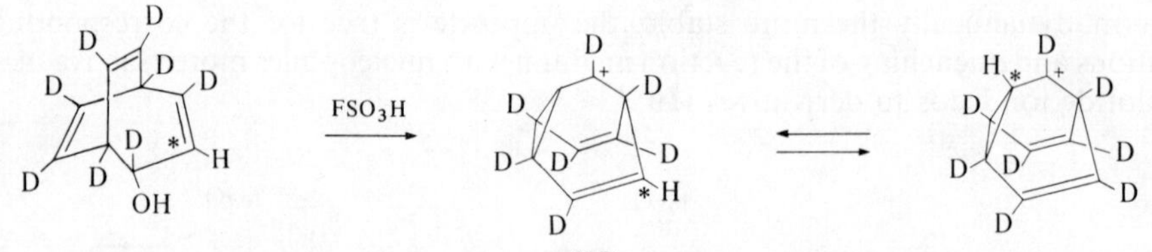

* indicates ^{13}C label

(12)

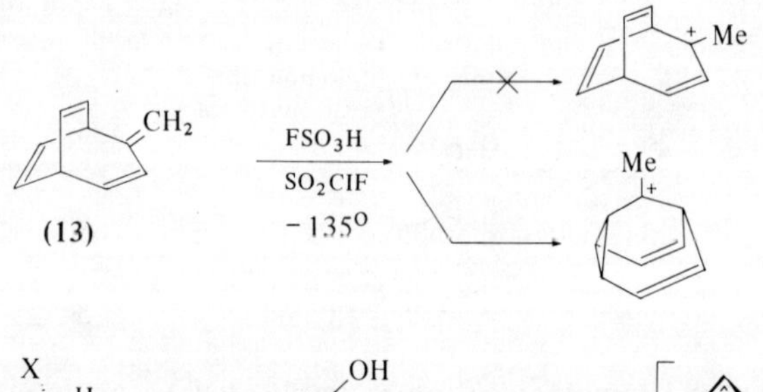

(13)

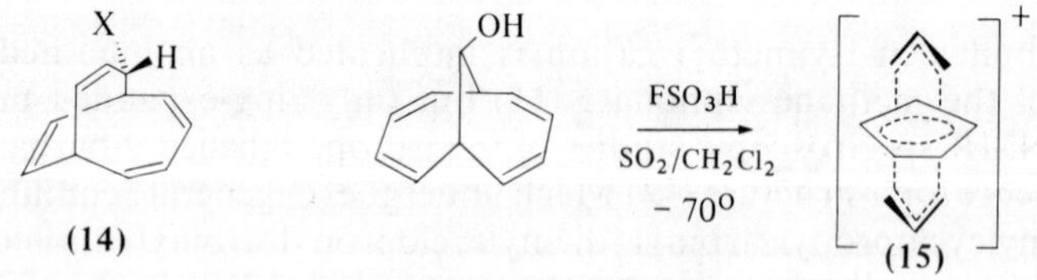

(14)

(15)

The stability of the 1-adamantyl cation has been ascribed to a favourable $(4n + 2)$ electron configuration arising from interaction of the σ-electrons from the β,γ-C—C and/or γ-C—H bonds and the carbocation centre.[19]

Solvolysis studies on 2-cyano-2-adamantyl trifluoromethanesulphonate (**16**) and the great instability observed for the cyanoprotoadamantyl trifluoromethanesulphonate (**17**) support the suggestion that an α-cyano carbocation is significantly more stable than its β-analogue as a result of back-donation of π-electron density from the cyano substituent to the cationic centre.[20]

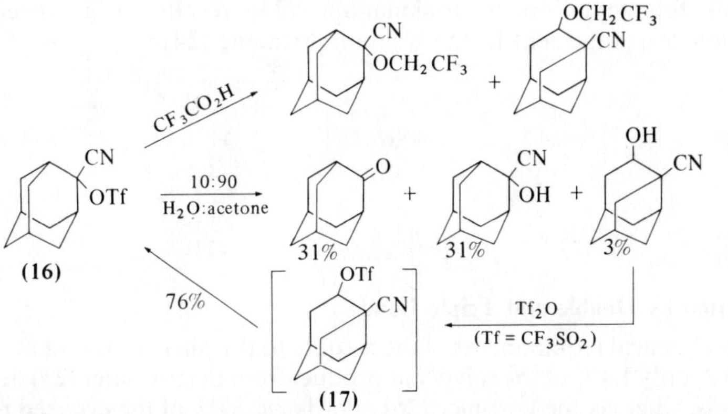

The results of a detailed product study of the solvolysis of precursors to the 4-methylprotoadamantyl cation (**18**), when taken in conjunction with previous data, provide strong evidence for a bridged structure (**19**). Hydrolysis of each of the esters (**20a**), (**21a**), and (**22a**) leads only to (**20b**) and (**21b**); none (<3 %) of the epimeric alcohol (**22b**) could be detected. ^{2}H-Labelling experiments demonstrate that formation of 1-methyladamantan-2-ol occurs with very high stereoselectivity; the pathway without rearrangement (from **21a**) proceeds with retention of configuration, whereas routes requiring rearrangement (from **20a** and **22a**) afford inversion at the migration origin.[21]

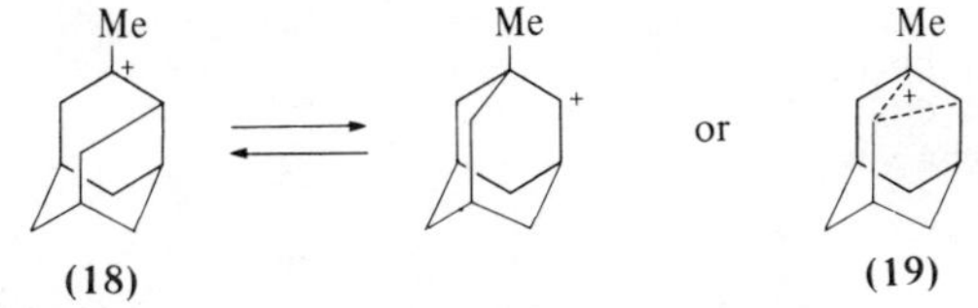

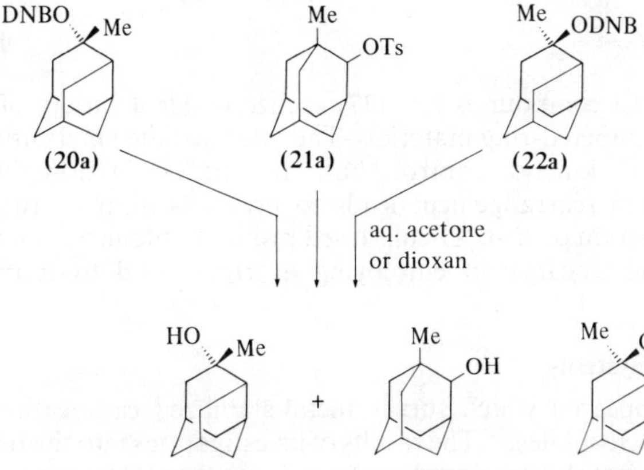

Schmidt fragmentation of basketanone (**23**) results in a novel one-step carbocation rearrangement to the bishomoprismane (**24**).[22]

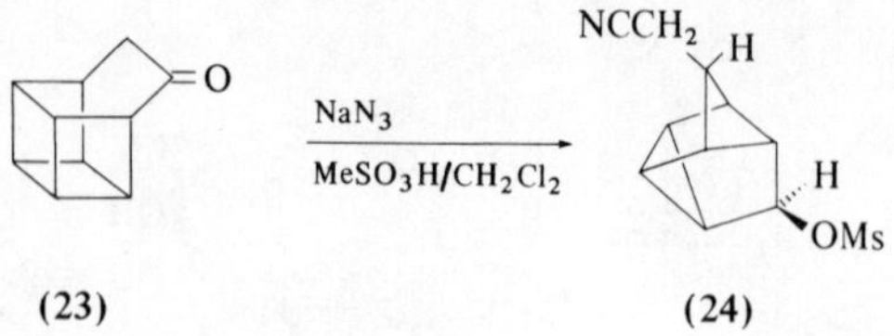

NaN_3, $MeSO_3H/CH_2Cl_2$

NCCH$_2$ H H OMs

(**23**) (**24**)

Participation by Double and Triple Bonds

The stereochemical requirements of the π-route to the phenyl cation have now been established; only 1.4% of the solvolysis product from the Z-isomer (**25**) derives from this process, whereas the E-isomer (**26**) affords *ca.* 25% of the cyclized material.[23]

A linear-free-energy study suggests that terpene condensation reactions using farnesyl pyrophosphate synthetase proceed by a stepwise pathway in which cleavage of the C—O bond of geranyl pyrophosphate leads to a highly structured ion-pair on the enzyme in which the bulk of the positive charge is localized on the primary carbon of the allyl cation, thus encouraging attack by the π-bond of isopentenyl pyrophosphate.[24,25]

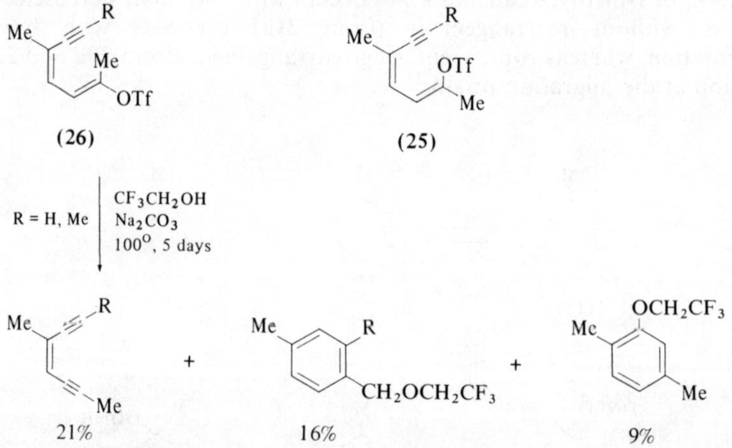

(**26**) (**25**)

R = H, Me; CF_3CH_2OH, Na_2CO_3, 100°, 5 days

21% + 16% ($CH_2OCH_2CF_3$) + 9% (OCH_2CF_3)

Polyunsaturated compounds like (**27**) cyclize under a variety of conditions to both 5- and 6-membered-ring materials. The intermediate vinyl cation (**28**) affords the products of kinetic control but, in media of low nucleophilicity, Wagner–Meerwein rearrangement leads to products from capture of (**29**). The significant proportion of *cis*-C/D-ring-fused product obtained from the cyclization of (**30**) mars the elegance of employing a triple bond to terminate polyene cyclizations.[26]

Organometallic Systems

Reviews have appeared which survey metal-stabilized carbocations[27] and their reactions with nucleophiles.[28] The results of investigations into the reactions of salts of tricarbonyl(η^5-cyclohexadienylium)iron cations (**31**) with a variety of

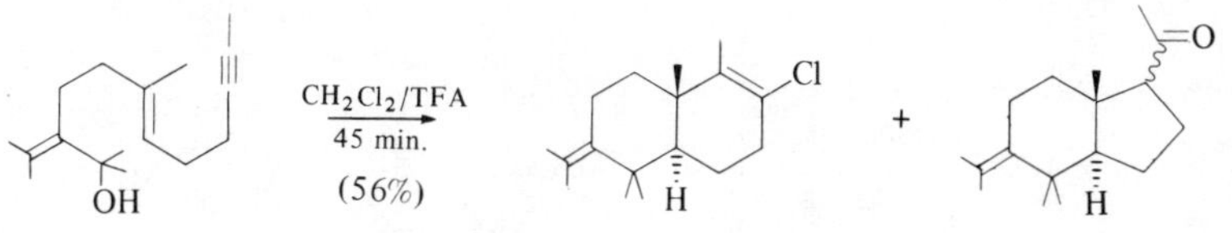

CH_2Cl_2/TFA, 45 min. (56%)

(27)

-78°	10	:	1
0°	55	:	45

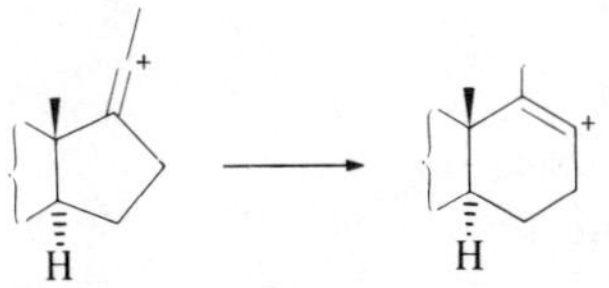

(28) (29) 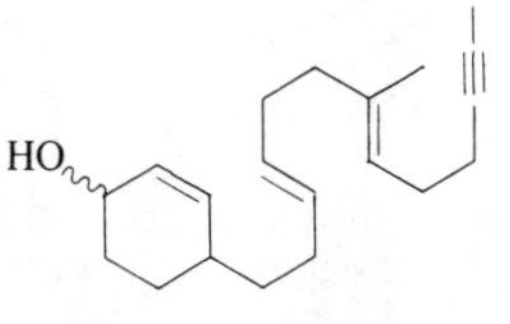 (30)

nucleophiles have been described. Reaction of (**31a**) with *p*-toluidine in acetonitrile involves a two-step pathway in which initial reversible addition to the dienyl ring is followed by rapid proton loss (assisted by solvent or amine).[29] Reaction with substituted pyridines is second order and the relative ease of addition of 2-methyl- and 2,6-dimethyl-pyridine demonstrates that steric effects on this process are non-additive.[30] Regioselectivity of the reduction of (**31b**) by borohydride reducing agents decreases at lower temperatures, but the opposite result is observed for 9-borabicyclononane, (**32a**) $>95\%$ being formed at $-78°$ in this case. It is suggested that the transition states for these reductions are early in terms of charge separation but do not resemble starting material in terms of orbital interactions contributing to the metal–ligand bond; it is this latter effect which influences regioselectivity.[31] The rates of reaction of pentane-2,4-dione with methyl-, methoxy-, or methoxycarbonyl-substituted salts (**31**) are affected not only by the nature of the substituent but also by its position.[32] The cations (**31c**) and (**31d**) are formed as transient intermediates in the reaction of trityl tetrafluoroborate with tricarbonyl[1- and 2-(trimethylsiloxy)-η^4-cyclohexadiene]iron complexes, but they are rapidly desilylated affording (**33**) and (**34**). The latter can be hydrolysed and reacts with nucleophiles in such a manner as to provide a synthetic route to 5-substituted tricarbonyl(2-hydroxy-1,3-cyclohexadiene)iron complexes (**35**).[33] Nucleophilic addition of malonate ion to tricarbonyl(1-ethyl-4-methoxycyclohexadiene)iron cation is a key step in the novel synthesis of a known precursor to aspidospermine.[34]

It has been noted that salts of tricarbonyl(cyclohexadienylium)metal cations (**36**) are the synthetic equivalents of spatially directed cations and thus have a particular value in asymmetric synthesis;[35] this has been exploited in a novel synthetic route to 6-keto-steroids.[36] An efficient route to (**37**), which is the synthetic equivalent of a *m*-methoxybenzene cation or the C(5) cation of cyclohex-2-enone, has been described.[37] It reacts with a variety of nucleophiles ranging from *tert*-butyllithium to ketones and enol trimethylsilyl ethers.[38] The anion of nitromethane reacts cleanly with (**31a**) to give tricarbonyl[5-*exo*-(nitromethyl)cyclohexa-1,3-diene]iron which is readily reduced to the aminomethyl derivative.[39] Nitromethylation is, however, inferior to cyanide addition as a route for nucleophilic formylation of tricarbonyl(η^5-cyclohexadienylium)iron salts, though the former does sometimes offer better regioselectivity.[40]

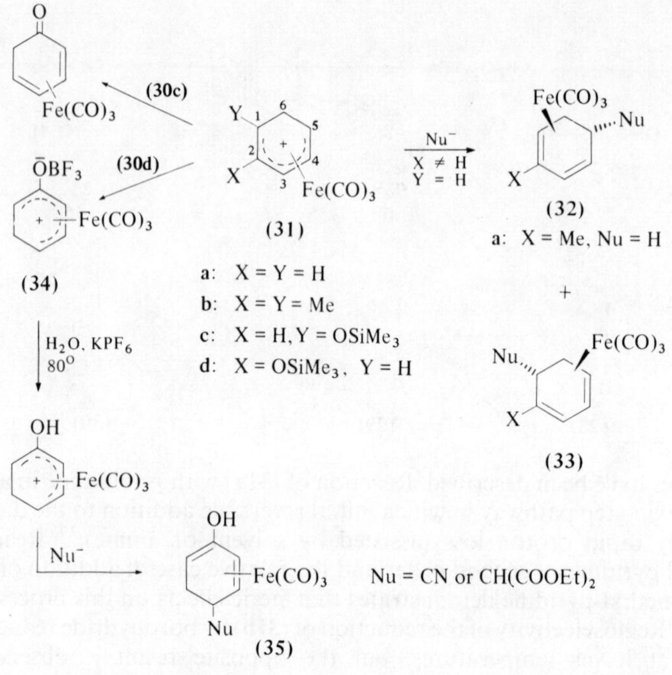

O

$Fe(CO)_3$

(30c)

$\bar{O}BF_3$

(30d)

+ $Fe(CO)_3$

(34)

Y 1 6 5 2 + 4 X 3 $Fe(CO)_3$

(31)

a: X = Y = H

b: X = Y = Me

c: X = H, Y = $OSiMe_3$

d: X = $OSiMe_3$, Y = H

Nu^-

X ≠ H

Y = H

$Fe(CO)_3$

Nu

X

(32)

a: X = Me, Nu = H

+

Nu

$Fe(CO)_3$

X

(33)

H_2O, KPF_6

80°

OH

$Fe(CO)_3$

Nu^-

OH

$Fe(CO)_3$

Nu

(35)

Nu = CN or $CH(COOEt)_2$

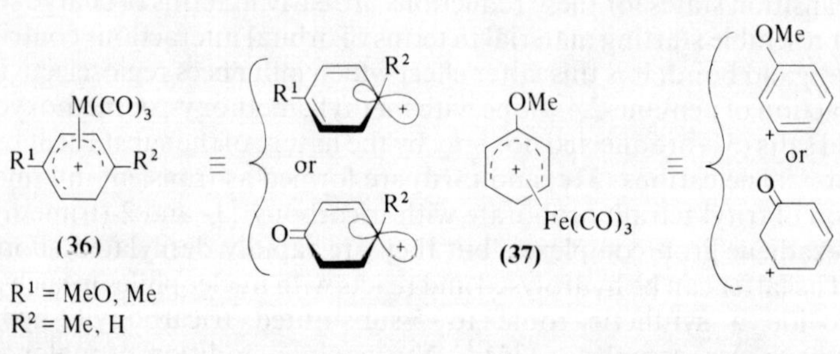

$M(CO)_3$

R^1 + R^2

(36)

≡

R^1 R^2 +

or

O R^2 +

R^1 = MeO, Me

R^2 = Me, H

OMe

+

$Fe(CO)_3$

(37)

≡

OMe

+

or

O

+

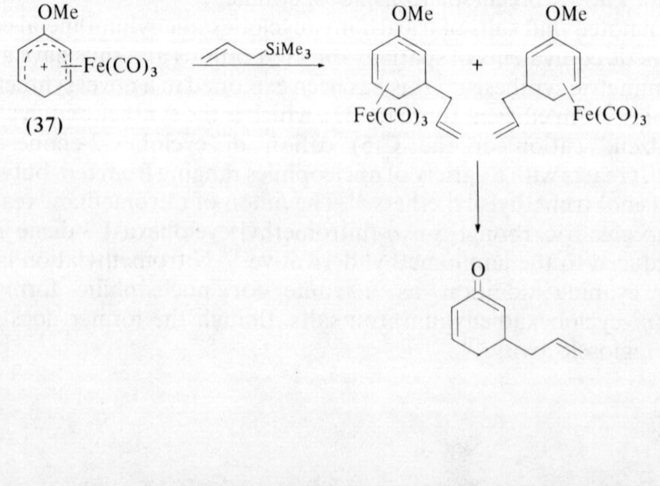

OMe

+ $Fe(CO)_3$

(37)

$SiMe_3$

OMe

$Fe(CO)_3$

+

OMe

$Fe(CO)_3$

O

The stereochemistry of addition of phosphines to iron-complexed (1-5-η-cycloheptadienylium) cations depends on the reaction conditions and solvent used; in dichloromethane, direct addition leads to 5-*exo*-substituted product, whereas in acetonitrile a red intermediate is observed and the 5-*endo*-isomer which is formed probably arises from a metal-assisted pathway.[41] Ligand substitution and nucleophilic reactivity of tricarbonyl(1-5-η-cyclooctadienylium)iron complexes has also been studied.[42]

Iron-stabilized carboxonium salts like $(C_5H_5)Fe(CO)_2[\overset{+}{C}(OEt)R]BF_4^-$ are readily prepared by reaction of $[(C_5H_5)Fe(CO)_2(\text{isobutene})]^+$ BF_4^- with monoalkylated acetylenes in $CH_2Cl_2/EtOH$ solutions. The acetylene complexes thus formed react with alcohols to give metal-stabilized carboxonium ion complexes which can be hydrolysed to acyliron complexes.[43] The synthesis and reactions of a series of cyclic and heterocyclic derivatives of dicarbonyl(cyclopentadienyl (3-ethoxy- and 3-dimethylamino-prop-2-enylidene)iron tetrafluoroborates have been reported. For example, (**38**) has been prepared and is readily converted into (**39**).[44]

Fe(CO)$_2$Cp; $Et_3O^+BF_4^-$; EtO; Fe(CO)$_2$Cp; Me_2NSiMe_3, $CH_2Cl_2/20^\circ$; Me_2N; Fe(CO)$_2$Cp

(**38**) (39)

The complexes $[(\text{benzene})_2M]^{2+}$ (M = Fe, Ru, Os) react rapidly and reversibly with tertiary phosphines to form cyclohexadienylium–phosphonium adducts $[(C_6H_6PR_3)(C_6H_6)M]^{2+}$; there is no evidence for addition to both benzene rings. Electrophilic reactivity for adduct formation is very metal-dependent (k_{rel} Fe:Ru:Os = 390:7:1), whereas the reverse process is only weakly metal-dependent and the reaction is thought to involve direct bimolecular attack on the arene with significant C—P bond formation in the transition state.[45]

The conversion of diferrocenylphenylmethanol into its carbocation is general-acid-catalysed in $H_2O:CH_3CN$. Comparison of this reaction with the conversion of tropyl alcohol into the tropylium ion reveals that the two processes are similar but have opposite solvent kinetic hydrogen isotope effects. It is suggested that cation formation transforms from a stepwise (protonation of OH then ionization) to a concerted process as cation stability increases.[46] Ion-pairing and initial-state conformation both appear to be important in determining the ease of proton elimination by N and O bases from 1-ferrocenylalkyl cations $Fc\overset{+}{C}(R^1)CH_2R^2$.[47]

The Ritter reaction can be greatly improved through the use of carbocationic intermediates stabilized by transition-metal residues like $Cr(CO)_3$, though excessive stabilization can inhibit reactivity. In these complexed systems, the reaction takes place with total stereochemical control.[48] The order of reactivity of various nucleophiles in water towards $[Mn(CO)_3(\eta\text{-}C_6H_6)]^+PF_6^-$ parallels that previously observed for addition to free carbocations.[49]

The stable cationic rhodium complex (**41a**) has been prepared and spectroscopic studies indicate that the bicyclic ligand is bound to the metal by η^2-ethylene and η^3-allyl bonds. There is little charge delocalization into phenyl or ferrocenyl substituents when these are attached (**41c**) to the exocyclic carbon atom.[50] Ionization of alcohols like (**40b**) and hydrolysis of cations (**41**) proceeds stereoselectively, the nucleophile attacking the cationic centre in the latter from the face opposite to that adjacent to the metal.[51]

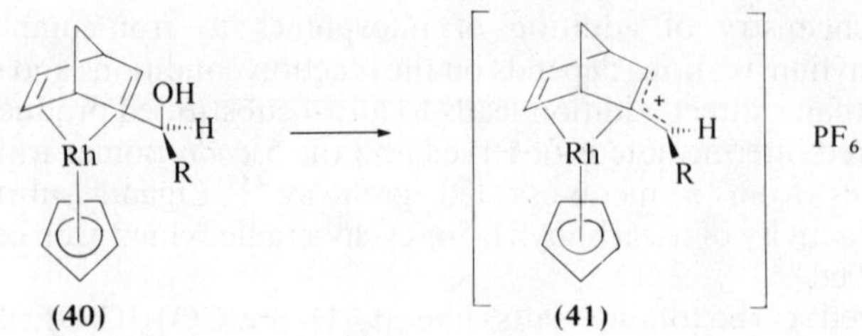

(40)

a: R = H
b: R = Me

(41)

a: R = H
b: R = Me
c: R = Ph or Fc

Stable Carbocations: their Structures and Reactions

A comparison of the effects of varying electron demand on the ^{13}C chemical shifts of the carbocation centres in aryl-substituted Coates' cation and substituted 2-adamantyl cations, relative to the corresponding 1-arylcyclopentyl cations, reveals that there is an inverse dependence of δ^{C^+} between classical and non-classical systems as a function of electron demand.[52] There is excellent correlation between substituent constants σ^{C^+} and $\Delta\delta^{C^+}$ ($\Delta\delta^{C^+} = \delta^{C^+}_{Ar} - \delta^{C^+}_{Ph}$) for a series of *meta*- and *para*-substituted 2-phenylbut-2-yl and 4-phenylhept-4-yl cations,[53] 3-phenylpent-3-yl and 2-phenyladamant-2-yl cations (where correlation with σ^+ is only fair),[54] and 1-phenylcyclohex-2-yl and 1-phenylcyclohept-1-yl cations.[55] From the latter work, it is suggested that variations observed in ρ^{C^+} for different ring sizes reflect differences in hyperconjugative and inductive contributions in the individual rings whereas changes in ρ^+ also include ground-state energy differences.[55] Linear correlations are also observed for a series of *meta*- and *para*-substituted 9-phenyl-*exo*- and *endo*-5,6-trimethylenenorborn-9-yl carbocations (**42**) and (**43**). In the latter, the chemical shift for C(7) is shifted upfield as Ar becomes more electron-withdrawing in contrast to zero or downfield shifts observed for the other carbons of the norbornyl skeleton; no such effect is observed in the spectra of the *exo*-isomers (**42**).[56] The ^{13}C chemical shifts of the methyl groups in *meta*- and *para*-substituted *tert*-cumyl cations do not correlate well with either σ^+ or σ^{C^+}. Accordingly, a new scale of substituent constants $\sigma(\alpha\text{-C}^+)$ has been established which it is hoped may be used to probe σ-bridging in arylnorbornyl cations.[57] From the variation of $^1J_{CH}$ for groups adjacent to the cationic centre in a series of benzylic carbocations, it has been found that the parameter ΔJ [$= {}^1J_{CH}$ (in the cation) $- {}^1J_{CH}$ (in a neutral model ketone)] varies linearly with the electron demand of the cationic centre as measured by σ^+ and $\sigma(\alpha\text{-C}^+)$.[58] The relationship between ^{11}B chemical shifts of trigonal boranes and $\delta^{^{13}C^+}$ of carbenium ions is more complex than previously thought; in the case of ferrocenyl derivatives, markedly different trends are observed, which reflect a fulvene-like structure for the substituted cyclopentadienyl ring in the carbocation.[59]

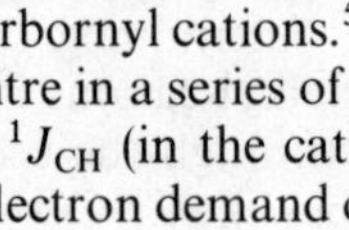

(42) 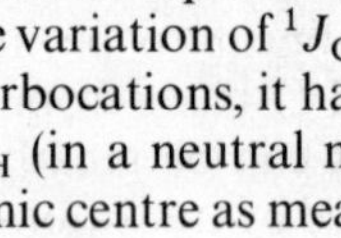 **(43)**

Reactions of alkylcarbenium ions in relation to isomerization and cracking of hydrocarbons have been reviewed.[60] A variable-temperature ^{13}C-NMR spectroscopic study of the *sec*-butyl cation in the solid state, generated by reaction of allyl halide with SbF_5, reveals an NMR spectrum similar to that obtained in solution. At the temperature of ion formation (−85°), a ^{13}C label is scrambled over all carbons, but the rate of this process is too slow for coalescence effects to be observed up to −60°; there is no evidence for a static $Me\overset{+}{C}HEt$ structure down to −190°.[61] The equilibrium isotope effect has been measured for equilibration of the 2,3-dimethylbut-2-yl cations (**44**) and varies from 1.0114 at −135° to 1.0197 at −62° in favour of the structure (**44b**) in which positive charge resides on the ^{13}C-labelled carbon.[62]

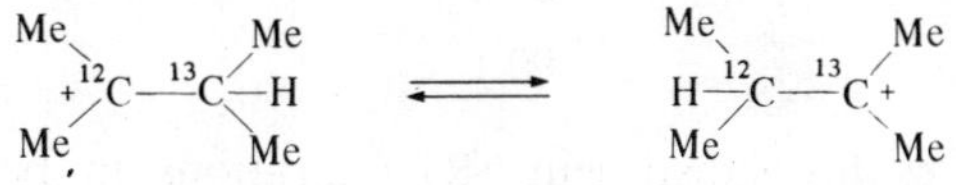

(**44a**) ⇌ (**44b**)

Full details have now appeared of the evidence supporting μ-hydrido bridging in 1,5- and 1,6-dimethylcyclododecyl cations; Saunders' "equilibrium deuterium isotope effect" criterion points to a resonance system.[63] Whilst these di-tertiary cations appear to exist in a symmetrical single-minimum potential energy well, there is evidence for an unsymmetrical cation μ-hydrido-bridged between a secondary and a tertiary site. Slight variations in the electron demand of the tertiary centre have a large effect on the structure of the hydrido bridge, and this study provides a model for a hydride-transfer process and suggests that there is a virtually flat potential for approach of a CH_2 group to a tertiary carbocation centre.[64]

Ionization of (**45a**) is immediately followed by ring-contraction to a primary cation, the ^{1}H- and ^{13}C-NMR spectra of which indicate unambiguously a bisected primary cyclopropylcarbinyl structure (**46**). In contrast, (**45b**) ionizes to (**47**).[65] The spiro[2.5]oct-4-yl cation (**48**) has been prepared; there is significant charge delocalization into the cyclopropyl ring.[66]

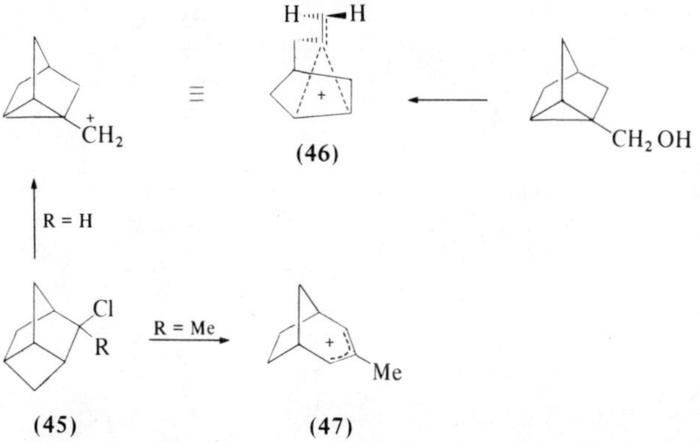

(**46**)

(**45**) (**47**)

a: R = H
b: R = Me

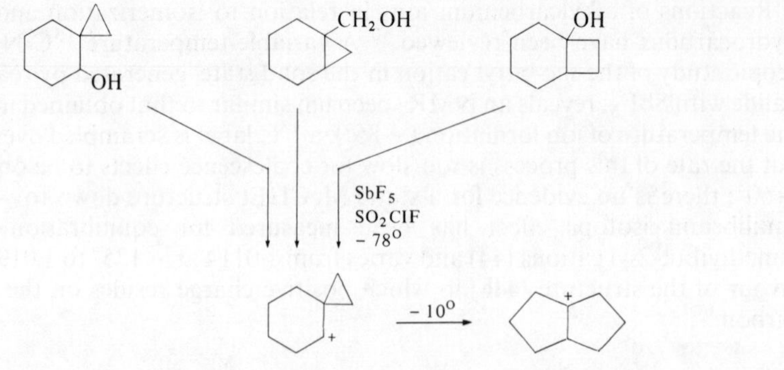

SbF$_5$
SO$_2$ClF
$-78°$
$-10°$
(48)

Activation energies for substituent (R) migrations in benzenium (**49**) and analogous naphthenium, pyrenium, and phenanthrenium ions have been calculated from bond energy, proton affinity, and ^{13}C-NMR data, and agree well with experiment.[67] Degenerate rearrangement of the tetramethylnaphthalenium ion (**50**) proceeds by successive 1,2-hydride shifts from C(1) to C(4).[68] Kinetic data for the degenerate isomerization of (**51**)[69] and (**52**)[70] have also been reported. An activation energy of 21.5 kcal mol^{-1} has been determined for rearrangement of (**53**) into (**54**): relative migrating abilities of R = Cl:Me:Et:Ph of 680:1:52:20 are found for (**53**).[71] From a study of the protonation of 2,4-di-, 3,4-di-, and 2,4,5-tri-alkylphenols and anisoles (R = Me or Et) in FSO$_3$H at $-50°$, it has been observed that

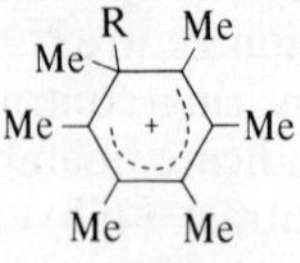

(49)
R = alkyl, aryl, Cl, Br

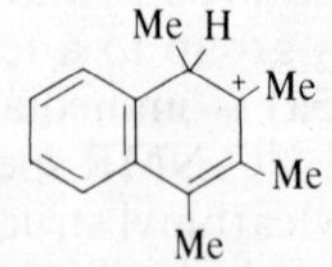

(50)

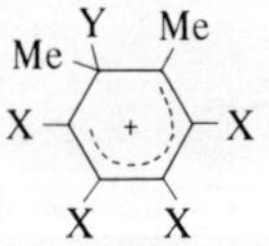

(51)
X = F, Cl
Y = Cl

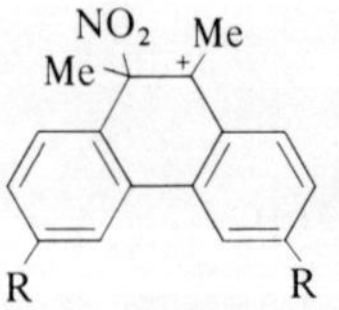

(52)
R = H, Me

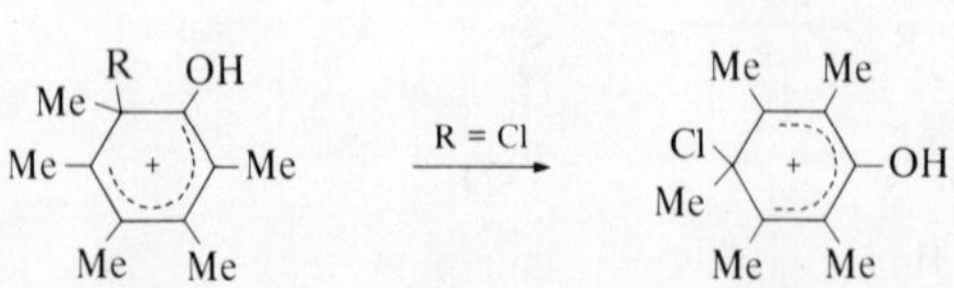

R = Cl

(53)

(54)

protonation *ipso* to a C(4)-ethyl substituent occurs much more readily than for the methyl analogue.[72] 1-Naphthol is protonated exclusively at C(4) but introduction of a 4-methyl substituent favours C(2) protonation; in contrast, protonation of 3,4-dimethyl-1-naphthol occurs exclusively at C(4). From these and related observations, it is suggested that steric strain is the predominant, but not exclusive, factor which determines the protonation site of these compounds.[73]

The activation energy for rotation about the aryl-oxygen bond in a series of (*p*-anisyl)$\overset{+}{C}$(R′)R cations has been used to estimate the relative electron-releasing ability of the substituents R, the sequence R = cyclopropyl > phenyl > methyl being observed for both secondary and tertiary cations.[74] A matrix photo-ionization study of benzyl bromide reveals photo-induced isomerization between the benzyl cation and the tropylium ion and is consistent with an equilibration between these two species at low internal energies in gas-phase studies.[75] The stabilization of benzylic cations through intramolecular participation by an *ortho*-nitro group has been reviewed.[76] 9-Ethylanthracene is oxidized to a dication in SbF_5/SO_2ClF at $-30°$ and this latter species is converted by subsequent addition of FSO_3H/SbF_5 into the 1-(9-anthracenyl)ethyl cation; the octamethylnaphthalene dication is not converted into a benzylic analogue under such conditions. Neither process could be reversed. Thus, whilst for secondary benzylic cations the monocation $\rightleftarrows$ dication equilibrium lies completely to the left, for primary benzylic cations it is shifted to the right but a high activation barrier prevents protonation of the monocation.[77] The bis(biphenylyl)carbenium ion has been prepared by reaction of bis(biphenylyl)chloromethane with $SbCl_5$ in CCl_4; it is stable and reacts as an electrophile or as a hydride-acceptor.[78]

Activation barriers for internal rotation in $(4\text{-}R_2NC_6H_4)_2\overset{+}{C}H$ (R = Me, Et) and $(4\text{-}Me_2NC_6H_4)_2\overset{+}{C}(C_6H_4\text{-}4\text{-}R')$ [R′ = MeO, $Ph_2P(O)$, Ph_2P, and Ph_2CH] have been measured: in the former $\Delta G^{\neq}_{rot}$ is *ca* 11 kcal mol^{-1}, whereas in the latter the C_6H_4R' ring exhibits essentially free rotation.[79] The reactions of the tris(2,4-dimethoxyphenyl)methyl cation,[80] of phenol red dimethyl ether (which is an intramolecular ion-pair)[81], and of the formally dinegatively charged electrophile bromophenol blue[81] with various nucleophiles in aqueous solution have been investigated. Medium effects on the kinetics of interaction of such amphoteric dyes with hydroxyl ion have also been studied.[82] Cationic micelles enhance and anionic micelles inhibit the reaction of setoglaucin $[(4\text{-}Me_2NC_6H_4)_2\overset{+}{C}(C_6H_4\text{-}2\text{-}Cl)Cl^-]$ with cyanide ion.[83] The effects of solvent dielectric constant and of ionic strength on the kinetics of the reaction of aminotriarylmethyl cations and their derivatives with hydroxide ion have been investigated.[84] Rate constants and activation energies for this process at various temperatures have been reported.[85] The Ritchie equation for rates of reaction of these organic cations with nucleophile–solvent systems does not well describe such processes with hydroxide and methoxide ion[86] and has been improved by inclusion of a parameter characteristic of the electrophilic reagent.[87,88] The kinetics for formation of the corresponding triarylmethanols in acetonitrile–water mixtures have been rationalized in terms of specific solvation and screening of the reaction centre.[89] The ionization ratios of four arylcarbinols in aqueous perchloric acid solutions have been measured and the concept of excess acidity has been used in determinations of pK_{R^+} values.[90]

It has been shown that $^3H/^1H$ discrimination cannot be used to identify the rate-determining step for the reduction of the cyclopropenium ion by sodium borohydride.[91] A broad maximum is observed for kinetic deuterium isotope effects for hydride transfer from triphenylmethane and 4,4′-dimethoxytriphenylmethane to a series of 9-arylfluoren-9-yl cations, the peak value occurring when the difference in pK_{R^+} between reactants and products is *ca.* zero; it is not clear whether this effect arises from quantum-mechanical tunnelling or simply from differences in transition-state zero-point energies.[92] It is found that second-order rate constants for one-electron reduction of hexa- and hepta-methyltropylium perchlorates are unexpectedly high, consistent with the conclusion that these ions are highly congested.[93] The ^{13}C-NMR spectra for all 17 methylated tropylium ions have been reported; chemical shifts of the ring carbons correlate linearly with charge densities calculated by the simple Hückel MO method. Deviations were noted for the penta-, hexa-, and hepta-methyl cations, reflecting out-of-plane distortion of the ring to relieve crowding.[94]

The BF_4^- salt of the 1,4-dihydro-1,4-ethenobenzotropylium cation (**55**) has been prepared; it is very stable (though pK_{R^+} measurements suggest that it is less so than tropylium itself) and there are intense charge-transfer bands in its UV–visible spectrum.[95] There is also evidence for charge-transfer interactions in (**56**), (**57**), and (**58**).[96] Whilst protonation of 2,3-benzo-6,7-homotropone in FSO_3H/SbF_5 leads to (**59**) in which there is complete delocalization, the ion (**60**) derived from *trans*-2,3-benzo-4,5:6,7-bishomotropone is best described as delocalized but non-aromatic.[97] Aromaticity in conjugated ions has been reviewed.[98]

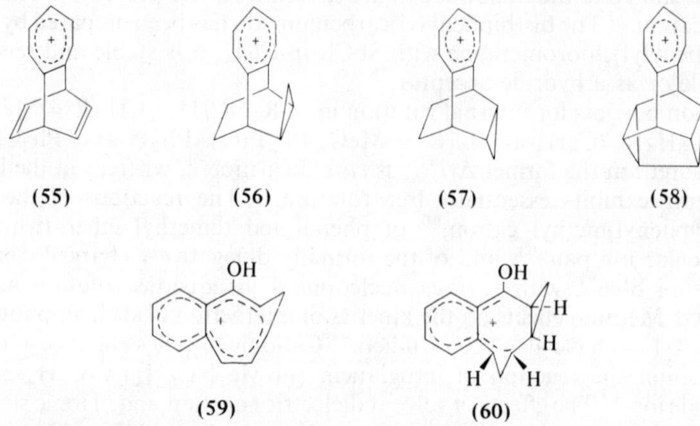

NMR spectra of a series of simple alkoxycarbenium ions have been recorded.[99] 1,3-Hydride shifts in methoxy-, ethoxy-, and propoxy-methyl cations have also been observed.[100,101] The rate-determining step in the formation of 1,3-dioxolenium cations by acid-catalysed decomposition of 2-alkoxy-1,3-dioxolanes is heterolysis of the exocyclic C—O bond.[102] Full details have now appeared of the preparation of dication ether salts from ketones.[103] Equilibrium constants have been measured for methyl transfer between alkoxycarbenium ions and ketones; it is found that carbon basicity measured by this method parallels proton basicity for the ketones under study.[104] Benzilic acid and diphenic acid form dipositive cations in superacid.[105] An X-ray diffraction analysis of tris(trichlorophosphazeno)carbenium hexachloroantimonate reveals that the cation has C_{3h} symmetry and that the C, N, P, and Cl (one per PCl_3 group) atoms lie approximately in a plane.[106] The cation (**61**) does not behave as a bridged dithia[13]annulene cation but the charge remains largely localized on the carbon flanked by the two thiophene rings.[107] An X-ray diffraction study of the BF_4^- salt of the 1*H*-thiophenium cation (**62**) shows that the bonding around sulphur is pyramidal and not planar in accord with theoretical predictions.[108] It has been suggested that multihapto bonding occurs between P^+ and the cyclopentadienyl unit in (**63**).[109] CNDO/2 calculations of the electronic spectrum provide evidence to support the formation of the carbocation of *meso*-methyleneporphyrin (**64**).[110] Whilst nitrenium ions are probably formed in the conversion of *N*-substituted hydroxylamines into Schiff bases with PCl_5, it has not proved possible to trap these intermediates by reaction with the solvent.[111]

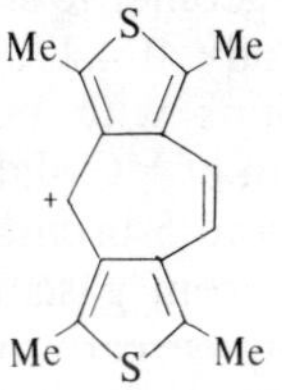

(**61**)

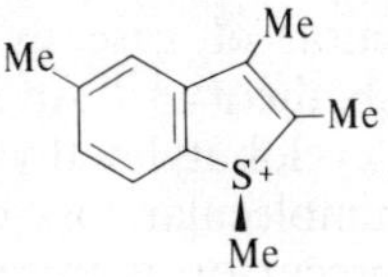

(**62**)

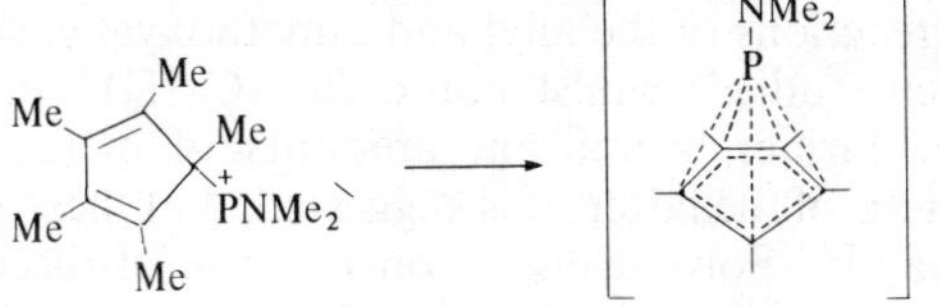

(**63**)

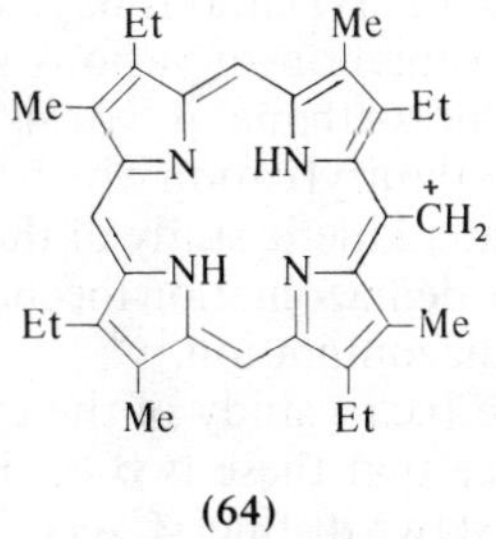

(**64**)

The formation of methyl-substituted acylium ions (in FSO_3H) and dialkoxycarbenium ions (in H_2SO_4) is found to be more exothermic than that of the corresponding phenyl-substituted ions, in contrast to the results expected simply from resonance effects.[112] 1,2-Migration of the axial substituent (either carbonyl or methyl) is observed in products derived from reaction of stereoisomers of 4-*tert*-butyl-1-methylcyclohexanecarbonyl cation with propyne.[113] Resonance Raman spectra of Wurster's cation in the solid state and solution have been reported.[114] Reviews have appeared concerned with charge distributions (NMR) in fluorine-containing carbocations,[115] the effect of structural factors and solvation effects on carbocation reactivity,[116] the isomerizations of carbocations formed in the presence of acid catalysts,[117] carbocation rearrangements[118–120] and the use of the Hammond postulate in the interpretation of their mechanisms,[121] and stable cationic intermediates as models for the study of electrophilic addition mechanisms.[122]

Other Reactions

The absolute heats of formation of simple alkyl cations in the gas phase have been measured.[123] The main product of the reaction of $\overset{+}{C}T_3$ with RCOOEt (R = H, Me, Et) in the gas phase is RCO_2CT_3 together with small amounts of $RCOCT_3$ CT_3COOEt, and CT_3OEt.[124] The electronic and geometric structure of $(C_2H_6)^+$ has been investigated by an ESR spectroscopic study at 4.2 K.[125] *O*-Alkylation is favoured in the reaction of gaseous ethyl cations with pentane-2,4-dione in accordance with h.s.a.b. theory.[126] An investigation of $^{13}C_2$-labelled isotopomeric bromocyclopentanes, cyclobutylmethyl bromide, and 5-bromopent-1-enes clearly demonstrates that unimolecular loss of ethylene from gaseous $(M - Br)^+$ ions generated from these precursors is preceded by complete carbon scrambling; a non-classical pyramidal ion is proposed as the intermediate for this process.[127] Direct evidence has been obtained from a radiolytic study for the existence of the cyclohexyl cation in the dilute gas phase; it has a lifetime $> 10^{-7}$ sec and readily rearranges into the methylcyclopentyl cation.[128]

The ion-molecule reactions of the allyl and 2-methoxyallyl cations with various enol ethers have been studied; whilst non-cyclic $(C_5H_5)^+$ products are formed preferentially by the former, significant amounts of methoxycyclopentadienyl cations arise in reactions of the latter; it is suggested, that these result from a 2 + 3-cycloaddition pathway.[129] Solvent effects on the rate of racemization of (−)-4-chlorobenzhydryl chloride,[130] and salt and solvent isotope effects on the rate of hydrolysis of *p*-methoxybenzal chloride and diphenyldichloromethane have been investigated.[131]

The formation and structure of aryl cations have been reviewed.[132] The effect of pressure on the rate of dediazoniation of various substituted benzenediazonium tetrafluoroborates in 1,2-dichloroethane is consistent with a rate-determining unimolecular fragmentation to the aryl cation which then rapidly reacts with solvent to form chloroarenes.[133] From a kinetic study of the effect of crown ethers on this process, it is concluded that the dediazoniation mechanism for the complexed cation is different from that of free diazonium ion.[134]

An ion-cyclotron-resonance (i.c.r.) study of the cyclopropenium and propargyl cations provides clear evidence that these two $(C_3H_3)^+$ ions have quite different chemical reactivities.[135] At least two distinct $(C_4H_4)^+$ structures are produced in the unimolecular fragmentation of $(C_6H_6)^+$ and $(C_5H_5N)^+$ and one of these has the but-3-en-1-yne structure.[136] An i.c.r. study shows that the methoxymethyl cation attacks 3-methoxyprop-1-ene and 3-(methylthio)prop-1-ene at the π-bond rather than at the nucleophilic hetero-atom. Dissociation of the ionic intermediate follows, in which fission of the methylene-oxygen bond of the reactant ion occurs.[137] Gaseous thiosulphonium ions thiolate neutral sulphides and 2-methoxypropene in a manner analogous to that observed in condensed phases.[138] Gaseous $CH_2{=}CHCH_2\overset{+}{S}iMe_2$ reacts with alcohols to give adducts which decompose by loss of C_3H_6 to give $Me_2\overset{+}{S}iOR$; a similar fragmentation occurs in reactions with ethers.[139]

$(Me_2SiT)^+$ has been generated by β-decay of Me_2SiT_2.[140] A comparison of the solvolytic reactivities of α-silicon- and α-alkynyl-stabilized vinyl cations, generated from triflates in aqueous ethanol, shows that the Me_3Si group has an accelerating effect and is stabilizing relative to H (but destabilizing relative to a *tert*-butyl substituent), consistent with unexpected α-C^+ stabilization by the silicon atom, whereas the α-ethynyl group is not (relative to a methyl group); it is suggested that this reflects the very great electron demand of the vinyl cation.[141] Tertiary alcohols with a γ-silyl group (**65**) generally undergo a simple carbocation rearrangement involving either a hydride or an alkyl(aryl) group migration, though in a cyclic system like (**66**) structural constraints lead only, as expected, to a hydride shift; such silicon-controlled carbocation rearrangements appear to be cleaner than the corresponding pinacol rearrangements.[142] *Ab initio* calculations on the reactions of trimethylsilyl cations with ambident nucleophiles like esters and thio-esters suggest that adducts arising from attack at the acyl hetero-atom are somewhat more stable than those from attack at the ethereal hetero-atom.[143]

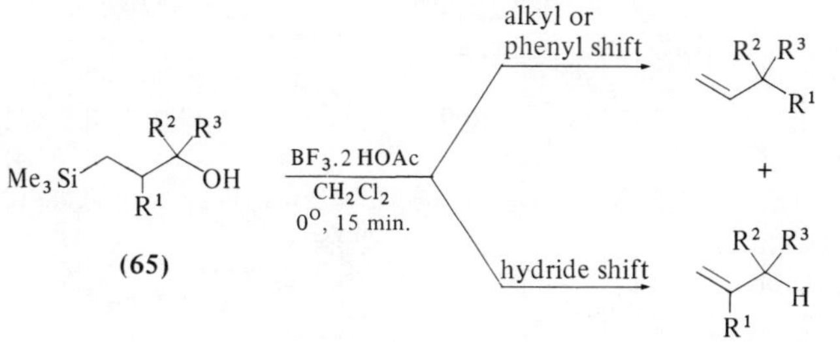

The hydrolysis of methyl vinyl selenides proceeds by a partially reversible initial slow protonation step to form a selenium-stabilized carbenium ion which is then captured by solvent affording an intermediate hemiselenoacetal; accumulation of this latter species and the relative stability of the cation are responsible for the partial reversibility of the first step.[144]

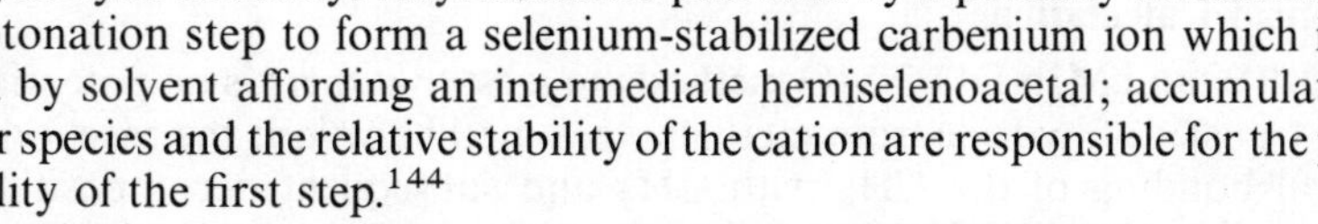

Carbocationic intermediates have been implicated in the reaction of triisobutylaluminium with 1,3,3-trimethylcyclopropene. It is suggested that attack by aluminium leads to a species with sufficient carbocationic character that electrocyclic ring-opening of this "cyclopropyl cation" readily occurs.[145]

A study of the kinetic energy release accompanying bromine atom loss from dissociative ionization of E- and Z-phenylvinyl bromides in the gas phase suggests that, even if these form a 2-phenylvinyl cation, such a species immediately isomerizes without activation energy to a 1-phenylvinyl cation or to the bridged ion (**67**).[146] The gas-phase reactions of the vinyl cation precursors (**68**)–(**71**) differ significantly from those observed under solvolysis conditions for the 2-phenylcyclobutenyl cation (**72**). This latter cation is relatively stable despite its lack of benzylic character, a result which points to C(2)—C(3) σ-bond stabilization as suggested by *ab initio* calculations.[147] The rates of acetolysis and bromide exchange for (**73**) and (**74**) are in accord with the reactivity–selectivity principle.[148] Degenerate rearrangements in triarylvinyl cations have been reviewed.[149]

Reviews have also appeared which are concerned with reactivity–selectivity in nucleophilic substitution reactions involving carbocations or ion pairs,[150] carbocations in ionic hydrogenation, hydrogenolysis, hydroalkylation, and alkylation of saturated hydrocarbons,[151] and the use of isotopes to investigate cationic reactions.[152]

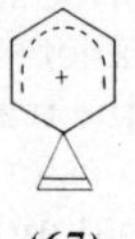

(**67**)

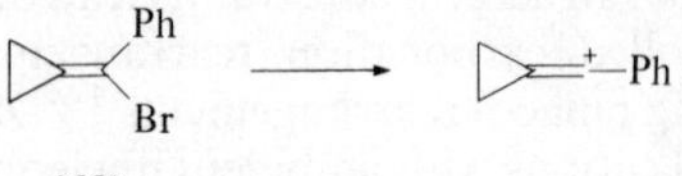

(**68**)

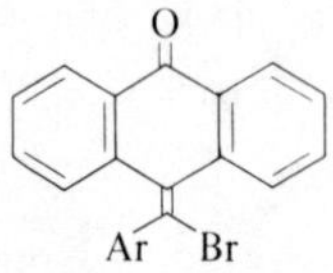

(**69**) (**72**)

(**73**)
Ar = anisyl, tolyl

Ph—≡—CH_2CH_2Br (**70**) ⟶ Ph—≡—$CH_2\overset{+}{C}H_2$

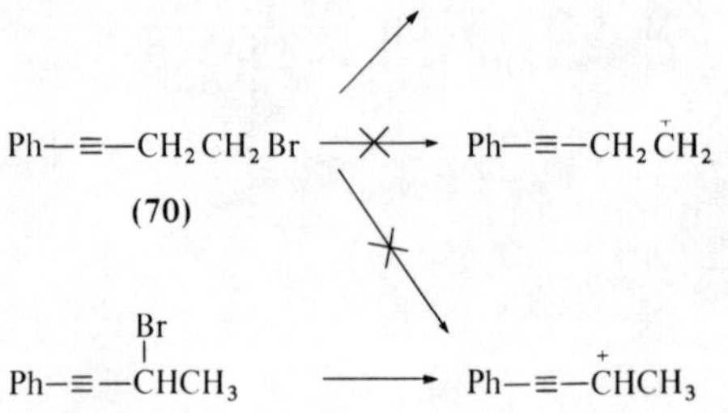

Ph—≡—CH(Br)CH_3 (**71**) ⟶ Ph—≡—$\overset{+}{C}HCH_3$

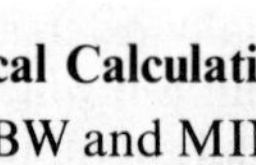

(**74**)
Ar = anisyl, phenyl
An = anisyl

Theoretical Calculations

CNDO/BW and MINDO/3 MO methods suggest that there is no potential barrier for the transfer of hydride from methane to a methyl cation; the reaction involves chemical bonding of the CH_4 with $\overset{+}{C}H_3$ and subsequent decomposition of the vibrationally excited $(C_2H_7)^+$ complex, the hydrogen atom serving as a conductor of electrons between the two.[153] An *ab initio* MO study (including electron correlation) of the structures and energetics of C_1 to C_3 carbocations leads to much better agreement with experimental data; bridged structures are predicted for the vinyl cation and $(C_2H_5)^+$, and protonation of methane and ethane is predicted to afford structures involving a three-centre bond between a carbenium ion and H_2.[154] Dissociation of the ethylene cation into $(C_2H_3)^+$ and $(C_2H_2)^+$ has also been investigated.[155] Hartree–Fock calculations indicate that $(HCC)^+$ has a triplet ground state with another triplet state only 7.6 kcal mol^{-1} higher in energy.[156] Elimination of hydrogen from the butyl cation proceeds through a tight pyramidal activated complex (structurally analogous to $\overset{+}{C}H_5$) which is associated with essentially quantitative dispersal of the reverse activation energy into translation.[157] The relative energies of localized and bridged structures for carbocations arising from protonation of alkenes of increasing chain-length and degree of branching have been calculated, and the influence of nucleophilic solvents on such structures has been simulated.[158]

Ab initio calculations have been carried out on small-ring boron–carbon compounds and their isoelectronic carbocations.[159] Magnetic susceptibility, proton magnetic shielding, and electron charge distribution in the cyclopropenium cation have been calculated and the results have been discussed in terms of the ring-current model.[160] The most stable cyclobutadiene dication structure is predicted to be non-planar (the planar form is 7.5 kcal mol^{-1} less stable) with a perpendicular methylenecyclopropene dication being somewhat less stable and the linear butatriene dication of still higher energy.[161] Homoaromatic stabilization of the cyclobut-3-enyl cation is apparently reinforced by σ non-classical effects.[162] A graph theoretical analysis of the rearrangement of the homotetrahedryl cation $(C_5H_5)^+$ has been reported.[163]

Calculations of spectroscopic parameters and the reactivity of arenium ions have been reviewed.[164] A theoretical study of the reaction of benzyl fluoride and *p*-methoxybenzyl fluoride with BF_3 suggests that stabilization by the methoxy group is primarily in the formation of the ion pair $(ArCH_2)^+ (BF_4)^-$ rather than in its dissociation, whereas solvent effects are important in the second step.[165]

Calculations of the stabilization energies of three α-cyanocarbenium ions relative to their parent cations suggest that the cyano group stabilizes a methyl cation but not the ethyl or propyl homologues.[166] Isocyano stabilization of carbenium ions is also predicted to be significant for primary species but the effect decreases with increasing cation size.[167]

It has been shown that the energy barriers for 1,2-hydride shifts in a range of carbenium ions are significantly influenced by the energy value and degree of localization of the LUMO; transition-state structures are in line with the Hammond postulate.[168] *Ab initio* MO theory has been used to study the intramolecular rearrangement of :CHMe to $CH_2{=}CH_2$ and $H_2\overset{+}{O}CH$: to $HO\overset{+}{C}H_2$.[169]

The electronic structures of immonium ions[170] and aryl and *N*-acetyl-*N*-arylnitrenium ions[171] have been investigated. A comparison of the stabilities of $H_3Si\overset{+}{C}H_2$ and $CH_3\overset{+}{S}iH_2$ suggests that the latter has significantly lower energy.[172] Neighbouring P or Cl atoms are much less efficient at stabilizing silicenium ions in comparison with their first-row counterparts, whereas S and O are about equally effective.[173]

The relative stabilities of various carbocations have been discussed in terms of charge distribution and hyperconjugation as compared to *s*- and *p*-orbital characteristics.[174] Quantum-chemical methods for evaluating carbon 1*s* level splittings and NMR parameters for cations,[175] and theoretical calculations and chemical reactivity[176] have been reviewed.

References

1 Barkhash, V. A., *Izv. Sib. Otd. Akad. Nauk SSSR, Ser. Khim. Nauk*, **1980,** 138; *Chem. Abs.*, **93,** 237978 (1980).

2 Lajunen, M., *Kem.-Kemi*, **7,** 643 (1980); *Chem. Abs.*, **94,** 173730 (1981).

3 Grob, C. A., Hanreich, R., and Waldner, A., *Tetrahedron Lett.*, **22,** 3231 (1981).

4 Cheremisin, A. A., and Schastnev, P. V., *Org. Magn. Reson.*, **14,** 327 (1980).

[5] Farçasiu, D., *J. Org. Chem.*, **46,** 223 (1981).
[6] Schleyer, P. von R., and Chandrasekhar, J., *J. Org. Chem.*, **46,** 225 (1981).
[7] Servis, K. L., and Shue, F.-F., *J. Am. Chem. Soc.*, **102,** 7233 (1980).
[8] Coxon, J. M., and Steel, P. J., *Aust. J. Chem.*, **33,** 2455 (1980).
[9] Creary, X., *J. Am. Chem. Soc.*, **103,** 2463 (1981).
[10] Lajunen, M., and Lyytikainen, H., *Acta Chem. Scand.*, **35A,** 139 (1981).
[11] Le Noble, W. J., Merbach, A. E., and Schulman, E. M., *J. Org. Chem.*, **46,** 3352 (1981).
[12] Grob, C. A., and Waldner, A., *Tetrahedron Lett.*, **22,** 3235 (1981).
[13] Schipper, P., de Haan, J. W., and Buck, H. M., *J. Chem. Soc., Perkin Trans. 2*, **1981,** 457.
[14] Birnberg, G. H., and Paquette, L. A., *J. Org. Chem.*, **45,** 5379 (1980).
[15] Ahlberg, P., Engdahl, C., and Jonsäll, G., *J. Am. Chem. Soc.*, **103,** 1583 (1981).
[16] Goldstein, M. J., Dinnocenzo, J. P., Ahlberg, P., Engdahl, C., Paquette, L. A., and Olah, G. A., *J. Org. Chem.*, **46,** 3751 (1981).
[17] Goldstein, M. J., Tomoda, S., Pressman, E. J., and Dodd, J. A., *J. Am. Chem. Soc.*, **103,** 6530 (1981).
[18] Fujise, Y., Yashima, H., Sato, T., and Ito, S., *Tetrahedron Lett.*, **22,** 1407 (1981).
[19] Sunko, D. E., *Croat. Chem. Acta*, **53,** 525 (1981); *Chem. Abs.*, **95,** 79789 (1981).
[20] Gassman, P. G., Saito, K., and Talley, J. J., *J. Am. Chem. Soc.*, **102,** 7613 (1980).
[21] Nordlander, J. E., and Haky, J. E., *J. Am. Chem. Soc.*, **103,** 1518 (1981).
[22] Mehta, G., and Srikrishna, A., *Indian J. Chem.*, **19B,** 997 (1980); *Chem. Abs.*, **94,** 208138 (1981).
[23] Hanack, M., and Holweger, W., *J. Chem. Soc., Chem. Commun.*, **1981,** 713.
[24] Poulter, C. D., Wiggins, P. L., and Le, A. T., *J. Am. Chem. Soc.*, **103,** 3926 (1981).
[25] Marsh, E. A., Gurria, G. M., and Poulter, C. D., *J. Am. Chem. Soc.*, **103,** 3927 (1981).
[26] Johnson, W. S., Ward, C. E., Boots, S. G., Gravestock, M. B., Markezich, R. L., McCarry, B. E., Okorie, D. A., and Parry, R. J., *J. Am. Chem. Soc.*, **103,** 88 (1981).
[27] Sokolov, V. I., *Izv. Sib. Otd. Akad. Nauk SSSR, Ser. Khim. Nauk*, **1980,** 63; *Chem. Abs.*, **93,** 237971 (1980).
[28] Pauson, P. L., *J. Organomet. Chem.*, **200,** 207 (1980).
[29] Kane-Maguire, L. A. P., Odiaka, T. I., and Williams, P. A., *J. Chem. Soc., Dalton Trans.*, **1981,** 200.
[30] Odiaka, T. I., and Kane-Maguire, L. A. P., *J. Chem. Soc., Dalton Trans.*, **1981,** 1162.
[31] Birch, A. J., and Stephenson, G. R., *J. Organomet. Chem.*, **218,** 91 (1981).
[32] Birch, A. J., Bogsanyi, D., and Kelly, L. F., *J. Organomet. Chem.*, **214,** C39 (1981).
[33] Effenberger, F., and Keil, M., *Tetrahedron Lett.*, **22,** 2151 (1981).
[34] Pearson, A. J., *Tetrahedron Lett.*, **22,** 4033 (1981).
[35] Birch, A. J., and Stephenson, G. R., *Tetrahedron Lett.*, **22,** 779 (1981).
[36] Mincione, E., Pearson, A. J., Bovicelli, P., Chandler, M., and Heywood, G. C., *Tetrahedron Lett.*, **22,** 2929 (1981).
[37] Birch, A. J., Kelly, L. F., and Thompson, D. J., *J. Chem. Soc., Perkin Trans. 1*, **1981,** 1006.
[38] Kelly, L. F., Dahler, P., Narula, A. S., and Birch, A. J., *Tetrahedron Lett.*, **22,** 1433 (1981).
[39] Johnson, B. F. G., Lewis, J., Parker, D. G., and Stephenson, G. R., *J. Organomet. Chem.*, **204,** 221 (1981).
[40] Pearson, A. J., and Chandler, M., *J. Organomet. Chem.*, **202,** 175 (1980).
[41] Brown, D. A., Glass, W. K., and Hussein, F. M., *J. Organomet. Chem.*, **218,** C15 (1981).
[42] Schiavon, G., and Paradisi, C., *J. Organomet. Chem.*, **210,** 247 (1981).
[43] Bates, D. J., Rosenblum, M., and Samuels, S. B., *J. Organomet. Chem.*, **209,** C55 (1981).
[44] Gompper, R., and Kottmair, E., *Tetrahedron Lett.*, **22,** 2865 (1981).
[45] Domaille, P. J., Ittel, S. D., Jesson, J. P., and Sweigart, D. A., *J. Organomet. Chem.*, **202,** 191 (1980).
[46] Bunton, C. A., Davoudzadeh, F., and Watts, W. E., *J. Am. Chem. Soc.*, **103,** 3855 (1981).
[47] Bunton, C. A., Carrasco, N., Davoudzadeh, F., and Watts, W. E., *J. Chem. Soc., Perkin Trans. 2*, **1981,** 924.
[48] Top, S., and Jaouen, G., *J. Org. Chem.*, **46,** 78 (1981).
[49] Evans, D. J., Kane-Maguire, L. A. P., and Sweigart, D. A., *J. Organomet. Chem.*, **215,** C27 (1981).
[50] Koridze, A. A., Chizhevsky, I. T., Petrovskii, P. V., Fedin, E. I., Kolobova, N. E., Vinogradova, L. E., Leites, L. A., Andrianov, V. G., and Struchkov, Yu. T., *J. Organomet. Chem.*, **206,** 373 (1981).
[51] Chizhevsky, I. T., Koridze, A. A., Bakhmutov, V. I., and Kolobova, N. E., *J. Organomet. Chem.*, **206,** 361 (1981).
[52] Farnum, D. G., and Clausen, T. P., *Tetrahedron Lett.*, **22,** 549 (1981).
[53] Brown, H. C., Periasamy, M., and Liu, K.-T., *J. Org. Chem.*, **46,** 1646 (1981).
[54] Kelly, D. P., Jenkins, M. J., and Mantello, R. A., *J. Org. Chem.*, **46,** 1650 (1981).
[55] Brown, H. C., and Periasamy, M., *J. Org. Chem.*, **46,** 3161 (1981).

[56] Brown, H. C., and Periasamy, M., *J. Org. Chem.*, **46,** 3166 (1981).
[57] Brown, H. C., Kelly, D. P., and Periasamy, M., *J. Org. Chem.*, **46,** 3170 (1981).
[58] Kelly, D. P., Farquharson, G. J., Giansiracusa, J. J., Jensen, W. A., Hügel, H. M., Porter, A. P., Rainbow, I. J., and Timewell, P. H., *J. Am. Chem. Soc.*, **103,** 3539 (1981).
[59] Wrackmeyer, B., *Z. Naturforsch.*, **35B,** 439 (1980); *Chem. Abs.*, **93,** 167026 (1980).
[60] Brouwer, D. M., *NATO Adv. Study Inst. Ser., Ser. E*, **39,** 137 (1980); *Chem. Abs.*, **94,** 29606 (1981).
[61] Myhre, P. C., and Yannoni, C. S., *J. Am. Chem. Soc.*, **103,** 230 (1981).
[62] Saunders, M., Kates, M. R., and Walker, G. E., *J. Am. Chem. Soc.*, **103,** 4623 (1981).
[63] Kirchen, R. P., Ranganayakulu, K., Rauk, A., Singh, B. P., and Sorensen, T. S., *J. Am. Chem. Soc.*, **103,** 588 (1981).
[64] Kirchen, R. P., Okazawa, N., Ranganayakulu, K., Rauk, A., and Sorensen, T. S., *J. Am. Chem. Soc.*, **103,** 597 (1981).
[65] Schmitz, L. R., and Sorensen, T. S., *Tetrahedron Lett.*, **22,** 1191 (1981).
[66] Olah, G. A., Fung, A. P., Rawdah, T. N., and Prakash, G. K. S., *J. Am. Chem. Soc.*, **103,** 4646 (1981).
[67] Borodkin, G. I., Koptyug, V. A., and Shubin, V. G., *Dokl. Akad. Nauk SSSR*, **255,** 587 (1980); *Chem. Abs.*, **94,** 120612 (1981).
[68] Morozov, S. V., Shakirov, M. M., and Shubin, V. G., *Zh. Org. Chim.*, **16,** 2103 (1980); *Chem. Abs.*, **94,** 83358 (1981).
[69] Loktev, V. F., Korchagina, D. V., and Shubin, V. G., *Izv. Sib. Otd. Akad. Nauk SSSR, Ser. Khim. Nauk*, **1981,** 112; *Chem. Abs.*, **95,** 23943 (1981).
[70] Loktev, V. F., Korchagina, D. V., and Shubin, V. G., *Izv. Sib. Otd. Akad. Nauk SSSR, Ser. Khim. Nauk*, **1980,** 86; *Chem. Abs.*, **94,** 191366 (1981).
[71] Berezina, R. N., Akimova, S. N., Korchagina, D. V., and Shubin, V. G., *Izv. Sib. Otd. Akad. Nauk SSSR, Ser. Khim. Nauk*, **1981,** 144; *Chem. Abs.*, **95,** 168203 (1981).
[72] Blackstock, S. M., Hartshorn, M. P., and Richards, K. E., *Aust. J. Chem.*, **33,** 2753 (1980).
[73] Blackstock, S. M., Hartshorn, M. P., and Richards, K. E., *Aust. J. Chem.*, **34,** 223 (1981).
[74] Jost, R., Sommer, J., Engdahl, C., and Ahlberg, P., *J. Am. Chem. Soc.*, **102,** 7663 (1980).
[75] Andrews, L., and Keelan, B. W., *J. Am. Chem. Soc.*, **103,** 99 (1981).
[76] Shabarov, Yu. S., Mochalov, S. S., and Daineko, V. I., *Izv. Sib. Otd. Akad. Nauk. SSSR, Ser. Khim. Nauk*, **1980,** 43; *Chem. Abs.*, **94,** 46256 (1981).
[77] Bodoev, N. V., Krysin, A. P., and Koptyug, V. A., *Zh. Org. Khim.*, **16,** 1002 (1980); *Chem. Abs.*, **93,** 185532 (1980).
[78] Volz, H., and Mayer, W. D., *Chem.-Ztg.*, **104,** 369 (1980); *Chem. Abs.*, **94,** 174057 (1981).
[79] Negrebetskii, V. V., Bychkov, N. N., and Stepanov, B. I., *Zh. Obshch. Khim.*, **50,** 2051 (1980); *Chem. Abs.*, **94,** 46655 (1981).
[80] Ritchie, C. D., Kamego, A. A., Virtanen, P. O. I., and Kubisty, C., *J. Org. Chem.*, **46,** 1957 (1981).
[81] Ritchie, C. D., and Hofelich, T. C., *J. Am. Chem. Aoc.*, **102,** 7039 (1980).
[82] Aleksandr, A. V., Sinev, V. V., and Ginzburg, O. F., *Zh. Org. Khim.*, **17,** 800 (1981); *Chem. Abs.*, **95,** 60943 (1981).
[83] Srivastava, S. K., and Katiyar, S. S., *Ber. Bunsenges. Phys. Chem.*, **84,** 1214 (1980); *Chem. Abs.*, **94,** 173892 (1981).
[84] Sinev, V. V., Aleksandr, A. V., and Pavlova, M. P., *Zh. Org. Khim.*, **16,** 1684 (1980); *Chem. Abs.*, **94,** 29963 (1981).
[85] Aleksandr, A. V., Sinev, V. V., Ginzburg, O. F., and Stepanova, T. F., *Zh. Org. Khim.*, **16,** 1688 (1980); *Chem. Abs.*, **94,** 29749 (1981).
[86] Sinev, V. V., and Smirnova, E. N., *Zh. Org. Khim.*, **16,** 1238 (1980); *Chem. Abs.*, **93,** 185547 (1980).
[87] Sinev, V. V., Kuznetsova, V. P., and Ginzburg, O. F., *Zh. Org. Khim.*, **17,** 978 (1981); *Chem. Abs.*, **95,** 114364 (1981).
[88] Sinev, V. V., Ginzburg, O. F., and Kuznetsova, V. P., *Zh. Org. Khim.*, **16,** 2435 (1980); *Chem. Abs.*, **94,** 64958 (1981).
[89] Sinev, V. V., Smirnova, E. N., and Ginzburg, O. F., *Zh. Org. Khim.*, **16,** 1234 (1980); *Chem. Abs.*, **93,** 185546 (1980).
[90] Dorion, F., and Gaboriaud, R., *J. Chim. Phys. Phys.-Chim. Biol.*, **77,** 1057 (1980); *Chem. Abs.*, **94,** 156093 (1981).
[91] Mata-Segreda, J. F., and Schowen, R. L., *J. Org. Chem.*, **46,** 644 (1981).
[92] Bethell, D., Hare, G. J., and Kearney, P. A., *J. Chem. Soc., Perkin Trans. 2*, **1981,** 684.
[93] Takeuchi, K., Kurosaki, T., Yokomichi, Y., Kimura, Y., Kubota, Y., Fujimoto, H., and Okamoto, K., *J. Chem. Soc., Perkin Trans. 2*, **1981,** 670.
[94] Takeuchi, K., Yokomichi, Y., Kubota, Y., and Okamoto, K., *Tetrahedron*, **36,** 2939 (1980).

[95] Nakazawa, T., Kubo, K., and Murata, I., *Angew. Chem. Int. Ed.*, **20,** 189 (1981).
[96] Nakazawa, T., Kubo, K., Okimoto, A., Segawa, J., and Murata, I., *Angew. Chem. Int. Ed.*, **20,** 813 (1981).
[97] El-Fayoumy, M. A. G., Bell, H. M., and Ogliaruso, M. A., *J. Org. Chem.*, **46,** 1603 (1981).
[98] Ilic, P., and Trinajstic, N., *Kem. Ind.*, **29,** 417 (1980); *Chem. Abs.*, **94,** 191049 (1981).
[99] Akhmatdinov, R. T., Kantor, E. A., Imashev, U. B., Yasman, Ya. B., Rakhmankulov, D. L., and Karakhanov, R. A., *Dokl. Akad. Nauk SSSR*, **252,** 622 (1980); *Chem. Abs.*, **93,** 220018 (1980).
[100] Akhmatdinov, R. T., Kantor, E. A., Karakhanov, R. A., and Rakhmankulov, D. L., *Zh. Org. Khim.*, **17,** 478 (1981); *Chem. Abs.*, **95,** 23942 (1981).
[101] Akhmatdinov, R. T., Kantor, E. A., Imashev, U. B., Yasman, Ya. B., and Rakhmankulov, D. L., *Zh. Org. Khim.*, **17,** 718 (1981); *Chem. Abs.*, **95,** 96639 (1981).
[102] Akhmatdinov, R. T., Chalova, O. B., Kantor, E. A., and Rakhmankulov, D. L., *Zh. Org. Khim.*, **16,** 962 (1980); *Chem. Abs.*, **93,** 185312 (1980).
[103] Stang, P. J., Maas, G., Smith, D. L., and McCloskey, J. A., *J. Am. Chem. Soc.*, **103,** 4837 (1981).
[104] Quirk, R. P., Gambill, C. R., and Thyvelikakath, G. X., *J. Org. Chem.*, **46,** 3181 (1981).
[105] Bali, A., Chaudhry, S. C., and Malhotra, K. C., *Acta Cienc. Indica, [Ser.] Chem.*, **6,** 136 (1980); *Chem. Abs.*, **94,** 191209 (1981).
[106] Mueller, U., *Z. Anorg. Allg. Chem.*, **463,** 117 (1980); *Chem. Abs.*, **93,** 238455 (1980).
[107] Brown, T. M., and Carruthers, W., *J. Chem. Soc., Perkin Trans. 1*, **1981,** 2904.
[108] Acheson, R. M., Prince, R. J., Procter, G., Wallis, J. D., and Watkin, D. J., *J. Chem. Soc., Perkin Trans. 2*, **1981,** 266.
[109] Baxter, S. G., Cowley, A. H., and Mehrotra, S. K., *J. Am. Chem. Soc.*, **103,** 5572 (1981).
[110] Mamaev, V. M., Gloriozov, I. P., and Ponomarev, G. V., *Zh. Struct. Khim.*, **21,** 170 (1980); *Chem. Abs.*, **95,** 23660 (1981).
[111] Stolyarov, B. V., and Krylov, A. I., *Zh. Org. Khim.*, **16,** 1802 (1980); *Chem. Abs.*, **94,** 46453 (1981).
[112] Larsen, J. W., Bouis, P. A., and Riddle, C. A., *J. Org. Chem.*, **45,** 4969 (1980).
[113] Kanishchev, M. I., Shchegolev, A. A., Smit, V. A., Caple, R., and Sepetov, N. F., *Izv. Akad. Nauk SSSR, Ser. Khim.*, **1980,** 2102; *Chem. Abs.*, **94,** 29894 (1981).
[114] Yoshimizu, N., Tajima, Y., Kamisuki, T., and Maeda, S., *Koen Yoshishu-Bunshi Kozo Sogo Toronkai*, **1979,** 558; *Chem. Abs.*, **93,** 167036 (1980).
[115] Shteingarts, V. D., *Izv. Sib. Otd. Akad. Nauk SSSR, Ser. Khim. Nauk*, **1980,** 53; *Chem. Abs.*, **94,** 3309 (1981).
[116] Ginzburg, O. F., and Sinev, V. V., *Izv. Sib. Otd. Akad. Nauk SSSR, Ser. Khim. Nauk*, **1980,** 72; *Chem. Abs.*, **93,** 237972 (1980).
[117] Lipovich, V. G., *Izv. Sib. Otd. Akad. Nauk SSSR, Ser. Khim. Nauk*, **1980,** 12; *Chem. Abs.*, **94,** 14614 (1981).
[118] Shubin, V. G., *Izv. Sib. Otd. Akad. Nauk SSSR, Ser. Khim. Nauk*, **1980,** 18; *Chem. Abs.*, **94,** 3308 (1981).
[119] Saunders, M., Chandrasekhar, J., and Schleyer, P. von R., *Org. Chem. (N.Y.)*, **42,** 1 (1980); *Chem. Abs.*, **94,** 3317 (1981).
[120] Koptyug, V. A., and Shubin, V. G., *Zh. Org. Khim.*, **16,** 1977 (1980); *Chem. Abs.*, **94,** 29608 (1981).
[121] Temnikova, T. I., *Izv. Sib. Otd. Akad. Nauk SSSR, Ser. Khim. Nauk*, **1980,** 5; *Chem. Abs.*, **94,** 14613 (1981).
[122] Smit, V. A., *Izv. Sib. Otd. Akad. Nauk SSSR, Ser. Khim. Nauk*, **1980,** 128; *Chem. Abs.*, **93,** 237977 (1980).
[123] Traeger, J. C., and McLoughlin, R. G., *J. Am. Chem. Soc.*, **103,** 3647 (1981).
[124] Nefedov, V. D., Sinotova, E. N., and Bermudez, R. K., *Zh. Org. Khim.*, **16,** 2281 (1980); *Chem. Abs.*, **94,** 29785 (1981).
[125] Iwasaki, M., Toriyama, K., and Nunome, K., *J. Am. Chem. Soc.*, **103,** 3591 (1981).
[126] Westdemiotis, C., and McLafferty, F. W., *Tetrahedron*, **37,** 3111 (1981).
[127] Franke, W., Schwarz, H., Thies, H., Chandrasekhar, J., Schleyer, P. von R., Hehre, W. J., Saunders, M., and Walker, G., *Chem. Ber.*, **114,** 2808 (1981).
[128] Attina, M., Cacace, F., and Giacomello, P., *J. Am. Chem. Soc.*, **103,** 4711 (1981).
[129] Van Tilborg, M. W. E. M., Van Doorn, R., and Nibbering, N. M. M., *Org. Mass Spectrom.*, **15,** 152 (1980).
[130] Hamuda, El M., Nikishova, N. G., Butin, K. P., Bundel, Yu. G., and Reutov, O. A., *Vestn. Mosk. Univ., Ser. 2: Khim.*, **21,** 589 (1980); *Chem. Abs.*, **94,** 120690 (1981).
[131] Vitullo, V. P., and Wilgis, F. P., *J. Am. Chem. Soc.*, **103,** 1982 (1981).
[132] Ambroz, H. B., and Kemp, T. J., *Wiad. Chem.*, **34,** 773 (1980); *Chem. Abs.*, **95,** 79352 (1981).

[133] Kuokkanen, T., *Finn. Chem. Lett.*, **1980,** 192; *Chem. Abs.*, **95,** 96446 (1981).
[134] Nakazumi, H., Szele, I., and Zollinger, H., *Tetrahedron Lett.*, **22,** 3053 (1981).
[135] Ausloos, P. J., and Lias, S. G., *J. Am. Chem. Soc.*, **103,** 6505 (1981).
[136] Ausloos, P. J., *J. Am. Chem. Soc.*, **103,** 3931 (1981).
[137] Kim, J. K., Bonicamp, J., and Caserio, M. C., *J. Org. Chem.*, **46,** 4236 (1981).
[138] Kim, J. K., Bonicamp, J., and Caserio, M. C., *J. Org. Chem.*, **46,** 4230 (1981).
[139] Blair, I. A., Trenerry, V. C., and Bowie, J. H., *Org. Mass Spectrom.*, **15,** 15 (1980).
[140] Nefedov, V. D., Kharitonov, N. P., Sinotova, E. N., Kochina, T. A., and Balakin, I. M., *Zh. Obshch. Khim.*, **50,** 2499 (1980); *Chem. Abs.*, **94,** 83297 (1981).
[141] Schiavelli, M. D., Jung, D. M., Vaden, A. K., Stang, P. J., Fisk, T. E., and Morrison, D. S., *J. Org. Chem.*, **46,** 92 (1981).
[142] Fleming, I., and Patel, S. K., *Tetrahedron Lett.*, **22,** 2321 (1981).
[143] Trenerry, V. C., Sheldon, J. C., and Bowie, J. H., *Nouveau J. Chim.*, **5,** 193 (1981).
[144] Hevesi, L., Piquard, J.-L., and Wautier, H., *J. Am. Chem. Soc.*, **103,** 870 (1981).
[145] Richey, H. G., Kubala, B., and Smith, M. A., *Tetrahedron Lett.*, **22,** 3471 (1981).
[146] Apeloig, Y., Franke, W., Rappoport, Z., Schwarz, H., and Stahl, D., *J. Am. Chem. Soc.*, **103,** 2589 (1981).
[147] Franke, W., and Schwarz, H., *J. Org. Chem.*, **46,** 2806 (1981).
[148] van Ginkel, F. I. M., Hartman, E. R., Lodder, G., Greenblatt, J., and Rappoport, Z., *J. Am. Chem. Soc.*, **102,** 7514 (1980).
[149] Lee, C. C., *Isot. Org. Chem.*, **5,** 1 (1980); *Chem. Abs.*, **94,** 46278 (1981).
[150] Dneprovskii, A. S., and Karavan, V. S., *Izv. Sib. Otd. Akad. Nauk SSSR, Ser. Khim. Nauk*, **1980,** 78; *Chem. Abs.*, **93,** 237973 (1980).
[151] Parnes, Z. N., *Izv. Sib. Otd. Akad. Nauk SSSR, Ser. Khim. Nauk*, **1980,** 27; *Chem. Abs.*, **94,** 3287 (1981).
[152] Buncel, E., and Lee, C. C., *Isotopes in Organic Chemistry, Vol. 5: Isotopes, in Cationic Reactions* (Eds. Buncel, E., and Lee, C. C.), Elsevier, Amsterdam, 1980.
[153] Pronin, A. F., and Holer, J., *Zh. Fiz. Khim.*, **55,** 635 (1981); *Chem. Abs.*, **94,** 208051 (1981).
[154] Raghavachari, K., Whiteside, R. A., Pople, J. A., and Schleyer, P. von R., *J. Am. Chem. Soc.*, **103,** 5649 (1981).
[155] Lorquet, J. C., Sannen, C., and Raşeev, G., *J. Am. Chem. Soc.*, **102,** 7976 (1980).
[156] Krishnan, R., Frisch, M. J., Whiteside, R. A., Pople, J. A., and Schleyer, P. von R., *J. Chem. Phys.*, **74,** 4213 (1981).
[157] Day, R. J., and Crooks, R. G., *Int. J. Mass Spectrom. Ion Phys.*, **35,** 293 (1980).
[158] Rauscher, H. J., Heidrich, D., Koehler, H. J., and Michel, D., *Theor. Chim. Acta*, **57,** 255 (1980).
[159] Krogh-Jespersen, K., Cremer, D., Dill, J. D., Pople, J. A., and Schleyer, P. von R., *J. Am. Chem. Soc.*, **103,** 2770 (1981).
[160] Lazzeretti, P., and Zanasi, R., *Chem. Phys. Lett.*, **80,** 533 (1981).
[161] Chandrasekhar, J., Schleyer, P. von R., and Krogh-Jespersen, K., *J. Comput. Chem.*, **2,** 356 (1981); *Chem. Abs.*, **95,** 168295 (1981).
[162] Wirth, D., and Bauld, N. L., *J. Comput. Chem.*, **1,** 189 (1980); *Chem. Abs.*, **93,** 238474 (1980).
[163] Randic, M., *Int. J. Quantum Chem., Quantum. Chem. Symp.*, **14,** 557 (1980); *Chem. Abs.*, **94,** 139056 (1981).
[164] Abronin, I. A., Chuvylkin, N. D., and Zhidomirov, G. M., *Izv. Sib. Otd. Akad. Nauk SSSR, Ser. Khim. Nauk*, **1980,** 96; *Chem. Abs.*, **93,** 237975 (1980).
[165] Decoret, C., Royer, J., and Dannenberg, J. J., *J. Org. Chem.*, **46,** 4074 (1981).
[166] Moffat, J. B., *Chem. Phys. Lett.*, **76,** 304 (1980).
[167] Moffat, J. B., *Tetrahedron Lett.*, **22,** 1001 (1981).
[168] Frenking, G., and Schwarz, H., *Z. Naturforsch.*, **36B,** 797 (1981).
[169] Nobes, R. H., Radom, L., and Rodwell, W. R., *Chem. Phys. Lett.*, **74,** 269 (1980).
[170] Bochvar, D. A., Gal'pern, E. G., Gambaryan, N. P., Pogrebnyak, A. A., Rybin, L. V., and Rybinskaya, M. I., *Arm. Khim. Zh.*, **34,** 370 (1981); *Chem. Abs.*, **95,** 149705 (1981).
[171] Ford, G. P., and Scribner, J. D., *J. Am. Chem. Soc.*, **103,** 4281 (1981).
[172] Hopkinson, A. C., and Lien, M. H., *J. Org. Chem.*, **46,** 998 (1981).
[173] Apeloig, Y., Godleski, S. A., Heacock, D. J., and McKelvey, J. M., *Tetrahedron Lett.*, **22,** 3297 (1981).
[174] Chen, C.-C., *Hua Hsueh Tung Pao*, **1981,** 60; *Chem. Abs.*, **94,** 156120 (1981).
[175] Cheremisin, A. A., and Schastnev, P. V., *Izv. Sib. Otd. Akad. Nauk SSSR, Ser. Khim. Nauk*, **1980,** 105; *Chem. Abs.*, **93,** 237981 (1980).
[176] Bodrikov, I. V., *Izv. Sib. Otd. Akad. Nauk SSSR, Ser. Khim. Nauk*, **1980,** 119; *Chem. Abs.*, **93,** 237976 (1980).

Organic Reaction Mechanisms 1981
Edited by A. C. Knipe and W. E. Watts

CHAPTER 9

Nucleophilic Aliphatic Substitution

J. SHORTER

Department of Chemistry, The University, Hull

Vinylic Systems . . . 323
Allylic and Various Unsaturated Systems . . . 325
Norbornyl and Closely Related Systems . . . 325
Miscellaneous Polycyclic Systems . . . 327
Epoxide Reactions . . . 328
Other Small Rings . . . 329
Substitution at Elements other than Carbon . . . 330
Substitution at Silicon and other Group IV Elements . . . 330
Substitution at Sulphur . . . 331
Intramolecular Substitution . . . 332
Anchimeric Assistance . . . 333
Ambident Nucleophiles . . . 334
Isotope Effects . . . 335
Gas-phase Reactions . . . 336
Radical Processes . . . 336
Solvent Effects . . . 337
Phase-transfer Catalysis and other Intermolecular Effects . . . 338
Structural Effects . . . 339
Correlation Analysis by Hammett, Brønsted, or Taft Equations . . . 340
Steric Effects . . . 341
Nucleophilicity . . . 342
Reactivity–Selectivity Principle (RSP) . . . 342
Theoretical Treatments . . . 343
S_N1 Reactions (Miscellaneous) . . . 343
S_N2 Reactions (Miscellaneous) . . . 344
Kinetic Studies (Miscellaneous) . . . 344
References . . . 354

Vinylic Systems

Rappoport's work has continued. In an extensive review he has discussed the question: "is the bimolecular substitution which involves nucleophilic attack on C_α a single- or a multi-step process?"[1] The main conclusions are that "for a perpendicular attack on C_α there is overwhelming evidence for a multi-step substitution of poor leaving groups. For good leaving groups, the evidence for this route is strong for strongly activated systems and highly suggestive for moderately activated ones." The single-step route is predicted "for very slightly activated systems but clear-cut experimental evidence is meager". Simultaneous exchange with radioactive Br^- and solvolysis of 9-(α-bromo-*p*-methoxybenzylidene)anthrone (**1a**), its *p*-methyl analogue (**1b**), and three α-anisyl-β,β-diarylvinyl bromides (**2**) proceed *via* free vinyl cations and there is evidence for the rare $S_N2(C^+)$ mechanism.[2] The solvolysis of (E)- and (Z)-1-anisyl-2-*p*-nitrophenylpropen-1-yl chlorides, (**3**) and (**4**), in various hydroxylic solvents and with various nucleophilic additives, proceeds in most cases *via* a cationoid species.[3]

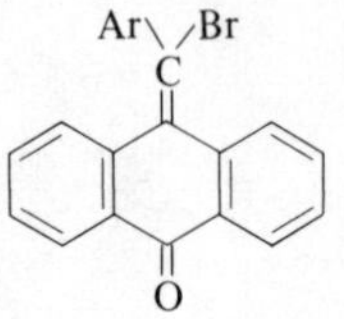

(**1**)

a: Ar = *p*-$MeOC_6H_4$
b: Ar = *p*-MeC_6H_4

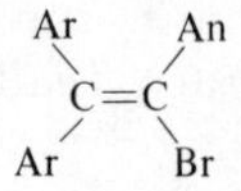

(**2**)

An = *p*-$MeOC_6H_4$

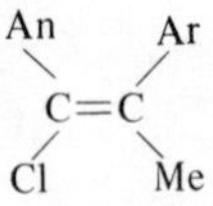

(**3**)

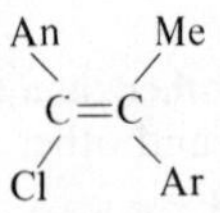

(**4**)

An = *p*-$MeOC_6H_4$, Ar = *p*-$O_2NC_6H_4$

Rearrangement studies with ^{14}C have continued in work on the solvolysis of triphenyl[2-^{14}C]vinyl and trianisyl[2-^{14}C]vinyl bromides in 70% acetic acid–30% water.[4] For the former substrate the extent of scrambling from C(2) to C(1) was unaffected by various nucleophilic additives, but for the latter substrate the extent of scrambling in the product was found to decrease with added AcO^-. These observations support the occurrence of the 1,2-phenyl shift in an ion-pair for the former case but in the free trianisylvinyl cation for the latter.

Kinetic data for the substitution reactions of $Cl_2C{=}CHSO_2R$ (R = Bu or Ph) with NaOMe and NaOEt and the observation of H–D exchange have indicated an elimination–addition mechanism.[5] On the other hand kinetic study of the reactions of various $RC_6H_4SO_2CH{=}CCl_2$ with various $R'C_6H_4SNa$ has indicated a direct substitution mechanism.[6] The small effect of the leaving group in the reactions of 4-$O_2NC_6H_4SO_2CH{=}CHX$ (X = Br, Cl) with piperidine has indicated a step-wise mechanism with loss of X as the non-rate-determining step.[7]

Methyl groups in the 6-position of cyclohexen-1-yl triflate facilitate solvolysis by anchimeric assistance.[8] The solvolytic reactivity of a number of silicon- and alkynyl-substituted vinyl triflates in aqueous ethanol has been investigated.[9] Other mechanistic studies on vinylic systems have involved the reactions of 3-bromo-3-formylacrylic derivatives with ammonia or piperidine,[10] the reactions of methyl β-chlorocinnamates with PhS^-,[11] and the reactions of polychlorobuta-1,3-dienes and polychlorobutenes with thiolates.[12]

Allylic and Various Unsaturated Systems

There seems to be unusually little to report for these systems this year, although the main lines go on.

There is continued interest in the stereochemistry and other aspects of the S_N2' mechanism. Theoretical treatment of the stereochemistry of the vinylogous S_N2' process, indicated schematically in (**5**), by orbital-symmetry methods indicates that its stereochemistry will be opposite to that of the S_N2' reaction, *i.e.* conditions which favour *syn*-S_N2' reactions of allylic systems should result in *anti*-S_N2' reaction in pentadienyl systems and *vice versa*.[13] The possibility of *anti*-S_N2' as well as S_N2 and *syn*-S_N2' reactions in nucleophilic attack on an allylic system has been discussed.[14] The allylation of sulphur-substituted active methylene compounds $RSCH_2Y$ (Y = COR, CO_2R, or CN) with allylic bromides in the presence of bases gives a mixture of S_N2 and S_N2' products.[15] The former product is selectively obtained in a non-polar solvent such as benzene.

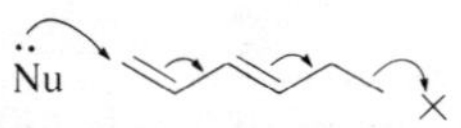

Nu X

(**5**)

The use of metal compounds to influence regio- or stereo-selectivity continues to excite interest. Copper dienolates derived from α,β-unsaturated acids undergo alkylation at the γ-carbon atom with high regioselectivity.[16] The factors affecting scope and regio- and stereo-selectivity have been extensively investigated. The factors influencing the regioselectivity of reactions involving organocuprate reagents and allyl acylates have been investigated in relation to work on prostaglandins.[17] Organocopper(I)-induced S_N2' reactions have been applied to the syntheses of various conjugated dienes including myrcene, and a rationale in terms of Pearson's HSAB principle has been given.[18]

Norbornyl and Closely Related Systems

Interest in these systems appears unabated, although there are no contributions from H. C. Brown to report. However, in September 1981, a conference at Bangor on *Carbocations* brought together most of the chemists who are prominent in the classical–non-classical ion controversy. This meeting will no doubt stimulate further publication in due course. In the meantime the major rôle in print has been played by C. A. Grob's group.[19–25] Most of their work has dealt with the solvolysis of 6-*exo*- or 6-*endo*-substituted 2-*exo*- or 2-*endo*-norbornyl tosylates, (**6**)–(**9**).

The solvolysis rates (80% v/v EtOH–H_2O) and products (70% v/v dioxan–H_2O) of reactions of about twenty members of each of the substrate types (**6**) and (**8**) have been determined.[19,20] In general the log*k* values correlate well with σ_I^q, but the sensitivity to the polar effect of 6-*exo*-*X* is much larger in the 2-*exo*-series (**6**) than in the 2-*endo*-series (**8**). Thus the 2-*exo*/2-*endo* rate ratio is 2388 for X = Bu^t and 0.37 for X = Br. The high sensitivity of the 2-*exo*-series indicates an unusually strong inductive interaction between C(6) and the incipient cationic centre at C(2). This interaction is envisaged as graded "1,3-bridging", which is enhanced by donor 6-X and reduced by acceptor 6-X. The product analyses are also explicable in terms of such bridging.

The solvolysis rates and products (solvents as above) have also been determined for about fifteen members of the series (**7**).[21,22] Correlation of log k with σ_I^q is somewhat restricted, deviations being shown when the 6-*endo*-substituent is nucleophilic (leading to *endo*-cyclization) or an *n*-electron donor (causing concerted fragmentation). Identical or different product mixtures are obtained from (**7**) or (**6**) with a given X depending on whether X is an electron donor or acceptor. The concept of graded, 1,3-bridging again provides adequate explanation. A limited series (**9**) has also been studied and the results seem to confirm that polar rather than steric effects control the relative rates of (**9**) and (**7**)[23] (*cf*. H. C. Brown's views).

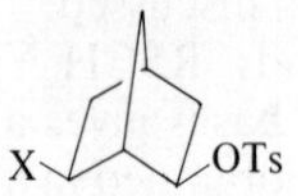

X OTs (**6**) X OTs (**7**) X OTs (**8**) X OTs (**9**)

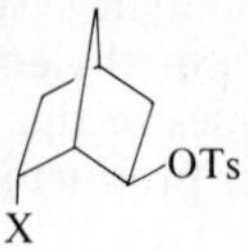

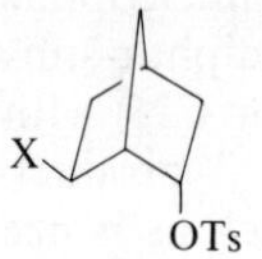

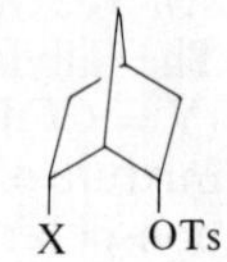

Grob has also studied the reactions of 2-*exo*- and 2-*endo*-norbornyl bromide in 90% ethanol with a large excess of KOH, and of 2-*exo*-norbornyl tosylate with excess of sodium thiophenolate in 2-methoxyethanol.[24] The reactions are first order in substrate and zero order with respect to HO^- or PhS^-, but the ratio of 1,2- and 1,3-elimination to *exo*-substitution depends strongly on the base–nucleophile concentrations; ion-pair intermediates are indicated. Finally, large variations in *exo*/*endo* rate ratios have been shown for several bicyclic tosylates which are epimers of the 2-norbornyl system and these have been ascribed to differential bridging strain accompanying the formation of intermediate cations.[25]

International groups associated with Schleyer have made two contributions.[26,27] A 5-cyano substituent decelerates the solvolysis of 2-*exo*-norbornyl brosylate by a factor of 1790, while a much smaller deceleration (by a factor of 24–90) is found for the *endo* secondary and for both *exo* and *endo* tertiary 2-norbornyl derivatives.[26] The results are considered to support the occurrence of σ-participation in the solvolysis of 2-*exo*-norbornyl brosylate. In a further communication, it is reported that tetracyclo[6.2.1.0^{2,6}.0^{5,10}]undec-3-yl derivatives (**10**) show typically high tertiary *exo*/*endo* rate ratios in solvolysis, but the ratios for secondary reactants are suppressed.[27] The behaviour of these constrained 2-norbornyl systems is held to support the theory of anchimeric assistance in the parent 2-*exo*-norbornyl solvolysis.

Steric effects in *endo,exo*-tetracyclo[6.2.1.1^{3,6}.0^{2,7}]dodecane systems have been studied with particular reference to the effect of 4-substituents on the solvolysis of epimeric 11-sulphonates.[28] The reaction of sodium phenoxide with 2-*exo*-norbornyl or 1-adamantyl (**11**) tosylate in THF shows good second-order kinetics, but deuterium scrambling and product studies show that the reactions proceed with rate-determining ionization.[29] The rate of solvolysis of tricyclo[3.2.1.0^{3,6}]octan-1-ylmethyl tosylate (**12**) differs only slightly from that of the parent norbornyl derivative, indicating that the cyclobutane ring has little effect.[30] An unusual long-range cyclopropyl participation has been detected in the solvolysis of 1-substituted *exo*-tricyclo[3.2.1.0^{2,4}]octanes.[31] Acetolyses of tricarbonyliron π-complexes of 2,3-dimethylidene-7-*anti*-norbornyl and 5,6-dimethylidene-2-*exo*-norbornyl esters have been studied.[32]

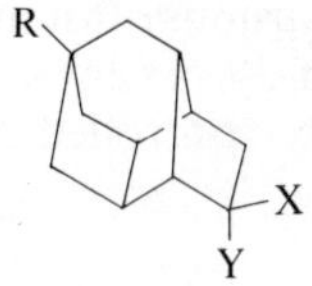

(10)

a: R = Y = H, X = OBs
b: R = X = H, Y = OBs
c: R = Me, Y = H, X = OTs
d: R = Me, X = H, Y = OTs
e: R = Y = H, X = OPNB
f: R = X = H, Y = OPNB
(PNB = $COC_6H_4NO_2$-*p*)

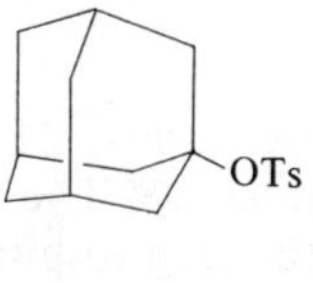

(11)

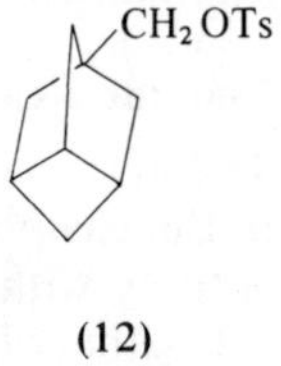

(12)

Miscellaneous Polycyclic Systems

Rates of solvolysis (60% propan-2-ol) have been measured for nine 1-adamantyl halides with 0, 2, or 3 methyl groups in the bridgehead positions.[33] The retarding effect of two Me groups is not uniform for the three halides (Cl, Br, or I) but that of three Me groups is. This indicates that the main effect of three Me groups is steric in nature. Kinetic data have been obtained for the hydrolysis of several brominated (fluoroalkyl)adamantanes.[34]

The solvolysis rate constants of (**13**)–(**15**), which are bridgehead chlorides with a strained bridgehead double bond, are in the ratio 214:105:15, the rate constant of the corresponding saturated compound being taken as unity.[35,36] These results indicate that homoallylic participation of the strained bridgehead double bond with the cationic centre at the opposite bridgehead position operates in the solvolysis of (**13**) and (**14**).

Product analyses for the acetolysis of the stereoisomeric 6-, 7-, and 8-bicyclo[3.3.0]oct-2-enyl tosylates show the occurrence of 1,2-hydride shifts and elimination solvolysis.[37] The solvolyses of the 3-*endo*- and 3-*exo*-3,5-dinitrobenzoates (**16a, b**) derived from tricyclo[4.2.0.0^{1,4}]octane strongly resemble those of the corresponding bicyclo[2.2.0]hexyl derivatives with regard both to rates and products.[38]

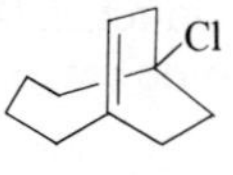

(13) 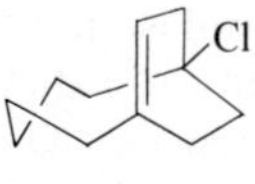 (14) 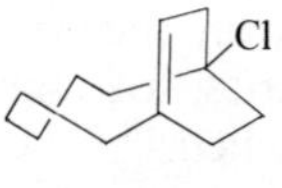 (15)

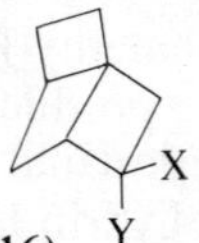

(16)

a: X = H, Y = ODNB
b: X = ODNB, Y = H
(DNB = 3,5 dinitrobenzoyl)

Other studies of polycyclic ring systems include nucleophilic substitution at tricyclo[4.4.1.0^{1,6}]undec-3-ene-11-diazonium ions,[39] solvolyses of *exo*-3,3-fluorenylidenetricyclo[3.2.1.0^{2,4}]oct-8-*anti*-yl tosylate and 6,6-diphenylbicyclo-[3.1.0]hex-3-*exo*-yl tosylate,[40] and the hydrolysis of *endo*-tetracyclo-[5.2.0.0^{2,4}0^{3,5}]non-8-en-6-yl *p*-nitrobenzoate.[41]

In the steroid field there have been studies of solvolytic ring-expansions of vitamin D_3,[42] and of the acetolysis of 4-, 6α-, and 6β-bromocholest-4-en-3-ones by silver acetate.[43] The latter reactions proceed *via* a stereoelectronically controlled S_N1 mechanism.

Epoxide Reactions

The steric course of the reaction of *trans*-[$^2H_2$]ethylene oxide with hydrogen halides in the gas phase has been examined.[44] The reactions with HCl or HBr take place entirely with *anti*-opening of the ring to give *erythro*-2-chloro- or -2-bromo-[1,2-$^2H_2$]ethanol. This finding contrasts with a mechanistic proposal, based on *ab initio* calculations, which involves concerted *syn*-opening. Reaction with HF yields only 5% 2-fluoroethanol; dioxan (37%), oligomers, and polymers are the main products.

Ab initio MO theory has been used to study the $C_2H_5O^+$ potential energy surface.[45] Thirteen structures (and many conformational variants) have been examined. The most stable isomer is the 1-hydroxyethyl cation. *O*-Corner-protonated oxiran is the only protonated oxiran structure which is at a minimum on the $C_2H_5O^+$ surface.

Kinetic and product studies of the ring-opening of propylene oxide have been carried out over the "entire" pH range in aqueous formate and aqueous phosphate buffers.[46] For the latter the observed catalysis was dissected into general and nucleophilic modes for $H_2PO_4^-$ and HPO_4^{2-}.

The susceptibility of propylene oxide to nucleophilic attack by primary amines is enhanced by the presence of polarizing groups such as OH or NH in the amine.[47] Rate studies for reactions of 2-substituted ethylamines with propylene oxide in THF, ethanol, or water show that while a 2-SH group may increase rates 15–25 fold, a 2-OH group only appears to be effective in aprotic solvents.

Rate constants have been measured for successive substitution of the amino H-atoms of ethylene diamine by propylene oxide.[48] Conformations of α-ethylene epoxides have been examined by the EHMO method and dipole-moment measurements.[49] Theoretical study (variation–perturbation method) of attack by aromatic or aliphatic amines shows a modification of the epoxide conformation. Factors influencing the reactions of *o*- and *p*-*tert*-hexylcresols with propylene oxide have been studied.[50]

A reinvestigation of the hydrolysis of 1,3-dichloropropene oxide has found the product to be 2-hydroxy-3-chloropropanal, which dimerizes to 2,5-dihydroxy-3,6-bis(chloromethyl)-1,4-dioxan.[51] α-Chloroacraldehyde, originally reported as the hydrolysis product, is actually from the primary thermal isomerization product, 2,3-dichloropropanal. The kinetics of the reaction of dicyanodiamide with epichlorohydrin have been studied.[52]

Structure–reactivity studies of epoxide reactions include reactions of phenoxy-methylphosphonic acid chloride with a series of asymmetric α-oxides[53] (application of Taft's σ^* parameter) and the polar effect of substituents in the reaction of oxides of substituted benzylidene acetophenones with morpholine in acetonitrile.[54]

The acidic ring-opening reactions of 1-ethynyl-1,2-epoxycyclohexane are almost completely regiospecific, giving mainly products from attack at the tertiary carbon.[55] The stereoselectivity is not completely *anti*, mixtures of *syn*- and *anti*-addition products being formed. The stabilizing effects of groups on an adjacent carbenium centre appear to lie in the order Ph > HC≡C > Me. The kinetics of the reaction of cyclohexene oxide with ammonia have been studied.[56]

Reaction of organolithium and organomagnesium reagents with γ,δ-epoxy ketones gives tetrahydrofuran derivatives by intramolecular cyclization of an intermediate addition alkoxide.[57] The cyclization is regioselective (the tetrahydropyran isomer is never formed), and the oxiran ring-opening is stereospecific, occurring with inversion. Addition of the organometallic reagent to the carbonyl group is, however, not stereoselective. Steric effects have been studied for the regioselective cyclization of 3,4-epoxy-alcohols to oxetans.[58] Influences of reagent and of solvent are very important. The structure and stereochemistry of the products produced by treatment of vinyloxirans with lithium dibutyl- or dimethyl-cuprate have been discussed.[59]

The catalytic influence of transition-metal compounds, *e.g.* $Mo(CO)_6$, $MoCl_5$, and $TaCl_5$, on the reaction of 1,2-epoxyoctane with ethanol at 100° in dioxan or chlorobenzene has been studied.[60] Various other epoxides and alcohols have also been examined and reactivities compared. A mechanism involving the co-ordination sphere of the transition metal was proposed. Epoxide ring-opening of various α,β-epoxy silanes by a variety of nucleophiles involves nucleophilic attack at the carbon α to silicon.[61]

In the area of steroidal epoxides, it has been shown that ring-opening of certain $4\beta,5\beta$-epoxides and $4\alpha,5\alpha$-epoxides, having a hydroxy or acetoxy group at C(6), with perchloric acid in THF, occurs regioselectively at C(4).[62] A 6β-*ax*-acetoxy group completely blocks ring-cleavage, 6α-*eq*-acetoxy and 6β-hydroxy groups retard reaction, and a 6α-*eq*-hydroxy group has no significant effect. Product analyses of the acid cleavage of steroidal $5\alpha,6\alpha$-epoxides[63] and of $3\alpha,4\alpha$- and $4\alpha,5\alpha$-epoxides[64] show that 19-OMe and 19-OAc substituents may exert marked participation, which may however be suppressed by external nucleophilic attack.

Other epoxide studies of biological interest include the kinetics and products of the hydrolysis of 3,4-dihydroprecocene 3,4-epoxide in aqueous solvents[65] and the enantioselectivity of microsomal epoxide hydrolase towards arene oxide substrates.[66]

Various studies of the reactions of phenyl glycidyl ether[67–69] and related compounds[70,71] with amines and other nucleophiles continue to be reported.

Other Small Rings

Cyclopropyl sulphides with halogen or a dimethylsulphonio group in the α-position react with a variety of nucleophiles to give 1-substituted cyclopropyl sulphides.[72] Thus the presence of an alkylthio group at the same ring carbon atom as the leaving group largely removes the difficulties often encountered with nucleophilic substitution at three-membered rings, *i.e.* low rates and ring-opening. The kinetics of solvolysis of 1-halogenocyclopropyl sulphides indicate an S_N1 mechanism, but methyl substituents in the ring promote ring-opening.[73] Mechanisms have been discussed for reactions of the bicyclic chloro-lactone (**17**) and related compounds.[74]

The ^{13}C-NMR spectra of fourteen β-cyclopropylidenic alcohols (**18**) and of the hydrolysis products of the tosylates of (**18a**) and (**18b**) have been recorded.[75] The analysis of the NMR data permits the structure and stereochemistry of the products to be unambiguously established. Nucleophilic ring-opening of cyclopropanimines has been studied.[76]

The hydrolyses of several episulphoxides (**19**) in aqueous mineral acids proceed by concurrent *A*-2 nucleophile-catalysed pathways, but some substrates show an *A*-1 mechanism under certain conditions.[77]

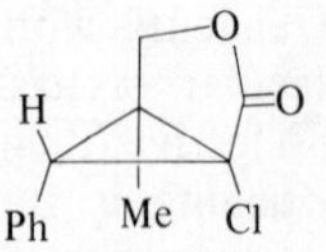

(17)

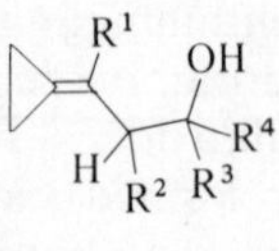

(18)

a: $R^1 = R^2 = R^3 = R^4 = H$
b: $R^1 = R^2 = R^4 = H$, $R^3 = Me$

MINDO/3 calculations indicate that cleavage of protonated ethylenimine (**20**) to the open carbocation involves a high activation barrier, which is lowered in the presence of Cl^-.[78] Reaction with Cl^- proceeds in one step by an S_N2 mechanism. The formation of *N*-(2-acetoxyalkyl)carbamates by acetolysis of 1-ethoxycarbonylaziridines (**21**) in cyclohexane is first order in aziridine and second order in acetic acid.[79] An *A*-2 mechanism is indicated, with fine details depending on the substituents in the substrate. Various mechanisms have been observed for acidic cleavage of 1-arenesulphonyl-2,2-dimethylaziridines.[80]

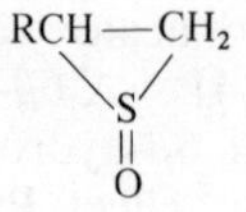

(19)

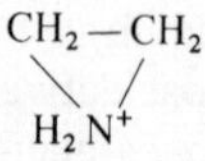

(20)

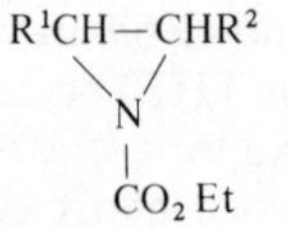

(21)

In a study of oxaziridines it has been found that nucleophilic attack occurs preferentially on the N-atom and decomposition into a carbonyl compound and an ylide occurs.[81] As the bulk of the ring substituents is increased, the reaction site shifts towards the O-atom, but the C-atom of the ring is completely inert towards nucleophiles. *cis*-Isomers react more rapidly than *trans*-isomers. Kinetic and product studies have been made for the hydrolytic cleavage of 2-benzyl-3,3-diethyloxaziridine.[82]

Substitution at Elements other than Carbon

Substitution at Silicon and other Group IV Elements

The reactions of (arylthio)trimethylsilanes with phenacyl bromide, giving aryl phenacyl sulphides and bromotrimethylsilane, show a remarkably large positive substituent effect ($\rho = +2.2$) and large negative entropy of activation.[83] The suggested mechanism involves a 5-co-ordinate silicon intermediate, which undergoes rate-limiting heterolysis of the Si—S bond.

The solvolysis of certain highly sterically hindered organosilicon perchlorates and iodides proceeds by an S_N1 mechanism involving anchimerically assisted ionization of the $Si-OClO_3$ or Si—I bond to give a methyl-bridged cation with unusual solvation requirements.[84,85]

Corriu has presented a lengthy review of the stereochemistry and mechanistic implications of nucleophilic displacement at silicon.[86] His group has continued experimental studies in reporting kinetic and stereochemical results for reactions between organolithiums or $LiAlH_4$ and some chiral organosilanes.[87] The results reveal the dominant influence of ion-pair dissociation, and thus of the electronic character of the nucleophile. In a study of the stereochemical behaviour of silicon-containing five-membered rings, nucleophilic displacements of intracyclic Si—O and Si—S bonds have been compared.[88] Analogies between silicon and phosphorus stereochemistry have been further explored in a study of the influence of the attacking anion on the stereochemistry of nucleophilic displacement at tetrahedral phosphorus.[89]

A study of the hydrolytic etherification of phenyltrichlorosilane by an H_2O—EtOH mixture has found that the rate-determining step is the reversible etherification of $PhSi(OEt)_2Cl$.[90] There have also been kinetic studies of the reactions of *p*-bromophenylsilane with azoles in aprotic solvents,[91] of the hydrolysis of 1-(2′,4′,6′-trimethylphenoxy)silatrane,[92] and of the reaction of disilazanes with *tert*-butanol.[93]

The stereochemistry of the nucleophilic substitution of various 1,2-dimethylgermacyclopentanes $MeC_4H_7Ge(Me)X$ (X = Cl, OR, SR, NR_2, or PR_2) by reducing agents ($LiAlH_4$ or $LiBH_4$), Grignard reagents, or organolithium compounds depends both on the nucleophile and on the leaving group, as for reactions of analogous Si compounds.[94]

Reaction pathways involving four-, five, and six-co-ordinate Sn(IV) have been derived from an examination of 186 crystal structures found from a search of the Cambridge Crystallographic Data Centre Database.[95] Facets of the S_N2 pathway both with inversion and with retention of configuration are revealed. A termolecular or S_N3 pathway is also well defined. Substitution at Ge(IV) and Pb(IV) is briefly discussed.

Substitution at Sulphur

Stable carbanions react with diaryl disulphides in aqueous solution by a direct displacement (S_N2) reaction to yield an arylthiol anion and the corresponding sulphide.[96] While the reaction of 1,3-dicarbonyl carbanions with 2,2′-dinitro-5,5′-dithiobenzoic acid (Ellman's reagent) is characterized by a Brønsted β_C value of 0.5, nitroalkane carbanions react 10^2–10^4 slower than 1,3-dicarbonyl carbanions of the same pK and have a β_C value of 0.95. The effect of changing the solvent from water to DMSO has also been studied.

For the hydrolysis of Ellman's reagent, rate constants have been measured as functions of $[HO^-]$ in water, micellar hexadecyltrimethylammonium bromide, and dioctadecyldimethylammonium chloride surfactant vesicles.[97] The micelles and particularly the surfactant vesicles cause enormous rate enhancement, which is attributed to highly increased concentrations of substrate and HO^- therein. The retardatory effect of anions on micellar catalysis of S—S bond cleavage in Ellman's reagent has been attributed to repulsion of HO^- from the micellar surface.[98]

The kinetics of the *S*-fluorination of arylsulphinates by perchloryl fluoride have been studied.[99]

Rate studies have been carried out for the reaction of substituted thiophen-2-sulphonyl chlorides and fluorides with anionic or neutral nucleophiles in water at 25°.[100] For some systems good Hammett correlations were obtained, while for others U-shaped plots were observed. The results are consistent with an S_N2-type mechanism, which can shift towards an S_N1 or an S_AN (addition–elimination) process depending on the nucleophile, on the ring substituent, and on leaving-group ability. The kinetics of hydrolysis, methanolysis, and ethanolysis of furan-2- and -3-, thiophen-2- and -3-, and benzene-sulphonyl chlorides have also been studied.[101] Various linear free energy relationships have been found; the mechanistic interpretation is similar to the above.

Intramolecular Substitution

Substantial review articles have appeared on effective molarities for intramolecular reactions[102] and on ring-closure reactions of bifunctional chain molecules.[103]

Product analyses, rates, and effective molarity (e.m.) values have been obtained for the formation of catechol polymethylene ethers by the intramolecular alkylation of *ortho*-ω-bromoalkoxyphenoxides (**22**) in DMSO–water (99:1 v/v).[104] Twelve

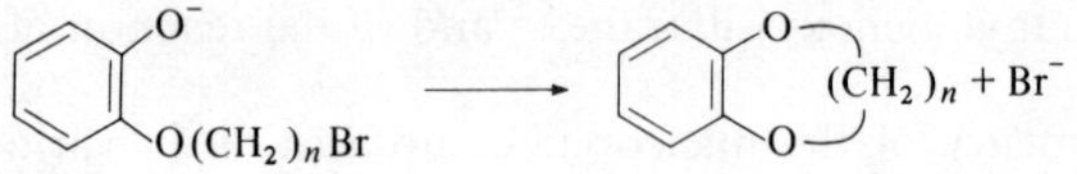

O^- / $O(CH_2)_nBr$ ⟶ $(CH_2)_n + Br^-$

(**22**)

ring sizes were studied in the range $n = 2$–28. Comparisons with analogous data for reaction in EtOH–water (75:25 v/v) have shown that e.m. values are largely independent of solvent, despite the large solvent effect on rate observed both in the cyclization and the related intermolecular model reaction. The same authors have also studied the kinetics of formation of 9-, 12-, 15-, 18-, 21-, 30-, and 48-membered benzo-crown ethers by intramolecular Williamson synthesis in DMSO–water (99:1 v/v).[105] Comparison with the data for formation of catechol polymethylene ethers throws light on the effect on ease of ring-closure of the replacement of methylene groups by oxygen atoms. The "template" catalytic effect of alkali and alkaline earth cations on the formation of benzo-18-crown-6 in methanol solution has also been studied.[106] Observed accelerations range from 13.2 for Cs^+ to 540 for Sr^{2+}.

In the solvolyses of stereoisomeric dienyl trifluoromethanesulphonates in trifluoroethanol (TFE), the E-isomers (**23**), but not the Z-isomers, rearrange to a remarkable extent to form the phenyl ethers (**25**) *via* the phenyl cations (**24**).[107]

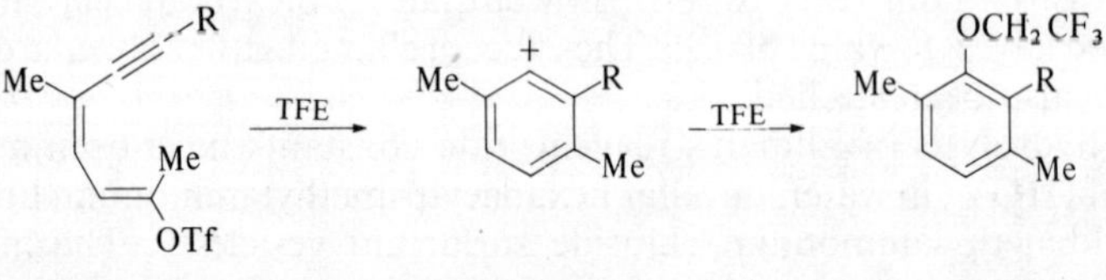

Me, R, Me, OTf —TFE→ Me, +, R, Me —TFE→ OCH_2CF_3, Me, R, Me

(**23**) (**24**) (**25**)

There has been an investigation[108] of the effect of chain-length, concentration, and solvent on the ring-closure of caesium ω-haloalkanoates (generated *in situ* from the acids and Cs_2CO_3) to form (macro)cyclic lactones (macrolides); DMF is the solvent giving the best yield, and caesium salts are far superior in this respect to those of other alkali metals, alkaline earth metals, silver, or thallium.

Asymmetric cyclizations of several chloroalkanols to give optically active oxirans have been studied.[109] The catalysts examined were optically active complexes of Co(II) (salicylidene). Mechanisms have been investigated by circular dichroism and absorption spectroscopy; the rôle of intermediates and stereochemistry have been discussed. Kinetics and stereochemistry of the iodo cyclization of several unsaturated alcohols have been studied.[110]

One-election reduction of 1-(4-biphenylyl)-ω-haloalkane by solvated electrons and the intramolecular reactions of the radical anion thus formed have been studied by pulse radiolysis.[111] The mechanism appears to involve "an intramolecular carbanionic displacement of the biphenyl radical anion on the halogen centre, which is a novel type of intramolecular S_N2 reaction". Chloro compounds have been examined in detail, and comparative study of some bromo and iodo derivatives has also been carried out.

For the oxidative cyclization of γ- and δ-hydroxy-olefins induced by TeO_2, a postulated mechanism involves an electrophilic attack by a solubilized tellurium species, followed by an intramolecular nucleophilic attack by a hydroxy group.[112]

The electrocyclic opening of cyclopropane derivatives containing internal nucleophiles provides a new route to vinyl-lactones, tetrahydropyrans, and tetrahydrofurans, shown schematically in (**26**).[113]

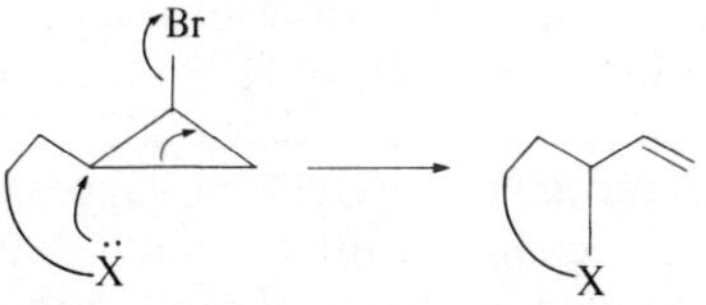

(26)

Ring-closure reactions involving various iodoisothiocyanates have been described, *e.g.* reactions of *trans*-1-iodo-2-isothiocyanatocyclohexane with carbon nucleophiles to form 2-substituted 2-thiazolines,[114] or with sulphur nucleophiles to form either 2-sulphur-substituted 2-thiazolines or thiazoline-2-thiones.[115]

The kinetics of the "intraresin reaction" (**27**) have been studied. The resin (P) was derived from chloromethylated polystyrene.[116]

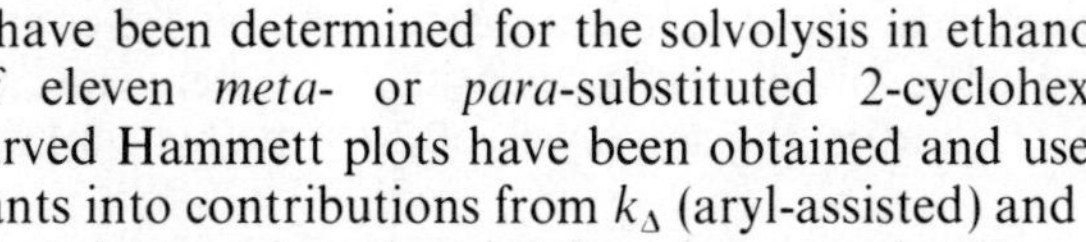

$$\textcircled{P}\begin{matrix}CH_2\overset{+}{N}Me_2Bu^n\overset{-}{O}Ac\\CH_2Cl\end{matrix}\longrightarrow\textcircled{P}\begin{matrix}CH_2\overset{+}{N}Me_2Bu^nCl^-\\CH_2OAc\end{matrix}$$

(27)

Anchimeric Assistance

Rate constants have been determined for the solvolysis in ethanol, acetic acid, or formic acid of eleven *meta*- or *para*-substituted 2-cyclohexyl-2-phenylethyl tosylates.[117] Curved Hammett plots have been obtained and used to analyse the total rate constants into contributions from k_Δ (aryl-assisted) and k_s (non-assisted) pathways. Aryl assistance is enhanced by electron-releasing substituents and decreased by electron-attracting substituents.

The course of the solvolysis of *syn*- and *anti*-*N*-chloro-1,4-dihydro-1,4-iminonaphthalenes, (**28a**) and (**28b**) respectively, below 0° is governed by the

(28a) 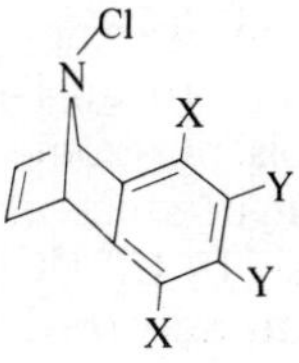 (28b)

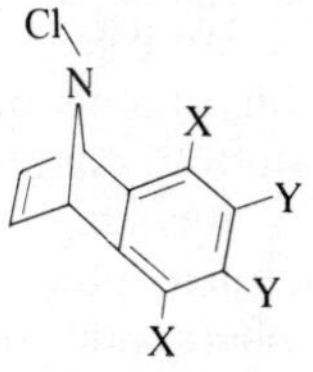

configuration of the chlorine, because the inversion at N is very slow.[118] Thus the rates of reaction of the *anti* compounds vary according to the ability of the substituents in the benzo ring to encourage benzo participation.

Rate and product studies have shown that pent-4-enyl *p*-nitrobenzenesulphonate undergoes insignificant solvolysis by the π-route in hexafluoropropan-2-ol.[119] However, in the solvolysis of 3-(cyclohex-1′-enyl)propyl *p*-nitrobenzenesulphonate (**29**) (i.e. a pent-4-enyl derivative) in fluoroalkanols, π-participation to give the perhydroinden-3a-yl (8-hydrindyl) cation (**30**) is highly significant.

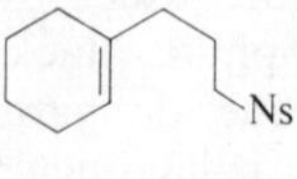

Ns

(29) 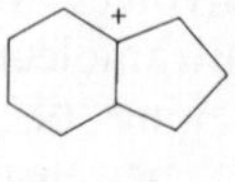 (30)

The solvolysis in aqueous ethanol of *syn*-(6-oxabicyclo[3.1.0]hex-3-yl)methyl *p*-bromobenzenesulphonate (**31**) is about 10^4–10^5 times faster than that of its *anti*-epimer (**32**) but, unexpectedly, of similar order to (**33**).[120] These results, along with studies of products and deuterium isotope effects, suggest that solvolysis of (**31**) and (**33**) proceeds by way of the common intermediate oxonium ion (**34**), involving participation of the non-bonded electrons of the O-atom.

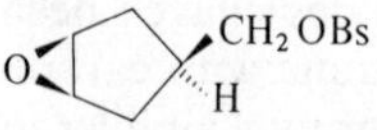

CH_2OBs H CH_2OBs OBs

(31)  (32) 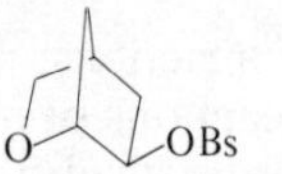 (33) 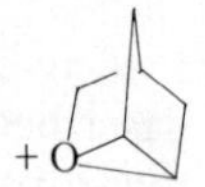 (34)

In the reaction of (s)-2-amino-3-cyclohexylpropionic acid with excess of sodium nitrite in poly(hydrogen fluoride)–pyridine, 2-fluoro-3-cyclohexylpropionic acid is formed with retention of configuration, which is ascribed to anchimeric assistance by the carboxyl group.[121] The dioxabicyclization of 3,4-dibromocyclopentyl hydroperoxide induced by silver trifluoroacetate involves preferential displacement of *cis*-3-Br.[122] This is probably assisted by the vicinal Br, with the formation of an intermediate *trans*-bromonium ion.

In the area of carbohydrate chemistry, anchimeric assistance has been variously found in the alkaline cleavage of aldofuranosides with *cis*-1,2-configurations,[123] in lactol ring-opening in the solvolysis of a glycoside,[124] in stereospecific synthesis of 2-glycosylaziridines,[125] and in the solvolysis of 2-*O*-(methylsulphonyl)fortimicin B derivatives.[126]

Ambident Nucleophiles

Regioselective carbon–carbon bond-forming reactions involving ambident nucleophiles have been reviewed (in Japanese).[127]

Studies of alumina-catalysed reactions of hydroxyarenes continue.[128,129] At 250–350° the reaction of phenol with excess of propan-1-ol gives >90% *C*-alkylation to form mono- to penta-*n*-propylphenols, plus some *O*-alkylation to give *n*-propyl aryl ethers.[128] It is suggested that the main products arise from an S_N2-type mechanism involving nucleophilic attack (variously by C(2), C(4), C(6), or O) of an adsorbed ambident phenoxide ion on C(1) of an adsorbed *n*-propoxide group.

Reaction with propan-2-ol has also been studied.[129] The main mechanism is analogous to the above, C(2) of the adsorbed isopropoxide group being involved, but minor *n*-propylation is ascribed to a side-reaction of S_N1 type.

In the reactions of 5-aryltetrazolide ions with CH_3I in EtOH–H_2O (1:1 *v*/*v*), the log (product ratio) for formation of *N*(1)- and *N*(2)-methyl isomers has been correlated with the σ^- constants for *para*-substituents in the 5-aryl group.[130] Resonance electron-withdrawing substituents deactivate *N*(1). Product ratios and second-order rate constants have also been determined for the reaction of 5-aryltetrazolide ions with Me_2SO_4 in acetonitrile.[131] Linear Hammett plots were obtained for rate constants and for product ratios.

A long series of papers on the reactivity and tautomerism of azolidines has continued with studies of the dual reactivity of the ambident anions of 2-aryliminothiazolidin-4-ones (**35**).[132–134] Rate constants and product ratios for exocyclic to endocyclic *N*-methylation have been determined for various alkylating agents and solvents, and appropriate correlation equations applied.[132,133] The work has included some studies with a salt of (**35**) as a solid phase.[134]

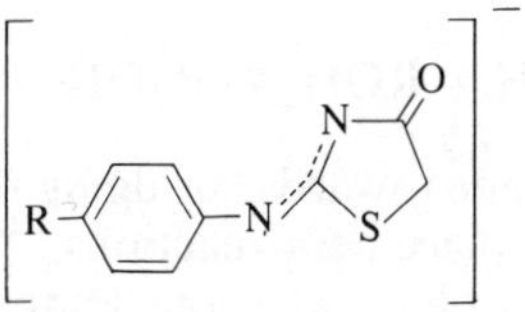

(35)

Isotope Effects

Measurement of α-, β- and γ-deuterium kinetic isotope effects (KIE) has been carried out for the solvolysis of octan-2-yl *p*-bromobenzenesulphonate in 88 % aqueous trifluoroethanol (TFE) and 83 % aqueous hexafluoroisopropyl alcohol (HFIPA).[135] Earlier studies employed 65 % aqueous ethanol as solvent and the isotope effects are now seen to increase in the order 65 % ethanol < 88 % TFE < 83 % HFIPA, corresponding to the decrease in the nucleophilic properties of the media. The results are discussed in relation to Winstein's mechanistic scheme and to the principal author's "unified mechanism".

The solvolysis of *p*-methoxybenzyl bromide in 80 % v/v dioxan–water involves intermediate carbocations which may be trapped.[136] This is shown by increases in rate and α-D KIE with increasing [ClO_4^-] and reductions in rate and α-D KIE with increasing concentrations of common-ion salt.

Model calculations of α-^{14}C and α-D KIE in the Menschutkin-type S_N2 reaction of benzyl arenesulphonates with *N*,*N*-dimethyl-*p*-toluidine have been carried out by use of the program BEBOVIB-IV.[137] A series of papers on model calculations of isotope effects has continued with an examination of secondary hydrogen isotope effects in the solvolysis of isopropyl halides.[138] A transition-state force field for S_N2 bromine-exchange reactions of a series of alkyl bromides has been developed.[139] Calculation of ^{14}C primary and (mostly) ^{2}H secondary KIE for these reactions gives reasonable agreement with experimental values for related reactions.

KIE have been studied for the alkylation of 5-phenyltetrazole by $(CD_3)_2SO_4$ in acetonitrile.[140]

Gas-phase Reactions

Work on gas-phase acid-induced nucleophilic displacement reactions has continued.[141–143] The existence and relative stability of gaseous three-membered cyclic butene-halonium ions has been studied by establishing the stereochemistry of the acid-induced displacement by nucleophiles such as H_2O and H_2S on several positively charged intermediates.[141] The formation of gaseous cyclic 2,3-butene-chloronium and -bromonium ions has been shown, but there is no evidence for the formation of stable 2,3-butene-fluoronium ions. The study has been extended to the quantitative evaluation of anchimeric assistance in the gas phase.[142] The order $OH \gg Br \geqslant Cl$ does not correspond to that observed in analogous solvolytic processes. The effects of the reaction environment on gas-phase proton-induced nucleophilic displacement at saturated carbon have also been studied.[143]

Product analyses and stereochemical studies of cationic S_N2 displacement on a protonated alcohol (equation 1) suggest that a simple backside displacement is involved.[144] On the other hand, evidence has been obtained for proton-bound dimers of simple aliphatic alcohols $[(R^1OH)H^+(R^2OH)]$ as intermediates in gas-phase reactions of the corresponding protonated alcohols with neutral alcohols.[145]

$$ROH + R\overset{+}{O}H_2 \rightarrow R_2\overset{+}{O}H + H_2O \tag{1}$$

A contribution has been made towards "bridging the gap between gas phase and solution" for nucleophilic displacement reactions.[146] The kinetics of reaction (2) have been studied in the gas phase at room temperature (by flowing-afterglow technique) for $n = 0$, 1, 2, and 3, and compared with results for aqueous solution. The reactivity drops progressively from $n = 0$ to $n = 3$ by about *four* orders of magnitude; a further drop of *twelve* orders of magnitude occurs from $n = 3$ to aqueous solution conditions.

$$HO^-(H_2O)_n + CH_3Br \rightarrow Br^-(H_2O)_n + CH_3OH \tag{2}$$

A qualitative study, using the virial theorem, has been made of the displacement of the transition state of symmetrical gaseous substitution reactions as a function of the nature of the atoms directly involved.[147]

Radical Processes

Nucleophilic substitutions through radical anions proceeding by chain mechanisms have been reviewed (in Hungarian).[148]

The mechanism of nucleophilic substitution *via* radical intermediates has been investigated by *ab initio* calculations.[149] The proposed mechanism involves electron transfer from an anion (or radical anion) to a neutral molecule and dissociation of the radical anion to an anion and a radical. The particular system studied involves the reactants $(CH_2NO_2)^-$ and $ClCH_2NO_2$, and the study shows that $(ClCH_2NO_2)^{\cdot-}$ cannot be considered as an intermediate species, since it dissociates spontaneously to Cl^- and CH_2NO_2. A comparative study of the reactions of substituted 4-nitrobenzylidene dichlorides ($ArCHCl_2$) with aqueous alcoholic alkali shows that chlorine substitution enhances the electron-transfer radical mechanism, while methyl substitution favours the solvolysis pathway.[150]

Reactions of 2-chloro-2-(4-pyridyl)propane with various nucleophiles have been studied.[151] With lithium propane-2-nitronate, four products result, and the formation of 2-nitro-3-(4-pyridyl)-2,3-dimethylbutane is suppressed by radical inhibitors. However, N_3^- and PhS^- do not show radical pathways with the above substrate.

The stereochemistry of photo-solvolysis of (R)-(+)-1-(3,5-dimethoxyphenyl)-ethyl acetate (with ^{18}O-labelling of the ether oxygen) in 50% v/v methanol–water and trifluoroethanol has been studied.[152]

Solvent Effects

Linear solvation energy relationships have been reviewed (in English[153] and in Japanese[154]). An elaborate statistical treatment has been carried out for the solvent dependence of Hammett ρ values of a wide variety of reactions, including various nucleophilic aliphatic substitutions.[155]

For the solvolyses of secondary sulphonates in aqueous ethanol and acetone, the tendency for Grunwald–Winstein correlation lines to show "dispersion" as between various binary mixed solvents has been attributed both to leaving-group effects and to variations in solvent nucleophilicity.[156] An alternative solvent ionizing-power parameter for tosylates Y_{OTs} and the solvent nucleophilicity parameter N_{OTs} have been evaluated for several aqueous–organic mixtures. In a related paper extensive kinetic data are presented for the solvolyses of secondary and tertiary alkyl sulphonates in fluorinated alcohols.[157] "Nucleophilic solvent assistance is a major determinant of the reactivity and solvation of carbocations, but its importance has frequently been underestimated."

Maxima are observed in Grunwald–Winstein plots for the solvolyses of methyl perchlorate in aqueous acetone and aqueous dioxan.[158] They are attributed to the low hydration tendency of the perchlorate anion. In a study of the solvolysis of methyl perchlorate in hydroxylic solvents, standard Gibbs energies of transfer, corrected for cavity or non-electrostatic effects, indicate that reaction takes place through a "looser-than-usual" S_N2 transition state, with a charge separation of about 0.52 units.[159]

Ultrasonic irradiation (80 kHz) increases the rate coefficients for the solvolysis of 2-chloro-2-methylpropane in aqueous ethanol by a factor of 1.5 at 10% v/v ethanol and 1.9 at 40% v/v ethanol.[160] This has been attributed to the disruption of hydrogen-bonded aggregates in the solvent.

The solvatochromic comparison method (Kamlet and Taft) has been used to analyse solvent effects on the unimolecular heterolytic decompositions of *tert*-butyl halides.[161] Reaction rates depend on the solvent dipolarity/polarizability parameter (π^*) and on solvent hydrogen-bond donor acidity (α), but the influence of solvent hydrogen-bond acceptor basicity (β) is essentially nil.

The effects of hydroxylic solvents on $\Delta G^{\neq}$ for the solvolyses of *tert*-butyl chloride and other alkyl halides have been dissected into initial-state and transition-state contributions.[162] Several inter-related papers have re-examined the activation parameters for the solvolysis of *tert*-butyl chloride and related compounds.[163–165] The previously reported dependence of heat capacity of activation on the mole fraction of ethanol in the aqueous ethanol solvent may be a consequence of the operation of the two-stage Albery–Robinson mechanism (equation 3) and the range of temperature over which the kinetic data are obtained; *i.e.* a new explanation has been devised for the curvature of the ln *k versus* T^{-1} plots.

$$RCl \underset{k_2}{\overset{k_1}{\rightleftharpoons}} R^+Cl^- \xrightarrow{k_3} \text{products} \tag{3}$$

The influence of solvent composition on activation parameters has been studied for the alkaline solvolysis of the dichloroacetate anion[166] and for the solvolysis of some alkyl fluorosulphates.[167] The remarkable range of activation parameters for the solvolysis of 1-adamantyl bromide and cyclohexyl bromide in *tert*-butyl alcohol–water mixtures can be explained in terms of the relative apparent molal properties of the *tert*-butyl alcohol and, in the case of cyclohexyl bromide, also of the partial molal properties of water.[168]

The kinetics have been studied of the solvolysis of, and the reaction of azide ion with, 1,7-dichloro-2,4,6-trinitro-2,4,6-triazaheptane in binary aqueous–organic solvents.[169,170] The methanolysis of various substrates in mixtures of methanol and acetonitrile has been studied.[171,172] Other investigations of solvent effects include studies of the solvolysis of methyl chloroformate and its thio analogue in mixtures of alcohols with water,[173] studies of the solvolysis of 9-(chloromethyl)anthracene[174] and of *tert*-cumyl chloride[175,176] in binary solvent mixtures, a study of the effect of the medium on the rate of racemization of (−)-4-chlorobenzhydryl chloride,[177] and a study of the reaction of benzyl chloride with methacrylic acid in the presence of tertiary amines.[178] Electrophilic assistance and solvation of transition states in S_N1 and S_N2 reactions have been reviewed (in Russian).[179]

The reaction of *N,N'*-dimethylimidazolidine-2-thione with methyl iodide follows second-order kinetics in polar solvents, but in less polar solvents the reverse reaction is significant.[180] Solvent effects on the forward rate constant are related linearly to those on the rate of reaction of tetramethylthiourea with methyl iodide. For the reactions of 1-benzyl-2,4,6-triphenylpyridinium cation with iodide ion, log *k* is linearly related to solvent dielectric constant, but for the reaction of the same substrate with piperidine a relationship to the Dimroth–Reichardt E_T value has been found.[181]

The rates for several S_N2 reactions and the activation parameters for Bu^nBr reacting with N_3^- in dimethylsulphoximide show that the latter behaves almost like a polar aprotic solvent, despite the presence of the NH group.[182]

Phase-transfer Catalysis and other Intermolecular Effects

In the displacement of the methanesulphonate group by a series of nucleophiles, anionic activation by the perhydrotribenzo[2.2.2.]cryptand (**36**) is practically the same as that observed in the presence of the less hindered [2.2.2,C_{14}]cryptand (**37**), both under phase-transfer conditions and in homogeneous chlorobenzene solution.[183] Methyl 2-pyridyl sulphoxide has been found to be an excellent phase-transfer catalyst for promoting S_N2 reactions of alkyl halides.[184] The kinetics of the reaction of benzyl chloride with solid NaOAc/NaOBz catalysed by a variety of

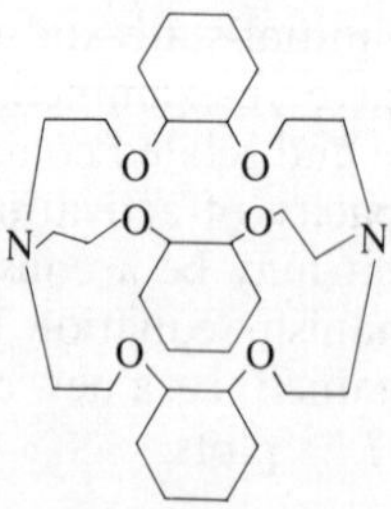

(CH$_2$)$_{13}$Me

(36)

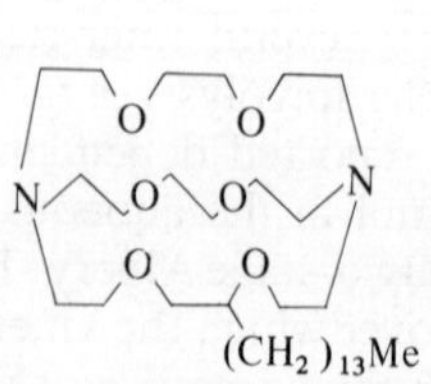

(37)

tertiary amines and quaternary ammonium salts as phase-transfer catalysts have been studied in a 10 cm internal diameter, mechanically agitated contactor.[185] Cetyldimethylbenzylammonium chloride was the most efficient of the catalysts used.

The water-insoluble compounds, bis(4-nitrophenyl) carbonate and *p*-chlorobenzhydryl chloride, are solvolysed in micelles very rapidly.[186] This indicates highly aqueous binding sites and supports a "porous cluster" micelle model, with water-filled regions where "guests" bind hydrophobically. Reactions of phenacyl bromide with Na salts of C_1–C_8 aliphatic carboxylic acids have been studied in acetone–water (72:28) in presence of varying concentrations of the cationic detergent cetyltrimethylammonium bromide (CTAB).[187] The second-order rate constant increases sharply with increasing [CTAB] above its critical micelle concentration.

A long series of papers on the chemistry of crown ethers has continued with a study of the activities of several immobilized benzo-18-crown-6 derivatives as triphase catalysts in the reaction between 1-bromo-octane in toluene and aqueous KCN.[188] The influence of various factors was examined. A series of oligo(oxyethylenes) containing 3–30 ethylene oxide residues and various polystyrene resin-supported analogues have been examined as phase-transfer catalysts in the reaction of solid PhOK with 1-bromobutane in toluene.[189] For both groups of catalysts the activity increases with the length of the oligo-ether chain. The rate of reaction of 1-bromo-octane with aqueous NaCN catalysed by insoluble polystyrene-bound benzyltri-*n*-butylphosphonium salts has been studied as a function of the method of mixing of the triphase system, catalyst particle size, degree of polymer cross-linking, solvent, and temperature.[190] A similar study has been made of the reaction of benzyl bromide with NaCN.[191]

Reactions of oriented organic molecules have been reviewed (in Japanese) in respect of control of regioselectivity[192] and control of stereoselectivity in a chiral environment.[193]

Structural Effects

In a recent treatment of the Menschutkin reaction, (summarized in *Organic Reaction Mechanisms, 1980*), a gross disparity was found between free energy and enthalpy behaviour on the one hand and that of entropies on the other. It was concluded that the transition state is "early" as far as bond-formation between C and N is concerned but "late" in terms of bond-rupture between the alkyl group and the leaving group, with solvent reorganization being nearly complete. These conclusions have now been criticized from several directions. The influences of substitution at the α-carbon atom, of solvent, and of substitution within the attacking anion have been studied for 'onium salt decompositions.[194] It has been proposed that the results are best explained in terms of a normal S_N2 transition state, rather than one with extensive bond-breaking and little bond-making. The earlier work has been criticized more fundamentally, however, by pointing out a fallacy in the discussion of the alleged "disparity" in thermodynamic quantities.[195] This discussion involves the comparison of entropies of activation $\Delta S^{\neq}$ with overall entropies of reaction ΔS^0. The numerical value of ΔS^0 for a Menschutkin reaction does not depend on the standard state chosen, but the value of $\Delta S^{\neq}$ *depends on the units used in the calculation of the rate constant.* Nothing can therefore be deduced from the ratio $\Delta S^{\neq}/\Delta S^0$ concerning the nature of the transition state. When correctly made, comparisons of $\Delta S^{\neq}$ and ΔS^0 show that transition states in Menschutkin reactions do not resemble the product dissociated ions, "a conclusion that is in agreement with all previous work in terms of free energy, enthalpy, and volume".

Comparison of the forward and reverse Menschutkin reactions of a number of nitrogen heterocycles indicates that the presence of a nitrogen lone pair in place of CH aids the forward reaction but does not have a commensurate retarding effect on the reverse process.[196]

Potential energy diagrams (More-O'Ferrall-type) have been constructed for the Doering–Zeiss mechanism of aliphatic nucleophilic substitution and for the conventional mechanism, in which rate-determining ionization proceeds without nucleophilic solvent assistance.[197] These diagrams, along with other evidence, predict variations in transition-state structure that are consistent with either of the above-mentioned mechanisms.

The products of the reaction of several (9-anthryl)arylmethyl chlorides or bromides and of related antimonate salts with nucleophiles have been determined under solvolytic conditions.[198] Notable effects of reaction conditions and of substituents on the reaction sites have been observed.

Correlation Analysis by Hammett, Brønsted, or Taft Equations

Second-order rate constants for the reactions of tetra-*n*-butylammonium arenesulphonates with methyl perchlorate in acetonitrile at 0.3° give a Hammett ρ value of -1.10 ± 0.04, almost the same as the ρ values previously reported for reactions with other powerful methylating agents.[199]

Pyridinium pK_a values and rates of quaternization of pyridines with ethyl iodide in various solvents have been used to evaluate the effect of substituents on the pyridine nitrogen reactivity.[200] The Brønsted and Hammett relations have been applied extensively. The influence of potential resonance acceptors is apparently due to their inductive effect alone, and the general question of the applicability of the dual substituent–parameter equation to substituent effects on nitrogen reactivity in pyridine has been discussed.

The Hammett σ values for *meta*- or *para*-substituted styryl groups at the 4-position of pyridine have been based on kinetic studies of the quaternization of substituted pyridines with methyl iodide.[201] Second-order rate constants have been determined for the *N*-alkylation of various 3-$RC_6H_4NH_2$ and of (3-NO_2-5-CO_2Me)$C_6H_3NH_2$ by 3-$R'C_6H_4CH_2Br$ in a nitrobenzene–DMSO mixture.[202] "Individual" and "crossed" Hammett correlations were obtained, and isoparametric points found, *i.e.* σ values for substituents in one reactant at which $\log k$ was independent of substituent in the other reactant.

The ρ value (-0.78) for the reaction of several (aryloxy)trimethylstannanes with benzyl bromide is in accord with bimolecular nucleophilic attack of the aryloxy-*O* on the benzyl-*C*, to give aryl benzyl ether and bromotrimethylstannane.[203]

The kinetics of ^{36}Cl isotopic exchange between Cl^- and *O,O*-diarylphosphorochloridates or *O,O*-diarylphosphorochloridothionates yield almost identical Hammett ρ values in correlation with σ^0.[204]

1-(*p*-Methoxybenzyl)- and 1-(2-furfuryl)-2,4,6-triphenylpyridiniums react with piperidine in chlorobenzene by S_N1 and S_N2 mechanisms, the proportion of S_N1 increasing with temperature, while other 1-(substituted benzyl) compounds show only S_N2;[205] Hammett treatments (σ and σ^+) have been used.

4′-Substituted 4-benzyloxybenzyl chlorides undergo methanolysis by the S_N1 mechanism (*cf.* simple benzyl chlorides) and new σ^+ values have been estimated for several 4′-$XC_6H_4CH_2O-$ groups.[206] Rate constants have been determined for the solvolysis of substituted 2-phenyl-2-chloropropanes (**38**), in 90% aqueous acetone, with X and Y variously Cl, Br, and H;[207] σ^+ constants for halocyclopropyl groups have been estimated. Hammett plots against σ or σ^+ were also involved in studies of substituent effects on the solvolysis of the chlorophthalide derivatives (**39**).[208]

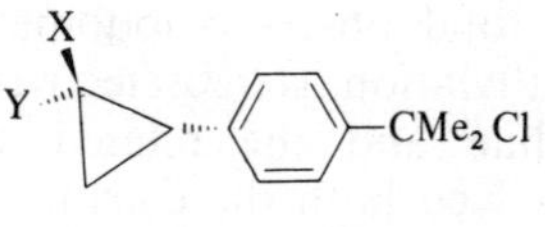

(38) 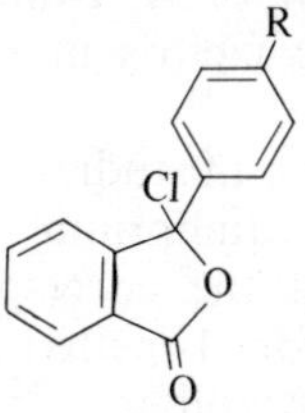 **(39)**

First-order rate constants have been measured for the hydrolyses of the Z- and E-isomers of *O*-methylbenzohydroximoyl chloride, PhCCl=NOMe, and also for reaction of several 4-substituted derivatives of the former.[209] For correlation with σ^+ the Hammett ρ value is -2.40, in accord with rate-determining ionization to a nitrilium ion. The solvolysis of twenty-five chloromethylated phenols has been studied kinetically (MeOH, 25°).[210] The results were complicated but limited correlation analysis gave highly negative ρ values, corresponding to an S_N1 mechanism.

The application of the "expanded" Hammett equation has been reviewed (in English).[211] Rate constants for the reactions of $RC_6H_4COCHBrPh$ with $R^1C_6H_4NH_2$ or $3\text{-}R^2C_6H_4O^-$ have been analysed through polylinear relationships.[212]

For the co-pyrolysis of 2-chloroethyl compounds with methanol on activated alumina, there is reasonably good correlation of log k with Taft's σ^* parameter, and the ρ^* value is $+0.49$, indicating nucleophilic attack by methanol on the substrate.[213] Taft's σ^* parameter has also been applied to the kinetics of the Arbuzov reaction in a series of fluoro-containing phosphites.[214]

When the equation $\log (k/k^0) = \rho^q\sigma_I^q$ is applied to the solvolysis of 3-substituted propyl bromides in EtOH–H_2O (4:1 v/v), there is good linear correlation except in cases where the 3-substituent exerts an anchimeric effect.[215] The reaction constant ρ^q is -0.12, typical for a nucleophilic solvent-assisted k_s process at a primary carbon atom. Series of substituted tertiary halides give considerably more negative ρ^q values, in accord with their reactions involving little nucleophilic solvent assistance.

Steric Effects

The rates of S_N2 reactions are subject to steric effects that can be measured quantitatively if bonding effects are held constant and polar effects are either negligible or can be allowed for.[216] Steric effects due to structural variation in the nucleophile can be correlated with Taft's E_s parameter.

Rate constants for the amine-catalysed cyclization of glycidyl-*N*-phenylcarbamate may be correlated with σ^* values of the amine substituents and the steric constants E_N of the amines.[217]

Differing behaviour of sulphonium as compared with phosphonium or ammonium salts in reverse Menschutkin reactions has been explained in terms of different steric hindrance around the cationic centre.[218]

N-Benzyl groups are transferred to piperidine from pyridinium ions in S_N1 and/or S_N2 processes.[219] Various steric effects have been observed, *e.g.* steric acceleration by α-phenyl groups is reduced by an adjacent β-methyl group but increased by constraining the phenyl group to near-planarity by a $(CH_2)_2$ chain. In the quaternization of *N*,*N*-dimethylbenzylamine and related compounds by alkyl halides, steric factors in both reactants and other structural factors influence reactivity.[220]

In the course of studies of steric effects in conformationally mobile systems, the ground-state equilibrium distribution, the total observed iodomethylation rate constant, and the corresponding iodomethylation stereoselectivity have been determined for 1-methyl-2-phenylpyrrolidine and the related 1-methyl-2-(2-alkylphenyl)pyrrolidines.[221] Discussion involved both the Curtin–Hammett and Winstein–Holness equations.

In the solvolysis of several 9-(*ortho*-substituted-benzyl)fluorene-9-trimethylammonium salts in ethanol, steric acceleration has been found to favour the loss of poor leaving groups.[222]

Nucleophilicity

Extrapolation of Brønsted-type plots shows that thiophenoxide ions are about 100–1000 times more nucleophilic towards alkyl halides in DMSO than fluorenyl carbanions and about 10^4–10^5 times more nucleophilic than 2-naphthoxide ions of the same basicity.[223]

Attempts have been made to correlate gas-phase proton affinities with S_N2 nucleophilicities in solution, and thereby to make a quantitative assessment of steric hindrance in S_N2 reactions.[224]

The phenylselenide anion is a potent nucleophile for S_N2-type cleavages of esters and lactones.[225] Its nucleophilicity can be greatly changed by varying the counter-ion and/or the extent of solvation of the anion.

Rate constants for the reactions of 2-chloropropanoic acid with aqueous amines have been discussed in terms of the Swain–Scott and Taft equations.[226] Relative nucleophilicities of several negative and neutral species towards the N=O group in butyl nitrite have been established.[227] Basicity and polarizability are both important factors. A study of nucleophilic selectivity ratios of model and clinical alkylating agents has been made by 4-(4′-nitrobenzyl)pyridine competition.[228]

CNDO/2 calculations for several S_N2 reactions indicate that reactivity is controlled by basicity and induced polarizability.[229]

Reactivity–Selectivity Principle (*RSP*)

A model for estimating the effect of substituents on transition-state structure in S_N2 reactions has been presented.[230] It is based on describing the reaction complex in terms of a wave-function built up from a linear combination of reactant configurations, and is applied to the S_N2 reactions of benzyl derivatives. The main conclusion is that within a limited family, substituent effects are likely to affect the transition state in an "anti-Hammond" manner, and only with substrates of widely differing structure and reactivity are "Hammond" effects likely to be seen.

Kinetic α-deuterium isotope effects have been determined for CH_3I/CD_3I reacting with 3-*X*-5-*Y*-substituted pyridines.[231] Considerable isotope effects, which vary with X and Y, were observed and these results indicate a gradual change in transition-state structure. This finding conflicts with the inference of constant structure from the constant selectivity observed in a series of Menschutkin reactions covering an enormous range of reactivity. Thus "selectivity is a relatively insensitive probe and frequently does show an invariance in cases in which the transition state does vary".

The RSP has been applied to solvolyses of 1-phenylneopentyl methanesulphonates in the presence of azide ion.[232] Carbocations and ion-pairs in nucleophilic substitution have been reviewed (in Russian) from the standpoint of reactivity–selectivity.[233]

Theoretical Treatments

Force-field calculations[139] have been made for non-potential-energy effects in the Br^--exchange reactions (Finkelstein reactions) of a series of simple alkyl bromides.[234] The effects on rate constants cover a range of about 30:1, neglecting contributions (if any) from variation in rotational symmetry numbers. The results were analysed in detail and compared with data from the literature. The treatment was extended to certain closely related reactions.

The stereochemistry of S_N2 reactions has been treated in terms of the HOMO of the pentaco-ordinate systems MH_5^-.[235] For M = C a large predominance of a D_{3h} structure with two elongated axial bonds over a C_s structure leads invariably to Walden inversion, while for M = Si a small predominance allows the possibility of retention; the influence of the nucleophile has also been considered.

A comparison of the results of experiments and of force-field calculations has been made for nucleophilic substitution and competing elimination in cyclohexane and steroid reactions.[236] The reactivity of indolizine, its derivatives, and azaindolizines towards nucleophilic ring-opening and recyclization to an isomer has been examined by a PMO method.[237] MO calculations have also been done in relation to the effect of aza substitution on the formation of anhydro bases of quinolylindoles, and the conclusions have been confirmed by experiment.[238]

S_N1 Reactions (Miscellaneous)

Rate ratios (H/α-CN and H/β-CN) have been determined for solvolyses of sulphonates of several non-cyclic aliphatic alcohols.[239] The β-CN group is far more rate-retarding than the α-CN group, and this is attributed to conjugative stabilization of the carbocation by α-CN.

The stereochemistry of solvolyses including 1,5-hydride shifts has been studied for 3-*tert*-butylcyclo-octyl compounds.[240] Solvolysis of the epimeric tosylates in various solvents occurs with >99% retention.

An elaborate study and interpretation of kinetic isotope effects in the solvolysis of neophyl arenesulphonates strongly confirms the view that these compounds react *via* the k_Δ pathway.[241]

Free energies of ionization of nine primary or secondary alkyl halides exceed slightly (by 7–13 kJ mol^{-1}) the activation energies for their hydrolysis.[242] It is therefore suggested that both ion-pair and S_N1 mechanisms are reasonable alternatives for these hydrolyses.

The hydrolysis of geranyl pyrophosphate (GPP) is catalysed by salts of Mn(II).[243] The most reactive species is GPP$[Mn(II)]_2^+$ at the optimum pH of 6.5–7, and the reaction involves C—O cleavage to a carbocation and $Mn_2P_2O_7$, which constitutes the leaving group.

The kinetics and products of the hydrolysis of RX (R = Et, Pr, Me_2CHCH_2; X = Cl, Br, I) in the presence of Ag^+, Hg^{2+}, or Tl^{3+} are in accord with an S_N1 mechanism.[244]

Rates of solvolysis of alkyl fluorides in sulphuric acid are in the sequence $Bu^tF > Pr^iF >$ ErF and the ratios $10^8:10^5:1$, a much greater spread of rates than obtains in the hydrolysis in water.[245] Various explanations have been suggested.

S_N2 Reactions (Miscellaneous)

The theoretical background of S_N2 reactions without inversion has been reviewed (in Korean).[246]

Several *p*- or *o*-nitrobenzoates have been found to undergo competing S_NAr and S_N2 reactions with azide, alkoxide, and thiophenoxide ions.[247]

Caesium carboxylates in dimethylformamide give clean S_N2 substitutions with a variety of substrates.[248]

Relative reactivities of trimethyltinsodium, as a nucleophilic reagent, towards various alkyl halides have been determined by competition experiments, with particular reference to the effects of structure and solvent.[249] Studies have been made of the reactions of Me_3SnM and of Ph_3SnM (M = Li, Na, K) with optically active 2-octyl tosylate, chloride, and bromide.[250] The observed stereochemical behaviour ranges from essentially complete inversion to predominant loss of stereochemical integrity.

Only anionic nucleophiles enter into substitution with bis(triphenylalkoxyphosphonium) salts, suggesting that prior association of Nuc^- with the $\overset{+}{O}PPh_3$ entity is important.[251] Sterically unconstrained bis(triphenylalkoxyphosphonium) salts, however, behave as if they are mono salts, and undergo substitution very slowly or not at all at room temperature.[252] The ion-pair cluster mechanism has been discussed in the light of these results.

A sensitive and reproducible competitive method has been devised for quantitative evaluation of the relative reactivities of alkylating agents.[253]

Solvolysis of *cis*- and *trans*-*S*-methylthiazolidinium phenylalanine cyclic dipeptides has S_N2 character.[254]

Kinetic Studies (Miscellaneous)

These have included the following: temperature and pressure effects on the methanolysis of 2-chloro-2-methylpropane and 3-chloro-3-ethylpentane;[255] dependence of rate constant of alkylation of carbanions by $Me(CH_2)_nCl$ on the value of *n*;[256] reaction of alkali salts of triphenylmethane with alkyl halides;[257] the reaction of alkyl halides with thiolate anions as a model reaction for protein alkylation;[258] the reaction of *p*-nitrobenzyl bromide with some α,β-unsaturated carboxylate ions;[259] the reactions of methyl iodide with $AgClO_4$ in nitrobenzene and nitromethane;[260] nucleophilic substitution in α-chlorocarboxylic acid amides;[261] the ionization of α-chlorodibenzyl sulphide studied by NMR spectroscopy;[262] the hydrolysis of *N*-phenylbenzohydrazonyl bromide;[263] the heat capacities of activation for displacements at primary and secondary carbon centres in water;[264] the acetolysis of *p*-anisylethyl tosylate, studied by ^{13}C-NMR spectroscopy;[265] the reaction of benzyl benzenesulphonate with pyridine in acetone under high pressure;[266] the thermolysis of sucrose in dimethyl sulphoxide solution;[267] further studies of the Acree relation, which treats the effect of ion-pairing;[268] the acid-catalysed decomposition of trialkyltriazines[269] and of diazomethane;[270] kinetics and thermodynamics in alkylation by alkoxy-phosphonium salts;[271] the triphenylphosphine–tetrachloromethane–alcohol reaction;[272] the nucleophilic substitution of thiamine analogues by thiolate ions, catalysed by HO^-;[273] and the formation of unsymmetrical dimethylhydrazine from monochloramine and dimethylamine.[274]

References

1 Rappoport, Z., *Acc. Chem. Res.*, **14**, 7 (1981).
2 van Ginkel, F. I. M., Hartman, E. R., Lodder, G., Greenblatt, J., and Rappoport, Z., *J. Am. Chem. Soc.*, **102**, 7514 (1980).
3 Rappoport, Z., and Pross, N., *J. Org. Chem.*, **45**, 4309 (1980).
4 Lee, C. C., Ko, E. C. F., and Rappoport, Z., *Can. J. Chem.*, **58**, 2369 (1980).
5 Shainyan, B. A., and Mirksova, A. N., *Zh. Org. Khim.*, **16**, 1797 (1980); *Chem. Abs.*, **94**, 46400 (1981).
6 Shainyan, B. A., and Mirskova, A. N., *Zh. Org. Khim.*, **16**, 2569 (1980); *Chem. Abs.*, **94**, 120502 (1981).
7 Popov, A. F., Kostenko, L. I., and Kravchenko, V. V., *Zh. Org. Khim.*, **16**, 1082 (1980); *Chem. Abs.*, **93**, 167161 (1980).
8 Garcia Martinez, A., Herrera Fernandez, A., and Martinez Alvarez, R., *An. Quim.*, **76C**, 87 (1980); *Chem. Abs.*, **94**, 83464 (1981).
9 Schiavelli, M. D., Jung, D. M., Vaden, A. K., Stang, P. J., Fisk, T. E., and Morrison, D. S., *J. Org. Chem.*, **46**, 92 (1981).
10 Farina, F., Martin, M. R., Martin, M. V., and Sanchez, F., *Rev. R. Acad. Cienc. Exactas, Fis. Nat. Madrid*, **73**, 599 (1979); *Chem. Abs.*, **93**, 220331 (1980).
11 Youssef, A.-H.A., Sharaf, S. M., El-Sadany, S. K., and Hamed, E. A., *J. Org. Chem.*, **46**, 3813 (1981).
12 Roedig, A., Ibis, C., and Zaby, G., *Chem. Ber.*, **114**, 684 (1981).
13 Murphy, W. S., and O'Mahony, B., *Tetrahedron Lett.*, **22**, 585 (1981).
14 Murahashi, S., and Tanigawa, Y., *Kagaku* (*Kyoto*), **35**, 925 (1980); *Chem. Abs.*, **95**, 24143 (1981).
15 Ono, N., Miyake, H., Tanabe, Y., Tanaka, K., and Kaji, A., *Chem. Lett.*, **1980**, 1365.
16 Savu, P. M., and Katzenellenbogen, J. A., *J. Org. Chem.*, **46**, 239 (1981).
17 Roberts, S. M., Woolley, G. T., and Newton, R. F., *J. Chem. Soc., Perkin Trans. 1*, **1981**, 1729.
18 Kleijn, H., Westmijze, H., Meijer, J., and Vermeer, P., *Recl. Trav. Chim. Pays-Bas*, **99**, 340 (1980).
19 Fischer, W., Grob, C. A., Hanreich, R., von Sprecher, G., and Waldner, A., *Helv. Chim. Acta*, **64**, 2298 (1981).
20 Grob, C. A., and Waldner, A., *Tetrahedron Lett.*, **22**, 3231 (1981).
21 Grob, C. A., Günther, B., and Hanreich, R., *Helv. Chim. Acta*, **64**, 2312 (1981).
22 Grob, C. A., Günther, B., Hanreich, R., and Waldner, A., *Tetrahedron Lett.*, **22**, 835 (1981).
23 Grob, C. A., Günther, B., and Hanreich, R., *Tetrahedron Lett.*, **22**, 1211 (1981).
24 Grob, C. A., and Lutz, E., *Helv. Chim. Acta*, **64**, 153 (1981).
25 Grob, C. A., and Waldner, A., *Tetrahedron Lett.*, **22**, 3235 (1981).
26 Apeloig, Y., Arad, D., Lenoir, D., and Schleyer, P. von R., *Tetrahedron Lett.*, **22**, 879 (1981).
27 Nordlander, J. E., Neff, J. R., Moore, W. B., Apeloig, Y., Arad, D., Godleski, S. A., and Schleyer, P. von R., *Tetrahedron Lett.*, **22**, 4921 (1981).
28 Lenoir, D., and Frank, R. M., *Chem. Ber.*, **114**, 3336 (1981).
29 Takeuchi, K., Kato, Y., Moriyama, T., and Okamoto, K., *Chem. Lett.*, **1981**, 935.
30 Luh, T.-Y., and Lei, K. L., *J. Chem. Soc., Chem. Commun.*, **1981**, 214.
31 Chenier, P. J., Kiland, P. J., Schmitt, G. D., and VanderWegen, P. G., *J. Org. Chem.*, **45**, 5413 (1980).
32 Barras, C., Bell, L. G., Roulet, R., and Vogel, P., *Tetrahedron Lett.*, **22**, 233 (1981).
33 Perkins, R., Tibbo, P., Reid, K., and Dornan, J., *Chem. Ind.* (*London*), **1981**, 571.
34 Nazarova, M. P., and Podkhalyuzin, A. T., *Zh. Org. Khim.*, **17**, 128 (1981); *Chem. Abs.*, **95**, 23750 (1981).
35 Sakai, Y., Ohtani, M., Tobe, Y., and Odaira, Y., *Tetrahedron Lett.*, **21**, 5025 (1980).
36 Sakai, Y., Toyotani, S., Ohtani, M., Matsumoto, M., Tobe, Y., and Odaira, Y., *Bull. Chem. Soc. Jpn.*, **54**, 1474 (1981).

[37] Nee, M., and Roberts, J. D., *J. Org. Chem.*, **46**, 67 (1981).
[38] Wiberg, K. B., Olli, L. K., Golembeski, N., and Adams, R. D., *J. Am. Chem. Soc.*, **102**, 7467 (1980).
[39] Engbert, T., and Kirmse, W., *Liebigs Ann. Chem.*, **1980**, 1689.
[40] Wilt, J. W., Kurek, J., and Roberts, W. N., *J. Org. Chem.*, **45**, 4243 (1980).
[41] Isobe, M., and Yano, K., *Bull. Chem. Soc. Jpn.*, **53**, 2929 (1980).
[42] Gerdes, J. M., Norman, A. W., and Okamura, W. H., *J. Org. Chem.*, **46**, 599 (1981).
[43] Koga, T., and Nogami, Y., *Tetrahedron Lett.*, **22**, 3075 (1981).
[44] Bellucci, G., Berti, G., Bianchini, R., Ingrosso, G., and Moroni, A., *J. Chem. Soc., Perkin Trans. 2*, **1981**, 1336.
[45] Nobes, R. H., Rodwell, W. R., Bouma, W. J., and Radom, L., *J. Am. Chem. Soc.*, **103**, 1913 (1981).
[46] Pocker, Y., Ronald, B. P., and Ferrin, L., *J. Am. Chem. Soc.*, **102**, 7725 (1980).
[47] Burfield, D. R., Khoo, T.-K., and Smithers, R. H., *J. Chem. Soc., Perkin Trans. 1*, **1981**, 8.
[48] Plucinski, J., and Prystasz, H., *Pol. J. Chem.*, **54**, 2201 (1980); *Chem. Abs.*, **95**, 96458 (1981).
[49] Lissillour, R., Etrillard, J., Sauleau, J., and Huet, J., *J. Chim. Phys. Phys.-Chim. Biol.*, **77**, 875 (1980); *Chem. Abs.*, **94**, 102688 (1981).
[50] Makarov, M. G., Shvets, V. F., Zavorotnyi, V. A., Korenev, K. D., and Belov, P. S., *Neftekhimiya*, **21**, 402 (1981); *Chem. Abs.*, **95**, 131885 (1981).
[51] Gold, B., and Leuschen, T., *J. Org. Chem.*, **46**, 1372 (1981).
[52] Nikolaev, A. F., Karkozov, V. G., and Morozov, A. G., *Zh. Prikl. Khim.* (*Leningrad*), **53**, 1799 (1980); *Chem. Abs.*, **94**, 29743 (1981).
[53] Khardin, A. P., Tuzhikov, O. I., V'yunov, K. A., Khokhlova, T. V., and Rygalov, L. N., *Zh. Obshch. Khim.*, **50**, 1733 (1980); *Chem. Abs.*, **93**, 238168 (1980).
[54] Semenova, S. N., El Sadani, S. K., Karavan, V. S., and Temnikova, T. I., *Zh. Org. Khim.*, **17**, 983 (1981); *Chem. Abs.*, **95**, 131878 (1981).
[55] Battistini, C., Crotti, P., and Macchia, F., *J. Org. Chem.*, **46**, 434 (1981).
[56] Cerveny, L., Marhoul, A., Berka, Z., and Ruzicka, V., *Sb. Vys. Sk. Chem.-Technol. Praze, Org. Chem. Technol.*, **C26**, 59 (1980); *Chem. Abs.*, **95**, 96467 (1981).
[57] Chastrette, M., and Axiotis, G. P., *J. Organomet. Chem.*, **206**, 139 (1981).
[58] Masamune, T., Sato, S., Abiko, A., Ono, M., and Murai, A., *Bull. Chem. Soc. Jpn.*, **53**, 2895 (1980).
[59] Ziegler, F. E., and Cady, M. A., *J. Org. Chem.*, **46**, 122 (1981).
[60] Lauterbach, G., Posselt, G., Schaefer, R., and Schnurpfeil, D., *J. Prakt. Chem.*, **323**, 101 (1980); *Chem. Abs.*, **95**, 60913 (1981).
[61] Davis, A. P., Hughes, G. J., Lowndes, P. R., Robbins, C. M., Thomas, E. J., and Whitham, G. H., *J. Chem. Soc., Perkin Trans. 1*, **1981**, 1934.
[62] Ishiguro, M., Saito, H., Hirano, Y., and Ikekawa, N., *J. Chem. Soc., Perkin Trans. 1*, **1980**, 2503.
[63] Kočovský, P., and Černý, V., *Collect. Czech. Chem. Commun.*, **45**, 3190 (1980).
[64] Kočovský, P., and Černý, V., *Collect. Czech. Chem. Commun.*, **45**, 3199 (1980).
[65] Hamnett, A. F., Ottridge, A. P., Pratt, G. E., Jennings, R. C., and Stott, K. M., *Pestic. Sci.*, **12**, 245 (1981); *Chem. Abs.*, **95**, 131882 (1981).
[66] Armstrong, R. N., Kedzierski, B., Levin, W., and Jerina, D. M., *J. Biol. Chem.*, **256**, 4726 (1981).
[67] Sorokin, M. F., Shode, L. G., Veslov, V. V., and Petrova, L. P., *Izv. Vyssh. Uchebn. Zaved., Khim. Khim. Tekhnol.*, **23**, 963 (1980); *Chem. Abs.*, **94**, 14752 (1981).
[68] Urbanski, J., and Iwanska, S., *Polimery* (*Warsaw*), **25**, 237 (1980); *Chem. Abs.*, **94**, 191184 (1981).
[69] Uejima, A., *Nippon Kagaku Kaishi*, **1981**, 399; *Chem. Abs.*, **94**, 191251 (1981).
[70] Sorokin, M. F., Shode, L. G., and Pavlyukov, S. A., *Deposited Doc.*, **1980**, VINITI 2030–80; *Chem. Abs.*, **95**, 96468 (1981).
[71] Sorokin, M. F., Shode, L. G., and Kuz'min, A. I., *Deposited Doc.*, **1980**, VINITI 328; *Chem. Abs.*, **95**, 6016 (1981).
[72] Jorritsma, R., Steinberg, H., and de Boer, Th.J., *Recl. Trav. Chim. Pays-Bas*, **100**, 184 (1981).
[73] Jorritsma, R., Steinberg, H., and de Boer, Th.J., *Recl. Trav. Chim. Pays-Bas*, **100**, 194 (1981).
[74] Hülskämper, L., and Weyerstahl, P., *Chem. Ber.*, **114**, 746 (1981).
[75] Fauré, R., Leandri, G., and Meou, A., *Can. J. Chem.*, **59**, 1089 (1981).
[76] Quast, H., and Frank, R., *Liebigs Ann. Chem.*, **1980**, 1939.
[77] Saleh, A., and Tillett, J. G., *J. Chem. Soc., Perkin Trans. 2*, **1981**, 132.
[78] Timofeeva, L. M., and Avakyan, V. G., *Izv. Akad. Nauk SSSR, Ser. Khim.*, **1980**, 1557; *Chem. Abs.*, **93**, 220082 (1980).
[79] Takeuchi, H., and Koyama, K., *J. Chem. Soc., Perkin Trans. 2*, **1981**, 121.
[80] Markov, V. I., and Danileiko, D. A., *Ukr. Khim. Zh.*, **47**, 80 (1981); *Chem. Abs.*, **94**, 155973 (1981).
[81] Hata, Y., and Watanabe, M., *J. Org. Chem.*, **46**, 610 (1981).
[82] Bhattacharjee, G., *Indian J. Chem.*, **19B**, 377 (1980); *Chem. Abs.*, **93**, 185327 (1980).

[83] Kozuka, S., Higashino, T., and Kitamura, T., *Bull. Chem. Soc. Jpn.*, **54**, 1420 (1981).
[84] Eaborn, C., and Mahmoud, F. M. S., *J. Chem. Soc., Perkin Trans. 2*, **1981**, 1309.
[85] Eaborn, C., and Mahmoud, F. M. S., *J. Chem. Soc., Chem. Commun.*, **1981**, 63.
[86] Corriu, R. J. P., and Guérin, C., *J. Organomet. Chem.*, **198**, 231 (1980).
[87] Corriu, R. J. P., and Guérin, C., *Tetrahedron*, **37**, 2467 (1981).
[88] Corriu, R. J. P., Fernandez, J. M., Guérin, C., and Kpoton, A., *Bull. Soc. Chim. Belg.*, **89**, 783 (1980).
[89] Corriu, R. J. P., Dutheil, J. P., and Lanneau, G., *Tetrahedron*, **37**, 3681 (1981).
[90] Kolobkov, V. S., and Utkin, O. V., *Zh. Prikl. Khim. (Leningrad)*, **54**, 671 (1981); *Chem. Abs.*, **95**, 79532 (1981).
[91] Ivanov, V. A., Saratov, I. E., and Reikhsfel'd, V. O., *Zh. Obshch. Khim.*, **50**, 1787 (1980); *Chem. Abs.*, **93**, 220093 (1980).
[92] Lukasiak, J., and Jamrogiewicz, Z., *Acta Chim. Acad. Sci. Hung.*, **105**, 19 (1980); *Chem. Abs.*, **94**, 191222 (1981).
[93] Sevast'yanova, I. V., Klebanskii, A. L., and Ponomarev, A. I. *Zh. Obshch. Khim.*, **50**, 2283 (1980); *Chem. Abs.*, **94**, 83232 (1981).
[94] Dubac, J., Cavezzan, J., Laporterie, A., and Mazerolles, P., *J. Organomet. Chem.*, **209**, 25 (1981).
[95] Britton, D., and Dunitz, J. D., *J. Am. Chem. Soc.*, **103**, 2971 (1981).
[96] Gilbert, H. F., *J. Am. Chem. Soc.*, **102**, 7059 (1980).
[97] Fendler, J. H., and Hinze, W. L., *J. Am. Chem. Soc.*, **103**, 5439 (1981).
[98] Hiramatsu, K., Hidaka, M., and Aoki, K., *Yukagaku*, **29**, 841 (1980); *Chem. Abs.*, **94**, 120465 (1981).
[99] Vigalok, I. V., Petrova, G. G., Lukashina, S. G., Vigalok, A. A., and Levin, Ya. A., *Zh. Org. Khim.*, **16**, 1442 (1980); *Chem. Abs.*, **94**, 3384 (1981).
[100] Arcoria, A., Ballistreri, F. P., Musumarra, G., and Tomaselli, G. A., *J. Chem. Soc., Perkin Trans. 2*, **1981**, 221.
[101] Ballistreri, F. P., Cantone, A., Maccarone, E., Tomaselli, G. A., and Tripolone, M., *J. Chem. Soc., Perkin Trans. 2*, **1981**, 438.
[102] Kirby, A. J., *Adv. Phys. Org. Chem.*, **17**, 183 (1980).
[103] Illuminati, G., and Mandolini, L., *Acc. Chem. Res.*, **14**, 95 (1981); *Chem. Abs.*, **94**, 173729 (1981).
[104] Cort, A. D., Illuminati, G., Mandolini, L., and Masci, B., *J. Chem. Soc., Perkin Trans. 2*, **1980**, 1774.
[105] Illuminati, G., Mandolini, L., and Masci, B., *J. Am. Chem. Soc.*, **103**, 4142 (1981).
[106] Ercolani, G., Mandolini, L., and Masci, B., *J. Am. Chem. Soc.*, **103**, 2780 (1981).
[107] Hanack, M., and Holweger, W., *J. Chem. Soc., Chem. Commun.*, **1981**, 713.
[108] Kruizinga, W. H., and Kellogg, R. M., *J. Am. Chem. Soc.*, **103**, 5183 (1981).
[109] Takeichi, T., Arihara, M., Ishimori, M., and Tsuruta, T., *Tetrahedron*, **36**, 3391 (1980).
[110] Gevaza, Yu. I., Kupchik, I. P., and Staninets, V. I., *Khim. Geterotsikl. Soedin.*, **1981**, 32; *Chem. Abs.*, **94**, 191215 (1981).
[111] Kigawa, H., Takamuku, S., Toki, S., Kimura, N., Takeda, S., Tsumori, K., and Sakurai, H., *J. Am. Chem. Soc.*, **103**, 5176 (1981).
[112] Bergman, J., and Engman, L., *J. Am. Chem. Soc.*, **103**, 5196 (1981)
[113] Danheiser, R. L., Morin, J. M., Yu, M., and Basak, A. *Tetrahedron Lett.*, **22**, 4205 (1981).
[114] Cambie, R. C., Chambers, D., Rutledge, P. S., and Woodgate, P. D., *J. Chem. Soc., Perkin Trans. 1*, **1981**, 40.
[115] Cambie, R. C., Mayer, G. D., Rutledge, P. S., and Woodgate, P. D., *J. Chem. Soc., Perkin Trans. 1*, **1981**, 52.
[116] Regen, D. L., and Bolikal, D., *J. Am. Chem. Soc.*, **103**, 5248 (1981).
[117] Krstić, V., and Muškatirović, M., *Liebigs Ann. Chem.*, **1980**, 1637.
[118] Durrant, M. L., and Malpass, J. R., *J. Chem. Soc., Chem. Commun.*, **1981**, 1028.
[119] Ferber, P. H., and Gream, G. E., *Aust. J. Chem.*, **34**, 1051 (1981).
[120] David, F., *J. Org. Chem.*, **46**, 3512 (1981).
[121] Faustini, F., De Munari, S., Panzeri, A., Villa, V., and Gandolfi, C. A., *Tetrahedron Lett.*, **22**, 4533 (1981).
[122] Bloodworth, A. J., and Eggelte, H. J., *Tetrahedron Lett.*, **22**, 169 (1981).
[123] Lönnberg, H., and Heikkinen, E., *Finn. Chem. Lett.*, **1980**, 181.
[124] Lee, D.-S., and Perlin, A. S., *Carbohydr. Res.*, **91**, C5 (1981).
[125] Tronchet, J. M. J., and Martin, O. R., *Carbohydr. Res.*, **96**, 167 (1981).
[126] Tadanier, J., Hallas, R., Martin, J. R., Cirovic, M., and Stanaszek, R. S., *Carbohydr. Res.*, **92**, 191 (1981).
[127] Ono, M., *Yukei Gosei Kagaku Kyokaishi*, **38**, 836 (1980); *Chem. Abs.*, **94**, 207901 (1981).
[128] Klemm, L. H., and Taylor, D. R., *J. Org. Chem.*, **45**, 4320 (1980).
[129] Klemm, L. H., and Taylor, D. R., *J. Org. Chem.*, **45**, 4326 (1980).

[130] Butler, R. N., Garvin. V. C., and McEvoy, T. M., *J. Chem. Res. (S)*, **1981**, 174.
[131] Ostrovskii, V. A., Shirobokov, I. Yu., and Koldobskii, G. I., *Zh. Org. Khim.*, **17**, 146 (1981); *Chem. Abs.*, **95**, 23988 (1981).
[132] Ramsh, S. M., Ginak, A. I., Basova, Yu. G., and Shamina, L. P., *Zh. Org. Khim.*, **17**, 846 (1981); *Chem. Abs.*, **95**, 96462 (1981).
[133] Ramsh, S. M., Ginak, A. I., Basova, Yu. G., and Shamina, L. P., *Zh. Org. Khim.*, **17**, 851 (1981); *Chem. Abs.*, **95**, 96463 (1981).
[134] Basova, Yu. G., Ramsh, S. M., and Gunak, A. I., *Zh. Org. Khim.*, **17**, 986 (1981); *Chem. Abs.*, **95**, 114365 (1981).
[135] Gregoriou, G., and Varveri, F. S., *J. Chem. Soc., Perkin Trans. 2*, **1981**, 985.
[136] Vitullo, V. P., Grabowski, J., and Sridharan, S., *J. Chem. Soc., Chem. Commun.*, **1981**, 737.
[137] Yamataka, H., and Ando, T., *J. Phys. Chem.*, **85**, 2281 (1981).
[138] McLennan, D. J., *J. Chem. Soc., Perkin Trans. 2*, **1981**, 1316.
[139] McKenna, J., Sims, L. B., and Williams, I. H., *J. Am. Chem. Soc.*, **103**, 268 (1981).
[140] Shirobokov, I. Yu, Ostrovskii, V. A., and Koldobskii, G. I., *Zh. Org. Khim.*, **16**, 2145 (1980); *Chem. Abs.*, **94**, 64779 (1981).
[141] Angelini, G., and Speranza, M., *J. Am. Chem. Soc.*, **103**, 3792 (1981).
[142] Angelini, G., and Speranza, M., *J. Am. Chem. Soc.*, **103**, 3800 (1981).
[143] Attina, M., Angelini, G., and Speranza, M., *Tetrahedron*, **37**, 1221 (1981).
[144] Hall, D. G., Gupta, C., and Morton, T. H., *J. Am. Chem. Soc.*, **103**, 2416 (1981).
[145] Bomse, D. S., and Beauchamp, J. L., *J. Am. Chem. Soc.*, **103**, 3292 (1981).
[146] Bohme, D. K., and Mackay, G. I., *J. Am. Chem. Soc.*, **103**, 978 (1981).
[147] Blaise, P., and Henri-Rousseau, O., *C.R. Hebd. Séances Acad. Sci., Sér. C*, **292**, 1429 (1981).
[148] Simig, G., *Magy. Kem. Lapja*, **35**, 347 (1980); *Chem. Abs.*, **94**, 64649 (1981).
[149] Bigot, B., Roux, D., and Salem, L., *J. Am. Chem. Soc.*, **103**, 5271 (1981).
[150] Goh, S. H., and Kam, T. S., *J. Chem. Soc., Perkin Trans. 1*, **1981**, 423.
[151] Feuer, H., Doty, J. K., and Kornblum, N., *J. Heterocycl. Chem.*, **18**, 783 (1981).
[152] Jaeger, D. A., and Angelos, G. H., *Tetrahedron Lett.*, **22**, 803 (1981).
[153] Kamlet, M. J., Abboud, J. L. M., and Taft, R. W., *Prog. Phys. Org. Chem.*, **13**, 485 (1981); *Chem. Abs.*, **95**, 41617 (1981).
[154] Inamoto, N., *Kagaku (Kyoto)*, **35**, 842 (1980); *Chem. Abs.*, **94**, 207933 (1981).
[155] Nummert, V., and Palm, V., *Org. React. (Tartu)*, **17**, 292, 331 (1980).
[156] Bentley, T. W., Bowen, C. T., Brown, H. C., and Chloupek, F. J., *J. Org. Chem.*, **46**, 38 (1981).
[157] Bentley, T. W., Bowen, C. T., Morten, D. H., and Schleyer, P. von R., *J. Am. Chem. Soc.*, **103**, 5466 (1981).
[158] Kevill, D. N., and Wang, A., *J. Chem. Soc., Chem. Commun.*, **1981**, 83.
[159] Abraham, M. H., Nasehzadeh, A., and Ramos, J. J. M., *Tetrahedron Lett.*, **22**, 1929 (1981).
[160] Lorimer, J. P., and Mason, T. J., *J. Chem. Soc., Chem. Commun.*, **1980**, 1135.
[161] Abraham, M. H., Taft, R. W., and Kamlet, M. J., *J. Org. Chem.*, **46**, 3053 (1981).
[162] Abraham, M. H., in *Advances in Solution Chemistry* (Eds. Bertini, I., Lunazzi, L., and Dei, A.), Plenum, New York, 1981, p. 341; *Chem. Abs.*, **95**, 79892 (1981).
[163] Blandamer, M. J., Burgess, J., Duce, P. P., Robertson, R. E., and Scott, J. M. W., *J. Chem. Soc., Chem. Commun.*, **1981**, 13.
[164] Blandamer, M. J., Burgess, J., Duce, P. P., Robertson, R. E., and Scott, J. M. W., *J. Chem. Soc., Faraday Trans., 1*, **77**, 1999 (1981).
[165] Blandamer, M. J., Burgess, J., Duce, P. P., Robertson, R. E., and Scott, J. M. W., *J. Chem. Soc., Perkin Trans. 2*, **1981**, 1157.
[166] Diefallah, E.-H.M., Ashy, M. A., and Baghlaf, A. O., *Can. J. Chem.*, **59**, 1208 (1981).
[167] Cafferata, L. F. R., Desvard, O. E., and Sicre, J. E., *J. Chem. Soc., Perkin Trans. 2*, **1981**, 940.
[168] Yang, Y.-C., and Fagley, T. F., *J. Am. Chem. Soc.*, **103**, 5849 (1981).
[169] Krauklis, I. V., Sinev, V. V., Tselinskii, I. V., and Starostin, B. S., *Zh. Org. Khim.*, **17**, 685 (1980); *Chem. Abs.*, **95**, 96459 (1981).
[170] Krauklis, I. V., Starostin, B. S., Sinev, V. V., and Tselinskii, I. V., *Zh. Org. Khim.*, **17**, 690 (1981); *Chem. Abs.*, **95**, 96460 (1981).
[171] Lee, H. W., La, S., and Lee, I., *Taehan Hwahakhoe Chi*, **24**, 115 (1980); *Chem. Abs.*, **94**, 29733 (1981).
[172] Lee, H. W., Song, K. B., Song, H. B., and Lee, I., *Kich'o Kwahak Yonguso Nonmunjip (Inha Taehakkyo)*, **2**, 27 (1981); *Chem. Abs.*, **95**, 131852 (1981).
[173] La, S., Koh, K. S., and Lee, I., *Taehan Hwahakhoe Chi*, **24**, 1 (1980); *Chem. Abs.*, **93**, 238144 (1980).
[174] Kim, W.-K., *Taehan Hwahakhoe Chi*, **24**, 413 (1980); *Chem. Abs.*, **94**, 138865 (1981).

[175] Amelichev, V. A., and Saidov, G. V., *Zh. Obshch. Khim.*, **51**, 187 (1981); *Chem. Abs.*, **94**, 191446 (1981).
[176] Amelichev, V. A., and Saidov, G. V., *Zh. Obshch. Khim.*, **51**, 1874 (1981); *Chem. Abs.*, **95**, 132120 (1981).
[177] Hamuda, E. M., Nikishova, N. G., Butin, K. P., Bundel, Yu. G., and Reutov, O. A., *Vestn. Mosk. Univ., Ser. 2: Khim.*, **21**, 589 (1980); *Chem. Abs.*, **94**, 120690 (1981).
[178] Charushnikov, K. A., Bulatov, M. A., and Surovtsev, L. G., *Izv. Akad. Nauk SSSR, Ser. Khim.*, **1980**, 1531; *Chem. Abs.*, **93**, 203580 (1980).
[179] Rudakov, E. S., *Izv. Sib. Otd. Akad. Nauk SSSR, Ser. Khim. Nauk*, **1980**, 152; *Chem. Abs.*, **93**, 237979 (1980).
[180] Kondo, Y., Yamada, T., and Kusabayashi, S., *J. Chem. Soc., Perkin Trans. 2*, **1981**, 414.
[181] Katritzky, A. R., Musumarra, G., Sakizadeh, K., and Misic-Vukovic, M., *J. Org. Chem.*, **46**, 3820 (1981).
[182] Furukawa, N., Takahashi, F., Yoshimura, T., Morita, H., and Oae, S., *J. Chem. Soc., Perkin Trans., 2*, **1981**, 432.
[183] Landini, D., Maia, A., Montanari, F., and Rolla, F., *J. Chem. Soc., Perkin Trans. 2*, **1981**, 821.
[184] Furukawa, N., Kishimoto, K., Ogawa, S., Kawai, T., Fujihara, H., and Oae, S., *Tetrahedron Lett.*, **22**, 4409 (1981).
[185] Yadav, G. D., and Sharma, M. M., *Ind. Eng. Chem. Process Des. Dev.*, **20**, 385 (1981); *Chem. Abs.*, **94**, 155964 (1981).
[186] Menger, F. M., Yoshinaga, H., Venkatasubban, K. S., and Das, A. R., *J. Org. Chem.*, **46**, 415 (1981).
[187] Mishra, S., Das, B. N., and Sinha, B. K., *Indian J. Chem.*, **20B**, 728 (1981); *Chem. Abs.*, **95**, 168067 (1981).
[188] van Zon, A., de Jong, F., and Onwezen, Y., *Recl. Trav. Chim. Pays-Bas*, **100**, 429 (1981).
[189] Heffernan, J. G., MacKenzie, W. M., and Sherrington, D. C., *J. Chem. Soc., Perkin Trans. 2*, **1981**, 514.
[190] Tomoi, M., and Ford, W. T., *J. Am. Chem. Soc.*, **103**, 3821 (1981).
[191] Tomoi, M., and Ford, W. T., *J. Am. Chem. Soc.*, **103**, 3828 (1981).
[192] Sugimoto, T., and Tanimoto, S., *Yuki Gosei Kagaku Kyokaishi*, **38**, 747 (1980); *Chem. Abs.*, **94**, 3316 (1981).
[193] Sugimoto, T., and Tanimoto, S., *Yuki Gosei Kagaku Kyokaishi*, **38**, 823 (1980); *Chem. Abs.*, **94**, 207921 (1981).
[194] Kevill, D. N., *J. Chem. Soc., Chem. Commun.*, **1981**, 421.
[195] Abraham, M. H., and Nasehzadeh, A., *J. Chem. Soc., Chem. Commun.*, **1981**, 905.
[196] Deady, L. W., *Aust. J. Chem.*, **34**, 163 (1981).
[197] Shafer, S. G., and Harris, J. M., *J. Org. Chem.*, **46**, 2164 (1981).
[198] Ogata, F., Takagi, M., Nojima, M., and Kusabayashi, S., *J. Am. Chem. Soc.*, **103**, 1145 (1981).
[199] Kevill, D. N., Lin, G. M. L., and Bahari, M. S., *J. Chem. Soc., Perkin Trans. 2*, **1981**, 49.
[200] Johnson, C. D., Rogert, I., and Taylor, P. G., *J. Chem. Soc., Perkin Trans. 2*, **1981**, 409.
[201] Ananthakrishnanadar, P., and Rajasekaran, K., *Indian J. Chem.*, **19B**, 324 (1980); *Chem. Abs.*, **94**, 14758 (1981).
[202] Litvinenko, L. M., Shpan'ko, I. V., and Korostylev, A. P., *Zh. Org. Khim.*, **17**, 972 (1981); *Chem. Abs.*, **95**, 96518 (1981).
[203] Kozuka, S., Kikuchi, A., and Yamaguchi, S., *Bull. Chem. Soc. Jpn.*, **54**, 307 (1981).
[204] Reimschüssel, W., Mikolajczyk, M., Ślebocka-Tilk, H., and Gajl, M., *Int. J. Chem. Kinet.*, **12**, 979 (1980).
[205] Katritzky, A. R., Musumarra, G., and Sakizadeh, K., *J. Org. Chem.*, **46**, 3831 (1981).
[206] Jorge, J. A. L., Kiyan, N. Z., Mujata, Y., and Miller, J., *J. Chem. Soc., Perkin Trans. 2*, **1981**, 100.
[207] Kulinkovich, O. G., Tishchenko, I. G., Reznikov, I. V., and Pap, A. A., *Zh. Org. Khim.*, **17**, 473 (1981); *Chem. Abs.*, **95**, 23790 (1981).
[208] Bhatt, M. V., and El Ashry, S. H., *Indian J. Chem.*, **19B**, 487 (1980); *Chem. Abs.*, **93**, 238529 (1980).
[209] Johnson, J. E., and Cornell, S. C., *J. Org. Chem.*, **45**, 4144 (1980).
[210] Stein, G., Boehmer, V., Lotz, W., and Kaemmerer, H., *Z. Naturforsch.*, **36B**, 231 (1981); *Chem. Abs.*, **94**, 208212 (1981).
[211] Zalewski, R. I., *Zanco*, **5A**, 187 (1979); *Chem. Abs.*, **94**, 14622 (1981).
[212] Istomin, B. I., Voronkov, M. G., and Zhdankovich, E. L., *Izv. Akad. Nauk SSSR, Ser. Khim.*, **1981**, 270; *Chem. Abs.*, **94**, 208038 (1981).
[213] Shinoda, K., *Bull. Chem. Soc. Jpn.*, **54**, 1236 (1981).
[214] Maslennikov, I. G., Lavrent'ev, A. N., Kuz'mina, N.Ya., and Kirichenko, L. N., *Zh. Obshch. Khim.*, **51**, 1567 (1981); *Chem. Abs.*, **95**, 168055 (1981).
[215] Grob, C. A., and Waldner, A., *Helv. Chim. Acta*, **63**, 2152 (1980).

[216] DeTar, D. F., *J. Org. Chem.*, **45**, 5174 (1980).
[217] Sorokin, M. F., Shode, L. G., Pavlyukov, S. A., and Onsova, L. A., *Zh. Org. Khim.*, **16**, 1241 (1980); *Chem. Abs.*, **93**, 185548 (1980).
[218] Landini, D., Maia, A., Montanari, F., Mikolajczyk, M., and Zatorski, A., *Nouv. J. Chim.*, **4**, 723 (1980).
[219] Katritzky, A. R., El-Mowafy, A. M., Musumarra, G., Sakizadeh, K., Sana-Ullah, C., El-Shafie, S. M. M., and Thind, S. S., *J. Org. Chem.* **46**, 3823 (1981).
[220] Teitelbaum, A. B., Kurguzova, A. M., Kudryavtseva, L. A., Bel'skii, V. E., and Ivanov, B. E., *Izv. Akad. Nauk SSSR, Ser. Khim.*, **1981**, 531; *Chem. Abs.*, **95**, 41787 (1981).
[221] Seeman, J. I., Secor, H. V., Hartung, H., and Galzerano, R., *J. Am. Chem. Soc.*, **102**, 7741 (1980).
[222] Pradhan, J., and Smith, P. J., *Can. J. Chem.*, **59**, 911 (1981).
[223] Bordwell, F. G., and Hughes, D. L., *J. Org. Chem.*, **46**, 3570 (1981).
[224] McManus, S. P., *J. Org. Chem.*, **46**, 635 (1981).
[225] Liotta, D., Sunay, U., Santiesteban, H., and Markiewicz, W., *J. Org. Chem.*, **46**, 2605 (1981).
[226] Ogata, Y., and Yamauchi, Y., *Kenkyu Hokoku–Asahi Garasu Kogyo Gijutsu Shoreikai*, **36**, 225 (1980); *Chem. Abs.*, **95**, 6311 (1981).
[227] Bhattacharjee, G., *Indian J. Chem.*, **20B**, 526 (1981); *Chem. Abs.*, **95**, 79602 (1981).
[228] Spears, C. P., *Mol. Pharmacol.*, **19**, 496 (1981); *Chem. Abs.*, **95**, 149454 (1981).
[229] Lee, B. S., Lee, I., and Yang, K., *Taehan Hwahakhoe Chi*, **25**, 145 (1981); *Chem. Abs.*, **95**, 168068 (1981).
[230] Pross, A., and Shaik, S. S., *J. Am. Chem. Soc.*, **103**, 3702 (1981).
[231] Harris, J. M., Paley, M. S., and Prasthofer, T. W., *J. Am. Chem. Soc.*, **103**, 5915 (1981).
[232] Nair, M. R., and Raveendran, G., *Indian J. Chem.*, **20B**, 258 (1981); *Chem. Abs.*, **95**, 60894 (1981).
[233] Dneprovskii, A. S., and Karavan, V. S., *Izv. Sib. Otd. Akad. Nauk SSSR, Ser. Khim. Nauk*, **1980**, 78; *Chem. Abs.*, **93**, 237973 (1980).
[234] Chalk, C. D., McKenna, J., and Williams, I. H., *J. Am. Chem. Soc.*, **103**, 272 (1981).
[235] Minot, C., *Nouv. J. Chim.*, **5**, 319 (1981).
[236] Gschwendtner, W., Hoppen, V., and Schneider, H.-J., *J. Chem. Res.*, **1981**, 96 (S), 1201 (M).
[237] Vysotskii, Yu. B., and Zemskii, B., *Khim. Geterosikl. Soedin.*, **1980**, 984; *Chem. Abs.*, **93**, 238535 (1980).
[238] Strupnikova, T. V., Lopatinskaya, Kh. Ya., Zemskii, B. P., Vysotskii, Yu. B., and Sagitullin, R. S., *Khim. Geterotskikl. Soedin.*, **1980**, 1365; *Chem. Abs.*, **94**, 64938 (1981).
[239] Gassman, P. G., and Saito, K., *Tetrahedron Lett.*, **22**, 1311 (1981).
[240] Schneider, H.-J., and Heiske, D., *J. Am. Chem. Soc.*, **103**, 3501 (1981).
[241] Ando, T., Kim, S.-G., Matsuda, K., Yamataka, H., Yukawa, Y., Fry, A., Lewis, D. E., Sims, L. B., and Wilson, J. C., *J. Am. Chem. Soc.*, **103**, 3505 (1981).
[242] Zamashchikov, V. V., *Dopov. Akad. Nauk. Ukr. RSR, Ser. B: Geol., Khim., Biol. Nauki*, **1981**, 58; *Chem. Abs.*, **95**, 114614 (1981).
[243] Vial, M. V., Rojas, C., Portilla, G., Chayet, L., Pérez, L. M., Cori, O., and Bunton, C. A., *Tetrahedron*, **37**, 2351 (1981).
[244] Zamashchikov, V. V., Rudakov, E. S., Litvinenko, S. L., and Uzhik, O. N., *Dokl. Akad. Nauk. SSSR*, **258**, 113 (1981); *Chem. Abs.*, **95**, 96495 (1981).
[245] Zamashchikov, V., Rudakov, E., and Bezbozhnaya, T., *Org. React. (Tartu)*, **17**, 162 (1980).
[246] Son, C. K., and Lee, I. K., *Hwahak Kwa Kongop Ui Chinbo*, **20**, 352 (1980); *Chem. Abs.*, **94**, 155835 (1981).
[247] Logue, M. W., and Han, B. H., *J. Org. Chem.*, **46**, 1638 (1981).
[248] Kruizinga, W. H., Strijtveen, B., and Kellogg, R. M., *J. Org. Chem.*, **46**, 4321 (1981).
[249] Kuivila, H. G., and Reeves, W. G., *Bull. Soc. Chim. Belg.*, **89**, 801 (1980).
[250] San Filippo, J., and Silbermann, J., *J. Am. Chem. Soc.*, **103**, 5588 (1981).
[251] Ramos, S., and Rosen, W., *Tetrahedron Lett.*, **22**, 35 (1981).
[252] Ramos, S., and Rosen, W., *J. Org. Chem.*, **46**, 3530 (1981).
[253] Bodor, N., Kaminski, J. J., Worley, S. D., and Gerson, S. H., *Z. Naturforsch.*, **35B**, 758 (1980); *Chem. Abs.*, **93**, 203543 (1980).
[254] Anteunis, M. J. O., Becu, C., Kolodziejczyk, A., and Liberek, B., *Bull. Soc. Chim. Belg.*, **90**, 785 (1981).
[255] Nunes Viana, C. A., and Calado, M. I. L. T., *Rev. Port. Quim.*, **21**, 139 (1979); *Chem. Abs.*, **95**, 60958 (1981).
[256] Solov'yanov, A. A., Karpyuk, A. D., Beletskaya, I. P., and Reutov, O. A., *Izv. Akad. Nauk SSSR, Ser. Khim.*, **1981**, 695; *Chem. Abs.*, **95**, 168008 (1981).
[257] Solov'yanov, A. A., Karpyuk, A. D., Beletskaya, I. P., and Reutov, O. A., *Zh. Org. Khim.*, **17**, 458 (1981); *Chem. Abs.*, **95**, 23789 (1981).

[258] Dahl, H. K., and McKinley-McKee, J. S., *Bioorg. Chem.*, **10**, 329 (1981).
[259] Chandrasekhar, J., *Indian J. Chem.*, **19B**, 327 (1980); *Chem. Abs.*, **94**, 3412 (1981).
[260] Zamashchikov, V. V., and Litvinenko, S. L., *Ukr. Khim. Zh.*, **47**, 728 (1981); *Chem. Abs.*, **95**, 114381 (1981).
[261] Mashevskaya, M. S., *Izv. Vyssh. Uchebn. Zaved., Khim. Khim. Tekhnol.*, **24**, 423 (1981); *Chem. Abs.*, **95**, 114340 (1981).
[262] Shimizu, A., Sakamaki, Y., Azuma, K., Kihara, H., Nakamura, N., and Oki, M., *Bull. Chem. Soc. Jpn.*, **54**, 2774 (1981).
[263] Kwon, K.-S., Sung, N.-D., and Park, M.-K., *Haksul Yonguchi–Chungnam Taehakkyo, Chayon Kwahak Yonguso*, **6**, 29 (1979); *Chem. Abs.*, **94**, 64764 (1981).
[264] Blandamer, M. J., Robertson, R. E., Golding, P. D., MacNeil, J. M., and Scott, J. M. W., *J. Am. Chem. Soc.*, **103**, 2415 (1981).
[265] Seki, Y., Fujio, M., Mishima, M., and Tsuno, Y., *Mem. Fac. Sci., Kyushu Univ., Ser. C*, **12**, 197 (1980); *Chem. Abs.*, **94**, 29760 (1981).
[266] Hwang, J.-U., Yoh, S.-D., and Jee, J.-G., *Taehan Hwahakhoe Chi*, **24**, 150 (1980); *Chem. Abs.*, **93**, 238143 (1980).
[267] Poncini, L., and Richards, G. N., *Carbohydr. Res.*, **87**, 209 (1980).
[268] Cayzergues, P., Georgoulis, C., and Mathieu, G., *J. Chim. Phys. Phys.-Chim. Biol.*, **77**, 401 (1980); *Chem. Abs.*, **93**, 185351 (1980).
[269] Sieh, D. H., and Michejda, C. J., *J. Am. Chem. Soc.*, **103**, 442 (1981).
[270] McGarrity, J. F., and Smyth, T., *J. Am. Chem. Soc.*, **102**, 7303 (1980).
[271] Lewis, E. S., and Colle, K. S., *J. Org. Chem.*, **46**, 4369 (1981).
[272] Slagle, J. D., Huang, T. T.-S., and Franzus, B., *J. Org. Chem.*, **46**, 3526 (1981).
[273] Zoltewicz, J. A., and Uray, G., *J. Am. Chem. Soc.*, **103**, 683 (1981).
[274] Delalu, H., Marchand, A., Ferriol, M., and Cohen-Adad, R., *J. Chim. Phys. Phys.-Chim. Biol.*, **78**, 247 (1981); *Chem. Abs.*, **95**, 41779 (1981).

Organic Reaction Mechanisms 1981
Edited by A. C. Knipe and W. E. Watts

CHAPTER 10

Carbanions and Electrophilic Aliphatic Substitution

I. Gosney

Department of Chemistry, University of Edinburgh

Carbanion Structure and Stability . . . 353
Aromaticity . . . 358
Reactions of Carbanions . . . 359
Enolates . . . 365
Dianions . . . 368
Anions α to Sulphur . . . 369
Other α-Hetero-substituted Anions . . . 371
Proton-transfer and Hydrogen Isotope Effects . . . 373
Absolute and Relative Acidity . . . 374
Electrophilic Aliphatic Substitution Reactions . . . 376
References . . . 377

Carbanion Structure and Stability

Comparison of calculated energy data for substituted cyclohexadienes with previously determined molecular electrostatic potentials for substituted cyclohexadienyl anions suggests that the thermodynamically favoured and kinetically favoured sites of protonation in the cyclohexadienyl anion frequently differ.[1] Peralkylcyclohexadienyllithium compounds (**1**), formed by addition of alkyllithiums, RLi (R = Bu^n, Bu^s, Bu^t), to 1,1,2,3,5,6-hexamethyl-4-methylenecyclohexa-2,5-diene,[2] exist as ion-paired aggregates containing conjugated anions bent out of the plane of the saturated ring carbon.[3] Contrary to current thought, it is proposed that protonation of dianion intermediates in the Birch reduction of common polynuclear hydrocarbons occurs at the site which produces the most stable monoanion, not necessarily the site of highest electron density.[4] Comparison of calculated with experimental proton affinities of a large number of anions shows that diffuse function augmented split-valence basis sets of relatively modest size give more reliable values.[5] A remarkable analogy exists between the zirconio–ethylene complex (**2**), with ZrCC angles of 75.9° (X-ray), and 1,2-dilithioethane (**3**), with LiCC angles of 73.3° (4-31G).[6] Vinyllithium, indicated by standard *ab initio* methods to have a classical geometry, is *ca.* 11 kcal mol^{-1} more stable than ethyllithium in a hydride-exchange reaction.[7] X-Ray structure determinations show that the Li_4 clusters in the tetramer units of crystalline ethyllithium tend to increase their symmetry at low temperature.[8] While the methyl stabilization energies of fifteen C_1–C_3 organolithium compounds, RLi, reflect (attentuation factor = 0.71) the methyl stabilization energies of the corresponding anions, R^-, deviations are

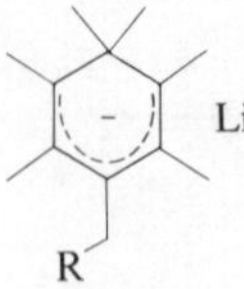

(1) 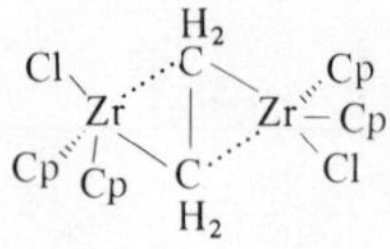 (2) (3)

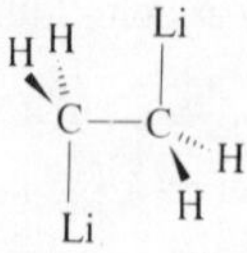

C_{2h}

observed when the degree of charge localization or delocalization is changed significantly by association with lithium.[9] Use of an improved reactor has raised hopes for structural characterization of gas-phase polylithium organic species, *e.g.* CLi_4, C_2Li_4, C_3Li_4, C_2Li_2, the focus of much intensive theoretical effort.[10] According to minimal basis *ab initio* calculations, monolithiated acetonitrile ($LiCH_2CN$) has three local singlet potential-energy surface minima, corresponding to three different structures.[11] $LiCH_2OH$ and $LiCH_2NH_2$ are suggested by *ab initio* molecular orbital theory to possess a number of isomeric forms; in the lowest energy structures, the hydroxy or amino group bridges the C—Li bond and extra stabilization results.[12] The stability of the dianions, $RCH(NPhM)CH_2Li$, prepared for the first time through a mercury–alkali-metal transmetallation process, decreases in the series M = Li > Na > K.[13] MNDO calculations indicate that the dimer of 2,6-dihydroxyphenyllithium (a model for the dimethoxy derivative) prefers a geometry with two planar tetracoordinate carbon atoms.[14] An X-ray structure determination shows that 2,6-dimethoxyphenyllithium forms a hexamer with crystallographic symmetry $\bar{3}$ with the lithium atoms clustered around the centre in the form of an octahedron.[15] Quantum-chemical calculation (CNDO/BW methods) of the potential surface of the reaction of fluoride anion with ethylene shows that, depending on the distance between reactants and the angle of F^- attack, two reactions are possible: *viz.* nucleophilic addition to give $FCH_2\bar{C}H_2$, and cleavage of a proton to give HF and $CH_2{=}\bar{C}H$.[16] ESR parameters for the tetrafluoroethylene anion point to a non-planar structure.[17]

A comprehensive review entitled "NMR of carbanions" has been published,[18] as well as a detailed ^{13}C-NMR study of the influence of substituents on the charge distributions in enolates and related carbanions.[19] The observation of two sharp ^{7}Li-NMR signals in the spectra of the "trapped" triple-ion cryptate (**4**) illustrates the usefulness of ^{7}Li-NMR spectroscopy as a simple probe for the detection of (Li^+, Li^+ triple-ion) complexes in solution.[20] From NMR studies at −60° in DMF or MeOH, the percentages of alkali-metal derivatives of 1,3-dicarbonyl compounds in the (E,Z) (**5a**) and (E,E) (**5b**) conformers increase with the size of M and on introduction of a

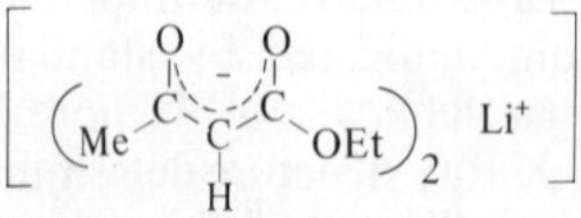

[Li^+⊂2·1·1]

(4) (5a) 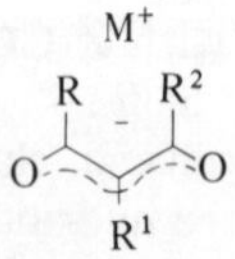 (5b)

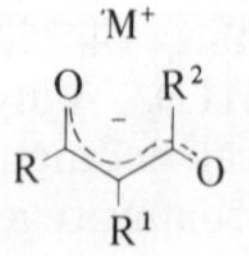

M = Li, Na, or K

methyl group at the central carbon atom, and decrease with increasing size of R(alkyl, Ph, or EtO) and R^2 [H, Me, or (un)substituted Ph].[21] For the Na, K, and Cs salts of $MeCOCH_2COOEt$, the Na and K salts of $(MeCO)_2CH_2$, and the Na salts of Me_3CCOCH_2CHO and MeCOCH(Me)COOEt in DMF, MeOH, and HMPA, the free ions prefer the (E,Z)-conformation whereas the ion pairs prefer the (Z,Z)-conformation.[22] Kinetic and isotope-effect data indicate that the (E,Z) → (E,E) conformational interconversion of 1,3-keto-aldehyde enolates in MeOH occurs *via* hindered rotation about the C—C double bond, whereas the (Z,Z) → (E,Z) interconversion takes place either *via* this mechanism (Li and Na salts of Me_3CCOCH_2CHO) or *via* intermediate formation of the dicarbonyl form (*C*-protonation-deprotonation as in the case of HCOCH(Na)COOEt).[23] Theoretical calculations predict that the "gas-phase" lithium enolate of acetaldehyde is chiral with a structure similar to that of a delocalized alkyllithium, although a planar structure with the lithium bound to oxygen is close in energy.[24] RO^- (R = Me_3C, Br, Et, Me, or H), generated by dissociative electron attachment from RONO, deprotonates Me_2CO in an ion cyclotron resonance spectrophotometer to give $MeCO\bar{C}H_2$ which is stable with respect to 1,3-hydrogen shifts.[25] Verification of dipole stabilization in the carbanions formed by loss of a proton from the methyl groups of methyl formate and *N*-methylformamide has come from *ab initio* SCF calculations, but differences between theory for the free anions and experiment is attributed to the effect of lithium-ion complexation.[26] Reaction pathways starting from different ester conformations are thought to be responsible for observed differences in stereoselectivities in base-promoted double- bond migrations in ethyl alk-2-enoates.[27] Anionic-induced *cis–trans* isomerization of dimethyl perfluoro(4-methylpent-2-ene)dioate takes place in a heterogenous system (undissolved potassium fluoride) and is promoted by the solvating effect of dipolar aprotic solvents such as DMSO, HMPA, or tetrahydrothiophene 1,1-dioxide.[28] A semi-empirical study of carbanions derived from cyanoacetic acid derivatives predicts carbanionic structures throughout, except in the case of the dinegative ion of cyanoacetamide which could be considered as originating from its aminoacetylenic tautomer.[29] Evidence from UV studies shows that delocalized and unfunctionalized carbanions can hydrogen-bond not only as the reported "free ions" in pure alcohols, but also as solvent-separated ion pairs in a medium of high dissociating power and cation-coordinating ability such as DMSO.[30] An ESR study of the ion pairs of the 9-fluorenone anion radical with 18-crown-6-ether-complexed Na^+ and K^+ ions indicates that an ethereal solvent molecule can solvate the crown-complexed cation on the opposite side of the anion with respect to the plane of the crown ether ring.[31] Interaction of 9-(*n*-propyl)fluorenyllithium with tertiary polyamines in aprotic solvents has been investigated by optical spectroscopy and the results explained in terms of externally complexed tight ion pairs and ligand-separated ion pairs, the latter complexes being much less soluble.[32] ^1H- and ^{13}C-NMR spectra of the Li, Na, and K salts of 2-ethylpyridine without and in the presence of crown ethers and cryptands indicate that the uncomplexed salts are present as contact ion pairs.[33] Physical measurements, primarily high-resolution NMR spectroscopy, support the contention that the allyl alkali-metal compounds are basically ionic in solution, but exist as intimate ion pairs or clusters.[34] Metallation of 2-methylbut-2-ene with two equivalents of *n*-butyllithium–TMEDA complex under argon at 25° gives initial formation of the linearly conjugated dianion (**6**), with facile and complete isomerization to the thermodynamically more stable cross-conjugated dianion

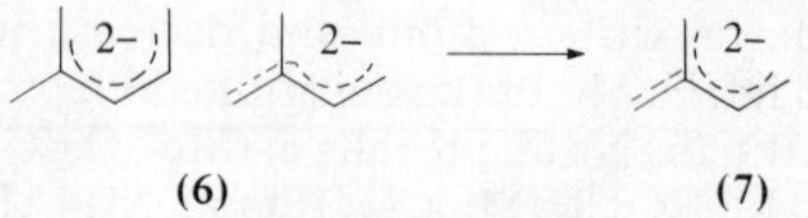

(6) (7)

(7).[35] CNDO/2 calculations of bond-lengths, total energies, atomic charges, and bond-orders of *cis*- and *trans*-crotyl anions and their complexes with lithium show that the *cis*- is more stable than the *trans*-configuration by 0.3 kcal mol^{-1} in the case of the free anion and by 31.2 kcal mol^{-1} in the case of the lithium complex.[36] Chemical trapping of the potassium salts of pentadienide, 2-methylpentadienide, or 2,4-dimethylpentadienide with chlorotrimethylsilane in THF produces preferentially (Z)-2,4-pentadienyltrimethylsilane or its methyl-substituted analogues, indicating that these salts have the "U"-shaped structure in solution; by contrast, trimethylsilylation of potassium pentadienide and 2-methylpentadiene in the crystalline state produces the (E)- and the (E),(Z)-isomers, respectively, suggesting that the geometry is drastically changed to "W"- or "S"-shaped by the medium effect.[37,38] Reaction of (E)- and (Z)-(1-iodoalkenyl)silanes with 2.1 equivalents of *tert*-butyllithium produces the synthetically important (E)- and (Z)-(1-lithioalk-1-enyl)silanes which isomerize at temperatures $> -70°$, reaching an (E):(Z) equilibrium distribution at 0° of 86:14 ($\pm 5\%$), respectively, irrespective of the size of the β-alkyl substituent.[39]

Rearrangements of carbanions have been reviewed.[40] An intermolecular transmetallation mechanism with deaggregation and ionization steps is proposed to explain the orders of reaction, parameters of activation, and solvent dependency for the rapid and quantitative vinyl-to-allyl anion rearrangement of (**8**) to (**9**).[41] The (Z)-isomer is the only observable form of 2-(2-methylphenyl)-1,2-diphenylvinyllithium (**10**) in THF solutions between $-70°$ and (temporarily) $+40°$,[42] whereupon it isomerizes to the benzyllithium derivative (**11**) which then cyclizes by a rarely encountered reversible attack on the *syn*-phenyl group to give a cyclohexadienide anion (**12**) containing a seven-membered ring;[43] (**12**) exhibits ^{1}H-NMR signal coalescence phenomena and rearranges under thermodynamic control to 1,2-diphenyl-1-indanide anion (**13**) (*cf*. related example with even better separability of the individual isomerization steps[44]). Vinylic deprotonation of β-substituted acrylic

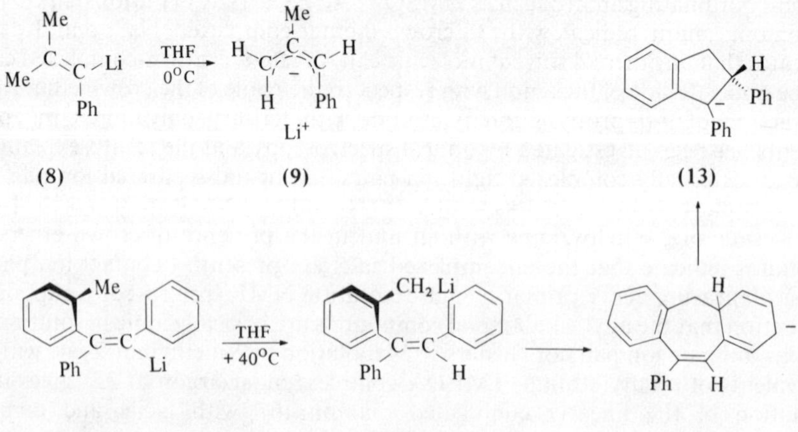

(8) THF, 0°C (9) (13)

(10) THF, +40°C (11) (12)

ester and nitrile derivatives occurs at either the α- or β-position depending on the nature of the substituent[45] (see also References 110 and 111). The isomerization of optically active secondary propargyl alcohols, $RCH(OH)C{\equiv}C(CH_2)_nMe$, to terminal acetylenic alcohols, $RCH(OH)(CH_2)_{n+1}C{\equiv}CH$, by potassium 3-aminopropylamide proceeds without loss of configuration at the hydroxylated carbon.[46] In the base-induced isomerization of 1,4-dialkoxybut-2-ynes, $ROCH_2C{\equiv}CCH_2OR$, to conjugated dienes of the type $ROCH{=}CHCH{=}CHOR$, the preferred formation of the (Z),(E)- and (Z),(Z)-isomers indicates that the intermediate allenic carbanion with (Z)-configuration is favoured above that with (E)-configuration, probably as a result of its higher kinetic stability.[47] The diacetylenic diamine, $Me_2NCH_2C{\equiv}C(CH_2)_3C{\equiv}CCH_2NMe_2$, rearranges instantaneously and quantitatively *via* its lithium derivative to a conjugated trienediamine [1,2-bis(β-dimethylaminovinyl)cyclopentene].[48] ^1H- and ^{13}C-NMR analysis of the deprotonation of π-(tricarbonylchromium)fluorene by excess of KH in THF shows that when the anion is produced at −20° or below, the $Cr(CO)_3$ group is bonded to one of the phenyl rings (η^6-anion), whereas ionization at room temperature produces solutions containing mainly the anion with the $Cr(CO)_3$ bonded to the cyclopentadienyl ring (η^5-anion) in equilibrium with the η^6-isomer.[49] The reversible isomerization of η^6- and η^5-fluorenylchromiumtricarbonyl anions in THF solution shifts towards the former on addition of 18-crown-6 ether or HMPA, but the most pronounced shift is observed upon replacement of an alkali metal by $Ph_3\overset{+}{P}Me$.[50–52]

A review entitled "Spectrophotometric investigations of arylmethyl carbanions" has been published.[53] Reactions of R^- ions, produced from the nucleophilic displacement reaction between MeO^- and Me_3SiR in an ion cyclotron resonance spectrometer, with methyl nitrite show that (*i*) the benzyl ion does not undergo benzylic hydrogen/phenyl hydrogen scrambling prior to reaction, (*ii*) the allyl ion reacts as a symmetrical species, and (*iii*) that hydrogen scrambling does not occur with the propargyl anion, and that $C_3H_3^-$ may either react as $HC{\equiv}\bar{C}H_2$ or $H\bar{C}{=}C{=}CH_2$.[54] Treatment of benzyl methyl ether with *n*-butyllithium–TMEDA in hexane leads to benzylic lithiation, without *ortho*-lithiation or a Wittig-type rearrangement.[55] The technique of multi-photon dissociation using a low-powered infra-red laser has been used to probe the mechanistic details of an ion-molecule reaction in the gas phase for the first time, showing that deprotonation of norbornadiene by MeO^- is accompanied by structural rearrangement to benzyl and cycloheptatrienyl anions.[56] Evidence from NMR and ESR spectroscopic studies indicates that the major product of the reaction between *tert*-butyllithium and trimesitylborane in THF is the boron-stabilized carbanion (**14**) and not a radical anion as previously reported.[57] The use of THF as solvent in metallations of 2,4-dimethyl-pyridine and -quinoline by strongly basic reagents promotes the formation of the 2-lithiomethyl reagents which isomerize to the thermodynamically more stable 4-lithiomethyl derivatives after relatively long reaction periods, or in the presence of amines, or an excess of the parent heterocycle.[58] From the effect of substituents on the rate of rotation around the C(8)—C(9) bond in 9-(1-naphthyl)methylenecyclooctatrienyl anions,[59] it is concluded that the process occurs through the transition state (**15**) with the negative charge located in the region of C(9) as indicated.[60]

The use of vibrational IR and Raman spectroscopy in structural studies of carbanions has been reviewed.[61] Russian workers have correlated values of ν_{CO} of anions containing ester, amide, and carboxylato groups with the C—O bond orders as calculated by the HMO method.[62] Formation kinetics for the growth of the anions of *trans*-stilbene, pyrene, and biphenyl in liquid methylamine using a 5 ns pulse (490, 410, and 490 nm, respectively) appear to be diffusion-controlled.[63]

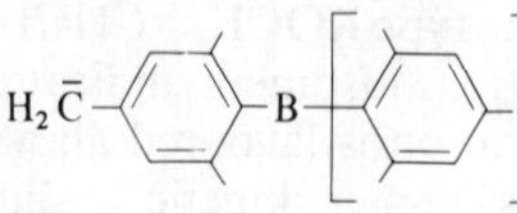

(14)

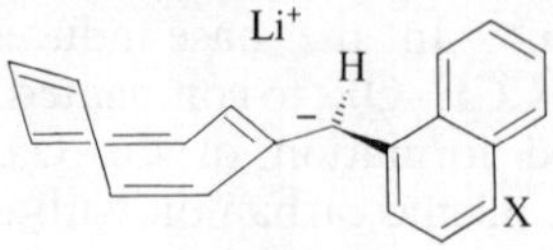

(15)

X = OMe, Me, H, or Cl

Aromaticity

8-Phenyl- and 8,9-diphenylbicyclo[5.2.0]nonatetraenyl anion (**16**), obtained in fairly stable solutions by base treatment of the corresponding bicyclo[5.2.0]nona-1,3,5,8-tetraenes, display properties characteristic of aromatic systems.[64] ^{1}H- and ^{13}C-NMR experiments as well as MO calculations give a consistent picture of the *s*-indacenyl dianion as a peripheral, delocalized, diatropic 14π-electron system similar to the 10π-electron indenyl anion system.[65] Arguments proferred in support of the anti-bicycloaromaticity of the 7-norbornadienyl anion appear to be no longer wholly valid following its efficient preparation from 7-chloronorbornadiene and lithium *p,p*-di-*tert*-butylbiphenyl and subsequent reaction with some electrophiles.[66] Contrary to a previous report of *endo*-stereoselection, electrophilic attack on bicyclodecatetraenyl anion (**17**) proceeds on the *exo*-side, an observation that is explained in terms of anti-bonding interaction between the canted π-lobes of the distorted allyl anion moiety, C(7)—C(8)—C(9), and the vacant orbital of the attacking species (alkyl halides, CO_2, or bulky trimethylsilyl chloride).[67] The ^{1}H-NMR spectra of carbanions obtained from simply 7-substituted cycloheptatrienes with KNH_2 in liquid ammonia are consistent with a non-planar pentadienyl-like structure having paratropic character.[68] A thorough analysis of the orbital interactions between the π-type molecular orbitals in a series of bicyclic carbanions, *e.g.* bicyclo[3.2.1]octa-3,6-dien-2-yl anion (**18**) reveals that homo-aromaticity and bicyclo-aromaticity are absent.[69,70] A review covering the aromaticity of conjugated ions has appeared[71] as has a theoretical method of calculating the maximum possible charge on an aromatic anion.[72] A combined Möbius–Hückel type of aromaticity in doubly-lithium-bridged $R_4C_4Li_2$ systems is indicated by molecular-orbital calculations.[73] The unusually extensive exocyclic "charge flight" from the cyclopropenide ring to the cyano group in cyanocyclopropenide ion is affirmed by MINDO/3 calculation, although it should be emphasized that the MINDO wave-function of this ion is not anti-aromatic in the absolute sense.[74] The structure of the 4-hydridopyridyl anion is confirmed by NMR spectroscopy to be planar with no 1,5-homo-aromatic overlap.[75] The diheteroanthracenide anions (**19a**), (**19b**), and (**19c**), prepared from the corresponding amines using KNH_2 in liquid ammonia, exhibit paratropicity, the magnitude of the induced ring current being greatest in the former and smallest in the latter.[76] The linearly benzannulated azocinnyl anion (**20**) distinctly lacks the type of π-frame stability normally associated with "aromatic" systems as evidenced by its rapid ring-closure to the dibenzobicyclic frame of the anion (**21**) on warming gradually to *ca.* 0°.[77]

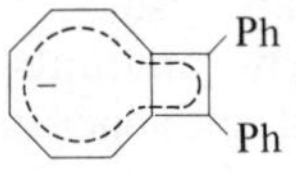

(16) 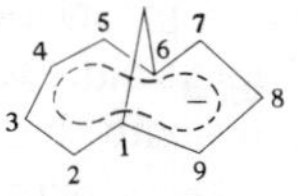 (17) 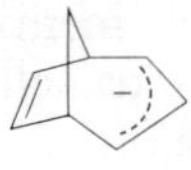 (18)

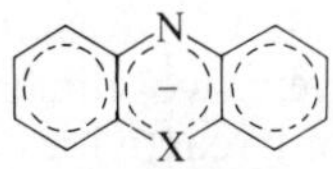

(19) a: X = NMe
b: X = O
c: X = S

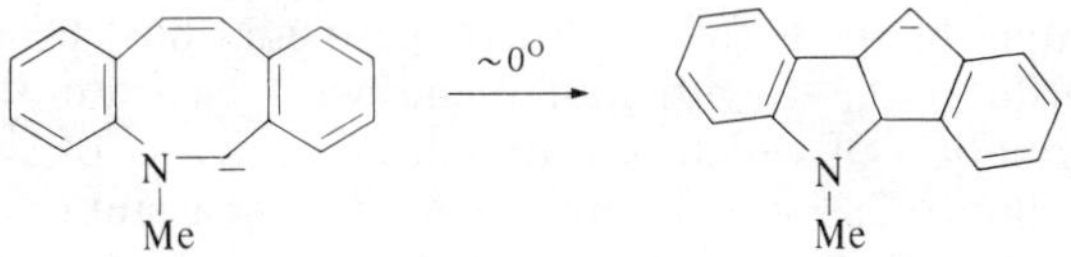

(20) ~0° (21)

Reactions of Carbanions

The dichotomous behaviour long observed in the condensation of active methylene compounds with nitrosobenzene seems to be resolved at last; kinetic evidence suggests that the stronger the CH-acid the more favourable is Sach's reaction (leading to azomethines), and conversely, the weaker the CH-acid the more probable Kröhnke's route (leading to nitrones) becomes.[78] Carbanions derived from active methylene compounds with pK_a 7–14 add regiospecifically to the pyridinium γ-position in *N*-(2,6-dimethyl-4-oxopyridin-1-yl)pyridinium tetrafluoroborate to yield 1,4-dihydro adducts which in some cases fragment to give 4-substituted pyridines, a class of compounds of which few examples are known.[79] Judicious choice of reaction conditions allows the addition of carbanions to unsymmetrical *p*-quinones at either carbonyl carbon.[80] Reaction of acyclic α-haloketones with malonate anion induces a Favorskii-type rearrangement leading to cyclopropanols.[81] Butadiene is transformed into diethyl 2,7-di(ethoxycarbonyl)-oct-4-enedioates and ethyl 2-ethoxycarbonyl-6-methoxyhex-4-enoates by reaction of π-4-chloro- and π-4-methoxy-butenylpalladium complexes, respectively, with malonate anion.[82] Reaction of 4-acetoxy-2-azetidinone with tertiary-stabilized carbanions affords carbon-substituted 2-azetidinones which can be transformed into 4-alkylidene-2-azetidinones.[83] Rate constants for the alkylation of Ph_3CLi, Ph_3CNa, fluoren-9-yl-lithium, and 9-cyanofluoren-9-ylcaesium by $Me(CH_2)_nCl$ ($n = 0$–5) in ether solvents pass through a minimum at $n = 2$.[84] In the alkylation of the free Ph_3C^- ion by haloalkanes in DME and THF, the rate exceeds that of the $Ph_3C^-\ Cs^+$ contact ion pair by three orders of magnitude and is 2–3 times greater than that of the solvent-separated $Ph_3C^-\ Li^+$ ion pair.[85] Reaction of benzaldehyde with Cl_3C^- in the presence of benzylquininium chloride as phase-transfer catalyst yields a 5.7% enantiomeric excess of (R)-$PhCH(OH)CCl_3$.[86] Addition of trichloromethyl anion, generated by cathodic reduction of carbon tetrachloride in the presence of chloroform, to electrophiles such as aldehydes or vinyl acetate occurs with extremely high current efficiency.[87] Reaction of tris(trimethylsilyl)-methyllithium with epoxides involves the transfer of a silyl group from carbon to oxygen in the alkoxide product, and in one case (styrene oxide) the intermediate so formed reacts to give a cyclopropane in what is a hitherto unobserved homologue of the Peterson reaction.[88] The facile conversion of tributyl(trimethylsilylmethyl)tin into Me_3SiCH_2Li by transmetallation has found further application in the synthesis of 2-oxoalkylsilanes (silylmethyl ketones) from carboxylic acids, esters, and chlorides.[89] Reaction of trimethylsilylisothiocyanate with catalytic amounts of tetrabutylammonium fluoride offers a facile and mild method of generating isothiocyanatomethanide, $\bar{C}H_2NCS$, which reacts with carbonyl compounds to give oxazilidinethiones in fair-to-good yields.[90] Oxazole alcohols are readily formed *via* the interaction of lithiomethyl isocyanide with lactones.[91]

The effect of the nature of the solvent on the kinetics of the reaction of butyllithium with *tert*-butyl peroxide has been investigated.[92] Reaction of perfluoro-2,3-dimethylbutadiene with caesium fluoride gives two dimers, the formation of which is rationalized in terms of stabilities and reactivities of the perfluoro-2,3-dimethylbutadiene with caesium fluoride gives two dimers, the cyanophenyl)propionitriles with base is dichotomous, giving 5-amino-11*H*-indeno[1,2-*c*]isoquinolines derived from both the less stable C(3)-carbanion, as well as the more stable C(2)-carbanion.[94] The matter of cyclopropyl anion ring-opening continues to provoke debate with a fresh suggestion that the reaction proceeds in a symmetry-forbidden manner.[95] 1-Lithio-1-vinylcyclopropanes, prepared from the corresponding methylseleno derivative and alkyllithiums, react with electrophiles (H_2O, RX, CO_2) exclusively on the cyclopropyl ring producing functionalized vinylcyclopropanes in good-to-high yields.[96] According to semi-empirical MNDO molecular-orbital calculations, the most favourable pathway for the recently demonstrated exchange between cyanide ion and the nitrile group of acetonitrile is a prototropic-type mechanism *via* a cyclopropane intermediate.[97] Regioselective oxetane formation from 3,4-epoxy-alcohols and alkoxide anions in aqueous DMSO is much less sensitive to steric hindrance at the attached oxirane ring than the corresponding reactions under anhydrous conditions (NaH in THF).[98] Adducts of carbanions derived from cyclic vinyl ethers to alkyl halides can be unmasked under extremely mild conditions to give hydroxyketones or dicarbonyl systems in generally high yield.[99] Heating of α,α'-tetra-substituted ethers with sodium amide at 225–300° leads to smooth cleavage of one of the α-substituents which is reduced in the process.[100] Mixing of a photo-responsive surfactant with a conventional cationic micelle produces rate constants for photo-irradiated proton abstraction from benzoin that are much larger than the rate constants for "dark" reactions, indicating that micellar catalysis can be controlled, in principle, by switching the light source on and off.[101]

Full experimental details have appeared of the utility of secondary α-keto-amide 2,4,6-triisopropylbenzenesulphonylhydrazones in the Shapiro reaction for the introduction of the acrylate functionality *via* the allenic dianions $[RCH{=}C{=}C(O^-)\bar{N}R']$,[102] which are versatile intermediates for the preparation of substituted 3-methylenetetrahydrofuran-2-ones[103] and 3-methyleneazetidin-3-one derivatives.[104] 2,4,6-Triisopropylbenzenesulphonylamidrazones, on treatment with *tert*-butyllithium and after decomposition of the initial dilithio species, yield α-lithioenamines which act as acyl anion equivalents.[105] A mechanism involving the direct displacement of an acyl anion is considered to be the most likely mechanism in the fluoride-catalysed conversion of acylsilanes into aldehydes and ketones.[106] The carbonylation of ArLi (Ar = Ph, *o*-anisyl) in the presence of alkyl bromides affords diaryl(alkyl)carbinols in good yields, apparently *via* the intermediacy of *benzoyllithium.*[107] 2,2-Difluorovinyllithium, prepared in quantitative yield from $CF_2{=}CH_2$ and *sec*-butyllithium at low temperature, reacts with carbonyl compounds to give the corresponding alcohols in excellent yield.[108] Trapping of vinyl carbanions derived by alkylation and subsequent decomposition of arenesulphonylhydrazone dianions with a variety of electrophiles results in an overall reaction sequence **(22)** → **(23)**.[109] Treatment of vinyl carbanions produced from ethyl *cis*- and *trans*-cinnamate with various electrophiles gives mostly β-substituted products, PhC(E)=CHCOOEt, having *trans*-geometry,[110] whereas the corresponding products in the case of the vinyl carbanions derived from *cis*-cinnamonitrile under comparable conditions are the C(α)-derivatives, PhCH=C(E)CN, with variable geometry determined by the solvating properties of the reaction medium.[111]

Complexes of 2,5- or 2,4-diphenyl-1,6,6$a\lambda^4$-trithiapentalene, anhydrous zinc or iron(III) chloride, and lithium serve as catalysts for the direct lithiation of olefins.[112] The products from the dimerization of β-alkyl-substituted acrylates in the presence of promoted potassium or sodium catalysts consist exclusively of the diesters of alkylidene-β-alkylglutaric acid which are formed by a mechanism involving metallation of the α-vinylic carbon, followed by addition at C(3) of a second monomeric molecule.[113] Dilithio species, obtained from treatment of $\Delta^{3,4}$-3-bromocyclohexenone ketals with two equivalents of *n*-butyllithium, act as "one-pot" equivalents of β-vinyl carbanions of cyclohexenones.[114] The phenol-derived 3-hydroxy-5-oxocyclopentenyl carbanion equivalent **(24)** reacts efficiently with various electrophiles to form substituted cyclopentenediol derivatives which can be converted into the corresponding 2-substituted 4-hydroxycyclopent-2-enones.[115]

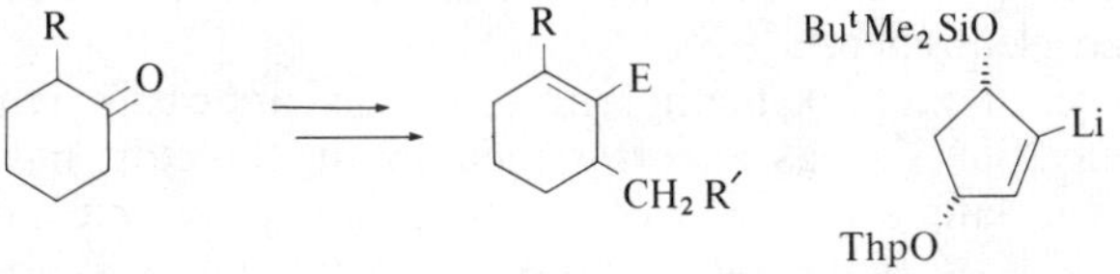

R, O, R, E, CH_2R', Bu^tMe_2SiO, Li, ThpO

(22) (23) (24)

3-Methylbut-2-enenitrile undergoes base-catalysed cyclo-dimerization to 3-amino-4-cyano-1,5,5-trimethylcyclohexa-1,3-diene by a mechanism thought to involve initial deprotonation to form an allylic anion, which adds to a second molecule in a Michael reaction at the γ-carbon atom, followed by a Thorpe–Ziegler cyclization.[116] α-Chloroallyllithium species, generated *in situ* by lithium diisopropylamide deprotonation of the corresponding allyl chloride, react with a variety of primary alkyl bromides exclusively at the α-position; the derived secondary allyl chloride intermediates are useful synthetic "building blocks" for the direct, facile synthesis of di-substituted olefinic sex pheromones.[117] 3-Silyloxydienyl anions of the type **(25)** present a convenient C_5-unit for both the construction and attachment of functionalized dienes **(26)** and dienophiles **(27)** *via* successive γ-alkylation and fluoride-promoted silyl ether cleavage.[118] Introduction of a sulphur moiety into **(25)** further increases the functionality of this C_5-unit so that **(26**; E = MeS) is a practical equivalent of the hypothetical β-deprotonated divinyl ketone **(28)**; a further advantage is that both enone functions can be liberated separately, thus allowing the regioselective poly-substitution of (**26**; E = MeS) by a range of different electrophiles and nucleophiles (see arrows).[119]

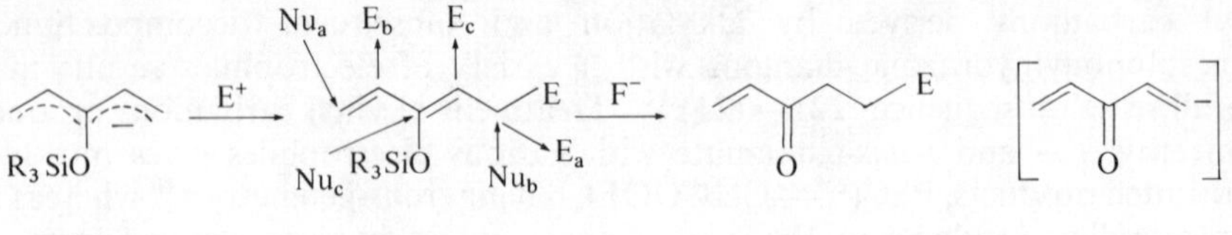

(25) (26) (27) (28)

Hydrolysis of α-lithiated (trimethylsilyloxy)butatrienes, generated by 1,4-elimination of methanol from 1-trimethylsilyloxy-4-methoxyalk-2-ynes with *tert*-butyllithium, yields separable mixtures of α-allenic aldehydes and allenyl trimethylsilyl ketones.[120] Reaction of allenyllithium reagents with carbon disulphide affords β,γ-unsaturated γ-dithiolactones *via* lithium alk-3-ynedithioates which can be intercepted by methyl iodide.[121] Sequential treatment of lithium chloropropargylide with thexylalkenylchloroboranes and aldehydes affords, depending on the reaction conditions, either 1,3-enynols or 1,2,4-trienols.[122] The overall migratory-aptitude order of alkyl groups in the iodination of lithium ethynyltrialkylborates is bicyclooctyl > *n*-butyl > cyclohexyl > *iso*-butyl > *sec*-butyl > thexyl.[123] Reaction of lithium phenylacetylide with the complex BuLi–Bu^{t}OK in a mixture of THF and hexane leads to the *ortho*-metallated phenylacetylide; subsequent reaction with methyl iodide, dimethyl disulphide, trimethylchlorosilane, and selenium completes an excellent synthetic procedure for the corresponding *ortho*-substituted phenylacetylene, not easily accessible in other ways, and benzoselenophen.[124]

The excellence of methoxymethyl not only as an easily removed phenol-protecting group, but also as an *ortho*-metallation director, has found further application in the latent synthesis of averufin, a key intermediate in aflatoxin biosynthesis.[125] The use of the *N*-(*tert*-butoxycarbonyl) group as an NH protecting group offers for the first time the possibility of functionalizing 1-unsubstituted pyrroles at the 2-position *via* lithiation (with lithium 2,2,6,6-tetramethyl-piperide).[126] When treated with *n*-butyllithium in THF at −70°, 3-[2-(2′-bromophenyl)ethyl]thiophene is metallated at C(2) in the thiophene moiety, apparently *via* the cyclic state (**29**).[127] Metallation of 3,5-dimethylisooxazole occurs selectively, first at the 5-position, and then at the 3-position.[128] 2-Fluoropyridine is lithiated regioselectively at the 3-position by lithium diisopropylamide at low temperatures without side-reactions (such as nucleophilic attack); subsequent reaction with aldehydes yields the corresponding fluorinated alcohols which can be selectively oxidized.[129] Lithiated *N*-pivaloyltetrahydroisoquinoline possesses a surprisingly high nucleophilicity, allowing alkylation at C(1) even with a primary alkyl *chloride*, a *secondary* alkyl iodide, or an easily deprotonated ketone; removal of the pivaloyl group then provides a convenient way of obtaining 1,1-di-substituted tetrahydroquinolines.[130] A key step in a new versatile regiospecific route to tetracyclic anthracyclinones entails the attachment of a nucleophilic C/D-ring synthon (**30**) to a suitably functionalized A-ring intermediate.[131] Metallation of 7-chloro-1-methyl-5-phenyl-1,3-dihydro-2*H*-1,4-benzodiazepin-2-one (diazepam) with two equivalents of lithium diisopropylamide in THF–hexane affords its monolithium salt in sufficient concentration to react with alkyl halides, aldehydes, ketones, and esters to give 3-substituted diazepam derivatives in good yields.[132] Other reactions of synthetic interest that have appeared this year include an improved preparation of 3-fluoroveratrole *via* lithiation of 3-fluoroanisole at the 2-position,[133] a new α-methylene ketone synthesis involving a highly regio-controlled and rapid lithiation of 3-methyl-4*H*-5,6-dihydro-1,2-oxazine,[134] the formation and *C*-alkylation of lithioimidazolines,[135] and the efficient side-chain metallation of tri-substituted oxazoles.[136]

Full details have appeared of the obtention of dipole-stabilized carbanions by lithiation of sterically hindered esters adjacent to oxygen, a key step in a reaction sequence that allows electrophilic substitution of the α-hydrogen of a primary alcohol.[137] Condensation of the novel dilithiated carbamate (**31**) with electrophiles

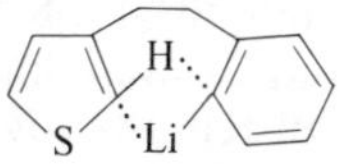

(29) 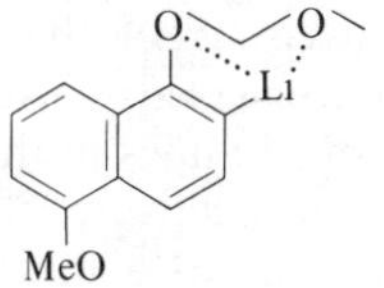 (30) 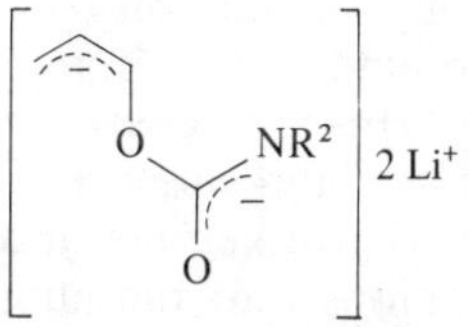 (31)

affords predominantly (z)-enol esters (γ-adducts) from which β-substituted carbonyl compounds can be obtained by acid hydrolysis.[138] Lithiation of *N*,*N*-dialkyl-2,2-diethylbutyramides adjacent to nitrogen, followed by addition to an aldehyde or ketone, an acid-driven nitrogen-to-oxygen rearrangement, and basic hydrolysis, provides secondary (α-lithioalkyl)alkylamine synthetic equivalents.[139] A significant enhancement in γ-regioselectivity is observed in alkylation reactions of metallated β,β-di-substituted unsaturated amides (see **32**) by conversion into their cuprates using cuprous iodide.[140] The reactions of α-bromoisobutyramides with the anions from amides and a thioamide afford oxazolin-4-ones and a thiazolidin-4-one, respectively; these are useful intermediates for the preparation of ester derivatives.[141]

This year sees the appearance of both a review[142] and an account[143] of current progress in the gas-phase chemistry of organic anions as studied by ion cyclotron resonance spectroscopy and the flowing afterglow technique, respectively. The gas-phase reactions of carbanions with triplet and singlet molecular oxygen have been studied using the flowing afterglow technique.[144] The reaction of a variety of carbanions with esters in the gas phase proceeds by an addition–elimination–deprotonation mechanism similar to that for the Claisen condensation in solution.[145] Allyl anion is observed to add to 1,1-dimethylsilacyclobutane in a flowing afterglow apparatus, giving adduct (**33**), the first silicon anion bearing *five carbon substituents*.[146]

Electron-transfer reactions of carbanions and related species have been reviewed.[147] A series of mechanistic steps is proposed for the reaction of phenylalanine dimethylamide with $PhNO_2$ and potassium *tert*-butoxide in *tert*-butyl alcohol at 50°, beginning with the one-electron oxidation of the α-amino carbanion, and proceeding through a ketimine and/or an enamine which is rapidly oxidized to the eventual products, potassium nitrobenzenide ($PhNO_2^{-}K^{+}$) and degradative fragments of the aminoamide.[148] Electron transfer from fluoren-9-yl anion causes facile reduction of 9-diazofluorene *via* radical-anion intermediates.[149] In DMSO solution electron transfer from 9-arylfluorenyl carbanions, 9-$ArFl^{-}$, to electron acceptors including $PhSO_2CH_2Br$, $PhSO_2CH_2I$, and $R_2C(NO_2)_2$, follows second-order kinetics and results in the formation of dimers, $(9\text{-}ArFl)_2$, in high yields.[150] The cyclic voltammogram of 1,3,5-tris(*p*-biphenyl)benzene in 0.2 M (M = mol dm^{-3}) TBAP–DME at −50° exhibits five reversible one-electron steps which may be evidence for the electrochemical form of mono-, di-, tri-, tetra-, and penta-anions of this hydrocarbon.[151]

In an article with the rather broad title "The search for new organometallic reagents for organic synthesis", several topics of interest are covered, including organo-element-group-assisted 1,3-anionic cycloadditions, element–halogen exchange in organic synthesis, element–lithium exchange in organic synthesis, new carbonyl olefination reagents, and hydrogen–lithium exchange in organometallic compounds.[152] A comprehensive account of the use of metalloids ($-SiMe_3$, $-SnR_3$) as protected carbanions has appeared.[153] Acylsilanes and *C*-stannylimines react as anion equivalents with organic halides in the presence of KF-18-crown-6-ether to give ketones and ketimines, respectively.[154] Prior conversion of classical carbanions into titanium reagents leads to a marked increase in their chemo-, diastereo-, and enantio-selectivity.[155] The trimethylsilylcopper reagent **(34)**,

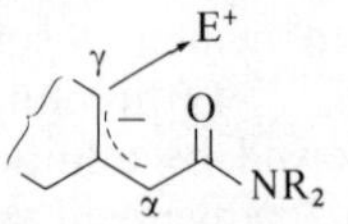

E^{+}, γ, O, NR_2, α

(32)

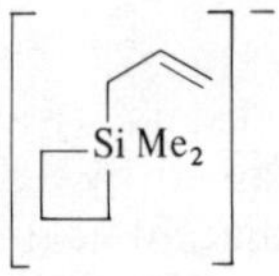

$[\text{(CH}_2\text{)}_3\text{SiMe}_2\text{CH}_2\text{CH=CH}_2]^{-}$ (Si Me_2)

(33)

$[Me_3SiCH{=}CHCH_2]^{-}$ $Cu(CN)Li^{+}$

(34)

prepared by mixing one equivalent of allyltrimethylsilyl anion with one equivalent of copper(I) cyanide at −78°, gives selective γ-substitution reactions with alkyl and acyl halides, and highly selective conjugate addition reactions with α,β-unsaturated esters and ketones; in contrast, α,β-unsaturated aldehydes give only 1,2-addition.[156] A combination of thermal stability (up to 100°) and high solubility in various organic solvents makes mesitylcopper(I) not only a useful metallation reagent, but also an effective "holding group" in mixed lithium cuprate reagents.[157] Reaction of lithium dialkyl cuprates with *S*-2-pyridyl thioates in the presence of oxygen affords carboxylic esters in high yields, whereas under nitrogen ketones are generally obtained.[158] Mixed organoboranes can be conveniently prepared in a single stage by treatment of a dialkylborane with various lithium dialkyl- or diaryl-cuprates.[159] Lithium trialkylalkynylborates react in a stereoselective fashion with benzo-1,3-dithiolium tetrafluoroborate to give vinylboranes which, on oxidation, yield protected 3-oxo-aldehydes and, on hydrolysis, give protected α,β-unsaturated aldehydes.[160] The reaction of the anion of a methylmethoxycarbenotungsten, *viz.* $(OC)_5W{=}C(OMe)\bar{C}H_2$, with enol ethers leads, after protonation, to new alkylidene complexes which result from the pericyclic addition of the two aforementioned species.[161]

Enolates

Substituent effects in the methylation of enolate anions have been studied,[162] and confirmation obtained from the combined effects of temperature, LiCl, crown ethers, [2.2.1]cryptand, and HMPA on *C/O* ratios that ion-aggregates are the true reactants in the methylation of lithioisobutyrone in weakly polar aprotic solvents.[163] Lithium enolates from pentan-3-one, cyclohexanone, acetophenone, and mesityl oxide react with acyclic and cyclic allylic acetates in a palladium-catalysed alkylation in fair-to-good yields under mild conditions; the substitution is shown to proceed with retention of configuration.[164] The reaction of Li, Na, and K enolates of ethyl acetoacetate with diethyl sulphate in THF at room temperature gives 60–70% *O*-alkylation when performed in the presence of one molar equivalent of solid HMPA; without HMPA the Li enolate does not react, the Na enolate gives only *C*-alkylation, while the K enolate leads to 90% *C*-alkylation, the reaction being quite slow in the two latter cases.[165] Slow addition of pre-cooled solutions of lithium enolates in THF to solutions of equimolar amounts of acyl chlorides in the same solvent at temperatures between -80 and $-100°$ furnishes 1,3-diketones in acceptable-to-good yields.[166] A drastic solvent effect occurs in the asymmetric alkylation of *exo*- and *endo*-acyloxynorbornanes *via* lithium enolates; in THF and in THF–HMPA (4:1) as solvent, products with inverse configuration are obtained with almost complete asymmetric induction in each case.[167] The formation of carbon–carbon bonds by condensation of enolates with hexachlorobutadiene takes place by at least two mechanisms: *viz.* direct addition–elimination or a secondary condensation with an intermediate 1-chloroalkyne.[168] The trianions of methyl 3,5-dioxohexanoate and *N,N*-dimethyl-3,5-dioxohexanoamide are highly nucleophilic at the 6-position, yielding terminal condensation products upon alkylation with benzyl chloride, aldol condensation with benzophenone, and acylation with methyl benzoate.[169] The dianion (**35**) of (s)-(−)-β-hydroxybutyrolactone reacts with alkyl halides to give exclusively *trans*-2-alkyl-3-hydroxylactones, and with aldehydes to give 2,3-*trans*-di-substituted lactones which exhibit *erythro*-selectivity in the newly formed aldol moiety.[170] Dienone dianion derivatives (**36**), accessible by double deprotonation of γ,δ-unsaturated (*i.e.* α-allylated) ketones and dithio-esters, display ω- or d^5-reactivity, condensing with benzophenone exclusively in the δ-position.[171] The alkylation of *O*-silylated dienolates of unsaturated aldehydes, ketones, and esters with the thiocarbocation electrophile, 1,3-dithenium tetrafluoroborate, shows useful γ-selectivity ($\gamma:\alpha > 5:1$), the γ-products being selectively protected 1,5-dicarbonyl compounds.[172] α-Alkylated products of lithium 3-trimethylsilylmethyl dienolate anions are readily transformed into 3-alkylated 2-methallylsilanes by pyrolysis.[173] On the basis of the concept of soft and hard acids and bases, a step-wise nucleophilic attack of the anion of β-keto-esters at the electrophilic centre of the 2- and 6-positions or the 2- and 3-positions of 1-substituted 3,5-dinitro-4-pyridones is proposed to explain the remarkable difference observed in the reaction pathways; *e.g.* diethyl sodio-3-oxopentanedioate gives 1-substituted 3,5-bis(ethoxycarbonyl)-4-pyridones and sodio-1,3-dinitropropan-2-one, whereas the corresponding reaction with ethyl sodioacetoacetate gives ethyl 4-hydroxy-3,5-dinitrobenzoate together with aminopyridine homologues.[174] Further work on the α-alkylation of dianions derived from β-hydroxy-esters has led to an enantioselective synthesis of 4,4- and 6,6-di-substituted cyclohex-2-en-1-ones (each with an enantiomeric excess of 86%).[175] The lithium enolate (**37**) of methyl 3-(dimethylamino)propionate serves as an acrylate anion equivalent for the synthesis of α-methylene esters, acids, and lactones.[176] Mono-enolate anions [E = RC(O$^-$)=CHR′] are easily oxidized with $XCMe_2NO_2$ (X = Cl, NO_2, or p-$MeC_6H_4SO_2$) leading to coupling [$ECMe_2NO_2$ and RCOC(R′)=CMe_2] and symmetrical dimerization (E–E) products, the competition between these processes, which proceed by free-radical chains, is determined by bimolecular reactions of $XCMe_2NO_2^{-\cdot}$ and not by reactions of free $O_2N\dot{C}Me_2$.[177]

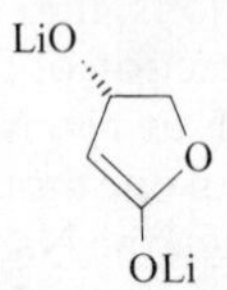

LiO OLi O Me₂N Li O OMe O⁻ R

(35) 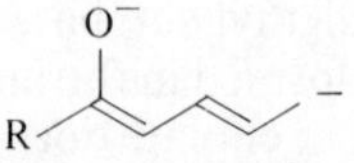 **(36)** 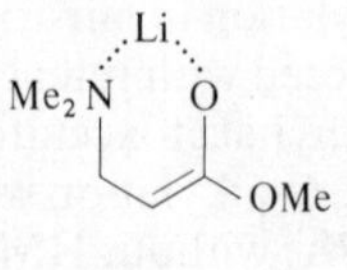 **(37)**

Full synthetic details have appeared of the enantioselective alkylation of ketones *via* chiral non-racemic lithioenamines;[178] in an accompanying paper, it is shown that (E)/(Z)-isomerization occurs for lithioenamines derived from acyclic and macrocyclic ketones, giving rise to optical antipodes with moderate-to-excellent enantiomeric purity.[179] By using a combination of aymmetric C—C coupling, chromatographic separation of diastereomers and their conversion into enantiomers (R)- and (S)-α-hydroxyketones and vicinol diols are formed in enantiomerically pure form from ketones, the chiral carbamoyllithium compound (**38**), and methyllithium.[180] Significant improvement of Cram's rule selectivity is realized in aldol condensations with chiral aldehydes by using the principle of double stereo-differentiation (double asymmetric induction).[181] Pure *erythro*-aldols such as (**40**) can be produced with high stereoselectivity (> 80:1) in a straightforward process starting from the reagent (**39**).[182] Condensation of ketone-derived enolates with phenylselenoacetaldehyde affords hydroxyselenides which are

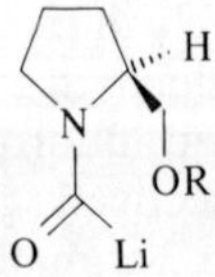

N H OR O Li O OSiMe₃ OH O Ph R

(38) 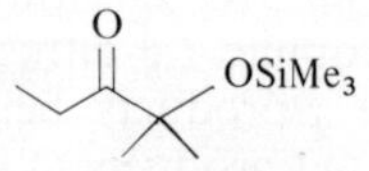 **(39)** 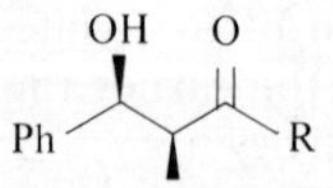 **(40)**

R = Me, Et, or Bu^n

easily converted into β,γ-unsaturated ketones by the action of methanesulphonyl chloride and triethylamine.[183] Aldol reaction of tris(diethylamino)sulphonium enolates (**41**) affords *erythro*-adducts in a stereoselective fashion which is independent of enolate geometry.[184] A solution to the recurring synthetic problem of introducing an α-substituent directly into α,β-unsaturated carbonyl compounds is provided by a sequence of reactions involving initial conjugate addition of organoaluminium reagents R_2AlX (X = SPh, SeMe), aldol condensation of the resulting aluminium enolate, and elimination of HX; the overall transformation is equivalent to the addition of aldehydes to the 1-acylethenyl anion equivalent

(**42**).[185] Triphenyltin enolates, prepared from lithium enolates and triphenyltin chloride, undergo a rapid aldol condensation with aldehydes, without the need for the presence of Lewis acids, to give predominantly the *erythro*-product regardless of the geometry of the starting enolates.[186] Crotyl boronates add to various aldehydes, yielding β-methylhomoallyl alcohols with a diastereoselectivity of $> 95\%$.[187] Full details of the generation of stereochemically homogeneous boron enolates and their stereoselective aldol condensations with aldehydes have been published.[188] The chiral 2-oxazolidone (**43**; R = H) serves as a useful, recyclable chiral auxiliary for carboxylic acids in highly enantioselective aldol condensations *via* the boron enolate derived from the corresponding *N*-propionylimide (**43**; R = COEt).[189] Zirconium enolates, *e.g.* (**44**), derived from prolinol amides, also exhibit excellent levels of asymmetric induction in the aldol process.[190] Boron enolates obtained from chiral ethyl ketones display remarkably high stereoselectivity in the aldol reaction,[191] as evidenced by the total synthesis of 6-deoxyerythronolide B, a monocyclic 14-membered lactone system with 10 asymmetric centres, *via* a sequence involving *four* highly stereoselective condensations.[192] The use of boron azaenolates, derived from achiral and chiral oxazolines, respectively, alters the stereochemical course of the aldol process from *threo*- to *erythro*-products with enantioselectivity of *ca.* 90% (77–88% e.e.) in the former and *ca.* 70–80% (40–60% e.e.) in the latter.[193]

The ability of α,β-unsaturated *sec*-thioamides to serve as Michael acceptors is greatly enhanced by selective *N*-alkylation of the derived monoanion;[194] by comparison, alkylation of *sec*-thioamide dianions (**45**) occurs selectively at the carbon α to the thiocarbonyl group.[195] Attempted preparation of a 2-aryloxycyclohex-2-enone has uncovered an unusual enolate-promoted 3,3-sigmatropic (Claisen) rearrangement.[196] Further work to delineate the scope and limitations of enolate anions as protecting groups for the selective reduction of dicarbonyl compounds has been published.[197] Reductive α-deoxygenation of α-acetoxy- or α-alkoxy-esters by naphthalenide anions provides a direct method for the generation of ester enolates; for example, in one application of this procedure, diethyl oxomalonate serves as a conjunctive reagent for stitching together an alkene and an alkyl halide with a malonyl group as linchpin.[198]

Rate constants for the α-epimerization of the rigid lactone (**46**) exhibit an unusually rare linear-free-energy correlation with the solvent donicity parameters, DN or B.[199] Base-catalysed homoketonization of a series of polycyclic

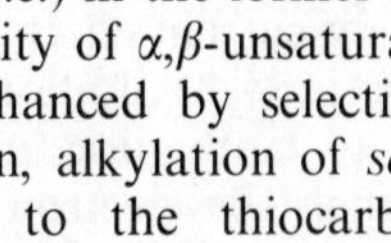

O^- $(Et_2N)_3S^+$; R^1–C(O$^-$)=CHR^2

(41)

[>C=$\bar{C}$–C(=O)R]

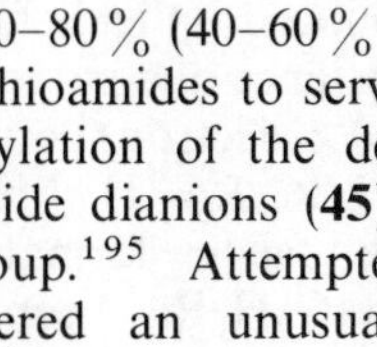

(42)

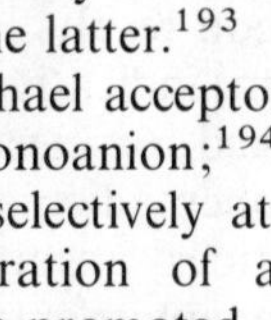

(43)

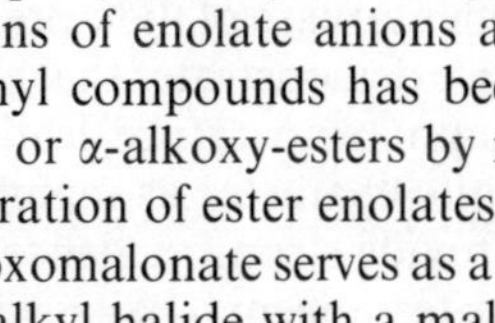

(44)

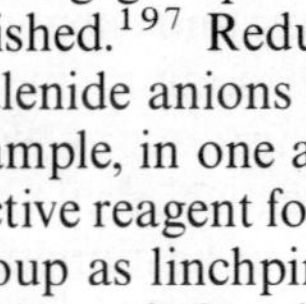

(45)

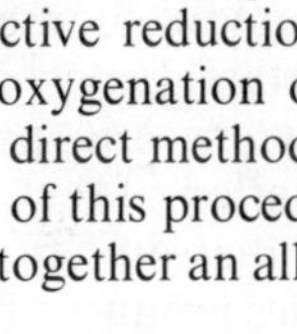

(46)

cyclopropanols has provided strong evidence for the intermediacy of their anions in hydrogen–deuterium exchange at α-methyl carbons in polycyclic ketones under homoenolization conditions.[200] Homoketonization of some readily prepared cyclopropoxides affords a new synthetic method for ring-expansion of [2.2.1] and [2.2.2] ring-systems.[201] Enolization of oxalacetate shows linear rate–buffer-concentration behaviour with tertiary amines, vitiating the prime evidence previously cited for a carbinolamine mechanism for the tautomerization.[202] A substantial enhancement in yield is observed in the reaction of enolizable ketones with the lithium salts of *N,N*-diethylbenzamide and *N,N*-diethyl-1-naphthamide when the hydrogen atoms α to the carbonyl group are replaced with deuterium.[203] Treatment of *N*-pivalylphenylalanine dimethylamide with potassium *tert*-butoxide in *tert*-butanol-*O-d* leads to hydrogen–deuterium exchange at the asymmetric α-carbon atom with a small, but reproducible excess of exchange *with retention of configuration*.[204]

Dianions

A comprehensive review entitled "Dianions and polyanions" has appeared.[205] Reduction of cyclohept[*f,g*]acenaphthylene (acepleiadylene) with lithium to its dianion results in a pronounced upfield shift of the ^{1}H-NMR resonances (by *ca.* 8 ppm) indicative of the existence of a paramagnetic ring-current effect, *i.e.* a vinyl-bridged 16π-electron (*i.e.* 4*n* π-electron) perimeter structure.[206] Addition of lithium cycloheptadienides (LiC_7H_8R; R = H, Me, or Bu^t) to lanthanide and actinide chlorides results in the facile formation of the 10π-electron aromatic cycloheptatrienide trianion ($C_7H_6R^{3-}$) complexed to the metal ion.[207] Calculated resonance energies of selected delocalized dicarbanions and higher delocalized carbanions show a good correlation with the experimental ease of their preparation and stabilities.[208] Spectroscopic and chemical evidence demonstrates that reduction of pyrene and perylene with sodium metal at room temperature leads to the formation of the respective tetra-anions with aromatic character as manifested by their diatropicity.[209] ^{13}C-Chemical-shift changes provide strong evidence that the species produced by reduction of 1,1-dimethyl-2,3,4,5-tetraphenyl-1-silacyclopentadiene with an *excess* of lithium metal in THF is the highly charged tetra-anion (**47**).[210] A convenient, low-cost synthesis of dipotassium cyclooctatetraenide is provided by the action of potassium on cycloocta-1,5-diene at 106–110° in the absence of solvent.[211]

Some new bond-delocalized dianions, named pseudo-oxocarbons, *e.g.* 2,1,3-bis-, 1,2,3-tris(dicyanomethylene)croconate salts, have been prepared.[212] Contrary to previous prediction, dianions of most monocyclic oxocarbons, together with all the neutral ones, are predicted by the graph theory of aromaticity to have very small resonance energies.[213] Qualitative analysis of electronic absorption and magnetic circular dichroism spectra of the oxocarbon dianions, $C_4O_4^{2-}$, $C_5O_5^{2-}$ and $C_6O_6^{2-}$, in terms of the perimeter model, and quantitatively by π-electron and semi-empirical all-valence calculations, indicates that many $n\pi^*$ states are present at relatively low energies but their identification in the experimental spectra is difficult.[214,215]

The dilithium reagent (**48**), derived from propargyl selenide, is a useful synthetic equivalent of acrolein dianion (see **49**), allowing the formation of two C—C bonds in a "one-pot" reaction; the methodology is exemplified by the synthesis of (±)-7-hydroxymyoporone.[216] In general, electrophilic reactions of the dianionic species,

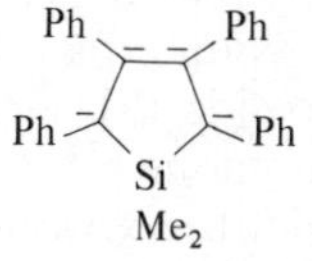

(47) 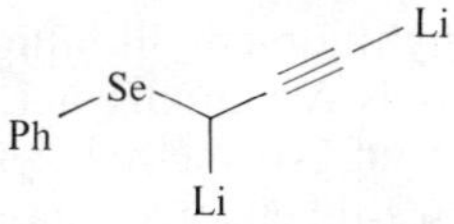 (48) 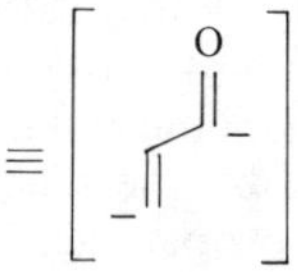 (49)

formed directly from furancarboxylic acids and lithium diisopropylamide, are very efficient with aldehydes and ketones, but not as satisfactory with alkyl halides and epoxides; they do not couple with allylic or benzylic halides, nor with nitriles or orthoesters.[217] Dimetallation of *N-tert*-butylmethacrylamide with *n*-butyllithium gives a useful reagent (equivalent to the dianion of methacrylic acid) for the facile preparation of α-methylene lactones and α-substituted acrylamides.[218] 1,1-Distannyl-1-alkenes, formed by hydrostannation of 1-stannyl-1-alkynes, give α-stannylvinyl anions (and in principle, vinyl dicarbanions) on treatment with methyllithium.[219] Dilithio-*S*-indacene, on reaction with an excess of chlorocarbene, produces *S*-benzvalenobenzobenzvalene, a double-valence isomer of anthracene.[220] Use of vicinal diester anions in a simple annelation procedure affords bicyclo[4.4.0]decan-2-one derivatives, which are valuable intermediates in sesquiterpene synthesis.[221] Acylation of multiple anions of poly-β-ketones by hydroxy- and alkoxy-benzoates yields tetraketones, which under mild basic conditions are cyclized in biomimetic processes to naturally occurring benzophenones and xanthones.[222] Double lithiation of 1-tosyltetrahydropyrimidine produces the α-metallated isocyanide (**50**), a novel reagent for nucleophilic 1,3-diaminopropylation (electrophilic addition at the metallated carbon atom followed by unmasking using known methods).[223] 9-Keto-9*H*-pyrrolo[1,2-*a*]indole is formed by the intramolecular cyclization of the dilithio derivatives generated from either 1-(2′-carboxyphenyl)pyrrole or 1-(2′-bromophenyl)pyrrole-2-carboxylic acid.[224]

Anions α to Sulphur

The rates of deprotonation of vinyl and allyl aryl sulphides with amide bases show that allyl sulphides are kinetically more acidic than allyl selenides, whilst vinyl sulphides are less acidic than vinyl selenides.[225] Treatment of the 1-oxides or 1,1-dioxides of *trans*-homoallylic (eight-to-ten)-membered cyclic sulphides with butyllithium in THF results in transannular addition of the α-thiocarbanion to the (E)-double bond with formation of bicyclic products.[226] Methylthio- and *tert*-butylthio-allyllithium react irreversibly with cyclopent-2-enone by α- and γ-1,2-addition in THF at −78°, whereas in the presence of one equivalent of HMPA, the major reaction pathway is α-1,4-addition with barely detectable γ-1,4-addition (*ca.* 5%) also taking place.[227] Interaction of lithium tris(phenylthio)methanide with trialkylboranes followed by oxidation allows the production of ketones or tertiary alcohols in good yields under mild conditions.[228] Phenylthiotrimethylsilylmethyllithium is a useful reagent for preparing aldehydes (by alkylation or by reaction with an aldehyde or ketone and hydrolysis of the resulting vinyl sulphide) and, regioselectively, α-phenylthioketones (by reaction with an ester).[229] Ketones are converted into α-sulphenylated aldehydes with addition of one carbon atom *via* reaction with methoxyphenylthiomethyllithium followed by rearrangement of the adducts.[230] Some fulvalenoids containing a 1,3-dithiepin counterpart have been synthesized by the reaction of 2-trimethylsilyl-1,3-dithiepinide (**51**) with di-*tert*-butyl-cyclopropenone, 2,3;6,7-dibenzotroponone, and fluorenone.[231]

From electron-density difference maps, 2-lithio-2-methyldithiane would seem to have a covalent, at most polarized, Li—C bond, whereas 2-lithio-2-phenyldithiane can be regarded as an example of a contact ion-pair complex.[232] In attempts to prepare 2-alkyl-2-trimethylsilylmethyl-1,3-dithianes, an unusual ring-cleavage is observed, giving access to 2-alkylthiovinylsilanes by a mechanism thought to involve carbanion formation α to silicon, followed by thiolate elimination and alkylation.[233] Dithianyl anions capable of carbocyclization *via* either direct or Michael addition to a carbonyl unit can be generated from 2-trimethylsilyl-1,3-dithianes.[234] When treated with an anionic base in the gas phase, 1,3-dithiane undergoes successive elimination reactions to form thiolates in competition with deprotonation at C(2), which is the sole solution-phase channel.[235] The lithium anion of the dithiane of (E)-2-methylbut-2-enal (**52**) serves as the functional equivalent of the thermodynamic enolate of methyl ethyl ketone in a highly regioselective Michael addition–alkylation sequence culminating in the syntheses of the pseudo-quaianolides, (±)-aromatin, and (±)-confertin.[236] The dianion from 2,2′-[1,2-ethanediylbis(thio)]bis-1,3-dithiane undergoes ring-opening with loss of two moles of ethylene to give 1,2-ethanebis(trithiocarbonic acid) dianion, which fragments further on alkylation with loss of alkyl halides, *e.g.* *n*-propyl bromide, producing dialkyl trithiocarbonates.[237] Bis(phenylthio)acetals can be lithiated with *n*-butyllithium–TMEDA complex in hexane and consequently alkylated.[238]

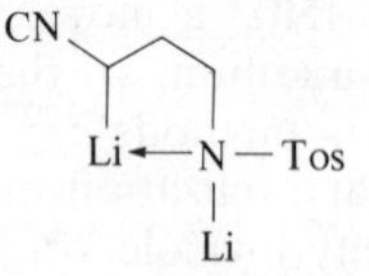

CN, Li←N—Tos, Li; S, S, SiMe$_3$; S, S

(50) 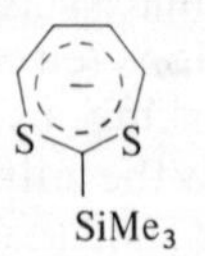 **(51)** 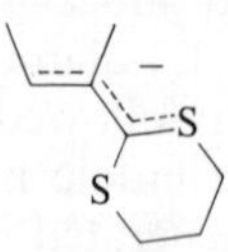 **(52)**

Electronic factors are thought to be responsible for the "remote" asymmetric induction observed in the formation of γ-products from the addition reactions of anions of aryl allyl sulphoxides to benzaldehyde.[239] Asymmetric syntheses of α-hydroxy-aldehydes with an α-C—H bond can be realized in quite satisfactory optical yield by performing a chiral *E*1 reaction on both aromatic and aliphatic aldehydes with the anion derived from (+)-(*S*)-*p*-tolyl *p*-tolylthiomethyl sulphoxide, a chiral formyl anion equivalent;[240] conjugate addition of the latter to cyclopentenone derivatives constitutes the key step in the asymmetric synthesis of 11-deoxy-*ent*-prostaglandin intermediates.[241] A sequence of reactions involving the stereoselective alkylation of alkenyl sulphoxides, reduction to the corresponding sulphides, followed by a (nickel–phosphine)-complex-catalysed coupling reaction with Grignard reagents, has culminated in a highly stereoselective synthesis of tri-substituted olefins.[242] 4,4-Di-substituted 1,4-dihydropyridines are accessible by intramolecular addition of a sulphoxide-stabilized carbanion to pyridines, followed by removal of the "handle" with Raney nickel.[243] In the chlorination of alkyl methyl sulphoxides (alkyl = ethyl, *iso*-butyl, or *iso*-propyl) with *N*-chlorosuccinimide, the corresponding 1-chloroalkyl methyl sulphoxide is formed predominantly, while addition of pyridine to the reaction medium reverses the regioselectivity to give an alkyl chloromethyl sulphoxide as a major product.[244] Other reactions of synthetic interest that have been studied this year include a new stereoselective approach to acyclic systems *via* condensation of α-lithiosulphinyl carbanions with aldehydes,[245] a concerted ring-opening reaction of cyclobutenes facilitated by arylsulphoxy and sulphonyl carbanion substituents,[246] and the conversion of pairs of diastereomeric sulphoximides and sulphilimines of known absolute configuration into the corresponding carbanions with a definite optical stability.[247]

Carbanions of alkyl *p*-tolyl sulphones react with carbon tetrahalides to give the corresponding polyhalogenated products in good yields.[248] Secondary alcohols of the general formula $R^1CH(OH)R^2$, are obtained in fair-to-good yield by condensation of lithium derivatives of sulphones of general formula, $PhSO_2CO_2R^1$, with trialkylboranes (BR^2_3) and subsequent oxidation with alkaline hydrogen peroxide.[249] Contrary to a previous report, $PhSO_2CH_2CN$ exhibits a one-electron reduction (two-electron reductive cleavage followed by deprotonation of $PhSO_2CH_2CN$ by $^-CH_2CN$).[250] The potential usefulness of sulphonyl-stabilized carbanions as acyl anion equivalents is further enhanced by the discovery that lithium derivatives of alkoxyalkyl phenyl sulphones are efficiently trapped with alkyl halides to provide monoalkylated products, which on hydrolysis under mild conditions yield aldehydes or ketones in moderate-to-good isolated yields.[251] In the lithiation of β-aminoalkyl sulphones, *opposite* diastereoselectivity is observed depending upon the type of amino group present.[252] Ambident lithium carbanions of allyl phenyl sulphones add α-1,4 to both acyclic and cyclic enones exclusively if HMPA is present in the medium; in contrast, α-1,2-addition to cyclic enones is the major kinetic course in the absence of HMPA, followed by rearrangement to the γ-1,4-adduct.[253] (Ethoxycarbonylcyclopropyl)carbonyl anions stabilized by phenylsulphonyl or phenylthio groups undergo "one-pot" nucleophilic ring-fission, alkylation, and elimination of the sulphur function to give substituted dienoates or dienedioates.[254] Kinetic data for the alkaline hydrolysis of aryl (methylsulphonyl)-methanesulphonate esters are consistent with a dissociative mechanism, *i.e.*, ionization of the substrate is followed by the slow unimolecular breakdown of the conjugate base to phenoxide ion and sulphene, but at high pH values a further ionization occurs to give a dicarbanion ($MeSO_2\overset{2-}{C}SO_2OAr$), which undergoes *E*1 elimination of phenolate ion to yield a sulphene anion in the rate-determining step.[255]

Other α-Hetero-substituted Anions

Attention is drawn to the new monodentate alkyl ligand (**53**); the key features of interest are (*i*) the presence of a chiral carbon at the ligating carbon atom, (*ii*) the existence of derived metal alkyls or dialkyls in enantiomeric or diastereomeric forms, respectively, and (*iii*) consequent mechanistic implications for the study of reactions involving metal–carbon bond-formation and/or -scission.[256] Direct formation of 2,5-substituted tetrahydrofurans occurs in the alkylation of lithiated α,β-unsaturated aldimines (**54**) using oxirane.[257] 1-Benzyl-4,6-diphenyl-2-pyridone is lithiated by lithium diisopropylamide at the methylene carbon to form a carbanion which reacts with various electrophiles to give the corresponding 1-(α-substituted-benzyl)-4,6-diphenyl-2-pyridones; interestingly, the anion derived from

1-(α-methylbenzyl)-4,6-diphenyl-2-pyridone rapidly rearranges to 2-methyl-2,5,7-triphenylazepin-3-one.[258] A transition state involving a planar dihydropyrazine anion (see **55**), one diastereotopic side of which is strongly shielded by the comparatively large isopropyl group, is proposed to explain the induction observed in the formation of (R)-amino-acids ($> 95\%$ enantiomeric excess) from cyclic bislactim ethers (dihydropyrazines) of *cyclo*-(L-Val-Gly) by successive metallation, alkylation, and hydrolysis.[259]

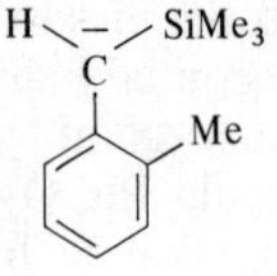

H, SiMe₃, C, Me — (**53**)   R¹, NR², Li⁺ — (**54**)   Br, H, C, R, H, Li⁺, OMe, H, N, H, Me, N, H Me OMe — (**55**)

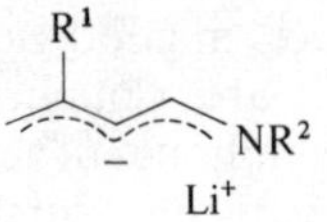

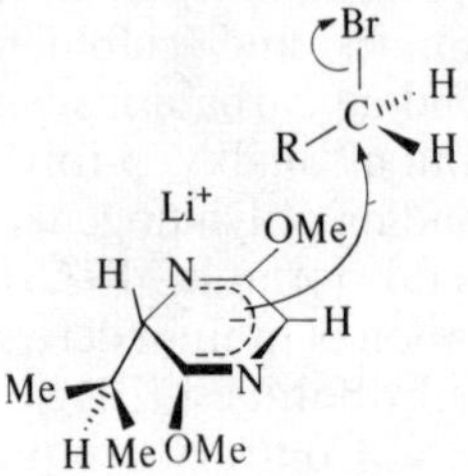

Carbanionic derivatives of methylphosphazenes react with mono-functional electrophiles to form phosphazenes carrying the groups PCH_2R [R = Me, Br, I, PhCO, COOH, $AsMe_2$, Me_3M (M = Si, Ge, or Sn)].[260] In the Wittig reaction between ketones and methylenetriphenylphosphorane in Et_2O, competitive enolization of the ketone is an important side-reaction that in many cases decreases the yield substantially, but which may be avoided by repeated additions of stoichimetric amounts of water (to regenerate the ketone) and of the phosphorane.[261] ω-Acyloxy-*n*-butylidenetriphenylphosphoranes give α-acyl-*n*-butylidenetriphenylphosphoranes by intermolecular condensation in *tert*-butanol, and 3,4-dihydro-(2*H*)-pyrans by intramolecular condensation in toluene.[262] By comparison, the corresponding reactions of ω-acyloxy-*n*-propylidenetriphenylphosphoranes lead to the formation of 2,3-dihydrofurans in toluene, and to cyclopropyl ketones (*via two* intermolecular condensations) in *tert*-butanol.[263] In the case of *O*-acyloxybenzylidenetriphenylphosphoranes, benzofurans are obtained in toluene and acylated products in *tert*-butanol.[264] Alkylation followed by elimination of triphenylphosphine makes the novel ylide anion (**56**) a synthetically useful source of a vinyl anion equivalent in the preparation of substituted fumarate esters.[265] Phosphinoylation, lithiation, reaction with electrophiles, and cleavage constitutes an efficient sequence for 1-alkylation of the isoquinoline nucleus.[266] Addition of electrophiles to the phosphoryl-stabilized allylic anion (**57**) produces products of γ-attack or α-attack depending upon the nature of the substituents R^1 and R^2 and the nature of the electrophile.[267] (Diphenylphosphino)methyllithium demonstrates an unusual preference toward deprotonation of 6,6-dimethylpentafulvene (to yield exclusively lithium isopropenylcyclopentadienide) both in homogeneous and heterogeneous media; only in hydrocarbon solvent is the product of the expected addition process obtained, and that in only very minor amounts.[268] Enol phosphates of the type $(R^1O)_2P(O)OC(R^2){=}CHSR^3$ are formed in good yields by a novel rearrangement of *S*-alkanoate *O,O*-dialkyl phosphate carbanions (see **58**) and subsequent trapping with alkyl halides.[269] The novel use of the phosphonate reagent $(MeO)_2P(O)CH(OR)COOMe$ (R = ethoxyethyl) as an acyl anion equivalent has resulted in a more efficient route to the enol lactam intermediate in cyctochalasin synthesis.[270]

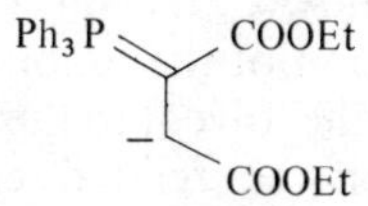

(56) 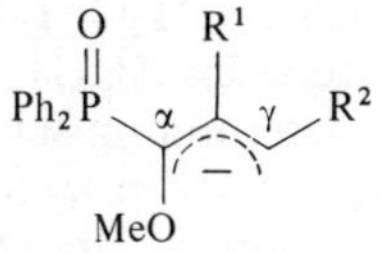 **(57)** 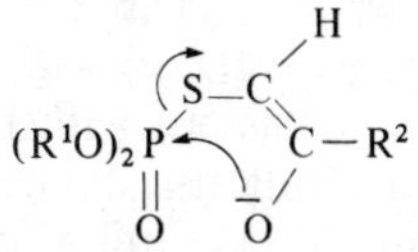 **(58)**

Proton-transfer and Hydrogen Isotope Effects

A review in Hungarian of hydrogen isotope effects in chemical reactions has appeared.[271] A reinvestigation of the kinetics of the proton-transfer reaction of 4-(nitrophenyl)nitromethane with amine bases in toluene solution gives a value of only 14 for the isotopic rates k_H/k_D;[272] this result is decisively against the unusually high value of 45 previously reported for this system, and it is argued that further experiments are desirable.[273] The kinetic isotope effect for proton-transfer reactions between ethyl nitroacetate and a series of oxygen bases has been calculated for the linear complex model.[274] The volume of activation for proton transfer from 2,4,6-trinitrotoluene to 1,8-diazabicyclo[5.4.0]undec-7-ene is larger than that to *N,N,N',N'*-tetramethylguanidine in acetonitrile and dichloromethane.[275] A classical mechanism is proposed for the proton transfer from 2-nitrohexafluoropropane to trioctylamine on the basis of the observed k_H/k_D ratio (> 6) and the linearity of the ln k *vs.* T^{-1} dependence.[276] Proton tunnelling appears to contribute to a significant extent to the proton transfer from 4-nitrophenylnitromethane to 2,7-dimethoxy-1,8-bis(dimethylamino)naphthalene in acetonitrile solvent.[277] The reliability of calculations on proton tunnelling has been critically examined.[278] For the reaction of tetranitromethane with hydroxide ion, the rate of formation of trinitromethanide ion increases in the presence of cationic micelles, an effect that is attributed mainly to a local concentration increase of reactant HO^- ions near the cationic micellar surface.[279]

Structure–reactivity correlations for nucleophilic reactions of stable carbanions (including those derived from nitroalkanes) with diaryl disulphides in aqueous solution are similar in many respects to those observed for proton-transfer reactions involving the same carbanions.[280] Although reactivity is practically zero at ambient temperature, gaseous CF_3H *does* undergo hydrogen isotope exchange in NaOD–D_2O solution at elevated temperature;[281] k_{obs} for hydroxide-ion-catalysed CF_3D exchange increases by a factor of 10^6 on changing from water to 70 mol % DMSO.[282] For nucleophilic reactions of various *gem*-difluoroalkenes of general structure PhC(R)=CF_2 (R = CF_3, CF_2Cl, CF_2CF_3, and CF_2H) with alkoxide ions in alcohol, solvent protonation of the carbanion intermediates is apparently slower than fluoride ion ejection from all groups studied other than trifluoromethyl.[283] A carbanion-mediated two-step mechanism with varying amounts of internal return is proposed for the alkoxide-promoted dehydrofluorination of $PhCH_2CH_2F$, $PhCH_2CHF_2$, $PhCH_2CF_3$, and $PhCH_2CF_2CF_3$ in *tert*-butanol and ethanol.[284] Side-chain nitroxylation of pentamethylbenzenes C_6Me_5R (R = Me, H, COMe, or $COCMe_3$) and of *m*-$H_2C_6Me_4$, *i.e.* formation of benzyl nitrate derivatives, gives a large primary hydrogen isotope effect, indicating that proton removal is partly involved in the rate-determining step.[285]

Analysis of the effects of substituents X in the cleavage of $XC_6H_4SiMe_3$ compounds by NaOH in 1/9 or 1/6 v/v H_2O–DMSO in the light of calculated energy differences, $(E_X^-) - (E_X)$, between $XC_6H_4^-$ and XC_6H_5 species suggests that they are consistent with rate-determining separation of the anions $XC_6H_4^-$ without electrophilic assistance by proton transfer from the solvent to the separating carbon atom;[286] a similar conclusion is drawn for the related cleavage of $R_3SiGePh_3$ with R = Me or Et in NaOMe–MeOH.[287] The absence of conjugative delocalization of charge in thienyl anions compared to benzyl-type anions is reflected in the significantly lower values of the product isotope effects obtained for the base cleavage of substituted 2-thienyltrimethylsilanes than for $XC_6H_4CH_2SiMe_3$ compounds.[288] The low product isotope effects of 1.2–1.3 for the base cleavage of 1-methyl-2-trimethylsilylbenzimidazole and 2-trimethylsilylbenzothiazole are attributed to the fact that the isotopic composition of the products is determined in a step involving transfer of hydrogen to a localized carbanion.[289] The compounds $PhCH_2SiMe_2(CH_2)_nOH$ ($n = 2$ or 3) are cleaved 0.75 and 95–135 times, respectively, as readily as $PhCH_2SiMe_3$ by NaOMe–MeOH; the greater reactivity of the compound with $n = 2$ may arise from intramolecular attack of the alkoxide centre on silicon in the anion $PhCH_2(SiMe_2)(CH_2)_3O^-$.[290] From rate measurements and solvent isotope effects, the mechanism of the cleavage of some $XC_6H_4CH{=}CHSnMe_3$ compounds in MeOH–MeONa is the same as that for the great majority of $RSnMe_3$ compounds studied, with the rate-determining step involving proton transfer from the solvent to the carbon atom of the C—Sn bond as it breaks as a result of the attack of the MeO^- ion at the tin atom.[291] Results from a related study involving the base cleavage of (dihalomethyl)trimethyltins in the presence of ammonia buffer support a mechanism in which free carbanions are separated in the rate-determining step (for reactions involving solvent conjugate base), and at the same time confirm the importance of electrophilic assistance of the solvent in the dominant process, which is catalysed by ammonia (general base).[292] In keeping with the existence of a difference in mechanism, there is a clear difference in behaviour between some benzyl-silicon and -tin compounds on base cleavage in various media.[293]

The intrinsic barrier for proton abstraction from cyclopentadiene by *tert*-butoxide in the gas phase as measured by pulsed ion cyclotron resonance spectroscopy is 10 kcal mol^{-1}, which is comparable to values obtained for proton transfer involving CH-acids and bases in solution.[294] Calculated thermodynamic parameters (proton affinities, free energies and entropies) for gas-phase proton-transfer reactions of a variety of molecules show good agreement with values obtained experimentally.[295]

Absolute and Relative Acidity

Equilibrium carbon acidities in solution have been reviewed.[296] Equilibrium acidity data for a series of esters, amides, and nitriles in DMSO follow the usual order of $GCH_2CN > GCH_2COOEt$ ($\Delta pK_a = 0.8$–2) and $GCH_2COOEt > GCH_2CONMe_2$ ($\Delta pK_a = 3.9$–4.5) where G is an acid-strengthening function.[297] Estimates derived from equilibrium acidity data in DMSO solution indicate that the aromatic stabilization energy of the cyclopentadienyl anion is about 24–27 kcal mol^{-1}.[298] Good agreement is found between equilibrium acidities of some phenalene hydrocarbons determined in the caesium cyclohexylamide–cyclohexylamine system and values obtained by a standard SCF-π approach with explicit incorporation of electron repulsions.[299] By measuring hydrogen isotope exchange of allylic hydrogens in propene using the aforementioned system, the solution ion-pair pK_a of propene is reliably estimated to be 43 ± 1.[300] Heats of deprotonation in solutions of the potassium salt of 1,3-diaminopropane in the parent amine as solvent show that it is over a million times more basic than the potassium salt of DMSO.[301] A reasonable correlation exists between protonic charge density and the kinetic acidities of carbon acids, *e.g.* nitroalkanes and fluoroalkanes, when hydrogen bonding in the contact complex plays a major part in determining the activation energy for the overall process of proton removal.[302] Double-zeta basis-set calculations for the carbanions $X\bar{C}H_2$ (X = H, Me, NH_2, OH, F, C≡CH, CH=CH_2, CHO, COMe, CN, or NO_2) give proton affinities over the range 449 (for $Me\bar{C}H_2$) to 355 kcal mol^{-1} (for $NO_2\bar{C}H_2$) with unsaturated anions having smaller affinities than saturated anions.[303] Deprotonation energies of MeR, RCH_2CN, $PhCH_2R$, Ph_2CHR, fluorene, and 9-cyanofluorene, calculated by the CNDO/2 method, show a linear correlation with p*K* values for each series; the CN group has an acidifying effect on CH_4 of *ca.* 27 log units.[304] According to standard self-consistent molecular-orbital calculations for systems XCH_3 offering competing sites of deprotonation, proton loss at CH_3 is generally favoured when X is an electropositive or unsaturated group, while deprotonation at X is favoured by electronegative substituents and by C≡CH.[305] Acidity differences (ΔpK) between mono-, bis-, and tris-(pentafluorophenyl)-methanes on the one hand, and mono-, di-, and tri-phenylmethanes on the other, correlate linearly with ^{19}F NMR chemical-shift differences ($\Delta\delta$) of the *para*-fluorine atom in going from the parent CH-acid to the carbanion.[306] Molecular-orbital *ab initio* calculations show that the small changes in hydrocarbon acidities or in the rates of nucleophilic substitution reactions, which are found when β-F is substituted by a β-CF_3 group, do not reflect the absence of fluorine hyperconjugation, but result from the stronger inductive effect of the CF_3 group relative to that of F, which compensates for its poorer hyperconjugation.[307] A theoretical comparison of ions formed by deprotonation and hydride removal from silane and methylsilane with their carbon analogues shows that silicon is more capable of accepting positive *and* (more surprisingly in terms of electronegativity) negative charge than carbon.[308] In molecules containing one silicon atom connected to carbon, with the exception of 1-silaallene, deprotonation occurs preferentially at silicon according to double-zeta-level calculations.[309] ^{13}C-NMR spin-lattice relaxation data obtained from solutions containing paramagnetic relaxation reagents, such as tris(acetylacetonato)-chromium, can be used for the determination of hydrogen-bond donor acidities of weak carbon acids.[310] From an examination of a number of $R\bar{C}$=O species by means of semi-empirical MNDO and *ab initio* molecular-orbital calculations with complete geometry optimization, it is concluded that bridgehead-substituted tertiary aldehydes such as 1-adamantyl- or 1-norbornyl-carboxaldehyde, aromatic aldehydes, and di-substituted formaldehydes should be best suited for experimental studies of proton abstraction from the CHO group.[311] Treatment of ketones, active methylene compounds, and acidic hydrocarbons with D_2O-treated alumina results in deuteriation of the compounds at acidic sites.[312]

Evaluation of the relative activity of the *C*-methyl groups of 2,4,6-trimethyl-1,3,5-triazine shows that the hydrogen–deuterium exchange rate is more than 2000 times slower than that of the 6-methyl group of 1,3,6-trimethyl-2,4-dioxo-1,3,5-triazine.[313] Silver(I) promotes hydrogen–deuterium exchange at the 2-methyl group of 2-methyl-6-nitrobenzothiazole in the presence of CD_3OD in DMSO.[314] Deuteriation under phase-transfer conditions offers a good method of obtaining specifically labelled thiazoles and certain other hetero-aromatic compounds with a deuterium content $> 90\%$.[315] Treatment of 4-methoxy-α-pyrones with butyllithium or lithium diisopropylamide causes rapid deprotonation at C(3) of the pyrone ring to give moderately stable vinylic carbanions.[316] In flavone and 4-methoxycoumarin, β-deprotonation by lithium diisopropylamide occurs readily and the resulting carbanions are easily carboxylated giving acids not previously accessible; in 2,6-dimethylchromone, β-deprotonation is kinetically favoured allowing 3-acylation to be achieved separately from the conventional acylation at the 2-methyl group.[317] Whereas tricarbonyl(η^6-1-methylindole)chromium(0) is lithiated at C(2), tricarbonyl(η^6-1-methyl-2-trimethylsilylindole)chromium(0) is lithiated predominantly at C(7).[318] Complexation of a $Cr(CO)_3$ unit to an aromatic hydrocarbon enhances the benzylic position towards attack by a base, allowing formation of a carbanion which reacts with carbonyl compounds to produce a complexed alcohol.[319]

A review article entitled "Gas-phase acidities of carbon acids" has appeared.[320] Gas-phase acidity values obtained for a series of fluorinated acetones lend some support to the hypothesis that fluorine substituents may play an ambiguous rôle in the stabilization of planar carbanions, and can in fact, have a destabilizing component due to the repulsive interaction between fluorine lone-pair electrons and the carbanion centre.[321] Under flowing afterglow conditions, D_2O serves as an exchange reagent for hydrogens which are much less acidic than those in water, *e.g.* it is possible to exchange not only the four allylic hydrogens in the 2-phenylallyl anion as shown in equation (1), but at least four of the five aryl hydrogens as well![322] Hydrogens attached to aromatic rings, to carbons adjacent to aromatic rings, and to carbons α to carbonyl groups are readily exchanged for deuterium under CI conditions when deuterium-labelled reagent gases are employed.[323]

$$C_6H_5C(\bar{C}H_2)=CH_2 \xrightarrow{D_2O} C_6HD_4C(\bar{C}D_2)=CD_2 \quad (1)$$

Electrophilic Aliphatic Substitution Reactions

The oxidative chlorination of ethanol by Cl_2 to give $CCl_3CH(OEt)_2$ proceeds *via* the formation of MeCHO (from EtOCl) as the primary intermediate; subsequent chlorination of MeCHO or $MeCH(OEt)_2$ in a sequence of steps then gives the product.[324] The rate of acid-catalysed bromination of 2,4,6-trimethylacetophenone in 50% aqueous acetic acid is first order in bromine, in marked contrast to acetophenone and other reported ketones at comparable concentrations.[325] A kinetic study of the bromination of some substituted acetophenones both in acidic and basic media has furnished evidence that the 3-Me group in 4-methoxy-3-methylacetophenone sterically enhances the resonance interaction of the MeO group with the aromatic ring.[326] Kinetics of iodination of 6-substituted 2-acetonaphthones and of 4-substituted 1-acetonaphthones, studied in pyridine–MeOH–H_2O mixtures (20:20:60 vol. ratio) at 25, 30, and 35° show a good Hammett correlation.[327] The kinetics and reaction mechanisms of iodination reactions have been reviewed.[328]

Thiourea makes a very efficient catalyst (better than SCN^- or Br^-) for the nitrosation of morpholine in acid solution.[329] The conversion of secondary amines into *N*-nitrosoamines in aqueous solution by dissolved NOCl gas (*ca.* six-fold excess) is more extensive in the presence of alkanolamines and probably proceeds *via* an intermediate allyl nitrite (generated from the alkanolamine and NOCl) which then nitrosates the secondary amine in a rate-determining step.[330] The nitrosation of 2-, 3-, and 4-methylaminopyridine in 0.002–5.00M perchloric acid proceeds mainly by the interaction of the nitrous acidium ion with the protonated form of these amines, whilst the nitrosation of their 1-oxide derivatives involves the simultaneous interaction of the nitrous acidium ion with the protonated and the free form of the amines.[331,332] *p*-Nitrosophenols (and structurally related compounds) catalyse the *N*-nitrosation of pyrrolidine and morpholine by a mechanism in which the nitrophenol reacts (*via* its quinone mono-oxime tautomer) with HNO_2 (or species in equilibrium with HNO_2) to produce an intermediate nitrosating agent which undergoes attack by the amine to produce the *N*-nitrosoamine and regenerate the catalyst.[333] Nitrolysis (*N*-nitration) of *N*,*N*-dialkylamides of carboxylic acids by $NO_2^+ BF_4^-$ in acetonitrile occurs by a first-order process, the rate-determining step being cleavage of the amide C—N bond in an initially-formed complex of reactants;[334–336] in the case of *N*-nitro-*N*-butylacetamide, the rate-determining step is the reaction of substrate with NO_2^+ regardless of substrate concentration.[337]

A review in Russian on $S_E1(N)$ processes in organometallic chemistry has appeared.[338] Replacement of mercury by deuterium in some organomercury halides by employing excess 1–2% Na–Hg in *ca.* 1M NaOD–D_2O as reducing agent occurs with complete regio- and stereo-specificity, giving support to the idea that the incorporation step is electrophilic cleavage of the C—Hg bond, probably in a sub-valent organomercury species.[339] On the basis of the stereochemistry of the products as well as the product distribution, it is highly likely that a 1,3-bridged mercurium ion intervenes in the oxymercuration of 3,5-dihydronoriceane which incorporates a bicyclo[2.1.0]pentane moiety in its cage structure.[340] A 1,3-bridging mechanism is also proposed for the corresponding bromination reaction in non-polar solvents in the dark.[341] The action of I_2/I^- in methanol on compounds $PhCH_2SnR_3$ leads only to cleavage of the benzyl group (as $PhCH_2I$) when R = Et, Pr^n, or Bu^n, whereas with $PhCH_2SnMe_3$, both benzyl and methyl groups are removed.[342]

References

1 Birch, A. J., Hinde, A. L., and Radom, L., *J. Am. Chem. Soc.*, **103,** 284 (1981).
2 Hallden-Abberton, M., Engelman, C., and Fraenkel, G., *J. Org. Chem.*, **46,** 538 (1981).
3 Fraenkel, G., and Hallden-Abberton, M. P., *J. Am. Chem. Soc.*, **103,** 5657 (1981).
4 Rabideau, P. W., Peters, N. K., and Huser, D. L., *J. Org. Chem.*, **46,** 1593 (1981).
5 Chandrasekhar, J., Andrade, J. G., and Schleyer, P., von R., *J. Am. Chem. Soc.*, **103,** 5609 (1981).
6 Kos, A. J., Jemmis, E. D., Schleyer, P. von R, Gleiter, R., Fischbach, U., and Pople, J. A., *J. Am. Chem. Soc.*, **103,** 4996 (1981).
7 Apeloig, Y., Clark, T., Kos, A. J., Jemmis, E. D., and Schleyer, P. von R., *Isr. J. Chem.*, **20,** 43 (1980).
8 Dietrich, H., *J. Organomet. Chem.*, **205,** 291 (1981).
9 Schleyer, P., von R., Chandrasekhar, J., Kos, A. J., Clark, T., and Spitznagel, G. W., *J. Chem. Soc., Chem. Commun.*, **1981,** 882.

[10] Shimp, L. A., Morrison, J. A., Gurak, J. A., Chinn, J. W., and Lagow, R. J., *J. Am. Chem. Soc.*, **103,** 5951 (1981).
[11] Moffat, J. B., *J. Chem. Soc., Chem. Commun.*, **1980,** 1108.
[12] Clark, T., Schleyer, P. von R., Houk, K. N., and Rondan, N. G., *J. Chem. Soc., Chem. Commun.*, **1981,** 579.
[13] Barluenga, J., Fañanás, F. J., and Yus, M., *J. Org. Chem.*, **46,** 1281 (1981).
[14] Chandrasekhar, J., and Schleyer, P. von R., *J. Chem. Soc., Chem. Commun.*, **1981,** 260.
[15] Dietrich, H., and Rewicki, D., *J. Organomet. Chem.*, **205,** 281 (1981).
[16] Shainyan, B. A., *Zh. Org. Khim.*, **16,** 1113 (1980); *Chem. Abs.*, **93,** 167423 (1980).
[17] Symons, M. C. R., *J. Chem. Res. (S)*, **1981,** 286.
[18] O'Brien, D. H., *Stud. Org. Chem. (Amsterdam)*, **5** (*Compr. Carbanion Chem., Pt. A*), 271 (1980); *Chem. Abs.*, **94,** 3328 (1981).
[19] Vogt, H.-H., and Gompper, R., *Chem. Ber.*, **114,** 2884 (1981).
[20] Cambillau, C., and Ourevitch, M., *J. Chem. Soc., Chem. Commun.*, **1981,** 996.
[21] Petrov, A. A., Esakov, S. M., and Ershov, B. A., *Zh. Org. Khim.*, **16,** 1576 (1980); *Chem. Abs.*, **94,** 15055 (1981).
[22] Petrov, A. A., Bizunok, S. N., and Ershov, B. A., *Zh. Org. Khim.*, **16,** 1582 (1980); *Chem. Abs.*, **94,** 15056 (1981).
[23] Petrov, A. A., Volosov, A. P., and Ershov, B. A., *Zh. Org. Khim.*, **16,** 1588 (1980); *Chem. Abs.*, **94,** 15057 (1981).
[24] Lynch, T. J., Newcomb, M., Bergbreiter, D. E., and Hall, M. B., *J. Org. Chem.*, **45,** 5005 (1980).
[25] Noest, A. J., and Nibbering, N. M. M., *Int. J. Mass Spectrom. Ion Phys.*, **34,** 383 (1980); *Chem. Abs.*, **94,** 14877 (1981).
[26] Rondan, N. G., Houk, K. N., Beak, P., Zajdel, W. J., Chandrasekhar, J., and Schleyer, P., von R., *J. Org. Chem.*, **46,** 4108 (1981).
[27] Krebs, E.-P., *Helv. Chim. Acta*, **64,** 1023 (1981).
[28] Paleta, O., Svoboda, J., Havlů, V., and Dědek, V., *Collect. Czech. Chem. Commun.*, **45,** 3360 (1980).
[29] Kaneti, J., Fattah, N. A., Binev, I., Radomirska, V., and Yukhnovski, I., *J. Mol. Struct.*, **68,** 11 (1980).
[30] Greifenstein, L. G., and Pagani, G. A., *J. Org. Chem.*, **46,** 3336 (1981).
[31] Nakamura, K., *Bull. Chem. Soc. Jpan.*, **53,** 2792 (1980).
[32] Helary, G., Lefevre-Jenot, L., Fontanille, M., and Smid, J., *J. Organomet. Chem.*, **205,** 139 (1981).
[33] Hogen-Esch, T. E., and Jenkins, W. L., *J. Am. Chem. Soc.*, **103,** 3666 (1981).
[34] Brownstein, S., Bywater, S., and Worsfold, D. J., *J. Organomet. Chem.*, **199,** 1 (1980).
[35] Mills, N. S., Shapiro, J., and Hollingsworth, M., *J. Am. Chem. Soc.*, **103,** 1263 (1981).
[36] Bondarenko, G. N., Ovchinnikov, A. A., Misurkin, I. A., Avakyan, V. G., and Dolgoplosk, B. A., *Dokl. Akad. Nauk SSSR*, **252,** 905 (1980); *Chem. Abs.*, **93,** 238520 (1980).
[37] Yasuda, H., Yamauchi, M., Ohnuma, Y., and Nakamura, A., *Bull. Chem. Soc. Jpn.*, **54,** 1481 (1981).
[38] Yasuda, H., Yamauchi, M., and Nakamura, A., *J. Organomet. Chem.*, **202,** C1 (1980).
[39] Zweifel, G., Murray, R. E., and On, H. P., *J. Org. Chem.*, **46,** 1292 (1981).
[40] Hunter, D. H., Stothers, J. B., and Warnhoff, E. W., *Org. Chem. (N.Y.)*, **1980,** 42 (*Rearrange. Ground Excited States, vl*), 391; *Chem. Abs.*, **93,** 219984 (1980).
[41] Knörr, R., and Lattke, E., *Chem. Ber.*, **114,** 2116 (1981).
[42] Knörr, R., Lattke, E., and Räpple, E., *Chem. Ber.*, **114,** 1581 (1981).
[43] Knörr, R., Lattke, E., Ruf, F., and Reissig, H.-U., *Chem. Ber.*, **114,** 1592 (1981).
[44] Lattke, E., and Knorr, R., *Chem. Ber.*, **114,** 1600 (1981).
[45] Schmidt, R. R., and Speer, H., *Tetrahedron Lett.*, **22,** 4259 (1981).
[46] Midland, M. M., Halterman, R. L., Brown, C. A., and Yamaichi, A., *Tetrahedron Lett.*, **22,** 4171 (1981).
[47] Van Rijn, P. E., Everhardus, R. H., Van der Ven, J., and Brandsma, L., *Recl. Trav. Chim. Pays-Bas*, **100,** 372 (1981).
[48] Epsztein, R., and Herman, B., *J. Chem. Soc., Chem. Commun.*, **1980,** 1250.
[49] Ceccon, A., Gambaro, A., Agostini, G., and Venzo, A., *J. Organomet. Chem.*, **217,** 79 (1981).
[50] Ustynyuk, N. A., Lokshin, B. V., Oprunenko, Yu. F., Roznyatovsky, V. A., Luzikov, Yu. N., and Ustynyuk, Yu. A., *J. Organomet. Chem.*, **202,** 279 (1980).
[51] Ustynyuk, N. A., Ustynyuk, Yu. A., and Nesmeyanov, A. N., *Dokl. Akad. Nauk SSSR*, **255,** 127 (1980); *Chem. Abs.*, **94,** 156051 (1981).
[52] Oprunenko, Yu. F., *Deposited Doc.*, **1979,** VINITI 3782, 161; *Chem. Abs.*, **94,** 156053 (1981).
[53] Buncel, E., and Balachandran, M., *Stud. Org. Chem. (Amsterdam)*, **5** (*Compr. Carbanion Chem., Pt. A*), 97 (1980); *Chem. Abs.*, **94,** 46263 (1981).

[54] Klass, G., Trenerry, V. C., Sheldon, J. C., and Bowie, J. H., *Aust. J. Chem.*, **34,** 519 (1981).
[55] Yeh, M. K., *J. Chem. Soc., Perkin Trans. 1*, **1981,** 1652.
[56] Wight, C. A., and Beauchamp, J. L., *J. Am. Chem. Soc.*, **103,** 6499 (1981).
[57] Ramsey, B. G., and Isabelle, L. M., *J. Org. Chem.*, **46,** 179 (1981).
[58] Kaiser, E. M., Thomas, W. R., Synos, T. E., McClure, J. R., Mansour, T. S., Garlich, J. R., and Chastain, J. E., *J. Organomet. Chem.*, **213,** 405 (1981).
[59] Staley, S. W., Dustman, C. K., and Linkowski, G. E., *J. Am. Chem. Soc.*, **103,** 1069 (1981).
[60] Staley, S. W., and Dustman, C. K., *J. Am. Chem. Soc.*, **103,** 4297 (1981).
[61] Corset, J., *Stud. Org. Chem. (Amsterdam)*, **5** (*Compr. Carbanion Chem., Pt. A*), 125 (1980); *Chem. Abs.*, **93,** 237988 (1980).
[62] Yukhnovski, I., Nazir, A. F., Sakhatchieva, M., Kaneti, J., and Binev, I., *Izv. Khim.*, **13,** 269 (1981); *Chem. Abs.*, **95,** 131790 (1981).
[63] Sami, A., *Radiochem. Radioanal. Lett.*, **43,** 319 (1980).
[64] Oda, M., Watabe, T., and Kawase, T., *Tetrahedron Lett.*, **22,** 4249 (1981).
[65] Edlund, U., Eliasson, B., Kowalewski, J., and Trogen, L., *J. Chem. Soc., Perkin Trans. 2*, **1981,** 1260.
[66] Stapersma, J., and Klumpp, G. W., *Tetrahedron*, **37,** 187 (1981).
[67] Takahashi, K., Takase, K., and Kagawa, T., *J. Am. Chem. Soc.*, **103,** 1186 (1981).
[68] Zwaard, A. W., and Kloosterziel, H., *Recl. Trav. Chim. Pays-Bas*, **100,** 126 (1981).
[69] Grutzner, J. B., and Jorgensen, W. L., *J. Am. Chem. Soc.*, **103,** 1372 (1981).
[70] Kaufmann, E., Mayr, J., Chandrasekhar, J., and Schleyer, P. von R., *J. Am. Chem. Soc.*, **103,** 1375 (1981).
[71] Ilic, P., and Trinajstic, N., *Kem. Ind.*, **29,** 417 (1980); *Chem. Abs.*, **94,** 191049 (1981).
[72] Kos, A. J., and Schleyer, P. von R., *J. Am. Chem. Soc.*, **102,** 7928 (1980).
[73] Tomilin, O. B., Stankevich, I. V., and Surin, N. G., *Izv. Akad. Nauk SSSR, Ser. Khim.*, **1980,** 1166; *Chem. Abs.*, **93,** 167387 (1980).
[74] Mitchell, R. D., and Bauld, N. L., *Isr. J. Chem.*, **20,** 319 (1980).
[75] Olah, G. A., and Hunadi, R. J., *J. Org. Chem.*, **46,** 715 (1981).
[76] Anastassiou, A. G., Kasmai, H. S., and Saadein, M. R., *Angew. Chem. Int. Ed.*, **20,** 115 (1981).
[77] Anastassiou, A. G., Kasmai, H. S., and Hauger, D., *J. Chem. Soc., Chem. Commun.*, **1981,** 647.
[78] Moskal, J., and Milart, P., *J. Chem. Res. (S)*, **1981,** 284.
[79] Sammes, M. P., Leung, C. W. F., and Katritzky, A. R., *J. Chem. Soc., Perkin Trans. 1*, **1981,** 2835.
[80] Liotta, D., Saindane, M., and Barnum, C., *J. Org. Chem.*, **46,** 3369 (1981).
[81] Sakai, T., Katayama, T., and Takeda, A., *J. Org. Chem.*, **46,** 2924 (1981).
[82] Åkermark, B., Ljungqvist, A., and Panunzio, M., *Tetrahedron Lett.*, **22,** 1055 (1981).
[83] Greengrass, C. W., and Hoople, D. W. T., *Tetrahedron Lett.*, **22,** 1161 (1981).
[84] Solov'yanov, A. A., Karpyuk, A. D., Beletskaya, I. P., and Reutov, O. A., *Izv. Akad. Nauk SSSR, Ser. Khim.*, **1981,** 695; *Chem. Abs.*, **95,** 168008 (1981).
[85] Solov'yanov, A. A., Karpyuk, A. D., Beletskaya, I. P., and Reutov, O. A., *Zh. Org. Khim.*, **17,** 458 (1981); *Chem. Abs.*, **95,** 23789 (1981).
[86] Julia, S., and Ginebreda, A., *Afinidad*, **37,** 194 (1980); *Chem. Abs.*, **94,** 14740 (1981).
[87] Shono, T., Ohmizu, H., Kawakami, S., Nakano, S., and Kise, N., *Tetrahedron Lett.*, **22,** 871 (1981).
[88] Fleming, I., and Floyd, C. D., *J. Chem. Soc., Perkin Trans. 1*, **1981,** 969.
[89] Seitz, D. E., and Zapata, A., *Synthesis*, **1981,** 557.
[90] Hirao, T., Yamada, A., Oshiro, Y., and Agawa, T., *Angew. Chem. Int. Ed.*, **20,** 126 (1981).
[91] Jacobi, P. A., Walker, D. G., and Odeh, I. M. A., *J. Org. Chem.*, **46,** 2065 (1981).
[92] Baryshnikov, Yu. N., Vesnovskaya, G. I., and Kaloshina, N. N., *Khimiya Elementoorgan Soedin (Gor'kii)*, **1979,** 36; *Chem. Abs.*, **93,** 185567 (1980).
[93] Chambers, R. D., Lindley, A. A., and Fielding, H. C., *J. Chem. Soc., Perkin Trans. 1*, **1981,** 939.
[94] Ando, K., Tokoroyama, T., and Kubota, T., *Bull. Chem. Soc. Jpn.*, **53,** 2885 (1980).
[95] Mulvaney, J. E., Londrigan, M. E., and Savage, D. J., *J. Org. Chem.*, **46,** 4592 (1981).
[96] Halazy, S., and Krief, A., *Tetrahedron Lett.*, **22,** 4341 (1981).
[97] Andrade, J. G., Clark, T., Chandrasekhar, J., and Schleyer, P. von R., *Tetrahedron Lett.*, **22,** 2957 (1981).
[98] Masamune, T., Sato, S., Abiko, A., Ono, M., and Murai, A., *Bull. Chem. Soc. Jpn.*, **53,** 2895 (1980).
[99] Boeckman, R. K., and Bruza, K. J., *Tetrahedron*, **37,** 3997 (1981).
[100] Perold, G. W., and Ourisson, G., *Tetrahedron Lett.*, **22,** 3783 (1981).
[101] Shinkai, S., Matsuo, K., Sato, M., Sone, T., and Manabe, O., *Tetrahedron Lett.*, **22,** 1409 (1981).
[102] Adlington, R. M., and Barrett, A. G. M., *Tetrahedron*, **37,** 3935 (1981).
[103] Adlington, R. M., and Barrett, A. G. M., *J. Chem. Soc., Chem. Commun.*, **1981,** 65; *J. Chem. Soc., Perkin Trans. 1*, **1981,** 2848.

[104] Adlington, R. M., Barrett, A. G. M., Quayle, P., Walker, A., and Betts, M. J., *J. Chem. Soc., Chem. Commun.*, **1981,** 404.
[105] Baldwin, J. E., and Bottaro, J. C., *J. Chem. Soc., Chem. Commun.*, **1981,** 1121.
[106] Schinzer, D., and Heathcock, C. H., *Tetrahedron Lett.*, **22,** 1881 (1981).
[107] Nudelman, N. S., and Vitale, A. A., *J. Org. Chem.*, **46,** 4625 (1981).
[108] Sauvêtre, R., and Normant, J. F., *Tetrahedron Lett.*, **22,** 957 (1981).
[109] Bond, F. T., and DiPietro, R. A., *J. Org. Chem.*, **46,** 1315 (1981).
[110] Feit, B. A., Melamed, U., Schmidt, R. R., and Speer, H., *J. Chem. Soc., Perkin Trans. 1*, **1981,** 1329.
[111] Feit, B. A., Melamed, U., Schmidt, R. R., and Speer, H., *Tetrahedron*, **37,** 2143 (1981).
[112] Bogdanovic, B., and Wermeckes, B., *Angew. Chem. Int. Ed.*, **20,** 684 (1981).
[113] Shabtai, J., Ney-Igner, E., and Pines, H., *J. Org. Chem.*, **46,** 3795 (1981).
[114] Shih, C., and Swenton, J. S., *Tetrahedron Lett.*, **22,** 4217 (1981).
[115] Gill, M., Bainton, H. P., and Rickards, R. W., *Tetrahedron Lett.*, **22,** 1437 (1981).
[116] Tucker, H., Golding, G., and Purvis, S. R., *Tetrahedron Lett.*, **22,** 1373 (1981).
[117] Macdonald, T. L., Narayanan, B. A., and O'Dell, D. E., *J. Org. Chem.*, **46,** 1504 (1981).
[118] Oppolzer, W., Snowden, R. L., and Simmons, D. P., *Helv. Chim. Acta*, **64,** 2002 (1981).
[119] Oppolzer, W., Snowden, R. L., and Briner, P. H., *Helv. Chim. Acta*, **64,** 2022 (1981).
[120] Visser, R. G., Brandsma, L., and Bos, H. J. T., *Tetrahedron Lett.*, **22,** 2827 (1981).
[121] Meijer, J., Ruitenberg, K., Westmijze, H., and Vermeer, P., *Synthesis*, **1981,** 551.
[122] Zweifel, G., and Pearson, N. R., *J. Org. Chem.*, **46,** 829 (1981).
[123] Slayden, S. W., *J. Org. Chem.*, **46,** 2311 (1981).
[124] Hommes, H., Verkruijsse, H. D., and Brandsma, L., *J. Chem. Soc., Chem. Commun.*, **1981,** 366.
[125] Townsend, C. A., and Bloom, L. M., *Tetrahedron Lett.*, **22,** 3923 (1981).
[126] Hasan, I., Marinelli, E. R., Lin, L. C., Fowler, F. W., and Levy, A. B., *J. Org. Chem.*, **46,** 157 (1981).
[127] Gronowitz, S., Stenhammar, K., and Svensson, L., *Heterocycles*, **15,** 947 (1981).
[128] Brunelle, D. J., *Tetrahedron Lett.*, **22,** 3699 (1981).
[129] Güngör, T., Marsais, F., and Queguiner, G., *J. Organomet. Chem.*, **215,** 139 (1981).
[130] Lohmann, J.-J., Seebach, D., Syfrig, M. A., and Yoshifuji, M., *Angew. Chem. Int. Ed.*, **20,** 128 (1981).
[131] Yadav, J., Corey, P., Hsu, C.-T., Perlman, K., and Sih, C. J., *Tetrahedron Lett.*, **22,** 819 (1981).
[132] Reitter, B. E., Sachdeva, Y. P., and Wolfe, J. F., *J. Org. Chem.*, **46,** 3945 (1981).
[133] Ladd, D. L., and Weinstock, J., *J. Org. Chem.*, **46,** 203 (1981).
[134] Lidor, R., and Shatzmiller, S., *J. Am. Chem. Soc.*, **103,** 5916 (1981).
[135] Anderson, M. W., Jones, R. C. F., and Saunders, J., *Tetrahedron Lett.*, **22,** 261 (1981).
[136] Lipshutz, B. H., and Hungate, R. W., *J. Org. Chem.*, **46,** 1410 (1981).
[137] Beak, P., and Carter, L. G., *J. Org. Chem.*, **46,** 2363 (1981).
[138] Hanko, R., and Hoppe, D., *Angew. Chem. Int. Ed.*, **20,** 127 (1981).
[139] Reitz, D. B., Beak, P., and Tse, A., *J. Org. Chem.*, **46,** 4316 (1981).
[140] Majewski, M., Mpango, G. B., Thomas, M. T., Wu, A., and Snieckus, V., *J. Org. Chem.*, **46,** 2029 (1981).
[141] Cavicchioni, P. S., Veronese, A. C., and D'Angeli, F., *J. Chem. Soc., Chem. Commun.*, **1981,** 416.
[142] Nibbering, N. M. M., *Recl. Trav. Chim. Pays-Bas*, **100,** 297 (1981).
[143] DePuy, C. H., and Bierbaum, V. M., *Acc. Chem. Res.*, **14,** 146 (1981).
[144] Bierbaum, V. M., Schmitt, R. J., and DePuy, C. H., *EHP, Environ. Health Perspect.*, **36,** 119 (1980); *Chem. Abs.*, **94,** 120558 (1981).
[145] Bartmess, J. E., Hays, R. L., and Caldwell, G., *J. Am. Chem. Soc.*, **103,** 1338 (1981).
[146] Sullivan, S. A., DePuy, C. H., and Damrauer, R., *J. Am. Chem. Soc.*, **103,** 480 (1981).
[147] Guthrie, R. D., *Stud. Org. Chem. (Amsterdam)*, **5** (*Compr. Carbanion Chem., Pt. A*), 197 (1980); *Chem. Abs.*, **94,** 3327 (1981).
[148] Guthrie, R. D., Hrovat, D. A., Prahl, F. G., and Swan, J., *J. Org. Chem.*, **46,** 498 (1981).
[149] Handoo, K. L., and Bahadur, Y., *Indian J. Chem.*, **20B,** 412 (1981); *Chem. Abs.*, **95,** 61085 (1981).
[150] Bordwell, F. G., and Clemens, A. H., *J. Org. Chem.*, **46,** 1035 (1981).
[151] Saji, T., and Aoyagui, S., *Chem. Lett.*, **1981,** 983.
[152] Kauffmann, T., *Top. Curr. Chem.*, **92,** 109 (1980).
[153] Andersen, N. H., McCrae, D. A., Grotjahn, D. B., Gabhe, S. Y., Theodore, L. J., Ippolito, R. M., and Sarkar, T. K., *Tetrahedron*, **37,** 4069 (1981).
[154] Degl'Innocenti, A., Pike, S., Walton, D. R. M., Seconi, G., Ricci, A., and Fiorenza, M., *J. Chem. Soc., Cem. Commun.*, **1980,** 1201.
[155] Reetz, M. T., Steinbach, R., Wenderoth, B., and Westermann, J., *Chem. Ind. (London)*, **1981,** 541.
[156] Corriu, R. J. P., Guerin, C., and M'Boula, J., *Tetrahedron Lett.*, **22,** 2985 (1981).
[157] Tsuda, T., Yazawa, T., Watanabe, K., Fujii, T., and Saegusa, T., *J. Org. Chem.*, **46,** 192 (1981).

[158] Kim, S., Lee, J. I., and Chung, B. Y., *J. Chem. Soc., Chem. Commun.*, **1981,** 1231.
[159] Whiteley, C. G., *J. Chem. Soc., Chem. Commun.*, **1981,** 5.
[160] Pelter, A., Rupani, P., and Stewart, P., *J. Chem. Soc., Chem. Commun.*, **1981,** 164.
[161] Rudler-Chauvin, M., and Rudler, H., *J. Organomet. Chem.*, **212,** 203 (1981).
[162] Gompper, R., and Vogt, H.-H., *Chem. Ber.*, **114,** 2866 (1981).
[163] Jackman, L. M., and Lange, B. C., *J. Am. Chem. Soc.*, **103,** 4494 (1981).
[164] Fiaud, J.-C., and Malleron, J.-L., *J. Chem. Soc., Chem. Commun.*, **1981,** 1159.
[165] Nee, G., Leroux, Y., and Seyden-Penne, J., *Tetrahedron*, **37,** 1541 (1981).
[166] Seebach, D., Weller, T., Protschuk, G., Beck, A. K., and Hoekstra, M. S., *Helv. Chim. Acta*, **64,** 716 (1981).
[167] Schmierer, R., Grotemeier, G., Helmchen, G., and Selim, A., *Angew. Chem. Int. Ed.*, **20,** 207 (1981).
[168] Kende, A. S., Fludzinski, P., and Hill, J. H., *J. Am. Chem. Soc.*, **103,** 2904 (1981).
[169] Hubbard, J. S., and Harris, T. M., *J. Org. Chem.*, **46,** 2566 (1981).
[170] Shieh, H.-M., and Prestwich, G. D., *J. Org. Chem.*, **46,** 4321 (1981).
[171] Seebach, D., and Pohmakotr, M., *Tetrahedron*, **37,** 4047 (1981).
[172] Paterson, I., and Price, L. G., *Tetrahedron Lett.*, **22,** 2833 (1981).
[173] Nishiyama, H., Itagaki, K., Takahashi, K., and Itoh, K., *Tetrahedron Lett.*, **22,** 1685 (1981).
[174] Matsumura, E., Ariga, M., and Tohda, Y., *Bull. Chem. Soc. Jpn.*, **53,** 2891 (1980).
[175] Fráter, G., *Tetrahedron Lett.*, **22,** 425 (1981).
[176] Yu, L.-C., and Helquist, P., *J. Org. Chem.*, **46,** 4536 (1981).
[177] Russell, G. A., Mudryk, B., and Jawdosiuk, M., *J. Am. Chem. Soc.*, **103,** 4610 (1981).
[178] Meyers, A. I., Williams, D. R., Erickson, G. W., White, S., and Druelinger, M., *J. Am. Chem. Soc.*, **103,** 3081 (1981).
[179] Meyers, A. I., Williams, D. R., White, S., and Erickson, G. W., *J. Am. Chem. Soc.*, **103,** 3088 (1981).
[180] Enders, D., and Lotter, H., *Angew. Chem. Int. Ed.*, **20,** 795 (1981).
[181] Heathcock, C. H., White, C. T., Morrison, J. J., and VanDerveer, D., *J. Org. Chem.*, **46,** 1296 (1981).
[182] White, C. T., and Heathcock, C. H., *J. Org. Chem.*, **46,** 191 (1981).
[183] Clive, D. L. J., and Russell, C. G., *J. Chem. Soc., Chem. Commun.*, **1981,** 434.
[184] Noyori, R., Nishida, I., and Sakata, J., *J. Am. Chem. Soc.*, **103,** 2106 (1981).
[185] Itoh, A., Ozawa, S., Oshima, K., and Nozaki, H., *Bull. Chem. Soc. Jpn.*, **54,** 274 (1981).
[186] Yamamoto, Y., Yatagai, H., and Maruyama, K., *J. Chem. Soc., Chem. Commun.*, **1981,** 162.
[187] Hoffmann, R. W., and Zeiss, H.-J., *J. Org. Chem.*, **46,** 1309 (1981).
[188] Evans, D. A., Nelson, J. V., Vogel, E., and Taber, T. R., *J. Am. Chem. Soc.*, **103,** 3099 (1981).
[189] Evans, D. A., Bartroli, J., and Shih, T. L., *J. Am. Chem. Soc.*, **103,** 2127 (1981).
[190] Evans, D. A., and McGee, L. R., *J. Am. Chem. Soc.*, **103,** 2876 (1981).
[191] Masamune, S., Choy, W., Kerdesky, F. A. J., and Imperiali, B., *J. Am. Chem. Soc.*, **103,** 1566 (1981).
[192] Masamune, S., Hirama, M., Mori, S., Ali, Sk. A., and Garvey, D. S., *J. Am. Chem. Soc.*, **103,** 1568 (1981).
[193] Meyers, A. I., and Yamamoto, Y., *J. Am. Chem. Soc.*, **103,** 4278 (1981).
[194] Tamaru, Y., Kagotani, M., and Yoshida, Z., *Tetrahedron Lett.*, **22,** 3409 (1981).
[195] Tamaru, Y., Kagotani, M., Furukawa, Y., Amino, Y., and Yoshida, Z., *Tetrahedron Lett.*, **22,** 3413 (1981).
[196] Schultz, A. G., and Napier, J., *J. Chem. Soc., Chem. Commun.*, **1981,** 224.
[197] Kraus, G. A., and Frazier, K., *J. Org. Chem.*, **45,** 4262 (1980).
[198] Pardo, S. N., Ghosh, S., and Salomon, R. G., *Tetrahedron Lett.*, **22,** 1885 (1981).
[199] Mulzer, J., and Zippel, M., *Tetrahedron Lett.*, **22,** 2165 (1981).
[200] Patel, V., and Stothers, J. B., *Can. J. Chem.*, **58,** 2728 (1980).
[201] Hoyano, Y., Patel, V., and Stothers, J. B., *Can. J. Chem.*, **58,** 2730 (1980).
[202] Emly, M., and Leussing, D. D., *J. Am. Chem. Soc.*, **103,** 628 (1981).
[203] Jacobs, S. A., Cortez, C., and Harvey, R. G., *J. Chem. Soc., Chem. Commun.*, **1981,** 1215.
[204] Guthrie, R. D., and Nicolas, E. C., *J. Am. Chem. Soc.*, **103,** 4637 (1981).
[205] Bates, R. B., *Stud. Org. Chem. (Amsterdam)*, **5** (*Compr. Carbanion Chem., Pt. A*), 1 (1980); *Chem. Abs.*, **93,** 237987 (1980).
[206] Becker, B. Ch., Huber, W., and Müllen, K., *J. Am. Chem. Soc.*, **102,** 7803 (1980).
[207] Miller, J. T., and Dekock, C. W., *J. Organomet. Chem.*, **216,** 39 (1981).
[208] Bates, R. B., Hess, B. A., Ogle, C. A., and Schaad, L. J., *J. Am. Chem. Soc.*, **103,** 5052 (1981).
[209] Minsky, A., Klein, J., and Rabinovitz, M., *J. Am. Chem. Soc.*, **103,** 4586 (1981).
[210] O'Brien, D. H., and Breeden, D. L., *J. Am. Chem. Soc.*, **103,** 3237 (1981).
[211] Evans, W. J., Wink, D. J., Wayda, A. L., and Little, D. A., *J. Org. Chem.*, **46,** 3925 (1981).
[212] Fatiadi, A. J., *J. Res. Natl. Bur. Stand.*, **85,** 73 (1980); *Chem. Abs.*, **93,** 238465 (1980).

[213] Aihara, J., *J. Am. Chem. Soc.*, **103,** 1633 (1981).
[214] West, R., Downing, J. W., Inagaki, S., and Michl, J., *J. Am. Chem. Soc.*, **103,** 5073 (1981).
[215] Michl, J., and West, R., *Oxocarbons*, **1980,** 79; *Chem. Abs.*, **95,** 41598 (1981).
[216] Reich, H. J., Shah, S. K., Gold, P. M., and Olson, R. E., *J. Am. Chem. Soc.*, **103,** 3112 (1981).
[217] Knight, D. W., and Nott, A. P., *J. Chem. Soc., Perkin Trans. 1*, **1981,** 1125.
[218] Fitt, J. J., and Gschwend, H. W., *J. Org. Chem.*, **45,** 4257 (1980).
[219] Amamria, A., and Mitchell, T. N., *J. Organomet. Chem.*, **210,** C17 (1981).
[220] Gandillon, G., Bianco, B., and Burger, U., *Tetrahedron Lett.*, **22,** 51 (1981).
[221] Bilyard, K. G., and Garratt, P. J., *Tetrahedron Lett.*, **22,** 1755 (1981).
[222] Sandifer, R. M., Bhattacharya, A. K., and Harris, T. M., *J. Org. Chem.*, **46,** 2260 (1981).
[223] Hoppe, I., and Schöllkopf, U., *Liebigs Ann. Chem.*, **1981,** 103.
[224] Cartoon, M. E. K., and Cheeseman, G. W. H., *J. Organomet. Chem.*, **212,** 1 (1981).
[225] Reich, H. J., and Willis, W. W., *J. Org. Chem.*, **45,** 5227 (1980).
[226] Ceré, V., Paolucci, C., Pollicino, S., Sandri, E., and Fava, A., *J. Chem. Soc., Chem. Commun.*, **1981,** 764.
[227] Binns, M. R., and Haynes, R. K., *J. Org. Chem.*, **46,** 3790 (1981).
[228] Pelter, A., and Rao, J. M., *J. Chem. Soc., Chem. Commun.*, **1981,** 1149.
[229] Ager, D. J., *Tetrahedron Lett.*, **22,** 2803 (1981).
[230] de Groot, A., and Jansen, B. J. M., *Tetrahedron Lett.*, **22,** 887 (1981).
[231] Sugihara, Y., Fujiyama, Y., and Murata, I., *Chem. Lett.*, **1980,** 1427.
[232] Amstutz, R., Dunitz, J. D., and Seebach, D., *Angew. Chem. Int. Ed.*, **20,** 465 (1981).
[233] Hase, T. A., and Lahtinen, L., *Tetrahedron Lett.*, **22,** 3285 (1981).
[234] Grotjahn, D. B., and Andersen, N. H., *J. Chem. Soc., Chem. Commun.*, **1981,** 306.
[235] Bartmess, J. E., Hays, R. L., Khatri, H. N., Misra, R. N., and Wilson, S. R., *J. Am. Chem. Soc.*, **103,** 4746 (1981).
[236] Ziegler, F. E., and Fang, J.-M., *J. Org. Chem.*, **46,** 825 (1981).
[237] Tanimoto, S., Oida, T., Hatanaka, K., and Sugimoto, T., *Tetrahedron Lett.*, **22,** 655 (1981).
[238] Ager, D. J., *Tetrahedron Lett.*, **21,** 4763 (1980).
[239] Antonjuk, D. J., Ridley, D. D., and Smal, M. A., *Aust. J. Chem.*, **33,** 2635 (1980).
[240] Colombo, L., Gennari, C., Scolastico, C., Guanti, G., and Narisano, E., *J. Chem. Soc., Perkin Trans. 1*, **1981,** 1278.
[241] Colombo, L., Gennari, C., Resnati, G., and Scolastico, C., *J. Chem. Soc., Perkin Trans. 1*, **1981,** 1284.
[242] Takei, H., Sugimura, H., Miura, M., and Okamura, H., *Chem. Lett.*, **1980,** 1209.
[243] Goldmann, S., *Angew. Chem. Int. Ed.*, **20,** 779 (1981).
[244] Ogura, K., Imaizumi, J., Iida, H., and Tsuchihashi, G., *Chem. Lett.*, **1980,** 1587.
[245] Williams, D. R., Phillips, J. G., and Huffman, J. C., *J. Org. Chem.*, **46,** 4103 (1981).
[246] Kametani, T., Tsubuki, M., Nemoto, H., and Suzuki, K., *J. Am. Chem. Soc.*, **103,** 1256 (1981).
[247] Annunziata, R., Cinquini, M., Colonna, S., and Cozzi, F., *J. Chem. Soc., Chem. Commun.*, **1981,** 1005.
[248] Kotake, H., Inomata, K., Kinoshita, H., Sakamoto, Y., and Kaneto, Y., *Bull. Chem. Soc. Jpn.*, **53,** 3027 (1980).
[249] Uguen, D., *Bull. Soc. Chim. Fr. II*, **1981,** 99.
[250] Bellamy, A. J., and MacKirdy, I. S., *J. Chem. Soc., Perkin Trans. 2*, **1981,** 1093.
[251] Tanaka, K., Matsui, S., and Kaji, A., *Bull. Chem. Soc. Jpn.*, **53,** 3619 (1980).
[252] Eisch, J. J., and Galle, J. E., *J. Org. Chem.*, **45,** 4534 (1980).
[253] Hirama, M., *Tetrahedron Lett.*, **22,** 1905 (1981).
[254] Tomažič, A., and Ghera, E., *Tetrahedron Lett.*, **22,** 4349 (1981).
[255] Thea, S., Guanti, G., and Williams, A., *J. Chem. Soc., Chem. Commun.*, **1981,** 535.
[256] Lappert, M. F., and Raston, C. L., *J. Chem. Soc., Chem. Commun.*, **1981,** 173.
[257] Takabe, K., Nagaoka, N., Endo, T., and Katagiri, T., *Chem. Ind. (London)*, **1981,** 540.
[258] Katritzky, A. R., Arrowsmith, J., bin Bahari, Z., Jayaram, C., Siddiqui, T., and Vassilatos, S., *J. Chem. Soc., Perkin Trans. 1*, **1980,** 2851.
[259] Schöllkopf, U., Groth, U., and Deng, C., *Angew. Chem. Int. Ed.*, **20,** 798 (1981).
[260] Gallicano, K. D., Oakley, R. T., Paddock, N. L., and Sharma, R. D., *Can. J. Chem.*, **59,** 2654 (1981).
[261] Aldercreutz, P., and Magnusson, G., *Acta Chem. Scand.*, **34B,** 647 (1980).
[262] Hercouet, A., and Le Corre, M., *Tetrahedron*, **37,** 2861 (1981).
[263] Hercouet, A., and Le Corre, M., *Tetrahedron*, **37,** 2855 (1981).
[264] Hercouet, A., and Le Corre, M., *Tetrahedron*, **37,** 2867 (1981).
[265] Cooke, M. P., *Tetrahedron Lett.*, **22,** 381 (1981).
[266] Seebach, D., and Yoshifuji, M., *Helv. Chim. Acta*, **64,** 643 (1981).
[267] Maleki, M., Miller, J. A., and Lever, O. W., *Tetrahedron Lett.*, **22,** 3789 (1981).
[268] Schore, N. E., and LaBelle, B. E., *J. Org. Chem.*, **46,** 2306 (1981).

[269] Sturtz, G., and Baboulene, M., *Tetrahedron*, **37**, 3067 (1981).
[270] Nakamura, E., *Tetrahedron Lett.*, **22**, 663 (1981).
[271] Simonyi, M., and Fitos, I., *Kem. Ujabb Eredmenyei*, **46**, 7 (1980); *Chem. Abs.*, **94**, 3315 (1981).
[272] Kresge, A. J., and Powell, M. J., *J. Am. Chem. Soc.*, **103**, 201 (1981).
[273] Caldin, E. F., Mateo, S., and Warrick, P., *J. Am. Chem. Soc.*, **103**, 202 (1981).
[274] German, E. D., and Kuznetsov, A. M., *J. Chem. Soc., Faraday Trans. 1*, **77**, 397 (1981).
[275] Sugimoto, N., Sasaki, M., and Osugi, J., *Bull. Chem. Soc. Jpn.*, **54**, 2598 (1981).
[276] Borodin, P. M., Golubev, N. S., Denisov, G. S., and Safarov, N. A., *React. Kinet. Catal. Lett.*, **16**, 1 (1981); *Chem. Abs.*, **95**, 96481 (1981).
[277] Leffek, K. T., and Pruszynski, P., *Can. J. Chem.*, **59**, 3034 (1981).
[278] Agresti, A., Bacci, M., and Ranfagni, A., *Chem. Phys. Lett.*, **79**, 100 (1981); *Chem. Abs.*, **95**, 24069 (1981).
[279] Harada, T., Nishikido, N., Moroi, Y., and Matuura, R., *Bull. Chem. Soc. Jpn.*, **54**, 2592 (1981).
[280] Gilbert, H. F., *J. Am. Chem. Soc.*, **102**, 7059 (1980).
[281] Symons, E. A., and Clermont, M. J., *J. Am. Chem. Soc.*, **103**, 3127 (1981).
[282] Symons, E. A., Clermont, M. J., and Coderre, L. A., *J. Am. Chem. Soc.*, **103**, 3131 (1981).
[283] Koch, H. F., Koch, J. G., Donovan, D. B., Toczko, A. G., and Kielbania, A. J., *J. Am. Chem. Soc.*, **103**, 5417 (1981).
[284] Koch, H. F., Tumas, W., and Knoll, R., *J. Am. Chem. Soc.*, **103**, 5423 (1981).
[285] Mishina, T., *J. Sci. Hiroshima Univ., Ser. A: Phys. Chem.*, **44**, 157 (1980); *Chem. Abs.*, **94**, 14732 (1981).
[286] Eaborn, C., Stamper, J. G., and Seconi, G., *J. Organomet. Chem.*, **204**, 27 (1981).
[287] Eaborn, C., and Mahmoud, F. M. S., *J. Organomet. Chem.*, **205**, 47 (1981).
[288] Seconi, G., Eaborn, C., and Stamper, J. G., *J. Organomet. Chem.*, **204**, 153 (1981).
[289] Seconi, G., and Eaborn, C., *J. Chem. Soc., Perkin Trans. 2*, **1981**, 1260.
[290] Eaborn, C., and Mahmoud, F. M. S., *J. Organomet. Chem.*, **209**, 13 (1981).
[291] Alvanipour, A., Eaborn, C., and Walton, D. R. M., *J. Organomet. Chem.*, **201**, 233 (1980).
[292] Stanczyk, W., and Chojnowski, J., *J. Organomet. Chem.*, **202**, 257 (1980).
[293] Eaborn, C., and Mahmoud, F. M. S., *J. Organomet. Chem.*, **206**, 49 (1981).
[294] Meyer, F. K., Pellerite, M. J., and Brauman, J. I., *Helv. Chim. Acta*, **64**, 1058 (1981).
[295] Koehler, H. J., *Wiss. Z.-Karl-Marx-Univ. Leipzig, Math.-Naturwiss. Reihe*, **28**, 625 (1979); *Chem. Abs.*, **93**, 238467 (1980).
[296] Streitwieser, A., Juaristi, E., and Nebenzahl, L. L., *Stud. Org. Chem. (Amsterdam)*, **5** (*Compr. Carbanion Chem., Pt. A*), 323 (1980); *Chem. Abs.*, **94**, 46264 (1981).
[297] Bordwell, F. G., and Fried, H. E., *J. Org. Chem.*, **46**, 4327 (1981).
[298] Bordwell, F. G., Drucker, G. E., and Fried, H. E., *J. Org. Chem.*, **46**, 632 (1981).
[299] Streitwieser, A., Word, J. M., Guibé, F., and Wright, J. S., *J. Org. Chem.*, **46**, 2588 (1981).
[300] Boerth, D. W., and Streitwieser, A., *J. Am. Chem. Soc.*, **103**, 6443 (1981).
[301] Arnett, E. M., and Venkatasubramaniam, K. G., *Tetrahedron Lett.*, **22**, 987 (1981).
[302] Slater, C. D., *J. Org. Chem.*, **46**, 2173 (1981).
[303] Hopkinson, A. C., and Llen, M. H., *Int. J. Quantum Chem.*, **18**, 1371 (1980).
[304] Petrov, E. S., and Ryaboi, V. M., *Zh. Org. Khim.*, **17**, 470 (1981); *Chem. Abs.*, **95**, 24043 (1981).
[305] Pross, A., DeFrees, D. J., Levi, B. A., Pollack, S. K., Radom, L., and Hehre, W. J., *J. Org. Chem.*, **46**, 1693 (1981).
[306] Vlasov, V. M., and Yakabson, G. G., *Izv. Sib. Otd. Akad. Nauk SSSR, Ser. Khim. Nauk*, **1980**, 133; *Chem. Abs.*, **94**, 46616 (1981).
[307] Apeloig, Y., *J. Chem. Soc., Chem. Commun.*, **1981**, 396.
[308] Hopkinson, A. C., and Lien, M. H., *J. Org. Chem.*, **46**, 998 (1981).
[309] Hopkinson, A. C., and Lien, M. H., *J. Organomet. Chem.*, **206**, 287 (1981).
[310] Holak, T. A., and Aksnes, D. W., *J. Chem. Soc., Chem. Commun.*, **1980**, 1237.
[311] Chandrasekhar, J., Andrade, J. G., and Schleyer, P., von R., *J. Am. Chem. Soc.*, **103**, 5612 (1981).
[312] Kabalka, G. W., Pagni, R. M., Bridwell, P., Walsh, E., and Hassaneen, H. M., *J. Org. Chem.*, **46**, 1513 (1981).
[313] Iorga, I., Csunderlik, C., Safta, M., and Ostrogovich, G., *Rev. Roum. Chim.*, **25**, 1351 (1980).
[314] Aresta, M., and Ciminale, F., *J. Chem. Soc., Dalton Trans.*, **1981**, 1520.
[315] Spillane, W. J., Kavanagh, P., Young, F., Dou, H. J.-M., and Metzger, J., *J. Chem. Soc., Perkin Trans. 1*, **1981**, 1763.
[316] Carpenter, T. A., Jenner, P. J., Leeper, F. J., and Staunton, J., *J. Chem. Soc., Chem. Commun.*, **1980**, 1227.
[317] Costa, A. M. B. S. R. C. S., Dean, F. M., Jones, M. A., Smith, D. A., and Varma, R. S., *J. Chem. Soc., Chem. Commun.*, **1980**, 1224.

[318] Nechvatal, G., Widdowson, D. A., and Williams, D. J., *J. Chem. Soc., Chem. Commun.*, **1981,** 1260.
[319] Brocard, J., Lebibi, J., and Couturier, D., *J. Chem. Soc., Chem. Commun.*, **1981,** 1264.
[320] Pellerite, M. J., and Brauman, J. I., *Stud. Org. Chem. (Amsterdam)*, **5** (*Compr. Carbanion Chem., Pt. A*), 55 (1980); *Chem. Abs.*, **94,** 29611 (1981).
[321] Faird, R., and McMahon, T. B., *Can. J. Chem.*, **58,** 2307 (1980).
[322] Squires, R. R., DePuy, C. H., and Bierbaum, V. M., *J. Am. Chem. Soc.*, **103,** 4256 (1981).
[323] Hunt, D. F., and Sethi, S. K., *J. Am. Chem. Soc.*, **102,** 6953(1980).
[324] Bermejo, J., and Martinez, A., *Publ. INCAR*, **1980,** 12; *Chem. Abs.*, **95,** 41905 (1981).
[325] Pinkus, A. G., and Gopalan, R., *J. Chem. Soc., Chem. Commun.*, **1981,** 1016.
[326] Ganapathy, K., and Ramanujam, M., *Indian J. Chem.*, **19B,** 901 (1980).
[327] Ananthakrishnanadar, P., and Gnanasekaran, C., *Indian J. Chem.*, **19A,** 646 (1980).
[328] Argentini, M., *EIR-Ber.*, **1979,** 393; *Chem. Abs.*, **93,** 237966 (1980).
[329] Meyer, T. A., and Williams, D. L. H., *J. Chem. Soc., Perkin Trans. 2*, **1981,** 361.
[330] Challis, B. C., and Shuker, D. E. G., *Food Cosmet. Toxicol.*, **18,** 283 (1980); *Chem. Abs.*, **93,** 238180 (1980).
[331] Kalatzis, E., and Papadopoulos, P., *J. Chem. Soc., Perkin Trans. 2*, **1981,** 239.
[332] Kalatzis, E., and Papadopoulos, P., *J. Chem. Soc., Perkin Trans. 2*, **1981,** 248.
[333] Davies, R., Massey, R. C., and McWeeny, D. J., *Food Chem.*, **6,** 115 (1980); *Chem. Abs.*, **94,** 208024 (1981).
[334] Andreev, S. A., Lebedev, B. A., Tselinskii, I. V., and Shokhor, I. N., *Zh. Org. Khim.*, **16,** 1353 (1980); *Chem. Abs.*, **93,** 238148 (1980).
[335] Andreev, S. A., Lebedev, B. A., and Tselinskii, I. V., *Zh. Org. Khim.*, **16,** 1360 (1980); *Chem. Abs.*, **93,** 238149 (1980).
[336] Andreev, S. A., Lebedev, B. A., and Tselinskii. I. V., *Zh. Org. Khim.*, **16,** 1374 (1980); *Chem. Abs.*, **93,** 238151 (1980).
[337] Andreev, S. A., Lebedev, B. A., and Tselinskii, I. V., *Zh. Org. Khim.*, **16,** 1370 (1980); *Chem. Abs.*, **93,** 238150 (1980).
[338] Reutov, O. A., *Izv. Akad. Nauk SSSR, Ser. Khim.*, **1980,** 2058; *Chem. Abs.*, **94,** 3322 (1981).
[339] Kitching, W., Atkins, A. R., Wickham, G., and Alberts, V., *J. Org. Chem.*, **46,** 563 (1981).
[340] Katsushima, T., Yamaguchi, R., Iemura, S., and Kawanisi, M., *Bull. Chem. Soc. Jpn.*, **53,** 3318 (1980).
[341] Katsushima, T., Yamaguchi, R., Iemura, S., and Kawanisi, M., *Bull. Chem. Soc. Jpn.*, **53,** 3324 (1980).
[342] Abraham, M. H., and Andonian-Haftvan, J., *Bull. Soc. Chim. Belg.*, **89,** 819 (1980).

Organic Reaction Mechanisms 1981
Edited by A. C. Knipe and W. E. Watts

CHAPTER 11

Elimination Reactions

A. F. HEGARTY

Chemistry Department, University College, Belfield, Dublin 4, Ireland

Stereochemistry, Orientation, and Isotope Effects in *E*2 Reactions . . . 385
The *ElcB* Mechanism 388
Pyrolytic Elimination Reactions 390
Esters and Carbamates 390
Decarboxylation and Related Reactions 392
Loss of N_2 394
Other Pyrolyses 395
Other Topics 396
Highly Strained Alkenes 396
Triple-bond Formation 397
Metal Ion Catalysis and Related Reactions 398
Silanes, Selenanes, and Related Substrates 399
Other Eliminations 401
References 404

Stereochemistry, Orientation, and Isotope Effects in *E*2 Reactions

In a major review of the mechanism of alkene formation, Bartsch and Zavada[1] have dealt with strong alkoxide base *E*2 eliminations. The effect of "free" (in the presence of crown ethers) and associated bases has been dissected and it is concluded that elimination stereochemistry depends on electrostatic effects between the attacking base and leaving-group and also on steric interactions which may tip the balance. A practically complete suppression of *syn*-elimination by the crown ether occurs with Bu^tOK–Bu^tOH when conformationally labile tosylates are used as substrates. By comparing[2] *anti*-elimination of a series of tosylates $RCH_2CHOTsC_5H_{11}$ and $RCHOTsCH_2C_5H_{11}$ (R = H, Me, Et, Pr^n, Pr^i, Bu^t) in the presence and absence of crowns it was concluded that the steric requirements of associated base species in the transition state *do not* exceed those of dissociated base; this also provides evidence against the "clump" base model for *syn*-elimination.

Orientation in *E*2 reactions is commonly found to be concentration-dependent. A systematic study[3] using 2-alkyl halides and tosylates as substrates with Bu^tOK–Bu^tOH as base–solvent system has shown that the percentage of terminal olefin and *trans*/*cis* ratio of internal alkene both increase with increasing Bu^tO^- in each case (solvent and substrate structure are less important). This is consistent with the explanation that competition occurs between elimination by associated (at high base concentration) and non-associated species; an alternative interpretation which attributes the change to solvation of the leaving-group cannot however be ruled out at this stage.

The same primary isotope effect is observed for the elimination of HX from 1-X-1,2-diphenylethanes in Bu^tOK–Bu^tOH in the presence and absence of crown ethers from which it is concluded that no variation in H—C bond-breaking occurs in the *E*2 reaction.[4]

Cyclic ethers, such as (ethylsulphonylmethyl)oxiran (**1**) undergo eliminative ring-fission *ca.* 2.5×10^6-fold more rapidly than acyclic models[5] such as (**2**). Cyclic ethers such as (**3**) also undergo elimination more rapidly than acyclic analogues but the ratios are small. For (**3**), a primary deuterium kinetic isotope effect of ~1 is observed but a value of 2.5 is observed for (**1**) suggesting an *E*2 or $(E1cB)_1$ mechanism. The *E*2 mechanism is ultimately preferred for (**1**) since the deprotonation rate (measured independently) is estimated to be far slower than ring-opening.

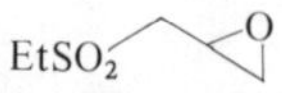

EtSO₂ (1) EtSO₂ OMe (2) EtSO₂ O (3)

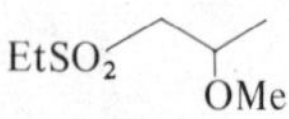

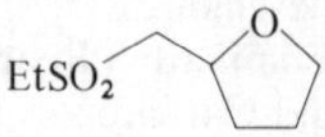

The chloro-alcohol (**4**) reacts intramolecularly in aqueous solution with barium hydroxide to give an oxetane whereas its isomer (**6**) forms unsaturated alcohols by elimination;[6] fragmentations occur from (**5**) and (**7**) in addition to elimination, and the individual rate parameters have been determined.

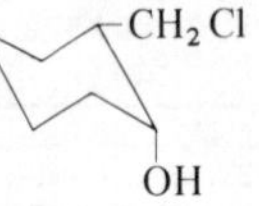

CH_2Cl OH (4) CH_2Cl OH (5) CH_2OH Cl (6) CH_2OH Cl (7)

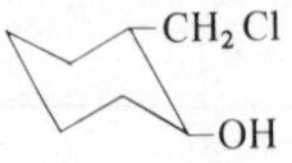

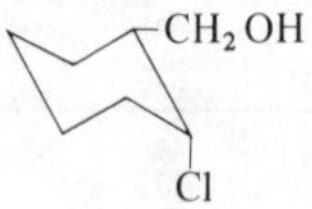

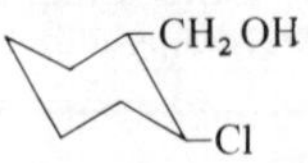

Sodium-alkoxide-activated $NaNH_2$ has a remarkable propensity for inducing *syn*-dehydrohalogenation.[7] For 1,2-dibromides the nature of the alkoxide has little effect on the debromination: dehydrobromination ratios but has a marked effect on reactivity. Best at producing *syn*-elimination are 2°-alkoxides and hindered bases (but a lower reactivity may be shown by these); at a minimum the alkoxide group should have *some* hydrophobic bulk.

Hofmann elimination from 1,1-dimethyl cyclic ammonium ions (**8**) is the major process observed upon reaction with NaOMe–MeOH, in competition with S_N2 displacements; elimination is the *only* ring-opening reaction observed with 5- and 6-membered substrates.[8] An interesting observation is that the α-methyl substituents provide a rate enhancement of 860–290-fold showing that the stereochemistry required in the transition state is more easily met in these series. For 2-aryl-1-phenylethylammonium ions with EtO^-–EtOH,[9] the primary isotope effect does not vary in a simple way with the pK_a of the leaving group. For example with *N*-methylpyrrolidine as leaving group k_H/k_D varies from 4.58 to 5.01 when the aryl group changes from p-MeC_6H_4 to p-ClC_6H_4 while the opposite variation (6.13 to

5.50) is observed when *N*-methylmorpholine is the leaving-group. It has been concluded that the proton is about half-transferred in the transition state.

As a mild alternative to Hofmann elimination, Katritzky[10] has suggested the use of the pentacyclic salt (**9**). The pyridinium salt (**9**) is obtained by reaction of the corresponding pyrylium salt (which is readily available) with primary amines; typical elimination conditions involve heating (**9**) at 150° with triphenylpyridine as non-nucleophilic base.

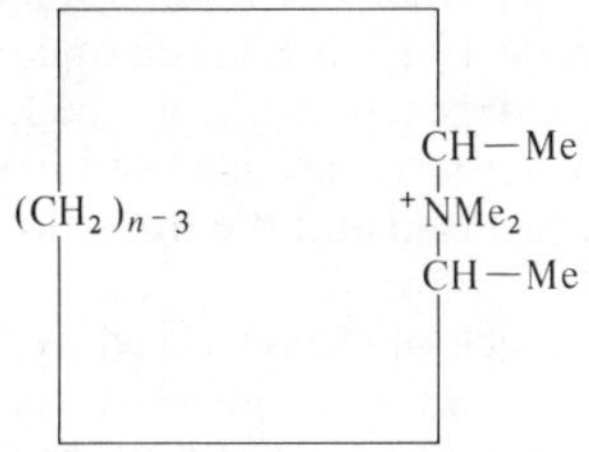

(**8**) 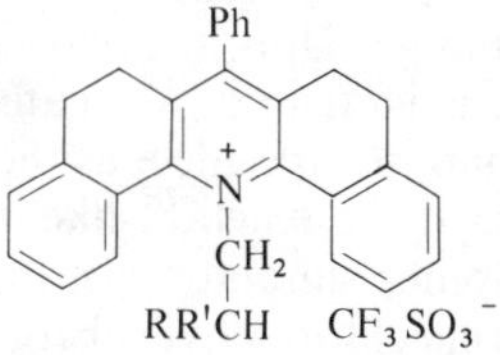 (**9**)

1-Phenyl-1-propyl- and 1-phenyl-2-propyl-trimethylammonium ions with a series of alkoxides give a variety of reactions including *syn*- and *anti*-β-elimination and α'–β-elimination (as well as Sommelet rearrangement and S_N2 displacements).[11] The low proportion of *syn*-elimination as well as high *trans*/*cis* alkene ratios (in *anti*-eliminations) observed in the presence of phenyl substituents has been ascribed to the lack of steric hindrance towards base approach.

Theoretical calculations on Hofmann eliminations of (2-phenylethyl)-trimethylammonium ions give poor agreement with experimental isotope effects when non-solvated models are used.[12] Differential solvation (which can be included in the calculations) largely overcomes these problems. In the absence of other factors, the extent to which negative charge is accumulated at the β-carbon in the transition state is solely a function of the leaving-group.

The controversy about the interpretation of the temperature dependence of isotope effects in terms of tunnelling, or otherwise, continues. Saunders[13] has reported the carbon isotope effects for the reaction of (2-phenylethyl-2-^{14}C)trimethylammonium ion with hydroxide; comparison with ^{13}C-isotope effects for the same reaction suggests that tunnelling *does* make a significant contribution to the observed isotope effects and the data fit Bell's theory of tunnelling. When the tunnelling correction is subtracted, the carbon isotope effects are small (or even inverse). The inverse carbon isotope effect, it is concluded, provides evidence that the symmetry of the transition state has been overstated. If this is so, then the large ρ value indicates that C—N bond cleavage lags well behind proton transfer in the transition state (and not necessarily that the extent of proton transfer is large).

The difficulty (arising largely from possible experimental errors) in accurately determining the tunnel correction (and thus the semi-classical primary isotope effect) has been emphasized.[14] Using 2-[(*p*-trifluoromethyl)-phenyl]ethyltrimethylammonium ions (**10**; $X = CF_3$) the k_H/k_D value at 50° in 35% DMSO is 7.25 and the error in the Q_{tH}/Q_{tD} (1.6–1.8) is 3–18%; the error in the pre-exponential term is even greater (10–50%) and consequently renders the interpretation less certain. As found previously there is an apparent maximum in both the semi-classical primary isotope effect and in the tunnelling correction (although this is less certain due to possible errors). The k_H/k_D value increases in the order H < *p*-Cl < *p*-CF_3 due either to a decreasing contribution of heavy atom motion or to changes in the extent of proton transfer. By increasing the DMSO content from 17% to 57%, a 10^3-fold increase in the rate of elimination is observed[15] for (**10**; X = H) while the k_H/k_D value goes through a maximum (of 5.21) at 34% DMSO. The maximum is quite noticeable since k_H/k_D is 3.22 at 17% DMSO and 3.83 at 57% DMSO. Nitrogen isotope effects do not show the same maxima and in fact just a small decrease (1.009 to 1.006) is observed when the DMSO fraction is increased; the Hammett ρ values show a small increase (3.11 to 3.38) on making the same change. Thus, the transition state becomes more reactant-like as the DMSO content is increased. A greater shift in transition-state structure is noted from the deuterium isotope effect as the basicity of HO^- is increased and the transition state becomes more reactant-like.

Computer simulation of the temperature dependence of the observed hydrogen isotope effect for dehydrobromination of (**11**; X = Br) can be achieved by assuming a mechanism involving internal return. However, measurement of both hydrogen and chlorine isotope effects[16] for (**11**; X = Cl) eliminates this possibility and provides firm evidence for an *E*2 mechanism. This diagnostic method of measuring a hydrogen isotope effect on a heavy-atom isotope effect need not be limited to leaving-groups and carbon atoms directly involved in the elimination could similarly be used. For example, for (**11**; X = I) $k_{^{12}C}/k_{^{14}C}$ values of 1.034 and 1.028 for the β-protium and β-deuterium compounds respectively were measured at 50° in MeONa–MeOH, supporting an *E*2 mechanism. Results for dehydrochlorination of (**12**) remain unexplained since the observed temperature independence of hydrogen isotope effects does not agree either with an *E*2 mechanism or with an *E*1*cB* reaction featuring internal return.

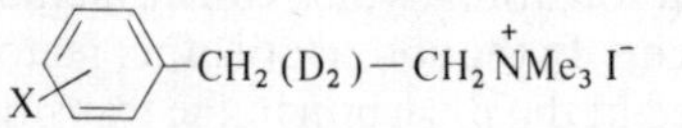

X–C_6H_4–$CH_2(D_2)$–$CH_2\overset{+}{N}Me_3$ I^- **(10)**

Ph–CH(CH$_3$)–*CH_2X **(11)**

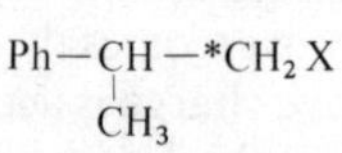

$ArCHXCF_2X$ **(12)**

The *E*1*cB* Mechanism

A large inverse solvent deuterium isotope effect is observed in the initial rates for the base-induced elimination from β-(*p*-nitrophenyl)ethylammonium ions (**13**) in aqueous acetohydroxamate buffers.[17] The curved buffer plots (concave downwards) observed have been quantitatively attributed to steady-state carbanion formation. The inverse solvent isotope effect is as large as 7.7; this arises since reprotonation of the carbanion (k_{-1}[BH]) competes with its unimolecular reaction (k_2) in H_2O while in D_2O under these conditions $k_2 > k_{-1}$[BD]; thus partitioning of the intermediate is dominated by k_{-1}[BH] in H_2O and not by k_{-1}[BD] in D_2O. The use of initial rates implies little H/D exchange in the substrate. The method provides a sensitive test for the *E*1*cB* mechanism whenever the proton-transfer step is not entirely rate-controlling.

Carbon leaving-groups are exceptionally poor in alkene-forming eliminations.[18] No correlation in nucleofugalities (leaving-group rank) is found with pK_a of the corresponding carbon acid using substrates of type (**14**; X = CN, Me; Y = CH_2Ph, NO_2, CN). As in other *E*1*cB* reactions the extent of cleavage of the leaving-group is small; delocalization in the forming anion is relatively unimportant in elimination (but not in pK_a measurements). By using "isobasic plots" it has been shown[19] that RO^- departs 5×10^3-fold faster than RS^- (of the same pK_a) from acetoacetates (**15a, b**). The major factor which contributes to the difference between thioalkoxide and alkoxide ions as leaving-groups is the pK_a difference (largely a solvation effect) while the ingrinsic nucleofugacities act in the opposite direction.

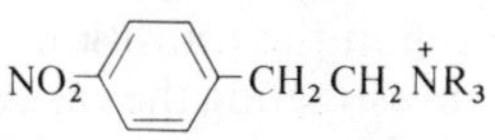

$NO_2C_6H_4CH_2CH_2\overset{+}{N}R_3$ **(13)**

$PhCO\bar{C}HCH(Ph)CX_2Y$ **(14)**

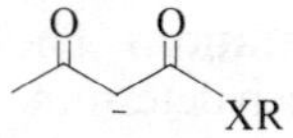

(15) a; X = O
b; X = S

An abrupt change in slope (−1.42 to +0.11) is noted for esters (**16**) in plots of log k against pK_a of HOR corresponding to a change-over from an $E1cB$ to a $B_{Ac}2$ mechanism.[20] The positive slope for the $B_{Ac}2$ mechanism (one previous example is noted) arises since two factors are involved: the substituent effect on the pK_a of the substrate and on the attack by HO^- on unionized substrate. The latter shows a "normal" negative slope ($\beta = -0.25$). The preferred conformation of the carbanion is not that required (**17**) for further reaction in which an increase in the electron density in the bonding orbital of the OR moiety leads to better vertical overlap and thus higher reaction rates. The kinetic advantage of the $E1cB$ over the $B_{Ac}2$ mechanism varies from 2×10^2 (**16**; X = H, Y = CN) to 8×10^3 (**16**; X = H, Y = CO_2Et) to 10^5 (**16**; X = H, Y = MeCO).

In spite of the accepted high acidity of thiolacetates (**18**) an insignificant amount of *S*-acetylcoenzyme A reacts *via* the E1*cB* pathway.[21] Both *S*-aryl and *S*-alkyl thiolacetates are correlated by the same $\beta_{l.g.}$ of −0.33 (unlike the oxygen analogues) and in general are one or two orders of magnitude less reactive than the oxygen esters.

S-Aryl- and *O*-aryl-thiocarbamates (**19**; X = S or O) react by the $E1cB$ pathway (to give isothiocyanate intermediates) and the Hammett ρ values obtained are, in general, close to those observed for analogous carbamates.[22] The hydrolysis of 2,4-

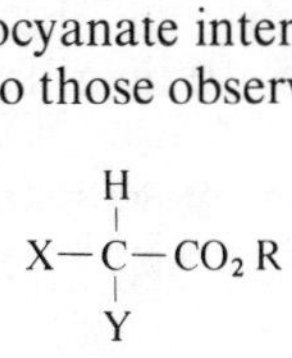

$X-CH(Y)-CO_2R$ **(16)**

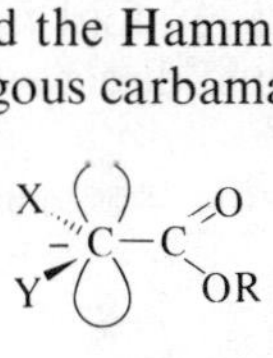

(17)

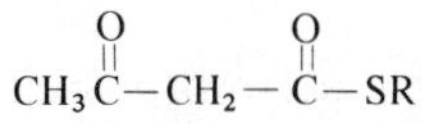

$CH_3C(=O)-CH_2-C(=O)-SR$ **(18)**

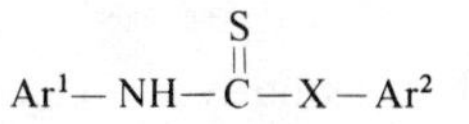

$Ar^1-NH-C(=S)-X-Ar^2$ **(19)**

dinitrophenyl arylmethanesulphonates (**20**) shows general base catalysis at low buffer concentrations but becomes non-linear (interpreted as a change in mechanism from rate-limiting proton transfer to breakdown of a sulphene–dinitrophenolate encounter complex) at high buffer concentrations.[23] This occurs since the stability of the anion (**21**) is reduced by the presence of the very good leaving-group. No deuterium exchange is observed at high buffer

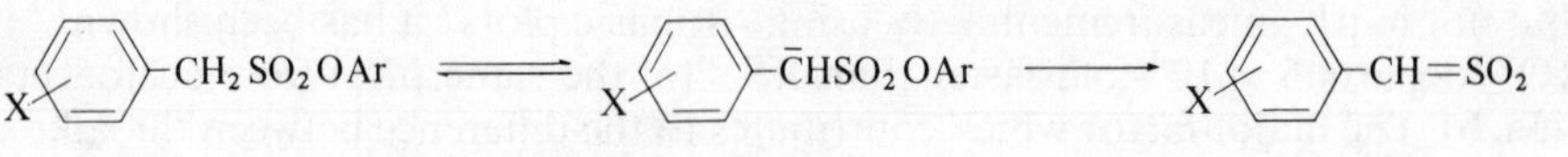

$X\text{-}C_6H_4\text{-}CH_2SO_2OAr \rightleftharpoons X\text{-}C_6H_4\text{-}\bar{C}HSO_2OAr \longrightarrow X\text{-}C_6H_4\text{-}CH{=}SO_2 + ArO^-$

(20) (21)

concentrations. The effective charges on carbon and oxygen in the transition state have been determined[24] (at low buffer concentrations) by investigating the variation in the rate of sulphene formation as X and Ar are changed. The change in effective charge on the oxygen of the phenoxy group is -0.4 unit, while the charge on carbon changes by -0.3 unit; the $-SO_2-$ group, on the other hand, undergoes no effective change. As the basicity of the leaving-group is decreased the mechanism of sulphene formation changes progressively from $(E1cB)_R$ to $(E1cB)_I$ to $E2$. When the $E2$ mechanism is operative[25] the transition state is unsymmetrical with differing amounts of C—H and S—O bond-cleavage in the transition state.

A dicarbanionic intermediate (**23**) has been implicated[26] in the alkaline hydrolysis of aryl methylsulphonylmethanesulphonate esters (**22**) at high hydroxide ion concentrations. Loss of ArO^- shows a high substituent effect ($\rho = +4.2$ from monoanion, $+3.8$ from **23**) consistent with the $E1cB$ mechanism. Substitution of one of the methylene hydrogens by a methyl group causes the reactivity of the monoanion to increase 500-fold while no reaction proportional to $[HO^-]$ is observed above the pK_a of (**22**). Although dicarbanions are synthetically useful, this study provides the first firm example in solution for such species from kinetic studies. A sulphene intermediate has been proposed in the reaction of phenols with $MeSO_2Cl$ in the presence of triethylamine;[27] the reaction is zero-order in phenol. Although the great majority of reactions of α-disulphones (**24**; R = alkyl) occur *via* an elimination–addition (sulphene) mechanism, bimolecular substitution is noted with good nucleophiles such as azide ion.[28] For (**24**; R = $PhCH_2$) only the $E1cB$ mechanism is observed, even with strong nucleophiles. For a given nucleophile elimination–addition is *ca.* 5–10-fold faster for (**24**) than for the corresponding alkanesulphonyl chloride while direct substitution is 5–10 times faster.

$$MeSO_2CH_2SO_2OAr \rightleftharpoons \text{monoanion} \rightleftharpoons MeSO_2\overset{2-}{C}SO_2OAr$$

(22) (23)

$$R{-}SO_2{-}SO_2{-}R$$

(24)

Base-catalysed oxidative cleavage of 1,1-bis-(*p*-chlorophenyl)-2,2,2-trichloroethanol to 4,4′-dichlorobenzophenone occurs *via* the expected $E1cB$ mechanism.[29]

Pyrolytic Elimination Reactions

Esters and Carbamates

A pulsed laser can be used to induce a thermal elimination to give a labile product; an excellent example[30] is provided by the reaction of cyclobutane acetate (**25**) on irradiation at 1078.6 cm^{-1} to yield cyclobutene (**26**). The activation energy for this process is 47 kcal mol^{-1} and (**26**) can be isolated using this technique, even though the activation energy for the further reaction of (**26**) to (**27**) is <33 kcal mol^{-1} (an allowed conrotatory ring-opening). The laser pulse heats the reactant for a very short time (10^{-7} s) and is followed by a cooling wave which quenches the reaction.

Pyrolysis of methyl 3-acetoxypropionate gives excellent first-order kinetics;[31] this and the activation energy observed suggest that grossly different results obtained previously were for a strongly surface-catalysed reaction. The direction of alkene formation from 4,4-dimethylpent-2-yl acetate is kinetically controlled by steric hindrance in the eclipsed *cis*-conformation.[32] The rate is increased by increasing steric crowding at the highly substituted carbon. The presence of $(CH_3)_2C{=}CH$ group at the β-carbon atom of ethyl acetate does not provide anchimeric assistance to pyrolytic elimination.[33] Rates of pyrolysis of several ethyl chloro-esters[34] are correlated by σ^* values ($\rho^* = 0.357$) showing that electron-withdrawing substituents at the acyl carbon enhance the rate of elimination. The effect of *ortho*-substituents has been investigated for the conversion of *o*-$RC_6H_4CH_2CH_2OAc$ to *o*-$RC_6H_4CH{=}CH_2$.[35]

Me O O (25) → (26) ----→ (27)

Thermal decomposition of fluorinated esters follows the normal elimination route to give an alkene and trifluoroacetic acid (no HF elimination is observed).[36] Trifluoroacetic acid is unstable at the pyrolysis temperatures, however, and other products (including those produced from **:**CF_2) result. Esters of terephthalic acid have been shown to undergo homolytic dissociation (*via* a six-membered transition state) in the first stage of pyrolysis.[37] The predominant reaction observed for gas-phase thermolysis of ^{18}O-labelled methyl acetate is the intramolecular oxygen to oxygen methyl-group migration.[38] Minor amounts of ketene (formed by 1,2-elimination of MeOH) have also been detected.

An unexpected decarbalkoxylation occurs when ethyl 2-ethoxynicotinate (**28**) is pyrolysed at 280° for 30 h, (**30**) being the main product.[39] At lower temperatures only O → N migration (formation of **29**) occurs. When the alkyl group is D-labelled, there is partial D-incorporation at C(5); a bimolecular reaction with the *N*-alkylpyridine acting as a nucleophile has been proposed.

The uses of *N,N*-disubstituted carbamates in lieu of simple esters has been probed;[40] the reaction is stereospecifically *syn* and *N,N*-diphenylcarbamates are recommended for synthesis since lower pyrolysis temperatures are required (although similar product distributions are observed as with acetates).

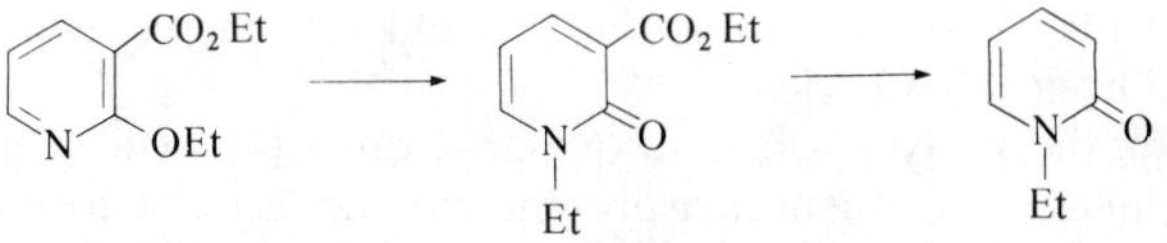

CO_2Et, N, OEt (28) → CO_2Et, N, O, Et (29) → N, O, Et (30)

Decarboxylation and Related Reactions

Decarboxylation of 6-nitrobenzisoxazole-3-carboxylate (**31**) continues to be used as a probe for micro-environment since the rate of this reaction is highly sensitive to solvent. In micro-emulsions the rate of decarboxylation ($3 \times 10^{-4} s^{-1}$) is similar to that previously observed in MeOH or in micellized CTABr (the value in pure water is $3 \times 10^{-6} s^{-1}$ under these conditions).[41] Aqueous aggregates such as CTABr micelles and dialkylammonium bilayer membranes give rate accelerations proportional to local hydrophobicities although the rigidity of the membrane is another factor which is important in determining its rôle as a catalyst.[42]

Decarboxylation of the potassium salt of 1-carbomethoxy-1-carboxycyclohexane is accelerated 10^4-fold by the presence of 18-crown-6 and this effect has been developed into a useful method for the decarboxylation of malonate esters.[43]

Previous studies on the decarboxylation of β-lactones (**32**) indicated complete retention of configuration in the alkene (**34**) formed *via* the zwitterion (**33**). Addition of acid has now been found[44] to increase the rate of alkene formation but as the acid concentration is increased the more stable *trans*-alkene (**36**) is formed. At high acid concentrations the reaction is again stereospecific, but this time for the formation of (**36**). Stabilization of the zwitterion (**33**) by proton transfer to (**35**), which is sufficiently long-lived to allow C—C bond rotation, seems to be indicated.

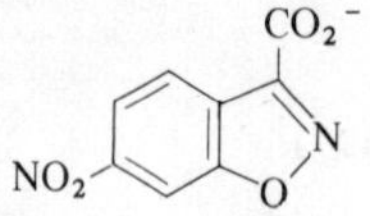

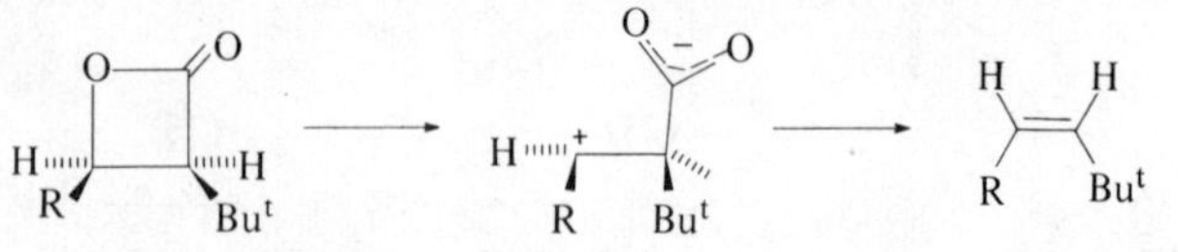

CO_2^- NO_2 O N (**31**) O O H R H Bu^t (**32**) → O O H + R Bu^t (**33**) → H H R Bu^t (**34**)

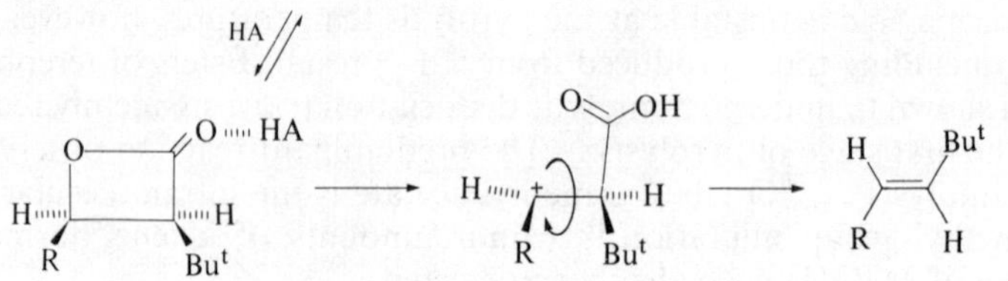

HA

O O ···· HA H R H Bu^t → O OH H + H R Bu^t (**35**) → H Bu^t R H (**36**)

Decarboxylation of *N*-alkyl-5-amino-1,3,4-oxadiazole-2-carboxylic acids (**37**) is unimolecular and occurs *via* a decarboxyprotonation mechanism; rate maxima are observed at *ca.* pH 0.[45] Decarboxylation of cinnamic acids by *Saccharomyces cerevisiae* has also been reported.[46]

Thermal decarbonylation is the principal reaction noted for (**38**) (and analogues) in hydrogen; dehydration and oligomerizations are parallel minor reactions.[47,48] Thermal decarbonylation of the caged ketone (**39**) to give the tricyclic triene (**40**) in quantitative yield is a particularly facile reaction.[49] The thermally labile C(1)—C(2) "long bond" (as shown by X-ray analysis) is thought to play a key rôle in enhancing the reactivity of (**39**); also notable is the 100-fold rate increase on replacing the groups R = Me by R = CO_2Me.

Flash vacuum thermolysis of 1,3-dioxolan-4-ones (**41**) yields aldehydes and ketones; when phenyl substituents are present epoxides are also formed (suggesting the intermediacy of a carbonyl ylid **42**).[50] Gas-phase thermolyses of thietan-1-oxide,

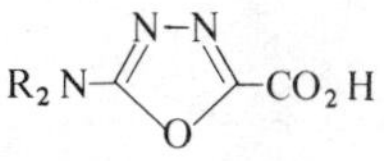

(37)

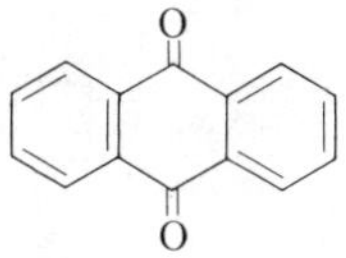

(38)

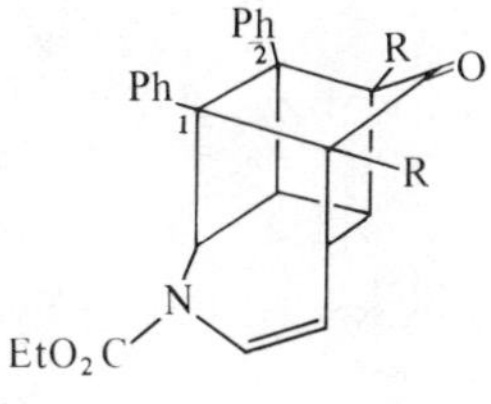

(39)

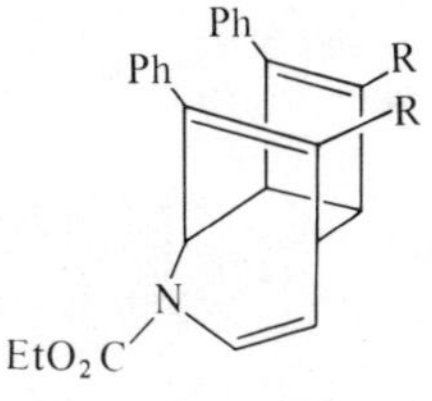

(40)

1,2-oxathiolan 2-oxide, and DMSO are believed to involve atomic oxygen extrusion and sulphoxide–sulphenate rearrangements.[51] Aromatic aldehydes react with *N*-methyl-*N*-sulphinylmethanamminium tetrafluoroborate with elimination of SO_2 to yield the iminium salts (**43**).[52] Aldimines are obtained in quantitative yield on heating (or treatment with acid) of the thiaziridine 1,1-dioxides (**44**),[53] while the

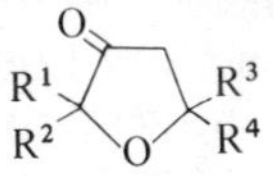

(41)

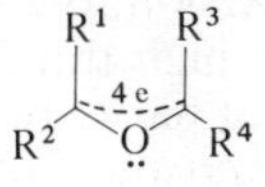

(42)

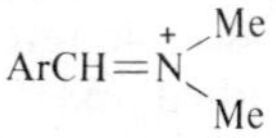

(43)

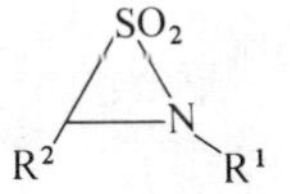

(44)

isomeric dihydro-1,3,4-thiadiazole 1,1-dioxides (**45**) and (**47**) yield the azine (**46**) and hydrazone (**48**), respectively.[54] Benzocyclobutenes (at low temperature) and alkenes (derived from [1,5]-hydride shift at high temperatures) are the major products formed by flash thermolysis of 1,3-dihydrobenzo[*c*]thiophen 2,2-dioxides.[55] Extrusion of SO from the sulphoxides (**49**) yields the thiophenes (**50**) but bulky R groups decrease the yield.[56] The rate-determining step in the reaction of benzoin with $SOCl_2$ is attack by the carbonyl on sulphur; extrusion of SO to give benzil occurs in a rapid subsequent step.[57] Methylsulphonium salts react with methyl-lithium to give desulphuration products, biradicals which may undergo ring-closure or [2,3]-sigmatropic rearrangement.[58] Thermolytic elimination of sulphur from α-halosulphides (**51**) to give alkenes[59] is analogous to the Ramberg–Bäcklund rearrangement of α-halosulphones.

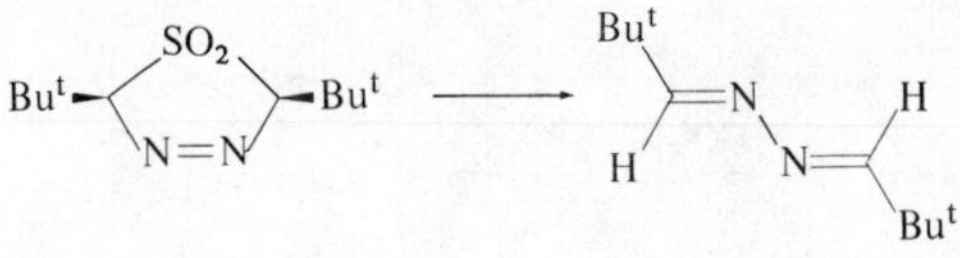

(45) (46)

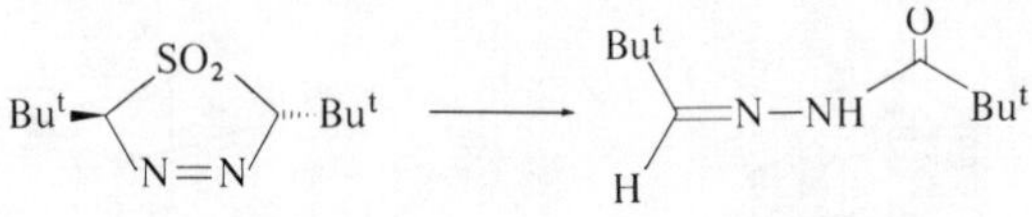

(47) (48)

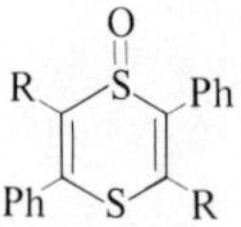

(49) 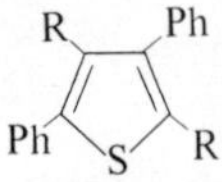 (50) 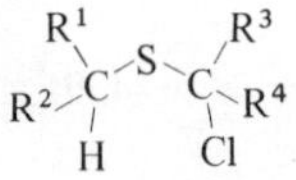 (51)

Loss of N_2

Thermolysis of (+)- and (−)-*trans*-3,5-diphenyl-1-pyrazoline (**52**) gives two competing processes: (*a*) cycloreversion to an azoalkane and (*b*) the formation of (**54**) by a "double retention" with high transfer of optical activity from (+)-(**52**) *via* the biradical (**53**).[60] The optical purity of (**54**) is higher in thermolysis than in photolysis and the degree of conservation of stereochemistry is the highest ever

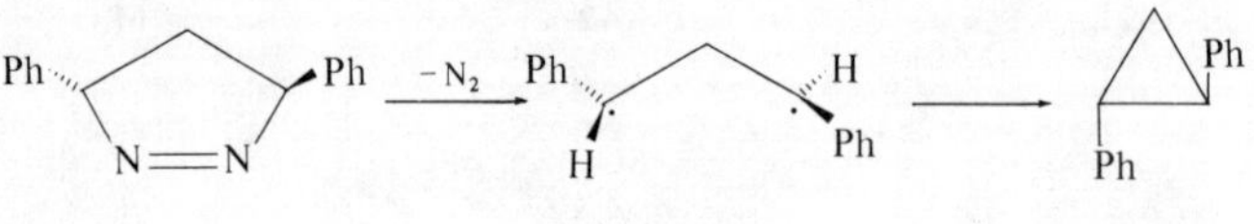

(52) (53) (54)

reported for the decomposition of a simple 1-pyrazoline. A disrotatory diazacyclobutene ring-opening has been reported for the first time in the conversion of 2,3-diazabicyclo[2.2.0]hex-2-ene (**55**) into a trimer of (**56**).[61] A compilation of thermal fragmentations of five-membered diazenes has been made and the extrusion of N_2 compared with loss of CO_2, CO, HCN, and similar fragments.[62] Electrocyclic ring-closure initiates the decomposition of *N*-nitrosodiphenylmethylimine (**57**) to benzophenone and nitogen.[63] Irradiation of 3,3,5,5-tetramethyl-1-pyrazoline with intense broad-band radiation gives 2,3-dimethylbutene and acetone azine by photocycloelimination.[64] The thiazolinones (**58**) decompose in solution to N_2, CO, and isopropylidene arylamine at 148–200°; a low substituent effect is observed (ρ for Ar substituents is −0.17) while the solvent effects show that the transition state is less polar than the ground state.[65] The adducts (**59**) (formed by [3 + 2]-cycloaddition from bis(trimethylsilyl)amino(trimethylsilylmethylene)phosphorane undergo loss of N_2 to give a λ^3-phosphorane.[66]

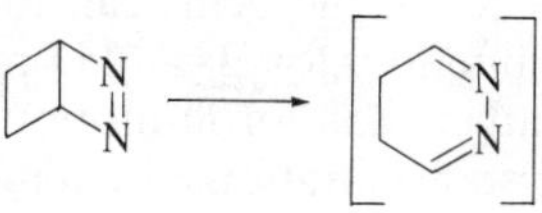

(55) (56)

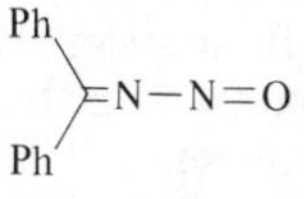

(57) 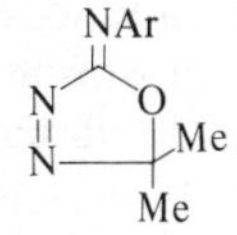 (58) 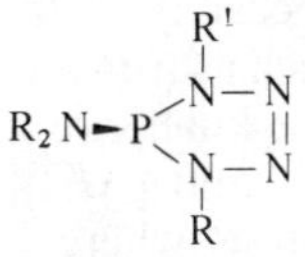 (59)

Other Pyrolyses

Gas-phase pyrolysis of 4-methyl-3,6-dihydro-2*H*-pyran to formaldehyde and 2-methylbuta-1,3-diene is a unimolecular process which occurs *via* a concerted transition state in which considerable charge separation does not occur.[67] The very-low-pressure pyrolysis technique has been used for the decomposition of cyclopentyl cyanide;[68] the main processes are HCN elimination to give cyclopentene and ring-fragmentation to give vinyl cyanide and propylene. The difference in activation energies for the ring-opening of cyclopentane and cyclopentyl cyanide is reasonably close to the estimated value for the cyano-group stabilization energy, supporting the assumption of a biradical mechanism.

The pyrolysis of 4-chlorobutan-2-one in the gas phase catalysed by HCl is complex, being of order 1.5 in the ketone; the ketone is 31-fold more reactive than ethyl chloride.[69] The effect of alkyl substitution on the pyrolysis of alkyl chlorides has been investigated,[70] using 2-chloro-2-methylbutane and 2-chloro-2,3-dimethylbutane, and a good correlation with σ^* obtained ($\rho^* = -4.75$) at 300°. The primary isotope effect (k_H/k_D) increases with decreasing pressure in the pyrolysis of 1-bromoethane-2-*d*, and the results compare favourably with theoretical calculations.[71] Arrhenius parameters have been determined for pyrolysis of neopentyl bromide,[72] which occurs by a simpler free-radical process than for the corresponding chloride. A study of free-radical chain dechlorination of chloroethanes in liquid triethylsilane has led to a determination of the absolute Arrhenius parameters for hydrogen abstraction from Et_3SiH.[73]

Anchimeric assistance by the neighbouring double bond is noted in the gas-phase dehydrohalogenation of 5-chloro-2-methylpent-2-ene and confirms that the three-membered conformation is the most favoured structure for participation.[74] MNDO-SCF MO calculations have been used to reproduce the kinetic isotope effects determined experimentally for dehydrochlorination of alkyl chlorides,[75] consistent with the concerted transition state.

While thermolysis of sulphoxides such as $C_6H_{11}CHDCH_2SOC_6H_5$ is characterized by a "normal" temperature dependence of the isotope effect consistent with a cyclic transition state, the closely related corresponding amine oxide ($C_6H_{11}CHDCH_2ONMe_2$) shows a temperature-independent k_H/k_D value. In an attempt to rationalize this difference Kwart and Brechbiel[76] point out that the lower activation energy observed for the thermolyses in DMSO compared to non-co-ordinating solvents is due to a change in mechanism rather than to the sequestering of H_2O as previously suggested. In the bent transition state implied by the five-membered ring the potential functions for H- and D-motions are the same but there is greater amplitude for H- rather than for D-motion. The weaker restoring forces associated with the transition-state vibrations result in an even greater amplitude for the H-motion than for D-. Several cases are noted with a bent transition state (for H-transfer) where $[\Delta E_a]_D^H < 0.1$ kcal. In dilute DMSO the amine oxide shows a variation in k_H/k_D with temperature and a linear transition state involving a change in mechanism (possibly to the solvent adduct **60**) has been proposed.

Selenoxides such as (**61**) undergo stereospecific *cis*-elimination of heating[77] in the range 40–85°. The kinetic results imply a tunnelling pathway for this β-elimination since $A_H/A_D < 0.7$ and $[\Delta E_a]_D^H >$ zero-point energy difference. This contrasts with sulphoxide pyrolysis where $[\Delta E_a]_D^H$ is 1.15 kcal mol^{-1}. Thus the lowering in the activation energy on going from a sulphoxide to a selenoxide is associated with a severe shortening of the separation of the O and C centres; due to the longer C—Se bond (compared to S) even the bulkiest substituents on Se would be mutually less repulsive.

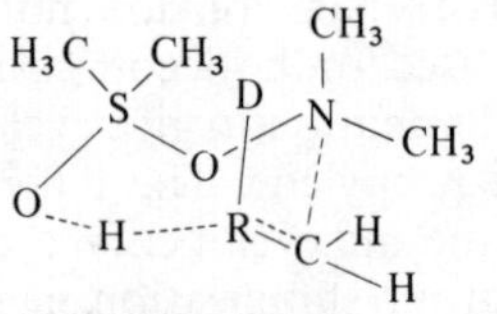

(**60**) 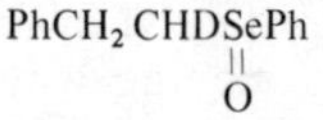 (**61**) $PhCH_2CHDSePh$ (=O)

Flash vacuum thermolysis of silylated thionocarboxylic acids gives ketenes rather than thioketenes;[78] gas-phase dechlorination of 1,1,2-trichloroethane is characterized by preferential formation of $CH_2ClCCl_2\cdot$ radicals;[79] very-low-pressure pyrolysis for the retroene decomposition of hepta-1,6-diene,[80] pyrolysis of *N,N*-disubstituted 5-methyl- and 5-chloro-benzimidazolines,[81] and gas-phase elimination of ethers induced by amide and hydroxide ions have been reported.[82]

Other Topics

Highly Strained Alkenes

The 1-chlorotricycloheptane (**62**) reacts with bulky lithium amides or Bu^{t}OK to give tricyclo[4.1.0.2^{2,7}]hept-1(7)-ene (**63**) as short-lived intermediate;[83] this has been trapped by reactive dienes such as anthracenes, furans, and spiro[2,4]hepta-4,6-diene giving propellanes. The existence of (**63**) as a distinct species has been established in an elegant series of competition experiments by using different bases and mixtures of trapping agents. Calculations show that the structure (**63**) is 11 kcal mol^{-1} more stable than the planar (**64**). Evidence that a chiral allene (**66**) is formed in the reaction of the 1-bromocyclohexene (**65**) with Bu^{t}OK has been presented.[84] The chirality of the intermediate was tested by using an optically active base, potassium menthoxide; of the four adducts isolated with 1,3-diphenylbenzo[*c*]furan, (**67**) showed a specific activity of $-1.3°$ which rapidly dropped to zero as the reaction temperature increased. Thus, (**65**) undergoes a concerted *cis*-elimination to give the chiral twisted allene (**66**) (which is the only possible chiral intermediate).

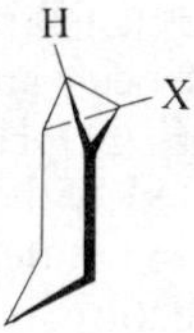

(62) 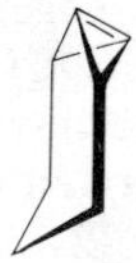 (63) 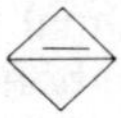 (64)

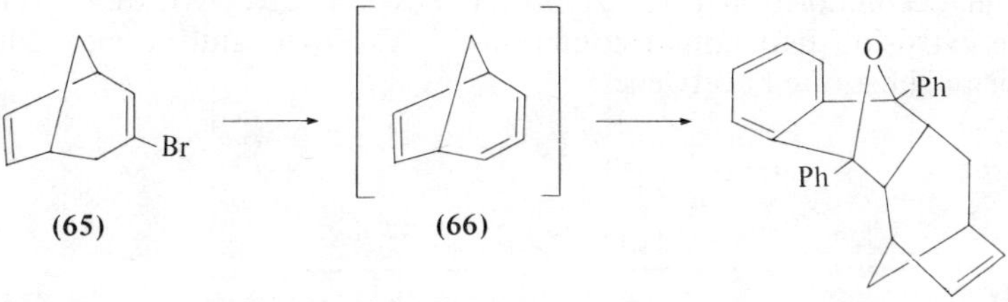

(65) (66) (67)

A "perpendicular olefin" does not intervene in the reaction of (**68**) in 96% aqueous HOAc buffered by NaOAc at 115°;[85] this has been shown using the doubly labelled substrate (**68**); the olefin (**69**) does not interconvert to (**70**) (*i.e.* the bridgehead olefins retain their configuration).

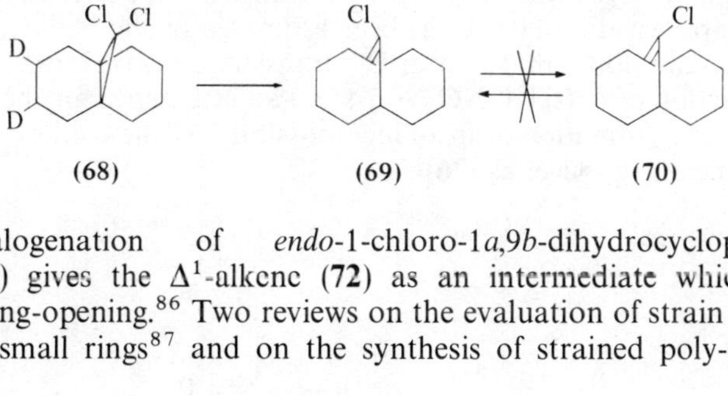

(68) (69) (70)

Dehydrohalogenation of *endo*-1-chloro-1*a*,9*b*-dihydrocyclopropa(*l*)phenanthrene (**71**) gives the Δ^1-alkene (**72**) as an intermediate which undergoes subsequent ring-opening.[86] Two reviews on the evaluation of strain effects on the reactivity of small rings[87] and on the synthesis of strained poly-cycles[88] have appeared.

Triple-bond Formation

N-Trifluoroacetylimidazole has been added to the formidable array of reagents which can be used to convert oximes to nitriles. The intermediacy of (**73**) has been proposed; this undergoes elimination catalysed by imidazole.[89] Fragmentation of benzoin-*O*-(carbamoyl)oximes depends on configuration; the nitrile is formed from

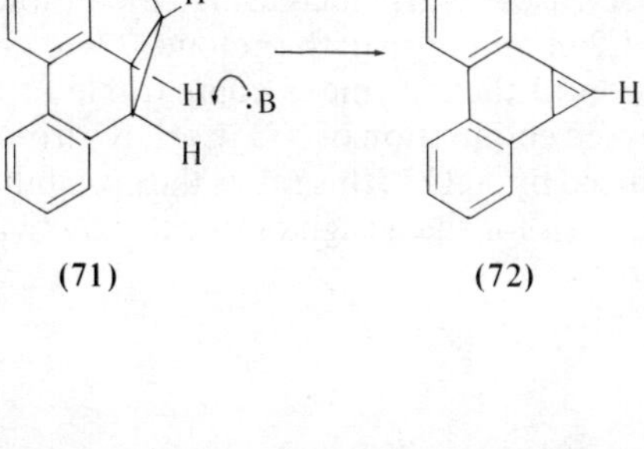

$RCH{=}NOCOCF_3$

(71) (72) (73)

the E-isomer in a concerted reaction whereas the (Z)-carbamate yields the rearranged isocyanide.[90] Another reagent which has been successfully used for dehydration of α-acetoxyoximes is DCC–DMSO–CF_3CO_2H (Moffat fragmentation).[91] The optimal yield of the nitrile is associated with an *anti*-periplanar arrangement of the bonds to be broken in the transition state.

The highly hindered 3,5-dichloro-2,4,6-trimethylbenzohydroximoyl chloride in methanol in the presence of HCl is characterized by rapid pre-equilibrium deprotonation followed by slow Cl^- loss;[92] the reverse reaction (addition of HCl to the nitrile oxide) has also been studied.

Thermal decomposition of 3-azidothiophens (**74**) gives rise unexpectedly to acetylene extrusion under mild conditions.[93] A ketone–anthracene adduct is a precursor of substituted acetylenes.[94]

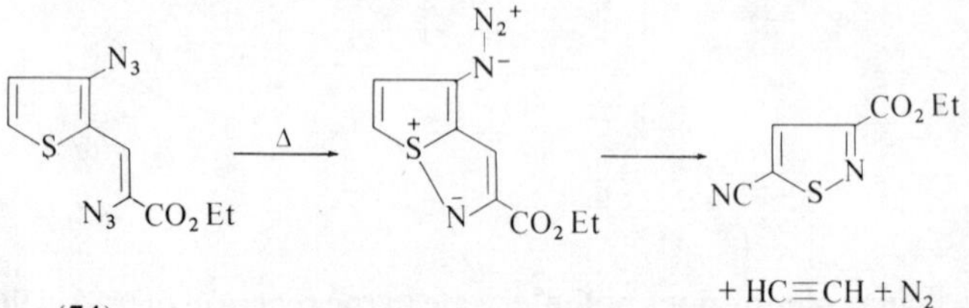

N_3 N_3 CO_2Et S $\xrightarrow{\Delta}$ N_2^+ N^- $\overset{+}{S}$ $\bar{N}$ CO_2Et $\longrightarrow$ CO_2Et NC S N

+ HC≡CH + N_2

(**74**)

The transition state for the elimination of NO_2^- from alkyl nitrites (RCH_2ONO_2) is shown to be cyclic, involving predominant proton tunnelling when base-catalysed and a cyclic non-linear proton transfer in the uncatalysed gas-phase process;[95] a barrier width of 0.9 Å has been estimated by the Bell–Caldin method. The simplest assumption is that a linear H^+-transfer occurs *via* tunnelling. Since the ρ value for reaction of $ArCH_2ONO_2$ is 3.4–3.9 an alternative for the base (RO^-)-catalysed process is formation of an adduct (possibly **75**); the saddle-point occurs at almost the same energy level as (**76**).

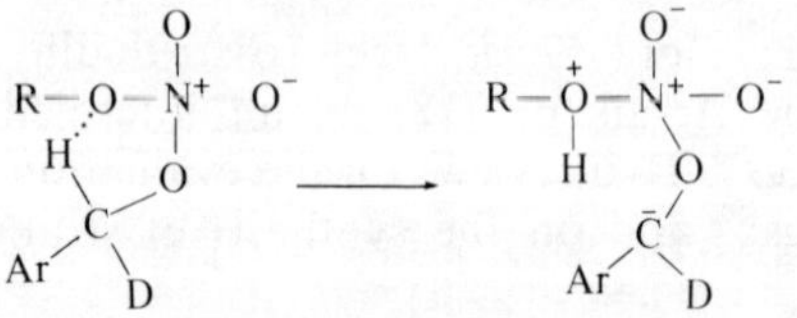

R–O–N^+(=O)(O$^-$)–O ··· H–C(Ar)(D) $\longrightarrow$ R–$\overset{+}{O}$(H)–N^+(O$^-$)(O$^-$)–O–$\bar{C}$(Ar)(D)

(**75**) (**76**)

Metal-Ion Catalysis and Related Reactions

Pyridoxal-catalysed β-elimination and dealdolization compete upon reaction with β-hydroxylglutamic acid as substrate.[96] Metal-ion catalysis is less effective than proton catalysis in the promotion of γ-decarboxylation and the relative effectiveness of the metal as a catalyst decreases with increasing metal-ion charge.[97] Carboxypeptidase A catalyses the 1,2-elimination of the ketone (**77**) to give (**78**). The k_{cat}/K_m value for the catalytic action of the enzyme is comparable to the second-order rate constant for HO^--catalysed elimination of $NO_2C_6H_4S^-$ from (**77**) and at least 10^5-fold greater than that induced by AcO^-; this raises the possibility that such elimination reactions can form the basis for the design of suicide inactivators of this enzyme.

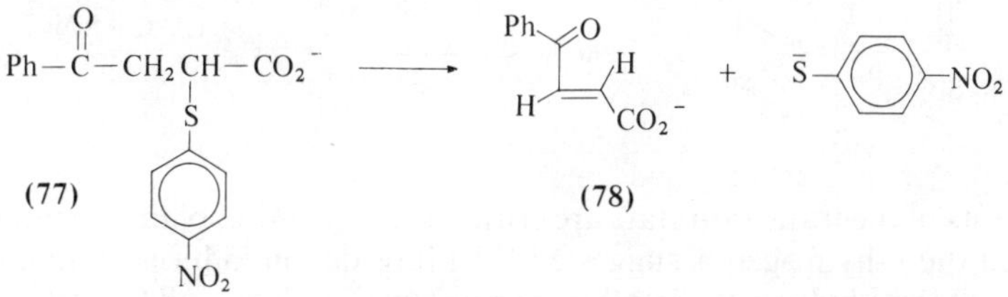

$$Ph-C(=O)-CH_2CH(S-C_6H_4-NO_2)-CO_2^- \longrightarrow PhC(=O)CH{=}CHCO_2^- + \bar{S}-C_6H_4-NO_2$$

(77) (78)

Vitamin B_{12} reacts with 1,1-dichloro-2,2-bis(*p*-chlorophenyl)ethane to form *trans*-4,4′-dichlorostilbene, a reaction which involves initial nucleophilic attack by Co(I) at C(l) with displacement of Cl^-;[98] the product-forming reaction is an elimination of cobalt chloride which proceeds *via* a carbene-type intermediate which rapidly rearranges to the alkene.

Formation of but-2-enes from the reaction of *threo*-3-deuterio-2-(trimethylstannyl)butane with trityl cation proceeds with ~99% *anti*-stereochemistry.[99] The proposed mechanism involves a hydride abstraction leading to a carbocation which is stabilized by σ–π conjugation and then subsequently loses a trimethyltin group.

Small isotope effects (1.2–1.5) have been observed in the dehydration of 2-methyl-3-phenylpropan-2-ol with anhydrous aluminium chloride.[100] Elimination of H_2 between dimethylaluminium hydride and benzylamine (at −63° in PhMe) is complex, involving several equilibria, as shown by kinetic studies.[101]

Silanes, Selenanes, and Related Substrates

The cycloelimination of silyl and sulphoxide groups has been compared with conventional cycloelimination of sulphoxides. The reaction is stereospecifically *syn* (to give either an alkene or alkyne); however, when there is a hydrogen α to the silyl group this is lost preferentially to give β-silylenones.[102] Protodesilylation of 1-dimethyl(phenyl)silyl-3-trimethylsilylpropene (**79**) and its isomeric propene (**81**) give the same 4:1 mixture of allylsilanes, showing that both of these reactions have a common intermediate, most probably the cation (**80**);[103] thus there is no measurable concertedness in the protodesilylation of allylsilanes of this type.

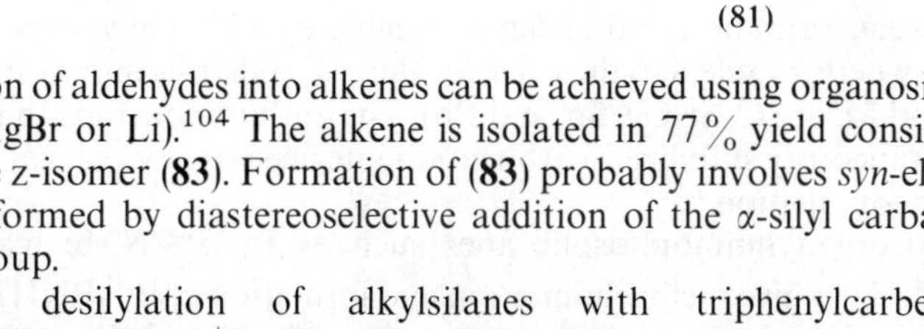

$$Me_3Si\text{–CH=CH–CH}_2\text{–}SiMe_2Ph \ (79) \xrightarrow{H^+} Me_3Si\text{–CH}_2\text{–}\overset{+}{C}H\text{–CH}_2\text{–}SiMe_2Ph \ (80) \leftarrow Me_3Si\text{–CH}_2\text{–CH=CH–}SiMe_2Ph \ (81)$$

(79) (80) (81)

Conversion of aldehydes into alkenes can be achieved using organosilicon allenes (**82**; M = MgBr or Li).[104] The alkene is isolated in 77% yield consisting mainly (10%) of the z-isomer (**83**). Formation of (**83**) probably involves *syn*-elimination of the adduct formed by diastereoselective addition of the α-silyl carbanion to the carbonyl group.

Oxidative desilylation of alkylsilanes with triphenylcarbenium ions ($RCH_2CH_2SiR_3 + Ph_3C^+ \rightarrow Ph_3CH + RCH{=}CH_2$) is highly sensitive to the nature of R. For example when R = Ph_2CH, the silane is inert. The steric

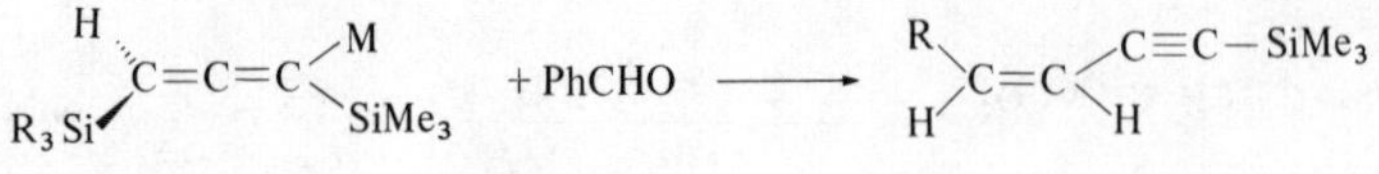

(82) + PhCHO ⟶ (83)

requirements of the transition state are critical and an anti-coplanar arrangement of silicon and the β-hydrogen is suggested.[105] Fluoride ion induces elimination of *p*-(trimethylsilyl)methylbenzyltrimethylammonium iodide (**84**) to give *p*-quinodimethane (**85**) which is isolated as the dimeric [2.2]paracyclophane (and proves a useful method for the synthesis of related furanophanes and thiophenophanes).[106]

Deoxysilylation of trimethylsilyl and bis-trimethylsilyl alcohols in the presence of acid is correlated ($\rho^+ = -11$) by σ^+ values of the substituent on the carbon carrying the leaving-group, suggesting a mechanism involving a conjugatively stabilized carbocation intermediate.[107] This represents one extreme; on the other hand, deoxymetallation of triphenyltin, triphenyl-lead, and iodomercury analogues exhibit characteristics of an *E*2 mechanism. Oxidation of β-amidoalkyl phenyl selenides with H_2O_2 leads to elimination and the formation of allylic amides.[108]

Wittig intermediates in alkene synthesis can be observed at $-78°$[109] in the absence of stabilizing substituents. The oxaphosphetanes can be reversed to aryl aldehyde and ylide; the formation of *cis*-alkenes is rationalized by the necessity to have ylide and aldehyde planes tilted towards an orthogonal arrangement to minimize steric interactions. The use of methyltriphenoxyphenonium iodide as an effective dehydration and dehalogenation agent has been probed using 1,3-dimethylimidazolidin-2-one as solvent.[110] In DMSO as solvent alcohol dehydration is hindered due to interaction between the solvent and the phenonium iodide.

The allylsilane (**86**) has been shown by deuterium-labelling studies to undergo protodesilylation initiated by protonation at C(3) of the allyl group. The position with regard to protonation is quite delicately balanced since closely related

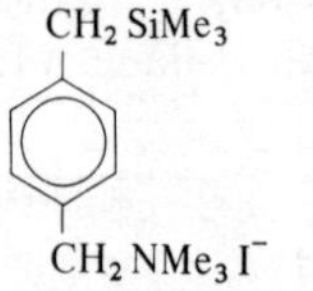

(84) 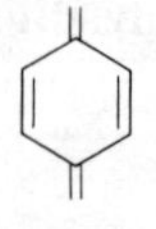 (85) 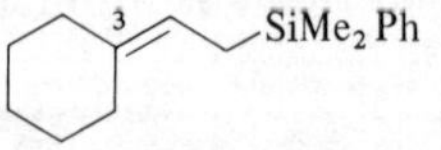 (86)

allylsilanes protonate at C(2) before undergoing a hydride shift.[111] The rate constants for the removal of hydride from $Me_3M(CH_2)_3M'Me_3$ using Ph_3C^+ (and ultimately alkene formation) have been correlated by a Hammett-type relationship against $\Sigma\sigma^+$, where σ^+ refers to the value for the $-CH_2MPh_3$ moiety for both metal groups (M and M′ are C, Si, Ge, Sn, and Pb). The authors conclude that both metal groups simultaneously stabilize a carbocation intermediate by vertical stabilization through σ–π conjugation.[112]

N-Methyltriorganyliminophosphoranes such as $Pr^i{}_3P{=}NMe$ react in liquid NH_3 with KNH_2 to give olefin elimination and formation of $Pr^i{}_2P\bar{N}HK^+$,[113] while tributyltin hydride reacts with *vic*-dinitro compounds to yield elimination products by a free-radical chain mechanism.[114]

Synthesis of exocyclic alkenes using *B*-(cycloalkylmethyl)-9BBN derivatives involves a facile reaction with an aldehyde in a key step (**87** → **88**); coupled with the use of lithium aluminium deuteride (LAD) this provides a useful synthesis of specifically labelled alkenes (*e.g.* **88**).[115]

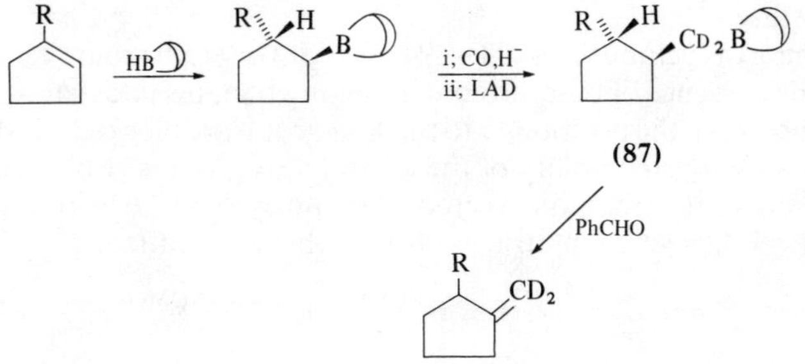

(87)

(88)

Other Eliminations

The use of isotope effects has been reviewed[116] as has phase-transfer catalysis.[117] The use of dimethylsulphoximide (DMSOI) as a solvent for eliminations has been examined. Rates of *E*2 reactions using MeO^- or Bu^tO^- are substantially higher (240-fold) than in Bu^tOH or MeOH, while the orientation observed for 2-substituted octanes is the same as observed in MeO^-–MeOH.[118]

Dehydration of 1,1-diadamantylmethyl carbinol in anhydrous acetic acid leads exclusively to the 1,1-bis(1-adamantyl)ethylene; the secondary deuterium isotope effect of 1.32 indicates that carbenium-ion formation is rate-determining. The cation can be trapped by the addition of water so that deprotonation of the carbenium ion can become kinetically important; under these conditions a primary isotope effect of 2.98 has been observed.[119]

Kinetic studies at high acidities indicate that the dication of (**89**) is the reactive species for the loss of hydroxylamines from 4-alkoxyimino-5,6-dihydro-6-alkoxyaminopyrimidin-2(1*H*)-ones.[120] Very weakly basic amines have the best combination of pK_a and reactivity for use as catalysts for the promotion of α-proton abstraction *via* iminium ion formation (*e.g.* the conversion of **90** to **92** via **91**). Using 1M hexafluoroisopropylamine as catalyst, formation of (**92**) *via* the iminium ion

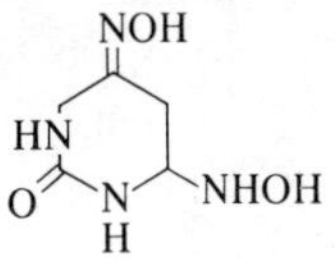

(89) 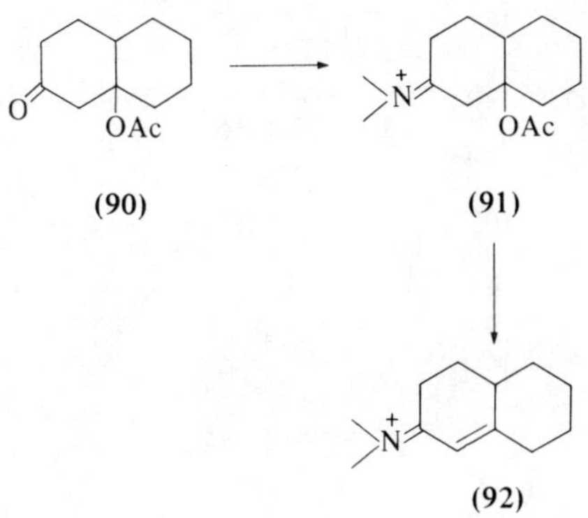 **(90)** **(91)**

(92)

route occurs > 10^4-fold more rapidly than direct elimination.[121] β-Elimination from *O*-phosphoserine, β-chloroalanine, and *S*-ethylcysteine is catalysed by pyridoxal in the presence of metal ions. A rate maximum at pD 8–9 is observed in D_2O; the metal ions typically show rate enhancements of the order of 10-fold.[122] Also reported is the reaction of ethanolamine *O*-sulphate with 4-aminobutyrate aminotransferase.[123]

Ring-opening of cyclobutenes (**93** → **94**; X = SO or SO_2) occurs very rapidly even at −30° in the presence of base, although the ring remains intact at −78°. No ring-opening occurs when the positions α to the XAr group are blocked by Me groups;[124] a key factor is clearly the ability of the arylsulphoxy and arylsulphonyl groups to stabilize the adjacent carbanion. Acetolysis of 6-bicyclo[3.2.0]hept-2-enyl mesylate (**95**) yields the 1,3,5-cycloheptatriene (**96**) *via* the carbocation.[125]

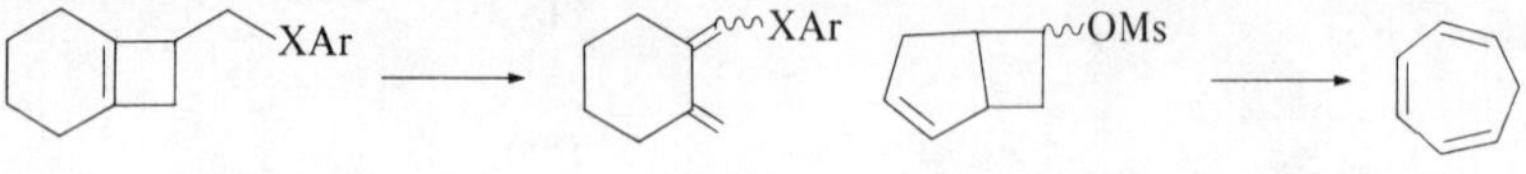

XAr → XAr; OMs →

(93) **(94)** **(95)** **(96)**

Also described are the coupling of α-halosulphones (Ramberg–Bäcklund homologation),[126] the elimination of sulpholanes,[127] and the oxidative cleavage of aliphatic and cyclic iodides.[128]

Although carbanions are poor leaving-groups in eliminations to give alkenes, reactions to give aldehydes and ketones are well known. A kinetic study of the β-nitro-alcohols (**97**) has revealed a specific base-catalysed carbonyl-forming elimination in which the slow step, involving C—C bond-cleavage, has a transition state resembling the carbanion.[129] A plot of log k_{HO^-} was correlated by the Jencks and Young modification of the Hammett equation with $\rho = +0.24$ and $\rho^r = -0.36$. β-Nitrosulphones and α-carbonylsulphones are cleaved to α,β-unsaturated ketones on treatment with one-electron-transfer agents.[130]

Potassium superoxide reacts in a regiospecific way to cause loss of H(4) and the 5-mesyloxy group (as well as the 1-mesyloxy group) from 2,3,4,6-tetra-*O*-benzyl-1,5-di-*O*-mesyl-D-glucitol. It has been suggested than a peroxy-anion intermediate is formed (by selective displacement of the 1-*O*-mesyl group) which results in intramolecular H(4) abstraction.[131]

When dehydrohalogenated by Bu^tO^- the 8,8-dichlorobicyclo[5,1,0]octa-2,4-diene (**98**) yields E- and Z-isomers of β-*t*-butoxystyrene (**100**). An intermediate on the

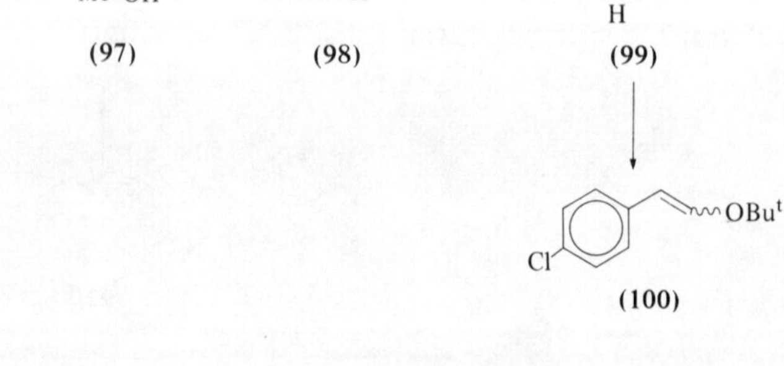

NO_2—C(Me)(Me)—C(Ar)(OH)—H

(97)

Cl, Cl, R, R → Cl, H, OBu^t, R, H

(98) **(99)**

↓

Cl, OBu^t

(100)

reaction pathway is thought to be the cyclobutene (**99**).[132] The cyclopropane (**98**) is formed from a bis-dichlorocarbene adduct of a cyclohexa-1,4-diene followed by regiospecific dehydrohalogenation by strong base.[133] Related reactions involving the trapping of adducts analogous to (**94**) by pyridine have also been described[134] as has the dehydrohalogenation of bromocyclopropanes by Bu^tO^- in DMSO (which suggest a *syn*-elimination mechanism).[135] DMSO has also been used as a dehydration agent for alcohols and diols; competing ether formation most likely occurs *via* a carbenium-ion intermediate.[136] The Kishner elimination using hydrazine can be used as a route to deoxygenated sugars and reaction conditions have been investigated to optimize yields.[137]

The trimethylammonium salt (**101**) undergoes both elimination and substitution reactions on solvolysis and provides one of the very few examples of *E*1 eliminations from a trimethylammonium salt. This is undoubtedly related to steric hindrance at the C—H bond since a 2,6-dimethyl-substituted substrate reacts *ca.* 100-fold more rapidly than the 2-methyl analogue.[138] Brønsted β values ranging from 0.32 to 0.45 have been determined for the ferrocenylalkyl cations $Fc\overset{+}{C}(Ph)CH_2Bu^t$ and $Fc_2\overset{+}{C}Me$ using amines as bases. Sterically hindered amines show negative deviations as do all oxygen bases. An apparent maximum is reached in the plot of primary deuterium isotope effects *vs.* pK_a. Deprotonation of the cation preferentially occurs in a direction *exo* to the ferrocenyl group; ion-pairing is also important in determining reactivity.[139] The solvolysis of 2-(trifluoromethyl)-2-propyl trifluoromethanesulphonate shows unusual behaviour, no doubt related to the strongly electron-withdrawing character of the CF_3 group. For example, for a range of solvents there is no correlation between the rate of alkene formation from $PrCH(CF_3)CH_2OTf$ and the rate of reaction of 2-adamantyl tosylate in the same solvent; however, substantial secondary isotope effects are observed. A mechanism involving rate-limiting solvent attack on an intimate ion-pair formed from the trifluoromethylsulphonate[140] has been suggested.

Addition of alkoxide ions to fluorinated alkenes $PhCR{=}CF_2$ leads to carbanions which lose F^- more rapidly than they are protonated by the solvent (ethanol); rates of fluoride ion elimination vary in the order $R = -CF_2H \gg -CF_2OR > -CF_2CF_3 \gg -CF_3$.[141]

Elimination of HBr from *erythro*-2′-hydroxychalcones (**102**) gives (E)- and (Z)-α-bromo-2′-hydroxychalcones (**103**) in a ratio of ~1:2 in water–ethanol (4:1) under neutral conditions. Both (E)- and (Z)-isomers cyclize to 3-bromoflavanone, the (Z)-isomer reaction 20-fold more rapidly. *syn*-Elimination is observed in preference to *trans*-elimination and a concerted *E*2 elimination (or *E*1 mechanism) can be discounted.[142]

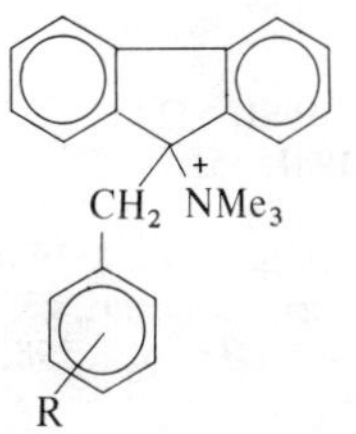

CH2 NMe3 (+), R

(101)

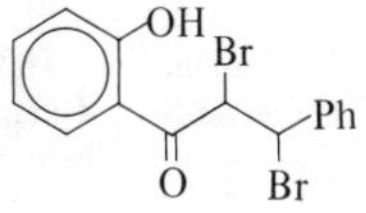

OH, Br, Ph, O, Br

(102)

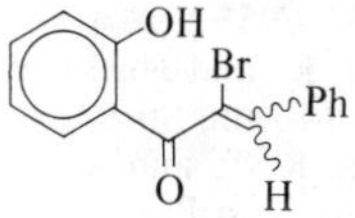

OH, Br, Ph, O, H

(103)

Also reported are force field calculations on the regiospecificity of steroid eliminations[143] and on the transition state for the *E*2 reaction between HO^- and simple chloroalkanes.[144] An *ab initio* study of H^- abstraction from alkyl-lithium compounds has been reported.[145]

Sulphur participation is a feature of the acetolysis of α-diazoketones $RSCHR'CH_2COCHN_2$ which competes with elimination.[146] Decomposition of α-hydroxyketones by metal hydroxides to give CH_2O involves rate-determining C—C bond-cleavage.[147] Reaction of $Me_2C{=}CHCH_2NR_2$ with iodine results in the formation of the iminium salt $Me_2C{=}CHCH{=}\overset{+}{N}R_2\,I^-$; the slow step has been assigned to the formation of a four-membered cycle.[148] 1,4-Dehydrochlorination of aryl (δ-chloroalkenyl) sulphones is first order in sulphone and in triethylamine and inhibited by alkyl group substitution.[149] Dehydrochlorination of DDT (using ammonia or methylamine) gives alkenes of high purity; the corresponding Hammett ρ values are thought to favour an *E*2 mechanism rather than the *E*1*cB* mechanism as proposed previously.[150] Also reported are the vapour-phase dehydrochlorination of 1,1,2,2-tetrachloroethane catalysed by chlorine,[151] base-catalysed dehydrochlorination of 9,10,12,13,15,16-hexachlorooctadecanoate,[152] the cyclodehydration of β-(*p*-toluidino)acrolein,[153] β-elimination of halogen atoms from diradicals formed from δ-haloketones,[154] elimination from carbanions formed by negative-ion spectroscopy from 1,3-dithianes and 1,3-dithiolanes,[155] and a theoretical calculation using SCF methods on the chemiluminescence of dioxetanones.[156]

References

1 Bartsch, R. A., and Zavada, J., *Chem. Rev.*, **80,** 453 (1980).
2 Pankova, M., and Zavada, J., *Collect. Czech. Chem. Commun.*, **45,** 3150 (1980).
3 Zavada, J., Pankova, M., Bartsch, R. A., and Bong Rae Cho, *Collect. Czech. Chem. Commun.*,**46,** 850 (1981).
4 Alunni, S., Perucci, P., and Ruzziconi, R., *Gazz. Chim. Ital.*, **110,** 261 (1980); *Chem. Abs.*, **93,** 238135 (1980).
5 Palmer, R. J., and Stirling, C. J. M., *J. Am. Chem. Soc.*, **102,** 7888 (1980).
6 Bartok, M., Felfoldi, K., and Bozoki-Bartok, G., *Helv. Chim. Acta*, **63,** 2173 (1980).
7 Ndebeka, G., Raynal, S., Caubere, P., and Bartsch, R. A., *J. Org. Chem.*, **45,** 5394 (1980).
8 Cospito, G., Illuminati, G., Lillocci, C., and Petride, H., *J. Org. Chem.*, **46,** 2944 (1981).
9 Wu, S.-L., and Smith, P. J., *Can. J. Chem.*, **59,** 3016 (1981).
10 Katritzky, A. R., and El-Mowafy, A. M., *J. Chem. Soc., Chem. Commun.*, **1981,** 96.
11 Machkova, Z., and Zavada, J., *Collect. Czech. Chem. Commun.*, **46,** 833 (1981).
12 Lewis, D. E., Sims, L. B., Yamataka, H., and McKenna, J., *J. Am. Chem. Soc.*, **102,** 7411 (1980).
13 Miller, D. J., Subramanian, R., and Saunders, W. H., *J. Am. Chem. Soc.*, **103,** 3519 (1981).
14 Miller, D. J., and Saunders, W. H., *J. Org. Chem.*, **46,** 4247 (1981).
15 Brown, K. C., Romano, F. J., and Saunders, W. H., *J. Org. Chem.*, **46,** 4242 (1981).
16 Koch, H. F., Koch, J. G., Tumas, W., McLennan, D. J., Dobson, B., and Lodder, G., *J. Am. Chem. Soc.*, **102,** 7955 (1980).
17 Keefe, J. R., and Jencks, W. P., *J. Am. Chem. Soc.*, **103,** 2457 (1981).
18 Varma, M., and Stirling, C. J. M., *J. Chem. Soc., Chem. Commun.*, **1981,** 553
19 Douglas, K. T., and Alborz, M., *J. Chem. Soc., Chem. Commun.*, **1981,** 551.
20 Inoue, M., and Bruice, T. C., *J. Chem. Soc., Chem. Commun.*, **1981,** 884.
21 Douglas, K. T., Yaggi, N. F., and Mervis, C. M., *J. Chem. Soc., Perkin Trans. 2*, **1981,** 171.
22 Minde, J., Balcarek, L., Silar, L., and Vecera, M., *Collect. Czech. Chem. Commun.*, **45,** 3130 (1980).
23 Thea, S. T., Kashefi-Naini, N., and Williams, A., *J. Chem. Soc., Perkin Trans. 2*, **1981,** 65.
24 Williams, A., and Thea, S., *J. Chem. Soc., Perkin Trans. 2*, **1981,** 72.
25 Thea, S., Harun, M. G., Kashefi-Naini, N., and Williams, A., *J. Chem. Soc., Perkin Trans. 2*, **1981,** 78.
26 Thea, S., Guanti, G., and Williams, A., *J. Chem. Soc., Chem. Commun.*, **1981,** 535.

[27] Bezrodnyi, V. P., Skrypnik, Yu. G., and Baranov, S. N., *Zh. Org. Khim.*, **16,** 1863 (1980); *Chem. Abs.*, **94,** 14772 (1981).
[28] Farng, L. O., and Kice, J. L., *J. Am. Chem. Soc.*, **103,** 1137 (1981).
[29] Nome, F., Erbs, W., and Correia, V. R., *J. Org. Chem.*, **46,** 3802 (1981).
[30] Nguyen, H. H., and Danen, W. C., *J. Am. Chem. Soc.*, **103,** 6253 (1981).
[31] Taylor, R., *J. Chem. Res. (S)*, **1981,** 250.
[32] Chuchani, G., and Dominguez, R. M., *Int. J. Chem. Kinet.*, **13,** 577 (1981).
[33] Chuchani, G., Martin, I., and Alonso, M. E., *J. Phys. Chem.*, **85,** 1241 (1981).
[34] Chuchani, G., Triana, J. L., Rotinov, A., and Carballo, D. F., *J. Phys. Chem.*, **85,** 1243 (1981).
[35] Rotinov, A., Carballo, D. F., Medina, J. D., and Chuchani, G., *React. Kinet. Catal. Lett.*, **13,**173 (1980); *Chem. Abs.*, **93,** 167375 (1980).
[36] Blake, P. G., and Shraydeh, B. F., *Int. J. Chem. Kinet.*, **13,** 463 (1981).
[37] Nair, T. D., *Thermochim. Acta*, **43,** 365 (1981); *Chem. Abs.*, **95,** 41997 (1981).
[38] Carlsen, L., Egsgaard, H., and Pagsberg, P., *J. Chem. Soc., Perkin Trans. 2*, **1981,** 1256.
[39] Newkome, G. R., Kohli, D. K., and Kawato, T., *J. Org. Chem.*, **45,** 4508 (1980).
[40] Atkinson, R. F., Balko, T. W., Westman, T. R., Sypniewski, G. C., Carmody, M. A., Pauler, C. T., Schade, C. L., Coulter, D. E., Phan, H. T., Barea, F., and Hassner, A., *J. Org. Chem.*, **46,** 2804 (1981).
[41] Bunton, C. A., and de Buzzaccarini, F., *J. Phys. Chem.*, **85,** 3139 (1981).
[42] Kunitake, T., Okahata, Y., Ando, R., Shinkai, S., and Hirakawa, S., *J. Am. Chem. Soc.*, **102,** 7877 (1980).
[43] Hunter, D. H., Patel, V., and Perry, R. A., *Can. J. Chem.*, **58,** 2271 (1980).
[44] Mulzer, J., and Zippel, M., *J. Chem. Soc., Chem. Commun.*, **1981,** 891.
[45] Noto, R., Buccheri, F., Consiglio, G., and Spinelli, D., *J. Chem. Soc., Perkin Trans. 2*, **1980,** 1627.
[46] Gramalica, P., Ranzi, B. M., and Manitto, P., *Bioorg. Chem.*, **10,** 14 (1981).
[47] Sakai, T., Hattori, M., and Yamane, N., *Sekiyu Gakkaishi*, **23,** 341 (1980); *Chem. Abs.*, **94,** 138966 (1981).
[48] Sakai, T., Hattori, M., and Yamane, N., *Sekiyu Gakkaishi*, **23,** 334 (1980); *Chem. Abs.*, **94,** 138965 (1981).
[49] Harano, K., Ban, T., Yasuda, M., Osawa, E., and Kanematsu, K., *J. Am. Chem. Soc.*, **103,** 2310 (1981).
[50] Cameron, T. B., El-Kabbani, F. M., and Pinnick, H. W., *J. Am. Chem. Soc.*, **103,** 5414 (1981).
[51] Carlsen, L., Egsgaard, H., and Harpp, D. N., *J. Chem. Soc., Perkin Trans. 2*, **1981,** 1166.
[52] Perez, M. A., Rossert, M., and Kresze, G., *Liebigs Ann. Chem.*, **1981,** 65.
[53] Quast, H., and Kees, F., *Chem. Ber.*, **114,** 774 (1981).
[54] Quast, H., and Kees, F., *Chem. Ber.*, **114,** 802 (1981).
[55] Durst, T., Lancaster, M., and Smith, D. J. H., *J. Chem. Soc., Perkin Trans. 1*, **1981,** 1846.
[56] Cajurel, C. L., and Vaidya, S. R., *Indian J. Chem.*, **19B,** 911 (1980); *Chem. Abs.*, **94,** 102378 (1981).
[57] Okumura, Y., Hase, N., Chikyu, H., Mori, N., and Suzuki, T., *Rep. Fac. Sci., Shizuoka Univ.*, **14,** 27 (1980); *Chem. Abs.*, **93,** 220059 (1980).
[58] Uyehara, T., Ohnuma, T., Saito, T., Kato, T., Yoshida, T., and Takahashi, K., *J. Chem. Soc., Chem. Commun.*, **1981,** 127.
[59] Pommelet, J.-C., Nyns, C., Lahousse, F., Merenyi, R., and Viehe, H. G., *Angew. Chem. Int. Ed.*, **20,** 585 (1981).
[60] Schneider, M. P., and Bippi, H., *J. Am. Chem. Soc.*, **102,** 7363 (1980).
[61] Wildi, E. A., VanEngen, D., and Carpenter, B. K., *J. Am. Chem. Soc.*, **102,** 7994 (1980).
[62] Paine, A. J., and Warkentin, J., *Can. J. Chem.*, **59,** 491 (1981).
[63] Liu, M. T. H., and Ibata, T., *Can. J. Chem.*, **59,** 559 (1981).
[64] Quast, H., and Fuss, A., *Angew. Chem. Int. Ed.*, **20,** 291 (1981).
[65] Cabelkova-Taguchi, L. M., and Warkentin, J., *Can. J. Chem.*, **59,** 100 (1981).
[66] Niecke, E., Schoeller, W. W., and Wildbredt, D.-A., *Angew. Chem. Int. Ed.*, **20,** 131 (1981).
[67] Frey, H. M., and Watts, H. P., *Int. J. Chem. Kinet.*, **13,** 729 (1981).
[68] King, K. D., and Goddard, R. D., *Int. J. Chem. Kinet.*, **13,** 755 (1981).
[69] Dominquez, R. M., and Chuchani, G., *Int. J. Chem. Kinet.*, **13,** 403 (1981).
[70] Chuchani, G., and Martin, I., *J. Phys. Chem.*, **84,** 3188 (1980).
[71] Kang, S. H., Yoo, H. S., and Jung, K.-H., *Bull. Korean Chem. Soc.*, **2,** 35 (1981); *Chem. Abs.*, **95,** 149502 (1981).
[72] Failes, R. L., Mollah, Y. M. A., and Shapiro, J. S., *Int. J. Chem. Kinet.*, **13,** 7 (1981).
[73] Aloni, R., Rajbenbach, L. A., and Horowitz, A., *Int. J. Chem. Kinet.*, **13,** 23 (1981).
[74] Chuchani, G., Martin, I., Alonso, M. E., and Jano, P., *Int. J. Chem. Kinet.*, **13,** 1 (1981).
[75] Rzepa, H. S., *J. Chem. Soc., Chem. Commun.*, **1981,** 939.
[76] Kwart, H., and Brechbiel, M., *J. Am. Chem. Soc.*, **103,** 4650 (1981).

[77] Kwart, L. D., Horgan, A. D., and Kwart, H., *J. Am. Chem. Soc.*, **103,** 1232 (1981).
[78] Carlsen, L., Egsgaard, H., Schaumann, E., Mrotzek, H., and Wolf-Rüdiger, K., *J. Chem. Soc., Perkin Trans. 2*, **1980,** 1557.
[79] Aver'yanov, V. A., Somov, G. V., and Dryndin, V. M., *Deposited Doc.*, **1980,** VINITI 582; *Chem. Abs.*, **94,** 120750 (1981).
[80] King, D., *J. Phys. Chem.*, **84,** 2517 (1980); Chem. Abs., **93,** 185478 (1980).
[81] Rao, C. V., Reddy, K., and Rao, N. V. S., *Indian J. Chem.*, **19B,** 967 (1980); *Chem. Abs.*, **94,** 174022 (1981).
[82] Deluy, C. H., and Bierbaum, V. M., *J. Am. Chem. Soc.*, **103,** 5034 (1981).
[83] Szeimies-Seebach, U., Schöffer, A., Römer, R., and Szeimies, G., *Chem. Ber.*, **114,** 1767 (1981).
[84] Balci, M., and Jones, W. M., *J. Am. Chem. Soc.*, **103,** 2874 (1981).
[85] Warner, P., and Palmer, R. F., *J. Am. Chem. Soc.*, **103,** 1584 (1981).
[86] Halton, B., and Officer, D. L., *Tetrahedron Lett.*, **22,** 3687 (1981).
[87] C. J. M. Stirling, *Isr. J. Chem.*, **21,** 111 (1981).
[88] Hirao, K., *Yakugaku Zasshi*, **100,** 473 (1980); *Chem. Abs.*, **93,** 203420 (1980).
[89] Keumi, T., Yamamoto, T., Saga, H., and Kitajima, H., *Bull. Chem. Soc. Jpn.*, **54,** 1579 (1981).
[90] Pfoertner, H.-K., Foricher, J., and Meister, W., *Helv. Chim. Acta*, **63,** 1908 (1980).
[91] Pfenninger, J., and Graf, W., *Helv. Chim. Acta*, **63,** 2338 (1980).
[92] Beltrame, P., Carniti, P., Gelli, G., and Loi, A., *Gazz. Chim. Ital.*, **110,** 603 (1980); *Chem. Abs.*, **94,** 173894 (1981).
[93] Moody, C. J., Rees, C. W., and Tsoi, S. C., *J. Chem. Soc., Chem. Commun.*, **1981,** 550.
[94] Tarnchompoo, B., Thebtaranonth, Y., Utamapanya, S., and Kasemsri, P., *Chem. Lett.*, **1981,** 1241.
[95] Kwart, H., George, T. J., Horgan, A. G., and Lin, Y. T., *J. Org. Chem.*, **46,** 1970 (1981).
[96] Tansumoto, K., and Martell, A. E., *J. Am. Chem. Soc.*, **103,** 6203 (1981).
[97] Nashed, N. T., and Kaiser, E. T., *J. Am. Chem. Soc.*, **103,** 3611 (1981).
[98] Nome, F., and Zanette, D., *Can. J. Chem.*, **58,** 2402 (1980).
[99] Hannon, S. J., and Traylor, T. G., *J. Org. Chem.*, **46,** 3645 (1981).
[100] Tsukurimichi, E., and Nohara, T., *Nippon Kagaku Kaishi*, **1981,** 1040; *Chem. Abs.*, **95,** 114351 (1981).
[101] Beachley, O. T., *Inorg. Chem.*, **20,** 2825 (1981); *Chem. Abs.*, **95,** 114328 (1981).
[102] Fleming, I., and Perry, D. A., *Tetrahedron Lett.*, **22,** 5095 (1981).
[103] Fleming, I., and Langley, J. A., *J. Chem. Soc., Perkin Trans. 1*, **1981,** 1421.
[104] Yamakado, Y., Ishiguro, M., Ikeda, N., and Yamamoto, H., *J. Am. Chem. Soc.*, **103,** 5568 (1981).
[105] Washburne, S. S., and Szendroi, R., *J. Org. Chem.*, **46,** 691 (1981).
[106] Ito, Y., Miyata, S., Nakatsuka, M., and Saegusa, T., *J. Org. Chem.*, **46,** 1043 (1981).
[107] Davis, D. D., and Jacocks, H. M., *J. Organomet. Chem.*, **206,** 33 (1981).
[108] Toshimitsu, A., Owada, H., Aoai, T., Uemura, S., and Okano, M., *J. Chem. Soc., Chem. Commun.*, **1981,** 546.
[109] Vedejs, E., Meier, G. P., and Snoble, K. A. J., *J. Am. Chem. Soc.*, **103,** 2823 (1981).
[110] Spangler, C. W., Kjell, D. P., Wellander, L. L., and Kinsella, M. A., *J. Chem. Soc., Perkin Trans. 1*, **1981,** 2287.
[111] Fleming, I., Marchi, D., and Patel, S. K., *J. Chem. Soc., Perkin Trans. 1*, **1981,** 2518.
[112] Traylor, T. G., and Koermer, G. S., *J. Org. Chem.*, **46,** 3651 (1981).
[113] Ross, B., Reetz, K. P., *Z. Anorg. Allg. Chem.*, **466,** 203 (1980); *Chem. Abs.*, **93,** 238510 (1980).
[114] Ono, N., Miyake, H., Tamura, R., Hamamoto, I., and Kaji, A., *Chem. Lett.*, **1981,** 1139.
[115] Brown, H. C., and Ford, T. M., *J. Org. Chem.*, **46,** 647 (1981).
[116] Simonyi, M., and Fitos, I., *Kem. Ujabb Eredmenyei*, **46,** 7 (1980); *Chem. Abs.*, **94,** 3315 (1981).
[117] Oda, R., *Hyomen*, **1981,** 234; *Chem. Abs.*, **95,** 131744 (1981).
[118] Furukawa, N., Takahashi, F., Yoshimura, T., Morita, H., and Oae, S., *J. Chem. Soc., Perkin Trans. 2*, **1981,** 432.
[119] Lomas, J. S., *J. Org. Chem.*, **46,** 412 (1981).
[120] Palling, D. J., Schalke, P. M., Atkins, P. J., and Hall, C. D., *J. Chem. Soc., Perkin Trans. 2*, **1981,** 113.
[121] Roberts, R. D., Ferran, H. E., Gula, M. J., and Spencer, T. A., *J. Am. Chem. Soc.*, **102,** 7054 (1980).
[122] Tatsumoto, K., Martell, A. E., and Motekaitis, R. J., *J. Am. Chem. Soc.*, **103,** 6197 (1981).
[123] Fowler, L. J., and John, R. A., *Biochem. J.*, **197,** 149 (1981).
[124] Kametani, T., Tsubuki, M., Nemoto, H., and Suzuki, K., *J. Am. Chem. Soc.*, **103,** 1256 (1981).
[125] Nee, M. J., Gorham, W. F., and Roberts, J. D., *J. Org. Chem.*, **46,** 1021 (1981).
[126] Burger, J. J., Chen, T. B. R. A., de Waard, E. R., and Huisman, H. O., *Tetrahedron*, **37,** 417 (1981).
[127] Bezmenova, T. E., *Katal. Sint. Org. Soedin. Sery*, **1979,** 191; *Chem. Abs.*, **93,** 238186 (1980).
[128] McCabe, P. H., de Jenga, C. I., and Stewart, A., *Tetrahedron Lett.*, **22,** 3679 (1981).

[129] Guanti, G., Petrillo, G., Thea, S., Cevasco, G., and Stirling, C. J. M., *Tetrahedron Lett.*, **21,** 4735 (1980).
[130] Ono, N., Tamura, R., Nakatsuka, T., Hayami, J., and Kaji, A., *Bull. Chem. Soc. Jpn.*, **53,** 3295 (1980).
[131] Rao, V. S., and Perlin, A. S., *Can. J. Chem.*, **59,** 333 (1981).
[132] Banwell, M. G., and Halton, B., *Aust. J. Chem.*, **33,** 2685 (1980).
[133] Banwell, M. G., and Halton, B., *Aust. J. Chem.*, **33,** 2673 (1980).
[134] Yagoub, A. K., and Iskander, G. M., *J. Chem. Soc., Perkin Trans. 1*, **1980,** 2405.
[135] Yakushkina, N. I., and Bolesov, I. G., *Zh. Org. Khim.*, **16,** 1335 (1980); *Chem. Abs.*, **93,** 185625 (1980).
[136] Molnar, A., and Bartok, M., *Helv. Chim. Acta*, **64,** 389 (1981).
[137] Wickremesinghe, L. K. G., and Slessor, K. N., *Can. J. Chem.*, **58,** 2628 (1980).
[138] Pradhan, J., and Smith, P. J., *Can. J. Chem.*, **59,** 911 (1981).
[139] Bunton, C. A., Carrasco, N., Davoudzadeh, F., and Watts, W. E., *J. Chem. Soc., Perkin Trans. 2*, **1981,** 924.
[140] Jansen, M. P., Koshy, K. M., Mangru, N. N., and Tidwell, T. T., *J. Am. Chem. Soc.*, **103,** 3863 (1981).
[141] Koch, H. F., Koch, J. G., Donovan, D. B., Toczko, A. G., and Kielbania, A. J., *J. Am. Chem. Soc.*, **103,** 5417 (1981).
[142] David, S. K., Main, L., and Old, K. B., *J. Chem. Soc., Perkin Trans. 2*, **1981,** 1367.
[143] Gschwendtner, W., Hoppen, V., and Schneider, H.-H., *J. Chem. Res.(S)*, **1981,** 96.
[144] Chalk, C. D., McKenna, J., Sims, L. B., and Williams, I. H., *J. Am. Chem. Soc.*, **103,** 281 (1981).
[145] Reetz, M. T., and Stephan, W., *J. Chem. Res.(S)*, **1981,** 44.
[146] Rosnati, V., and Saba, A., *Gazz. Chim. Ital.*, **110,** 211 (1980); *Chem. Abs.*, **93,** 203566 (1980).
[147] Morozov, A. A., and Levanevskii, O. E., *Kinet. Katal.*, **21,** 954 (1980); *Chem. Abs.*, **93,** 238445 (1980).
[148] Gevaza, Yu. I., Degurko, T. A., and Staninets, V. I., *Ukr. Khim. (Russ. Ed.)*, **46,** 760 (1980); *Chem. Abs.*, **93,** 220166 (1980).
[149] Tanaskov, M. M., Stadnichuk, M. D., and Kekisheva, L. V., *Zh. Obshch. Khim.*, **50,** 1738 (1980); *Chem. Abs.*, **94,** 3387 (1981).
[150] MacLaury, M. R., *Polym. Prepr., Am. Chem. Soc., Div. Polym. Chem.*, **20,** 361 (1979); *Chem. Abs.*, **95,** 41776 (1981).
[151] Venkateshwar, S., and Rao, M. B., *Indian J. Technol.*, **19,** 106 (1981); *Chem. Abs.*, **95,** 114394 (1981).
[152] Ketola, M. R., *Fette, Seifen, Anstrichm.*, **83,** 106 (1981); *Chem. Abs.*, **95,** 23787 (1981).
[153] Tamura, S., and Todoriki, R., *Chem. Pharm. Bull.*, **28,** 3401 (1980); *Chem. Abs.*, **94,** 155947 (1981).
[154] Wagner, P. J., Lindstrom, M. J., Sedon, J. H., and Ward, D. R., *J. Am. Chem. Soc.*, **103,** 3842 (1981).
[155] Bartmess, J. E., Hays, R. L., Khatri, H. N., Misra, R. N., and Wilson, S. R., *J. Am. Chem. Soc.*, **103,** 4746 (1981).
[156] Schmidt, S. P., Vincent, M. A., Dykstra, C. E., and Schuster, G. B., *J. Am. Chem. Soc.*, **103,** 1292 (1981).

Organic Reaction Mechanisms 1981
Edited by A. C. Knipe and W. E. Watts

CHAPTER 12

Addition Reactions: Polar Addition

R. M. PATON

Department of Chemistry, University of Edinburgh

Electrophilic Additions 409
Halogen and Related Additions 409
Addition of Hydrogen Halide 414
Hydration and Related Reactions 415
Miscellaneous Electrophilic Additions 416
Nucleophilic Additions 421
References 426

Additions to fluoroalkenes[1] and the reactions of acetylene in superbasic media[2] have been reviewed, as have asymmetric additions to α,β-unsaturated carboxylic acids.[3] The mechanisms of the reactions of ketenes, allenes, and related compounds have also been discussed.[4] In order to rationalize the stereoselectivities found for polar additions and cycloadditions a general rule has been proposed[5] which states that "attack of a reagent at an unsaturated site occurs such as to minimize anti-bonding secondary orbital interactions between the critical frontier molecular orbital of the reagent and those of the vicinal bonds." Houk and his coworkers have also noted[6] that pyramidalization towards tetrahedral geometry for alkenes such as norbornene is a molecular distortion which parallels the stereoselectivity of their addition reactions.

Electrophilic Additions

A detailed analysis has been made[7] of the relative reactivities of carbon–carbon double and triple bonds towards electrophiles. The formation and transformation of stable cationoid intermediates have been examined[8] as models for the study of electrophilic addition mechanisms.

Halogen and Related Additions

The mechanisms of electrophilic additions of halogens to multiple bonds have been reviewed.[9]

Acetyl hypofluorite, generated by passing fluorine through HOAc–$CFCl_3$, undergoes regiospecific *syn*-addition to cinnamate esters to yield a mixture of *threo*- and *erythro*- $PhCH(OAc)CHFCO_2R$.[10] $CsSO_4F$ also proves to be a mild but effective reagent for the fluorination of alkenes.[11]

The presence of electron-withdrawing substituents is found[12] to retard the addition of FCl to olefins. A mechanism involving a donor–acceptor complex and a cyclic transition state has been suggested[13,14] for the chlorophenoxylation of 3-substituted propenes; the kinetics of the corresponding chloroalkoxylations have also been examined.[15] The addition of $ClN(SO_2F)_2$ to *cis*- and *trans*-MeCH=CHMe is stereospecifically *trans*, but its reaction with *cis*-CHF=CHF is non-stereospecific;[16] addition of $HN(SO_2F)_2$ to these alkenes follows Markownikoff's rule. The chlorination of methyl 2,3-diphenyl-2-cyclopropenecarboxylate yields the *cis*-adduct (**1**).[17] *Ab initio* calculations indicate[18] that $C_2H_4Cl^+$ is unlikely to be an intermediate in the gas-phase addition of Cl_2 to ethene. The chlorination of but-2-enes using $CuCl_2$ giving mostly *anti*-adducts proceeds *via* a chloronium ion intermediate.[19]

Chloro- and bromo-fluorination of E- and Z-$ClCH_2CR{=}CHCl$ in liquid HF by *N*-bromosuccinimide, hexachloromelamine, and bromine proceeds regiospecifically *via* a bromonium ion intermediate.[20] *syn*-Collapse of ion-pair intermediates accounts for the substantial amounts of *cis*-addition products obtained from the reactions of Br_2 and BrCl with indene; in contrast only *trans*-dihalides are obtained from indenone suggesting that the neighbouring carbonyl group stabilizes the bridged bromium ion.[21] The existence and relative stability of cyclic 2,3-butenechloronium and bromonium ions in the gas phase has been demonstrated by establishing the stereochemistry of acid-induced nucleophilic displacements on 2,3-dihalobutanes;[22] in accord with theoretical predictions no evidence could be found for the formation of stable 2,3-butenefluoronium ions. The bromination of the cyclohexene double bond in diazatetracyclic alkenes, *e.g.* (**2**), occurs in a highly regio- and stereo-selective manner; as the main product and rate-determining step was found to be attack by the nucleophile the usual Ad_E2 mechanism was discounted.[23]

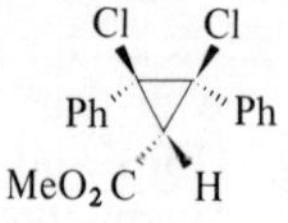

(**1**) 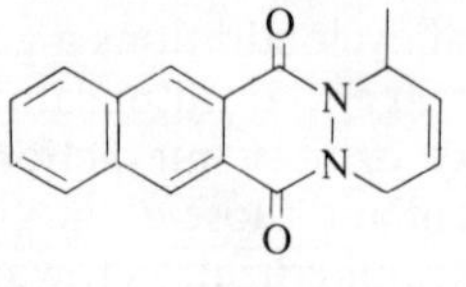 (**2**)

The relative rates of bromination of methyl-substituted alkenes have been measured[24] and are found to increase in the order: $CH_2{=}CH_2 < MeCH{=}CH_2 < Me_2C{=}CH_2 < Me_2C{=}CH\text{-}Me < Me_2C{=}CMe_2$; their π-bond orders and ionization potentials decrease in this same order. Substituent effects on the rates of bromination of styrenes and enol ethers have also been studied[25] and have been shown to be consistent with a carbocation-like transition state; in contrast it is bromonium-like for non-conjugated alkenes. Weakly bridged halonium ions in equilibrium with open carbocations have been invoked[26] to explain the degrees of stereoselectivity observed for the halogenation of *cis*- and *trans*-3-methylpenta-1,3-diene. For the bromination of 2-bromonorbornene using Br_2–CCl_4, raising the temperature from 0°C to 77°C appears to induce a change in reaction mechanism from ionic to radical.[27] The stereochemical assignment for the adducts resulting from the reaction of bromine with olefinic bonds in bicyclic systems have been assigned by NMR spectroscopy.[28] To account for the exclusive formation of (**6**; R = CN, CO_2Me) from the bromination of (**3**) a mechanism has been proposed[29] which involves transannular participation of the *endo*-nitro group giving (**4**), followed by methanolysis and deprotonation to yield (**5**), with subsequent bromination and demethylation.

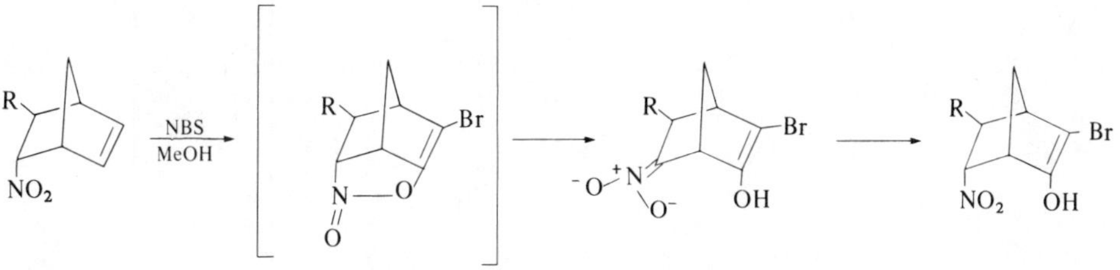

(3) (4) (5) (6)

For the bromination of cyclohexene in the presence of pyridine a mechanistic scheme has been proposed[30] which involves not only direct reaction with bromine *via* a bromonium ion, but also nucleophilic attack on an alkene–bromine charge-transfer complex by both Br_3^- and the complex between Br_2 and pyridine. Likewise, comparison of the kinetics and product distributions for the reactions of buta-1,3-diene with molecular bromine, with the pyridine–Br_2 charge-transfer complex, and with $Bu_4\overset{+}{N}\,Br_3^-$ strongly suggest that both of the latter reagents act as independent nucleophiles rather than as sources of Br_2.[31] The steric course of bromoepoxide formation from the reaction of bromine with E- and Z-pent-3-en-2-ols is critically dependent on the conditions used.[32] Hypobromous acid with 10β-vinylcholestanes not only yields products resulting from neighbouring-group participation but also from rearrangement of the bromonium ion intermediate (Scheme 1).[33] For the addition of HOBr to 5-cholestenes the previously reported 19-substituent participation is found[34] to be modified by the presence of an acetoxy group in the 3-position. The role of 19-substituents on the corresponding reaction for 3-cholestenes has also been investigated.[35]

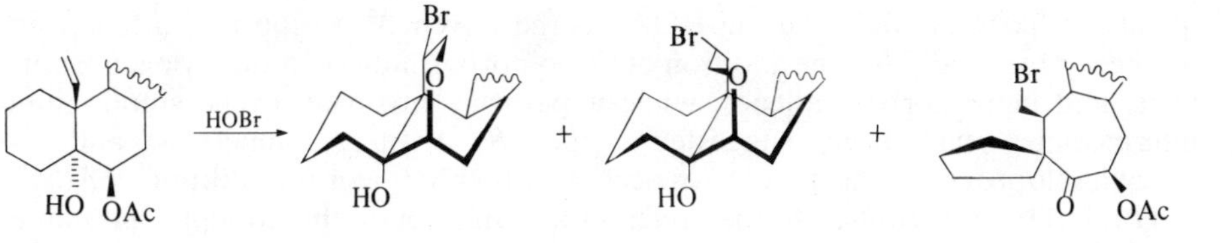

SCHEME 1

A molecule-induced homolytic radical mechanism has been put forward[36] to explain the formation of only 1,2-diiodoalkanes from the reaction of terminal alkynes with I_2–MeOH; on the other hand solvent incorporation is found when $AgNO_3$ is present, and an ionic pathway has been invoked in this case. The kinetics and thermodynamics of keto–enol tautomerism for simple carbonyl compounds have been investigated[37] by measuring their rates of bromination and/or iodination at very low halogen concentrations. Treatment of but-3-enyl acetate with I_2–AgNCO followed by NH_3 yields (**7**) and not (**8**) as previously reported;[38] a mechanism involving iodonium and dioxolenium ions has been proposed (Scheme 2).[39] An iodonium ion intermediate also accounts for the formation of both iodothiocyanato and iodoisothiocyanato vicinal adducts from E-hex-3-ene and I_2–KSCN.[40]

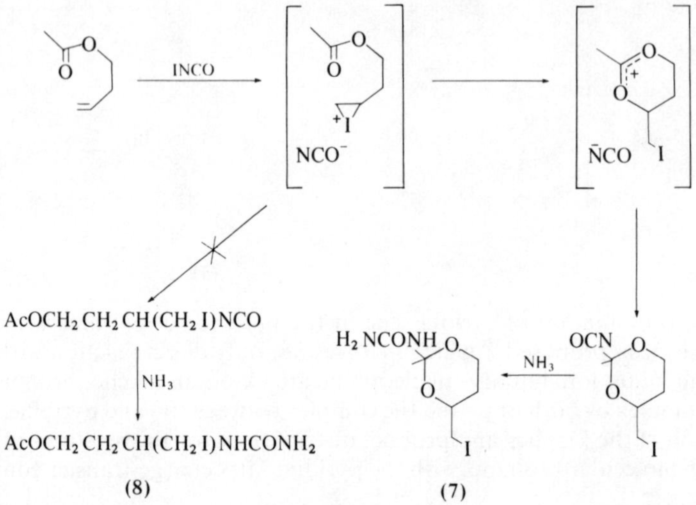

SCHEME 2

The role of episulphonium ions in the reactions of alkenes with sulphenyl derivatives has been reviewed.[41] Quantum-mechanical calculations indicate that the polarizability of the π-electrons is the rate-determining factor for the additions of halogens, alkanesulphinyl halides, and peracids to non-conjugated alkenes.[42] Kinetically controlled addition[43] of benzenesulphenyl chloride to $PhCH{=}CHCH_2COX$ (X = OH, OMe, NHPh, Cl) gives $PhCHClCH(SPh)CH_2COX$, which isomerizes in dry solvents to $PhCH(SPh)CHClCH_2COX$. The effective electrophilicity of 2,4-dinitrobenzenesulphenyl chloride (**9**) is increased by the addition of strong electrolytes (e.g. alkali metal perchlorates) and by using HOAc as the solvent.[44] An ion-pair mechanism has been proposed[45] for the addition of (**9**) to norbornadiene, 5-methylenenorbornene, and benzonorbornadienes, while a pathway involving cyclic sulphonium intermediates has been suggested[46] for (**9**) with 3,3-dimethyl- and 1-methylcyclopropene. The high regioselectivity observed[47] for the addition of ArSCl to (**10**) has been attributed to the preferential formation of the *exo*-episulphonium ion, followed by nucleophilic attack by Cl^- at the less hindered C(3) site (Scheme 3).

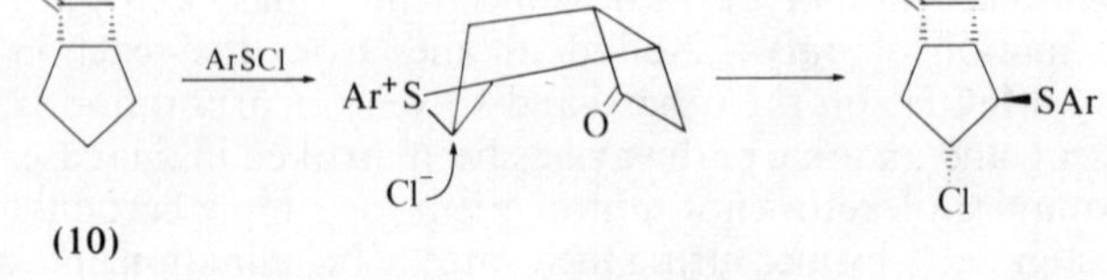

SCHEME 3

The significance of steric and electronic effects in the reaction of ArSCl with allenes has been examined;[48] an Ad_E2 mechanism has been proposed involving rate-determining formation of an alkylidenethiiranium ion intermediate by S_N2 attack at sulphur. The unusual formation of chlorodisulphides, instead of the expected chlorosulphides, from the reaction of Ph_3CSCl with norbornene has been attributed[49] to addition of a second molecule of Ph_3CSCl to the first-formed thiirane intermediate. Crystal structure determination[50] for (**11**) provides unequivocal proof for the earlier hypothesis[51] that attack of ArSCl occurs *exo* to the endocyclic double bond of (**12**). The formation of a mixture of (**13**) and (**14**) from the reaction of RSCl with (**15**) can be rationalized[52] in terms of initial formation of episulphonium ions at the 3,4- and 4,5-double bonds respectively, the subsequent cyclizations resulting from interaction with the vinyl group. The reactivities of $(MeO)_2P(O)SCl$ and PhSCl in their addition reactions with vinylsilanes have been compared;[53] for the latter the proposed transition state was of the ion-pair type, while for the former it resembled a cyclic α-sulphurane. Sulphenyl trihaloacetates ($RSOCOCX_3$; X = F,Cl), generated as transient intermediates from the corresponding thiosulphinates and $(CX_3CO)_2O$, undergo stereospecific *trans*-addition to olefins to give β-trihaloacetoxysulphides with Markownikoff regioselectivity.[54] Linear free energy relationship analysis for the addition of ArSSCN to styrenes indicates two kinetically distinct Ad_E2 mechanisms.[55] Treatment of olefins with thiocyanogen bromide, prepared from Br_2 and TlSCN in wet chloroform, affords *vic*-bromothiocyanates *via* an ionic pathway involving nucleophilic attack by Br^- on an *S*-cyanothiiranium cation.[56]

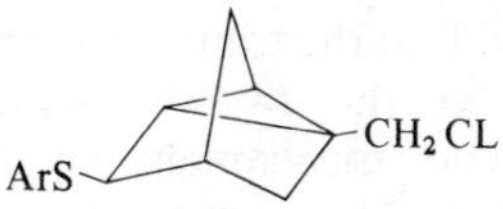

(**11**) 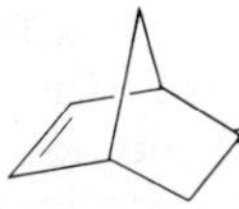 (**12**)

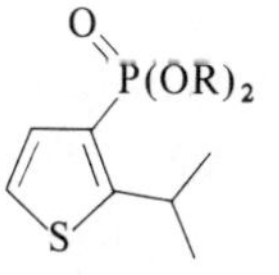

(**13**) 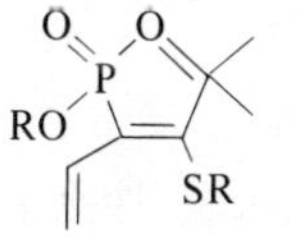 (**14**) 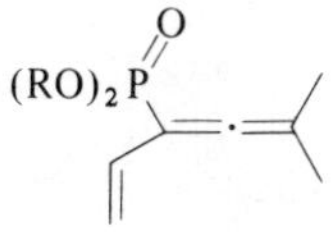 (**15**)

Anti-stereospecificity is observed when PhSeCl adds to propargyl alcohols; the regioselectivity of the process depends on the nature of the substituents geminal to the alcoholic moiety, bulky groups favouring *anti*-Markownikoff products under conditions of kinetic control.[57] A mechanism involving rate-limiting formation of alkylideneepiseleniranium ions and/or alkylideneepiselenuranes, followed by isomerization prior to the product-determining transition state has been proposed[58] to account for the pronounced steric effects observed for the addition of ArSeBr to allenes. Phenylselenolactonization and etherification continue to attract attention[59] as a means of achieving highly stereoselective ring-closures, the value of the technique being illustrated by the synthesis of stable and biologically active prostacyclins. The selenolactone (**16**), derived from homogeranic acid (**17**), undergoes acid-catalysed rearrangement to give the bicyclic lactone (**18**);[60] the seleniranium ions (**19**) and (**20**) are believed to be key intermediates. The phenylselenolactonization of (**17**) occurs stereospecifically.[61]

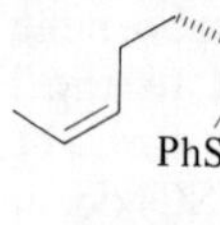

(16) 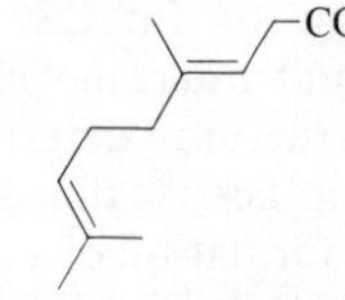 (17) 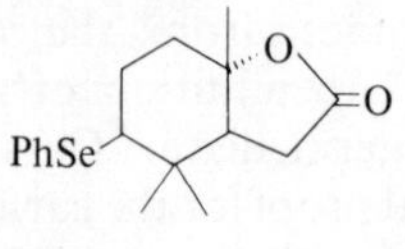 (18)

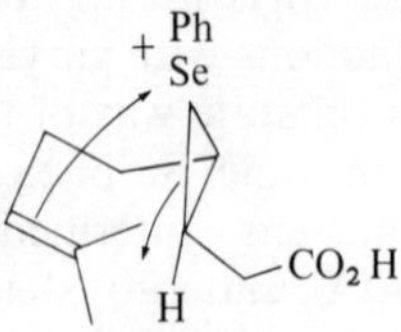

(19) 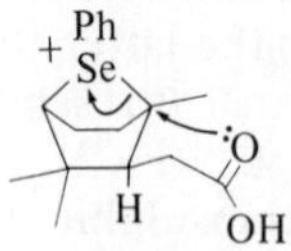 (20)

Addition of $PhSeSO_2Ar$ to olefins in the presence of $BF_3 \cdot Et_2O$ produces β-phenylselenosulphones with Markownikoff regioselectivity *via* a stereospecific *anti*-addition;[62] in the absence of the catalyst *anti*-Markownikoff adducts are formed by way of a non-stereospecific free-radical pathway. Results of a kinetic study[63] of the reaction of PhSeSCN with styrenes suggest that the transition state has a more open ion-like structure than that involved in the rate-determining step for the corresponding addition of PhSeCl. The mechanism of the formation of bis(phenylseleno) cyclic ethers [e.g. (**21**) and (**22**)] from treatment of dienes such as *cis,cis*-cycloocta-1,5-diene with PhSeCN–$CuCl_2$ in alcohols has been explained;[64] initial alkoxyselenation of one double bond is followed by transannular attack of the resulting alkoxy group on the episelenonium ion generated at the other double bond by addition of a second molecule of PhSeCN.

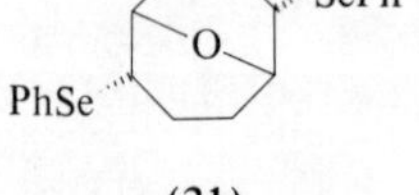

(21) 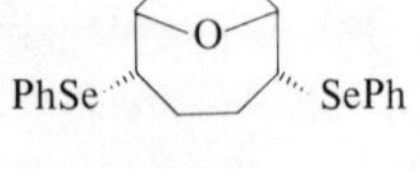 (22)

Addition of Hydrogen Halide

The regio- and stereo-selectivity observed for hydrochlorination of mono- and di-methyl derivatives of cyclohex-3-ene-1-carboxylic acid has been interpreted[65] in terms of conformational effects in the intermediate cations. For the $SnCl_4$-catalysed addition of HCl to cyclohexene at low temperatures (143–293 K) a multi-step process with at least two competing mechanisms has been proposed[66] involving complexation of the HCl with the alkene and with the catalyst. The regioselectivity found for the addition of HBr to hept-1-ene has been explained[67,68] by invoking a 2:1 HBr–olefin complex. The kinetics of the hydrochlorination of acetylene and vinylacetylene in non-aqueous solutions of CuCl have also been examined.[69,70] For the hydrochlorination of acetylene catalysed by Hg(II)–amine complexes a chloride ion displaced from the inner co-ordination sphere plays an active role.[71] The reaction of HCl with 1,2-alkadienephosphonate esters affords cyclized products such as (**23**).[72] The reactions of trimethylsilylcyclopropanes with $TiCl_4$ and HCl leads to ring-cleavage, the electrophile becoming attached to the silyl-substituted carbon except where ring-strain factors become too large;[73] there is thus some similarity to the behaviour of vinylsilanes.

Hydration and Related Reactions

The reactivity of unsaturated bridging bicyclic hydrocarbons towards the addition of carboxylic acids is dependent on ring strain.[74] Treatment of vinylnorbornane with acetic acid results mainly in ring-expansion to the bicyclo[3.2.1]octane skeleton, whereas with 5-vinylnorbornene reaction occurs exclusively at the endocyclic double bond;[75] the corresponding reaction for the vinyltricycloheptane (**24**) affords (**26**) *via* the exocyclic adduct (**25**). The formation of acetophenone from

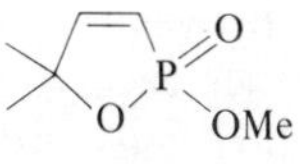

(23) 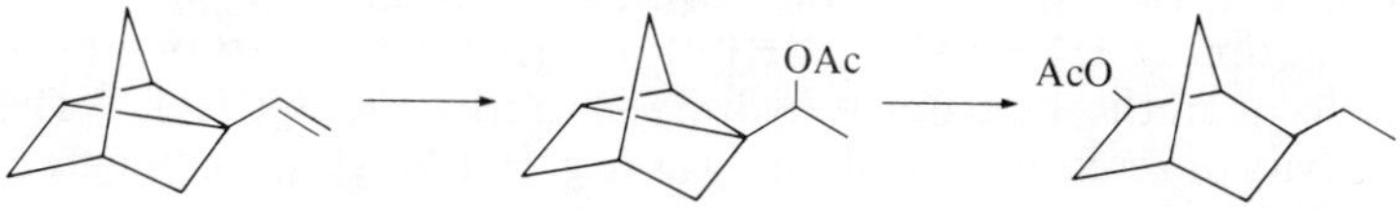 (24) (25) (26)

phenylacetylene in the presence of acetic acid proceeds *via* α-acetoxystyrene as an intermediate.[76] The rate constants for the reaction of $Me_2C{=}CHMe$ with aqueous formic acid suggest an *A*1 mechanism.[77] Kinetic and spectral studies indicate that the acetoxylation of ethylene in HOAc with PdX_4^{2-} (X = halide) catalysts proceeds mainly by formation of a binuclear π-complex between the alkene and Pd, isomerization of this π-complex by O_2 or ONO^- to a σ-complex, and heterolysis of the Pd—C bond with participation of H_2O.[78] 1,2-Addition products predominate for the diacetoxylation of conjugated dienes with $Tl(OAc)_3$–HOAc.[79] Protoadamantane (**27**) with CF_3CO_2H affords (**28**), the rearrangement accompanying the addition involving only a simple Wagner–Meerwein shift;[80] the isolation (using CF_3CO_2D) of four diastereoisomeric 2-adamantyl-4-*d* trifluoroacetates with mainly *exo*-attachment of the electrophilic deuterium indicates the involvement of an olefin–CF_3CO_2H π-complex rather than a carbocation as the first and key intermediate. The addition of Cl_2, HOCl, and H_2O_2 to cyclohexene-3-carboxylic acids in polar solvents gives *trans*-diaxial products.[81] The isomerization of *cis*-1,2-diarylethylenes catalyzed by methanesulphonic acid, is believed[82] to proceed *via* a proton addition–elimination mechanism. The stereoselective acid-catalysed cyclization of the β-hydroxyselenide (**29**) derived from nerolidol affords the bicyclic ether (**30**).[83]

A detailed kinetic analysis has been performed[84] on the hydrolysis of $Ph_2C{=}C(NO_2)_2$ to Ph_2CO and $CH_2(NO_2)_2$, and the results compared with those

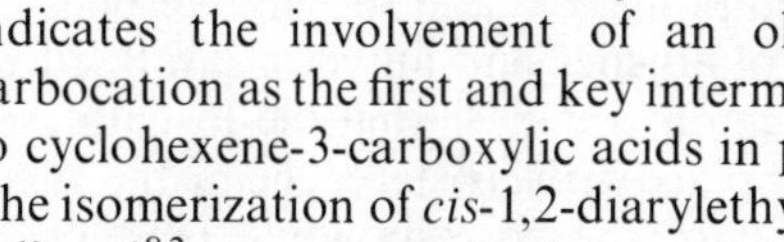

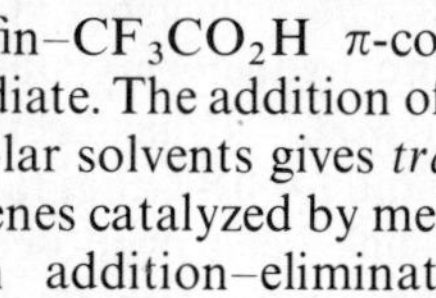

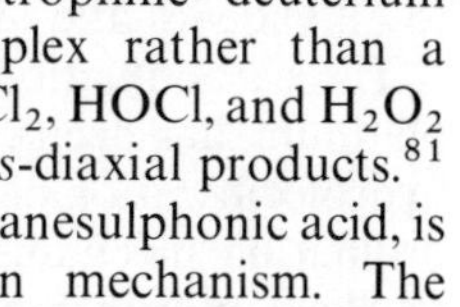

CF_3CO_2H

(27) (28) (29) (30)

for the corresponding reaction with benzylidene Meldrum's acid. MO calculations for the hydration and hydrohalogenation of olefins suggest[85] that the reaction rate and product distribution are related to the polarizability of the π-bond. For the hydrolysis at low pH of enaminoketones the mechanism involves rapid protonation at carbon, followed by slow hydration;[86] as the pH increases the first step becomes rate-determining. MINDO/3 calculations performed on the protonation of ethylene by H^+ and H_3O^+ indicate that proton solvation reduces the heat of the exothermic reaction and changes the mechanism;[87] electrophilic attack is accompanied by desolvation of the proton, this process occurring after the transition state. A kinetic study of the hydrolysis of chalcone has shown[88] that at pH < 12 the rate-limiting step involves attack by H_2O on the C=C, whereas at pH 13.5 the rate depends only on HO^-. For the conversion of allene and propyne to acetone in aqueous H_2SO_4, solvent isotope effects and the dependence of the rates on acidity are consistent with the Ad_E2 mechanism of rate-determining protonation of the terminal carbon leading to the prop-2-enyl cation as intermediate in each case, followed by hydration to the enol and subsequent tautomerization.[89] Linear Taft plots are obtained for the hydration of HC≡CCH=CHOR to MeCOCH=CHOR; this step proves to be faster than the subsequent hydrolysis to $MeCOCH_2CHO$.[90] The higher catalytic activity of heteropoly-acid catalysts (e.g. $H_3PW_{12}O_{40}$), compared with H_2SO_4 and $HClO_4$, for the hydration of phenylacetylene has been attributed[91] to the cooperative action of the heteropoly-anion.

Miscellaneous Electrophilic Additions

The regiospecificity found for the acetylation of fulvenes by MeI–CO in the presence of $Co_2(CO)_8$ and a phase-transfer catalyst has been rationalized[92] in terms of the steric requirements of the acetylfulvene–$Co(CO)_3$ complex intermediate. A phase-transfer catalyst has also been used[93] to achieve the hydroxyacetylation of allenes using $AcCo(CO)_4$. A study[94] of the $ZnCl_2$-catalysed reaction of propargyl chlorides with acyclic 1,3-dienes has confirmed the hypothesis[95] that such Lewis acid-catalysed additions of alkyl halides (RX) to olefins are only possible when the initial carbocation R^+ is better stabilized than the cation RCC^+ formed in the addition step. Thiazolium salts catalyse the addition of aldehydes to phenyl vinyl sulphone,[96] to alkylidene- and arylidene-β-dicarbonyl compounds,[97] and to vinyl ketones.[98]

The carbonylation and hydroformylation of alkenes and alkynes, catalysed by platinum, has been reviewed.[99] It has been demonstrated[100] that the Pt-catalysed hydroformylation of Z- and E-but-2-enes takes place with *cis*-stereochemistry. A co-ordination mechanism has been proposed[101] for the corresponding reaction of ethylene in toluene containing $(Ph_3P)_2Rh(CO)Cl$. The asymmetric induction resulting from the Rh-catalysed hydroformylation of vinyl esters in the presence of chiral phosphine ligands has been investigated.[102] With bis-olefinic amine derivatives there are cyclized, and both mono- and bis-hydroformylated products, depending on the substrate and catalyst.[103] $HCo(CO)_4$ undergoes *cis*-addition to the more hindered face of 1,2-diphenyl-3-(methoxycarbonyl)cyclopropene.[104] Polymer-bound ruthenium catalysts for the hydroformylation of olefins have been examined,[105] and their effectiveness has been compared to the homogeneous action of $(Ph_3P)_2Ru(CO)_3$. Molecular hydrogen promotes the regioselective hydrocarboxylation of propene using CO–MeOH–$[PdCl_2(PPh_3)_2]$.[106] The main factors governing the isomeric composition of the product of the hydrocarbomethoxylation of hex-1-ene are the relative rates of isomerization of the starting olefin.[107]

SCF MO calculations suggest[108] that for the hydroalumination of acetylene there is a symmetrical π-complex as intermediate. In marked contrast with other carbometallation reactions, the Zr-catalysed carboalumination of propargyl and homopropargyl derivatives containing OH, $OSiMe_2Bu^t$, SPh, and I substituents displays an unexpectedly high regioselectivity.[109] Evidence has been presented[110] for carboalumination in the Zr-catalysed reactions of terminal alkynes with organoalanes.

The observed regioselectivities and kinetic data indicate[111] that similar mechanisms are operative for the hydrosilylation of alkenes and alkynes in the presence of homogeneous and supported phosphonium and ammonium salts containing a Pt anion. Intervention of a π-allylic palladium species is believed[112] to account for the enantioselectivity observed for hydrosilylation of styrene in the presence of menthyldiphenylphosphine. The existence of intermediate complexes containing oxygen has been demonstrated[113] for the $(Ph_3P)_3RuCl_2$-catalysed addition of $(RO)_3SiH$ to alkenes in the presence of oxygen. The rate constants for the reactions of RSiHMePh with hept-1-ene are related to σ^* and E_S parameters,[114] and also to the stretching vibrational frequency and NMR chemical shift of the Si—H. The catalytic activity of metal carbonyl clusters for the hydrosilylation of $Me_3SiCH{=}CH_2$ decreases in the order: $Rh_4(CO)_{12} \gg [Co(CO)_4]_2Sn > Co_2(CO)_8 > Co_4(CO)_{12} > [Co(CO)_4]Sn \simeq [Co(CO)_4]_2SnCl_2 > Rh_6(CO)_{16} \simeq Co_6(CO)_{16}$; the hydrido form of the metal carbonyl is the active species.[115,116] Treatment of allyl alcohol with R_3SiH in the presence of ion exchangers gives both dehydrocondensation and hydrosilylation products, the latter being favoured by increased temperature.[117] The isolation of the same ratio of allylsilanes, (**31**) and (**32**), from protodesilylation of both (**33**) and (**34**) shows that the cationic intermediate (**35**) is involved in both processes (Scheme 4).[118]

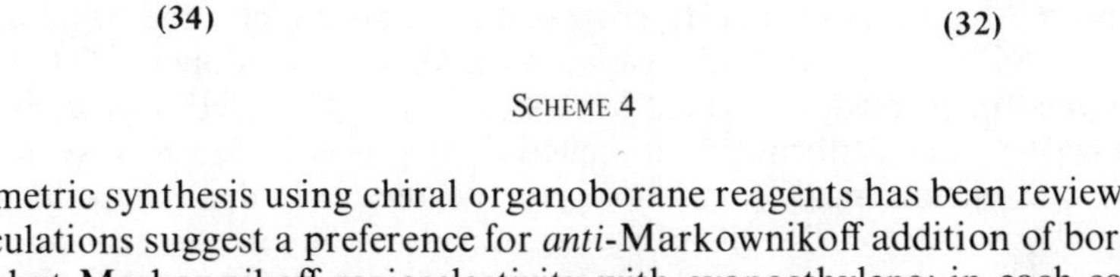

Me_3Si—(33)—$SiMe_2Ph$ $\xrightarrow{H^+}$ (35) $\xleftarrow{H^+}$ Me_3Si—(34)—$SiMe_2Ph$

(35) $\xrightarrow{-H^+}$ (31) $SiMe_2Ph$; (35) $\xrightarrow{-H^+}$ (32) Me_3Si

(**33**) (**34**) (**35**) (**31**) (**32**)

SCHEME 4

Asymmetric synthesis using chiral organoborane reagents has been reviewed.[119] MO calculations suggest a preference for *anti*-Markownikoff addition of borane to propene but Markownikoff regioselectivity with cyanoethylene; in each case the reaction pathway involves π-complex formation between the reactants, and subsequent deformation through an asymmetic four-centre transition state to yield the organoborane product in a donation–back-donation mechanism;[120] the regioselectivities arise from electronic effects induced by the electron-donating or electron-withdrawing substituents, steric factors being negligible. A three-centre transition state for the hydroboration of ethylene and acetylene is indicated by a semi-empirical MO study.[121] The identification by ^{1}H- and ^{11}B-NMR spectroscopy of (**36**) as the product of the reaction of isoprene with (**37**) has led to a mechanism being proposed [122] which involves (**38**) and (**39**) as intermediates (Scheme 5). ^{11}B-NMR has also been used to establish[123] 3-methyl-1-boracyclopentane as an intermediate in the corresponding reaction with BH_3–THF and BH_3–Me_2S. An extensive study[124] of the hydroboration of alkenes by 9-borabicyclo[3.3.1]nonane (9-BBN) has shown that substituents have little or no effect when separated from the olefinic centre by three or more methylene groups, but that both the reaction rate and product distribution are influenced as the functional group approaches the reaction site. The large steric requirements of 9-BBN are believed[125] to be responsible for the high regioselectivities found in its reaction with styrenes, the β-phenylethyl adduct being preferred to the extent of at least 39:1. The hypothesis that the hydroboration of alkenes by $(9\text{-BBN})_2$ proceeds through prior dissociation of the dimer into monomer (9-BBN), followed by reaction of the monomer with the alkene, has been confirmed by a kinetic investigation[126] in which the disappearance of the infrared absorption due to the B—H—B linkages of $(9\text{-BBN})_2$ was monitored quantitatively. Dilongifolylborane, synthesized from (+)-longifolene, has proved to be an effective reagent for asymmetric reaction with prochiral olefins.[127] The 1-pyrrolylborane–THF complex has also been used for the hydroboration of alkenes.[128]

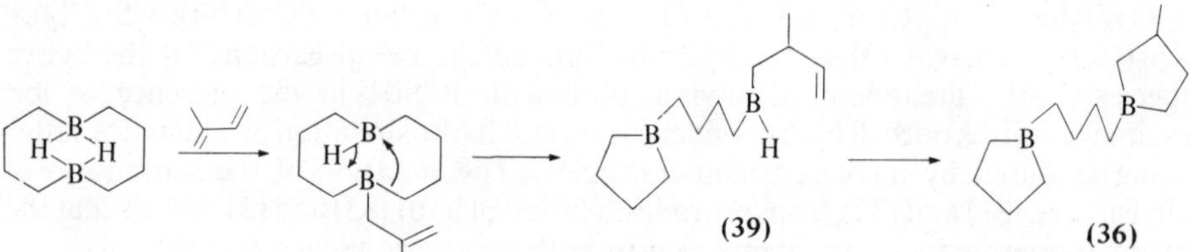

(37) (38) (39) (36)

SCHEME 5

In a new approach to examining the mechanisms of alkene bromination and mercuration Fukuzumi and Kochi[129] have evaluated steric effects directly from the electron donor-acceptor complexes of the alkenes with Br_2 and Hg(II)salts; after taking the steric term into account a striking similarity has been found between the activation processes for the two reactions, and this is attributed to a common mechanism. Mercuric salts add to ethyne, methyl propiolate, and hex-1-yne in the presence of SCN^- to yield β-thiocyanatoalkenylmercuric adducts;[130] the isomeric isothiocyanato product is formed with hex-3-yne. The methoxymercuration of styrenes[131,132] and stilbenes[131] is believed to proceed *via* an unsymmetrically bridged cationic intermediate, with varying degrees of bridging depending on the nature of the substrate. The site of mercuration of 5-vinylnorbornene and (**40**) has been found[133] to depend on the polarity of the solvent, HOAc and THF favouring reaction at the endocyclic double bond, whereas in MeOH and THF–H_2O addition occurs at the exocyclic group. Acetoxymercuration of (**41**) by $Hg(OAc)_2$–HOAc–KBr affords both *cis-endo-* and *cis-exo-*adducts, the transition state being stabilized by the bridging and aromatic ring π-systems.[134] The hydration of mono- and 1,2-disubstituted alkenes *via* solvomercuration–demercuration proceeds with very high (>99%) Markownikoff regioselectivity.[135] The presence of an anionic surfactant in the hydroxymercuration of non-conjugated dienes such as limonene leads to selective monofunctionalization at the exocyclic double bond;[136] on further reaction the cyclic ether (**42**) is formed in preference to the diol (**43**), due to strong stabilization of the cationic mercury intermediate by the head groups of the surfactant; the influence of the water, which would otherwise have solvated the

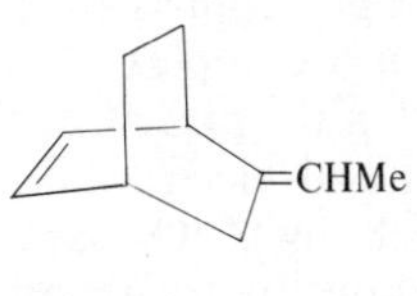

=CHMe F F F F O OH HO

(40) 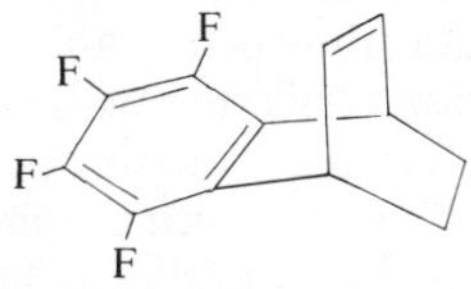 **(41)** 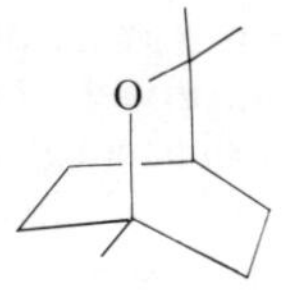 **(42)** 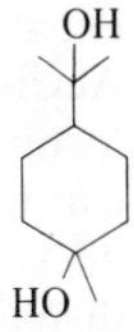 **(43)**

covalent complex, is thereby decreased. The oxymercuration–demercuration of a variety of alkenes bearing halogen, oxygen, and sulphur substituents has also been examined.[137,138] Treatment of olefins with aniline and HgO–HBF_4 affords aminomercurials,[139] which are stable at -10°C and suitable for subsequent hydroxylation or alkoxylation. With prop-2-ynol primary aromatic amines in the presence of $Hg(OAc)_2$ give $H_2C{=}C(NHAr)CH_2OH$, which subsequently undergo allylic oxidation and amination to yield *N*,*N*′-diarylpropane-1,2-diimines.[140] The aminomercuration–demercuration process has also been applied[141] to allylamines and allyl sulphides. Intramolecular aminomercuration–demercuration of the ε,ζ-unsaturated amine (**44**) to (**45**) provides the key step in a stereoselective synthesis of optically active Prosopis alkaloids.[142] Markownikoff amidation of olefins is achieved by an amidomercuration–demercuration process involving treatment of the olefin with the amide in the presence of anhydrous $Hg(NO_2)_2$, followed by *in situ* reduction by $NaBH_4$.[143] A similar sulphonamidomercuration has also been reported.[144] The intramolecular amidomercuration–demercuration of (**46**) to (**47**) occurs with much greater stereoselectivity[145] than the corresponding aminomercuration–demercuration.[146] The cyclizations of (**48**) and (**49**) have also been described.[147]

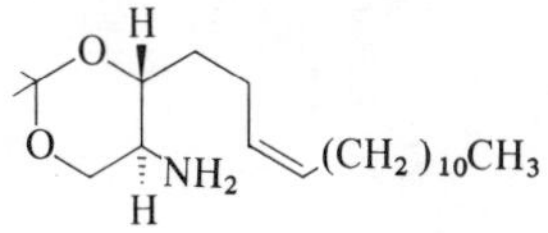

H O O NH_2 H $(CH_2)_{10}CH_3$ H O O N H H H $(CH_2)_{11}CH_3$

(44) 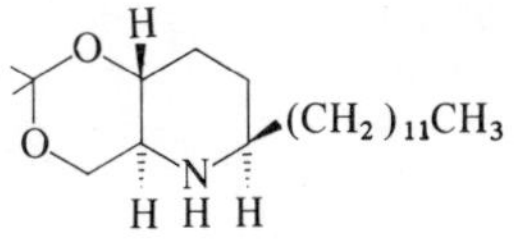 **(45)**

HN CO R N CO R OCOR O NMe $^+CH_2$ OCOR OMe $^+$CHOMe

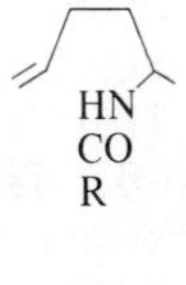

(46) 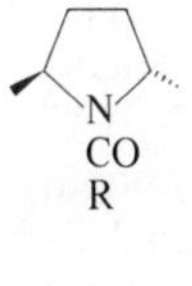 **(47)** 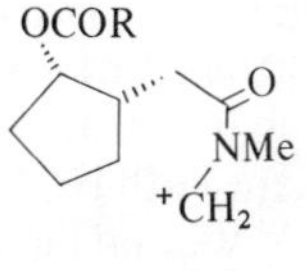 **(48)** 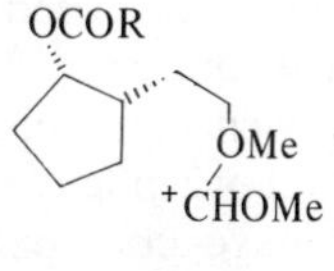 **(49)**

The preferred formation of (**50**) rather than its regioisomer from the hydrostannation of (**51**) by Bu_3SnH is believed[148] to result from the co-ordinative effect of the lone-pair electrons of the neighbouring nitrogen; for (**52**) the presence of the hydroxyl group reverses the direction of addition. The *syn/anti* ratios observed[149] for the perepoxidation of 11-isopropylidenedibenzonorbornadiene correlate well with the relative abilities of the internally competing aromatic rings to enter into homoaromatic charge delocalization; stronger electrophiles (e.g. $AcCl–AlCl_3$) approach the aromatic ring better able to form a π-complex. The formation of enol acetates by reaction of alkynes with acetic acid in the presence of catalytic amounts of silver salts has been attributed[150] to initial electrophilic addition of Ag^+ to give (**53**), followed by nucleophilic attack by AcO^- and displacement of Ag^+ by H^+. The inversion of olefins brought about by treatment with $TeCl_4$ and Na_2S involves *cis*-chlorotelluration followed by *trans*-dechlorotelluration.[151]

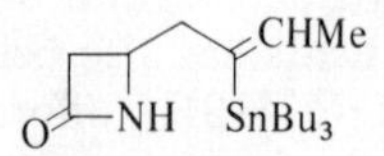

(**50**) 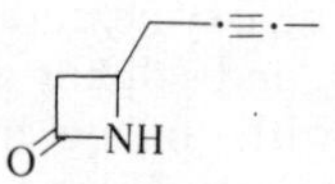 (**51**) 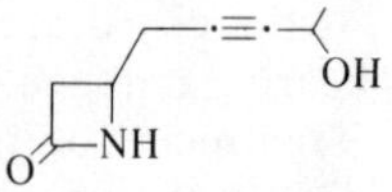 (**52**) 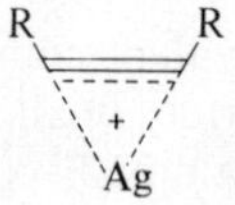 (**53**)

The initiation of polyene cyclization using 3-methylcyclohexenone has been discussed.[152] The ability of thallium(III) salts to activate alkene double bonds to nucleophilic attack has allowed the cyclization of *o*-prenylphenols to be achieved.[153] The naturally occurring bromine-containing terpenoid (**54**) has been synthesized from the prenyl precursor (**55**) by treatment with (**56**), which causes concerted cyclization in a biomimetic fashion.[154] The rate data for the addition of chlorosulphonyl isocyanate to 6-substituted benzonorbornadienes are consistent with either a concerted addition or a process which involves a carbocation intermediate; a mechanism has been proposed[155] which proceeds through initial generation of a π-complex, which decays to the carbocation in a product-determining step. Nitration of the enol acetates of substituted cyclohexanones with

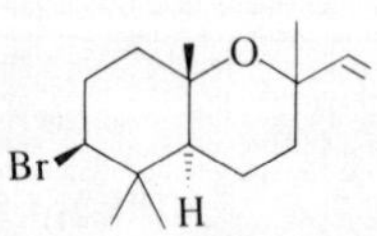

(**54**) 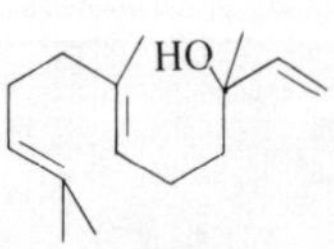 (**55**) 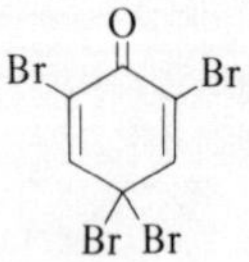 (**56**)

$HNO_3–Ac_2O$ leads to *cis*- and *trans*-substituted 2-nitrocyclohexanones in a kinetically controlled process.[156] The reactivity of 3-phenoxycyclohexene towards electrophiles has been investigated.[157] UV data and the effect of substituents on the initial rate indicate[158,159] that the Markownikoff addition of propylene sulphide to arylacetylenes proceeds *via* a charge-transfer complex. The main products from the treatment of 2-phenylnorbornene with $HCHO–Pr^i{}_2NH_2{}^+Cl^-–HCO_2H$ are formed *via* a Prins reaction pathway.[160] Vicinal bis-tosylation of olefins by $PhIOH^+TsO^-$ is believed[161] to involve electrophilic addition, followed by dehydroxylation and nucleophilic displacement of iodobenzene by a second tosylate anion. The complex

[$C_5H_5Fe(CO)_5/CH_2{=}C(OEt)CO_2Et$] acts as an α-acrylic ester cation equivalent in its reaction with cyclohexanone lithium enolate.[162] The facile oligomerization of alkenes in the presence of $Pd(MeCN)_4^{2+}$ may also be attributed[163] to the formation of incipient carbocations which form on complexation to the strongly electrophilic catalyst. The reaction of $ZrCp_2$[isoprene] with simple alkenes proceeds with highly regioselective carbon–carbon bond formation between C(2) of the alkene and C(4) of the isoprene;[164] in the presence of excess isoprene tail-to-tail dimerization products are obtained. Neighbouring-group participation by the benzamido moiety in the addition of electrophiles to $PhCONHCH_2C{\equiv}CH$ results in the formation of 3-methyleneoxazolium salts.[165]

Nucleophilic Additions

Mechanistic and synthetic aspects of Michael additions have been the subject of several reviews,[166–168] and the significance of the addition–elimination pathway in nucleophilic vinyl substitutions has been assessed.[169] A detailed MO study has been made[170] of the relationship between the electronic structure of olefins complexed to transition metals and their reactivity towards nucleophiles. MO treatment has also been applied[171] to the regioselectivity of additions to α-enones. The formation of $CF_3CF(CO_2Me)CF{=}CFCO_2Me$ from treatment of $CFCl{=}CFCO_2Me$ with KF has been attributed[172] to substitution of the Cl by F *via* a nucleophilic addition–elimination pathway, followed by dimerization of the resulting trifluoropropenoate ($CF_2{=}CFCO_2Me$). An addition–elimination mechanism explains[173] the formation of *cis*(95%)- and *trans*(5%)-$N{\equiv}CCF{=}CFCO_2Me$ from nucleophilic vinyl substitution of $CF_2{=}CFCO_2Me$ with KCN. MO calculations indicate[174] that for the reaction of fluoride with ethylene there are two pathways possible, depending on the distance between the reactants and the angle of F^- attack; either nucleophilic addition to give $FCH_2CH_2^-$, or proton abstraction yielding HF and $CH_2{=}CH^-$. Conformational analysis based on the conservation of the torsion angle and the maximum overlap of orbitals has been used to explain the products of the hydrocyanation of keto-3Δ′-steroids; generally the cyano group is introduced in the axial position under kinetic control,[175,176] but when there is a methyl substituent at the β-position of the α-enone the preferred position is equatorial. A kinetic study of the hydrocyanation of various α-enones has been undertaken.[177] Deuterium labelling shows that the nickel-catalysed addition of HCN to olefins occurs in a *cis*-fashion.[178] The formation of 2-cyano-1,2-dihydro addition products from the reaction of cyanide ion with quaternary quinolinium salts suggests[179] that the usual designation of CN^- as a soft nucleophile may need refinement.

The kinetics of the addition of NaOR in ROH to various *gem*-difluoroalkenes indicate[180] that protonation of the carbanion intermediate by the solvent is generally slower than ejection of the fluoride ion. The kinetics of the Michael addition of pentaerithritol to acrolein and the reverse process have been established.[181] Further examples of base-catalysed cyclization of tricyclic olefinic alcohols to yield ethers have been reported; in each case the reactivity of the C=C double bond has been attributed[182] to relief of ground-state strain. A nucleophilic addition–elimination mechanism has been proposed[82] for the isomerization of *cis*-1,2-diarylethylenes catalysed by $KOBu^t$. The hydroxyurea anion ($H_2NCONHO^-$) under basic conditions undergoes 1,4-addition to α-acetylenic esters giving both *cis*- and *trans*-ethylenic β-ureidoxy-esters.[183] An addition–elimination mechanism involving a carbanion intermediate is operative for substitution by ArS^- of Cl^- in β,β-dichlorovinyl sulphones;[184] this contrasts with the corresponding reaction with sodium alkoxides which proceeds *via* an E–Ad_N pathway. For the Et_3N-catalysed addition of alkane thiols to acrylic and maleic acid derivatives there is a linear relationship between the rate constant and the π-charge on the β-carbon of the carbon–carbon double bond.[185] Alkyl thiols readily add to divinyl sulphoxides in the presence of catalytic amounts of KOH.[186] Intramolecular hydrogen bonding enhances the rate of the Michael addition of ArSH to a series of cyclopentenones.[187] The stoichiometry of the addition of thiols to indene has been determined.[188] The asymmetric addition of thioglycolic acid to β-nitrostyrene catalysed by cinchona alkaloids is due to interaction between the carboxyl group of the addend and the quinuclidine nitrogen of the catalyst.[189] High degrees of enantiomeric selectivity have been achieved[190,191] for the Michael addition of thiols to cyclohexenones using chiral bases as catalysts; a tight transition-state complex composed of thiol, amine, and catalyst has been proposed.[191] Asymmetric induction for this reaction has also been achieved[192] using chiral phase-transfer catalysts derived from cinchona and ephedra alkaloids; the presence of a hydroxy group β to the onium function is found to be essential. Reaction of thiols with the esters and amides of 1-methylpyridinium-3,5-dicarboxylic acid salts affords a mixture of 1,2- and 1,4-adducts.[193] A cyclic transition state (**57**) has been invoked to explain the regio- and stereo-specificity observed[194] for the addition of Et_2AlSPh to vinyl oxiranes. Thiol addition to steroid α,β-unsaturated carbonyl groups is considered[195] to be sufficiently favourable to be significant in steroid–protein interactions. Michael addition of $RCS_2^-\,MgBr^+$ to α-enones affords β-ketodithio-esters.[196] The cyclofunctionalization of allylic and homoallylic alcohol carbonates is highly regioselective yielding five- and six-membered rings respectively.[197] The addition of arenesulphinic acids to aryl acrylates involves an ionic mechanism in which there is rate-determining nucleophilic attack by the sulphur atom on the β-carbon of the acrylate, the products being formed *via* a charged cyclic intermediate.[198] Alkyne-ω-sulphenic acids (**58**; $n = 1$–4) cyclize regiospecifically to yield 2-methylenethiacycloalkane *S*-oxides (**59**); regioselective intermolecular addition of sulphenic acids to both unactivated and activated alkynes is also reported.[199] A rare example of facile addition of a nucleophile to an unactivated carbon–carbon double bond is provided[200] by the intramolecular reactions of the α-lithiosulphoxides (**60**), the product being the transannular adduct (**61**).

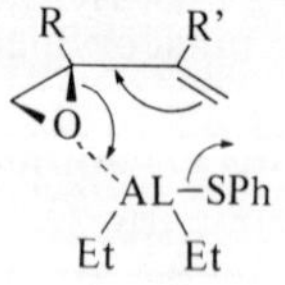

(**57**) 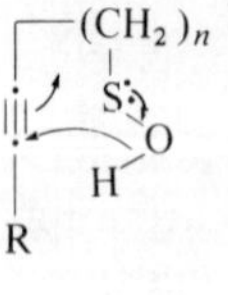 (**58**) 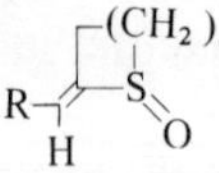 (**59**)

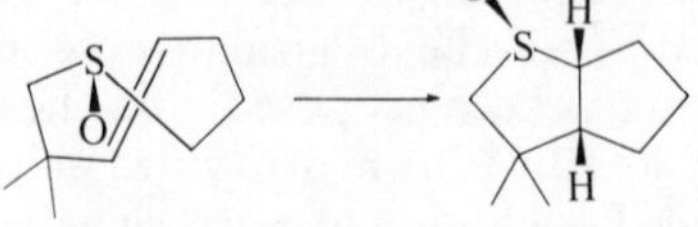

(**60**) (**61**)

The stereochemistry of the adducts of allylmagnesium halides to hydroxynorbornenes indicate[201] that during the addition process the allyl group is associated with the metallated hydroxyl group; likewise, with 3-(hydroxymethyl)cyclopropenes reaction occurs at the side of the double bond nearest the hydroxyl.[202] The rôle of the OH group in the addition of BuLi to hydroxynorbornenes has also been considered.[203] From a study of the reaction of allylic Grignard reagents with allylic and propargylic amines it has been concluded[204] that tertiary amino and metallated primary and secondary amino groups can assist the addition of organomagnesium compounds to alkenes and alkynes. Similarly metallated phenolic groups, even though relatively remote from the double bond, facilitate the addition of allylmagnesium halides to *o*-allylphenol.[205] Disubstituted acetylenes react with Bu^tMgX in the presence of Cp_2TiCl_2 to afford *trans*-alkenyl Grignard reagents selectively.[206] The proportion of 1,3-addition products found for the reaction of Bu^+MgCl with ethyl *trans*-cinnamates correlates with the ^{13}C-NMR chemical shifts, suggesting[207] that, although it is a radical process, polar effects determine the regioselectivity and that the *tert*-butyl radical has nucleophilic character. Me_3SiCH_2MgCl undergoes 1,4-addition to α-enones to give γ-silanes,[208] in contrast to the corresponding process for Me_3SiCH_2Li which gives 1,2-adducts.

Nucleophilic addition to the alkene double bond is believed[209] to be the first step in the formation of both 3-oxathian-1,1-dioxides and cyanocyclopropanes from the reaction of vinyl sulphones with α-lithionitriles. Hexamethylphosphoramide (HMPA) promotes α-1,4-addition of α-alkylthio-, α-phenylthio-, and α-phenylseleno-allyl lithium reagents to cyclopent-2-enone at the expense of α- and γ-1,2-addition.[210] Lithium alkyls (RLi) and Grignard reagents (RMgX) readily undergo 1,4-addition to $Me_2NCH_2C{\equiv}CCH{=}CHCH_2Y$ ($Y = OH, OMe, NMe_2$) to yield α,β-difunctional allenes ($Me_2NCH_2CH{=}C{=}CHCHRCH_2Y$);[211] with $MeOCH_2C{\equiv}CCH{=}CHCH_2Y$ 1,4-addition and substitution occurs giving $RCH_2CH{=}C{=}CHCHRCH_2Y$,[212] but reaction with $HOCH_2C{\equiv}CCH{=}CHCH_2Y$ yields either allenes or dienes depending on the nature of the metallic group.[213] α,β-Unsaturated amides undergo tandem conjugate addition–α-alkylation with strong (RLi, RMgX) and weak [$(RS)_2CHLi$] nucleophiles.[214,215] The reactivities of $Bu^tC(OM){=}CHR$, where M = Li, MgBr, $SiMe_3$ and R = H, Me, have been compared in their 1,2- and 1,4-addition to chalcone.[216] Under kinetic control *cis*-enolates of β,β-diphenylpropiomesitylene are formed from PhM (M = Li, MgBr, Na) and α-enones, but the *trans*-isomers result from the ketone itself.[217] *cis*-Decalone adducts result from the reaction of lithiated arylacetonitriles with 2-octalones,[218] while with cyclohex-2-enones in THF–HMPA at −70°C δ-ketoalkylnitriles generally result from 1,4-addition;[219] a reagent-like transition state has been proposed,[219] with the nucleophile approaching the alkene group by the less hindered face and the CN directed towards the carbonyl group. The lithium enolate of ethyl dithianecarboxylate undergoes 1,4-addition to cinnamaldehyde and crotonaldehyde.[220] With alkylidenemalonates and alkylideneacetoacetates, lithium trialkylalkynylborates undergo unusual Michael additions involving migration of alkyl groups from boron to carbon yielding alkenylboranes;[221] the process is not stereospecific. Lithium carbanions of allyl phenyl sulphones add α-1,4 to cyclic enones in the presence of HMPA, but α-1,2-addition is the major kinetic course otherwise.[222] Conjugate addition to cyclopent-2-enone with asymmetric induction has been achieved[223] using *p*-tolyl *p*-tolylithiomethyl sulphoxide as a chiral formyl anion equivalent. Cyclohex-2-enones will undergo conjugate addition with $[EtOCHMeOCMeCN]^- Li^+$ even when substituted in the 3-position.[224] The ability of α,β-unsaturated secondary thioamides to act as Michael acceptors has been examined.[225]

The formation of E-Bu^{t}CH=C=C=CHBu^t from treatment of Z-Bu^{t}CH=CBrC≡CH with Bu^{t}Cu involves *cis*-addition to the triple bond followed by *anti*-elimination of CuBr from the resulting adduct.[226] Results of a stopped-flow UV and IR study of the conjugate addition of Me_2CuLi to α-enones suggest[227] that the mechanism involves equilibrium between the reactants and an intermediate complex, which may unimolecularly rearrange to a trialkylcopper(III) species in which copper is bound to the β-carbon of the lithium enolate; this is followed by a reductive elimination process, involving the copper ligands, to yield MeCu and the β-methyllithium enolate product. The ability of 2-(phenylselenyl)enones to undergo alkylation by lithium dialkylcuprates, at the 3-position, is a key step in the synthesis of 2,3-dialkylenones such as *cis*-jasmone.[228] The regio- and stereo-chemistry of dialkylcuprate additions to alkylideneoxiranes has also been examined.[229] Dialkenylcuprates, generated from dialkylcuprates and acetylene, react with α,β-unsaturated sulphones to give *cis*-γ,δ-unsaturated sulphones which are useful precursors of *cis*-olefins.[230] Moderate degrees of enantiomeric selectivity ($<34\%$) have been reported[231] for the 1,4-addition to chalcone of heterocuprates, prepared from chiral amino-alcohols, imino-alcohols, and imino-phenols; higher selectivity has been observed with α,β-ethylenic sulphoxides.[232] The use of RCu–$AlCl_3$ rather than RCu allows conjugate addition to γ-acetoxy-α,β-unsaturated ketones to be accomplished[233] without the formation of α,β- or β,γ-enone by-products. 1,4-Methylation predominates over 1,2-addition to a much greater extent for $Me_5Cu_3Li_2$ reacting with α-enals than for Me_2CuLi.[234] Halomagnesium diorganocuprates ($XMgR_2Cu$) derived from Grignard reagents and CuI undergo conjugate addition with α,β-unsaturated esters and ketones thus providing an alternative to R_2CuLi.[235] Secondary dialkylcuprates react with α,β-ethylenic aldehydes to give 1,2- and 1,4-adducts, the ratio of the products depending on the nature of the metal;[236] $R_2CuMgCl$ in THF gives the greatest proportion of 1,4-adduct. The stereochemical course of the conjugate addition of (trimethylstannyl)cuprate reagents to α,β-acetylenic esters is dependent on the nature of the reagent as well as the structure of the substrate;[237] under kinetic control $PhS(Me_3Sn)CuLi$ gives only *trans*-β-trimethylstannyl α,β-olefinic esters, whereas the corresponding *cis*-isomers are formed exclusively under thermodynamically controlled conditions.[238] $RCu(Me_2S)\cdot MgBr_2$, prepared from RMgBr and the complex of CuBr and Me_2S, undergoes stereoselective addition to $MeO_2CC{\equiv}CCO_2Me$ to give 2-substituted maleates.[239] A mixture of Cu_4MgMe_6 and Cu_6MgMe_8 in the presence of $MgBr_2$ is thought to be involved in the formation of PhCMe=CH_2 and E-PhCH=CHMe from the reaction of the Normant reagent (MeMgBr and CuBr) with phenylacetylene.[240] $(PhMe_2Si)_2CuLi$ readily adds to α-enones to give β-silylcarbonyl compounds,[241] while with terminal alkynes the products result from *syn*-addition of the $PhMe_2Si$ and the copper, with the silyl group attached predominantly at the end carbon atom.[242] Organoargentates, [RAgBr]MgCl, add regiospecifically to diynes such as HC≡CC≡CPh to give RCH=C(AgBr)C≡CPh; in contrast the corresponding cuprate, [RCuBr]MgCl, reacts with low selectivity.[243]

Reaction conditions influence the stereochemistry of the Michael adducts resulting from the $NaNH_2$-catalysed reaction of $PhCH_2CONR_2$ with $PhCH{=}CHCO_2Me$ or $PhCH{=}CHCONH_2$.[244,245] The combination of fluoride ion catalysis and the use of high pressure promotes Michael additions which are normally sluggish;[246] e.g. $PhCH(CO_2Et)_2$ to $PhCH{=}CHCOPh$. 1,4-Addition of nucleophiles to 2-methylene-3-alkoxycyclohexenones gives only β-alkylated 2-cyclohexenones due to simultaneous elimination of the β-alkoxy group;[247] a second 1,4-conjugate addition to the product then allows α- and β-disubstituted ketones to be generated. The kinetics and deuterium isotope effects have been studied[248] for the phenylation of alkenes by $Pd(OAc)_2$ in benzene; a π-complex intermediate has been proposed. The rate-limiting step for the Pd-catalysed addition of arylmercury compounds to α-enones is the formation of the arylpalladium σ-complex.[249] Oxazolin-5-ones with bulky substituents at C(2) undergo Et_3N-catalysed addition to activated olefins exclusively at C(4).[250] The direction of Michael additions to the unsymmetrical enedione (**62**) is under product development control in the emerging enolate intermediates.[251] The regioselectivity observed[252] for the addition of carboxylic acid dianions to α-enals and α-enones is dependent on their substitution pattern, the complexing ability of the counter-ion, and the Lewis basicity of the solvent. The stereoselectivity of Michael additions to β-substituted α-methylene-lactams (e.g. **63**) is dependent on the β-substituent and the nature of the donor,[253] but intramolecular reactions are not observed. Chiral crowns complexed to potassium bases catalyse, with high turnover numbers, the Michael additions of $PhCHRCO_2Me$ ($R = Me, H$) and 2-methoxycarbonylindan-1-one to methyl vinyl ketone to give products of 60–99% optical purity;[254] the configurational bias can be rationalized on the basis of steric effects. The product resulting from the condensation of chalcone with ethyl cyanoacetate has been identified[255] as (**64**) and not as (**65**) as previously reported;[256] the reaction pathway involves a double Michael addition yielding (**66**), which then undergoes an aldol-type condensation.

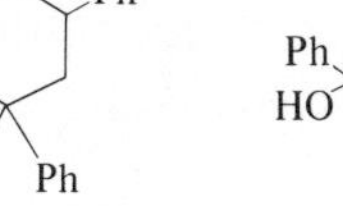

(**62**)

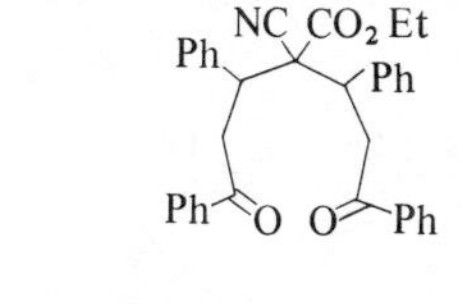

(**63**)

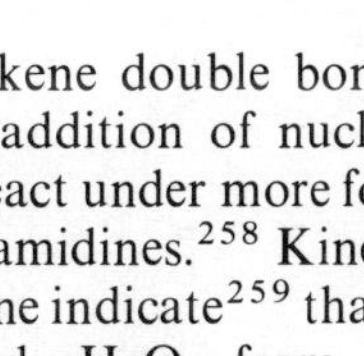

(**64**)

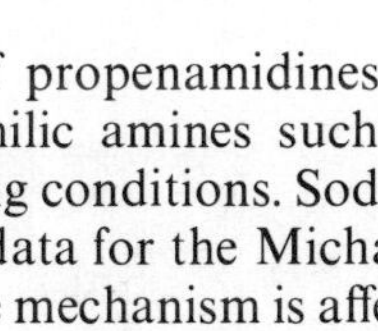

(**65**)

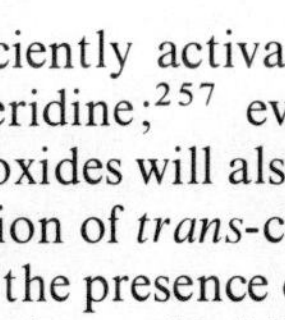

(**66**)

The alkene double bond of propenamidines is sufficiently activated to allow Michael addition of nucleophilic amines such as piperidine;[257] even aromatic amines react under more forcing conditions. Sodium alkoxides will also add to give 3-alkoxyamidines.[258] Kinetic data for the Michael reaction of *trans*-chalcone with pyrrolidine indicate[259] that the mechanism is affected by the presence or absence of CO_2 and H_2O from the atmosphere. Dialkylamines (R_2NH) add to

$CH_2=C(OMe)CMe{=}CH_2$ to give $MeC(OMe)=CMeCH_2NH_2$ and $CH_2{=}C(OMe)CHMeCH_2NR_2$, but $R_2NCH_2C(OMe){=}CMe_2$ is not formed.[260] The direction of amine attack on alkenes co-ordinated to Pt yielding 2-ammonioethanide complexes is determined by electronic rather than steric factors,[261] unless the latter are very large. Electron-deficient olefins undergo nucleophilic addition with anilines in the presence of Pd(II) catalysts.[262] The ene reaction between 2-pyridone and $MeO_2CC{\equiv}CCO_2Me$ predominates over 1,4-addition and appears to be charge-controlled rather than frontier-orbital-controlled, with bond formation taking place after the transition state.[263]

Product distribution studies and SCF *ab initio* MO calculations for substituted benzo- and naphtho-quinones show that the preferred site of nucleophilic attack is that position having the largest LUMO coefficient unless a donor group is directly attached;[264] FMO theory thus parallels resonance theory arguments used to explain regioselectivity. It has been proposed[265] on the basis of MINDO/3 calculations that HNO_2 acts as a nucleophile in its reactions with olefins, and that the addition to nitroethylene should give an *anti*-Markownikoff product due to the nature of the LUMO of the alkene. Michael addition of enamines to nitro-olefins affords γ-nitroketones in high (>90%) diastereomeric yields;[266] a topological rule for carbon–carbon bond-forming processes between pro-chiral centres has thereby been developed. The oxazepine (**67**) reacts asymmetrically with 1-nitrocyclohexene in the presence of base and crown ether catalysts to afford the optically active adduct (**68**).[267] Formation of the Diels–Alder and pseudo-Michael adducts from the reaction of furan with acetylenic aldehydes is believed[268] to involve a common intermediate (**69**), formed by electrophilic attack of the protonated aldehyde $RC{\equiv}\overset{+}{C}HOH$ on the furan. Only 1,4-adducts are observed[269] for the BF_3-catalysed reaction between 2-methylfuran and α,β-unsaturated carbonyl compounds, even in the case of hindered enones. The reaction of $H_2C{=}C(NMePh)_2$ with 1,4-benzoquinones proceeds by way of the zwitterion (**70**), which is subsequently oxidized to (**71**);[270] this contrasts with the behaviour of $H_2C{=}C(OR)_2$ which gives 1:2 adducts under similar conditions.[271] Nucleophilic attack by Me_2PPh on ethene complexed to ruthenium(II) has been reported.[272] Hammett $\rho\sigma$ analysis for the hydrostannylation of arylacetylenes by Me_3SnLi in THF indicates[273] a nucleophilic addition mechanism with a polar transition state in the rate-determining step. Addition of nucleophiles to nitrilium ions proceeds stereospecifically.[274] CsF in the presence of $Si(OR)_4$ acts as an efficient catalyst for

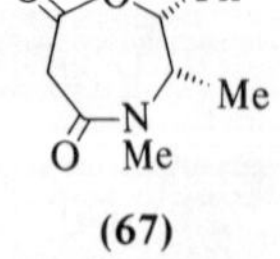

(**67**) (**68**)

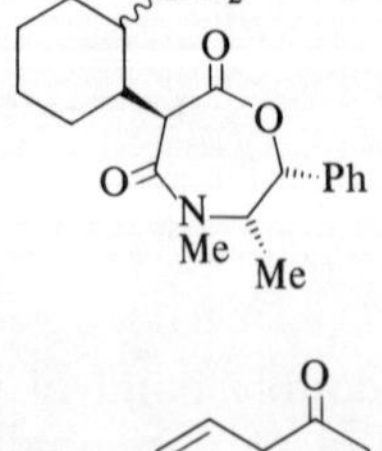

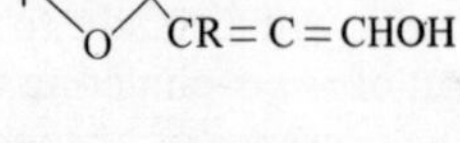

(**69**) 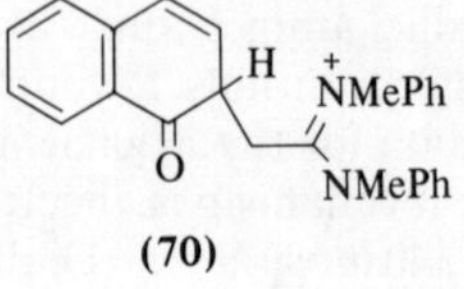 (**70**) 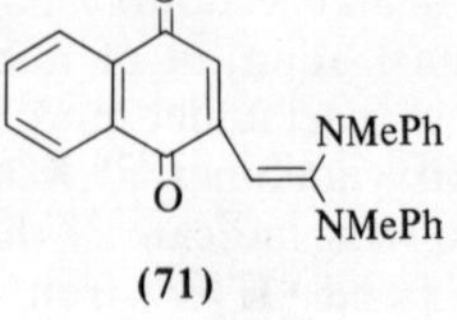 (**71**)

Michael additions to α,β-unsaturated carbonyl compounds thus allowing the direct reaction of relatively unreactive monoketones;[275] the $Si(OR)_4$ is believed to have two rôles in the process: generation of the base at the surface of CsF, and trapping the enolate to give a silyl enol ether which immediately reacts *in situ*.

References

[1] Ishikawa, N., Inukai, K., and Muramatsu, H., *Kagaku Sosetsu*, **27**, 135 (1980); *Chem. Abs.*, **94**, 3284 (1981).

[2] Trofimov, B. A., *Usp. Khim.*, **50**, 248 (1981); *Chem. Abs.*, **94**, 207925 (1981).

[3] Tomioka, K., and Koga, K., *Kagaku No Ryoiki*, **34**, 920 (1980); *Chem. Abs.*, **94**, 138728 (1981).

[4] Blake, P., *Chem. Ketenes, Allenes Relat. Compd.*, **1**, 309 (1980).

[5] Caramella, P., Rondan, N. G., Paddon-Row, M. N., and Houk, K. N., *J. Am. Chem. Soc.*, **103**, 2438 (1981).

[6] Rondan, N. G., Paddon-Row, M. N., Caramella, P., and Houk, K. N., *J. Am. Chem. Soc.*, **103**, 2436 (1981).

[7] Melloni, G., Modena, G., and Tonellato, U., *Acc. Chem. Res.*, **14**, 227 (1981).

[8] Smit, V. A., *Izv. Sib. Otd. Akad. Nauk SSSR, Ser. Khim. Nauk*, **1980**, 128; *Chem. Abs.*, **93**, 237977 (1980).

[9] Vyunov, K. A., and Ginak, A. I., *Usp. Khim.*, **50**, 273 (1981); *Chem. Abs.*, **94**, 207926 (1981).

[10] Rozen, S., Lerman, O., and Kol, M., *J. Chem. Soc., Chem. Commun.*, **1981**, 443.

[11] Stavber, S., and Zupan, M., *J. Chem. Soc., Chem. Commun.*, **1981**, 795.

[12] Boguslavskaya, L. S., Chuvatkin, N. N., Panteleeva, I. Yu., Ternovskoi, L. A., and Krom, E. N., *Zh. Org. Khim.*, **16**, 2525 (1980); *Chem. Abs.*, **94**, 138858 (1981).

[13] Dmitrieva, O. M., Vyunov, K. A., Ginak, A. I., and Sochilin, E. G., *Zh. Org. Khim.*, **17**, 58 (1981); *Chem. Abs.*, **95**, 23748 (1981).

[14] Dmitrieva, O. M., Vyunov, K. A., Ginak, A. I., and Sochilin, E. G., *Zh. Org. Khim.*, **17**, 63 (1981); *Chem. Abs.*, **95**, 23749 (1981).

[15] Larionova, L. A., Vyunov, K. A., and Ginak, A. I., *Zh. Org. Khim*, **17**, 294 (1981); *Chem. Abs.*, **95**, 41759 (1981).

[16] Colburn, C. B., Hill, W. E., and Verma, R. D., *J. Fluorine Chem.*, **17**, 75 (1981).

[17] Grigorova, T. N., and Komendantov, M. I., *Zh. Org. Khim.*, **17**, 317 (1981); *Chem. Abs.*, **95**, 6355 (1981).

[18] Kochanski, E., *Quantum Theory Chem. React.*, **2**, 177 (1981); *Chem. Abs.*, **95**, 42052 (1981).

[19] Misono, M., Takeyasu, H., and Yoneda, Y., *Nippon Kagaku Kaishi*, **1980**, 1844; *Chem. Abs.*, **94**, 102704 (1981).

[20] Melnikova, N. B., Yasman, Y. B., Buguslavskaya, L. S., and Kartashov, V. R., *Zh. Org. Khim.*, **16**, 2026 (1980); *Chem. Abs.*, **94**, 208009 (1981).

[21] Heasley, V. L., Arnold, S., Carter, T. L., Yaeger, D. B., Gipe, B. T., and Shellhamer, D. F., *J. Org. Chem.*, **45**, 5150 (1980).

[22] Angelini, G., and Speranza, M., *J. Am. Chem. Soc.*, **103**, 3792 (1981).

[23] Cano, M. C. Gomez Contreras, F., and Sanz, A. M., *J. Heterocycl. Chem.*, **17**, 1265 (1980).

[24] Akhmetkarimov, K. A., and Ayapbergenov, K. A., *Deposited Doc.*, **1979**, VINITI 1807; *Chem. Abs.*, **93**, 167185 (1980).

[25] Bienvenue-Goetz, E., and Dubois, J.-E., *J. Am. Chem. Soc.*, **103**, 5388 (1981).

[26] Heasley, G. E., Smith, D. A., Smith, J. N., Heasley, V. L., and Shellhamer, D. F., *J. Org. Chem.*, **45**, 5206 (1980).

[27] Gassman, P. G., and Gennick, I., *J. Org. Chem.*, **45**, 5211 (1980).

[28] Singh, N., and Verma, S. M., *Indian J. Chem.*, **19B**, 290 (1980).

[29] Blom, N. F., Edwards, D. M. F., Field, J. S., and Michael, J. P., *J. Chem. Soc., Chem. Commun.*, **1980**, 1240.

[30] Bellucci, G., Berti, G., Bianchini, R., Ingrosso, G., and Ambrosetti, R., *J. Am. Chem. Soc.*, **102**, 7480 (1980).

[31] Bellucci, G., Besti, G., Bianchini, R., Ingrosso, G., and Yates, K., *J. Org. Chem.*, **46**, 2315 (1981).

[32] Midland, M. M., and Halterman, R. L., *J. Org. Chem.*, **46**, 1227 (1981).

[33] Kocovsky, P., and Turecek, F., *Tetrahedron Lett.*, **22**, 2699 (1981).

[34] Kocovsky, P., and Cerny, V., *Collect. Czech. Chem. Commun.*, **45**, 3023 (1980).

[35] Kocovsky, P., and Cerny, V., *Collect. Czech. Chem. Commun.*, **45**, 3030 (1980).

[36] Heasley, V. L., Shellhamer, D. F., Heasley, L. E., and Yaeger, D. B., *J. Org. Chem.*, **45**, 4649 (1980).
[37] Dubois, J.-E., El-Alaoui, M., and Toullec, J., *J. Am. Chem. Soc.*, **103**, 5393 (1981).
[38] von Rohrscheidt, C., and Kohn, H., *Tetrahedron Lett.*, **1979**, 215.
[39] Kohn, H., Bean, M. B., von Rohrscheidt, C., Willcott, M. R., and Warnhoff, E. W., *Tetrahedron*, **37**, 3195 (1981).
[40] Cambie, R. C., Chambers, D., Rutledge, P. S., Woodgate, P. D., and Woodgate, S. D., *J. Chem. Soc., Perkin Trans. 1*, **1981**, 33.
[41] Smit, V. A., Zefirov, N. S., and Bodrikov, I. V., *Org. Sulphur Chem., Invited Lect. Int. Symp., 9th*, **1980**, 159; *Chem. Abs.*, **95**, 149330 (1981).
[42] Chang, C.-C., Liu, H.-H., and Liu, J.-C., *K'o Hsueh T'ung Pao*, **25**, 539 (1980); *Chem. Abs.*, **94**, 14921 (1981).
[43] Paciute, N., Rasteikiene, L., and Knunyants, I. L., *Izv. Akad. Nauk SSSR, Ser. Khim.*, **1980**, 2363; *Chem. Abs.*, **94**, 29780 (1981).
[44] Andreeva, L. A., *Deposited Doc., VINITI*, **3782**, 227 (1979); *Chem. Abs.*, **94**, 83481 (1981).
[45] Akhmedova, R. S., *Deposited Doc., VINITI*, **3782**, 180 (1979); *Chem. Abs.*, **94**, 102396 (1981).
[46] Kartashov, V. R., Akimkina, N. F., Skorobogatova, E. V., and Trub, E. P., *Kinet. Katal.*, **21**, 1330 (1980); *Chem. Abs.*, **94**, 46422 (1981).
[47] Bigg, M. G., Roberts, S. M., and Suschitzky, H., *J. Chem. Soc., Perkin Trans. 1*, **1981**, 926.
[48] Garratt, D. G., and Beaulieu, P. L., *Can. J. Chem.*, **58**, 2737 (1980).
[49] Majewski, J. M., and Zakrzewski, J., *Tetrahedron Lett.*, **22**, 3659 (1981).
[50] Przybylska, M., and Garratt, D. G., *Can. J. Chem.*, **59**, 658 (1981).
[51] Garratt, D. G., Beaulieu, P. L., and Morisset, V. M., *Can. J. Chem.*, **58**, 1021 (1980).
[52] Angelov, C. M., and Vachkov, K. V., *Tetrahedron Lett.*, **22**, 2517 (1981).
[53] Kutyrev, G. A., Vinokurov, A. I., Istomin, B. I., Cherkasov, R. A., and Pudovik, A. N., *Zh. Obshch. Khim.*, **51**, 1003 (1981); *Chem. Abs.*, **95**, 96502 (1981).
[54] Morishita, T., Furukawa, N., and Oae, S., *Tetrahedron*, **37**, 2539 (1981).
[55] Bodrikov, I. V., Chumakov, L. V., Pryadilova, A. N., Zefirov, N. S., and Smit, V. A., *Dokl. Akad. Nauk SSSR*, **251**, 1402 (1980); *Chem. Abs.*, **93**, 203561 (1980).
[56] Cambie, R. C., Larsen, D. S., Rutledge, P. S., and Woodgate, P. D., *J. Chem. Soc., Perkin Trans. 1*, **1981**, 58.
[57] Garratt, D. G., Beaulieu, P. L., and Morisset, V. M., *Can. J. Chem.*, **59**, 927 (1981).
[58] Garratt, D. G., Beaulieu, P. L., Morisset, V. M., and Ujjainwalla, M., *Can. J. Chem.*, **58**, 2745 (1980).
[59] Nicolaou, K. C., *Tetrahedron*, **37**, 4097 (1981).
[60] Rouessac, F., and Zamarlik, H., *Tetrahedron Lett.*, **22**, 2643 (1981).
[61] Rouessac, A., Rouessac, F., and Zamarlik, H., *Tetrahedron Lett.*, **22**, 2641 (1981).
[62] Back, T. G., and Collins, S., *J. Org. Chem.*, **46**, 3249 (1981).
[63] Parr, W. J. E., and Crafts, R. C., *Tetrahedron Lett.*, **22**, 1371 (1981).
[64] Toshimitsu, A., Aoai, T., Uemura, S., and Okano, M., *J. Org. Chem.*, **46**, 3021 (1981).
[65] Ismailov, A. G., Gashimov, G. A., and Rustamov, M. A., *Azerb. Khim. Zh.*, **1979**, 58; *Chem. Abs.*, **94**, 46653 (1981).
[66] Rostovshchikova, T. N., *Deposited Doc., VINITI*, **3784**, 502 (1979); *Chem. Abs.*, **94**, 138869 (1981).
[67] Sergeev, G. B., Leenson, I. A., and Tyurina, L. A., *Kinet. Katal.*, **21**, 1137 (1980); *Chem. Abs.*, **94**, 29782 (1981).
[68] Sergeev, G. B., Leenson, I. A., and Tyurina, L. A., *Kinet. Katal.*, **21**, 1130 (1980); *Chem. Abs.*, **94**, 64769 (1981).
[69] Gasparyan, L. A., Manukyan, T. K., Kazazyan, S. S., Galoyan, M. G., Oganesyan, I. I., Tarkhanyan, A. S., and Karapetyan, N. G., *Arm. Khim. Zh.*, **34**, 683 (1981); *Chem. Abs.*, **95**, 168070 (1981).
[70] Karapetyan, N. G., Tarkhanyan, A. S., Gasparyan, L. A., and Galoyan, M. G., *Arm. Khim. Zh.*, **34**, 707 (1981); *Chem. Abs.*, **95**, 168071 (1981).
[71] Balyatinskaya, L. N., Milyaev, Yu. F., Khorishko, S. A., and Zelenova, G. P., *Koord. Khim.*, **6**, 724 (1980); *Chem. Abs.*, **93**, 167147 (1980).
[72] Mikhailova, T. S., Zakharov, V. I., Ignatev, V. M., Ionin, B. I., and Petrov, A. A., *Zh. Obshch. Khim.*, **50**, 1690 (1980); *Chem. Abs.*, **93**, 238167 (1980).
[73] Daniels, R. G., and Paquette, L. A., *J. Org. Chem.*, **46**, 2901 (1981).
[74] Bobyleva, A. A., Belikova, N. A., Dzhigirkhanova, A. V., and Plate, A. F., *Zh. Org. Khim.*, **16**, 915 (1980); *Chem. Abs.*, **93**, 167159 (1980).
[75] Bobyleva, A. A., Belikova, N. A., Kalinichenko, A. N., Baryshnikov, A. T., Dubitskaya, N. F., Pehk, T., Lippmaa, E., and Plate, A. F., *Zh. Org. Khim.*, **16**, 1645 (1980); *Chem. Abs.*, **94**, 29747 (1981).
[76] Montheard, J.-P., Camps, M., and Benzaid, A., *Chem. Lett.*, **1981**, 523.
[77] Makhin, A. A., Frolov, A. F., and Aronovich, K. A., *Kinet. Katal.*, **22**, 344 (1981); *Chem. Abs.*, **95**,

79557 (1981).
[78] Devekki, A. V., Koshelev, Yu. N., and Mushenko, D. V., *Zh. Org. Khim.*, **16,** 2518 (1980); *Chem. Abs.*, **94,** 102411 (1981).
[79] Uemura, S., Miyoshi, H., Tabata, A., and Okano, M., *Tetrahedron*, **37,** 291 (1981).
[80] Nordlander, J. E., Haky, J. E., and Landino, J. P., *J. Am. Chem. Soc.*, **102,** 7487 (1980).
[81] Ismailov, A. G., Gashimov, G. A., and Rustamov, M. A., *Dokl. Akad. Nauk Az. SSR*, **36,** 45 (1980); *Chem. Abs.*, **94,** 102712 (1981).
[82] Maccarone, E., Mamo, A., Perrini, G., and Torre, M., *J. Chem. Soc., Perkin Trans. 2*, **1981,** 324.
[83] Kametani, T., *Tetrahedron Lett.*, **22,** 3653 (1981).
[84] Bernasconi, C. F., Carré, D. J., and Kanavarioti, A., *J. Am. Chem. Soc.*, **103,** 4850 (1981).
[85] Chang, C.-C., Liu, H.-H., and Liu, L.-C., *Ko Hsueh Tung Pao*, **25,** 886 (1980); *Chem. Abs.*, **94,** 102607 (1981).
[86] Kiselev, S. S., Polievktov, M. K., and Granik, V. G., *Khim. Geterotsikl. Soedin.*, **1981,** 352; *Chem. Abs.*, **95,** 6071 (1981).
[87] Sordo, T., Arumi, M., and Bertran, J., *J. Chem. Soc., Perkin Trans. 2*, **1981,** 708.
[88] Pack, M.-K., *Haksul Yonguchi-Chungnam Tae Hakkyo, Chayon Kwahak Yonguso*, **6,** 85 (1979); *Chem. Abs.*, **94,** 46415 (1981).
[89] Cramer, P., and Tidwell, T. T., *J. Org. Chem.*, **46,** 2683 (1981).
[90] Vavilova, A. N., Trofimov, B. A., Volkov, A. N., and Keiko, V. V., *Zh. Org. Khim.*, **17,** 926 (1981); *Chem. Abs.*, **95,** 114363 (1981).
[91] Matsuo, K., Urabe, K., and Izumi, Y., *Chem. Lett.*, **1981,** 1315.
[92] Alper, H., and Laycock, D. E., *Tetrahedron Lett.*, **22,** 33 (1981).
[93] Gambarotta, S., and Alper, H., *J. Org. Chem.*, **46,** 2142 (1981).
[94] Mayr, H., and Klein, H., *J. Org. Chem.*, **46,** 4097 (1981).
[95] Mayr, H., *Angew. Chem. Int. Ed.*, **20,** 184 (1981).
[96] Stetter, H., and Bender, H.-J., *Chem. Ber.*, **114,** 1226 (1981).
[97] Stetter, H., and Jonas, F., *Chem. Ber.*, **114,** 564 (1981).
[98] Stetter, H., and Mertens, A., *Chem. Ber.*, **114,** 2479 (1981).
[99] Yamanaka, T., *Kagaku Kogyo*, **32,** 116 (1981); *Chem. Abs.*, **94,** 173736 (1981).
[100] Haelg, P., Consiglio, G., and Pino, P., *Helv. Chim. Acta*, **64,** 1865 (1981).
[101] Polievka, M., Uhlar, L., and Macho, V., *Petrochemia*, **20,** 33 (1980); *Chem. Abs.*, **93,** 238160 (1980).
[102] Hobbs, C. F., and Knowles, W. S., *J. Org. Chem.*, **46,** 4422 (1981).
[103] Garst, M. E., and Lukton, D., *J. Org. Chem.*, **46,** 4433 (1981).
[104] Nalesnik, T. E., Fish, J. G., Horgan, S. W., and Orchin, M., *J. Org. Chem.*, **46,** 1987 (1981).
[105] Pittman, C. U., and Wilemon, G. M., *J. Org. Chem.*, **46,** 1901 (1981).
[106] Cavinato, G., and Toniolo, L., *Congr. Naz. Chim. Inorg. [ATTI], 12th*, **1979,** 430; *Chem. Abs.*, **95,** 41799 (1981).
[107] Katsnelson, M. G., and Kashina, V. V., *Zh. Prikl. Khim.*, **54,** 1201 (1981); *Chem. Abs.*, **95,** 114367 (1981).
[108] Gropen, O., and Haaland, A., *Acta Chem. Scand.*, **35A,** 305 (1981).
[109] Rand, C. L., Van Horn, D. E., Moore, M. W., and Negishi, E.-i., *J. Org. Chem.*, **46,** 4096 (1981).
[110] Yoshida, T., and Negishi, E.-i., *J. Am. Chem. Soc.*, **103,** 4985 (1981).
[111] Brevko, V. S., Skvortsov, N. K., and Reikhsfeld, V. O., *Zh. Obshch. Khim.*, **51,** 411 (1981); *Chem. Abs.*, **95,** 6352 (1981).
[112] Yamamoto, K., Kiso, Y., Ito, R., Tamao, K., and Kumanda, M., *J. Organomet. Chem.*, **210,** 9 (1981).
[113] Marciniec, B., and Gulinski, J., *J. Mol. Catal.*, **10,** 123 (1980).
[114] Zaslavskaya, T. N., Filippov, N. A., and Reikhsfeld, V. O., *Zh. Obshch. Khim.*, **51,** 107 (1981); *Chem. Abs.*, **94,** 191208 (1981).
[115] Magomedov, G. K., Druzhkova, G. V., Syrkin, V. G., and Shkolnik, O. V., *Koord. Khim.*, **6,** 767 (1980); *Chem. Abs.*, **93,** 167148 (1980).
[116] Magomedov, G. K., and Shkolnik, O. V., *Zh. Obshch. Khim.*, **50,** 1103 (1980); *Chem. Abs.*, **93,** 167153 (1980).
[117] Zaslavskaya, T. N., Reikhsfeld, V. O., Filippov, N. A., and Shornik, N. A., *Zh. Obshch. Khim.*, **50,** 2478 (1980); *Chem. Abs.*, **94,** 83244 (1981).
[118] Fleming, I., and Langley, J. A., *J. Chem. Soc., Perkin Trans. 1*, **1981,** 1421.
[119] Brown, H. C., Prabhakar, K. J., and Mandal, A. K., *Tetrahedron*, **37,** 3547 (1981).
[120] Graham, G. D., Freilich, S. C., and Lipscomb, W. N., *J. Am. Chem. Soc.*, **103,** 2546 (1981).
[121] Datta, M. K., Dasgupta, S., and Datta, R., *Indian J. Chem.*, **19A,** 611 (1980); *Chem. Abs.*, **93,** 238589 (1980).
[122] Contreras, R., and Wrackmeyer, B., *J. Organomet. Chem.*, **205,** 15 (1981).

[123] Contreras, R., and Wrackmeyer, B., *Z. Naturforsch., B: Anorg. Chem., Org. Chem.*, **35**, 1229 (1980).
[124] Brown, H. C., and Chen, J. C., *J. Org. Chem.*, **46**, 3978 (1981).
[125] Vishwakarma, L. C., and Fry, A., *J. Org. Chem.*, **45**, 5306 (1980).
[126] Wang, K. K., and Brown, H. C., *J. Org. Chem.*, **45**, 5303 (1980).
[127] Jadhav, P. K., and Brown, H. C., *J. Org. Chem.*, **46**, 2988 (1981).
[128] Anez, M., Uribe, G., Mendoza, L., and Contreras, R., *Synthesis*, **1981**, 214.
[129] Fukuzumi, S., and Kochi, J. K., *J. Am. Chem. Soc.*, **103**, 2783 (1981).
[130] Giffard, M., and Cousseau, J., *J. Organomet. Chem.*, **201**, Cl (1980).
[131] Bassetti, M., Floris, B., and Illuminati, G., *J. Organomet. Chem.*, **202**, 351 (1980).
[132] Lewis, A., and Azoro, J., *J. Org. Chem.*, **46**, 1764 (1981).
[133] Lermontov, S. A., Belikova, N. A., Shornyakova, T. G., Pehk, T., Lippmaa, E., and Plate, A., *Zh. Org. Khim.*, **16**, 2322 (1980); *Chem. Abs.*, **94**, 83233 (1981).
[134] Kartashov, V. R., Povelikina, L. N., Barkhash, V. A., and Bodrikov, I. V., *Dokl. Akad. Nauk SSSR*, **252**, 883 (1980); *Chem. Abs.*, **94**, 15033 (1981).
[135] Brown, H. C., Geoghegan, P. J., and Kurek, J. T., *J. Org. Chem.*, **46**, 3810 (1981).
[136] Link, C. M., Jansen, D. K., and Sukenik, C. N., *J. Am. Chem. Soc.*, **102**, 7798 (1980).
[137] Brown, H. C., and Lynch, G. J., *J. Org. Chem.*, **46**, 930 (1981).
[138] Brown, H. C., and Lynch, G. J., *J. Org. Chem.*, **46**, 531 (1981).
[139] Barluenga, J., Alonso-Cires, L., and Asensio, G., *Synthesis*, **1981**, 376.
[140] Barluenga, J., Aznar, F., and Liz, R., *J. Chem. Soc., Chem. Commun.*, **1981**, 1181.
[141] Barluenga, J., Jimenez, C., Najera, C., and Yus, M., *Synthesis*, **1981**, 201.
[142] Saitoh, Y., Moriyama, Y., Hirota, H., Takahashi, T., and Khuong-Huu, Q., *Bull. Chem. Soc. Jpn*, **54**, 488 (1981).
[143] Barluenga, J., Jimenez, C., Najera, C., and Yus, M., *J. Chem. Soc., Chem. Commun.*, **1981**, 670.
[144] Barluenga, J., Jiminez, C., Najera, C., and Yus, M., *J. Chem. Soc., Chem. Commun.*, **1981**, 1178.
[145] Harding, K. E., and Burks, S. R., *J. Org. Chem.*, **46**, 3920 (1981).
[146] Perie, J. J., Laval, J. P., Roussel, J., and Lattes, A., *Tetrahedron*, **28**, 675 (1972).
[147] Li, T.-T., Lesko, P., Ellison, R. H., Subramanian, N., and Fried, J. H., *J. Org. Chem.*, **46**, 111 (1981).
[148] Nishida, A., Shibasaki, M., and Ikegami, S., *Tetrahedron Lett.*, **22**, 4819 (1981).
[149] Paquette, L. A., Klinger, F., and Hertel, L. W., *J. Org. Chem.*, **46**, 4403 (1981).
[150] Ishino, Y., Nishiguchi, I., Nakao, S., and Hirashima, T., *Chem. Lett.*, **1981**, 641.
[151] Bäckvall, J.-E., and Engman, L., *Tetrahedron Lett.*, **22**, 1919 (1981).
[152] Sutherland, J. K., *Chem. Soc. Rev.*, **9**, 265 (1980).
[153] Yamada, Y., Nakamura, S., Iguchi, K., and Hosaka, K., *Tetrahedron Lett.*, **22**, 1355 (1981).
[154] Kato, T., Ishii, K., Ichinose, I., Nakai, Y., and Kumagai, T., *J. Chem. Soc., Chem. Commun.*, **1980**, 1106.
[155] Mazzocchi, P. H., and Harrison, A. M., *Isr. J. Chem.*, **21**, 164 (1981).
[156] Ozbal, H., and Zajac, W. W., *J. Org. Chem.*, **46**, 3082 (1981).
[157] Marcincal-Lefebvre, A., and Depreux, P., *J. Chem. Res. (S)*, **1981**, 60.
[158] Hasegawa, H., Nishibayashi, H., and Takatsudo, S., *Kenkyu Hokoku — Asahi Garasu Kogyo Gijutsu Shoreikai*, **37**, 305 (1980); *Chem. Abs.*, **95**, 23778 (1981).
[159] Hasegawa, H., Nishibayashi, H., and Takatsudo, S. *Waseda Daigaku Rikogaku Kenkyusho Hokuko*, **1980**, 56; *Chem. Abs.*, **95**, 149472 (1981).
[160] Manninen, K., and Parhi, S., *Acta Chem. Scand.*, **35B**, 45 (1981).
[161] Koser, G. F., Rebrovic, L., and Wettach, R. H., *J. Org. Chem.*, **46**, 4324 (1981).
[162] Chang, T. C. T., and Rosenblum, M., *J. Org. Chem.*, **46**, 4626 (1981).
[163] Sen, A., and Lai, T.-W., *J. Am. Chem. Soc.*, **103**, 4627 (1981).
[164] Yasuda, H., Kajihara, Y., Nagasuna, K., Mashima, K., and Nakamura, A., *Chem. Lett.*, **1981**, 719.
[165] Capozzi, G., Caristi, C., Gattuso, M., and Stagno d'Alcontres, G., *Tetrahedron Lett.*, **22**, 3325 (1981).
[166] Toma, S., *Chem. Listy*, **74**, 589 (1980).
[167] Stefanovskii, Y., and Viteva, L., *Organometalliques Fonct. Ambidents, Recl. Commun., Colloq. Fr.-Bulg.*, **1980**, 100: *Chem. Abs.*, **95**, 60733 (1981).
[168] Toma, S., *Chem. Listy.*, **75**, 1 (1981); *Chem. Abs.*, **94**, 173714 (1981).
[169] Rappoport, Z., *Acc. Chem. Res.*, **14**, 7 (1981).
[170] Eisenstein, O., and Hoffmann, R., *J. Am. Chem. Soc.*, **103**, 4308 (1981).
[171] Rover, J., *Quantum Theory Chem. React.*, **2**, 169 (1981); *Chem. Abs.*, **94**, 207931 (1981).
[172] Svoboda, J., Paleta, O., and Dedek, V., *Collect. Czech. Chem. Commun.*, **46**, 1272 (1981).
[173] Svoboda, J., Paleta, O., and Dedek, V., *Collect. Czech. Chem. Commun.*, **46**, 1495 (1981).
[174] Shainyan, B. A., *Zh. Org. Khim.*, **16**, 1113 (1980); *Chem. Abs.*, **93**, 167423 (1980).
[175] Agami, C., Fadlallah, M., and Levisalles, J., *Tetrahedron*, **37**, 903 (1981).

[176] Agami, C., Fadlallah, M., and Levisalles, J., *Tetrahedron*, **37**, 909 (1981).
[177] Agami, C., Fadlallah, M., Levisalles, J., and Cayzergues, P., *Tetrahedron*, **37**, 3723 (1981).
[178] Bäckvall, J.-E., and Andell, O. S., *J. Chem. Soc., Chem. Commun.*, **1981**, 1098.
[179] Sidorov, E. O., Matern, A. I., and Chupakhim, O. N., *Zh. Org. Khim.*, **17**, 418 (1981); *Chem. Abs.*, **95**, 6035 (1981).
[180] Koch, H. F., Koch, J. G., Donovan, D. B., Toczko, A. G., and Kielbania, A.-J., *J. Am. Chem. Soc.*, **103**, 5417 (1981).
[181] Guba, G., Koudelka, L., and Kostkova, V., *Petrochemia*, **20**, 114 (1980); *Chem. Abs.*, **95**, 114323 (1981).
[182] Tombo, G. M. R., Pfund, R. A., and Ganter, C., *Helv. Chim. Acta*, **64**, 813 (1981).
[183] Lassalvy, C., Petrus, C., and Petrus, F., *Can. J. Chem.*, **59**, 175 (1981).
[184] Shainyan, B. A., and Mirskova, A. N., *Zh. Org. Khim.*, **16**, 2569 (1980).
[185] Tanchuk, Yu. V., Gunko, V. M., Roev, L. M., and Komienko, A. A., *Teor. Eksp. Khim.*, **16**, 609 (1980); *Chem. Abs.*, **94**, 46593 (1981).
[186] Trofimov, B. A., Gusarova, N. K., Efremova, G. G., Amosova, S. V., Kletsko, F. P., Vlasova, N. N., and Voronkov, M. G., *Zh. Org. Khim.*, **16**, 2538 (1980).
[187] Wilson, S. R., *Bioorg. Chem.*, **9**, 212 (1980).
[188] Szmant, H. H., and Baeza, H. J., *J. Org. Chem.*, **45**, 4902 (1980).
[189] Kobayashi, N., and Iwai, K., *J. Org. Chem.*, **46**, 1823 (1981).
[190] Mukaiyama, T., Ikegawa, A., and Suzuki, K., *Chem. Lett.*, **1981**, 165.
[191] Hiemstra, H., and Wynberg, H., *J. Am. Chem. Soc.*, **103**, 417 (1981).
[192] Colonna, S., Re, A., and Wynberg, H., *J. Chem. Soc., Perkin Trans. 1*, **1981**, 939.
[193] Piepers, O., and Kellogg, R. M., *J. Chem. Soc., Chem. Commun.*, **1980**, 1147.
[194] Yasuda, A., Takahashi, M., and Takaya, H., *Tetrahedron Lett.*, **22**, 2413 (1981).
[195] Fahey, R. C., Myers, P. A., and Di Stefano, D. L., *Bioorg. Chem.*, **9**, 293 (1980).
[196] Ramachandran, J., Ramadas, S. R., and Pillai, C. N., *Org. Prep. Proced. Int.*, **13**, 71 (1981).
[197] Cardillo, G., Orena, M., Porzi, G., and Sandri, S., *J. Chem. Soc., Chem. Commun.*, **1981**, 465.
[198] Matsuda, I., Akiyama, K., Kobayashi, S., Katada, T., and Mizuta, M., *Nippon Kagaku Kaishi*, **1981**, 405; *Chem. Abs.*, **95**, 6058 (1981).
[199] Bell, R., Cottam, P. D., Davies, J., and Jones, D. N., *J. Chem. Soc., Perkin Trans. 1*, **1981**, 2106.
[200] Cere, V., Paolucci, C., Pollicino, S., Sandri, E., and Fava, A., *J. Chem. Soc., Chem. Commun.*, **1981**, 764.
[201] Richey, H. G., and Wilkins, C. W., *J. Org. Chem.*, **45**, 5027 (1980).
[202] Richey, H. G., and Bension, R. M., *J. Org. Chem.*, **45**, 5036 (1980).
[203] Richey, H. G., Wilkins, C. W., and Bension, R. M., *J. Org. Chem.*, **45**, 5042 (1980).
[204] Richey, H. G., Moses, L. M., Domalski, M. S., Erickson, W. F., and Heyn, A. S., *J. Org. Chem.*, **46**, 3773 (1981).
[205] Richey, H. G., and Domalski, M. S., *J. Org. Chem.*, **46**, 3780 (1981).
[206] Sato, F., Ishikawa, H., and Sato, M., *Tetrahedron Lett.*, **22**, 85 (1981).
[207] Jalander, L., *Acta Chem. Scand.*, **35B**, 419 (1981).
[208] Taylor, R. T., and Galloway, J. G., *J. Organomet. Chem.*, **220**, 295 (1981).
[209] Agawa, T., Yoshida, Y., Komatsu, M., and Ohshiro, Y., *J. Chem. Soc., Perkin Trans. 1*, **1981**, 751.
[210] Binns, M. R., and Haynes, R. K., *J. Org. Chem.*, **46**, 3790 (1981).
[211] Mesnard, D., Charpentier, J.-P., and Miginiac, L., *J. Organomet. Chem.*, **214**, 15 (1981).
[212] Mesnard, D., Charpentier, J.-P., and Miginiac, L., *J. Organomet. Chem.*, **214**, 23 (1981).
[213] Mesnard, D., Charpentier, J.-P., and Miginiac, L., *J. Organomet. Chem.*, **214**, 135 (1981).
[214] Mpango, G. B., Mahalanabis, K. K., Mahdavi-Damghani, Z., and Snieckus, V., *Tetrahedron Lett.*, **21**, 4823 (1980).
[215] Mpango, G. B., and Snieckus, V., *Tetrahedron Lett.*, **21**, 4827 (1980).
[216] Meyer, R., Gorrichon, L., Bertrand, J., and Maroni, P., *Organometalliques Fonct. Ambidents, Recl. Commun., Colloq. Fr.-Bulg.*, **1980**, 224; *Chem. Abs.*, **95**, 114352 (1981).
[217] Ignatova, E., and Pozharliev, J., *Organometalliques Fonct. Ambidents, Recl. Commun., Colloq. Fr.-Bulg.*, **1980**, 233; *Chem. Abs.*, **95**, 114650 (1981).
[218] Roux-Schmitt, M.-C., Seyden-Penne, J., Baddeley, G. V., and Wenkert, E., *Tetrahedron Lett.*, **22**, 2171. (1981).
[219] Roux, M. C., Wartski, L., and Seyden-Penne, J., *Tetrahedron*, **37**, 1927 (1981).
[220] Braun, M., and Esdar, M., *Chem. Ber.*, **114**, 2924 (1981).
[221] Pelter, A., and Rao, J. M., *Tetrahedron Lett.*, **22**, 797 (1981).
[222] Hirama, M., *Tetrahedron Lett.*, **22**, 1905 (1981).
[223] Colombo, L., Gennari, C., Resnati, G., and Scolastico, C., *Synthesis*, **1981**, 74.
[224] Seuron, N., Wartski, L., and Seyden-Penne, J., *Tetrahedron Lett.*, **22**, 2175 (1981).

225 Tamaru, Y., Kagotani, M., and Yoshida, Z., *Tetrahedron Lett.*, **22**, 3409 (1981).
226 Tigchelaar, M., Kleijn, H., Elsevier, C. J., Meijer, J., and Vermeer, P., *Tetrahedron Lett.*, **22**, 2237 (1981).
227 Krauss, S. R., and Smith, S. G., *J. Am. Chem. Soc.*, **103**, 141 (1981).
228 Liotta, D., Barnum, C. S., and Saindane, M., *J. Org. Chem.*, **46**, 4301 (1981).
229 Ziegler, F. E., and Cady, M. A., *J. Org. Chem.*, **46**, 122 (1981).
230 De Chirico, G., Fiandanese, V., Marchese, G., Naso, F., and Sciacovelli, O., *J. Chem. Soc., Chem. Commun.*, **1981**, 523.
231 Huché, M., Berlan, J., Pourcelot, G., and Cresson, P., *Tetrahedron Lett.*, **22**, 1329 (1981).
232 Posner, G. H., Mallamo, J. P., and Miura, K., *J. Am. Chem. Soc.*, **103**, 2886 (1981).
233 Ibuka, T., Minakata, H., Mitsui, Y., Kinoshita, K., and Kawami, Y., *J. Chem. Soc., Chem. Commun.*, **1980**, 1193.
234 Clive, D. L. J., Farina, V., and Beaulieu, P., *J. Chem. Soc., Chem. Commun.*, **1981**, 643.
235 Rahman, M. T., Saha, S. L., and Hansson, A.-T., *J. Organomet. Chem.*, **199**, 9 (1980).
236 Chuit, C., Foulon, J. P., and Normant, J. F., *Tetrahedron*, **37**, 1385 (1981).
237 Piers, E., Chong, J. M., and Morton, H. E., *Tetrahedron Lett.*, **22**, 4905 (1981).
238 Piers, E., and Morton, H. E., *J. Org. Chem.*, **45**, 4263 (1980).
239 Nishiyama, H., Sasaki, M., and Itoh, K., *Chem. Lett.*, **1981**, 905.
240 Ashby, E. C., Smith, R. S., and Goel, A. B., *J. Organomet. Chem.*, **215**, Cl (1981).
241 Ager, D. J., Fleming, I., and Patel, S. K., *J. Chem. Soc., Perkin Trans. 1*, **1981**, 2520.
242 Fleming, I., Newton, T. W., and Roessler, F., *J. Chem. Soc., Perkin Trans. 1*, **1981**, 2527.
243 Kleijn, H., Tigchelaar, M., Meijer, J., and Vermeer, P., *Recl. Trav. Chim. Pays Bas*, **100**, 337 (1981).
244 Stefanovsky, Y. N., and Viteva, L., *Monatsh. Chem.*, **111**, 1287 (1980).
245 Stefanovsky, Y. N.. and Viteva, L. Z., *Monatsh. Chem.*, **112**, 125 (1981).
246 Matsumoto, K., *Angew. Chem. Int. Ed.*, **20**, 770 (1981).
247 Takahashi, T., Hori, K., and Tsuji, J., *Tetrahedron Lett.*, **22**, 119 (1981).
248 Kozhevnikov, I. V., and Sharapova, S. T., *React. Kinet. Catal. Lett.*, **15**, 49 (1980); *Chem. Abs.*, **94**, 138872 (1981).
249 Cacchi, S., Misiti, D., and Palmieri, G., *Tetrahedron*, **37**, 2941 (1981).
250 Wegmann, H., and Steglich, W., *Chem. Ber.*, **114**, 2580 (1981).
251 Danishefsky, S., and Kahn, M., *Tetrahedron Lett.*, **22**, 485 (1981).
252 Mulzer, J., Brüntrup, G., Hartz, G., Kühl, U., Blaschek, U., and Böhrer, G., *Chem. Ber.*, **114**, 3701 (1981).
253 Ernst, J., Ottow, E., Recker, H.-G., and Winterfeldt, E., *Chem. Ber.*, **114**, 1907 (1981).
254 Cram, D. J., and Sogah, G. D. Y., *J. Chem. Soc., Chem. Commun.*, **1981**, 625.
255 Soto, J. L., Seoane, C., Mansilla, A. M., and Pardo, M. C., *Tetrahedron Lett.*, **22**, 4845 (1981).
256 Degny, E., Zard, S. Z., Pastor, R., and Cambon, A., *Tetrahedron Lett.*, **22**, 2169 (1981).
257 Fuks, R., and Van Den Bril, M., *Tetrahedron*, **37**, 2895 (1981).
258 Van Den Bril, M., and Fuks, R., *Tetrahedron*, **37**, 2905 (1981).
259 Kinatowski, S., Grabarkiewicz-Szczesna, J., and Kostecki, M., *Pol. J. Chem.*, **54**, 1679 (1980); *Chem. Abs.*, **95**, 23755 (1981).
260 Kobesheva, N. I., Frolkov, A. I., Kheruze, Y. I., Chistokletov, V. N., and Petrov., A. A., *Zh. Org. Khim.*, **17**, 260 (1981); *Chem. Abs.*, **94**, 208035 (1981).
261 Green, M., Sarhan, J. K. K., and Al-Najjar, I. M., *J. Chem. Soc., Dalton Trans.*, **1981**, 1565.
262 Bozell, J. J., and Hegedus, L. S., *J. Org. Chem.*, **46**, 2561 (1981).
263 El-Basil, S., and Said, M., *Indian J. Chem.*, **19B**, 1071 (1980); *Chem. Abs.*, **95**, 23976 (1981).
264 Rozeboom, M. D., Tegmo-Larsson, I.-M., and Houk, K. N., *J. Org. Chem.*, **46**, 2338 (1981).
265 Faustov, V. I., Shevelev, S. A., Anikin, N. A., Yufit, S. S., and Fainzilberg, A. A., *Izv. Akad. Nauk SSSR, Ser. Khim.*, **1980**, 2220; *Chem. Abs.*, **94**, 46668 (1981).
266 Seebach, D., and Golinski, J., *Helv. Chim. Acta*, **64**, 1413 (1981).
267 Takeda, T., Hoshiko, T., and Mukaiyama, T., *Chem. Lett.*, **1981**, 797.
268 Gorgues, A., Simon, A., Le Coq, A., and Corre, F., *Tetrahedron Lett.*, **22**, 625 (1981).
269 ApSimon, J., Srinivasan, V. S., L'Abbe, M. R., and Seguin, R., *Heterocycles*, **15**, 1079 (1981).
270 Cameron, D. W., Feutrill, G. I., and Thiel, J. M., *Aust. J. Chem.*, **34**, 453 (1981).
271 Cameron, D. W., Crossley, M. J., Feutrill, G. I., and Griffiths, P. G., *Aust. J. Chem.*, **31**, 1335 (1978).
272 Stephenson, M., and Mawby, R. J., *J. Chem. Soc., Dalton Trans.*, **1981**, 2112.
273 Klyuchinskii, S. A., Zubova, T. P., Zavgorodnii, V. S., and Petrov, A. A., *Zh. Obshch. Khim.*, **50**, 1752 (1980); *Chem. Abs.*, **93**, 238170 (1980).
274 Hegarty, A. F., *Acc. Chem. Res.*, **13**, 448 (1980).
275 Boyer, J., Corriu, R. J. P., Perz, R., and Reye, C., *J. Chem. Soc., Chem. Commun.*, **1981**, 122.

Organic Reaction Mechanisms 1981
Edited by A. C. Knipe and W. E. Watts

CHAPTER 13

Addition Reactions: Cycloaddition

I. R. DUNKIN

Department of Pure and Applied Chemistry, University of Strathclyde

Introduction 433
2 + 2-Cycloaddition 433
2 + 3-Cycloaddition 437
2 + 4-Cycloaddition 443
Miscellaneous Cycloadditions 450
References 453

Introduction

The trend, noted last year,[1] for work on cycloaddition reactions to be devoted more and more towards synthetic goals has been maintained. Syntheses involving cycloadditions can now exploit the accrued knowledge and understanding of site-, regio-, and stereo-selectivity which have come from the mechanistic studies of earlier years. This, perhaps, is a sign that these reactions have come of age, and reviews of steroid synthesis *via* intramolecular cycloadditions,[2] and cycloadditions of singlet oxygen[3] exemplify recent applications.

Amongst reports of mechanistic studies, there have been several with rather general aims. Huisgen, in presenting a PMO approach to a number of problems, has discussed the available means of discriminating between concerted and non-concerted reactions. These include the enormous rate increases brought about by extra cyano groups in some of the Diels–Alder reactions of cyanoethylene, and the effects of non-equivalent π-orbital extension in norbornene.[4] A semi-empirical MO method has been used to calculate potential energy surfaces for three cycloreversions: (*i*) cyclopentyl anion to allyl anion and ethylene, (*ii*) cyclohexene to butadiene and ethylene, and (*iii*) cyclohept-4-enyl cation to butadiene and allyl cation.[5] Major differences were found in the pathways of these reactions, and diradical character was found only for the fragmentation of the six-membered ring. Site-selectivity in the cycloaddition reactions of substituted ketenimines has been discussed in the light of MO assignments of their PE spectra,[6] and site-, regio-, and stereo-selectivities in the cycloadditions of 7-isopropylidenebenzonorbornadiene have been examined.[7]

2 + 2-Cycloaddition

Both semi-empirical and *ab initio* techniques have been used to explore portions of the excited singlet and triplet hypersurfaces for the cycloaddition of ethene.[8] The

2s + 2s excited singlet state surfaces correspond to the ideas derived from orbital symmetry rules, while the addition of triplet ethene to ground-state ethene resembles, in important respects, the addition of methyl radical to ethene. Semi-empirical calculations have also been carried out for the cycloreversions of cyclobutane and several substituted cyclobutanes.[9] The reaction pathway with the lowest activation energy is non-concerted, proceeding *via* a biradicaloid, non-zwitterionic intermediate. For the unsubstituted cyclobutane, the two transition states are approximately equal in energy, but for the substituted molecules, the breakdown of the intermediate is much slower than its formation. Photo-cycloadditions of coumarin and carbostyril to $MeOCH{=}CH_2$ have also been studied using the intermolecular orbital theory of Salem.[10] The results indicate that the bonds close in a concerted asymmetric manner.

Product distributions have been compared for the photo-dimerization of ethyl cinnamate both in the liquid phase and in glasses at 77–193 K.[11] Six of the eleven possible dimers were identified in the products, but the proportions of each vary widely with the conditions and are thought to reflect the configurations of reactant pairs in the glasses. Aryl vinyl ethers undergo a photo-induced 2 + 2-cycloaddition with alkyl vinyl ethers in the presence of an electron acceptor such as 1,4-dicyanobenzene.[12] Among the products, the *cis* head-to-head cycloadduct predominates, but no explanation of this selectivity has so far been proposed. The specific formation of an *anti* head-to-tail photo-dimer of 1,3-dimethyl-5-chlorouracil has also been reported.[13]

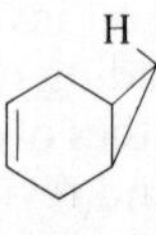

H N_2^+

(1)

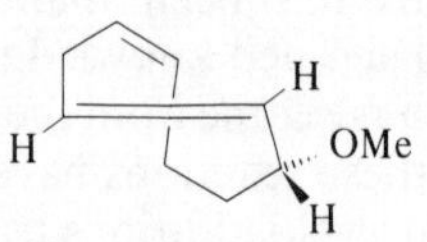

H H OMe H

(2)

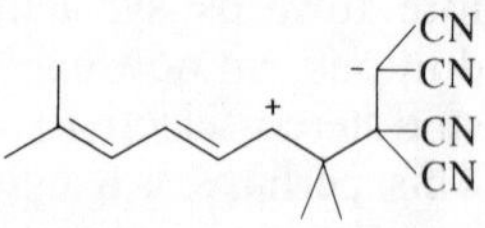

CN CN CN CN

(3)

The 1(Z),4(E)-cycloheptadiene (**2**) has been generated from (**1**) in weakly alkaline methanol. It undergoes *trans–cis* isomerization in competition with dimerization, and the main dimer has been shown by an X-ray structure determination to correspond to the ${}_{\pi}2_s + {}_{\pi}2_a$ cycloadduct of the *trans*-portions of two enantiomeric molecules of (**2**), with head-to-head connection.[14] Thermal cycloadditions involving tetracyanoethene often occur *via* zwitterionic intermediates.[1] The additional rôle of donor–acceptor complexes in these reactions has also been investigated in a number of laboratories. Rate constants for the cycloadditions of a series of compounds, 4-$RC_6H_4XCH{=}CH_2$ (R = MeO, Me, H, Cl, Br, NO_2; X = O, S, Se), with TCNE have been measured and good linear correlations have been found between log *k* and $1/\lambda$ for the corresponding charge-transfer band.[15] For the thio-ether series, rate constants were also determined in ten solvents. The selectivity of TCNE in a given solvent was defined as the slope of the line obtained by plotting log *k vs.* log *k* in PhCl. No correlation was found between the selectivity and solvent polarity parameters. On the other hand, a good linear correlation between the selectivity and solvation enthalpies of TCNE suggests that the selectivity is determined by the acceptor properties of TCNE in a given solvent.[16] Kinetic measurements for the high-pressure cycloaddition of enol ethers and TCNE, followed spectrophotometrically by the disappearance of the donor–acceptor complex, have led to the conclusion that the complex lies on the path to the zwitterion and final cycloadduct.[17] The reaction between TCNE and 2,7-dimethylocta-2,4,6-triene gives initially a blue charge-transfer complex and then a mixture of the terminal 2 + 2- and 2 + 4-cycloadducts.[18] The 2 + 2-adduct is favoured by polar solvents, implicating the zwitterion (**3**) as an intermediate. The ability of a diene portion of the molecule to assume a planar, or nearly planar, cisoid conformation must also be an important factor in the competition between 2 + 2- and 2 + 4-cycloaddition, however. Thermal 2 + 2-cycloadditions of vinyl cations with various cycloalkenes have been reported,[19] and the central bond in bicyclo[2.1.0]pentane adds to electron-deficient alkenes under the action of bis(acrylonitrile)nickel(0).[20]

The first observation of a 1,4-biradical in the Paterno–Büchi reaction has been made.[21] This was achieved for the photo-cycloaddition of benzophenone and dioxene, which was followed by absorption spectroscopy with a time-resolution of 25 ps. The biradical underwent unimolecular decay with a rate constant of $6.3 \times 10^8\ s^{-1}$, and studies both with and without quenchers led to the proposal of a mechanism involving initial formation of a charge-transfer contact ion-pair between triplet benzophenone and the alkene, followed by collapse to a triplet biradical, intersystem crossing, and finally ring-closure. The unimolecular photochemistry of a series of 2-allylcycloalkanones (*e.g.* **4**) has been investigated.[22] Oxetanes (*e.g.* **5** and **6**; R = H) are produced, with the cross-cycloadduct (e.g. **6**; R = H) predominating in each case. This is similar to the behaviour of hex-5-enyl radicals, which cyclize preferentially through a five-membered ring transition state. The sterically more hindered enone (**7**), in contrast, gives oxetane (**5**; R = Me) preferentially, thus resembling 5-methylhex-5-enyl radicals, which cyclize through a six-membered ring transition state. The reaction kinetics suggest that an exciplex between the excited singlet carbonyl group and the ground state of the olefinic bond is formed first. This subsequently reverts to the ground state or proceeds to a 1,4-biradical. Substituted *N*-methylphthalimides (**8**) undergo photochemical $2\sigma + 2\pi$ cycloaddition with

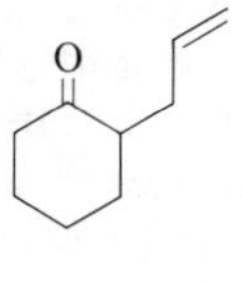

(**4**) 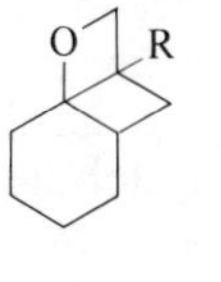 (**5**) 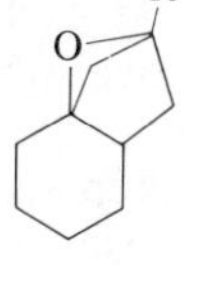 (**6**) 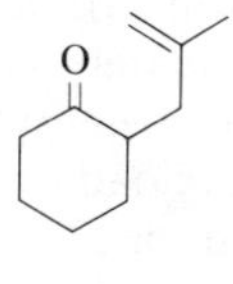 (**7**)

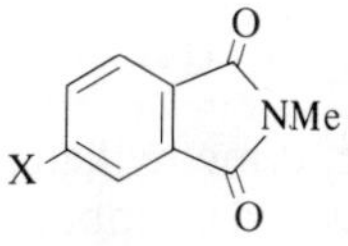

(**8**) 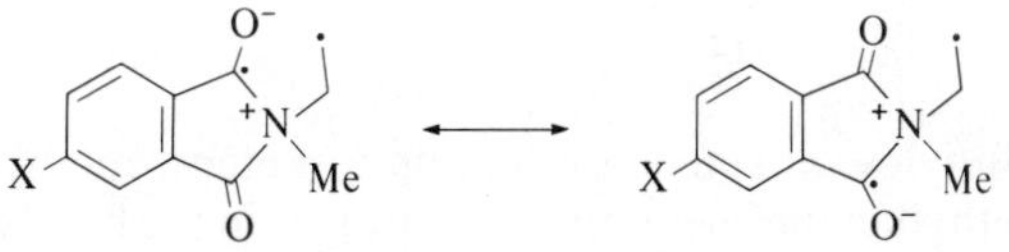 (**9**)

ethene and hex-1-ene, which may involve either of the amide C—N bonds.[23] The product distributions, but not the overall yields, depend greatly on the aryl substituent. Previous work had limited the possible mechanisms essentially to two, *viz.* a concerted $2\sigma + 2\pi$ reaction, or the intermediate formation of biradicals (**9**). The observed product distributions are inconsistent with results from radical-anion-trapping experiments, but consistent with predictions for a concerted reaction. 3-Phenyl-1,2-benzisothiazole (**10**) undergoes photochemical 2 + 2-addition with alkenes in a regio- and stereo-specific manner.[24] Thus the reaction with z-2-butene yields (**11a**), and that with ethyl vinyl ether (**11b**). The efficiency of the reaction is increased by polar solvents, and the process does not appear to involve the triplet benzisothiazole; a favoured mechanism has not yet emerged. Isolation of the 2 + 2-adduct (**12**) of singlet oxygen and cycloheptatriene completes the set of possible mono-cycloadducts from these species,[25] and has permitted the observation that (**12**) is the precursor of benzaldehyde but not of the 2 + 6-cycloadduct in the overall reaction. Evidence has been presented that 6,6-dimethyl-6-silafulvene undergoes 2 + 2-cycloaddition to benzaldehyde, yielding an unstable adduct (**13**).[26]

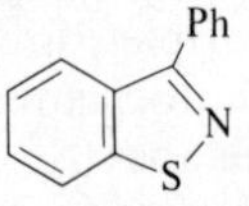

(10) 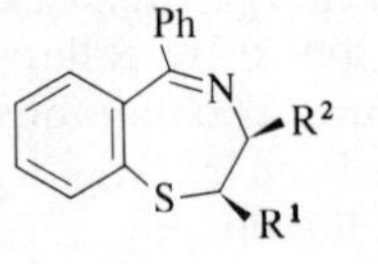 (11) 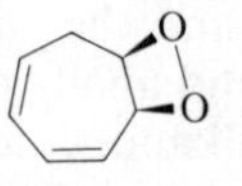 (12) 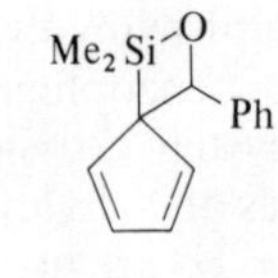 (13)

a: $R^1 = R^2 = Me$

b: $R^1 = OEt$, $R^2 = H$

The recent suggestion[27] that 2 + 2-cycloadditions of allenes are more likely to proceed by a ${}_\pi2_s + ({}_\pi2_s + {}_\pi2_s)$ mechanism rather than the alternative ${}_\pi2_s + {}_\pi2_a$ pathway has received some experimental verification. Predictions for 1,1-dimethylallene are that reaction of the least substituted C=C bond should be strongly favoured by the former but not by the latter mechanism. The reaction between 1,1-dimethylallene and maleic anhydride gives exclusively the product arising from 2 + 2-addition across the least substituted C=C bond.[28] This behaviour contrasts with that of 1-ethylallene, in which both C=C bonds are approximately equally reactive, and is taken as evidence against the ${}_\pi2_s + {}_\pi2_a$ pathway. Methyl buta-2,3-dienoate undergoes $EtAlCl_2$-catalysed stereospecific 2 + 2-cycloaddition with alkenes, to give cyclobutylideneacetates.[29] For these reactions, a ${}_\pi2_s + {}_\pi2_a$ mechanism involving the complex (**14**) has been proposed.

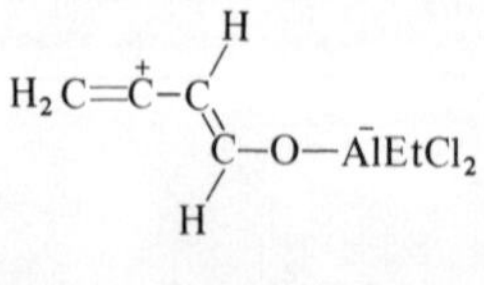

(14) 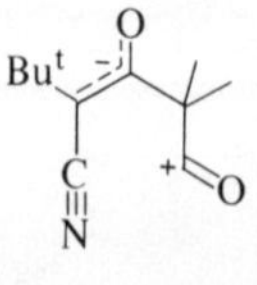 (15) 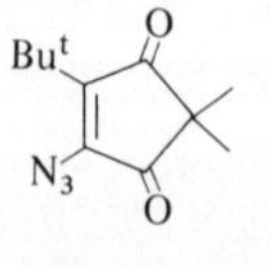 (16)

tert-Butylcyanoketene cycloadds to ketene, methylketene, dimethylketene, and ethylmethylketene, yielding 2-oxetanones in the first two cases and cyclobutane-1,3-diones in the last two.[30] These results are accommodated by a mechanism in which initial head-to-tail bond formation gives rise to a zwitterionic intermediate such as (**15**), which subsequently undergoes ring-closure by C—C or C—O bond formation. The zwitterions have also been generated by thermolysis of appropriate azidocyclopentenediones such as (**16**), and are found to ring-close in the same way independently of their method of generation. *N*-Arylketenimines, $(CF_2)_2C{=}C{=}NC_6H_4R$ (R = H, 4-MeO, 2-Me, 4-Me, 4-CO_2Et), undergo 2 + 2-cycloadditions with PhC≡CH, EtOCH=CH_2, and isonitriles, which are found to be assisted by electron-attracting and inhibited by electron-donating substituents (R).[31] A two-step polar mechanism is indicated by the substituent effect. A 2 + 2-cycloaddition of dimethylketenimines and SO_2 has been reported,[32] and the tetramethylketiminium ion (**17**) has been shown to be sufficiently reactive to undergo 2 + 2-cycloaddition even with α,β-unsaturated carbonyl compounds, a reactivity attributed to its exceptionally low-lying LUMO.[33] The 2 + 2-additions of arylsulphonylisothiocyanates to vinyl ethers exhibit characteristics of a two-step reaction with zwitteronic intermediates, but have an unexpectedly slight solvent effect.[34] It has been suggested that the two reactants approach orthogonally to give a non-planar zwitterion, in which rotation about single bonds competes with ring-closure.

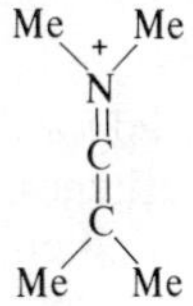

Me Me N⁺ C C Me Me

(**17**)

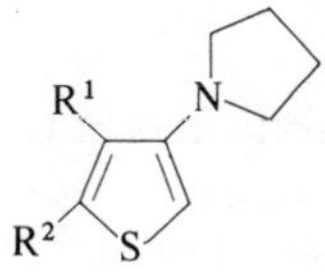

R^1 R^2 N S

(**18**)

N CO_2Me R^1 CO_2Me R^2 S

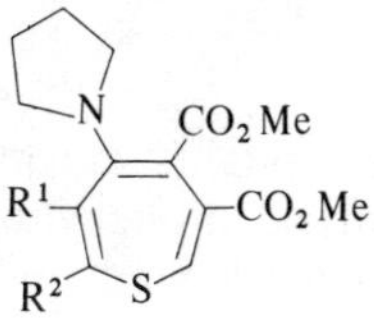

(**19**)

The influence of $MgBr_2$ on the cycloadditions of ynamines with cycloalkenones has been investigated.[35] Cyclohexenones react at the carbonyl group in the presence of $MgBr_2$ but at the C=C bond without catalysts, whereas cyclopentenones react at the C=C bond with or without $MgBr_2$. No explanation of this observation has so far been advanced. 2 + 2-Cycloadditions of 3-(1-pyrrolidinyl)thiophenes (**18**) and electron-deficient alkynes occur in apolar solvents, yielding, ultimately, substituted benzenes *via* thiepins such as (**19**).[36] In polar solvents, an alternative route involving zwitterions is followed.

Synthetic applications of the 2 + 2-cycloadditions of α,β-unsaturated carbonyl compounds[37] and of halogenated ketenes[38] have been reviewed.

2 + 3-Cycloaddition

Several reviews covering mainly synthetic aspects of 2 + 3-cycloadditions have appeared;[39–41] one of these[41] includes a short discussion of anionic 2 + 3-cycloadditions. Mass-spectrometric evidence has been presented for cationic 2 + 3-addition between substituted allyl cations and alkenes.[42] The use of 1,3-dipoles as trapping agents for the transient species, thiophene sulphone, has been described.[43]

The reaction between benzyne and a variety of substituted thiophenes has been investigated.[44] The products are benzo[*b*]thiophenes, arising from loss of an acetylene moiety from an intermediate adduct. Contrary to earlier proposals, it appears that adjacent α- and β-carbons of the thiophenes are either retained as C(2) and C(3) of the product, or lost. The most likely mechanism, therefore, involves

2 + 3-cycloaddition of benzyne as shown in Scheme 1. Rate constants have been determined for the reactions of TCNE with a range of substituted cycloheptatriene complexes of tricarbonyliron.[45] In most cases, 2 + 3-cycloaddition of TCNE at the uncoordinated double bond of the complex occurs. The reaction rates are first-order in both TCNE and the complex, and exhibit only a small solvent effect, ruling out a polar intermediate. The mechanism is most probably a concerted one, and the preference for 2 + 3-cycloaddition as observed has been confirmed by FMO theory.[46] 2 + 3-Cycloadditions of 1-methylenecyclopropanes to alkenes under the action of palladium catalysis[47,48] and of a π-allylruthenium complex to alkynes[49] have also been described.

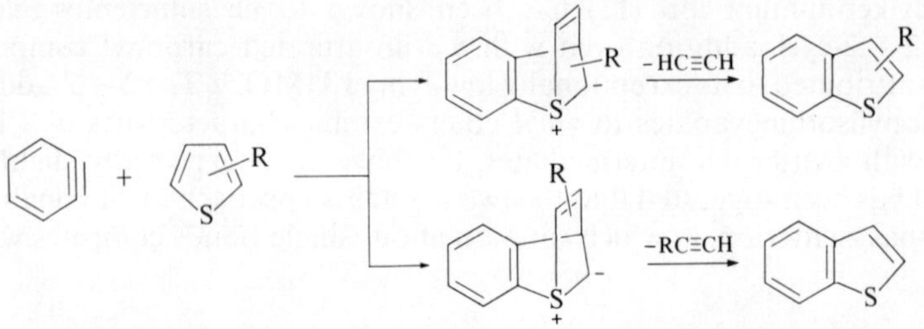

SCHEME 1

Rate constants and activation parameters have been measured for the 1,3-dipolar addition of diphenyldiazomethane to cyano-substituted ethenes.[50] Reactivity appears to depend on both the acceptor properties of the alkene and on the localization energy. *N*-Allyldiazoacetamides (**20**; $R^1 = Pr^n$ or Ph) and *N*-allyldiazomalonamide esters (**20**; $R^1 = CO_2Et$) undergo regio- and stereo-specific intramolecular 2 + 3-cycloadditions, yielding products derived from the intermediate adducts (**21**).[51] In accord with qualitative appraisals of HOMO and LUMO energies, the diazocetamides were found to be much less reactive than the diazomalonamide esters. *tert*-Butyldiazomethane reacts with SO_2 to give a mixture of the thiirane 1,1-dioxide (**22**) and the diastereomeric dihydrothiadiazole 1,1-dioxides (**23**).[52] An increase in solvent polarity favours the formation of (**22**) over (**23**), but has virtually no effect on the diastereomer ratio *cis*-(**23**):*trans*-(**23**). The ratio (**22**):(**23**) did not depend on whether the intermediate sulphene (Bu^tCHSO_2) was generated from the diazo compound and SO_2 or independently from 2,2-dimethylpropanesulphonyl chloride and triethylamine. It thus seems that (**22**) arises from a zwitterion ($Bu^t\bar{C}HSO_2CHBu^t\overset{+}{N}_2$), and (**23**) arises from concerted 2 + 3-addition. The dichotomous behaviour of sulphenes towards diazoalkanes can be attributed to the existence of two low-lying sulphene MOs only one of which has π-symmetry. A 2 + 3-cycloaddition between diazoalkanes and thioketene *S*-oxides has also been reported.[53]

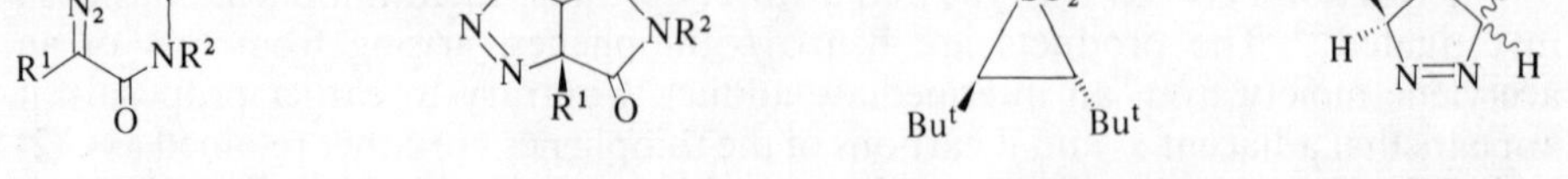

(20) (21) (22) (23)

p-Nitrobenzenesulphonyl azide adds to alkylidenecycloalkanes with the electrophilic end of the dipole bonding to the least substituted carbon atom.[54] The resulting triazolines have not, however, been isolated. With isopropylidenecycloalkanes, where the double bond is fully substituted, the regiospecificity is lost. 2-Isopropylidenenorbornane reacts to give a surprising proportion of the product from *endo*-attack ($k_{exo}/k_{endo} = 4$), confirming the importance of electronic as well as steric factors in these reactions. The bulkier azide, 1-azidoadamantane, has been found to add to a range of norbornene derivatives to give only *exo* 2 + 3-adducts.[55] 1-Azidoadamantane also adds to alkynes and electron-poor alkenes, with regioselectivity depending on both steric and FMO interactions.

p-Chlorobenzonitrile oxide reacts only at the C=P bond of the diazaphosphole (**24a**), yielding only that regioisomer in which the oxygen atom is bound to the carbon atom of (**24a**).[56] The diazaarsole (**24b**), on the other hand, reacts at the C=As bond to give both regioisomers; the kinetic product corresponds to the single regioisomer obtained from (**24a**), and the other isomer is the thermodynamic product. Diacylfuroxans (**25**) decompose, when heated in solution, by either of two pathways, depending on the substituents, R.[57] Thus, uncrowded diacylfuroxans rearrange to α-acyloximinonitrile oxides, while bulky substituents promote fragmentation into two "half-molecule" acyl nitrile oxides, $R(CO)C{\equiv}\overset{+}{N}{-}O^{-}$. The acyl nitrile oxides may be trapped by dipolarophiles and exhibit regioselectivity according with LUMO(nitrile oxide)–HOMO(dipolarophile) control. Calculations have shown that, compared with alkyl and aryl nitrile oxides, the acyl group has only a small effect on the HOMO, but lowers the LUMO energy substantially (~ 3 eV). The aryl group, therefore, narrows the HOMO–LUMO gap in the nitrile oxide itself, and this may account for the high dimerization rates for the acyl compounds.

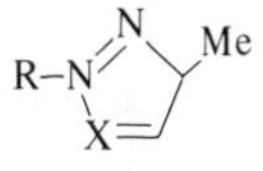

(**24**) a: R = MeCO, X = P
b: R = Ph, X = As

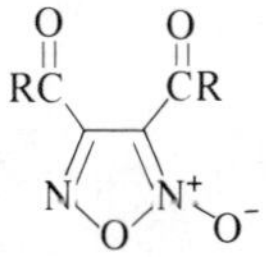

(**25**)

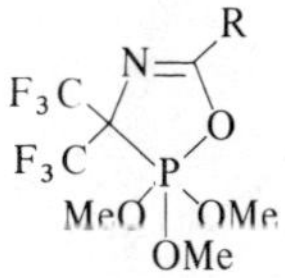

(**26**)

2 + 3-Additions of phenylglyoxylonitrile oxide, $PhC(O)C{\equiv}\overset{+}{N}{-}O^{-}$, to several bicyclic dienes have been examined, with the clear importance of steric factors emerging.[58] A thorough investigation of the reactions of benzonitrile *N*-phenylimide with α,β-unsaturated ketones has been carried out.[59] Hitherto, FMO theory had failed to give a satisfactory account of regioselectivity in these reactions. The results indicate that the reactions are controlled by HOMO(nitrile imine)–LUMO(dipolarophile) interactions, but also lead to the conclusion that the largest HOMO coefficient is that of the carbon atom in the nitrile imine, contrary to widely held belief. This conclusion has also received support from *ab initio* MO calculations for $HC{\equiv}\overset{+}{N}{-}\overset{-}{N}H$. Regioselectivity in the reactions with alkynones suggests the importance of secondary orbital interactions between the nitrile imine and the carbon atom of the carbonyl group; this interaction is important in the

alkynes, owing to their linear geometry, but is almost negligible in the alkenes. *o*-Allylbenzonitrile *N*-phenylimide, generated photochemically, has been shown to undergo an intramolecular 1,3-dipolar addition,[60] and the additions of nitrile imines to 1,2-dibenzoylethenes have been described.[61] Thermolysis of (**26**; R = Me_3C, Ph, 4-FC_6H_4) generates the corresponding nitrile ylids, $(F_3C)_2\bar{C}-\overset{+}{N}\equiv CR$, each of which gives two regioisomeric 2 + 3-cycloadducts and an open-chain product with 1,1-diphenylethene.[62] The product ratio depends on the polarity of the solvent. The same nitrile ylids also give 2 + 3-adducts with a variety of aza-arenes.[63] Nitrile ylids have also been generated by the photolysis of 3-(*N*-methylanilino)-2*H*-azirines and trapped by dipolarophiles.[64]

PMO theory has been applied to the 2 + 3-cycloadditions of *C*-benzoyl-*N*-phenylnitrone and furan derivatives.[65] Ionization potentials of the furan derivatives were determined from energies of the corresponding charge-transfer complexes with TCNE, and show a good correlation with second-order rate constants for the cycloadditions. Taking the ionization potentials as estimates of HOMO energies leads to the conclusion that the reactions are controlled by LUMO-(nitrone)–HOMO(furan) interactions. Rate constants for the cycloadditions of the same nitrone to aryl vinyl ethers, 4-$RC_6H_4OCH{=}CH_2$, and to (**27**; R = MeO, Me, H, Cl, Br, NO_2) have also been measured.[66] Reactions of the ethers are governed by HOMO(dipole)–LUMO(dipolarophile) interactions, while those of (**27**) are of the reverse type. A stereochemical study of the reaction between pent-1-ene and 3,4,5,6-tetrahydropyridine-1-oxides, such as (**28**), has shown that the resulting 2 + 3-adducts arise mainly or exclusively from axial *exo*-addition.[67] The 3*H*-pyrrol-3-one (**29**) reacts with dimethyl acetylenedicarboxylate to give a mixture of products, all apparently derived from an intermediate 2 + 3-adduct;[68] nitrone cycloadditions as routes to *β*-lactams have been examined,[69] and the cycloadditions of cycloimmonium ylids (**30**) to diphenylcyclopropene have been described.[70]

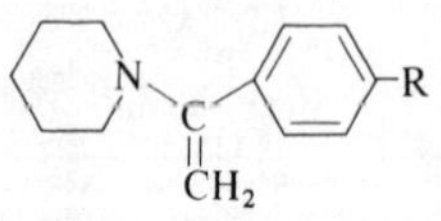

(27) 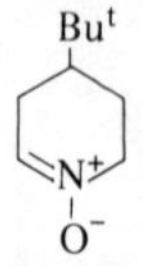 (28) 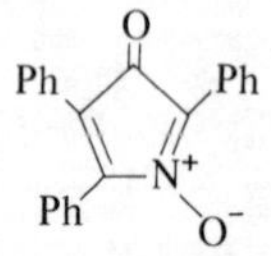 (29) 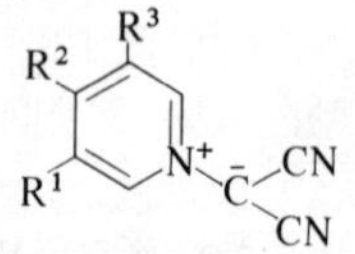 (30)

Site selectivity in the reactions of various 1,3-dipoles with norbornadiene derivatives (**31a–c**) has been investigated.[71] Aryl azides, and benzonitrile oxides attack (**31a**) preferentially at the electron-poor C=C bond, but attack (**31b**) and (**31c**) at the unsubstituted C=C bond. 2-Diazopropane and *C*-phenyl-*N*-methylnitrone, on the other hand, attack all three bicyclic compounds only at the electron-poor C=C bond, and in a stereospecific (*exo*) manner. The triazoline (**32**) loses N_2 on thermolysis, giving the *α*-cyaniminocarbene (**33**), which behaves as a 1,3-dipole towards MeCN and benzene.[72] This behaviour contrasts with that of *α*-oxocarbenes, which usually undergo Wolff rearrangement and give only minimal yields of 1,3-dipolar adducts.

A novel photochemical 2 + 3-cycloaddition has been discovered for the *α*,*β*-acetylenic ketone (**34**) and alkenes, such as tetramethylethene.[73] The product vinylhydrofurans (*e.g.* **36**) are accompanied by propynyloxetanes as by-products.

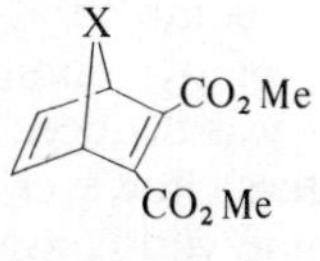

(31) a: X = O
b: X = CH_2
c: X = $Me_2C{=}C$

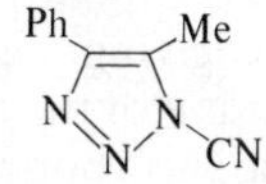

(32)

(33)

The reaction with 1,1-dimethylethene shows no regiospecificity, and those with E- and Z-2-butene, no stereospecificity. A mechanism involving an intermediate biradical (**35**) (Scheme 2) accounts for both types of product and also the loss of stereochemistry. In the photochemical and thermal generation of 1,3-diaryl carbonyl ylids from 2,3-diaryloxiranes and in the subsequent reaction of the ylids with dipolarophiles, the stereochemistry of the dipolarophiles is preserved in the cycloadducts.[74] This is consistent with a concerted addition process, but solvent effects, steric hindrance and secondary orbital overlap may all play a part in determining the product distribution.

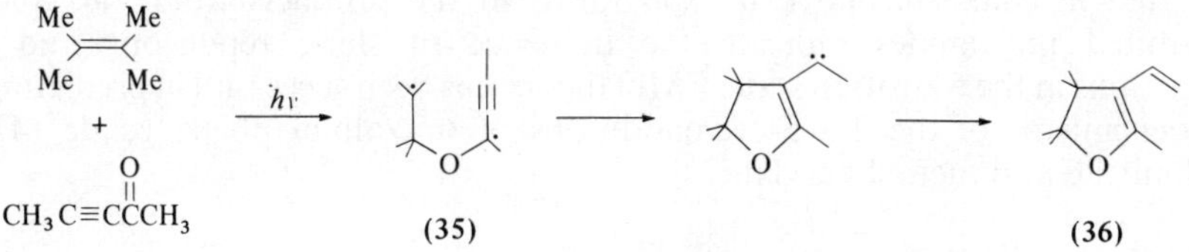

$CH_3C{\equiv}CCOCH_3$ (34) (35) (36)

SCHEME 2

Thioketenes effect C=N cleavage in the aminoazirine (**37**; R^3 = Me), yielding the 2 + 3-adducts (**38**) or ketenimines (**39**), depending on the substituents (Scheme 3).[75] In contrast, the phenyl-substituted aminoazirine (**37**; R^3 = Ph) undergoes 1,2-bond cleavage under similar conditions, yielding 2-thiazolin-4-ones after hydrolysis of (**40**). A similar competition between 1,2- and 1,3-ring-opening of aminoazirines occurs with carbodiimides, which in certain cases can enter into cycloaddition *via*

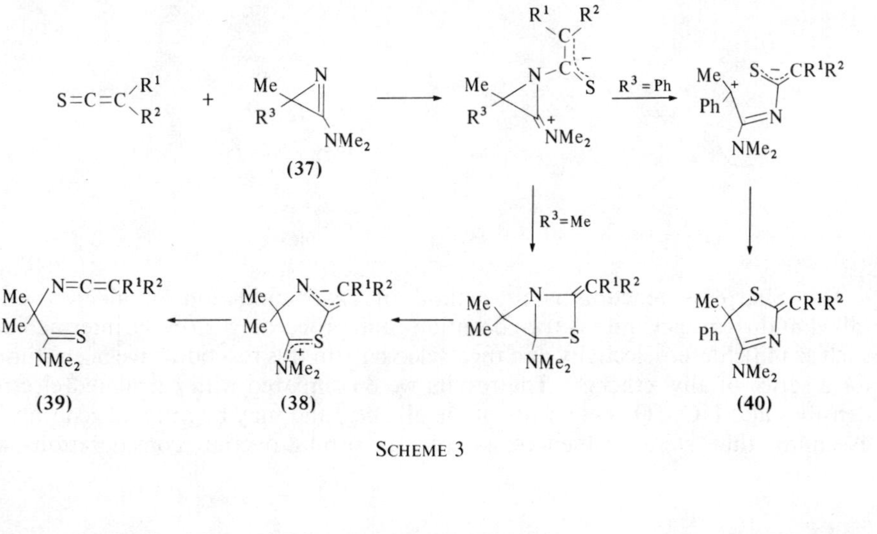

(37) (38) (39) (40)

SCHEME 3

either C=N bond.[76] 1-Benzyl-Z- and -E-2,3-diphenylaziridines combine at 110° with dipolarophiles *via* azomethine ylids, giving high yields of pyrrolidines.[77] Stereospecific conrotation for the ring-opening of the aziridines was deduced from the structure of the cycloadducts. 1-Ethoxycarbonyl analogues, however, give diastereomeric pyrrolidines with dimethyl fumarate. Azomethine ylids have also been generated by thermolysis of 2-cyano-3-phenyl-*N*-alkylaziridines,[78] and the azomethine imine (**41**) has been reported to undergo an intramolecular 2 + 3-addition, in which steric interactions apparently determine the product stereochemistry.[79]

Mesoionic 1,3-oxazolium-5-olates (**42**) react with endocyclic double bonds in 6-substituted and 6,6-disubstituted fulvenes, yielding pyrrofulvenes after loss of CO_2.[80,81] In one instance, the intermediate was trapped by dimethyl acetylenedicarboxylate as the adduct (**43**). 2,3,4,5-Tetraarylfulvenes react with (**42**) at the exocyclic double bond.[82] The allyloxy mesoion (**44**) and its propargyloxy analogue both undergo intramolecular 2 + 3-addition; in the former case, the adduct (**45**) was isolated in excellent yield, whereas in the latter case, the adduct decomposed by loss of COS.[83] The mesoion (**46**) reacts with a variety of dipolarophiles in a stereospecific manner.[84] The stereospecificity suggests a concerted mechanism, and *exo–endo* ratios in the products from (**46**) and *N*-substituted maleimides indicate the influence of steric repulsions and π–π interactions in the transition state. FMO theory has been successful in predicting the stereochemistry of the 2 + 3-cycloadditions of triazolium phenacyclide (**47**) to acrylonitrile and methyl acrylate.[85]

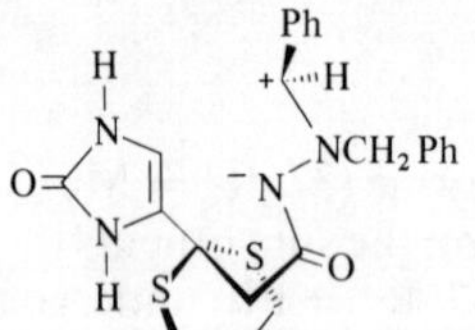

(**41**) 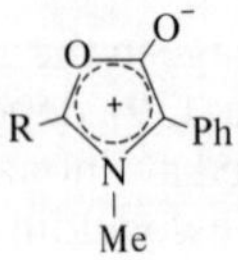 (**42**) 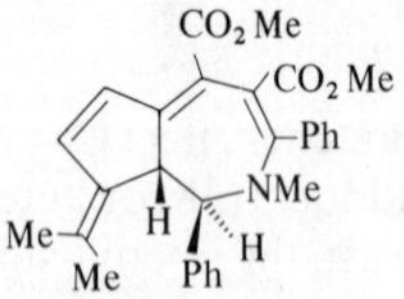 (**43**)

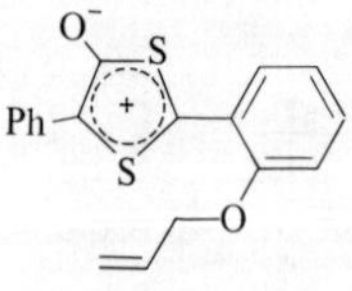

(**44**) 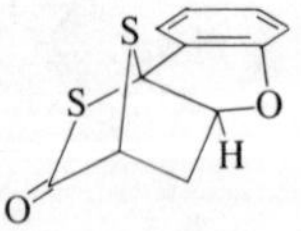 (**45**) 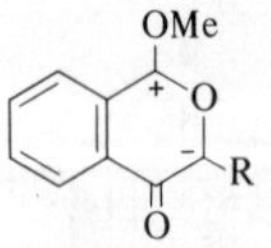 (**46**) 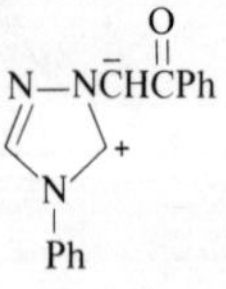 (**47**)

The Sharpless reaction is a method of *cis*-oxamination of alkenes using (alkylimido)- or (acylimido)-trioxoosmium, and proceeding *via* cyclic intermediates such as (**48**). Stereoselectivity and regioselectivity in this reaction have been studied for a series of allyl ethers.[86] The results were compared with calculated electron densities and HOMO coefficients of the alkenes, and may be rationalized only by assuming that steric influences as well as orbital-overlap considerations are important. Thermal 2 + 3-cycloadditions of cyclopropanone acetals to TCNE yield 2,2,3,3-tetracyanocyclopentanone acetals, presumably *via* zwitterionic intermediates.[87] Both the fluorenylidene-*N*-(*p*-toluenesulphonyl)sulphimide (**49**) and bis(*p*-toluenesulphonyl)sulphur diimide (TsN=S=NTs) behave as 1,3-dipoles towards fulvenes.[88]

2 + 4-Cycloaddition

Limitations of PMO treatments of Diels–Alder reactions have been pointed out in a theoretical study of eleven 2 + 4-cycloadditions.[89] All interactions within the π-electron approximation were considered, and reactions were classified, according to the nature of their transition states, as either overlap-controlled or closed-shell repulsion-controlled. For the latter type of reactions, involving hetero-dienophiles, the use of the frontier orbital approximation as an index of reactivity is essentially meaningless. Even the classification of Diels–Alder reactions as "normal" or "inverse" in electron demand has been shown to be problematical.[90] In the reaction between 9,10-dimethylanthracene and typical cyclic dienophiles reactivity decreases in the order maleic anhydride > maleic thioanhydride > maleimide; LUMO energies from MINDO/3 calculations for these species are −0.63, −0.87, and −0.47 eV, respectively. Analogous behaviour is observed for the chlorinated derivatives of the dienophiles. In both series, the observations can be rationalized by the FMO model only if, besides the orbital energies, the differing frontier-orbital densities in the dienophile and the electrophilicity of the diene are also taken into account. Second-order rate constants and regioselectivity have been determined for the Diels–Alder reactions of 90 aromatic hydrocarbons with maleic anhydride.[91] The results were correlated with various reactivity indices, including second-order perturbation energies and Polansky indices, the latter being a quantification of Clar's idea of benzenoid character. All the models gave some correlation with the data, but the Polansky indices were the least satisfactory.

Rate measurements including activation parameters have been reported for the liquid-phase Diels–Alder reaction of cyclopentadiene and norbornene, and also for the reverse reaction in the gas phase.[92] 4-Substituted styrenes add to the di(methoxycarbonyl)cyclopentadienone (**50a**) at rates that correlate well with Okamoto–Brown σ_p^+ constants ($\rho = -0.941$), indicating a Diels–Alder reaction with inverse electron demand.[93] Rate constants for the analogous reactions of the diethylcyclopentadienone (**50b**), in contrast, indicate a reaction with approximately neutral electron demand. 2,5-Diaryl-2,4-*o*,*o*-biphenylenecyclopentadienones (**51**) undergo Diels–Alder reactions with *N*-arylmaleimides, maleic anhydride, acrylonitrile, and cyanoethenes, the rate constants for which give linear Hammett

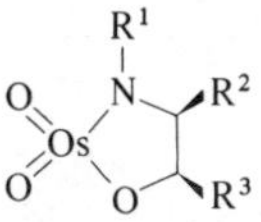

R^1 N R^2 O Os O O R^3

(**48**)

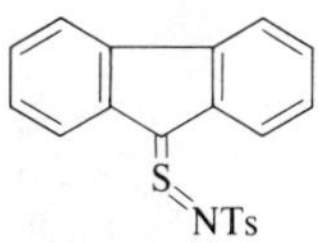

S NTs

(**49**)

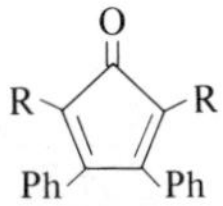

O R R Ph Ph

(**50**) a: $R = CO_2Me$
b: R = Et

relationships with negative ρ values.[94,95] Additionally, the ρ values for the reactions of the cyano-substituted dienophiles correlate well with the electron affinities of the nitriles. These observations suggest the cycloadditions are of diene donor–dienophile acceptor type; but in these reactions and also those of tetraphenylcyclopentadienone with cyanoethenes,[96] localization energies are also thought to be a factor determining reactivity. Kinetic data, including Hammett plots, for hexabromo- and 5,5-dimethoxytetrabromo-cyclopentadiene (**52a** and **52b**) indicate that, in their Diels–Alder reactions with *N*-arylmaleimides and 4-phenyl-1,2,4-triazoline-3,5-dione, the hexabromo compound (**52a**) acts as an electron acceptor and the dimethoxy analogue (**52b**) as a donor.[97] Kinetic results have been presented for Diels–Alder reactions of hexachlorocyclopentadiene with a variety of dienophiles.[98–100] Similar studies of the cycloadditions of (**53**) with 4-substituted styrenes also gave good correlations between rate constants and σ_p^+ parameters ($\rho = -0.84$), again indicating LUMO (diene)–HOMO(dienophile) control.[101]

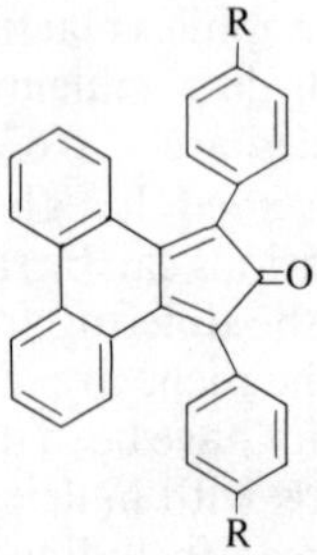

R R Br Br O SO$_2$ Cl Cl Br Br Cl Cl

(51) 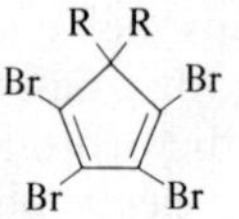 (52) a: R = Br b: R = OMe (53)

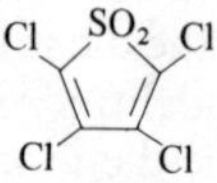

With acyclic dienes, the conformation of the diene is an additional factor determining reactivity towards dienophiles. A series of methoxybutadienes, having different electron-donating properties and conformational preferences, have been found to add to β,β-dicyanostyrenes with rate constants that correlate with σ^+ parameters, yielding plots with positive ρ values.[102] Conformational effects seem to be less important in these reactions than frontier-orbital interactions. The reactions of 1-(acylamino)-1,3-dienes with various dienophiles have been studied:[103] carbamates (**54**) react even with poor dienophiles such as cyclohex-2-enone, and in the case of unsymmetrical dienophiles a very high degree of regioselectivity obtains. Regioselectivity in the Diels–Alder reactions of electron-rich dienes and quinones may be rationalized on the basis of PMO theory, provided that secondary orbital interactions are invoked.[104,105] Studies of selectivity in the Diels–Alder reactions of diquinones with cyclopentadiene and quadricyclane,[106] of quinone acetals with 1-substituted isobenzofurans,[107] and of (alk-3-enyl)cyclopentadienes[108] have been described.

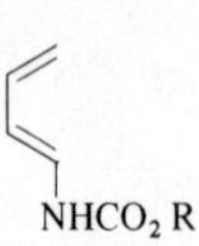

NHCO$_2$R R H H R H H Me

(54) 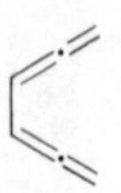 (55) 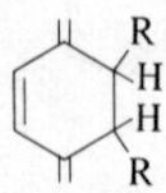 (56) 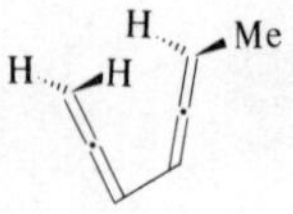 (57)

Conjugated bis-allenes such as (**55**) give only 2 + 4 adducts (*e.g.* **56**) with maleic anhydride, maleimides, quinones, azo-group-containing dienophiles, and acyclic dienophiles.[109] Maleic anhydride attacks the asymmetic bis-allene (**57**) only from the less shielded side, *anti* to the methyl group. The tricyclic dienes (**58**) and (**59**) react with various dienophiles, yielding only the products of *endo*-attack, whereas (**60**) yields two isomers in each case, with the *exo*-product preferred.[110] It is claimed that the stereospecificity shown by (**58**) and (**59**) cannot be due to steric factors, since *endo*-approach is likely to be more hindered than *exo*-approach. Instead, mixing of σ- and π-orbitals in the diene, leading to a tilting of the diene orbitals, is postulated to minimize antibonding interactions on the *endo*-faces of (**58**) and (**59**) relative to the *exo*-faces. The premise that steric factors would indeed favour *exo*-attack has, however, been challenged.[111] The oxa analogue (**61**) reacts with maleic anhydride and dimethyl acetylenedicarboxylate to give only the *endo*-adducts. Even at room temperature, the reactants and adducts are equilibrated, and no trace of the *exo*-isomers can be detected. Thus, for (**61**), and very probably for (**58**) and (**59**) also, the *endo*-adduct is thermodynamically favoured. If the Bell–Evans–Polyani principle is followed, the more stable isomer would also be formed faster. Stereospecificity in the Diels–Alder reactions of (**58**), (**59**), and (**61**) may still, therefore, be attributable to straightforward steric influences. A preliminary report has appeared on the effects of differential geminal substitution (X, Y) in dienones (**62**) on face-selectivity in Diels–Alder reactions of these compounds.[112] Steric and secondary orbital effects have been invoked to rationalize high degrees of stereo- and regio-selectivity in Diels–Alder reactions of 6-acetoxy-2,6-dimethylcyclohexa-2,4-dienone.[113]

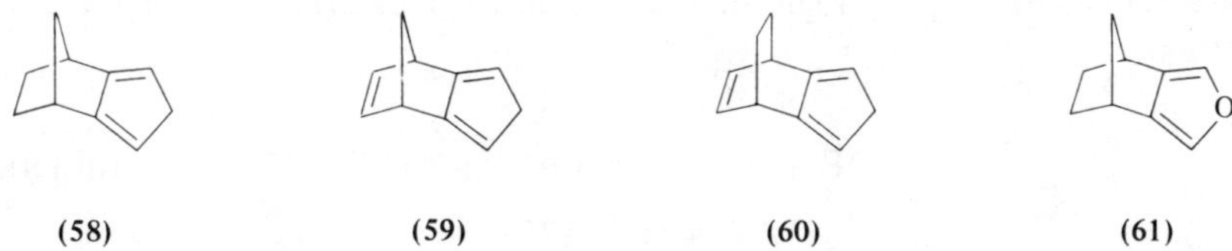

(**58**) (**59**) (**60**) (**61**)

The rôle of charge-transfer complexes in 2 + 4-cycloadditions has been the subject of several papers. Formation constants for the complexes between TCNE and a range of phenyl-substituted ethenes have been measured, and are found to follow the order *trans*-stilbene > styrene > α-methylstyrene > *cis*-stilbene > 1,1-diphenylethene.[114] Both electronic and steric effects are operative. α-Methylstyrene and 1,1-diphenylethene undergo thermal cycloaddition with TCNE faster than styrene, but the stilbenes do not react with TCNE. The reactivity is best explained in terms of localization energies, and clearly does not correlate with the stabilities of the charge-transfer complexes. Kinetic measurements on the reaction between TCNE and 1,2- and 9-substituted anthracenes have permitted the separation of the formation constants (K) for the intermediate complexes from the overall rate constant (k_2^{obs}).[115] A relatively small range of values for K indicates a very small charge-transfer development in the complexes. The transition state leading from the complexes to the adducts, however, shows a marked degree of charge development, particularly for the 9-substituted derivatives ($\rho = -7$). Substituents in the outer rings influence the rate constants in a way that gives a good correlation with composite substituent constants ($\sigma_m^+ + \sigma_p^+$), with a ρ value of -3. Such a composite correlation supports the postulate of concerted attack by the incoming dienophile.

Solvent effects on the rate of reaction of TCNE with anthracene correlate better with solvation of the charge-transfer complex than with solvation of reactants or product.[116] Phencyclone (**51**; R = H) undergoes Diels–Alder reaction with electron-rich dienophiles *via* charge-transfer complexes, which seem to play an important part in determining the regiochemistry and stereochemistry of the adducts.[117]

The effects of catalysts upon Diels–Alder reactions have also recently received much attention. Infrared data indicate that at 75–165 K in the presence of $TiCl_4$, cyclopentadiene forms a complex with methyl acrylate.[118] On warming, this complex is converted into the bicyclic 2 + 4-adduct. $AlCl_3$ catalysis lowers by 20–25 kJ mol^{-1} the activation energies for Diels–Alder reactions of buta-1,3-diene and isoprene with acrylonitrile and methacrylonitrile.[119] In this connection, it has been suggested that increases in the acceptor strengths of dienophiles on complexation with $AlCl_3$ are not sufficient to explain the observed catalytic acceleration.[120] Changes in the C=C bond energies of complexed dienophiles are small, and cannot cause 10^5–10^6-fold increases in reactivity. A change, during reaction, of the interaction of $AlCl_3$ with a functional group may be the primary catalytic mechanism. Inherently inefficient Diels–Alder reactions involving approximately neutral or electron-rich dienophiles may be accelerated by conversion of the dienophiles into the corresponding radical cations—highly electron-deficient species. An example of such acceleration has been reported for the 2 + 4-dimerization of cyclohexa-1,3-diene.[121] Without catalysts, 30 % dimerization is achieved after 20 hours at 200°; in the presence of 5–10 mol % of the stable radical-cation salt, bis(*p*-bromophenyl)aminium hexachlorostibinate, dimerization occurs to the extent of 70 % within 15 minutes at 0°. The following reaction scheme has been proposed:

$$DP + Ar_3N^{+\cdot} \rightleftharpoons DP^{+\cdot} + Ar_3N \qquad \text{Initiation}$$

$$\left.\begin{aligned} DP^{+\cdot} + D &\rightarrow A^{+\cdot} \\ A^{+\cdot} + DP &\rightarrow A + DP^{+\cdot} \end{aligned}\right\} \quad \text{Propagation}$$

(D = diene; DP = dienophile; A = adduct)

Several interesting examples of hydrophobic catalysis have been discovered.[122] The reaction between cyclopentadiene and butenone is 700-times faster in water than in 2,2,4-trimethylpentane, and is further increased by the presence of LiCl. Most strikingly, the reaction between 9-(hydroxymethyl)anthracene and *N*-ethylmaleimide is faster in non-polar solvents than in polar solvents, with the exception of water, in which the reaction is very fast. Only a hydrophobic effect seems capable of explaining this behaviour. Models suggest that a reacting pair made up from a molecule of cyclopentadiene and a molecule of butenone or of acrylonitrile should fit together into the hydrophobic cavity of β-cyclodextrin. Both reactions are even faster in water when β-cyclodextrin is added, but are slowed by the addition of α-cyclodextrin. In the latter case, inibition occurs because cyclopentadiene on its own fits into the cavity, but there is no room for the dienophile as well. Rate enhancement of Diels–Alder reactions by an increase of solvent viscosity has also been discussed, with particular reference to enzymic catalysis.[123]

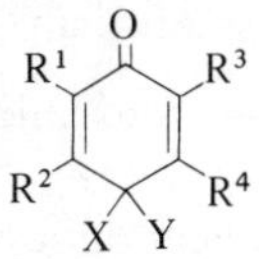

(62) 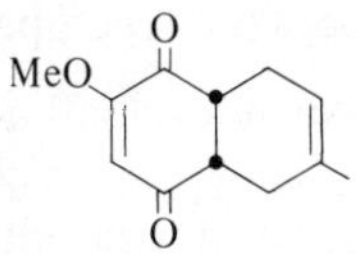 (63) 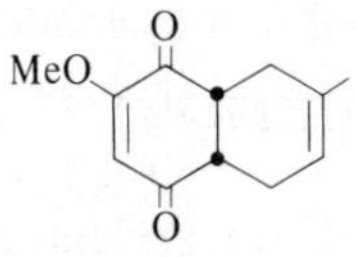 (64)

In addition to increasing overall rates, catalysts may influence regio- and stereoselectivity. Poor regioselectivity in the Diels–Alder reactions of 2-methoxy-5-methylbenzoquinone with alkyl-substituted dienes can be improved and directed to favour either isomeric adduct by the use of stannic chloride or boron trifluoride as catalyst.[124] With isoprene, for example, the uncatalysed reaction at 100° gives at 1:1 mixture of regioisomers (**63**) and (**64**); in the presence of $SnCl_4$ at 16°, (**64**) is favoured by more than 20:1, while in the presence of BF_3 at 0° (**63**) is favoured by 2.4:1. Lewis acid-catalysed intramolecular Diels–Alder reactions of (**65**; R = H or Pr^i) give exclusively the *trans*-fused adducts (**66**; R = H or Pr^i),[125] whereas the uncatalysed reactions give mixtures of *cis*- and *trans*-adducts.[126] Regioselectivity in the nickel-catalysed cyclodimerization of isoprene,[127] and stereoselectivity in the Lewis acid-catalysed reaction of cyclopentadiene and mesityl oxide[128] have also been examined.

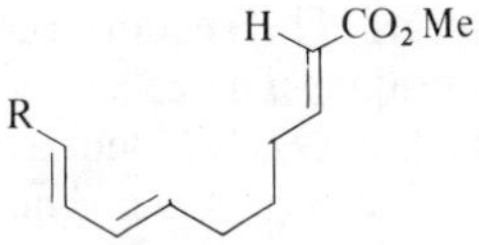

(65) 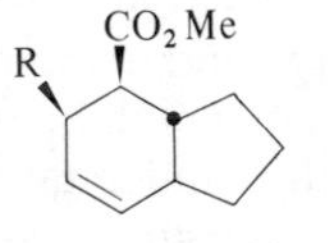 (66) 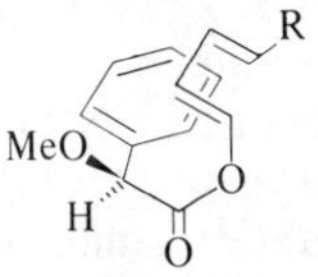 (67a) 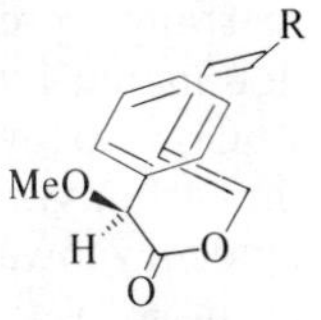 (67b)

There have been two reports on the effects of Lewis acid catalysts on asymmetric induction in Diels–Alder reactions.[129,130] A π-stacking model has been proposed to account for asymmetric induction in Diels–Alder reactions of dienes with phenyl groups on asymmetric carbon atoms.[131] This is illustrated by structures (**67a**) and (**67b**), which represent two conformations of such a diene. They differ in that (**67a**) suffers from an unfavourable steric interaction between the methoxyl group and the diene moiety, whereas (**67b**) does not and is therefore favoured. The aromatic ring thus acts as a shielding group, directing the incoming dienophile to one of the two enantiotopic faces of the diene. The reaction between (**67**; R = H) and acrolein, for example, yields a mixture of the two diastereomeric adducts in the ratio 82:18, and the reaction with juglone is virtually specific, yielding only one diastereomer. This model of the directing effect receives further support from the fact that, when the diene has a similar phenyl-bearing group at both ends [**67**; R = (*S*)-*O*-methylmandeloxy), both sides of the diene are shielded by π-stacking and the molecule is completely inert towards dienophiles. The principle of co-operativity in asymmetric induction has been used as a test for concertedness in the Diels–Alder reaction of chiral fumarates and 1,3-diphenylisobenzofuran.[132] The results provide strong support for a concerted mechanism.

Cationic 2 + 4-cycloadditions between cyclopentadiene and an allyl cation ($Me_2C{=}\overset{+}{C}HMe$)[133,134] and between cycloalka-1,3-diene and allenyl cations ($ArC{=}\overset{+}{C}{-}CMe_2$)[135] have been observed. In both instances, monocyclic adducts are also formed, and alkylallenyl cations ($RC{=}\overset{+}{C}{-}CMe_2$; R = alkyl or H) give 3 + 4-cycloadducts rather than 2 + 4-adducts.

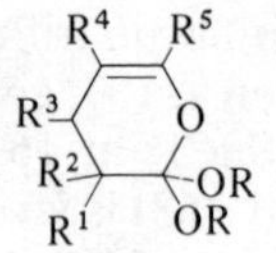

(68) 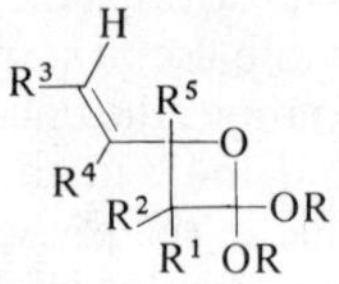 (69) 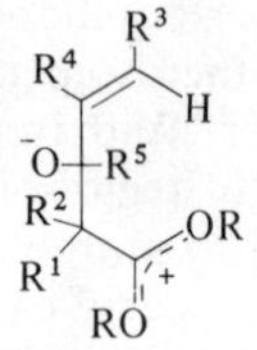 (70)

Ketene acetals [$R^1R^2C{=}C(OR)_2$] and enones ($R^3CH{=}CR^4COR^5$) undergo a 2 + 4-cycloaddition catalysed by $ZnCl_2$, yielding 2,2-dialkoxy-3,4-dihydropyrans (**68**).[136] At low temperatures, the reaction follows a 2 + 2-pathway, giving oxetanes (**69**), which can sometimes be isolated. At higher temperatures, the oxetanes rearrange, presumably *via* zwitterions (**70**), to the dihydropyrans. The reaction is dominated by HOMO(ketene acetal)–LUMO(enone) interactions. The effect of $ZnCl_2$ is to lower the enone LUMO energy and to increase the LUMO coefficient at the carbonyl carbon atom to a far greater extent than that at the β-carbon atom. The tendency towards oxetane formation is thus understandable. The zwitterion (**71**), an example of the supposed intermediates in the 2 + 4-cycloadditions of dichloroketene and *N*,*N*-disubstituted enaminones, has actually been isolated.[137] Simultaneous thermolysis of the 2-diazo-1,3-diketone, $(4\text{-}MeOC_6H_4CO)_2CN_2$, and the phosphinyldiazoalkane, $Ph_2P(O)C(N_2)Ph$, leads to the hitherto unknown ring-system (**72**; R = $4\text{-}MeOC_6H_4$) *via* 2 + 4-cycloaddition of the corresponding acylketene and $Ph_2C{=}P(O)Ph$.[138] Analogous reactants do not give the same type of product, but since the reaction depends on the half-lives of reactants and intermediates, this may not be surprising. An FMO study of Diels–Alder reactions of alkenes and *o*-quinone methides led to the conclusion that LUMO-(diene)–HOMO(alkene) interactions dominate.[139] 2 + 4-Adducts of tetrachloro-*o*-benzoquinone and ketene tautomers (**73**) of 6-oxo-6*H*-1,3-diazin-1-ium-4-olates have been obtained,[140] as have 2 + 4-adducts of ethyl vinyl ether and oxoindolin-3-ylidene derivatives β,β-disubstituted with electron-withdrawing groups.[141]

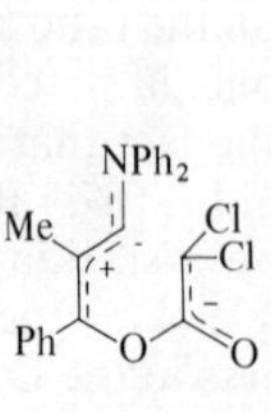

(71) 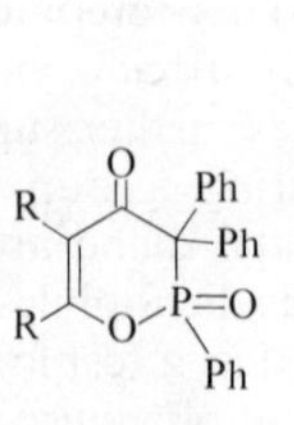 (72) 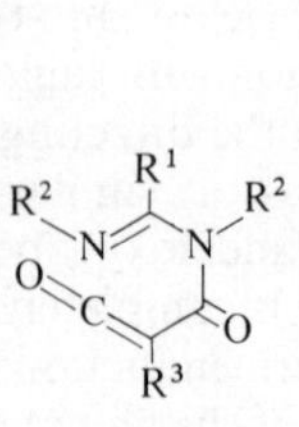 (73)

Dimethyl tetrathiooxalate (**74**; X = S) reacts with dienes as a 2π-component and with alkenes and acetylenes as a 4π-component.[142] *O,O*-Dimethyl dithioxolate (**74**; X = O) similarly acts as a 2π-component with dienes and also with quadricyclane, but acts as a 4π-component only with strained double bonds such as those in norbornenes, cyclopropenes, and benzvalene.[143] 2-Furylthiones and 2-thienylthiones also behave as hetero-dienes.[144] Chloroalkyl phenyl sulphides (**75**) lose chloride ion in the presence of $SnCl_4$ and then may undergo polar $2 + 4^+$-cycloadditions with styrene, *trans*-stilbene, and phenylacetylene, yielding thiochroman and thiochromene derivatives.[145]

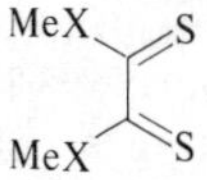

MeX S MeX S (**74**) S R Cl (**75**) Cl Cl Cl Cl N Cl (**76**) Cl Cl Cl Cl N Cl (**77**)

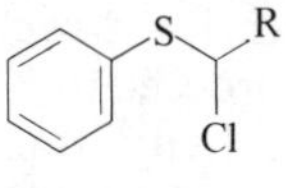

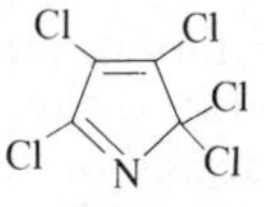

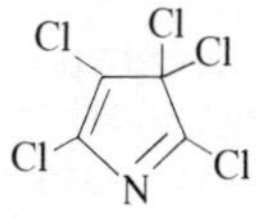

Pentachloroazacyclopentadiene exists in two tautomeric forms (**76**) and (**77**); (**76**) behaves as a dienophile but not a diene, and (**77**) as a diene but not a dienophile.[146] In reactions with dienes, either pathway may be followed: cyclopentadiene reacts only with tautomer (**76**) and cyclohexa-1,3-diene and acyclic dienes such as *trans*-piperylene only with tautomer (**77**). The reactions most likely involve dominant interaction of the LUMO of (**76**) or (**77**) with the HOMO of the diene. Ionization potentials of cyclopentadiene, cyclohexadiene, and *trans*-piperylene are very similar (8.55, 8.40, 8.56 eV, respectively), so that the LUMO–HOMO energy gap cannot be the cause of the variation in the cycloadditions. Efficiency of orbital overlap in conjunction with steric factors is thought to be the determining factor. Reissert salts[147] and an intermediate from thermolysis of *N*-acyl-*O*-acetyl-*N*-alkylhydroxylamines[148] have provided other examples of azadienes that undergo 2 + 4-cycloadditions.

Hetero-dienophiles that have received recent attention include α- and β-halonitrosoalkenes, which may act as either 2π- or 4π- components,[149] *N*-sulphinylimonium ions ($Me_2\overset{+}{N}{=}S{=}O$), which react with dienes by addition across the N=S bond[150], and intermediate silaimines (RR′Si=NR; R′ = Bu^t or Ph) which react with 2-methylprop-2-enal to give the 2 + 4-adducts (**78**).[151] Further studies[27] of the reactions of propelladienes with 4-substituted 1,2,4-triazoline-3,5-diones[152–154] and with nitrosoarenes[155] have been published.

Photochemical 2 + 4-cycloadditions have received relatively little attention and include, in the past year, only the photo-addition of maleic anhydride and chrysene,[156] the photo-dimerization of α-phenylmaleimide, which gives both 2 + 2- and 2 + 4-dimers,[157] and photo-dimerization of 1,1-diphenylethene or cross-cycloaddition of 1,1-diphenylethene to 2-methylpropene.[158] Synthetic applications of both thermal and photochemical 2 + 4-cycloadditions have been reviewed.[159]

Finally, 2 + 4-cycloadditions have been used to trap a variety of interesting intermediates, including 1-ethoxyisobenzofurans from thermolysis of 1,1-diethoxyphthalans,[160,161] dehydrothiophenes from thermolysis of thiophene-dicarboxylic anhydrides,[162] and cyclopentadienone from reactions of 4-tosylcyclopent-2-en-1-one.[163] New cycloadducts (**79**) of 1,2-di-*tert*-butyl-3,4,5,6-tetramethylbenzocyclobutadiene with TCNE (R = CN) and *trans*-dicyanoethylene (R = H) have also been reported.[164]

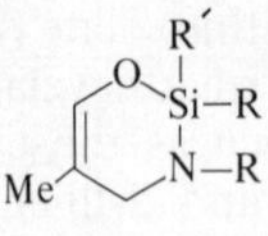

(78) 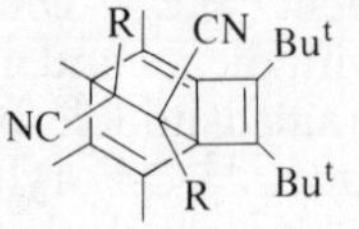 (79) 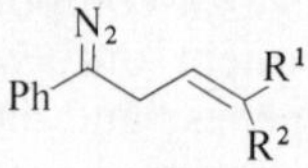 (80) 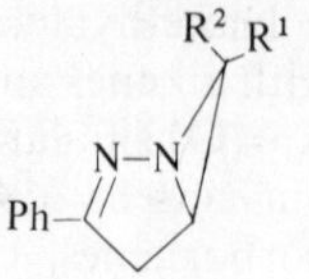 (81)

Miscellaneous Cycloadditions

Further reports[165,166] on the intramolecular 1 + 2-cycloadditions of allyldiazomethanes (**80**) have shown that the reactions proceed to 1,2-diazabicyclo[3.1.0]hex-2-enes (**81**) with a high degree of stereoselectivity, thus supporting a concerted mechanism. Nitrilimines (**82**) are thought to give the 1 + 2-adducts (**83**) with similar stereoselectivity, but the two product stereoisomers are found to equilibrate rapidly *via* a benzodiazepine.[167] A 1 + 2-cycloaddition of sulphur across the P=C bond of methylene thioxophosphoranes has been reported,[168] and 1 + 4-adducts such as (**84**) are formed in a reversible reaction between phenyl isocyanide and 1,5-diaryl-4-benzoylpyrrole-2,3-diones.[169]

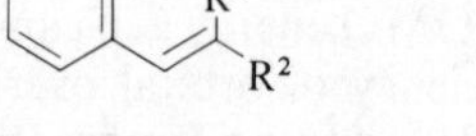

(82) 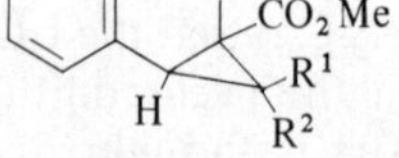 (83) 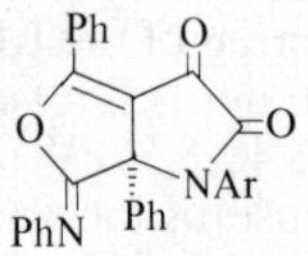 (84)

Diazoazoles (**85**) usually react with ylids to give 1 + 7-cycloadducts, but with acyltriphenylphosphonium methylids, 2 + 7-adducts are obtained instead.[170] Fidecene derivatives such as (**86**) can react with dienophiles in a variety of ways; in particular (**86**) adds TCNE at the positions indicated, to give a cycloadduct that, at least formally, is a symmetry-allowed $_{\pi}2_s + _{\pi}16_s$ product.[171]

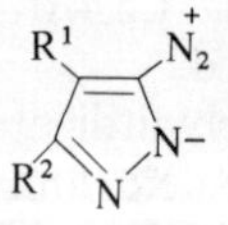

(85) 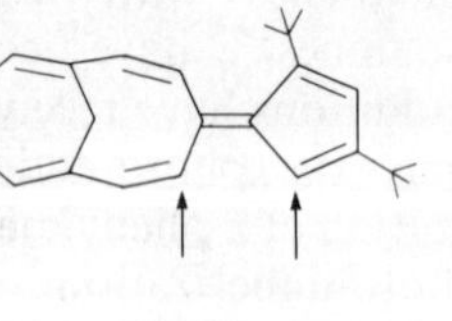 (86) 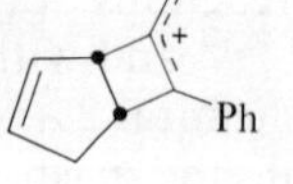 (87)

Allyl cations, including (**87**), tend to undergo 3 + 4-addition to conjugated dienes,[172,173] whereas allenyl cations $R'C{=}\overset{+}{C}{-}CR^2R^3$, may undergo either 2 + 4- or 3 + 4-cycloaddition with dienes and either 2 + 2- or 2 + 3-cycloaddition with monoenes.[174] The interception of an acyclic intermediate in the latter type of reaction speaks for a step-wise mechanism. Oxyallyl cations react with conjugated dienes[175] and furan[176–178] to give 3 + 4-cycloadducts. More complex reactions, however, have been observed between oxyallyl cations and tropones (3 + 8-cycloaddition), and 6-(dimethylamino)fulvene (3 + 6-cycloaddition).[179] These reactions follow second-order kinetics and have a low sensitivity to solvent polarity, suggesting a concerted mechanism. Complexation with tricarbonyliron diverts the reaction between 8-(4-chlorophenyl)-8-azaheptafulvene and oxyallyl cations from a 3 + 8- to a 2 + 3-pathway.[180]

4 + 4-Cycloadducts have been obtained from reactions of 1,3-diazolium-4-olates and a 1-oxa analogue with tetrachloro-*o*-benzoquinone and an *o*-benzoquinone diimine.[181] The photo-dimerization of *N*-carboxyalkyl-2-pyridones is affected by micellar interactions.[182] Novel 4 + 6-cycloadducts of the cyclopentadienone (**50a**) with cyclooctatetraene[183] and with *N*-(ethoxycarbonyl)azepine[184] have been reported; the azepine derivative (**88**) spontaneously dimerizes in a 4 + 6-*exo-anti* manner.[185] The anion (**89**), derived from addition of 2-lithiopyridine to 2,2′-bipyridyl, dimerizes to a stable dianion in a 4 + 6-cycloaddition.[186] 8,8-Dimethylisobenzofulvene (**90**) spontaneously forms an *endo* and an *exo* 8 + 10-dimer.[187] In principle, there are 32 different possible modes of dimerization, but PMO calculations correctly favour the two observed products.

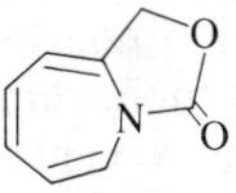

(88) 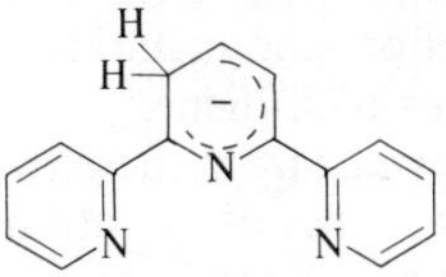 **(89)** 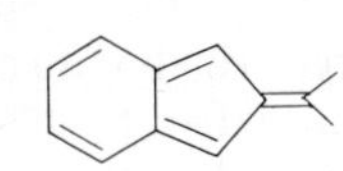 **(90)**

Cycloadditions of triazoline-3,5-diones to bicyclo[3.2.1]octa-2,6-diene[188] and to *exo*- and *endo*-tricyclo[3.2.1.0^{2,4}]oct-6-enes[189] give mixtures of rearranged and unrearranged adducts, best rationalized by assumption of zwitterionic intermediates. The "inside–outside" bicyclo[*n*.2.2]alkadienes (**91**; $n = 6$–8) react with *N*-phenyltriazolinedione to give predominantly the rearranged cycloadducts (**92**; $n = 6$–8).[190] This type of process is unprecedented, but may also be rationalized on the basis of a zwitterionic pathway (**93**). Nevertheless, such a route seems to be ruled out by a very low dependence of the rate on solvent polarity; indeed the reaction is faster in benzene than in acetonitrile. Polar cycloadditions of triazolinediones to heptalene[191] and to heptafulvene[192] have also been discussed.

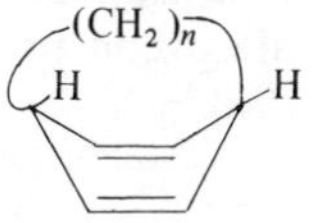

(91) 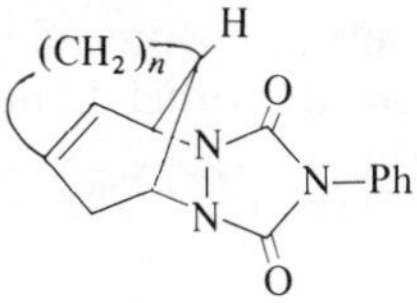 **(92)** 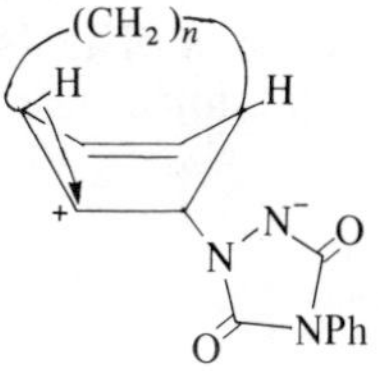 **(93)**

Gilbert has reviewed his own extensive work on inter- and intra-molecular photo-cycloadditions of alkenes to aromatic compounds,[193] and a much less accessible review of photo-cycloadditions of aromatic nitriles has appeared.[194] The regio- and stereo-selectivities of *ortho*- and *meta*-photo-adducts of alkenes to benzene and mono-substituted benzenes have been examined.[195] The *meta*-adducts usually exhibit high degrees of selectivity, but for the *ortho*-adducts there seems to be no simple underlying feature which controls the regio- and stereo-selectivities of the reaction. The *meta*-photo-cycloaddition of cyclopentene to *o*-methylanisole probably involves an exciplex.[196] Photo-cycloadditions of benzene with 1,2-, 1,3- and 1,4-dienes have been reported.[197] Allene reacts mainly in the 1,2-*para*-mode; the conformationally fixed *cis*-diene, 1,2-dimethylenecyclohexane, undergoes selective 1,4-*para*- and 1,4-*meta*-addition, but more flexible 1,3-dienes give complex mixtures; penta-1,4-diene and cyclohexa-1,4-diene yield 1,2-*meta*-cycloadducts. Photo-reactions of benzenes with furans or cyclic 1,3-dienes give 1,4–1,4-photo-adducts as the major products in most cases, but *meta*- and *ortho*-adducts may also be formed.[198] Photo-cycloadditions of 9-cyanoanthracene with α-terpinene and 1-methoxycyclohexa-1,3-diene also give 2 + 4- and 4 + 4-adducts, probably *via* two distinct biradicals in each case.[199] Photo-cycloaddition of cyclohexa-1,3-diene to a variety of polynuclear arenes has been investigated.[200] The results were found to be difficult to explain in terms of Woodward–Hoffmann or derivative theories, and it was suggested that local nodal patterns in the FMOs about the reacting point may be of importance. The 4 + 4-photo-dimerization of anthracene is impeded not enhanced by the presence of hexa-2,4-diene.[201] This observation contrasts with earlier reports that an exciplex formed from these two species led to the dimer faster than alternative routes. Measurements of quantum yields of fluorescence, photo-reaction, and internal deactivation have led one author to conclude that biradical intermediates do not take part in the photo-cycloadditions of α,ω-(bis-9-anthryl)-*n*-alkanes,[202] but this view has been vigorously criticized.[203] Photo-cycloadditions of cycloheptatriene to substituted anthracenes[204] and of cyclopentene to C_6F_5OR (R = Me, Et, Me_2CH, Me_3C)[205] have been reported, and the photo-cycloaddition of alkenes to arenes has found an application in the synthesis of (±)-α-cedrene.[206]

Cycloadditions of three or more addends seem most readily brought about by metal catalysts. Recent examples include the nickel-catalysed cyclotrimerization of alkynes,[207] palladium-catalysed cyclo-oligomerizations of 3,3-dialkylcyclo-propenes,[208] and cobalt-catalysed 2 + 2 + 2-cycloadditions of enediynes,[209] α,ω-enynes to alkynes,[210] and (trimethylsilyl)acetylene to carbon monoxide.[211] The most entertaining chemistry in this area, however, concerns compounds such as (**94**) and (**95**) that contain sterically fixed norbornadiene or quadricyclane systems.[212] In line with expectations based on models, the [3]-series of compounds (*e.g.* **94**; X = CH_2 or O) undergo only intracyclic processes, *viz.* the photochemical conversion of a norbornadiene moiety into a quadricyclane and the thermal reversal of this reaction. The [5]-series, on the other hand, prefer intercyclic processes; heating (**95**), for example, produces the $_{\pi}2 + _{\sigma}2 + _{\sigma}2$ adduct (**96**).

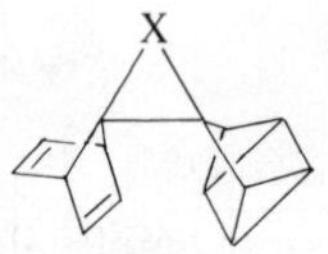

X

(**94**)

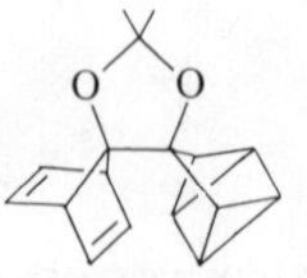

O O

(**95**)

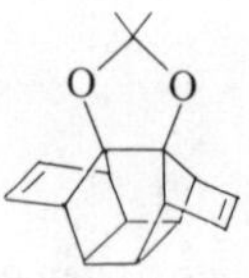

O O

(**96**)

References

[1] *Org. Reaction Mech.*, **1980,** 471.

[2] Kametani, T., and Nemoto, H., *Tetrahedron*, **37,** 3 (1981).

[3] Wasserman, H. H., and Ives, J. L., *Tetrahedron*, **37,** 1825 (1981).

[4] Huisgen, R., *Pure Appl. Chem.*, **53,** 171 (1981).

[5] Jug, K., and Dwivedi, C. P. D., *Theor. Chim. Acta*, **59,** 357 (1981); *Chem. Abs.*, **95,** 96778 (1981).

[6] Fernando, B., Bottoni, A., Battaglia, A., Distefano, G., and Dondoni, A., *Z. Naturforsch.*, **35A,** 521 (1980); *Chem. Abs.*, **93,** 185234 (1980).

[7] Sasaki, T., Hayakawa, K., Manabe, T., and Nishida, S., *J. Am. Chem. Soc.*, **103,** 565 (1981).

[8] Kassab, E., Evleth, E. M., Dannenberg, J. J., and Rayez, J. C., *Chem. Phys.*, **52,** 151 (1980); *Chem. Abs.*, **95,** 96783 (1981).

[9] Jug, K., and Mueller, P. L., *Theor. Chim. Acta*, **59,** 365 (1981); *Chem. Abs.*, **95,** 96779 (1981).

[10] Chadha, R., and Ray, N. K., *Proc. Indian Acad. Sci.*, [*Ser.*]*: Chem. Sci.*, **89,** 539 (1980); *Chem. Abs.*, **94,** 191459 (1981).

[11] Egerton, P. L., Hyde, E. M., Trigg, J., Payne, A., Beynon, P., Mijovic, M. V., and Reiser, A., *J. Am. Chem. Soc.*, **103,** 3859 (1981).

[12] Mizuno, K., Ueda, H., and Otsuji, Y., *Chem. Lett.*, **1981,** 1237.

[13] Morrison, H., Izuno, C., Byrn, S., and McKenzie, A., *Photochem. Photobiol.*, **34,** 107 (1981); *Chem. Abs.*, **95,** 149530 (1981).

[14] Jendralla, H., *Chem. Ber.*, **113,** 3557 (1980).

[15] Solomonov, B. N., Arkhireeva, I. A., and Konovalov, A. I., *Zh. Org. Khim.*, **16,** 1670 (1980); *Chem. Abs.*, **94,** 29748 (1981).

[16] Solomonov, B. N., Arkhireeva, I. A., and Konovalov, A. I., *Zh. Org. Khim.*, **16,** 1666 (1980); *Chem. Abs.*, **94,** 14959 (1981).

[17] Sasaki, M., Tsuzuki, H., and Osugi, J., *J. Chem. Soc., Perkin Trans. 2*, **1980,** 1596.

[18] Josey, A. D., *Angew. Chem. Int. Ed.*, **20,** 686 (1981).

[19] Hanack, M., Harder, I., and Bofinger, K.-R., *Tetrahedron Lett.*, **22,** 553 (1981).

[20] Suzuki, T., Kumagai, Y., Yamakawa, M., and Noyori, R., *J. Org. Chem.*, **46,** 2846 (1981).

[21] Freilich, S. C., and Peters, K. S., *J. Am. Chem. Soc.*, **103,** 6255 (1981).

[22] Kossanyi, J., Jost, P., Furth, B., Daccord, G., and Chaquin, P., *J. Chem. Res. (S)*, **1980,** 368.

[23] Mazzocchi, P. H., Khachik, F., Wilson, P., and Highet, R., *J. Am. Chem. Soc.*, **103,** 6498 (1981).

[24] Sindler-Kulyk, M., Neckers, D. C., and Blount, J. R., *Tetrahedron*, **37,** 3377 (1981).

[25] Adam, W., and Rebollo, H., *Tetrahedron Lett.*, **22,** 3049 (1981).

[26] Barton, T. J., Burns, G. T., Arnold, E. V., and Clardy, J., *Tetrahedron Lett.*, **22,** 7 (1981).

[27] *Org. Reaction Mech.*, **1979,** 445.

[28] Pasto, D. J., *Tetrahedron Lett.*, **21,** 4787 (1980).

[29] Snider, B. B., and Spindell, D. K., *J. Org. Chem.*, **45,** 5017 (1980).

[30] Moore, H. W., and Wilbur, D. S., *J. Org. Chem.*, **45,** 4483 (1980).

[31] Gam Garyan, N. P., Avetisyan, E. A., Del'tsova, D. P., and Safronova, Z. V., *Arm. Khim. Zh.*, **34,** 380 (1981); *Chem. Abs.*, **95,** 149483 (1981).

[32] Dondoni, A., Giorgianni, P., Battaglia, A., and Andreetti, G. D., *J. Chem. Soc., Chem. Commun.*, **1981,** 350.

[33] Heine, H.-G., and Hartmann, W., *Angew. Chem. Int. Ed.*, **20,** 782 (1981).

[34] Schaumann, E., Bäuch, H.-G., and Adiwidjaja, G., *Angew. Chem. Int. Ed.*, **20,** 613 (1981).

[35] Ficini, J., Krief, A., Guingant, A., and Desmaele, D., *Tetrahedron Lett.*, **22,** 725 (1981).

[36] Reinhoudt, D. N., Geevers, J., Trompenaars, W. P., Harkema, S., and van Hummel, G. J., *J. Org. Chem.*, **46,** 424 (1981).

[37] Baldwin, S. W., *Org. Photochem.*, **5,** 123 (1981); *Chem. Abs.*, **94,** 207939 (1981).

[38] Brady, W. T., *Tetrahedron*, **37,** 2949 (1981).

[39] Carrié, R., *Heterocycles*, **14,** 1529 (1980); *Chem. Abs.*, **94,** 3289 (1981).

[40] Galishev, V. A., Chistokletov, V. N., and Petrov, A. A., *Usp. Khim.*, **49,** 1801 (1980); *Chem. Abs.*, **94,** 14585 (1981).

[41] Kauffmann, T., *Top. Curr. Chem.*, **92,** 109 (1980).

[42] Van Tilborg, M. W. E. M., Van Doorn, R., and Nibbering, N. M. M., *Mass Spectrom.*, **15,** 152 (1980); *Chem. Abs.*, **94,** 29722 (1981).

[43] Bened, A., Durand, R., Pioch, D., Geneste, P., Declercq., J. P., Germain, G., Rambaud, J., and Roques, R., *J. Org. Chem.*, **46,** 3502 (1981).

[44] Del Mazza, D., and Reinecke, M. G., *J. Chem. Soc., Chem. Commun.*, **1981,** 124.

[45] Chopra, S. K., Hynes, M. J., and McArdle, P., *J. Chem. Soc., Dalton Trans.*, **1981,** 586.

[46] Chopra, S. K., Moran, G., and McArdle, P., *J. Organomet. Chem.*, **214,** C36 (1981).
[47] Binger, P., and Schuchardt, U., *Chem. Ber.*, **114,** 3313 (1981).
[48] Binger, P., and Germer, A., *Chem. Ber.*, **114,** 3325 (1981).
[49] Lutsenko, Z. L., Kisin, A. V., Kuznetsova, M. G., Bezrukova, A. A., Khandkarova, V. S., and Rubezhov, A. Z., *Inorg. Chim. Acta.*, **53,** L28 (1981).
[50] Samuilov, Ya. D., Movchan, A. I., and Konovalov, A. I., *Zh. Org. Khim.*, **17,** 1205 (1981); *Chem. Abs.*, **95,** 149497 (1981).
[51] Sturm, H., Ongania, K.-H., Daly, J. J., and Klötzer, W., *Chem. Ber.*, **114,** 190 (1981).
[52] Quast, H., and Kees, F., *Chem. Ber.*, **114,** 787 (1981).
[53] Schaumann, E., Behr, H., Adiwidjaja, G., Tangerman, A., Lammerink, B. H. M., and Zwanenburg, B., *Tetrahedron*, **37,** 219 (1981).
[54] McManus, S. P., Ortiz, M., and Abramovitch, R. A., *J. Org. Chem.*, **46,** 336 (1981).
[55] Sasaki, T., Eguchi, S., Yamaguchi, M., and Esaki, T., *J. Org. Chem.*, **46,** 1800 (1981).
[56] Carrié, R., Yeung Lam Ko, Y. Y. C., DeSarlo, F., and Brandi, A., *J. Chem. Soc., Chem. Commun.*, **1981,** 1131.
[57] Brittelli, D. R., and Boswell, G. A., *J. Org. Chem.*, **46,** 316 (1981).
[58] Umano, K., Mizone, S., Tokisato, K., and Inoue, H., *Tetrahedron Lett.*, **22,** 73 (1981).
[59] Bianchi, G., Gandolfi, R., and De Micheli, C., *J. Chem. Res. (S)*, **1981,** 6.
[60] Meier, H., Heinzelmann, W., and Heimgartner, H., *Chimia*, **34,** 506 (1980); *Chem. Abs.*, **94,** 191252 (1981).
[61] Oida, T., Shimizu, T., Hayashi, Y., and Teramura, K., *Bull. Chem. Soc. Jpn.*, **54,** 1429 (1981).
[62] Burger, K., Goth, H., and Roth, W. D., *Z. Naturforsch.*, **35B,** 1426 (1980); *Chem. Abs.*, **94,** 120463 (1981).
[63] Burger, K., Tremmel, S., Roth, W.-D., and Goth, H., *J. Heterocycl. Chem.*, **18,** 247 (1981).
[64] Dietliker, K., Stegmann, W., and Heimgartner, H., *Heterocycles*, **14,** 929 (1980); *Chem. Abs.*, **93,** 238239 (1980).
[65] Fišera, L., Gaplovský, A., Timpe, H.-J., and Kováč, J., *Collect. Czech. Chem. Commun.*, **46,** 1504 (1981).
[66] Samuilov, Ya. D., Solov'eva, S. E., and Konovalov, A. I., *Zh. Org. Khim.*, **16,** 1228 (1980); *Chem. Abs.*, **93,** 185323 (1980).
[67] Gössinger, E., *Monatsh. Chem.*, **112,** 1017 (1981).
[68] Freeman, J. P., and Haddadin, M. J., *J. Org. Chem.*, **45,** 4898 (1980).
[69] Padwa, A., Koehler, K. F., and Rodriguez, A., *J. Am. Chem. Soc.*, **103,** 4974 (1981).
[70] Matsumoto, K., and Uchida, T., *J. Chem. Soc., Perkin Trans. 1*, **1981,** 73.
[71] Cristina, D., De Amici, M., De Micheli, C., and Gandolfi, R., *Tetrahedron*, **37,** 1349 (1981).
[72] Danion, D., Arnold, B., and Regitz, M., *Angew. Chem. Int. Ed.*, **20,** 113 (1981).
[73] Hussain, S., and Agosta, W. C., *Tetrahedron*, **37,** 3301 (1981).
[74] Wong, J. P. K., Fahmi, A. A., Griffin, G. W., and Bhacca, N. S., *Tetrahedron*, **37,** 3345 (1981).
[75] Schaumann, E., Grabley, S., Grabley, F.-F., Kausch, E., and Adiwidjaja, G., *Liebigs Ann. Chem.*, **1981,** 277.
[76] Schaumann, E., and Grabley, S., *Liebigs Ann. Chem.*, **1981,** 290.
[77] Huisgen, R., Matsumoto, K., and Ross, C. H., *Heterocycles*, **15,** 1131 (1981); *Chem. Abs.*, **95,** 6026 (1981).
[78] Merah, B., and Texier, F., *Bull. Soc. Chim. Fr.II*, **1980,** 552.
[79] Jacobi, P. A., Brownstein, A., Martinelli, M., and Grozinger, K., *J. Am. Chem. Soc.*, **103,** 239 (1981).
[80] Friedrichsen, W., and Schröer, W.-D., *Liebigs Ann. Chem.*, **1981,** 476.
[81] Friedrichsen, W., Schröer, W.-D., and Debaerdemaeker, T., *Liebigs Ann. Chem.*, **1981,** 491.
[82] Debaerdemaeker, T., Schröer, W.-D., and Friedrichsen, W., *Liebigs Ann. Chem.*, **1981,** 502.
[83] Gotthardt, H., and Huss, O. M., *Liebigs Ann. Chem.*, **1981,** 347.
[84] Ibata, T., Jitsuhiro, K., and Tsubokura, Y., *Bull. Chem. Soc. Jpn.*, **54,** 240 (1981).
[85] Surpateanu, G., Luchian, C., Constantinescu, M., Barboiu, V., and Petrovanu, M., *Bul. Inst. Politeh. Iasi, Sect. 2: Chim. Ing. Chim.*, **26,** 97 (1980); *Chem. Abs.*, **95,** 149812 (1981).
[86] Friege, H., Friege, H., and Dyong, I., *Chem. Ber.*, **114,** 1822 (1981).
[87] Wiering, P. G., and Steinberg, H., *J. Org. Chem.*, **46,** 1663 (1981).
[88] Saito, T., Musashi, T., and Motoki, S., *Bull. Chem. Soc., Jpn.*, **53,** 3377 (1980).
[89] El-Enany, M., El-Fattah, B. A., and El-Basil, S., *Indian J. Chem.*, **20B,** 592 (1981); *Chem. Abs.*, **95,** 149718 (1981).
[90] Fleischhauer, J., Asaad, A. N., Schleker, W., and Scharf, H.-D., *Liebigs Ann. Chem.*, **1981,** 306.
[91] Hess, B. A., Schaad, L. J., Herndon, W. C., Biermann, D., and Schmidt, W., *Tetrahedron*, **37,** 2983 (1981).

[92] Moll, K. K., Ramhold, K., Zimmermann, G., and Pohle, C., *J. Prakt. Chem.*, **322,** 1048 (1980); *Chem. Abs.*, **94,** 174017 (1981).
[93] Mori, M., Hayamizu, A., and Kanematsu, K., *J. Chem. Soc., Perkin Trans. 1*, **1981,** 1259.
[94] Samuilov, Ya. D., Nurullina, R. L., Uryadova, L. F., and Konovalov, A. I., *Zh. Org. Khim.*, **16,** 1858 (1980); *Chem. Abs.*, **94,** 29778 (1981).
[95] Samuilov, Ya. D., Bukharov, S. V., and Konovalov, A. I., *Zh. Org. Khim.*, **17,** 121 (1981); *Chem. Abs.*, **95,** 23986 (1981).
[96] Samuilov, Ya. D., Bukharov, S. V., and Konovalov, A. I., *Zh. Obshch. Khim.*, **51,** 1368 (1981); *Chem. Abs.*, **95,** 149499 (1981).
[97] Mustafaev, A. M., Adigezalov, N. R., Kiselev, V. D., Konovalov, A. I., and Guseinov, M. M., *Zh. Org. Khim.*, **16,** 2549 (1980); *Chem. Abs.*, **94,** 155957 (1981).
[98] Musaeva, N. F., Salakhov, M. S., and Umaeva, V. S., *Zh. Org. Khim.*, **17,** 116 (1981); *Chem. Abs.*, **95,** 23985 (1981).
[99] Musaeva, N. F., Salakhov, M. S., Aleskerov, A. A., Suleimanov, S. N., and Kopylova, T. A., *Zh. Org. Khim.*, **16,** 1376 (1980); *Chem. Abs.*, **94,** 3383 (1981).
[100] Salakhov, M. S., Musaeva, N. F., Israfilov, A. I., Gasanov, A. A., and Kopylova, T. A., *Dokl. Akad. Nauk Az. SSR*, **36,** 62 (1980); *Chem. Abs.*, **94,** 208033 (1981).
[101] Kanematsu, K., Harano, K., and Dantsuji, H., *Heterocycles*, **16,** 1145 (1981); *Chem. Abs.*, **95,** 114358 (1981).
[102] Broekhuis, A. A., Scheeren, J. W., and Nivard, R. J. F., *Recl. Trav. Chim. Pays-Bas*, **100,** 143 (1981).
[103] Overman, L. E., Freerks, R. L., Petty, C. B., Clizbe, L. A., Ono, R. K., Taylor, G. F., and Jessup, P. J., *J. Am. Chem. Soc.*, **103,** 2816 (1981).
[104] Tegmo-Larsson, I.-M., Rozeboom, M. D., and Houk, K. N., *Tetrahedron Lett.*, **22,** 2043 (1981).
[105] Tegmo-Larsson, I.-M., Rozeboom, M. D., Rondan, N. G., and Houk, K. N., *Tetrahedron Lett.*, **22,** 2047 (1981).
[106] Yoshino, S., Hayakawa, K., and Kanematsu, K., *J. Org. Chem.*, **46,** 3841 (1981).
[107] Warrener, R. N., Hammer, B. C., and Russell, R. A., *J. Chem. Soc., Chem. Commun.*, **1981,** 942.
[108] Snowden, R. L., *Tetrahedron Lett.*, **22,** 97 (1981).
[109] Schön, G., and Hopf, H., *Liebigs Ann. Chem.*, **1981,** 165.
[110] Böhm, M. C., Carr, R. V. C., Gleiter, R., and Paquette, L. A., *J. Am. Chem. Soc.*, **102,** 7218 (1980).
[111] Haganbuch, J.-P., Vogel, P., Pinkerton, A. A., and Schwarzenbach, D., *Helv. Chim. Acta*, **64,** 1818 (1981).
[112] Liotta, D., Saindane, M., and Barnum, C., *J. Am. Chem. Soc.*, **103,** 3224 (1981).
[113] Auksi, H., and Yates, P., *Can. J. Chem.*, **59,** 2510 (1981).
[114] Uosaki, Y., Nakahara, M., and Osugi, J., *Bull. Chem. Soc. Jpn.*, **54,** 2569 (1981).
[115] Roberts, R. M. G., and Yavari, F., *Tetrahedron*, **37,** 2657 (1981).
[116] Kiselev, V. D., Mavrin, G. V., and Konovalov, A. I., *Zh. Org. Khim.*, **16,** 1435 (1980); *Chem. Abs.*, **93,** 238515 (1980).
[117] Yasuda, M., Harano, K., and Kanematsu, K., *J. Org. Chem.*, **46,** 3836 (1981).
[118] Sergeev, G. B., Komarov, V. S., and Putyatin, A. A., *Vestn. Mosk. Univ. Ser. 2: Khim.*, **21,** 349 (1980); *Chem. Abs.*, **94,** 14737 (1981).
[119] Efendiev, E. Kh., Rasulbekova, T. I., and Mekhtiev, S. I., *Azerb. Khim. Zh.*, **1980,** 103; *Chem. Abs.*, **94,** 120508 (1981).
[120] Kiselev, B. D., Konovalov, A. I., Mavrin, G. V., Korshin, E. E., and Dimukhametov, M. N., *Deposited Doc.*, **1980,** *SPSTL 601 khp-D80*; *Chem. Abs.*, **95,** 168065 (1981).
[121] Bellville, D. J., Wirth, D. D., and Bauld, N. L., *J. Am. Chem. Soc.*, **103,** 718 (1981).
[122] Rideout, D. C., and Breslow, R., *J. Am. Chem. Soc.*, **102,** 7816 (1980).
[123] Firestone, R. A., and Vitale, M. A., *J. Org. Chem.*, **46,** 2160 (1981).
[124] Tou, J. S., and Reusch, W., *J. Org. Chem.*, **45,** 5012 (1980).
[125] Roush, W. R., and Gillis, H. R., *J. Org. Chem.*, **45,** 4267 (1980).
[126] Roush, W. R., Ko, A. I., and Gillis, H. R., *J. Org. Chem.*, **45,** 4264 (1980).
[127] van Leeuwen, P. W. N. M., and Roobeek, C. F., *Tetrahedron*, **37,** 1973 (1981).
[128] Bachner, J., Huber, U., and Buchbauer, G., *Monatsh. Chem.*, **112,** 679 (1981).
[129] Oppolzer, W., Kurth, M., Reichlin, D., and Moffatt, F., *Tetrahedron Lett.*, **22,** 2545 (1981).
[130] Helmchen, G., and Schmierer, R., *Angew. Chem. Int. Ed.*, **20,** 205 (1981).
[131] Trost, B. M., O'Krongly, D., and Belletire, J. L., *J. Am. Chem. Soc.*, **102,** 7595 (1980).
[132] Tolbert, L. M., and Ali, M. B., *J. Am. Chem. Soc.*, **103,** 2104 (1981).
[133] Vathke-Ernst, H., and Hoffmann, H. M. R., *Chem. Ber.*, **114,** 1464 (1981).
[134] Hoffmann, H. M. R., and Vathke-Ernst, H., *Chem. Ber.*, **114,** 2208 (1981).
[135] Mayr, H., and Schütz, F., *Tetrahedron Lett.*, **22,** 925 (1981).

[136] Bakker, C. G., Scheeren, J. W., and Nivard, R. J. F., *Recl. Trav. Chim. Pays-Bas*, **100,** 13 (1981).
[137] Bargagna, A., Schenone, P., Bondavalli, F., and Longobardi, M., *J. Heterocycl. Chem.*, **17,** 1201 (1980).
[138] Capuano, L., and Tammer, T., *Chem. Ber.*, **114,** 456 (1981).
[139] Lanteri, P., Longeray, R., and Royer, J., *J. Chem. Res.*, (*S*), **1981,** 168.
[140] Friedrichsen, W., Schmidt, R., van Hummel, G. J., and van der Ham, D. M. W., *Liebigs Ann. Chem.*, **1981,** 521.
[141] Righetti, P. P., Gamba, A., Tacconi, G., and Desimoni, G., *Tetrahedron*, **37,** 1779 (1981).
[142] Hartke, K., Quante, J., and Kämpchen, T., *Liebigs Ann. Chem.*, **1980,** 1482.
[143] Hartke, K., Henssen, G., and Kissel, T., *Liebigs Ann. Chem.*, **1980,** 1665.
[144] Ohmura, H., and Motoki, S., *Chem. Lett.*, **1981,** 235.
[145] Tamura, Y., Ishiyama, K., Mizuki, Y., Maeda, H., and Ishibashi, H., *Tetrahedron Lett.*, **22,** 3773 (1981).
[146] Rammash, B. K., Gladstone, C. M., and Wong, J. L., *J. Org. Chem.*, **46,** 3036 (1981).
[147] McEwen, W. E., Hernandez, M. A., Ling, C.-F., Marmugi, E., Padronaggio, R. M., Zepp, C. M., and Lubinkowski, J. J., *J. Org. Chem.*, **46,** 1656 (1981).
[148] Cheng, Y.-S., Fowler, F. W., and Lupo, A. T., *J. Am. Chem. Soc.*, **103,** 2090 (1981).
[149] Francotte, E., Merényi, R., Vandenbulcke-Coyette, B., and Viehe, H.-G., *Helv. Chim. Acta*, **64,** 1208 (1981).
[150] Kresze, G., and Rössert, M., *Liebigs Ann. Chem.*, **1981,** 58.
[151] Neemann, J., and Klingebiel, U., *Liebigs Ann. Chem.*, **1980,** 1978.
[152] Peled, M., and Ginsburg, D., *Tetrahedron*, **37,** 151 (1981).
[153] Peled, M., and Ginsburg, D., *Tetrahedron*, **37,** 161 (1981).
[154] Böhm, M. C., and Gleiter, R., *Tetrahedron*, **36,** 3209 (1980).
[155] Ashkenazi, P., Gleiter, R., von Philipsborn, W., Bigler, P., and Ginsburg, D., *Tetrahedron*, **37,** 127 (1981).
[156] Karpf, H., Polansky, O. E., and Zander, M., *Monatsh. Chem.*, **112,** 659 (1981).
[157] Ichimura, K., Watanabe, S., Ueno, K., and Ochi, H., *Nippon Kagaku Kaishi*, **1980,** 846; *Chem. Abs.*, **93,** 185480 (1980).
[158] Arnold, D. R., Borg, R. M., and Albini, A., *J. Chem. Soc., Chem. Commun.*, **1981,** 138.
[159] Wagner-Jauregg, T., *Synthesis*, **1980,** 769; *Chem. Abs.*, **94,** 102212 (1981).
[160] Contreras, L., and MacLean, D. B., *Can. J. Chem.*, **58,** 2573 (1980).
[161] Contreras, L., and MacLean, D. B., *Can. J. Chem.*, **58,** 2580 (1980).
[162] Reinecke, M. G., Newsom, J. G., and Chen, L.-J., *J. Am. Chem. Soc.*, **103,** 2760 (1981).
[163] Gaviña, F., Costero, A. M., Gil, P., Palazón, B., and Luis, S. V., *J. Am. Chem. Soc.*, **103,** 1797 (1981).
[164] Straub, H., and Miller, B., *Z. Naturforsch.*, **35B,** 1484 (1980); *Chem. Abs.*, **94,** 155940 (1981).
[165] Miyashi, T., Fujii, Y., Nishizawa, Y., and Mukai, T., *J. Am. Chem. Soc.*, **103,** 725 (1981).
[166] Padwa, A., and Rodriguez, A., *Tetrahedron Lett.*, **22,** 187 (1981).
[167] Padwa, A., and Nahm, S., *J. Org. Chem.*, **46,** 1402 (1981).
[168] Niecke, E., and Wildbredt, D.-A., *J. Chem. Soc., Chem. Commun.*, **1981,** 72.
[169] Kollenz, G., Ott, W., Ziegler, E., Peters, K., von Schnering, H. G., and Quast, H., *Liebigs Ann. Chem.*, **1980,** 1801.
[170] Ege, G., and Gilbert, K., *J. Heterocycl. Chem.*, **18,** 675 (1981).
[171] Knothe, L., Prinzbach, H., and Hädicke, E., *Chem. Ber.*, **114,** 1656 (1981).
[172] Hoffmann, H. M. R., and Vathke-Ernst, H., *Chem. Ber.*, **114,** 2898 (1981).
[173] Mayr, H., Wilhelm, E., and Kaliba, C., *J. Chem. Soc., Chem., Commun.*, **1981,** 683.
[174] Mayr, H., Seitz, B., and Halberstadt-Kausch, I. K., *J. Org. Chem.*, **46,** 1041 (1981).
[175] Hoffmann, H. M. R., and Matthei, J., *Chem. Ber.*, **113,** 3837 (1980).
[176] Föhlisch, B., Gottstein, W., Herter, R., and Wanner, I., *J. Chem. Res.* (*S*), **1981,** 246.
[177] Föhlisch, B., and Gottstein, W., *J. Chem. Res.* (*S*), **1981,** 94.
[178] Mann, J., and Usmani, A. A., *J. Chem. Soc., Chem. Commun.*, **1980,** 1119.
[179] Ishizu, T., Mori, M., and Kanematsu, K., *J. Org. Chem.*, **46,** 526 (1981).
[180] Ishizu, T., Harano, K., Yasuda, M., and Kanematsu, K., *J. Org. Chem.*, **46,** 3630 (1981).
[181] Friedrichsen, W., Schröer, W.-D., and Debaerdemaeker, T., *Liebigs Ann. Chem.*, **1980,** 1850.
[182] Nakamura, Y., Kato, T., and Morita, Y., *Tetrahedron Lett.*, **22,** 1025 (1981).
[183] Yasuda, M., Harano, K., and Kanematsu, K., *J. Am. Chem. Soc.*, **103,** 3120 (1981).
[184] Harano, K., Yasuda, M., Ban, T., and Kanematsu, K., *J. Org. Chem.*, **45,** 4455 (1980).
[185] Meth-Cohn, O., and Rhouati, S., *J. Chem. Soc., Chem. Commun.*, **1980,** 1161.
[186] Newkome, G. R., Hager, D. C., and Fronczek, F. R., *J. Chem. Soc., Chem. Commun.*, **1981,** 858.

[187] Warrener, R. N., Paddon-Row, M. N., Russell, R. A., and Watson, P. L., *Aust. J. Chem.*, **34,** 397 (1981).
[188] Adam, W., and De Lucchi, O., *Tetrahedron Lett.*, **22,** 3501 (1981).
[189] Erden, I., *Chem. Lett.*, **1981,** 263.
[190] Gassman, P. G., and Hoye, R. C., *J. Am. Chem. Soc.*, **103,** 2496 (1981).
[191] Horn, K. A., Browne, A. R., and Paquette, L. A., *J. Org. Chem.*, **45,** 5381 (1980).
[192] Erden, I., and Kaufmann, D., *Tetrahedron Lett.*, **22,** 215 (1981).
[193] Gilbert, A., *Pure Appl. Chem.*, **52,** 2669 (1980).
[194] Sakurai, H., and Pac, C., *Mem. Inst. Sci. Ind. Res., Osaka Univ.*, **37,** 59 (1980); *Chem. Abs.*, **93,** 219974 (1980).
[195] Gilbert, A., and Yianni, P., *Tetrahedron*, **37,** 3275 (1981).
[196] Jans, A. W. H., and Cornelisse, J., *Recl. Trav. Chim. Pays-Bas*, **100,** 213 (1981).
[197] Berridge, J. C., Forrester, J., Foulger, B. E., and Gilbert, A., *J. Chem. Soc., Perkin Trans. 1*, **1980,** 2425.
[198] Cantrell, T. S., *J. Org. Chem.*, **46,** 2674 (1981).
[199] Kaupp, G., and Schmitt, D., *Chem. Ber.*, **114,** 1567 (1981).
[200] Yang, N. C., Masnovi, J., Chiang, W., Wang, T., Shou, H., and Yang, D.-D.H., *Tetrahedron*, **37,** 3285 (1981).
[201] Kaupp, G., and Teufel, E., *Chem. Ber.*, **113,** 3669 (1980).
[202] Ferguson, J., *Chem. Phys. Lett.*, **76,** 398 (1980); *Chem. Abs.*, **94,** 102449 (1981).
[203] Jones, G., Bergmark, W. R., and Halpern, A. M., *Chem. Phys. Lett.*, **76,** 403 (1980); *Chem. Abs.*, **94,** 102450 (1981).
[204] Kondo, H., Mori, M., and Kanematsu, K., *J. Org. Chem.*, **45,** 5273 (1980).
[205] Sket, B., and Zupan, M., *Croat. Chem. Acta*, **52,** 387 (1979); *Chem. Abs.*, **93,** 220304 (1980).
[206] Wender, P. A., and Howbert, J. J., *J. Am. Chem. Soc.*, **103,** 688 (1981).
[207] Bicev, P., Furlani, A., and Russo, M. V., *Gazz. Chim. Ital.*, **110,** 25 (1980); *Chem. Abs.*, **93,** 238120 (1980).
[208] Binger, P., and Schuchardt, U., *Chem. Ber.*, **114,** 1649 (1981).
[209] Gadek, T. R., and Vollhardt, K. P. C., *Angew. Chem. Int. Ed.*, **20,** 802 (1981).
[210] Chang, C.-A., King, J. A., and Vollhardt, K. P. C., *J. Chem. Soc., Chem. Commun.*, **1981,** 53.
[211] Gesing, E. R. F., Tane, J. P., and Vollhardt, K. P. C., *Angew. Chem. Int. Ed.*, **19,** 1023 (1980).
[212] Prinzbach, H., Weidmann, K., Trah, S., and Knothe, L., *Tetrahedron Lett.*, **22,** 2541 (1981).

Organic Reaction Mechanisms 1981
Edited by A. C. Knipe and W. E. Watts

CHAPTER 14

Molecular Rearrangements

A. W. MURRAY

Department of Chemistry, University of Dundee

Aromatic Rearrangements . . . 460
Benzene Derivatives . . . 460
Heterocyclic Derivatives . . . 465
Cyclohexadiene Derivatives . . . 469
Sigmatropic Rearrangements . . . 472
[3,3]-Migrations . . . 472
[2,3]-Migrations . . . 480
[1,3]-Migrations . . . 483
[1,5]-Migrations . . . 485
Miscellaneous . . . 488
Electrocyclic Reactions . . . 490
Anionic Rearrangements . . . 495
Cationic Rearrangements . . . 500
Rearrangements in Polycyclic Systems . . . 506
Rearrangements in Natural-product Systems . . . 509
Miscellaneous . . . 513
Rearrangements involving Electron-deficient Hetero-atoms . . . 514
Metal-catalysed Rearrangements . . . 516
Rearrangements Involving Ring-opening and Ring-closure . . . 519
Three-membered Rings . . . 519
Four-membered Rings . . . 526
Five-membered Rings . . . 528
Six-membered and Larger Rings . . . 531
Isomerizations . . . 533
Tautomerism . . . 537
Acyl and Related Migrations . . . 538
Addendum . . . 539
Photochemical Studies . . . 539
References . . . 540

Aromatic Rearrangements

Benzene Derivatives

A review that outlines a theoretical approach to rearrangements,[1] a general account of benzene rearrangements,[2] and an analysis of the use of isotopes in the mechanistic studies of aromatic cationic rearrangements,[3] have appeared. A quantitative description of 1,2-migrations in benzenonium and phenanthrenium derivatives has been reported[4] and the energetics of a 1,2-proton shift in protonated benzene have been calculated[5] by the MINDO/3 method. Degenerate 1,2-hydrogen atom shifts in 1*H*-1,2-dimethyl-3,4,5,6-tetrachloro- and -tetrafluoro-benzenonium ions have been studied,[6] kinetic parameters have been determined for the rearrangements of dimethylpentachloro- and dimethylchlorotetrafluoro-benzenonium ions,[7] for the conversion of the 1-chloro-1,3,4,5,6-pentamethyl-2-hydroxybenzenonium ion into the 1-chloro-1,2,3,5,6-pentamethyl-4-hydroxy-benzenonium ion,[8] and for the rearrangement of *β*-nitro-substituted arenonium ions.[9] In this latter rearrangement, intermediate species having the nitrogen atom and one oxygen atom of the nitro group attached to adjacent carbon atoms are proposed to intervene. NMR data have indicated[10] that the degenerate rearrangement of the 1*H*-1,2,3,4-tetramethylnaphthalenonium ion proceeds by successive hydride shifts from C(1) to C(2), C(3), and C(4). The free energies of activation of the isomerization of protonated polymethylnaphthalenes have been calculated,[11] and the first report of naphthyl shifts on aromatic rings, *via* naphthalenonium ions has appeared.[12] A novel ring-expansion of anilides (**1**; X = R′CO) to 1*H*-azepines (**3**) *via* benzenonium ions (**2**) has been reported.[13]

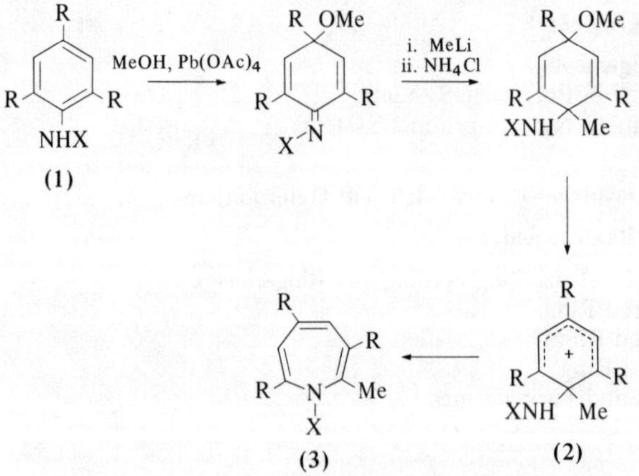

MeOH, $Pb(OAc)_4$

i. MeLi
ii. NH_4Cl

(1) (2) (3)

Evidence has been presented[14] for a concerted mechanism for the benzidine rearrangement; an interesting result, suggesting the intramolecular nature of the triazene rearrangement, has been obtained[15] on the basis of the effect of added *N*,*N*-dimethylaniline on the *ortho*-rearrangement of 1,3-bis(4-methylphenyl)triazene, and a detailed study[16] has shown that the experimental observations for the diazoamino rearrangement can only be explained by postulating two mechanisms, namely, a competitive complete back-reaction from the diazonium compound to the amine and the diazonium ion, and an attack of amine on a molecular complex formed between the diazonium ion and the amine.

Selective *ortho*-Wallach rearrangements of azoxybenzene–$SbCl_5$ complexes have been reported.[17] Comparative analysis of the kinetics of the Fischer–Hepp rearrangement of mono- and di-arylnitrosoamines has shown that rearrangement to the *C*-nitroso compound proceeds by an intermolecular mechanism involving acid-catalysed denitrosation and subsequent *C*-nitrosation.[18] With monoarylnitrosoamines, the rate-determining step is *C*-nitrosation, but with diarylnitrosoamines, it is denitrosation. The kinetics of the hydrohalide-catalysed rearrangement have also been determined.[19] A kinetic study of the Bamberger rearrangement of phenylhydroxylamines to *p*-aminophenols in aqueous H_2SO_4 has indicated that the reaction occurs by an S_N1 mechanism and that the elimination of water from $ArNH\overset{+}{O}H_2$ is rate-determining,[20] while *N*-alkyl-*N*-arylhydroxylamines have been shown to react with sulphuric acid in methanol to form nitrenium ions which combine with methoxide ion to give *o*- and *p*-$MeOC_6H_4NHR$.[21] Kinetic results are in agreement with a cyclic process, see (**4**) to (**5**), for the mechanism of the thermal rearrangement of *N*-aryl cyclic amine oxides to *O*-arylhydroxylamines.[22] A comprehensive analysis[23] of the rates for the trifluoroacetic-acid-catalysed rearrangement of *N*-bromo-4-chloroacetanilide to 2-bromo-4-chloroacetanilide has revealed that the rearrangement involves the intramolecular migration of bromine in at least three mechanistic steps. Thermolysis of *N*-(α-phenethyl)- and *N*-(β-phenethyl)-aniline results in the migration of the phenethyl group to the aniline nucleus in a process which has been interpreted[24] in terms of a free-radical mechanism.

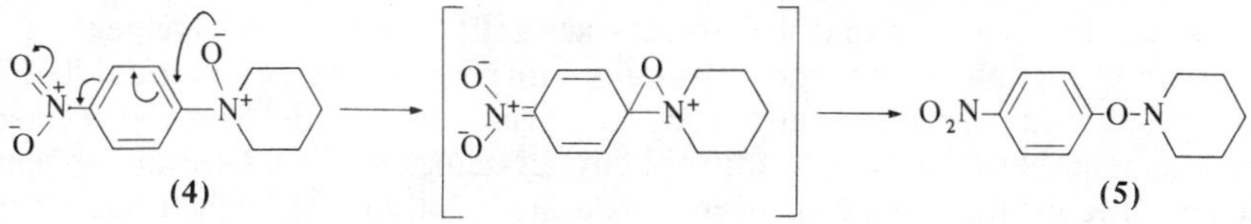

(4) (5)

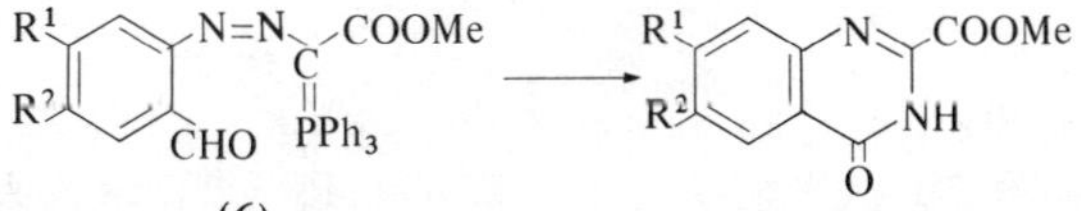

(6)

Arylazomethylenetriphenylphosphoranes (**6**) have been shown to afford quinazolin-4-one derivatives *via* a facile thermally promoted rearrangement involving N=N bond-cleavage and elimination of triphenylphosphine,[25] while *N*-aryl(triphenylphosphoranylidene)ketenimines (**7**) appear, on melting, to dimerize by [4 + 2]cycloaddition, proton migration, and Smiles rearrangement to yield [4-(diarylamino)-3(diphenylphosphino)-2-quinolylmethylene]triphenylphosphoranes (**8**); see Scheme 1.[26] A mechanism involving a radical ion-pair and a spiro-type Meisenheimer complex has been proposed[27,28] for the photo-induced intramolecular conversion of a series of 1-(*p*-nitrophenoxy)-ω-anilinoalkanes into *N*-(*p*-nitrophenyl)-ω-anilinoalkan-1-ols. A Smiles rearrangement has been utilized in the preparation of 3,5-diisopropylbenzoic acid.[29] The Smiles rearrangement of 2-acyloxy-5-nitrophenoxides has been described[30] as a possible route to tribenzo[*beh*]trioxonins, and the above-named rearrangement is considered to be involved in a recent conversion of 3-substituted 2-aminobenzenethiols into

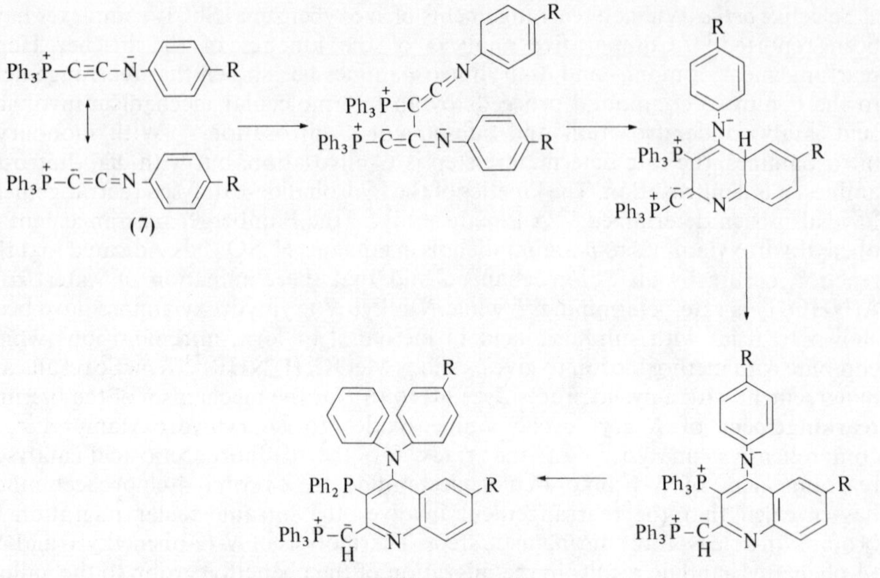

$Ph_3\overset{+}{P}-C\equiv C-\overset{-}{N}-C_6H_4-R \leftrightarrow Ph_3\overset{+}{P}-\overset{-}{C}=C=N-C_6H_4-R$

(7)

(8) SCHEME 1

phenothiazines.[31] The thermal rearrangement of (phenylthio)bromouracils (**9**) has been shown to yield pyrimidobenzothiazines (**10**).[32] The rearrangement of (ω-hydroxyalkyl)methyl (*p*-nitrophenyl)sulphonium perchlorates in aqueous alkali has been investigated[33] and a kinetic study has been carried out[34] on the reversible rearrangement between *S*-(2,4-dinitrophenyl)cysteine and *N*-(2,4-dinitrophenyl)-cysteine. Interestingly, it has been postulated[35] that the lichen metabolite, canesolide, arises by catabolism of its congeneric depsidone by a Smiles rearrangement.

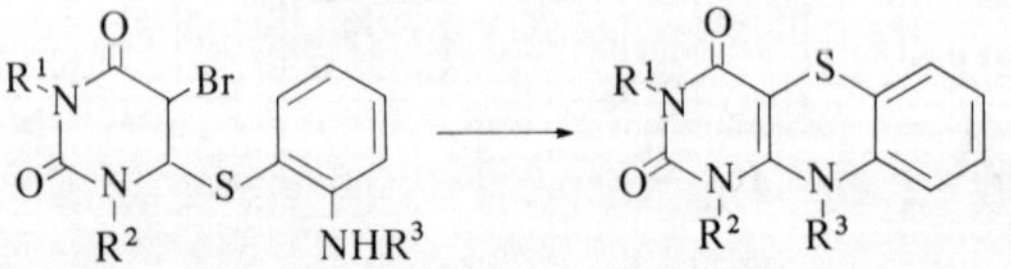

(9) (10)

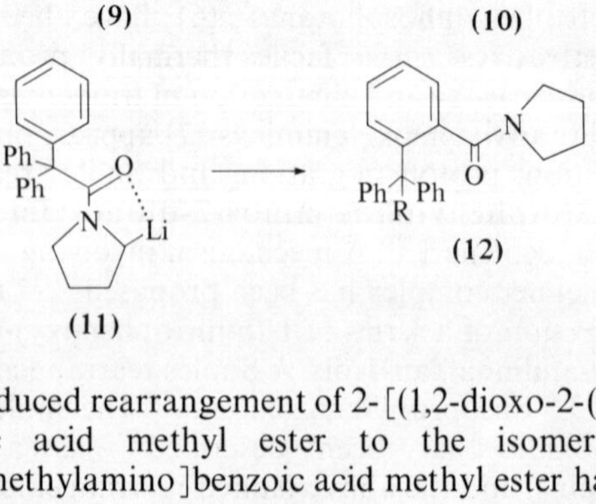

(11) (12)

An efficient base-induced rearrangement of 2-[(1,2-dioxo-2-(methylamino)ethyl-phenylamino]benzoic acid methyl ester to the isomeric 2-[(1,2-dioxo-2-(phenylamino)ethyl)methylamino]benzoic acid methyl ester has been described.[36] It is suggested that this novel rearrangement proceeds through a spiro intermediate with benzoate acting as a Michael receptor. Metalated triphenylacetamides, such as (**11**), have been shown to rearrange to (**12**; R = Li,H) by a process which involves a 1,3-shift of the carbamoyl group to the *ortho*-position of the benzene ring.[37] A high-yield conversion of phenols into 2-hydroxyphenylphosphonates (**14**) *via* base-catalysed rearrangement of diethyl phenyl phosphates, (**13**; R = Et) has been achieved[38] in the manner shown in Scheme 2. A study has been made of the mechanism of cyclization of *o*- and *m*-chloro-α-phenylcinnamic acids to phenanthrene derivatives[39] under the influence of potassium amide in liquid ammonia. It seems that, on exposure to the reagent, the substrates isomerize through ion-radicals. If an *ortho*-halogen is present, it is lost to give the phenyl radical which cyclizes to phenanthrene, while *meta*-halogenated reactants form benzyne intermediates which undergo ring-closure with the carboxy function. It has been demonstrated[40] that, although the three trichlorobenzenes fail to participate in the base-catalysed "halogen dance", 1,2,3,5- and 1,2,4,5-tetrachlorobenzenes do undergo this degenerate rearrangement.

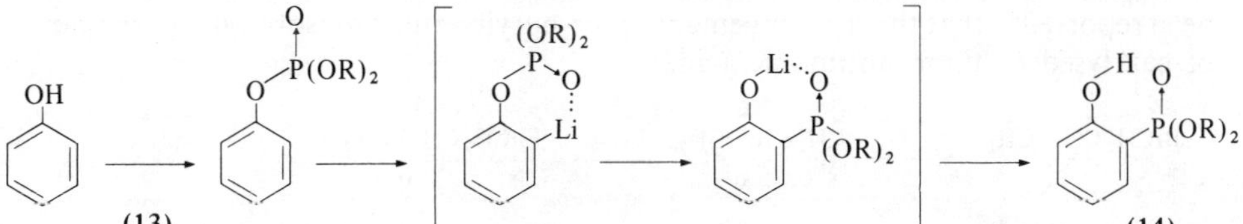

(**13**) (**14**)

SCHEME 2

Interestingly, the dibromodinitro compound (**15**), formed by nitration of 4,6-dibromo-2,5-dimethylphenol with fuming nitric acid in acetic acid, has been shown[41] to rearrange further in halocarbon solvents to the α-diketone (**16**). A possible rearrangement mechanism is outlined in Scheme 3. The same authors have likewise found[42] that dinitration of 2,4,5-tribromo-3,6-dimethylphenol (**17**) gives either the tribromodinitro compound (**18**) or its acyloin rearrangement product (**20**) depending on the reaction conditions. Formation of this cyclopentenol stereoisomer in the acyloin rearrangement of hydroxy-ketone (**18**) requires the rearrangement to occur in the half-chair conformation (**19**). Again, depending on reaction conditions, bromination of 3,3-bis(*p*-methoxyphenyl)propionic acid has been shown to give

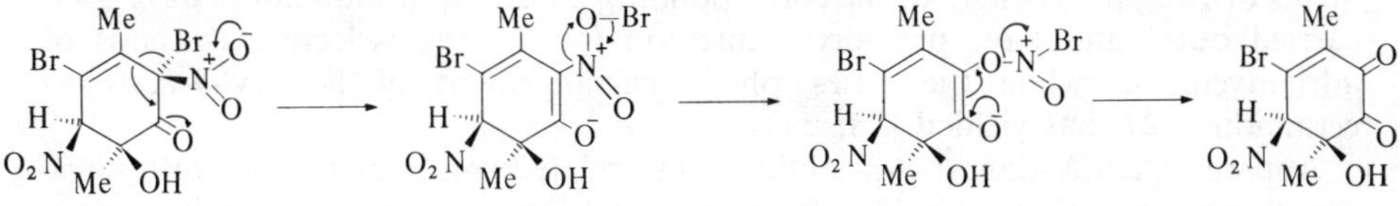

(**15**) (**16**)

SCHEME 3

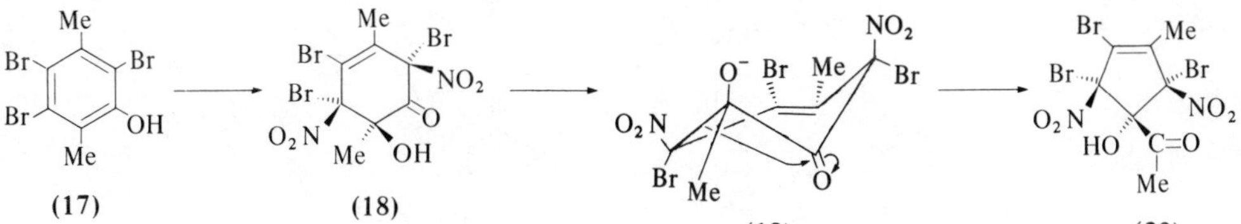

(**17**) (**18**) (**19**) (**20**)

both rearrangement and bromination products,[43] while the nature of the dienones derived from further bromination of the di- and poly-bromophenols obtained from *o*-, *m*-, and *p*-cresols have been reinvestigated[44] in an attempt to rationalize the course and regiospecificity of their rearrangements to ring-substituted polybromophenols.

The Friedel–Crafts reaction of 1-benzenesulphonyl-2-(bromomethyl)ethylenimine-2-^{14}C (**21**) and benzene was shown to yield 3,3-diphenyl-1-benzenesulphonamidopropane-2-^{14}C (**26**).[45] These data, coupled with the results of an earlier tracer study, have led to a revised pathway for the process. The first step is taken as the generation of carbocation (**22**) and alkylation of benzene to form 1-benzenesulphonyl-2-benzylethylenimine (**23**). Next, under the influence of $AlCl_3$, the ring in (**23**) opens to give carbocation (**24**) which by hydride shift forms the more stable benzylic ion (**25**). This probably serves as the immediate precursor of the final product (**26**). Results obtained on the alkylation of toluene with but-1-ene to give *sec*-butyltoluene, indicate[46] that the initially formed *meta*- and *para*-products isomerize mainly by an intermolecular mechanism to the *meta*-isomer, while it has been reported[47] that the rearrangement of *sec*-butylbenzene to isobutylbenzene can be catalysed by fluoroantimonic acid.

$BrCH_2\overset{*}{C}H$—CH_2 (N–SO_2Ph) (21) ⟶ $\overset{+}{C}H_2CH$—CH_2 (N–SO_2Ph) (22) $\xrightarrow{C_6H_6}$ $PhCH_2CH$—CH_2 (N–SO_2Ph) (23) $\xrightarrow{AlCl_3}$ $PhCH_2\overset{+}{C}H$—CH_2–$\bar{N}(AlCl_3)SO_2Ph$ (24) ⟶ $Ph\overset{+}{C}HCH_2CH_2NHSO_2Ph$ (25) $\xrightarrow{C_6H_6}$ $Ph_2CH\overset{*}{C}H_2CH_2NHSO_2Ph$ (26)

A comparative applied study of the Fries rearrangement has been made.[48] This rearrangement of the benzoates of 3-hydroxy-6,7-dimethoxycoumarins has been reported,[49] and the synthesis of orcinoylhydroquinones has been achieved by photochemical Fries rearrangement of their esters.[50] The photo-Fries rearrangements of 1-naphthyl esters to the corresponding 2-acylnaphthoquinones have been carried out[51] and have provided a method for the regioselective synthesis of adriamycinone, while the Fries photo-rearrangement of 5-acetyl-9-acetoxyceramidine (**27**) has yielded (**28**).[52]

Apparently, *N*,*N*-dimethyl-*p*-toluidine and related compounds react with nitric acid in aqueous sulphuric acid to form the *ipso*-Wheland intermediate with a nitro

NAc, OAc (27) ⟶ N, Ac, O (28)

group at the 4-position. However, where the *ortho*-positions are unsubstituted, the formation of the *ipso*-Wheland intermediate is followed by a 1,3-rearrangement to give the 2-nitro product. This rearrangement appears to be mainly intramolecular, but studies of isotopic exchange in concurrent rearrangements show that there is also an intermolecular component and this is considered to involve the cation radical ($ArNMe_2^{+\bullet}$) and nitrogen dioxide.[53] *Ipso*-attack at the 1-position of the naphthalene nucleus, followed by rearrangement, has also been invoked[54] to account for the cyclization of 1-naphthylbutanols to tetrahydrophenanthrene in the presence of $BF_3 \cdot OEt_2$.

A mechanism involving synchronous rearrangement of protonated isopropylbenzene hydroperoxide in a four-membered cyclic transition state has been proposed[55] for the decomposition of the hydroperoxide to phenol. Other rearrangements which can be classified as rearrangements of benzene derivatives include the alumina-catalysed rearrangement of α-chloroalkyl aryl sulphides to α,ω-bis(arylthio)alkanes[56] in a unique aryl coupling which compares interestingly with the Ullman procedure, the phase-transfer-catalysed isomerization of allylbenzene to *trans*- and *cis*-β-methylstyrenes,[57] and the thermal rearrangement of 2-halogeno-1,1-diphenylethane derivatives to stilbene compounds.[58]

Heterocyclic Derivatives

The photochemical rearrangements of five-membered ring heterocycles[59] and the rearrangements of indazole, indoxazene, and furazan oxides[60] have been reviewed. An account of pyridine-ring nucleophilic recyclizations, which describes[61] a new enamine rearrangement of pyridinium salts to anilines, has appeared.

The thermal rearrangements of 2,6-bis(pyridylmethylamino)pyridines into 3,5-bis(pyridylmethyl-2,6-diamino)pyridines have been described,[62] as have the thermal and photochemical rearrangements of *N*-substituted 2-pyridones to 3- and 5-substituted 2-pyridones.[63] The authors favour a radical N—O cleavage and recombination pathway for this latter rearrangement. A convenient synthesis of 1,2-oxazines (**30**) has been achieved[64] by photolysis of 2-azidopyridine 1-oxides (**29**). The photolysis leads to nitrogen elimination and ring-opening followed by recyclization to 6-cyano-1,2-oxazines (**30**), which compounds rearrange further under the influence of heat to yield 2-cyano-1-hydroxypyrroles (**31**). New rearrangements have been described[65] in the reaction of 4-chloro-2,2-dimethyl-2*H*-1,3-benzoxazine with substituted pyridine *N*-oxides, while the thermolysis of 2-vinylpiperidine and 6-vinyl-1,2,5,6-tetrahydropyridine *N*-oxides appears to proceed[66] by a Meisenheimer-type rearrangement to give the corresponding 1,2-oxazepine derivatives. The thermal rearrangements of pyridine *N*-arylimines have been studied[67] by differential scanning calorimetry. Novel 1*H*-1,3-benzodiazepines have been synthesized[68] by a photo-induced two-step rearrangement of 1-substituted isoquinoline *N*-acylimides; 3*H*-1,3-benzodiazepines have been obtained[69] from the photolysis of quinoline *N*-acylimides, and a new mechanism has

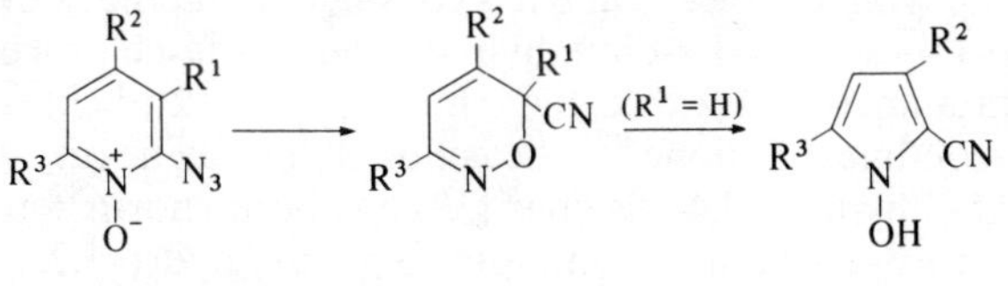

(29) (30) (31)

been proposed[70] for the dihydroisoquinoline rearrangement. A novel oxidative rearrangement of 11-methyl-6-oxo-3,4-dihydro-2*H*,6*H*-1,3-thiazino [3,2-*b*]isoquinoline (**32**), resulting in ring-enlargement to give (**33**), and the further conversion of (**33**) into (**34**), has been reported.[71]

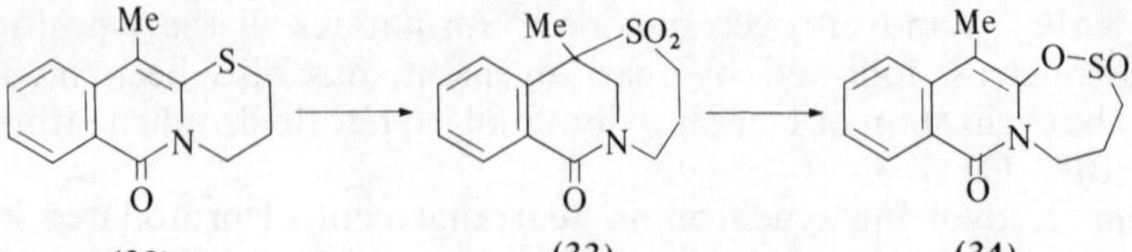

(**32**) (**33**) (**34**)

4,9-Dihydro-9-oxo-1*H*-*v*-triazolo[4,5-*b*]quinolines have been prepared[72] by the polyphosphoric-acid-catalysed Dimroth rearrangement of 5-arylamino-*v*-triazole-4-carboxylic acids. A Dimroth rearrangement is the key step in a three-step synthesis of 2,4,6-tris(alkylamino)pyridine-3-carbonitriles from *N*-alkylcyanoacetamides.[73] A Dimroth rearrangement of benzofuro[3,2-*d*]pyrimidines has been reported,[74] and this rearrangement has also been shown to be involved in a novel purine-to-1-deazapurine transformation which leads to the synthesis of 1-deazaadenosine derivatives.[75]

A new synthesis of 4-formylimidazoles has been achieved[76] by treating 5-aminopyrimidines with *N*-alkyl(or aryl)imidoyl chlorides in the presence of phosphorus oxychloride. A novel one-step pyrimidine-to-pyridine transformation outlined in Scheme 4, has been devised and adapted to the synthesis of pyrido[2,3-*d*]pyrimidines,[77] while angular annelated imidazo[1,2-*c*]- and pyrimido[1,2-*c*]-quinazolines have been rearranged to linear annelated pyrimido[2,1-*b*]- and imidazo[2,1-*b*]-quinazolines.[78] Allylimidazo[1,5-*a*]-1,3,5-triazinones have been synthesized from 5-allyl-4,5-dihydropyrimidin-4-ones by a cyclization–rearrangement that appears to proceed through 5-allylguanines as transient intermediates.[79] The thermal rearrangements of perfluoropyridazines and perfluoroalkylpyridazines to pyrimidines are now regarded as occurring by free-radical-promoted formation of valence isomers on the basis of recent experiments.[80]

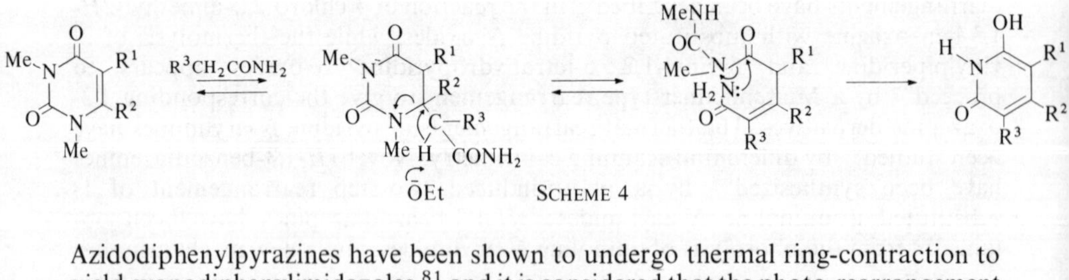

$R^3CH_2CONH_2$

SCHEME 4

Azidodiphenylpyrazines have been shown to undergo thermal ring-contraction to yield cyanodiphenylimidazoles,[81] and it is considered that the photo-rearrangement of 5,6-dichloro-1,4-dihydro-1,4-dimethylpyrazine-2,3-dione proceeds reversibly by way of a simultaneous 1,2-chlorine shift to give 2-(dichloromethylene)-1,3-dimethylimidazolidine-4,5-dione.[82] The action of hydrazine on 3-methylmercapto-5-oxo-4-aryl-4,5-dihydro-1,2,4-triazines (**35**) has been shown[83] to lead by a novel rearrangement to 3-arylamino-4-amino-5-oxo-4,5-dihydro-1,2,4-triazines (**36**). A reasonable pathway for the process is shown in Scheme 5.

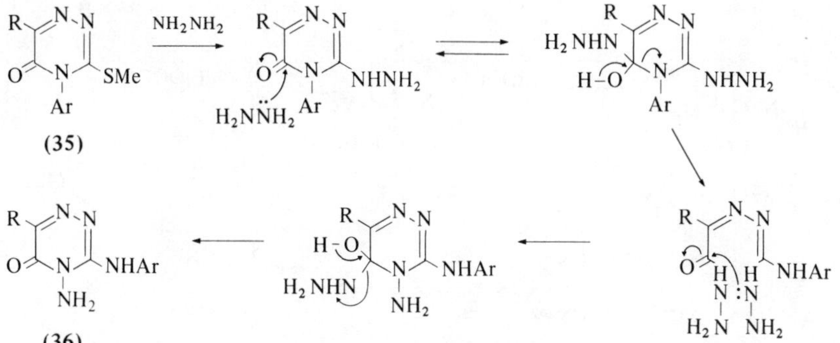

(35)

(36)

SCHEME 5

2-(Arylsulphinyl)- and 2-(alkylsulphinyl)-pyrroles have been found to undergo a remarkably facile acid-promoted rearrangement to the isomeric 3-sulphinyl derivatives.[84] This transposition appears to be, at least in part, an intermolecular process. A free radical mechanism has been proposed[85] for the light-induced N-to-O and N-to-C(4) benzyl migration in 3-pyrazolin-5-ones (**37**). 1-Benzyl- and 1-benzhydryl-3,5-dimethylpyrazoles have been shown to rearrange to pyrimidines in the presence of sodamide,[86] and 2*H*-1,2,3-triazole-4-carboxylic acids have been prepared by ring-transformation of 4,4-diazido-2-pyrazolin-5-ones.[87] The thermally labile spiro-3*H*-pyrazoles (**39**), obtained by the reaction of diazoindenothiophenes (**38**; X = S, Y = CH and X = CH, Y = S) with acetylene dicarboxylates, rearrange into the corresponding 3*H*-pyrazoles (**40**) *via* a Van Alphen–Hüttel rearrangement and migration of an ester group; see Scheme 6.[88,89]

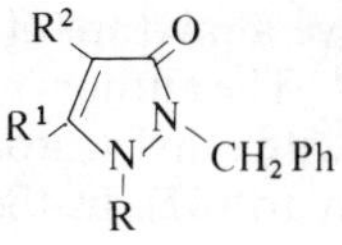

(37)

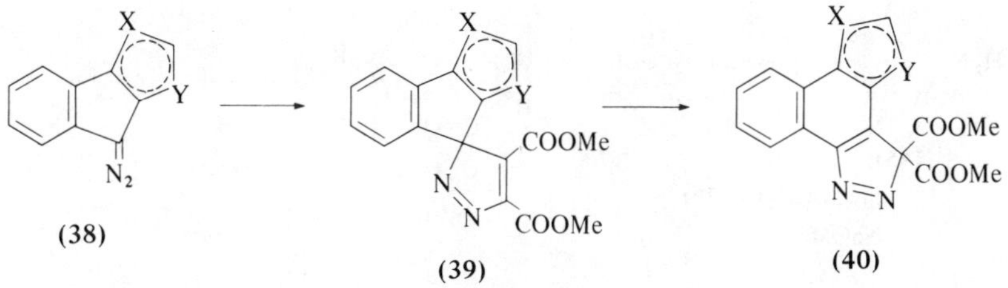

(38)

(39)

(40)

SCHEME 6

2*H*-Thiopyrans and dihydro-2*H*-thiopyrans have been established[90] as useful synthons for thiophenes. The novel and efficient photochemical rearrangement of 1,3-thiazines (**41**) to the fused cyclopropathiazolines (**42**) has been regarded[91] as a new version of the di-π-methane rearrangement where the sulphur hetero-atom replaces a C=C bond (see Scheme 7), while the novel heterocyclic system, 1,4-dithiino[2,3-*c*:6,5-*c*′]diisothiazole (**44**), has been synthesized by a sulphur

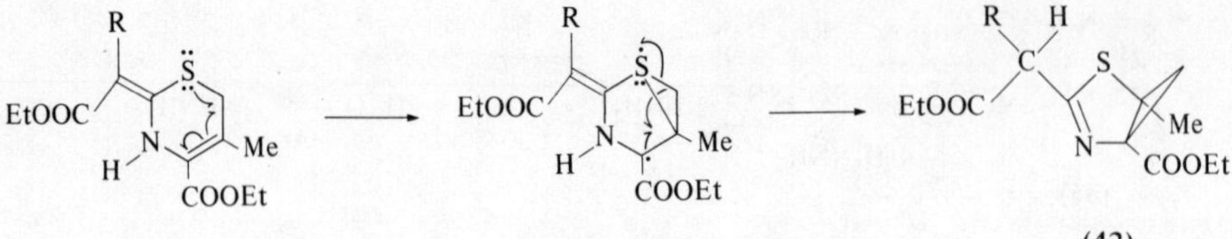

(41)

SCHEME 7

(42)

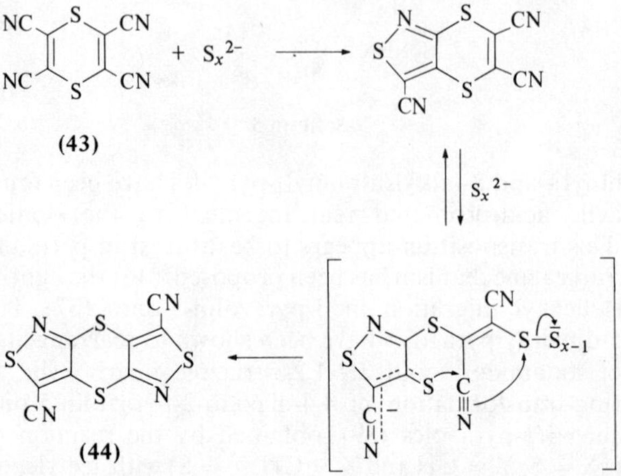

(43)

(44)

insertion–rearrangement reaction of 1,4-dithiin-2,3,5,6-tetracarbonitrile (**43**), as illustrated.[92] 2-Amino-5-substituted-1,3,4-thiadiazoles (**45**) have been shown to react with ethyl acetoacetate to give a mixture of thiadiazolopyrimidines (**47**) and (**49**), depending on the conditions.[93] The authors conclude that, in the presence of *p*-toluenesulphonic acid, (**45**) reacts with ethyl acetoacetate on the ketonic function to give (**46**) which cyclizes exclusively to (**47**). In the presence of sodium methoxide, however, nucleophilic attack on the carboethoxy group yields the intermediate (**48**) which cyclizes through partial rearrangement to a mixture of (**47**) and (**49**).

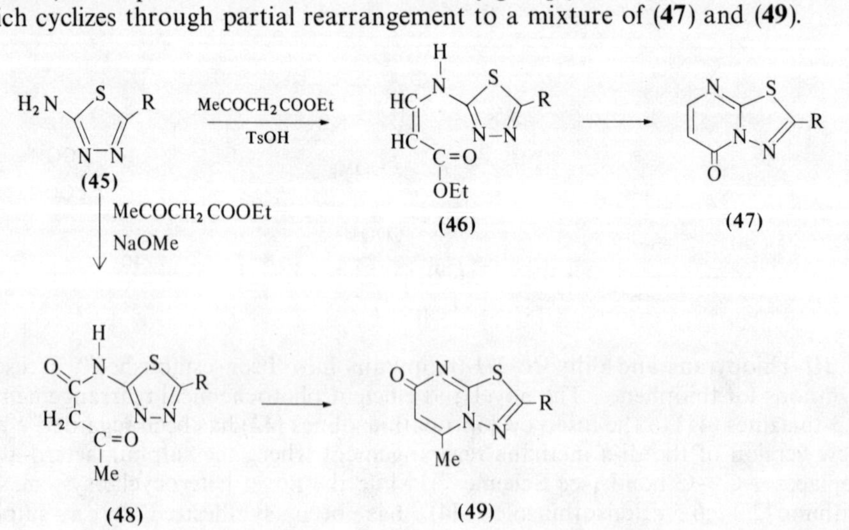

(45) (46) (47)

(48) (49)

When heated under reflux in ethanol containing small amounts of HCl, 1,2,4-oxadiazoles (**50**) have been found to rearrange to the corresponding isoxazoles (**51**).[94] Substituent effects[95] and kinetics[96] of the base-catalysed rearrangement of the arylhydrazones of 3-benzoyl-5-phenyl-1,2,4-oxadiazole into 2-aryl-4-benzylamino-5-phenyl-1,2,3-triazoles have been investigated and evidence for general base catalysis in this rearrangement has been presented.[97]

(**50**) (**51**)

Further heterocyclic rearrangements include the rearrangement of *N*-methylisoindolo[1,2-*b*][3]benzazepinium ions into *N*-methyldibenzo[*a,g*]quinolizinium (i.e. *N*-methylberbinium) ions,[98] a transformation which appears to have wide applicability for the synthesis of alkaloids, the sodium hypochlorite-promoted rearrangement of *N*-chloroindoles to their respective 3*H*-indoles,[99] the conversion of (pyrrolyl-2)-2-oxadiazol-5-one into 1,4-dioxo-1,2,3,4-tetrahydropyrrolo[1,2-*d*]-1,2,4-triazine,[100] the interconversion of isothiazolethioacetamides and 1,3-dithietanecarboxamides,[101] and the ring-rearrangement of 1,2-dihydro-4*H*-3,1-benzothiazine-2,4-dithione with *o*-phenylenediamine to yield 5,6-dihydrobenzimidazo[1,2-*c*]quinazolin-6-thione.[102]

Cyclohexadiene Derivatives

The photochemical transformations of enones[103] and conjugated cyclic dienones[104] have been reviewed. The regioselectivity of both the type A cyclohexadienone photochemistry and the dark type A zwitterion rearrangement have been studied,[105] and the evidence found to favour an *n*-π^* triplet as the bridging and reacting species in the type A dienone rearrangement, although it is noted[106] that, in selected cases, not all of the mechanism proceeds *via* the type A zwitterion. An unambiguous determination of the stereochemistry of the electrocyclic ring-closure and the subsequent [1,4]-sigmatropic shift in the photochemical rearrangement of cyclohexa-2,5-dienones to bicyclo[3.1.0]hex-3-en-2-ones, has been presented.[107] Cyclohexadienones of the type (**52**) under UV irradiation have been shown to yield products resulting from profound structural reorganization of the substrate.[108] The authors postulate that all products start with zwitterionic intermediates of the type already postulated in the photolysis of similar dienones, and further experimental work has established[109] the intermediacy of zwitterions in this reaction. Irradiation of the *n*-π^* band of the dichlorocarbene adducts of cyclohexa-2,4-dienones (**53**; R = Cl) has led[110] to ring-expansion and 1,2-chlorine migration to yield (**54**), whereas the dihydrogen analogue (**53**; R = H) has given the vinylcyclopropane–cyclopentene rearrangement product (**55**)—presumably a consequence of the absence of a leaving group or migrating group in the cyclopropane ring.

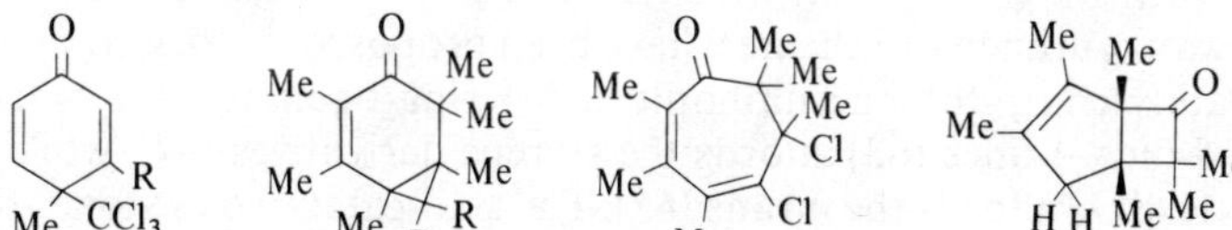

(**52**) (**53**) (**54**) (**55**)

A study of the effect of substituents on the kinetics and direction of dienone–phenol rearrangements has been carried out.[111] Kinetic studies[112,113] have enabled the relative importance of the various mechanisms for the dienone–phenol rearrangement of bicyclic cyclohexa-2,5-dienones of the 4*a*-alkyl-5,6,7,8-tetrahydronaphthalen-2(4*aH*)-one type (**56**) to be assessed. These mechanisms involve either a direct alkyl shift from C(4*a*) to C(4) and deprotonation to give a 4-alkyl-5,6,7,8-tetrahydro-2-naphthol, or a shift of the 4*a*,5-bond from C(4*a*) to C(8*a*) with formation of a spiran intermediate and further ring migration from C(8*a*) to C(1) to form a 4-alkyl-5,6,7,8-tetrahydro-1-naphthol. A further mechanism, for which there is little strong evidence, involves an alkyl shift from one angular position, C(4*a*), to the other, C(8*a*), followed by further ring-migrations *via* a spiro intermediate to give an alternative route to the 4-alkyltetrahydro-2-naphthol product. Interestingly, *N*-4-methoxybenzyl- and *N*-3,4-dimethoxybenzyl-aminoacetonitriles with both 2-aryl and *N*-alkyl substituents, for example (**57**), have been shown to undergo *O*-demethylation and rearrangement with elimination of ammonia in concentrated H_2SO_4 to yield 1-aryl-*N*-alkylcyclohepta[*c*]pyrrol-6(2*H*)-ones (**59**).[114] A possible mechanism involves the spiro intermediate (**58**), as shown in Scheme 8. Hydroxy ketone (**60**) has been shown[115] to undergo dehydration with partial rearrangement to yield (**61**) and (**62**); the fist-formed dienone (**61**) suffers, on heating, a reversible migration of the 19-methyl group.

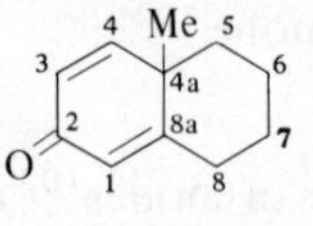

(**56**)

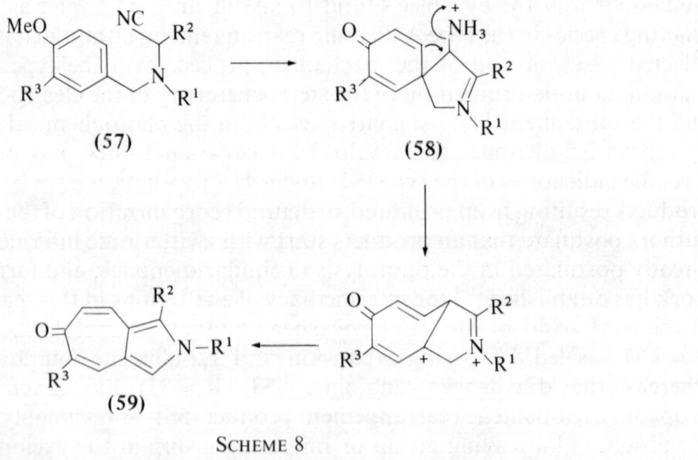

(**57**) (**58**) (**59**)

SCHEME 8

The pathways outlined in Scheme 9 have been proposed[116] to account for the fact that acid-catalysed (*p*-toluenesulphonic acid) ring-opening of 3,5-dialkylspiro-[2,5]octa-2,5-dien-4-ones (**63**) affords the styrene derivatives (**64**), while treatment with Ac_2O–H_2SO_4 affords the furans (**65**). The acid-catalysed dienone–phenol and dienol–benzene rearrangements of eupodienone-1 and its derivatives[117] have been shown to produce dibenzocyclooctene derivatives resulting from alkyl rather than

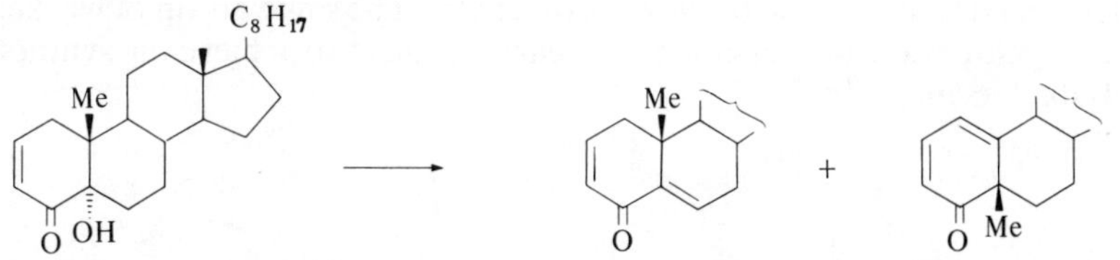

(60) (61) (62)

aryl group migration, while the unusual dibenzazonine structure of protostephanine has been synthesized[118] by a route analogous to its biosynthetic pathway.

Kinetic data, deuterium isotope effects, and ^{13}C-NMR experiments have all indicated that proton migrations between *ortho*-atoms in various 2,5-dihydroxy-1,4-benzoquinones do not occur simultaneously and involve both intermolecular and intramolecular mechanisms.[119] Recent work[120] has provided evidence to suggest that cyclohexenone derivatives may be intermediates in the acyl-pyrone–benzene rearrangement, while in the presence of hydrohalic acids, anthra[1,9-*cd*]isoxazol-6-ones have been converted into bromo-substituted aminoanthraquinones by a mechanism involving rate-determining attack of halide ion on protonated substrate.[121]

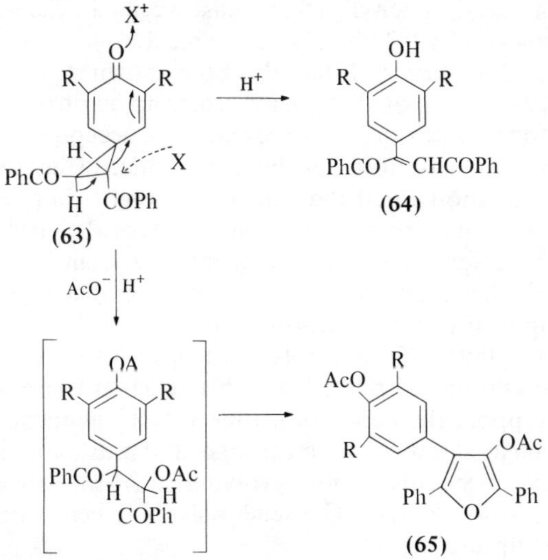

(63) (64) (65)

SCHEME 9

Various rearrangement products, in addition to the usual dioxetane scission product, have been identified[122] from the acid-catalysed decomposition of the *exo*-1,2-dioxetane derivative from 1-isopropylidene-4,4-diphenylcyclohexa-2,5-diene.

1-Benzyl-4-*tert*-butyl-4-methoxy-1,4-dihydroarsenine has been reported[123] to undergo a 1,4-dihydroarsenine–arsenine rearrangement to yield 2-benzyl-4-*tert*-butylarsenine on treatment with $BF_3 \cdot OEt_2$, while a study of the 1,4-dihydrophosphorin (**66**) to λ^5-phosphorin (**68**) rearrangement has indicated that direct acid-catalysed 1,4-migration of the methoxy group probably takes place in the case of the *trans*-isomers, while 1-alkylphosphorinium cations (**67**) are proposed as intermediates in the case of the *cis*-isomers.[124] The same group of workers have used the cyclohexadienol–arsabenzene rearrangement to achieve the synthesis of 2-aryl-4-alkylarsenines.[125]

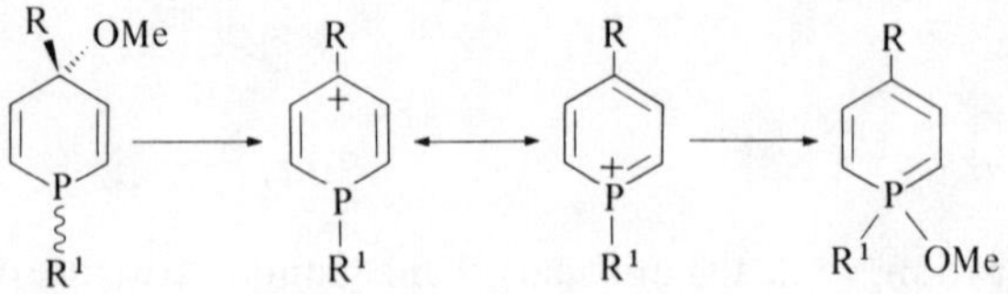

(66) **(67)** **(68)**

Sigmatropic Rearrangements

[3,3]-*Migrations*

The syntheses of medium-ring compounds *via* sigmatropic rearrangements have been reviewed,[126] and a review[127] has appeared on the stereospecificity in intramolecular asymmetric rearrangements, especially [3,3]-, [2,3]-, and [1,2]-sigmatropic rearrangements; it includes reference to a number of syntheses of natural products using these reactions. An interesting study of the Claisen rearrangement of allyl phenyl ether has verified the corollary of the vibration–activation theory,[128] namely that typical bond-making reactions ought to go faster at high viscosities.[129] It has also been shown[130] that electron-releasing substituents at the 3-position of allyl phenyl ethers appear to favour rearrangement of the allyl group to the 6-position, whereas electron acceptors favour migration to the 2-position. A mechanistic study of the light-induced Claisen rearrangement of aryl benzyl ethers has shown that the reaction proceeds *via* a radical pathway.[131] The thermolysis of silicon and germanium analogues of allyl aryl ethers has resulted in an aromatic Claisen-type rearrangement *via* transient metal π-bonded intermediates, and the generation of new oxametallacyclohexanes by intramolecular trapping of the metal–carbon double bond.[132] Although the reported conversion of 4-phenylbut-1-ene to *o*-tolylpropenes has been cited as a precedent in mechanistic interpretations of related possible "carbon-Claisen" rearrangements, recent results[133] support the conclusion that a [3,3]-sigmatropic rearrangement step in the all-carbon analogue of this rearrangement must be a high-energy process, and that alternative mechanisms not involving [3,3]-sigmatropic shifts must be considered for possible "carbon-Claisen" rearrangements not activated by a cyclopropyl bond rupture.

New Lewis acids generated by reacting alkylaluminium halides with Brønsted acids have been used[134] to catalyse Claisen rearrangements, and organoaluminium compounds have been found effective in promoting the aliphatic Claisen rearrangement of allyl vinyl ether derivatives.[135]

The observed deuterium isotope effect in the thermal Claisen rearrangement of phenyl propargyl ether has indicated[136] that bond-cleavage is taking place in advance of bond-formation. The thermal rearrangement of propargyl 3,5-dimethylpyrid-4-yl ether (**69**) has been shown[137] to afford 3,7-dimethyl-5-azaindan-2-one (**70**), which was subsequently transformed into the racemic actinidine (**71**). Naphthofurans have been prepared[138] by the Claisen rearrangement of some allyl β-naphthyl ethers and subsequent cyclization of the products. A new general route

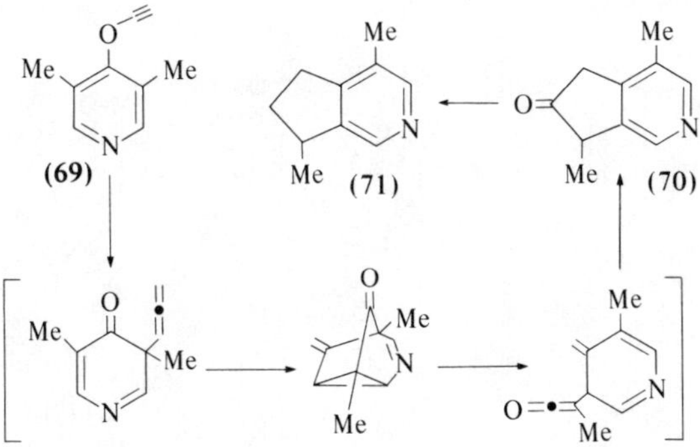

(69) (71) (70)

to γ,δ-unsaturated α,α-dichloro ketones from allyl 2,2,2-trichloroethyl ethers, *via* the [3,3]-sigmatropic rearrangement of intermediary 2,2-dichlorovinyl ethers, has been reported.[139] Trimethyl methoxyorthoacetate (**72**) has been shown to participate[140] as the orthoester partner in the Claisen rearrangement producing α-methoxy γ,δ-unsaturated methyl esters (**73**) (see Scheme 10), while the orthoester Claisen rearrangement (**74**) → (**75**) has been used[141] as the key step in the synthesis of both enantiomers of methyl (E)-tetradeca-2,4,5-trienoate. A total synthesis has been described[142] of (−)- and (+)-nonactic acids from D-mannose and D-gulono-γ-lactone, respectively, which utilizes the enolate Claisen rearrangement and serves to define a boat-like transition state for the rearrangement in these heterocyclic systems. An evaluation has been made[143] of the Claisen rearrangement of cyclohex-2-enols for the stereoselective construction of a terpene synthon (**76**), and a recently

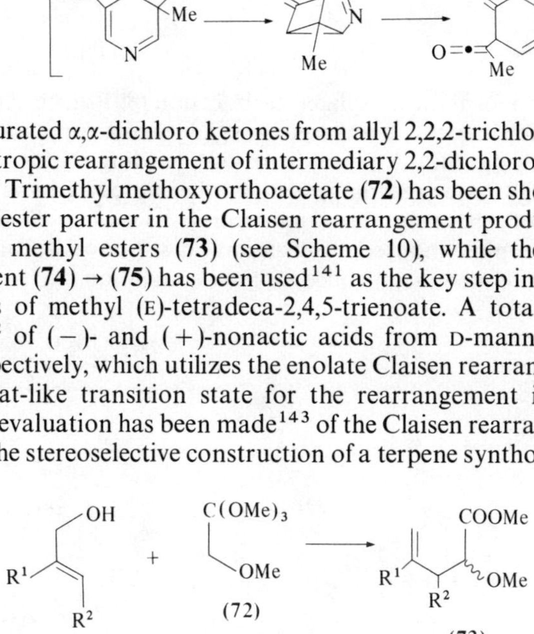

(72) (73)

SCHEME 10

$$\text{RCHC}\equiv\text{CH (OH)} \xrightarrow[\text{EtCOOH, } \Delta]{\text{MeC(OEt)}_3} \text{H(R)C=C=C(H)CH}_2\text{COOEt}$$

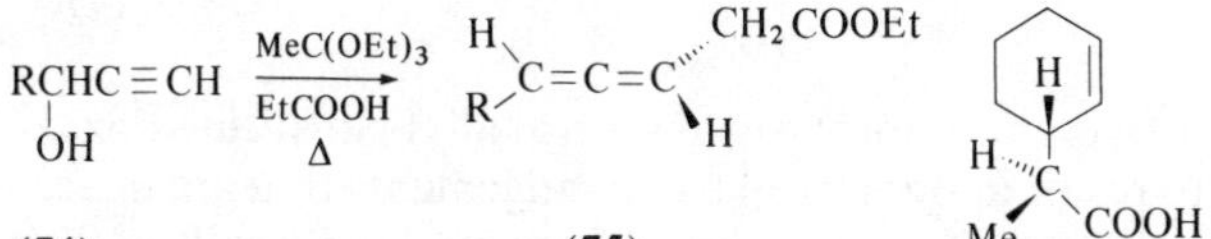

(74) (75) (76)

devised Pd(II)-catalysed [3,3]-sigmatropic rearrangement of allylic acetates[144] has permitted the facile transfer of chirality with complete stereochemical control. In particular, the methodology has offered a mild, general solution to the problem of controlling stereochemistry at C(15) of rigid bicyclo[2.2.1]heptane intermediates along the pathway to prostaglandins. The base-catalysed rearrangement of the 2-aryloxycyclohex-2-enone (**77**) to the enone (**79**) is considered to occur by an enolate-promoted [3,3]-sigmatropic rearrangement of the enolate aryl ether (**78**).[145] The diacetate (**82**), obtained during the aromatization of Diels–Alder adducts derived from 1-benzylisobenzofuran, has been accounted for[146] by postulating breakage of

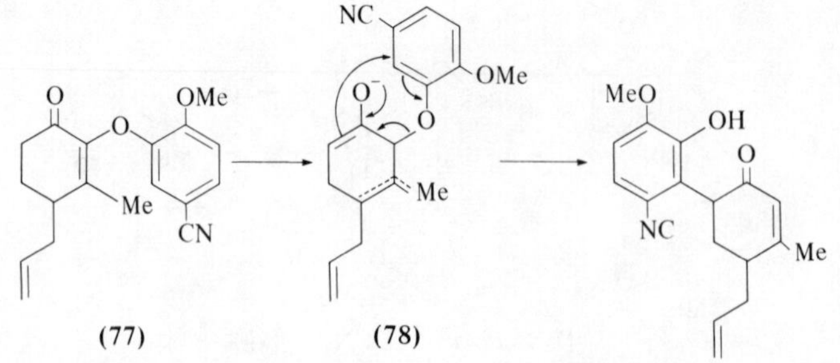

(77) (78) (79)

the 1,4-epoxy bridge to form the acetylated carbocation (**80**) which can give rise to the diacetate (**81**), which in turn can undergo two [3,3]-sigmatropic rearrangements of the aryl groups to yield (**82**). A novel diene (**83**), yielding Diels–Alder adducts capable of undergoing Claisen rearrangement, has been prepared and utilized as a potential synthon for the rubradirin antibiotics,[147] and a tandem Claisen rearrangement–nucleophile-induced cyclization has shown promise as an important technique for stereo-controlled lactone annulation.[148]

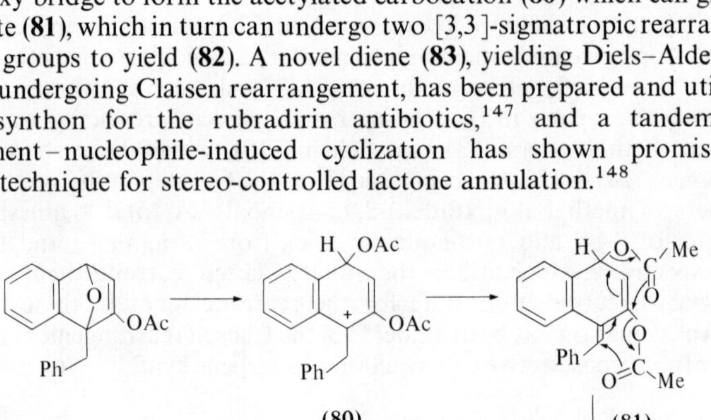

(80) (81) (82) (83)

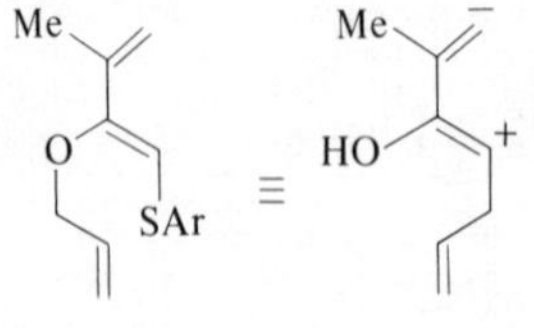

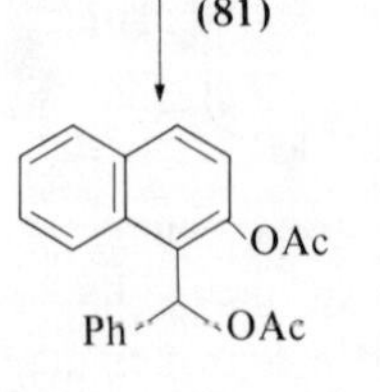

The thermolysis and photolysis of several allyl-substituted oxazolin-5-ones have been shown to result in sigmatropic rearrangements that are in accord with orbital-symmetry predictions,[149] while the [3,3]-sigmatropic shift of *N*-benzoyl-2-aza-3-oxabicyclo[2.2.1]heptane (**84**), and its derivatives, has provided[150] a novel route to unusual bicyclic heterocycles (**85**). The thermal rearrangement of *N*-(substituted-allyl)-1-naphthylamines has been examined[151] and the results have been shown to accord with deductions made about the *para*-rearrangement of the corresponding amido-Claisen rearrangements. A kinetic study[152] of the thermal isomerization of *N*-pyridino-2-vinylaziridines to pyridoazepines shows substituent effects completely analogous to the related benzenic *O*-Claisen rearrangement. This is an argument in favour of a similar transition state for both reactions, so it looks as if the ring-expansion of *N*-aryl-2-vinylaziridines should be described as a concerted [3,3]-sigmatropic rearrangement. Trichloroacetimidates from carbohydrates with an endo- or an exo-cyclic allylic alcohol grouping have been shown[153] to rearrange

(84) (85)

under thermal conditions with the formation of unbranched or R—C—N branched amino-sugars. Of the pseudo-ureas examined to date, those derived from *N*-cyano-*N*-methylaniline (**86**; R^3 = H,Ar) have been found to be the intermediates of choice for preparing 2-pyridones with hydrogens or aryl substituents at C(6).[154] The first step in the process is thought to be a thermal [3,3]-sigmatropic rearrangement of (**86**) to yield allene (**87**). Subsequent tautomerization could then produce the (*Z,E*)-diene (**88**), and a second tautomerization (**88**) → (**89**) would then occur *via* an intramolecular [1,5]-sigmatropic hydrogen migration, and finally lead after ring-closure to substituted 2-pyridones (**90**) which have R^3 = H or aryl; see Scheme 11. A new highly stereoselective method for preparing acylamino-1,3-dienes, which depends upon a [3,3]-sigmatropic rearrangement of propargylic trichloro-acetimidates has been unearthed,[155] and a novel method for stereoselective piperidine annulation involving the cyclization and subsequent [3,3]-sigmatropic rearrangement of (5,5-dimethoxypentyl)but-3-enylamines has been devised.[156]

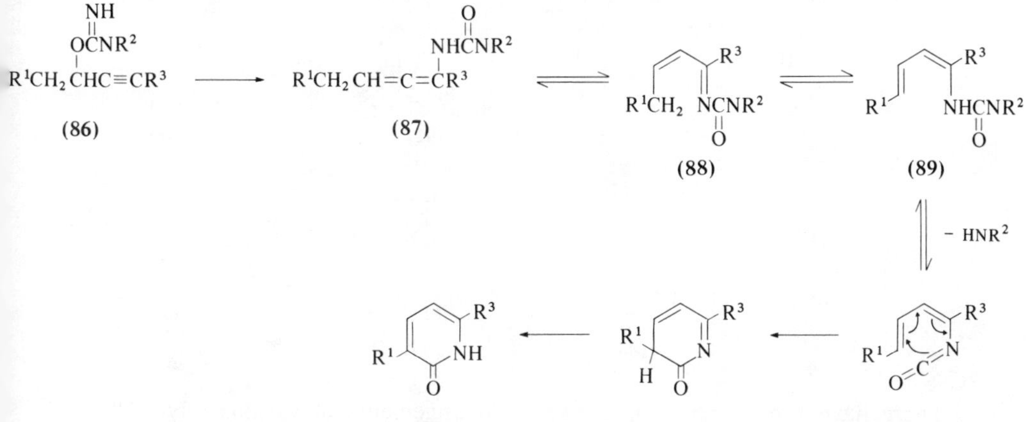

$R^1CH_2CH(OC(=NH)NR^2)C{\equiv}CR^3$ (86) → $R^1CH_2CH{=}C{=}C(NHC(=O)NR^2)R^3$ (87) ⇌ (88) ⇌ (89) → $-HNR^2$ → ... → (90)

SCHEME 11

A new versatile method for the stereo-controlled synthesis of conjugated (E,E)-dienamides, which relies upon the acid-catalysed or thermal ynamine-Claisen rearrangement of readily accessible (arylthio)-ynamines, has been described.[157] The S → N rearrangement of *S*-allylthioimidates to *N*-allylthioamides has been efficiently catalysed by Pd(II) salts.[158] A paper has appeared[159] describing the stereoselective generation of (Z)-enolates of thioamides and their application in thio-Claisen rearrangements. Unique thermal rearrangements of allyl 2,2-dichlorovinyl and 1,2,2-trichlorovinyl sulphides (**91**; X = H and Cl) have been reported.[160] Both substrates are thought to undergo a [3,3]-sigmatropic rearrangement to produce a chlorinated thiocarbonyl compound (**92**) which can give the observed products, 1,2-dichloro-penta-1,4-dienes (**93**), the 2,3-dichloro-5-chloromethyl-4,5-dihydrothiophene (**94**), and 3,5,6-trichloro-3,4-dihydro-2*H*-thiopyran (**95**), as shown. The cyclization-induced rearrangement mechanism proposed to account for nucleophilic catalysis of the thio-Claisen rearrangement has been tested[161] by application of two criteria, *viz.* the secondary deuterium kinetic isotope effect at the β (side-chain) carbon in allyl phenyl sulphide and the substituent rate effect. The observed results can be construed to support the previously validated mechanism of nucleophilic triggering of the sigmatropic rearrangement. Rate constants and activation parameters have been determined[162] for the [3,3]-sigmatropic rearrangements of allyl and but-2-enyl 1-naphthyl sulphides and allyl (2-benzothienyl) sulphide;[163] 4,5,6,7-tetrafluoro-2,3-dihydro-2-methyl-1-benzothiophene and analogues have been prepared[164] from pentafluorophenyl prop-2-enyl sulphide by a reaction that appears to proceed through homolytic fission of the aliphatic C—F bond of the Claisen rearrangement product. Claisen rearrangements of diallyldixanthogen and di(but-2-enyl)dixanthogen have been studied.[165]

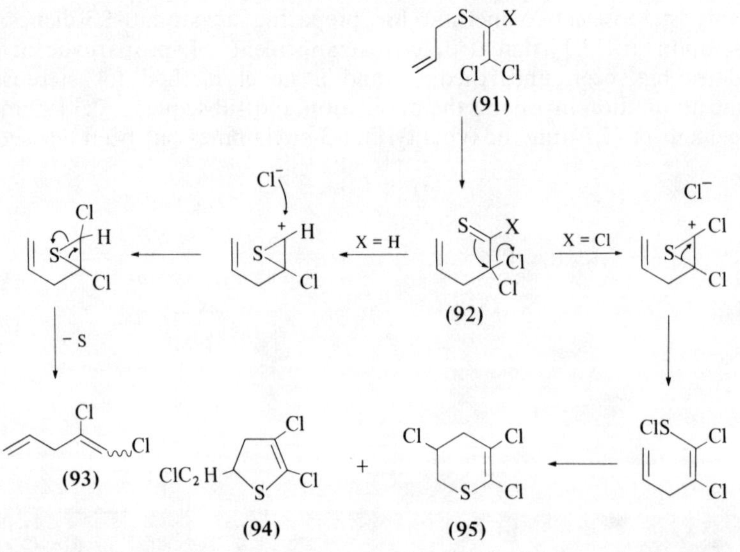

(91)

X = H   X = Cl

(92)

− S

(93) ClC_2H (94)   +   (95)

There have been reports of Claisen rearrangements of various allyl-,[166,167] allyloxy-,[168] and prenyloxy-[169] coumarins, of 7-(3′-methylbut-2′-enyloxy)-flavones and -isoflavones,[170] and of 1,4-bis(allyloxy)-[171] and 1,4-bis(prop-2′-enyloxy)-anthraquinones.[172] The reaction has been used as a key step in the syntheses of 25-hydroxycholecalciferol-26,23-lactone[173], aphidicolin,[174] and secodine,[175] and the Claisen–Eschenmoser reaction of codeines has been used to prepare thebaines and oripavines.[176] Claisen rearrangements have also been observed in the acid-catalysed reaction of 1-prenylindoles and 3-prenylindolenines,[177] during the thermal reaction of feniculin,[178] on heating α-acetylenic alcohols substituted with Δ'-pyrroline,[179] and in the thermolysis of γ-chloroallyl xanthates.[180] A product study of the Claisen rearrangement of a series of allyloxy unsaturated tetrahydroazepinones has been carried out.[181]

An interesting demonstration that the tandem Cope–Claisen rearrangement can be employed for the preparation of (E,E)-cyclodeca-1,6-dienes has been presented.[182] The transformation appears to occur in a concerted manner through chair-like cyclodecadiene conformations since the product produced is not the thermodynamically most stable isomer.

The rearrangements of unsaturated phosphonate esters have been examined systematically.[183] The results show that α-hydroxyalk-2-enyl phosphonates undergo the Claisen orthoester rearrangement on heating with orthoesters, and their arylsulphenates undergo [2,3]-sigmatropic rearrangement to give 3-arylsulphinylalk-1-enyl phosphonates. The addition of allyloxide anion to the central carbon of the allene $Me_2C{=}C{=}CHP(O)(OEt)_2$ is followed by a rapid Claisen rearrangement of the resulting allylic carbanion to give the two possible β-ketoalkyl phosphonates. Allenic phosphonates of the general formula $R^1R^2C{=}CH(CH_2)_nCR^3{=}C{=}CHP(O)(OEt)_2$, on the other hand, have been shown to undergo Cope rearrangement to give isomeric dienes.

Reactions of acetylenic substrates that require the triple bond to become part of an intramolecular pericyclic transition state (reactions such as the acetylenic oxy-Cope reaction) have been discussed.[184] Barbaralane-2,6-dicarbonitrile and its bromo derivative have been prepared and used to study the degenerate Cope rearrangement in substituted tricyclenes.[185] The ionization reactions of 9-chloro-9-methoxy-*endo*-tricyclo[4.2.1.0^{2,5}]nona-3,7-diene are attended by a facile Cope rearrangement to the tautomeric 7-methoxy-*endo*-tricyclo[4.3.0.0^{2,5}]nona-3,9-dienyl cation, the conclusion being that the σ-bond involved in the Cope rearrangement provides stabilization of the incipient cation in the tricyclic system.[186] A multi-step process with one of the steps involving a [3,3]-sigmatropic Cope rearrangement has been invoked[187] to account for the thermal rearrangement of tricyclo[4.2.2.0^{2,5}]decatriene derivatives, and a study of the conversion of tricyclo[4.2.2.2^{2,5}]dodeca-1,5-diene (**96**) to 2,5-dimethylenetricyclo[4.2.2.0^{1,6}]decane (**97**)[188] has provided a unique opportunity to examine an essentially degenerate Cope rearrangement, for which the less favoured boat activated complex is required to obtain a measure of the interaction energy between the double bonds in (**96**). Substituent rate effects on the thermal isomerization of bicyclo[3.2.0]hepta-2,6-diene suggest[189] a preference for disrotatory ring-opening to give cycloheptatriene and a small preference for conrotatory ring-opening to give a *cis,trans,cis*-cycloheptatriene which gives the product of [3,3]-shift.

(**96**) (**97**)

The hitherto unsuspected rôle of η^3-allylic intermediates in the Pd(II)-catalysed Cope rearrangement of acyclic 1,5-dienes has been revealed.[190] It has been found[191] that acid-catalysed Cope rearrangement of 2-acyl-1,5-dienes occurs very rapidly at room temperature, a simple example being the conversion of (**98**) into (**99**), while the thermal rearrangement of divinylcyclopropanes to cycloheptadienes has been explained[192] by invoking epimerization of the *trans*-divinylcyclopropane followed by a Cope rearrangement. The hetero-aromatic Cope rearrangements of 1-pyridyl-, 1-furyl-, and 1-thienyl-2-vinylcyclopropanes have been reported;[193] in all cases the (Z)-vinylcyclopropanes undergo the rearrangement faster than their (E)-isomers.

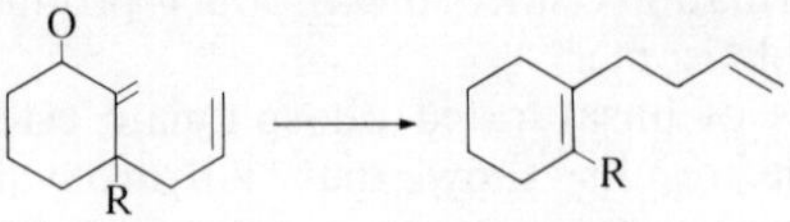

(98) (99)

New methods for the synthesis of macrocycles from relatively simple precursors are always attractive, and recently it has been found[194] that the (mainly *trans*) bis(butadienyl)cyclohexanol (**100**) gives the 14-membered trienone (**101**) in 90% yield upon treatment with KH at room temperature. The authors point out[195] that this reaction could occur by two [3,3]-sigmatropic rearrangements or by a single (unprecedented) [5,5]-sigmatropic rearrangement. Both processes could be expected to benefit from the usual rate acceleration of the oxy-Cope type. Bicyclic bridgehead olefins with a carbonyl group included in the largest bridge, e.g. (**103**), have been synthesized[196] by the oxy-Cope rearrangement of easily accessible spirocyclic precursors (**102**) (see Scheme 12), while a model study for the synthesis of germacranes, which utilizes a consecutive oxy-Cope cyclobutene ring-opening rearrangement has been described;[197] see (**104**) → (**105**).

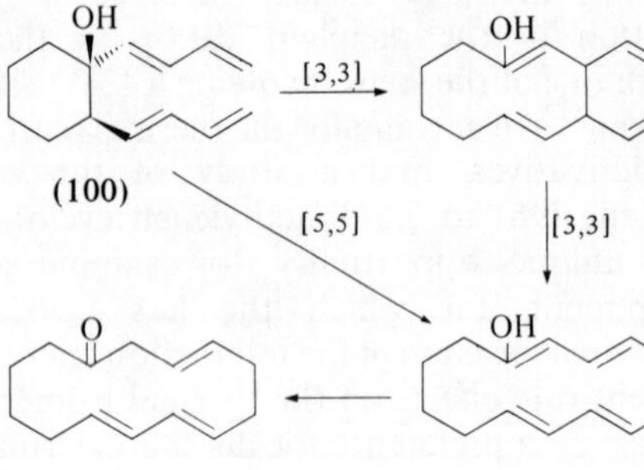

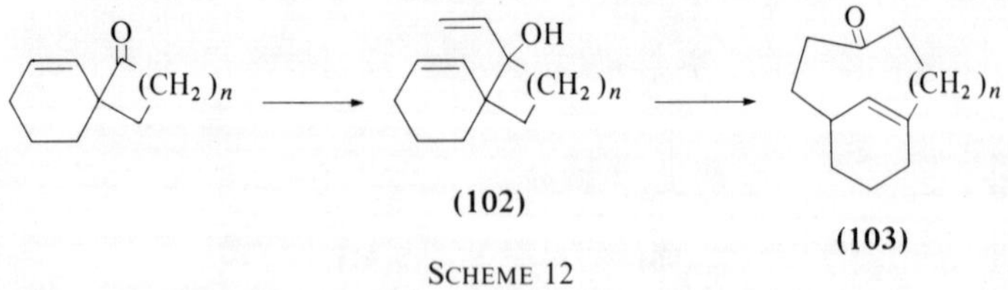

(100) [3,3] [5,5] [3,3] (101)

(102) (103)

SCHEME 12

The behaviour of 1,2-divinylcycloalkanols in the oxy-Cope rearrangement has been examined,[198] and a recent study has been demonstrated that the anionic oxy-Cope rearrangement of tertiary *all-cis*-2,4,7-cyclononatrienyl alkoxides becomes progressively less favoured relative to [1,3]-hydrogen shifts, as charge separation is enhanced.[199] It seems that such systems find it kinetically more feasible to undergo intramolecular abstraction of a doubly allylic ring-proton when the basicity of the

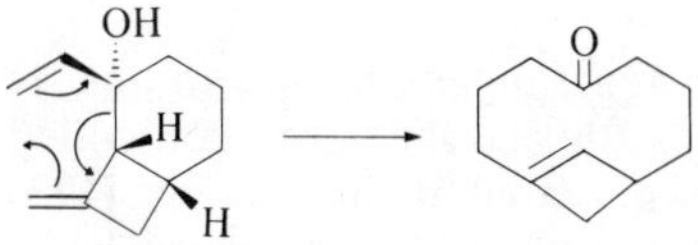

(104) (105)

alkoxide anion becomes sufficiently heightened by charge separation. An approach to the synthesis of pseudoguianolides based on an oxy-Cope rearrangement,[200] and a facile stereo-controlled synthesis of (±)-estrone employing the trimethylsilylcyanohydrin-Cope rearrangement,[201] have been reported.

Treatment of carbinolamide (**106**) with acid is considered to generate the *N*-acyliminium ion (**107**) which rearranges to ion (**108**) presumably *via* a 2-aza-Cope process,[202] while the formation of dienes (**111**; $R^3 = CH_2OCOH$) from formic acid treatment of ethoxylactams (**109**) has also been rationalized[203] as an example of a 2-aza-Cope rearrangement *via* an α-acyliminium ion (**110**). A recent synthetic application of the above rearrangement includes a stereoselective synthesis of 9*a*-arylhydrolilolidines,[204] which can be utilized to prepare *Aspidosperma* alkaloids.

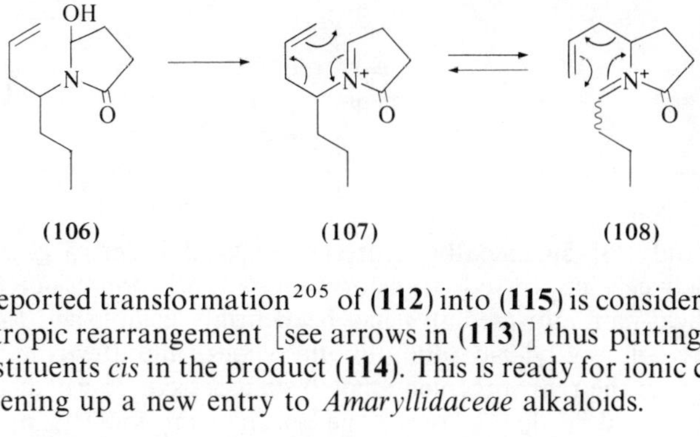

(106) (107) (108)

The recently reported transformation[205] of (**112**) into (**115**) is considered to involve a [3,3]-sigmatropic rearrangement [see arrows in (**113**)] thus putting the aryl and hydrogen substituents *cis* in the product (**114**). This is ready for ionic cyclization to (**115**), thus opening up a new entry to *Amaryllidaceae* alkaloids.

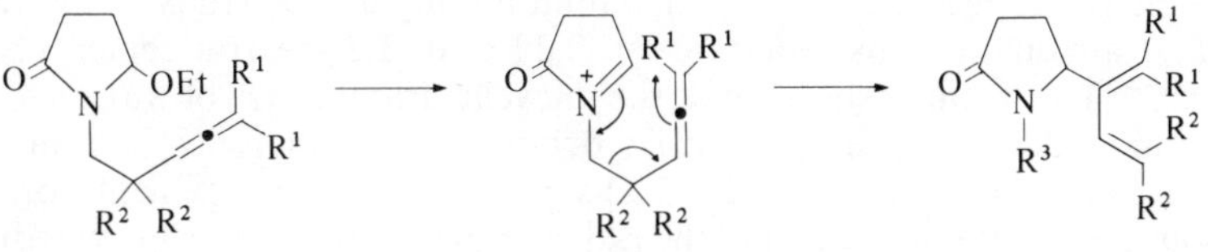

(109) (110) (111)

A recent report[206] on the properties of the $\pi(p\text{-}p)$ multiple bonds in phospha-alkenes and phospha-alkynes indicates that these compounds are more similar to olefins than to phosphorus ylides. Indeed, (E,Z)-isomerism occurs, and a pericyclic reaction of the Cope type has been observed with compounds such as (**116**; X = NPh). A reaction constituting a phospha-Cope rearrangement and proceeding by way of an intermediate monomeric metaphosphonate (**118**) is thought to be involved[207] in the thermal rearrangement of sodium allylvinylphosphinate (**117**) to sodium hydrogen pent-4-enephosphonate (**119**).

[2,3]-*Migrations*

The [2,3]-sigmatropic rearrangement of anions (**121**), available from easily prepared ethers (**120**), has allowed the stereo-controlled synthesis of β,γ-unsaturated aldehydes (**122**) and conjugated dienoic acids (**123**),[208] while the [2,3]-Wittig rearrangement of unsymmetrical bis-allylic ethers has opened up a facile method for the regio- and stereo-selective synthesis of 1,5-dien-3-ols.[209] Imino-ethers have been converted[210] into α-amino ketones by way of an imino-Wittig rearrangement, and optically active (4β,15α)-18-norkaur-16-ene-15-carboxylic acids have been prepared[211] from (4β)-18-norkaur-16-en-3-one by means of a potassium *tert*-butoxide-induced [2,3]-sigmatropic rearrangement.

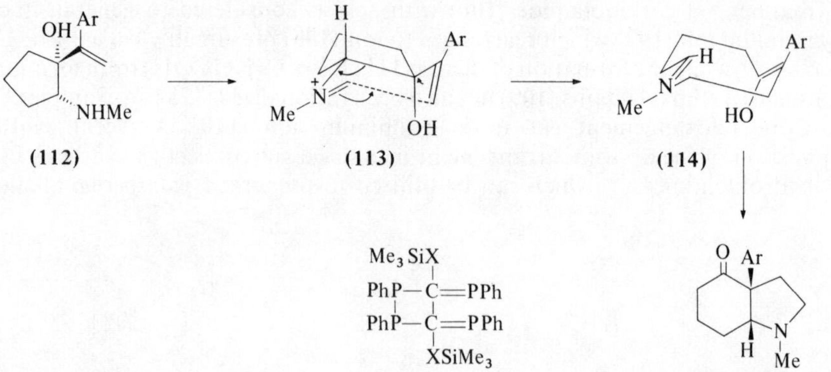

(112) **(113)** **(114)**

(116) **(115)**

Through study of specifically deuteriated *N*-allyl derivatives, quantitative distinction between concerted transformations and non-concerted processes involving radical-pair intermediates has been firmly established for the thermal rearrangements of *N*-allyl-2-oxidoanilinium ylides, and the rôle of cyclohexadienone species in the rearrangements has been assessed.[212] Further evidence has also been presented[213] for a two-stage mechanism for the apparent [3,2]-sigmatropic rearrangement of propynylammonium ylides. A study of the effects of torsional strain on the transition states of [3,2]- and [1,2]-rearrangements has been initiated.[214] The results show that the bicyclic framework of various bicyclic ammonium ylides can prevent correct orbital alignment for π-bonding in the transition state for a concerted [3,2]-sigmatropic rearrangement, or in the vinylogous nitroxide component of the radical-pair required as an intermediate in a [1,2]-sigmatropic rearrangement. Recent results[215] show that allylic ammonium and sulphonium ylides may be isolated as relatively stable compounds when the normal "allowed" [3,2]-rearrangement mode is inhibited by steric effects. Such inhibition has provided clear evidence for steric requirements in the transition states

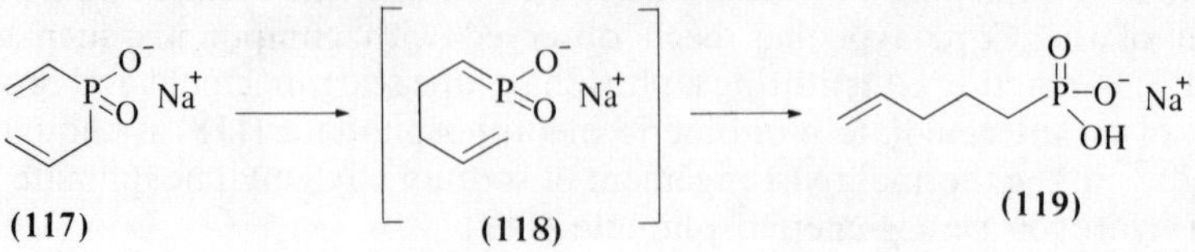

(117) **(118)** **(119)**

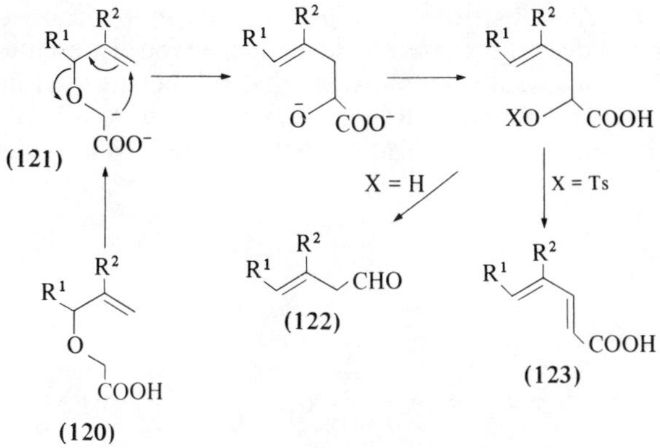

(121)

(122)

(123)

(120)

of [3,2]-sigmatropic rearrangements that are consistent with a reaction mechanism involving simultaneous making and breaking of the two σ-bonds that characterize the reaction.

Dihydrobenzoisothiazoles have been prepared by intramolecular rearrangement of the ylides of the diaminooxosulphonium salts,[216] and a spontaneous [2,3]-sigmatropic rearrangement of *S*-arylthiosulphonium ylides (**125**) has been utilized[217] to obtain *ortho*-(methylthiomethyl) thiophenols (**126**) in the reaction of arylsulphenyl chlorides (**124**; X = H, Me, Cl) with (methylthiomethyl)trimethylsilane. Allylic oxosulphonium ylides (**127**) have been found to undergo a [3,2]-ring rearrangement at room temperature or below to give the sulphoxides (**128**) which readily lose methanesulphinic acid on heating to give the dihydropyridines (**130**), presumably by way of the intermediate azatrienes (**129**).[218] An approach that appears to be uniquely suitable for the synthesis of short-bridged betweenanenes has been uncovered.[219] The crucial step in the synthesis involves a stereospecific [2,3]-sigmatropic rearrangement of *cis*- and *trans*-isomers of spirocyclic sulphonium ylides; thus, derivatives of *cis*- and *trans*-15-thiabicyclo[10.7.0]nonadec-1(12)-enes (**133**) and (**134**) have been prepared from (**131**) and (**132**), respectively. It has been shown by the same authors that the geometry of the cyclic homoallylic sulphides produced by [2,3]-sigmatropic ring-enlargement of cyclic sulphonium ylides is primarily determined by configurational and conformational factors.[220] Interestingly (E,E)- and (2Z,6E)-hedycaryol phenyl sulphides have been converted by [2,3]-sigmatropy of their sulphoxides to the same allyl alcohol which in turn has afforded a stereospecific transformation to cadinane derivatives.[221]

The gas-phase thermolyses of thietan-1-oxide and 1,2-oxathiolan-2-oxide have

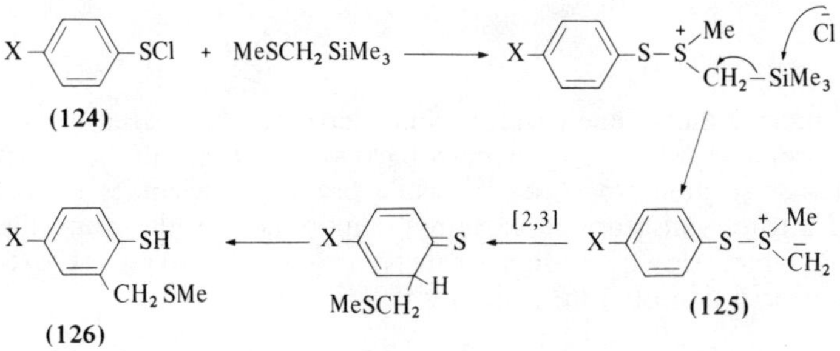

(124)

(126)

(125)

been rationalized[222] in terms of atomic oxygen extrusion and sulphoxide–sulphenate rearrangement; a general synthesis of β-hydroxy-α-methylene-γ-butyrolactones based on the sulphoxide–sulphenate rearrangement has been presented,[223] and it has been established[224] that the allyl-sulphoxide–allyl-sulphenate rearrangement of allyl sulphoxides $RS(O)CH_2CR^1{=}CR^2R^3$ (R = aryl; R^1 *trans* to R^3), proceeds *via* two diastereoisomeric transition states leading to the different enantiomeric forms of an allyl alcohol.

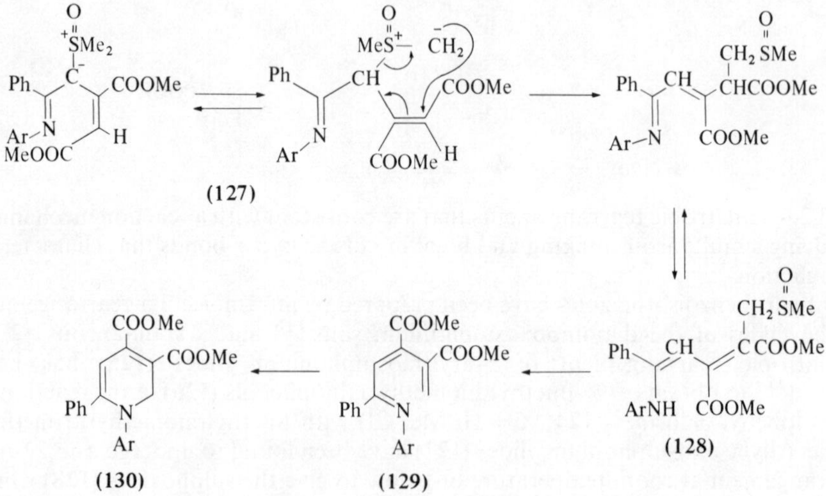

(127)

(128)

(129)

(130)

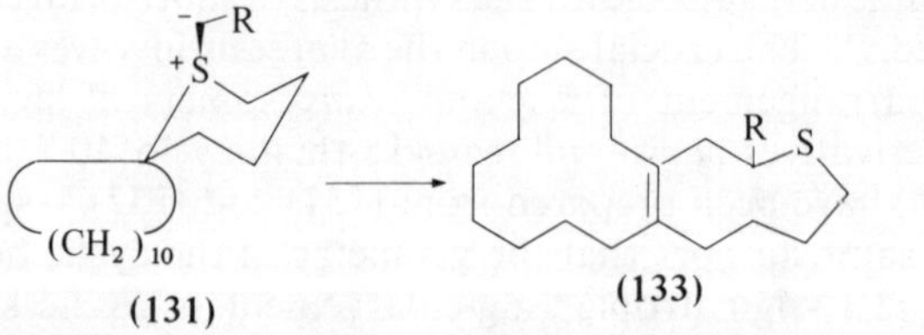

(131)

(133)

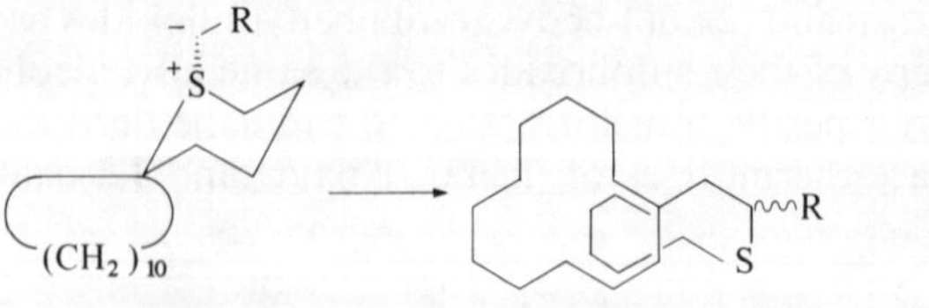

(132)

(134)

Functionalized alkylidene cyclopropanes have been prepared from 1-seleno-1-vinylcyclopropanes using [2,3]-sigmatropic rearrangements of their corresponding selenoxides or selenonium ylides,[225] while propargyl selenides (**135**) have been converted into α,β-unsaturated carbonyl compounds by the route illustrated in Scheme 13, namely slow [2,3]-sigmatropic rearrangement to give (**136**) followed by a rapid isomerization of (**136**) to (**137**).[226]

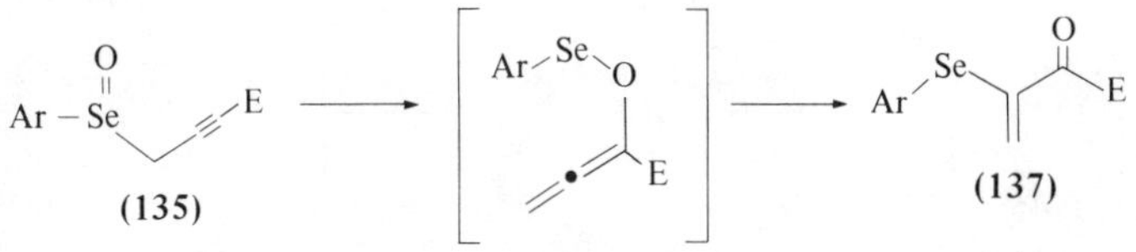

(135) (136) (137)

SCHEME 13

[1,3]-*Migrations*

Ab initio MO theory has been used[227] to study the [1,3]-sigmatropic hydrogen tautomerization in propene and formic acid and the conversion of $CH_2{=}CHOH$ into MeCHO. The degenerate 1,3-shift rearrangements of allylamine and allylborane have been mapped out by the PRDDO method;[228] apparently the former is neither pericyclic nor pseudo-pericyclic in character, but the latter conforms to the pseudo-pericyclic concept. MNDO SCF-MO calculations have suggested[229] that the uncatalysed rearrangements of *O*-acyl imidates (**138**) to the corresponding imides (**139**) are (1,3)-concerted reactions involving synchronous C—N bond-formation and C—O bond-cleavage. On the other hand, the transition states for [1,3]-acyl migrations in compounds such as (**140**), which have no participating lone pair of electrons, have been predicted as having significantly different properties.

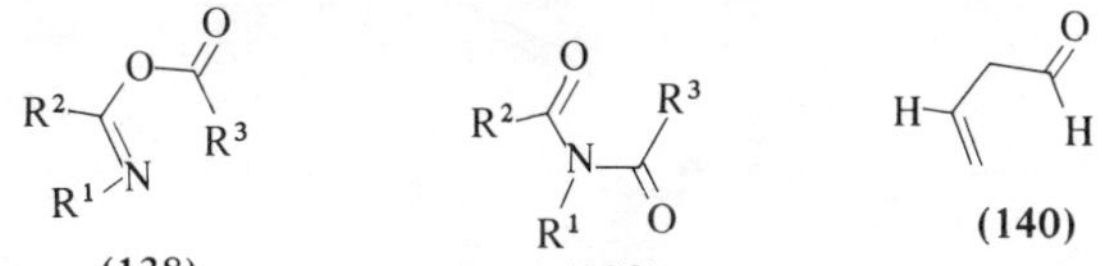

(138) (139) (140)

The photochemical behaviour of a series of alkyl-substituted hexa-1,5-dienes has been studied.[230] Competing allylcyclopropane formation and rearrangement to isomeric hexa-1,5-dienes *via* competing [1,2]- and [1,3]-sigmatropic allyl shifts have been observed. In contrast with acyclic dienes, cycloocta-1,5-diene afforded products from competing [3,2]-, [1,3]-, and [3,3]-allyl shifts. The divergent behaviour in this case apparently arises because appropriate interaction between the two double bonds can occur only in the boat conformation. The photochemical rearrangements of several arylalkyl-substituted indenes has been depicted[231] as proceeding *via* an electrocyclic ring-closure, a [1,3]-sigmatropic shift, ring-opening to an isoindene, and a subsequent [1,5]-sigmatropic hydrogen shift. It has been shown[232] that bicyclic [3.2.*n*]dienones which differ from one another only in the presence or absence of an oxygen atom as the "*n*" bridge, show very different photo-isomerization behaviour. Thus, irradiation of hexamethyl-8-oxabicyclo[3.2.1]octa-dienone (**141**) gives *endo*-6-acetylpentamethylbicyclo[3.1.0]hexenone (**142**) probably *via* an initial [1,3]-sigmatropic rearrangement. In contrast, hexamethyl-bicyclo[3.2.0]heptadienone (**143**) photo-isomerizes *via* an initial [3,3]-sigmatropic rearrangement, the isolated product being the relatively stable cyclopentadiene–ketene (**144**). A study of the photochemical transformations of the non-dissociating, substituted cyclopentadienone dimers (**145**) has shown that all rearrangements consist of cleavage of the weak, doubly allylic C(6)–C(7) bond and appropriate bond-formation by [1,3]-shifts.[233] Thermal transformations occur by [3,3]-shifts. Quite different results were obtained from the irradiation of "mixed dimers" and

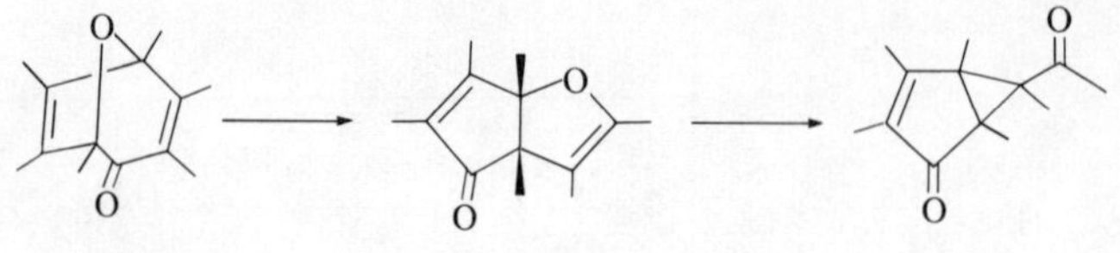

(141) (142)

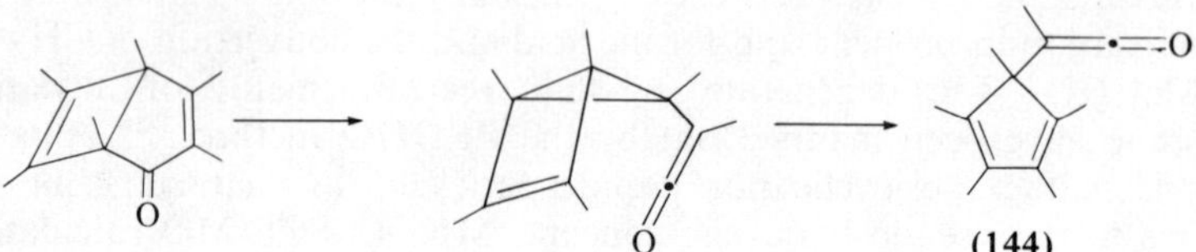

(143) (144)

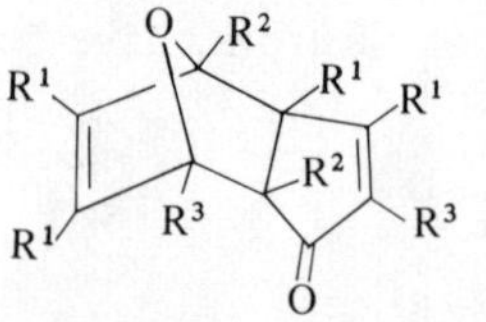

(145)

stereoelectronic arguments have been put forward to rationalize the behaviour in a general framework of this class of compounds.[234] An unusually facile 1,3-allylic rearrangement of an apparently novel type has been observed in tetraarylated dihydrosemibullvalene derivatives,[235] and it has been shown[236] that the pyrolysis of 1,3-bis(dideuteriomethylene)cyclopentane results in a first-order degenerate rearrangement in which products of [1,3]- and [3,3]-shift are formed in a 2:1 ratio respectively, thus indicating the intermediacy of an effectively orthogonal 2,2′-bis(allylmethane) biradical. A recently reported alkoxy-accelerated rearrangement of vinylcyclobutanes (**146**) has provided[237] a powerful new annulation approach to six-membered carbocyclic compounds (**147**). The stereochemical course of the reaction can be rationalized in terms of a concerted [1,3]-sigmatropic rearrangement, although the results may also be compatible with non-concerted mechanisms involving diradical or allylic-anion–aldehyde intermediates.

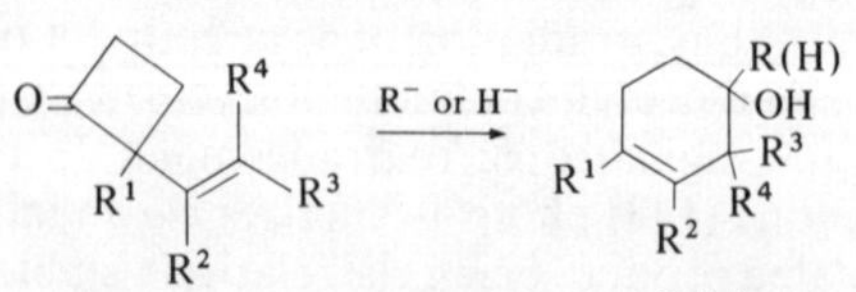

R^- or H^-

(146) (147)

The quantitative isomerization has been reported[238] of 2,2-bis(ethylthio)-3,3-dimethylpent-4-enal to 2,2-bis(ethylthio)-5-methylhex-4-enal in a rearrangement that can formally be regarded as a [1,3]-sigmatropic shift. A [4 + 2]-cycloaddition followed by an allylic rearrangement has been proposed[239] to account for the formation of 2-(1-pyrrolidinyl) (piperidino)-2*H*-thiopyrans (**149**; X = H,D) from the reaction of enaminothiones (**148**) with electrophilic acetylenes; see Scheme 14. Treatment of 3-(trimethylsilyl) allylic alcohols (**150**) with phenylsulphenyl chloride has been shown[240] to yield unstable allylic sulphoxides (**151**) which rearrange and are then hydrolysed to unsaturated aldehydes (**152**) in a process that is analogous to a sila-Pummerer rearrangement. This rearrangement has been utilized[241] as the key step in the total synthesis of optically active Prelog–Djerassi lactonic acid. The [1,3]-migration of a trimethylsilyl group has also been observed in aminofluorosilanes.[242] A study of the mechanism of the [1,3]-sigmatropic shift in η^5-$C_5H_5Fe(CO)(L)(\eta^1\text{-allyl})$ complexes has produced results that are in accord with a radical-chain process.[243]

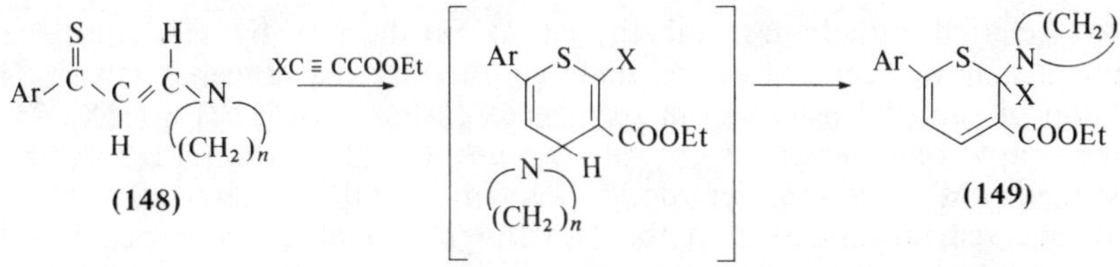

(148) (149)

SCHEME 14

[1,5]-*Migrations*

A review has appeared[244] covering [1,5]-shifts of hydrogen and substituted groups, primarily in cyclic unsaturated systems. It has been shown that the [1,5]-shifts of vinyl and related groups occur by a mechanism in which σ-bond making is almost complete before bond-breaking begins. However, in the rearrangement of *cis*-9,10-dihydronaphthalenes a discrete intermediate appears to be involved.[245]

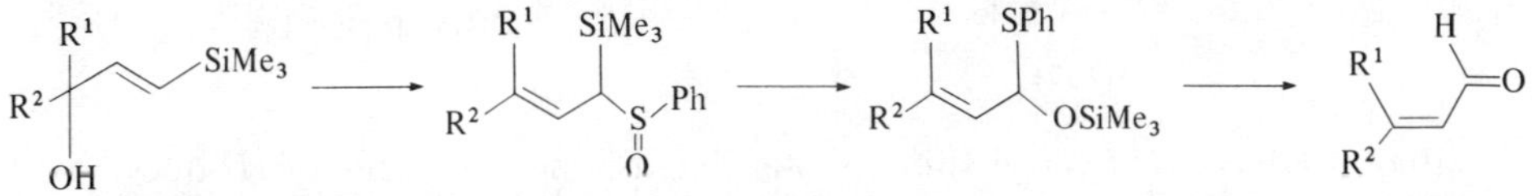

(150) (151) (152)

A new reaction, the formation of an ene-allenic compound by a [1,5]-sigmatropic hydrogen shift, has been found[246] in the photolysis of (Z,E)-5-phenylhepta-1,3,5-triene. Penta-1,5-diene oxygen shifts have been noted in the thermal reactions of (Z)-2,4,5,5-tetrachloro-3-(arylthio)penta-2,4-dienals,[247] (Z)-2,4,5,5-tetra-chloro-3-organylthio- and (Z,Z)-2,3,4,5-tetrachloro-5-organylthio-penta-2,4-diene-thioates,[248] and the thermal transformations of chloro- and organylthio-substituted (2Z,4Z)-penta-2,4-dien-als and -ones to (2Z,4Z)-penta-2,4-dienethioates[249] by a 1,5-chlorine shift have been examined. Pyrolysis of polyfluorocyclohexadienes at 480° over new quartz or pyrex glass has been shown to yield products mainly derived from [1,5]-sigmatropic migrations of fluorine.[250] A [1,5]-homodienyl hydrogen shift has been invoked[251] to account for the stereoselective rearrangement of hexadienols (**153**) into cyclopropylacetaldehydes (**154**), and siloxyvinylcyclopropanes, such as (**155**), have been shown to rearrange cleanly to their respective cyclopentyl enol silyl ethers (**156**) by a route which can be regarded as a homo-1,5-hydrogen shift.[252] In a process without literature precedent, trialkylsilyl ethers of

(153) (154)

(155) (156)

1,1-dioxygenated butadienes, substituted at position 3 by electron-releasing substituents, have been shown to undergo thermal rearrangement with [1,5]-migration of the silyl group from oxygen to carbon; see (**157**) → (**158**).[253] The relatively rapid rearrangement of γ-silyl-α,β-unsaturated carbonyl compounds to siloxybutadienes is best understood[254] as an orbital-symmetry-allowed [1,5]-sigmatropic rearrangement, and the [1,5]-migration of a silyl group has been contrived as the key step in a recent total synthesis of (±)-prostaglandin E_2 methyl ester.[255] The thermolysis of certain *trans*-1-isopropenyl-4-methylenespiro[2.*n*]-alkanes ($n = 4,5$) has been observed[256] to cause quantitative rearrangement to 1-methylene-2-(3′-methylbut-2′-enyl)cycloalkanes by a route that can be described as a sigmatropic [1,5]-hydrogen shift and concomitant cyclopropane bond-cleavage.

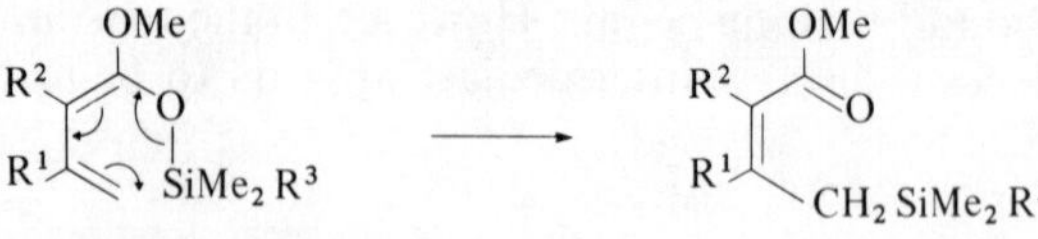

(157) (158)

Photo-induced [1,5]-acyl shifts of 3-acylated 3-methyloxepin-2(3*H*)-ones have been reported,[257] as has an uncatalysed thermal sigmatropic [1,5]-shift of the acyl group of 5-acyl-5-methylcyclohexa-1,3-dienes.[258] The uncatalysed skeletal isomerization of 5,5-disubstituted cyclohexa-1,3-dienes has been investigated[259] with the aim of establishing the extent to which sigmatropic [1,5]-shifts of hydrocarbon groups are participating in these reactions. A reasonable mechanism for the thermal rearrangement of 5,6-benzotricyclo[3.2.0.0^{2,7}]hept-5-ene (**159**) to 2-vinylindene (**161**) is *via* the intermediary isoindene (**160**) formed by an intramolecular retrocyclic Diels–Alder reaction;[260] a subsequent [1,5]-hydrogen shift in (**160**) would afford the more stable, aromatized (**161**). The photolysis and flash thermolysis of 1,3-dihydrobenzo[*c*]thiophene 2,2-dioxides (**162**) have been studied.[261] It appears that

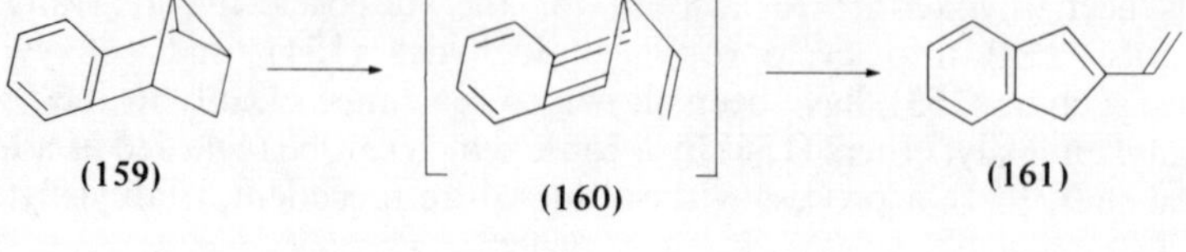

(159) (160) (161)

the reaction involves extrusion of SO_2 to yield (*E*)-*o*-quinodimethanes (**163**) which ring-close to afford the benzocyclobutenes (**164**). At higher temperatures, (Z)-*o*-quinodimethanes (**165**) predominate and these undergo a [1,5]-hydride shift to yield *o*-methylstyrenes (**166**). The migratory aptitudes of aryl migrations in the ground and electronically excited states have been determined by studying the sigmatropic rearrangements of 1,1-diarylindenes.[262] The thermal, photochemical, and electron-transfer-induced rearrangements of 1,1′-spirobiindenes have afforded[263,264] the isomeric 7*H*-benzo[*c*]fluorenes by sigmatropic shift of the alkenyl group followed by hydrogen migration or hydrogen abstraction and protonation of the fluorenyl anion. An efficient stereoselective synthesis of 2,3-dihydroindoles *via* 1,5-electrocyclization of imines to dihydroindole 3,3-spiroimides has been contrived.[265] A series of imidoylketenimines has been prepared and most of them have shown a readiness to produce 1,2-dihydropyrimidines by way of a [1,5]-hydrogen shift.[266] Conjugated triene diamines (**168**) have been generated[267] by isomerization of the lithio derivatives of α,ω-diaminononadiynes (**167**) in a cyclization that very much depends on the nature of the *N*-substituents. The postulated mechanism for this reaction involves the cyclization of the first-formed allenic carbanion followed by a [1,5]-sigmatropic hydrogen shift from one side-chain to the other. Apparently, complexation of 3*H*-1,2-diazepines with diiron nonacarbonyl much reduces the rate of the [1,5]-sigmatropic hydrogen shift across the diazepine ring,[268] while in the light-induced rearrangement of thieno[3,2-*c*][1λ^4,2]thiazines (**169**) into thieno[3,2-*b*]pyrroles (**171**), it is interesting to note that the product arises by exclusive migration of the sulphur substituent in intermediate (**170**), in preference to migration of the ester group.[269]

R SO2 (**162**) → R (**163**) ⇌ R (**164**)

heat

R¹ Me (**166**) ← R¹ H (**165**)

[1,5]-Sigmatropic shifts are also involved in the photo-cyclization of 1,4-diarylbutenynes to aryl-substituted aromatics,[270] in the thermal conversion of 7*b*-methyl-7*bH*-cyclopent[*cd*]indene into its 2*aH*-isomer,[271] in the photolytic rearrangement of (1s)-*exo*-1-(methoxycarbonylmethyl)-3-*tert*-butyl-1*a*,7*b*-dihydro-1*H*-cyclopropa[*a*]naphthalene into (7s)-7-(methoxycarbonylmethyl)-5-*tert*-butyl-7*H*-benzocycloheptene,[272] in the photolytic transformation of photosantonin into

NR^1_2 NR^2_2 (**167**) → NR^1_2 NR^2_2 (**168**)

neophotosantonin,[273] and in the high-temperature conversion of 1-phenylphospholes into 2*H*-phospholes.[274]

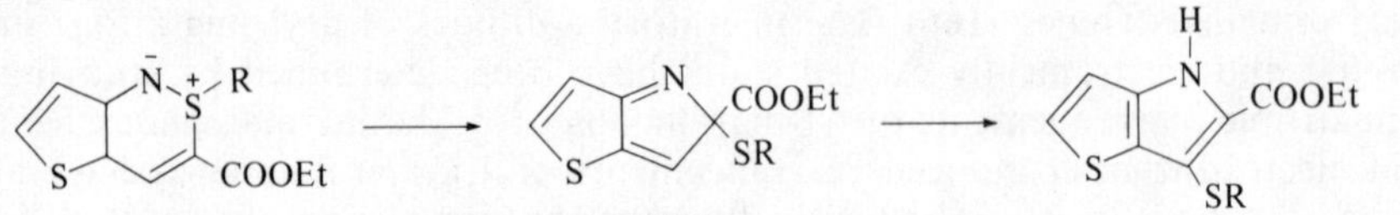

(169) **(170)** **(171)**

Miscellaneous

MNDO calculations on the [1,2]- and [1,3]-shift mechanisms for both SiH_3 and hydrogen shifts in silylcyclopentadienes have been investigated[275] and it has been established that the [1,2]-shift mechanism is the lowest energy pathway in each case. An intriguing [1,2]-silyl migration in an allylsilane system is considered to be responsible for the photo-induced isomerization of a 2,3-benzo-7,7,8,8-tetramethyl-7,8-disilabicyclo[2.2.2]octa-2,5-diene to 3,4-benzo-6,6,7,7-tetramethyl-6,7-disilatricyclo[$3.3.0.0^{2,8}$]octane;[276] indeed, a number of 1,4-dihydronaphthalene derivatives bridged by a $SiMe_2XSiMe_2$ group (X = O,CH_2 *etc.*) at the 1,4-positions have been photolysed to give isomers resulting from [1,2]-silyl migration.[277] A concerted [1,2]-sigmatropic rearrangement has been visualized[278] to account for the formation of 3-oxo-1-pyrroline-1-oxides (**173**) on heating 4-oxo-5,6-dihydro-1,2,4*H*-oxazines (**172**), and a novel [1,2]-shift, this time intermolecular,

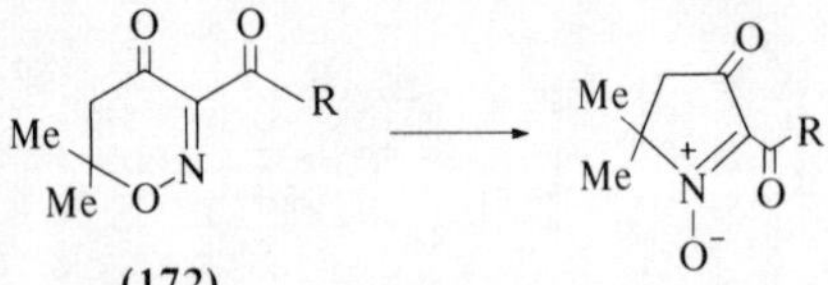

(172) **(173)**

has been invoked[279] to explain the conversion of 2-*tert*-butyl-3-methyl-1-phenylbenzothiazolinium ylides into 2-*tert*-butyl-3-methyl-2-phenylbenzothiazolines. The same group has found[280] that 2-alkyl (or aryl)-1-benzoyl-3,4-dihydro-2-thianaphthalenes (**174**) undergo a novel intermolecular [1,4]-rearrangement to afford enol ethers of the type (**175**). The minimum-energy path of dual intramolecular [1,4]-proton transfer has been calculated[281] by the MINDO/3 method for HO_2CCO_2H, HOCH=C(OH)CHO, squaric acid, and 2,5-dihydroxy-*p*-benzoquinone; interestingly, in no case was a synchronous shift of the two migrating protons found to be the most favoured path. Products arising from a [1,4]-shift of the thio-ether group were obtained[282] from acetolysis of 4-(arylthio)- and 4-(aralkylthio)-1-diazobutan-2-ones — normal substitution being only a secondary pathway. The products (**177**; *e.g.* R = CH_2Ph, (E)-$CH_2CH{=}CHPh$ of a

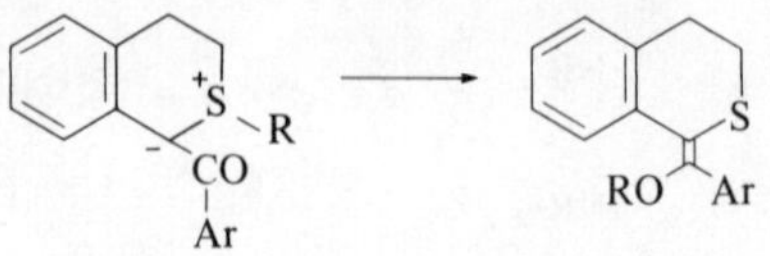

(174) **(175)**

[1,4]-rearrangement have been identified from the thermolysis of 2-oxidoanilinium ylides (**176**),[283] and further rearrangements of these ylides have been identified.[284] Dyotropic [1,4]-rearrangements of trimethylsilyl groups in bis-imidazolines have been studied,[285] and the addition products (**178**) of trialkylsilyl cyanide with conjugated aldehydes react as their anions (**179**) with carbonyl compounds to yield products (**180**) that result from a [1,4]-oxygen-to-oxygen silyl rearrangement.[286] A remarkable oxygen-to-oxygen migration of a trimethylsilyl group has also been noted[287] in the intermediate cyclodecadienones observed upon the thermolytic rearrangement of 2,5-bis(trimethylsiloxy)tricyclo[4.4.0.0^{2,5}]decan-7-ones into hydroazulenes.

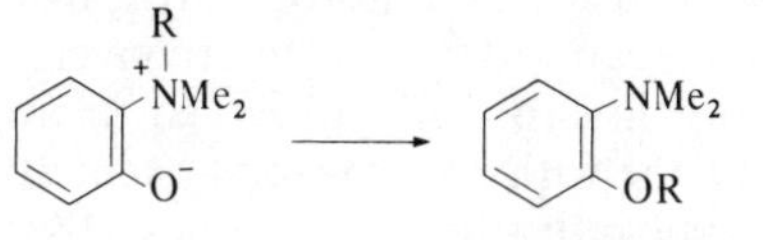

(**176**) (**177**)

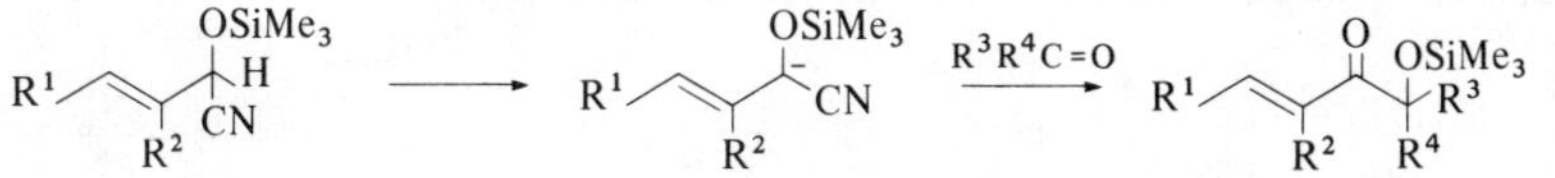

(**178**) (**179**) (**180**)

Fluorescence quenching and sensitization experiments have indicated[288] that the [1,7]-sigmatropic shift of substituted cycloheptatrienes occurs in the singlet state. A model has been proposed[289] which accounts for the regioselectivities and periselectivities of [1,7]-shifts and electro-cyclizations of substituted cycloheptatrienes from singlet excited states as a function of substituent, and it further shows why donor substituents promote electro-cyclizations while acceptors promote sigmatropic shifts. *cis,cis,cis*-Cycloundeca-1,3,5-triene has been found[290] to undergo a thermal [1,7]-hydrogen migration to yield the *trans,cis,cis*-isomer.

The first example of the cationic [3,4]-sigmatropic reaction has been achieved[291] by reaction of 1,3-dibromo-3-methyl-hept-6-en-2-one and diiron nonacarbonyl. Contrary to the formally similar Claisen rearrangement of allyl phenyl ethers, the thermal rearrangements of 1-allyloxy-λ^5-phosphorin derivatives (**181**) have been shown to afford in the irreversible step a "forbidden" [3,5]-allyl shift to yield 4-allyl-1,4-dihydro-λ^5-phosphorin derivatives (**182**). This is followed by a Cope rearrangement to give 2-allyl-1,2-dihydro-λ^5-phosphorins (**183**) which at high temperatures undergo an intramolecular [4 + 2]-cycloaddition to produce tricyclics (**184**).[292]

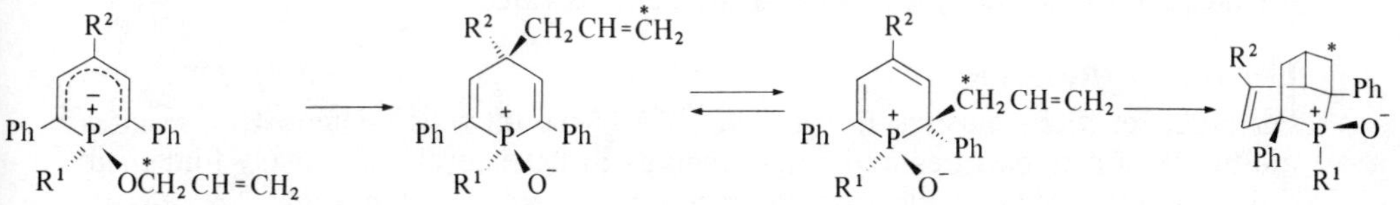

(**181**) (**182**) (**183**) (**184**)

The behaviour of a number of arylhydrazones of aromatic carbonyl compounds and arylaliphatic ketones has been described,[293] and it has been shown that, in addition to products of the expected Fischer indole synthesis, biphenyl derivatives are produced by way of a [5,5]-sigmatropic rearrangement; in the case of the 4-aminoacetophenone-2,6-dimethylphenylhydrazone, however, a completely new [3,5]-sigmatropic rearrangement occurs. Diaryl ethers, whose formation can be attributed to an unprecedented [5,7]-sigmatropic rearrangement, have also been identified in these reactions.[294]

An intramolecular, thermoneutral $[_{\sigma}2_s + _{\sigma}2_s + _{\pi}2_s]$ dyotropic transfer of hydrogen has been observed to compete with cycloreversion in the *endo,endo*-11-oxatetracyclo$[6.2.1.1^{3,6}.0^{2,7}]$dodecenes.[295] *cis*-2′-Trifluoromethylspiro(acenaphthene-1,7′-norcaradiene-2-one has been reported[296] to interconvert with its 3′-trifluoromethyl isomer, this being the first directly observed example of a degenerate rearrangement for a norcaradiene with a spiro ring-system attached at C(7). It has been suggested[297] that cyano-substituted pyrazoles (**185**; R = D) transpose photochemically into imidazoles (**186**) and (**187**) by two concurrent paths (see Scheme 15), namely, a 1,5-interchange, probably by 2,5-bonding to a diazabicyclopentene which isomerizes by nitrogen "walk" before rearomatization and 2,3-interchange, probably *via* an intermediate azirine.

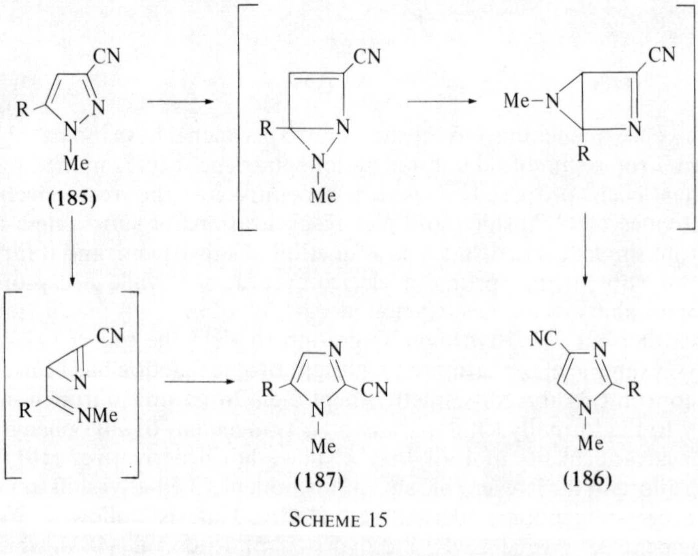

(185) (187) (186)

SCHEME 15

Finally, a suggestion of a possible concerted mechanism for the thermal rearrangement of 2,2-difluorovinylcyclopropane to 4,4-difluorocyclopentene has been put forward[298] as a reasonable explanation for a very puzzling result, although it has certainly not been unambiguously demonstrated.

Electrocyclic Reactions

Pericyclic reactions have been reviewed.[299] An account which demonstrates that a multitude of ring-closures and ring-openings in heterocyclic chemistry find their ordering principle in 1,5-electrocyclic reactions has appeared in print,[300] as has an article which presents a systematic account of thermal and photochemical cyclizations of hetero-substituted hexa-1,3,5-trienes into five-membered rings;[301] most of these latter reactions can be viewed as pericyclic reactions. The photochemical electrocyclic reaction mechanisms of a variety of hetero-atom conjugated systems have recently been elucidated by a unique concept.[302] The proposed classification of the reaction, which is based upon the number and types of electrons involved in the reaction centres, takes into account the participation of the lone-pair electrons and gives the correct description of the electron behaviour during the electrocyclic reaction process.

The photochemical electrocyclic reaction paths of *s-cis*-acrylaldehyde to oxetene have been examined[303] on the basis of the potential surfaces obtained by the MINDO/3 CI calculation, while the barriers on the MINDO/3 reaction paths for π and σ approaches to concerted cycloaddition of methylene to *s-cis*-buta-1,3-diene have been analysed[304] by a new energy-decomposition scheme and compared to the results of published extended-Hückel calculations.

A study of some concerted, intramolecular, pericyclic reactions of acetylenic compounds has shown[305] that acetylenic bonds can readily participate in reactions which require considerable distortion from the normal *sp*-bond angle and that they undergo retro-ene reactions more readily than their olefinic analogues. Moreover, retro-ene reactions of acetylenic and olefinic substrates have been shown[306] to differ in the stereochemical requirements of their respective transition states; the reactions of olefins proceed *via* a chair-like (or boat-like) configuration while those of acetylenes require planarity. By using the technique of low-pressure pyrolysis to investigate the decomposition of pent-1-yne, the process has been shown to proceed predominantly *via* a molecular retro-ene pathway to yield allene and ethylene.[307] The first example of transfer of axial chirality into central chirality has been reported[308] during a reaction of hetero-ene synthesis performed with optically active γ-allenic aldehyde (**188**); see Scheme 16. The pyrolysis of the 4-mercaptoazetidin-2-one (**189**) has provided a new γ-lactam (**191**) stereospecifically, *via* an ene reaction of an intermediate thioaldehyde (**190**).[309] 3β-Methyl-*A*-nor-5β-cholestan-5-ol-6-one has been obtained[310] from 3β-tosyloxy-5α-cholestane-5,6β-diol through an ene reaction of an unsaturated acyloin, and an expedient and highly

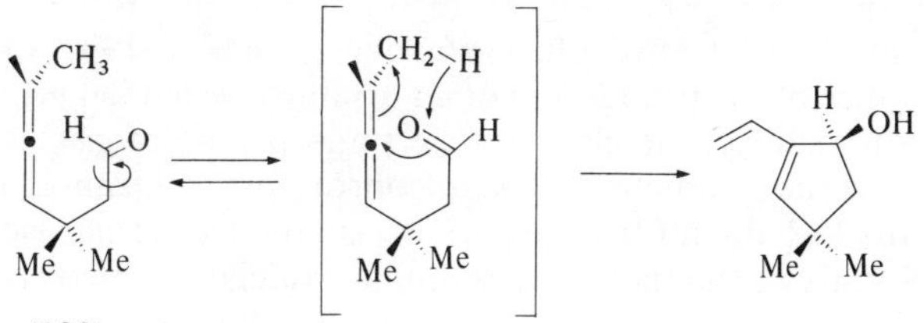

(188)

SCHEME 16

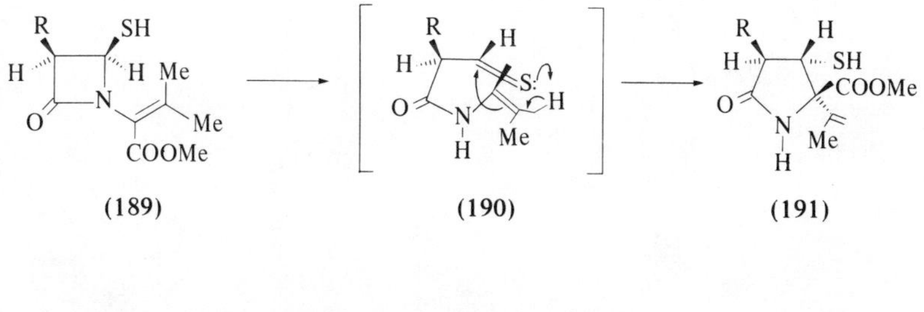

(189) (190) (191)

stereoselective intramolecular ene reaction has furnished a useful synthesis of (±)-modhephene.[311] A novel regiochemical effect associated with the intramolecular ene reaction of tetra-substituted cyclopropene derivatives has been described,[312] and the vinylaziridine (**192**) has been shown[313] to react as expected with electrophilic acetylenes and olefins to produce azepine derivatives. However, with β-nitrostyrene, a novel rearrangement has been shown to occur, presumably *via* an ene reaction (see Scheme 17) to form (**193**). A kinetic study[314] of the thermal ring-opening of two *N*-cyclohexyl-2-cyanoaziridines has established that replacement of a hydrogen at the C(3) position by a phenyl group enhances the electrocyclic reaction.

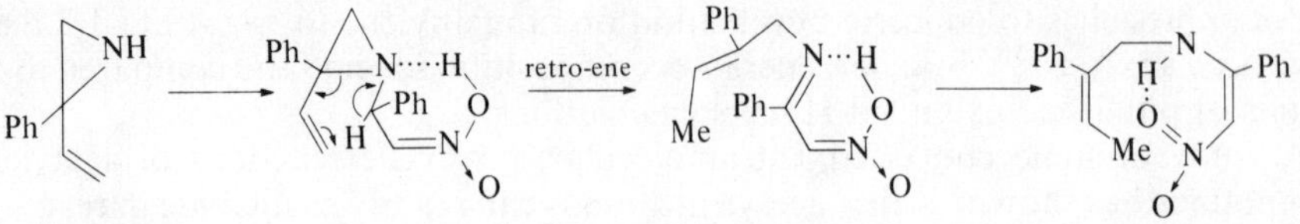

(**192**) retro-ene (**193**)

SCHEME 17

Recent observations[315] have reinforced the conclusion that the dominant pathway for double rotation in the thermally induced 1,2-dimethylspiropentane geometric isomerization, is disrotatory. Irradiation of (E)-4-(2′,4′,4′-trimethylbicyclo-[4.1.0]hept-2′-en-3′-yl)but-3-en-2-one yields (E)- and (Z)-4-(2′,7′,7′-trimethylbicyclo[3.2.0]hept-2′-en-1′-yl)but-3-en-2-one by an electrocyclic process involving C(ε)—C(ζ) cleavage,[316] and the thermally induced ring-expansion reactions of several butadienyloxiranes (**194**) have been described;[317] a multi-step sequence (see Scheme 18) has been proposed with 2-oxaheptatrienyl dipoles (**195**) as intermediates. These undergo electrocyclization with participation of either six or eight electrons leading to dihydrofurans and dihydrooxepins, respectively. In an analogous manner, the eight-electron cyclization of the conjugated carbonyl ylide (**197**), generated by thermolysis of butadienyloxirane (**196**), takes place in the theoretically expected conrotatory fashion yielding the *cis*-2,3-dihydrooxepin (**198**).[318] An electrocyclic opening of cyclopropane derivatives containing internal nucleophilic groups has been used[319] to provide a new route to vinyllactones, tetrahydropyrans, and tetrahydrofurans as illustrated by (**199**) → (**200**).

The first instance of the production of an orbitally forbidden product by a laser-induced reaction, namely, the electrocyclic ring-opening of *cis*-3,4-dichlorocyclobutene to produce the symmetry-allowed (conrotatory) *cis,trans*-1,4-dichlorobuta-1,3-diene and the forbidden (disrotatory) *cis,cis*- and *trans,trans*-isomers, has been reported.[320] A study of the thermally promoted valence isomerization of *cis*-fused

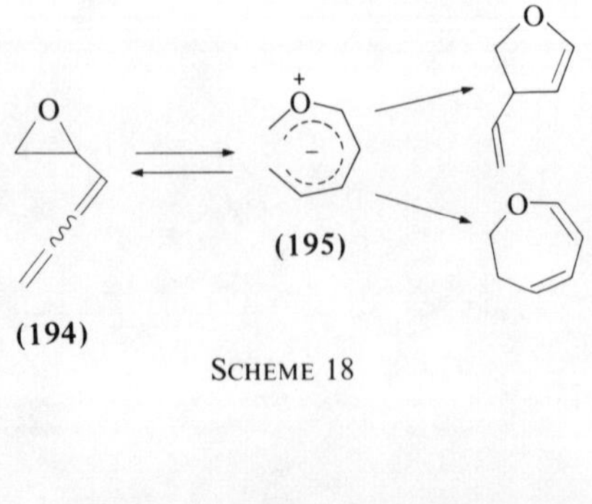

(**194**) (**195**)

SCHEME 18

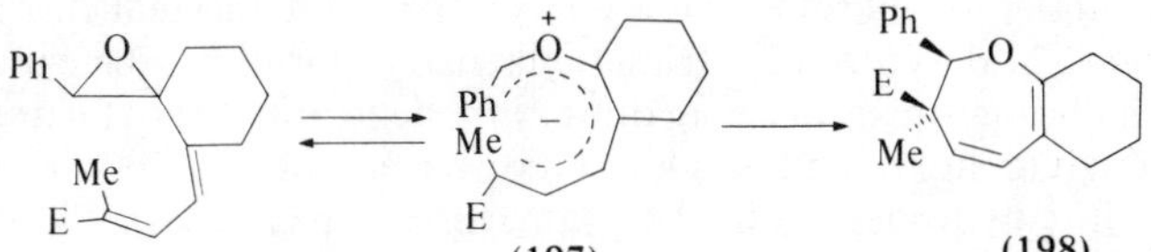

(196) (197) (198)

(199) (200)

bicyclic cyclobutenes has shown[321] that, when free of steric constraint, these compounds yield products that arise from both orbital-symmetry-allowed conrotatory openings of the labile σ-bond in the cyclobutene ring. Moreover, in the case of the related vinylcyclobutenes, one conrotatory mode produces a stable *cis,trans,cis* monocyclic triene whereas fission in the other allowed sense affords a thermally labile *trans,cis,cis* monocyclic triene which, in turn, cyclizes to yield a cyclohexadiene derivative. The possible extension of the Woodward–Hoffmann rules to the isomerization of heterocyclobutenes has now been confirmed[322] by the finding that 2-(*N,N*-diethylcarbamoyl)-2,4-dimethyl-3-phenyl-2,3-dihydroazete-1-oxide isomerizes on heating to *N*-[1-(*N,N*-diethylcarbamoyl)ethylidene]-1-phenylprop-1-en-2-amine-*N*-oxide, thus showing that ring-opening has taken place in a conrotatory mode in line with the isomerization of cyclobutenes.

A kinetic study of the thermal isomerization of fulvene into benzene has indicated[323] that the reaction proceeds by a concerted mechanism with an activated complex as shown (**201**). It has been demonstrated[324] that the vinylogous sesquifulvalene (**202**) undergoes thermal 14-electron electrocyclization with ease and perispecifically. However, due to the steric constraints imposed by the system, the process is considered to take place exclusively in the "symmetry-forbidden" conrotatory sense to give *trans*-10*a*,10*b* dihydrophenazulene (**203**) which is rapidly isomerized into the benzenoid 1,8- and 3,8-dihydrocyclohept[*e*]indenes (**204**) and (**205**). (E)-α-3-Furylethylidene(isopropylidene)succinic anhydride (**206**) has been

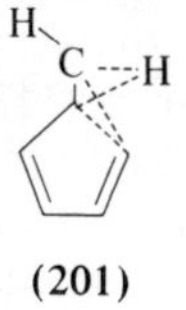

(201)

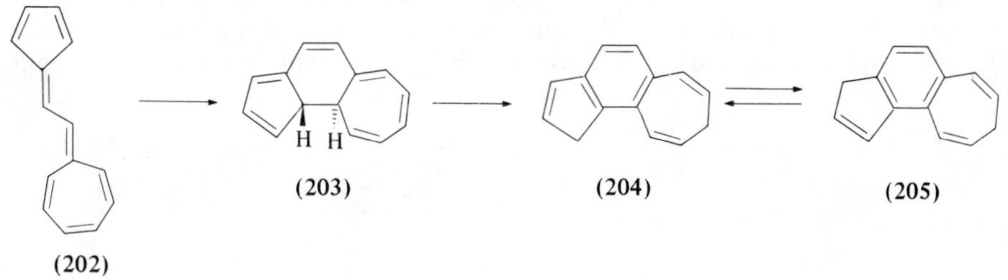

(202) (203) (204) (205)

shown[325] to undergo photochemical conrotatory and thermal disrotatory ring-closure to give 7,7*a*-dihydro-4,7,7-trimethylbenzo[*b*]furan-5,6-dicarboxylic anhydride (**207**), which ring-opens in a conrotatory mode to afford the starting product on exposure to white light, and is slowly converted into (**208**) on heating by a suprafacial [1,5]-hydrogen shift. The same group has shown[326] that (E)-α-2,5-dimethyl-3-furylethylidene(alkylidene)succinic anhydrides undergo reversible photochemical conrotatory electrocyclic reactions to give 7,7*a*-dihydrobenzo[*b*]-furan-5,6-dicarboxylic anhydrides. Interesting thermal and photochemical rearrangements involving the adducts from benzyne and substituted 2,1,3-benzoselenadiazoles have been reported.[327] The authors envisage an electrocyclic rearrangement involving the side-chain of the 5-(1,2-benzoselenazol-3-yl)pentadienonitrile derivatives (**209**) to give the intermediate (**210**) which can then rearrange further to give 2-(2-pyridyl)phenyl selenocyanates (**211**), probably by an ionic mechanism.

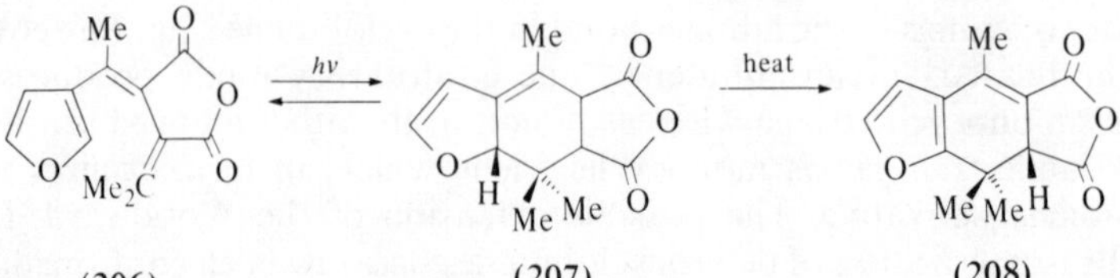

(206) (207) (208)

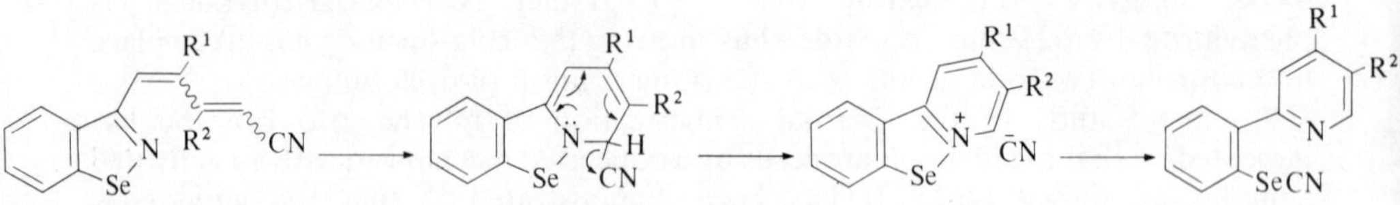

(209) (210) (211)

Intramolecular photo-addition of the 1,7-diene system in 4-(1,5-dimethylhex-4-enyl)cyclohex-2-enone yielded, *inter alia*, two ketones containing the tricyclo-[5.3.1.0^{5,11}]undecan-8-one system.[328] A kinetic study has been made of cyclization and cycloreversion reactions of aryl-substituted cycloheptatrienes,[329] and the initial product (**212**) from the dimerization of cycloocta-1,5-dien-3-yne has been readily converted by electrocyclic reaction into (**213**) and (**214**) with the former valence isomer preponderating in the equilibrium.[330] A preliminary communication on the acid-catalysed pericyclization of benzo[9]annulenone has appeared[331] and the results rationalized by assuming that the necessary driving force is supplied by the development of a π-destabilized system. Recent experimental evidence has supported[332] a vinylketene mechanism for both the photolysis and pyrolysis of bicyclic δ-thia-α,β-unsaturated ketones. The unexpected amount of sulphide

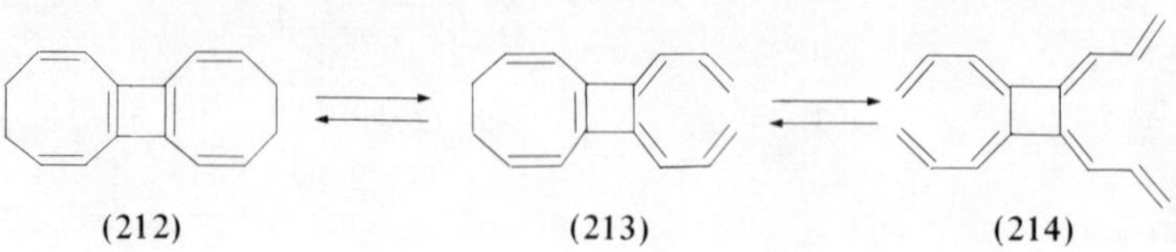

(212) (213) (214)

rearrangement observed to take place on reaction of butenyl methyl sulphides with methanesulphenyl reagents has been rationalized[333] by postulating that the initial methylthiolation step involves reversible attack at sulphur followed by pericyclic interconversion of the thiosulphonium ions so formed.

It has been suggested[334] that the rearrangement of the symmetrical triepoxide, trispiro{tricyclo[3.3.3.$0^{1,5}$]undecane-2,2′:8,2″:9,2‴-tris[oxirane]} to the tris-ether, 2,5,14-trioxahexacyclo[5.5.2.1.2.$^{4,10}0^{4,17}0^{10,17}$]heptadecane proceeds *via* an unprecedented $[2\sigma + 2\sigma + 2\sigma]$ to $[2\sigma + 2\sigma + 2\sigma]$ electrocyclic rearrangement, while an electrocyclic ring-opening and recyclization has been postulated[335] to occur on irradiation of the 4*a*,10:9,9*a*-diepoxide of 9,10-diphenylanthracene.

Anionic Rearrangements

The rearrangements of carbanions[336] and the Stevens rearrangement[337] have been reviewed. A MINDO/3 investigation of the photochemical rearrangement of the *aci*-nitrate anion $CH_2N(O)O^-$ to $HCONHO^-$ has indicated[338] that, in the first step, C—N bond rotation is accompanied by oxygen migration to form a bridged structure, (**215**), or (**216**). A recent study[339] has shown that, although *N*-acetyl-*N*-nitrosophenylalanine thermally decomposes in the expected manner, its carboxylate anion undergoes a series of facile rearrangements initiated by carboxylate attack to give 2-hydroxy-3-phenylpropanoic acid.

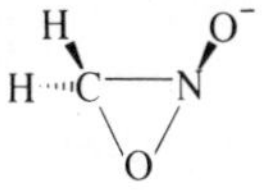

(**215**) 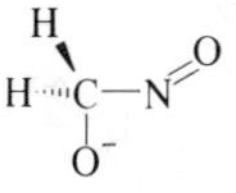 (**216**) 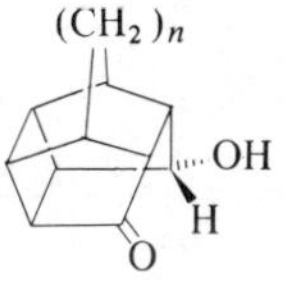 (**217**)

The intramolecular hydride rearrangements of a number of homologous hydroxyketones of the type (**217**) have been described,[340] while the mild base-promoted conversion of *tert-cis,cis*-cyclonona-2,4-dienols to bicyclo[4.3.1]nona-2,4-dienes has been reconciled with the generation and antarafacial cyclization of helical carbanions under extraordinarily mild conditions.[341] Anionic rearrangements have been invoked to explain the efficient five-step assembly of a novel pentacyclic C_{15}-quinane system from cyclopentadiene and naphthoquinone by way of an intermediate hexacyclic propellane.[342] A new route to 3-alkyl-2,5-di-*tert*-butylcyclopenta-2,4-dienones (**219**) from 4-alkyl-2,6-di-*tert*-butylphenols has been devised.[343] It involves the base-catalysed reaction of 5-acylmethyl-2,5-di-*tert*-butyl-4-oxacyclopent-2-enones (**218**) which are selectively derived from the phenols in three steps; a possible pathway is represented in Scheme 19.

2-Hydroxyalkyl-2-phenylsulphonylcycloalkanones have been shown to undergo base-catalysed isomerization to phenylsulphonylactones with incorporation of the side-chain into the ring.[344] The reaction of several dibenzylidene sulphamides (**220**) with various lithium amides has been shown to yield *N*′-substituted N^2-benzylsulphamoylbenzamidines (**221**) *via* an intramolecular hydride-transfer reaction.[345] Hydride transfer to a tetrazine molecule has been invoked[346] to explain the formation of 4-hydroxy-3-(α-hydroxybenzyl)-4-phenyl-2-pyrazolin-5-one on reaction of 3,6-bis(α-hydroxybenzyl)-*s*-tetrazine with methanolic potassium hydroxide. No conclusive evidence has been obtained concerning the mechanism of

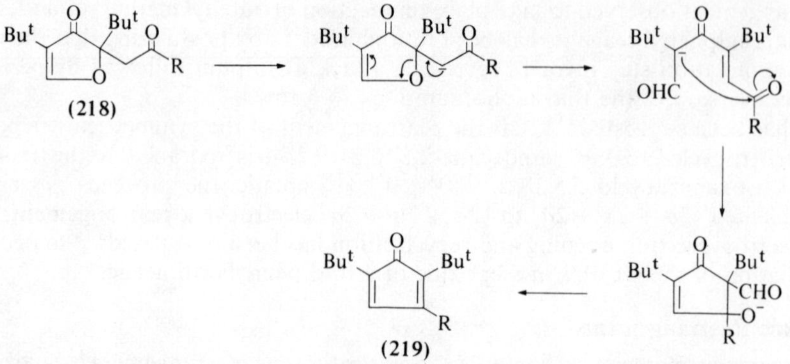

(218)

(219)

SCHEME 19

the facile base-catalysed rearrangement of the diacetylenic diamine (**222**) into the conjugate diaminotriene (**223**), but it seems reasonable to assume the pathway shown in Scheme 20 for the process.[347] (Dialkylamino)allenes have been prepared by the base-catalysed isomerization of prop-2-ynylamines.[348]

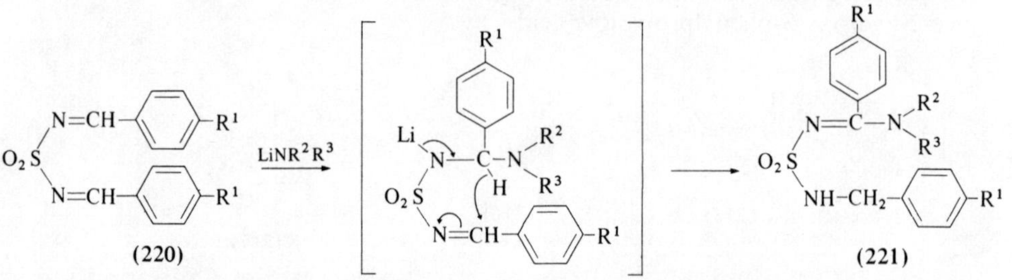

(220)

(221)

Instances of potassium-fluoride-induced Stevens rearrangements have been described.[349,350] The same workers have shown[351] that the Stevens rearrangement of quatenary ammonium salts bearing an (alkoxycarbonyl)methyl and cyanomethyl group is accompanied by allylic rearrangement of the cyanomethyl group to yield conjugated dienic amino-esters, and they have used the Stevens rearrangement to synthesize α-(dialkylamino)ketones with allenic and vinylallenic structures.[352] The

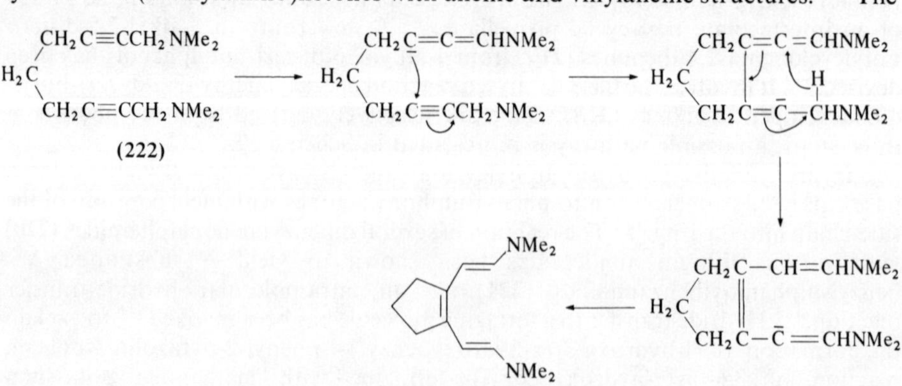

(222)

SCHEME 20 (223)

best analogy of the reactivity of ammonium ylides of the type $PhCH_2\overset{+}{N}Me_2\overset{-}{C}HCOPh$ has been shown[353] to be that of the corresponding sulphonium ylides, in that in both cases, [1,2]- or [3,2]-rearrangements occur readily when a suitable migrating group is attached to the 'onium centre; most of the other reactions are formally alkylidene-transfer reactions. Stevens rearrangement products have been obtained[354] on base treatment of 3,5-dioxopiperazinium salts. A reinvestigation of the Stevens rearrangement of 1-benzyl-1,3,4-trimethyl-1,2,5,6-tetrahydropyridinium salts has been effected.[355] The thermal rearrangements of the conjugate bases (**224**) of *N*-benzyl open-chain analogues of Reissert compounds have been shown to afford products formally derived from two competing [1,2]-rearrangement–elimination reactions.[356] The products of the major sequence of reactions are deoxybenzoins and benzonitriles, while those of the minor sequence are α-benzamidostilbenes, and the results are compatible with a radical-pair mechanism involving initial homolysis of the conjugate base (**224**) followed by two possible modes of radical recombination. 2,3,4,5,6,7-Hexahydro-1*H*-2-benzazonines (**225**) have been prepared[357] by a sodamide–liquid-ammonia-induced Sommelet–Hauser rearrangement. The Meisenheimer rearrangement of phenylbis-(trifluoromethyl)amine oxide, $Ph\overset{+}{N}(CF_3)_2O^-$, to $PhN(OCF_3)CF_3$ has been described.[358]

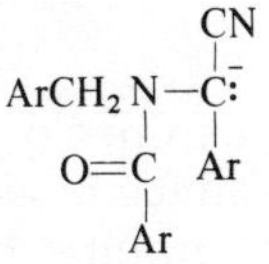

(224) 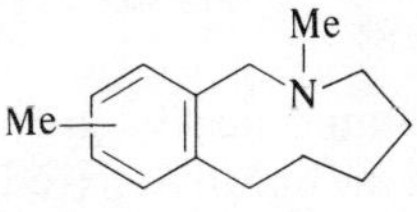 **(225)**

Kinetic details have been established[359] for the apparent vinyl-to-allyl anion rearrangement of 2-methyl-1-phenylpropen-1-yllithium, and the results of the rearrangement of isomeric linear decyn-1-ols by reaction with the sodium salt of 1,3-diaminopropane have been found to be consistent with a general mechanism in which a series of reversible 1,3-proton shifts effect the migration of the triple bond along the methylene chain.[360]

The base-induced conversion of the [8 + 2] π-cycloadducts (**226**), from azaheptafulvenes and mono-substituted ketones, into (z)-α-substituted cinnamamides (**228**) has been presumed to proceed through an intermediate

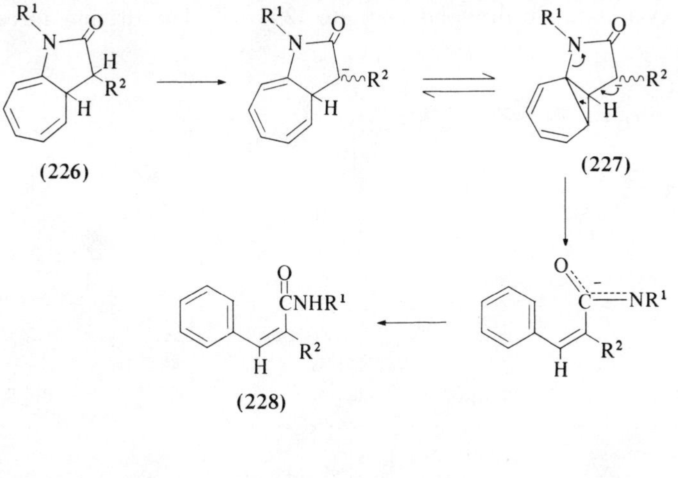

(226) **(227)**

(228)

norcaradiene (**227**);[361] a follow-up paper[362] has reported that the cycloadducts from azaheptafulvenes and sulphenes, as well as from tropones and arylsulphones, rearrange in a similar manner upon oxidation of their corresponding α-sulphonyl anions, to give 1,2-disubstituted indoles or 2-arylbenzofurans, respectively. Pyrazolo[2,3-*a*]pyrimidines (**230**) have been obtained from 2-ethynylpyridinium *N*-imides (**229**) on treatment with base.[363] A possible mechanism is shown, although none of the intermediates have as yet been isolated.

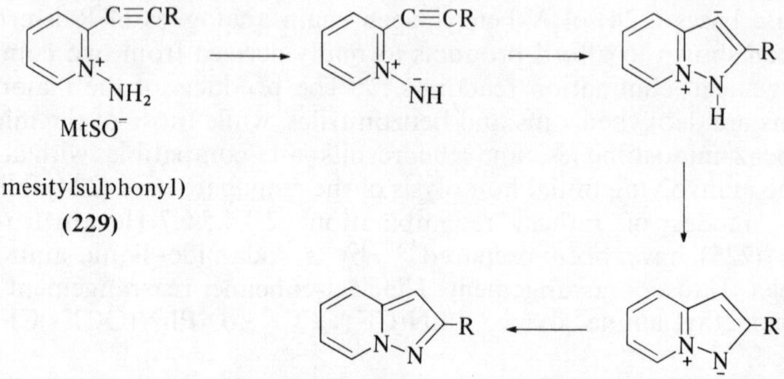

C≡CR, N^+, NH_2, $MtSO^-$

(MtS≡mesitylsulphonyl)

(**229**)

(**230**)

An interesting 1,4-anionic silyl group migration from oxygen to carbon has been reported to occur in the anions derived from the adducts of substituted acroleins and trimethylsilyl cyanide,[364] while transfer of a silyl group from carbon to oxygen is invoked[365] to account for the reaction of tris(trimethylsilyl)methyllithium with a variety of carbon electrophiles; see (**231**; R = $SiMe_3$) → (**232**; R = $SiMe_3$).

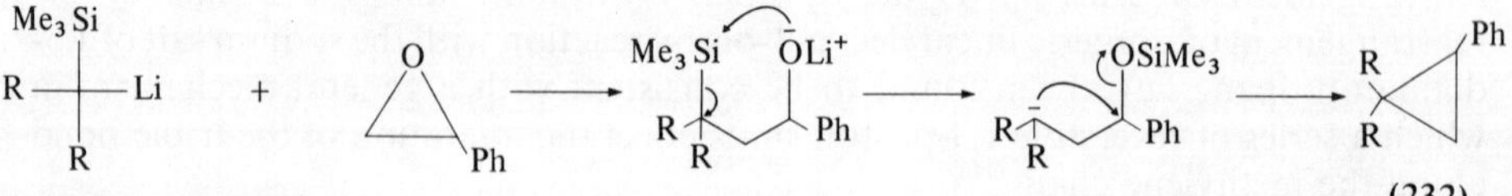

(**231**) (**232**)

Diethyl 1-hydroxy-3-phenyl-1-(trimethylsilyl)propylphosphonate (**233**) has been shown to undergo a silicon–tin rearrangement with tributylin alkoxides to afford the 1-silyloxy-1-stannylpropylphosphonate (**234**).[366] The driving force for this

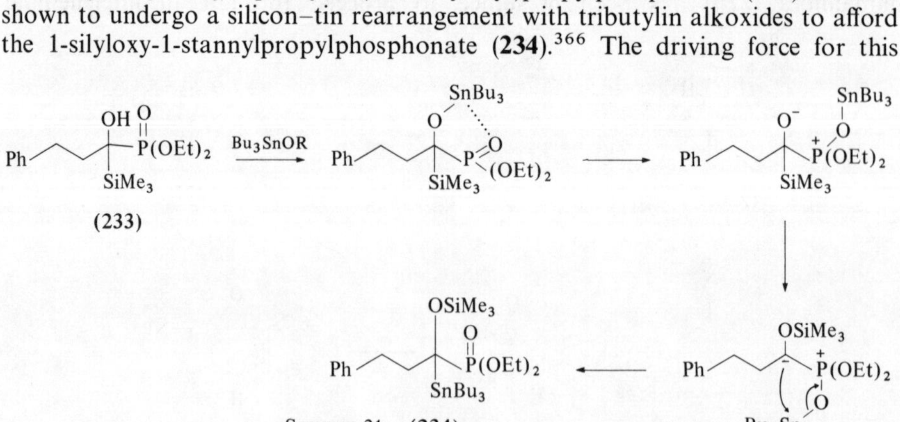

Bu_3SnOR

(**233**)

SCHEME 21 (**234**)

reaction arises from coordination of the stannyl group to the α-phosphonyl oxygen and the strong affinity of the silyl group for the resulting oxide ion; see Scheme 21. Thermal and fluoride-ion-promoted phosphoryl rearrangements of diethyl α-benzoyl-α-[(trimethylsilyl)oxy]benzylphosphonate and analogues have been investigated,[367] and the postulated mechanism for this reaction has provided further evidence that the Perkow and Kukhtin–Ramirez reactions proceed *via* an initial attack of phosphorus on carbonyl carbon. An interesting rearrangement involving the intramolecular cycloaddition reaction of phosphorus ylides (**235**) has given rise to 8,13-methano-5*H*-dibenzo[*b*,*e*]phosphonin derivatives (**236**);[368] see Scheme 22. An examination of the stability of *S*-alkanoate-*O*,*O*-dialkyl phosphate carbanions has unearthed an intriguing enol-phosphate–phosphorothiolate rearrangement.[369]

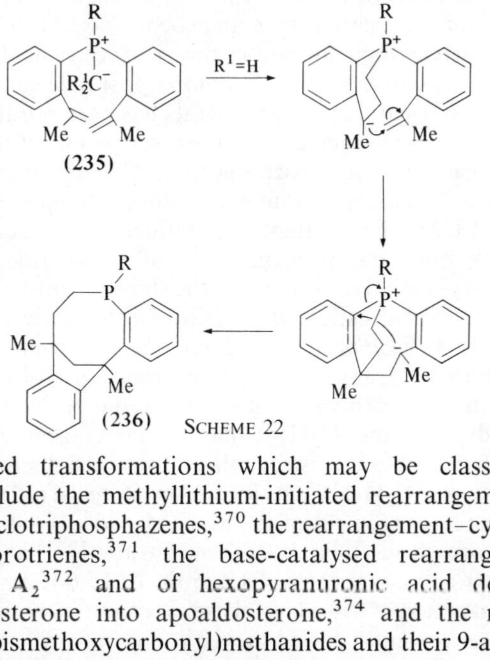

(235) $\xrightarrow{R^1=H}$

(236)

SCHEME 22

Recently reported transformations which may be classified as anionic rearrangements include the methyllithium-initiated rearrangement of propynyl- and propadienyl-cyclotriphosphazenes,[370] the rearrangement–cyclization of cross-conjugated perfluorotrienes,[371] the base-catalysed rearrangements of 3-*O*-demethylfortimicin A_2[372] and of hexopyranuronic acid derivatives,[373] the conversion of aldosterone into apoaldosterone,[374] and the rearrangement of thioxanthen-10-io(bismethoxycarbonyl)methanides and their 9-alkyl derivatives to the corresponding 9-[bis(methoxycarbonyl)methyl]thioxanthens.[375]

Oxyallyl has been computed[376] to close in a disrotatory manner to cyclopropanone with no energy barrier, and is thus excluded as a possible intermediate in the Favorskii rearrangement. An interesting reaction of acyclic α-haloketones with malonate anion represents the first example of a carbanion-induced Favorskii-type rearrangement and also provides a new synthetic route to a variety of cyclopropanols.[377] The first example of the Favorskii-type rearrangement of α-monochloroketimines has been reported.[378] The regiospecific opening of the intermediate cyclopropylideneamines has been shown to parallel the opening of cyclopropanones. The report of a selective Favorskii rearrangement in a macrocyclic ring has appeared.[379] 2-Bromo-5-chlorooctan-4-one has been shown to undergo Favorskii rearrangement to give a mixture of methyl (E)- and (Z)-2-propylpent-2-enoates,[380] and the same rearrangement has been used to prepare bicyclo[2.2.0]hexane-1-carboxylic acid from 7-oxobicyclo[2.2.1]heptane-1-carbonyl chloride.[381] The Favorskii rearrangement has been used[382] in a versatile stereoselective synthesis of 3-(2,2-dihalovinyl)-2,2-dimethylcyclopropanecarboxylic acids which have been further used[383] in the synthesis of pyrethroids. The reaction of 4,8,8-tribromobicyclo[5.1.0]oct-4-en-3-ones under Favorskii conditions has been shown to yield both rearranged, ring-contracted products and saturated bicyclic ketals and epoxides.[384] Pentaprismane has been synthesized[385] by way of a reaction that is analogous to a Favorskii ring-contraction, while methyl 3β-hydroxy-16α-methyl-5α-pregn-9(11)-en-21-oate arose as an unexpected product from an unusual Favorskii rearrangement of 17ζ-bromo-16α-methyl-20-oxo-5α-pregn-9(11)-en-3β-yl acetates.[386]

Cationic Rearrangements

Reviews of rearrangements of various types of carbocations have appeared.[387–390] The characteristics of carbocation rearrangements have been reviewed[391] and the isomerizations of carbocations formed in the presence of acid catalysts have been discussed.[392] An account of how the Hammond postulate can be used to interpret the mechanisms of carbocation rearrangements has been produced,[393] and reviews of new methods for ring-enlargement,[394] the mechanisms of the rearrangement of peroxides,[395] and gas-phase ion rearrangements[396] have appeared.

It has been shown[397] that for various carbocations the energy value and degree of localization of the LUMO have a significant influence on the energy barriers of 1,2-hydrogen shifts. A study has been made[398] of 1,3-hydride shifts in methoxy-, ethoxy-, and propoxy-methyl cations, while the stereochemistry of transannular 1,5-hydride shifts have been investigated[399] using 3-*tert*-butylcyclooctyl compounds. MINDO/3 calculations have been carried out[400] to determine the structures of the transition states in the rearrangement and decomposition of formamide molecular cations, and the symmetry properties of a graph representing the rearrangement of the homotetrahedryl cation, $C_5H_5^+$, have been considered.[401] A paper has appeared[402] which describes the use of collision-induced dissociation spectra in the structural elucidation of $C_4H_5X^+$ ions formed by chlorine or hydroxyl migration in a McLafferty-type rearrangement of the molecular ion of ω-functionalized allenes and their isomeric acetylenes. It has been demonstrated[403] that processes revealed in the chemical ionization mass spectra of a number of 3,5,6-triaryl-4-nitrocyclohex-1-enes involve the formation of triarylcyclohexenyl cations. The observed specificity of carbocation rearrangements during acylation of propyne with 4-*tert*-butyl-1-methylcyclohexane cation stereoisomers has confirmed[404] the rôle of orbital control in carbocation rearrangements. Rearrangements of halotoluene parent cations[405] and phenylalkene cations[406] have been examined in solid argon. Studies of the alkylaluminium-chloride-induced cyclization of unsaturated aldehydes and ketones, which indicate the advantages of these reagents in Lewis acid-initiated reactions, have been reported.[407] With such reagents, proton-catalysed reactions apparently do not occur and the alkyl group can enter into the reaction in a synthetically useful manner.

A study of the kinetic isotope effect in the solvolysis of neophyl arenesulphonates has been reported[408] and the experimental results have been compared with those calculated for various transition-state models. The usefulness of this combined experimental–theoretical approach has been extended to rearrangement reactions. The nature of σ-participation as a composite of two formally conflicting modes, namely hyperconjugation and bridging, has been discussed on the basis of results obtained from a study of the kinetic isotope effect upon the solvolysis of neopentyl-type arenesulphonates.[409] A study of deuterium isotope effects for migrating and non-migrating groups in the solvolysis of neopentyl-type esters has indicated[410] that the products of solvolysis are predominantly rearranged products resulting from methyl migration. The results are consistent with a mechanism that involves neighbouring-group participation during ionization. A Wagner–Meerwein rearrangement of the carbon skeleton has been observed in the gas-phase pyrolysis of neopentyl bromide[411] and unambiguous proof for alkoxycarbonyl group migration in Wagner–Meerwein rearrangements of β-hydroxy-esters has been presented.[412] It has been shown[413] that the reaction of alkyl halides with sodium iodide in acetone can yield coupled or rearranged products *via* cationic intermediates, and a ^{14}C-labelling study has indicated[414] that the isomerization of 1(2)-bromo-2(1)-acetoxypropanes through the 1,2-migration of bromine proceeds at least partially *via* solvent-separated ion-pairs. Relative migratory aptitudes of various alkyl groups have been determined in the iodine-induced rearrangement of lithium ethynyltrialkylborates.[415] The reaction intermediate in the acid-catalysed rearrangement of indolines (**237**) to the 2,3-disubstituted indoles (**239**) appears to be the carbocation (**238**) which then undergoes attack on the C(2) substituent having the highest migratory aptitude.[416] Isotopic-scrambling studies have been carried out on the acetolysis of labelled trianisylvinyl bromide,[417] 1,2-phenyl shifts in the 1,2-dianisyl-2-phenylvinyl cation have been observed during the decomposition of *cis*- or *trans*-1,2-dianisyl-2-phenyl[2-^{14}C]vinylphenyltriazenes in acetic acid,[418] and a simple synthetic approach to α-arylalkanoic acids has been accomplished[419] by inducing a novel 1,2-aryl rearrangement. A 1,2-phenyl shift to the double bond of a vinyl cation has been observed[420] on reaction of 3-phenylpropyne with HCl, 1,1-difluoro-2-phenylethanes have been conveniently prepared[421] by deamination of 2-fluoro-2-phenylethylamines with concomitant 1,2-phenyl migration, and substituent effects and migratory aptitudes in the fluorination–rearrangement of 1,1-diarylethenes with aryliodine(III) difluorides have proved consistent[422] with an electrophilic addition of (*p*-iodophenyl)acetic acid difluoride to the diarylethene, followed by a rearrangement step involving a phenonium ion.

OH R² R¹ N R³ → H + R² R¹ N R³ → R² R¹ N R³

(237) (238) (239)

A pinacol rearrangement coupled with the contraction of a tetrahydropyran ring has been observed.[423] The intrinsic migration aptitudes of alkyl groups in the pinacol rearrangement of substituted *cis*-2-tosyloxycyclopentanols have been deduced[424] and a kinetic study has been undertaken[425] to provide information about the pinacol rearrangement of secondary *vic*-alkanediols. Pinacol rearrangements have been observed[426] in the mass spectrometry of spontaneous and collision-induced fragmentations of protonated aldehydes and ketones, and collision-induced dissociation and mass-analysed ion-kinetic-energy spectrometry has been used[427] to identify pinacol rearrangements and related reactions in the gas phase. A method has been described[428] which involves a pinacol-like shift of a vinyl group, which permits a one-step conversion of 2-chlorocycloalkanones into differentiated 1,2-divinylcycloalkanols, 2-vinylcycloalkanones, or 2-vinylcycloalkanols. Interestingly, tertiary alcohols with a γ-silyl group, namely (**240**), have been found[429] to undergo a simple acid-catalysed pinacol-like rearrangement, yielding a single alkene product (**241**). This silicon-controlled carbocation rearrangement is cleaner than the corresponding pinacol rearrangement, and when γ-silyl-β-phenylthio-alcohols are treated with acid it has been shown[430] that the strategically placed silyl group encourages the rearrangement of the phenylthio group, both from a secondary migration origin to a secondary migration terminus and from a secondary migration origin to a tertiary migration terminus.

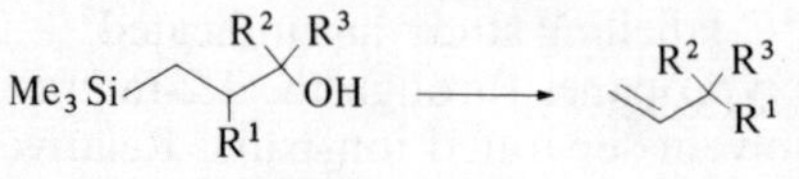

Me_3Si … R^2 R^3 OH R^1 (240) → R^2 R^3 R^1 (241)

Isotopic labelling studies of the thermal rearrangement of phenyloxirane (**242**) to phenylethanal (**243**) have suggested[431] that the reaction occurs by a 1,2-hydrogen shift as in path (*a*) and/or path (*b*). An investigation[432] of migratory aptitudes in the Criegee rearrangement has supported a transition state (**244**) rather than (**245**). The formation of acyclic products in the reaction of triisobutylaluminium and 1,3,3-trimethylcyclopropene has been ascribed to a carbocation rearrangement[433] which supports the earlier proposed mechanism for the carbalumination of alkenes. The first unambiguous observation of a "bisected" primary cyclopropylcarbinyl cation has been reported,[434] the cation being formed by ring-contraction of a C_8 tricyclic framework. Although totally unexpected from a bond-energy–steric-strain standpoint, this observation underlines the large resonance stabilization associated with "bisected" cyclopropylcarbinyl cations. Although a further example of the use of oxy and thio substituents to direct the course of carbocation rearrangements in general, and cyclopropylcarbinyl rearrangements in particular, has been observed in the deamination of phenylthio- and 1-oxy-substituted chrysanthemylamines, the kinetic stabilizing effects of these hetero-atom substituents on competing processes

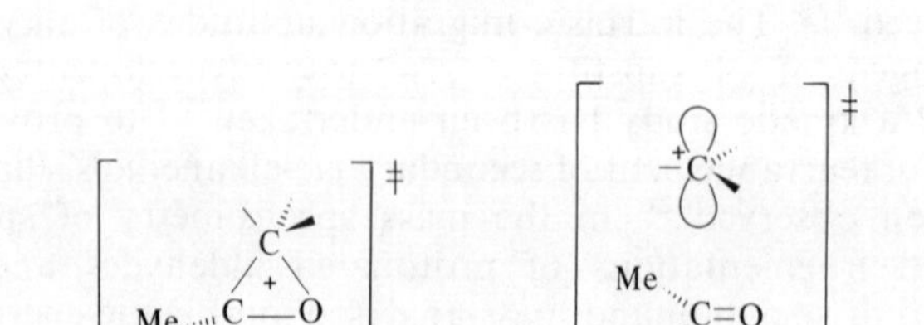

PhCH–CH₂ (O bridged) (242) —(a)→ ·O–CH₂ / ·PhCH–CH₂ → $PhCH_2CHO$ (243)

(242) —(b)→ O^- / $\overset{+}{PhCH}$–CH₂ → (243)

[Me⋯C–O with $\overset{+}{C}$ bridged]‡ (244)

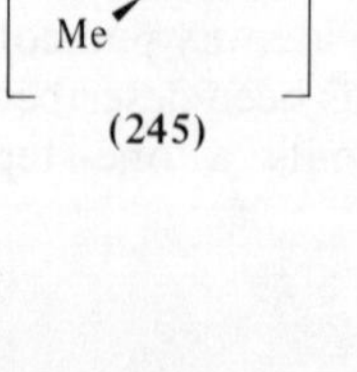

[$-\overset{+}{C}$ over Me, Me C=O]‡ (245)

have been found to be less than expected.[435] The formation of methyl 1-methylcyclopropyl ketone (**248**) from the oxythallation of the 1,3-diene (**246**) with thallium(III) nitrate has provided good evidence that the transformation is presumably derived from a cyclopropylmethyl cation (**247**) after a vinyl migration.[436]

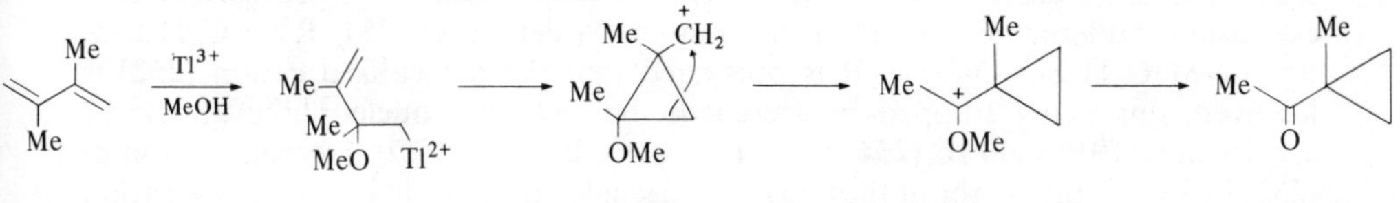

(**246**) (**247**) (**248**)

The spiro[2,5]oct-4-yl cation has been found to rearrange irreversibly on warming to the thermodynamically more stable, rapidly equilibrating bicyclo-[3.3.0]oct-1-yl cation.[437] It has been established[438] that the amount of rearrangement to four-membered and three-membered ring-compounds from the solvolysis of pent-3-ynyl triflate is strongly dependent upon the nucleophilicity of the solvent, while the favoured formation of phenyl cyclopropyl ketone from gas-phase dissociative ionization of various isomeric phenyl-substituted cyclopropylidene methyl and cyclobuten-1-yl cations has been accepted[439] as direct evidence for the unusual stability of the intermediate phenyl-substituted cyclopropylidenemethyl cation. A cationic mechanism has been proposed[440] for the cyclization of aryl cyclopropyl ketones to 1-tetralones, and 2,3,6,7-dibenzo-9-oxabicyclo[3.3.1]nona-2,6-dienes (**250**) have been generated by treatment of phenylacetaldehydes (**249**) with fluorosulphuric acid.[441] A reasonable rationale of the reaction is shown in Scheme 23.

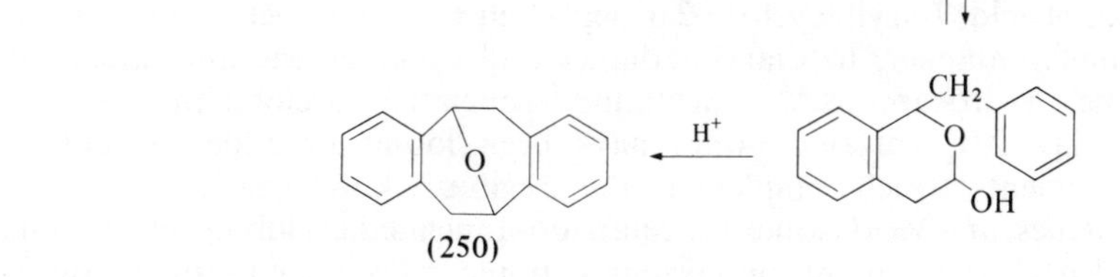

(**249**) (**250**)

SCHEME 23

Data obtained from a study of the mechanism of acid-catalysed opening of the cyclopropane ring in bicyclo[3.1.0]hexan-3-ols[442] are consistent with the view that the reaction is initiated at the cyclopropane ring, rather than at the hydroxy group, although the latter is subsequently lost during formation of the 1-methylcyclopentenium ion. In this study it was impossible to deduce a detailed mechanism because of the plane of symmetry possessed by all the compounds under scrutiny, but in a further study of thujan-3-ols,[443] substrates which lack a plane of symmetry, it was observed that the initial ring-opening involved at least two mechanisms yielding a carbocation and an olefin. Subsequent reaction, possibly by elimination and olefin shift, would give the 2,3-dimethyl-4-isopropylcyclopent-2-enol which could rapidly ionize to the observed ion. 1,4-Dihydroxy-*A*-homo-19-nor-9,10-secocholesta-5,7-diene (**254**; R = R′ = H) has been obtained[444] by solvolytic ring-expansion (buffered acetolysis) of the vitamin D_3 derivative (**251**; $R^1 = C_6H_5CO$, $R^2 = p\text{-}MeC_6H_4SO_2OCH_2$). It is presumed that the delocalized cation (**252**) is involved, this being trapped by stereospecific *endo*-face nucleophilic attack by acetate at C(10) to afford (**254**; $R^1 = C_6H_5CO$, $R = CH_3CO$). Stereomutation of (**252**) by 180° rotation about the C(5)—C(6) single bond would lead to a new cation (**253**) which by analogous *endo*-face acetate attack at C(10) could account for the minor product (**254**; $R^1 = CH_3CO$, $R = C_6H_5CO$).

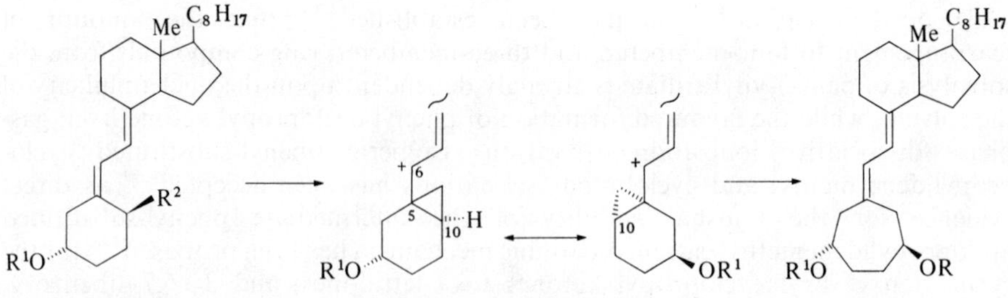

(**251**) (**252**) (**253**) (**254**)

The tendency of 3-substituted 2,2-dimethylcyclobutanol systems to undergo ring-contraction preferentially is thought to be due to a concerted ionization–rearrangement step (**255**) → (**256**),[445] in which the 1,2-bond cleavage and the backside attack at C(3) is facilitated by the geometry of the favoured 1,3-diequatorial conformation of (**255**), as shown. Cyclobutanones derived from 1-naphthyl ketones and cycloalka-1,3-dienes have been observed to undergo acid-catalysed rearrangements, under a variety of conditions, to yield tetracyclic ketones which can be further converted into chrysenes and substituted cyclopentenophenanthrenes. These rearrangements can be rationalized[446] according to Scheme 24. In the presence of acid, 7-arylbicyclo[3.2.0]hept-2-en-6-ols have been found to rearrange, in a similar manner, to diarylmethanes and cyclopenteno-annelated polycyclic aromatic hydrocarbons.[447] Acetamido-annelated cyclobutenes such as 2-azabicyclo[3.2.0]hept-2-en-3-ones have been found to undergo acid-catalysed rearrangements, which complement the previously known cationic conversions of cyclobutenes, and yield isomeric acetamido-annelated cyclobutenes.[448] It has been concluded[449] that, in intramolecular cationic π-cyclizations of cyclohex-2-enol derivatives, the first-formed intermediate is the allylic cation.

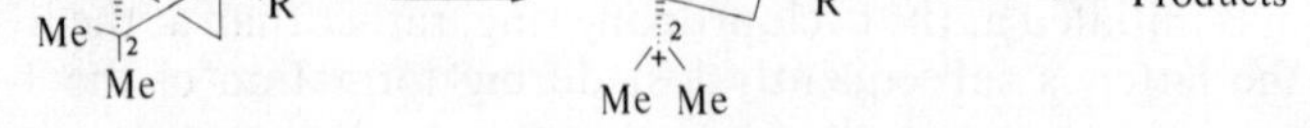

(**255**) → (**256**) → Products

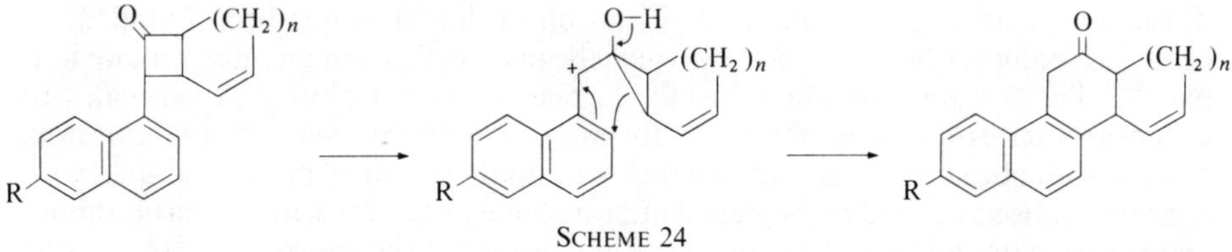

SCHEME 24

It has been demonstrated[450] that δ-lactones are converted by hot formic acid into either pendant or spiro-γ-lactones, depending on their constitution and stereochemistry. The nature of the products obtained from the thermolysis of α-azidochalcone has indicated[451] that at least part of the pathway proceeds by a cleavage mechanism which allows for an intermediate acyclic cation to react with nucleophilic species. It has been observed[452] that thioketals react with IN_3 to afford α-azidosulphides which on treatment with trifluoroacetic acid lead smoothly to amides in which the more substituted atom migrates preferentially. In an attempted total synthesis of (±)-chelidonine from the diazoketone (**257**), migration of an aromatic ring was observed[453] in a reaction which mechanistically can be viewed as proceeding through a spirocyclic cation (**258**). The Lewis-acid-promoted decomposition of unsaturated α-diazoketones has been used[454] in an efficient approach to simple and annulated cyclopentenones. The feasibility of employing the diazoketone functionality for the initiation of polyolefinic cationic cyclizations has been explored.[455] These have been shown to proceed in a non-stereospecific fashion, a result which is believed to reflect the step-wise nature of the cyclization process.[456] A thietanone has been established[457] as the main product of both thermal and photochemical treatment of the α-diazoketone of 2,2,5,5-tetramethylthiolane-3,4-dione.

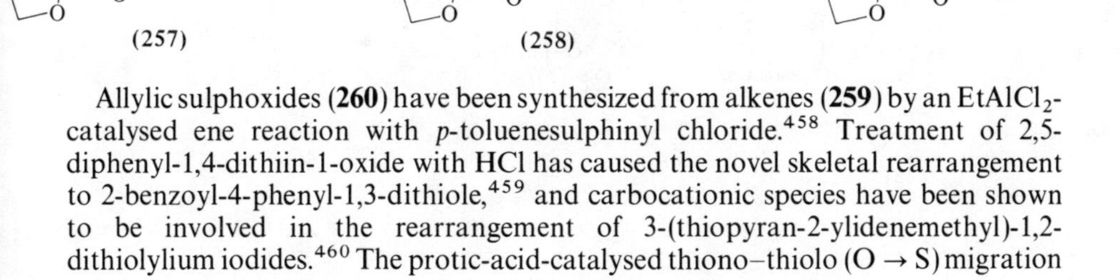

(257) (258)

Allylic sulphoxides (**260**) have been synthesized from alkenes (**259**) by an $EtAlCl_2$-catalysed ene reaction with *p*-toluenesulphinyl chloride.[458] Treatment of 2,5-diphenyl-1,4-dithiin-1-oxide with HCl has caused the novel skeletal rearrangement to 2-benzoyl-4-phenyl-1,3-dithiole,[459] and carbocationic species have been shown to be involved in the rearrangement of 3-(thiopyran-2-ylidenemethyl)-1,2-dithiolylium iodides.[460] The protic-acid-catalysed thiono–thiolo (O → S) migration

(259) $EtAlCl_3^-$ $EtAlCl_3^-$ (260)

of secondary alkyl groups in trialkyl phosphorothionates has been found[461] to occur in a complex fashion, and to involve the intermediacy of an intermediate ion-pair.[462] The trifluoroacetic acid-catalysed rearrangement of dialkyl xanthates to dithiocarbonates has been shown[463] to progress with inversion of configuration. Finally, a kinetic study[464] has shown that the initial reaction of primary amines with pyrylium cations to yield ring-opened intermediates, is fast for strongly basic amines and is base-catalysed for weak amines. Ring-closure of the intermediate to give the pyridinium derivative is subject to steric and electronic hindrance and is acid-catalysed.

Rearrangements in Polycyclic Systems

The syntheses and reactions of strained, polycyclic cage compounds have been reviewed,[465] and a detailed account of the importance of the rearrangements of benzvalene and its derivatives in the synthesis of organic compounds has appeared.[466] Copper(II) chloride and thorium(IV) nitrate have been used to induce the Wagner–Meerwein rearrangement of camphene.[467] An example of the norbornene–norpinene rearrangement has been reported[468] on decomposition of norborn-5-ene-2-diazonium ions, while the composition of the methylnorbornan-2-*exo*-ols obtained on deamination of 3-, 5-, 6-, and 7-methylnorbornane-2-diazonium ions has indicated virtually complete Wagner–Meerwein equilibration of the cationic intermediates.[469] The effect of a 5-methyl group on the rearrangement rates of *endo*- and *exo*-norborn-5-en-2-ols has been studied.[470]

The rearrangement of the bicyclo[2.2.2]octane (**261**) to a bicyclo[3.2.1]octane system (**263**) has been proposed[471] to involve oxidative decarboxylation of a single carboxylic acid group to give a carbocation (**262**) that undergoes rearrangement *via* a 1,2-acyl migration (see Scheme 25), while addition of acetic acid to 2-vinylbicyclo[2.2.1]heptane has resulted[472] mainly in ring-expansion to the bicyclo[3.2.1]octane skeleton and to a lesser extent direct addition with a hydride shift and a Wagner–Meerwein rearrangement. Some unexpected aspects of the chemistry of homoisodrin have been explained by postulating a route which necessitates the intervention of a 1,3-hydride shift.[473] A kinetic study of the rearrangement of benzo[*b*]bicyclo[3.1.0]hex-2-en-4-yl derivatives to naphthalenium ions has been made[474] and an MO study has shown[475] that this eight-electron process proceeds either photochemically or thermally, depending upon the

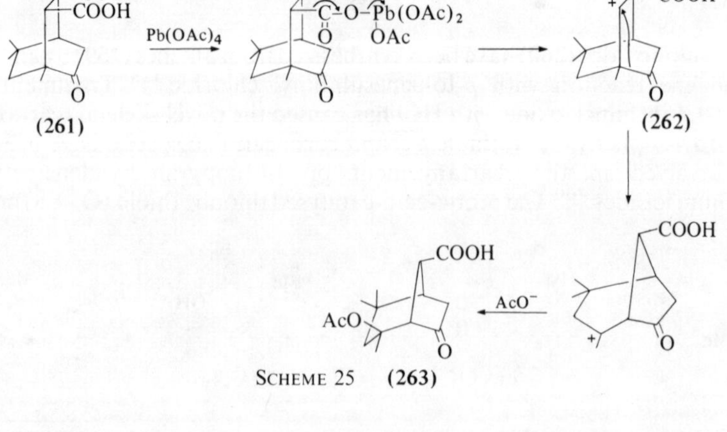

COOH COOH Pb(OAc)$_4$ C(=O)–O–Pb(OAc)$_2$ OAc O (**261**) (**262**) AcO$^-$ AcO (**263**)

SCHEME 25

substituents. Apparently methyl substituents lower the energy barrier to ring-cleavage of the bicyclic cation. The structural features which determine the rearrangement pathway of the bicyclic cations generated from 1- and 3-methyl-, and 1,3- and 1,7-dimethyl-2-arylbicyclo[2.2.1]heptan-2-ols have been discussed.[476] Unexpected differences in the $AlBr_3$-catalysed rearrangement behaviour of 1,2-*endo*-trimethylenenorbornane (**264**) and its 1,2-*exo*-isomer (**265**) have been interpreted.[477] Thus, isotopic-labelling studies have indicated that reversible abstraction of the tertiary 2-*endo*-hydride in (**265**) does not occur. Instead rearrangement to (**266**) seems to be favoured. 2-Cyclopentyladamantane has been observed[478] to undergo skeletal rearrangement to the isomeric methyl-1,2-butanoadamantanes in the presence of $AlBr_3$, while the same catalyst, in the presence of aromatic substrates, has been used to convert 1-azidoamantane into 3-aryl-4-azahomoadamantane.[479] Interestingly, it has been shown[480] that the use of anhydrous hexafluoroisopropanol as solvent enhances the carbocation rearrangements during solvolysis of 2-adamantyl tosylate. Thionyl chloride treatment of protoadamantan-5-*endo*-ol (**267**) has yielded the chloride (**268**).[481] A simple Wagner–Meerwein rearrangement has been shown to attend the addition of neat trifluoroacetic acid to protoadamantane,[482] while recent studies[483] on the solvolysis of 1-methyl-2-adamantyl tosylate, 4-methyl-*exo*- and 4-methyl-*endo*-4-protoadamantyl 3,5-dinitrobenzoate have strongly complemented previous evidence for an intermediate bridged cation (**269**). A recent study[484] has indicated that the adamantane rearrangement of the [3.3.2]propellane ring-system follows the route (**270**) → (**271**) → (**272**) → (**273**), while the remarkable distinction in reactivity in the acid-catalysed rearrangement between [5.3.2]- and [5.4.2]-propella-ε-lactones and the corresponding δ-lactones has been attributed[485] to the effect of lactone ring-size rather than to the steric factors of the third ring. An $AlCl_3$-catalysed

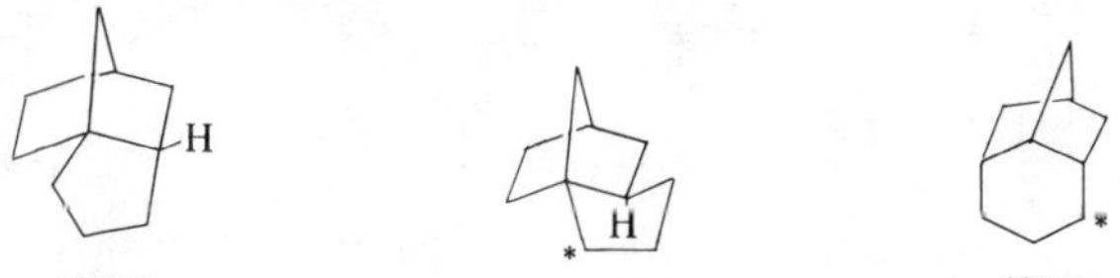

H

(264) (265) (266)

HO

Cl

Me

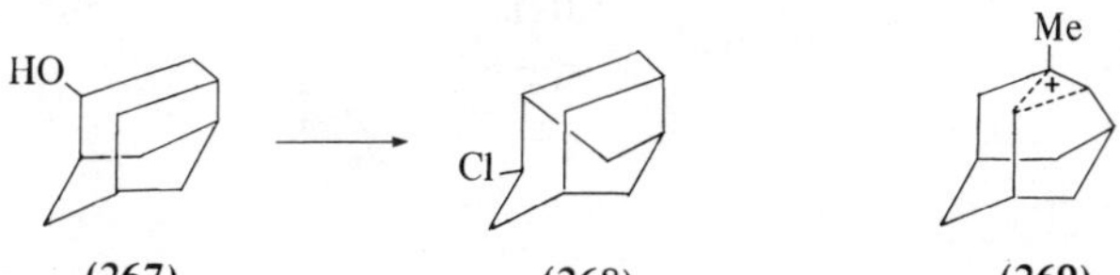

(267) (268) (269)

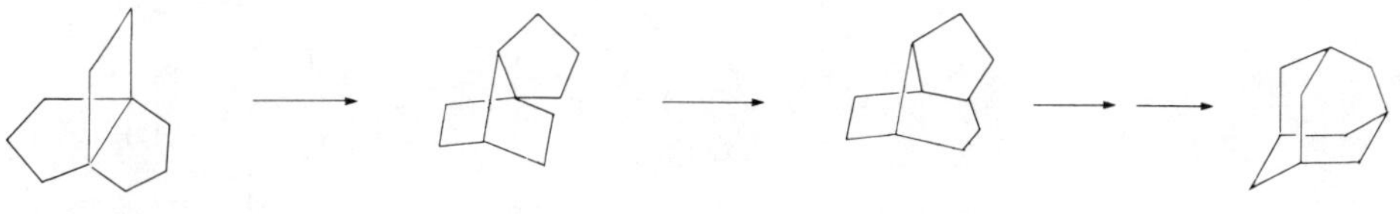

(270) (271) (272) (273)

rearrangement is involved in a recent efficient synthesis of triamantane.[486] The acid-catalysed rearrangement of *anti*-tricyclo[4.2.1.1^{2,5}]deca-3,7-diene-9,10-diol (**274**) to (**275**) has been shown to proceed[487] through concomitant electrophilic and nucleophilic attack on an alkene by an incipient carbocation and the remaining alcohol function. It has been shown[488] that entry to fused and spiro ring-frames can be regulated by the selective cleavage, under appropriate conditions, of interior or exterior cyclopropane bonds in tricyclic ketones of the type (**276**). Application of this strategy to the synthesis of two spiro sesquiterpenes, chamigrene and acorenone B, has been described and a mechanism involving opening of a cyclopropylcarbinyl cation either *via* a bicyclobutonium ion or by a step-wise pathway which entails a subsequent 1,2-alkyl shift has been considered as consistent with these results. Functionalized bicyclo[3.2.1]octa-2,6-dienes have been obtained[489] from tricyclo-[4.2.2.0^{2,5}]deca-3,7-dien-9-ones *via* a Schmidt fragmentation and a carbocation rearrangement. Novel one-step rearrangements of pentacyclo[5.3.0.0^{2,5}0^{3,9}.0^{4,8}]decan-6-one to *exo*-2-methanesulphonoxy-9-cyanobrend-4-one,[490] and pentacyclo[5.4.0.0^{2,6}0^{3,10}.0^{5,9}]undecane-8,11-dione to tetracyclo[4.3.1.0^{4,9}]-decane and 3,7-ethenotricyclo[3.3.0.0^{3,17}]octane ring-systems,[491] have been reported. Analogous treatment (sodium azide in methanesulphonic acid) of basketanone has resulted in a novel one-step carbocation rearrangement to the tetracyclo[3.3.0.0^{2,8}0^{3,6}]octyl system.[492] Moreover, rapid acid-catalysed re-arrangements of *seco*-basketanones have been reported.[493] A simple new intramolecular *C*-alkylation rearrangement of *β,γ*-unsaturated *α′*-diazoketones has been utilized[494] in the incorporation of a functionalized pentalene moiety into a variety of systems; see Scheme 26.

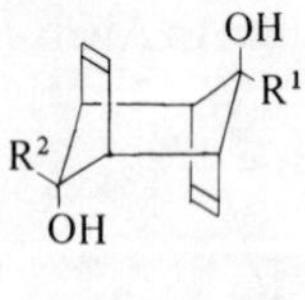

(274) 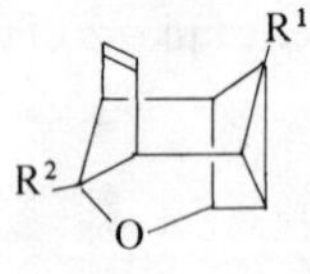 (275) 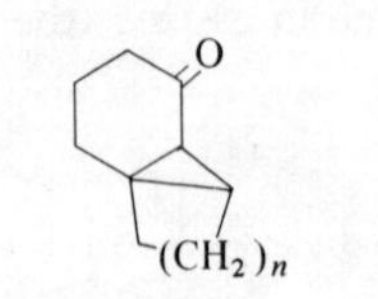 (276)

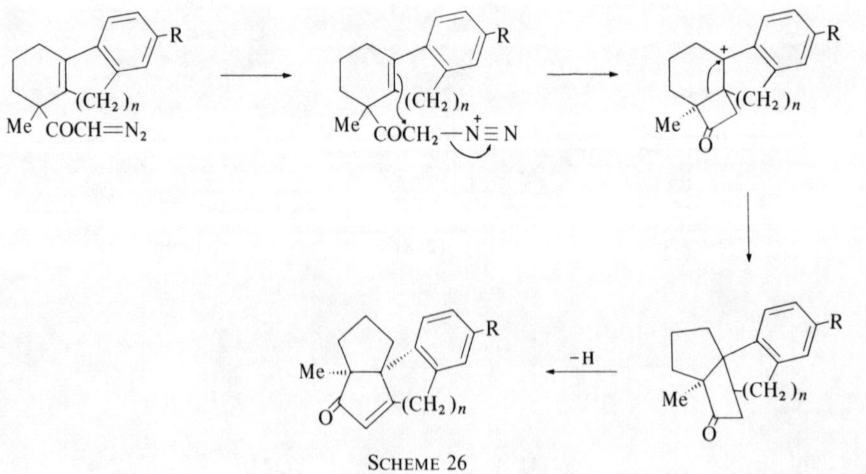

SCHEME 26

A product-analysis study[495] on the solvolysis of a series of 2-*exo*-chloroazabicycloalkanes has shown that exclusively rearranged products are formed in all cases from solvent capture of the corresponding cyclo-immonium cations. A kinetic study on the same system[496] has established that anchimeric assistance by the nitrogen bridge is evidently not important in the rate-controlling step of these rearrangements. The skeletal rearrangement observed on photochemical reaction of methanol with 10*H*-azepino[1,2-*a*]indoles **(277)** has been considered as proceeding by the pathway illustrated in Scheme 27; the crucial stage for the production of the cyclobutane **(279)** is a Wagner–Meerwein rearrangement of **(278)**.[497]

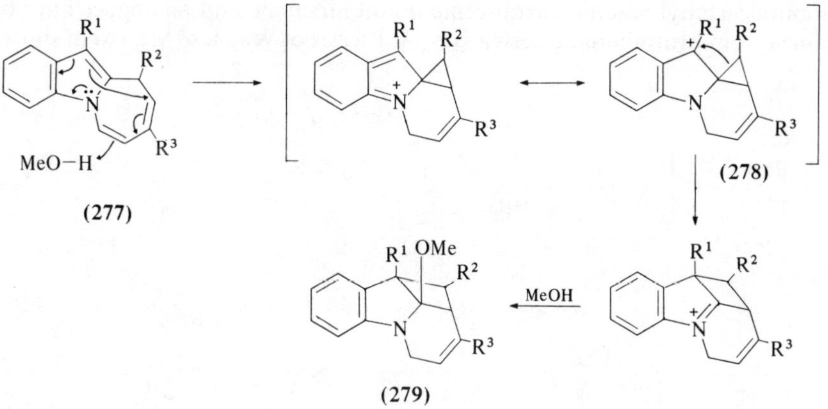

(277)

(278)

MeOH

(279)

SCHEME 27

It has been demonstrated[498] that the addition of phenylsulphenium ion to hexamethyl(Dewar benzene) first involves addition from the *endo*-side to produce a thiiranium ion. At higher temperatures, this ion readily undergoes rearrangement which results in degeneracy of five of the six methyl groups. Such a process is considered to be in accord either with rearrangement *via* a bicyclo[2.1.1]hexenyl cation or *via* a hypervalent ion formed by capping a pentamethylcyclopentadienyl ring with a hexa-coordinate methyl-bearing carbon atom from one side and a hexa-coordinate phenyl-bearing sulphur from the other side. Finally, the super-acid-induced skeletal rearrangement of *β*-hydroxyselenide derivatives of bicyclo-[4.2.1]nonatriene, namely **(280)**, has yielded selenides **(281)**.[499]

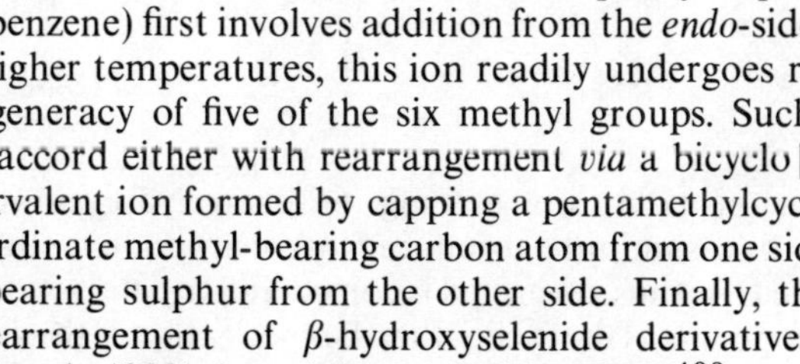

p-MeC_6H_4Se~~~CRR^1OH

(280)

(281)

Rearrangements in Natural-product Systems

These studies are classified thus because they involve rearrangements of natural-product systems that probably proceed *via* carbocations.

Acid-catalysed rearrangements of longifolene have been reviewed[500] and a recent detailed account of the terpenoid metabolites of mushrooms describes a number of rearrangements of these compounds.[501] The *A*-homograyanotoxane ring-system (**284**) has been constructed[502] by thermolysis of benzocyclobutene (**282**; X = SBu) followed by a Wagner–Meerwein rearrangement of the resulting kaurane type of compound (**283**; X = CH_2OH, Y = R = H). A series of remarkable rearrangement reactions involving (1β,9β)-1-methoxypicras-11-en-16-one and related derivatives have been described.[503] Thus, treatment of epoxide (**285**) with $LiAlH_4$ yielded (**286**) in a process found to be general in the context of the quassin series. A novel rearrangement of angularly fused cyclobutanones[504] has opened up a route to the C_{20}-gibberellins. The sesquiterpene isocomene has been synthesized[505] *via* a cyclobutylcarbinyl ketone rearrangement and hirsutene and analogues have been obtained from protoilludane derivatives by a series of Wagner–Meerwein shifts.[506]

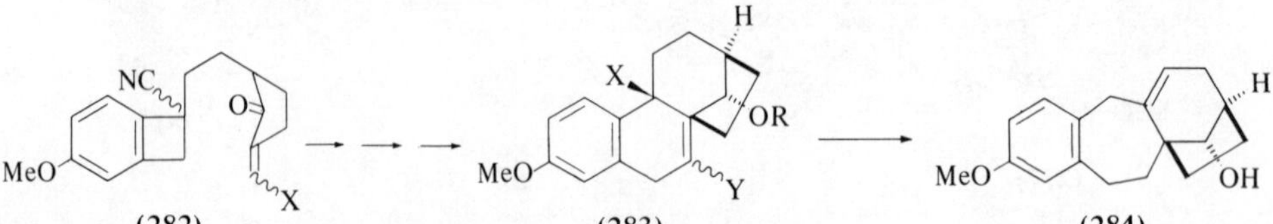

(**282**) (**283**) (**284**)

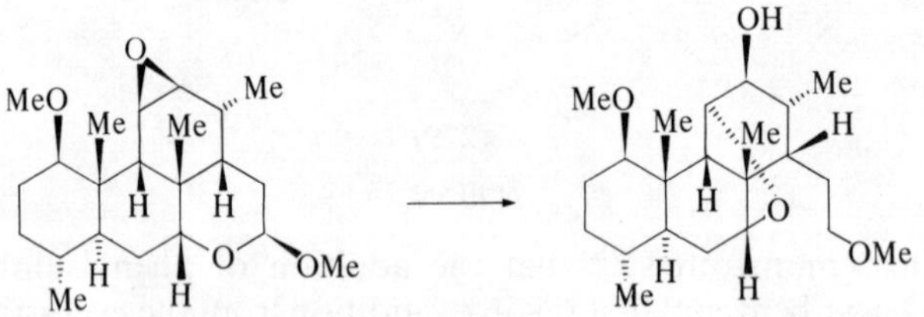

(**285**) (**286**)

The formation of (**288**) on heating 6β-hydroxyaplysistatin (**287**) has been rationalized[507] by assuming heat-induced loss of bromide, methyl migration, and loss of a proton. The HBr thus generated is considered to catalyse dehydration of the 6β-OH group and opening of the cyclic ether. Isomerization of the double bonds followed by dehydration would lead to (**288**). An unexpected change of mechanism has been observed[508] in the Lewis acid-catalysed perezone–pipitzol transformation, and an attempt has been made[509] to rationalize the formation of products from acid and base treatment of himachalene derivatives. The route shown in Scheme 28 has been invoked[510] to account for the DDQ-induced rearrangements of paulownin (**289**;

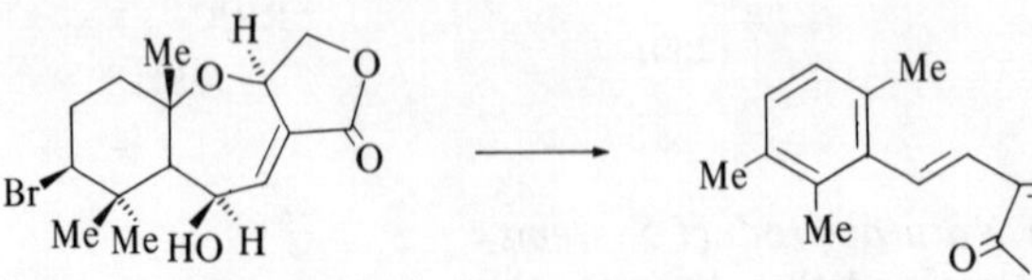

(**287**) (**288**)

Ar = 3,4-methylenedioxyphenol) and gmelinol (**289**; Ar = 3,4-dimethoxyphenyl) into 4-pyrone derivatives (**290**).

The mechanism of the isomerization of Δ^5-3-keto-steroid by the enzyme *Pseudomonas testosteroni* has been reinvestigated with androst-5-ene,3,17-dione.[511] A determination of the label distribution in polydeuteriated isoholamine, resulting from D_2SO_4-catalysed rearrangement of holamine, has been carried out[512] and on the basis of the study a mechanism involving formation of a carbocation at C(5) followed by migration of the methyl group from C(10) to C(5) has been proposed for the backbone rearrangement. Further backbone rearrangements of the steroid and triterpene skeleton have been reported. Thus, it has been concluded that the individual steps in the backbone rearrangement of cholest-5-ene occur relatively slowly compared with the carbocation $\rightleftharpoons$ olefin interconversion at the various backbone sites, while the subsequent equilibration at C(20) also proceeds in a manner which allows intervention of both $\Delta^{20(21)}$ and $\Delta^{20(22)}$ as well as $\Delta^{17(20)}$ olefinic intermediates.[513] A backbone rearrangement of 4,4-dimethyl-6α,20α-epoxy(5α)androstane with BCl_3 has been reported.[514] Wagner–Meerwein rearrangement of *D*:*A*-friedooleananes[515] and lupanes[516] have been described, *D*:*A*-friedo-18β-lup-19-en-3-one has been transformed into *D*:*B*-friedo-18β,19α*H*-lup-5-en-3β-ol *via* a backbone rearrangement of 3β,4β-epoxy-*D*:*A*-friedo-18β,19α-lupane,[517] and a study has been made[518] of solvent effects on the $BF_3 \cdot OEt_2$-catalysed rearrangements of the above epoxide and its 3α,4α-isomer. The structures of the acid-induced rearrangement products of 16-keto-friedel-3-ene have been determined[519] and rearrangements of the carbon skeleton, and methyl migration from C(20) to C(19), have been confirmed in the biosynthesis of oleanene- and ursene-type triterpenes in tissue cultures.[520]

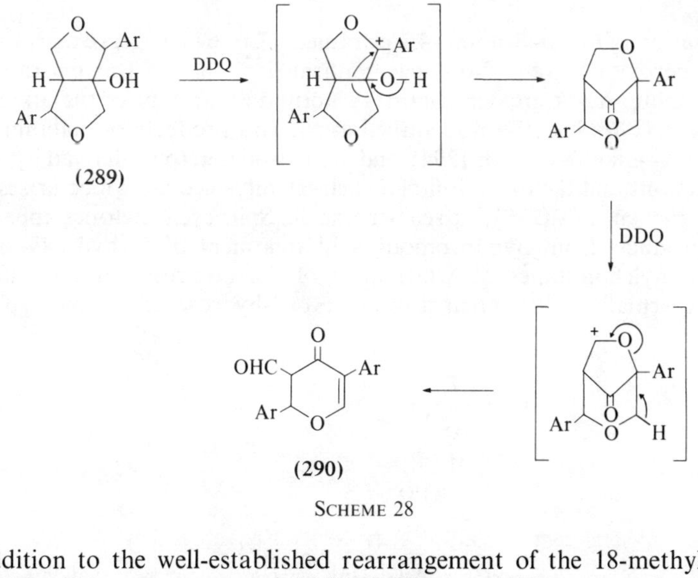

(289) —DDQ→ [intermediate] → … —DDQ→ [intermediate] → (290)

SCHEME 28

In addition to the well-established rearrangement of the 18-methyl group to C(17), $BF_3 \cdot OEt_2$ treatment of 11β,17β-dihydroxy-17α-propynyl-androsta-1,4-dien-3-one (**291**) was found to yield (**292**) in non-nucleophilic solvents.[521] Treatment of (**292**) with $SnCl_4$ in Ac_2O, followed by mild saponification caused extensive rearrangement to (**293**). This latter rearrangement most likely occurs through an intermediate vinyl cation of the type indicated in Scheme 29, followed by vinylidene–cyclobutene rearrangement. 3β,19-Epoxy-3α-methoxy-steroids have been converted in a highly stereoselective manner into the corresponding 3α-methoxy-19-oxo compounds[522] in the presence of $BF_3 \cdot OEt_2$, while the same catalyst has effected the rearrangement of 8α,9α-epoxy-12,12-ethylenedioxypodocarpan-16-ol acetate into a novel *C*-seco,*B*-aromatic system.[523]

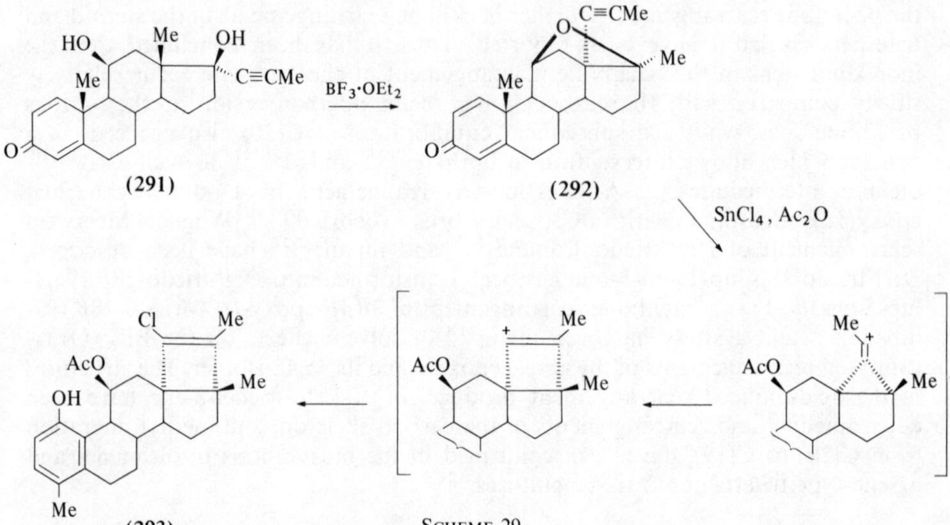

(291) $\xrightarrow{BF_3 \cdot OEt_2}$ (292) $\xrightarrow{SnCl_4, Ac_2O}$ [intermediates] → (293)

SCHEME 29

The formation of 6-oxa-*B*-homo-4,7-estradien-17-one on Criegee rearrangement of 5α-hydroperoxy-6-estren-17-one has confirmed[524] that, in this rearrangement, the alkenyl group has a greater migratory aptitude than that of the alkyl group. Spiro-lactones (**296**), have been identified as the end-products of thallium acetate treatment of 3,4-diketo-steroids (**294**), and are considered to be derived by *A* and *B* ring-contractions and lactonization of a carboxy intermediate which arises from a thallium adduct such as (**295**);[525] see Scheme 30. Spirocyclic ketones appear to be the main products from hypobromous acid treatment of 5α-hydroxy- and 5α-alkoxy-10β-vinylcholestanes,[526] while most of the observed products from the photo- and thermally-induced rearrangements of 3-hydroxy-Δ^5-steroid hypoiodites

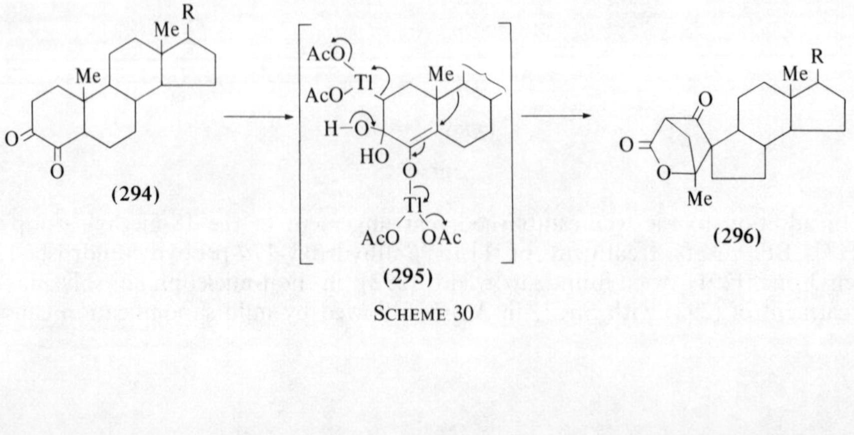

(294) → [(295)] → (296)

SCHEME 30

in the presence of mercury(II) oxide–iodine have been accounted for by assuming the intermediacy of an intermediate allyl radical resulting from β-scission of a 3β-oxyl radical.[527]

Oxacarbene (**298**) has been invoked as an intermediate[528] in the unusual photochemical rearrangement of 1,2:5,6-di-*O*-isopropylidene-α-D-ribo-3-hexulofuranose (**297**) to (**299**), and the fate of the intermediate benzoxonium ions in the rearrangements of derivatives of galactopyranosides,[529] and manno-, altro-, and iodo-pyranosides[530] has been studied in some detail. Skeletal rearrangements that possibly proceed by intermediate carbocations have been reported during the deamination of 4-amino-4-des(oxymethylene)anhydrolycoctonam[531] and on the acetylation of the *trans*-erythrinane ring-system.[532]

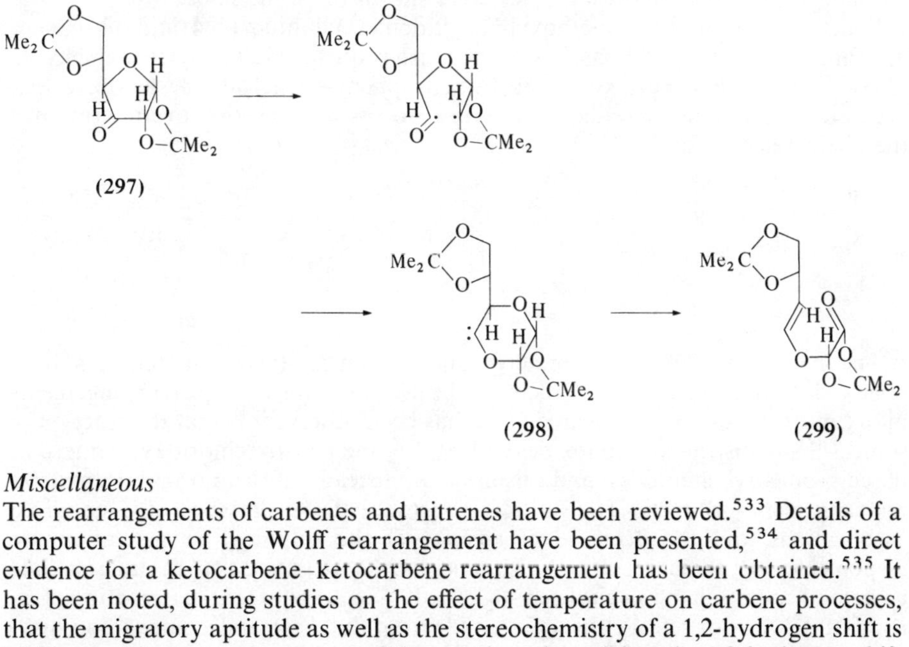

(**297**)

(**298**) (**299**)

Miscellaneous

The rearrangements of carbenes and nitrenes have been reviewed.[533] Details of a computer study of the Wolff rearrangement have been presented,[534] and direct evidence for a ketocarbene–ketocarbene rearrangement has been obtained.[535] It has been noted, during studies on the effect of temperature on carbene processes, that the migratory aptitude as well as the stereochemistry of a 1,2-hydrogen shift is highly sensitive to temperature and the reaction phase. Thus the 1,2-hydrogen shift of $Ph\ddot{C}(CH_2R)$ in a rigid matrix produces the (Z)-olefin predominantly, particularly as the R group becomes more bulky.[536] These results have been interpreted in terms of special matrix effects on the population of carbene conformations. A facile homologation of angular α-diazomethyl ketones by a photo-induced Wolff rearrangement has been reported,[537] and details of an experimental test of an alternative hypothesis for isotope-position scrambling in the photochemical Wolff rearrangement of azibenzil, *i.e.* $PhCOC(N_2)Ph$, have been put forward.[538] Tricyclo[4.2.0.0^{1,4}]octane-3-carboxylic acid[539] and its esters[540] ("broken window compounds") have been prepared by diazoketone ring-contraction of tricyclo-[5.2.0.0^{1,5}]nonan-4-one, while carbene (**301**) and phosphene (**302**) have been proposed[541] as intermediates in the rearrangement of diazomethylenebis(diphenylphosphane oxide) (**300**) to the phosphinic acid (**303**).

An investigation of the rearrangements of *N*-substituted hydroxylamines as a potential source of nitrenium ions has been undertaken,[542] and the rearrangements

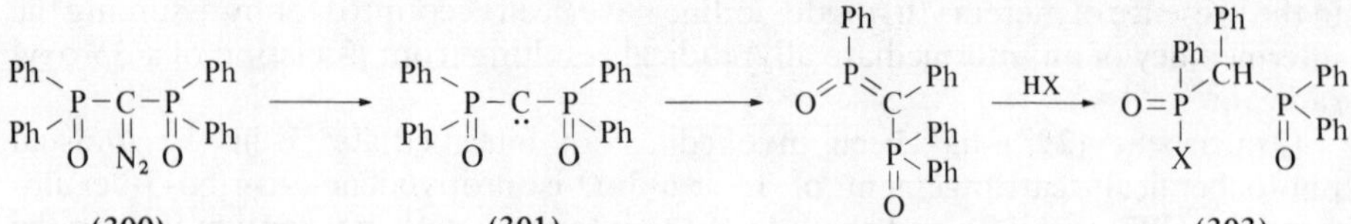

(300) (301) (302) (303)

of *N*-nitroso- and *N*-nitro-sulphonamides to the corresponding diazo- and diazoxy-esters have been studied with the intention of comparing these reactions with the rearrangements of the *N*-nitroso- and *N*-nitro-carboxamides of primary alkylamines. The study showed that the counter-ion had little effect on the mechanisms of these processes which were shown to be the same.[543]

Reaction of ketones with methoxyphenylthiomethyllithium (**304**) and subsequent rearrangements of the adducts, has opened up a route to α-sulphenylated aldehydes (**305**),[544] and a number of alkenyl-substituted α-phenylseleno-β-ketoesters have been cyclized in the presence of acidic catalysts in a process involving the novel migration of the phenylseleno moiety.[545]

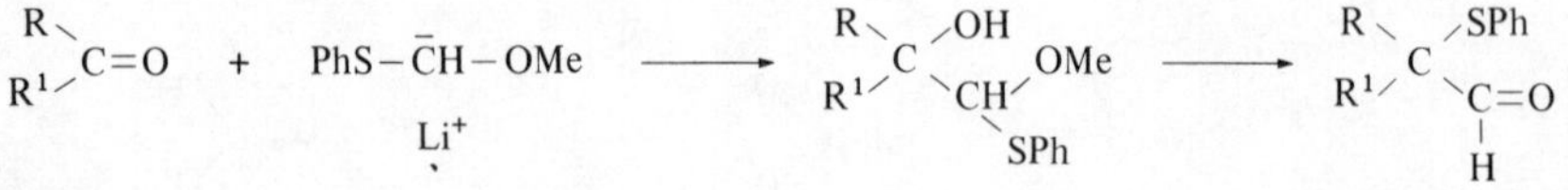

(304) (305)

The mechanism of the Pummerer and Pummerer-type reactions has been reviewed[546] and a review which includes the use of Pummerer-type rearrangements in organic syntheses has appeared.[547] It has been observed[548] that the reaction of some β-thia-substituted sulphoxides with $SOCl_2$ and benzoyl chloride give mixtures of chloromethyl sulphides and disulphides instead of the normal Pummerer rearrangement products. The authors suggest that the presence of *S*-stabilized carbocations and the decrease of basicity of the sulphinyl oxygen atom are responsible for these results. A Pummerer-type reaction *via* an intermediate acylaminosulphonium ion has been suggested[549] to account for the formation of *N*-*tert*-butyl-2-(acyloxymethylthio)benzamide from the thermal decomposition of *O*-acyl-*N*-[2-(methylthio)benzoyl]-*N*-*tert*-butylhydroxylamines. A new synthetic route to 4-phenylthio-4-butanolide derivatives makes use of the Pummerer rearrangement.[550]

Rearrangements Involving Electron-deficient Hetero-atoms

The heterolytic cleavage of the N—O bond of 6-methylhept-5-en-2-one oxime under a variety of conditions has led to two types of cationic cyclization reactions, *viz*. a direct and a rearrangement cyclization, both of which produce Δ′-pyrrolines. The latter process, (**306**) to (**308**), is considered to occur[551] by way of an intermediate nitrilium ion (**307**) formed by Beckmann rearrangement of the oxime. Beckmann rearrangements carried out in alkaline media in the presence of cyanide anions have been described,[552] and trimethylsilyl polyphosphate, conveniently prepared from P_2O_5 and hexamethyldisiloxane, has proved to be a useful reagent for initiating the rearrangement.[553] A report of the Beckmann rearrangement of cyclohexanone oxime in the gas phase has appeared,[554] and a study of catalyst deactivation in the

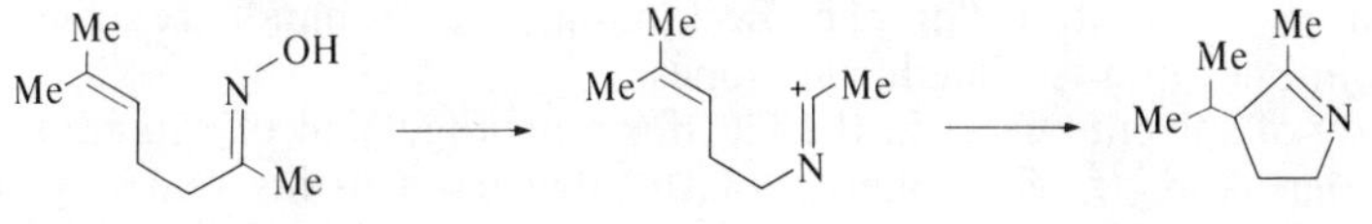

(306) (307) (308)

vapour-phase rearrangement of the same oxime has been initiated.[555] The photo-Beckmann rearrangement of steroidal β,γ-unsaturated ketoximes,[556] and the Beckmann rearrangements of α-substituted β-aryl-α,β-unsaturated ketoximes,[557] 3β-chloro-19-nor-5-methyl-5β-cholest-9(10)-en-6-one oxime,[558] tetronic acid ketoxime derivatives,[559] and 2-(*N*-alkoxycarbonyl)hydroxyamino-1-tetralone oximes[560] have been described; under Beckmann rearrangement conditions, the oximes of 4-aryl-4,5-dihydro-6*H*-cyclopenta[*b*]thiophen-6-ones have been shown to yield 4-aryl-4,5-dihydro-6*H*-thieno[2,3-*c*]pyridine-7-ones.[561] The key step in the synthesis of *B*-homo-5-azaandrostane-3β,17β-diol by a multi-stage testosterone transformation was the Beckmann rearrangement of oxosecoandrostanoate oxime,[562] while the same rearrangement was utilized in the syntheses of 3β-acetoxy-7α-aza-*B*-homo-22*a*-spirost-5(6)-en-7-one[563] and of 7,8-dihydroxy-1,2,3,4,5,6-hexahydro-2,6-methano-3-benzazocin-1-ol.[564] The possibility that the Beckmann rearrangement of bridging oximes of the triptycene series can proceed by migration of either the tripticenyl or the alkyltripticenyl group has been substantiated,[565] and a critical review of nitrogen insertion reactions of bridged bicyclic ketones leading to bridged bicyclic lactams has appeared.[566] Although the photo-reaction of 6-*exo*-methoxy-1,5,6-trimethyl- or 6-*exo*-methoxy-1,3,5,6-tetramethyltricyclo[3.2.1.0^{2,7}]-oct-3-en-8-one oxime has been shown to undergo aza-di-π-methane rearrangement leading to 7-*endo*-methoxy-2,6,7-trimethyl- or 7-*endo*-methoxy-1,2,6,7-tetramethyltetracyclo[3.3.0.0^{2,8}0^{4,6}]octan-3-one oxime, the one-carbon ring-enlarged 5,8-dimethyl-9-methylenetricyclo[3.3.1.0^{2,8}]non-3-en-6-one oxime undergoes photo-Beckmann rearrangement to afford 5-aza-2,6-dimethyl-10-methylenetricyclo[4.3.1.0^{2,9}]dec-7-en-4-one.[567]

The migration of C(2) and/or C(4) to oxygen or nitrogen in the Beckmann, Schmidt, and Baeyer–Villiger reactions of 3-keto-steroids has been studied[568] with the aid of ^{13}C-NMR spectroscopy, and several ring-expansion reactions, previously reported to give only one product, have been found to yield both possible migration products. A recent review[569] of methods for effecting oxygen-insertion reactions of bridged bicyclic ketones has stressed the versatility of the Baeyer–Villiger rearrangement with organic peroxides, or basic hydrogen peroxide. Interestingly, the Beckmann rearrangement of 5α- and 5β-spirostanones (**309**) yielded azahomospirostanones (**310**), whereas their Schmidt reaction afforded tetrazolospirostans (**311**).[570] Subtle differences in bond migration have also been

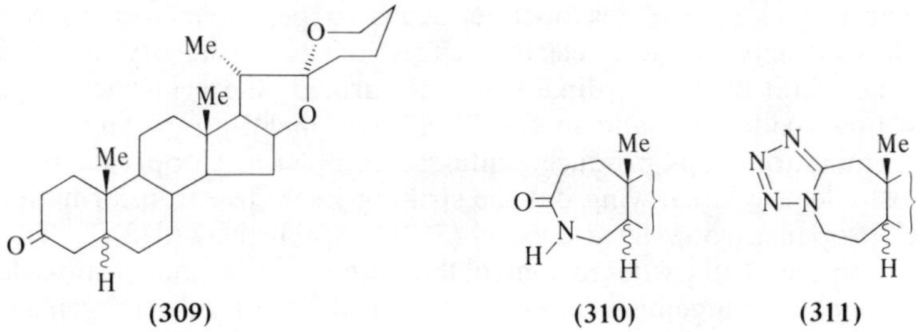

(309) (310) (311)

observed[571] in a study of the Beckmann and Schmidt rearrangements of isoquinolinones and isothiachromanones.

A study of solvent effects in the Curtius rearrangement of cinnamoyl azide has been undertaken.[572] The kinetics of the thermal Curtius rearrangement of 2-benzofuroyl azide and related compounds have been studied by infrared spectroscopy,[573] and it has been established[574] that acyclic dialkylphosphinic azides share with phosphetinic azides the ability to undergo Curtius-like rearrangement on irradiation in protic solvents. The Curtius rearrangements of acyl amino acid and peptide azides,[575] and of pinanoyl, pinonoyl, and pinoyl azides,[576] have been examined, and the Curtius rearrangement has been used in the syntheses of the 6-dialkylamino derivatives of 1,3-oxazin-2-ones,[577] and of 1,2-oxazoles bearing a fused heterocyclic ring.[578] The same rearrangement has also been utilized as a key step in the syntheses of 2,3-dihydroxybenzoic acid-(carboxyl-^{14}C),[579] 6-phenylaminoimidazo[1,2-*a*]pyridine-2-carbamate,[580] and 5-substituted pyrimidine-2,4-diones.[581] The Hofmann rearrangement of 3-carbamoyl-5-methylisoxazole to 3-amino-5-methylisoxazole has been reported[582] and a recently described synthesis of N^2-protected L-2,3-diaminopropanoic acid[583] makes use of a Hofmann rearrangement as the crucial step.

Metal-catalysed Rearrangements

Intramolecular metallotropic rearrangements[584] and rearrangements in coordination complexes[585] have been reviewed, and the rôles of transition-metal complexes in catalysis, as observed in recent studies on the kinetics and mechanisms of homogeneous catalytic hydrogenations, have been discussed.[586] A review has also appeared on various mechanistic aspects of coenzyme B_{12}-dependent rearrangements,[587] while conclusive evidence has been cited[588] to confirm that 1,2-bond-shift rearrangements catalysed by coenzyme B_{12} are not unique but occur widely in heterogeneous transition-metal catalysis and in homogeneous organometallic reactions as well. An instance of vitamin B_{12} acting as catalyst for a non-enzymic carbon-skeleton model rearrangement has been observed,[589] and the radical cleavage of the carbon–cobalt bond of 1-phenyl-2-oxocyclopentylmethyl cobaloxime and 1-ethoxycarbonyl-2-oxocyclopentylmethyl cobaloxime gave only 3-phenylcyclohex-2-enone and 3-ethoxycarbonylcyclohex-2-enone, respectively, in a rearrangement that is considered[590] to be a reasonable model for the ester migration mediated by coenzyme B_{12}. The recently reported[591] formation of (cyclopentadienylcobalt)cyclononatetraene represents the first isolation of a σ,π-bonded intermediate in a metal-catalysed ring-expansion, and its formation can only be accounted for by invoking pathways other than that of the classic Woodward–Hoffmann type.

It has been demonstrated[592] that transition-metal-catalysed dehydrogenative transannular ring-closure of cyclooctatetraenes to pentalenes occurs when the metal–cyclooctatrienyl moiety electrocyclizes in a disrotatory manner thus rendering the metal atom coordinatively unsaturated; the bridgehead hydrogen atoms can now oxidatively add to the 16-electron metal atoms and subsequent reductive elimination steps produce pentalene complexes. Copper catalysts have proved most effective in carrying out the strikingly selective transformation of β-alkoxycyclopropanecarboxylic acid esters (**312**) to vinyl ethers (**313**).[593] The use of early transition-metal alkoxides to control the stereo-, regio-, and chemo-selectivity of epoxy-alcohol rearrangements has been described.[594] A rigid arrangement of the metal–epoxy-alcohol complex as in (**314**) provides an attractive rationale for the results observed.

trans-Tetrahydrofuran-3-carbaldehydes have been prepared[595] by ruthenium-catalysed isomerization of 4,7-dihydro-1,3-dioxepines and subsequent Lewis-acid-catalysed 1,3-alkyl migration, while a mechanism involving (η^3-vinylcarbene)iron complexes and a novel 1,4-oxygen shift reaction has been proposed[596] to account for the thermal rearrangement of (tricarbonyl)(η^4-5,6-dimethoxy-4-methoxycarbonyl-2-pyrone)iron to (tricarbonyl) (η^4-5,6-dimethoxy-3-methoxycarbonyl-2-pyrone)iron. Further mechanistic details on the fluxional behaviour of (η^1-cyclopentadienyl)iron complexes have appeared[597] and rearrangements of (1,3-dithiol-2-ylidene)iron complexes into hetero-metallacycles have been recorded.[598]

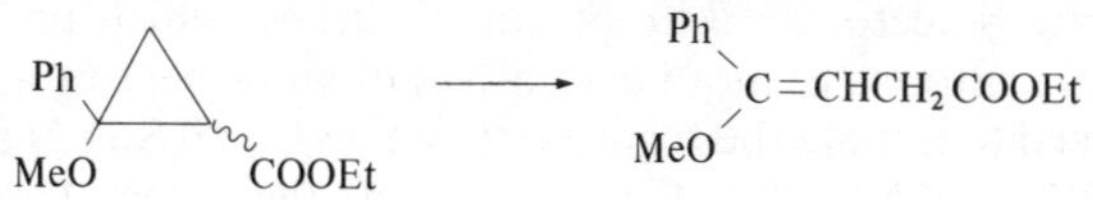

Ph, MeO, COOEt (**312**) ⟶ Ph(MeO)C=CHCH$_2$COOEt (**313**); (**314**): Ln, M, :O:, O, :OR, R, H

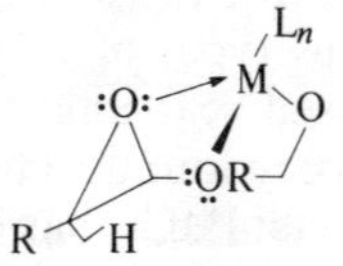

A study of the reversible isomerization of η^6- and η^5-indenyl and -fluorenyl transition-metal complexes by IR and NMR spectroscopy has been made,[599] while the reaction of [Mo(η-MeC$_2$Me)(SC$_6$H$_4$—*o*-SC$_6$H$_5$)(η-C$_5$H$_5$)] with 3,3-dimethyl-cyclopropene has led to displacement of the acetylene, unexpected migration of the phenyl group from sulphur to a carbon atom of a ring-opened cyclopropene, and formation of the η^3-allylic complex, [Mo(η^3-*syn*-1-Ph-3,3-Me$_2$C$_3$H$_2$)(1,2-C$_6$H$_4$S$_2$)(η-C$_5$H$_5$)].[600] The skeletal isomerization of 3,3-dimethylbut-1-ene on molybdenum(VI) oxide has been observed.[601] The mechanism of double-bond shifts in η-cyclopentadienyl(η-3-methoxycarbonylcyclohexa-1,4-diene)rhodium(I) has been investigated[602] and selective *cis*-isomerization of pent-1-ene has been achieved[603] using Ni(I)-triphenylphosphine complexes. Selective *cis*-isomerization of α-aryl-β-methyl-substituted cinnamonitriles has been induced by using Raney nickel or europium powder in an alcohol.[604] Anchimeric assistance by a transition metal in the homolysis of a C—O bond has been postulated[605] for the thermal rearrangement of methyl zirconocenyldiphenylmethyl ether. Mechanistic aspects of the nickel-catalysed addition reaction of alkenylzirconium reagents to cyclopropyl ketones have been outlined,[606] and the formation of an iridiumethyl complex by methyl migration to a coordinated methylene group has been achieved.[607] This latter result suggests that alkyl migration to a carbene ligand may be a general and useful method for forming carbon–carbon bonds under very mild conditions. A mechanism involving α- and β-elimination reactions promoted by metal–metal bond-rupture has been put forward[608] in an attempt to explain the interesting rearrangement of a tungsten complex to a tungstole-tungsten derivative.

Evidence has been presented[609] for the formation of incipient carbocations as intermediates in the novel catalytic transformations of alkenes by tetrakis-(acetonitrile)palladium ditetrafluoroborate. The Pd(0)-catalysed epiperoxide to hydroxyenone conversion (**315**) → (**318**), has been interpreted[610] in terms of a Pd(0)/Pd(II) exchange mechanism as outlined in Scheme 31. Oxidative addition of the O—O bond of (**315**) to the zerovalent Pd atom produces the cyclic structure (**316**), and subsequent β-elimination of a PdH element, giving (**317**), followed by

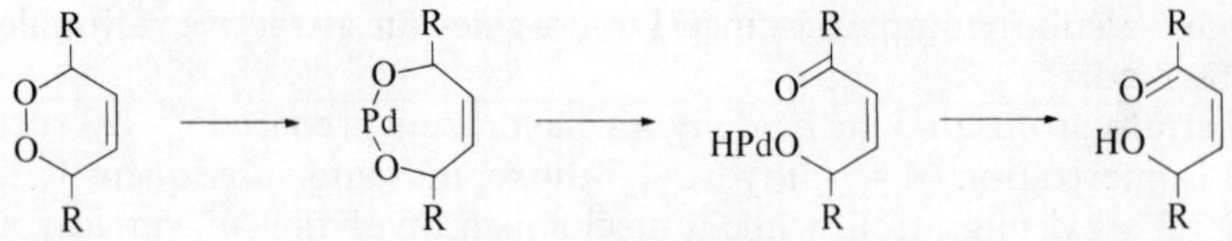

(315) (316) SCHEME 31 (317) (318)

reductive elimination of Pd(0) species leads ultimately to the hydroxyenone (**318**). An interesting communication[611] has attempted to present a mechanistic insight into the stereo- and regio-chemistry of the palladium-catalysed isomerization of allyl vinyl ethers such as (**319**) into cyclopentanones (**320**), when these compounds are known to rearrange to cycloheptenones (**321**) under thermolysis conditions. Various types of α-cyanoallylic acetates have been converted,[612] in a palladium-catalysed rearrangement, into γ-acetoxy-α,β-unsaturated nitriles, which products have been found to be useful synthons of furan derivatives often present in natural products; $PdCl_2$ has been used to initiate the regioselective S → C and S → N allylic rearrangements of *S*-allylthioimidates.[613] Complexes of the type L_2PdCl_2 (L = MeCN, EtCN, PhCN) in benzene or chloroform have been found to act as homogeneous catalysts in the room-temperature *cis–trans* isomerization of ethyl chrysanthemate and chrysanthemic acid,[614] and 3-acetoxy-1,4-dienes have been rapidly converted into 1-acetoxy-2,4-dienes in a stereo- and regio-selective manner using a Pd(II) catalyst.[615] An attempt to elucidate the stereochemical features of the 1,3-chloropalladation of the bicyclic methylenecyclopropanes, *cis*- and *trans*-9-methylenebicyclo[6.1.0]nonane, has indicated[616] a preference for disrotatory opening of the cyclopropane ring. Novel fluxional rearrangements in which a dinuclear platinum moiety commutes between pairs of sulphur atoms in an intramolecular manner have been observed[617] on warming a solution of the *s*-trithian complex $[(Me_3PtCl)_2(\overline{SCH_2SCH_2SCH_2})]$. It has been further noted that, in order to account for the above intramolecular 1,3-shifts, a change in ring conformation from boat to chair is almost certainly involved. The photo-assisted isomerization of carbon–carbon double bonds has been effected with high efficiency in the presence of platinum complexes,[618] and a "cyclopropanoid" mechanism has been proposed for the isomerization of alkanes on platinum-supported catalysts.[619]

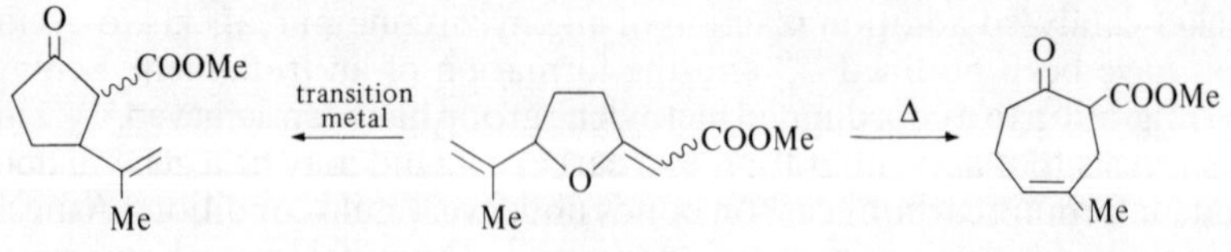

(320) (319) (321)

Cyclopropene derivatives have been rapidly isomerized to indenes by trace amounts of silver ion, but it has been recorded[620] that the silver-induced reactions of these cyclic alkenes differ dramatically from their well-established photochemical rearrangements in that quite different carbon–carbon bonds are cleaved in the two processes. The silver-catalysed rearrangement of dehydrolinalool-1,2-$^{14}C_2$ to citral-1,2-$^{14}C_2$ is a key step in the recently reported synthesis of 13-*cis*-retinoic acid-6,7-^{14}C.[621] Benzonorbornadienes have been readily produced by silver(I)-induced rearrangement of the benzotricyclo[3.2.0.0^{2,7}]heptene ring-system,[622] and treatment of pentacyclo[5.3.1.0.^{2,6}0.^{3,5}0^{4,9}]undecane (**322**) with a catalytic amount of silver tetrafluoroborate has resulted in a novel skeletal rearrangement to afford tetracyclo[5.3.1.0.^{2,6}0^{3,9}]undec-4-ene (**323**). This rearrangement has clearly demonstrated[623] that both ring-strain and suitable geometry are equally important in silver(I)-ion-catalysed rearrangements of strained polycyclic molecules. Silver-ion catalysis has been used[624] to convert tricyclic fulvenes (**324**) and (**326**) into sesquifulvalene (**325**) and the vinylogous sesquifulvalene (**327**), respectively, and 1,4- and 1,5-cyano migrations have been observed to occur on peroxydisulphate–silver-ion oxidation of cyanohydrins.[625]

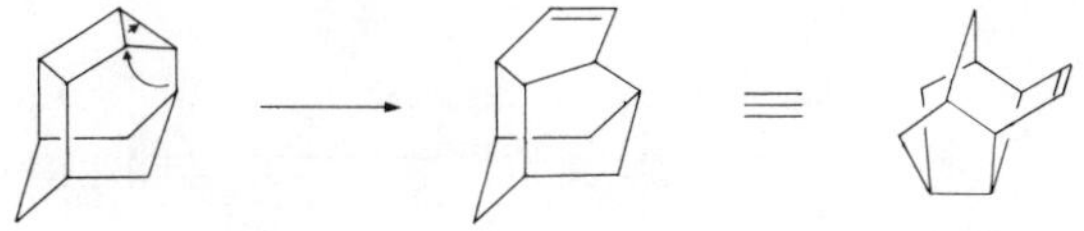

(**322**) (**323**)

(**324**) (**325**)

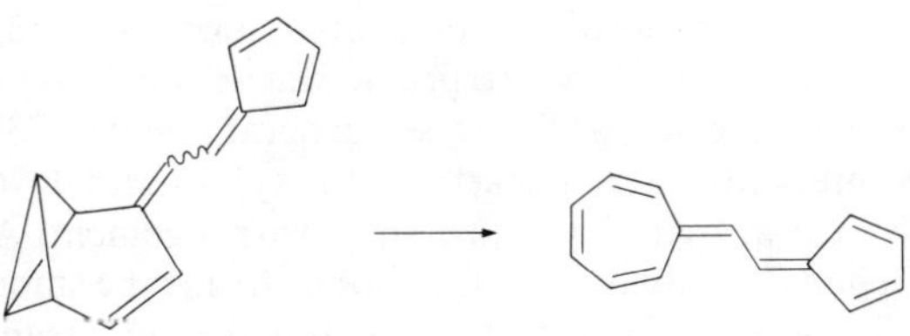

(**326**) (**327**)

Permethylated linear polysilanes, $Me(Me_2Si)_nMe$ (n = 4–10 and 12), have been found to undergo skeletal rearrangement to give branched isomers when treated with a catalytic amount of $AlCl_3$ in boiling benzene,[626] and observations on the nature of polylithium organic compounds and their rearrangements have appeared in print.[627] Finally, tri-*n*-butyltin hydride reduction of an α,β-epoxy-*O*-thiocarbonylimidazolide derivative of an alcohol has led, *via* oxiran ring-opening, to the formation of an allylic alkoxy radical[628] whose rearrangement constitutes a useful alternative to the Wharton reaction (treatment of the epoxide of an α,β-unsaturated ketone with hydrazine).

Rearrangements Involving Ring-opening and Ring-closure

Three-membered Rings

Photochemical rearrangements involving three-membered rings have been reviewed[629] and a review of the thermal rearrangements of *gem*-difluorocyclopropanes[630] has drawn attention to the dramatic and completely unprecedented effects of geminal fluorine substituents on the reactivity of the ring. A detailed investigation has been made of the thermal rearrangement of 1,1-difluorospiropentane (**328**) to the methylenecyclobutane (**329**) and it has been concluded[631] that the rearrangement proceeds by two consecutive processes, path (*a*) and path (*b*). Further attempts have been made to investigate the effect of a second-geminal fluorine substituent on the extrusion of :CF_2 relative to rearrangement;[632] indeed, a systematic study[633] of the effect of increasing *gem*-difluoro substitution on the thermal chemistry of spiropentane has produced results that are completely consistent with the current understanding concerning the effect of fluorine substituents on the thermodynamic and kinetic stability of a cyclopropane ring.

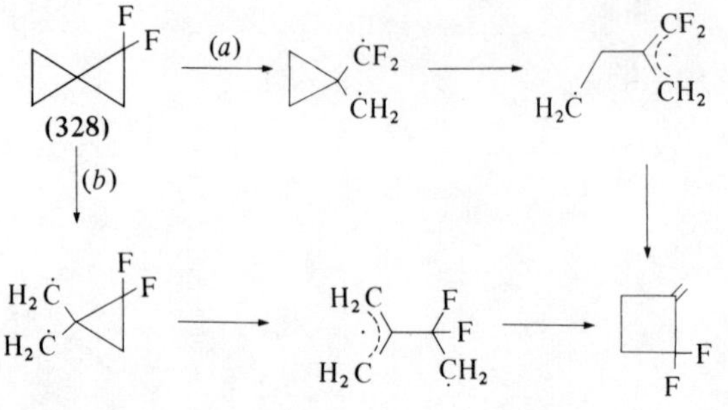

(328)

(329)

An alkoxy-accelerated vinylcyclopropane rearrangement has provided a stereoselective method for the conversion of 1,3-dienes into cyclopentene derivatives.[634] The intramolecular cyclopropanation–rearrangement sequence of dienic diazoketones has been shown to provide facile access to bicyclo[3.3.0]octanes which are of value as terpene synthons, see general scheme (**330**) → (**331**);[635] while the regiospecific preparation of 2-(carbomethoxy)-4-methylcyclohept-4-enone has been achieved[636] *via* a divinylcyclopropane rearrangement. A biradical pathway involving cyclorpopyl cleavage of the bond from the more substituted side, subsequent cyclization of the resultant biradical, and aromatization, has been proposed to account for the rearrangements of phenyl-substituted spirocyclopropaneanthrones and related compounds to dihydroaceanthrones.[637] The stereoisomerization of aryl-substituted cyclopropanes *via* trimethylene radical anions has been reported,[638] and the photo-sensitized electron-transfer-induced rearrangements of several 3-phenyl-substituted cyclopropenes to indenes have been studied.[639] Tetrachlorocyclopropene has been shown to react with azide anion to yield 1,2,3-trichloropropenyl azide which rapidly transforms into trichloroacrylonitrile.[640] A number of examples of the cyclopropenylcarbinyl–allenyl rearrangement, which illustrate a new synthetic route to vinylallenic alcohols from cyclopropenes, have been reported.[641]

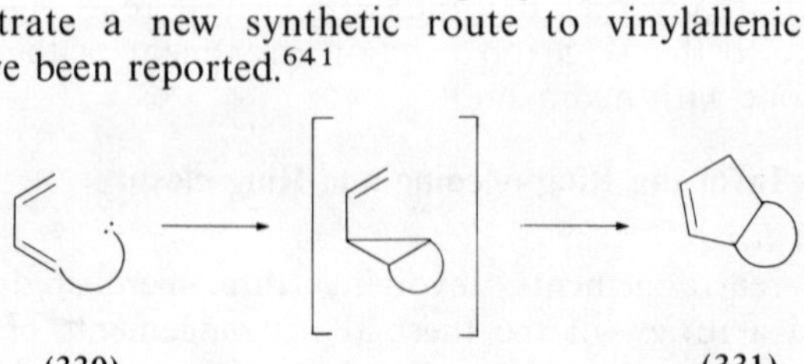

(330) (331)

The observed thermal rearrangement of tricyclo[4.1.0.0^{2,7}]hept-1(7)ene (**332**) into cumulene (**335**), in which two bonds on opposite sides of the bicyclo[1.1.0]but-1(3)-ene moiety in (**332**) are broken, is forbidden as a synchronous reaction according to the Woodward–Hoffmann rules. On the other hand, a step-wise isomerization of (**332**) to (**335**), with the diradical (**333**) or the carbene (**334**) as an intermediate, has not been ruled out.[642] The thermal conversions of spirobicyclo-[2.1.0]pentane-5,2′-methylenecyclopropanes have been investigated.[643] At temperatures around 80°, a double epimerization (bridge flip), initiated by cleavage of the bridge bond, is encountered. Above 120°, isomeric 6- and 7-methylenebicyclo-[3.2.0]hept-1-enes are produced. Polycyclic methylenecyclobutanes have been obtained unexpectedly from thermal rearrangement of cyclopropanic 9,10-ethano-9,10-dihydroanthracenes,[644] while labelling studies have shown that the rearrangement of 4-acetoxytricyclo[4.1.0.0^{2,7}]hept-4-en-3-one (**336**) into 1-acetoxybicyclo-[3.2.0]hepta-3,6-dien-2-one (**337**) probably involves initial C(1)—C(6) bond-cleavage followed by construction of the cyclobutene ring *via* C(2)—C(7) cleavage and then C(4)—C(7) bonding.[645]

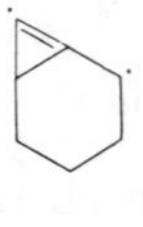

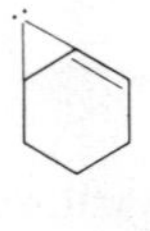

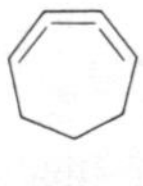

(332) **(333)** **(334)** **(335)**

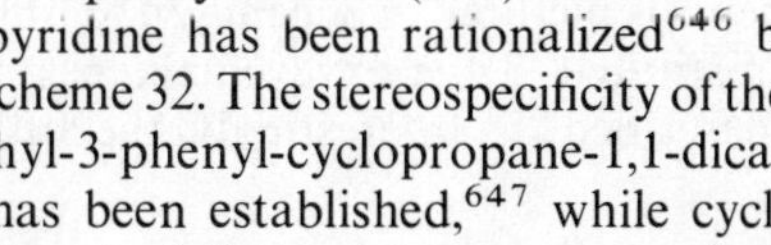

(H)D OAc **(336)** (D)H OAc **(337)**

The isolation of (−)-(R)-3-fluoro-2-phenylpropionic acid (**339**) as the only product from the reaction of phenylalanine (**338**) with an excess of sodium nitrite in polyhydrogen-fluoride–pyridine has been rationalized[646] by assuming a process such as that outlined in Scheme 32. The stereospecificity of the rearrangement of 2,3-diphenyl- and 2-methyl-3-phenyl-cyclopropane-1,1-dicarboxylic acids into carboxybutyrolactones has been established,[647] while cyclopropanecarboxylates having a carbonyl substituent at the cyclopropane α carbon have been converted into γ-lactones in the presence of bis(trimethylsilyl) sulphate.[648] 1-(Arylthio)cyclopropanecarboxaldehydes have been used[649] in a secoalkylation sequence which creates a versatile functionalized chain at the α position of a carbonyl group. The first example of a Lewis-acid-assisted cyclopropyl ketone to dihydrofuran rearrangement, and more importantly the first such transformation to occur with complete retention of cyclopropane stereochemistry, has been reported.[650]

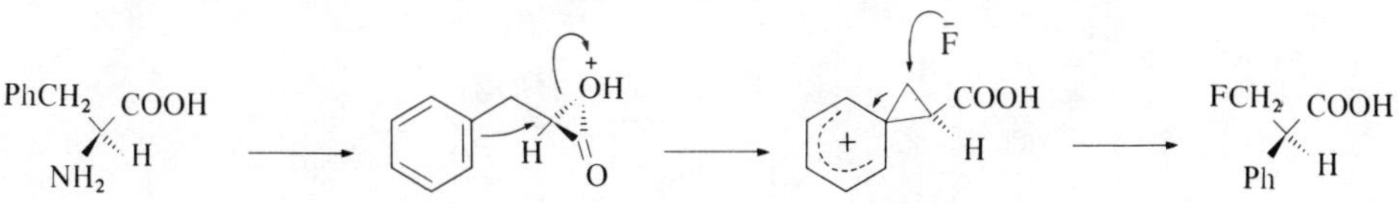

$PhCH_2$ COOH H NH_2 **(338)** → OH H O → F COOH H → FCH_2 COOH Ph H **(339)**

SCHEME 32

It has been observed[651] that the thermal rearrangement of 1,3-dichloro-6-ethoxycarbonyl-2-thiabicyclo[3.1.0]hex-3-ene (**340**) does not follow the anticipated course but leads instead to ethyl 2,4-dichloro-5-hydroxy-6-methylbenzoate (**341**). The formation of this product is difficult to rationalize, but a plausible scheme is shown in Scheme 33. A spirocyclopropylpyrrolium ion has been postulated[652] as an intermediate in the thionyl chloride-induced transformation of 2-hydroxyethylpyrroles into 2-haloethylpyrroles.

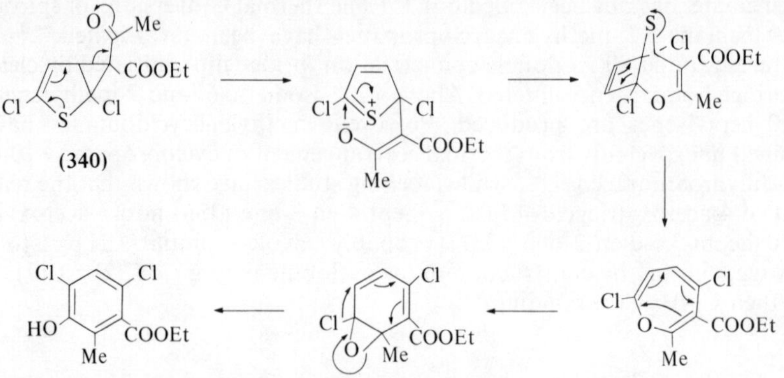

(340) (341)

SCHEME 33

A series of papers on the photo-induced cleavage of γ,δ-epoxyenones[653–656] and of conjugated 5,6-epoxytrienes[657] has appeared, and the oxidation of 5-acyl-4-oxopyran-2-carboxylic acid esters to novel epoxypyrones and their subsequent rearrangement to the previously inaccessible 6-substituted 5-hydroxy- or 5-chloropyrones, has been described.[658] An interesting epoxyketone–furan rearrangement has been reported,[659] and epoxy cyclopropyl ketones (**342**) have been found to undergo two consecutive rearrangement reactions with BF_3 as catalyst. It is considered that, in the first step, an acyl migration takes place to yield a β-keto-aldehyde (**343**) which further rearranges to an enol ester (**344**). A rationale for the two rearrangements is pictured in Scheme 34.[660] Thermolytic and photolytic ring-expansions of substituted 3,4-epoxycycloalkenes into benzene and furan derivatives have been described.[661] Interestingly, the authors propose a concerted

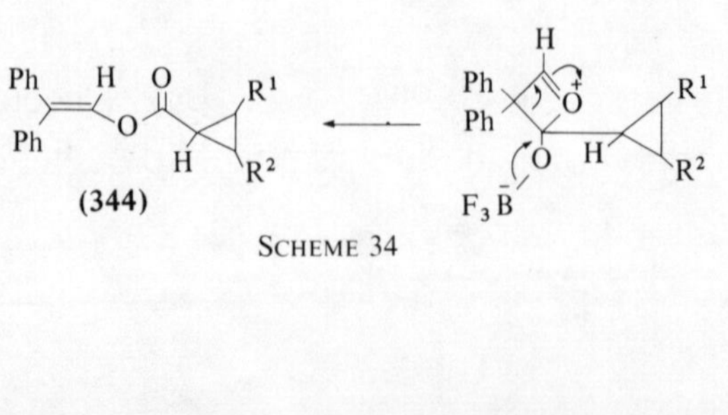

(342) (343) (344)

SCHEME 34

[1,3]-carbon migration for the light-induced formation of the furans. Benzocyclobutenols have been prepared[662] by metallation of *o*-halostyrene oxides, and the pathways shown in Scheme 35 have been proposed[663] to account for the base-catalysed rearrangements of aurone epoxides (**345**) to flavonols (**346**) and to aroylcoumaran-3-ones (**347**). In the presence of methanolic potassium borohydride, 3-hydroperoxy-1,3,6-trimethyl-2-phenylindolines have been observed to rearrange into the corresponding 3-hydroxy-6-methoxymethylindolines[664] by a route that possibly involves epoxide ring-formation and subsequent opening.

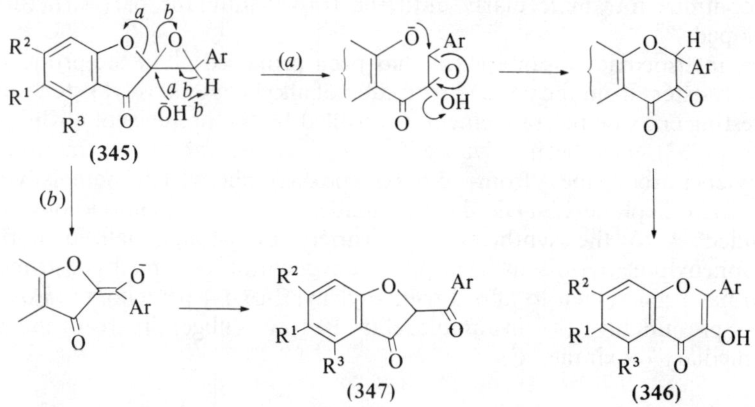

(345) (347) (346)

SCHEME 35

The kinetics of the gas-phase decomposition of 2,3-epoxy-1,1,1-trifluoropropane to 1,1,1-trifluoropropanone and 3,3,3-trifluoropropanal have been studied,[665] and a comparison of this reaction with the decomposition of 2,3-epoxypropane has demonstrated that substitution of CF_3 for CH_3 in the oxirane ring apparently strengthens the C—O bond adjacent to the substituent but has little effect on the C—O bond opposite the substituent. The acid-catalysed isomerization of epoxides containing tertiary ring carbon atoms has resulted[666] in the selective formation of the corresponding alcohols when the reaction is carried out in chloroform in the presence of small quantities of H_2SO_4. The observation of a 1,2-hydride shift in both the peracid oxidation and microsomal metabolism of the terminal acetylene, [1-^2H]biphenylacetylene, has established[667] that acetylenic moieties are in fact oxidatively metabolized. This study has further demonstrated that this oxidative process involves reaction of activated oxygen with the π-bonds rather than with the terminal C—H bond of the acetylene; see Scheme 36.

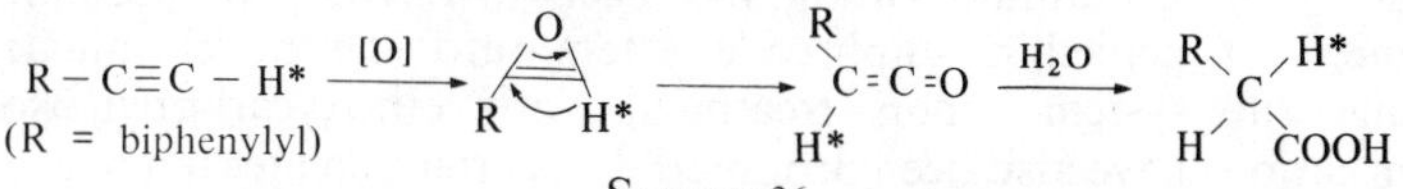

R−C≡C−H* (R = biphenylyl) →[O] → R(H*)C=C=O →H_2O → RCH(H*)COOH

SCHEME 36

A remarkable rearrangement of a [3.3.3]propellane triepoxide (**348**), which provides direct access to the first known C_{17}-hexaquinane (**351**), has been encountered.[668] This unusually facile conversion, which requires the unprecedented rupture of an equivalent bond in three different epoxide rings, is considered to be made possible by the unique proximity factors contained within the system. As illustrated in (**349**) and (**350**), Lewis-acid-promoted opening of one epoxide ring so

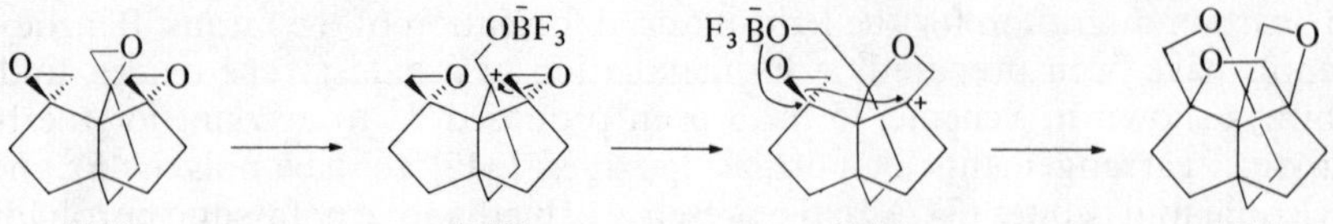

(348) (349) (350) (351)

as to induce cationic character at a more highly substituted cyclopentyl carbon can induce neighbouring-group participation by a second oxygen atom. This process can continue intramolecularly until the trioxatriquinane part structure is fully developed.[669]

The intermediacy of an epoxide has been proposed[670] in the production of α-hydroxy-esters from the peracid oxidation of alkyl trimethylsilyl ketene acetals. An interesting epoxide rearrangement, controlled by the presence of a silyl group, see **(352)** → **(353)**, has been advanced[671] to account for the formation of a 5-methylenenorbornane from 5,6-*exo*-epoxy-*N*-phenyl-1-trimethylsilylmethylbicyclo[2.2.1]heptane-2,3-*endo*-dicarboximide, and this methodology has been extended[672] to the synthesis of a variety of 7-functionalized norbornenes. Tetraphenylporphyrin-sensitized photo-oxygenation of 2-methyl-5-trimethylsilylfuran has been shown to afford trimethylsilyl 2-oxo-4-pentenoate in quantitative yield, presumably *via* intramolecular Baeyer–Villiger rearrangement of an intermediary dioxirane.[673]

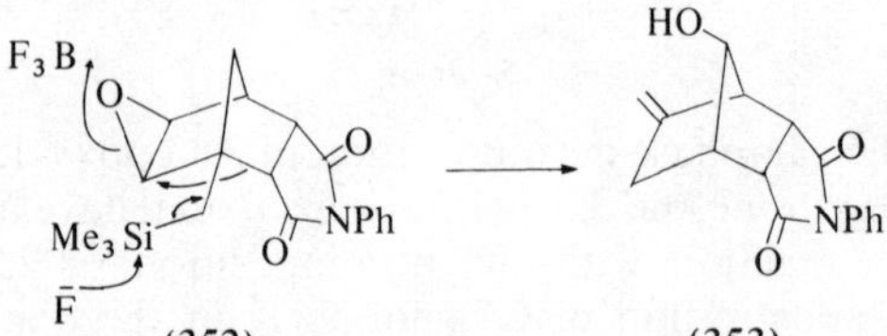

(352) (353)

A kinetic analysis has been carried out for the interconversion of oxirenes or thiirenes and their open-chain valence isomers.[674] The finding of a new rearrangement of alkyl (*N*-alkylamido)methyl sulphides, promoted by strong bases, has been described.[675] The reaction induces migration of the alkyl group linked to the sulphur to the amide carbonyl, thus yielding the corresponding ketone, and it is thought to proceed by thiiran formation from a β-keto-thiolate, and subsequent ring-opening with extrusion of sulphur. A thiiranium intermediate has been postulated[676] in the conversion of 2β-iodomethyl-2α-methylpenams into 3β-iodo-3α-methylcephams, while a mechanism involving C(5)—S cleavage and the intermediacy of a thiiranium cation has been invoked[677] to account for the rearrangement of penicillin sulphoxide esters into the novel imidazo[5,1-*c*][1,4]thiazine ring-system, upon treatment with ethoxycarbonyl isocyanate. Thiiranium cations have also been proposed[678] as intermediates in the acetate-ion-promoted rearrangement of 3-bromothiochroman-4-one into a dimeric spirothioindoxyl and thence to vinylenethioindigo.

An *ab initio* SCF-CI analysis[679] of the photochemical conversion of 1-amino-2-cyanoethylene into imidazole has shown that the formation of an azirine intermediate represents the most favourable reaction path. An MO picture of the possible reaction pathway of photoisomerization of isoxazoles to oxazoles through azirine intermediates has been presented.[680]

Base treatment of amidoxime-*O*-sulphonates, such as (**354**), afforded 2-amino-1-azirines (**355**) instead of the expected 3-amino-2-pyrazolin-5-ones in a transformation that can be viewed[681] as a new variant of the Neber rearrangement. On treatment with an excess of NaOMe, (**355**) underwent a novel rearrangement to the 2-amino-5-imidazolone (**357**), a process for which the intermediacy of the bicyclo[2.1.0] compound (**356**) seems reasonable.

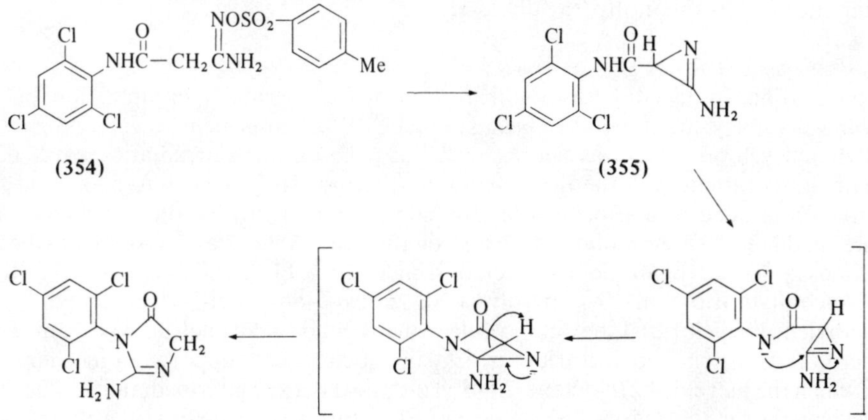

(**354**) (**355**)

(**357**) (**356**)

2-Imino-3-acylaziridines are assumed to be intermediates in the spontaneous rearrangement of 3-iminoisoxazolines to 2-iminooxazolines.[682] The products observed in a recent reinvestigation of the Prévost reaction of *N*-allylmorpholine with silver benzoate and iodine have been rationalized[683] by assuming the initial formation of an aziridinium ion from participation of the tertiary amino group, while the reversible rearrangements of 3-azawurtzitanes (**358**) and 3(4 → 5)abeo-3-azawurtzitanes (**359**) have likewise been described as proceeding by neighbouring-group participation of the N(3) atom.[684] Aziridinium cations are also invoked[685] to account for the completely stereospecific thermal rearrangement of (S)-2-chloromethyl-1-ethylpyrrolidine to (R)-3-chloro-1-ethylpiperidine, and in the indenobenzazepine–spirobenzylisoquinoline rearrangement used in a recent synthesis of (±)-raddeanine.[686] The halide-ion-catalysed rearrangement of *N*-acyl-2(S)-benzylaziridines to oxazolines has been shown[687] to proceed quantitatively, and a highly regiospecific and stereoselective rearrangement of *N*-ethoxycarbonyl-α-chloroaziridines to *N*-ethoxycarbonyl-2-chloro-2-alkenylamines has been reported.[688] A new type of rearrangement involving conrotatory cleavage of the C—C bond of the aziridine ring and the intermediacy of an azomethine ylide, has been

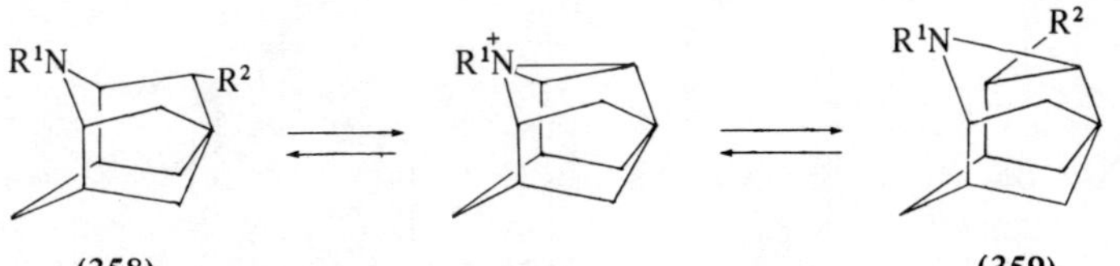

(**358**) (**359**)

described[689] for the thermal reaction of vinylaziridines, and firm evidence has been furnished[690] for the participation of vinyl nitrene intermediates in the thermal rearrangement of 2*H*-azirines into indoles. The results obtained[691] from the photolysis and thermolysis of 3-*n*-butyl-3-phenyldiazirine are consistent with the fact that the oxidation products are derived mainly from a common intermediate, *viz.* 1-diazo-1-phenylpentane. Finally, diazaphosphiridine oxides are proposed as intermediates in the alkoxide-induced rearrangement of *N*-chlorophosphonic diamides to phosphonohydrazidic esters.[692,693]

Four-membered Rings

Constitutional and configurational aspects of the thermal reorganization of 1-cyano-2-vinylcyclobutane have been studied.[694] Rearrangement to *cis*- and *trans*-allylvinylcyclopropanes was established[695] as the outcome of irradiation of *cis*-1,3-divinylcyclobutane. On the other hand, its thermolysis yielded butadiene and 4-vinylcyclohexene in a process which is considered to involve diradical (**360**) as intermediate.[696] A homolytic cleavage of the vinylcyclobutane system has been proposed[697] to explain the unexpected isomerization of the 5-methyl group to the *endo*-configuration in the pyrolysis of 1-*exo*-5-dimethyl-2-[(carbomethoxy)-methylene]bicyclo[2.1.1]hexane, while kinetic and calorimetric work on the thermolysis of *syn*- and *anti*-tricyclo[4.2.0.0^{2,5}]octane has supported a mechanism in which the bicyclo[4.2.0]octane-2,5-diyl diradical is a key intermediate.[698] The full mechanism for the thermolysis is discussed in terms of a likely potential energy hypersurface for C_8H_{12} hydrocarbons. The variable-encounter method has been used[699] to study transients in the thermal isomerization of cyclobutene to buta-1,3-diene, and it has been argued[700] that the mechanism of the facile rearrangement of spiro[benzocyclobutene-1,7′-cyclohepta-1′,3′,5′-triene] (**361**) to 9*a*,10-dihydro-benz[*a*]azulene (**363**) must involve opening of the benzocyclobutene moiety of (**361**) to give an *o*-xylylene intermediate (**362**) which possesses considerable diradical character. Benzocyclobutenes have also been used as starting materials in a simple synthesis of the basic skeletons of some naturally occurring terpenes possessing lactone moieties.[701] Thus, thermolysis of 1,2-dihydro-6-methoxybenzocyclobuten-1-ylbut-3-enoate (**364**) in a sealed tube at 180–200° afforded the lactone (**367**). The proposed route is outlined in Scheme 37. The fact that only the *cis*-product (**367**) was obtained can most reasonably be explained by assuming that the first-formed product (**365**) of kinetic control is isomerized to (**366**) and that recyclization then occurs to produce the thermodynamically more stable *cis*-lactone (**367**), Rearrangement of tetrakis(phenylimino)cyclobutane to the blue indigo dianil

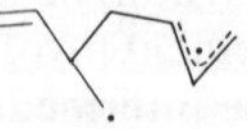

(**360**)

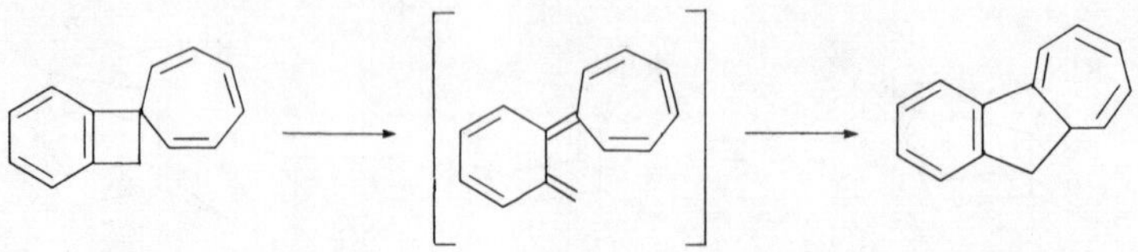

(**361**) (**362**) (**363**)

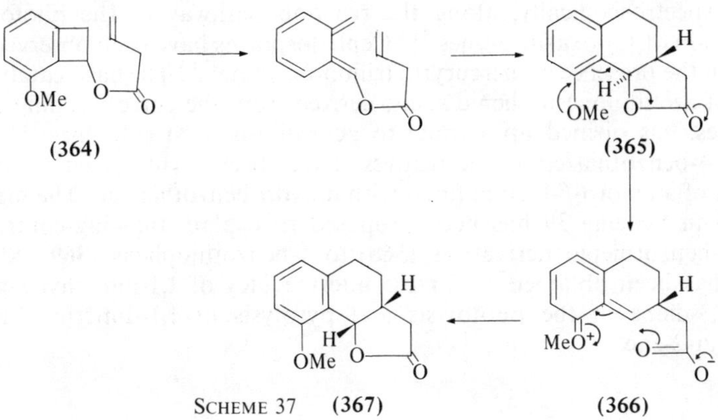

SCHEME 37

involves initial cleavage of a carbon–carbon bond in the cyclobutane ring followed by a two-fold intramolecular, electrophilic, aromatic substitution.[702]

A new method for the chain-expansion of functionalized alkanones, a rearrangement reaction that proceeds by way of a four-membered intermediate, has been reported,[703] and thiolactones have been established as the main products from the photolysis of dispiro-substituted 3-thioxocyclobutanones.[704] An unusual cyclobutanone ring-cleavage in tricarbonyl[(η-2,3,4,5)-9,9-diphenylbicyclo[5.2.0]-nona-2,4-dien-8-one]iron has been induced by catalytic amounts of acid and base.[705] A preliminary kinetic study has been set up[706] in an attempt to differentiate between Criegee rearrangement (path *a*) and dioxetane rearrangement (path *b*) mechanisms (see Scheme 38) for decomposition of α,β-unsaturated hydroperoxides, and further examination of a series of 3-(hydroperoxy)indolenines has favoured a mechanism involving base-catalysed decomposition of the dioxetane for their rearrangement.[707] The effect of cyclic substituents on the thermolysis of 3,4-dialkyl-1,2-dioxetanes has been studied,[708] and a 1,2-dioxetane species has been postulated[709] as an intermediate in the photo-oxidation of 1-(9-phenanthryl)-4-phenylbut-1-en-3-yne.

The $AlCl_3$-catalysed rearrangement of 1-phenyl-β-lactams to dihydroquinolinones has been reported,[710] as has the base-catalysed ring-expansion of 7,7-disbustituted and 7-substituted 2-azabicyclo[3.2.0]heptane-3,4-diones to dihydroazatropolones.[711] Two oxaazabicyclo[2.2.0]hexenone intermediates have been

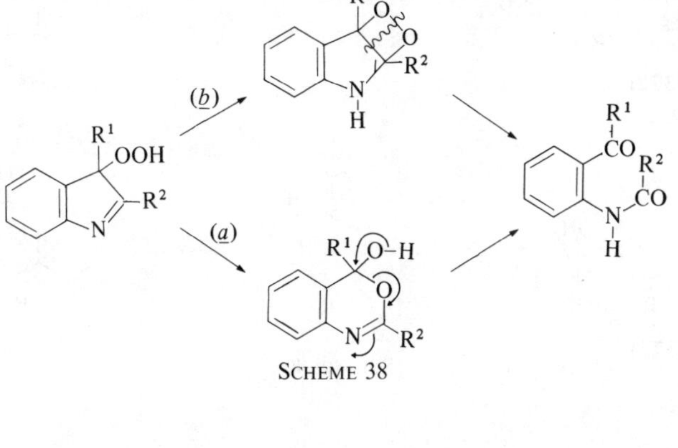

SCHEME 38

observed spectroscopically, along the reaction pathway of the photochemical isomerization of 1,3-oxazin-6-ones.[712] Cephalosporins have been observed to yield oxazines in the presence of mercury(II) trifluoroacetate.[713] The base-catalysed ring-opening of azetidinonesulphenic acids, derived from the corresponding penicillin sulphoxides, has opened up a route to penicillin-derived sulphines,[714] and 4,5-dihydro-1,4-benzothiazepine derivatives have been synthesized[715] *via* ring-expansion of a chloro-β-lactam fused with dihydrobenzothiazine. The mechanism illustrated in Scheme 39 has been proposed to explain the ring-contraction of various 1-benzothiepin derivatives (**368**) to 1-benzothiophens (**369**).[716] Finally, evidence has been obtained[717] for the intermediacy of 1,1-dimethyl-2-phenyl-1-silabuta-1,3-diene in the photolysis and pyrolysis of 1,1-dimethyl-2-phenyl-1-silacyclobut-2-ene.

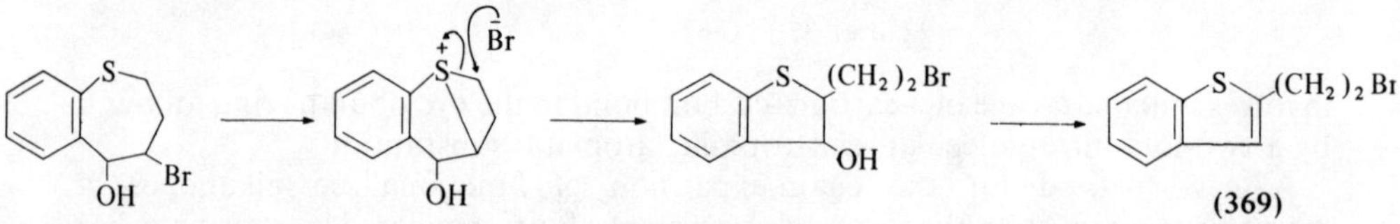

(368) (369)

SCHEME 39

Five-membered Rings

Treatment of 4-ylidenebutenolides (**370**) with sodium methoxide has resulted in rearrangement to the corresponding cyclopentene-1,3-diones (**371**),[718] and this methodology has been applied in a synthesis of calythrone, while the base-catalysed nitrosation of 1-indanones (**372**) has produced, exclusively, 2-hydroxyiso-carbostyrils (**373**) by a novel ring-expansion process; see Scheme 40.[719] The thermal

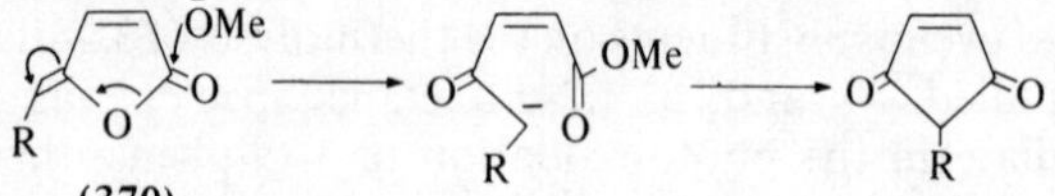

(370) (371)

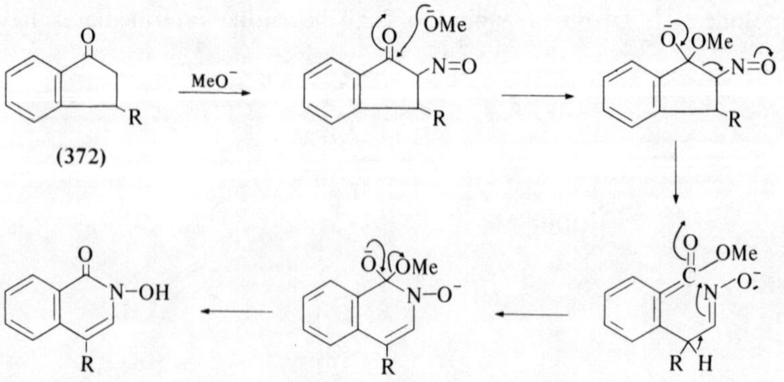

(372) (373)

SCHEME 40

rearrangement of the *endo*-peroxides of 2,5-dimethylfurans of the type (**374**) has led to the corresponding diepoxides (**375**) which provide a convenient entry to the synthesis of compounds which are structurally related to crotepoxide.[720] The addition of singlet oxygen to 2,5-furo-18-crown-6 (**376**) has yielded a crown ozonide (**377**) which rearranged quantitatively to a macrocyclic lactone (**378**) in a process that offers possibilities for an easy introduction of the γ-keto-α,β-unsaturated ester unit into macrocycles.[721] Anomalous ozonolysis products, resulting from cleavage of the carbon–carbon single bond adjacent to the carbon–carbon double bond, have been observed[722] on addition of singlet oxygen to methoxymethylfurans. The thermal rearrangement of a furan endoperoxide to an enol ester by way of an intramolecular Baeyer–Villiger rearrangement has been recorded.[723]

Interestingly, *N*-acylation of *N*-hydroxy-2-azabicyclo[2.2.1]hept-2-ene has caused rearrangement to oxa-azabicyclic compounds by insertion of the hydroxalamine oxygen into the heterocyclic system.[724] During the diene reaction of *N*-unsubstituted and *N*-monosubstituted enamine thiones with maleic anhydride, the primary adduct (**379**) was found to rearrange to the bicyclic compound (**380**); see Scheme 41.[725] The photochemical rearrangement of 2-aminopyrrolin-5-ones to aminocyclopropyl isocyanates has been described,[726] and thermolysis or photolysis of 4-azidopyrrolin-2-ones has resulted in their ring-contraction to 3-cyanoazetidin-2-ones.[727] Evidence has been presented to establish zwitterionic intermediates as the penultimate precursors to the azetidin-2-ones. *N*-Acylimines and enamides have been obtained from the gas-phase pyrolysis of 4-alkyl-2-oxazolin-5-ones,[728] and an interesting oxazole–triamide rearrangement has been utilized[729] in a general

(**374**) (**375**)

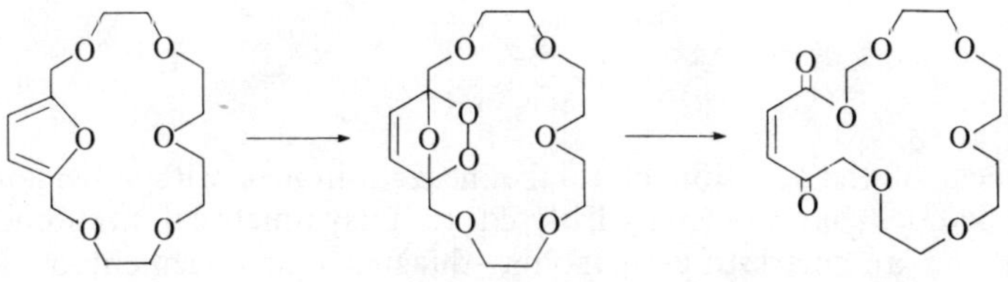

(**376**) (**377**) (**378**)

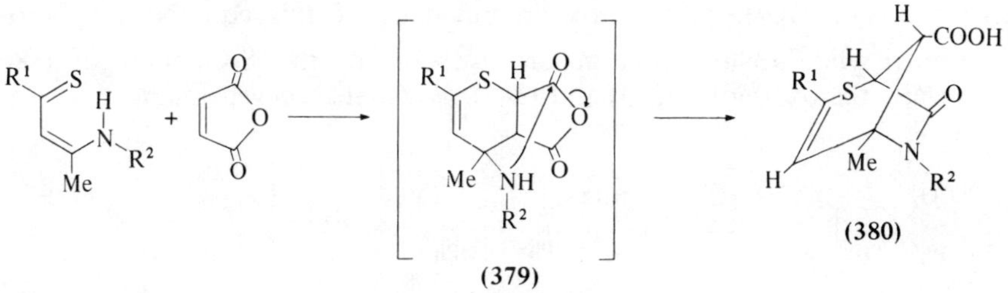

(**379**) (**380**)

SCHEME 41

synthesis of lactones ranging from five- and six-membered units to macrocyclic systems. A surprisingly selective thermal isomerization has been observed[730] in the case of the bicyclic diazene *N*-oxides (**381**). At 130° only the C—NO bond and not the C—N bond is opened and there is ample evidence for the participation of a diradical of type (**382**) in the process. An extension of this work involving a study of the thermolysis of 7,8-diazabicyclo[4.2.2]deca-2,4,7-triene *N*-oxide has thrown to light the first example of a *cis*-diazen–isodiazen *N*-oxide isomerization.[731] The addition of *p*-nitrobenzenesulphonyl azide to a series of exocyclic olefins has been observed[732] to give ring-expanded sulphonimides by way of unstable triazoline intermediates.

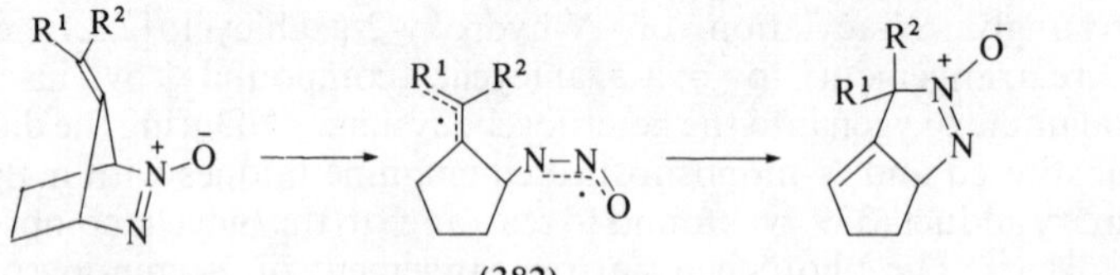

(381) (382)

A convenient synthesis of 4-(2-pyridyl)imidazolidine-2-thiones (**384**) from imidazo[1,5-*a*]pyridine-3-thione (**383**) has been developed[733] by treating the latter with amines and carbonyl compounds. A rationalization of the process is illustrated in Scheme 42. Acylation of α-arylhydrazono-2,5-dihydro-2-methyl-2-thiazoleacetic esters apparently occurs at the sulphur atom with concomitant opening of the ring system to give 3[(z)-2(acylthio)vinylimino]-2-(arylhydrazono)butanoic esters.[734]

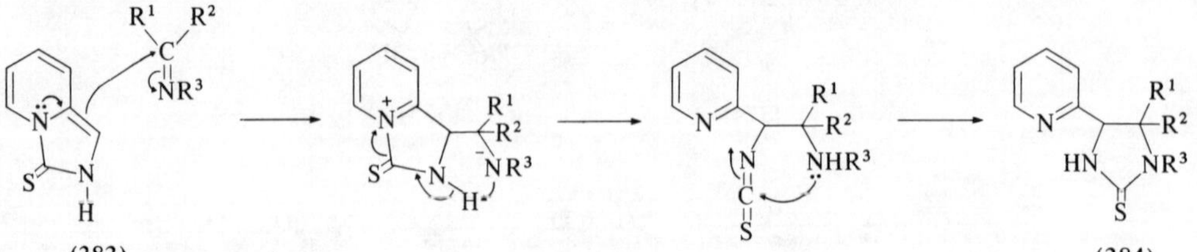

(383) (384)

SCHEME 42

The products of the reaction of thiazolineazetidinones with alkyl and aralkyl sulphenyl iodides have been hydrolysed to unsymmetrical azetidinone disulphides,[735] and an interesting thiazoline–thiazine rearrangement of methylenethiazolines has been observed[736] on storage of the latter compounds. A new molecular rearrangement of 1,3-oxathioles has been described.[737] Kinetic experiments have revealed that the intermediate of this equilibration is the thiocarbonyl ylide (**386**); rotation or inversion of the $\overset{+}{S}$-$\overset{-}{C}$ bond of (**386**) interconverts the oxathioles (**385**) and (**387**), *via* 6π-electrocyclization.

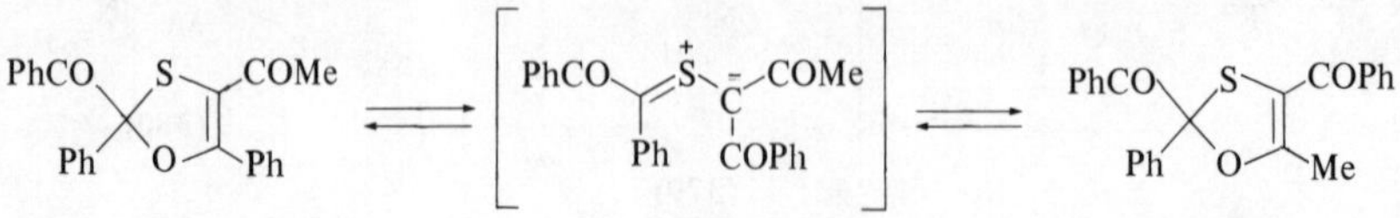

(385) (386) (387)

Other rearrangements that are considered to involve ring-opening of a five-membered ring include the flash thermolysis of methylenephthalide and 3-methylenecoumaran-2-one,[738] the novel rearrangement of a 6-ethoxycarbonyl-2,7,8-trioxoerythrinan to an isoquinolino-α-pyridone derivative,[739] the rearrangement of 2-iminofuran-3-ones,[740] the isomerization of 5-iminothiazolidine-2-thiones to 5-methyliminothiazolidine-2-thiones,[741] the rearrangement of 1,3,2-oxazaphospholidine-2-thiones to 1,3,2-thiazaphospholidine-2-ones,[742] the rearrangement of oxazaphospholines into diazaphospholanones,[743] and the imide–amide rearrangement of cyclic imidophosphorous compounds in the presence of $BF_3 \cdot OEt_2$.[744] A mechanism involving a transient pentavalent silicon anion has been proposed[745] to account for the first reported thermal isomerization of cyclic β-chloro-β-vinylsilicon compounds.

Six-membered and Larger Rings

An interesting review of singlet oxygen in organic syntheses includes a section on the rearrangement of endoperoxides to diepoxides.[746]

A novel type of rearrangement has been observed[747] on treating cyclic β-dicarbonyl compounds in strong acid media. Thus 5,5-dimethylcyclohexane-1,3-dione and its derivatives were shown to yield 5,6-dihydropyran-4-ones on irradiation in FSO_3H or H_2SO_4, although the 2,2-dimethyl derivative, lacking hydrogen on C(2), failed to afford products under the same conditions. A similar type of isomerization was also observed with thiolane-2,4-diones, and a mechanism which involves 1,2-bond fission of the protonated species (see Scheme 43) has been proposed for the process. 2*H*-Pyran intermediates have been established as intermediates in the (Z,E)-isomerizations of *S*-alkyl and *S*-aryl 2,3,4,5-tetrachloro-2,4-pentadienethioates;[748] 2-dialkylaminomethylene tetrahydropyrans have been rearranged into 2-dialkylaminocyclohexanones,[749] and the transformation of 2-amino-4*H*-pyrans into the corresponding 2-pyridones has been achieved when the pyrans are allowed to react with nitrosylsulphuric acid in acetic acid. The reaction has been explained[750] by assuming the formation of an open-chain intermediate which arises from nucleophilic attack of water on the protonated pyran ring. A new route to 3,1-benzoxazepines has involved the thermal rearrangement of 1-azidoisochromenes,[751] and 2,3-dihydrobenzo-γ-pyrone oximes have been transformed into isomeric Δ^2-isoxazoline derivatives by treatment with trifluoroacetic acid.[752] Treatment of *N*-benzyloxycarbonylspectinomycin (**388**) with weakly acidic catalysts has led[753] to the formation of a lactone (**390**) *via* a stereo-controlled tertiary α-ketol rearrangement; see (**389**). A new photo-oxidation of a cyclic ketone has been exemplified[754] in a one-step transformation of a 7-oxo-steroid,

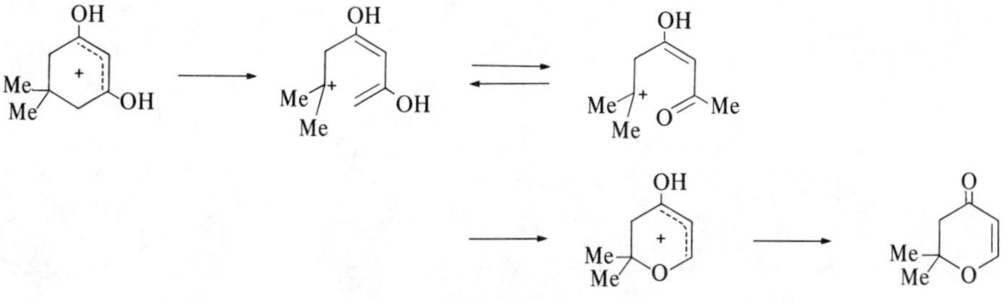

Scheme 43

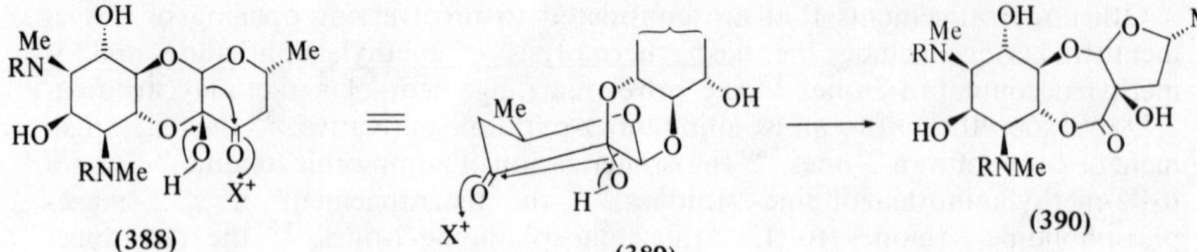

(388) (389) (390)

namely 3α,5-cyclo-5α-cholestan-7-one, into a 7-oxa-steroid, 3α,5-cyclo-7-oxa-5α-cholestane, and a novel rearrangement of the adduct from CS-epoxide, *i.e.* α-(*o*-chlorophenyl)-*β*,*β*-dicyanostyrene oxide, and dioxan-2-hydroperoxide has been described.[755] A reductive rearrangement of 4-phenyl-1,3-dioxans to 2-phenylbutane-1,4-diols has been observed[756] upon treatment of the former with sodium–potassium alloy. A number of different methyl-substituted compounds of the new acetal system (**392**) have been synthesized from 4-(5-methyl-2-furyl)alkanols (**391**) and have been rearranged to ketones (**393**).[757]

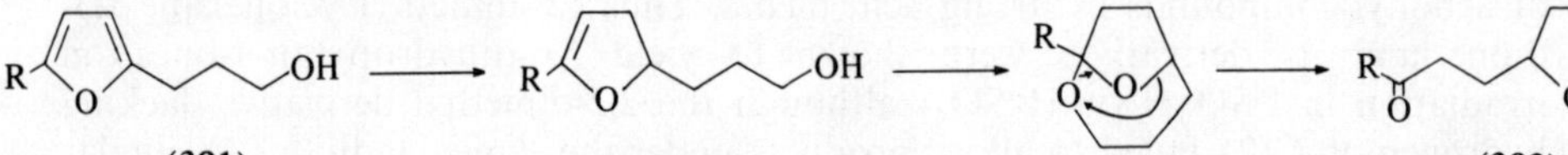

(391) (392) (393)

Diacylfuroxans have been prepared by the reaction of α-ethynyl acetates (**394**) with nitrosyl-fluoride–nitrosonium-tetrafluoroborate. Although no mechanistic studies have been carried out on the reaction, the observed formation of 3,4-bis[(acetoxycycloalkyl)carbonyl]furoxans (**396**) can be rationalized[758] by invoking the intermediacy of a nitroacetylene (**395**) which dimerizes with rearrangement; see

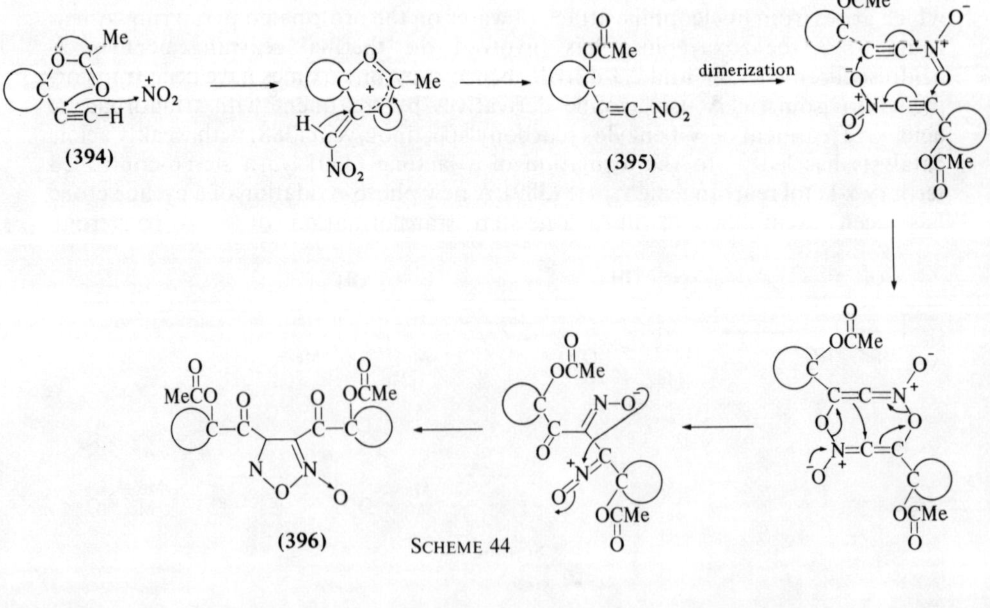

(394) (395) dimerization (396)

SCHEME 44

Scheme 44. The opening of an antiaromatic oxazine ring has been postulated[759] to account for the formation of indolones from the reaction of quinoneimine *N*-oxides with acetylenedicarboxylate esters. The course of the diastereoselective rearrangement of cyclic semi-aminals with α-mercapto groups, for example (**397**), to the racemic mixture of 2-azaspiro[4,4]nonenes (**398**) has been discussed,[760] and an interesting ring-expansion of 4-oxothiochroman-1-io(bismethoxycarbonyl)methanides (**399**) to tetrahydro-1-benzothiepin-5-ones (**400**) has been described.[761]

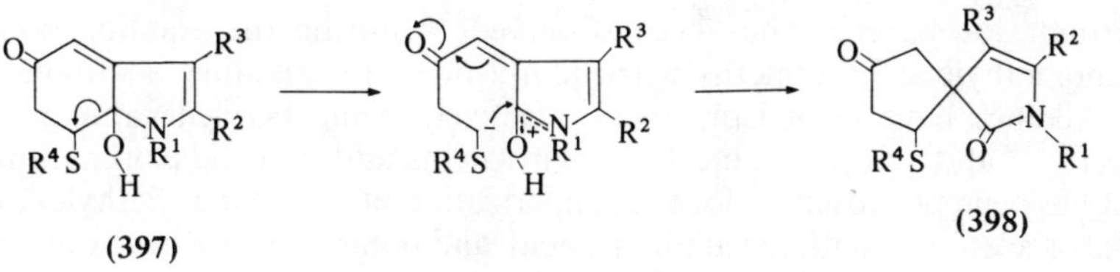

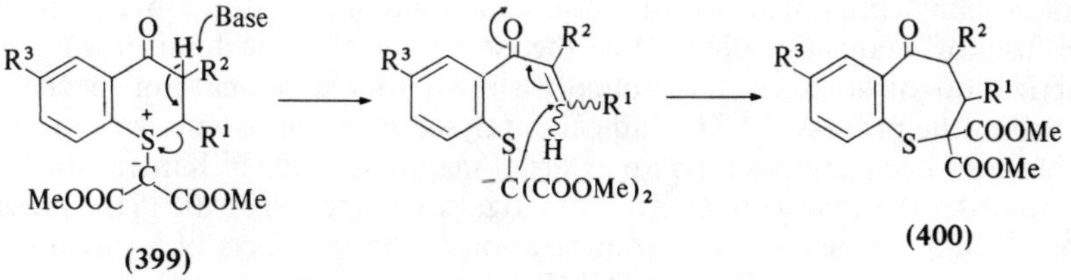

(**397**) (**398**)

(**399**) (**400**)

Observations that constitute strong support for the intermediacy of 1,4-dehydrobenzene biradicals in the thermal rearrangement of diethynyl olefins have been presented.[762] Further rearrangements that proceed by the opening of six-membered rings (or larger) include the thermal rearrangement of alkoxycyclohexenetricarbonitriles into dicyanooxocyclohexenecarboxylate esters,[763] the thermal and photochemical rearrangements of 1,4-dithiin sulphoxides,[764] the acid-catalysed rearrangements of hetisine,[765] and the acid-catalysed rearrangements of the *N*-(3-oxo-1-butenyl) derivatives of hexahydrocyclohept[*b*]indole and hexahydrocyclooct[*b*]indole.[766] The effects of electron-withdrawing substituents on the aspidospermane–eburnane rearrangement of indole alkaloids have been studied.[767]

Isomerizations

The application of two-dimensional ^{13}C-NMR spectroscopy to the elucidation of chemical-exchange networks has been proposed, and the utility of ^{13}C chemical-exchange maps has been demonstrated by application to systems involving conformational rearrangements, bond shifts, and solvation-exchange processes.[768] *Ab initio* calculations on the aminonitrene → *trans*-diimide rearrangement have been reported,[769] where it has been shown that a bimolecular hydrogen exchange is much easier than a unimolecular 1,2-hydrogen shift. *Ab initio* calculations have also been carried out on the cyanamide–isocyanamide rearrangement,[770] on the rearrangement paths of ketene, hydroxyacetylene, formylmethylene, and oxirene,[771] and *ab initio* MO theory has been used to study the intramolecular rearrangement of ethylidene to ethylene and of the oxoniomethylene cation to the hydroxymethyl cation.[772] A systematic theoretical study has also been carried out on the new degenerate rearrangement between the radical cations of formaldehyde and hydroxymethylene.[773] The thermally induced rearrangement of methyl acetate in the gas phase has been studied.[774] Apparently the predominant reaction is intramolecular oxygen-to-oxygen methyl migration and this isomerization has been discussed in terms of vibrational excitation of specific in-plane bonding modes observed for the methyl acetate molecule. The kinetics of the *cis–trans* isomerization of methyl nitrite have been examined.[775,776]

The *cis–trans* isomerization of olefins has been reviewed,[777] and a *cis–trans* equilibrium isotope effect has been observed[778] during the equilibration of 5-arylcyanomethylene-3,4-dimethylpyrrolidin-2-ones in alkaline methanolic solution. Alkenes have been isomerized effectively using *p*-xylene and phenol as sensitizers,[779] and a step-wise mechanism involving addition and elimination of the catalyst has been proposed[780] for the isomerization of *cis*-1,2-diarylethylenes in the presence of selenium, methanesulphonic acid, and potassium *tert*-butoxide. Radical ions or charge-transfer complexes have been proposed as participating in the geometrical isomerization of nitrostilbenes in the presence of aromatic amines,[781] and of α-phenylcinnamic acids on potassium amide-liquid ammonia treatment,[782] while use of magnetic effects for the study of the mechanism of *trans–cis* isomerization of stilbenes has actually shown that a radical-ion mechanism is involved in the process.[783] The radical-catalysed *cis–trans* isomerization of maleic acid has also been catalysed by an external magnetic field.[784] Kinetic studies have been carried out on the geometrical isomerization of methyl oleate in the presence of iodine,[785] and on the *cis–trans* isomerization of the enol form of formylnitroacetic acid,[786] while kinetic and dynamic NMR spectroscopic data have indicated that the *cis–trans* isomerization of α-nitro-*β*-*p*-tolylaminoacrylic acid ester proceeds by thermal rotation about the carbon–carbon double bond and by N—H bond-cleavage.[787] More recent studies[788] have shown that in general the *cis–trans* isomerization of α-nitro-*β*-arylaminoacrylic esters occurs both by a thermal mechanism and by ionization of the N—H bond. The isomerization of *cis*- to *trans*-polyacetylene has been examined[789] by ^{13}C magic-angle NMR spectroscopy. Interestingly, the major primary reaction on photolysis of *cis*-octa-1,2,6-triene appears to be *cis–trans* isomerization.[790] The thermal isomerization of a series of cyclic conjugated *cis,trans,cis*-trienes has been examined[791] and the results shown to support a structure–reactivity correlation between the size of the ring and the type of product observed. Studies on the *cis–trans* photo-isomerization of cyclooctene have revealed[792] that methyl benzoate and some other aromatic esters can non-vertically promote simple alkenes to a twisted alkene in the excited singlet state *via* a singlet exciplex. Such results have prompted the use[793] of methyl benzoate as a singlet sensitizer for cycloocta-1,5-diene, where intramolecular competition towards singlet sensitization and transannular interaction between the two double bonds is expected.

Detailed studies of the catalytic and non-catalytic *trans–cis* isomerization of pyridine-2-aldoximes have been made.[794,795] An investigation of the acid-catalysed isomerization of (E)-*O*-methylbenzohydroximoyl chloride has supported a mechanism which involves nucleophilic attack on the conjugate acid to yield a tetrahedral intermediate which then undergoes stereo-mutation,[796] while a novel isomerization of oximes to α- and α′-acyloxyenimides has been described.[797] Free energies of activation have been calculated for isomerization about the C=N bond in a series of 2,6-dimethyl-4-aryliminopyrans and their hydro and metho salts.[798]

The effects of temperature, solvent, and aryl substituent on the energies of activation have indicated that the metho salts isomerize by C=N bond rotation, the imines by inversion, or a mechanism intermediate between inversion and rotation, and the hydro salts by a previously unreported deprotonation–imine-isomerization–reprotonation mechanism. A *cis–trans* photo-isomerization of the C=N double bond has been shown to be responsible for the photochromic properties observed with the 2-arylhydrazones of 2-substituted 1,2-diketones.[799]

The stereoisomers of *N*-(2,3-dihydro-2-oxobenzooxazol-3-yl)- and *N*-(1,2-dihydro-2-oxoquinolin-1-yl)-*N*-1-methylallylarenesulphenamides have been interconverted by formal rotation around their N—N (chiral) axes.[800] The photochemical *cis–trans* isomerization of eight-membered cyclic azo compounds such as 1,2-diazacyclooct-1-ene has been described,[801] while the photo-control of polypeptide helix sense by *cis–trans* isomerism of side-chain azobenzene moieties has been achieved.[802] Interestingly, this photo-induced reversible inversion in chirality might permit photo-control in chiral recognition or chiral catalytic reactions when binding or catalytic functional groups are incorporated into the polypeptide sequence. A detailed study of the isomerization of bridgehead *cis*-1,2-diazenes has shown[803] that they can isomerize thermally to *trans*, decompose to nitrogen, or undergo both reactions simultaneously. The general shape of the potential-energy curve for isomerization has also been deduced. The kinetics and mechanism of the thermal stereoisomerization of 2,3-diaryl-1-phthalimidoazimines have been investigated.[804]

The efficient and selective isomerization of 2,3-dimethyl- and 2-methyl-buta-1,3-diene to the corresponding cyclobutenes has been achieved[805] *via* infrared multiphoton excitation, the *gem*-dimethyl effect in a Grignard-reagent cyclization–cleavage rearrangement has been discussed,[806] the perfluorobutadiene to perfluorocyclobutene isomerization has been studied,[807] highly diluted in argon and behind reflected waves in a single-pulse shock tube, and the thermal isomerizations of heptafluorocyclohexadienes under various reaction conditions have been described.[808] The steric effect on the cycloheptatriene–norcaradiene equilibrium has been studied in simple 7-alkyl-7-cyanocycloheptatrienes[809] and it has been observed that the norcaradiene component increases with the bulkiness of the 7-alkyl group. In much the same way, introduction of a *tert*-butyl group into the 2-, 3-, 2,4-, or 2,5-positions of 7-*tert*-butyl-7-cyanocyclohepta-1,3,5-trienes dramatically shifted the equilibrium to the norcaradiene, showing the importance of non-bonded interactions in this equilibrium.[810] A quantitative kinetic assessment of the effect of non-vicinal *tert*-butyl groups on [8]annulene ring-inversion and bond-shifting barriers has been made,[811] and it has been established[812] that the substitution pattern of 7,8-di-*tert*-butylbicyclo[4.2.0]octa-2,4,7-triene is unique in that it best minimizes steric interaction between the peripheral substituents and allows the less stable [8]annulene valence tautomer to be thermodynamically favoured. An unusual equilibrium between 1,4- and 1,6-di-*tert*-butylcyclooctatetraenes has been observed.[813] Labelling studies[814] on the azulene–naphthalene rearrangement have supported the recently proposed vinylidene–norcaradiene mechanism, although it is clear from a more recent labelling study[815] of the same isomerization that no single pathway yet proposed for the process can, by itself, adequately account for the observed results.

Quantitative data on the thermal conversion of ground-state fulvene into benzene have been reported.[816] The isomerization of Dewar benzenes involving electron donor–acceptor exciplexes has been studied,[817] as has the the thermo-reversible photo-valence isomerization of 9-*tert*-butylanthracene into its 9,10-Dewar isomer.[818] An interesting case of reversal of stability has been reported[819], where crowding of very bulky groups has made the Dewar-benzene isomer more stable than the partner with an intact benzene ring. The photo-isomerization of bicyclo[3.2.2]nona-2,6,8-triene to two homologues of semibullvalene has been compared[820] with the photo-reactions of its dihydro derivatives. This study has afforded new knowledge about the energy surface of the excited states of C_9H_{10} hydrocarbons.

A new type of isomerism, translational isomerism, has been observed[821] in compounds made up of three rings in which the central ring contains two bulky groups. The lateral rings are either arranged next to each other or are separated by the bulky groups. The effect of substituents on the activation energy of the thermal *cis–trans* isomerization of substituted cyclopropanes has confirmed[822] the biradical nature of the transition state. Indeed, the low barrier of activation energy (9 kJ mol^{-1}) observed for the isomerization of the tetraphenyl-substituted derivative has been attributed to merostabilization of the 1,3-biradical. The kinetics of the thermal skeletal inversions of bicyclo[2.1.0]pentane and methylbicyclo-[2.1.0]pentanes have been assessed[823] in terms of electronic and steric subsituent effects on ground-state bicyclopentanes and transition states leading to planar cyclopentane-1,3-diyl diradicals. The photochemical reactions of 6,7-dicyano- and 6,7-bis(methoxycarbony)-3-phenylsulphonyl-(2,4-*exo*)-3-azatricyclo-[3.2.1.0^{2,4}]oct-6-ene, *e.g.* (**401**), have been shown to afford the corresponding tetracyclic compounds (**402**) by a skeletal isomerization.[824] These results appear to demonstrate that the bent C(2)–C(4) bond directed towards the *endo* side is necessary for the photochemical isomerization of these tricyclic aziridines. On the basis of kinetic considerations[825] the gas-phase thermal isomerization of aminomethylisoxazoles is thought to proceed by way of a [1,3]-sigmatropic shift and the intermediacy of 1-azirines. The equilibration of β-lactones of the type (**403**)/(**405**) has been used[826] as a model for studying the position of equilibrium of *cis–trans*-disubstituted four-membered rings as a function of the substituents. The mutual rearrangement is considered to proceed *via* a contact ion-pair (**404**). Chrysene 3,4-oxide has been found to racemize[827] thermally with an activation energy of 25.2 kcal mol^{-1}, a value consistent with an earlier perturbational MO calculation

R, R, 4, 2, NSO_2Ph → R, R, NSO_2Ph

(**401**) (**402**)

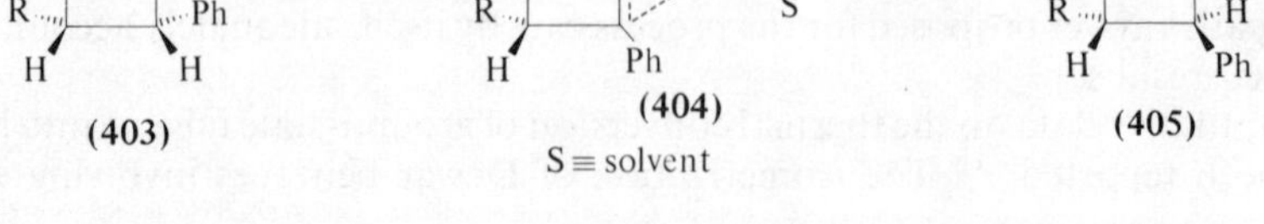

(**403**) ⇌ (**404**) ⇌ (**405**)

S ≡ solvent

which predicted an intermediate oxepine mechanism. A novel thermal isomerization reaction of penems has been described,[828] the kinetics of isomerization of 2-alkyl-5-hydroxy-1,3-dioxanes and 2-alkyl-4-(hydroxymethyl)-1,3-dioxolanes in aqueous solution of HCl have been studied,[829] and it has been proposed[830] that the *cis–trans* isomerization of η^5-cyclopentadienyl(dimethylphenylphosphine)dicyanocobaltacyclopentane proceeds *via* a polytopal rearrangement.

"Tailor-made" dicarboxylic acids and dihydroxynaphthalene–triethylamine complexes have been used[831] successfully to catalyse the isomerization of Δ-5-cholestenone. The exceptionally rapid thermal isomerization[832] of *B*-(3-hexyl)-bis(2,5-dimethylcyclohexyl)borane has opened up a sterically enhanced, highly efficient, synthetic route for the conversion of internal acyclic olefins into terminal olefins and their derivatives. The isomerization of optically active propargyl alcohols to terminal acetylenes has been achieved[833] without loss of configuration at the hydroxy centre by using potassium 3-aminopropylamide.

The dissociative permutational isomerization of a persulphurane has been reported.[834] Dynamic ^{31}P-NMR spectroscopic investigations have provided evidence[835] of phosphorus permutation isomerism in 1,3,2-benzoxazaphospholines, and NMR spectroscopic studies of some caged oxyazaphosphoranes have helped to delineate[836] those features which control rates of intramolecular ligand reorganization of phosphoranes. A quantitative study[837] has been made of the major contributions to the activation barrier of multiple pseudo-rotation processes occurring in a homogeneous series of monocyclic phosphoranes with an oxaphospholene ring. A novel phosphorus isomerization has been reported,[838] *viz.* the valence isomerization of an imino(methylene)phosphorane to a 1,2-λ^3-azaphosphiridine. It has also been demonstrated that the thiocyanate-ion- and amine-catalysed thiocyanidate–isothiocyanidate isomerization, $\rangle$P(O)SCN → $\rangle$P(O)NCS, occurs stereospecifically with inversion of configuration at phosphorus. This result has been rationalized[839] by postulating a phosphorane intermediate, formed by nucleophilic attack of SCN^- on phosphorus, in which thiocyanate and isothiocyanate groups occupy apical positions.

Tautomerism

The usefulness of the mechanistic approach to the molecular design of new tautomeric systems has been outlined in a recent review,[840] and a classification of tautomeric processes with respect to the types of transformation of corresponding multi-centred systems has been undertaken on the basis of a formallogical approach developed for reactions with cyclic electron transfer.[841] The CNDO/2 method has been used[842] to rationalize the participation of a water molecule in the tautomeric interconversion of the lactone form to the lactim form of 2-pyridone in aqueous solution, and a theoretical treatment of the tautomerism of some substituted pyridines has been presented.[843]

High-resolution ^{13}C-NMR spectroscopy has provided a powerful tool for studying annular tautomerism of pyrazoles and imidazoles in the solid state.[844] A dynamic ^{1}H-NMR spectroscopic investigation of the imino-phosphorane–benzoxazaphospholine tautomerism has been described,[845] and ^{15}N-NMR spectroscopy has been used to detect the tetrazolo[1,5-*a*]pyrimidine-2-azidopyrimidine equilibrium.[846] The porphyrinogen to corphin tautomerization has been mediated[847] by transition-metal ions such as Ni^{2+} and also by Mg^{2+}. A kinetic study has been initiated[848] on the ketimine–enamine tautomerism of 2-ethoxycarbonylmethylene-4-methyl-3-oxo-1,2,3,4-tetrahydroquinoxaline, while 4-ethoxycarbonylmethylene-4,5-dihydrotetrazolo[1,5-*a*]quinoxaline and its α-substituted derivatives have also been shown to display ketimine–enamine tautomerism.[849]

The primary dependence of 7-aryl-2,5-di-*tert*-butylcyclohepta-1,3,5-triene valence tautomerism on the π acceptor strength of the 7-aryl substituent has been demonstrated,[850] a study of the tautomerism of 2- and 4-ethoxycarbonylthiolan-3-ones[851] has implied the stereochemical effects of ring-substitution, and ring-chain tautomerism has been confirmed[852] in the 2-trimethylsiloxyphenylisocyanate–trimethylsilylbenzoxazolone system. A mechanism involving an N–O silanotropic reversible rearrangement in which the equilibrium is displaced toward the lactim form has been established for the process.

Acyl and Related Migrations

When 2-acylalkane-1,4-diones are heated to temperatures above 120°C, acyl-group migration occurs with the formation of 3-acylalkane-1,4-diones.[853] In the presence of acid, 2-alkyl-2-vinylcyclobutanones have been observed[854] to rearrange mainly by a 1,2-acyl migration to give cyclopentenones (see Scheme 45) whereas 2-vinylcyclobutanones, lacking 2-alkyl substituents, have been shown to produce cyclohexenones by a 1,3-acyl migration (see Scheme 46). 1,3-Acyl shifts have also been prominent in the photolytic conversion of a number of β,γ-unsaturated ketones into β,γ-unsaturated spiro-ketones,[855] while irradiation of 3-oxo-$\Delta^{5(10)}$-steroids in alcoholic solvents has produced[856] products of both 1,2- and 1,3-acyl shifts.

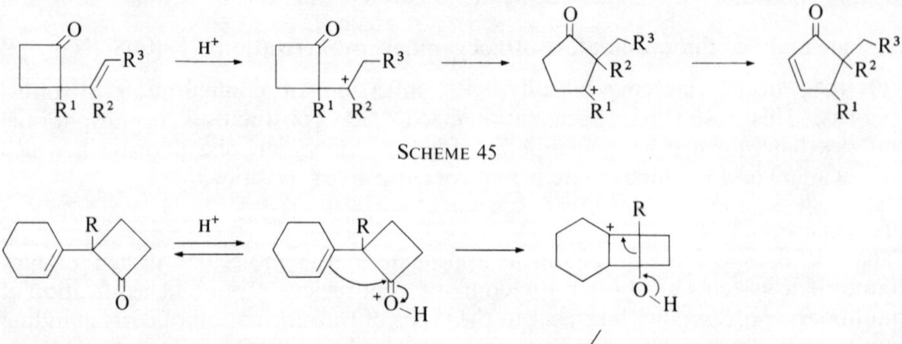

SCHEME 45

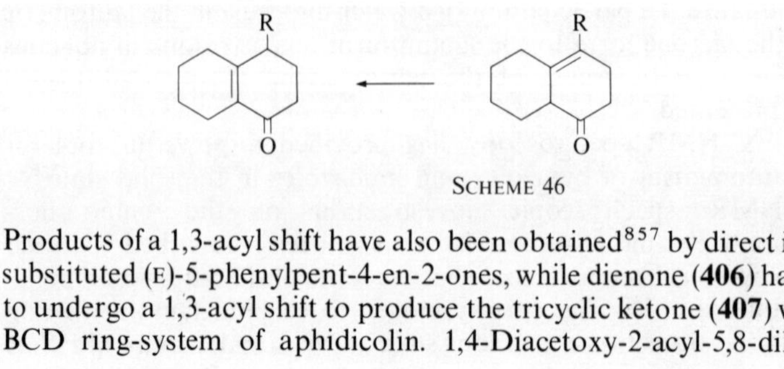

SCHEME 46

Products of a 1,3-acyl shift have also been obtained[857] by direct irradiation of alkyl-substituted (E)-5-phenylpent-4-en-2-ones, while dienone (**406**) has been observed[858] to undergo a 1,3-acyl shift to produce the tricyclic ketone (**407**) which embodies the BCD ring-system of aphidicolin. 1,4-Diacetoxy-2-acyl-5,8-dihydronaphthalenes

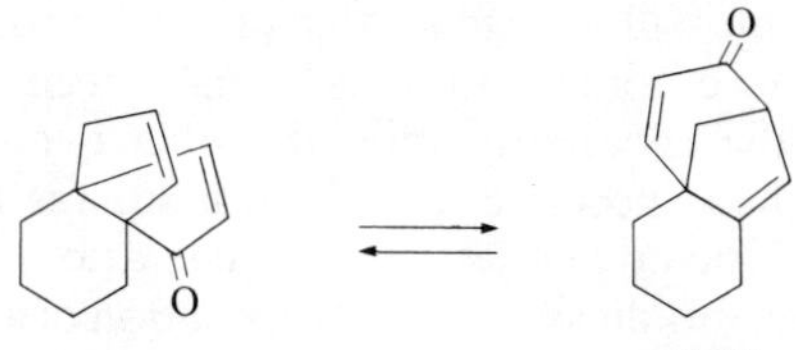

(406) (407)

have been prepared[859] by treatment of 4*a*-acyl-4*a*,5,8,8*a*-tetrahydro-1,4-naphthoquinones with boiling Ac_2O; [1,5]- and [1,2]-acetyl shifts have been observed[860] in the Diels–Alder adducts of 2-acetyl-6-methyl-1,4-benzoquinone, and a similar base-induced benzoyl migration has been described[861] for the Diels–Alder adducts of benzoyl-1,4-benzoquinones.

N-Alkyl-2-acylpyrroles have been converted into *N*-alkyl-3-acylpyrroles by strong anhydrous acid,[862] and the mechanism of acyl-group migration in the pyrrole ring of indole has been investigated.[863] The known photo-rearrangement of *N*-acetylcarbazole yielding carbazole ketones has also been observed[864] for the *N*-acetyl and *N*-(2-naphthoyl) derivatives of 5*H*-benzo[*b*]carbazole. A new example of a nitrogen-to-carbon acyl migration has been described[865] in the rearrangement of *N*-acylspirocycloalkylbenzothiazolines (**408**) to ω-(benzothiazolyl)alkyl aryl (or alkyl) ketones (**409**). 2′-*N*-Acetylfortimicins have been formed by acyl migration of 4-*N*-acylfortimicins[866] and an unusual $N \rightarrow N'$ acid-catalysed migration of the acyl group has been observed[867] in 5-aminomethyl-4-oxamoylamino-1,2,3-triazoles. $N \rightarrow O$ and $O \rightarrow N$ acyl migrations have been observed[868] in 3-amino-1,2,3-triphenylpropanols and rate constants have been reported[869] for the intramolecular *O*,*N*-acyl-transfer reactions of 2-amino-*N*-benzyl-*N*-[2-(acyloxy)-4-nitrobenzyl]-acetamides and analogues. 4-Arylisothiazol-3-yl *O*-thio-esters have been synthesized[870] by utilizing the *S*-thio-ester to *O*-thio-ester rearrangement of 4-aryl-5-thioxo-3-isothiazolin-3-yl *S*-thio-esters induced by acylation reagents, peracid, or *N*-bromosuccinimide. The thiol-ester–thioxo-ester rearrangements of 3-acylthio-4-aryl-3-isothiazoline-5-ones induced by diazoalkane, alkyl iodide, or triethyloxonium tetrafluoroborate have been reported.[871]

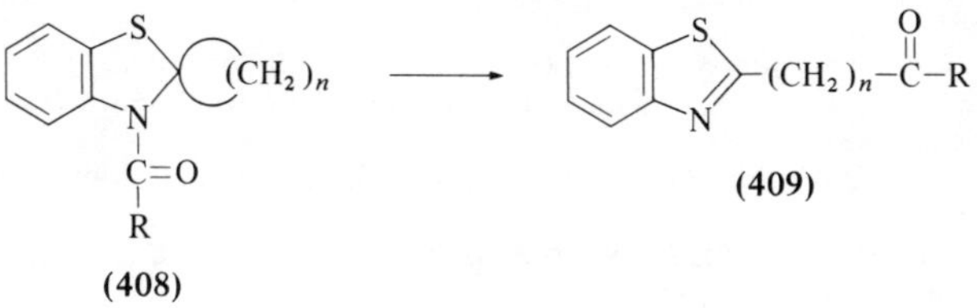

(408) (409)

Addendum

Photochemical Studies

The photochemical rearrangements of trienes have been reviewed[872] and a review has appeared on the di-π-methane rearrangement.[873] Allylbenzene and its derivatives bearing methyl groups on the aromatic ring or at C(1) of the allyl chain have been found[874] to produce cyclopropylbenzenes *via* a triplet di-π-methane rearrangement, when irradiated in MeCN–Me_2CO, while the di-π-methane rearrangements of simple allylbenzene derivatives both with π-donor substituents[875] and with π-acceptor substituents[876] have been described. A paper has appeared[877] describing the apparently anomalous behaviour of polar substituents on the benzene ring in the regioselectivity of di-π-methane rearrangements of 5,6-benzo-2-azabicyclo[2.2.2]octadienones. The generation of 2,3-benzobicyclo-[2.2.1]hept-2-ene-5,7-diyl diradicals *via* denitrogenation of appropriate azoalkanes has been described,[878] and their mechanistic implications in the di-π-methane rearrangement of benzonorbornadienenes have been discussed.[879] The di-π-methane photo-rearrangement of 6-methylenetricyclo[3.2.1.0^{2,7}]oct-3-en-8-*endo*-ol derivatives[880] and the aza-di-π-methane rearrangement of the corresponding 8-one oxime derivatives[881] have been described. The benzoylnaphthobarrelene system has been observed to undergo a di-π-methane type process to produce 1-benzoylnaphtho[*de*-2,3,4]tricyclo[4.3.0.0^{2,9}]nona-3,7-dienes,[882] and a di-π-methane rearrangement has been utilized[883] in a synthesis of the carbon skeleton of taylorione. Deuterium labelling results from the photolysis of 3,3-dimethylbicyclo-[2.2.2]octa-5,7-dien-2-one have shown[884] that the product, 4,4-dimethyltricyclo-[3.3.0.0^{2,8}]oct-6-en-3-one, is formed in a di-π-methane rearrangement rather than in an oxa-di-π-methane process. A reversible di-π-methane rearrangement of bicyclic organosilicon compounds has been reported.[885]

The effect of alkyl substituents on the photo-rearrangement of cyclohex-2-en-1-ones to bicyclo[3.1.0]hexan-2-ones has been investigated,[886] kinetically distinct triplets have been observed[887] in the photo-rearrangement of 2-phenylcyclohexanone to *cis*- and *trans*-6-phenylhex-5-enals, and a study has been made[888] of the UV irradiation of 4-(3′,7′,7′-trimethyl-2′-oxabicyclo[3.2.0]hept-3′-ene-1′-yl)but-3-en-2-one. The photochemical rearrangements of flavonols[889] and of 3-methyl-3-(1′-naphthyl)but-1-ene[890] have been described, and the liquid- and solid-state photolysis of *N*-nitroso-*N*-alkyl amino-acids has been shown to cause rearrangement with concurrent decarboxylation to amidoximes. This reaction is thought to proceed by an aminium radical.[891]

References

1 Epiotis, N. D., Shaik, S., and Zander, W., *Org. Chem.* (*N.Y.*) (Rearrangements Ground Excited States, part 2), **42**, 1 (1980).

2 Bryce-Smith, D., and Gilbert, A., Org. Chem. (*N.Y.*) (Rearrangements Ground Excited States, part 3), **42**, 349 (1980).

3 Williams, D. L. H., and Buncel, E., *Isot. Org. Chem.* (Isot. Cationic React.), **5**, 147 (1980).

4 Borodkin, G. I., Koptyug, A. V., and Shubin, V. G., *Dokl. Akad. Nauk SSSR*, **255**, 587 (1980); *Chem. Abs.*, **94**, 120612 (1981).

5 Abronin, I. A., and Zhidomirov, G. M., *Izv. Akad. Nauk SSSR, Ser. Khim.*, **1981**, 1752; *Chem. Abs.*, **95**, 169327 (1981).

6 Loktev, V. F., and Shubin, V. G., *Izv. Sib. Otd. Akad. Nauk SSSR, Ser. Khim. Nauk*, **1981**, 108; *Chem. Abs.*, **95**, 41984 (1981).

7 Loktev, V. F., Korchagina, D. V., and Shubin, V. G., *Izv. Sib. Otd. Akad. Nauk SSSR, Ser. Khim. Nauk*, **1981**, 112; *Chem. Abs.*, **95**, 23943 (1981).

8 Berezina, R. N., Akimova, S. N., Korchagina, D. V., and Shubin, V. G., *Izv. Sib. Otd. Akad. Nauk SSSR, Ser. Khim. Nauk*, **1981**, 144; *Chem. Abs.*, **95**, 168203 (1981).

9 Loktev, V. F., Korchagina, D. V., and Shubin, V. G., *Izv. Sib. Otd. Akad. Nauk SSSR, Ser. Khim. Nauk*, **1980**, 86; *Chem. Abs.*, **94**, 191366 (1981).

10 Morozov, S. V., Shakirov, M. M., and Shubin, V. G., *Zh. Org. Khim.*, **16**, 2103 (1980); *Chem. Abs.*, **94**, 83358 (1981).

11 Rodionov, V. I., Shakirov, M. M., Isaev, I. S., and Koptyug, V. A., *Zh. Org. Khim.*, **16**, 1515 (1980); *Chem. Abs.*, **94**, 14878 (1981).

12 Werstiuk, N. H., and Timmins, G., *Can. J. Chem.*, **59**, 3218 (1981).

[13] Eckhardt, H. H., Hege, D., Massa, W., Perst, H., and Schmidt, R., *Angew. Chem. Int. Ed.*, **20**, 699 (1981).
[14] Shine, H. J., Zmuda, H., Park, K. H., Kwart, H., Horgan, A. G., Collins, C., and Maxwell, B. E., *J. Am. Chem. Soc.*, **103**, 955 (1981).
[15] Ogata, Y., Nakagawa, Y., and Inaishi, M., *Bull. Chem. Soc. Jpn.*, **54**, 2853 (1981).
[16] Penton, J. R., and Zollinger, H., *Helv. Chim. Acta*, **64**, 1728 (1981).
[17] Yamamoto, J., Nishigaki, Y., Umezu, M., and Matsuura, T., *Tetrahedron*, **36**, 3177 (1980).
[18] Nikitenkova, L. P., Karakotov, S. D., Strepikheev, Y. A., Kashemirov, B. A., and Naumova, I. I., *Zh. Obshch. Khim.*, **51**, 1633 (1981); *Chem. Abs.*, **95**, 149642 (1981).
[19] Nikitenko, L. P., Karakotov, S. D., Naumova, I. I., and Strepikheev, Y. A., *Zh. Obshch. Khim.*, **51**, 1382 (1981); *Chem. Abs.*, **95**, 149641 (1981).
[20] Sone, T., Tokuda, Y., Sakai, T., Shinkai, S., and Manabe, O., *J. Chem. Soc., Perkin Trans. 2*, **1981**, 298.
[21] Krylov, A. I., and Stolyarov, B. V., *Zh. Org. Khim.*, **17**, 811 (1981); *Chem. Abs.*, **95**, 79758 (1981).
[22] Khuthier, A.-H., Al-Kazzaz, A.-K. S., Al-Rawi, J. M. A., and Al-Iraqi, M. A., *J. Org. Chem.*, **46**, 3634 (1981).
[23] Golding, P. D., Reddy, S., Scott, J. M. W., White, V. A., and Winter, J. G., *Can. J. Chem.*, **59**, 839 (1981).
[24] Badr, M. Z. A., Abdel-Rahman, A. E., El-Sherief, H. A. H., and Aly, M. M., Indian J. Chem., **20B**, 422 (1981).
[25] Alemagna, A., Buttero, P. D., Licandro, E., and Maiorana, S., *J. Chem. Soc., Chem. Commun.*, **1981**, 894.
[26] Bestmann, H. J., Schmid, G., Böhme, R., Wilhelm, E., and Burzlaff, H., *Chem. Ber.*, **113**, 3937 (1980).
[27] Yokoyama, K., Nakagaki, R., Nakamura, J., Nagakura, S., and Mutai, K., *Koen Yoshishu-Bunshi Kozo Soyo Toronkai*, **1979**, 236; *Chem. Abs.*, **93**, 220114 (1980).
[28] Mutai, K., and Kobayashi, K., *Bull. Chem. Soc. Jpn.*, **54**, 462 (1981).
[29] Alvarez, A. J., Armesto, D., and Perez-Ossorio, R., *An. Quim.*, **76C**, 100 (1980); *Chem. Abs.*, **94**, 156484 (1981).
[30] Ramsden, C. A., *J. Chem. Soc., Perkin Trans. 1*, **1981**, 2456.
[31] Gupta, R. R., Ojha, K. G., and Kumar, M., *J. Heterocycl. Chem.*, **17**, 1325 (1980).
[32] Maki, Y., Sako, M., Tanabe, M., and Suzuki, M., *Synthesis*, **1981**, 462.
[33] Irie, T., and Tanida, H., *J. Org. Chem.*, **45**, 4961 (1980).
[34] Kondo, H., Moriuchi, F., and Sunamoto, J., *J. Org. Chem.*, **46**, 1333 (1981).
[35] Sala, T., Sargent, M. V., and Elix, J. A., *J. Chem. Soc., Perkin Trans. 1*, **1981**, 849.
[36] Peet, N. P., Sunder, S., and Barbuch, R. J., *J. Heterocycl. Chem.*, **17**, 1513 (1980).
[37] Wykypiel, W., Lohmann, J.-J., and Seebach, D., *Helv. Chim. Acta*, **64**, 1337 (1981).
[38] Melvin, L. S., *Tetrahedron Lett.*, **22**, 3375 (1981).
[39] Kessar, S. V., Nadir, U. K., Pahwa, P. S., Singh, P., and Gupta, Y. P., *Indian J. Chem.*, **20B**, 7 (1981).
[40] Mach, M. H., and Bunnett, J. F., *J. Org. Chem.*, **45**, 4660 (1980).
[41] Bates, P. A., Gray, M. J., Hartshorn, M. P., Ing, H. T., Richards, K. E., and Robinson, W. T., *Tetrahedron Lett.*, **22**, 1279 (1981).
[42] Bates, P. A., Ditzel, E. J., Hartshorn, M. P., Ing, H. T., Richards, K. E., and Robinson, W. T., *Tetrahedron Lett.*, **22**, 2325 (1981).
[43] Abdou, S. E., and Sakla, A. B., *Curr. Sci.*, **49**, 728 (1980); *Chem. Abs.*, **94**, 64745 (1981).
[44] Brittain, J. M., de la Mare, P. B. D., and Newman, P. A., *J. Chem. Soc., Perkin Trans., 2*, **1981**, 32.
[45] Gensler, W. J., and Dheer, S. K., *J. Org. Chem.*, **46**, 4051 (1981).
[46] Kubsrestha, G. N., Shanker, U., Pathania, B. S., Negi, J., and Bhattacharyya, K. K., *Indian J. Chem.*, **20B**, 298 (1981).
[47] Roberts, R. M., *Rev. Roum. Chim.*, **25**, 687 (1980); *Chem. Abs.*, **94**, 64867 (1981).
[48] Hocking, M. B., *J. Chem. Technol. Biotechnol.*, **30**, 626 (1980).
[49] Box, V. G. S., and Jackson, Y. A., *Heterocycles*, **14**, 1265 (1980).
[50] Lewis, J. R., and Paul, J. G., *J. Chem. Soc., Perkin Trans. 1*, **1981**, 770.
[51] Crouse, D. J., Hurlbut, S. L., and Wheeler, D. M. S., *J. Org. Chem.*, **46**, 374 (1981).
[52] Shishkina, R. P., Ektova, L. V., Korchagin, P. V., and Fokin, E. P., *Izv. Sib. Otd. Akad. Nauk SSSR, Ser. Khim. Nauk*, **1981**, 112; *Chem. Abs.*, **95**, 80689 (1981).
[53] Al-Omran, F., Fujiwara, K., Giffney, J. C., Ridd, J. H., and Robinson, S. R., *J. Chem. Soc., Perkin Trans. 2*, **1981**, 518.
[54] Jackson, A. H., Shannon, P. V. R., and Taylor, P. W., *J. Chem. Soc., Perkin Trans. 2*, **1981**, 286.
[55] Kozhevnikov, I. V., and Kulikov, S. M., *Kinet. Katal.*, **22**, 956 (1981); *Chem. Abs.*, **95**, 168238 (1981).
[56] Jigajinni, V. B., and Wightman, R. H., *J. Chem. Soc., Chem. Commun.*, **1981**, 87.
[57] Halpern, M., Yonowich-Weiss, M., Sasson, Y., and Rabinovitz, M., *Tetrahedron Lett.*, **22**, 703 (1981).

[58] Arcoleo, A., Natoli, M. C., Fontana, G., Giammona, G., and Cicero, G., *Chem. Ind.* (*London*), **1981**, 471.
[59] Padwa, A., *Org. Chem.* (*N.Y.*) (Rearrangements Ground Excited States, part 3), **42**, 501 (1980).
[60] Boulton, A. J., *Bull. Soc. Chim. Belg.*, **90**, 645 (1981).
[61] Kost, A. N., Gromov, S. P., and Sagitullin, R. S., *Tetrahedron*, **37**, 3423 (1981).
[62] Kowalski, P., and Czuba, W., *Pol. J. Chem.*, **54**, 1185 (1980); *Chem. Abs.*, **103**, 46520 (1981).
[63] Katritzky, A. R., Chapman, A. V., Cook, M. J., and Millet, G. H., *J. Chem. Soc., Perkin Trans. 1*, **1980**, 2743.
[64] Abramavitch, R. A., and Dupuy, C., *J. Chem. Soc., Chem. Commun.*, **1981**, 36.
[65] Wachi, K., and Terada, A., *Chem. Pharm. Bull.*, **28**, 3020 (1980).
[66] Tsuchiya, T., and Sashida, H., *Heterocycles*, **14**, 1925 (1980).
[67] Bird, C. W., Partridge, I. J., and Wong, C. K., *Tetrahedron*, **37**, 1011 (1981).
[68] Tsuchiya, T., Enkaku, M., and Okajima, S., *Chem. Pharm. Bull.*, **28**, 2602 (1980).
[69] Tsuchiya, T., Okajima, S., Enkaku, M., and Kurita, J., *J. Chem. Soc., Chem. Commun.*, **1981**, 211.
[70] Knabe, J., and Heckmann, R., *Arch. Pharm.* (*Weinheim, Ger.*), **313**, 1033 (1980).
[71] Kubo, K., Ito, N., Isomura, Y., and Murakami, M., *Chem. Pharm. Bull.*, **28**, 1131 (1980).
[72] Buckle, D. R., *J. Chem. Res.* (*S*), **1980**, 308.
[73] Bornatsch, W., Reel, H., and Schündehütte, K.-H., *Chem. Ber.*, **114**, 937 (1981).
[74] Sangapure, S. S., and Agasimundin, Y. S., *Indian J. Chem.*, **19B**, 115 (1980).
[75] Inoue, H., Takada, S., Tanigawa, S., and Ueda, T., *Heterocycles*, **15**, 1049 (1981).
[76] Gönczi, C., Swistun, Z., and van der Plas, H. C., *J. Org. Chem.*, **46**, 608 (1981).
[77] Hirota, K., Kitade, Y., Senda, S., Halat, M. J., Watanabe, K. A., and Fox, J. J., *J. Org. Chem.*, **46**, 846 (1981).
[78] Leistner, S., Wagner, G., and Hentschel, K., *Z. Chem.*, **20**, 143 (1980).
[79] Holtwick, J. B., and Leonard, N. J., *J. Org. Chem.*, **46**, 3681 (1981).
[80] Chambers, R. D., Musgrave, W. K. R., and Sargent, C. R., *J. Chem. Soc., Perkin Trans. 1*, **1981**, 1071.
[81] Ota, A., Watanabe, T., Nishiyama, J., Uehara, K., and Hirate, R., *Heterocycles*, **14**, 1963 (1980).
[82] Wamhoff, H., and Kleimann, W., *J. Chem. Soc., Chem. Commun.*, **1981**, 743.
[83] Ibrahim, Y. A., Eid, M. M., and Abdel-Hady, S. A. L., *J. Heterocycl. Chem.*, **17**, 1733 (1980).
[84] Carmona, O., Greenhouse, R., Landeros, R., and Muchowski, J. M., *J. Org. Chem.*, **45**, 5336 (1980).
[85] Singh, G., Singh, D., and Ram, R. N., *Tetrahedron Lett.*, **22**, 2213 (1981).
[86] Tertov, B. A., and Bogachev, Y. G., *Khim. Geterotsikl. Soedin.*, **1981**, 119; *Chem. Abs.*, **94**, 208798 (1981).
[87] Weber, G., Mann, G., Wilde, H., and Hauptmann, S., *Z. Chem.*, **20**, 437 (1980).
[88] Mataka, S., Takahashi, K., Oshima, T., and Tashiro, M., *Chem. Lett.*, **1980**, 915.
[89] Mataka, S., Oshima, T., and Tashiro, M., *J. Org. Chem.*, **46**, 3960 (1981).
[90] Praefcke, K., and Weichsel, C., *Justus Liebigs Ann. Chem.*, **1980**, 1604.
[91] Hitchcock, P. B., McCabe, R. W., Young, D. W., and Davies, G. M., *J. Chem. Soc., Chem. Commun.*, **1981**, 608.
[92] Vladuchick, S. A., Fukunaga, T., Simmons, H. E., and Webster, O. W., *J. Org. Chem.*, **45**, 5122 (1980).
[93] Kornis, G., Marks, P. J., and Chidester, C. G., *J. Org. Chem.*, **45**, 4860 (1980).
[94] Ronsisvalle, G., Guerrera, F., and Siracusa, M. A., *Tetrahedron*, **37**, 1415 (1981).
[95] Frenna, V., Vivona, N., Corrao, A., Consiglio, G., and Spinelli, D., *J. Chem. Res.* (*S*), **1981**, 308.
[96] Frenna, V., Vivona, N., Spinelli, D., and Consiglio, G., *J. Heterocycl. Chem.*, **18**, 723 (1981).
[97] Frenna, V., Vivona, N., Consiglio, G., Corrao, A., and Spinelli, D., *J. Chem. Soc., Perkin Trans. 2*, **1981**, 1325.
[98] Barili, P. L., Fiaschi, R., Napolitano, E., Pistelli, L., Scartoni, V., and Marsili, A., *J. Chem. Soc., Perkin Trans. 1*, **1981**, 1654.
[99] De Rosa, M., Carbognani, L., and Febres, A., *J. Org. Chem.*, **46**, 2054 (1981).
[100] Lancelot, J. C., Maume, D., and Robba, M., *J. Heterocycl. Chem.*, **18**, 743 (1981).
[101] Iwanami, M., Maeda, T., Fujimoto, M., Nagano, Y., Nagano, N., Yamazaki, A., Shibanuma, T., Tamazawa, K., and Yano, K., *Chem. Pharm. Bull.*, **28**, 2629 (1980).
[102] Leistner, S., Wagner, G., and Strohscheidt, T., *Pharmazie*, **35**, 293 (1980); *Chem. Abs.*, **94**, 47255 (1981).
[103] Schuster, D. I., *Org. Chem.* (*N.Y.*) (Rearrangements Ground Excited States, part 3), **42**, 167 (1980).
[104] Schaffner, K., and Demuth, M., *Org. Chem.* (*N.Y.*) (Rearrangements Ground Excited States, part 3), **42**, 281 (1980).
[105] Zimmerman, H. E., and Pasteris, R. J., *J. Org. Chem.*, **45**, 4864 (1980).
[106] Zimmerman, H. E., and Pasteris, R. J., *J. Org. Chem.*, **45**, 4876 (1980).

[107] Schuster, D. I., Prabhu, K. V., Smith, K. J., van der Veen, J. M., and Fujiwara, H., *Tetrahedron*, **36**, 3495 (1980).
[108] Schuster, D. I., Smith, K. J., Brisimitzakis, A. C., van der Veen, J. M., and Gruska, R. P., *J. Org. Chem.*, **46**, 473 (1981).
[109] Schuster, D. I., and Liu, K.-C., *Tetrahedron*, **37**, 3329 (1981).
[110] Hart, H., and Weiner, M., *Tetrahedron Lett.*, **1981**, 3115.
[111] Koptyug, V. A., and Buraev, V. I., *Zh. Org. Khim.*, **16**, 1882 (1980); *Chem. Abs.*, **94**, 14891 (1981).
[112] Waring, A. J., Zaidi, J. H., and Pilkington, J. W., *J. Chem. Soc., Perkin Trans. 1*, **1981**, 1454.
[113] Waring, A. J., Zaidi, J. H., and Pilkington, J. W., *J. Chem. Soc., Perkin Trans. 2*, **1981**, 935.
[114] Waigh, R. D., *J. Chem. Soc., Chem. Commun.*, **1980**, 1164.
[115] Hanna, R., and Kodeih, M., *Tetrahedron Lett.*, **22**, 5051 (1981).
[116] Ward, R. S., and Thatcher, K. S., *Tetrahedron Lett.*, **22**, 4831 (1981).
[117] Bowden, B. F., Read, R. W., and Taylor, W. C., *Aust. J. Chem.*, **34**, 799 (1981).
[118] Battersby, A. R., Bhatnagar, A. K., Hackett, P., Thornber, C. W., and Staunton, J., *J. Chem. Soc., Perkin Trans. 1*, **1981**, 2002.
[119] Bren, V. A., Chernoivanov, V. A., Konstantinovskii, L. E., Nivorozhkin, L. E., Zhdanov, Y. A., and Minkin, V. I., *Dokl. Akad. Nauk SSSR*, **251**, 1129 (1980); *Chem. Abs.*, **93**, 185466 (1980).
[120] Eiden, F., and Leister, H. P., *Arch. Pharm. (Weinheim, Ger.)*, **313**, 972 (1980).
[121] Gornostaev, L. M., and Sakilidi, V. T., *Khim. Geterotsikl. Soedin.*, **1980**, 1471; *Chem. Abs.*, **94**, 120471 (1981).
[122] Kawamoto, A., Uda, H., and Harada, N., *Bull. Chem. Soc. Jpn.*, **53**, 3279 (1980).
[123] Maerkl, G., and Liebl, R., *Justus Liebigs Ann. Chem.*, **1980**, 2095.
[124] Maerkl, G., Baier, H., Liebl, R., and Stephenson, D. S., *Justus Liebigs Ann. Chem.*, **1980**, 870.
[125] Maerkl, G., Liebl, R., and Baier, H., *Justus Liebigs Ann. Chem.*, **1981**, 1610.
[126] Nakai, T., ad Mikarni, K., *Yuki Gosei Kagaku Kyokaishi*, **38**, 381 (1980); *Chem. Abs.*, **94**, 46807 (1981).
[127] Hiroi, K., *Annu. Rep. Tohoku Coll. Pharm.*, **1980**, 1; *Chem. Abs.*, **95**, 149345 (1981).
[128] Firestone, R. A., and Christensen, B. G., *Tetrahedron Lett.*, **1973**, 389.
[129] Firestone, R. A., and Vitale, M. A., *J. Org. Chem.*, **46**, 2160 (1981).
[130] Bruce, J. M., and Roshan-Ali, Y., *J. Chem. Soc., Perkin Trans. 1*, **1981**, 2677.
[131] Timpe, H. J., Dietrich, R., Boeckelmann, J., Friedel, I., Boegel, H., and Haucke, G., *Collect. Czech. Chem. Commun.*, **46**, 219 (1981.
[132] Bertrand, G., Mazerolles, P., and Ancelle, J., *Tetrahedron*, **37**, 2459 (1981).
[133] Newcomb, M., and Vieta, R. S., *J. Org. Chem.*, **45**, 4793 (1980).
[134] Snider, B. B., Rodini, D. J., Karras, M., Kirk, T. C., Deutsch, E. A., Cordova, R., and Price, R. T., *Tetrahedron*, **37**, 3927 (1981).
[135] Takai, K., Mori, I., Oshima, K., and Nozaki, H., *Tetrahedron Lett.*, **22**, 3985 (1981).
[136] Al-Sader, B. H., and Al-Asadi, Z. A. K., *Iraqi J. Sci.*, **21**, 338 (1980); *Chem. Abs.*, **94**, 102532 (1981).
[137] Nitta, M., Sekiguchi, A., and Koba, H., *Chem. Lett.*, **1981**, 933.
[138] Prajer-Janczewska, L., and Wroblewski, J., *Pol. J. Chem.*, **54**, 1431 (1980); *Chem. Abs.*, **95**, 24669 (1981).
[139] Morimoto, T., and Sekiya, M., *Synthesis*, **1981**, 308.
[140] Daub, G. W., Teramura, D. H., Bryant, K. E., and Burch, M. T., *J. Org. Chem.*, **46**, 1485 (1981).
[141] Mori, K., Nukada, T., and Ebata, T., *Tetrahedron*, **37**, 1343 (1981).
[142] Ireland, R. E., and Vevert, J.-P., *J. Org. Chem.*, **45**, 4259 (1980).
[143] Bartlett, P. A., and Pizzo, C. F., *J. Org. Chem.*, **46**, 3896 (1981).
[144] Grieco, P. A., Takigawa, T., Bongers, S. L., and Tanaka, H., *J. Am. Chem. Soc.*, **102**, 7587 (1980).
[145] Schultz, A. G., and Napier, J., *J. Chem. Soc., Chem. Commun.*, **1981**, 224.
[146] Smith, J. G., and Chu, N. G., *J. Org. Chem.*, **46**, 4083 (1981).
[147] Kozikowski, A. P., Sugiyama, K., and Huie, E., *Tetrahedron Lett.*, **22**, 3381 (1981).
[148] Paquette, L. A., Annis, G. D., Schostarez, H., and Blount, J. F., *J. Org. Chem.*, **46**, 3768 (1981).
[149] Padwa, A., Akiba, M., Cohen, L. A., and MacDonald, J. G., *Tetrahedron Lett.*, **22**, 2435 (1981).
[150] Ranganathan, D., Ranganathan, S., and Rao, C. B., *Tetrahedron*, **37**, 637 (1981).
[151] Inada, S., and Kurata, R.-i., *Bull. Chem. Soc. Jpn.*, **54**, 1581 (1981).
[152] Figeys, H. P., and Jammar, R., *Tetrahedron Lett.*, **22**, 637 (1981).
[153] Dyong, I., Weigand, J., and Merten, H., *Tetrahedron Lett.*, **22**, 2965 (1981).
[154] Overman, L. E., and Roos, J. P., *J. Org. Chem.*, **46**, 811 (1981).
[155] Overman, L. E., Clizbe, L. A., Freerks, R. L., and Marlowe, C. K., *J. Am. Chem. Soc.*, **103**, 2807 (1981).

[156] Overman, L. E., and Yokomatsu, T., *J. Org. Chem.*, **45**, 5229 (1980).
[157] Nakai, T., Setoi, H., and Kageyama, Y., *Tetrahedron Lett.*, **22**, 4097 (1981).
[158] Tamaru, Y., Kagotani, M., and Yoshida, Z.-i., *J. Org. Chem.*, **45**, 5221 (1980).
[159] Tamaru, Y., Harada, T., Nishi, S.-i., Mizutani, M., Hioki, T., and Yoshida, Z.-i., *J. Am. Chem. Soc.*, **102**, 7806 (1980).
[160] Nagashima, E., Suzuki, K., and Sekiya, M., *Tetrahedron Lett.*, **22**, 2587 (1981).
[161] Kwart, H., Miles, W. H., Horgan, A. G., and Kwart, L. D., *J. Am. Chem. Soc.*, **103**, 1757 (1981).
[162] Aukharieva, R. G., Danilova, T. A., Anisimov, A. V., and Viktorova, E. A., *Khim. Geterotsikl. Soedin.*, **1981**, 611; *Chem. Abs.*, **95**, 96643 (1981).
[163] Anisimov, A. V., Babaitsev, V. S., Grobovenko, S. Y., Danilova, T. A., and Viktorova, E. A., *Khim. Geterotsikl. Soedin.*, **1981**, 615; *Chem. Abs.*, **95**, 114489 (1981).
[164] Brooke, G. M., and Wallis, D. I., *J. Chem. Soc., Perkin Trans. 1*, **1981**, 1659.
[165] Fielding, L. J., Porter, Q. N., and Winter, G., *Aust. J. Chem.*, **33**, 2615 (1980).
[166] Banjeree, S. K., Gupta, B. D., and Atal, C. K., *Phytochemistry*, **19**, 1256 (1980).
[167] Singh, K., Banerjee, S. K., and Atal, C. K., *Indian J. Chem.*, **20B**, 108, (1981).
[168] Patra, A., Mukhopadhyay, A. K., and Mitra, A. K., *Indian J. Chem.*, **17B**, 638 (1979).
[169] Shah, R. R., and Trivedi, K. N., *Indian J. Chem.*, **20B**, 210 (1981).
[170] Raju, K. V. S., Sudha, K., and Srimannorayana, G., *Indian J. Chem.*, **19B**, 866 (1980).
[171] Cambie, R. C., Zhen-Dong, H., Noall, W. I., Rutledge, P. S., and Woodgate, P. D., *Aust. J. Chem.*, **34**, 819 (1981).
[172] Cambie, R. C., Dunlop, M. G., Noall, W. I., Rutledge, P. S., and Woodgate, P. D., *Aust. J. Chem.*, **34**, 1079 (1981).
[173] Morris, D. S., Williams, D. H., and Norris, A. F., *J. Org. Chem.*, **46**, 3422 (1981).
[174] McMurry, J. E., Andrus, A., Ksander, G. M., Musser, J. H., and Johnson, M. A., *Tetrahedron*, **37** (Supplement to Issue 9), 319 (1981).
[175] Raucher, S., Macdonald, J. E., and Lawrence, R. F., *J. Am. Chem. Soc.*, **103**, 2419 (1981).
[176] Fleischhacker, W., and Richter, B., *Chem. Ber.*, **113**, 3866 (1980).
[177] Baird, K. J., Grundon, M. F., Harrison, D. M., and Magee, M. G., *Heterocycles*, **15**, 713 (1981).
[178] Okely, H. M., and Grundon, M. F., *J. Chem. Soc., Perkin Trans. 1*, **1981**, 897.
[179] Dannhardt, G., and Obergrusberger, R., *Arch. Pharm. (Weinheim, Ger.,)*, **313**, 858 (1980).
[180] Schulze, K., Richter, F., and Muehlstaedt, M., *Z. Chem.*, **20**, 297 (1980).
[181] Mooney, B. A., Prager, R. H., and Ward, A. D., *Aust. J. Chem.*, **33**, 2717 (1980).
[182] Raucher, S., Burks, J. E., Hwang, K.-J., and Svedberg, D. P., *J. Am. Chem. Soc.*, **103**, 1853 (1981).
[183] Cooper, D., and Trippett, S., *J. Chem. Soc., Perkin Trans. 1*, **1981**, 2127.
[184] Viola, A., Collins, J. J., and Filipp, N., *Tetrahedron*, **37**, 3765 (1981).
[185] Quast, H., Gorlach, Y., and Stawitz, J., *Angew. Chem. Int. Ed.*, **20**, 93 (1981).
[186] Schipper, P., de Haan, J. W., and Buck, H. M., *J. Chem. Soc., Perkin Trans. 2*, **1981**, 457.
[187] Dinulescu, I. G., Georgescu, E. G., Stanescu, L., and Avram, M., *Tetrahedron*, **37** (Supplement to Issue 9), 55 (1981).
[188] Wiberg, K. B., Matturro, M., and Adams, R., *J. Am. Chem. Soc.*, **103**, 1600 (1981).
[189] Otterbacher, E. W., and Gajewski, J. J., *J. Am. Chem. Soc.*, **103**, 5862 (1981).
[190] Hamilton, R., Mitchell, T. R. B., and Rooney, J. J., *J. Chem. Soc., Chem. Commun.*, **1981**, 456.
[191] Dauben, W. G., and Chollet, A., *Tetrahedron Lett.*, **22**, 1583 (1981).
[192] Brulé, D., Chalchat, J.-C., Garry, R.-P., Lacroix, B., Michet, A., and Vessière, R., *Bull. Soc. Chim. Fr. II*, **1981**, 57.
[193] Maas, G., and Hummel, C., *Chem. Ber.*, **113**, 3679 (1980).
[194] Wender, P. A., and Sieburth, S. McN., *Tetrahedron Lett.*, **22**, 2471 (1981).
[195] Wender, P. A., Sieburth, S. McN., Petraitis, J. J., and Singh, S. K., *Tetrahedron*, **37**, 3967 (1981).
[196] Levine, S. G., and McDaniel, R. L., *J. Org. Chem.*, **46**, 2199 (1981).
[197] Schreiber, S. L., and Santini, C., *Tetrahedron Lett.*, **22**, 4651 (1981).
[198] Kato, T., Kondo, H., Nishino, M., Tanaka, M., Hata, G., and Miyake, A., *Bull. Chem. Soc. Jpn.*, **53**, 2958 (1980).
[199] Crouse, G. D., and Paquette, L. A., *Tetrahedron Lett.*, **22**, 3167 (1981).
[200] Tice, C. M., and Heathcock, C. H., *J. Org. Chem.*, **46**, 9 (1981).
[201] Ziegler, F. E., and Wang, T.-F., *Tetrahedron Lett.*, **22**, 1179 (1981).
[202] Hart, D. J., and Tsai, Y.-M., *Tetrahedron Lett.*, **22**, 1567 (1981).
[203] Nossin, P. M. M., and Speckamp, W. N., *Tetrahedron Lett.*, **22**, 3289 (1981).
[204] Overman, L. E., Sworin, M., Bass, L. S., and Clardy, J., *Tetrahedron*, **37**, 4041 (1981).
[205] Overman, L. E., and Mendelson, L. T., *J. Am. Chem. Soc.*, **103**, 5579 (1981).

[206] Appel, R., Knoll, F., and Ruppert, I., *Angew. Chem. Int. Ed.*, **20**, 731 (1981).
[207] Loewus, D. I., *J. Am. Chem. Soc.*, **103**, 2292 (1981).
[208] Nakai, T., Mikami, K., Taya, S., Kimura, Y., and Mimura, T., *Tetrahedron Lett.*, **22**, 69 (1981).
[209] Nakai, T., Mikami, K., Taya, S., and Fujita, Y., *J. Am. Chem. Soc.*, **103**, 6492 (1981).
[210] Katritzky, A. R., and Ponkshe, N. K., *Tetrahedron Lett.*, **22**, 1215 (1981).
[211] Cossey, A. L., Mander, L. N., and Turner, J. V., *Aust. J. Chem.*, **33**, 2061 (1980).
[212] Ollis, W. D., Somanathan, R., and Sutherland, I. O., *J. Chem. Soc., Chem. Commun.*, **1981**, 573.
[213] Mageswaran, S., Ollis, W. D., Southam, D. A., Sutherland, I. O., and Thebtaranonth, Y., *J. Chem. Soc., Perkin Trans. 1*, **1981**, 1969.
[214] Ollis, W. D., Sutherland, I. O., and Theblaranonth, Y., *J. Chem. Soc., Perkin Trans. 1*, **1981**, 1963.
[215] Mageswaran, S., Ollis, W. D., and Sutherland, I. O., *J. Chem. Soc., Perkin Trans. 1*, **1981**, 1953.
[216] Okuma, K., Higuchi, N., Ota, H., and Kobayashi, M., *Chem. Lett.*, **1980**, 1503.
[217] Gassman, P. G., and Miura, T., *Tetrahedron Lett.*, **22**, 4787 (1981).
[218] Faragher, R., Gilchrist, T. L., and Southon, I. W., *J. Chem. Soc., Perkin Trans. 1*, **1981**, 2352.
[219] Ceré, V., Paolucci, C., Pollicino, S., Sandri, E., and Fava, A., *J. Org. Chem.*, **46**, 486 (1981).
[220] Ceré, V., Paolucci, C., Pollicino, S., Sandri, E., and Fava, A., *J. Org. Chem.*, **46**, 3315 (1981).
[221] Kodama, M., Shimada, K., Takahashi, T., Kabuto, C., and Itô, S., *Tetrahedron Lett.*, **22**, 4271 (1981).
[222] Carlsen, L., Egsgaard, H., and Harpp, D. N., *J. Chem. Soc., Perkin Trans. 2*, **1981**, 1166.
[223] Corbet, J.-P., and Benezra, C., *J. Org. Chem.*, **46**, 1141 (1981).
[224] Hoffmann, R. W., *Org. Sulphur Chem., Invited Lect. Int. Symp. 9th*, **1980**, 69; *Chem. Abs.*, **95**, 149820 (1981).
[225] Halazy, S., and Krief, A., *Tetrahedron Lett.*, **1981**, 2135.
[226] Reich, H. J., Shah, S. K., Gold, P. M., and Olson, R. E., *J. Am. Chem. Soc.*, **103**, 3112 (1981).
[227] Nodwell, W. R., Bouma, W. J., and Radom, L., *Int. J. Quantum Chem.*, **18**, 107 (1980); *Chem. Abs.*, **94**, 14926 (1981).
[228] Henriksen, U., Snyder, J. P., and Halgren, T. A., *J. Org. Chem.*, **46**, 3767 (1981).
[229] Rzepa, H. S., *Tetrahedron*, **37**, 3107 (1981).
[230] Manning, T. D. R., and Kropp, P. J., *J. Am. Chem. Soc.*, **103**, 889 (1981).
[231] Padwa, A., Goldstein, S., Loza, R., and Pulwer, M., *J. Org. Chem.*, **46**, 1858 (1981).
[232] Hart, H., Chen, S.-M., and Nitta, M., *Tetrahedron*, **37**, 3323 (1981).
[233] Fuchs, B., Pasternak, M., and Pazhenchevsky, B., *J. Org. Chem.*, **46**, 2017 (1981).
[234] Fuchs, B., and Pasternak, M., *Tetrahedron*, **37**, 2501 (1981).
[235] Taylor, G. E., Mackenzie, K., and Fray, G. I., *Tetrahedron Lett.*, **21**, 4935 (1980).
[236] Gajewski, J. J., and Salazar, J. D. C., *J. Am. Chem. Soc.*, **103**, 4145 (1981).
[237] Danheiser, R. L., Martinez-Davila, C., and Sard, H., *Tetrahedron*, **37**, 3943 (1981).
[238] Bates, G. S., and Ramaswamy, S., *Can. J. Chem.*, **59**, 3120 (1981).
[239] Rasmussen, J. B., Shabana, R., and Lawesson, S.-O., *Tetrahedron*, **37**, 3693 (1981).
[240] Cutting, I., and Parsons, P. J., *Tetrahedron Lett.*, **22**, 2021 (1981).
[241] Isobe, M., Ichikawa, Y., and Goto, T., *Tetrahedron Lett.*, **22**, 4287 (1981).
[242] Klingebiel, U., and Neemann, J., *Z. Naturforsch.*, **35B**, 1155 (1980); *Chem. Abs.*, **94**, 30058 (1981).
[243] Rosenblum, M., and Waterman, P., *J. Organomet. Chem.*, **206**, 197 (1981).
[244] Mironov, V. A., Fedorovich, A. D., and Akhrem, A. A., *Usp. Khim.*, **50**, 1272 (1981); *Chem. Abs.*, **95**, 114226 (1981).
[245] Alder, R. W., and Grimme, W., *Tetrahedron*, **37**, 1809 (1981).
[246] Courtot, P., and Salaun, J.-Y., *Tetrahedron*, **36**, 3187 (1980).
[247] Roedig, A., and Goepfert, H., *Chem. Ber.*, **114**, 165 (1981).
[248] Roedig, A., Fleischmann, K., Frank, F., and Rettenberger, R., *Justus Liebigs Ann. Chem.*, **1981**, 858.
[249] Roedig, A., and Goepfert, H., *Chem. Ber.*, **114**, 3625 (1981).
[250] Burdon, J., Childs, A., Parsons, I. W., and Rimmington, T. W., *J. Fluorine Chem.*, **18**, 75 (1981).
[251] Klärner, F.-G., Rüngeler, W., and Maifield, W., *Angew. Chem. Int. Ed.*, **20**, 595 (1981).
[252] Trost, B. M., and Scudder, P. H., *J. Org. Chem.*, **46**, 506 (1981).
[253] Anderson, G., Cameron, D. W., Feutrill, G. I., and Read, R. W., *Tetrahedron Lett.*, **22**, 4347 (1981).
[254] Casey, C. P., Jones, C. R., and Tukada, H., *J. Org. Chem.*, **46**, 2089 (1981).
[255] Howard, C., Newton, R. F., Reynolds, D. P., and Roberts, S. M., *J. Chem. Soc., Perkin Trans. 1*, **1981**, 2049.
[256] Sarel, S., Schlossman, A., and Langbeheim, M., *Tetrahedron Lett.*, **22**, 691 (1981).
[257] Hoshi, N., and Uda, H., *J. Chem. Soc., Chem. Commun.*, **1981**, 1163.
[258] Schiess, P., Dinkel, R., and Funfschilling, P., *Helv. Chim. Acta*, **64**, 787 (1981).
[259] Schiess, P., and Dinkel, R., *Helv. Chim. Acta*, **64**, 801 (1981).

[260] Adam, W., and De Lucchi, O., *J. Org. Chem.*, **45**, 4167 (1980).
[261] Durst, T., Lancaster, M., and Smith, D. J. H., *J. Chem. Soc., Perkin Trans. 1*, **1981**, 1846.
[262] Manning, C., McClory, M. R., and McCullough, J. J., *J. Org. Chem.*, **46**, 919 (1981).
[263] Kiesele, H., *Tetrahedron Lett.*, **22**, 1097 (1981).
[264] Imajo, S., Shingu, K., Kuritani, H., and Kato, A., *Tetrahedron Lett.*, **22**, 107 (1981).
[265] Speckamp, W. N., Veenstra, S. J., Dijkink, J., and Fortgens, R., *J. Am. Chem. Soc.*, **103**, 4643 (1981).
[266] Goerdeler, J., Lindner, C., and Zander, F., *Chem. Ber.*, **114**, 536 (1981).
[267] Epsztein, R., and Le Goff, N., *Tetrahedron Lett.*, **22**, 1965 (1981).
[268] Argo, C. B., and Sharp, J. T., *Tetrahedron Lett.*, **22**, 353 (1981).
[269] Moody, C. J., Rees, C. W., Tsoi, S. C., and Williams, D. J., *J. Chem. Soc., Chem. Commun.*, **1981**, 927.
[270] van Arendonk, R. J. F. M., de Violet, P. F., and Laarhoven, W. H., *Recl. Trav. Chim. Pays-Bas*, **100**, 256 (1981).
[271] Gilchrist, T. L., Tuddenham, D., McCague, R., Moody, C. J., and Rees, C. W., *J. Chem. Soc., Chem. Commun.*, **1981**, 657.
[272] Kato, M., Takatoku, K., Ito, S., Funakura, M., and Miwa, T., *Bull. Chem. Soc. Jpn.*, **53**, 3648 (1980).
[273] Burgstahler, A. W., *J. Org. Chem.*, **46**, 1741 (1981).
[274] Mathay, F., Mercier, F., Charrier, C., Fischer, J., and Mitschler, A., *J. Am. Chem. Soc.*, **103**, 4595 (1981).
[275] Cuthbertson, A. F., and Glidewell, C., *J. Organomet. Chem.*, **221**, 19 (1981).
[276] Nakadaira, Y., Otsuka, T., and Sakurai, H., *Tetrahedron Lett.*, **22**, 2417 (1981).
[277] Nakadaira, Y., Otsuka, T., and Sakurai, H., *Tetrahedron Lett.*, **22**, 2421 (1981).
[278] Deshayes, C., and Gelin, S., *Tetrahedron Lett.*, **22**, 2557 (1981).
[279] Hori, M., Kataoka, T., Shimizu, H., and Ueda, N., *Tetrahedron Lett.*, **22**, 3071 (1981).
[280] Hori, M., Kataoka, T., Shimizu, H., and Tomoto, A., *Tetrahedron Lett.*, **22**, 3629 (1981).
[281] Simkin, B. Y., Golyanskii, B. V., and Minkin, V. I., *Zh. Org. Khim.*, **17**, 3 (1981); *Chem. Abs.*, **95**, 23982 (1981).
[282] Rosnati, V., and Saba, A., *Gazz. Chim. Ital.*, **110**, 211 (1980); *Chem. Abs.*, **93**, 203566 (1980).
[283] Ollis, W. D., Sutherland, I. O., and Thebtaranonth, Y., *J. Chem. Soc., Perkin Trans. 1*, **1981**, 1981.
[284] Ollis, W. D., Somanathan, R., and Sutherland, I. O., *J. Chem. Soc., Perkin Trans. 1*, **1981**, 2930.
[285] Bren, V. A., Chernoivanov, V. A., Borisenko, N. I., Konstantinovskii, L. E., Bren, Z. V., and Minkin, V. I., *Dokl. Akad. Nauk SSSR*, **255**, 107 (1980); *Chem. Abs.*, **94**, 102546 (1981).
[286] Hünig, S., and Öller, M., *Chem. Ber.*, **114**, 959 (1981).
[287] Audenaert, F., and Vandewalle, M., *Tetrahedron Lett.*, **22**, 4521 (1981).
[288] Abraham, W., Csongar, C., and Kreysig, D., *J. Prakt. Chem.*, **323**, 420 (1981).
[289] Tezuka, T., Kikuchi, O., Houk, K. N., Paddon-Row, M. N., Santiago, C. M., Rondan, N. G., Williams, J. C., and Gandour, R. W., *J. Am. Chem. Soc.*, **103**, 1367 (1981).
[290] Dauben, W. G., and Michno, D. M., *Tetrahedron*, **37**, 3263 (1981).
[291] Nishizawa, M., and Noyori, R., *Bull. Chem. Soc. Jpn.*, **54**, 2233 (1981).
[292] Dimroth, K., Schaffer, O., and Weierhauser, G., *Chem. Ber.*, **114**, 1752 (1981).
[293] Fusco, R., and Sannicolò, F., *J. Org. Chem.*, **46**, 83 (1981).
[294] Fusco, R., and Sannicolò, F., *J. Org. Chem.*, **46**, 90 (1981).
[295] Hagenbuch, J.-P., Stampfli, B., and Vogel, P., *J. Am. Chem. Soc.*, **103**, 3934 (1981).
[296] Brown, P. J. N., Cadogan, J. I. G., Gosney, I., and Johnstone, A., *J. Chem. Soc., Chem. Commun.*, **1981**, 1035.
[297] Barltrop, J. A., Day, A. C., Mack, A. G., Shahrisa, A., and Wakamatsu, S., *J. Chem. Soc., Chem. Commun.*, **1981**, 604.
[298] Dolbier, W. R., Al-Sadar, B. H., Sellers, S. F., and Koroniak, H., *J. Am. Chem. Soc.*, **103**, 2138 (1981).
[299] Bushby, R. J., *Annu. Rep. Prog. Chem.*, **76B**, 45 (1980).
[300] Huisgen, R., *Angew. Chem. Int. Ed.*, **19**, 947 (1980).
[301] George, M. V., Mitra, A., and Sukumaran, K. B., *Angew. Chem. Int. Ed.*, **19**, 973 (1980).
[302] Kikuchi, O., *Tetrahedron Lett.*, **22**, 859 (1981).
[303] Kikuchi, O., Kubota, H., and Suzuki, K., *Bull. Chem. Soc. Jpn.*, **54**, 1126 (1981).
[304] Bauld, N. L., and Wirth, D., *J. Comput. Chem.*, **2**, 1 (1981); *Chem. Abs.*, **94**, 191500 (1981).
[305] Viola, A., Collins, J. J., and Filipp, N., *Prepr. Div. Pet. Chem., Am. Chem. Soc.*, **21**, 206 (1979); *Chem. Abs.*, **94**, 156058 (1981).
[306] Viola, A., and Collins, J. J., *J. Chem. Soc., Chem. Commun.*, **1980**, 1247.
[307] King, K. D., *Int. J. Chem. Kinet.*, **13**, 245 (1981).
[308] Bertrand, M., Roumestant, M. L., and Sylvestre-Panthet, P., *Tetrahedron Lett.*, **22**, 3589 (1981).
[309] Baldwin, J. E., Jung, M., and Kitchin, J., *J. Chem. Soc., Chem. Commun.*, **1981**, 578.

[310] Brown, D. S., Foster, R. W. G., Marples, B. A., and Mason, K. G., *Tetrahedron Lett.*, **21**, 5057 (1980).
[311] Oppolzer, W., and Bättig, K., *Helv. Chim. Acta*, **64**, 2489 (1981).
[312] Padwa, A., and Ricker, W. F., *J. Am. Chem. Soc.*, **103**, 1859 (1981).
[313] Hassner, A., D'Costa, R., McPhail, A. T., and Butler, W., *Tetrahedron Lett.*, **22**, 3691 (1981).
[314] Derdour, A., and Texier, F., *C. R. Hebd. Seances Acad. Sci.*, **291C**, 101 (1980).
[315] Gajewski, J. J., and Chang, M. J., *J. Am. Chem. Soc.*, **102**, 7542 (1980).
[316] Ishii, K., Frei, B., Wolf, H. R., and Jeger, O., *Helv. Chim. Acta*, **64**, 1235 (1981).
[317] Eberbach, W., and Trostmann, U., *Chem. Ber.*, **114**, 2979 (1981).
[318] Eberbach, W., Hädicke, E., and Trostmann, U., *Tetrahedron Lett.*, **22**, 4953 (1981).
[319] Danheiser, R. L., Morin, J. M., Yu. M., and Basak, A., *Tetrahedron Lett.*, **22**, 4205 (1981).
[320] Mao, C.-R., Presser, N., John, L.-S., Moriarty, R. M., and Gordon, R. J., *J. Am. Chem. Soc.*, **103**, 2105 (1981).
[321] Dauben, W. G., and Michno, D. M., *J. Am. Chem. Soc.*, **103**, 2284 (1981).
[322] Pennings, M. L. M., Reinhoudt, D. N., Harkema, S., and van Hummel, G. J., *J. Am. Chem. Soc.*, **102**, 7570 (1980).
[323] Gaynor, B. J., Gilbert, R. G., King, K. D., and Harman, P. J., *Aust. J. Chem.*, **34**, 449 (1981).
[324] Prinzbach, H., Bingmann, H., Beck, A., Hunkler, D., Sauter, H., and Hädicke, E., *Chem. Ber.*, **114**, 1697 (1981).
[325] Heller, H. G., and Oliver, S., *J. Chem. Soc., Perkin Trans. 1*, **1981**, 197.
[326] Darcy, P. J., Heller, H. G., Strydom, P. J., and Whittall, J. *J. Chem. Soc., Perkin Trans. 1*, **1981**, 202.
[327] Bryce, M. R., Reynolds, C. D., Hanson, P., and Vernon, J. M., *J. Chem. Soc., Perkin Trans. 1*, **1981**, 607.
[328] Croft, K. D., Ghisalberti, E. L., Jefferies, P. R., Stuart, A. D., Raston, C. L., and White, A. H., *J. Chem. Soc., Perkin Trans. 2*, **1981**, 1473.
[329] Abraham, W., Paulick, W., and Kreysig, D., *J. Prakt. Chem.*, **323**, 427 (1981); *Chem. Abs.*, **95**, 114498 (1981).
[330] Meier, H., Echter, T., and Zimmer, O., *Angew. Chem. Int. Ed.*, **20**, 865 (1981).
[331] Anastassion, A. G., Kasmai, H. S., and Saadein, M. R., *Helv. Chim. Acta*, **64**, 617 (1981).
[332] Miyashi, T., Suto, N., Yamaki, T., and Mukai, T., *Tetrahedron Lett.*, **22**, 4421 (1981).
[333] Kim, J. K., and Caserio, M. C., *Tetrahedron Lett.*, **22**, 4159 (1981).
[334] Benner, S. A., Maggio, J. E., and Simmons, H. E., *J. Am. Chem. Soc.*, **103**, 1581 (1981).
[335] Rigaudy, J., Scribe, P., and Breliere, C., *Tetrahedron*, **37**, 2585 (1981).
[336] Hunter, D. H., Stothers, J. B., and Warnhoff, E. W., *Org. Chem.* (*N.Y.*) (Rearrangements Ground Excited States, part 1), **42**, 391 (1980).
[337] Pant, J., and Joshi, B. C., *Indian, J. Chem. Educ.*, **7**, 11 (1980); *Chem. Abs.*, **94**, 29597 (1981).
[338] Kikuchi, O., Tanaka, S., Kanekiyo, T., Naruchi, K., and Yamada, K., *Kogakubu Kenkyu Hokoku* (*Chiba Daigaku*), **32**, 53 (1981); *Chem. Abs.*, **95**, 42122 (1981).
[339] Chow, Y. L., and Polo, J., *J. Chem. Soc., Chem. Commun.*, **1981**, 297.
[340] Craze, G.-A., and Watt, I., *J. Chem. Soc., Perkin Trans. 2*, **1981**, 175.
[341] Paquette, L. A., and Crouse, G. D., *J. Am. Chem. Soc.*, **103**, 6235 (1981).
[342] Mehta, G., Rao, K. S., Bhadbhade, M. M., and Venkatesan, K., *J. Chem. Soc., Chem. Commun.*, **1981**, 755.
[343] Nishinaga, A., Nakamura, K., and Matsuura, T., *Tetrahedron Lett.*, **22**, 3261 (1981).
[344] Bhat, V., and Cookson, R. C., *J. Chem. Soc., Chem. Commun.*, **1981**, 1123.
[345] Knollmüller, M., and Kosma, P., *Monatsh. Chem.*, **112**, 489 (1981).
[346] Hunter, D., and Neilson, D. G., *J. Chem. Soc., Perkin Trans. 1*, **1981**, 1371.
[347] Epsztein, R., and Herman, B., *J. Chem. Soc., Chem. Commun.*, **1980**, 1250.
[348] Verkruijsse, H. D., Bos, H. J. T., de Noten, L. J., and Brandsma, L., *Recl. Trav. Chim. Pays-Bas*, **100**, 244 (1981).
[349] Kocharyan, S. T., Razina, T. L., and Babayan, A. T., *Arm. Khim. Zh.*, **33**, 684 (1980); *Chem. Abs.*, **94**, 174552 (1981).
[350] Kocharyan, S. T., Razina, T. L., and Babayan, A. T., *Arm. Khim. Zh.*, **34**, 409 (1981); *Chem. Abs.*, **95**, 168724 (1981).
[351] Kocharyan, S. T., Grigoryan, V. V., and Babayan, A. T., *Arm. Khim. Zh.*, **34**, 223 (1981); *Chem. Abs.*, **95**, 132213 (1981).
[352] Kocharyan, S. T., Voskanyan, V. S., Grigoryan, V. V., and Babayan, A. T., *Arm. Khim. Zh.*, **34**, 429 (1981); *Chem. Abs.*, **95**, 168468 (1981).
[353] Jemison, R. W., Mageswaran, S., Ollis, W. D., Sutherland, I. O., and Thebtaranonth, Y., *J. Chem. Soc., Perkin Trans. 1*, **1981**, 1154.

[354] Tanaka, T., Kubota, K., Watanabe, Y., and Kawamura, A., *Nippon Kagaku Kaishi*, **1980**, 600; *Chem. Abs.*, **93**, 168224 (1980).
[355] Bosch, J., and Rubiralta, M., *J. Heterocycl. Chem.*, **18**, 485 (1981).
[356] Stamegna, A. P., and McEwen, W. E., *J. Org. Chem.*, **46**, 1653 (1981).
[357] Benattar, A., Barbry, D., Hasiak, B., and Couturier, D., *J. Heterocycl. Chem.*, **18**, 63 (1981).
[358] Banks, R. E., Brown, A. K., Haszeldine, R. N., Kenny, A., and Tipping, A. E., *J. Fluorine Chem.*, **17**, 85 (1981).
[359] Knorr, R., and Lattke, E., *Chem. Ber.*, **114**, 2116 (1981).
[360] Macaulay, S. R., *Can. J. Chem.*, **58**, 2567 (1980).
[361] Dean, B. D., and Truce, W. E., *J. Org. Chem.*, **45**, 5429 (1980).
[362] Dean, B. D., and Truce, W. E., *J. Org. Chem.*, **46**, 3575 (1981).
[363] Tsuchuja, T., and Sashida, H., *J. Chem. Soc., Chem. Commun.*, **1980**, 1109.
[364] Hünig, S., and Öller, M., *Chem. Ber.*, **113**, 3803 (1980).
[365] Fleming, I., and Floyd, C. D., *J. Chem. Soc., Perkin Trans. 1*, **1981**, 969.
[366] Sekine, M., Kume, A., and Hata, T., *J. Chem. Soc., Chem. Commun.*, **1981**, 969.
[367] Sekine, M., Nakajima, M., and Hata, T., *J. Org. Chem.*, **46**, 4030 (1981).
[368] Hellwinkel, D., Krapp, W., and Sheldrick, W. S., *Chem. Ber.*, **114**, 1786 (1981).
[369] Sturtz, G., and Baboulene, M., *Tetrahedron*, **37**, 3067 (1981).
[370] Allcock, H. R., Harris, P. J., and Nissan, R. A., *J. Am. Chem. Soc.*, **103**, 2256 (1981).
[371] Ter-Gabrielyan, V. G., Gambaryan, P. N., Lur'e, E. P., and Knunyants, I. L., *Dokl. Akad. Nauk SSSR*, **254**, 898 (1980); *Chem. Abs.*, **94**, 120597 (1981).
[372] Lartey, P. A., and Grampovnik, D. J., *J. Antibiot.*, **33**, 1071 (1980).
[373] Kiss, J., and Arnold, W., *Helv. Chim. Acta*, **64**, 1566 (1981).
[374] Mattox, V. R., Carpenter, P. C., and Graf, E., *J. Steroid Biochem.*, **14**, 19 (1981).
[375] Tamura, Y., Mukai, C., Nakajima, N., Ikeda, M., and Kido, M., *J. Chem. Soc., Perkin Trans. 1*, **1981**, 212.
[376] Schaad, L. J., Hess, B. A., and Zahradník, R., *J. Org. Chem.*, **46**, 1909 (1981).
[377] Sakai, T., Katayama, T., and Takeda, A., *J. Org. Chem.*, **46**, 2924 (1981).
[378] De Kimpe, N., Verhé, R., De Buyck, L., Moëns, L., and Schamp, N., *Tetrahedron Lett.*, **22**, 1837 (1981).
[379] Abad, A., Arnó, M., Pedro, J. R., and Seoane, E., *Tetrahedron Lett.*, **22**, 1733 (1981).
[380] Abad, A., Arnó, M., Pedro, J. R., and Seoane, E., *Chem. Ind. (London)*, **1981**, 157.
[381] Kirmse, W., and Sandkühler, P., *Justus Liebigs Ann. Chem.*, **1981**, 1394.
[382] Martin, P., Grueter, H., and Bellus, D., *Pestic. Sci.*, **11**, 141 (1980); *Chem. Abs.*, **94**, 102873 (1981).
[383] Grueter, H., Bissig, P., Martin, P., Flueck, V., and Gsell, L., *Pestic. Sci.*, **11**, 148 (1980); *Chem. Abs.*, **94**, 102874 (1981).
[384] Khawad, I. E., and Iskander, G. M., *Rev. Roum. Chim.*, **25**, 1321 (1980).
[385] Eaton, P. E., Or, Y. S., and Branca, S. J., *J. Am. Chem. Soc.*, **103**, 2134 (1981).
[386] Logan, R. T., Roy, R. G., and Woods, G. F., *J. Chem. Soc., Perkin Trans. 1*, **1981**, 2631.
[387] Shubin, V. G., *Izv. Sib. Otd. Akad. Nauk SSSR*, **1980**, 18; *Chem. Abs.*, **94**, 3308 (1981).
[388] Saunders, M., Chandrasekhar, J., and Schleyer, P. von R., *Org. Chem.* (*N.Y.*) (Rearrangements Ground Excited States, part 1), **42**, 1 (1980); *Chem. Abs.*, **94**, 3317 (1981).
[389] Lee, C. C., *Isot. Org. Chem.* (Isot. Cationic React.), **5**, 1 (1980); *Chem. Abs.*, **94**, 46278 (1981).
[390] Shchegolev, A. A., and Kanishchev, M. I., *Usp. Khim.*, **50**, 1046 (1981); *Chem. Abs.*, **95**, 114225 (1981).
[391] Koptyug, V. A., and Shubin, V. G., *Zh. Org. Khim.*, **16**, 1977 (1980); *Chem. Abs.*, **94**, 29608 (1981).
[392] Lipovich, V. G., *Izv. Sib. Otd. Akad. Nauk SSSR, Ser. Khim. Nauk*, **1980**, 12; *Chem. Abs.*, **94**, 14614 (1981).
[393] Temnikova, T. I., *Izv. Sib. Otd. Akad. Nauk SSSR, Ser. Khim. Nauk*, **1980**, 5; *Chem. Abs.*, **94**, 14613 (1981).
[394] Heimgartner, H., *Chimia*, **34**, 333 (1980); *Chem. Abs.*, **94**, 3285 (1981).
[395] Yablokov, V. A., *Usp. Khim.*, **49**, 1711 (1980); *Chem. Abs.*, **94**, 3324 (1981).
[396] Bowen, R. D., and Williams, D. H., *Org. Chem.* (*N.Y.*) (Rearrangements Ground Excited States, part 1), **42**, 55 (1980); *Chem. Abs.*, **94**, 3318 (1981).
[397] Frenking, G., and Schwarz, H., *Z. Naturforsch.*, **36B**, 797 (1981); *Chem. Abs.*, **95**, 168269 (1981).
[398] Akhmatdinov, R. T., Kantor, E. A., Karakhanov, R. A., and Rakhmankulov, D. L., *Zh. Org. Khim.*, **17**, 478 (1981); *Chem. Abs.*, **95**, 23942 (1981).
[399] Schneider, H.-J., and Heiske, D., *J. Am. Chem. Soc.*, **103**, 3501 (1981).
[400] Bews, J. R., and Glidewell, C., *J. Mol. Struct.*, **67**, 151 (1980).
[401] Randic, M., *Int. J. Quantum. Chem.*, **14**, 557 (1980); *Chem. Abs.*, **94**, 139056 (1981).

402 Arseniyadis, S., Guenot, P., Goré, J., and Carrié, R., *Tetrahedron Lett.*, **22**, 2251 (1981).
403 Das, K. G., and Vairamani, M., *Adv. Mass Spectrom.*, **8A**, 691 (1980); *Chem. Abs.*, **94**, 191353 (1981).
404 Kanishchev, M. I., Shchegolev, A. A., Smit, V. A., Caple, R., and Sepetov, N. F., *Izv. Akad. Nauk SSSR, Ser. Khim.*, **1980**, 2102; *Chem. Abs.*, **94**, 29894 (1981).
405 Keelan, B. W., and Andrews, L., *J. Am. Chem. Soc.*, **103**, 822 (1981).
406 Andrews, L., Harvey, J. A., Kelsall, B. J., and Duffey, D. C., *J. Am. Chem. Soc.*, **103**, 6415 (1981).
407 Karras, M., and Snider, B. B., *J. Am. Chem. Soc.*, **102**, 7951 (1980).
408 Ando, T., Kim, S.-G., Matsuda, K., Yamataka, H., Yukawa, Y., Fry, A., Lewis, D. E., Sims, L. B., and Wilson, J. C., *J. Am. Chem. Soc.*, **103**, 3505 (1981).
409 Ando, T., Yamataka, H., Morisaki, H., Yamawaki, J., Kuramochi, J., and Yukawa, Y., *J. Am. Chem. Soc.*, **103**, 430 (1981).
410 Shiner, V. J., and Tai, J. J., *J. Am. Chem. Soc.*, **103**, 436 (1981).
411 Failes, R. L., Mollah, Y. M. A., and Shapiro, J. S., *Int. J. Chem. Kinet.*, **13**, 7 (1980).
412 Berner, D., Dahn, H., and Vogel, P., *Helv. Chim. Acta*, **63**, 2538 (1980).
413 Smith, W. B., and Branum, G. D., *Tetrahedron Lett.*, **22**, 2055 (1981).
414 Brusova, G. P., Gopius, E. D., Smolina, T. A., and Reutov, O. A., *Dokl. Akad. Nauk SSSR*, **253**, 349 (1980); *Chem. Abs.*, **94**, 3516 (1981).
415 Slayden, S. W., *J. Org. Chem.*, **46**, 2311 (1981).
416 Berti, C., Greci, L., and Poloni, M., *J. Chem. Soc., Perkin Trans. 1*, **1981**, 1610.
417 Lee, C. C., Obafemi, C. A., Quail, J. W., and Rappoport, Z., *Can. J. Chem.*, **59**, 2342 (1981).
418 Lee, C. C., and Obafemi, C. A., *Can. J. Chem.*, **59**, 1636 (1981).
419 Tsuchihashi, G.-i., Kitajima, K., and Mitawmura, S., *Tetrahedron Lett.*, **22**, 4305 (1981).
420 Vittinghoff, K., and Griesbaum, K., *Tetrahedron Lett.*, **22**, 1889 (1981).
421 Wade, T. N., *J. Chem. Res. (S)*, **1980**, 388.
422 Patrick, T. B., Scheibel, J. J., Hall, W. E., and Lee, Y. H., *J. Org. Chem.*, **45**, 4492 (1980).
423 Gevorkyan, A. A., Kazaryan, P. I., and Khizantsyan, N. M., *Arm. Khim. Zh.*, **34**, 435 (1981); *Chem. Abs.*, **95**, 168886 (1981).
424 Wistuba, E., and Rüchardt, C., *Tetrahedron Lett.*, **22**, 4069 (1981).
425 Herlihy, K. P., *Aust. J. Chem.*, **34**, 107 (1981).
426 Audier, H. E., Flammang, R., Maquestiau, A., and Millet, A., *Nouveau J. Chim.*, **4**, 531 (1980).
427 Maquestiau, A., Flammang, R., Flammang-Barbieux, M., Mispreuve, H., Howe, I., and Beynon, J. H., *Adv. Mass. Spectrom.*, **8A**, 698 (1980); *Chem. Abs.*, **94**, 46521 (1981).
428 Holt, D. A., *Tetrahedron Lett.*, **22**, 2243 (1981).
429 Fleming, I., and Patel, S. K., *Tetrahedron Lett.*, **22**, 2321 (1981).
430 Fleming, I., Paterson, I., and Pearce, A., *J. Chem. Soc., Perkin Trans. 1*, **1981**, 256.
431 Roberts, R. M., and Elrod, L. W., *J. Org. Chem.*, **46**, 3732 (1981).
432 Wistuba, E., and Rüchardt, C., *Tetrahedron Lett.*, **22**, 3389 (1981).
433 Richey, H. G., Kubala, B., and Smith, M. A., *Tetrahedron Lett.*, **22**, 3471 (1981).
434 Schmitz, L. R., and Sorensen, T. S., *Tetrahedron Lett.*, **22**, 1191 (1981).
435 Van Cantfort, C. K., and Coates, R. M., *J. Org. Chem.*, **46**, 4331 (1981).
436 Murakami, N., and Nishida, S., *Chem. Lett.*, **1981**, 997.
437 Olah, G. A., Fung, A. P., Rawdah, T. N., and Prakash, G. K., *J. Am. Chem. Soc.*, **103**, 4646 (1981).
438 Hanack, M., Collins, C. J., Stutz, H., and Benjamin, B. M., *J. Am. Chem. Soc.*, **103**, 2356 (1981).
439 Franke, W., and Schwarz, H., *J. Org. Chem.*, **46**, 2806 (1981).
440 Murphy, W. S., and Wattanasin, S., *J. Chem. Soc., Perkins Trans. 1*, **1981**, 2920.
441 Kagan, J., Agdeppa, D. A., Chang, A. I., Chen, S.-A., Harmata, M. A., Melnick, B., Patel, G., Poorker, C., Singh, S. P., Watson, W. H., Chen, J. S., and Zabel, V., *J. Org. Chem.*, **46**, 2916 (1981).
442 Rees, J. C., and Whittaker, D., *J. Chem. Soc., Perkin Trans. 2*, **1981**, 948.
443 Rees, J. C., and Whittaker, D., *J. Chem. Soc., Perkin Trans. 2*, **1981**, 953.
444 Gerdes, J. M., Norman, A. W., and Okamura, W. H., *J. Org. Chem.*, **46**, 599 (1981).
445 Karpf, M., and Djerassi, C., *J. Am. Chem. Soc.*, **103**, 302 (1981).
446 Lee-Ruff, E., Hopkinson, A. C., and Dao, L. H., *Can. J. Chem.*, **59**, 1675 (1981).
447 Lee-Ruff, E., Hopkinson, A. C., Kazarians-Moghaddam, H., Duperrouzel, P., Gupta, B., and Katz, M., *Tetrahedron Lett.*, **22**, 4917 (1981).
448 Becker, H.-D., and Gustafsson, K., *J. Org. Chem.*, **46**, 4077 (1981).
449 Ladika, M., Bregovec, I., and Sunko, D. E., *J. Am. Chem. Soc.*, **103**, 1285 (1981).
450 Edward, J. T., Cooke, E., and Paradellis, T. C., *Can. J. Chem.*, **59**, 597 (1981).
451 Belinka, B. A., Hassner, J., and Hendler, J. M., *J. Org. Chem.*, **46**, 631 (1981).
452 Trost, B. M., Vaultier, M., and Santiago, M. L., *J. Am. Chem. Soc.*, **102**, 7929 (1980).

453 Cushman, M., Choong, T.-C., Valko, J. T., and Koleck, M. P., *J. Org. Chem.*, **45**, 5067 (1980).
454 Smith, A. B., Toder, B. H., Branca, S. J., and Dieter, R. K., *J. Am. Chem. Soc.*, **103**, 1996 (1981).
455 Smith, A. B., and Dieter, R. K., *J. Am. Chem. Soc.*, **103**, 2009 (1981).
456 Smith, A. B., and Dieter, R. K., *J. Am. Chem. Soc.*, **103**, 2017 (1981).
457 Bolster, J., and Kellogg, R. M., *J. Org. Chem.*, **45**, 4804 (1980).
458 Snider, B. B., *J. Org. Chem.*, **46**, 3155 (1981).
459 Kobayashi, K., and Mutai, K., *Chem. Lett.*, **1981**, 1105.
460 Sauve, J. P., and Lozac'h, N., *Bull. Soc. Chim. Fr. II*, **1980**, 577.
461 Bruzik, K., and Stec, W. J., *J. Org. Chem.*, **46**, 1618 (1981).
462 Bruzik, K., and Stec, W. J., *J. Org. Chem.*, **46**, 1625 (1981).
463 Fichtner, M. W., and Haley, N. F., *J. Org. Chem.*, **46**, 3141 (1981).
464 Katritzky, A. R., and Manzo, R. H., *J. Chem. Soc., Perkin Trans. 2*, **1981**, 571.
465 Hirao, K., *Yakugaku Zasshi*, **100**, 473 (1980); *Chem. Abs.*, **93**, 203420 (1980).
466 Christl, M., *Angew. Chem. Int. Ed.*, **20**, 529 (1981).
467 Thappa, R. K., Taneja, S. C., Dhar, K. L., and Atal, C. K., *Indian Perfum.*, **23**, 172 (1979); *Chem. Abs.*, **93**, 204861 (1980).
468 Kirmse, W., Knöpfel, N., Loosen, K., Siegfried, R., and Wroblowsky, H.-J., *Chem. Ber.*, **114**, 1187 (1981).
469 Kirmse, W., Hartmann, M., Siegfried, R., Wroblowsky, H.-J., Zang, B., and Zellmer, V., *Chem. Ber.*, **114**, 1793 (1981).
470 Lajunen, M., and Lyytikainen, H., *Acta Chem. Scand.*, **35A**, 131 (1981).
471 Yates, P., and Langford, G. E., *Can. J. Chem.*, **59**, 344 (1981).
472 Bobyleva, A. A., Belikova, N. A., Kalinichenko, A. N., Baryshnikov, A. T., Dubitskaya, N. F., Pehk, T., Lippmaa, E., and Plate, A. F., *Zh. Org. Khim.*, **16**, 1645 (1980); *Chem. Abs.*, **94**, 29747 (1981).
473 Bird, C. W., and Yeung, C. Y. A., *Chem. Ind. (London)*, **1981**, 95.
474 Razus, A. C., Wertheimer, V., Glatz, A. M., Arvay, Z., and Badea, F., *Rev. Roum. Chim.*, **26**, 457 (1981).
475 Badea, F., Glatz, A. M., and Razus, A. C., *Rev. Roum. Chim.*, **26**, 629 (1981).
476 Coxon, J. M., and Steel, P. J., *Aust. J. Chem.*, **33**, 2455 (1980).
477 Farcasin, M., Hagaman, E. W., Wenkert, E., and Schleyer, P. von R., *Tetrahedron Lett.*, **22**, 1501 (1981).
478 Bagrii, E. I., Frid, T. Yu., and Sanin, P. I., *Neftekhimiya*, **20**, 812 (1980); *Chem. Abs.*, **94**, 156400 (1981).
479 Margosian, D., and Kovacic, P., *J. Org. Chem.*, **46**, 877 (1981).
480 Allard, B., Casadevall, A., Casadevall, E., and Largean, C., *Nouveau J. Chim.*, **4**, 539 (1980).
481 Jäggi, F. J., and Ganter, C., *Helv. Chim. Acta*, **63**, 2087 (1980).
482 Nordlander, J. E., Haky, J. E., and Landino, J. P., *J. Am. Chem. Soc.*, **102**, 7487 (1980).
483 Nordlander, J. E., and Haky, J. E., *J. Am. Chem. Soc.*, **103**, 1518 (1981).
484 Tobe, Y., Terashima, K., Sakai, Y., and Odaira, Y., *J. Am. Chem. Soc.*, **103**, 2307 (1981).
485 Kakinchi, K., Tsugaru, T., Tobe, Y., and Odaira, Y., *J. Org. Chem.*, **46**, 4204 (1981).
486 Hollowood, F. S., McKervey, M. A., Hamilton, R., and Rooney, J. J., *J. Org. Chem.*, **45**, 4954 (1980).
487 Doecke, C. W., and Garratt, P. J., *Tetrahedron Lett.*, **22**, 1051 (1981).
488 Ruppert, J. F., and White, J. D., *J. Am. Chem. Soc.*, **103**, 1808 (1981).
489 Mehta, G., and Srikrishna, A., *J. Org. Chem.*, **46**, 1730 (1981).
490 Mehta, G., Pandey, P. N., Usha, R., and Venkatesan, K., *Indian J. Chem.*, **20B**, 177 (1981).
491 Mehta, G., Chaudhury, B., Singh, V. K., Usha, R., Varughese, K. I., and Venkatesan, K., *Indian J. Chem.*, **20B**, 186 (1981).
492 Mehta, G., and Srikrishna, A., *Indian J. Chem.*, **19B**, 997 (1980).
493 Klunder, A. J. H., van Seters, A. J. C., Buza, M., and Zwanenburg, B., *Tetrahedron*, **37**, 1601 (1981).
494 Satyanarayana, G. O. S. V., Kanjilal, P. R., and Ghatak, U. R., *J. Chem. Soc., Chem. Commun.*, **1981**, 746.
495 Bastable, J. W., Cooper, A. J., Dunkin, I. R., Hobson, J. D., and Riddell, W. D., *J. Chem. Soc., Perkin Trans. 1*, **1981**, 1339.
496 Bastable, J. W., Dunkin, I. R., and Hobson, J. D., *J. Chem. Soc., Perkin Trans. 1*, **1981**, 1346.
497 Hayes, P. C., and Jones, G., *Tetrahedron Lett.*, **22**, 3897 (1981).
498 Bolster, J. M., Hogeveen, H., Kellogg, R. M., and Zwart, L., *J. Am. Chem. Soc.*, **103**, 3955 (1981).
499 Gillissen, H. M. J., Schipper, P., and Buck, H. M., *Recl. Trav. Chim. Pays-Bas*, **99**, 346 (1980).
500 Dev, S., *Acc. Chem. Res.*, **14**, 82 (1981).
501 Ayer, W. A., and Browne, L. M., *Tetrahedron*, **37**, 2199 (1981).
502 Kametani, T., Matsumoto, H., Honda, T., and Fukumoto, K., *Tetrahedron Lett.*, **22**, 2379 (1981).

[503] Grieco, P. A., Ferriño, S., Vidari, G., Huffman, J. C., and Williams, E., *Tetrahedron Lett.*, **22**, 1071 (1981).
[504] Ghatak, U. R., Ghosh, S., and Sanyal, B., *J. Chem. Soc., Perkin Trans. 1*, **1980**, 2881.
[505] Pirrung, M. C., *J. Am. Chem. Soc.*, **103**, 82 (1981).
[506] Hayano, K., Ohfune, Y., Shirahama, H., and Matsumoto, T., *Helv. Chim. Acta*, **64**, 1347 (1981).
[507] Capon, R., Ghisalberti, E. L., Jefferies, P. R., and Skelton, B. W., *Tetrahedron*, **37**, 1613 (1981).
[508] Sánchez, I. H., Yáñez, R., Enriquez, R., and Joseph-Nathan, P., *J. Org. Chem.*, **46**, 2818 (1981).
[509] Harref, A. B., Bernardini, A., Fkih-Tetouani, S., Jacquier, R., and Viallefont, P., *J. Chem. Res.*, (*S*), **1981**, 372.
[510] Ward, R. S., Pelter, A., Jack, I. R., Satyanarayana, P., Rao, S. V. G., and Subrahmanyam, P., *Tetrahedron Lett.*, **22**, 4111 (1981).
[511] Viger, A., Coustal, S., and Marquet, A., *J. Am. Chem. Soc.*, **103**, 451 (1981).
[512] Frappier, F., Hull, W. E., and Lukacs, G., *J. Org. Chem.*, **46**, 4314 (1981).
[513] Akporiaye, D. E., Farrant, R. D., and Kirk, D. N., *J. Chem. Res.* (*S*), **1981**, 210.
[514] Fetizon, M., and Sozzi, G., *Tetrahedron*, **37**, 61 (1981).
[515] Anjaneyulu, A. S. R., and Rao, M. N., *Indian J. Chem.*, **19B**, 634 (1980).
[516] Anjaneyulu, A. S. R., Rao, M. N., Sree, A., and Murty, V. S., *Indian J. Chem.*, **19B**, 735 (1980).
[517] Yokoyama, Y., Moriyama, Y., Tsuyuki, T., Takahashi, T., Itai, A., and Iitaka, Y., *Bull. Chem. Soc. Jpn.*, **53**, 2971 (1980).
[518] Yokoyama, Y., Moriyama, Y., Tsuyuki, T., and Takahashi, T., *Bull. Chem. Soc. Jpn.*, **54**, 234 (1981).
[519] Kikuchi, T., Niwa, M., Takayama, M., Yokoi, T., and Shingu, T., *Chem. Pharm. Bull.*, **28**, 1999 (1980).
[520] Seo, S., Tomita, Y., and Tori, K., *J. Am. Chem. Soc.*, **103**, 2075 (1981).
[521] Teutsch, G., Lang, C., Smolik, R., Mornon, J. P., and Delettre, J., *Tetrahedron Lett.*, **22**, 327 (1981).
[522] Acklin, G., and Graf, W., *Helv. Chim. Acta*, **63**, 2342 (1980).
[523] Nakano, T., Haces, A., Martin, A., and Rojas, A., *J. Chem. Soc., Perkin Trans. 1*, **1981**, 2075.
[524] Peters, J. A. M., Van Vliet, N. P., and Zeelen, F. J., *Recl. Trav. Chim. Pays-Bas*, **100**, 226 (1981).
[525] Maione, A. M., Romeo, A., Cerrini, S., Fedeli, W., and Mazza, F., *Tetrahedron*, **37**, 1407 (1981).
[526] Kočovský, P., and Tureček, F., *Tetrahedron Lett.*, **22**, 2699 (1981).
[527] Suginome, H., Furusaki, A., Kato, K., Maeda, N., and Yonebayashi, F., *J. Chem. Soc., Perkin Trans. 1*, **1981**, 236.
[528] Binkley, R. W., and Jarrell, H. F., *J. Org. Chem.*, **46**, 4564 (1981).
[529] Jacobsen, S., and Mols, O., *Acta Chem. Scand.*, **35B**, 163 (1981).
[530] Jacobsen, S., and Mols, O., *Acta Chem. Scand.*, **35B**, 169 (1981).
[531] Edwards, O. E., *Can. J. Chem.*, **59**, 3039 (1981).
[532] Janssen, H.-W., Mohr, S., and Mondon, A., *Chem. Ber.*, **114**, 2158 (1981).
[533] Jones, W. M., *Org. Chem.* (*N.Y.*) (Rearrangements Ground Excited States, part 1), **42**, 95 (1980).
[534] Bargon, J., Tanaka, K., and Yoshimine, M., *Comput. Methods Chem.*, (*Proc. Int. Symp.*), **1979**, 239.
[535] Tomioka, H., Okuno, H., Kondo, S., and Izawa, Y., *J. Am. Chem. Soc.*, **102**, 7123 (1980).
[536] Tomioka, H., Ueda, H., Kondo, S., and Izawa, Y., *J. Am. Chem. Soc.*, **102**, 7817 (1980).
[537] Sinha, G., and Ghatak, U. R., *Indian J. Chem.*, **20B**, 411 (1981).
[538] Blaustein, M. A., and Berson, J. A., *Tetrahedron Lett.*, **22**, 1081 (1981).
[539] Wiberg, K. B., Olli, L. K., Golembeski, N., and Adams, R. D., *J. Am. Chem. Soc.*, **102**, 7467 (1980).
[540] Wolff, S., and Agosta, W. C., *J. Chem. Soc., Chem. Commun.*, **1981**, 118.
[541] Regitz, M., Bennyarto, F., and Heydt, H., *Justus Liebigs Ann. Chem.*, **1981**, 1044.
[542] Stolyarov, B. V., and Krylov, A. I., *Zh. Org. Khim.*, **16**, 1802 (1980); *Chem. Abs.*, **94**, 46453 (1981).
[543] White, E. H., Lewis, C. P., Ribi, M. A., and Ryan, T. J., *J. Org. Chem.*, **46**, 552 (1981).
[544] de Groot, A., and Jansen, B. J. M., *Tetrahedron Lett.*, **22**, 887 (1981).
[545] Jackson, W. P., Ley, S. V., and Morton, J. A., *Tetrahedron Lett.*, **22**, 2601 (1981).
[546] Oae, S., and Numata, T., *Isot. Org. Chem.* (Isot. Cationic React.), **5**, 45 (1980); *Chem. Abs.*, **94**, 46279 (1981).
[547] Warren, S., *Chem. Ind.* (*London*), **1980**, 824.
[548] Wladislaw, B., Marzorati, L., and Andrade, M. A. C., *An. Acad. Bras. Cienc.*, **52**, 11 (1980); *Chem. Abs.*, **94**, 29885 (1981).
[549] Uchida, Y., Kobayashi, Y., and Kozuka, S., *Bull. Chem. Soc. Jpn.*, **54**, 1781 (1981).
[550] Watanabe, M., Nakamori, S., Hasegawa, H., Shirai, K., and Kumamoto, T., *Bull. Chem. Soc. Jpn.*, **54**, 817 (1981).
[551] Gawley, R. E., Termine, E. J., and Onan, K. D., *J. Chem. Soc., Chem. Commun.*, **1981**, 568.
[552] Aginskii, V. N., Lyamin, I. A., and Chebotarev, O. V., *Zh. Vses. Khim. O-va.*, **25**, 597 (1980); *Chem. Abs.*, **94**, 46922 (1981).

553 Imamoto, T., Yokoyama, H., and Yokoyama, M., *Tetrahedron Lett.*, **22**, 1803 (1981).
554 Esteban, S., and Marinas, J. M., *Afinidad*, **38**, 19 (1981); *Chem. Abs.*, **94**, 174270 (1981).
555 Costa, A., Deya, P. M., Sinisterra, J. V., and Marinas, J. M., *Afinidad*, **38**, 225 (1981); *Chem. Abs.*, **95**, 132026 (1981).
556 Suginome, H., Maeda, N., Takahashi, Y., and Miyata, N., *Bull. Chem. Soc. Jpn.*, **54**, 846 (1981).
557 Zielinski, W., *Pol. J. Chem.*, **54**, 745 (1980); *Chem. Abs.*, **94**, 64863 (1981).
558 Shafuillah, G. M. A., *J. Indian Chem. Soc.*, **57**, 762 (1980).
559 Gelin, S., *J. Heterocycl. Chem.*, **18**, 535 (1981).
560 Kiseleva, N. I., Baskakov, Y. A., Faddeeva, M. I., and Putsykin, Y. G., *Zh. Org. Khim.*, **17**, 343 (1981); *Chem. Abs.*, **95**, 42725 (1981).
561 Maffrand, J. P., Boigegrain, R., Courregelongue, J., Ferrand, G., and Frehel, D., *J. Heterocycl. Chem.*, **18**, 727 (1981).
562 Rodewald, W. J., and Zaworska, A., *Pol. J. Chem.*, **54**, 1147 (1980).
563 Singh, A. K., and Dhar, D. N., *Z. Naturforsch.*, **35B**, 1575 (1980); *Chem. Abs.*, **95**, 25395 (1981).
564 Ito, K., and Oka, Y., *Chem. Pharm. Bull.*, **28**, 2862 (1980).
565 Brunovlenskaya, I. I., Alekseeva, A. B., and Skvarchenko, V. R., *Zh. Org. Khim.*, **16**, 2141 (1980); *Chem. Abs.*, **95**, 24626 (1981).
566 Krow, G. R., *Tetrahedron*, **37**, 1283 (1981).
567 Nitta, M., Kasahara, I., and Kobayashi, T., *Bull. Chem. Soc. Jpn.*, **54**, 1275 (1981).
568 Dave, V., Stothers, J. B., and Warnhoff, E. W., *Can. J. Chem.*, **58**, 2666 (1980).
569 Krow, G. R., *Tetrahedron*, **37**, 2697 (1981).
570 Irismetov, M. P., Goryaev, M. I., and Alibaeva, K. A., *Izv. Akad. Nauk Kaz. SSR, Ser. Khim.*, **1980**, 45; *Chem. Abs.*, **95**, 7586 (1981).
571 Venugopal, V. K., Rao, N., Rahman, M. F., Bhalerao, U. T., and Thyagarajan, G., *Indian J. Chem.*, **20B**, 156 (1981).
572 Iskander, M. L., *Z. Phys. Chem.* (*Leipzig*), **262**, 76 (1981); *Chem. Abs.*, **94**, 156135 (1981).
573 Saikachi, H., Kitagawa, T., Nasu, A., and Sasaki, H., *Chem. Pharm. Bull.*, **29**, 237 (1981); *Chem. Abs.*, **95**, 23930 (1981).
574 Harger, M. J. P., and Stephen, M. A., *J. Chem. Soc., Perkin Trans. 1*, **1981**, 736.
575 Okada, Y., Tsuda, Y., and Yagyu, M., *Chem. Pharm. Bull.*, **28**, 2254 (1980).
576 Avotins, F., Gudriniece, E., Bizdena, E. O., and Zaitseva, V. E., *Latv. PSR Zinat. Akad. Vestis, Kim. Ser.*, **1981**, 229; *Chem. Abs.*, **95**, 97141 (1981).
577 Bayder, A. E., and Boyd, G. V., *J. Chem. Soc., Perkin Trans. 1*, **1981**, 2871.
578 Chantegrel, B., and Gelin, S., *Synthesis*, **1981**, 315.
579 Sundaram, M. G., *J. Labelled Compd. Radiopharm.*, **18**, 489 (1981).
580 Peterson, L. H., Douglas, A. W., and Tolman, R. L., *J. Heterocycl. Chem.*, **18**, 659 (1981).
581 Bisagni, E., Lhoste, J. M., and Nguyen, C. H., *J. Heterocycl. Chem.*, **18**, 755 (1981).
582 Li, Y.-G., Ma, L.-T., and Wang, X., *Yao Hsueh Hsueh Pao*, **16**, 230 (1981); *Chem. Abs.*, **95**, 150510 (1981).
583 Michinori, W., Kitajima, Y., and Izumuja, N., *Synthesis*, **1981**, 266.
584 Tanaka, T., *Kagaku Sosetsu*, **32**, 135 (1981); *Chem. Abs.*, **95**, 167899 (1981).
585 Jackson, W. G., and Sargeson, A. M., *Org. Chem.* (*N.Y.*) (Rearrangements Ground Excited States, part 2), **42**, 273 (1980).
586 Halpern, J., *Inorg. Chim. Acta*, **50**, 11 (1981).
587 Halpern, J., *Adv. Chem. Ser.*, **1980**, 191.
588 Hamilton, R., Mitchell, T. R. B., McIlgorm, E. A., Rooney, J. J., and McKervey, M. A., *J. Chem. Soc., Chem. Commun.*, **1981**, 686.
589 Scott, A. I., Kang, J., Dowd, P., and Trivedi, B. K., *Bioorg. Chem.* **9**, 227 (1980); *Chem. Abs.*, **93**, 167331 (1980).
590 Tada, M., Miura, K., Okabe, M., Seki, S., and Mizukami, H., *Chem. Lett.*, **1981**, 33.
591 Beer, H. R., Bigler, P., von Philipsborn, W., and Salzer, A., *Inorg. Chim. Acta*, **53**, L49 (1981).
592 McGlinchey, M. J., *Inorg. Chim. Acta*, **49**, 125 (1981).
593 Doyle, M. P., and van Leusen, D., *J. Am. Chem. Soc.*, **103**, 5917 (1981).
594 Morgans, D. J., Sharpless, K. B., and Traynor, S. G., *J. Am. Chem. Soc.*, **103**, 462 (1981).
595 Suzuki, H., Yashima, H., Hirose, T., Takahashi, M., Moro-oka, Y., and Ikawa, T., *Tetrahedron Lett.*, **21**, 4927 (1980).
596 Kafuku, K., Kinoshita, K., and Nakatsu, K., *J. Chem. Soc., Chem. Commun.*, **1981**, 22.
597 Fabian, B. D., and Labinger, J. A., *J. Organomet. Chem.*, **204**, 387 (1981).
598 Le Bozec, H., Gorgues, A., and Dixneuf, P. H., *Inorg. Chem.*, **20**, 2486 (1981).

599 Ustynyuk, N. A., Lokshin, B. V., Oprunenko, Y. F., Roznyatovsky, V. A., Luzikov, Y. N., and Ustynyuk, Y. A., *J. Organomet. Chem.*, **202**, 279 (1980).
600 Green, M., Norman, N. C., and Orpen, A. G., *J. Organomet. Chem.*, **221**, C11 (1981).
601 Halasz, I., and Gati, G., *Magy. Kem. Foly.*, **87**, 125 (1981); *Chem. Abs.*, **95**, 114478 (1981).
602 Drew, M. G. B., Regan, C. M., and Nelson, S. M., *J. Chem. Soc., Dalton Trans.*, **1981**, 1034.
603 Kanai, H., Kushi, K., Sakanoue, K., and Kishimoto, N., *Bull. Chem. Soc. Jpn.*, **53**, 2711 (1980).
604 Prashad, M., Seth, M., and Bhaduri, A. P., *Indian J. Chem.*, **19B**, 393 (1980).
605 Erker, G., and Rosenfeldt, F., *Tetrahedron Lett.*, **22**, 1379 (1981).
606 Loots, M. J., Dayrit, F. M., and Schwartz, J., *Bull. Soc. Chim. Belg.*, **89**, 897 (1980).
607 Thorn, D. L., and Tulip, T. H., *J. Am. Chem. Soc.*, **103**, 5984 (1981).
608 Levisalles, J., Rose-Munch, F., Rudler, H., Daran, J.-C., and Jeannin, Y., *J. Chem. Soc., Chem. Commun.*, **1981**, 1057.
609 Sen, A., and Lai, T.-W., *J. Am. Chem. Soc.*, **103**, 4627 (1981).
610 Suzuki, M., Oda, Y., and Noyori, R., *Tetrahedron Lett.*, **22**, 4413 (1981).
611 Trost, B. M., and Runge, T. A., *J. Am. Chem. Soc.*, **103**, 2485 (1981).
612 Mandai, T., Hashio, S., Goto, J., and Kawada, M., *Tetrahedron Lett.*, **22**, 2187 (1981).
613 Tamaru, Y., Kagotani, M., and Yoshida, Z., *Tetrahedron Lett.*, **22**, 4245 (1981).
614 Williams, J. L., and Rettig, M. F., *Tetrahedron Lett.*, **22**, 385 (1981).
615 Golding, B. T., and Pierpoint, C., *J. Chem. Soc., Chem. Commun.*, **1981**, 1030.
616 Clemens, P. R., Hughes, R. P., and Margerum, L. D., *J. Am. Chem. Soc.*, **103**, 2428 (1981).
617 Abel, E. W., Booth, M., Orrell, K. G., and Pring, G. M., *J. Chem. Soc., Chem. Commun.*, **1981**, 29.
618 Courtot, P., Pichon, R., and Salaün, J.-Y., *J. Chem. Soc., Chem. Commun.*, **1981**, 542.
619 Guo, X.-X., Tao, L.-X., Xu, Y.-D., Ruan, Z.-K., Huang, J.-S., Shi, Y.-Z., Wang, X.-M., and Pan, L.-R., *Ts'ui Hua Hsueh Pao*, **1**, 119 (1980); *Chem. Abs.*, **94**, 120613 (1981).
620 Padwa, A., Blacklock, T. J., and Loza, R., *J. Am. Chem. Soc.*, **103**, 2404 (1981).
621 Muccino, R. R., and Wasiowich, C. A., *J. Labelled Cmpd. Radiopharm.*, **17**, 463 (1980).
622 Padwa, A., and Rieker, W. F., *Tetrahedron Lett.*, **22**, 1487 (1981).
623 Katsushima, T., Yamaguchi, R., Kawanisi, M., and Osawa, E., *Bull. Chem. Soc. Jpn.*, **53**, 3313 (1980).
624 Bingmann, H., Beck, A., Fritz, H., and Prinzbach, H., *Chem. Ber.*, **114**, 1679 (1981).
625 Ogibin, Y. M., Velibekova, D. S., Troyanskii, E. I., and Nikishin, G. I., *Izv. Akad. Nauk SSSR, Ser. Khim.*, **1981**, 633; *Chem. Abs.*, **95**, 61096 (1981).
626 Ishikawa, M., Iyoda, J., Ikeda, H., Kotake, K., Hashimoto, T., and Kumada, M., *J. Am. Chem. Soc.*, **103**, 4845 (1981).
627 Shimp, L. A., Morrison, J. A., Gurak, J. A., Chinn, J. W., and Lagow, R. J., *J. Am. Chem. Soc.*, **103**, 5951 (1981).
628 Barton, D. H. R., Motherwell, R. S. H., and Motherwell, W. B., *J. Chem. Soc., Perkin Trans. 1*, **1981**, 2363.
629 Nastasi, M., and Streith, J., *Org. Chem. (N.Y.)* (Rearrangements Ground Excited States, part 3), **42**, 445 (1980); *Chem. Abs.*, **94**, 102255 (1981).
630 Dolbier, W. R., *Acc. Chem. Res.*, **14**, 195 (1981).
631 Dolbier, W. R., Sellers, S. F., Al-Sader, B. H., and Elsheimer, S., *J. Am. Chem. Soc.*, **103**, 715 (1981).
632 Dolbier, W. R., Sellers, S. F., Al-Sader, B. H., and Fielder, T. H., *J. Am. Chem. Soc.*, **103**, 717 (1981).
633 Dolbier, W. R., Sellers, S. F., and Smart, B. E., *Tetrahedron Lett.*, **22**, 2953 (1981).
634 Danheiser, R. L., Martinez-Davila, C., Auchus, R. J., and Kadonaga, J. T., *J. Am. Chem. Soc.*, **103**, 2443 (1981).
635 Hudlicky, T., Koszyk, F. F., Kutchan, T. M., and Sheth, J. P., *J. Org. Chem.*, **45**, 5020 (1980).
636 Marino, J. P., and Ferro, M. P., *J. Org. Chem.*, **46**, 1912 (1981).
637 Hirakawa, K., and Nosaka, T., *J. Chem. Soc., Perkin Trans. 1*, **1980**, 2835.
638 Boche, G., and Wintermayr, H., *Angew. Chem. Int. Ed.*, **20**, 874 (1981).
639 Padwa, A., Chou, C. S., and Rieker, W. F., *J. Org. Chem.*, **45**, 4555 (1980).
640 Dehmlow, E. V., and Naser-ud-Din, *Chem. Ber.*, **114**, 1546 (1981).
641 Moiseenkov, A. M., Czeskis, B. A., Rudashevskaya, T. Y., Nesmeyanova, O. A., and Semenovsky, A. V., *Tetrahedron Lett.*, **22**, 151 (1981).
642 Zoch, H.-H., Szeimies, G., Römer, R., and Schmitt, R., *Angew. Chem. Int. Ed.*, **20**, 877 (1981).
643 Lokensgard, D. M., Dougherty, D. A., Hilinski, E. F., and Berson, J. A., *Proc. Natl. Acad. Sci. U.S.A.*, **77**, 3090 (1980); *Chem. Abs.*, **94**, 14876 (1981).
644 Lasne, M.-C., and Ripoll, J.-L., *Bull. Soc. Chim. Fr. (II)*, **1981**, 340.
645 Sugihara, Y., Yamato, A., and Murata, I., *Tetrahedron Lett.*, **22**, 3257 (1981).
646 Faustini, F., De Munari, S., and Panzeri, A., *Tetrahedron Lett.*, **22**, 4533 (1981).

[647] Mandel'shtam, T. V., Kolesova, S. V., Polina, T. V., Solomentsev, V. V., and Osmolovskaya, N. S., *Zh. Org. Khim.*, **16**, 1186 (1980); *Chem. Abs.*, **93**, 185624 (1980).
[648] Morizawa, Y., Hiyama, T., and Nozaki, H., *Tetrahedron Lett.*, **22**, 2297 (1981).
[649] Trost, B. M., and Jungheim, L. N., *J. Am. Chem. Soc.*, **102**, 7910 (1980).
[650] Alonso, M. E., and Morales, A., *J. Org. Chem.*, **45**, 4530 (1980).
[651] Gillespie, R. J., Murray-Rust, J., Murray-Rust, P., and Porter, A. E. A., *Tetrahedron*, **37**, 743 (1981).
[652] Smith, K. M., Martynenko, Z., and Tabba, H. D., *Tetrahedron Lett.*, **22**, 1291 (1981).
[653] Murato, K., Wolf, H. R., and Jeger, O., *Helv. Chim. Acta*, **63**, 2212 (1980).
[654] Nakamura, N., Schweizer, W. B., Frei, B., Wolf, H. R., and Jeger, O., *Helv. Chim. Acta*, **63**, 2230 (1980).
[655] Tsutsumi, K., and Wolf, H. R., *Helv. Chim. Acta*, **63**, 2370 (1980).
[656] de Weck, G., and Wolf, H. R., *Helv. Chim. Acta*, **64**, 224 (1981).
[657] Alder, A. P., Wolf, H. R., and Jeger, O., *Helv. Chim. Acta*, **64**, 198 (1981).
[658] Ross, W. J., Todd, A., Clark, B. P., Morgan, S. E., and Baldwin, J. E., *Tetrahedron Lett.*, **22**, 2207 (1981).
[659] Cormier, R. A., and Francis, M. D., *Synth. Commun.*, **11**, 365 (1981).
[660] Smeets, F. L. M., Thigs, L., and Zwanenburg, B., *Tetrahedron*, **36**, 3269 (1980).
[661] Eberbach, W., and Carré, J. C., *Chem. Ber.*, **114**, 1027 (1981).
[662] Akgün, E., Glinski, M. B., Dhawan, K. L., and Durst, T., *J. Org. Chem.*, **46**, 2730 (1981).
[663] Brady, B. A., Geoghegan, M., McMurtrey, K. D., and O'Sullivan, W. I., *J. Chem. Soc., Perkin Trans. 1*, **1981**, 119.
[664] Amsterdamsky, C., and Rigaudy, J., *Tetrahedron Lett.*, **22**, 1403 (1981).
[665] Flowers, M. C., and Honeyman, M. R., *J. Chem. Soc., Faraday Trans. 1*, **76**, 2290 (1980).
[666] van Zon, A., and Huis, R., *Recl. Trav. Chim. Pays-Bas*, **100**, 425 (1981).
[667] de Montellano, P. R. O., and Kunze, K. L., *J. Am. Chem. Soc.*, **102**, 7373 (1980).
[668] Simmons, H. E., and Maggio, J. E., *Tetrahedron Lett.*, **22**, 287 (1981).
[669] Paquette, L. A., and Vazeux, M., *Tetrahedron Lett.*, **22**, 291 (1981).
[670] Rubottom, G. M., and Marrero, R., *Synth. Commun.*, **11**, 505 (1981).
[671] Fleming, I., and Williams, R. V., *J. Chem. Soc., Perkin Trans. 1*, **1981**, 684.
[672] Fleming, I., and Michael, J. P., *J. Chem. Soc., Perkin Trans. 1*, **1981**, 1549.
[673] Adam, W., and Rodriguez, A., *Tetrahedron Lett.*, **22**, 3505 (1981).
[674] Meier, H., and Kolshorn, H., *Z. Naturforsch.*, **35B**, 1040 (1980); *Chem. Abs.*, **94**, 3515 (1981).
[675] Kurebayashi, Y., Ishikawa, Y., Terao, Y., and Sekiya, M., *Tetrahedron Lett.*, **22**, 123 (1981).
[676] Micetich, R. G., Fortier, R. A., and Wolfert, P., *Can. J. Chem.*, **59**, 1018 (1981).
[677] Nudelman, A., Haram, T. E., and Shakked, Z., *J. Org. Chem.*, **46**, 3026 (1981).
[678] Cox, P. J., MacKenzie, N. E., and Thomson, R. H., *Tetrahedron Lett.*, **22**, 2221 (1981).
[679] Bigot, B., and Roux, D., *J. Org. Chem.*, **46**, 2872 (1981).
[680] Tanaka, H., Osamura, Y., Matsushita, T., and Nishimoto, K., *Bull. Chem. Soc. Jpn.*, **54**, 1293 (1981).
[681] Hyatt, J. A., *J. Org. Chem.*, **46**, 3953 (1981).
[682] De Sarlo, F., Guarna, A., Mascagni, P., Carrié, R., and Guenot, P., *J. Chem. Soc., Perkin Trans. 1*, **1981**, 1367.
[683] Cambie, R. C., Hayward, R. C., Jurlina, J. L., Rutledge, P. S., and Woodgate, P. D., *Aust. J. Chem.*, **34**, 1349 (1981).
[684] Klaus, R. O., and Ganter, C., *Helv. Chim. Acta*, **63**, 2559 (1980).
[685] Hammer, C. F., and Weber, J. D., *Tetrahedron*, **37**, 2173 (1981).
[686] Blaskó, G., Murugesan, N., Freyer, A. J., Gula, D. J., Sener, B., and Shamma, M., *Tetrahedron Lett.*, **22**, 3139 (1981).
[687] Bates, G. S., and Varelas, M. A., *Can. J. Chem.*, **58**, 2562 (1980).
[688] Pellacani, L., Persia, F., and Tardella, P. A., *Tetrahedron Lett.*, **21**, 4967 (1980).
[689] Hasegawa, H., Suzuki, M., Arai, N., Fujita, M., and Watanabe, K., *Kenkyu Hokoku-Asahi Garasu Kogyo Gijutsu Shoreikai*, **37**, 295 (1980); *Chem. Abs.*, **95**, 41979 (1981).
[690] Isomura, K., Ayabe, G.-I., Hatano, S., and Taniguchi, H., *J. Chem. Soc., Chem. Commun.*, **1980**, 1252.
[691] Liu, M. T. H., Palmer, G. E., and Chishti, N. H., *J. Chem. Soc., Perkin Trans. 2*, **1981**, 53.
[692] Quast, H., Heuschmann, M., and Abdel-Rahman, M. O., *Justus Liebigs Ann. Chem.*, **1981**, 943.
[693] Quast, H., and Heuschmann, M., *Justus Liebigs Ann. Chem.*, **1981**, 967.
[694] Doering, W. von E., and Mastrocola, A. R., *Tetrahedron*, **37**, (Supplement to Issue 9), 329 (1981).
[695] Trautmann, W., and Musso, H., *Chem. Ber.*, **114**, 982 (1981).
[696] Schwarz, W., Trautmann, W., and Musso, H., *Chem. Ber.*, **114**, 990 (1981).

[697] Gibson, T., *J. Org. Chem.*, **46**, 1073 (1981).
[698] Walsh, R., Martin, H.-D., Kunze, M., Oftring, A., and Beckhaus, H.-D., *J. Chem. Soc., Perkin Trans. 2*, **1981**, 1076.
[699] Wolters, F. C., Flowers, M. C., and Rabinovitch, B. S., *J. Phys. Chem.*, **85**, 589 (1981).
[700] O'Leary, M. A., Richardson, G. W., and Wege, D., *Tetrahedron*, **37**, 813 (1981).
[701] Kametani, T., Honda, T., Matsumoto, H., and Fukumoto, K., *J. Chem. Soc., Perkin Trans. 1*, **1981**, 1383.
[702] Bestmann, H. J., Wilhelm, E., and Schmid, G., *Angew. Chem. Int. Ed.*, **19**, 1012 (1980).
[703] Lorenzi-Riatsch, A., Nakashita, Y., and Hesse, M., *Helv. Chim. Acta*, **64**, 1854 (1981).
[704] Kimura, K., Fukuda, Y., Negoro, T., and Odaira, Y., *Bull. Chem. Soc. Jpn.*, **54**, 1901 (1981).
[705] Goldschmidt, Z., and Antebi, S., *J. Organomet. Chem.*, **206**, C1 (1981).
[706] Muto, S., and Bruice, T. C., *J. Am. Chem. Soc.*, **102**, 7379 (1980).
[707] McCapra, F., and Long, P. V., *Tetrahedron Lett.*, **22**, 3009 (1981).
[708] Baumstark, A. L., and Wilson, C. E., *Tetrahedron Lett.*, **22**, 4363 (1981).
[709] van Arendonk, R. J. F. M., and Laarhoven, W. H., *Recl. Trav. Chim. Pays-Bas*, **100**, 263 (1981).
[710] Ongania, K.-H., and Hohenlohe-Oehringen, K., *Chem. Ber.*, **114**, 1203 (1981).
[711] Sano, T., Horiguchi, Y., and Tsuda, Y., *Heterocycles*, **16**, 355 (1981); *Chem. Abs.*, **95**, 41973 (1981).
[712] de Mayo, P., Weedon, A. C., and Zabel, R. W., *Can. J. Chem.*, **59**, 2328 (1981).
[713] Gunda, T. E., and Eneback, C., *Acta Chem. Scand.*, **34B**, 299 (1980).
[714] Baldwin, J. E., Herchen, S. R., Schulz, G., Falshaw, C. P., and King, T. J., *J. Am. Chem. Soc.*, **102**, 7815 (1980).
[715] Fodor, L., Szabó, J., Bernáth, G., Párkányi, L., and Sohár, P., *Tetrahedron Lett.*, **22**, 5077 (1981).
[716] Chatterjee, A., Sen, B., and Chatterjee, S. K., *J. Chem. Soc., Perkin Trans. 1*, **1981**, 1707.
[717] Tzeng, D., Fong, R. H., Soysa, H. S. D., and Weber, W. P., *J. Organomet. Chem.*, **219**, 153 (1981).
[718] Clemo, N. G., Gedge, D. R., and Pattenden, G., *J. Chem. Soc., Perkin Trans. 1*, **1981**, 1448.
[719] Chatterjea, J. N., Bhakta, C., Sinha, A. K., Jha, H. C., and Zilliken, F., *Justus Liebigs Ann. Chem.*, **1981**, 52.
[720] Graziano, M. L., Iesce, M. R., and Scarpati, R., *J. Chem. Soc., Chem. Commun.*, **1981**, 720.
[721] Feringa, B. L., *Tetrahedron Lett.*, **22**, 1443 (1981).
[722] Feringa, B. L., and Butselaar, R. J., *Tetrahedron Lett.*, **22**, 1447 (1981).
[723] Adam, W., and Rodriguez, A., *Tetrahedron Lett.*, **22**, 3509 (1981).
[724] Heesing, A., and Herdering, W., *Tetrahedron Lett.*, **22**, 4675 (1981).
[725] Adiwidjaja, G., Proll, T., and Walter, W., *Tetrahedron Lett.*, **22**, 3175 (1981).
[726] Swanson, B. J., Crockett, G. C., and Koch, T. H., *J. Org. Chem.*, **46**, 1082 (1981).
[727] Moore, H. W., Hernandez, L., Kunert, D. M., Mercer, F., and Sing, A., *J. Am. Chem. Soc.*, **103**, 1769 (1981).
[728] Jendrzejewski, S., and Steglich, W., *Chem. Ber.*, **114**, 1337 (1981).
[729] Wasserman, H. H., Gambale, R. J., and Pulwer, M. J., *Tetrahedron Lett.*, **22**, 1737 (1981).
[730] Olsen, H., and Oth, J. F. M., *Angew. Chem. Int. Ed.*, **20**, 983 (1981).
[731] Olsen, H., *Angew. Chem. Int. Ed.*, **20**, 984 (1981).
[732] McManus, S. P., Ortiz, M., and Abramovitch, R. A., *J. Org. Chem.*, **46**, 336 (1981).
[733] Collington, E. W., Middlemiss, D., Panchal, T. A., and Wilson, D. R., *Tetrahedron Lett.*, **22**, 3675 (1981).
[734] Gasteiger, J., and Strauss, U., *Chem. Ber.*, **114**, 2336 (1981).
[735] Micetich, R. G., Fortier, R. A., and Liew, J., *Can. J. Chem.*, **59**, 1020 (1981).
[736] Tsoi, L. A., Kalkabaeva, L. T., Cholpankulova, S. T., Ryskieva, G. A., and Salimbaeva, A. D., *Izv. Akad. Nauk Kaz. SSR, Ser. Khim.*, **1980**, 73; *Chem. Abs.*, **94**, 175014 (1981).
[737] Oka, K., Dobashi, A., and Hara, S., *J. Am. Chem. Soc.*, **103**, 2757 (1981).
[738] Bloch, R., and Orvane, P., *Tetrahedron Lett.*, **22**, 3597 (1981).
[739] Tsuda, Y., Sakai, Y., and Sano, T., *Heterocycles*, **15**, 1097 (1981).
[740] Capuano, L., and Tammer, T., *Chem. Ber.*, **114**, 456 (1981).
[741] Chubb, F. L., and Edward, J. T., *Can. J. Chem.*, **59**, 2724 (1981).
[742] Hall, C. R., and Williams, N. E., *Tetrahedron Lett.*, **21**, 4959 (1980).
[743] Gololobov, Y. G., and Nesterova, L. I., *Zh. Obshch. Khim.*, **50**, 683 (1980); *Chem. Abs.*, **93**, 186250 (1980).
[744] Gilyarov, V. A., Tikhonina, N. A., Shcherbina, T. M., and Kabachnik, M. I., *Zh. Obshch. Khim.*, **50**, 1438 (1980); *Chem. Abs.*, **94**, 3994 (1981).
[745] Manuel, G., Bertrand, G., Mazerolles, P., and Ancelle, J., *J. Organomet. Chem.*, **212**, 311 (1981).

[746] Wasserman, H. H., and Ives, J. L., *Tetrahedron*, **37**, 1825 (1981).
[747] Saito, K., Yuki, H., Ohyama, T., Nakane, R., Nagumo, K., and Sato, T., *Can. J. Chem.*, **59**, 1717 (1981).
[748] Roedig, A., and Fleischmann, K., *Justus Liebigs Ann. Chem.*, **1980**, 1960.
[749] Karakhanov, R. A., Vartanyan, M. M., and Apandiev, R. B., *Izv. Akad. Nauk SSSR, Ser. Khim.*, **1981**, 694; *Chem. Abs.*, **95**, 97526 (1981).
[750] Seoane, C., Soto, J. L., Zamorano, P., and Quinteiro, M., *J. Heterocycl. Chem.*, **18**, 309 (1981).
[751] Le Roux, J. P., Desbene, P. L., and Cherton, J. C., *J. Heterocycl. Chem.*, **18**, 847 (1981).
[752] Witczak, Z., *Heterocycles*, **14**, 1319 (1980).
[753] Hanessian, S., and Roy, R., *Tetrahedron Lett.*, **22**, 1005 (1981).
[754] Suginome, H., and Shea, C.-M., *Bull. Chem. Soc. Jpn.*, **53**, 3387 (1980).
[755] Harrison, J. M., and Inch, T. D., *Tetrahedron Lett.*, **22**, 679 (1981).
[756] Bailey, W. F., and Cioffi, E. A., *J. Chem. Soc., Chem. Commun.*, **1981**, 155.
[757] Francke, W., and Reith, W., *Tetrahedron Lett.*, **22**, 2029 (1981).
[758] Brittelli, D. R., and Boswell, G. A., *J. Org. Chem.*, **46**, 312 (1981).
[759] Damavandy, J. A., and Jones, R. A. Y., *J. Chem. Soc., Perkin Trans. 1*, **1981**, 712.
[760] Kuckländer, U., Edoho, E. J., Rinus, O., and Massa, W., *Chem. Ber.*, **113**, 3405 (1980).
[761] Tamura, Y., Takebe, Y., Mukai, C., and Ikeda, M., *J. Chem. Soc., Perkin Trans. 1*, **1981**, 2978.
[762] Lockhart, T. P., Comita, P. B., and Bergman, R. G., *J. Am. Chem. Soc.*, **103**, 4082 (1981).
[763] Adamczyk, M., Mirek, J., and Mokrosz, M., *Synthesis*, **1980**, 916.
[764] Kobayashi, K., and Mutai, K., *Tetrahedron Lett.*, **22**, 5201 (1981).
[765] Pelletier, S. W., Mody, N. V., Finer-Moore, J., Ateya, A.-M. M., and Schramm, L. C., *J. Chem. Soc., Chem. Commun.*, **1981**, 327.
[766] Teuber, H.-J., Gholami, A., and Reinehr, U., *Justus Liebigs Ann. Chem.*, **1981**, 152.
[767] Lewin, G., Rolland, Y., and Poisson, J., *Heterocycles*, **14**, 1915 (1980).
[768] Huang, Y., Macura, S., and Ernst, R. R., *J. Am. Chem. Soc.*, **103**, 5327 (1981).
[769] Kemper, M. J. H., and Buck, H. M., *Can. J. Chem.*, **59**, 3044 (1981).
[770] Vincent, M. A., and Dykstra, C. E., *J. Chem. Phys.*, **73**, 3838 (1980).
[771] Tanaka, K., and Yoshimine, M., *J. Am. Chem. Soc.*, **102**, 7655 (1980).
[772] Nobes, R. H., Radom, L., and Rodwell, W. R., *Chem. Phys. Lett.*, **74**, 269 (1980).
[773] Osamura, Y., Goddard, J. D., Schaefer, H. F., and Kim, K. S., *J. Chem. Phys.*, **74**, 617 (1981).
[774] Carlesen, L., Egsgaard, H., and Pagsberg, P., *J. Chem. Soc., Perkin Trans. 2*, **1981**, 1256.
[775] Bauer, S. H., and True, N. S., *J. Phys. Chem.*, **84**, 2507 (1980).
[776] Bauer, S. H., *J. Phys. Chem.*, **85**, 1276 (1981).
[777] Saltiel, J., and Charlton, J. L., *Org. Chem. (N.Y.)* (Rearrangements Ground Excited States, part 3), **42**, 25 (1980).
[778] Ribó, J. M., and Vallés, A., *J. Chem. Soc., Chem. Commun.*, **1981**, 205.
[779] Snyder, J. J., Tise, F. P., Davis, R. D., and Kropp, P. J., *J. Org. Chem.*, **46**, 3609 (1981).
[780] Maccarone, E., Mamo, A., Perrini, G., and Torre, M., *J. Chem. Soc., Perkin Trans. 2*, **1981**, 324.
[781] Todres, Z. V., Kuryaeva, T. T., Ryzhova, G. L., and Kursanov, D. N., *Izv. Akad. Nauk SSSR, Ser. Khim.*, **1980**, 1602; *Chem. Abs.*, **93**, 220200 1980.
[782] Kessar, S. V., Nadir, U. K., Narula, S., Kumar, P., and Mohammad, T., *Indian J. Chem.*, **20B**, 4 (1981).
[783] Leshina, T. V., Belyaeva, S. G., Mar'yasova, V. I., Sagdeev, R. Z., and Molin, Y. N., *Dokl. Akad. Nauk SSSR*, **255**, 141 (1980); *Chem. Abs.*, **94**, 102547 (1981).
[784] Ponomarev, V. S., and Stepukhovich, A. D., *Zh. Fiz. Khim.*, **54**, 1769 (1980); *Chem. Abs.*, **93**, 203747 (1980).
[785] Kubota, Y., *Yukagaku*, **30**, 85 (1981); *Chem. Abs.*, **94**, 191372 (1981).
[786] Bakhmutov, V. I., Babievskii, K. K., Belikov, V. M., and Fedin, E. I., *Izv. Akad. Nauk SSSR, Ser. Khim.*, **1980**, 2143; *Chem. Abs.*, **94**, 29895 (1981).
[787] Bakhmutov, V. I., and Kochetkov, K. A., *Izv. Akad. Nauk SSSR, Ser. Khim.*, **1980**, 1157; *Chem. Abs.*, **93**, 203738 (1980).
[788] Bakhmutov, V. I., and Fedin, E. I., *Izv. Akad. Nauk SSSR, Ser. Khim.*, **1980**, 2036; *Chem. Abs.*, **94**, 29893 (1981).
[789] Gibson, H. W., Pochan, J. M., and Kaplan, S., *J. Am. Chem. Soc.*, **103**, 4619 (1981).
[790] Karan, H. I., *J. Org. Chem.*, **46**, 2186 (1981).
[791] Dauben, W. G., Michno, D. M., and Olsen, E. G., *J. Org. Chem.*, **46**, 687 (1981).
[792] Inone, Y., Takamuku, S., Kunitomi, Y., and Sakurai, H., *J. Chem. Soc., Perkin Trans. 2*, **1980**, 1672.
[793] Goto, S., Takamuku, S., Sakurai, H., Inone, Y., and Hakushi, T., *J. Chem. Soc., Perkin Trans. 2*, **1980**, 1678.

[794] Ivanenko, A. G., Somin, I. N., and Kuznetsov, S. G., *Zh. Org. Khim.*, **16**, 1457 (1980); *Chem. Abs.*, **94**, 3513 (1981).
[795] Ivanenko, A. G., Somin, I. N., and Kuznetsov, S. G., *Zh. Org. Khim.*, **16**, 2258 (1980); *Chem. Abs.*, **94**, 64880 (1981).
[796] Johnson, J. E., Silk, N. M., Nalley, E. A., and Arfan, M., *J. Org. Chem.*, **46**, 546 (1981).
[797] Reddy, G. S., and Bhatt, M. V., *Indian J. Chem.*, **19B**, 213 (1980).
[798] Sammes, M. P., *J. Chem. Soc., Perkin Trans. 2*, **1981**, 1501.
[799] Pichon, R., Le Saint, J., and Courtot, P., *Tetrahedron*, **37**, 1517 (1981).
[800] Atkinson, R. S., and Judkins, B. D., *J. Chem. Soc., Perkin Trans. 2*, **1981**, 509.
[801] Overberger, C. G., and Chi, M.-S., *J. Org. Chem.*, **46**, 303 (1981).
[802] Ueno, A., Takahashi, K., Anzai, J.-i., and Osa, T., *J. Am. Chem. Soc.*, **103**, 6410 (1981).
[803] Chae, W.-K., Baughman, S. A., Engel, P. S., Bruch, M., Özmeral, C., Szilagyi, S., and Timberlake, J. W., *J. Am. Chem. Soc.*, **103**, 4824 (1981).
[804] Hoesch, L., *Helv. Chim. Acta*, **64**, 38 (1981).
[805] Buechele, J. L., Weitz, E., and Lewis, F. D., *J. Am. Chem. Soc.*, **103**, 3588 (1981).
[806] Hill, E. A., Link, D. C., and Donndelinger, P., *J. Org. Chem.*, **46**, 1177 (1981).
[807] Lifshitz, A., and Kahana, P., *J. Phys. Chem.*, **85**, 2827 (1981).
[808] Feast, W. J., and Morland, J. B., *J. Fluorine Chem.*, **18**, 57 (1981).
[809] Takahashi, K., Takase, K., and Toda, H., *Chem. Lett.*, **1981**, 979.
[810] Takeuchi, K., Arima, M., and Okamoto, K., *Tetrahedron Lett.*, **22**, 3081 (1981).
[811] Paquette, L. A., Hanzawa, Y., McCullough, K. J., Tagle, B., Swenson, W., and Clardy, J., *J. Am. Chem. Soc.*, **103**, 2262 (1981).
[812] Hanzawa, Y., and Paquette, L. A., *J. Am. Chem. Soc.*, **103**, 2269 (1981).
[813] Lyttle, M. H., Streitwieser, A., and Kluttz, R. Q., *J. Am. Chem. Soc.*, **103**, 3232 (1981).
[814] Zeller, K. P., and Wentrup, C., *Z. Naturforsch.*, **36B**, 852 (1981); *Chem. Abs.*, **95**, 114523 (1981).
[815] Scott, L. T., and Kirms, M. A., *J. Am. Chem. Soc.*, **103**, 5875 (1981).
[816] Gaynor, B. J., Gilbert, R. G., King, K. D., and Harman, P. J., *Aust. J. Chem.*, **34**, 449 (1981).
[817] Jones, G., and Chiang, S.-H., *Tetrahedron*, **37**, 3397 (1981).
[818] Dreeskamp, H., Jahn, B., and Pabst, J., *Z. Naturforsch.*, **36A**, 665 (1981); *Chem. Abs.*, **95**, 96655 (1981).
[819] Maier, G., and Schneider, K.-A., *Angew. Chem. Int. Ed.*, **19**, 1022 (1980).
[820] Kumagai, T., Shimizu, K., Tsuruta, H., and Mukai, T., *Tetrahedron Lett.*, **22**, 4965 (1981).
[821] Schill, G., Rissler, K., Fritz, H., and Vetter, W., *Angew. Chem. Int. Ed.*, **20**, 187 (1981).
[822] Arnold, D. R., and Yoshida, M., *J. Chem. Soc., Chem. Commun.*, **1981**, 1203.
[823] Baldwin, J. E., and Ollerenshaw, J., *J. Org. Chem.*, **46**, 2116 (1981).
[824] Umano, K., Koura, H., and Inoue, H., *Bull. Chem. Soc. Jpn.*, **54**, 2827 (1981).
[825] Pérez, J. D., de Díaz, R. G., and Yranzo, G. I., *J. Org. Chem.*, **46**, 3505 (1981).
[826] Mulzer, J., and Zippel, M., *Angew. Chem. Int. Ed.*, **20**, 399 (1981).
[827] Boyd, D. R., Burnett, M. G., and Greene, R. M. E., *J. Chem. Soc., Chem. Commun.*, **1981**, 838.
[828] Girijavallabhan, V. M., Ganguly, A. K., McCombie, S. W., Pinto, P., and Rizvi, R., *Tetrahedron Lett.*, **22**, 3485 (1981).
[829] Sokolowski, A., and Burczyk, B., *J. Prakt. Chem.*, **323**, 63 (1981); *Chem. Abs.*, **95**, 41823 (1981).
[830] Wakatsuki, Y., and Yamazaki, H., *J. Chem. Soc., Chem. Commun.*, **1980**, 1270.
[831] Fauve, A., and Kergomard, A., *Tetrahedron*, **37**, 1697 (1981).
[832] Brown, H. C., Racherla, U. S., and Taniguchi, H., *J. Org. Chem.*, **46**, 4313 (1981).
[833] Midland, M. M., Halterman, R. L., Brown, C. A., and Yamaichi, A., *Tetrahedron Lett.*, **22**, 4171 (1981).
[834] Michalak, R. S., and Martin, J. C., *J. Am. Chem. Soc.*, **103**, 214 (1981).
[835] Scheffler, K., Burmester, A., Haller, R., and Stegmann, H. B., *Chem. Ber.*, **114**, 23 (1981).
[836] Denney, D. B., Denney, D. Z., Gavrilovic, D. M., Hammond, P. J., Huang, C., and Tseng, K.-S., *J. Am. Chem. Soc.*, **102**, 7072 (1980).
[837] Buono, G., and Llinas, J. R., *J. Am. Chem. Soc.*, **103**, 4532 (1981).
[838] Niecke, E., Seyer, A., and Wildbredt, D.-A., *Angew. Chem. Int. Ed.*, **20**, 675 (1981).
[839] Łopusiński, A., Łuczak, L., Michalski, J., Kabachnik, M. M., and Moriyama, M., *Tetrahedron*, **37**, 2011 (1981).
[840] Minkin, V. I., Olekhnovich, L. P., and Zhdanov, Y. A., *Acc. Chem. Res.*, **14**, 210 (1981).
[841] Zefirov, N. S., and Trach, S. S., *Chem. Scr.*, **15**, 4 (1980); *Chem. Abs.*, **93**, 167417 (1980).
[842] Lledós, A., and Bertrán, J., *Tetrahedron Lett.*, **22**, 775 (1981).
[843] El-Basil, S., and Hilal, R., *Indian J. Chem.*, **19B**, 898 (1980).

844 Elguero, J., Fruchier, A., and Pellegrin, V., *J. Chem. Soc., Chem. Commun.*, **1981**, 1207.
845 Stegmann, H. B., Haller, R., Burmester, A., and Scheffler, K., *Chem. Ber.*, **114**, 14 (1981).
846 Hull, W. E., Künstlinger, M., and Breitmaier, E., *Angew. Chem. Int. Ed.*, **19**, 924 (1980).
847 Johansen, J. E., Piermattie, V., Angst, C., Diener, E., Kratky, C., and Eschenmoser, A., *Angew. Chem. Int. Ed.*, **20**, 261 (1981).
848 Toman, J., Štěrba, V., and Klicnar, J., *Collect. Czech. Chem. Commun.*, **46**, 2104 (1981).
849 Klicnar, J., and Toman, J., *Collect. Czech. Chem. Commun.*, **46**, 2110 (1981).
850 Takeuchi, K., Fujimoto, H., and Okamoto, K., *Tetrahedron Lett.*, **22**, 4981 (1981).
851 Duus, F., *Tetrahedron*, **37**, 2633 (1981).
852 Kuznetsova, M. G., Mironova, N. V., Kisin, A. V., Kozyukov, V. P., Nikitin, V. S., and Alekseev, N. V., *Zh. Obshch. Khim.*, **51**, 1096 (1981); *Chem. Abs.*, **95**, 150748 (1981).
853 Stetter, H., and Jonas, F., *Tetrahedron Lett.*, **22**, 4945 (1981).
854 Matz, J. R., and Cohen, T., *Tetrahedron Lett.*, **22**, 2459 (1981).
855 Williams, J. R., and Sarkisian, G. M., *J. Org. Chem.*, **45**, 5088 (1980).
856 Williams, J. R., and Abdel-Magid, A., *Tetrahedron*, **37**, 1675 (1981).
857 van Der Weerdt, A. J. A., and Cerfontain, H., *Tetrahedron*, **37**, 2121 (1981).
858 Cargill, R. L., Bushey, D. F., Dalton, J. R., Prasad, R. S., and Dyer, R. D., *J. Org. Chem.*, **46**, 3389 (1981).
859 Ahmad, F. B. H., Bruce, J. M., Khalafy, J., Pejanović, V., Sabetian, K., and Watt, I., *J. Chem. Soc., Chem. Commun.*, **1981**, 166.
860 Ahmad, F. B. H., Bruce, J. M., Khalafy, J., and Sabetian, K., *J. Chem. Soc., Chem. Commun.*, **1981**, 169.
861 Al-Hamdany, R., Bruce, J. M., Pardasani, R. T., and Watt, I., *J. Chem. Soc., Chem. Commun.*, **1981**, 171.
862 Carson, J. R., and Davis, N. M., *J. Org. Chem.*, **46**, 839 (1981).
863 Budylin, V. A., Matveeva, E. D., and Kost, A. N., *Khim. Geterotsikl. Soedin*, **1980**, 1235; *Chem. Abs.*, **94**, 29899 (1981).
864 Zander, M., *Chem. Ber.*, **114**, 2665 (1981).
865 Liso, G., Trapani, G., Reho, A., and Latrofa, A., *Tetrahedron Lett.*, **22**, 1641 (1981).
866 Shirahata, K., Iida, T., Sato, M., and Mochida, K., *Carbohydr. Res.*, **92**, 168 (1981).
867 Albert, A., *J. Chem. Soc., Perkin Trans. 1*, **1981**, 887.
868 Lyapova, M. J., Pojarlieff, I. G., and Kurtev, B. J., *J. Chem. Res. (S)*, **1981**, 351.
869 Kemp, D. S., Kerkman, D. J., Leung, S.-L., and Hanson, G., *J. Org. Chem.*, **46**, 490 (1981).
870 Kawamura, E., Abe, N., and Iori, M., *J. Chem. Soc., Perkin Trans. 1*, **1981**, 1401.
871 Nishiwaki, T., Kawamura, E., Abe, N., and Iori, M., *J. Chem. Soc., Perkin Trans. 1*, **1980**, 2693.
872 Dauben, W. G., McInnis, E. L., and Michno, D. M., *Org. Chem. (N.Y.)* (Rearrangements Ground Excited States, part 3), **42**, 91 (1980).
873 Zimmerman, H. E., *Org. Chem. (N.Y.)* (Rearrangements Ground Excited States, part 3), **42**, 131 (1980).
874 Fasel, J. P., and Hansen, H. J., *Chimia*, **35**, 9 (1981); *Chem. Abs.*, **94**, 174005 (1981).
875 Jolidon, S., and Hansen, H. J., *Chimia*, **35**, 49 (1981); *Chem. Abs.*, **94**, 191374 (1981).
876 Dietliker, K., and Hansen, H. J., *Chimia*, **35**, 52 (1981); *Chem. Abs.*, **94**, 191375 (1981).
877 Kuzuya, M., Mano, E., Ishikawa, M., Okuda, T., and Hart, H., *Tetrahedron Lett.*, **22**, 1613 (1981).
878 Adam, W., and De Lucchi, O., *Angew. Chem. Int. Ed.*, **20**, 400 (1981).
879 Adam, W., and De Lucchi, O., *J. Org. Chem.*, **46**, 4133 (1981).
880 Nitta, M., Inoue, O., and Kasahara, I., *Waseda Daigaku Rikogaku Kenkyusho Hokoku*, **1981**, 27; *Chem. Abs.*, **95**, 168207 (1981).
881 Nitta, M., Inone, O., and Uchida, T., *Waseda Daigaku Rikogaku Kenkyusho Hokoku*, **1981**, 63; *Chem. Abs.*, **95**, 149635 (1981).
882 Demuth, M., Amrein, W., Bender, C. O., Braslavsky, S. E., Burger, U., George, M. V., Lemmer, D., and Schaffner, K., *Tetrahedron*, **37**, 3245 (1981).
883 Pattenden, G., and Whybrow, D., *J. Chem. Soc., Perkin Trans. 1*, **1981**, 1046.
884 Luibrand, R. T., Broline, B. M., Charles, K. A., and Drues, R. W., *J. Org. Chem.*, **46**, 1874 (1981).
885 Rich, J. D., Drahnak, T. J., West, R., and Michl, J., *J. Organomet. Chem.*, **212**, C1 (1981).
886 Schuster, D. I., and Rao, M., *J. Org. Chem.*, **46**, 1515 (1981).
887 Wagner, P. J., and Stratton, T. J., *Tetrahedron*, **37**, 3317 (1981).
888 Murato, K., Wolf, H. R., and Jeger, O., *Helv. Chim. Acta*, **64**, 215 (1981).
889 Yokoe, I., Higuchi, K., Shirataki, Y., and Komatsu, M., *Chem. Pharm. Bull.*, **29**, 894 (1981).
890 Zimmerman, H. E., and Blinn, J. R., *Tetrahedron*, **37**, 3237 (1981).
891 Chow, Y. L., Horning, D. P., and Polo, J., *Can. J. Chem.*, **58**, 2477 (1980).

Author Index

In this index bold figures relate to chapter numbers, roman figures are reference numbers.

Aalstad, B., **3**, 696
Abad, A., **14**, 379, 380
Abatjoglou, A. G., **4**, 208
Abboud, J. L. M., **9**, 153
Abdallah, A. A., **1**, 177; **5**, 25
Abdel Halim, F. M., **2**, 87
Abdel' razek, F. M., **7**, 76
Abdel-Hady, S. A. L., **14**, 83
Abdel-Magid, A., **14**, 856
Abdel-Rahman, A. E., **14**, 24
Abdel-Rahman, M. O., **14**, 692
Abdelkader, M., **3**, 569
Abdou, S. E., **14**, 43
Abdul-Malik, N. F., **4**, 306
Abdullabekov, I. M., **6**, 68
Abdullaev, F. Z., **4**, 468
Abdullin, M. I., **4**, 520
Abe, H., **6**, 90
Abe, K., **1**, 114
Abe, N., **14**, 870, 871
Abe, S., **3**, 516
Abel, E. W., **14**, 617
Abeles, R. H., **4**, 758
Abeywickrema, R. S., **3**, 739
Abiko, A., **9**, 58; **10**, 98
Abou-Donia, M. B., **3**, 880
Abou-Elenien, G., **3**, 335
Abraham, M. H., **9**, 159, 161, 162, 195; **10**, 342
Abraham, W., **14**, 288, 329
Abramovitch, D. A., **3**, 507
Abramovitch, R. A., **3**, 507; **5**, 127, 128; **7**, 148; **13**, 54; **14**, 64, 732
Abronin, I. A., **7**, 5, 6; **8**, 164; **14**, 5
Abuin, E., **3**, 320
Abul'khanov, A. G., **3**, 670, 673
Accrombessi, G. C., **4**, 695
Acharya, R. C., **4**, 8
Acheson, R. M., **8**, 108
Achmatowicz, O., **1**, 136
Ackermann, P., **3**, 827
Acklin, G., **14**, 522
Ács, G., **3**, 202, 204, 270
Adachi, H., **3**, 152, 237
Adam, F. A., **1**, 47
Adam, W., **3**, 102, 103, 108, 562, 563; **4**, 414, 415, 423; **13**, 25, 188; **14**, 260, 673, 723, 878, 879
Adamčíkova, L., **3**, 742; **4**, 108
Adamczyk, M., **14**, 763
Adams, R., **14**, 188
Adams, R. D., **9**, 38; **14**, 539
Addink, R., **4**, 728
Adeleke, B. B., **3**, 44, 1177
Adembri, G., **3**, 577
Adeniji, S. A., **4**, 286
Adigezalov, N. R., **13**, 97
Adiwidjaja, G., **13**, 34, 53, 75; **14**, 725
Adlington, R. M., **10**, 102–104
Adolph, H. G., **6**, 60
Adrian, G., **3**, 687, 688; **4**, 44
Aerts, G. M., **1**, 30
Afanas'ev, I. B., **3**, 272
Agabekow, V. E., **4**, 462, 513, 515, 519, 521
Agaev, F. M., **4**, 468
Agaeva, S. I., **4**, 364
Agafonova, A. S., **6**, 143
Agami, C., **12**, 175–177
Agarwal, N. L., **6**, 62, 63
Agarwal, S. C., **4**, 192, 193
Agasimundin, Y. S., **14**, 74
Agawa, T., **4**, 661; **10**, 90; **12**, 209
Agdeppa, D. A., **14**, 441
Ager, D. J., **10**, 229, 238; **12**, 241
Aginskii, V. N., **14**, 552
Ağirbaŝ, H., **3**, 5, 1136
Agisheva, S. A., **3**, 334
Agmon, N., **3**, 4
Agosta, W. C., **3**, 438; **13**, 73; **14**, 540
Agostini, G., **10**, 49
Agrawal, V. B., **4**, 193
Agresti, A., **10**, 278
Agropovich, J. W., **4**, 279
Ah-Kow, G., **6**, 151
Ahern, E. P., **1**, 45
Ahern, M. F., **6**, 6
Ahlberg, A., **3**, 776
Ahlber, P., **3**, 291; **8**, 15, 16, 74
Ahmad, A. M., **2**, 94
Ahmad, F., **4**, 45, 57
Ahmad, F. B. H., **14**, 859, 860
Ahmed, A. M., **7**, 127
Ahmed, F., **4**, 58
Ahmed, M. G., **1**, 64
Ahmed, S. A., **1**, 64
Ahonkhai, S. I., **3**, 343
Ahrens, W., **3**, 64
Aihara, J., **7**, 4; **10**, 213
Airoldi, G., **4**, 450
Aizawa, T., **6**, 138
Akaba, R., **3**, 889; **4**, 464
Åkermark, B., **10**, 82
Akgün, E., **14**, 662
Akhmatdinov, R. T., **1**, 11; **2**, 8; **8**, 99–102; **14**, 398
Akhmedova, R. S., **12**, 45
Akhmetkarimov, K. A., **12**, 24
Akhrem, A. A., **14**, 244
Akhtar, M., **4**, 742
Akiba, K., **6**, 108
Akiba, M., **14**, 149
Akimkina, N. F., **12**, 46
Akimoto, H., **3**, 412
Akimova, S. N., **8**, 71; **14**, 8
Akiyama, K., **3**, 53, 250; **12**, 198
Akiyama, S. K., **4**, 760
Akkerman, O. S., **3**, 854, 864
Akporiaye, D. E., **14**, 513
Aksnes, D. W., **10**, 310

Aksnes, G., **2**, 207
Akutagawa, K., **4**, 217
Al Akeel, N. Y., **3**, 322
Al'Ainen, S. A., **3**, 959
Al-Asadi, Z. A. K., **14**, 136
Al-Hamdany, R., **14**, 861
Al-Iraqi, M. A., **14**, 22
Al-Lohedan, H., **2**, 208, 209
Al-Najjar, I. M., **12**, 261
Al-Omran, F., **7**, 62; **14**, 53
Al-Rawi, J. M. A., **14**, 22
Al-Sader, B. H., **3**, 486, 1117–1119, 1121; **14**, 136, 298, 631, 632
Al-kazzaz, A.-K. S., **14**, 22
AlAruri, A. D. A., **6**, 135, 137
Alanso, R. A., **3**, 801
Alazard, J.-P., **4**, 362
Albert, A., **14**, 867
Alberti, A., **3**, 45, 77, 81, 420, 1049
Alberts, V., **10**, 339
Albery, W. J., **3**, 727
Albini, A., **3**, 894, 899; **4**, 419; **5**, 85; **13**, 158
Alborz, M., **2**, 38a; **11**, 19
Albright, T. A., **6**, 81
Alcaide, B., **4**, 550
Alder, A. P., **14**, 657
Alder, R. W., **3**, 19, 943; **6**, 27; **14**, 245
Aldercreutz, P., **10**, 261
Aldred, S. E., **7**, 86
Alegria, A., **3**, 678
Alekhina, N. N., **6**, 143
Aleksandr, A. V., **8**, 82, 84, 85
Aleksandrov, A. L., **3**, 197–199; **4**, 522, 523
Aleksandrov, E. N., **3**, 396; **4**, 384,385
Aleksandrov, G. G., **6**, 119
Aleksandrov, Yu., **3**, 93, 94
Alekseev, N. V., **14**, 852
Alekseeva, A. B., **14**, 565
Alemagna, A., **14**, 24, 25
Alender, J., **5**, 83
Aleskerov, A. A., **13**, 99
Alexandrou, N. E., **5**, 26
Alexiou, M., **6**, 65
Alfassi, Z. B., **3**, 280–282
Ali, M. B., **13**, 132
Ali, Sk. A., **10**, 192
Ali-Zade, N. I., **4**, 365
Alibaeva, K. A., **14**, 570
Alibhai, M., **2**, 9
Alimardanov, Kh. M., **4**, 492
Allara, D. L., **3**, 1100
Allard, B., **14**, 480
Allcock, H. R., **14**, 370
Allen, A. O., **3**, 1048
Allen, D. W., **1**, 98, 99
Allen, R. W., **5**, 133
Allevi, P., **4**, 312
Alma, N. C. M., **3**, 1164
Almgren, C. W., **3**, 889a
Alnajjar, M. S., **3**, 834
Aloni, R., **3**, 351; **11**, 73
Alonso, M. E., **2**, 133; **11**, 33, 74; **14**, 650
Alonso, R. A., **3**, 804; **6**, 19, 21
Alonso-Cires, L., **4**, 64; **12**, 139
Alper, H., **4**, 648, 655; **12**, 92, 93
Alunni, S., **11**, 4
Alvanipour, A., **10**, 291
Alvarez Diaz, M., **4**, 506
Alvarez, A. J., **14**, 29
Alvarez, E., **4**, 379
Alvernhe, G., **7**, 148
Alves, K. B., **1**, 76
Aly, M. M., **14**, 24
Amamria, A., **10**, 219
Amaratunga, S., **4**, 648
Amarla, L., **1**, 76
Amato, A., **4**, 257
Ambrosetti, R., **12**, 30
Ambroz, H. B., **8**, 132
Amdrianov, V. G., **8**, 50
Amelichev, V. A., **9**, 175, 176
Ameta, S. C., **4**, 38, 40
Amila, M. F., **2**, 94, 95
Amino, Y., **10**, 195
Aminov, S. N., **3**, 387
Amira, M. F., **7**, 127
Amita, F., **6**, 140
Amosova, S. V., **12**, 186
Amrein, W., **14**, 882
Amrich, M. J., **3**, 206
Amsterdamsky, C., **14**, 664
Amstutz, R., **10**, 232
Anan, F. K., **4**, 772
Anand, B. H., **4**, 159
Anandan, S., **1**, 176; **4**, 247
Anandasundaresan, P., **7**, 24
Ananthakrishnanadar, P., **1**, 151; **2**, 52; **4**, 554; **9**, 201; **10**, 327
Anastasia, M., **4**, 312
Anastassiou, A. G., **10**, 76, 77; **14**, 331
Anastassopoulou, J. D., **3**, 625
Ancelle, J., **14**, 132, 745
Anciaux, A. J., **5**, 112
Andell, O. S., **12**, 178
Andersen, N. H., **10**, 153, 234
Anderson, D. R., **3**, 617
Anderson, G., **14**, 253
Anderson, G. H., **5**, 135
Anderson, L., **1**, 20
Anderson, M. W., **10**, 135
Ando, K., **10**, 94
Ando, R., **2**, 205; **11**, 42
Ando, T., **2**, 29; **3**, 354; **4**, 600; **9**, 137, 241; **14**, 408, 409
Ando, W., **3**, 120; **4**, 391, 421; **5**, 27, 98, 99, 152
Ando, Y., **6**, 142
Andoh, T., **7**, 78
Andonian-Haftvan, J., **10**, 342
Andrade, J. G., **4**, 69; **10**, 5, 97, 311
Andrade, M. A. C., **14**, 548
Andreetti, G. D., **13**, 32
Andreev, N. A., **2**, 380
Andreev, S. A., **2**, 148, 149, 151, 152; **10**, 334–337
Andreeva, L. A., **12**, 44
Andrejevic, V., **4**, 100, 101
Andreou, A. D., **7**, 126, 128
Andresen, P., **4**, 383
Andrews, L., **3**, 970; **8**, 75; **14**, 405, 406
Andrews, M. A., **4**, 173
Andrews, P. E., **3**, 777
Andrus, A., **14**, 174
Andruzzi, R., **3**, 779
Anez, M., **12**, 128
Angeletakis, C. N., **2**, 448; **4**, 349, 350
Angelini, G., **9**, 141–143; **12**, 22
Angelis, C. T., **3**, 1060; **4**, 378
Angelos, G. H., **9**, 152
Angelov, C. M., **12**, 52
Angert, J. L., **3**, 604
Angiolini, L., **4**, 543
Angst, C., **14**, 847
Anikin, N. A., **12**, 265
Anisimov, A. V., **14**, 162, 163
Anjaneyulu, A. S. R., **14**, 515, 516
Ankner, K., **3**, 786
Annis, G. D., **14**, 148
Annunziata, R., **10**, 247
Anoardi, L., **2**, 231
Ansari, A. H., **4**, 22
Antebi, S., **14**, 705
Anteunis, M. J. O., **9**, 254
Antonioletti, R., **4**, 5
Antonjuk, D. J., **10**, 239
Antonova, V. V., **4**, 488
Anzai, J.-i., **14**, 802
Aoai, T., **4**, 353; **11**, 108; **12**, 64
Aoki, K., **9**, 98
Aoki, M., **6**, 138
Aoyagui, S., **10**, 151

Aoyama, H., **3**, 1083
Aoyama, Y., **4**, 706
ApSimon, J., **12**, 269
Apandiev, R. B., **14**, 749
Apatu, J. O., **3**, 541; **4**, 329; **7**, 143
Apeloig, Y., **8**, 146, 173; **9**, 26, 27; **10**, 7, 307
Apoussidis, T., **3**, 87, 89
Appel, R., **14**, 206
Appelman, E. H., **7**, 19
Applequist, D. E., **3**, 406, 660
Apte, D. V., **7**, 114
Arad, D., **9**, 26, 27
Arai, I., **4**, 696
Arai, M., **4**, 285; **7**, 78
Arai, N., **14**, 689
Araya, D., **3**, 614
Archer, W. J., **7**, 44–46
Arcoleo, A., **14**, 58
Arcoria, A., **2**, 444; **9**, 100
Arduengo, A. J., **3**, 678
Aresta, M., **10**, 314
Arfan, M., **1**, 72; **14**, 796
Argentini, M., **1**, 146; **7**, 38; **10**, 328
Argo, C. B., **14**, 268
Aribike, D. S., **3**, 1109
Ariga, M., **6**, 122, 123; **10**, 174
Arihara, M., **9**, 109
Arikata, S., **5**, 51
Ariko, N. G., **3**, 117, 157, 333; **4**, 471, 501
Arima, M., **14**, 810
Arjona, O., **1**, 130
Arkhireeva, I. A., **13**, 15, 16
Armand, J., **4**, 645
Armesto, D., **14**, 29
Armstead, C. R., **4**, 28
Armstrong, R. N., **9**, 66
Arnaud, R., **3**, 360, 513
Arnett, E. M., **10**, 301
Arno, M., **14**, 379, 380; **13**, 072
Arnold, D. R., **3**, 41, 42, 750, 903; **13**, 158; **14**, 822
Arnold, E., **4**, 411
Arnold, E. V., **13**, 26
Arnold, S., **12**, 21
Arnold, W., **14**, 373
Arnone, C., **6**, 94, 95
Aronovich, K. A., **12**, 77
Arous-Chtara, R., **1**, 61
Arrowsmith, J., **10**, 258
Arseniyadis, S., **14**, 402
Arsent'ev, S. D., **3**, 154
Artamkina, G. A., **6**, 148, 154
Arthur, C. D., **7**, 19
Arthur, N. L., **3**, 268, 276
Artym, I. I., **3**, 142
Aruga, T., **3**, 781
Arumi, M., **7**, 43; **12**, 87
Arutyunov, V. S., **3**, 396; **4**, 384, 385
Arvay, Z., **14**, 474
Arzamaskova, L. N., **4**, 18, 19
Asaad, A. N., **13**, 90
Asaki, M., **6**, 58
Asami, M., **4**, 594
Asao, M., **4**, 733
Asao, T., **6**, 92
Asensio, G., **4**, 64; **12**, 139
Ash, D. K., **4**, 662
Ashby, E. C., **1**, 129; **3**, 836, 838–848, 855, 857; **4**, 529–533; **12**, 240
Ashe, A. J., **7**, 12
Ashkenazi, P., **13**, 155
Ashy, M. A., **9**, 166
Aslam, M., **2**, 49–51
Asmus, K.-D., **3**, 342, 877, 878, 1139, 1140; **4**, 347
Atal, C. K., **14**, 166, 167, 467
Atamanyuk, V. Yu., **3**, 725, 979; **4**, 13, 168
Atanasow, P., **4**, 319
Ateya, A.-M., **14**, 765
Atherton, N. M., **3**, 1015, 1016; **4**, 639
Atkins, A. R., **10**, 339
Atkins, P. J., **11**, 120
Atkinson, R. F., **11**, 40
Atkinson, R. S., **4**, 103; **5**, 66, 69, 129; **14**, 800
Atmaram, S., **3**, 443
Attina, M., **8**, 128; **9**, 143
Aubry, J. M., **4**, 377
Auchus, R. J., **14**, 634
Audenaert, F., **14**, 287
Audier, H. E., **14**, 426
Audley, D. C., **3**, 315
Audo, D., **1**, 105
Aukharieva, R. G., **14**, 162
Auksi, H., **13**, 113
Aumann, R., **5**, 154
Aunir, D., **3**, 1171
Aurich, G., **3**, 39
Aurich, H. G., **3**, 587, 610–612
Ausloos, P., **7**, 49
Ausloos, P. J., **8**, 135, 136
Auterhoff, **6**, 155
Avakyan, V. G., **9**, 78; **10**, 36
Aver'yanov, S. F., **3**, 699
Aver'yanov, V. A., **3**, 286, 287, 289, 305, 529; **11**, 79
Aveta, R., **6**, 96, 167
Avetisyan, E. A., **13**, 31
Avico, U., **4**, 257
Avotins, F., **14**, 576
Avram, M., **14**, 187
Awad, S. B., **4**, 306
Axiotis, G. P., **9**, 57
Ayabe, G.-I., **5**, 124; **14**, 690
Ayala, A. D., **3**, 421
Ayapbergenov, K. A., **12**, 24
Ayer, W. A., **14**, 501
Ayyangar, N. R., **5**, 118
Azarko, V. A., **4**, 521
Aznar, F., **4**, 65; **12**, 140
Azogu, C. I., **5**, 108
Azoro, J., **12**, 132
Azran, J., **4**, 615
Azuma, K., **9**, 262

Baba, N., **4**, 698, 708
Babaitsev, V. S., **14**, 163
Baban, J. A., **3**, 676
Babayan, A. T., **14**, 349–352
Babcock, B. W., **3**, 464
Babievskii, K. K., **14**, 786
Babin, V. N., **3**, 640, 768–770
Babler, J. H., **4**, 266, 571
Baboulene, M., **10**, 269; **14**, 369
Bacaloglu, I., **2**, 113, 127
Bacaloglu, R., **2**, 113, 126–128
Bacci, M., **10**, 278
Bachi, M. D., **3**, 440, 441
Bachner, J., **13**, 128
Bachur, N. R., **3**, 777
Baciocchi, E., **3**, 683; **4**, 88
Back, T. G., **3**, 434; **4**, 202–204; **12**, 62
Bäckvall, J.-E., **12**, 151, 178
Badasyan, G. V., **4**, 320, 322
Baddeley, G. V., **12**, 218
Badea, F., **14**, 474, 475
Badr, M. Z. A., **14**, 24
Bae, D.-H., **3**, 211
Baert, F., **6**, 52
Baeza, H. J., **12**, 188
Bagavant, G., **2**, 30
Bagdasar'yan, Kh., **3**, 229
Baghal-vayjooee, M. H., **3**, 232
Baghlaf, A. O., **9**, 166
Bagrii, E. I., **14**, 478
Bahadur, Y., **10**, 149
Bhari, M. S., **9**, 199
Bahnemann, D., **3**, 878, 1139
Baier, H., **14**, 124, 125
Bailey, W. F., **4**, 629; **14**, 756
Baillie, P. J., **7**, 80
Bainton, H. P., **10**, 115
Baiocchi, C., **4**, 59
Bairamov, F. G., **4**, 364
Baird, K. J., **14**, 177
Baird, M. S., **5**, 17
Baj, S., **4**, 483

Bakac, A., **4**, 53, 62
Bakalo, L. A., **2**, 158
Bakalova, E. P., **7**, 113
Baker, D. S., **3**, 111
Baker, G. L., **4**, 673, 675
Baker, S. R., **1**, 111; **3**, 709
Bakhmutov, V. I., **8**, 51; **14**, 786–788
Bakker, C. G., **13**, 136
Bakore, G. V., **4**, 220, 221
Bakos, D., **3**, 332
Bakuzis, M. L. F., **4**, 231
Bakuzis, P., **4**, 231
Bal, B. S., **4**, 216
Balaban, A. T., **3**, 59, 551
Balabanov, E., **7**, 8
Balachandran, M., **10**, 53
Balakin, I. M., **8**, 140
Balasubramanian, P., **5**, 113
Balasubramanian, V., **4**, 89; **7**, 24
Balber, J. H., **4**, 612, 616
Balcarek, L., **11**, 22
Balcárek, P., **2**, 122
Balci, M., **3**, 104; **11**, 84
Baldea, I., **4**, 20
Baldwin, J. E., **3**, 497, 1114; **6**, 80; **10**, 105; **14**, 309, 658, 714, 823
Baldwin, R. R., **3**, 160
Baldwin, S. W., **13**, 37
Bali, A., **8**, 105
Baliah, V., **2**, 43, 44; **4**, 11, 12
Balko, T. W., **11**, 40
Ball, M., **6**, 144
Ball, S. S., **3**, 1061; **4**, 374, 726
Ballester, M., **6**, 61
Balli, H., **5**, 19
Ballistreri, F. P., **2**, 433, 444; **9**, 100, 101
Ballod, A. P., **3**, 265
Balou, D., **7**, 27
Balyatinskaya, L. N., **12**, 71
Bambal, R. B., **5**, 118
Bamkole, T. O., **6**, 43
Ban, T., **11**, 49; **13**, 184
Banas, B., **4**, 21, 684
Bancroft, E. E., **3**, 974
Bandow, H., **3**, 412
Banerjee, S. K., **14**, 167
Banerjee, T. K., **4**, 688
Banerji, K. K., **4**, 232, 233
Banjeree, S. K., **14**, 166
Banjoko, O., **2**, 439; **6**, 41, 42
Banks, B. E. C., **2**, 365
Banks, R. E., **3**, 618, 619 ,621; **14**, 358
Bansal, N. L., **2**, 22
Banthiya, U. S., **4**, 77
Banwell, M. G., **11**, 132, 133
Banyard, S. A., **5**, 16
Bao Iglesias, M., **4**, 506
Baranov, S. N., **11**, 27
Barbaric, S., **2**, 197
Barbarin, F., **3**, 596
Barbaro, G., **3**, 77
Barboiu, V., **13**, 85
Barbry, D., **14**, 357
Barbuch, R. J., **14**, 36
Barcelo, J., **6**, 10
Barclay, L. R. C., **3**, 176; **4**, 446, 447
Barea, F., **11**, 40
Bargagna, A., **13**, 137
Bargar, T., **6**, 22
Bargon, J., **14**, 534
Barili, P. L., **14**, 98
Barillier, D., **4**, 281
Barkalov, I. M., **3**, 405
Barker, G. C., **3**, 542
Barker, M. W., **4**, 129
Barker, P. E., **3**, 80
Barkhash, V. A., **8**, 1; **12**, 134
Barlet, R., **5**, 48
Barlett, P. A., **14**, 143
Barltrop, J. A., **14**, 297
Barluenga, J., **4**, 64, 65; **10**, 13; **12**, 139–141, 143, 144
Barner, U., **3**, 1101
Barnum, C., **10**, 80; **12**, 228; **13**, 112
Barradas, R. G., **3**, 1050
Barras, C., **9**, 32
Barrett, A. G., **4**, 630, 632
Barrett, A. G. M., **10**, 102–104
Barry, I., **3**, 1127
Barsheva, N. S., **4**, 105
Bartels, H. M., **3**, 383
Bartetzko, R., **3**, 963
Bartl, A., **3**, 56, 58
Bartlett, P. D., **3**, 1080
Bartmess, J. E., **10**, 145, 235; **11**, 155
Bartnik, R., **7**, 148
Bartocci, C., **3**, 316
Bartok, M., **11**, 6, 136
Bartolini, G., **4**, 265
Barton, B. D., **3**, 51, 1112, 1123
Barton, D. H. R., **3**, 470–474; **4**, 107, 205, 209, 229, 630, 632; **14**, 628
Barton, D. L., **3**, 756
Barton, T. J., **3**, 492; **5**, 151; **13**, 26
Bartroli, J., **1**, 91; **10**, 189
Bartsch, R. A., **6**, 5; **11**, 1, 3, 7
Baruch, G., **3**, 278, 346, 377
Baryshnikov, A. T., **12**, 75; **14**, 472
Baryshnikov, Yu. N., **3**, 94; **10**, 92
Barzaghi, M., **3**, 1003
Basak, A., **9**, 113; **14**, 319
Basak, S., **5**, 34
Basco, N., **3**, 152, 237
Basiulis, D. I., **4**, 75
Baskakov, Y. A., **14**, 560
Basova, Yu. G., **9**, 132–134
Bass, L. S., **14**, 204
Basset, J. M., **5**, 155
Bassetti, M., **12**, 131
Bastable, J. W., **14**, 495, 496
Bastos, M. P., **1**, 76
Basu, S., **1**, 118
Basu, S. K., **4**, 72
Baswani, V. S., **4**, 45, 57, 58
Bates, D. J., **8**, 43
Bates, G. S., **14**, 238, 687
Bates, P. A., **14**, 41, 42
Bates, R. B., **10**, 205, 208
Batog, A. E., **4**, 460
Battaglia, A., **3**, 77; **13**, 6, 32
Battersby, A. R., **14**, 118
Battig, K., **14**, 311
Battistini, C., **9**, 55
Battye, P. J., **2**, 117
Batyrbaev, N. A., **3**, 273, 274, 325, 345
Bäuch, H.-G., **13**, 34
Bauer, C., **5**, 157
Bauer, C.-A., **2**, 281
Bauer, H., **3**, 74
Bauer, S. H., **14**, 775, 776
Baughman, S. A., **14**, 803
Baulch, D. L., **3**, 581
Bauld, N. L., **3**, 977; **8**, 162; **10**, 74; **13**, 121;**14**, 304
Baumann. H., **3**, 186
Baumberger, R. S., **3**, 684
Baumstark, A. L., **3**, 107; **4**, 307, 727; **14**, 708
Bawa, J. S., **4**, 480
Baxter, S. G., **8**, 109
Bayder, A. E., **14**, 577
Bayes, K. D., **3**, 267
Bayles, K. D., **3**, 161
Bazanov, A. G., **3**, 818, 882
Bazanova, G. V., **4**, 170, 171; **7**, 84, 85
Bazhenov, B. N., **2**, 33, 344
Bazhin, N. M., **3**, 932
Beachley, O. T., **11**, 101
Beadle, J. R., **6**, 6
Beak, P., **10**, 26, 137, 139
Bean, M. B., **12**, 39
Bearden, R., **3**, 1046

Beauchamp. J. L., **9**, 145; **10**, 56
Beaulieu, N., **1**, 4
Beaulieu, P., **12**, 234
Beaulieu, P. L., **12**, 48, 51, 57, 58
Bechara, E. J. H., **3**, 106
Beck, A., **14**, 324, 624
Beck, A. K., **10**, 166
Becker, A. R., **4**, 768
Becker, B. Ch., **10**, 206
Becker, H.-D., **14**, 448
Becker, H. G. O., **3**, 218, 222
Becker, N., **4**, 317
Beckhaus, H.-D., **14**, 698
Beckwith, A. L. J., **3**, 16, 181, 224, 225, 324, 439, 484, 497; **6**, 13
Becu, Chr., **9**, 254
Beddard, C., **3**, 1074
Bedford, C. D., **5**, 115
Bedin, M. P., **3**, 882
Bee, L. K., **7**, 21
Beebe, T. R., **3**, 711; **4**, 230
Beenackers, A. A. C. M., **7**, 97
Beer, H. R., **14**, 591
Begley, M. J., **3**, 709
Bégué, J.-P., **3**, 147
Begunov, A. V., **2**, 345
Behar, D., **3**, 518, 1027–1029
Behera, G. B., **4**, 95
Behr, H., **13**, 53
Behrens, G., **3**, 967
Beileryan, N. M., **3**, 139; **4**, 184, 189, 505, 507
Bekhazi, M., **5**, 28
Bel'skii, V. E., **2**, 211, 339, 353; **9**, 220
Belenkii, L. I., **7**, 6
Beletskaya, I. P., **2**, 255; **3**, 658, 659, 689, 690, 717; **4**, 435; **6**, 148, 154; **9**, 256, 257; **10**, 84, 85
Belevskii, V. N., **3**, 636
Belikov, V. M., **14**, 786
Belikova, N. A., **12**, 74, 75, 133; **14**, 472
Belinka, B. A., **14**, 451
Bell, H. M., **8**, 97
Bell, L. G., **9**, 32
Bell, R., **12**, 199
Bell, T. N., **3**, 279, 1134; **6**, 170
Bellamy, A. J., **10**, 250
Bellamy, F., **4**, 412
Beller, J. D., **4**, 214
Belletire, J. L., **13**, 131
Belli, A., **3**, 692, 719, 720; **4**, 183
Bellucci, G., **9**, 44; **12**, 30, 31
Bellus, D., **14**, 382
Bellville, D. J., **13**, 121
Belostotskaya, I. S., **3**, 499, 867
Belous, N. P., **4**, 474
Belousov, V. M., **4**, 121
Belousov, Yu. A., **3**, 640, 768–770
Belousova, I. A., **2**, 443
Belov, P. S., **7**, 122; **9**, 50
Beltrame, P., **4**, 525; **11**, 92
Beltrame, P. L., **2**, 112; **4**, 525
Beltrami, H., **6**, 107
Belville, D. J., **3**, 977
Belyaeva, S. G., **3**, 1038, 1039; **14**, 783
Belyalov, R. U., **2**, 382
Benati, L., **3**, 449, 572, 978; **5**, 130, 131
Benattar, A., **14**, 357
Bender, C. O., **14**, 882
Bender, H.-J., **12**, 96
Bender, M. L., **2**, 278, 279
Bened, A., **13**, 43
Benedict, J. H., **6**, 100
Benezra, C., **14**, 223
Benjamin, B. M., **6**, 32; **14**, 438
Benko, J., **2**, 35, 36
Benkovic, S. J., **2**, 393, 399, 400
Benner, S. A., **14**, 334
Bennett, J. E., **3**, 155
Bennyarto, F., **5**, 101; **14**, 541
Benoiton, N. L., **2**, 162
Bensadoun, N., **3**, 454
Bension, R. M., **12**, 202, 203
Benson, S. W., **3**, 232, 303, 450, 1104
Bentley, T. W., **9**, 156, 157
Bentrude, W. G., **3**, 195; **4**, 524
Benz, R. C., **3**, 969
Benzaid, A., **12**, 76
Berbee, R. M., **6**, 115
Berberova, N. T., **3**, 528
Bercaw, J. E., **5**, 158
Berces, T., **3**, 365
Berdutin, A. Ya., **2**, 72
Berenblyum, A. S., **3**, 200; **4**, 455
Berends, W., **4**, 728
Berenschot, D. R., **4**, 225
Berezin, I. V., **2**, 174; **4**, 73, 162, 163
Berezina, R. N., **8**, 71; **14**, 8
Bergbreiter, D. E., **1**, 84; **10**, 24
Bergler, H. U., **3**, 194, 1056
Bergman, J., **4**, 210; **9**, 112
Bergman, R. G., **3**, 566; **14**, 762
Bergmann, K., **4**, 963
Bergmark, W. R., **13**, 203
Bergon, M., **2**, 118
Bergsma, J. P., **3**, 896
Berka, Z., **9**, 56
Berlan, J., **12**, 231
Berlin, K. D., **6**, 53
Bermejo, J., **10**, 324
Bermudez, R. K., **8**, 124
Bernadskii, A. A., **5**, 140
Bernardi, F., **3**, 77, 998
Bernardini, A., **14**, 509
Bernardon, C., **1**, 132
Bernasconi, C. F., **12**, 84
Bernashevskii, N. V., **4**, 292
Bernáth, G., **14**, 715
Berndt, A., **3**, 64, 65, 958, 985
Berner, D., **14**, 412
Berner, J., **2**, 121
Berneth, H., **3**, 937
Bernier, J. L., **6**, 52
Berridge, J. C., **13**, 197
Berson, J. A., **3**, 553, 554; **5**, 93; **14**, 538, 643
Berti, C., **3**, 606, 623, 779; **14**, 416
Berti, G., **9**, 44; **12**, 30
Bertrán, J., **7**, 43; **12**, 87; **14**, 842
Bertrand, G., **14**, 132, 745
Bertrand, J., **12**, 216
Bertrand, M., **14**, 308
Bespalova, A. M., **4**, 488
Besse, J., **3**, 214, 216
Best, W. M., **6**, 174
Besti, G., **12**, 31
Bestmann, H. J., **14**, 26, 702
Betancor, C., **4**, 379
Beth, L. D., **6**, 99
Bethell, D., **3**, 1054, 1055; **8**, 92
Bettinetti, G. F., **4**, 419; **5**, 85
Betts, M. J., **10**, 104
Beugelmans, R., **6**, 23
Beulen, J., **3**, 145
Bevlen, J., **3**, 254
Bews, J. R., **14**, 400
Beyer, M., **7**, 111
Beynon, J. H., **14**, 427
Beynon, P., **13**, 11
Beyrich, J., **4**, 481
Bezbozhnaya, T., **9**, 245
Bezmenova, T. E., **2**, 120; **11**, 127
Bezrodnyi, V., **2**, 436; **11**, 27
Bezrukova, A. A., **13**, 49
Bhacca, N. S., **13**, 74
Bhadbhade, M. M., **14**, 342
Bhadoria, G. P. S., **1**, 145
Bhaduri, A. P., **14**, 604
Bhakta, C., **14**, 719
Bhaleran, U. T., **2**, 429

Bhalerao, U. T., **14**, 571
Bhat, V., **14**, 344
Bhatia, I., **4**, 233
Bhatnagar, A. K., **14**, 118
Bhatt, M. V., **2**, 105, 192, 193; **3**, 691; **4**, 191, 259; **9**, 208; **14**, 797
Bhattacharjee, G., **2**, 451; **4**, 261–264; **9**, 82, 227
Bhattacharya, A. K., **10**, 222
Bhattacharya, B., **4**, 165
Bhattacharyya, K. K., **4**, 480; **7**, 110; **14**, 46
Bhattacharyya, S. N., **3**, 1143
Bialer, M., **4**, 302
Bianchi, G., **13**, 59
Bianchini, R., **9**, 44; **12**, 30,31
Bianco, B., **5**, 110; **10**, 220
Bicev, P., **13**, 207
Bickelhaupt, F., **3**, 854, 864
Bida, G., **3**, 706
Biehl, E. R., **6**, 175
Bielski, B. H. J., **3**, 1048
Bienvenue-Goetz, E., **12**, 25
Bierbaum, V. M., **10**, 143, 144, 322; **11**, 82
Biermann, D., **13**, 91
Biesiada, K. A., **4**, 650; **5**, 61
Bigg, M. G., **12**, 47
Bigler, P., **13**, 155; **14**, 591
Bignozzi, C. A., **3**, 316
Bigot, B., **3**, 813; **9**, 149; **14**, 679
Billingham, N. C., **3**, 5, 1136
Billups, W. E., **3**, 565; **5**, 23
Bilyard, K. G., **10**, 221
Binev, I., **10**, 29, 62
Binger, P., **13**, 47, 48, 208
Bingmann, H., **14**, 324, 624
Binkley, R. W., **14**, 528
Binns, M. R., **10**, 227; **12**, 210
Bippi, H., **3**, 560; **11**, 60
Birch, A. J., **4**, 618; **6**, 129, 130; **8**, 31, 32, 37, 38; **10**, 1
Birchall, J. M., **3**, 408, 619
Bird, C. W., **14**, 67, 473
Birnberg, G. H., **8**, 14
Bisagni, E., **14**, 581
Bissig, P., **14**, 383
Biswas, M. K., **2**, 252
Bizdena, E. O., **14**, 576
Bizunok, S. N., **10**, 22
Bizzigotti, G. O., **2**, 219
Black, D. St. C., **3**, 778
Blackburn, E. V., **3**, 824
Blackburn, G. M., **2**, 386
Blackburn, I. V., **3**, 300
Blacklock, T. J., **14**, 620
Blackstock, S. M., **8**, 72, 73
Blair, I. A., **8**, 139
Blaise, P., **9**, 147
Blake, P., **1**, 165; **12**, 4
Blake, P. G., **11**, 36
Blanc, J., **4**, 3, 609
Blandamer, M. J., **9**, 163–165, 264
Blaschek, U., **12**, 252
Blaskó, G., **14**, 686
Blau, K., **3**, 178
Blaustein, M. A., **5**, 93; **14**, 538
Bleeker, I. P., **3**, 504
Bleicher, W., **4**, 693
Bleisch, S., **3**, 55–58
Blinn, J. R., **14**, 890
Blinov, N. N., **4**, 299
Bloch, R., **2**, 132; **3**, 1130; **14**, 738
Blom, N. F., **12**, 29
Bloodworth, A. J., **3**, 103, 111; **4**, 601; **9**, 122
Bloom, L. M., **10**, 125
Blount, H. N., **3**, 974
Blount, J. F., **14**, 148
Blount, J. R., **13**, 24
Blucher, W., **7**, 59
Bludssus, W., **5**, 43
Blum, J., **4**, 615
Boar, R. B., **4**, 630
Bobillier, C., **4**, 327
Bobkov, V. N., **4**, 32
Bobrowski, K., **3**, 1153, 1154
Bobyleva, A. A., **12**, 74, 75; **14**, 472
Boche, G., **3**, 1033; **14**, 638
Bochkarev, V. N., **5**, 140
Bochvar, D. A., **8**, 170
Bock, H., **3**, 953, 957, 1002; **5**, 105
Bodoev, N. V., **8**, 77
Bodor, A., **1**, 67
Bodor, N., **2**, 188; **9**, 253
Bodrikov, I. V., **8**, 176; **12**, 41, 55, 134
Boeckelmann, J., **14**, 131
Boeckman, R. K., **10**, 99
Boegel, H., **14**, 131
Boehmer, V., **9**, 210
Boekelheide, V., **3**, 991
Boeriu, C., **2**, 113
Boersma, J., **4**, 699
Boerth, D. W., **10**, 300
Bofinger, K.-R., **13**, 19
Bogachev, Y. G., **14**, 86
Bogatskii, A. V., **1**, 14
Bogdanovic, B., **10**, 112
Bógel, H., **3**, 1069
Bogsanyi, D., **8**, 32
Boguslavskaya, L. S., **12**, 12
Bóhm, M. C., **4**, 412; **13**, 110, 154
Bohme, D. K., **9**, 146
Bohme, R., **14**, 26
Bohonek, J., **2**, 5
Bóhrer, G., **12**, 252
Boigegrain, R., **14**, 561
Boiko, T. S., **4**, 142
Boiko, V. N., **6**, 145, 146
Boiveaut, A., **3**, 423
Bokadia, M. M., **4**, 106
Boldt, P., **3**, 383
Bolesov, I. G., **11**, 135
Bolikal, D., **9**, 116
Bolster, J., **14**, 457
Bolster, J. M., **14**, 498
Bolton, J. R., **3**, 1174
Bolton, R., **3**, 517
Bolze, R., **3**, 65, 958
Bomse, D. S., **9**, 145
Bond, F. T., **10**, 109
Bondarenko, G. N., **10**, 36
Bondarenko, L. I., **2**, 166, 167
Bondarenko, T. G., **4**, 496
Bondavalli, F., **13**, 137
Boneva, M., **4**, 469
Bong Rae Cho., **11**, 3
Bongers, S. L., **14**, 144
Bonicamp, J., **8**, 137, 138
Bonini, B. F., **3**, 45
Bonner, T. G., **1**, 15
Bonvino, V., **7**, 118
Booth, B. L., **7**, 131
Booth M., **14**, 617
Boots, S. G., **8**, 26
Borauskaite, A., **7**, 14
Borden, W. T., **3**, 550, 946; **5**, 11, 88
Bordwell, F. G., **3**, 821; **9**, 223; **10**, 150, 297, 298
Boreham, C. J., **2**, 244
Borg, R. M., **13**, 158
Borghese, A., **2**, 455
Borhani, K. J., **3**, 1053
Borisenko, N. I., **14**, 285
Borisov, I. M., **4**, 516, 517
Born, L., **3**, 570
Bornatsch, W., **14**, 73
Borodaev, S. V., **7**, 99
Borodin, P. M., **10**, 276
Borodkin, G. I., **8**, 67; **14**, 4
Bortolini, O., **4**, 119, 123
Bos, H. J. T., **10**, 120; **14**, 348
Bosch, J., **14**, 355
Bosnich, B., **4**, 670
Boswell, G. A., **13**, 57; **14**, 758
Bothe, E., **1**, 167; **2**, 71; **3**, 967; **4**, 503
Bottaro, J. C., **10**, 105

Bottoni, A., **13**, 6
Bottura, G., **3**, 542
Bouab, O., **2**, 6
Bougeard, P., **3**, 653, 654, 869
Bougrois, J.-L., **3**, 446
Bouis, P. A., **8**, 112
Boulton, A. J., **14**, 60
Bouma, W. J., **9**, 45; **14**, 227
Bourgeois, J., **3**, 146
Bourque, R. A., **3**, 1093
Bovicelli, P., **8**, 36
Bowden, B. F., **14**, 117
Bowen, C. T., **9**, 156, 157
Bowen, R. D., **14**, 396
Bowers, J. C., **3**, 855
Bowers, P. G., **6**, 2
Bowie, J. H., **8**, 139, 143; **10**, 54
Bowman, W. R., **3**, 812, 814
Box, V. G. S., **14**, 49
Boyd, D. R., **14**, 827
Boyd, G. V., **14**, 577
Boyd, R. J., **3**, 994
Boyer, B., **1**, 65
Boyer, J., **4**, 607; **12**, 275
Boyer, J. H., **4**, 359
Boyer, M., **3**,54
Boyer, M. P., **3**, 603
Bozell, J. J., **12**, 262
Bozoki-Bartok G., **11**,, 6
Bracken, C., **3**, 908
Bradley, S. M., **2**, 308
Brady, B. A., **14**, 663
Brady, W. T., **1**, 168; **13**, 38
Braga, A. L., **4**, 104
Bráhler, G., **3**, 953
Brailovskii, S. M., **2**, 160
Branca, S. J., **14**, 385, 454
Branchini, B. R., **2**, 283
Brand, J. C., **3**, 62, 667
Brandi, A., **13**, 56
Brandina, L. I., **2**, 171
Brandsma, L., **10**, 47, 120, 124; **14**, 348
Brandström, A., **7**, 100
Branum, G. D., **14**, 413
Brarda, M. T., **3**, 404
Braslavsky, S. E., **14**, 882
Brauer, H. D., **3**, 975
Brault, D., **3**, 789
Brauman, J. I., **3**, 1052; **10**, 294, 320
Braun, M., **12**, 220
Bravo-Zhivotovskii, D. A., **3**, 1023
Brayer, G. D., **2**, 281
Brazina, I. O., **3**, 255
Breazu, D., **1**, 67
Brechbiel, M., **11**, 76
Brede, O., **3**, 222
Breeden, D. L., **10**, 210
Bregadze, V. I., **3**, 668
Bregovec, I., **14**, 449
Breitenbach, L. P., **3**, 150, 164, 304, 306; **4**, 510
Breitmaier, E., **14**, 846
Breliere, C., **14**, 335
Bren, V. A., **14**, 119, 285
Bren, Z. V., **14**, 285
Brenner, D. G., **2**, 306
Breslow, R., **2**, 239; **13**, 122
Brettle, R., **4**, 628
Breuckmann, R., **3**, 573
Brevnova, T. N., **3**, 128; **4**, 356
Brewster, A. G., **4**, 205
Bricca, C. E., **4**, 335
Brich, A. J., **8**, 35
Bridwell, P., **10**, 312
Brilev, V.V., **4**, 471
Brindley, P. B., **3**, 196
Briner, P. H., **10**, 119
Bringmann, G., **3**, 473, 474
Brinker, U. H., **5**, 102, 103
Brisimitzakis, A. C., **14**, 108
Brittain, J. M., **7**, 28; **14**, 44
Brittelli, D. R., **13**, 57; **14**, 758
Britton, D., **9**, 95
Britton, H. G., **2**, 409
Brocard, J., **10**, 319
Brocklehurst, K., **2**, 295
Brocksom, T. J., **4**, 104
Brody, R. S., **2**, 398
Broekhuis, A. A., **13**, 102
Broline, B. M., **14**, 884
Brooke, D. N., **6**, 160, 161
Brooke, G. M., **3**, 445; **14**, 164
Brossmer, R., **1**, 23
Brouwer, D. M., **8**, 60
Brovkina, G. V., **2**, 111
Brovko, V. S., **12**, 111
Brown, A. K., **3**, 618; **14**, 358
Brown, C., **1**, 143
Brown, C. A., **10**, 46; **14**, 833
Brown, D. A., **8**, 41
Brown, D. S., **14**, 310
Brown, H. C., **4**, 578; **8**, 53, 55–57; **9**, 156; **11**, 115; **12**, 119, 124, 126, 127, 135, 137, 138; **14**, 832
Brown, J. M., **2**, 225; **4**, 667
Brown, K. C., **7**, 152; **11**, 15
Brown, P. J. N., **14**, 296
Brown, R. F. C., **5**, 82, 117
Brown, R. S., **2**, 250
Brown, S. B., **4**, 752
Brown, S. P., **7**, 115
Brown, T. M., **8**, 107
Browne, A. R., **13**, 191
Browne, L. M., **14**, 501
Brownstein, A., **13**, 79
Brownstein, S., **10**, 34
Broxterman, Q. B., **3**, 945
Broxton, T. J., **2**, 38, 215, 216; **6**, 73
Bru, N., **5**, 10
Bruce, J. M., **4**, 48; **14**, 130, 859–861
Bruce, M. R., **7**, 134
Bruch, M., **14**, 803
Brudnik, B. M., **3**, 188; **4**, 298
Bruice, T. C., **3**, 892, 1061; **4**, 374, 721, 723, 724, 726; **11**, 20; **14**, 706
Brule, D., **14**, 192
Brun, P., **3**, 454
Brunelle, D. J., **10**, 128
Brunet, J.-J., **3**, 802, 803; **4**, 596, 597; **6**, 25
Brunner, H., **3**, 69
Brunovlenskaya, I. I., **14**, 565
Brunton, G., **3**, 155
Brüntrup, G., **12**, 252
Brusova, G. P., **14**, 414
Bruza, K. J., **10**, 99
Bruzik, K., **2**, 359; **14**, 461, 462
Bryant, D. R., **4**, 208
Bryant, F. R., **2**, 393, 399, 400
Bryant, K. E., **14**, 140
Bryce, M. R., **14**, 327
Bryce-Smith, D., **14**, 2
Bub, G. K., **4**, 128
Bubani, B., **3**, 542
Bubnov, N. N., **3**, 499, 528, 661, 668, 672
Buccheri, F., **11**, 45
Buchachenko, A. L., **3**, 123, 600
Buchanan, J. G., **6**, 114
Buchbauer, G., **13**, 128
Buchman, O., **4**, 615
Buchwald, S. L., **2**, 401
Buck, H. M., **2**, 328–330; **3**, 664–666; **5**, 104; **8**, 13; **14**, 186, 499, 769
Buckingham, D. A., **2**, 244, 370
Buckle, D. R., **14**, 72
Budilova, I. Yu., **2**, 337
Budylin, V. A., **7**, 15; **14**, 863
Budzelaar, P. H., **4**, 699
Buechele, J. L., **14**, 805
Buguslavakaya, L. S., **12**, 20
Bukharov, S. V., **13**, 95, 96
Bulatov, M. A., **9**, 178
Bulbulian, R. V., **7**, 126, 128
Bulman-Page, P. C., **1**, 18
Bumbure, G., **4**, 653
Bunce, N. J., **3**, 896

Buncel, E., **6**, 134, 163; **8**, 152; **10**, 53; **14**, 3
Bundel, Yu. G., **6**, 49, 50; **8**, 130; **9**, 177
Bunnett, J. F., **3**, 800; **6**, 18, 38–40, 67, 176; **14**, 40
Bunting, J. W., **4**, 701, 714; **6**, 169
Bunton, C. A., **2**, 100, 198, 208, 209, 214, 346, 347; **6**, 72; **8**, 46, 47; **9**, 243; **11**, 41, 139
Buono, G., **14**, 837
Buraev, V. I., **14**, 111
Burch, M. T., **14**, 140
Burczyk, B., **14**, 829
Burden, A. G., **2**, 49
Burdon, J., **14**, 250
Burenko, S. N., **3**, 690
Burfield, D. R., **9**, 47
Burger, J. J., **11**, 126
Burger, K., **3**, 408; **13**, 62, 63
Burger, U, **3**, 752; **5**, 67, 110; **10**, 220; **14**, 882
Burgess, J., **9**, 163–165
Burgstahler, A. W., **14**, 273
Burka, L. T., **4**, 355, 746
Burks, J. E., **14**, 182
Burks, S. R., **12**, 145
Burmester, A., **14**, 835, 845
Burnett, M. G., **14**, 827
Burns, G. T., **13**, 26
Burns, J. M., **3**, 870
Burrington, J. D., **4**, 457
Burt, R. A., **1**, 171
Burton, D. E., **7**, 34
Burton, D. J., **3**, 483
Burton, G. W., **3**, 189, 190; **4**, 444, 445
Burton, R. M., **6**, 99
Burzlaff, H., **14**, 26
Buscemi, S., **2**, 53
Buschhaus, H.-U., **3**, 87, 89
Buse, C. T., **1**, 103
Bushby, R. J., **14**, 299
Bushey, D. F., **14**, 858
Buss, A. D., **1**, 92
Buss, V., **4**, 548
Butcher, J. A., **5**, 53
Butin, K. P., **8**, 130; **9**, 177
Butler, A. R., **2**, 114
Butler, D., **3**, 565
Butler, J. E., **3**, 261
Butler, R., **4**, 396
Butler, R. N., **1**, 78; **4**, 76; **9**, 130
Butler, S. C., **3**, 527
Butler, W., **14**, 313
Butovskata, G. V., **4**, 519
Butselaar, R. J., **4**, 416; **14**, 722
Buttafava, A., **3**, 27, 37, 1152
Buttero, P. D., **14**, 25
Buttrill, S. E., **7**, 61
Buynak, J. D., **3**, 565
Buza, M., **14**, 493
Buzzaccarini, F., **6**, 72
Bychlov, N. N., **8**, 79
Byers, L. D., **2**, 57, 194
Byers, M., **7**, 25
Bykova, N. V., **4**, 477
Byrce, M. R., **6**, 171
Byrn, S., **13**, 13
Bywater, S., **10**, 34

Cabelkova-Taguchi, L. M., **11**, 65
Cabiddu, S., **3**, 420
Cacace, F., **7**, 104; **8**, 128
Cacchi, S., **12**, 249
Cacioli, P., **6**, 80
Cadman, P., **3**, 349
Cadogan, J. I. G., **3**, 15; **14**, 296
Cady, M. A., **9**, 59; **12**, 229
Cafferata, L. F. R., **2**, 424; **3**, 91; **9**, 167
Cajurel, C. L., **11**, 56
Calado, M. I. L. T., **9**, 255
Calas, R., **3**, 609
Calder, M. R., **4**, 742
Caldin, E. F., **10**, 273
Caldwell, G., **10**, 145
Calmon, J.-P., **2**, 118
Caluwe, P., **4**, 538
Calvaruso, G., **4**, 93, 94
Calvert, J. G., **3**, 166
Calvo, K. C., **2**, 277
Camaioni, D. M., **3**, 493
Cambie, R, C., **3**, 865; **9**, 114, 115; **12**, 40, 56; **14**, 171, 683
Cambillau, C., **10**, 20
Cambon, A., **12**, 256
Camerini, E., **4**, 66
Cameron, D. W., **12**, 270, 271; **14**, 253
Cameron, T. B., **11**, 50
Cameron, T. S., **3**, 903
Camilleri, P., **3**, 341
Campagnole, M., **3**, 146
Campana, C. F., **4**, 335
Camparini, A., **3**, 577
Campbell, I. M., **3**, 315
Campbell, W. H., **4**, 335
Camps, F., **4**, 303
Camps, M., **12**, 76
Cano, M. C., **12**, 23
Canosa, C. E., **3**, 397, 398
Canosa-Mas, C. E., **5**, 16
Cantacuzene, D., **3**, 374
Cantone, A., **2**, 433; **9**, 101
Cantrell, T. S., **13**, 198
Caple, R., **8**, 113; **14**, 404
Capon, B., **1**, 137, 164; **2**, 2, 3
Capon, R., **14**, 507
Caporiccio, G., **3**, 37, 1152
Caporusso, A. M., **4**, 546,
Capozzi, G., **12**, 165
Capretto de Castillo, M. E., **7**, 96
Caproiu, M. T., **3**, 59
Capuano, L., **13**, 138; **14**, 740
Caraballo, D. F., **2**, 135
Caramella, P., **1**, 127
Carassiti, V., **3**, 316
Carballeira, N., **3**, 563
Carballo, D. F., **11**, 34, 35
Carbini, M., **4**, 475
Carbognani, L., **14**, 99
Carde, R. N., **5**, 122
Cardillo, G., **12**, 197
Cargill, R. L., **14**, 858
Caristi, C., **12**, 165
Carlsen, L., **11**, 38, 51, 78; **14**, 222, 774
Carlsen, P. H. J., **4**, 152
Carlsen, R., **2**, 131
Carmella, P., **12**, 5, 6
Carmody, M. A., **11**, 40
Carmona, O., **14**, 84
Carniti, P., **4**, 525; **11**, 92
Caronna, T., **3**, 911
Carpenter, B. K., **6**, 81; **11**, 61
Carpenter, P. C., **14**, 374
Carpenter, T. A., **10**, 316
Carpentier, J.-M., **2**, 79
Carr, R. V. C., **4**, 284, 411; **13**, 110
Carr, R. W., **3**, 1065
Carrasco, N., **8**, 47; **11**, 139
Carré, D. J., **12**, 84
Carré, J. C., **14**, 661
Carré, R., **13**, 39, 56; **14**, 402, 682
Carrington, T., **3**, 234
Carriur, M., **3**, 780
Carruthers, W., **8**, 107
Carsky, P., **6**, 5
Carson, J. R., **14**, 862
Cartano, A. V., **6**, 38
Carter, L. G., **10**, 137
Carter, T. L., **1**, 152; **12**, 21
Cartoon, M. E. K., **10**, 224
Carver, D. R., **3**, 807; **6**, 20
Casadevall, A., **14**, 480
Casadevall, E., **14**, 480
Casagrande, F., **3**, 720
Casanova, J., **3**, 454
Caserio, M. C., **2**, 130; **8**, 137,

138; **14**, 333
Casewit, C., **5**, 136
Casey, C. P., **14**, 254
Casini, G., **7**, 118
Caspi, E., **4**, 738, 739
Casserly, E. W., **5**, 24
Cassidy, J. F., **2**, 117
Castaner, J., **6**, 61
Castedo, L., **3**, 627; **4**, 180, 641
Castellano, A., **3**, 33
Castellijns, M. M. C. F., **2**, 329
Castlehano, A. L., **3**, 20
Castro, C., **2**, 68
Castro, E. A., **2**, 68
Caswell, M., **2**, 179
Catteau, J. P., **3**, 33
Catus, W. B., **4**, 744
Caubere, P., **3**, 802, 803; **4**, 596, 597; **6**, 25; **11**, 7
Causey, J. G., **3**, 32
Cava, M. P., **3**, 876
Cavasino, F. P., **4**, 93, 94
Cavazza, M., **6**, 147
Cavezzan, J., **9**, 94
Cavicchioni, P. S., **10**, 141
Cavinato, G., **12**, 106
Cayzergues, P., **9**, 268; **12**, 177
Ceccherelli, P., **4**, 408
Ceccon, A., **10**, 49
Celina, M., **3**, 1016
Ceré, V., **10**, 226; **12**, 200; **14**, 219, 220
Cerfontain, H., **7**, 47, 94; **14**, 857
Černý, V., **9**, 63, 64; **12**, 34, 35
Cerrini, S., **14**, 525
Cerveny, L., **1**, 12; **9**, 56
Cesca, S., **3**, 27
Cevasco, G., **2**, 62; **11**, 129
Chadha, R., **13**, 10
Chae, W.-K., **14**, 803
Chaichit, N., **5**, 82
Chaigneau, M., **4**, 110
Chaintreau, A., **3**, 687, 688; **4**, 44
Chakrabarti, J., **2**, 24
Chalchat, J.-C., **14**, 192
Chalk, C. D., **9**, 234; **11**, 144
Challis, B. C., **10**, 330
Chaloner, P. A., **4**, 667
Chalova, O. B., **1**, 11; **2**, 8; **8**, 102
Chambers, D., **9**, 114; **12**, 40
Chambers, R. D., **3**, 355, 384, 993; **10**, 93; **14**, 80
Chan, W.-T., **7**, 12
Chandalia, S. B., **4**, 323, 332
Chandler, M., **8**, 36, 40
Chandrasekhar, J., **8**, 6, 119, 127, 161; **10**, 5, 9, 14, 26, 70, 97, 311; **14**, 388
Chandrinos, J. D., **3**, 625
Chang, A. I., **14**, 441
Chang, C.-A., **13**, 210
Chang, C.-C., **12**, 42, 85
Chang, C. K., **3**, 964; **4**, 743, 748
Chang, C. T., **2**, 183
Chang, J. S., **3**, 21
Chang, M. H., **3**, 559
Chang, M. J., **3**, 1116; **5**, 55; **14**, 315
Chang, P. C., **5**, 108
Chang, S.-C., **5**, 81
Chang, T. C. T., **12**, 162
Chang, Y. H., **5**, 40
Chantegrel, B., **14**, 578
Chapat, J.-P., **6**, 116
Chapman, A. V., **14**, 63
Chapman, D. C., **3**, 541; **4**, 329; **7**, 143
Chapman, N. B., **2**, 49–51
Chaquin, P., **13**, 22
Charl, S., **5**, 3
Charles, K. A., **14**, 884
Charlton, J. L., **14**, 777
Charnas, R. L., **2**, 305, 308, 309
Charpentier, J.-P., **12**, 211, 212, 213
Charrier, C., **14**, 274
Charton, M., **2**, 49–51
Charushnikov, K. A., **9**, 178
Chastain, J. E., **10**, 58
Chastrette, M., **9**, 57
Chatgilialoglu, C., **3**, 352, 584, 585, 586, 601
Chatrousse, A.-P., **6**, 134
Chatterjea, J. N., **14**, 719
Chatterjee, A., **4**, 621; **14**, 716
Chatterjee, S. K., **4**, 621; **14**, 716
Chatterjee, U., **4**, 165, 166
Chattopadhyay, S., **4**, 23
Chaturvedi, R. K., **2**, 179
Chatzopoulos, M., **3**, 447
Chaudhry, S. C., **8**, 105
Chaudhury, B., **14**, 491
Chauhan, K. S., **4**, 201
Chawla, R., **6**, 24
Chayet, L., **9**, 243
Chdrasekharan, J., **9**, 259
Chebotarev, N. F., **3**, 400
Chebotarev, O. V., **14**, 552
Chechulina, I. N., **3**, 999
Cheeseman, G,. W. H., **10**, 224
Chekir, K., **4**, 645
Chekulaev, V. P., **3**, 543
Chelibanov, V. P., **3**, 240
Chen, C.-C., **8**, 174
Chen, F. M. F., **2**, 162
Chen, J. C., **12**, 124
Chen, J. S., **14**, 441
Chen, K. S., **3**, 44, 1176
Chen, L.-J., **13**, 162
Chen, S.-A., **14**, 441
Chen, S.-M., **14**, 232
Chen, T. B. R. A., **11**, 126
Chenets, V. V., **7**, 112, 113
Cheng, Y.-S., **13**, 148
Chenier, J. H. B., **3**, 169
Chenier, P. J., **9**, 31
Chepaikin, E. G., **4**, 689
Cheremisin, A. A., **8**, 4, 175
Cherkasov, A. V., **7**, 5
Cherkasov, R. A., **2**, 384; **12**, 53
Cherkasov, V. K., **3**, 97
Cherkaspva, V. A., **1**, 66
Chernikov, V. V., **4**, 482
Chernoivanov, V. A., **14**, 119, 285
Chernyshova, T. M., **3**, 1040
Cherton, J. C., **14**, 751
Chervinskii, K. A., **4**, 511
Chesnokova, T. A., **3**, 128
Chevrier, B., **4**, 749
Chi, M.-S., **14**, 801
Chiang, S.-H., **14**, 817
Chiang, V., **1**, 171
Chiang, W., **13**, 200
Chiang, Y., **1**, 172
Chiari, G., **6**, 113
Chiavarelli, S., **4**, 257
Chichagov, V. N., **4**, 496
Chicu, A., **2**, 126, 127
Childester, C. G., **14**, 93
Chignell, C. F., **3**, 644
Chihi, A., **5**, 148
Chikamatsu, H., **4**, 733–735
Chikhacheva, I. P., **3**, 140
Chikvaidze, N., **3**, 265
Chikyu, H., **4**, 175; **11**, 57
Childers, W. E., **4**, 216
Childs, A., **14**, 250
Chimishkyan, A. L., **2**, 159
Chin, A., **3**, 227; **6**, 11
Chin, J., **2**, 251
Chinn, J. W., **10**, 10; **14**, 627
Chinnasamy, P., **4**, 622
Chiorboli, C., **3**, 316
Chiriac, A., **2**, 262
Chirkova, L. I., **2**, 158
Chisaka, A., **3**, 501
Chishti, N. H., **4**, 357; **14**, 691
Chistokletov, V. N., **12**, 260; **13**, 40

Chistyakov, A. N., **4**, 487
Chiu, F.-T., **5**, 40
Chiu, T.-K., **3**, 1174
Chiusoli, G. P., **3**, 520
Chizhevsky, I. T., **8**, 50, 51
Chizhov, O. S., **3**, 386, 721, 722
Chkhubianishvili, N. G., **3**, 272
Chkubo, K., **2**, 224
Chloupek, F. J., **9**, 156
Chodak, I., **3**, 332
Choi, H.-D., **7**, 133
Choi, H. K. J., **3**, 279; **6**, 170
Chojnowski, J., **10**, 292
Chollet, A., **14**, 191
Cholpankulova, S. T., **14**, 736
Cholvas, V., **3**, 122, 123
Chong, J. M., **12**, 237
Choo, K. Y., **3**, 450
Choong, T.-C., **14**, 453
Chopa, A. B., **3**, 421
Chopra, S. K., **13**, 45, 46
Chorn, T. A., **4**, 90
Chottard, J.-C., **4**, 749
Chou, C. S., **3**, 917; **14**, 639
Chou, F. E., **3**, 777
Chou, Y., **1**, 36
Choubani, S., **3**, 293
Chow, M.-F., **3**, 1159, 1160; **4**, 393
Chow, Y. L., **3**, 1074; **14**, 339, 891
Chowdry, H. C., **4**, 38, 40
Chowdury, A. K., **3**, 1051
Choy, W., **1**, 90; **10**, 191
Chrisope, D. R., **4**, 307, 727
Christensen, B. G., **14**, 128
Christensen, S. B., **5**, 79
Christl, M., **14**, 466
Chu, N. G., **14**, 146
Chubarov, G. A., **2**, 93, 111
Chubb, F. L., **14**, 741
Chuchani, G., **2**, 133, 135, 136; **11**, 32–35, 69, 70, 74
Chuikova, T. V., **7**, 132
Chuit, C., **12**, 236
Chumakov, L. V., **12**, 55
Chung, B. Y., **10**, 158
Chung, C.-J., **3**, 1159, 1160
Chung, S. K., **3**, 820
Chupakhim, O. N., **12**, 179
Church, D,. F., **3**, 301
Churkina, T. D., **3**, 428, 429, 468
Chuvatkin, N. N., **12**, 12
Chuvylkin, N. D., **3**, 31; **8**, 164
Cicero, G., **14**, 58
Cidda, C., **6**, 98
Cieplak, A. S., **1**, 125
Ciminale, F., **3**, 43, 1049; **10**, 314
Cingolani, G. M., **7**, 118
Cinquini, M., **10**, 247
Cioffi, E. A., **4**, 629; **14**, 756
Cipiciani, A., **2**, 100, 214
Ciranni, G., **7**, 104
Ciranni-Signoretti, E., **4**, 257
Cirinna, C., **2**, 326
Cirovic, M., **9**, 126
Citterio, A., **3**, 17, 226, 692, 719, 720, 788; **4**, 183; **6**, 30
Citterio, H., **3**, 524
Ciuffarin, E., **2**, 425
Ciuhandu, G., **2**, 261, 262
Clancy, M. G., **5**, 123
Clardy, J., **4**, 411; **13**, 26; **14**, 204, 811
Clark, B. C., **4**, 407
Clark, B. P., **14**, 658
Clark, C. R., **2**, 370
Clark, G. R., **6**, 82
Clark, H. R., **6**, 99c
Clark, T., **3**, 980; **10**, 7, 9, 12, 97
Clarke, S. J., **4**, 590
Clasesson, S., **3**, 1087
Claspy, P. C., **3**, 969
Clausen, T. P., **8**, 52
Cleland, W. W., **4**,, 755, 762, 766
Clemens, A. H., **3**, 821; **10**, 150
Clemens, P. R., **14**, 616
Clemo, N. G., **14**, 718
Clennan, E. L., **3**, 889a
Clerici, A., **3**, 758; **4**, 642
Clermont, M. J., **10**, 281, 282
Cliff, C. J., **5**, 122
Clive, D. L. J., **10**, 183; **12**, 234
Clizbe, L. A., **13**, 103; **14**, 155
Closs, G. L., **3**, 1172, 1173
Clouet, F. L., **7**, 134
Coates, R. M., **14**, 435
Cobbledick, R. E., **7**, 68
Coblens, K. E., **2**, 37
Coderre, J. A., **2**, 397
Coderre, L. A., **10**, 282
Cohen, G. H., **2**, 271
Cohen, J. B., **2**, 312
Cohen, L. A., **14**, 149
Cohen, P. G., **3**, 340
Cohen, S. G., **2**, 312; **3**, 340, 929
Cohen, T., **14**, 854
Cohen-Adad, R., **9**, 274
Cohn, M., **2** 404
Colburn, A., **3**, 581
Colburn, C. B., **12**, 16
Coleman, M. C., **3**, 172
Coll, J., **4**, 303
Colle, K. S., **9**, 271
Collington, E. W., **14**, 733
Collins, C. J., **6**, 32; **14**, 14, 438
Collins, J. J., **14**, 184, 305, 306
Collins, S., **3**, 434; **4**, 203; **12**, 62
Colman, R., **5**, 107
Colombo, L., **10**, 240, 241; **12**, 223
Colonna, F. P., **3**, 45
Colonna, M., **3**, 536, 607; **7**, 81
Colonna, S., **10**, 247; **12**, 192
Colter, A. K., **3**, 730
Colussi, A. J., **3**, 232
Combes, D., **1**, 32
Comita, P. B., **3**, 566; **14**, 762
Compton, D. A. C., **3**, 585
Compton, R. G., **3**, 727
Concannon, C., **3**, 403
Conlin, R. T., **5**, 150
Connors, K. A., **2**, 69
Consiglio, G., **2**, 53; **6**, 94, 95; **11**, 45; **12**, 100; **14**, 95–97
Constant, G., **2**, 110
Constantinescu, M., **13**, 85
Contreras, L., **13**, 160, 161
Contreras, R., **12**, 122, 123, 128
Cook, M. J., **14**, 63
Cook, P. F., **4**, 755, 762, 766
Cooke, E., **14**, 450
Cooke, **10**, 265
Cookson, R. C., **14**, 344
Coon, M. J., **4**, 754
Cooper, A. J., **14**, 495
Cooper, C. S., **1**, 37
Cooper, D., **14**, 183
Cooper, R., **3**, 159
Cooperman, B. S., **2**, 412
Corbet, J. P., **14**, 223
Corbett, B. R., **4**, 611
Corbett, J. F., **7**, 152
Cordes, E. H., **1**, 31
Cordova, R., **14**, 134
Corey, P., **10**, 131
Corfield, G. C., **6**, 160
Cori, O., **9**, 243
Corina, D. L., **4**, 742
Cormier, R. A., **14**, 659
Cornelisse, J., **3**, 893; **4**, 425; **13**, 196
Cornell, S. C., **9**, 209
Corrao, A., **14**, 95, 97
Corre, F., **12**, 268
Correia, V. R., **11**, 29
Corriu, R. J. P., **2**, 355, 357; **4**, 607; **9**, 86–89; **10**, 156; **12**, 275
Corset, J., **10**, 61
Cort, A. D., **9**, 104
Cortez, C., **10**, 203
Corti, C., **3**, 37, 1152
Cosa, F., **3**, 369

Cosemans, J., **3**, 254
Cospito, G., **11**, 8
Cossey, A. L., **14**, 211
Cossy, J., **3**, 1092
Costa, A., **14**, 555
Costa, A. M. B. S. R. C. S., **10**, 317
Costa, Bizzarri, P., **4**, 543
Costero, A. M., **13**, 163
Cotarcă, L., **2**, 127
Cottam, P. D., **12**, 199
Couffignal, R., **4**, 649
Coulter, D. E., **11**, 40
Coulter, G. A., **3**, 593
Counotte-Potman, A, **6**, 132, 133
Courregelongue, J., **14**, 561
Courtneidge, J. L., **4**, 601
Courtot, P., **14**, 246, 618, 799
Cousseau, J., **12**, 130
Coustal, S., **14**, 511
Coutouli-Argyropoulou, E., **5**, 26
Coutts, I. G., **3**, 702
Couture, A., **3**, 1082
Couturier, D., **3**, 687, 688; **4**, 44; **10**, 319; **14**, 357
Cowan, D. A., **4**, 340
Cowell, A. B., **3**, 620
Cowley, A. H., **8**, 109
Cox, A., **7**, 90
Cox, M. M., **2**, 63, 65
Cox, P. J., **14**, 678
Cox, R. A., **2**, 89–92
Cox, S. D., **4**, 681
Coxon, J. M., **3**, 635; **8**, 8; **14**, 476
Cozzi, F., **10**, 247
Crabtree, R. H., **4**, 681
Crafts, R. H., **12**, 63
Craig, D. J., **2**, 187
Cram, D. J., **12**, 254
Cramer, P., **12**, 89
Crampton, M. R., **6**, 135–137, 150, 160–162
Crans, D. C., **3**, 549
Crawford, R. J., **3**, 559
Craze, G.-A., **14**, 340
Creary, X., **8**, 9
Creaser, I. I., **2**, 247
Creber, K. A. M., **3**, 79
Creed, D., **4**, 429
Cremaschi, P., **3**, 1042
Cremer, D., **4**, 275–278; **8**, 159
Cresson, P., **12**, 231
Criosy-Delcey, M., **4**, 315
Cristina, D., **13**, 71
Crivello, J. V., **4**, 172; **7**, 57
Crockett, G. C., **14**, 726
Croft, K. D., **14**, 328
Crombie, L., **1**, 111
Crombie, L. J., **3**, 709
Crooks, P. A., **4**, 340
Crooks, R. G., **8**, 157
Cross, R. L., **2**, 385
Crossley, M. J., **12**, 271
Crotti, P., **9**, 55
Crouse, D. J., **4**, 72; **14**, 51
Crouse, G. D., **14**, 199, 341
Crow, W. D., **5**, 116
Csizmadia, I. G., **3**, 28
Csongar, C., **14**, 288
Csunderlik, C., **10**, 313
Cueto, O., **4**, 423
Cugnon de Sévricourt, M., **7**, 11
Cullis, P. M., **2**, 364
Cumming, J. B., **3**, 159
Curci, R., **4**, 309
Curini, M., **4**, 408
Curran, D. P., **4**, 181
Cushman, M., **14**, 453
Cuthbertson, A. F., **14**, 275
Cutting, I., **14**, 240
Cvetanovic, R. J., **4**, 387, 388
Czarnowski, J., **3**, 98
Czepluch, H., **3**, 587
Czeskis, B. A., **14**, 641
Czuba, W., **14**, 62
Czytko, M. P., **4**, 128

D'Alba, F., **4**, 81
D'Angeli, F., **10**, 141
D'Auria, M., **4**, 5
D'Costa, R., **14**, 313
D'Incan, E., **4**, 572, 573
Dabestani, R., **4**, 633
Dabi, S., **2**, 156, 157
Dabral, R. P., **4**, 480
Daccord, G., **13**, 22
Dadali, P. A., **2**, 443
Dadoun, H., **4**, 362
Dahl, H. K., **9**, 258
Dahl, D., **2**, 378
Dahler, P., **8**, 38
Dahlmann, J., **4**, 456
Dahn, H., **14**, 412
Dailly, B. N., **3**, 517
Daineko, V. I., **8**, 76
Dalton,, C. E., **4**, 28
Dalton, H., **4**, 741
Dalton, J. C., **3**, 1093
Dalton, J. R., **14**, 858
Daly, J. J., **13**, 51
Damani, L. A., **4**, 340
Damavandy, J. A., **14**, 759
Damgaard, S. E., **4**, 756, 759
Damin, B., **4**, 236
Dammel, R., **5**, 105
Dammers, E., **3**, 987
Damrauer, R., **10**, 146
Dams, R., **4**, 643
Dan, S., **3**, 567
Danen, W. C., **11**, 30
Danheiser, R. L., **9**, 113; **14**, 237, 319, 634
Daniel, T., **4**, 429
Daniels, L., **4**, 711
Daniels, R. G., **12**, 73
Danikeiko, D. A., **9**, 80
Danilova, T. A., **14**, 162, 163
Danion, D., **5**, 96; **13**, 72
Danishefsky, S., **12**, 251
Dannenberg, J. J., **8**, 165; **13**, 8
Dannhardt, G., **14**, 179
Danov, S. M., **2**, 93, 111
Dantsuji, H., **13**, 101
Dao, L. H., **14**, 446
Daran, J.-C., **14**, 608
Darcy, P. J., **14**, 326
Darcy, R., **3**, 590
Darensbourg, D. J., **4**, 598
Darensbourg, M. Y., **4**, 598
Darling, P., **4**, 6
Darmanjan, A. P., **3**, 1087
Darmograi, M. I., **3**, 131
Darnauer, S. M., **2**, 191
Darwent, J. R., **3**, 926
Das, A. R., **9**, 186
Das, B. N., **9**, 187
Das, K. G., **14**, 403
Das, P. K., **3**, 330, 922
Dasgupta, S., **12**, 121
Dash, A. C., **1**, 56, 57
Dash, B., **1**, 56, 57
Dash, M. N., **4**, 8
Dass Gupta, B., **3**, 653
Dassanayake, N. L., **7**, 148
Datta, M. K., **12**, 121
Datta, R., **12**, 121
Daub, G. W., **14**, 140
Dauben, W. G., **14**, 191, 290, 321, 791, 872
Dauksas, V., **7**, 14
Daum, U., **4**, 361
Dauplaise, D., **3**, 953
Dave, V., **14**, 568
Davenport, J. E., **3**, 539
Davenport, K. G., **1**, 84
Davey, P. N., **4**, 592
David, F., **9**, 120
David, L. F., **3**, 276
David, S. K., **11**, 142
Davidson, E. R., **3**, 550, 946; **5**, 11, 88
Davidson, I. M. T., **3**, 350, 1135; **5**, 147
Davidson, P. A., **4**, 28
Davidson, R. S., **3**, 933

Davies, A. G., **3**, 12, 48, 95, 239, 422, 485, 986; **6**, 101
Davies, D. R., **2**, 271
Davies, G. L. O., **2**, 24
Davies, G. M., **14**, 91
Davies, H. C., **4**, 767
Davies, J., **12**, 199
Davies, R., **10**, 333
Davis, A. P., **3**, 497; **9**, 61
Davis, D. D., **11**, 107
Davis, F. A., **4**, 306
Davis, M. S., **3**, 925; **4**, 743
Davis, N. M., **14**, 862
Davis, R., **3**, 745
Davis, R. D., **14**, 779
Davoudzadeh, F., **8**, 46, 47; **11**, 139
Day, A. C., **14**, 297
Day, R. J., **8**, 157
Day, R. O., **2**, 454
Dayrit, F. M., **14**, 606
Dchmidt, S. O., **3**, 109
De Amici, M., **13**, 71
De Bruyne, C. K., **1**, 30
De Buyck, L., **14**, 378
De Chirico, G., **12**, 230
De Fusco, A., **6**, 165
De Fusco, A. A., **6**, 78
De Kimpe, N., **14**, 378
De Konig, A. J., **4**, 699
De Lucchi, O., **3**, 562, 563; **4**, 423; **13**, 188; **14**, 260, 878, 879
De Maoy, P., **3**, 1171
De Maria, P., **1**, 138; **6**, 48
De Micheli, C., **13**, 59, 71
De Munari, S., **9**, 121; **14**, 646
De Rosa, M., **5**, 44; **14**, 99
De Sarlo, F., **13**, 56; **14**, 682
DeFrees, D. J., **10**, 305
DePriest, R. N., **3**, 836, 840, 841, 843, 845, 846; **4**, 529, 531, 533
DePuy, C. H., **10**, 143, 144, 146, 322
DeTar, D. F., **2**, 25, 274, 275; **9**, 216
DeWolf, B., **3**, 1095
Deady, L. W., **2**, 67; **9**, 196
Dean, B. D., **14**, 361, 362
Dean, F. M., **10**, 317
Debaerdemaeker, T., **13**, 81, 82, 181
Debal, A., **1**, 101
Deberitz, J., **4**, 99
Deberly, A., **1**, 132
Declercq, J. P., **13**, 43
Decoreille, A., **4**, 716
Decoret, C., **8**, 165
Dědek, V., **10**, 28; **12**, 172, 173
Dedeoglu, E., **7**, 54
Deganello, S., **2**, 317
Degl'Innocenti, A., **10**, 154
Degny, E., **12**, 256
Degtyarev, A. N., **3**, 668
Degtyareva, T. G., **4**, 516
Deguchi, Y., **2**, 181; **3**, 995, 996
Degurko, T. A., **11**, 148
Dehmlow, E. V., **14**, 640
Dehmlow, E. W., **5**, 74
Deicha, C., **6**, 166
Deiko, S. A., **4**, 73
Dekker, E. E., **1**, 44
Dekock, C. W., **10**, 207
Del Mazza, D., **6**, 172; **13**, 44
Del'tsova, D. P., **13**, 31
Delalu, H., **9**, 274
Delettre, J., **14**, 521
Delgado Diaz, S., **4**, 506
Dell'Erba, C., **2**, 62; **6**, 118
Della, W. S., **3**, 739
Dellacoletta, B. A., **2**, 190
Delpuech, J. J., **1**, 25
Deluy, C. H., **11**, 82
Dem'yanov, P. I., **3**, 689
Demmin, T. R., **4**, 50, 51
Demonceau, A., **5**, 72, 112
Demou, P. C., **2**, 397
Demuth, M., **14**, 4, 882
Demyanov, P. I., **3**, 690
Deng, C., **10**, 259
Deniau, J., **3**, 869
Denis, J. N., **4**, 659
Denisov, E. T., **3**, 167, 1132; **4**, 438, 496, 517
Denisov, G. S., **10**, 276
Denisovich, L. I., **3**, 1040
Denney, D. B., **2**, 331; **14**, 836
Denney, D. Z., **2**, 331; **5**, 52; **14**, 836
Deodhar, K. D., **7**, 114
Depew, M. C., **3**, 1177
Depreux, P., **12**, 157
Derdour, A., **14**, 314
Dereza, N. A., **7**, 124
Dersch, W., **3**, 612
Derstunagova, K. A., **2**, 353
Dervan, P. B., **5**, 13
Desauvage, S., **1**, 170
Desbene, P. L., **14**, 751
Descotes, G., **4**, 667; **5**, 63
Deshayes, C., **14**, 278
Desimoni, G., **13**, 141
Deslongchamps, P., **1**, 4, 5, 7; **4**, 8
Desmaele, D., **13**, 35
Desmeules, P., **2**, 403
Despax, B., **3**, 33
Desrut, M., **4**, 730
Dessouki, A. M., **1**, 167; **2**, 74
Desvard, O. E., **2**, 424; **9**, 167
Deuschle, E., **3**, 39
Deutsch, E. A., **14**, 134
Dev, B. R., **4**, 227
Dev, S., **14**, 500
Devekki, A. V., **12**, 78
Dewar, M. J. S., **4**, 650; **5**, 61
Dey, A. K., **7**, 39
Deya, P. M., **14**, 555
Dhar, D. N., **14**, 563
Dhar, K. L., **14**, 467
Dhavale, D. D., **6**, 66
Dhawan, K. L., **14**, 662
Dheer, S. K., **14**, 45
Di Furia, F., **4**, 119, 123
Di Stefano, D. L., **12**, 195
DiCosimo, R., **4**, 711
DiPietro, R. A., **10**, 109
Diaz Rodriguez, F., **4**, 506
Diaz, E. D., **3**, 824
Diaz, S., **3**, 328
Dick, Y. M., **6**, 2
Dickinson, I. L., **3**, 1007
Dickinson, R. A., **1**, 4
Dickson, J., **3**, 1060; **4**, 378
Diefallah, E., **2**, 104
Diefallah, E.-H. M., **9**, 166
Diener, E., **14**, 847
Dienys, G. J., **6**, 105
Dieter, R. K., **5**, 14; **14**, 454–456
Dietliker, K., **14**, 876
Dietrich, H., **10**, 8, 15
Dietrich, R., **14**, 131
Dignam, K. J., **1**, 45
Digurov, N. C., **4**, 467
Dijkink, J., **14**, 265
Dijkstra, B. W., **2**, 314, 315
Dill, J. D., **8**, 159
Dimitrijevic, D., **2**, 54
Dimitrov, D., **4**, 466
Dimitrov, V., **4**, 456
Dimmel, D. R., **3**, 464
Dimroth, K., **3**, 83, 84; **14**, 292
Dimukhametov, M. N., **13**, 120
Dinçtürk, S., **3**, 5, 1136, 1137
Ding, J.-Y., **3**, 105
Dinkel, R., **14**, 258, 259
Dinnocenzo, J. P., **8**, 16
Dinse, H. D., **2**, 48
Dinulescu, I. G., **14**, 187
Dinya, Z., **3**, 45
Distefano, G., **13**, 6
Ditzel, E. J., **14**, 42
Dixit, A. S., **1**, 38
Dixneuf, P. H., **14**, 598
Dixon, B. G., **3**, 129

Dixon, D. A., **3**, 552
Djerassi, C., **14**, 445
Dmitrieva, N. G., **4**, 10
Dmitrieva, D. M., **12**, 13, 14
Dmitrieva, O. P., **4**, 515
Dneptrovskii, A. S., **3**, 290; **8**, 150; **9**, 233
Dobaeva, N. M., **3**, 796
Dobashi, A., **14**, 737
Dobrov, I. V., **3**, 391
Dobson, B., **11**, 16
Dobson, B. C., **3**, 219
Dodd, J. A., **8**, 17
Dodd, J. R., **3**, 1034
Doddi, G., **6**, 96, 167
Dodonov, V. A., **3**, 97
Doecke, C. W., **14**, 487
Doering, W. von E., **14**, 694
Dohrmann, J. K., **3**, 236, 246
Doi, J. T., **4**, 212–214
Dolbier, W. R., **3**, 486, 1117–1121; **14**, 298, 630–633
Dolgoplosk, B. A., **10**, 36
Domaille, P. J., **8**, 45
Domalski, M. S., **12**, 204, 205
Dominguez, R. M., **2**, 136; **11**, 32, 69
Domschke, G., **3**, 55–58
Donady, J., **3**, 513
Dondoni, A., **3**, 77; **13**, 6, 32
Donets, G. I., **4**, 518
Donndelinger, P., **14**, 806
Donovan, D. B., **10**, 283; **11**, 141; **12**, 180
Dören, K., **3**, 88
Dorfman, Ya. A., **4**, 486
Dorion, F., **8**, 90
Dornan, J., **9**, 33
Dostovalova, V. I., **3**, 283, 378, 428, 481
Doty, J. K., **9**, 151
Dou, H. J.-M., **10**, 315
Douady, J., **3**, 360
Doubleday, C., **3**, 1165
Dougherty, D. A., **14**, 643
Douglas, A. W., **14**, 580
Douglas, K. T., **1**, 143; **2**, 38a, 77; **11**, 19, 21
Doull, J., **3**, 1007
Douzou, P., **2**, 265
Dowd, P., **14**, 589
Downing, J. W., **10**, 214
Doyle, D. J., **3**, 1071
Doyle, M. P., **14**, 593
Dragolov, V. V., **2**, 159
Drahnak, T. J., **14**, 885
Drakesmith, F. G., **3**, 993
Draney, D., **3**, 594
Draper, M. R., **7**, 52
Drapier, J., **1**, 54
Dreeskamp, H., **14**, 818
Dreher, E.-L., **3**, 213
Dreiding, A. S., **5**, 64
Drenth, J., **2**, 314, 315
Drenth, W., **2**, 228
Drew, M. G. B., **1**, 49, 50; **14**, 602
Drozd, V. N., **6**, 119, 149
Drucker, G. E., **10**, 298
Druelinger, M., **10**, 178
Drues, R. W., **14**, 884
Druet, L. M., **2**, 91
Drupp, P. V., **2**, 172
Druzhkova, G. V., **12**, 115
Dryer, F. L., **3**, 1103
Dryndin, V. M., **3**, 305; **11**, 79
Dubac, J., **4**, 406; **9**, 94
Dubey, R. C., **4**, 111
Dubitskaya, A. F., **12**, 75; **14**, 472
Dubois, J.-E., **1**, 140; **7**, 27; **12**, 25, 37
Dubrovina, I. V., **3**, 396
Duce, P. P., **9**, 163–165
Duddy, N. W., **2**, 38
Duesler, E. N., **2**, 415
Duffey, D. C., **3**, 970; **14**, 406
Duggal, S. K., **3**, 610, 611
Dumitreanu, A., **2**, 261, 262
Dumont, C., **1**, 105
Dumpis, M., **7**, 125
Dunaway-Mariano, D., **2**, 413
Dunbar, R. C., **3**, 969
Dunham, W. R., **3**, 962
Dunitz, J. D., **9**, 95; **10**, 232
Dunkin, I. R., **5**, 17; **14**, 495, 496
Dunlop, M. G., **14**, 172
Dunning, T. H., **3**, 552
Dunoguès, J., **3**, 609
Duong, K. N. V., **3**, 869
Dupas, G., **4**, 716
Duperrouzel, P., **14**, 447
Dupuy, C., **14**, 64
Durand, R., **13**, 43
Durban, P. C., **3**, 205, 266
Durbut, P., **4**, 120
Durochar, G., **3**, 934
Dürr, H., **5**, 21
Durrant, M. L., **9**, 118
Durst, T., **11**, 55; 14, 261, 662
Dushkin, A. V., **3**, 18
Dustman, C. K., **10**, 59, 60
Dutheil, J. P., **2**, 355, 357; **9**, 89
Dutka, F., **2**, 139, 141, 143
Dutly, A., **7**, 54
Dutsch, H.-R., **3**, 271
Duus, F., **14**, 851
Duynstee, E. F. J., **3**, 144, 145, 254
Dwivedi, C. P. D., **13**, 5
Dyadyusha, G. G., **3**, 357, 370, 436, 533
Dyer, J. C., **4**, 651; **5**, 132
Dyer, R. D., **14**, 858
Dykstra, C. E., **3**, 109; **11**, 156; **14**, 770
Dyong, I., **13**, 86; **14**, 153
Dzheiranishvili, M. S., **3**, 272
Dzhigirkhanova, A. V., **12**, 74
Dzierzkiewicz, H., **6**, 89

Eaborn, C., **1**, 59, 60; **7**, 150; **9**, 84; **10**, 286–291, 293
Eades, R. A., **3**, 552
Easton, C. J., **3**, 324
Easton, N. R., **5**, 133
Eastwood, F. W., **5**, 82, 117
Eaton, D. R., **3**, 595
Eaton, J. T., **5**, 133
Eaton, P. E., **14**, 385
Ebata, T., **3**, 162, 163; **14**, 141
Eberbach, W., **14**, 317, 318, 661
Eberhardt, E., **3**, 544
Eberson, L., **3**, 547, 548, 693, 792; **6**, 28, 29
Ebina, F., **4**, 748
Ebine, S., **7**, 35
Echter, T., **14**, 330
Ecker, P., **3**, 972
Eckes, H., **5**, 100
Eckhardt, H. H., **14**, 13
Eckstein, F., **2**, 394
Edelson, D., **4**, 80
Edlaud, U., **10**, 65
Edoho, E. J., **14**, 760
Edward, J. T., **14**, 450, 741
Edwards, D. M. F., **12**, 29
Edwards, J. O., **3**, 706, 732; **4**, 182, 309, 337
Edwards, O. E., **14**, 531
Edwige, C., **5**, 155
Efendiev, E. Kh., **13**, 119
Effenberger, F., **8**, 33
Effio, A. A., **3**, 490
Efremova, E. P., **3**, 140
Efremova, G. G., **12**, 186
Egashira, N., **3**, 1005
Egberink, H., **1**, 68
Ege, G., **5**, 29; **13**, 170
Egerton, P. L., **13**, 11
Eggelte, H. J., **9**, 122
Egorochkin, A. N., **3**, 874
Egorov, M. P., **6**, 148

Egsgaard, H., **2**, 131; **11**, 38, 51, 78; **14**, 222, 774
Eguchi, S., **13**, 55
Ehrenberg, A., **3**, 962
Ei'nitskii,, A. P., **3**, 135
Eichler, K., **5**, 58
Eid, M. M., **14**, 83
Eiden, F., **14**, 120
Eiki, T., **2**, 234
Einstein, F. W. B., **7**, 68
Eisch, J. J., **10**, 252
Eisenberg, W. C., **4**, 396
Eisenstein, O., **2**, 73; **3**, 551; **12**, 170
Ektova, L. V., **14**, 52
Ekwenchi, M. M., **3**, 646
El Ashry, S. H., **2**, 105, 192; **9**, 208
El Sadani, S. K., **9**, 54
El Seoud, M. I., **2**, 201
El Seoud, O. A., **1**, 113; **2**, 201
El Soueni, A., **3**, 372, 373
El-Alaoui, M., **1**, 140; **12**, 37
El-Bahaie, **2**, 182, 184
El-Basil, S., **12**, 263; **13**, 89; **14**, 843
El-Fattah, B. A., **13**, 89
El-Fayoumy, M. A. G., **8**, 97
El-Kabbani, F. M., **11**, 50
El-Kashef, H., **7**, 11
El-Mowafy, A. M., **9**, 219; **11**, 10
El-Nahas, H. M., **5**, 25
El-Sadany, S. K., **9**, 11
El-Shafei, A. K., **1**, 47
El-Shafie, S. M. M., **9**, 219
El-Sheikh, M. I., **6**, 175
El-Sheikh, Y. M., **2**, 21
El-Sherief, H. A. H., **14**, 24
El-Sukkary, M. M. A., **4**, 375
El-Zaru, R., **4**, 143
Elguero, J., **14**, 844
Eliasson, B., **10**, 65
Eling, T. E., **3**, 880
Elinson, M. N., **3**, 385, 386
Eliseeva, G. D., **2**, 32, 33, 341, 342
Elix, J. A., **14**, 35
Elkasaby, M. A., **5**, 119
Elkind, J. L., **2**, 312
Elliott, R. C., **2**, 225
Ellis, M. D., **5**, 16
Ellison, R. H., **12**, 147
Ellul, R., **3**, 341
Elrod, J. P., **2**, 58
Elrod, L. W., **14**, 431
Elsemongy, M. M., **2**, 94, 95; **7**, 127
Elsenbaumer, R. L., **4**, 551
Elsevier, C. J., **12**, 226
Elsheimer, S., **3**, 1117, 1118; **14**, 631
Emanuel, N. M., **4**, 432
Emel'yanov, V. I., **3**, 287
Emly, M., **1**, 117; **10**, 202
Emokpae, T. A., **6**, 51
Encina, M. V., **3**, 328, 1078
Encinas, M. V., **3**, 330, 1079
Enders, D., **10**, 180
Endo, H., **1**, 114
Endo, M., **3**, 230
Endo, T., **10**, 257
Eneback, C., **14**, 713
Engbert, T., **9**, 39
Engberts, J. B. F. N., **2**, 426, 428; **3**, 67, 504
Engdahl, C., **8**, 15, 16, 74
Engel, P. S., **3**, 201; **14**, 803
Engelhardt, U., **4**, 341
Engelman, C., **10**, 2
England, D. C., **1**, 169
Engman, L., **4**, 210; **9**, 112; **12**, 151
Enkaku, M., **14**, 68, 69
Enkelmann, V., **3**, 948
Enokida, R., **5**, 144
Enriqez, R., **14**, 508
Epiotis, N. D., **14**, 1
Epsztein, R., **10**, 48; **14**, 267, 347
Erbs, W., **11**, 29
Ercolani, G., **9**, 106
Erden, I., **13**, 189, 192
Erickson, A. S., **3**, 825
Erickson, G. W., **1**, 108; **10**, 178, 179
Erickson, W. F., **12**, 204
Erker, G., **3**, 506; **14**, 605
Erkine, R. W., **1**, 126
Ermakov, Yu. I., **4**, 18, 19, 161, 463
Ernst, J., **12**, 253
Ernst, R. R., **14**, 768
Ernstbrunner, E. E., **3**, 728
Erokhina, N. G., **3**, 286
Errede, L. A., **2**, 191
Ershov, B. A., **10**, 21–23
Ershov, V. V., **3**, 499, 867
Esaki, T., **13**, 55
Esakov, S. M., **10**, 21
Escale, R., **6**, 116
Eschenmoser, A., **14**, 847
Esdar, M., **12**, 220
Espenson, J. H., **3**, 851; **4**, 53, 62
Esser, M. L., **3**, 144
Esteban, S., **14**, 554
Estrina, G. Ya., **3**, 115, 334
Etrillard, J., **9**, 49
Etter, M. C., **2**, 191
Evans, B. L. B., **2**, 294
Evans, B. S., **3**, 344
Evans, D. A., **1**, 91, 92, 94; **10**, 188–190
Evans, D. J., **8**, 49
Evans, J. C., **3**, 663
Evans, S. A., **4**, 651; **5**, 132
Evans, W. J., **10**, 211
Everhardus, R. H., **10**, 47
Evleth, E. M., **13**, 8
Evmenenko, N. P., **4**, 485
Evteeva, N. M., **4**, 514
Exner, O., **2**, 76

Fabian, B. D., **14**, 597
Fabian, J., **3**, 57
Fabre, C., **3**, 596
Facsko, O., **2**, 70
Faddeeva, M. I., **14**, 560
Fadlallah, M., **12**, 175–177
Fadnavis, N. W., **2**, 30
Fagley, T. F., **9**, 168
Fahey, D. R., **4**, 680
Fahey, R. C., **12**, 195
Fahmi, A. A., **13**, 74
Failes, R. L., **11**, 72; **14**, 411
Fainzilberg, A. A., **12**, 265
Faird, R., **10**, 321
Fajer, F., **4**, 743
Fajer, J., **3**, 925, 964
Falshaw, C. P., **14**, 714
Falsig, M., **3**, 761, 783–785
Fananás, F. J., **10**, 13
Fang, J-M., **10**, 236
Farage, V. J., **4**, 79, 83
Faragher, R., **14**, 218
Farah, J. P. S., **1**, 113
Farcasin, M., **14**, 477
Farcasiu, D., **8**, 5
Farina, F., **9**, 10
Farina, V., **12**, 234
Farmer, T., **2**, 304
Farng, L. O., **2**, 449; **11**, 28
Farnum, D. G., **8**, 52
Farquharson, G. J., **8**, 58
Farrall, M. J., **4**, 6
Farrant, R. D., **14**, 513
Fasani, E., **3**, 899
Fasel, J. P., **14**, 874
Fatiadi, A. J., **10**, 212
Fattah, N. A., **10**, 29
Faucitano Martinotti, F., **3**, 27, 37, 1152
Faucitano, A., **3**, 27, 37, 1152
Fauré, R., **9**, 75
Faustini, F., **9**, 121; **14**, 646
Faustov, V. I., **12**, 265

Fauve, A., **4**, 732; **14**, 831
Fava, A., **10**, 226; **12**, 200; **14**, 219, 220
Favre, A., **2**, 245
Feast, W. J., **14**, 808
Febres, A., **14**, 99
Fedeli, W., **14**, 525
Fedevich, M. D., **4**, 358
Fedin, E. I., **8**, 50; **14**, 786, 788
Fedorchuk, E. S., **7**, 98
Fedorov, S. B., **2**, 211, 353
Fedorova, T. V., **3**, 265
Fedorova, V. A., **3**, 131
Fedorovich, A. D., **14**, 244
Fedorynski, M., **5**, 53
Fedyainov, N. V., **3**, 808, 809; **6**, 26
Fedyanin, N. P., **2**, 138
Feher, F. J., **6**, 100
Feit, B. A., **10**, 110, 111
Feld, W. A., **5**, 41
Feldhues, M., **3**, 134
Feldman, D. S., **2**, 317
Feldman, L., **3**, 280–282
Felföldi, K., **11**, 6
Felix, C. C., **3**, 1011, 1013
Feller, D., **3**, 946; **5**, 11, 88
Fendler, J. H., **9**, 97
Fendrich, G., **4**, 758
Ferappi, M., **7**, 118
Ferber, P. H., **9**, 119
Ferguson, J., **13**, 202
Feringa, B., **3**, 701; **4**, 137
Feringa, B. L., **4**, 416, 417; **14**, 721, 722
Fernandez Gonzalez, J., **4**, 506
Fernandez de la Pradilla, **4**, 550
Fernandez, J. M., **9**, 88
Fernando, B., **13**, 6
Fernando, D. R., **2**, 215, 216
Ferradini, C., **4**, 377
Ferran, H. E., **11**, 121
Ferrand, G., **14**, 561
Ferreira, B., **4**, 104
Ferrin, L., **9**, 46
Ferrino, S., **14**, 503
Ferriol, M., **9**, 274
Ferro, M. P., **14**, 636
Fetizon, M., **14**, 514
Feutrill, G. I., **12**, 270, 271; **14**, 253
Fever, H., **9**, 151
Fey, L., **1**, 67
Fiandanese, V., **12**, 230
Fiaschi, R., **14**, 98
Fiaud, J. C., **1**, 88; **10**, 164
Fichtner, M. W., **14**, 463
Ficini, J., **13**, 35
Fiecchi, A., **4**, 312
Fiege, H., **4**, 502
Field, J. S., **12**, 29
Fielder, T. H., **3**, 1117, 1119; **14**, 632
Fielding, H. C., **3**, 355; **10**, 93
Fielding, L. J., **14**, 165
Fife, T. H., **1**, 1, 34; **2**, 176, 248
Figeys, H. P., **14**, 152
Filby, W. G., **3**, 1085
Filipp, N., **14**, 184, 305
Filippov, A. P., **4**, 121
Filippov, M. P., **7**, 36
Filippov, N. A., **12**, 114, 117
Filliatre, C., **3**, 99, 100, 125, 146
Filseth, S. V., **3**, 234
Findeis, M. A., **3**, 98
Finer-Moore, J., **14**, 765
Fini, A., **1**, 138
Fink, A. L., **2**, 266
Finkel'shtein, B. L., **2**, 31–33
Finlayson, W. L., **2**, 67
Fiorentino, M., **4**, 309
Fiorenza, M., **1**, 59; **10**, 154
Fiorini, M., **4**, 671
Firestone, R. A., **3**, 7; **13**, 123; **14**, 128, 129
Fischbach, U., **10**, 6
Fischer, A., **7**, 68, 70, 73
Fischer, E. O., **5**, 154
Fischer, H., **3**, 69, 70, 236, 244, 271, 1066
Fischer, J., **14**, 274
Fischer, W., **9**, 19
Fischer, P., **5**, 50
Fišera, L., **13**, 65
Fish, J. G., **12**, 104
Fisher, J., **2**, 308
Fisk, T. E., **8**, 141; **9**, 9
Fitos, I., **10**, 271; **11**, 116
Fitt, J. J., **10**, 218
Fitts, S. W., **2**, 413
Fitzjohn, S., **4**, 48
Fkih-Tetouani, S., **14**, 509
Flammang, R., **14**, 426, 427
Flammang-Barbieux, M., **14**, 427
Flay, R. B., **2**, 325
Fleet, G. W. J., **4**, 590–592
Fleischhacker, W., **14**, 176
Fleischhauer, J., **13**, 90
Fleischmann, K., **14**, 248, 748
Fleming, I., **8**, 142; **10**, 88; **11**, 102, 103, 111; **12**, 118, 241, 242; **14**, 365, 429, 430, 671, 672
Fleury, D., **1**, 82
Fleury, M. B., **1**, 82
Florio, S., **2**, 102
Floris, B., **12**, 131
Floss, H. G., **4**, 740
Flowers, M. C., **14**, 665, 699
Floyd, C. D., **10**, 88; **14**, 365
Fludzinski, P., **10**, 168
Flueck, V., **14**, 383
Fodor, G., **2**, 129
Fodor, L., **14**, 715
Foergeteg, S., **3**, 365
Föhlisch, B., **13**, 176, 177
Fojtík, A., **3**, 218
Fokin, E. P., **14**, 52
Foldwhite, H., **3**, 82
Fong, R. H., **14**, 717
Font Freide, J. J. H. M., **2**, 332
Fontan, F., **3**, 806
Fontana, G., **14**, 58
Fontanille, M., **10**, 32
Foote, C. S., **3**, 891; **4**, 395, 422, 426
Forcellese, M. L., **4**, 66
Forch, B. E., **3**, 1058
Ford, G. P., **8**, 171
Ford, T. M., **11**, 115
Ford, W. J., **9**, 191
Ford, W. T., **9**, 190
Foresti, E., **6**, 48
Forestiere, A., **4**, 236
Foricher, J., **11**, 90
Forlani, L., **6**, 48, 97
Forman, A., **3**, 925
Fornasier, R., **2**, 231
Forrester, A. R., **3**, 68, 366, 443, 622, 723; **7**, 25
Forrester, J., **13**, 197
Forster, M., **3**, 931, 949
Fortgens, R., **14**, 265
Fortier, R. A., **14**, 676, 735
Fossey, J., **3**, 148, 480, 523
Foster, B., **5**, 18
Foster, B. A., **2**, 408
Foster, R. W. G., **14**, 310
Foulet, G., **3**, 307
Foulger, B. E., **13**, 197
Foulon, J. P., **12**, 236
Four, P., **4**, 603
Fowler, F. W., **10**, 126; **13**, 148
Fowler, L. J., **11**, 123
Fox, J. J., **14**, 77
Foxall, J., **3**, 1175
Fraenkel, G., **10**, 2, 3
Franchi, V., **3**, 524; **6**, 30
Francis, D. J., **2**, 244
Francis, M. D., **14**, 659
Francisco, J. S., **3**, 98
Francke, W., **14**, 757
Francotte, E., **13**, 149
Frank, F., **14**, 248

Frank, M. S., **4**, 134, 135
Frank, R., **9**, 76
Frank, R. M., **9**, 28
Franke, W., **8**, 127, 146, 147; **14**, 439
Frankel, E. N., **3**, 171
Frankfater, A., **2**, 288
Franklin, R. C., **2**, 199; **3**, 192
Franz, J. A., **3**, 493
Franzi, R., **3**, 645
Franzus, B., **9**, 272
Frappier, F., **14**, 512
Fráter, G., **10**, 175
Fray, G. I., **14**, 235
Frazier, K., **10**, 197
Frazier, K. A., **4**, 537, 605
Frazier, R. H., **3**, 708, 715; **4**, 130
Frechet, J. M. J., **4**, 6
Freeman, F., **2**, 448; **4**, 28, 349, 350
Freeman, J. P., **13**, 68
Freerks, R. L., **13**, 103; **14**, 155
Frehel, D., **14**, 561
Frei, B., **3**, 1070; **5**, 30; **14**, 316, 654
Freidlina, R. Kh., **3**, 255, 283, 375, 378–380, 478, 481, 633
Freilich, S. C., **3**, 580; **12**, 120; **13**, 21
Freire, R., **4**, 379
Frenking, G., **8**, 168; **14**, 397
Frenna, V., **6**, 94; **14**, 95–97
Frey, H. M., **5**, 16; **11**, 67
Frey, P. A., **2**, 393, 398
Freyer, A. J., **14**, 686
Frid, T. Y., **14**, 478
Fridman, A. L., **3**, 811
Fridovich, I., **4**, 761
Fridovich, S. E., **4**, 380, 761
Friebolin, H., **1**, 23; **3**, 96
Fried, H. E., **4**, 722; **10**, 297, 298
Fried, J. H., **12**, 147
Friedel, I., **14**, 131
Friedman, R. M., **4**, 666
Friedrich, M. S., **4**, 457
Friedrichsen, W., **13**, 80–82, 140, 181
Friege, H., **13**, 86
Frisch, M. J., **5**, 89; **8**, 156
Frishin, Yu. A., **3**, 18
Fritschel, S. J., **4**, 673, 675
Fritz, H., **14**, 624, 821
Fritz, H. P., **3**, 972
Frolkov, A. I., **12**, 260
Frolov, A. F., **12**, 77
Fromageot, P., **4**, 676
Fronczek, F. R., **13**, 186
Frost, J. W., **3**, 626; **4**, 725
Fruchier, A., **14**, 844
Fry, A., **9**, 241; **12**, 125; **14**, 408
Fry, A. J., **3**, 510, 786
Fry, J. L., **4**, 604
Fuchs, B., **14**, 233, 234
Fuchs, K. P., **3**, 182
Fueno, T., **1**, 166; **2**, 78; **4**, 401, 402
Fugedi, P., **4**, 540
Fujihara, H., **3**, 886; **9**, 184
Fujii, T., **4**, 528, 734; **10**, 157
Fujii, Y., **13**, 165
Fujimoto, H., **3**, 738; **8**, 93; **14**, 850
Fujio, M., **9**, 265
Fujise, Y., **8**, 18
Fujita, E., **1**, 102
Fujita, H., **3**, 866, 995, 996
Fujita, I., **3**, 925
Fujita, M., **14**, 689
Fujita, S., **3**, 1146
Fujita, Y., **14**, 209
Fujitsu, H., **4**, 690
Fujiwara, K., **7**, 53, 62, 77, 78; **14**, 53
Fujiyama, Y., **10**, 231
Fukata, G., **3**, 700; **4**, 138; **7**, 67
Fukawa, H., **4**, 613
Fukazawa, Y., **3**, 990
Fuke, S., **3**, 886
Fuks, R., **12**, 257, 258
Fukuda, Y., **5**, 31; **7**, 67; **14**, 704
Fukui, K., **3**, 230
Fukumoto, K., **14**, 502, 701
Fukunaga, T., **14**, 92
Fukuya, K., **2**, 227
Fukuzumi, S., **12**, 129
Fulara, J., **3**, 1096
Fuller, S. E., **3**, 292
Fuller, T. S. S., **4**, 602
Funakura, M., **14**, 272
Funanashi, T., **3**, 124
Funfschilling, P., **14**, 258
Fung, A. P., **4**, 328; **7**, 56, 58, 145; **8**, 66; **14**, 437
Fung, C. W., **7**, 142
Furlani, A., **13**, 207
Furth, B., **13**, 22
Furukawa, N., **4**, 217; **9**, 182, 184; **11**, 118; **12**, 54
Furukawa, Y., **10**, 195
Furusaki, A., **3**, 448, 465; **14**, 527
Furuta, K., **3**, 515, 1094
Furuta, N., **3**, 1168
Fusco, R., **14**, 293, 294
Fuselier, C. O., **4**, 28
Fushimi, M., **4**, 698
Fuss, A., **11**, 64
Fututa, H., **3**, 923

Gabay, Z., **6**, 109
Gabe, E. J., **3**, 189; **4**, 444
Gabel, R. A., **6**, 79
Gabhe, S. Y., **10**, 153
Gaboriaud, R., **8**, 90
Gabrielyan, S. M., **4**, 320, 322
Gaca, J., **6**, 89
Gadek, T. R., **13**, 209
Gaevskii, V. F., **4**, 485
Gaffney, J. S., **3**, 260
Gagarina, A. B., **3**, 177, 410
Gagliardi, L., **4**, 257
Gaicomelli, G., **4**, 547
Gajda, T. M., **4**, 524
Gajewski, J. J., **3**, 9, 556, 561, 1116; **14**, 189, 236, 315
Gajl, M., **2**, 381; **9**, 204
Gal'pern, E. G., **8**, 170
Gal, D., **3**, 156, 256
Galat, V. F., **4**, 478
Galimova, L. G., **3**, 174; **4**, 294
Galishev, V. A., **13**, 40
Galle, J. E., **10**, 252
Gallery, P. S., **4**, 534
Galli, C., **3**, 217; **6**, 14
Galliani, G., **3**, 729, 1063; **4**, 102, 369; **6**, 36
Gallicano, K. D., **10**, 260
Gallopo, A. R., **4**, 182
Galloway, J. G., **12**, 208
Gallucci, R. R., **1**, 153
Galoyan, M. G., **12**, 69, 70
Galvez, C., **3**, 806
Galzerano, R., **9**, 221
Gam Garyan, N. P., **13**, 31
Gamba, A., **3**, 1002; **13**, 141
Gambale, R. J., **14**, 729
Gambaro, A., **10**, 49
Gambarotta, S., **12**, 93
Gambaryan, N. P., **8**, 170
Gambaryan, P. N., **14**, 371
Gambie, R., **14**, 172
Gambill, C. R., **8**, 104
Gammill, R. B., **4**, 575
Ganapathy, K., **1**, 141; **10**, 326
Gancarz, R. A., **4**, 354
Gandillon, G., **5**, 67, 110; **10**, 220
Gandolfi, C. A., **9**, 121
Gandolfi, R., **13**, 59, 71
Gandour, R. W., **14**, 289
Ganem, B., **2**, 37; **4**, 562
Ganesan, R., **7**, 7, 31, 33

Ganguly, A. K., **14**, 828
Gans, P., **3**, 965
Ganter, C., **12**, 182; **14**, 481, 684
Gaoni, Y., **6**, 127
Gaplovsky, A., **13**, 65
Garapon, J., **4**, 236
Garbis, S. J., **6**, 113
Garcia, J. L., **6**, 82
Garcia-Martinez, A., **9**, 8
Gardner, R. D., **2**, 83
Gardrat, C., **3**, 393
Gareil, M., **3**, 1062; **6**, 35
Gargarina, A. B., **4**, 514
Garlich, J. R., **10**, 58
Garner, A. W., **5**, 83
Garratt, D. G., **12**, 48, 50, 51, 57, 58
Garratt, P. J., **7**, 21; **10**, 221; **14**, 487
Garrett, D. L., **6**, 99
Garry, R.-P., **14**, 192
Garst, M. E., **12**, 103
Garver, J. C., **1**, 20
Garvey, D. S., **10**, 192
Garvin, V. C., **9**, 130
Gasanov, A. A., **13**, 100
Gasanov, R. G., **3**, 429, 468, 482, 588, 633
Gasanova, L. M., **4**, 493
Gase, R. A., **4**, 709, 715
Gaset, E., **2**, 110
Gashimob, A. G., **4**, 468
Gashimov, G. A., **12**, 65, 81
Gaspard, S., **3**, 927
Gasparyan, L. A., **12**, 69, 70
Gassman, P. G., **8**, 20; **9**, 239; **12**, 27; **13**, 190; **14**, 217
Gasteiger, J., **14**, 734
Gatehouse, B. M., **5**, 82
Gati, G., **14**, 601
Gattuso, M., **12**, 165
Gaudemar, M., **1**, 61
Gaudemer, A., **3**, 869; **4**, 461
Gaviña, F., **13**, 163
Gavrilovic, D. M., **14**, 836
Gawali, B. B., **2**, 429
Gawley, R. E., **1**, 75; **14**, 551
Gaynor, B. J., **14**, 323, 816
Gazizov, T. Kh., **2**, 382
Gdanski, R. D., **3**, 406, 660
Gebeyehu, G., **4**, 211
Gedge, D. R., **14**, 718
Gedge, S., **2**, 5
Geevers, J., **13**, 36
Geise, H. J., **4**, 643
Geletii, Yu. V., **4**, 495
Gelin, S., **14**, 278, 559, 578
Gellerstedt, G., **4**, 326
Gelli, G., **11**, 92
Gemal, A. L., **4**, 588
Geneste, P., **3**; 1099; **4**, 695; **13**, 43
Gennari, C., **10**, 240, 241; **12**, 223
Gennick, I., **12**, 27
Gensler, W. J., **14**, 45
Geoffroy, M., **3**, 598, 645
Geoghegan, M., **14**, 663
Geoghegan, P. J., **12**, 135
Geol, A. B., **4**, 531
George, M. V., **14**, 301, 882
George, T. J., **2**, 453; **11**, 95
Georgescu, E. G., **14**, 187
Georgoulis, C., **9**, 268
Geppert, J. T., **7**, 69
Gerdes, J. M., **9**, 42; **14**, 444
Gerlt, J. A., **2**, 397
Germain, J. P., **3**, 596
Germain, G., **13**, 43
German, E. D., **10**, 274
German, L. S., **3**, 375
Germer, A., **13**, 48
Gerratt, J., **3**, 235
Gerson, F., **3**, 942, 963, 989, 991
Gerson, S. H., **9**, 253
Gesing, E. R. F., **13**, 211
Gesson, J.-P., **7**, 146
Getoff, N., **3**, 1141
Gevaza, Yu. I., **9**, 110; **11**, 148
Gevorkyan, A. A., **14**, 423
Gevorkyan, M. G., **4**, 189
Gey, C., **5**, 48
Gey, E., **3**, 57
Ghatak, U. R., **14**, 494, 504, 537
Ghera, E., **10**, 254
Ghisalberti, E. L., **14**, 328, 507
Ghobt-Sharif, J., **1**, 143
Gholami, A., **14**, 766
Ghoneim, F. B., **1**, 149
Ghosh, A. K., **1**, 164; **2**, 2, 3
Ghosh, S., **10**, 198; **14**, 504
Giacomelli, G., **4**, 546
Giacomello, P., **7**, 104; **8**, 128
Giada, K., **3**, 195
Giada, T. M., **3**, 195
Giammona, G., **14**, 58
Giannotti, C., **3**, 927
Giansiracusa, J. J., **8**, 58
Gibbs, H. W., **7**, 51
Gibson, B., **6**, 136
Gibson, H. W., **14**, 789
Gibson, T., **14**, 697
Gibson, T. L., **7**, 102
Giddings, M. R., **4**, 555
Gierke, W., **3**, 74
Giese, B., **3**, 361; **5**, 46
Giffard, M., **12**, 130
Giffney, J. C., **7**, 62; **14**, 53
Gil, P., **13**, 163
Gilbert, A., **13**, 193, 195, 197; **14**, 2
Gilbert, B. C., **3**, 318, 319, 321, 451, 476, 477, 495, 546, 635, 968, 1175
Gilbert, H. F., **9**, 96; **10**, 280
Gilbert, K., **5**, 29; **13**, 170
Gilbert, R. G., **14**, 323, 816
Gilbro, T., **3**, 1045
Gilchrist, T. L., **14**, 218, 271
Giles, J. R. M., **3**, 669, 849, 986
Giles, R. G. F., **4**, 90
Gill, M., **3**, 443; **10**, 115
Gillen, C. J., **2**, 56
Gillespie, R. J., **14**, 651
Gillet, B., **1**, 25
Gillies, C. W., **4**, 279
Gillis, H. R., **13**, 125, 126
Gillissen, H. M. J., **14**, 499
Gilow, H. M., **7**, 34
Gilyarov, V. A., **3**, 672; **14**, 744
Ginak, A. I., **9**, 132, 133; **12**, 9, 13–15
Ginebreda, A., **1**, 134; **10**, 86
Gineitite, V., **7**, 14
Gingerich, S. B., **4**, 335
Ginsburg, D., **4**, 404; **13**, 152, 153, 155
Ginzburg, O. F., **8**, 82, 85, 87–89, 116
Giongo, G. M., **4**, 671
Giordano, C., **3**, 692, 719, 720; **4**, 183
Giorgianni, P., **13**, 32
Gipe, B. T., **12**, 21
Gipe, D. E., **1**, 152
Girard, P., **4**, 649
Girijavallabhan, V. M., **14**, 828
Girling, R. B., **3**, 728
Gisler, M., **6**, 71
Gitis, B. S., **6**, 54, 55, 139, 143, 152, 153, 156–159
Giuffrè, L., **4**, 450
Gladstone, C. M., **13**, 146
Glass, W. K., **8**, 41
Glassman, I., **3**, 1103
Glatt, H. H., **2**, 113
Glatz, A. M., **14**, 474, 475
Glaz, A. I., **6**, 139, 153, 156, 157
Gleicher, G. J., **3**, 353
Gleiter, R., **3**, 963; **4**, 412; **10**, 6; **13**, 110, 154, 155
Glick, J., **2**, 317
Glidewell, C., **14**, 275, 400

Glinski, M. B., **14**, 662
Gloriozov, I. P., **8**, 110
Glover, S. A., **3**, 537
Glukhova, S. S., **3**, 1129
Glukhovtsev, V. G., **3**, 101, 327, 462, 463
Gnanapragasam, N. S., **7**, 26
Gnanasekaran, C., **1**, 151; **10**, 327
Göbl, M., **3**, 877, 1140; **4**, 347
Goddard, J. D., **5**, 139; **14**, 773
Goddard, R. D., **11**, 68
Godfrey, C. R., **4**, 630
Godleski, S. A., **8**, 173; **9**, 27
Godovikov, N. N., **3**, 668
Godoy, J., **4**, 146
Godreau, P. V., **2**, 145
Goel, A. B., **1**, 129; **3**, 836, 838–848, 857; **4**, 529, 530, 532, 533; **12**, 240
Goemann, M., **4**, 72
Goepfert, H., **14**, 247, 249
Goerdeler, J., **14**, 266
Gogan, N. J., **3**, 1007
Gogins, K. A. Z., **3**, 697; **4**, 14, 131
Goh, S. H., **3**, 810; **9**, 150
Going, R., **1**, 153
Gokel, G. W., **6**, 6
Gol'danskii, V. I., **3**, 405
Gol'dfarb, E. I., **3**, 1072
Gold, B., **9**, 51
Gold, P. M., **4**, 575; **10**, 216; **14**, 226
Gold, V., **3**, 787
Goldberg, V. M., **3**, 156
Golden, D. M., **3**, 50
Golding, B. T., **3**, 1073; **4**, 741; **14**, 615
Golding, G., **10**, 116
Golding, P., **6**, 160
Golding, P. D., **9**, 264; **14**, 23
Golding, S. L., **3**, 537
Goldman, P., **6**, 117
Goldmann, S., **10**, 243
Goldschmidt, J. M. E., **6**, 109
Goldschmidt, Z., **14**, 705
Goldstein,, M. J., **8**, 16, 17
Goldstein, S., **14**, 231
Golembeski, N., **9**, 38; **14**, 539
Golinski, J., **12**, 266
Golodets, G. I., **4**, 527
Gololobov, Y. G., **14**, 743
Gololobov, Yu. G., **2**, 337
Golopolosova, T. V., **6**, 139, 156–158
Golubev, N. S., **10**, 276
Golyanskii, B. V., **14**, 281
Golzke, V., **3**, 126
Gomes, B., **4**, 758
Gomez Contreras, F., **12**, 23
Gomez, J., **3**, 1082
Gomez, R. R., **3**, 105
Gompper, R., **1**, 107; **8**, 44; **10**, 19, 162
Goncharov, A. N., **2**, 170
Gonczi, C., **6**, 125; **14**, 76
Gonen, Y., **3**, 348
Gonzo, E. E., **7**, 96
Goodin, J. W., **3**, 933
Goodman, P. D., **6**, 9; **7**, 89, 90
Goosen, A., **1**, 13; **3**, 537
Gopala Rao, B. V., **4**, 273
Gopalakrishnan, G., **4**, 235
Gopalan, R., **1**, 147, 176; **4**, 247; **10**, 325
Gopinathan, M. B., **4**, 259
Gopius, E. D., **14**, 414
Gorbachevskaya, K. R., **3**, 142
Gorbatenko, N. G., **2**, 421, 422
Gorbatov, V. V., **3**, 93
Gordon, M. S., **5**, 146
Gordon, R. J., **14**, 320
Gordon, S., **3**, 159
Gore, J., **14**, 402
Gore, P. H., **6**, 104; **7**, 44, 126, 128
Gorenstein, D. G., **2**, 333
Gorgues, A., **12**, 268; **14**, 598
Gorham, W. F., **11**, 125
Gorlach, Y., **14**, 185
Gorman, A. A., **4**, 389
Gornostaev, L. M., **14**, 121
Gorrichon, L., **12**, 216
Goryaev, M. I., **14**, 570
Gosney, A. P., **7**, 83
Gosney, I., **14**, 296
Goss, L. P., **3**, 261
Gössinger, E., **13**, 67
Goth, H., **13**, 62, 63
Goto, J., **14**, 612
Goto, S., **14**, 793
Goto, T., **14**, 241
Gotoh, T., **3**, 918
Gotthardt, H., **13**, 83
Gottstein, W., **13**, 176, 177
Goure, W. F., **5**, 151
Goya, S., **2**, 241
Graafland, T., **2**, 426, 428
Grabarkiewicz-Szczesna, J., **12**, 259
Grabley, F.-F., **13**, 75
Grabley, S., **13**, 75, 76
Grabner, E. W., **3**, 975
Grabowski, J., **9**, 136
Grady, S. R., **1**, 44
Graf, E., **14**, 374
Graf, F., **3**, 1008
Graf, W., **11**, 91; **14**, 522
Gragerov, I. P., **4**, 271, 272, 652
Graham, D. H., **2**, 373
Graham, G. D., **12**, 120
Gramain, J.-C., **3**, 1088, 1089
Gramalica, P., **11**, 46
Grampovnik, D. J., **14**, 372
Grandbois, E. R., **4**, 544, 556
Grande, H. J., **3**, 962
Granik, V. G., **12**, 86
Granoth, J., **2**, 406
Grasse, P. B., **5**, 6
Grasselli, R. K., **4**, 457
Grassy, G., **6**, 116
Grate, J. H., **3**, 869a
Graubaum, H., **2**, 113
Grava, I., **1**, 66
Graves, D. P., **3**, 464
Gravestock, M. B., **8**, 26
Gray, M. J., **14**, 41
Grayson, B. T., **2**, 178
Graziano, M. L., **3**, 193; **14**, 720
Grdina, M. J., **4**, 392
Gream, G. E., **9**, 119
Greci, L., **3**, 536, 606, 607, 623, 779; **7**, 81; **14**, 416
Green, G. D. J., **2**, 286
Green, I. R., **4**, 90
Green, M., **12**, 261; **14**, 600
Green, R. C., **1**, 152
Greenberg, F. H., **6**, 127
Greenblatt, J., **8**, 148; **9**, 2
Greene, R. M. E., **14**, 827
Greengrass, C. W., **10**, 83
Greenhouse, R., **14**, 84
Gregor, R. B., **4**, 666
Gregoriou, G., **9**, 135
Greifenstein, L. G., **10**, 30
Grein, K., **3**, 988
Gren, A. I., **3**, 325
Gribble, G. W., **5**, 133
Grieco, P. A., **14**, 144, 503
Griesbaum, K., **14**, 420
Grieve, D. Mc. A., **2**, 2
Griffin, B. W., **3**, 121
Griffin, G. W., **13**, 74
Griffith, D. W., **2**, 318
Griffiths, D. V., **4**, 660
Griffiths, P. G., **12**, 271
Griggs, R., **4**, 148
Grigg, C. G., **2**, 225
Grigor'ev, N. A., **3**, 375
Grigoreva, E. N., **7**, 122
Grigorova, T. N., **12**, 17
Grigoryan, V. V., **14**, 351, 352
Griller, D., **3**, 20, 38, 48, 66,

329, 469, 489, 671; **5**, 5
Grimme, W., **3**, 573; **14**, 245
Grimshaw, C. E., **4**, 755
Grimshaw, J., **3**, 1064
Grishin, D. F., **3**, 97
Grishin, O. M., **4**, 142
Grishina, O. N., **2**, 380
Grisola, S., **2**, 409
Gritsan, N. P., **3**, 932
Grob, C. A., **8**, 3, 12; **9**, 19–25, 215
Grobovenko, S. Y., **14**, 163
Gromov, S. P., **6**, 120; **14**, 61
Gronowitz, S., **7**, 13; **10**, 127
Gropen, O., **12**, 108
Grossi, L., **3**, 29, 339, 424
Grossman, W. E. L., **3**, 728
Grosz, K.-P., **3**, 212
Grotemeier, G., **10**, 167
Groth, U., **10**, 259
Grotjahn, D. B., **10**, 153, 234
Groves, I. F., **3**, 745
Grozinger, K., **13**, 79
Grubbs, E. J., **3**, 502
Grueter, H., **14**, 382, 383
Grundon, M. F., **4**, 559; **14**, 177, 178
Grüner, H., **5**, 19
Grutzner, J. B., **10**, 69
Gschwend, H. W., **10**, 218
Gschwendtner, W., **9**, 236; **11**, 143
Gsell, L., **14**, 383
Gu, C., **4**, 426
Guanti, G., **2**, 62, 420; **6**, 18; **10**, 240, 255; **11**, 26, 129
Guarna, A., **14**, 682
Guba, G., **12**, 181
Gudimenko, Yu. I., **4**, 462
Gudriniece, E., **14**, 576
Guengerich, F. R., **4**, 746
Guenot, P., **14**, 402, 682
Guérin, C., **9**, 86–88; **10**, 156
Guerra, M., **3**, 45, 77, 420, 998
Guerrera, F., **14**, 94
Guest, M. F., **3**, 22
Guglielmetti, G., **3**, 720
Guha, D., **4**, 9
Guibé, F., **4**, 603; **10**, 299
Guida, W. C., **4**, 585
Guillaume, A., **3**, 298
Guingant, A., **13**, 35
Gukasyan, T. T., **4**, 184
Gula, D. J., **14**, 686
Gula, M. J., **11**, 121
Gülačar, F. O., **3**, 766
Gulinski, J., **12**, 113
Gumenyuk, V. V., **3**, 640, 768–770
Gumerova, V. K., **3**, 187
Gump, C. A., **3**, 312
Gunak, A. I., **9**, 134
Gunda, T. E., **14**, 713
Güngör, T., **10**, 129
Gunko, V. M., **12**, 185
Gunter, M. J., **4**, 747
Günther, B., **9**, 21–23
Gunther, K., **3**, 1085
Guo, W., **5**, 52, 53
Guo, X.-X., **14**, 619
Gupa, C., **9**, 144
Gupta, B., **14**, 447
Gupta, B. D., **14**, 166
Gupta, H. L., **4**, 40
Gupta, J. C., **4**, 194
Gupta, K. C., **4**, 144
Gupta, K. D., **4**, 9
Gupta, R. R., **14**, 31
Gupta, S. C., **4**, 225, 226
Gupta, V. K., **4**, 186–188, 194, 195, 261–263
Gupta, Y. K., **4**, 77
Gupta, Y. P., **14**, 39
Gurak, J. A., **10**, 10; **14**, 627
Gurria, G. M., **8**, 25
Gurudutt, K. N., **4**, 637
Gurumurthy, R., **2**, 43, 44
Gusarova, N. K., **12**, 186
Guseinov, M. M., **13**, 97
Gusevskaya, E. V., **4**, 463
Gust, S. L., **3**, 1050
Gustafsson, K., **14**, 448
Gutbrod, H. D., **1**, 19
Guthrie, J. P., **1**, 110; **2**, 196a
Guthrie, R. D., **3**, 879; **10**, 147, 148, 204
Gutman, D., **3**, 650
Gynane, M. J. S., **3**, 82, 86

Haake, P., **2**, 351, 352, 372
Haaland, A., **12**, 108
Habashi, F., **3**, 597
Haber, J., **4**, 441
Haberfield, P., **5**, 44
Habib, A., **3**, 139
Habon, I., **4**, 252
Habraken, C. L., **6**, 115
Haces, A., **14**, 523
Hackenberger, A., **5**, 21
Hacker, N., **5**, 17
Hackett, P., **14**, 118
Haddadin, M. J., **13**, 68
Hädicke, E., **3**, 207; **13**, 171; **14**, 318, 324
Hadju, J., **2**, 144
Haelg, P., **12**, 100
Häfelinger, G., **7**, 111
Hagaman, E. W., **14**, 477
Haganbuch, J.-P., **13**, 111; **14**, 295
Hageman, H. J., **3**, 252
Hager, D. C., **13**, 186
Hair, M. L., **3**, 639; **4**, 394
Häjek, M., **3**, 743, 744
Hakushi, T., **3**, 558; **14**, 793
Haky, J. E., **8**, 21; **12**, 80; **14**, 482, 483
Halasz, I., **14**, 601
Halat, M. J., **6**, 124; **14**, 77
Halazy, S., **10**, 96; **14**, 225
Halberstadt-Kausch, I. K., **13**, 174
Haley, N. F., **14**, 463
Halgren, T. A., **14**, 228
Hall, C. D., **11**, 120
Hall, C. R., **2**, 356, 374, 375; **14**, 742
Hall, D. G., **9**, 144
Hall, H. K., **3**, 569
Hall, M. B., **10**, 24
Hall, M. C., **3**, 336, 415
Hall, W. E., **14**, 422
Hallas, R., **4**, 581; **9**, 126
Hallden-Abberton, M. P., **10**, 2, 3
Haller, R., **14**, 835, 845
Halpern, A. M., **13**, 203
Halpern, J., **3**, 527; **4**, 665; **14**, 586, 587
Halpern, M., **14**, 57
Halterman, R. L., **10**, 46; **12**, 32; **14**, 833
Halton, B., **11**, 86, 132, 133
Hamad, H. J., **1**, 66
Hamada, H., **4**, 712
Hamada, Y., **5**, 152
Hamaguchi, M., **6**, 163
Hamamoto, I., **3**, 828; **11**, 114
Hamblin, M. R., **3**, 702
Hamed, E. A., **9**, 11
Hamel, P., **7**, 130
Hamerlinck, J. H. H., **3**, 664–666
Hamilton, R., **14**, 190, 486, 588
Hamilton, S. E., **2**, 276
Hammer, B. C., **13**, 107
Hammer, C. F., **14**, 685
Hammerich, O., **3**, 776, 1037
Hammerum, S., **3**, 971
Hammes, G. G., **4**, 760
Hammond, P. J., **2**, 331; **14**, 836
Hamnett, A. F., **9**, 65
Hamon, M., **4**, 110
Hamuda, E. M., **8**, 130; **9**, 177
Han, B. H., **6**, 47; **9**, 247

Hanack, M., **8**, 23; **9**, 107; **13**, 19; **14**, 438
Hanada, T., **2**, 200
Hanamara, M., **5**, 149
Handa, V. K., **3**, 786
Handel, H., **6**, 69
Handlir, K., **4**, 7
Handoo, K. L., **4**, 634; **10**, 149
Handoo, S. K., **4**, 634
Haneda, A., **4**, 631
Hanessian, S., **2**, 245; **14**, 753
Hanhela, P. J., **3**, 737
Hanko, R., **10**, 138
Hanna, R., **14**, 115
Hannah, D. J., **3**, 862
Hannon, S. J., **4**, 268; **11**, 99
Hanotier, J., **4**, 479
Hanotier-Bridoux, M., **4**, 479
Hanreich, R., **8**, 3; **9**, 19, 21–23
Hansen, D. E., **2**, 392
Hansen, H. J., **14**, 874–876
Hanson, G., **2**, 180; **14**, 869
Hanson, L. K., **3**, 925, 964; **4**, 743
Hanson, P., **3**, 10, 960; **14**, 327
Hansson, A.-T., **12**, 235
Hanzawa, Y., **14**, 811, 812
Hara, S., **1**, 81; **14**, 737
Harada, K., **4**, 566, 679
Harada, N., **14**, 122
Harada, T., **1**, 109; **5**, 76; **10**, 279; **14**, 159
Harakal, M. E., **4**, 306
Haram, T. E., **14**, 677
Harano, K., **11**, 49; **13**, 101, 117, 180, 183, 184
Harbour, J. R., **3**, 639; **4**, 394
Hardell, H. L., **4**, 326
Harder, I., **13**, 19
Harding, K. E., **12**, 145
Harding, L. J., **3**, 962
Harding, P. J. C., **4**, 591, 592
Hardman, K. D., **2**, 299
Hardman, M. J., **1**, 43; **4**, 771
Hare, G. J., **8**, 92
Harger, M. J. P., **2**, 324, 379; **14**, 574
Harkema, S., **13**, 36
Harkema, S., **14**, 322
Harland, J. J., **2**, 454
Harlow, R. L., **3**, 708, 715; **4**, 130
Harman, P. J., **14**, 323, 816
Harmata, M. A., **14**, 441
Harpp, D. N., **4**, 662; **11**, 51; **14**, 222
Harref, A. B., **14**, 509
Harrer, W., **3**, 74, 76
Harris, J. F., **3**, 651
Harris, J. M., **9**, 197, 231
Harris, P. J., **14**, 370
Harris, T. M., **10**, 169, 222
Harrison, A. M., **12**, 155
Harrison, D. M., **14**, 177
Harrison, J. F., **5**, 2
Harrison, J. J., **6**, 82; **7**, 37
Harrison, J. M., **14**, 755
Harron, J., **1**, 2
Harrowfield, J. M., **2**, 247, 369
Hart, D. J., **14**, 202
Hart, H., **4**, 640; **14**, 110, 232, 877
Hartke, K., **13**, 142, 143
Hartman, E. R., **8**, 148; **9**, 2
Hartman, R. F., **4**, 153
Hartmann, M., **14**, 469
Hartmann, W., **3**, 570; **13**, 33
Hartshorn, M. P., **7**, 74, 75; **8**, 72, 73; **14**, 41, 42
Hartung, H., **9**, 221
Hartz, G., **12**, 252
Harun, M. G., **2**, 419; **11**, 25
Haruna, M., **4**, 206
Harvey, J. A., **3**, 970; **14**, 406
Harvey, K., **3**, 835
Harvey, R. G., **4**, 316; **10**, 203
Hasan, F. B., **2**, 312
Hasan, I., **10**, 126
Hasan, M., **2**, 23
Hase, N., **4**, 175; **11**, 57
Hase, T. A., **10**, 233
Hasegawa, A., **3**, 981, 982
Hasegawa, H., **12**, 158, 159; **14**, 550, 689
Hashem, K. E., **4**, 655
Hashida, Y., **7**, 91
Hashihama, M., **7**, 66
Hashimoto, T., **2**, 327; **14**, 626
Hashio, S., **14**, 612
Hasiak, B., **14**, 357
Hassan, A. F., **1**, 138
Hassaneen, H. M., **10**, 312
Hassner, A., **11**, 40; **14**, 313
Hassner, J., **14**, 451
Haszeldine, R. N., **3**, 408, 618, 619, 621; **14**, 358
Hata, G., **14**, 198
Hata, T., **1**, 158; **14**, 366, 367
Hata, Y., **4**, 310; **9**, 81
Hatanaka, K., **10**, 237
Hatano, S., **5**, 124; **14**, 690
Hattori, M., **11**, 47, 48
Hattori, T., **3**, 872
Haucke, G., **3**, 1069; **14**, 131
Hauger, D., **10**, 77
Hauptmann, S., **14**, 87
Hautman, D. J., **3**, 1103
Havlu, V., **10**, 28
Hawari, J. A. A., **3**, 422, 485
Hawkins, D., **5**, 83
Hawley, M. D., **3**, 1053
Hayakawa, K., **13**, 7, 106
Hayami, J., **3**, 829; **6**, 58; **11**, 130
Hayamizu, A., **13**, 93
Hayano, K., **14**, 506
Hayashi, H., **3**, 521, 579
Hayashi, M., **3**, 981, 982
Hayashi, N., **4**, 324
Hayashi, T., **4**, 674
Hayashi, Y., **13**, 61
Hayes, D. M., **2**, 270
Hayes, G. F., **6**, 160
Hayes, P. C., **5**, 122; **14**, 497
Haynes, R. K., **10**, 227; **12**, 210
Hays, R. L., **10**, 145, 235; **11**, 155
Hayward, R. C., **3**, 865; **14**, 683
Hazle, M. A. S., **4**, 457
Heacock, D. J., **8**, 173
Heaney, H., **3**, 541; **4**, 329; **7**, 143
Heasley, G. E., **3**, 409; **12**, 26
Heasley, L. E., **3**, 409; **12**, 36
Heasley, V. L., **1**, 152; **3**, 409; **12**, 21, 26, 36
Heathcock, C. H., **1**, 85, 87, 103; **10**, 106, 181, 182; **14**, 200
Heck, R. F., **4**, 614
Heckmann, R., **14**, 70
Heesing, A., **14**, 724
Heffernan, J. G., **9**, 189
Hegarty, A. F., **1**, 45; **12**, 274
Hege, D., **14**, 13
Hegedic, D., **3**, 138
Hegedus, L. S., **12**, 262
Hehre, W. J., **8**, 127; **10**, 305
Heicklen, J., **3**, 309
Heide, W., **3**, 83, 84
Heideman, W., **2**, 394
Heidrich, D., **8**, 158
Heikkinen, E., **1**, 35; **9**, 123
Heimgartner, H., **13**, 64; **14**, 394
Heine, H.-G., **13**, 33
Heinrich, J.-M., **3**, 612
Heinzelmann, W., **13**, 60
Heiske, D., **9**, 240; **14**, 399
Heistand, R. H., **4**, 750
Helary, G., **10**, 32
Heller, H. G., **14**, 325, 326
Helling, D., **3**, 1018
Hellwinkel, D., **14**, 368
Helmchen, G., **2**, 225; **10**, 167; **13**, 130
Helquist, P., **10**, 176

Helsby, P., **3**, 794; **7**, 63
Hemmerich, P., **3**, 962
Henderson, G. N., **7**, 68, 70, 73
Henderson, J., **3**, 622
Henderson, P. A., **2**, 285
Hendi, S. B., **5**, 128
Hendler, J. M., **14**, 451
Hendry, D. G., **3**, 539; **4**, 434
Heneghan, S. P., **3**, 303
Henichart, J. P., **6**, 52
Henri-Rousseau, O., **9**, 147
Henriksen, U., **14**, 228
Henry, P. M., **4**, 125
Henssen, G., **13**, 143
Hentschel, K., **14**, 78
Hepburn, S. P., **3**, 622
Herbrandson, H. F., **6**, 100
Herchen, S. R., **14**, 714
Nercouet, A., **10**, 262–264
Herdering, W., **14**, 724
Heremans, K. A. H., **4**, 35
Herlihy, K. P., **14**, 425
Herman, B., **10**, 48; **14**, 347
Herman, H. H., **4**, 765
Hermkens, P. H. H., **3**, 666
Hermolin, J., **3**, 247–249, 251, 773
Hernandez, L., **14**, 727
Hernandez, M. A., **3**, 575; **13**, 147
Hernansky, C., **2**, 350
Herndon, W. C., **3**, 514; **13**, 91
Herold, B. J., **3**, 1015, 1016; **4**, 639
Herrera Fernandez, A., **9**, 8
Herrmann, R., **7**, 135
Herrmann, W. A., **5**, 157
Herron, J. T., **3**, 382; **4**, 280
Hershberger, J., **3**, 435, 873
Hertel, L. W., **12**, 149
Herter, R., **13**, 176
Herz, W., **4**, 409
Hesabi, M. M., **5**, 123
Hess, B. A., **10**, 208; **13**, 91; **14**, 376
Hess, U., **3**, 740
Hesse, M., **14**, 703
Hester, R. E., **3**, 728, 931, 949, 950, 961
Heuschmann, M., **14**, 692, 693
Heutzenroeder, K., **2**, 121
Hevesi, L., **1**, 170; **8**, 144
Hewgill, F. R., **3**, 698, 718
Heydt, H., **5**, 101; **14**, 541
Heydtmann, H., **5**, 58
Heyn, A. S., **12**, 204
Heywood, G. C., **8**, 36
Hiaro, K., **1**, 124
Hiatt, R. R., **3**, 90
Hickmott, P. W., **1**, 64
Hidai, M., **4**, 31
Hidaka, I., **7**, 82
Hidaka, M., **9**, 98
Hiemstra, H., **12**, 191
Higashino, T., **9**, 83
Higgins, R., **3**, 292; **4**, 741
Highet, R., **13**, 23
Higuchi, K., **14**, 889
Higuchi, N., **14**, 216
Hii, P., **3**, 711; **4**, 230
Hikage, R., **6**, 142
Hilal, R., **14**, 843
Hilbers, C. W., **3**, 1164
Hilinski, E. F., **3**, 554; **14**, 643
Hill, E. A., **14**, 806
Hill, J. H., **10**, 168
Hill, R. K., **4**, 736
Hill, W. E., **12**, 16
Hillebrand, W., **3**, 212
Hiller, K.-O., **3**, 877, 1140; **4**, 347
Hillgartner, H., **3**, 46
Hillhouse, J. H., **2**, 431
Hinde, A. L., **4**, 618; **6**, 129, 130
Hine, J., **1**, 36, 39; **2**, 86
Hinze, W. L., **9**, 97
Hioki, T., **14**, 159
Hiraga, S., **4**, 366
Hirai, H., **2**, 376; **5**, 111
Hirakawa, K., **14**, 637
Hirakawa, S., **2**, 204, 205; **11**, 42
Hirakubo, K., **3**, 517
Hirama, M., **10**, 192, 253; **12**, 222
Hiramatsu, K., **9**, 98
Hirano, M., **4**, 126
Hirano, Y., **9**, 62
Hirao, A., **4**, 558, 560
Hirao, K., **11**, 88; **14**, 465
Hirao, T., **4**, 661; **10**, 90
Hirashima, T., **12**, 150
Hirata, F., **5**, 31
Hirate, R., **14**, 81
Hirayama, F., **2**, 241
Hiroi, K., **14**, 127
Hirose, T., **14**, 595
Hirota, H., **12**, 142
Hirota, K., **6**, 124; **14**, 77
Hirota, M., **1**, 114
Hirsch, J., **4**, 539
Hirschon, A. S., **4**, 212, 214
Hirst, J., **6**, 43, 51
Hisaoka, M., **3**, 183; **4**, 453
Hitchcock, P. B., **14**, 91
Hitomi, T., **3**, 866
Hiyama, H., **7**, 92
Hiyama, T., **1**, 97; **14**, 648
Ho, T.-I., **3**, 904
Hobbs, C. F., **12**, 102
Hobson, J. D., **14**, 495, 496
Hochstein, P., **3**, 176
Hocki, T., **1**, 109
Hocking, M. B., **14**, 48
Hodges, M. L., **5**, 83
Hoeft, E., **4**, 456
Hoekstra, M. S., **10**, 166
Hoesch, L., **5**, 37, 38, 64; **14**, 804
Hofelich, T. C., **8**, 81
Hoffman, S. J., **4**, 744
Hoffmann, H. M. R., **13**, 133, 134, 172, 175
Hoffmann, M. Z., **3**, 1138
Hoffmann, R., **3**, 551; **12**, 170
Hoffmann, R. W., **5**, 20; **10**, 187; **14**, 224
Hofle, G., **1**, 28
Hogen-Esch, T. E., **10**, 33
Hogeveen, H., **3**, 945; **14**, 498
Hogg, J. L., **2**, 41, 96, 99
Hohenlohe-Dehringen, K., **14**, 710
Höhlein, P., **3**, 610, 611
Hol, W. G. H., **2**, 315
Holak, T. A., **10**, 310
Holba, V., **2**, 35, 36
Holcomb, W. D., **5**, 127
Holecek, J., **4**, 7
Holer, J., **8**, 153
Holland, H. L., **4**, 361
Hollingsworth, M., **10**, 35
Hollinshead, D. M., **4**, 630
Hollocher, T. C., **4**, 717, 718, 720
Hollowood, F. S., **14**, 486
Holmes, R. R., **2**, 454
Holt, D. A., **14**, 428
Holton, D. M., **3**, 1012
Holtwick, J. B., **14**, 79
Holweger, W., **8**, 23; **9**, 107
Holy, N. L., **4**, 692
Hombacj, H.-P., **6**, 32
Hommes, H., **10**, 124
Honda, T., **14**, 502, 701
Honeyman, M. R., **14**, 665
Hong, Y. S., **2**, 346, 347
Honzatko, R. B., **2**, 299
Hooper, D., **5**, 2
Hoople, D. W. T., **10**, 83
Hoornaert, C., **3**, 440, 441
Hoornaerts, M. T., **1**, 54
Hopf, H., **13**, 109
Hopkinson, A. C., **3**, 28; **8**, 172; **10**, 303, 308, 309; **14**, 446, 447

Hoppe, D., **10**, 138
Hoppe, I., **10**, 223
Hoppen, V., **9**, 236; **11**, 143
Horgan, A. D., **11**, 77
Horgan, A. G., **2**, 453; **11**, 95; **14**, 14, 161
Horgan, S. W., **12**, 104
Hori, F., **3**, 1005
Hori, K., **12**, 247
Hori, M., **6**, 173; **14**, 279, 280
Horiguchi, Y., **14**, 711
Horii, T., **3**, 516
Horino, H., **6**, 92
Horiuchi, S., **2**, 181
Horn, K. A., **13**, 191
Horner, L., **5**, 105
Horning, D. P., **14**, 891
Horowitz, A., **3**, 278, 346, 348, 351, 377; **11**, 73
Horton, H. R., **2**, 294
Horvath, I., **3**, 270; **4**, 508
Hosaka, K., **4**, 67; **12**, 153
Hosako, R., **2**, 222
Hoshi, N., **14**, 257
Hoshiko, T., **12**, 267
Hoshino, M., **3**, 567; **7**, 149
Hosomi, A., **1**, 160
Houk, K. N., **1**, 127; **3**, 24; **4**, 401, 402; **5**, 2, 52; **10**, 12, 26; **12**, 5, 6, 264; **13**, 104; **14**, 289
Houssin, R., **6**, 52
Hoving, H., **2**, 402
Howard, C., **14**, 255
Howard, J. A., **3**, 155, 169, 329; **4**, 442
Howard, S. I., **4**, 544, 556
Howbert, J. J., **13**, 206
Howe, I., **14**, 427
Hoyano, Y., **10**, 201
Hoye, R. C., **13**, 190
Hoyermann, K., **3**, 242, 302, 311
Hrovat, D. A., **3**, 879; **10**, 148
Hsu, C.-T., **10**, 131
Hu, H., **4**, 225, 226
Huang, C., **4**, 359; **14**, 836
Huang, T. T.-S., **9**, 272
Huang, Y., **14**, 768
Huang, J.-S., **14**, 619
Hubbard, J. S., **3**, 807; **6**, 20; **10**, 169
Huber, U., **13**, 128
Huber, W., **10**, 206
Hubert, A. J., **5**, 72, 112
Huche, M., **12**, 231
Hudéc, J., **4**, 555
Hudlicky, T., **14**, 635
Hudson, A., **3**, 80–82
Hudson, R. F., **2**, 187
Hudson, R. L., **3**, 1044
Huet, J., **9**, 49
Huffman, J. C., **10**, 245; **14**, 503
Hugel, H. M., **8**, 58
Huggenberger, C., **3**, 236, 244
Hughes, D. L., **9**, 223
Hughes, G. J., **9**, 61
Hughes, R. A., **3**, 619
Hughes, R. P., **14**, 616
Huguet, J., **2**, 250
Huhn, D., **3**, 740
Huhn, P., **3**, 202, 204, 270
Huie, E., **14**, 147
Huie, R. E., **3**, 382; **4**, 280
Huis, R., **14**, 666
Huisgen, R., **13**, 4, 77; **14**, 300
Huisman, H. O., **11**, 126
Hull, W. E., **14**, 512, 846
Hulskamper, L., **9**, 74
Humffray, A. A., **4**, 224
Hummel, C., **14**, 193
Humphreys, R. W. R., **3**, 707, 750; **4**, 132
Humphries, H., **4**, 753
Hunadi, R. J., **6**, 131; **10**, 75
Hundal, A. S., **6**, 104
Hung, M.-H., **3**, 227; **6**, 11
Hungate, R. W., **10**, 136
Hunig, S., **14**, 286, 364
Hunig, S. M., **3**, 937
Hunkler, D., **14**, 324
Hunt, D. F., **10**, 323
Hunter, D., **14**, 346
Hunter, D. H., **10**, 40; **11**, 43; **14**, 336
Hurlbut, S. L., **14**, 51
Hurst, A. F., **3**, 604
Hurst, J. R., **3**, 143
Huschens, R., **3**, 96
Huser, D. L., **4**, 619; **10**, 4
Huskey, W. P., **2**, 41, 96, 99
Huss, O. M., **13**, 83
Hussain, I., **2**, 114
Hussain, S., **13**, 73
Hussein, F. M., **8**, 41
Hutchings, G. J., **2**, 365
Hutchins, J. E. C., **2**, 176
Hutchins, M. G. K., **5**, 39
Hutchins, R. O., **4**, 584
Hüttermann, J., **3**, 1144
Hutton, R. S., **5**, 3, 8
Huygens, A. V., **7**, 141
Hwang, J.-U., **9**, 266
Hwang, K.-J., **14**, 182
Hwang, T. L., **5**, 142, 143
Hwang, W. S., **3**, 733
Hyatt, J. A., **14**, 681
Hyde, E. M., **13**, 11
Hynes, M. J., **13**, 45

Iacobucci, G. A., **4**, 407
Ibata, T., **11**, 63; **13**, 84
Ibis, C., **9**, 12
Ibrahim, M. H., **6**, 106
Ibrahim, Y. A., **14**, 83
Ibuka, T., **12**, 233
Ichihara, M., **2**, 29
Ichihashi, Y., **7**, 22
Ichikawa, Y., **14**, 241
Ichimura, K., **13**, 157
Ichinose, I., **12**, 154
Icli, S., **3**, 419
Iemura, S., **10**, 340, 341
Iesce, M. R., **14**, 720
Igeta, H., **3**, 1124
Ignat'ev, N. V., **6**, 145, 146
Ignatenko, A. V., **3**, 101
Ignatev, V. M., **12**, 72
Ignatova, E., **12**, 217
Iguchi, K., **4**, 67; **12**, 153
Ihama, M., **4**, 705
Ihara, H., **2**, 349
Ihara, M., **6**, 91
Ihara, Y., **2**, 222, 223
Iida, H., **10**, 244
Iida, K., **4**, 664
Iida, T., **14**, 866
Iino, M., **3**, 503
Iitaka, Y., **4**, 372; **14**, 517
Ijjima, S., **1**, 97
Ikariya, T., **4**, 677
Ikawa, T., **14**, 595
Ikeda, H., **14**, 626
Ikeda, M., **14**, 375, 761
Ikeda, N., **11**, 104
Ikeda, Y., **3**, 1169
Ikegami, S., **12**, 148
Ikegami, Y., **3**, 53, 248, 250
Ikegawa, A., **12**, 190
Ikekawa, N., **4**, 549; **9**, 62
Ikeno, M., **5**, 152
Il'chenko, A. Ya., **6**, 145
Il'in, Yu. V., **3**, 101, 327, 462, 463
Il'ina, O. M., **2**, 211
Ilic, P., **8**, 98; **10**, 71
Illarionova, L. V., **6**, 159
Illuminati, G., **6**, 96; **9**, 103–105; **11**, 8; **12**, 131
Imai, T., **3**, 576; **7**, 129
Imaizumi, J., **10**, 244
Imajo, S., **14**, 264
Imamoto, T., **1**, 180; **14**, 553
Imamura, S., **4**, 52
Imashev, U. B., **3**, 115, 187, 188, 273, 274, 334, 456–458,

1129; **4**, 298; **8**, 1, 99
Imberger, H. E., **4**, 224
Imperiali, B., **1**, 90; **10**, 191
Inada, S., **14**, 151
Inagaki, S., **10**, 214
Inagki, N., **3**, 75
Inaishi, M., **14**, 15
Inamoto, N., **3**, 656; **9**, 154
Inanaga, J., **6**, 57
Inbar, S., **3**, 929
Inbasekaran, M. N., **7**, 148
Inch, T. D., **2**, 356, 374; **14**, 755
Indukumari, P. V., **4**, 352
Ing, H. T., **7**, 74; **14**, 41, 42
Ingold, C. F., **4**, 27, 30
Ingold, K. U., **3**, 25, 48, 66, 176, 189, 190, 238, 323, 338, 352, 490, 491, 584–586, 601; **4**, 444–447
Ingrosso, G., **12**, 30, 31
Ino, M., **3**, 678a
Inomata, K., **10**, 248
Inone, O., **14**, 881
Inone, Y., **14**, 792, 793
Inoue, H., **13**, 58; **14**, 75, 824
Inoue, K., **3**, 887; **4**, 31. 397
Inoue, M., **11**, 20
Inoue, O., **14**, 880
Inoue, S., **1**, 120; **2**, 240, 242
Inoue, T., **2**, 200
Inoue, Y., **3**, 558
Inouye, Y., **2**, 226; **3**, 483; **4**, 698, 708
Inove, Y., **3**, 1083
Inoyatova, D. A., **6**, 111
Insam, N., **6**, 167
Int, J., **8**, 163
Invergo, B. J., **4**, 266, 571, 616
Ioganov, K. M., **2**, 169
Ionin, B. I., **12**, 72
Iorga, I., **10**, 313
Iori, M., **14**, 870, 871
Ip, D. P., **7**, 19
Ippolito, R. M., **10**, 153
Ireland, R. E., **14**, 142
Irie, T., **6**, 77; **14**, 33
Irikawa, H., **3**, 68, 366
Irismetov, M. P., **14**, 570
Irving, E. M., **4**, 590
Irwin, R. S., **3**, 310; **4**, 388
Isaacs, N. S., **4**, 35
Isabelle, L. M., **10**, 57
Isac, D., **6**, 3
Isaev, I. S., **14**, 11
Isaeva, S. A., **4**, 647
Isayama, S., **3**, 448, 465
Ise, N., **2**, 210
Isham, W. J., **3**, 960
Ishibashi, H., **7**, 133; **13**, 145
Ishiguro, M., **1**, 100; **4**, 549; **9**, 62; **11**, 104
Ishihara, E., **3**, 354
Ishihara, T., **4**, 600
Ishii, H., **1**, 77; **7**, 137, 138
Ishii, K., **12**, 154; **14**, 316
Ishii, M., **3**, 901
Ishikawa, H., **4**, 595; **12**, 206
Ishikawa, M., **5**, 144, 145; **14**, 626, 877
Ishikawa, N., **12**, 1
Ishikawa, R., **2**, 181
Ishikawa, Y., **14**, 675
Ishimori, M., **9**, 109
Ishino, Y., **12**, 150
Ishiwatari, T., **2**, 210
Ishiyama, K., **13**, 145
Ishizu, K., **3**, 71, 599, 954, 995
Ishizu, T., **13**, 179, 180
Iskander, G. M., **11**, 134; **14**, 384
Iskander, M. L., **14**, 572
Ismail, N., **3**, 335
Ismailov, A. G., **12**, 65, 81
Isobe, M., **9**, 41; **14**, 241
Isola, M., **2**, 425
Isomura, K., **5**, 124; **14**, 690
Isomura, Y., **14**, 71
Israel, M., **4**, 658
Israfilov, A. I., **13**, 100
Israfifilov, A. I., **13**, 100
Issa, I. M., **2**, 104
Issler, S. L., **3**, 639; **4**, 394
Istomin, B. I., **2**, 31–33, 242, 340, 341, 343, 344; **7**, 112; **9**, 212; **12**, 53
Itagaki, K., **10**, 173
Itai, A., **14**, 517
Ito, I., **4**, 41
Ito, K., **4**, 206; **14**, 564
Ito, N., **14**, 71
Ito, O., **3**, 431, 432, 781
Ito, R., **12**, 112
Ito, S., **3**, 902, 921, 990; **4**, 145; **8**, 18; **14**, 221, 272
Ito, T., **6**, 70
Ito, Y., **3**, 578, 579; **4**, 290; **11**, 106
Itoh, A., **10**, 185
Itoh, K., **10**, 173; **12**, 239
Itoh, R., **5**, 15
Itsuno, S., **4**, 557, 558, 560
Ittel, S. D., **8**, 45
Ivakhnenko, E. P., **3**, 528
Ivakhov, V. N., **3**, 1125
Ivanchev, S. S., **3**, 142
Ivanenko, A. G., **14**, 794, 795
Ivanov, A. M., **4**, 511
Ivanov, A. V., **6**, 54, 55, 143
Ivanov, B. E., **2**, 211, 339, 353; **9**, 220
Ivanov, S., **4**, 456, 469
Ivanov, V. A., **9**, 91
Ivanov, V. L., **3**, 959
Ivanov, L. P., **1**, 66
Ivanova, L. V., **3**, 633
Ivanova, N. S., **2**, 164
Ivanova, T. M., **6**, 37
Ivanova, V., **4**, 689
Ivanyutina, Z. M., **4**, 501
Ives, J. L., **4**, 390; **13**, 3; **14**, 746
Ivin, K. J., **4**, 458
Iwai, K., **12**, 189
Iwaizumi, M., **3**, 990
Iwakuma, T., **4**, 561
Iwamoto, H., **7**, 88
Iwamura, H., **3**, 861, 1017, 1169; **5**, 120
Iwanami, M., **14**, 101
Iwanska, S., **9**, 68
Iwasa, A., **7**, 82
Iwasa, K., **4**, 622
Iwasaki, M., **3**, 941; **8**, 125
Iwasaki, S., **4**, 381
Iwata, C., **4**, 623
Iwata, M., **1**, 42
Iwunre, M., **2**, 199
Iyanagi, T., **3**, 714; **4**, 772
Iyangar, D. S., **2**, 429
Iyer, R. S., **3**, 647, 648
Iyoda, J., **14**, 626
Iyoda, T., **4**, 333
Izawa, Y., **5**, 78, 90, 92, 94; **6**, 70; **14**, 535, 536
Izumi, Y., **12**, 91
Izumuja, N., **14**, 583
Izuno, C., **13**, 13

Jack, I. R., **14**, 510
Jackman, L. M., **3**, 819; **10**, 163
Jackson, A. H., **7**, 117; **14**, 54
Jackson, R. A., **3**, 5, 80, 655, 1136, 1137
Jackson, W. G., **14**, 585
Jackson, W. P., **14**, 545
Jackson, Y. A., **14**, 49
Jacobi, P. A., **10**, 91; **13**, 79
Jacobi, S. A., **3**, 492
Jacobs, M., **4**, 393
Jacobs, S. A., **10**, 203
Jacobsen, S., **14**, 529, 530
Jacobson, R. A., **5**, 81
Jacobus, J. O., **5**, 83
Jacocks, H. M., **11**, 107
Jacox, M. E., **5**, 70

Jacquesy, J. C., **7**, 30, 146
Jacquier, R., **14**, 509
Jadhav, M. B., **4**, 239–241
Jadhav, P. K., **4**, 578; **12**, 127
Jaeger, D. A., **9**, 152
Jagannadham, V., **1**, 58
Jagannadhaswamy, B., **4**, 465
Jagdale, M. H., **2**, 66; **4**, 199, 200
Jagdmann, G. E., **4**, 69
Jager, V., **4**, 548
Jäggi, F. J., **14**, 481
Jahn, B., **14**, 818
Jain, P. S., **3**, 510
Jaky, M., **4**, 29
Jalander, L., **3**, 860; **12**, 207
James, D. G. L., **3**, 152
James, F. C., **3**, 279; **6**, 170
James, M. N. G., **2**, 281
James, S. L. T., **2**, 296, 297
Jamison, W. C. L., **4**, 583
Jammar, R., **14**, 152
Jamrogiewicz, Z., **9**, 92
Janardhana, C., **7**, 42
Janata, E., **3**, 1150
Jander, J., **4**, 223
Janjic, D., **4**, 79, 83
Jano, P., **11**, 74
Jans, A. W. H., **13**, 196
Jansen, B. J. M., **10**, 230; **14**, 544
Jansen, D. K., **12**, 136
Jansen, M. P., **11**, 140
Janssen, H.-W., **14**, 532
January, J. R., **3**, 1053
Janzen, E. G., **3**, 593, 630
Jaouen, G., **8**, 48
Jaquesey, J.-C., **7**, 29
Jarrar, A. A., **4**, 143
Jarrell, H. F., **14**, 528
Jarvest, R. L., **2**, 363, 364, 388, 389, 396
Jarvis, B. B., **3**, 1133
Jasinskas, L. L., **6**, 105
Jastrzebski, J. T. B. H., **3**, 60
Jawdosiuk, M., **3**, 815, 816; **10**, 177
Jayakumar, A., **1**, 112
Jayaram, C., **10**, 258
Jayathirtha Rao, V., **4**, 291
Jeannin, Y., **14**, 608
Jee, J.-G., **9**, 266
Jefferies, P. R., **14**, 328, 507
Jefferson, J., **3**, 618
Jefford, C. W., **4**, 399, 400, 403
Jeger, O., **3**, 1070; **5**, 30; **14**, 316, 653, 654, 657, 888
Jehlicka, V., **2**, 76
Jelinski, L. W., **2**, 283
Jemison, R. W., **14**, 353
Jemmis, E. D., **10**, 6, 7
Jempty, T. C., **3**, 697; **4**, 131
Jencks, W. P., **2**, 63, 65, 263; **11**, 17
Jendralla, H., **13**, 14
Jendrzejewski, S., **14**, 728
Jenkins, J. A., **4**, 289
Jenkins, M. J., **8**, 54
Jenkins, W. L., **10**, 33
Jenner, G., **1**, 135
Jenner, P. J., **10**, 316
Jennings, P. W., **4**, 335
Jennings, R. C., **9**, 65
Jensen, R. K., **4**, 497
Jensen, W. A., **8**, 58
Jerina, D. M., **4**, 315; **9**, 66
Jesson, J. P., **8**, 45
Jessup, P. J., **13**, 103
Jha, H. C., **14**, 719
Jiang, Z.-Q., **3**, 592
Jigajinni, V. B., **14**, 56
Jimenez, C., **12**, 141, 143
Jiminez, C., **12**, 144
Jin, S.-S., **1**, 150
Jirkovský, J., **3**, 218
Jitsuhiro, K., **13**, 84
Jitsukawa, K., **4**, 122
Jo, S., **7**, 136
Jo, T., **3**, 1169
Johansen, J. E., **14**, 847
Johansson, E., **4**, 448
John, D. I., **3**, 496, 657; **4**, 599
John, I. L., **7**, 25
John, L.-S., **14**, 320
John, R. A., **11**, 123
Johnson, B. F. G., **8**, 39
Johnson, C. D., **9**, 200
Johnson, F. A., **2**, 289–291
Johnson, J. E., **1**, 72; **9**, 209; **14**, 796
Johnson, M., **3**, 869
Johnson, M. D., **3**, 653, 654
Johnson, M. N., **4**, 28
Johnson, M. W., **4**, 154; **7**, 69
Johnson, W. S., **8**, 26
Johnston, E. R., **2**, 84, 85
Johnston, L. J., **3**, 1171
Johnston, M. A., **14**, 174
Johnstone, A., **14**, 296
Johnstone, L., **3**, 778
Jolidon, S., **14**, 875
Jonas, F., **12**, 97; **14**, 853
Jones, B. B., **4**, 407
Jones, C. R., **3**, 883; **14**, 254
Jones, D. N., **12**, 199
Jones, D. R., **2**, 369
Jones, E. O., **2**, 317
Jones, G., **5**, 18, 122; **13**, 203; **14**, 497, 817
Jones, M., **5**, 60
Jones, M. A., **10**, 317
Jones, N. K., **4**, 28
Jones, R. A. Y., **14**, 759
Jones, R. C. F., **10**, 135
Jones, W. G. M., **3**, 384
Jones, W. M., **5**, 86; **11**, 84; **14**, 533
Jongsma, S. J., **3**, 893; **4**, 425
Jonsäll, G. J., **8**, 15
Jönsson, L., **3**, 547, 548; **6**, 28, 29
Joo, W.-C., **4**, 620
Jordan, F., **2**, 272
Jorge, J. A. L., **2**, 107; **9**, 206
Jorgensen, W. L., **10**, 69
Jorritsma, R., **5**, 106; **9**, 72, 73
Joseph, N., **7**, 26
Joseph-Nathan, P., **14**, 508
Josey, A. D., **13**, 18
Joshi, A., **3**, 602, 643
Joshi, B. C., **4**, 77; **14**, 337
Joshua, C. P., **4**, 352
Jost, P., **13**, 22
Jost, P. C., **3**, 598
Jost, R., **8**, 74
Jouannetaud, M.-P., **7**, 29, 30, 146
Jouin, P., **4**, 707
Joukhader, L., **4**, 630
Juan, A., **4**, 405
Juaristi, E., **10**, 296
Judkins, B. D., **5**, 66; **14**, 800
Jug, K., **13**, 5, 9
Julia, L., **6**, 61
Julia, S., **1**, 134; **10**, 86
Jullien, R., **3**, 1047
Jun, H. W., **1**, 63
Jung, D. M., **8**, 141; **9**, 9
Jung, K.-H., **11**, 71
Jung, M., **14**, 309
Jungheim, L. N., **14**, 649
Jurlina, J. L., **3**, 865; **14**, 683
Jutz, C., **3**, 989
Juznetsova, M. G., **13**, 49

Kabachnik, M. I., **3**, 499, 668, 672, 867; **14**, 744
Kabachnik,, M. M., **14**, 839
Kabąlka, G. W., **4**, 577; **10**, 312
Kabuto, C., **14**, 221
Kacher, M. L., **4**, 426
Kachhwaha, O. P., **4**, 139
Kachurin, O. I., **7**, 8, 98, 123, 124
Kada, R., **6**, 112

Kadam, S. D., **2**, 66
Kadam, S. R., **3**, 757; **4**, 624, 625
Kadentsev, V. I., **3**, 386, 721, 722
Kadonaga, J. T., **14**, 634
Kaemmerer, H., **9**, 210
Kafuku, K., **14**, 596
Kagan, H. B., **4**, 649, 676
Kagan, J., **14**, 441
Kagawa, T., **10**, 67
Kageyama, Y., **14**, 157
Kaghazchi, T., **7**, 79
Kagi, A., **3**, 831
Kagotani, M., **10**, 194, 195; **12**, 225; **14**, 158, 613
Kahan, B. H., **4**, 98
Kahana, P., **14**, 807
Kahn, M., **12**, 251
Kaim, W., **3**, 951, 952, 1002, 1057
Kaiser, E. M., **10**, 58
Kaiser, E. T., **2**, 301, 406; **4**, 722; **11**, 97
Kaiser, E. W., **4**, 509
Kaiser, T., **3**, 1066
Kaji, A., **3**, 828, 829; **4**, 700; **9**, 15; **10**, 251; **11**, 114, 130
Kajihara, Y., **12**, 164
Kakinchi, K., **14**, 485
Kakushima, M., **7**, 130
Kalabina, A. V., **2**, 340–342
Kalashnikov, S. M., **3**, 326, 455–459, 1129
Kalatzis, E., **10**, 331, 332
Kalbus, M., **1**, 156
Kalck, P., **2**, 110
Kalfus, K., **2**, 34
Kalhorn, T., **4**, 768
Kaliba, C., **13**, 173
Kalinenko, R. A., **3**, 1110
Kalinichenko, A. N., **12**, 75; **14**, 472
Kalk, K. H., **2**, 314, 315
Kalkabaeva, L. T., **14**, 736
Kaloshina, N. N., **3**, 94; **10**, 92
Kalyanaraman, B., **3**, 644, 880, 1041
Kalyanasundaram, -K., **3**, 926
Kam, T. S., **3**, 810; **9**, 150
Kamalov, G. L., **1**, 14; **4**, 320, 322
Kamego, C. D., **8**, 80
Kametani, T., **6**, 91; **10**, 246; **11**, 124; **12**, 83; **13**, 2; **14**, 502, 701
Kamimori, M., **3**, 624
Kaminski, J. J., **9**, 253
Kaminskii, A. Ya., **6**, 139, 156–159
Kamishimoto, M., **4**, 719
Kamisuki, T., **8**, 114
Kamiya, T., **3**, 686
Kamiya, Y., **4**, 282, 283
Kamlet, M. J., **9**, 153, 161
Kamounah, F. S., **7**, 44, 126
Kampars, V., **3**, 220; **4**, 653
Kämpchen, T., **13**, 142
Kan, C. S., **3**, 166
Kanai, H., **14**, 603
Kanaoka, V., **3**, 519
Kanavarioti, A., **12**, 84
Kanayama, M., **3**, 163
Kanda, N., **4**, 333
Kandror, I. I., **3**, 255
Kane-Maguire, L. A. P., **8**, 29, 30, 49
Kaneda, K., **4**, 122
Kanehire, K., **4**, 674
Kanekiyo, T., **14**, 338
Kanematsu, K., **11**, 49; **13**, 93, 101, 106, 117, 179, 180, 183, 184, 204
Kaneti, J., **10**, 29, 62
Kaneto, Y., **10**, 248
Kanfer, S., **4**, 393
Kang, H. K., **2**, 434, 435
Kang, J., **14**, 589
Kang, K.-T., **3**, 656
Kang, S. H., **11**, 71
Kanishchev, M. I., **8**, 113; **14**, 390, 404
Kanjilal, P. R., **14**, 494
Kannan, N., **2**, 52; **4**, 554
Kannappan, V., **7**, 7, 33
Kanno, T., **3**, 183; **4**, 453
Kanoktanaporn, S., **6**, 126
Kansak, B. C., **4**, 198
Kantlehner, W., **1**, 19
Kantor, E. A., **1**, 11; **8**, 99–102; **14**, 398
Kantov, E. A., **2**, 8
Kaplan, P., **4**, 85
Kaplan, S., **14**, 789
Kaplunov, M. G., **3**, 347
Kapoor, R. C., **4**, 139
Kappi, R., **1**, 35
Kaptein, R., **3**, 1164, 1166
Karakhanov, R. A., **3**, 317, 345, 459, 467; **8**, 99, 100; **14**, 398, 749
Karakotov, S. D., **7**, 87; **14**, 18, 19
Karan, H. I., **14**, 790
Karapetyan, N. G., **12**, 69, 70
Karasevich, E. I., **4**, 495
Karavan, V. S., **8**, 150; **9**, 54, 233
Karchefski, E. M., **4**, 28
Karger-Kocsis, J., **3**, 1000
Karkozov, V. G., **9**, 52
Karpf, H., **13**, 156
Karpf, M., **14**, 445
Karpova, S. G., **4**, 299
Karpova, V. V., **7**, 71, 72
Karpyuk, A. D., **9**, 256, 257; **10**, 84, 85
Karras, M., **14**, 134, 407
Karshalukov, K., **4**, 469
Kartasheva, Z. S., **3**, 177, 410
Kartashov, V. R., **12**, 20, 46, 134
Karukstis, J., **3**, 35
Karukstis, K. K., **3**, 34
Kasahara, I., **14**, 567, 880
Kasaikina, O. T., **3**, 177, 410
Kasatochkin, A. N., **3**, 290
Kasemsri, P., **11**, 94
Kashefi-Naini, N., **2**, 417, 419; **11**, 23, 25
Kashemirov, B. A., **14**, 18
Kashina, V. V., **12**, 107
Kashirskii, V. F., **4**, 467
Kasmai, H. S., **10**, 76, 77; **14**, 331
Kassab, E., **13**, 8
Kassem, A. A., **1**, 177
Kasukhim, L. F., **2**, 337
Katada, T., **12**, 198
Katagiri, T., **10**, 257
Kataoka, T., **14**, 280
Katayama, C., **3**, 448, 465
Katayama, S., **5**, 145
Katayama, T., **10**, 81; **14**, 377
Kates, M. R., **8**, 62
Kateva, I., **4**, 456
Katıyar, S. S., **8**, 83
Kato, H., **3**, 515, 545, 1094; **4**, 287
Kato, K., **14**, 527
Kato, M., **3**, 918; **14**, 272
Kato, N., **3**, 990
Kato, S., **7**, 148
Kato, T., **3**, 571; **4**, 41; **11**, 58; **12**, 154; **13**, 182; **14**, 198
Kato, Y., **9**, 29
Katoaka, T., **6**, 173
Katritzky, A. R., **6**, 106, 107, 121; **9**, 181, 205, 219; **10**, 79, 258; **11**, 10; **14**, 63, 210, 464
Katsin, M. I., **3**, 710; **4**, 196
Katsnelson, M. G., **12**, 107
Katsuki, T., **4**, 152, 314; **6**, 57
Katsushima, T., **3**, 296; **10**, 340, 341; **14**, 623
Katz, M., **2**, 405; **14**, 447
Katz, T. J., **5**, 156

Katzenellenbogen, J. A., **2**, 284; **6**, 12; **9**, 16
Kauffmann, T., **10**, 152; **13**, 41
Kaufmann, D., **13**, 192
Kaufmann, E., **10**, 70
Kaufmann, K., **3**, 1097
Kaupp, G., **3**, 574; **13**, 199, 201
Kaur, H., **3**, 631
Kausch, E., **13**, 75
Kaushik, D., **4**, 221
Kaválek, J., **2**, 184
Kavanagh, P., **10**, 315
Kaverinskii, V. A., **4**, 455
Kavetskaya, O. I., **4**, 162
Kawabata, N., **3**, 863
Kawabe, H., **2**, 229
Kawada, M., **14**, 612
Kawada, S., **2**, 234
Kawada, Y., **3**, 861, 1169
Kawaguchi, T., **3**, 1124
Kawai, T., **9**, 184
Kawakami, S., **10**, 87
Kawakaru, T., **4**, 324
Kawami, Y., **12**, 233
Kawamoto, A., **14**, 122
Kawamura, A., **14**, 354
Kawamura, E., **14**, 870, 871
Kawamura, S., **3**, 516
Kawanisi, M., **3**, 296, 866; **10**, 340, 341; **14**, 623
Kawao, S., **1**, 166
Kawasaki, M., **2**, 253
Kawase, T., **10**, 64
Kawashima, T., **6**, 123
Kawato, T., **11**, 39
Kawazura, H., **3**, 992
Kayser, M. M., **2**, 73; **4**, 563
Kazakova, V. M., **3**, 999; **4**, 606
Kazarians-Moghaddam, H., **14**, 447
Kazaryan, P. I., **14**, 423
Kazazyan, S. S., **12**, 69
Keana, J. F. W., **3**, 598
Kearney, P. A., **8**, 92
Kedzierski, B., **9**, 66
Keefe, J. R., **11**, 17
Keelan, B. W., **8**, 75; **14**, 405
Keene, F. R., **2**, 247
Kees, F., **11**, 53, 54; **13**, 52
Keiko, V. V., **12**, 90
Keil, M., **8**, 33
Keilich, G., **1**, 23
Keith, C., **2**, 317
Kekisheva, L. V., **11**, 149
Kellogg, R. M., **4**, 656, 707; **9**, 108, 248; **12**, 193; **14**, 457, 498
Kelly, D. P., **8**, 54, 57, 58
Kelly, K. P., **4**, 173
Kelly, L. F., **8**, 32, 37, 38
Kelly, N., **3**, 384
Kelsall, B. J., **3**, 970; **14**, 406
Kemal, C., **2**, 307
Kemp, D. S., **2**, 180; **14**, 869
Kemp, T. J., **3**, 726, 884, 1073; **6**, 9; **7**, 89, 90; **8**, 132
Kemper, M. J. H., **5**, 104; **14**, 769
Kende, A. S., **10**, 168
Kenigsberg, T. P., **3**, 157
Kenley, R. A., **1**, 71; **3**, 539
Kenny, A., **14**, 358
Kenyon, G. L., **2**, 403
Kerdesky, F. A. J., **1**, 90; **10**, 191
Kergomard, A., **4**, 730, 732; **14**, 831
Kerkman, D. J., **2**, 180; **14**, 869
Kerr, I. S., **3**, 727
Kerr, J. A., **4**, 286
Kerr, R. G., **4**, 203
Kessar, S. V., **6**, 24; **14**, 39, 782
Ketola, M. R., **11**, 152
Keumi, T., **4**, 328; **7**, 145; **11**, 89
Keute, J. S., **3**, 617
Kevill, D. N., **9**, 158, 194, 199
Khabashesku, V. N., **5**, 97
Khachatryan, L. A., **3**, 154
Khachik, F., **3**, 909; **13**, 23
Khait, I., **3**, 1167
Khal'kin, V. A., **6**, 64
Khalafy, J., **14**, 859, 860
Khalil-Ur-Rahman, J., **6**, 42
Khan, M. M. T., **4**, 683
Khan, N., **2**, 23
Khandkarova, V. S., **13**, 49
Khanna, R. K., **6**, 6
Khante, R. N., **4**, 323, 332
Khardin, A. P., **9**, 53
Kharicheva, E. M., **5**, 57
Kharitonov, N. P., **8**, 140
Khatri, H. N., **10**, 235; **11**, 155
Khawad, I. E., **14**, 384
Kheidorov, V. P., **3**, 93
Kheifets, V. I., **2**, 171
Khelevin, R. N., **7**, 95
Kheruze, Y. I., **12**, 260
Khidekel, M. L., **4**, 689
Khinkova, M., **4**, 469
Khitrin, S. V., **2**, 147
Khizantsyan, N. M., **14**, 423
Khodak, A. A., **3**, 672
Khodaskar, S. N., **2**, 66
Khodzhaev, O. M., **4**, 482
Khokhlov, P. S., **2**, 383
Khokhlova, T. V., **9**, 53
Khokhryakova, N. A., **3**, 333
Khoo, T.-K., **9**, 47
Khorishko, S. A., **12**, 71
Khromov-Borisov, N. V., **7**, 125
Khudyakov, I. V., **3**, 252, 735, 1087
Khuong-Hnu, Q., **12**, 142
Khurana, J. M., **3**, 837
Khuthier, A.-H., **14**, 22
Kice, J. L., **2**, 447, 449; **4**, 354; **11**, 28
Kido, M., **14**, 375
Kidokoro, H., **7**, 35
Kielbania, A. J., **10**, 283; **11**, 141; **12**, 180
Kiesele, H., **14**, 263
Kigawa, H., **3**, 1030; **9**, 111
Kihara, H., **9**, 262
Kihara, K., **2**, 302
Kii, N., **4**, 122
Kiji, J., **7**, 22
Kikuchi, A., **9**, 203
Kikuchi, H., **3**, 294
Kikuchi, J., **4**, 706
Kikuchi, M., **3**, 649
Kikuchi, O., **3**, 73; **14**, 289, 302, 303, 338
Kikuchi, T., **14**, 519
Kikukawa, K., **6**, 15, 16
Kiland, P. J., **9**, 31
Kilbourn, M. R., **6**, 12
Kim, C.-K., **4**, 620
Kim, J. D., **4**, 587
Kim, J. F., **2**, 431
Kim, J. K., **2**, 130; **8**, 137, 138
Kim, J. M., **14**, 333
Kim, K. S., **14**, 773
Kim, S., **10**, 158
Kim, S.-G., **9**, 241; **14**, 408
Kim, S. O., **4**, 593
Kim, S. S., **3**, 381, 475
Kim, W.-K., **9**, 174
Kim, Y. H., **4**, 258, 370, 371
Kimura, K., **1**, 97; **5**, 31; **14**, 704
Kimura, M., **4**, 136, 218
Kimura, N., **3**, 1030; **9**, 111
Kimura, Y., **2**, 237; **3**, 738; **8**, 93; **14**, 208
Kinatowski, S., **12**, 259
King, D., **11**, 80
King, D. M., **3**, 318, 319, 476, 477
King, J. A., **13**, 210
King, K. D., **3**, 49, 1105–1107; **11**, 68; **14**, 307, 323, 816
King, R. S. L. M., **2**, 86
King, T. E., **4**, 764

King, T. J., **3**, 366; **6**, 26; **14**, 714
Kingsbury, C. A., **3**, 594
Kinoshita, H., **10**, 248
Kinoshita, K., **12**, 233; **14**, 596
Kinsella, M. A., **11**, 110
Kinzel, E., **5**, 32
Kinzelmann, P., **1**, 23
Kiplov, V. M., **4**, 606
Kiprianova, L. A., **4**, 271, 272, 652
Kira, M., **3**, 85, 526
Kirby, A. J., **2**, 428
Kirchen, R. P., **8**, 63, 64
Kirichenko, A. I., **2**, 166, 167
Kirichenko, L. N., **9**, 214
Kirilicheva, V. G., **3**, 253
Kirilov, M., **1**, 53
Kirino, Y., **3**, 591
Kirk, D. N., **14**, 513
Kirk, T. C., **14**, 134
Kirms, M. A., **14**, 815
Krimse, W., **5**, 77; **9**, 39; **14**, 381, 468, 469
Kirowa-Eisner, E., **3**, 773
Kirsch, J. F., **1**, 29
Kirsch, L. J., **3**, 153
Kirshen, N. A., **1**, 71
Kirste, B., **3**, 74, 76, 988
Kiryukhin, D. P., **3**, 405
Kisabayashi, S., **9**, 198
Kise, N., **10**, 87
Kiselev, B. D., **13**, 120
Kiselev, V. D., **13**, 97, 116
Kiseleva, L. N., **3**, 283, 481
Kiseleva, N. I., **14**, 560
Kiseliv, S. S., **12**, 86
Kisfaludy, L., **2**, 143
Kishimoto, K., **9**, 184
Kishimoto, N., **14**, 603
Kishimoto, S., **7**, 92
Kishore, G. M., **4**, 763
Kishoyan, V. S., **4**, 189
Kisin, A. V., **13**, 49; **14**, 852
Kiso, Y., **12**, 112
Kispert, L. D., **3**, 36, 1041
Kiss, J., **14**, 373
Kissel, T., **13**, 143
Kitade, Y., **6**, 124; **14**, 77
Kitagawa, T., **14**, 573
Kitahara, A., **2**, 200
Kitahara, S., **7**, 92
Kitaigorodskii, A. N., **7**, 147
Kitajima, H., **11**, 89
Kitajima, K., **14**, 419
Kitajima, Y., **14**, 583
Kitamura, T., **3**, 411, 919; **9**, 83
Kitao, T., **4**, 430
Kitchin, J., **14**, 309
Kitching, W., **3**, 835; **10**, 339
Kivganova, V. I., **4**, 516
Kiyan, N. Z., **2**, 107; **9**, 206
Kjell, D. P., **11**, 110
Klages, C.-P., **3**, 1018
Klapper, M. H., **2**, 277
Klarner, F.-G., **14**, 251
Klass, G., **10**, 54
Klaus, R. O., **14**, 684
Klausner, A. E., **2**, 91
Klebanskii; A. L., **9**, 93
Kleier, D. A., **3**, 552
Kleijn, H., **9**, 18; **12**, 226, 243
Klein, H., **12**, 94
Klein, J., **5**, 75; **10**, 209
Klemimann, W., **14**, 82
Klemm, L. H., **7**, 120, 121; **9**, 128, 129
Klerks, J. M., **3**, 60
Kletsko, F. P., **12**, 186
Klicnar, J., **14**, 848, 849
Klimentov, B. N., **5**, 140
Klimov, E. S., **3**, 253
Klingebiel, U., **13**, 151; **14**, 242
Klingelhöfer, H.-G., **3**, 610, 611
Klinger, F., **12**, 149
Klingler, R. J., **3**, 733
Klinman, J. P., **4**, 753
Kloosterziel, H., **10**, 68
Klotz, H., **2**, 39
Klötzer, W., **13**, 51
Kluge, H., **5**, 134
Kluger, R., **2**, 251
Klumpp, G. W., **5**, 59; **10**, 66
Klunder, A. J. H., **14**, 493
Klusik, H., **3**, 985
Kluttz, R. Q., **14**, 813
Klyuchinskii, S. A., **12**, 273
Klyuchkinskii, A. I., **4**, 518
Knabe, J., **14**, 70
Knight, D. W., **10**, 217
Knight, W. B., **2**, 413
Knipe, A. C., **2**, 56; **6**, 83
Knoll, F., **14**, 206
Knoll, H., **3**, 362, 364, 365, 1127
Knoll, R., **10**, 284
Knollmuller, M., **1**, 62; **14**, 345
Knoot, P. A., **3**, 303
Knöpfel, N., **14**, 468
Knopp, J. A., **2**, 294
Knoppova, V., **6**, 112
Knörr, R., **10**, 41–44; **14**, 359
Knothe, L., **13**, 171, 212
Knowles, J. R., **2**, 305–309, 392, 401
Knowles, W., **4**, 666
Knowles, W. S., **12**, 102
Knox, C. V., **1**, 49
Knunyants, I. L., **12**, 43; **14**, 371
Knyazev, V. N., **6**, 119, 149
Ko, A. I., **13**, 126
Ko, E. C. F., **9**, 4
Koba, H., **14**, 137
Kobayashi, H., **7**, 88
Kobayashi, K., **3**, 866; **4**, 421; **6**, 75; **14**, 28, 459, 764
Kobayashi, M., **3**, 1168; **14**, 216
Kobayashi, N., **12**, 189
Kobayashi, S., **2**, 319, 327; **3**, 919; **12**, 198
Kobayashi, T., **14**, 567
Kobayashi, Y., **14**, 549
Kobesheva, N. I., **12**, 260
Kobrin, P. A., **2**, 84
Kobrina, L. S., **3**, 704
Koch, H. F., **10**, 283, 284; **11**, 16, 141; **12**, 180
Koch, J. G., **10**, 283; **11**, 16, 141; **12**, 180
Koch, T. H., **3**, 617, 870, 920; **14**, 726
Kochanski, E., **12**, 18
Kocharyan, S. T., **14**, 349–352
Kochetkiov, K. A., **14**, 787
Kochetkova, N. S., **3**, 769
Kochi, J. K., **3**, 2, 678, 733; **12**, 129
Kochin, V. A., **7**, 10
Kochina, T. A., **8**, 140
Kočovský, P., **9**, 63, 64; **12**, 33–35; **14**, 526
Kodama, M., **14**, 221
Kodeih, M., **14**, 115
Koehler, H. J., **3**, 362; **8**, 158; **10**, 295
Koehler, K. F., **13**, 69
Koenig, K., **4**, 666
Koermer, G. S., **4**, 269; **11**, 112
Koga, T., **9**, 43
Kogan, G. O., **6**, 49, 50
Kogure, T., **4**, 668
Koh, K. S., **2**, 109; **9**, 173
Kohda, A., **3**, 681; **4**, 133
Kohli, D. K., **11**, 39
Kohn, H., **12**, 38, 39
Koizumi, N., **4**, 549
Koizumi, T., **2**, 376
Kohima, M., **3**, 888; **4**, 459
Kojo, S., **3**, 433
Kok, D. M., **3**, 945
Kokars, V., **3**, 220; **4**, 653
Kol, M., **12**, 10
Koldobskii, G. I., **9**, 131, 140
Koleck, M. P., **14**, 453
Kolesnikov, V. A., **2**, 111

Kolesova, S. V., **14**, 647
Kolhe, J. N., **3**, 757; **4**, 624, 625
Kollenz, G., **13**, 169
Kollman, P. A., **2**, 270
Kolobkov, V. S., **9**, 90
Kolobova, N. E., **8**, 50, 51
Kolobova, T. A., **1**, 66
Kolodziejczyk, A., **9**, 254
Kolovskii, V. B., **3**, 400
Kolshorn, H., **14**, 674
Koltzenburg, G., **3**, 967
Komadel, P., **2**, 36
Komarov, V. S., **13**, 118
Komatenkova, N. G., **3**, 999
Komatsu, K., **3**, 944
Komatsu, M., **12**, 209; **14**, 889
Komendantov, M. I., **12**, 17
Komerov, V. S., **3**, 240
Komienko, A. A., **12**, 185
Komin, A. P., **3**, 807; **6**, 20
Komissarov, V. D., **3**, 677, 1131, 1132; **4**, 294–296
Komissarova, I. N., **4**, 298
Kömives, T., **2**, 139, 141, 143
Komiyama, M., **2**, 240, 242; **5**, 111
Kon-No, I., **2**, 200
Konar, A., **7**, 13
Kondo, H., **2**, 236; **6**, 74; **13**, 204; **14**, 33, 198
Kondo, S., **5**, 90, 92; **14**, 535, 536
Kondo, Y., **4**, 218; **9**, 180
Kondratenko, V. I., **2**, 173
Kondratov, S. A., **6**, 88
König, L., **5**, 103
Konishevskaya, G. A., **4**, 121
Konishi, H., **7**, 22
Kono, K., **6**, 16
Konovalov, A. I., **13**, 15, 16, 50, 66, 94–97, 116, 120
Konovatenko, V. V., **3**, 142
Konstantinovskii, L. E., **14**, 119, 285
Kontsova, L. V., **2**, 72
Koo, I. S., **2**, 432, 435
Koob, R. D., **3**, 1071
Kooyman, E. C., **7**, 141
Kopecky, K. R., **3**, 105, 336, 415
Köppel, B., **5**, 37
Koppes, W. M., **6**, 60
Koptyug, V. A., **8**, 67, 77, 120; **14**, 4, 11, 111, 391
Kopylov, V. M., **3**, 999
Kopylova, T. A., **13**, 99, 100
Korcek, S., **4**, 497
Korchagin, P. V., **14**, 52
Korchagina, D. V., **8**, 69; **14**, 7–9
Korenev, K. D., **9**, 50
Koridze, A. A., **3**, 1040; **8**, 50, 51
Koritskii, A. T., **3**, 1151
Kornblum, N., **3**, 825, 827; **9**, 151
Kornis, G., **14**, 93
Korochagina, D. V., **8**, 70, 71
Korodi, T., **2**, 128
Korolev, V. A., **5**, 97
Koroni, H., **3**, 486
Koroniak, H., **3**, 1121; **14**, 298
Koros, E., **4**, 252
Korostylev, A. P., **9**, 202
Korotkova, N. P., **4**, 297
Korshin, E. E., **13**, 120
Korth, H.-G., **3**, 47
Korzeniowski, S. H., **6**, 6
Korzhilova, O. I., **2**, 170
Korzun, N. V., **3**, 1108
Kos, A. J., **10**, 6, 7, 9, 72
Koser, G. F., **12**, 161
Koshechko, V. G., **3**, 725, 979; **4**, 13, 168
Koshelev, V. I., **7**, 99
Koshelev, Yu. N., **12**, 78
Kosheleva, L. M., **3**, 289
Koshkin, L. V., **3**, 885
Koshy, K. M., **11**, 140
Kosma, P., **1**, 62; **14**, 345
Kosmacheva, T. G., **4**, 513
Kosobutskii, V. S., **3**, 460
Kosolapov, V. T., **2**, 169
Kosower, E., **3**, 773
Kosower, E. M., **3**, 247–249, 251
Kossanyi, J., **13**, 22
Kossiakoff, A. A., **2**, 273
Kost, A. N., **6**, 120; **7**, 15; **14**, 61, 863
Kostecki, M., **12**, 259
Kostenko, L. I., **9**, 7
Kostikov, R. R., **5**, 57
Kostkova, V., **12**, 181
Koszyk, F. F., **14**, 635
Kotake, H., **10**, 248
Kotake, K., **14**, 626
Kotake, Y., **3**, 589
Kotlyar, S. A., **1**, 14
Kotowski, S., **3**, 975
Kotsuki, H., **4**, 413
Kottmair, E., **8**, 44
Kotyk, M., **2**, 184
Koudelka, L., **12**, 181
Koudstaal, H., **7**, 18
Koulkes-Pujo, A. M., **3**, 313
Koura, H., **14**, 824
Kovac, J., **6**, 112
Kovác, P., **4**, 539
Kovach, I. M., **2**, 58
Kovacic, P., **3**, 1122; **14**, 479
Kováčv, J., **13**, 65
Kovalenko, A. S., **4**, 82
Kovalev, M. D., **3**, 1151
Kovar, K.-A., **6**, 155
Kovbuz, M. A., **3**, 142
Kovtonyuk, V. N., **3**, 704
Kovtun, G. A., **3**, 200; **4**, 455
Kowalewski, J., **10**, 65
Kowalsi, P., **14**, 62
Kowari, K., **3**, 394
Koya, N., **3**, 230
Koyama, K., **5**, 36; **9**, 79
Kozhevnikov, I. V., **4**, 499, 500; **12**, 248; **14**, 55
Kozikowski, A. P., **4**, 118; **14**, 147
Kozima, S., **3**, 866
Kozliner, M. Z., **3**, 240
Kozlov, S. N., **3**, 396; **4**, 384, 385
Kozlov, V. A., **2**, 260
Kozlowska-Gramsz, E., **5**, 62, 63
Kozlowski, K., **6**, 89
Kozuka, S., **9**, 83, 203; **14**, 549
Kozyukov, V. P., **14**, 852
Kpoton, A., **9**, 88
Kraeutler, B., **3**, 1157, 1159
Kraft, G. A., **2**, 284
Krapcho, A. P., **4**, 657
Krapp, W., **14**, 368
Kratky, C., **14**, 847
Krauklis, I. V., **9**, 169, 170
Kraus, G. A., **4**, 537, 605; **10**, 197
Krauss, S. R., **12**, 227
Krausz, P., **3**, 927
Kravchenko, V. V., **9**, 7
Krebs, A., **3**, 108, 942
Krebs, E.-P., **10**, 27
Kreevoy, M. M., **1**, 10; **4**, 570
Kresge, A. J., **1**, 171, 172; **10**, 272
Kress, A. O., **5**, 128
Kresze, G., **11**, 52; **13**, 150
Kreutter, N. M., **3**, 312, 538
Kreysig, D., **14**, 288, 329
Kriby, A. J., **9**, 102
Krief, A., **4**, 659; **10**, 96; **13**, 35; **14**, 225
Krieg, R., **5**, 134
Krieger, R. L., **3**, 510
Krip, I. M., **4**, 518
Krisanova, L. D., **3**, 198; **4**, 522

Krishna, B., **4**, 158
Krishnan, R., **5**, 89; **8**, 156
Kristol, D. S., **2**, 39
Krivopalov, V. P., **6**, 102, 103
Krivska, M., **1**, 12
Krochman, D. E., **4**, 28
Krogh-Jespersen, K., **5**, 47; **8**, 159, 161
Kröhn, H., **3**, 246
Krohnke, C., **3**, 948
Krom, E. N., **12**, 12
Kropf, H., **3**, 113; **6**, 144
Kropp, P. J., **5**, 49; **14**, 230, 779
Krow, G. R., **4**, 300; **14**, 566, 569
Krstić, V., **9**, 117
Krudy, G. A., **2**, 362
Kruglaya, O. A., **3**, 1023
Kruglova, N. V., **3**, 378–380
Kruizinga, W. H., **9**, 108, 248
Kruk, I., **4**, 431
Krupenskii, V. I., **4**, 54, 55
Krylov, A. I., **8**, 111; **14**, 21, 542
Krysin, A. P., **8**, 77
Kryuchkova, L. V., **3**, 694
Ksander, G. M., **14**, 174
Kbáček, P., **3**, 771
Kubala, B., **8**, 145; **14**, 433
Kubisty, C., **8**, 80
Kubo, K., **8**, 95, 96; **14**, 71
Kubota, H., **14**, 303
Kubota, K., **14**, 354
Kubota, S., **3**, 53, 250
Kubota, T., **10**, 94
Kubota, Y., **3**, 738; **8**, 93, 94
Kubsrestha, G. N., **14**, 46
Kucher, R. V., **3**, 130, 191; **4**, 512
Kuckländer, U., **14**, 760
Kucybala, Z., **6**, 89
Kudesia, V., **4**, 43
Kudesia, V. P., **1**, 174; **4**, 219, 222
Kudo, T., **2**, 269
Kudryashova, N. I., **7**, 125
Kudryavtseva, L. A., **2**, 211, 339, 353; **9**, 220
Kühl, U., **12**, 252
Kuivila, H. G., **3**, 833, 834; **9**, 249
Kukalenko, S. S., **2**, 160
Kukhar, V. P., **3**, 659
Kukota, Y., **14**, 785
Kulagin, N. I., **2**, 160
Kulicki, Z., **4**, 437, 483
Kulikov, S. M., **14**, 55
Kulikova, T. V., **2**, 164
Kulinkovich, O. G., **9**, 207
Kulsrestha, G. N., **7**, 110
Kumada, M., **5**, 144, 145; **14**, 626
Kumagai, T., **12**, 154; **14**, 820
Kumagai, Y., **13**, 20
Kumamoto, T., **14**, 550
Kumanda, M., **12**, 112
Kumar, A., **3**, 452; **4**, 56, 195, 264
Kumar, M., **14**, 31
Kumar, P., **6**, 24; **14**, 782
Kumar, S., **4**, 45, 654
Kume, A., **14**, 366
Kumova, A. A., **3**, 265
Kunanec, S. M., **3**, 630
Kunert, D. M., **14**, 727
Kunikiyo, N., **2**, 223
Kunimasa, T., **2**, 223
Kunitake, T., **2**, 146, 202, 204, 205, 349; **11**, 42
Kunitomi, Y., **14**, 792
Künstlinger, M., **14**, 846
Kuntz, R. R., **3**, 1075
Kunze, K. L., **4**, 745, 770; **14**, 667
Kunze, M., **14**, 698
Kuokkanen, T., **6**, 1; **8**, 133
Kupchik, I. P., **9**, 110
Kuppy, T., **2**, 288
Kuramochi, J., **14**, 409
Kuramoto, N., **4**, 430
Kuramshin, E. M., **3**, 115, 187, 188, 334; **4**, 298
Kurata, R.-I., **14**, 151
Kurauchi, M., **1**, 180
Kurebayashi, Y., **14**, 675
Kurek, J., **9**, 40
Kurek, J. T., **12**, 135
Kurenkova, V. M., **7**, 48a
Kurguzova, A. M., **2**, 339; **9**, 220
Kuriacose, J. C., **4**, 113, 114, 685
Kurita, J., **14**, 69
Kuritani, H., **14**, 264
Kuritsyn, L. U., **2**, 163
Kurnersova, E. P., **4**, 527
Kuroda, H., **4**, 712
Kuroda, K., **4**, 420, 454, 719
Kuroda, Y., **1**, 178; **2**, 243
Kuroki, N., **2**, 222, 223
Kurosaki, T., **3**, 738; **8**, 93
Kurosawa, H., **3**, 872
Kurosawa, K., **3**, 716
Kurreck, H., **3**, 74, 76, 988
Kursanov, D. N., **3**, 1040; **14**, 781
Kurskii, Yu. A., **3**, 94
Kurtev, B. J., **14**, 868
Kurth, M., **13**, 129
Kurts, A. L., **6**, 49, 50
Kurusu, Y., **4**, 178
Kuryaeva, T. T., **14**, 781
Kurz, M. E., **3**, 895
Kusabayashi, S., **4**, 288; **9**, 180
Kusano, Y., **1**, 144; **4**, 729
Kuse, T., **4**, 52
Kushi, K., **14**, 603
Kutchan, T. M., **14**, 635
Kutskii, Yu. B., **3**, 94
Kutyrev, G. A., **2**, 384; **12**, 53
Kutzer, J. C., **3**, 920
Kuwabara, M., **3**, 1067, 1068, 1145
Kuwajima, I., **4**, 207
Kuwata, K., **3**, 589
Kuz'min, A. I., **9**, 71
Kuz'min, A. K., **4**, 487
Kuz'min, M. G., **3**, 959
Kuz'min, V. A., **3**, 228, 252, 1087
Kuz'mina, N. Ya., **9**, 214
Kuznetsov, A. M., **10**, 274
Kuznetsov, S. G., **14**, 794, 795
Kuznetsov, V. A., **3**, 874
Kuznetsova, M. G., **14**, 852
Kuznetsova, N. I., **4**, 161
Kuznetsova, V. P., **8**, 97, 88
Kuzuhara, H., **1**, 42
Kuzuya, M., **14**, 877
Kwak, S. T., **4**, 587
Kwan, T., **3**, 591
Kwantes, P. M., **5**, 59
Kwart, H., **2**, 453; **11**, 76, 77, 95; **14**, 14, 161
Kwart, L. D., **11**, 77; **14**, 161
Kwon, K.-S., **9**, 263

L'Abbe, M. M., **12**, 269
La, S., **2**, 108, 109; **9**, 171, 173
La, S., Koh, K. S., **1**, 17
LaBelle, B. E., **10**, 268
Laarhoven, W. H., **14**, 270, 709
Labeish, N. N., **5**, 57
Labinger, J. A., **14**, 597
Lablache-Combier, A., **3**, 33
Lace, D. A., **2**, 279
Lacroix, B., **14**, 192
Ladatko, A. A., **3**, 52
Ladd, D. L., **10**, 133
Ladika, M., **5**, 33; **14**, 449
Lafont, D., **4**, 667
Lagerström, P., **7**, 100
Lagow, R. J., **10**, 10; **14**, 627
Lahousse, F., **11**, 59
Lahtinen, L., **10**, 233
Lai, C. C., **3**, 730

Lai, T.-W., **12**, 163; **14**, 609
Lai, Y.-H., **6**, 84
Lajunen, M., **8**, 2, 10; **14**, 470
Lakshmi, V., **4**, 243, 251, 254, 255
Lakshmi, V., **4**, 243, 251, 254
Lakshmikantham, M. V., **3**, 876
Lakshmikanthan, N., **1**, 142
Laland, E., **3**, 393
Lam, W. Y., **2**, 414, 415
Lamartina, L., **4**, 24
Lamaty, G., **1**, 65; **2**, 6
Lambrecht, G., **2**, 313
Lambrechts, H. J. A., **7**, 94
Lammerink, B. H. M., **13**, 53
Lammertsma, K., **7**, 48
Lamotte, G., **3**, 474
Lampe, J., **1**, 87, 103
Lancaster, M., **11**, 55; **14**, 261
Lancelot, J. C., **14**, 100
Landeros, R., **14**, 84
Landheer, I., **4**, 404
Landini, D., **9**, 183, 218
Landino, J. P., **12**, 80; **14**, 482
Landis, M. E., **4**, 307
Landolt, R. G., **3**, 604
Landry, D. W., **5**, 56
Lane, S. M., **2**, 179
Lang, C., **14**, 521
Langbeheim, M., **14**, 256
Lange, B. C., **3**, 819; **10**, 163
Lange, M., **4**, 749
Langford, G. E., **14**, 471
Langley, J. A., **11**, 103; **12**, 118
Lanneau, G., **9**, 89
Lanneau, G. F., **2**, 355, 357
Lanteri, P., **13**, 139
Lapachev, V. V., **1**, 162; **6**, 102, 103
Lapin, S., **3**, 895
Laporterie, A., **4**, 406; **9**, 94
Lappert, M. F., **3**, 82, 86, 609; **10**, 256
Lapshin, S. A., **2**, 443
Lapshova, A. A., **3**, 317, 466, 467
Larchevêque, M., **1**, 101
Lardicci, L., **4**, 546, 547
Largean, C., **14**, 480
Larionova, L. A., **12**, 15
Larsen, D. S., **12**, 56
Larsen, J. W., **8**, 112
Lartey, P. A., **14**, 372
Lasker, J. M., **3**, 880
Lasne, M.-C., **14**, 644
Lasperas, M., **1**, 128, 131
Lassalvy, C., **12**, 183
Latif, F., **2**, 23
Latowski, T., **3**, 1096
Latrofa, A., **4**, 177; **14**, 865
Lattes, A., **3**, 33; **5**, 155; **12**, 146
Lattke, **10**, 41–44; **14**, 359
Lau, K. S. Y., **4**, 75
Lau, W., **3**, 678
Lauffer, R. B., **4**, 750
Lauterbach, G., **9**, 60
Laval, J. P., **5**, 155; **12**, 146
Laverdet, G., **3**, 307
Laviron, E., **3**, 976
Lavrent'ev, A. N., **9**, 214
Lawesson, S.-O., **14**, 239
Lawrence, G. W., **6**, 60
Lawrence, R. F., **14**, 175
Laxerera, A. M., **3**, 135
Laycock, D. E., **12**, 92
Lazana, R. L. R., **3**, 1016
Lazzeretti, P., **8**, 160
Le Bozec, H., **14**, 598
Le Bras, G., **3**, 307
Le Coq, A., **12**, 268
Le Corre, M., **10**, 262–264
Le Goff, N., **14**, 267
Le Houllier, C. S., **5**, 133
Le Noble, W. J., **8**, 11
Le Page, Y., **3**, 189; **4**, 444
Le Roux, J. P., **14**, 751
Le Saint, J., **14**, 799
Le, A. T., **8**, 24
LeGoaller, R., **5**, 48
LeGoff, E., **4**, 336
Leandri, G., **9**, 75
Lebedev, B. A., **2**, 148, 149, 151, 152; **10**, 334–337
Lebedev, N. N., **4**, 115, 116, 467
Lebedev, V. L., **3**, 31
Lebibi, J., **10**, 319
Ledon, H. J., **4**, 120
Lee, B. C., **2**, 440
Lee, B. H., **2**, 103
Lee, B. S., **9**, 229
Lee, C., **4**, 37
Lee, C. C., **5**, 108; **8**, 149, 152; **9**, 5; **14**, 389, 417, 418
Lee, D.-S., **9**, 124
Lee, D. G., **4**, 25, 26, 33, 34, 36
Lee, F. S. C., **3**, 649
Lee, H., **4**, 616
Lee, H. W., **2**, 438; **9**, 171, 172
Lee, I., **1**, 17; **2**, 82, 108, 109, 432, 434, 435, 438, 440; **9**, 171–173, 229
Lee, I. K., **9**, 246
Lee, J. I., **10**, 158
Lee, J. K., **2**, 434
Lee, K. H., **3**, 517
Lee, K. S., **3**, 747
Lee, S. J., **5**, 156
Lee, W.-B., **5**, 46
Lee, Y.-S., **2**, 213
Lee, Y. H., **14**, 422
Lee, Y. T., **4**, 542
Lee-Ruff, E., **14**, 446, 447
Leenson, I. A., **12**, 67, 68
Leeper, F. J., **10**, 316
Leete, E., **4**, 736
Lefevre-Jenot, L., **10**, 32
Leffek, K. T., **10**, 277
Lefort, D., **3**, 137, 147, 148, 523
Lehnig, M., **3**, 88, 89
Lehr, G. F., **3**, 1163
Lei, K. L., **9**, 30
Lei, X.-C., **3**, 132
Leichenko, A. A., **4**, 489
Leigh, W. J., **3**, 41, 42, 750
Leister, H. P., **14**, 120
Leister, S., **14**, 102
Leistner, S., **14**, 78
Leites, L. A., **8**, 50
Lemetais, P., **2**, 79
Lemieux, R. U., **4**, 27
Lemke, F., **2**, 40
Lemke, T. L., **1**, 80
Lemmer, D., **14**, 882
Lempert, K., **3**, 822
Lemura, S., **3**, 296
Lendzian, F., **3**, 1001
Leng, J. L., **2**, 102
Lenney, P. W., **3**, 581
Lenoir, D., **9**, 26, 28
Leonard, N. J., **14**, 79
Leone-Bay, A., **3**, 908
Leoni, P., **2**, 425
Leopold, A., **6**, 6
Lerman, O., **7**, 20; **12**, 10
Lerman, Z. A., **6**, 54, 55
Lermontov, S. A., **12**, 133
Leroux, Y., **1**, 106; **10**, 165
Leroy, G., **1**, 70; **3**, 6
Leshina, T. V., **3**, 18, 1038, 1039; **14**, 783
Lesko, P., **12**, 147
Lessard, J., **3**, 66, 413
Leuenberger, C., **5**, 64
Leung, C. W. F., **10**, 79
Leung, K. H. W., **3**, 631
Leung, S.-L., **2**, 180; **14**, 869
Leuschen, T., **9**, 51
Leuschner, R., **3**, 246
Leussing, D. D., **10**, 202
Leussing, D. L., **1**, 117
Levanevskii, D. E., **11**, 147
Lever, O. W., **10**, 267
Levey, G., **3**, 706, 732; **4**, 337
Levi, B. A., **10**, 305

Levin, M., **3**, 247–249
Levin, P. P., **3**, 228, 252
Levin, W., **9**, 66
Levin, Ya. A., **2**, 450; **3**, 670, 673, 1072; **9**, 99
Levine, S. G., **14**, 196
Levine, S. Z., **3**, 260
Levine-Pinto, H., **4**, 676
Levisalles, J., **12**, 175–177; **14**, 608
Levy, A. B., **10**, 126
Lewin, G., **14**, 767
Lewis, A., **12**, 132
Lewis, C. P., **14**, 543
Lewis, D., **1**, 15
Lewis, D. E., **9**, 241; **11**, 12; **14**, 408
Lewis, E. S., **9**, 271
Lewis, F. D., **3**, 904, 906; **14**, 805
Lewis, I. C., **3**, 180
Lewis, J., **8**, 39
Lewis, J. R., **3**, 703; **14**, 50
Lewis, M., **2**, 299
Lewis, R. J., **3**, 960
Lewis, S. D., **2**, 289–291
Ley, S. V., **1**, 18; **4**, 205, 209, 229; **14**, 545
Lhomme, M.-F., **3**, 1088
Lhoste, J. M., **14**, 581
Li, G.-Fu., **4**, 305
Li, T.-T., **12**, 147
Li, Y.-G., **14**, 582
Liao, S.-T., **2**, 447
Lias, S. G., **7**, 49; **8**, 135
Liberek, B., **9**, 254
Licandro, E., **14**, 25
Lichszteld, K., **4**, 431
Lidor, R., **10**, 134
Liebl, R., **14**, 123–125
Liebman, J. F., **5**, 2
Lien, M. H., **3**, 28; **8**, 172; **10**, 308, 309
Lienhard, G. E., **2**, 282
Liew, J., **14**, 735
Lifshitz, A., **14**, 807
Ligon, W. V., **2**, 190
Likforman, J., **4**, 110
Likholobov, V. A., **4**, 161, 463
Likic, U., **3**, 597
Lillocci, C., **2**, 371; **11**, 8
Lin, C.-L., **2**, 241
Lin, G. M. L., **9**, 199
Lin, L. C., **10**, 126
Lin, L. P., **5**, 24
Lin, M. C., **3**, 261
Lin, M. S., **7**, 105
Lin, S.-F., **2**, 69
Lin, T.-Y., **4**, 15
Lin, Y. T., **2**, 453; **11**, 95
Linda, P., **2**, 100, 214
Linden, M., **4**, 63
Lindfors, E. L., **4**, 326
Lindley, A. A., **3**, 355; **10**, 93
Lindley, J. M., **5**, 84
Lindner, C., **14**, 266
Lindoy, L. F., **2**, 369
Lindsay, D., **3**, 48
Lindstrom, M. J., **3**, 582; **11**, 154
Ling, C.-F., **3**, 575; **13**, 147
Link, C. M., **12**, 136
Link, D. C., **14**, 806
Linkowski, G. E., **10**, 59
Linnartz, T., **3**, 145, 254
Linschitz, H., **3**, 929
Lion, Y., **3**, 1067, 1068, 1145
Liotta, D., **9**, 225; **10**, 80; **12**, 228; **13**, 112
Lipatova, T. E., **2**, 158
Lipovich, V. G., **7**, 103, 112, 113; **8**, 117; **14**, 392
Lippmaa, E., **12**, 075, 133; **14**, 472
Lipscher, J., **3**, 244
Lipscomb, W. N., **2**, 299; **12**, 120
Lipshutz, B. H., **10**, 136
Lipták, A., **4**, 539, 540
Lisitsyn, V. N., **6**, 86, 87
Liso, G., **3**, 43; **4**, 177; **14**, 865
Lissi, E., **3**, 320, 328, 614, 1078
Lissillour, R., **9**, 49
Little, D. A., **10**, 211
Little, R. D., **3**, 564
Litvinenko, L. M., **2**, 106, 142, 166, 167, 170, 442, 443; **6**, 56; **9**, 202
Litvinenko, S. L., **9**, 244, 260
Litvinov, V., **7**, 13
Litvintsev, I. Yu., **4**, 115, 116, 318
Liu, C. S., **5**, 142, 143
Liu, G., **4**, 176
Liu, H.-H., **12**, 42, 85
Liu, J.-C., **12**, 42
Liu, J. H., **4**, 1
Liu, K.-C., **3**, 488, 583; **14**, 109
Liu, K.-T., **8**, 53
Liu, L.-C., **12**, 85
Liu, M. T. H., **4**, 357; **11**, 63; **14**, 691
Liu, R.-Z., **7**, 3
Liu, X.-H., **7**, 3
Liu, Y.-C., **3**, 132, 592
Liz, R., **4**, 65; **12**, 140
Ljungqvist, A., **10**, 82
Lledós, A., **14**, 842
Llen, M. H., **10**, 303
Llinas, J. R., **14**, 837
Lloyd, D., **3**, 772
Lloyd, R. V., **3**, 30, 32
Lockhart, T. P., **3**, 566; **14**, 76?
Lockwood, P. A., **3**, 105
Locsei, V., **2**, 155
Lodder, G., **8**, 148; **9**, 2; **11**, 16
Loechler, E. L., **4**, 717, 718, 720
Loehe, J. R., **6**, 100
Loehr, R., **3**, 1101; **4**, 386
Loewus, D. I., **2**, 323; **14**, 207
Löfas, S., **3**, 291
Loftfield, N. S., **3**, 242
Loftsson, T., **2**, 188
Logan, R. T., **14**, 386
Logue, M. W., **6**, 47; **9**, 247
Logutov, V. I., **2**, 111
Loh, J.-P., **4**, 646
Lohmann, J.-J., **10**, 130; **14**, 37
Loi, A., **11**, 92
Loiá, M. C. **3**, 1015; **4**, 639
Lokensgard, D. M., **14**, 643
Lokhov, R. E., **4**, 342
Lokshin, B. V., **10**, 50; **14**, 599
Loktev, V., **14**, 6
Loktev, V. F., **8**, 69, 70; **14**, 7
Lolkema, J. S., **2**, 402
Lomas, J. S., **11**, 119
Londrigan, M. E., **10**, 95
Long, P. V., **14**, 707
Longeray, R., **13**, 139
Longobardi, M., **13**, 137
Lönnberg, H., **1**, 35; **9**, 123
Loosen, K., **5**, 77; **14**, 468
Loots, M. J., **14**, 606
Lopatinskaya, Kh. Ya., **9**, 238
Lopatinskii, V. P., **2**, 55
Lopez, J., **3**, 942, 989, 991
Lopez, S., **4**, 405
Lopez-Mardomingo, C., **4**, 550
Łopusiński, A., **14**, 839
Lorand, J. P., **1**, 8
Lorenzi-Riatsch, A., **14**, 703
Lorimer, J. P., **9**, 160
Lorquet, J. C., **8**, 155
Löser, U., **3**, 257
Lossing, F. P., **3**, 38
Loth, K., **3**, 1008
Lotter, H., **10**, 180
Lotto, C. P., **2**, 84
Lotz, W., **9**, 210
Lotze, M., **5**, 20
Loupy, A., **1**, 123; **4**, 572, 573
Low, M., **2**, 143
Lowe, G., **2**, 363, 364, 387–390, 396, 409
Lowndes, P. R., **9**, 61

Lowry, S. A., **3**, 604
Loza, R., **14**, 231, 620
Lozac'h, N., **14**, 460
Lu, H.-L., **3**, 132
Lu, N.-C., **4**, 305
Lu, Y.-Z., **4**, 305
Lübbe, F., **3**, 212
Lubinkowski, J. J., **3**, 575; **13**, 147
Lubuzh, E. D., **3**, 721, 722
Lucaciu, N., **2**, 155
Luche, J.-L., **4**, 588
Luchian, C., **13**, 85
Lucivjansky, P., **4**, 108
Łuczak, L., **14**, 839
Lüdersdorf, R., **3**, 1167
Lugade, A. G., **5**, 118
Lugovoi, Yu. M., **3**, 388–390, 418
Luh, T.-Y., **3**, 747; **9**, 30
Lui, M. T. H., **3**, 210
Luibrand, R. T., **14**, 884
Luis, S. V., **13**, 163
Luisi, P., **2**, 197
Luk'yanenko, L. V., **3**, 130
Lukacs, G., **14**, 512
Lukacs, J., **3**, 156
Lukashina, G. S., **2**, 450
Lukashina, S. G., **9**, 99
Lukasiak, J., **9**, 92
Lukton, D., **12**, 103
Lumma, W. C., **1**, 133
Lunazzi, L., **3**, 29, 339, 424
Lund, A., **3**, 1031, 1045
Lund, H., **3**, 761, 783–785, 830
Lund, L., **3**, 780
Luntz, A. C., **4**, 383
Lupo, A. T., **13**, 148
Lur'e, B. A., **3**, 1125
Lur'e, E. P., **14**, 371
Lusztyk, J., **3**, 239, 986
Lutsenko, Z. L., **13**, 49
Lutsyk, A. I., **4**, 17
Lutz, E., **9**, 24
Luzikov, Y. N., **10**, 50; **14**, 599
Luzinchi, X., **4**, 362
Lyamin, I. A., **14**, 552
Lyapina, N. Sh., **3**, 253
Lyapova, M. J., **14**, 868
Lyatifov, I. R., **3**, 770
Lychkin, I. P., **2**, 72
Lynch, G. J., **12**, 137, 138
Lynch, T. J., **10**, 24
Lysenko, D. L., **3**, 200; **4**, 455
Lytle, F. W., **4**, 666
Lyttle, M. H., **14**, 813
Lyubimova, T. B., **2**, 171
Lyytikainen, H., **8**, 10; **14**, 470
M'Boula, J., **10**, 156
Mansuri, M. M., **7**, 21
Ma, L.-T., **14**, 582
Maas, G., **4**, 179; **8**, 103; **14**, 193
MacAlpine, D. K., **3**, 605
MacBride, J. A. H., **6**, 126
MacDonald, T. L., **10**, 117
MacKenzie, N. E., **14**, 678
MacKenzie, W. M., **9**, 189
MacKirdy, I. S., **10**, 250
MacLaury, M. R., **11**, 150
MacLean, D. B., **13**, 160, 161
MacNeil, J. M., **9**, 264
MacNeil, P. A., **4**, 670
Macaulay, S. R., **14**, 360
Maccarone, E., **1**, 46; **2**, 433; **9**, 101; **12**, 82; **14**, 780
Macchia, F., **9**, 55
Maccioni, A., **3**, 81
Macdonald, J. E., **14**, 175
Macdonald, J. G., **14**, 149
Macdonald, T. L., **3**, 453; **4**, 355
Mach, M. H., **6**, 67; **14**, 40
Macháček, S., **2**, 182
Machkova, Z., **11**, 11
Macho, V., **12**, 101
Mack, A. G., **14**, 297
Mackay, G. I., **9**, 146
Mackay, R. A., **2**, 350
Mackenzie, K., **14**, 235
Macomber, R. S., **2**, 362
Macpherson, M. T., **3**, 161
Macura, S., **14**, 768
Madhavan, V., **3**, 548a
Maeda, H., **7**, 065; **13**, 145
Maeda, N., **3**, 448, 465; **14**, 527, 556
Maeda, S., **8**, 114
Maeda, T., **14**, 101
Maerkl, G., **14**, 123–125
Maeshima, T., **3**, 886
Maffrand, J. P., **14**, 561
Magaretha, P., **3**, 766
Magaril, R. Z., **3**, 1108
Magee, M. G., **14**, 177
Mager, H., **2**, 27, 28
Mageswaran, S., **14**, 213, 215, 353
Maggio, J. E., **14**, 334, 668
Maggiora, G. M., **1**, 115
Magnusson, G., **10**, 261
Magomedov, G. K., **12**, 115, 116
Mahadevappa, D. S., **4**, 238–241
Mahalanabis, K. K., **12**, 214
Mahapatro, D. D., **4**, 2
Mahapatro, R. C., **4**, 228
Mahapatro, S. N., **4**, 1, 334, 339
Mahdavi-Damghani, Z., **12**, 214
Maheshwari, M. K., **4**, 194
Mahmoud, F. M. S., **9**, 84, 85; **10**, 287, 290, 293
Mahmoud, M., **2**, 104
Mahmound, M. R., **1**, 47
Mahoney, L. R., **4**, 497
Maia, A., **4**, 572; **9**, 183, 218
Maialo, F., **3**, 1032
Maier, G., **14**, 819
Maier, J. P., **3**, 936
Maier, W. F., **4**, 693
Maifield, W., **14**, 251
Maillard, B., **3**, 25, 99, 100, 125, 146, 238, 491
Maillard, P., **3**, 927
Main, L., **7**, 51; **11**, 42
Maiolo, F., **3**, 534
Maione, A. M., **14**, 525
Maiorana, S., **14**, 25
Maiti, S., **4**, 165, 166
Maitlis, P. M., **4**, 610
Majer, E., **3**, 728
Majer, J., **4**, 71
Majestic, V. K., **6**, 113
Majewski, J. M., **12**, 49
Majewski, M., **10**, 140
Majima, T., **3**, 910, 913
Majumdar, D. K., **2**, 252
Majur, Y., **3**, 697
Makani, S., **7**, 29
Makarov, I. G., **3**, 999; **4**, 606
Makarov, M. G., **9**, 50
Makarova, R. A., **2**, 441
Maker, P. D., **3**, 150, 164, 185, 304, 306; **4**, 510
Makhin, A. A., **12**, 77
Makhon'kov, D. I., **3**, 689, 690; **4**, 435
Maki, Y., **14**, 32
Makimoto, S., **2**, 206
Makino, R., **4**, 772
Maksimenko, N. N., **2**, 437
Malatesta, V., **3**, 323, 338, 584
Maldotti, A., **3**, 316
Maleeva, N., **2**, 446
Malek, F., **3**, 655
Málek, J., **3**, 743, 744
Maleki, M., **10**, 267
Maletin, Yu. A., **3**, 416
Malhotra, K. C., **8**, 105
Malik, L., **3**, 123, 859, 1035
Malinowski, M., **4**, 643
Malisheva, N. A., **3**, 499
Malkani, R. K., **4**, 221

Malkov, **3**, 356, 392
Mallamo, J. P., **12**, 232
Malleron, J.-L., **10**, 164
Malleron, J. C., **1**, 88
Mallick, N., **3**, 179
Malpass, J. R., **4**, 103; **5**, 69, 129; **9**, 118
Malthouse, J. P. G., **2**, 295
Mamaev, V. M., **8**, 110
Mamaev, V. P., **1**, 162; **6**, 102, 103
Mamed'yarov, G. M., **4**, 365
Mametsuka, H., **3**, 501
Mamo, A., **1**, 46; **12**, 82; **14**, 780
Manabe, O., **1**, 144; **4**, 703, 712, 729; **7**, 92; **10**, 101; **14**, 20
Manabe, T., **13**, 7
Mandai, T., **14**, 612
Mandal, A. K., **4**, 578; **12**, 119
Mandel'shtam, T. V., **5**, 57; **14**, 647
Mander, L. N., **4**, 747; **14**, 211
Mandolini, L., **3**, 683; **4**, 88; **9**, 103–106
Manelis, G. B., **3**, 1126
Manglik, A., **4**, 222
Manglik, A. K., **7**, 54
Mangru, N. N., **11**, 140
Manifand, C., **3**, 99, 100
Manikyamba, P., **4**, 244
Manion-Schilling, M. L., **5**, 8
Manitto, P., **11**, 46
Mann, G., **14**, 87
Mann, J., **13**, 178
Manninen, K., **12**, 160
Manning, C., **14**, 262
Manning, T. D. R., **14**, 230
Mano, E., **14**, 877
Manser, G. E., **1**, 71
Mansilla, A. M., **12**, 255
Mansour, T. S., **10**, 58
Mansuy, D., **4**, 749
Mantashyan, A. A., **3**, 154; **4**, 498
Mantello, R. A., **8**, 54
Mantsch, H. H., **3**, 585
Mantulo, A. P., **2**, 173
Manuel, G., **14**, 745
Manukyan, T. K., **12**, 69
Manzo, R. H., **6**, 121; **14**, 464
Mao, C.-R., **14**, 320
Maquestiau, A., **14**, 426, 427
Mar'yasova, V. I., **3**, 1038; **14**, 783
March, G., **3**, 5, 1136
Marchand, A., **9**, 274
Marchese, G., **12**, 230
Marchetti, L., **3**, 779
Marchi, D., **11**, 111
Marchini, P., **4**, 177
Marchon, J. C., **3**, 965
Marcincal-Lefebvre, A., **12**, 157
Marciniec, B., **12**, 113
Mardoyan, V. A., **3**, 165
Marecek, J. F., **2**, 366
Mareda, J., **5**, 67
Margerum, L. D., **14**, 616
Margosian, D., **14**, 479
Marhoul, A., **1**, 12; **9**, 56
Mariano, **3**, 487, 907, 908
Maricich, T. J., **4**, 349
Marinas, J. M., **14**, 554, 555
Marinelli, E. R., **10**, 126
Marino, J. P., **14**, 636
Mark, F., **3**, 1141
Markaryan, Sh. A., **3**, 139
Markezich, R. L., **8**, 26
Markiewicz, W., **9**, 225
Markov, V. I., **9**, 80
Markowitz, M., **4**, 584
Marks, A., **6**, 175
Marks, P. J., **14**, 93
Marlier, J. F., **2**, 15, 399
Marlowe, C. K., **14**, 155
Marmugi, E., **3**, 575; **13**, 147
Maroni, P., **1**, 81; **12**, 216
Maroulis, A. J., **3**, 903
Marples, B. A., **14**, 310
Marquet, A., **14**, 511
Marrero, R., **14**, 670
Marriott, P. R., **3**, 20, 329, 469, 671
Marsais, F., **10**, 129
Marsh, E. A., **8**, 25
Marshall, P. D. R., **3**, 321, 451
Marshall, R. M., **3**, 112, 205, 266, 397, 398
Marsili, A., **14**, 98
Martell, A. E., **1**, 40, 41; **11**, 96, 122
Martem'yanov, V. S., **4**, 516, 517, 520
Martemyanov, V. S., **3**, 167
Martens, F. M., **3**, 774
Martin, A., **4**, 379; **14**, 523
Martin, D., **2**, 113
Martin, G., **2**, 119
Martin, H.-D., **14**, 698
Martin, I., **2**, 133; **11**, 33, 70, 74
Martin, J. C., **2**, 414, 415; **3**, 678; **14**, 834
Martin, J. R., **4**, 581; **9**, 126
Martin, M. R., **9**, 10
Martin, M. V., **9**, 10
Martin, N. H., **4**, 399, 400
Martin, O. R., **9**, 125
Martin, P., **14**, 382, 383
Martin, R., **3**, 423, 1128
Martin, V. S., **4**, 152
Martin, V. V., **3**, 600
Martinelli, M., **13**, 79
Martinez Alvarez, R., **9**, 8
Martinez, A., **10**, 324
Martinez, R. I., **3**, 382; **4**, 280
Martinez-Davila, C., **14**, 237, 634
Martinez-Utrilla, R., **3**, 1091
Martins, L. J. A., **3**, 884
Martinson, P., **3**, 221
Marton, A., **2**, 139, 141, 143
Martynenko, Z., **14**, 652
Martynov, A. V., **3**, 430
Maruthamuthu, M., **1**, 142
Maruthamuthu, P., **3**, 705
Maruyama, K., **1**, 96, 97; **3**, 852, 923
Marx, J., **3**, 223
Maryanoff, B. E., **4**, 576
Maryasova, V. I., **3**, 1039
Marzorati, L., **14**, 548
Masai, M., **4**, 52
Masamune, S., **1**, 90; **10**, 191, 192
Masamune, T., **9**, 58; **10**, 98
Mascagni, P., **14**, 682
Masci, B., **9**, 104–106
Maselko, J., **4**, 253
Mashevskaya, M. S., **9**, 261
Mashima, K., **12**, 164
Maslennikov, I. G., **9**, 214
Maslennikov, S. I., **3**, 174
Masloch, B., **3**, 877, 1140; **4**, 347
Masnovi, J., **13**, 200
Mason, K. G., **14**, 310
Mason, R. P., **3**, 644, 880, 1041
Mason, T. J., **9**, 160
Massa, W., **14**, 13, 760
Massey, R. C., **10**, 333
Mastrocola, A. R., **14**, 694
Masuda, S., **3**, 608, 634
Masumoto, K., **3**, 256
Masunaga, T., **4**, 661
Masuyama, Y., **4**, 178
Mat'aŝova, E., **3**, 859
Mat'tsev, A. K., **5**, 97
Mata-Segreda, J. F., **2**, 20; **4**, 567; **8**, 91
Mataka, S., **14**, 88, 89
Mateo, S., **10**, 273
Materikova, R. B., **3**, 769, 770
Matern, A. I., **12**, 179
Mathay, F., **14**, 274
Mathieu, G., **9**, 268

Mathis, R., **4**, 716
Mathlouthi, M., **1**, 32
Mathur, P. C., **4**, 654
Matsuda, I., **12**, 198
Matsuda, K., **9**, 241; **14**, 408
Matsuda, M., **3**, 431, 432, 678a, 713, 781
Matsuda, T., **6**, 15, 16
Matsudo, M., **3**, 503
Matsui, K., **6**, 142; **7**, 91
Matsui, S., **10**, 251
Matsumaru, N., **7**, 149
Matsumoto, H., **14**, 502, 701
Matsumoto, K., **2**, 230; **12**, 246; **13**, 70, 77
Matsumoto, M., **4**, 145, 420, 454; **9**, 36
Matsumoto, S., **4**, 372
Matsumoto, T., **14**, 506
Matsumoto, Y., **2**, 220, 226
Matsumura, E., **4**, 690; **6**, 122, 123; **10**, 174
Matsuo, K., **10**, 101; **12**, 91
Matsuo, M., **4**, 372
Matsuo, N., **3**, 501
Matsuoka, H., **6**, 108
Matsushima, K., **2**, 234
Matsushita, T., **14**, 680
Matsuura, M., **2**, 367
Matsuura, T., **3**, 578, 579, 902, 921; **4**, 290, 397, 398, 402, 413, 528; **14**, 17, 343
Matsuuts, T., **3**, 887
Matta, M. S., **2**, 285
Mattay, J., **3**, 1161
Matteson, D. S., **2**, 282
Matthei, J., **13**, 175
Matthews, J. I., **3**, 350
Matthias, M., **5**, 134
Mattox, V. R., **14**, 374
Matturro, M., **14**, 188
Matturro, M. G., **3**, 1115
Matuura, R., **10**, 279
Matveev, K. I., **4**, 499, 500
Matveeva, E. D., **14**, 863
Matz, J. R., **14**, 854
Maul, R., **5**, 19
Maume, D., **14**, 100
Mavrin, G. V., **13**, 116, 120
Mawby, R. J., **12**, 272
Maxwell, B. E., **6**, 32; **14**, 14,
May, D. D., **3**, 72, 331
May, S. W., **4**, 765
Mayanna, S. M., **2**, 80; **4**, 237
Maybury, P. C., **4**, 697
Mayer, B., **5**, 141
Mayer, G. D., **9**, 115
Mayer, R., **3**, 55–58
Mayer, W., **3**, 570
Mayer, W. D., **8**, 78
Mayr, H., **12**, 94, 95; **13**, 135, 173, 174
Mayr, J., **10**, 70
Mazerolles, P., **4**, 406; **9**, 94; **14**, 132, 745
Mazur, M. R., **3**, 553
Mazur, Y., **4**, 131
Mazza, F., **14**, 525
Mazzanti, G., **3**, 45
Mazzocchi, P. H., **3**, 909; **12**, 155; **13**, 23
Mbarak, M. S., **4**, 143
McArdle, P., **13**, 45, 46
McAskill, N. A., **3**, 1155
McClory, M. R., **14**, 262
McCabe, P. H., **11**, 128
McCabe, R. W., **14**, 91
McCague, R., **14**, 271
McCann, M., **1**, 49
McCapra, F., **14**, 707
McCarry, B. E., **8**, 26
McCleery, D. G., **4**, 559
McClelland, C. W., **1**, 13; **3**, 537
McClelland, R. A., **1**, 2, 33; **2**, 5, 9, 12, 17, 326
McCloskey, J. A., **8**, 103
McClure, J. R., **10**, 58
McCombie, S. W., **14**, 828
McComsey, D. F., **4**, 576
McCormick, J. P., **3**, 756; **4**, 154
McCracken, S., **2**, 407
McCrae, D. A., **10**, 153
McCullough, J. J., **14**, 262
McCullough, K. J., **14**, 811
McDaniel, R. L., **14**, 196
McDonald, R. N., **3**, 1051, 1053
McDonald, R. S., **1**, 122
McEvoy, T. M., **9**, 130
McEwen, C. N., **3**, 367, 368
McEwen, W. E., **3**, 505, 575; **13**, 147; **14**, 356
McGarrity, J. F., **9**, 270
McGaw, B. A., **4**, 736
McGee, L. R., **1**, 92; **10**, 190
McGhie, J. F., **4**, 630
McGlinchey, M. J., **14**, 592
McGuiness, S. J., **6**, 83
McGuinness, R., **3**, 110
McIlgorm, E. A., **14**, 588
McInnis, E. L., **14**, 872
McKamey, D., **6**, 32
McKelvey, J. M., **8**, 173
McKelvey, R. D., **3**, 464
McKenna, J., **9**, 139, 234; **11**, 12, 144
McKenzie, A., **13**, 13
McKervey, M. A., **14**, 486, 588
McKillop, A., **4**, 69
McKinley-McKee, J. S., **9**, 258
McLafferty, F. W., **8**, 126
McLean, E. A., **7**, 115
McLennan, D. J., **9**, 138; **11**, 16
McLoughlin, R. G., **8**, 123
McMahon, P. E., **1**, 119
McMahon, R. J., **4**, 547
McMahon, T. B., **10**, 321
McManus, S. P., **9**, 224; **13**, 54; **14**, 732
McMillen, D. A., **3**, 598
McMillen, D. F., **3**, 50
McMurry, J. E., **14**, 174
McMurtrey, K. D., **14**, 663
McNab, H., **5**, 116
McNelis, E., **4**, 211
McPaulissen, L., **2**, 328
McPhail, A. T., **3**, 173; **14**, 313
McRobbie, I. M., **5**, 84
McWeeny, D. J., **10**, 333
Medelson, L. T., **14**, 205
Medina, J. D., **11**, 35
Medzhidov, A. A., **4**, 492
Meerholz, C. A., **4**, 209
Megremis, T. L., **3**, 736
Mehdi, S., **2**, 397
Mehnert, R., **3**, 222
Mehrotra, K. N., **1**, 52; **5**, 73, 95
Mehrotra, R. N., **4**, 47
Mehrotra, S. K., **8**, 109
Mehta, G., **8**, 22; **14**, 342, 489–492
Mehta, M., **4**, 47
Mehta, S. B., **4**, 16
Meidar, D., **7**, 101
Meier, G. P., **11**, 109
Meier, H., **1**, 79; **13**, 60; **14**, 330, 674
Meighey, E. A., **2**, 407
Meijer, E. W., **4**, 449
Meijer, J., **9**, 18; **10**, 121; **12**, 226, 243
Meijs, G. F., **3**, 224, 225; **6**, 13
Meinwald, J., **3**, 953; **4**, 160
Meister, A., **2**, 318
Meister, W., **11**, 90
Meixner, J., **3**, 361
Mekhtiev, D. S., **4**, 468
Mekhtiev, S. I., **13**, 119
Mel'nikov, A. I., **6**, 153
Melamed, U., **10**, 110, 111
Melichercik, M., **4**, 68, 86
Melloni, G., **12**, 7
Melnick, B., **14**, 441

Melnikova, N. B., **12**, 20
Melvin, S., **14**, 38
Mendenhall, G. D., **4**, 289
Mendoza, L., **12**, 128
Menger, F. M., **4**, 37; **9**, 186
Menghani, G. D., **4**, 159
Mentasti, E., **4**, 59
Meou, A., **9**, 75
Merah, B., **13**, 78
Merbach, A. E., **8**, 11
Merca, E., **2**, 70, 71
Mercer, F., **14**, 727
Mercier, F., **14**, 274
Merényi, R., **3**, 40; **11**, 59; **13**, 149
Merten, H., **14**, 153
Mertens, A., **12**, 98
Mervis, C. M., **2**, 77; **11**, 21
Mesnard, D., **12**, 211, 212, 213
Messeguer, A., **4**, 303
Messing, I., **3**, 234
Meth-Cohn, O., **5**, 84, 123, 125, 126; **13**, 185
Metzger, A., **3**, 989
Metzger, J., **10**, 315
Meunier, B., **4**, 568
Mews, R., **5**, 43
Meyer, D., **4**, 676
Meyer, F. K., **10**, 294
Meyer, G., **2**, 217, 218
Meyer, R., **12**, 216
Meyer, T. A., **10**, 329
Meyers, A. I., **1**, 108; **6**, 79; **10**, 178, 179, 193
Miazus, Z. K., **3**, 158
Micetich, R. G., **14**, 676, 735
Michael, J. P., **12**, 29; **14**, 672
Michael, J. V., **3**, 308, 314
Michalak, R. S., **14**, 834
Michalski, J., **14**, 839
Micheau, J. C., **3**, 33
Michejda, C. J., **9**, 269
Michel, D., **8**, 158
Michet, A., **14**, 192
Michinori, W., **14**, 583
Michl, J., **10**, 214, 215; **14**, 885
Michno, D. M., **14**, 290, 321, 791, 872
Michon, J., **3**, 596
Midden, W. R., **2**, 86
Middlemiss, D., **14**, 733
Midiwo, J. O., **3**, 1133
Midland, M. M., **10**, 46; **12**, 32; **14**, 833
Miginiac, L., **12**, 211–213
Migita, T., **3**, 297
Migliara, O., **4**, 24
Mihailovic, M. L., **4**, 100, 101
Mihelich, E. D., **3**, 172
Mijovic, M. V., **13**, 11
Mikami, K., **14**, 208, 209
Mikarni, K., **14**, 126
Mikhailova, T. S., **12**, 72
Mikhailovskaya, T. N., **4**, 511
Mikolajczyk, M., **2**, 381; **9**, 204, 218
Milaev, A. G., **3**, 11
Milart, P., **10**, 78
Mildvan, A. S., **2**, 403, 406
Miles, W. H., **14**, 161
Mill, T., **4**, 434
Miller, A. L., **6**, 99
Miller, B., **13**, 164
Miller, D. J., **11**, 13, 14
Miller, J., **2**, 107; **9**, 206
Miller, J. A., **10**, 267
Miller, J. K., **7**, 153
Miller, J. T., **10**, 207
Miller, L. L., **3**, 530, 531, 697; **4**, 14, 131; **6**, 33
Miller, R. E., **3**, 1113
Miller, R. J., **3**, 1173
Miller, V. B., **3**, 168
Millet, A., **14**, 426
Millet, G. H., **14**, 63
Mills, N. S., **10**, 35
Mills, S., **4**, 579
Milyaev, Yu. F., **12**, 71
Mimoun, H., **4**, 363
Mimura, T., **14**, 208
Min'ko, L. A., **4**, 151
Minakata, H., **12**, 233
Minamoto, K., **4**, 566
Minato, T., **4**, 136
Mincione, E., **4**, 66
Minde, J., **11**, 22
Mindl, J., **2**, 122, 123
Minicone, E., **8**, 36
Minisci, F., **3**, 17, 511, 524, 692, 719, 788; **4**, 183; **6**, 17, 30
Minkin, V. I., **3**, 52; **5**, 1; **14**, 119, 281, 285, 840
Minoli, **4**, 419; **5**, 085
Minot, C., **9**, 235
Minsker, K. S., **4**, 520
Minyaev, R. M., **5**, 1
Miranda, M. A., **3**, 1091
Mirao, A., **4**, 557
Mirek, J., **14**, 763
Miri, A. Y., **7**, 126
Mirksova, A. N., **9**, 5
Mironov, V. A., **14**, 244
Mironova, N. V., **14**, 852
Mirskova, A. N., **3**, 430; **9**, 6; **12**, 184
Mishima, M., **9**, 265
Mishina, T., **7**, 66, 82; **10**, 285
Mishra, J. P., **2**, 22
Mishra, K. K., **4**, 274
Mishra, S., **9**, 187
Mishra, S. P., **3**, 663; **6**, 34
Misic-Vukovic, M., **2**, 54; **9**, 181
Misiti, D., **12**, 249
Misono, M., **12**, 19
Mispreuve, H., **14**, 427
Misra, H. P., **4**, 761
Misra, P. C., **1**, 175; **4**, 149
Misra, R. N., **10**, 235; **11**, 155
Misra, V. D., **4**, 144
Missol, U., **3**, 119
Misurkin, I. A., **10**, 36
Mitawmura, S., **14**, 419
Mitchell, P. R. K., **4**, 90
Mitchell, R. D., **10**, 74
Mitchell, T. N., **10**, 219
Mitchell, T. R. B., **4**, 148; **14**, 190, 588
Mitchenko, E. S., **2**, 445
Mitra, A., **14**, 301
Mitra, A. K., **14**, 168
Mitschler, A., **14**, 274
Mitskevich, N. I., **3**, 117, 157, 333; **4**, 462, 471, 501, 513, 515, 519, 521
Mitsuhashi, T., **3**, 294
Mitsui, Y., **12**, 233
Mitsunobu, O., **4**, 174
Mitsuru, A., **7**, 77
Mittal, A. K., **4**, 195
Mitzner, R., **2**, 40, 336
Mitzutani, M., **14**, 159
Miura, K., **3**, 1169; **12**, 232; **14**, 590
Miura, M., **4**, 288; **10**, 242
Miura, T., **14**, 217
Miwa, T., **14**, 272
Miyake, A., **14**, 198
Miyake, H., **3**, 828; **9**, 15; **11**, 114
Miyake, T., **4**, 595
Miyano, T., **4**, 215
Miyashi, T., **13**, 165; **14**, 332
Miyashita, K., **4**, 623
Miyashita, S., **3**, 713
Miyata, N., **14**, 556
Miyata, S., **11**, 106
Miyata, T., **4**, 285
Miyata, Y., **2**, 107
Miyauchi, Y., **4**, 705
Miyazaki, H., **4**, 421
Miyoshi, H., **4**, 70; **12**, 79
Mizoguchi, I., **3**, 376
Mizone, S., **13**, 58
Mizukami, H., **14**, 590
Mizuki, Y., **13**, 145

Mizuno, H., **3**, 503, 678a
Mizuno, K., **3**, 901, 912, 916; **13**, 12
Mizuta, M., **12**, 198
Mizutani, M., **1**, 109
Mkhitaryan, R. P., **4**, 184
Moad, G., **3**, 133, 484
Möbius, K., **3**, 1001
Mochalov, S. S., **7**, 71, 72, 76; **8**, 76
Mochida, I., **4**, 690; **7**, 119
Mochida, K., **3**, 13, 85; **14**, 866
Mochizuki, H., **4**, 557, 558
Mock, G. B., **5**, 60
Mock, W. L., **1**, 163
Modena, G., **4**, 119, 123; **12**, 7
Modro, T. A., **2**, 88, 91, 373
Modryi, E. N., **4**, 87
Mody, N. V., **14**, 765
Moens, L., **14**, 378
Moffat, J. B., **8**, 166, 167; **10**, 11
Moffatt, F., **13**, 129
Moger, G., **3**, 632
Mogilyanskii, A. I., **2**, 160
Mohammad, T., **14**, 782
Mohanraj, S., **4**, 409
Mohiuddin, R., **4**, 683
Mohr, S., **14**, 532
Möhrie, H., **4**, 63
Moillet, J. S., **3**, 355
Moir, R. Y., **6**, 134
Moisa, C., **2**, 47
Moiseenkov, A. M., **14**, 641
Moiseev, I. I., **3**, 200; **4**, 164, 455, 484
Mojelsky, T. W., **3**, 741
Mokoki, S., **13**, 88
Mokrosz, M., **14**, 763
Molin, Yu. N., **3**, 18, 1038, 1039; **14**, 783
Moll, K. K., **13**, 92
Mollah, Y. A., **11**, 72
Mollah, Y. M. A., **14**, 411
Molnar, A., **11**, 136
Mols, O., **14**, 529, 530
Mondon, A., **14**, 532
Mondon, M., **3**, 413
Monig, J., **3**, 219
Monsan, P., **1**, 32
Monsky, A., **10**, 209
Montague, D. C., **3**, 581
Montanari, F., **9**, 183, 218
Montaudon, E., **3**, 393
Montevecchi, P. C., **3**, 449, 572, 978; **5**, 130, 131
Montgomery, S. H., **1**, 87
Montheard, J. P., **3**, 447; **12**, 76
Montoneri, E., **4**, 450
Moodie, R. B., **2**, 117, 153; **7**, 51, 54, 80
Moody, C. J., **5**, 137, 138; **11**, 93; **14**, 269, 271
Moody, W., **1**, 22
Mooney, B. A., **14**, 181
Moore, H. W., **13**, 30; **14**, 727
Moore, M. W., **12**, 109
Moore, P., **3**, 726
Moore, R. E., **4**, 265
Moore, W. B., **9**, 27
Mootz, D., **4**, 63
Morales, A., **14**, 650
Moran, G., **13**, 46
Morand, P., **4**, 563
Moravskii, A. P., **3**, 317; **4**, 495
Moreau, C., **1**, 65; **2**, 6
Moreau, J. L., **1**, 61
Moreau, M., **3**, 313
Morehouse, K. M., **3**, 1043, 1046, 1147
Morgan, S. E., **14**, 658
Morgans, D. J., **14**, 594
Morganti, G., **6**, 147
Morgat, J.-L., **4**, 676
Mori, I., **14**, 135
Mori, K., **4**, 731; **14**, 141
Mori, **2**, 234; **13**, 93, 179, 204
Mori, N., **4**, 175; **11**, 57
Mori, S., **10**, 192
Moriarty, R. M., **4**, 225, 226; **14**, 320
Morigami, Y., **3**, 858
Morild, E., **2**, 207
Morimoto, H., **4**, 324
Morimoto, T., **4**, 126; **14**, 139
Morin, J. M., **9**, 113; **14**, 319
Morin, L., **4**, 281
Morinaga, T., **3**, 294
Morine, G. H., **3**, 1075
Morisaki, H., **14**, 409
Morishita, T., **12**, 54
Morisset, V. M., **12**, 51, 57, 58
Morita, H., **9**, 182; **11**, 118
Morita, Y., **13**, 182
Moriuchi, F., **6**, 74; **14**, 34
Moriyama, M., **14**, 839
Moriyama, T., **3**, 944; **9**, 29
Moriyama, Y., **12**, 142; **14**, 517, 518
Morizawa, Y., **14**, 648
Morkovnik, A. S., **3**, 796, 797, 938
Morland, J. B., **14**, 808
Mornon, J. P., **14**, 521
Moro-oka, Y., **14**, 595
Moroi, Y., **10**, 279
Morokuma, K., **5**, 149
Moroni, A., **9**, 44
Morosi, G., **3**, 1042
Morozov, A. A., **11**, 147
Morozov, A. G., **9**, 52
Morozov, A. I., **2**, 423
Morozov, S. V., **8**, 68; **14**, 10
Morozova, V. Ya., **2**, 171
Morris, D. F. C., **6**, 104; **7**, 128
Morris, D. S., **14**, 173
Morris, G. J., **1**, 78; **4**, 76
Morrison, D. S., **8**, 141; **9**, 9
Morrison, H., **13**, 13
Morrison, J. A., **10**, 10; **14**, 627
Morrison, J. D., **4**, 544, 551, 556; **7**, 9, 109
Morrison, J. J., **1**, 85; **10**, 181
Morrison, M. A., **3**, 425
Morrocchi, S., **3**, 911
Morsi, S. E., **3**, 139
Morten, D. H., **9**, 157
Morton, H. E., **12**, 237, 238
Morton, J. A., **1**, 18; **14**, 545
Morton, J. R., **3**, 603
Morton, T. H., **9**, 144
Morvillo, A., **3**, 850
Morzycki, J. W., **4**, 229
Moses, L. M., **12**, 204
Mosher, H. S., **4**, 551
Moshkina, R. I., **4**, 490, 491
Moskal, J., **10**, 78
Moss, R. A., **2**, 213, 219; **5**, 47, 52–55
Mossoba, M. M., **3**, 642
Motekaitis, R. J., **1**, 40; **11**, 122
Motherwell, R. S. H., **3**, 472, 474
Motherwell, W. B., **3**, 470–474; **4**, 107, 229; **14**, 628
Motoki, S., **13**, 144
Motov, S. A., **3**, 140
Motoyama, N., **3**, 634
Motsarev, G. V., **3**, 289
Mousa, M. A., **2**, 104
Movchan, A. I., **13**, 50
Movrocíková, M., **3**, 522
Movsumzade, M. M., **6**, 68
Mozhaeva, T. Ya., **6**, 119, 149
Mpango, G. B., **10**, 140; **12**, 214, 215
Mracec, M., **6**, 2
Mrotzek, H., **11**, 78
Muceino, R. R., **14**, 621
Muchowski, J. M., **14**, 84
Mudryk, B., **3**, 815–817; **10**, 177
Muehlstaedt, M., **14**, 180
Mueller, P., **4**, 3
Mueller, P. H., **5**, 2
Mueller, P. L., **13**, 9

Mueller, P. W., **4**, 765
Mueller, U., **8**, 106
Muetterties, E. L., **4**, 124
Muggleton, B., **3**, 485
Mugnoli, A., **6**, 118
Muhlstadt, M., **1**, 69
Mujata, Y., **9**, 206
Mujica, C., **3**, 320
Mukai, C., **14**, 375, 761
Mukai, K., **3**, 71, 75, 599, 954
Mukai, T., **3**, 914; **13**, 165; **14**, 332, 820
Mukaiyama, T., **4**, 594; **12**, 190, 267
Mukherjee, J., **4**, 232
Mukhopadhyay, A. K., **14**, 168
Mulac, W. A., **3**, 159
Mulai, T., **4**, 674
Mulenko, S. A., **3**, 243
Müllen, K., **10**, 206
Müller, F., **3**, 962
Muller, G. W., **3**, 564
Muller, P., **4**, 146, 327, 609
Müller, U., **3**, 178
Müller-Litz, W., **2**, 115
Mulvaney, J. E., **10**, 95
Mulzer, J., **10**, 199; **11**, 44; **12**, 252; **14**, 826
Mumamoto, T., **3**, 297
Munegumi, T., **4**, 679
Munjal, R. C., **5**, 53
Munteanu, L., **4**, 20
Murahashi, S., **9**, 14
Murahashi, S.-I., **4**, 564
Murai, A., **9**, 58; **10**, 98
Murai, H., **3**, 557
Murai, O., **3**, 141
Murakami, H., **4**, 735
Murakami, M., **14**, 71
Murakami, N., **14**, 436
Murakami, Y., **2**, 227, 230; **4**, 706; **7**, 137, 138
Muramatsu, H., **12**, 1
Muramatsu, I., **4**, 696
Murata, I., **8**, 95, 96; **10**, 231; **14**, 645
Murata, S., **1**, 81
Murato, K., **14**, 653, 888
Murphy, D., **3**, 1012
Murphy, W. S., **3**, 755; **4**, 638; **7**, 116; **9**, 13; **14**, 440
Murray, K. S., **4**, 747
Murray, M. B., **3**, 621
Murray, R. E., **10**, 39
Murray, R. W., **4**, 396
Murray-Rust, P., **14**, 651
Murthy, N. K., **4**, 246
Murty, V. S., **14**, 516
Murugesan, N., **14**, 686
Muruyama, K., **10**, 186
Musaev, M. R., **4**, 492
Musaeva, N. F., **13**, 98–100
Musashi, T., **13**, 88
Muscio, O. T., **6**, 99
Musgrave, W. K. R., **3**, 384, 993; **14**, 80
Mushenko, D. V., **4**, 494; **12**, 78
Mushenko, V. D., **3**, 175; **4**, 494
Muškatirović, M., **9**, 117
Musker, W. K., **4**, 212–214
Muslin, D. V., **3**, 253
Musser, J. L., **14**, 174
Musso, H., **3**, 568; **14**, 695, 696
Mustafaev, A. M., **13**, 97
Musumarra, G., **2**, 444; **9**, 100, 181, 205, 219
Muszkat, K. A., **3**, 1167
Muszkat, L., **3**, 1152a
Mutai, K., **6**, 75, 76; **14**, 27, 28, 459, 764
Mutin, R., **5**, 155
Muto, S., **3**, 892; **4**, 723; **14**, 706
Myers, P. A., **12**, 195
Myhre, P. C., **7**, 69; **8**, 61
Myshkin, V. E., **3**, 356, 391, 392, 418

Nabiullina, Z. V., **4**, 517
Nacsa, A., **3**, 365
Nader, F. W., **5**, 29
Nadir, U. K., **14**, 39, 782
Nadjo, L., **3**, 784, 1047
Nadvornik, M., **4**, 7
Nagai, K., **7**, 65
Nagai, T., **1**, 179; **5**, 51
Nagakura, S., **3**, 521; **6**, 76; **14**, 27
Nagamatsu, H., **2**, 78
Nagano, N., **14**, 101
Nagano, Y., **14**, 101
Nagaoka, N., **10**, 257
Nagaoka, S., **4**, 764
Nagarajan, K., **7**, 24
Nagase, S., **3**, 394; **4**, 288; **5**, 149
Nagashima, E., **14**, 160
Nagata, R., **4**, 398
Nagayama, A., **3**, 982
Naggi, N. F., **2**, 77
Nagiev, M., **4**, 365
Nagiev, T. M., **4**, 364, 493
Nagieva, Z. M., **4**, 364
Nagira, K., **6**, 15, 16
Nagori, R. R., **4**, 47
Nagubandi, S., **2**, 129
Nagumo, K., **14**, 747
Nagusuna, K., **12**, 164
Nahm, S., **13**, 167
Naidu, H. M. K., **4**, 239–241
Naimushin, A. I., **3**, 31
Nair, M., **5**, 156
Nair, M. R., **9**, 232
Nair, T. D., **11**, 37
Nair, V., **1**, 37
Naito, I., **3**, 919
Najera, C., **12**, 141, 143, 144
Nakabayashi, T., **3**, 516
Nakadaira, Y., **14**, 276, 277
Nakagaki, R., **6**, 76; **14**, 27
Nakagawa, K., **5**, 144, 145
Nakagawa, S., **2**, 269
Nakagawa, Y., **2**, 226; **14**, 15
Nakahama, S., **4**, 557, 560
Nakahara, M., **13**, 114
Nakahara, S., **4**, 558
Nakai, T., **14**, 126, 157, 208, 209
Nakai, Y., **12**, 154
Nakajima, M., **14**, 367
Nakajima, N., **4**, 424; **14**, 375
Nakamori, S., **14**, 550
Nakamura, A., **10**, 37, 38; **12**, 164
Nakamura, E., **10**, 270
Nakamura, J., **6**, 76; **14**, 27
Nakamura, K., **2**, 200; **3**, 1022; **4**, 126; **10**, 31; **14**, 343
Nakamura, N., **3**, 1070; **5**, 30; **9**, 262; **14**, 654
Nakamura, S., **4**, 67; **12**, 153
Nakamura, Y., **13**, 182
Nakane, R., **14**, 747
Nakanishi, H., **4**, 421
Nakanishi, W., **3**, 1169
Nakano, A., **2**, 227, 230
Nakano, S., **10**, 87
Nakano, T., **14**, 523
Nakao, S., **12**, 150
Nakashima, E., **2**, 181
Nakashima, N., **2**, 146
Nakashima, Y., **3**, 256
Nakashita, Y., **14**, 703
Nakasone, A., **3**, 910
Nakata, T., **4**, 586
Nakatsu, K., **14**, 596
Nakatsuji, T., **4**, 734
Nakatsuka, M., **11**, 106
Nakatsuka, T., **3**, 829; **11**, 130
Nakausa, R., **3**, 61
Nakayama, J., **3**, 567; **7**, 149
Nakayama, M., **3**, 297
Nakazaki, M., **4**, 733–735
Nakazawa, H., **3**, 777
Nakazawa, T., **8**, 95, 96

Nakazumi, H., **6**, 4; **8**, 134
Nalbandyan, A. B., **3**, 165; **4**, 490, 491
Nalepa, C. J., **3**, 201
Nalesnik, T. E., **12**, 104
Nalley, E. A., **1**, 72; **14**, 796
Nametkin, N. S., **3**, 1110
Nand, K. C., **4**, 197
Nango, M., **2**, 222, 223
Nanjan, M. J., **7**, 7, 33
Nanni, E. J., **3**, 1059–1061; **4**, 374, 376, 378
Napier, J., **10**, 196; **14**, 145
Napier, R. J., **3**, 723
Napolitano, E., **14**, 98
Narang, S. C., **3**, 793; **7**, 56, 58, 60
Narasimhan, K., **5**, 113
Narasimhan, N., **4**, 355
Narayanan, B. A., **10**, 117
Nardelli, M., **3**, 520
Narisano, E., **10**, 240
Narita, N., **3**, 120
Naruchi, K., **14**, 338
Narukawa, Y., **2**, 327
Narula, A. S., **4**, 311; **8**, 38
Narula, S., **14**, 782
Nasehzadeh, A., **9**, 159, 195
Naser-ud-Din, **5**, 74; **14**, 640
Nashed, N. T., **2**, 301; **11**, 97
Naso, F., **12**, 230
Nastasi, M., **14**, 629
Nasu, A., **14**, 573
Nath, N., **4**, 111
Natoli, M. C., **14**, 58
Naumann, W., **3**, 222
Naumova, I. I., **7**, 87; **14**, 18, 19
Nava, D. F., **3**, 308, 314
Nava, M. E., **4**, 767
Navazio, G., **4**, 475
Nayak, R. N., **4**, 346
Nayar, B. M. S., **4**, 534
Nazaretyan, V. P., **3**, 375
Nazarova, M. P., **9**, 34
Nazin, G. M., **3**, 1126
Nazir, A. F., **10**, 62
Nazran, A. S., **3**, 209, 494
Ndebeka, G., **11**, 7
Nebenzahl, L. L., **10**, 296
Nechitailo, L. G., **3**, 191; **4**, 512
Nechvatal, G., **10**, 318
Neckers, D. C., **3**, 1076; **13**, 24
Nedelec, J. Y., **3**, 137, 148, 480
Nee, G., **1**, 106; **10**, 165
Nee, M. J., **9**, 37; **11**, 125
Neelakantan, P., **2**, 429
Neemann, J., **13**, 151; **14**, 242
Nefedov, O. M., **5**, 97
Nefedov, V. A., **3**, 694
Nefedov, V. D., **8**, 124, 140
Nefedov, V. P., **3**, 670
Neff, J. R., **9**, 27
Neff, W. E., **3**, 171
Neftekhimiya, **4**, 496
Negareche, M., **3**, 54
Negi, J., **7**, 110; **14**, 46
Negi, S. C., **4**, 233
Negishi, E.-i., **12**, 109, 110
Negoiță, N., **3**, 59
Negoro, T., **5**, 31; **14**, 704
Negrebetskii, V. V., **8**, 79
Neilands, O., **3**, 220; **4**, 653
Neilson, D. G., **14**, 346
Nelsen, S. F., **3**, 889, 939, 966; **4**, 464
Nelson, D. J., **4**, 650; **5**, 61
Nelson, J. V., **1**, 91; **10**, 188
Nelson, S. M., **1**, 49, 50; **14**, 602
Nemer, M. J., **4**, 344
Nemes, I., **3**, 637
Nemirova, L. I., **2**, 173
Nemoto, F., **3**, 954
Nemoto, H., **10**, 246; **11**, 124; **13**, 2
Neri, G., **1**, 48
Nesmeyanov, A. N., **10**, 51
Nesmeyanova, O. A., **14**, 641
Nesterova, L. I., **14**, 743
Nészmelyi, A., **4**, 539
Neta, P., **3**, 518, 789, 1027–1029, 1143
Neto, G. O., **1**, 76
Neuenschwander, K., **4**, 605
Neugebauer, F. A., **3**, 69, 70, 958
Neumann, R. C., **3**, 206
Neumann, W. P., **3**, 46, 87, 89; **5**, 141
Newcomb, M., **1**, 84; **10**, 24; **14**, 133
Newcombe, P. J., **5**, 109
Newitt, P. J., **3**, 268
Newkome, G. R., **6**, 113; **11**, 39; **13**, 186
Newman, P. A., **7**, 28; **14**, 44
Newsom, J. G., **13**, 162
Newton, R. F., **9**, 17; **14**, 255
Newton, T. W., **12**, 242
Ney-Igner, E., **10**, 113
Ng, J. S., **6**, 12
Ng, L. K., **4**, 553
Nguyen, C. H., **14**, 581
Nguyen, H. H., **11**, 30
Nguyen, M.-T., **1**, 70
Nguyen, T. T., **3**, 49, 1106, 1107
Nhan, D. D., **6**, 64
Niazyan, O. M., **3**, 154
Nibbering, N. M. M., **1**, 154; **8**, 129; **10**, 25, 142; **13**, 42
Nicholas, E. S., **2**, 145
Nicholas, P. E., **3**, 1133
Nicolaou, K. C., **12**, 59
Nicolas, E. C., **10**, 204
Nicole, D. J., **1**, 25
Nicovich, J. M., **3**, 538, 540
Niecke, E., **13**, 168; **14**, 838
Niederer, P., **3**, 213
Nieke, E., **11**, 66
Nieuwpoort, W. C., **2**, 426; **3**, 67
Nigam, A., **3**, 837
Niki, E., **4**, 282, 283
Niki, H., **3**, 150, 164, 185, 304, 306; **4**, 510
Nikishin, G. I., **3**, 101, 327, 385, 386, 463, 710, 721, 722; **4**, 196; **14**, 625
Nikishova, N. G., **8**, 130; **9**, 177
Nikitenko, L. P., **14**, 19
Nikitenkova, L. P., **7**, 87; **14**, 18
Nikitin, N. R., **6**, 143, 152, 153
Nikitin, V. S., **14**, 852
Nikolaev, A. F., **9**, 52
Nikoleav, A. I., **3**, 174
Nikonava, L. A., **4**, 647
Nill, G., **2**, 225
Nilsson, G., **3**, 1031, 1045
Ninomiya, Y., **2**, 226
Nip, W. S., **4**, 387
Nishi, S., **1**, 109
Nishi, S.-I., **14**, 159
Nishibayashi, H., **12**, 158, 159
Nishida, A., **12**, 148
Nishida, I., **1**, 95; **10**, 184
Nishida, K., **2**, 181
Nishida, S., **3**, 576; **13**, 7; **14**, 436
Nishigaki, Y., **14**, 17
Nishiguchi, I., **12**, 150
Nishijima, Y., **3**, 918
Nishijimi, M., **3**, 900
Nishikido, N., **10**, 279
Nishimoto, K., **14**, 680
Nishimura, H., **3**, 579
Nishinaga, A., **4**, 528; **14**, 343
Nishino, J., **4**, 366
Nishino, M., **4**, 735; **14**, 198
Nishio, T., **4**, 424
Nishiwaki, T., **14**, 871
Nishiyama, H., **10**, 173; **12**, 239
Nishiyama, J., **14**, 81

Nishiyama, T., **3**, 944
Nishizawa, M., **4**, 545; **14**, 291
Nishizawa, Y., **13**, 165
Nissan, R. A., **14**, 370
Nitta, M., **14**, 137, 232, 567, 880, 881
Nitteberg, E. B., **4**, 473
Nivard, R. J. F., **13**, 102, 136
Nivorozhkin, L, E., **14**, 119
Niwa, M., **4**, 74; **14**, 519
Noall, W. I., **14**, 171, 172
Nobes, R. H., **5**, 87; **8**, 169; **9**, 45; **14**, 772
Noda, H., **4**, 74
Nodwell, W. R., **14**, 227
Noels, A. F., **5**, 72, 112
Noest, A. J., **1**, 154; **10**, 25
Nogami, K., **3**, 432
Nogami, Y., **9**, 43
Noguchi, H., **4**, 31
Noguchi, Y., **3**, 488, 583
Nohara, D., **3**, 358, 359
Nohara, T., **11**, 100
Nojima, M., **4**, 288; **9**, 198
Nolte, R. J. M., **2**, 228
Nome, F., **3**, 868; **5**, 159; **11**, 29, 98
Nomoto, S., **4**, 679
Nona, S. N., **3**, 408, 619
Nonaka, Y., **2**, 302
Noori, G. F. M., **7**, 131
Nordeen, J., **1**, 152
Nordlander, J. E., **8**, 21; **9**, 27; **12**, 80; **14**, 482, 483
Norman, A. W., **9**, 42; **14**, 444
Norman, N. C., **14**, 600
Norman, R. O. C., **3**, 292, 321, 451, 495, 546, 635, 968, 1175
Normant, J. F., **10**, 108; **12**, 236
Norokhatka, D. A., *2*, 171
Norris, A. F., **14**, 173
Norris, A. R., **6**, 134, 163
Norris, R. K., **3**, 826; **5**, 109
Norseev, Yu. V., **6**, 64
Nortey, S. O., **4**, 576
Nosaka, T., **14**, 637
Nossin, P. M. M., **14**, 203
Nosyreva, O. V., **4**, 460
Noto, R., **2**, 53; **6**, 94, 95; **11**, 45
Nott, A. P., **10**, 217
Nouguier, R., **5**, 155
Noureldin, N. A., **4**, 36; **5**, 119
Nov, E., **3**, 875
Novi, M., **6**, 118
Novikov, S. S., **3**, 811
Novikova, I. A., **3**, 867
Novroćik, J., **3**, 522
Nowicki, A. W., **4**, 410
Noyori, R., **1**, 81; **4**, 545; **10**, 184; **13**, 20; **14**, 291, 610
Nozaki, H., **1**, 97; **4**, 147; **10**, 185; **14**, 135, 648
Npyori, R., **1**, 95
Nudelman, A., **14**, 677
Nudelman, N. S., **6**, 45, 46; **10**, 107
Nukada, T., **14**, 141
Numata, T., **3**, 714; **14**, 546
Nummert, V., **9**, 155
Nunes Viana, C. A., **9**, 255
Nunome, K., **3**, 941; **8**, 125
Nurullina, R. L., **13**, 94
Nutaitis, C. F., **4**, 580
Nwaedozie, J. M., **6**, 51
Nwokogu, G., **4**, 640
Nyman, R., **2**, 154
Nyns, C., **11**, 59
O'Brien, D. H., **10**, 18, 210
O'Connor, C. J., **2**, 233
O'Connor, D. E., **3**, 172
O'Dell, D. E., **3**, 453; **10**, 117
O'Donohue, A. M., **1**, 78; **4**, 76
O'Krongly, D., **13**, 131
O'Leary, M. A., **5**, 91, 114; **14**, 700
O'Leary, M. H., **2**, 15, 258, 259
O'Leary, S., **2**, 196a
O'Mahony, B., **9**, 13
O'Neal, H. E., **3**, 110
O'Sullivan, W. I., **14**, 663
Oae, S., **3**, 120, 714; **4**, 217, 258, 348, 370, 371, 664; **9**, 182, 184; **11**, 118; **12**, 54; **14**, 546
Oakley, R. T., **10**, 260
Obafemi, C. A., **3**, 874a; **14**, 417, 418
Obana, K., **6**, 142
Obana, M., **4**, 677
Obaza, J., **4**, 580
Oberti, R., **3**, 899
Obi, K., **3**, 162, 163
Obmelyukhina, T. N., **2**, 93
Ochai, M., **1**, 102
Ochi, H., **13**, 157
Oda, J., **4**, 698, 708
Oda, M., **10**, 64
Oda, R., **11**, 117
Oda, Y., **14**, 610
Odaira, Y., **5**, 31; **9**, 35, 36; **14**, 484, 485, 704
Odeh, I. M. A., **10**, 91
Odiaka, T. I., **8**, 29, 30
Oduwole, D. A., **3**, 1148
Oesch, F., **4**, 663
Oexler, E. V., **3**, 369
Officer, D. L., **11**, 86
Oftring, A., **14**, 698
Oganesyan, I. I., **12**, 69
Ogashiwa, S., **3**, 992
Ogata, F., **9**, 198
Ogata, Y., **3**, 149, 184, 515, 545, 682, 1077, 1086, 1094, 1098; **4**, 218, 287, 330, 331, 360, 367, 427, 452; **7**, 144; **9**, 226; **14**, 15
Ogawa, J., **3**, 916
Ogawa, S., **4**, 41; **9**, 184
Ogawa, Y., **4**, 381
Ogibin, Y. M., **14**, 625
Ogibin, Yu. N., **3**, 385, 386, 710, 721, 722; **4**, 196
Ogilby, P. R., **4**, 395
Ogilvie, K. K., **4**, 344
Ogiya, N., **3**, 192
Ogle, C. A., **10**, 208
Ogliaruso, M. A., **8**, 97
Ogryzlo, E. A., **3**, 267
Ogunye, A. F., **3**, 1109
Ogura, K., **10**, 244
Ohfune, Y., **14**, 506
Ohgai, M., **7**, 119
Ohkatsu, Y., **4**, 440
Ohkubo, K., **2**, 226; **3**, 256; **4**, 678
Ohkuma, T., **3**, 591
Ohmizu, H., **10**, 87
Ohmura, H., **13**, 144
Ohno, A., **3**, 871; **4**, 702, 704, 715
Ohno, T., **4**, 330; **7**, 144
Ohnuma, T., **3**, 571; **11**, 58
Ohnuma, Y., **10**, 37
Ohsawa, A., **3**, 1124
Ohsawa, T., **4**, 631; **6**, 91
Ohshiro, Y., **4**, 661; **12**, 209
Ohta, H., **2**, 224
Ohta, T., **3**, 376; **7**, 151
Ohtani, E., **3**, 354; **4**, 600
Ohtani, M., **9**, 35, 36
Ohwa, M., **4**, 558
Ohya-Nishiguchi, H., **3**, 995
Ohyama, T., **14**, 747
Oida, T., **10**, 237; **13**, 61
Oishi, T., **4**, 586, 631
Ojha, K. G., **14**, 31
Ojima, I., **4**, 668
Oka, K., **14**, 737
Oka, S., **3**, 871; **4**, 127, 702, 704, 713, 715; **6**, 93
Oka, Y., **14**, 564
Okabe, M., **3**, 498; **14**, 590
Okada, H., **3**, 872
Okada, T., **3**, 686

Okada, Y., **14**, 575
Okahata, Y., **2**, 202, 205, 349; **11**, 42
Okajima, S., **14**, 68, 69
Okamoto, H., **2**, 236
Okamoto, K., **3**, 141, 738, 944; **8**, 93, 94; **9**, 29; **14**, 810, 850
Okamoto, T., **4**, 127; **6**, 93; **7**, 151
Okamura, H., **10**, 242
Okamura, W. H., **9**, 42; **14**, 444
Okano, M., **4**, 70, 353; **7**, 136; **11**, 108; **12**, 64, 79
Okano, T., **7**, 22
Okawara, M., **4**, 215
Okazaki, K., **3**, 395, 399
Okazaki, R., **3**, 656
Okazawa, N., **8**, 64
Okely, H. M., **14**, 178
Okhlobystin, O. Yu., **3**, 11, 52, 528, 796, 797, 938
Oki, M., **3**, 294; **4**, 4; **9**, 262
Okimoto, A., **8**, 96
Okimoto, K., **1**, 158
Okorie, D. A., **8**, 26
Oku, A., **3**, 746; **5**, 76
Oku, J. I., **1**, 120
Okubo, M., **3**, 852, 858
Okubo, T., **2**, 210
Okuda, M., **3**, 412
Okuda, S., **4**, 381
Okuda, T., **14**, 877
Okuma, K., **14**, 216
Okumura, K., **1**, 178
Okumura, Y., **4**, 175; **11**, 57
Okuno, H., **5**, 78, 92, 94; **14**, 535
Okura, I., **3**, 775
Okuyama, T., **1**, 166; **2**, 78
Okwuiwe, R., **2**, 439
Olah, G. A., **3**, 793; **4**, 328; **6**, 131; **7**, 56, 58, 60, 101, 134, 145; **8**, 16, 66; **10**, 75; **14**, 437
Olah, J. A., **3**, 793; **7**, 58, 60
Old, K. B., **11**, 142
Oleinik, N. M., **2**, 142
Olekhnovich, L. P., **14**, 840
Olieman, C., **7**, 18
Oliva, C., **3**, 1003
Olive, J. L., **3**, 1099; **4**, 695
Oliver, S., **14**, 325
Oller, M., **14**, 286, 364
Ollerenshaw, J., **3**, 1114; **14**, 823
Olli, L. K., **9**, 38; **14**, 539
Ollis, W. D., **3**, 508, 509; **14**, 212–215, 283, 284, 353
Olmstead, M. M., **4**, 214
Olsen, E. G., **14**, 791
Olsen, H., **14**, 730, 731
Olson, R. E., **10**, 216; **14**, 226
Olszowy, H., **3**, 835
Omote, Y., **3**, 1083; **4**, 424
On, H. P., **10**, 39
Onan, K. D., **1**, 75; **14**, 551
Ongania, K.-H., **13**, 51; **14**, 710
Ono, A., **4**, 713
Ono, M., **4**, 710; **9**, 58, 127; **10**, 98
Ono, N., **3**, 828, 829, 831; **4**, 700; **6**, 58; **9**, 15; **11**, 114, 130
Ono, R. K., **13**, 103
Ono, S., **2**, 221
Onosova, L. A., **9**, 217
Onwezen, Y., **9**, 188
Onyido, I., **6**, 43
Ooi, N. S., **1**, 73
Opeida, I. A., **3**, 191; **4**, 460, 478, 512
Oppenheimer, N. J., **4**, 762
Oppolzer, W., **10**, 118, 119; **13**, 129; **14**, 311
Oprunenko, Yu. F., **10**, 50, 52; **14**, 599
Or, Y. S., **14**, 385
Orchin, M., **12**, 104
Orena, M., **12**, 197
Orfanopoulos, M., **4**, 392
Orlova, G. V., **6**, 86
Orlova, T. E., **4**, 520
Ornstein, P. L., **4**, 617
Orpen, A. G., **14**, 600
Orrell, K. G., **14**, 617
Ortiz de Montellano, P. R., **4**, 745, 770
Ortiz, M., **13**, 54; **14**, 732
Orvane, P., **2**, 132; **3**, 1130; **14**, 738
Osa, T., **14**, 802
Osakada, K., **4**, 677
Osamura, Y., **14**, 680, 773
Osawa, E., **11**, 49; **14**, 623
Osawa, T., **3**, 498
Oshima, K., **4**, 147; **10**, 185; **14**, 135
Oshima, T., **1**, 179; **5**, 51; **14**, 88, 89
Oshiro, Y., **10**, 90
Osina, O. I., **7**, 132
Osmolovskaya, N. S., **14**, 647
Ostah, N. A., **5**, 147
Ostanina, V. A., **4**, 472
Ostrizhko, F. N., **2**, 46
Ostrogovich, G., **10**, 313
Ostrovskii, V. A., **9**, 131, 140
Osugi, J., **6**, 140, 141; **10**, 275; **13**, 17, 114
Osuka, A., **6**, 90; **7**, 65, 82
Ota, A., **14**, 81
Ota, H., **14**, 216
Ota, T., **3**, 608
Otagiri, M., **2**, 241
Oth, J. F. M., **14**, 730
Otiono, P., **6**, 41
Otroshchenko, O. S,., **6**, 111
Otsuji, Y., **3**, 901, 912, 916; **13**, 12
Otsuka, T., **14**, 276, 277
Otsuki, J., **3**, 923
Ott, R. A., **4**, 604
Ott, W., **13**, 169
Otterbacher, E. W., **14**, 189
Ottow, E., **12**, 253
Ottridge, A. P., **9**, 65
Otutakowski, J., **4**, 125
Ouannes, C., **4**, 338
Ourevitch, M., **10**, 20
Ourisson, G., **10**, 100
Ovchinnikov, A. A., **10**, 36
Ovchinnikov, V. I.,, **3**, 118; **4**, 470, 476
Ovchinnikova, L. V., **2**, 55
Overberger, C. G., **14**, 801
Overill, R. E., **3**, 22
Overman, L. E., **13**, 103; **14**, 154–156, 204, 205
Owada, H., **4**, 353; **11**, 108
Owen, H. L., **3**, 349
Oyama, K., **2**, 302
Ozaslan, N., **3**, 655
Ozawa, S., **10**, 185
Özbal, H., **12**, 156
Özmeral, C., **5**, 83; **14**, 803

Pabst, J., **14**, 818
Pac, C., **3**, 897, 898, 910, 913; **4**, 705; **13**, 194
Pac, X., **4**, 565
Pacansky, J., **3**, 21, 23, 26
Pace, M. D., **3**, 36
Pacey, P. D., **3**, 1102
Paciute, N., **12**, 43
Pack, M.-K., **12**, 88
Packer, J. E., **3**, 219
Paddock, N. L., **10**, 260
Paddon-Row, M. N., **1**, 127; **3**, 24; **12**, 5, 6; **13**, 187; **14**, 289
Padmanabha, J., **4**, 256; **7**, 40, 41
Padronaggio, R. M., **3**, 575; **13**, 147
Padwa, A., **3**, 917; **5**, 68; **13**, 69, 166, 167; **14**, 59, 149, 231, 312, 620, 639

Paessun, R., **5**, 41
Pagani Zecchini, G., **4**, 541
Pagani, G. A., **10**, 30
Page, M. I., **2**, 264; **7**, 83
Paglialunga Paradisi, M., **4**, 541
Pagni, R. M., **3**, 615, 1034; **10**, 312
Pagsberg, P., **2**, 131; **11**, 38; **14**, 774
Pahwa, P. S., **14**, 39
Pai, B. R., **4**, 235
Pai, Y. M., **5**, 143
Paillous, N., **3**, 33
Paine, A. J., **11**, 62
Pal, S., **4**, 261–264
Palacios, S. M., **3**, 804; **6**, 19
Palazón, B., **13**, 163
Paleta, O., **10**, 28; **12**, 172, 173
Paley, M. S., **9**, 231
Palla, G., **3**, 525
Palleros, D., **6**, 46
Palling, D. J., **11**, 120
Palm, V., **9**, 155
Palmer, G. E., **4**, 357; **14**, 691
Palmer, R. F., **11**, 85
Palmer, R. J., **11**, 5
Palmieri, G., **12**, 249
Paltenghi, R., **3**, 267
Pan, L.-R., **14**, 619
Panchal, T. A., **14**, 733
Panchenko, Yu. V., **3**, 131
Panda, A. K., **4**, 334, 339
Panda, H. P., **4**, 150, 260
Pande, P. N., **4**, 40
Pandey, P. N., **14**, 490
Pandit, U. K., **4**, 709, 715
Panetta, C. A., **1**, 38
Panigrahi, G. P., **4**, 2, 334, 339, 346
Pankova, M., **11**, 2, 3
Panov, E. P., **2**, 437
Panov, V. B., **3**, 52
Panshin, Yu. A., **2**, 46
Pant, J., **14**, 337
Panteleeva, I. Yu., **12**, 12
Panunzio, M., **10**, 82
Panzeri, A., **9**, 121; **14**, 646
Paolucci, C., **10**, 226; **12**, 200; **14**, 219, 220
Pap, A. A., **9**, 207
Papadopoulos, M., **1**, 135
Papadopoulos, P., **10**, 331, 332
Pape, R. F., **2**, 155
Paquer, D., **4**, 281
Paquette, L. A., **4**, 284, 411, 412; **8**, 14, 16; **12**, 73, 149; **13**, 110, 191; **14**, 148, 199, 341, 669, 811, 812
Paradellis, T. C., **14**, 450
Paradisi, C., **8**, 42
Paraskevopoulos, G., **3**, 310
Pardasani, R. T., **4**, 48; **14**, 861
Pardini, V. L., **3**, 748
Pardo, M. C., **12**, 255
Pardo, S. N., **10**, 198
Parhi, S., **12**, 160
Parimala, N., **4**, 251
Park, K. H., **14**, 14
Park, M.-K., **9**, 263
Park, W. S., **3**, 847
Párkányi, C., **3**, 14
Párkányi, L., **14**, 715
Parker, D. G., **8**, 39
Parker, G. R., **1**, 80
Parker, R. C., **2**, 39
Parker, V. D., **3**, 684, 685, 696, 762–765, 776, 799, 1036, 1037, 1054, 1055
Parkes, D. A., **3**, 153, 363
Parkey, D., **4**, 667
Parkhomenko, P. I., **2**, 120
Parmelee, W. P., **3**, 966
Parnes, Z. N., **8**, 151
Parr, W. J. E., **12**, 63
Parry, R. J., **8**, 26
Parsons, I. W., **14**, 250
Parsons, P. J., **14**, 240
Partridge, I. J., **14**, 67
Pasquini, M. A., **6**, 69
Pass, M. C., **3**, 698
Pasteris, R. J., **14**, 105, 106
Pasternak, M., **14**, 233, 234
Pasto, D. J., **3**, 425, 426; **13**, 28
Pastor, R., **12**, 256
Pastukhova, G. L., **2**, 93
Pastushenko, E. V., **3**, 326, 455–457, 459
Paswa, A., **14**, 622
Patal, K. B., **3**, 231
Patel, G., **2**, 249, 326; **14**, 441
Patel, S. K., **8**, 142; **11**, 111, **12**, 241; **14**, 429
Patel, V., **10**, 200, 201; **11**, 43
Pater, R. H., **4**, 309
Paterson, I., **10**, 172; **14**, 430
Pathania, B. S., **7**, 110; **14**, 46
Pati, S. C., **4**, 227, 228
Pati, S. N., **4**, 78
Patil, N., **4**, 200
Patnaik, L. N., **3**, 179
Patra, A., **14**, 168
Patra, M., **1**, 57
Patrick, R., **3**, 233
Patrick, T. B., **2**, 285; **14**, 422
Pattenden, G., **14**, 718, 883
Patterson, J. F., **2**, 96
Patterson, J. R., **3**, 1007
Paul, D. B., **3**, 737
Paul, H., **3**, 1066
Paul, J. G., **3**, 703; **14**, 50
Pauler, C. T., **11**, 40
Paulick, W., **14**, 329
Pauson, P. L., **8**, 28
Pavia, A. A., **1**, 26; **4**, 695
Pavlov, G. P., **4**, 477
Pavlova, E. K., **3**, 731
Pavlova, M. P., **8**, 84
Pavlyukov, S. A., **9**, 70, 217
Payne, A., **13**, 11
Payne, W. A., **3**, 314
Pazhenchevsky, B., **14**, 233
Pearce, A., **14**, 430
Pearlman, P. M., **4**, 589
Pearson, A. J., **8**, 34, 36, 40
Pearson, N. R., **10**, 122
Pedersen, J. A., **3**, 1014
Pedro, J. R., **14**, 379, 380
Pedulli, G. F., **3**, 45, 77, 81, 420, 998, 1049
Peet, N. P., **14**, 36
Pegg, W. J., **4**, 658
Pehk, T., **12**, 75, 133; **14**, 472
Pejanovic, V., **14**, 859
Peled, M., **13**, 152, 153
Pelikan, P., **3**, 1035
Pelizzi, C., **3**, 520
Pellacani, L., **14**, 688
Pellegrin, V., **14**, 844
Pellegrini, J. P., **7**, 37
Pellerite, M. J., **3**, 1052; **10**, 294, 320
Pelletier, S. W., **14**, 765
Pellicciari, R., **4**, 408
Pellino, A. M., **1**, 34
Pelter, A., **4**, 579; **10**, 160, 228; **12**, 221; **14**, 510
Pemberton, J. E., **3**, 974
Pemberton, S. M., **3**, 787
Pen'kovskii, V. V., **3**, 370, 417
Penchatsharum, V. S., **7**, 24
Peñéñory, A. B., **3**, 805
Pennings, M. L. M., **14**, 322
Penton, J. R., **14**, 16
Pepe, N., **7**, 106
Pérez, J. D., **14**, 825
Perez, L. A., **5**, 47
Pérez, L. M., **9**, 243
Perez, M. A., **11**, 52
Perez-Ossorio, R., **1**, 128, 130; **4**, 550; **14**, 29
Perez-Rubalcaba, A., **1**, 128, 130, 131
Pergold, G. W., **10**, 100
Periasamy, M., **8**, 53, 55–57
Pericas, M. A., **4**, 303
Peri, J. J., **12**, 146

Perkins, C. A., **3**, 678
Perkins, K. A., **3**, 1134
Perkins, M. J., **3**, 628, 631, 652
Perkins, P. G., **3**, 1134
Perkins, R., **9**, 33
Perlin, A. S., **9**, 124; **11**, 131
Perlman, K., **10**, 131
Perrée-Fauvet, M., **4**, 461
Perrin, C. L., **2**, 84, 85
Perrini, G., **1**, 46; **12**, 82; **14**, 780
Perry, D. A., **11**, 102
Perry, R. A., **11**, 43
Persia, F., **14**, 688
Perst, H., **14**, 13
Perucci, P., **11**, 4
Perumal, P. T., **3**, 691
Perumal, S. I., **4**, 129
Perz, R., **12**, 275
Pete, J.-P., **3**, 1092
Péter, A., **3**, 202, 204, 270
Peter, R., **1**, 86
Peters, J. A. M., **14**, 524
Peters, K., **13**, 169
Peters, K. S., **3**, 580, 928, 930; **13**, 21
Peters, N. K., **4**, 619; **10**, 4
Peterson, L. H., **14**, 580
Petreitis, J. J., **14**, 195
Petrenko, O. P., **6**, 102, 103
Petride, H., **11**, 8
Petrillo, G., **2**, 62; **11**, 129
Petrov, A. A., **10**, 21–23; **12**, 72, 260, 273; **13**, 40
Petrov, E. S., **10**, 304
Petrov, L. V., **3**, 114
Petrova, G. C., **9**, 99
Petrova, I., **1**, 53
Petrova, L. P., **9**, 67
Petrova, R. G., **3**, 428, 429, 468
Petrovanu, M., **13**, 85
Petrovskii, P. V., **8**, 50
Petrus, C., **12**, 183
Petrus, F., **12**, 183
Petrusi, S., **4**, 24
Petryaev, E. P., **3**, 460, 461
Petsko, G. A., **2**, 266
Petsom, A., **3**, 830
Petty, C. B., **13**, 103
Pez, R., **4**, 607
Pfeiffer, J. M., **5**, 82
Pfenninger, J., **11**, 91
Pfoertner, H.-K., **11**, 90
Pfund, R. A., **12**, 182
Phan, H. T., **11**, 40
Phelan, K. G., **4**, 169
Phelps, D. J., **2**, 145
Philippi, K., **4**, 317
Phillipou, G., **3**, 439
Phillips, J. G., **10**, 245
Phillips, R. S., **4**, 765
Piancatelli, G., **4**, 5
Piazza, G. J., **2**, 259
Pibulik, I. I., **6**, 8
Pichon, R., **14**, 618, 799
Pickering, I. A., **3**, 160
Pickett, J. E., **4**, 418
Pieniazrk, M., **3**, 277
Pienta, N. J., **5**, 49
Piepers, O., **12**, 193
Piermattie, V., **14**, 847
Pierpoint, C., **14**, 615
Pierre, J. L., **6**, 69
Piers, E., **12**, 237, 238
Pietra, F., **6**, 147
Pietroni, B. R., **7**, 118
Pietruszko, R., **4**, 757
Pike, S., **10**, 154
Pikh, Z. G., **4**, 358
Pilkington, J. W., **14**, 112, 113
Pillai, C. N., **12**, 196
Pillai, G. C., **4**, 113, 114
Pillay, M. K., **4**, 42
Pilling, M. J., **3**, 161, 233
Pincus, M., **2**, 267
Pineda, N., **3**, 321, 451
Pines, H., **10**, 113
Pinkerton, A. A., **13**, 111
Pinkus, A. G., **1**, 147, 155; **10**, 325
Pinnick, H. W., **7**, 115; **11**, 50
Pinnick, J. W., **4**, 216
Pino, P., **12**, 100
Pinot de Moira, P., **6**, 9; **7**, 89, 90
Pinson, J., **3**, 1062; **4**, 645; **6**, 35
Pinto, P., **14**, 828
Pioch, D., **13**, 43
Piper, J., **4**, 739
Piper, J. U., **4**, 738
Piquard, J. L., **8**, 144
Pirazzini, G., **1**, 59
Pirgo, M. D., **6**, 56
Pirrung, M., **1**, 87, 103; **14**, 505
Piskunova, Zh. P., **2**, 142
Pistelli, L., **14**, 98
Pittman, C. U., **12**, 105
Pizzo, C. F., **14**, 143
Placucci, G., **3**, 29, 339, 424
Plancherel, D., **3**, 595
Plarnbeck, J. A., **3**, 741
Plate, A. F., **12**, 74, 75, 133; **14**, 472
Plato, M., **3**, 1001
Platt, K. L., **4**, 663
Plattner, G., **3**, 963
Platz, M. S., **5**, 4, 7
Plucinski, J., **9**, 48
Plumb, I. C., **3**, 269
Plumet, J., **4**, 550
Plusquellec, D., **2**, 119
Plust, S. J., **6**, 100
Pochan, J. M., **14**, 789
Pochat, F., **3**, 40
Pocker, Y., **9**, 46
Podda, G., **3**, 81
Podesta, J. C., **3**, 421
Podkhalyuzin, A. T., **9**, 34
Podoplelov, A. V., **3**, 18
Pogrebnyak, A. A., **8**, 170
Pohjonen, M. L., **3**, 262
Pohle, C., **13**, 92
Pohmakotr, M., **10**, 171
Poisson, J., **14**, 767
Pojarlieff, I. G., **14**, 868
Pokhodenko, V. D., **3**, 725 979; **4**, 168
Poladyan, E. A., **4**, 498
Polansky, O. E., **13**, 156
Poletaeva, N. K., **4**, 10
Polgár, L. **2**, 272
Poliakoff, M., **5**, 17
Polievka, M., **12**, 101
Polievktov, M. K., **12**, 86
Polina, T. V., **14**, 647
Polis, J., **1**, 66
Politzer, I. R., **2**, 283
Polivanov, A. N., **5**, 140
Pollack, S. K., **10**, 305
Pollicino, S., **10**, 226; **12**, 200; **14**, 219, 220
Polniaszek, R. P., **5**, 49
Polo, J., **14**, 339, 891
Poloni, M., **3**, 536, 606, 607; **7**, 81; **14**, 416
Polonov, V. M., **2**, 31
Polyak, S. S., **4**, 490, 491
Pomares, O., **2**, 6
Pommelet, J.-C., **11**, 59
Poncin, G., **2**, 119
Poncini, L., **9**, 267
Ponkshe, N. K., **14**, 210
Ponomarchuk, N. P., **2**, 337
Ponomarev, A. I., **9**, 93
Ponomarev, G. V., **8**, 110
Ponomarev, V. S., **3**, 1170; **14**, 784
Pons, J. M., **3**, 751; **4**, 644
Ponticelli, F., **3**, 577
Poonia, N. S., **1**, 112
Poorker, C., **14**, 441
Popkova, I. A., **2**, 260
Pople, J. A., **5**, 89; **8**, 154, 156, 159; **10**, 6
Popov, A. A., **4**, 299
Popov, A. F., **2**, 142, 445; **6**,

56; **9**, 7
Popova, R. S., **6**, 56
Poppl, R. F., **7**, 96
Poraicu, M., **2**, 70, 71
Porta, O., **3**, 511, 758; **4**, 642; **6**, 17
Porte, A. L., **3**, 605
Porter, A. E. A., **3**, 474; **14**, 651
Porter, A. J., **2**, 233
Porter, A. P., **8**, 58
Porter, J. D., **3**, 1050
Porter, N. A., **3**, 173; **4**, 380
Porter, Q. N., **14**, 165
Portilla, G., **9**, 243
Porwal, P. K., **1**, 112
Porzi, G., **12**, 197
Posner, G. H., **12**, 232
Pospiril, J., **4**, 443
Posselt, G., **9**, 60
Potekhin, V. M., **3**, 116, 118; **4**, 470, 476
Pothier, N., **1**, 5, 7; **4**, 608
Pottage, C., **2**, 356
Potter, B. V. L., **2**, 363, 364, 387, 388, 390
Potter, J. P., **2**, 12
Potter, P. C., **4**, 139
Potzinger, P., **3**, 341
Pouet, M.-J., **1**, 159; **6**, 151, 164
Poulin, J.-C., **4**, 676
Poulter, C. D., **8**, 24, 25
Poupart, M. A., **2**, 245
Pourcelot, G., **12**, 231
Povelikina, L. N., **12**, 134
Powell, D. R., **5**, 81
Powell, M., **4**, 109
Powell, M. J., **10**, 272
Power, P. P., **3**, 82
Powers, E. J., **3**, 861b
Powers, E. L., **3**, 3
Powis, G., **3**, 1142
Pozdeeva, N. N., **3**, 167
Pozdnyakovitch, Yu. V., **7**, 132
Pozharliev, J., **12**, 217
Prabhakar, K. J., **12**, 119
Prabhu, K. V., **14**, 107
Pradella, G., **6**, 48
Pradhan, J., **9**, 222; **11**, 138
Pradhan, S. K., **3**, 757; **4**, 624–626
Praefcke, K., **3**, 1167
Prager, R. H., **14**, 181
Pragst, F., **3**, 782
Praharaj, S., **1**, 56
Prahl, F. G., **3**, 879; **10**, 148
Prajer-Janczweska, L., **14**, 138
Prakash, B. S. J., **2**, 80
Prakash, G. K., **8**, 66; **14**, 437
Praofake, K., **14**, 90
Prasad, G., **1**, 52; **5**, 95
Prasad, H. S., **3**, 843; **4**, 533
Prasad, M., **4**, 198
Prasad, R. S., **14**, 858
Prashad, M., **14**, 604
Prasthofer, T. W., **9**, 231
Pratt, G. E., **9**, 65
Pravednikov, A. N., **3**, 140
Predieri, G., **3**, 520
Presser, N., **14**, 320
Pressman, E. J., **8**, 17
Preston, K. F., **3**, 603
Prestwich, G. D., **1**, 89; **10**, 170
Price, L. G., **10**, 172
Price, R. T., **14**, 134
Prikhod'ko, P. L., **3**, 999
Prince, R. J., **8**, 108
Pring, G. M., **14**, 617
Prinzbach, H., **13**, 171, 212; **14**, 324, 624
Pritzhow, W., **3**, 178, 182
Proctor, G., **3**, 497; **8**, 108
Prokof'ev, A. I., **3**, 499, 867
Prokopiou, P. A., **4**, 630, 632
Prokudin, V. G., **3**, 1126
Proll, T., **14**, 725
Pronin, A. F., **8**, 153
Proskuryakov, V. A., **3**, 175; **4**, 297, 487
Pross, A., **7**, 1; **9**, 230; **10**, 305
Pross, N., **9**, 3
Prosteanu, N., **2**, 155; **6**, 3
Protschuk, G., **10**, 166
Proudfoot, G. M., **3**, 718
Prousek, J., **3**, 823
Pruchnik, F.m, **4**, 684
Pruszynski, P., **10**, 277
Pryadilova, A. N., **12**, 55
Pryor, W. A., **3**, 1, 301
Prystasz, H., **9**, 48
Przybylska, M., **12**, 50
Przystas, T. J., **1**, 1; **2**, 248
Pshezhetskii, S. Ya., **3**, 400–402
Pterova, G. G., **2**, 450
Pucheault, J., **4**, 377
Pudovik, A. N., **2**, 382, 384; **12**, 53
Puga, A., **3**, 627; **4**, 180
Pukhal'skaya, G. V., **3**, 400–402
Pulwer, M., **14**, 231, 729
Pünkt, J., **3**, 1009
Purdy, J. R., **5**, 71
Purohit, P. C., **3**, 1084
Purohit, R., **4**, 38
Purvaneckas, G., **7**, 14
Purvis, S. R., **10**, 116
Putsykin, Y. G., **14**, 560
Putyatin, A. A., **13**, 118
Putyrskaya, G. V., **3**, 117
Pyatnitskii, Yu. I., **4**, 527
Pytela, O., **2**, 125

Qaiser, A., **4**, 197
Quail, J. W., **14**, 417
Quan, C., **2**, 346, 347
Quante, J., **13**, 142
Quast, H., **9**, 76; **11**, 53, 54, 64; **13**, 52, 169; **14**, 185, 692, 693
Quayle, P., **10**, 104
Quazzani-Chahdi, L., **3**, 1089
Que, L., **4**, 750
Queguiner, G., **4**, 716; **10**, 129
Quick, G. R., **3**, 726
Quigley, F. R., **4**, 740
Quintans, M. T., **3**, 91
Quinteiro, M., **14**, 750
Quirk, J. M., **4**, 681
Quirk, R. P., **8**, 104
Quiroga, M. L., **1**, 128, 130
Quiroga, O. D., **7**, 96
Quiroga-Feijoo, M. L., **1**, 131

Raber, D. J., **4**, 585
Rabideau, P. W., **4**, 619; **10**, 4
Rabinovitch, B. S., **14**, 699
Rabinovitz, M., **10**, 209; **14**, 57
Racheria, U. S., **14**, 832
Rackham, D. M., **2**, 24
Radhakrishnamurti, P. S., **1**, 175; **4**, 78, 149, 150, 234 , 249, 250, 260; **7**, 23, 32, 42
Radhakrishnan, T. V., **3**, 757; **4**, 624
Radner, F., **3**, 792
Radojkovic-Velickovic, M. **2**, 54
Radom, L., **2**, 81; **5**, 87; **6**, 129, 130; **8**, 169; **9**, 45; **10**, 1, 305; **14**, 227, 772
Radomirska, V., **10**, 29
Rafikova, V. S., **3**, 158
Raghavachari, K., **8**, 154
Raghavan, N. V., **3**, 1156
Raguel, B., **1**, 66
Rahil, J., **2**, 352, 372
Rahman, M. F., **14**, 571
Rahman, M. T., **12**, 235
Rai, J. P., **4**, 158
Raid, Y., **5**, 25
Raimondi, M., **3**, 235
Rainbow, I. J., **8**, 58
Rajananda, V., **4**, 752
Rajanna, K. C., **2**, 257
Rajaram, J., **4**, 113, 114, 685

Rajasekaran, K., **9**, 201
Rajbenbach, L. A., **3**, 346, 348, 351; **11**, 73
Raju, K. V. S., **14**, 170
Rak, L. V., **2**, 116
Rakhmankulov, D. L., **1**, 11; **2**, 8, **3**, 31, 115, 156, 187, 188, 273, 274, 317, 325–327, 334, 345, 455, 457, 458, 463, 466, 467, 1129; **4**, 298; **8**, 99–102; **14**, 398
Rakintzis, N. Th., **3**, 625
Rall, G. J. H., **4**, 69
Ram, R. N., **3**, 500; **14**, 85
Ramachandran, J., **12**, 196
Ramadas, S. R., **12**, 196
Ramaiah, A. K., **4**, 242
Ramalingam, K., **4**, 112; **6**, 53
Ramalingam, V., **4**, 248
Ramamurthy, V., **4**, 291, 428
Raman, P. S., **1**, 121
Ramanujam, M., **1**, 141; **10**, 326
Ramaswamy, S., **14**, 238
Ramazanov, O. M., **3**, 459
Rambaud, J., **13**, 43
Ramesh, S., **4**, 1
Ramesh, V., **4**, 428
Ramhold, K., **13**, 92
Ramirez, D., **3**, 121
Ramirez, F., **2**, 366
Rammash, B. K., **13**, 146
Ramnath, N., **4**, 428
Ramos, J. J. M., **9**, 159
Ramos, S., **9**, 251, 252
Ramperti, M., **3**, 226
Ramsden, C. A., **14**, 30
Ramsey, B. G., **10**, 57
Ramsh, S. M., **9**, 132–134
Rand, C. L., **12**, 109
Randic, M., **8**, 163; **14**, 401
Random, L., **4**, 618; **7**, 1
Ranfagni, A., **10**, 278
Ranganathan, D., **14**, 150
Ranganathan, S., **14**, 150
Ranganayakulu, K., **8**, 63, 64
Range, G., **3**, 126
Ranikaga, R., **4**, 700
Rantenstrauch, V., **3**, 752
Ranzi, B. M., **11**, 46
Rao, A. S., **4**, 117
Rao, B. D. N., **2**, 404
Rao, B. M., **4**, 96
Rao, B. V. D., **7**, 32
Rao, C. B., **14**, 150
Rao, C. V., **1**, 51; **11**, 81
Rao, Ch. N., **4**, 92
Rao, G. V., **4**, 92
Rao, J. M., **10**, 228; **12**, 221
Rao, K. S., **14**, 342
Rao, M., **14**, 886
Rao, M. A., **4**, 97
Rao, M. B., **11**, 151
Rao, M. D. P., **4**, 256; **7**, 40, 41
Rao, M. N., **14**, 515, 516
Rao, M. P., **4**, 155, 156
Rao, M. V., **2**, 193
Rao, N., **14**, 571
Rao, N. V. S., **1**, 51; **11**, 81
Rao, P. J. P., **4**, 60
Rao, P. R., **4**, 246
Rao, P. V. K., **4**, 134, 135, 242
Rao, R. B., **1**, 6
Rao, R. T., **1**, 58
Rao, S. N., **2**, 429
Rao, S. V. G., **14**, 510
Rao, T. N., **1**, 74; **4**, 46, 60, 97, 155, 156,
Rao, V. S., **11**, 131
Rao, V. V., **1**, 74; **4**, 46
Rao, Y. R., **2**, 256
Räpple, E., **10**, 42
Rappoport, Z., **8**, 146, 148; **9**, 1–4; **12**, 169; **14**, 417
Raşeev, G., **8**, 155
Raskova, O. G., **4**, 10
Rasmussen, J. B., **14**, 239
Rasmussen, J. K., **3**, 680; **4**, 190
Rassat, A., **3**, 596
Rasteikiene, L., **12**, 43
Rastetter, W. H., **3**, 626; **4**, 725
Rastogi, M., **4**, 274
Raston, C. L., **10**, 256; **14**, 328
Rasulbekova, T. I., **13**, 119
Ratajczak, E., **3**, 277
Rathi, U. K., **4**, 201
Raucher, S., **14**, 175, 182
Rauk, A., **8**, 63, 64
Rault, S., **7**, 11
Rauscher, H. J., **8**, 158
Rautenstrauch, V., **4**, 627
Raveendran, G., **9**, 232
Ravichandran, R., **5**, 42
Ravindranath, B., **4**, 637
Ravishankara, A. R., **3**, 312, 538, 540
Rawat, D. S., **4**, 480
Rawdah, T. N., **8**, 66; **14**, 437
Ray, N. K., **13**, 10
Raychaudhuri, S. R., **4**, 621
Rayez, J. C., **13**, 8
Raynal, S., **11**, 7
Raynier, B., **3**, 444
Razina, T. L., **14**, 349, 350
Razus, A. C., **14**, 474, 475
Razuvaev, G. A., **3**, 97, 128, 253, 874; **4**, 356
Re, A., **12**, 192
Read, C. M., **4**, 205
Read, R. W., **14**, 117, 253
Reading, C., **2**, 304
Rebollo, H., **13**, 25
Rebrovic, L., **12**, 161
Recker, H.-G., **12**, 253
Reddy, B. S. R., **4**, 458
Reddy, C. P., **1**, 6
Reddy, C. S., **4**, 254, 255
Reddy, G. S., **14**, 797
Reddy, K., **11**, 81
Reddy, K. K., **1**, 51
Reddy, S., **14**, 23
Reed, C. A., **3**, 965
Reed, D. W., **3**, 300, 741
Reed, L. E., **5**, 24
Reel, H., **14**, 73
Rees, C. W., **5**, 137, 138; **11**, 93; **14**, 269, 271
Rees, D. C., **2**, 299
Rees, J. C., **14**, 442, 443
Reetz, K. P., **11**, 113
Reetz, M. T., **1**, 86, 104; **10**, 155; **11**, 145
Reeves, W. G., **9**, 249
Reffy, J., **3**, 1000
Regan, C. M., **14**, 602
Regen, D. L., **9**, 116
Regitz, M., **5**, 96, 100, 101; **13**, 72; **14**, 541
Regnard, J. R., **3**, 965
Reho, A., **14**, 865
Rehorek, D., **3**, 223, 629, 1009
Reich, H. J., **10**, 216, 225; **14**, 226
Reich, K. P., **4**, 223
Reichlin, D., **13**, 129
Reid, A. H., **4**, 682
Reid, K., **9**, 33
Reiffen, M., **5**, 20
Reikhsfel'd, V. O., **9**, 91; **12**, 111, 114, 117
Reimann, B., **3**, 341
Reimschüssel, W., **2**, 381; **9**, 204
Reinders, J. H., **3**, 962
Reinecke, M. G., **6**, 172; **13**, 44, 162
Reinehr, U., **14**, 766
Reinhoudt, D. N., **13**, 36; **14**, 322
Reinking, P., **3**, 711; **4**, 230
Reiser, A., **13**, 11
Reiser, J., **3**, 335
Reiss, J. A., **6**, 80
Reissig, H.-U., **10**, 43
Reith, W., **14**, 757
Reitter, B. E., **10**, 132

Reitz, D. B., **10**, 139
Rekers, J. W., **3**, 112
Rempel, T. D., **1**, 152
Renard, M. F., **4**, 730
Rendell, R. W., **3**, 384
Rensch, R., **3**, 96
Resch, J. F., **4**, 160
Resnati, G., **10**, 241; **12**, 223
Rettenberger, R., **14**, 248
Rettig, M. F., **14**, 614
Reuman, M., **6**, 79
Reusch, W., **13**, 124
Reutov, O. A., **6**, 148, 154; **8**, 130; **9**, 177, 256, 257; **10**, 84, 85, 338; **14**, 414
Revinski, I. F., **4**, 321
Rewicki, D., **10**, 15
Réyé, C., **4**, 607; **12**, 275
Reynolds, C. D., **14**, 327
Reynolds, D. P., **14**, 255
Rexnikov, I. V., **9**, 207
Rhee, S.-W., **4**, 736
Rhouati, S., **5**, 125, 126; **13**, 185
Riad, Y., **1**, 177
Ribi, M. A., **14**, 543
Ribo, J., **3**, 557
Ribó, J. M., **14**, 778
Ricci, A., **1**, 59; **10**, 154
Rich, J. D., **14**, 885
Richard, C., **3**, 423, 1128
Richards, G. N., **1**, 22; **9**, 267
Richards, K. E., **7**, 74, 75; **8**, 72, 73; **14**, 41, 42
Richardson, G. D., **3**, 814
Richardson, G. W., **5**, 114; **14**, 700
Richardson, P. F., **3**, 964
Richardson, W. H., **3**, 110
Richarz, W., **4**, 481
Richelme, S., **3**, 609
Richey, H. G., **8**, 145; **12**, 201–205; **14**, 433
Richman, J. E., **2**, 325
Richter, B., **14**, 176
Richter, F., **14**, 180
Richters, V. E. M., **2**, 228
Rickards, R. W., **10**, 115
Ricker, W. F., **14**, 312
Rico, I., **3**, 374
Ridd, J. H., **3**, 794, 795; **7**, 52, 62–64; **14**, 53
Riddell, W. D., **14**, 495
Riddle, C. A., **8**, 112
Rideout, D. C., **13**, 122
Ridley, D. D., **10**, 239
Riecker, W. F., **3**, 917
Riederer, H., **3**, 1144
Rieger, A. L., **3**, 732; **4**, 337
Rieker, A., **3**, 213
Rieker, W. F., **14**, 622, 639
Riemland, E., **4**, 361
Riera, J., **6**, 61
Riesz, P., **3**, 642, 1067, 1068, 1145
Rigaudy, J., **4**, 377; **6**, 10; **14**, 335, 664
Riggs, N. V., **2**, 81
Righetti, P. P., **13**, 141
Riko, E. A., **4**, 318
Riley, D. P., **4**, 672
Riley, P. I., **3**, 86, 609
Rimbault, C. G., **4**, 403
Rimmington, T. W., **14**, 250
Rindone, B., **3**, 729, 1063; **4**, 102, 369; **6**, 36
Rinus, O., **14**, 760
Ripoll, J.-L., **14**, 644
Risley, J. M., **2**, 16 452
Rissler, K., **14**, 821
Rist, G. H., **3**, 1166
Ritchie, C. D., **8**, 80, 81
Rittberg, B. R., **3**, 1122
Ritzer, J., **5**, 102
Riva, M., **3**, 758; **4**, 642
Rivière, P., **3**, 86, 609
Rivière-Baudet, M., **3**, 86, 609
Rizvi, R., **14**, 828
Rizzardo, E., **3**, 133
Robaugh, D. A., **3**, 1111, 1112
Robba, M., **7**, 11; **14**, 100
Robbins, C. M., **9**, 61
Roberts, B. P., **3**, 62, 662, 667, 669, 674–676, 849
Roberts, C., **3**, 479
Roberts, I., **9**, 200
Roberts, J. D., **5**, 136; **9**, 37; **11**, 125
Roberts, J. L., **4**, 373
Roberts, J. T., **3**, 1122
Roberts, N. K., **4**, 670
Roberts, R. D., **11**, 121
Roberts, R. M., **7**, 102; **14**, 47, 431
Roberts, R. M. G., **7**, 140, 142; **13**, 115
Roberts, S. M., **9**, 17; **12**, 47; **14**, 255
Roberts, W. N., **9**, 40
Robertson, P. J., **4**, 761
Robertson, R. E., **9**, 163–165, 264
Robillard, G. T., **2**, 402
Robinson, C. H., **4**, 89
Robinson, S. R., **7**, 62; **14**, 53
Robinson, W. T., **7**, 74; **14**, 41, 42
Rocek, J., **4**, 1
Rockenbauer, A., **3**, 632, 637
Rodewald, W. J., **14**, 562
Rodgers, M. A. J., **3**, 3; **4**, 389
Rodini, D. J., **14**, 134
Rodionov, V. I., **14**, 11
Rodriguez, A., **4**, 414, 415; **5**, 68; **13**, 69, 166; **14**, 673, 723
Rodwell, W. R., **5**, 87; **8**, 169; **9**, 45; **14**, 772
Roe, A. N., **3**, 173
Roedig, A., **9**, 12; **14**, 247–249, 748
Roessler, F., **12**, 242
Roev, L. M., **12**, 185
Rogers, D., **3**, 231, 263
Rogers, G. F., **3**, 233
Rogers, J., **5**, 18
Rogic, M. M., **4**, 50, 51
Roginskii, V. A., **3**, 168
Rohde, R., **3**, 257
Rohlfes, W., **6**, 155
Rojas, A., **14**, 523
Rojas, C., **9**, 243
Rokach, J., **7**, 130
Rokicki, A., **4**, 598
Rokstad Odd, A., **4**, 473
Rol'nik, L. Z., **3**, 326, 455–457
Rol, C., **4**, 88
Rol, E., **3**, 683
Rolla, F., **4**, 574; **9**, 183
Rolland, Y., **14**, 767
Romanenko, A. V., **4**, 18, 19
Romano, F. J., **11**, 15
Romanovich, L. B., **4**, 490, 491
Romeo, A., **14**, 525
Römer, R., **11**, 83; **14**, 642
Romsted, L. S., **2**, 208, 346, 347
Ronald, B. P., **9**, 46
Ronaniuk, P. J., **2**, 394
Rondan, N. G., **1**, 127; **5**, 2, 52; **10**, 12, 26; **12**, 5, 6; **13**, 105; **14**, 289
Ronlán, A., **3**, 696
Ronsisvalle, G., **14**, 94
Roof, A. A. M., **3**, 1080
Rooney, J. J., **4**, 458; **14**, 190, 486, 588
Roos, J. P., **14**, 154
Roques, R., **13**, 43
Ros, F., **3**, 817; **6**, 61
Rose, S. D., **4**, 153
Rose-Munch, F., **14**, 608
Roseboom, M. D., **13**, 104
Rosen, W., **9**, 251, 252
Rosenberg, S., **1**, 29
Rosenblum, M., **8**, 43; **12**, 162; **14**, 243
Rosenfeld, M. N., **4**, 205

Rosenfeldt, F., **3**, 506; **14**, 605
Rosenheim, L. D., **3**, 712
Rosenthal, I., **3**, 642
Rosevear, P. R., **2**, 403
Roshan-Ali, Y., **14**, 130
Rosnati, V., **11**, 146; **14**, 282
Ross, B., **11**, 113
Ross, C. H., **13**, 77
Ross, D. S., **3**, 791; **7**, 59, 61
Ross, W. J., **14**, 658
Rosser, R., **4**, 579
Rössert, M., **11**, 52; **13**, 150
Rossi, R. A., **3**, 801, 804, 805; **6**, 19, 21
Rossiter, B. E., **4**, 314
Rostovschikova, T. N., **12**, 66
Roswell, D. F., **2**, 283
Roth, B. D., **4**, 605
Roth, H. D., **3**, 915, 947; **5**, 3, 8
Roth, P., **3**, 1101; **4**, 386
Roth, W.-D., **13**, 63
Roth, W. R., **3**, 573
Rothwell, E. J., **3**, 1020
Rotinov, A., **2**, 135; **11**, 34, 35
Rotondo, E., **1**, 48
Rotschova, J., **4**, 443
Rouessac, A., **12**, 61
Rouessac, F., **12**, 60, 61
Roulet, R., **9**, 32
Roullier, L., **3**, 748, 976
Roumestant, M. L., **14**, 308
Roush, W. R., **13**, 125, 126
Roussel, J., **12**, 146
Roussel, M., **4**, 363
Roussi, G., **6**, 23
Rout, M. K., **3**, 179
Rout, S. P., **3**, 179
Routledge, P. J., **6**, 150
Roux, D., **3**, 813; **9**, 149; **14**, 679
Roux, M. C., **12**, 219
Roux-Schmitt, M. C., **1**, 123; **12**, 218
Rowan, D. D., **1**, 5, 7; **4**, 608
Rowe, J. E., **2**, 215, 216
Rowell, R., **2**, 333
Rowland, F. S., **3**, 403, 647–649
Rowlett, R., **3**, 1041
Roy, R., **14**, 753
Roy, R. G., **14**, 386
Royer, G. P., **2**, 195
Royer, J., **8**, 165; **12**, 171; **13**, 139
Rozario, A. P., **2**, 187
Rozeboom, M. D., **12**, 264; **13**, 105
Rozen, S., **7**, 20; **12**, 10
Roznyatovsky, V. A., **10**, 50; **14**, 599
Ruan, Z.-K., **14**, 619
Rubezhov, A. Z., **13**, 49
Rubio, M. A., **3**, 614
Rubio, V., **2**, 409
Rubiralta, M., **14**, 355
Rubottom, G. M., **14**, 670
Rubtsov, V. L., **3**, 168
Rubtsova, I. I., **4**, 115
Rüchardt, C., **3**, 126, 207; **14**, 424, 432
Rudakov, E. S., **4**, 17, 151, 433; **9**, 179, 244, 245
Rudashevskaya, T. Y., **14**, 641
Rudat, M. A., **3**, 367, 368
Rudenko, A. P., **3**, 699; **4**, 105
Rudler, H., **10**, 161; **14**, 608
Rudler-Chauvin, M., **10**, 161
Ruf, F., **10**, 43
Ruffini, B., **4**, 66
Rüger, W., **3**, 942
Ruitenberg, K., **10**, 121
Ruiz, R. P., **3**, 161
Rumanskii, B. L., **3**, 668
Runge, T. A., **14**, 611
Rungeler, W., **14**, 251
Ruo, T. C.-S., **3**, 298
Rupani, P., **10**, 160
Ruppert, I., **14**, 206
Ruppert, J. F., **14**, 488
Rusling, J. R., **3**, 753
Russ, M., **4**, 636
Russell, C. G., **10**, 183
Russell, G. A., **3**, 435, 815–817, 873, 1017; **10**, 177
Russell, R. A., **4**, 313; **13**, 107, 187
Russo, M. V., **13**, 207
Rustamova, M. A., **12**, 65, 81
Rutkovskii, G. V., **2**, 345
Rutledge, P. S., **3**, 865; **9**, 114, 115;n **12**, 40, 56; **14**, 171, 172, 683
Rutter, K., **1**, 15
Ruzicka, V., **1**, 12; 9, 56
Ruzsicska, B., **6**, 170
Ruzziconi, R., **11**, 4
Ryabikova, V. N., **4**, 10
Ryaboi, V. M., **10**, 304
Ryabokobylko, Yu. S., **3**, 694
Ryabov, A. D., **4**, 73, 162
Ryan, K. R., **3**, 269
Ryan, T. J., **14**, 543
Ryang, H.-S., **4**, 422
Rybachenko, V. I., **2**, 441
Rybakova, M. V., **2**, 120
Rybakova, N. A., **3**, 283, 481
Rybin, L. V., **8**, 170
Rybinskaya, M. I., **8**, 170
Rycroft, D. S., **1**, 137
Rygalov, L. N., **9**, 53
Rykova, L. A., **4**, 271, 272, 652
Rys, P., **7**, 54
Ryskieva, G. A., **14**, 736
Ryzhkina, I. S., **3**, 673
Ryzhova, G. L., **14**, 781
Rzepa, H. S., **11**, 75; **14**, 229

Saá, J. M., **3**, 627; **4**, 180, 641
Saadein, M. R., **10**, 76; **14**, 331
Saba, A., **3**, 1003; **11**, 146; **14**, 282
Sabesan, A., **1**, 155
Sabetian, K., **14**, 859, 860
Saburi, M., **4**, 677
Sachdeba, Y. P., **10**, 132
Sadekov, I. D., **3**, 52
Sadhu, K. M., **2**, 282
Sadovnikov, A. I., **2**, 163
Sadovskii, Yu. S., **2**, 142
Sadowski, C. M., **3**, 234
Saed, A. A. H., **1**, 55
Saegusa, T., **2**, 319, 327; **10**, 157; **11**, 106
Safarik, I., **3**, 646
Safarov, A. S., **4**, 492
Safarov, N. A., **10**, 276
Safiullin, R. L., **3**, 677, 1131, 1132
Safronova, Z. V., **13**, 31
Safta, M., **10**, 313
Saga, H., **11**, 89
Sagdeev, R. Z., **3**, 18, 1038, 1039; **14**, 783
Sagitullin, R. S., **6**, 120; **9**, 238; **14**, 61
Saha, C. R., **4**, 688
Saha, S. L., **12**, 235
Sahu, N. C., **4**, 234
Said, M., **12**, 263
Said, Z., **2**, 18
Saidov, G. V., **9**, 175, 176
Saikachi, H., **14**, 573
Saindane, M., **10**, 80; **12**, 228; **13**, 112
Saini, M. S., **2**, 401
Sainsbury, M., **3**, 512; **4**, 109
Saiprakash, P. K., **2**, 256, 257; **4**, 92
Saito, H., **9**, 62
Saito, I., **3**, 887, 902, 921; **4**, 397, 398, 402, 413
Saito, K., **4**, 282; **8**, 20; **9**, 239; **14**, 747
Saito, R., **4**, 4
Saito, T., **3**, 571; **11**, 58; **13**, 88

Saitoh, Y., **12**, 142
Saji, T., **10**, 151
Sajus, L., **4**, 436
Sakaguchi, Y., **3**, 521
Sakai, T., **3**, 358, 359; **10**, 81; **11**, 47, 48; **14**, 20, 377
Sakai, Y., **5**, 31; **9**, 35, 36; **14**, 484, 739
Sakamaki, Y., **9**, 262
Sakamoto, Y., **10**, 248
Sakanoue, K., **14**, 603
Sakata, J., **1**, 95; **10**, 184
Sakata, Y., **2**, 243
Sakhatchieva, M., **10**, 62
Sakilidi, V. T., **14**, 121
Sakizadeh, K., **9**, 181, 205, 219
Sakla, A. B., **14**, 43
Sako, M., **14**, 32
Sakodinskaya, I. K., **4**, 162
Saksena, S. C., **4**, 198
Sakuragi, H., **3**, 183, 624, 888; **4**, 453, 459
Sakurai, H., **1**, 160; **3**, 85, 526, 897, 898, 910, 913, 1030; **4**, 565, 705; **9**, 111; **13**, 194; **14**, 277, 792, 793
Sakurai, T., **4**, 421
Sakwrai, H., **14**, 276
Sala, T., **14**, 35
Salakhov, M. S., **13**, 98–100
Salaün, J.-Y., **14**, 246, 618
Salazar, J. Del C., **3**, 561; **14**, 236
Saleh, A., **9**, 77
Salem, L., **3**, 813; **9**, 149
Salem, S. M., **2**, 87
Salickaite, L. P., **6**, 105
Salimbaeva, A. D., **14**, 736
Salimov, R. M., **3**, 769
Salomoichenko, T., **2**, 442
Salomon, R. G., **10**, 198
Saltiel, J., **14**, 777
Salunkhe, M. M., **2**, 66
Salzer, A., **14**, 591
Sami, A., **10**, 63
Sammes, M. P., **6**, 106, 107; **10**, 79; **14**, 798
Sammons, D., **2**, 393
Samodumov, S. A., **4**, 472
Samskog, P. O., **3**, 1031, 1045
Samuels, S. B., **8**, 43
Samuilov, Ya. D., **13**, 50, 66, 94–96
San Clemente, M. R., **4**, 526
San Filippo, J., **9**, 250
Sana, M., **1**, 70
Sana-Ullah, C., **9**, 219
Sanchez, F., **9**, 10
Sánchez, I. H., **14**, 508
Sandall, J. P. B., **3**, 794, 795; **7**, 63, 64
Sander, S. P., **3**, 151
Sandifer, R. M., **10**, 222
Sandilya, R. K., **4**, 185
Sandkuhler, P., **14**, 381
Sandri, E., **10**, 226; **12**, 200; **14**, 219, 220
Sandri, S., **12**, 197
Sangapure, S. S., **14**, 74
Sanghvi, Y. S., **4**, 117
Sangster, D. F., **3**, 1155
Sanin, P. I., **14**, 478
Sankpal, S. G., **4**, 199, 200
Sannen, C., **8**, 155
Sannicolo, F., **14**, 293, 294
Sano, S., **3**, 433
Sano, T., **14**, 711, 739
Sansom, P. J., **2**, 153
Santamaria, J., **3**, 890; **4**, 368
Santaniello, E., **4**, 569
Santappa, M., **2**, 254
Santelli, M., **3**, 751; **4**, 644
Santiago, C. M., **14**, 289
Santiago, M. L., **14**, 452
Santiesteban, H., **9**, 225
Santini, C., **14**, 197
Santoro, R. J., **3**, 1103
Sanyal, B., **14**, 504
Sanz, A. M., **12**, 23
Sapunov, V. N., **4**, 115, 116, 318
Saran, N. K., **4**, 8
Sarangi, L. D., **4**, 249, 250
Saratov, I. E., **9**, 91
Sard, H., **14**, 237
Sarel, S., **14**, 256
Sargent, C. R., **3**, 993; **14**, 80
Sargent, M. V., **14**, 35
Sargeson, A. M., **2**, 244, 247, 369; **14**, 585
Sarhan, J. K. K., **12**, 261
Sarkanen, K. V., **4**, 526
Sarkar, T. K., **10**, 153
Sarkisian, G. M., **14**, 855
Sarma, Y. R., **2**, 257
Sarussi, S. J., **4**, 612
Sasaki, H., **14**, 573
Sasaki, K., **1**, 160
Sasaki, M., **3**, 634; **6**, 140, 141; **10**, 275; **12**, 239; **13**, 17
Sasaki, T., **4**, 451, 566; **13**, 7, 55
Sasamoto, K., **3**, 820
Sashida, H., **14**, 66, 363
Sasmal, B. M., **7**, 23
Sass, V. P., **3**, 136, 275, 371
Sasson, Y., **14**, 57
Satanova, R. B., **3**, 1110
Satchell, D. P. N., **2**, 249
Satchell, R. S., **2**, 249
Satgé, J., **3**, 609
Satiloane, B. P., **3**, 300
Sato, F., **1**, 97; **4**, 595; **12**, 206
Sato, K., **3**, 914
Sato, M., **1**, 97; **4**, 595; **7**, 35; **10**, 101; **12**, 206; **14**, 866
Sato, S., **3**, 394, 395, 399, 411; **9**, 58; **10**, 98
Sato, T., **3**, 681; **4**, 133; **5**, 99; **8**, 18; **14**, 747
Satoh, S., **3**, 767
Satterthwait, A. C., **1**, 157; **2**, 322
Satyamurthy, N., **6**, 53
Satyanarayana, G. O. S., **14**, 494
Satyanarayana, N., **1**, 148
Satyanarayana, P., **4**, 273; **14**, 510
Satyanarayana, P. V. V., **4**, 11, 12
Sau, A. C., **2**, 454
Sauerwald, M., **1**, 104
Sauleau, J., **9**, 49
Saunders, J., **10**, 135
Saunders, J. K., **1**, 7
Saunders, M., **8**, 62, 119, 127; **14**, 388
Saunders, W. H., **11**, 13–15
Sauer, H., **3**, 867a
Sauter, H., **14**, 324
Sauve, J. P., **14**, 460
Sauvêtre, R., **10**, 108
Savage, C. M., **3**, 150, 164, 185, 304, 306; **4**, 810
Savage, D. J., **10**, 95
Savage, E. B., **5**, 156
Savéant, J. M., **3**, 784, 1062; **6**, 35
Savelli, G., **2**, 100, 214
Savelova, V. A., **2**, 442, 443, 446
Savinova, L. N., **6**, 139, 156–159
Savranskaya, R. L., **1**, 14
Savu, P. M., **9**, 16
Sawada, H., **3**, 354; **4**, 600
Sawada, M., **2**, 29
Sawaki, Y., **3**, 149, 545, 1086; **4**, 287, 330, 360, 427, 452; **7**, 144
Sawaku, Y., **3**, 184
Sawayanagi, M., **3**, 248
Sawyer, D. T., **3**, 1059–1061; **4**, 373, 374, 376, 378
Sawyer, J. A., **5**, 49
Saxena, L. K., **4**, 192

Sayer, J. M., **4**, 315
Sbriziolo, C., **4**, 93, 94
Scaiano, J. C., **3**, 92, 329, 330, 352, 674, 1079, 1081; **5**, 5
Scala, A., **4**, 312
Scalzi, F. V., **3**, 1122
Scapini, G., **4**, 543
Scardellato, C., **4**, 123
Scarpati, R., **3**, 193; **14**, 720
Scartoni, V., **14**, 98
Scatton, M. J., **3**, 196
Scettri, A., **4**, 5
Schaad, L. J., **10**, 208; **13**, 91; **14**, 376
Schade, C. L., **11**, 40
Schädler, H.-D., **5**, 134
Schaefer, C. G., **3**, 930
Schaefer, G., **5**, 50
Schaefer, H. F., **5**, 139; **14**, 773
Schaefer, R., **9**, 60
Schaefer-Ridder, M., **4**, 341
Schäfer, H. J., **3**, 134
Schafer, K., **3**, 342
Schäfer, W., **6**, 62, 63
Schaffer, O., **14**, 292
Schaffner, K., **14**, 104, 882
Schalke, P. M., **11**, 120
Schamp, N., **14**, 378
Scharf, H.-D., **13**, 90
Schastnev, P. V., **8**, 4, 175
Schaumann, E., **11**, 78; **13**, 34, 53, 75, 76
Scheek, R. M., **3**, 1164
Scheeren, J. W., **13**, 102, 136
Scheffler, K., **3**, 194, 867a, 1056; **14**, 835, 845
Scheibel, J. J., **14**, 422
Schellekens, R., **3**, 144, 254
Schellekens, S. J., **3**, 145
Schenone, P., **13**, 137
Schepp, H., **5**, 19
Scheraga, H. A., **2**, 267
Scherzer, K., **3**, 257, 1127
Schiavelli, M. D., **8**, 141; **9**, 9
Schiavon, G., **8**, 42
Schiess, P., **14**, 258, 259
Schill, G., **14**, 821
Schilling, M. L. M., **3**, 915, 947, 1162
Schinzer, D., **10**, 106
Schipper, P., **2**, 329; **3**, 664–666; **8**, 13; **14**, 186, 499
Schleich, T., **3**, 1164
Schleker, W., **13**, 90
Schleyer, P. von R., **4**, 693; **5**, 88; **8**, 6, 119, 127, 154, 156, 159, 161; **9**, 26, 27, 157; **10**, 5–7, 9, 12, 14, 26, 70, 72, 97, 311; **14**, 388, 477
Schlossman, A., **14**, 256
Schmalsteig, H., **3**, 108
Schmid, G., **14**, 26, 702
Schmidt, A. H., **4**, 636
Schmidt, J. L., **3**, 896
Schmidt, R., **13**, 140; **14**, 13
Schmidt, R. R., **10**, 45, 110, 111
Schmidt, S. P., **11**, 156
Schmidt, W., **13**, 91
Schmidt-Renner, W., **3**, 178
Schmierer, R., **10**, 167; **13**, 130
Schmiesing, R. J., **4**, 118
Schmir, G. L., **2**, 179
Schmitt, D., **3**, 574; **13**, 199
Schmitt, G. D., **9**, 31
Schmitt, R., **14**, 642
Schmitt, R. J., **3**, 791; **7**, 61; **10**, 144
Schmittel, M., **3**, 207
Schmitz, L. R., **8**, 65; **14**, 434
Schnabel, W., **3**, 919
Schnaut, R., **3**, 64, 65, 958
Schneider, H.-H., **11**, 143
Schneider, H.-J., **4**, 317; **9**, 236, 240; **14**, 399
Schneider, K.-A., **14**, 819
Schneider, M. P., **3**, 560; **11**, 60
Schnurpfeil, D., **9**, 60
Schoeller, W., **5**, 45
Schoeller, W. W., **11**, 66
Schöffer, A., **11**, 83
Schofield, K., **7**, 50, 51, 54, 80
Schöllkopf, U., **10**, 223, 259
Schomburg, D., **5**, 56
Schön, G., **13**, 109
Schore, N. E., **10**, 268
Schostarez, H., **14**, 148
Schowen, R. L., **1**, 115; **2**, 58; **4**, 597; **8**, 91
Schrade, R., **3**, 867a
Schramm, L. C., **14**, 765
Schrauzer, G. N., **3**, 869a
Schreiber, S. L., **14**, 197
Schreifels, J. A., **4**, 697
Schröer, W.-D., **13**, 80–82, 181
Schroth, W., **5**, 134
Schubert, I., **1**, 69
Schubert, K., **3**, 74
Schubert, W., **3**, 26
Schuchardt, U., **13**, 47, 208
Schuler, P., **3**, 867a
Schuler, R. H., **3**, 548a, 1150
Schulman, E. M., **8**, 11
Schulte-Frohlinde, D., **1**, 167; **2**, 74; **3**, 967; **4**, 503
Schulten, W., **3**, 46
Schultz, A. G., **5**, 42; **10**, 196; **14**, 145
Schultz, P. G., **5**, 13
Schultz, R. A., **4**, 580
Schulz, A., **3**, 207
Schulz, G., **14**, 714
Schulz, M., **3**, 119
Schulz, W., **3**, 957
Schulze, B., **1**, 69
Schulze, K., **14**, 180
Schumacher, H. J., **3**, 98
Schumachers, L., **3**, 573
Schundehutte, K.-H., **14**, 73
Schuster, D. I., **14**, 103, 107–109, 886
Schuster, G. B., **3**, 109, 129, 143, 1097; **5**, 6; **11**, 156
Schuster, P., **1**, 118
Schütz, F., **13**, 135
Schwab, J. M., **4**, 301
Schwab, W., **4**, 548
Schwartz, J., **4**, 691; **14**, 606
Schwarz, H., **3**, 940; **8**, 127, 146, 147, 168; **14**, 397, 439
Schwarz, H. A., **3**, 1048
Schwarz, U., **3**, 214
Schwarz, W., **3**, 213, 215, 568; **14**, 696
Schwarzenbach, D., **3**, 597; **13**, 111
Schweizer, W. B., **5**, 30; **14**, 654
Sciacovelli, O., **12**, 230
Scolastico, C., **10**, 240, 241; **12**, 223
Scott, J. M. W., **9**, 163–165, 264; **14**, 23
Scott, L. T., **14**, 815
Scribe, P., **14**, 335
Scribner, J. D., **8**, 171
Scrimin, P., **4**, 123
Scriven, E. F. V., **5**, 107
Scudder, P. H., **14**, 252
Sealy, R. C., **3**, 1011, 1013
Seaone, F., **2**, 280
Searle, G. H., **2**, 246
Sebastian, C. F., **4**, 33, 34
Sebedio, J. L., **3**, 523
Seconi, G., **1**, 59, 60; **7**, 150; **10**, 154, 286, 288, 289
Secor, H. V., **9**, 221
Sedon, J. H., **3**, 582; **11**, 154
Sedshaw, W., **3**, 178
Seebach, D., **10**, 130, 166, 171, 232, 266; **12**, 266; **14**, 37
Seeger, D. E., **3**, 554
Seely, M. J., **3**, 1122
Seeman, J. I., **9**, 221
Seetz, J. W. F. L., **3**, 854, 864
Segawa, J., **8**, 96
Seguin, R., **12**, 269
Seidman, D. A., **4**, 657

Seitz, B., **13**, 174
Seitz, D. E., **10**, 89
Seiyama, T., **7**, 119
Seki, K., **3**, 519
Seki, M., **4**, 708
Seki, S., **3**, 294; **14**, 590
Seki, Y., **9**, 265
Sekiguchi, A., **5**, 27, 98, 99; **14**, 137
Sekiguchi, S., **6**, 39, 40, 138, 142
Sekine, M., **1**, 158; **14**, 366, 367
Sekiya, M., **4**, 613; **14**, 139, 160, 675
Selby, K., **3**, 322
Selim, A., **10**, 167
Selim, E. A., **1**, 149
Selivanov, B. A., **7**, 132
Selivanov, N. T., **3**, 175, 264
Sellers, S. F., **3**, 486, 1117–1121; **14**, 298, 631–633
Selvaraj, K., **4**, 112
Selvarajan, N., **3**, 1156
Selwitz, C. M., **7**, 37
Semenov, V. V., **4**, 356
Semenova, S. N., **9**, 54
Semenovsky, A. V., **14**, 641
Semenyuk, G. V., **2**, 106, 168
Semmelhack, M. F., **6**, 22, 82
Sen Gupta, K. K., **4**, 165
Sen, A., **12**, 163; **14**, 609
Sen, B., **14**, 716
Sen, P. K., **4**, 165
Sen, S., **1**, 112
Senda, S., **6**, 124; **14**, 77
Sendega, R. V., **2**, 421–423
Sener, B., **14**, 686
Sengupta, K. K., **4**, 166
Senthilnathan, V. P., **4**, 112; **5**, 4, 7
Seo, S., **14**, 520
Seoane, C., **12**, 255
Seoane, E., **14**, 380
Seone, C., **14**, 750
Seone, E., **14**, 379
Sepetov, N. F., **8**, 113; **14**, 404
Serebryanskaya, A. I., **7**, 48a
Seree de Roch, I., **4**, 436
Serelis, A. K., **3**, 439
Sergeev, G. B., **3**, 853; **12**, 67, 68; **13**, 118
Sergeeva, O. R., **7**, 112
Serguchev, Yu. A., **2**, 255; **3**, 717
Serov, S. I., **3**, 136, 275, 371
Serravalle, G., **4**, 81
Serravalle, M., **3**, 788
Servé, M. P., **5**, 41
Servis, K. L., **8**, 7
Seshadri, R., **4**, 658
Sessions, R. B., **3**, 943
Seth, M., **14**, 604
Sethi, S. K., **10**, 323
Sethuram, B., **1**, 58, 74; **4**, 46, 60, 97, 155, 156
Setoi, H., **14**, 157
Seuron, N., **12**, 224
Sevast'yanova, I. V., **9**, 93
Sevenster, A. J. L., **3**, 1006
Sevilla, M. D., **3**, 1004, 1043, 1046, 1147
Seyden-Penne, J., **1**, 106, 123; **10**, 165; **12**, 218, 219, 224
Seyer, A., **14**, 838
Sha, R. C., **2**, 19
Shabana, R., **14**, 239
Shabanov, A. L., **6**, 68
Shabanov, A. V., **3**, 874
Shabarov, Yu. S., **7**, 71, 72, 76; **8**, 76
Shabtai, J., **10**, 113
Shadyro, O. I., **3**, 460, 461
Shafer, J. A., **2**, 289–291
Shafer, S. G., **9**, 197
Shafig, Y. E., **7**, 46
Shafikov, N. Ya., **4**, 295, 296
Shafuillah, G. M. A., **14**, 558
Shah, R. C., **3**, 538
Shah, R. R., **14**, 169
Shah, S. K., **10**, 216; **14**, 226
Shahrisa, A., **14**, 297
Shaik, S., **14**, 1
Shail, S. S., **9**, 230
Shainyan, B. A., **9**, 5, 6; **10**, 16; **12**, 174, 184
Shakhel'dyan, A. R., **6**, 55
Shakhel'dyan, I. V., **6**, 55, 153
Shakhtakjtinskii, T. N., **4**, 482
Shakirov, M. M., **8**, 68; **14**, 10, 11
Shakked, Z., **14**, 677
Shaller, H. E. A., **2**, 193
Shames, S. L., **2**, 57, 194
Shamina, L. P., **9**, 132, 133
Shamma, M., **4**, 622; **14**, 686
Shanker, R., **4**, 220
Shanker, U., **7**, 110; **14**, 46
Shanmuganathan, S., **2**, 254
Shannon, P., **4**, 724
Shannon, P. V. R., **7**, 117; **14**, 54
Shanthalakshmy, P., **4**, 685
Shapiro, J., **10**, 35
Shapiro, J. S., **11**, 72; **14**, 411
Shapiro, S., **4**, 738, 739
Sharaf, S. M., **9**, 11
Sharapova, S. T., **12**, 248
Sharma, A., **4**, 144
Sharma, K. K., **3**, 994
Sharma, M. M., **9**, 185
Sharma, N. K., **4**, 219
Sharma, R. D., **10**, 260
Sharma, R. G., **4**, 185
Sharma, S. K., **4**, 222
Sharma, T. C., **4**, 106
Sharon, M., **6**, 66
Sharp, J. T., **7**, 153; **14**, 268
Sharpless, K. B., **4**, 152, 314; **14**, 594
Sharykin, V. G., **4**, 115, 116
Shatenshtein, A. I., **7**, 48a
Shatzmiller, S., **10**, 134
Shaw, C. F., **4**, 61
Shaw, E., **2**, 286
Shaw, J. H., **3**, 166
Shaw, R., **3**, 1100
Shchedrinskaya, T. V., **4**, 489
Shchegolev, A. A., **8**, 113; **14**, 390, 404
Shcherbina, F. F., **4**, 474
Shcherbina, T. M., **14**, 744
Shcherbinin, M. B., **3**, 818, 882
Shchupak, G. M., **6**, 145
Shea, C.-M., **14**, 754
Sheena, H. H., **3**, 1073
Sheiko, S. G., **2**, 430, 445
Shein, S. M., **6**, 37, 88
Sheka, I. A., **3**, 416
Sheldon, J. C., **8**, 143; **10**, 54
Sheldon, R. A., **3**, 2; **4**, 304
Sheldrick, W. S., **14**, 368
Shellhamer, D. F., **1**, 152; **3**, 409; **12**, 21, 26, 36
Shelton, S. R., **4**, 92
Shen, D. Z., **1**, 83
Shen, M., **5**, 42
Shendrik, A. N., **4**, 478
Shepard, R., **5**, 9
Shepherd, P. T., **4**, 721
Sheppard, A., **3**, 397
Sheppard, W. A., **6**, 6
Sheradsky, T., **3**, 875; **5**, 65
Shereshovets, V. V., **4**, 294–296
Sherrington, D. C., **9**, 189
Sherstyuk, V. P., **4**, 13
Sheth, J. P., **14**, 635
Shetty, R. V., **3**, 630
Shevchuk, I. M., **3**, 543
Shevelev, S. A., **12**, 265
Shevlin, P. B., **4**, 650, 682; **5**, 61
Shevyreva, E. V., **4**, 467
Shi, Y.-Z., **14**, 619
Shiau, C. C., **5**, 142
Shibanuma, T., **14**, 101
Shibasaki, M., **12**, 148
Shibib, S. M., **4**, 628

Shieh, H-M., **10**, 170
Shieh, H. M., **1**, 89
Shigalevskii, V. A., **7**, 36
Shih, C., **10**, 114
Shih, T. L., **1**, 91; **10**, 189
Shik, G. L., **4**, 482
Shilov, A. E., **4**, 647
Shimada, K., **14**, 221
Shimanouchi, N., **2**, 253
Shimazu, K., **6**, 70
Shimidzu, T., **4**, 333
Shimizu, H., **6**, 173; **14**, 279, 280
Shimizu, K., **1**, 27; **14**, 820
Shimizu, M., **4**, 207
Shimizu, T., **2**, 181; **4**, 528; **13**, 61
Shimomura, M., **2**, 202, 204
Shimp, L. A., **10**, 10; **14**, 627
Shindo, H., **7**, 133
Shine, H. J., **3**, 211, 973; **14**, 14
Shiner, V. J., **14**, 410
Shingu, K., **14**, 264
Shingu, T., **14**, 519
Shinizu, A., **9**, 262
Shinkai, S., **1**, 144; **2**, 146, 204, 205; **4**, 703, 712, 729; **10**, 101; **11**, 42; **14**, 20
Shinoda, K., **9**, 213
Shio, T., **3**, 871; **4**, 702
Shirahama, H., **14**, 506
Shirahama, S., **4**, 690
Shirahata, K., **14**, 866
Shirai, K., **14**, 550
Shirataki, Y., **14**, 889
Shirobokov, I. Yu., **9**, 131, 140
Shiskhkina, R. P., **14**, 52
Shkolnik, O. V., **12**, 115, 116
Shmelev, V. A., **2**, 171
Shode, L. G., **9**, 67, 70, 71, 217
Shoenberger, D. C., **4**, 585
Shoikhet, A. A., **3**, 347
Shokhor, I. N., **10**, 334; **2**, 148
Shono, T., **10**, 87
Shopov, D., **4**, 689
Shornik, N. A., **12**, 117
Short, E. L., **7**, 128
Shorter, J., **2**, 49–51
Shosenji, H., **2**, 221
Shostenko, A. G., **3**, 356, 387–392, 418
Shou, H., **13**, 200
Shpan'ko, I. V., **2**, 170; **9**, 202
Shraydeh, B. F., **11**, 36
Shteingarts, V. D., **7**, 132; **8**, 115
Shteinman, A. A., **4**, 495
Shubin, V. G., **8**, 67–71, 118, 120; **14**, 4, 6–10, 387, 391
Shudo, K., **7**, 151
Shue, F.-F., **8**, 7
Shuker, D. E. G., **10**, 330
Shukla, R. K., **4**, 158
Shul'pin, G. B., **7**, 147
Shumate, R. E., **4**, 672
Shumskii, F. Z., **2**, 171
Shuvalov, V. F., **3**, 317
Shved, E. N., **6**, 56
Shvets, V. F., **3**, 287, 288, 529; **9**, 50
Shyamasundar, N., **4**, 538
Sibi, M. P., **5**, 133
Sibley, C. E., **1**, 122
Sicre, J. E., **2**, 424; **9**, 167
Sidahmed, I. M., **2**, 87
Siddiqui, T., **10**, 258
Sidorov, E. D., **12**, 179
Sidot, C., **3**, 802, 803; **4**, 597; **6**, 25
Sieburth, S. M[c]N., **14**, 194, 195
Siegfried, R., **14**, 468, 469
Siegfriedt, K.-H., **6**, 144
Sieh, D. H., **9**, 269
Sielecki, A. R., **2**, 281
Sienv, V. V., **8**, 87
Sievert, R., **3**, 242, 302, 311
Sigler, P. B., **2**, 317
Sih, C. J., **10**, 131
Sik, R. H., **3**, 644
Sikkar, R., **3**, 221
Šilar, L., **2**, 122; **11**, 22
Silbermann, J., **9**, 250
Silgar, S. G., **4**, 754
Šilhavý, P., **3**, 743, 744
Silk, N. M., **14**, 796
Silkiňa, N. N., **5**, 140
Sillion, B., **4**, 236
Silver, M. S., **2**, 296, 297
Silverman, R. B., **4**, 744, 751
Silverton, E. W., **2**, 271
Sim, G. A., **3**, 605
Simandi, L., **4**, 29
Simanenko, Yu. S., **2**, 443
Simig, G., **3**, 798, 822; **9**, 148
Simkin, B. Y., **14**, 281
Simmons, D. P., **10**, 118
Simmons, H. E., **6**, 128; **14**, 92, 334, 668
Simmons, W., **3**, 889a
Simon, A., **12**, 268
Simon, J., **5**, 9
Simon, J. D., **3**, 928
Simon, P., **3**, 632, 637
Simon, R., **3**, 833
Simon, Z., **2**, 262; **6**, 3
Simoncsits, A., **2**, 377
Simonet, J., **3**, 780
Simonetta, M., **3**, 235, 1042
Simonnin, M. P., **1**, 159; **6**, 151, 164
Simonov, M. A., **3**, 191; **4**, 512
Simonyi, M., **10**, 271; **11**, 116
Simpson, J. T., **3**, 904, 906
Sims, L. B., **9**, 139, 141; **11**, 12, 144; **14**, 408
Sindhuatmadja, S., **4**, 701, 714; **6**, 169
Sindler-Kulyk, M., **3**, 1076; **13**, 24
Sinev, V. V., **8**, 82, 84–86, 88, 89, 116; **9**, 169, 170
Sing, A., **14**, 727
Singer, L. S., **3**, 180
Singh, A. K., **14**, 563
Singh, B., **4**, 141, 157, 158
Singh, B. B., **4**, 141, 157
Singh, B. P., **8**, 63
Singh, D., **3**, 500; **14**, 85
Singh, G., **1**, 52; **3**, 500; **14**, 85
Singh, H., **3**, 616; **4**, 198, 201
Singh, K., **3**, 675; **14**, 167
Singh, K. P., **4**, 185
Singh, L., **2**, 19
Singh, M., **4**, 193
Singh, N., **12**, 28
Singh, P., **6**, 24, 176; **14**, 39
Singh, P. R., **3**, 837
Singh, R. N., **4**, 140
Singh, R. P., **4**, 157
Singh, R. T., **2**, 19
Singh, S., **4**, 141
Singh, S. B., **5**, 73
Singh, S. K., **14**, 195
Singh, S. M., **1**, 6
Singh, S. P., **14**, 441
Singh, V., **4**, 106
Singh, V. K., **14**, 491
Singhal, S. K., **4**, 185
Singleton, D. L., **3**, 310; **4**, 387, 388
Singleton, K. A., **3**, 497
Sinha, A., **2**, 86
Sinha, A. K., **14**, 719
Sinha, B. K., **4**, 95; **9**, 187
Sinha, B. P., **4**, 139, 274
Sinha, G., **14**, 537
Sinisterra, J. V., **14**, 555
Sinotova, E. N., **8**, 124, 140
Sinou, D., **4**, 667, 669
Sinovich, I. D., **4**, 477
Sipershtein, I. N., **4**, 84
Siracusa, M. A., **14**, 94
Siranson, R., **3**, 908
Sirazhitdinova, D. S., **3**, 387
Sircar, J. K., **7**, 39
Sitzmann, E. V., **3**, 1172
Sitzmann, M. E., **6**, 60

Sivakamasundari, S., **7**, 31
Sivarajah, K., **3**, 880
Sjawelski, R. J., **2**, 292
Sjölin, L., **2**, 411
Skarzewski, J., **3**, 679
Skell, P. S., **3**, 72, 331
Skelton, B. W., **14**, 507
Sket, B., **7**, 17; **13**, 205
Skibida, I. P., **3**, 158
Skinner, G. B., **3**, 231, 263
Skinner, K. L., **5**, 69
Skornyakova, T. G., **12**, 133
Skorobogatova, E. V., **12**, 46
Skramstad, J., **4**, 448
Skrypnik, Yu., **2**, 436
Skrypnik, Yu. G., **11**, 27
Skurko, M. R., **3**, 459
Skvarchenko, V. R., **14**, 565
Skvortsov, N. K., **12**, 111
Slagle, I. R., **3**, 650
Slagle, J. D., **9**, 272
Slater, C. D., **10**, 302
Slavinskii, N. V., **3**, 101
Slayden, S. W., **10**, 123; **14**, 415
Slayton, R. I., **5**, 133
Ślebocka-Tilk, H., **2**, 381; **9**, 204
Sleiter, G., **6**, 98
Slessor, K. N., **11**, 137
Sliskovic, D. R., **5**, 18
Slocum, G. H., **3**, 1097
Sluma, H.-D., **5**, 77
Slyusarenko, T. F., **5**, 140
Smal, M. A., **10**, 239
Smart, B. E., **3**, 1117; **14**, 633
Smeets, F. L. M., **14**, 660
Smetannikov, Yu. V., **3**, 390
Smid, J., **10**, 32
Smidt, J., **3**, 987
Smirnov, V. V., **3**, 853
Smirnova, E. N., **8**, 86, 89
Smirnova, T. I., **4**, 488
Smit, C. J., **3**, 759
Smit, V. A., **8**, 113, 122; **12**, 8, 41, 55; **14**, 404
Smith, A. B., **5**, 14; **14**, 454–456
Smith, A. K., **5**, 18
Smith, C. Z., **3**, 754
Smith, D. A., **10**, 317; **12**, 26
Smith, D. J. H., **11**, 55; **14**, 261
Smith, D. L., **8**, 103
Smith, F. X., **4**, 580
Smith, G. F., **3**, 833
Smith, G. M., **2**, 144; **7**, 130
Smith, H. K., **3**, 680; **4**, 190
Smith, I. G., **3**, 1149
Smith, J. G., **14**, 146
Smith, J. N., **12**, 26
Smith, J. R. L., **3**, 292
Smith, K. J., **14**, 107, 108
Smith, K. M., **3**, 925; **14**, 652
Smith, L., **4**, 767
Smith, L. A., **6**, 39
Smith, M. A., **8**, 145; **14**, 433
Smith, M. B., **4**, 535
Smith, P., **3**, 34, 35
Smith, P. A. S., **5**, 115
Smith, P. J., **9**, 222; **11**, 9, 138
Smith, R. A., **4**, 662
Smith, R. A. J., **3**, 862
Smith, R. J., **3**, 615
Smith, R. S., **12**, 240
Smith, S. G., **4**, 547; **12**, 227
Smith, W. B., **3**, 299; **14**, 413
Smithers, R. H., **9**, 47
Smolik, R., **14**, 521
Smolina, T. A., **14**, 414
Smyth, T., **2**, 251; **9**, 270
Smyth, T. A., **7**, 68
Smyth-King, R. J., **3**, 826
Snell, E. E., **4**, 763
Snelson, A., **4**, 396
Snider, B. B., **13**, 29; **14**, 134, 407, 458
Snieckus, V., **10**, 140; **12**, 214, 215
Snoble, K. A. J., **11**, 109
Snowden, R. L., **10**, 118, 119; **13**, 108
Snyder, J. J., **14**, 779
Snyder, J. P., **3**, 549
So, Y.-H., **3**, 530, 531; **6**, 33
Sobchak, Yu., **4**, 121
Sobieray, D. M., **4**, 575
Sochilin, E. G., **12**, 13, 14
Sogah, G. D. Y., **12**, 251
Sogomonyan, B. M., **4**, 505, 507
Soháni, S. V., **3**, 757; **4**, 624, 626
Sohar, P., **14**, 715
Sohrabi, M., **7**, 79
Sokolenko, V. N., **2**, 172
Sokolov, S. V., **3**, 136, 275
Sokolov, V. I., **8**, 27
Sokolova, G. D., **2**, 383
Sokolowski, A., **14**, 829
Solar, S., **3**, 1141
Solar, W., **3**, 1141
Solodar, S. L., **7**, 10
Solodovnikov, S. P., **3**, 499, 640, 661, 668, 672, 768
Sololov, S. V., **3**, 371
Solomatin, G. G., **7**, 36
Solomentsev, V. V., **14**, 647
Solomon, D,. H., **3**, 133
Solomonov, B. N., **13**, 15, 16
Solov'eva, S. E., **13**, 66
Solov'yanov, A. A., **9**, 256, 257; **10**, 84, 85
Solyanikov, V. M., **3**, 114; **4**, 496
Somanathan, R., **3**, 508; **14**, 212, 284
Somani, R., **1**, 33
Somayaji, V., **2**, 105, 192
Somin, I. N., **14**, 794, 795
Sommer, J., **8**, 74
Somov, G. V., **3**, 305; **11**, 79
Son, C. K., **9**, 246
Sonawane, H. R., **3**, 1084
Sone, T., **10**, 101; **14**, 20
Song, B. S., **4**, 542
Song, H. B., **2**, 434, 438; **9**, 172
Song, K. B., **2**, 438; **9**, 172
Sonoda, A., **4**, 564
Sonoda, T., **7**, 88
Sopchik, A. E., **3**, 195; **4**, 524
Sopher, D. W., **3**, 760
Soppe-Mbang, H., **3**, 353
Sorba, J., **3**, 148, 523
Sordo, T., **7**, 43; **12**, 87
Sorensen, T. S., **8**, 63–65; **14**, 434
Sorgi, K. L., **4**, 118
Sorokin, M. F., **9**, 67, 70, 71, 217
Sorrell, T. M., **4**, 589
Sosonkin, I. M., **3**, 808, 809; **6**, 26
Soto, H., **3**, 1078
Soto, J. L., **12**, 255; **14**, 750
Soto, R., **2**, 218
Soumillion, J. Ph., **3**, 1095
Southam, D. A., **14**, 213
Southon, I. W., **14**, 218
Sowadski, J. M., **2**, 408
Soysa, H. S. D., **14**, 717
Sozzani, P., **4**, 569
Sozzi, G., **14**, 514
Sopagnolo, P., **3**, 572, 978; **5**, 130, 131
Spaite, D. W., **1**, 152
Spangler, C. W., **11**, 110
Sparapani, C., **7**, 107
Spasskaya, R. I., **2**, 147
Spears, C. P., **9**, 228
Speckamp, W. N., **14**, 203, 265
Speer, H., **10**, 45, 110, 111
Speier, G., **4**, 49, 325, 375
Spencer, S. A., **2**, 273
Spencer, T. A., **11**, 121
Speranza, M., **7**, 106–108; **9**, 141–143; **12**, 22
Spillane, W. J., **10**, 315

Spindell, D. K., **13**, 29
Spinelli, D., **2**, 53; **6**, 94, 95; **11**, 45; **14**, 95–97
Spiro, V., **4**, 24
Spitznagel, G. W., **10**, 9
Springs, B., **2**, 412
Sproat, B. S., **2**, 409
Squires, R. R., **1**, 9; **10**, 322
Sree, A., **14**, 516
Sridharan, S., **9**, 136
Srikrishna, A., **8**, 22; **14**, 489, 492
Srimannorayana, G., **14**, 170
Srinivasan, N. S., **4**, 26
Srinivasan, R., **1**, 139
Srinivasan, S., **4**, 248
Srinivasan, V. S., **12**, 269
Srivastava, S. K., **8**, 83
Srivastava, S., **5**, 34
Srivastava, S. P., **4**, 185–188, 194, 195, 261–264
Srivastava, S. S., **4**, 654
Stachell, R. S., **2**, 154
Stadnichuk, M. D., **11**, 149
Stagno d'Alcontres, G., **12**, 165
Stahl, D., **8**, 146
Stahl-Lariviere, H., **3**, 1047
Staley, S. W., **10**, 59, 60
Stamegna, A. P., **3**, 505; **14**, 356
Stamper, J. G., **10**, 286, 288
Stampfli, B., **14**, 295
Stanaszek, R. S., **4**, 581; **9**, 126
Stanczyk, W., **10**, 292
Stanescu, L., **14**, 187
Stang, P. J., **4**, 179; **5**, 33, 35, 79, 80, 135; **8**, 103, 141; **9**, 9
Stange, A., **3**, 471
Staninets, V. I., **3**, 357, 370, 436, 533, 658, 659; **9**, 110; **11**, 148
Stankevich, A. I., **3**, 135
Stankevich, G. S., **6**, 87
Stankevich, I. V., **10**, 73
Stanney, K., **7**, 9, 109
Stapersma, J., **10**, 66
Starchevskii, M. K., **4**, 164, 484
Staricco, E. H., **3**, 369, 404
Stark, C. J., **4**, 308
Starostin, B. S., **9**, 169, 170
Staroverova, N. V., **3**, 286
Stashina, G. A., **3**, 466
Stasko, A., **3**, 56, 123, 859, 1035
Staunton, J., **10**, 316; **14**, 118
Staurt, A. D., **14**, 328
Stavber, S., **12**, 11
Stavinoha, J. L., **3**, 487, 907, 908
Stavrova, S. D., **3**, 140
Stawitz, J., **14**, 185
Stec, W., **2**, 359
Stec, W. J., **14**, 461, 462
Stec, Z., **4**, 437
Stedman, G., **4**, 169
Steel, P. J., **8**, 8; **14**, 476
Steenken, S., **3**, 330, 1146
Stefanovsky, Y. N., **12**, 167, 244, 245
Stegel, F., **6**, 96, 167
Steglich, W., **12**, 250; **14**, 728
Stegmann, H. B., **3**, 194, 867a, 1056; **14**, 835, 845
Stegmann, W., **13**, 64
Steichen, D. S., **3**, 891
Stein, G., **9**, 210
Stein, R. L., **1**, 31
Stein, S. E., **3**, 51, 1111–1113, 1123
Stein, U., **3**, 957
Steinbach, K., **5**, 20
Steinbach, R., **10**, 155
Steinbeck, K., **5**, 75
Steinberg, H., **5**, 106; **9**, 72, 73; **13**, 87
Steinfeld, J. I., **3**, 98
Stekhova, S. A., **1**, 162
Steliou, K., **2**, 245; **4**, 69
Stella, L., **3**, 40, 444, 446
Stenhammer, K., **10**, 127
Stepanov, B. I., **8**, 79
Stepanova, T. F., **8**, 85
Stephen, M. A., **2**, 324; **14**, 574
Stephen, W., **11**, 145
Stephens, C. W., **3**, 619
Stephenson, D. S., **14**, 124
Stephenson, G. R., **8**, 31, 35, 39
Stephenson, L. M., **4**, 392
Stephenson, M., **12**, 272
Stephenson, M. T., **3**, 973
Stepukhovich, A. D., **3**, 1170; **14**, 784
Šteřba, V., **2**, 13, 182, 184; **14**, 848
Sternbach, D. D., **4**, 583
Sternson, L. A., **4**, 768
Stetter, H., **12**, 96–98; **14**, 853
Stevenson, G. R., **3**, 1058
Stewart, A., **11**, 128
Stewart, P., **10**, 160
Stewart, R., **1**, 139; **4**, 552, 553
Stick, R. V., **4**, 602
Stief, L. J., **3**, 308, 314
Stille, J. K., **4**, 673, 675
Stille, J. R., **4**, 675
Stirling, C. J. M., **2**, 102; **11**, 5, 18, 87, 129
Stob, S., **3**, 1164
Stobie, A., **4**, 107; **6**, 114
Stock, L. M., **3**, 227; **6**, 11; **7**, 55, 139
Stockburn, W. A., **3**, 960
Stojanowa-Antoszczyn, M., **4**, 319
Stolyarov, B. V., **8**, 111; **14**, 21, 542
Stolze, K., **3**, 194
Stone, A. J., **1**, 126
Stone, J. A., **7**, 105
Stone, P. G., **3**, 340
Storm, D. R., **2**, 394
Stothers, J. B., **10**, 40, 200, 201; **14**, 336, 568
Stotskii, A. A., **4**, 170, 171; **7**, 84; 85
Stott, K. M., **9**, 65
Stradlund, G., **7**, 100
Strasak, M., **4**, 71
Stratton, T. J., **14**, 887
Straub, H., **13**, 164
Strauss, M. J., **6**, 78, 164, 165
Strauss, U., **14**, 734
Strausz, O. P., **3**, 279, 557, 646; **6**, 170
Streith, J., **14**, 629
Streitwieser, A., **10**, 296, 299, 300; **14**, 813
Strepikheev, Yu. A., **7**, 87; **14**, 18, 19
Strichkov, Yu. T., **8**, 50
Strijtveen, B., **9**, 248
Strizhakova, N. G., **3**, 416
Strobel, M.-P., **4**, 281
Strogov, G. N., **3**, 808, 809; **6**, 26
Strohscheidt, T., **14**, 102
Strom, J. G., **1**, 63
Struchkov, Yu. T., **6**, 119
Strum, H., **13**, 51
Strupnikova, T. V., **9**, 238
Strydom, P. J., **14**, 326
Stults, B. R., **4**, 666
Stumpe, R. W., **6**, 7; **7**, 93
Sturtz, G., **10**, 269; **14**, 369
Stutz, H., **14**, 438
Su, T. T., **2**, 183
Suarez, E., **4**, 379
Suau, R., **3**, 627; **4**, 180, 641
Subra, R., **3**, 360, 513
Subrahmanyam, P., **14**, 510
Subramanian, N., **3**, 705; **12**, 147
Subramanian, P. K., **6**, 53
Subramanian, P. S., **4**, 248
Subramanian, R., **11**, 13
Subramanian, T. V., **4**, 465
Suckling, C. J., **7**, 16

Sucrow, W., **3**, 212
Suda, H., **2**, 229
Suda, M., **3**, 427
Sudha, K., **14**, 170
Suehiro, T., **3**, 61, 634
Suenaga, R., **3**, 858
Sufiname, H., **3**, 448
Sugahara, K., **2**, 224; **4**, 678
Sugawara, K., **3**, 394, 395, 399
Sugawara, T., **3**, 1169; **5**, 120
Suggs, J. W., **4**, 681; **5**, 8
Sugihara, Y., **10**, 231; **14**, 645
Sugimori, A., **3**, 900
Sugimoto, N., **10**, 275
Sugimoto, T., **2**, 196; **7**, 136; **9**, 192, 193; **10**, 237
Sugimura, H., **10**, 242
Suginome, H., **3**, 465, 767, 1090; **14**, 527, 556, 754
Sugiyama, K., **14**, 147
Suh, J., **2**, 103
Sukenik, C. N., **12**, 136
Sukhareva, I. P., **2**, 159
Sukhorukhova, N. A., **2**, 340
Sukhorukov, Yu. I., **2**, 31
Sukumaran, K. B., **14**, 301
Sul'man, E. M., **4**, 686
Suleimanov, S. N., **13**, 99
Suleman, N. K., **3**, 1017
Sullivan, S. A., **10**, 146
Sultan, L., **3**, 833
Sulzer, J., **2**, 123
Sumegi, L., **3**, 637
Summers, S. T., **4**, 577
Sunamoto, J., **2**, 236; **6**, 74; **14**, 34
Sunay, U., **9**, 225
Sundaram, E. V., **1**, 148; **4**, 98, 243 246, 251, 254, 255
Sundaram, M. G., **14**, 579
Sundberg, R. J., **4**, 582
Sunder, S., **14**, 36
Sundin, S. E., **4**, 526
Sung, D. D., **2**, 165
Sung, N.-D., **9**, 263
Sunko, D. E., **8**, 19; **14**, 449
Supp, M., **1**, 23
Suprun, V. Ya., **4**, 87
Surbeck, J.-P., **3**, 766
Suresh, D. D., **4**, 457
Surikova, T. P., **7**, 76
Surin, N. G., **10**, 73
Surkov, V. D., **3**, 811
Surovtsev, L. G., **9**, 178
Surpateanu, G., **13**, 85
Suryanarayana, D., **3**, 1004, 1147
Surzur, J. M., **3**, 444, 446
Suschitzky, H., **5**, 107; **12**, 47
Sustmann, R., **3**, 47
Susu, A. A., **3**, 1109
Suszka, A., **4**, 351
Sutcliffe, R., **3**, 66, 95; **6**, 101
Sutherland, I. O., **3**, 508, 509; **14**, 212–215, 283, 284, 353
Sutherland, J. K., **12**, 152
Sutherland, R. G., **5**, 108
Suto, N., **14**, 332
Sutthivaiyakit, S., **4**, 148
Sutton, J., **3**, 313
Sutyagin, V. M., **2**, 55
Suvorou, B. V., **4**, 486
Suzuki, A., **1**, 81
Suzuki, H., **6**, 90; **7**, 65, 66, 82; **14**, 595
Suzuki, K., **2**, 206; **4**, 613; **10**, 246; **11**, 124; **12**, 190; **14**, 160, 303
Suzuki, M., **3**, 297; **14**, 32, 610, 689
Suzuki, N., **6**, 70
Suzuki, T., **11**, 57; **13**, 20
Suzuki, Y., **4**, 420
Svedberg, D. P., **14**, 182
Svensson, L., **10**, 127
Svingen, B. A., **3**, 1142
Svoboda, J., **10**, 28; **12**, 172, 173
Svoboda, P., **2**, 125
Swamy, R., **4**, 238
Swan, J., **3**, 879; **10**, 148
Swanson, B. J., **3**, 920; **14**, 726
Swartz, S., **3**, 1043, 1046
Swartz, W. E., **4**, 697
Sweany, R., **3**, 527
Sweigart, D. A., **8**, 45, 49
Swenson, W., **14**, 811
Swenton, J. S., **10**, 114
Swerdloff, M. D., **4**, 50
Swern, D., **5**, 39
Swidler, R., **1**, 71
Swiger, R. T., **3**, 827
Swistun, Z., **6**, 125; **14**, 76
Sworin, M., **14**, 201
Syfrig, M. A., **10**, 130
Sylvestre-Panthet, P., **14**, 308
Symons, E. A., **10**, 281, 282
Symons, M. C. R., **3**, 812, 935, 943, 984, 997, 1024, 1144, 1149, 1175; **6**, 34; **10**, 17
Synder, J. P., **14**, 228
Synos, T. E., **10**, 58
Sypniewski, G. C., **11**, 40
Syrkin, V. G., **12**, 115
Syroezhko, A. M., **4**, 297
Sytilin, M. S., **2**, 423
Szabö, J., **14**, 715
Szamosi, J., **2**, 366
Szczygielska-Nowosielska, **2**, 245
Szeimies, G., **5**, 32; **11**, 83; **14**, 642
Szeimies-Seebach, U., **11**, 83
Szele, I., **6**, 4; **8**, 134
Szendroi, R., **4**, 267; **11**, 105
Szilagyi, S., **14**, 803
Szirovica, L., **3**, 203
Szmant, H. H., **12**, 188
Szuki, T., **4**, 175
Szymoniak, J., **1**, 136

Taba, K. M., **7**, 12
Tabata, A., **4**, 70; **12**, 79
Tabata, Y., **3**, 885a
Tabba, H. D., **14**, 652
Taber, T. R., **1**, 91, 93, 94; **10**, 188
Tabner, B. J., **3**, 1019–1021
Tabner, B. L., **3**, 1006
Tabushi, I., **2**, 237, 243
Tacconi, G., **13**, 141
Tada, M., **3**, 498; **14**, 590
Tadanier, J., **9**, 126
Taft, R. W., **9**, 153, 161
Tagaki, W., **2**, 234
Tagawa, S., **3**, 885a
Tagaya, H., **3**, 781
Tagle, B., **14**, 811
Tai, J. J., **14**, 410
Tajima, Y., **8**, 114
Takabe, K., **10**, 257
Takada, S., **14**, 75
Takadate, A., **2**, 241
Takagaki, T., **4**, 631
Takagi, H., **2**, 376
Takagi, K., **3**, 1077, 1098; **4**, 367
Takagi, M., **9**, 198
Takahashi, F., **9**, 182; **11**, 118
Takahashi, K., **3**, 571; **10**, 67, 173; **11**, 58; **14**, 88, 802, 809
Takahashi, M., **12**, 194; **14**, 595
Takahashi, T., **12**, 142, 247; **14**, 221, 517, 518
Takahashi, Y., **14**, 556
Takai, K., **4**, 147; **14**, 135
Talamuku, S., **3**, 1030; **9**, 111; **14**, 792, 793
Takarabe, K., **2**, 202
Takase, K., **10**, 67; **14**, 809
Takashita, K., **7**, 119
Takata, T., **4**, 258, 348, 370, 371, 664
Takatoku, K., **14**, 272
Takatsudo, S., **12**, 158, 159
Takaya, H., **12**, 194

Takayama, M., **14**, 519
Takayuki, A., **7**, 77
Takebe, Y., **14**, 761
Takeda, A., **10**, 81; **14**, 377
Takeda, M., **4**, 561
Takeda, S., **3**, 1030; **9**, 111
Takeda, T., **12**, 267
Takei, H., **10**, 242
Takeichi, T., **9**, 109
Takeshita, H., **3**, 501
Takeshita, K., **4**, 690
Takeuchi, H., **5**, 36; **9**, 79
Takeuchi, K., **3**, 141, 738; **8**, 93, 94; **9**, 29; **14**, 810, 850
Takeyasu, H., **12**, 19
Takigawa, T., **14**, 144
Takiguchi, H., **3**, 298
Takimoto, S., **6**, 57
Takisawa, N., **6**, 140, 141
Takita, Y., **3**, 1005
Talley, J. J., **3**, 881; **4**, 270; **8**, 20
Tam, S. W., **3**, 747
Tam, W., **6**, 84
Tamaki, K., **4**, 366
Tameo, K., **12**, 112
Tamaru, Y., **1**, 109; **10**, 194, 195; **12**, 225; **14**, 158, 159, 613
Tamazawa, K., **14**, 101
Tamblyn, W. H., **3**, 733
Tamborski, C., **3**, 620
Tammer, T., **13**, 138; **14**, 740
Tamura, M., **7**, 53, 77, 78
Tamura, R., **3**, 828, 829, 831; **4**, 700; **6**, 58; **11**, 114, 130
Tamura, S., **11**, 153
Tamura, T., **3**, 598
Tamura, Y., **7**, 133; **13**, 145; **14**, 375, 761
Tan, R. Y., **4**, 313
Tanabe, A., **2**, 229
Tanabe, F., **7**, 91
Tanabe, M., **14**, 32
Tanabe, Y., **9**, 15
Tanaka, H., **3**, 992; **14**, 144, 680
Tanaka, I., **3**, 162, 163
Tanaka, K., **9**, 15; **10**, 251; **14**, 534, 771
Tanaka, M., **14**, 198
Tanaka, S., **14**, 338
Tanaka, T., **4**, 586; **14**, 354, 584
Tanaskov, M. M., **11**, 149
Tanchuk, Yu. V., **12**, 185
Tandanier, J., **4**, 581
Tane, J. P., **13**, 211
Taneja, S. C., **14**, 467
Tang, F. Y., **3**, 301
Tang, R. H., **3**, 301
Tangerman, A., **13**, 53
Tanida, H., **6**, 77; **14**, 33
Tanigawa, S., **14**, 75
Tanigawa, Y., **9**, 14
Taniguchi, H., **3**, 919; **5**, 124; **14**, 690, 832
Taniguchi, Y., **2**, 206
Tanikaga, R., **3**, 831
Tanimoto, M., **3**, 863
Tanimoto, S., **2**, 196; **7**, 136; **9**, 192, 193; **10**, 237
Tanimoto, Y., **3**, 1160
Tanka, H., **3**, 608
Tanner, D. D., **3**, 298, 300, 741, 824
Tansumoto, K., **11**, 96
Tao, L.-X., **14**, 619
Tapodi, H. P., **4**, 220
Taraban'ko, V. E., **4**, 499, 500
Taran, P. N., **2**, 116
Tarasov, A. K., **2**, 169
Tarasova, N. P., **3**, 388–390
Tarassova, N. P., **3**, 100
Tardella, P. A., **14**, 688
Tarkhanyan, A. S., **12**, 69, 70
Tarnchompoo, B., **11**, 94
Tartakovskaya, L. M., **4**, 606
Tarygina, L. K., **3**, 694
Taschner, M. J., **4**, 605
Tashiro, M., **3**, 700; **4**, 138; **7**, 67; **14**, 88, 89
Tashiro, T., **3**, 61
Tatsumoto, K., **1**, 40, 41; **11**, 122
Taya, S., **14**, 208, 209
Taylor, D. R., **7**, 120, 121; **9**, 128, 129
Taylor, E. C., **4**, 69
Taylor, G. E., **14**, 235
Taylor, G. F., **13**, 103
Taylor, J. A., **4**, 741
Taylor, P. G., **7**, 80; **9**, 200
Taylor, P. W., **7**, 117; **14**, 54
Taylor, R., **2**, 134; **7**, 44–46; **11**, 31
Taylor, R. T., **12**, 208
Taylor, W. C., **14**, 117
Tchapla, A., **3**, 596
Tebby, J. C., **4**, 660
Tedder, J. M., **3**, 284, 372, 373, 616; **7**, 9, 109
Tedeschi, P., **3**, 577
Teeinga, H., **3**, 67
Tegmo-Larsson, I.-M., **12**, 264; **13**, 104, 105
Teitel'boim, M. A., **3**, 347
Teitelbaum, A. B., **9**, 220
Temnikova, T. I., **8**, 121; **9**, 54; **14**, 393
Tempesti, E., **4**, 450
Teo, K. C., **4**, 552, 553
Teodorovic, A. V., **4**, 100, 101
Teok, I., **3**, 762
Ter-Gabrielyan, V. G., **14**, 371
Terada, A., **14**, 65
Terada, I., **4**, 678
Terada, O., **2**, 236
Teramura, D. H., **14**, 140
Teramura, K., **13**, 61
Teranishi, S., **4**, 122
Terao, Y., **14**, 675
Terashima, K., **14**, 484
Terashima, M., **3**, 519
Tercio, J., **4**, 104
Terent'ev, A. B., **3**, 478
Termine, E. J., **1**, 75; **14**, 551
Ternovskoi, L. A., **12**, 12
Terpigorev, A. N., **3**, 818
Terpki, M. O., **4**, 614
Terrier, F., **1**, 159; **6**, 151, 164–166
Tertov, B. A., **14**, 86
Testaferri, L., **3**, 534, 1032, 1049
Teuber, H.-J., **14**, 766
Teufel, E., **13**, 201
Teulade, J.-C., **6**, 116
Teutsch, G., **14**, 521
Texier, F., **13**, 78; **14**, 314
Teyssie, P., **1**, 54
Teyssié, P. J., **5**, 72, 112
Tezuka, T., **3**, 120; **14**, 289
Thac, T. D., **3**, 147
Thaker, V. B., **3**, 757; **4**, 624
Thal, C., **4**, 338
Thappa, R. K., **14**, 467
Tharunaraj, G. V., **2**, 52
Thatcher, K. S., **14**, 116
Thea, S., **2**, 62, 417–420; **10**, 255; **11**, 23–26, 129
Thebtaranonth, Y., **3**, 509; **6**, 82; **11**, 94; **14**, 213, 283, 353
Theodore, L. J., **10**, 153
Thiel, J. M., **12**, 270
Thies, H., **8**, 127
Thigs, L., **14**, 660
Thind, S. S., **9**, 219
Thirumalai Perumal, P., **4**, 191
Thirunavukkarasu, A., **4**, 42
Thomas, C. B., **3**, 318, 319, 476, 477
Thomas, D. R., **5**, 107
Thomas, E. J., **3**, 496, 657; **4**, 599; **9**, 61
Thomas, G. J., **7**, 153
Thomas, M. T., **10**, 140

Thomas, W. R., **10**, 58
Thommen, W., **3**, 752; **4**, 627
Thompson, D. J., **8**, 37
Thompson, I., **7**, 153
Thompson, R. L., **3**, 538, 540
Thompson, R. S., **7**, 75
Thomson, C., **3**, 1010
Thomson, R., **3**, 366
Thomson, R. H., **3**, 443, 723, 1014; **7**, 25; **14**, 678
Thorn, D. L., **14**, 607
Thornber, C. W., **14**, 118
Thorsen, A., **4**, 746
Thorsen, P. T., **4**, 570
Threlkel, R. S., **5**, 158
Thuan, N. K., **3**, 775
Thurst, B. A., **5**, 71
Thyagarajan, G., **14**, 571
Thyvelikakath, G. X., **8**, 104
Tibbo, P., **9**, 33
Tidwell, T. T., **1**, 172; **11**, 140; **12**, 89
Tiecco, M., **3**, 532, 534, 1032, 1049; **6**, 31
Tigchelaar, M., **12**, 226, 243
Tikhonina, N. A., **14**, 744
Tikhonova, L. P., **4**, 82
Tikoo, P. K., **4**, 140
Tillett, J. G., **2**, 18; **9**, 77
Timberlake, J. W., **5**, 83; **14**, 803
Timewell, P. H., **8**, 58
Timmins, G., **14**, 12
Timofeeva, L. M., **9**, 78
Timpe, H.-J., **3**, 186; **13**, 65; **14**, 131
Tingoli, M., **3**, 534, 1032
Tinner, U., **3**, 851
Tipping, A. E., **14**, 358
Tise, F. P., **14**, 779
Tischchenko, I. G., **4**, 321; **9**, 207
Tishchenko, N. A., **4**, 17, 151
Tissot, P., **3**, 766
Titakchuk, T. A., **3**, 265
Titov, E. V., **2**, 441
Titz, M., **3**, 522
Tkáč, A., **3**, 122, 123, 859, 1035
Tkacheva, G. A., **4**, 496
Tkacheva, G. D., **4**, 486
Tkacheva, O. P., **3**, 600
Tmenov, D. V., **4**, 474
Tobe, Y., **5**, 31; **9**, 35, 36; **14**, 484, 485
Tobin, P. S., **4**, 72
Toczko, A. G., **10**, 283; **11**, 141; **12**, 180
Toda, H., **14**, 809
Todd, A., **14**, 658
Toder, B. H., **14**, 454
Todesco, P., **6**, 97
Todoriki, R., **11**, 153
Todres, Z. V., **3**, 1040; **14**, 781
Tohda, Y., **6**, 122, 123; **10**, 174
Toi, H., **4**, 564
Tojo, G., **4**, 641
Tokamaru, K., **4**, 453
Toki, S., **3**, 1030; **9**, 111
Tokisato, K., **13**, 58
Tokoroyama, T., **10**, 94
Tokuda, K., **2**, 224
Tokuda, M., **3**, 767
Tokuda, Y., **14**, 20
Tokumaru, K., **3**, 183, 624, 888; **4**, 459
Tolbert, L. M., **13**, 132
Tolman, R. L., **14**, 580
Toma, S., **12**, 166, 168
Toman, J., **14**, 848, 849
Tomaselli, G. A., **2**, 433, 444; **9**, 100, 101
Tomasik, P., **3**, 507
Tomasik, W., **4**, 154
Tomasz, J., **2**, 377
Tomaszweski, J. E., **4**, 551
Tomažič, A., **10**, 254
Tombo, G. M. R., **12**, 182
Tomilenko, E. I., **3**, 658, 659
Tomilin, O. B., **10**, 73
Tomioka, H., **4**, 147; **5**, 78, 90, 92, 94; **14**, 535, 536
Tomioka, K., **12**, 3
Tomita, K., **4**, 623
Tomita, Y., **14**, 520
Tomizawa, K., **3**, 515, 682, 1077, 1094; **4**, 330, 331, 367; **7**, 144
Tomoda, S., **8**, 17
Tomoi, M., **9**, 190, 191
Tomoto, A., **14**, 280
Tomoto, N., **6**, 142
Tondeur, J.-J., **2**, 455
Tonellato, U., **2**, 231; **12**, 7
Tongpenyai, N., **4**, 148
Toniolo, L., **12**, 106
Top, S., **8**, 48
Topor, M. G., **3**, 1065
Tor, Y., **7**, 20
Tordo, P., **3**, 54
Tori, K., **14**, 520
Toriyama, K., **3**, 941; **8**, 125
Toromanova-Petrova, P., **3**, 337
Torre, M., **12**, 82; **14**, 780
Torres, M., **3**, 557
Toshimitsu, A., **4**, 353; **11**, 108; **12**, 64
Tou, J. S., **13**, 124
Touchard, D., **3**, 413
Toullec, J., **1**, 140; **12**, 37
Touzot, P., **7**, 11
Tovstokhat'ko, F. I., **3**, 118; **4**, 470
Townsend, C. A., **10**, 125
Townson, M., **3**, 5, 1136
Toyotani, S., **9**, 36
Trach, S. S., **14**, 841
Traeger, J. C., **8**, 123
Traficante, D. D., **2**, 397
Trah, S., **13**, 212
Trainor, G. L., **2**, 239
Tramontini, M., **4**, 543
Trapani, G., **3**, 43; **4**, 177; **14**, 865
Trautmann, W., **3**, 568; **14**, 695, 696
Traylor, T. G., **4**, 268, 269; **11**, 99, 112
Traynham, J. G., **3**, 285
Traynor, S. G., **14**, 594
Trazza, A., **3**, 779
Treger, Yu. A., **3**, 287; **4**, 320, 322
Treichel, P. M., **3**, 712
Treindl, L., **4**, 68, 85, 86, 108
Tremelling, M. J., **3**, 800; **6**, 18
Tremmel, S., **13**, 63
Trenerry, V. C., **8**, 139, 143; **10**, 54
Tret'yakov, V. P., **4**, 151
Trevor, P. L., **3**, 50
Triana, J. L., **2**, 135; **11**, 34
Tribunescu, P., **2**, 70, 71
Trice, C. M., **14**, 200
Triebe, F. M., **3**, 1053
Trigg, J., **13**, 11
Trill, H., **3**, 47
Trimm, D. L., **4**, 439
Trinajstic, N., **8**, 98; **10**, 71
Tripolone, M., **2**, 433; **9**, 101
Trippett, S., **2**, 332; **14**, 183
Trivedi, B. K., **14**, 589
Trivedi, K. N., **14**, 169
Trofimov, B. A., **12**, 2, 90, 186
Trogen, L., **10**, 65
Troin, Y., **3**, 1089
Troisi, L., **4**, 309
Trompenaars, W. P., **13**, 36
Tronchet, J. M. J., **3**, 597; **9**, 125
Troostwijk, C. B., **4**, 707
Trost, B. M., **4**, 176, 181, 617; **13**, 131; **14**, 252, 452, 611, 649
Tröster, G., **4**, 63
Trostmann, U., **14**, 317, 318
Troyanskii, E. I., **3**, 710, 721,

722; **4**, 196; **14**, 625
Trub, E. P., **12**, 46
Truce, W. E., **14**, 361, 362
True, N. S., **14**, 775
Truong, T. K., **2**, 113
Trushkova, L. V., **3**, 1108
Tsai, Y.-M., **14**, 202
Tse, A., **10**, 139
Tse, M. W., **3**, 485
Tselinskii, I. V., **2**, 148, 149, 151, 152; **3**, 818, 882; **9**, 169, 170; **10**, 334–337
Tseng, K.-S., **14**, 836
Tsitini-Tsamis, M., **4**, 110
Tsivunin, V. S., **1**, 173
Tsoi, L. A., **14**, 736
Tsoi, S. C., **5**, 137, 138; **11**, 93; **14**, 269
Tsou, H. R., **1**, 163
Tsubokura, Y., **13**, 84
Tsubuki, M., **10**, 246; **11**, 124
Tsuchihashi, G., **10**, 244; **14**, 419
Tsuchiya, T., **14**, 66, 68, 69, 363
Tsuda, T., **10**, 157
Tsuda, Y., **14**, 575, 711, 739
Tsugaru, T., **14**, 485
Tsuhako, A., **4**, 178
Tsuji, A., **2**, 181
Tsuji, H., **3**, 746
Tsuji, J., **12**, 247
Tsukamoto, Y., **4**, 360
Tsukurimichi, E., **11**, 100
Tsumori, K., **3**, 1030; **9**, 111
Tsunashima, S., **3**, 411
Tsuno, T., **4**, 703
Tsuno, Y., **2**, 29; **9**, 265
Tsuruta, H., **14**, 820
Tsuruta, T., **9**, 109
Tsuruya, S., **4**, 52
Tsutsumi, K., **14**, 655
Tsuyuki, T., **14**, 517, 518
Tsuzuki, H., **13**, 17
Tsuzuki, N., **3**, 954
Tsvetkov, O. N., **7**, 122
Tucek, E., **2**, 48
Tucker, H., **10**, 116
Tuddenham, D., **14**, 271
Tudos, F., **3**, 637
Tukada, H., **14**, 254
Tulip, T. H., **14**, 607
Tully, F. P., **3**, 538
Tumanskii, B. L., **3**, 672
Tumas, W., **10**, 284; **11**, 16
Tundo, A., **3**, 572; **5**, 130
Turco, A., **3**, 850
Turecek, F., **12**, 33; **14**, 526
Turner, A. B., **4**, 410
Turner, E. S., **3**, 652
Turner, J. J., **5**, 17
Turner, J. V., **14**, 211
Turovskii, A. A., **3**, 130
Turro, N. J., **3**, 734, 1157–1161, 1163; **4**, 393; **5**, 53
Tutubalina, V. P., **4**, 345
Tuulmets, A., **1**, 156
Tuzhikov, O. I., **9**, 53
Tuzikov, A. B., **6**, 154
Tuzovskaya, S. A., **2**, 55
Tyeklar, Z., **4**, 49, 325
Tyman, J., **6**, 65
Tyrrell, N. D., **3**, 496, 657; **4**, 599
Tyupalo, N. F., **4**, 292
Tyurina, L. A., **12**, 67, 68
Tzeng, D., **3**, 749; **4**, 635; **5**, 153; **14**, 717

Uchida, T., **3**, 1090; **13**, 70; **14**, 881
Uchida, Y., **4**, 31; **14**, 549
Uchimura, S., **3**, 981
Uda, H., **14**, 122, 257
Udrenaite, E., **7**, 14
Ueda, H., **3**, 912; **5**, 90; **13**, 12; **14**, 536
Ueda, K., **3**, 681; **4**, 133
Ueda, N., **6**, 173; **14**, 279
Ueda, T., **14**, 75
Ueda, Y., **3**, 558
Uehara, K., **14**, 81
Uejima, A., **9**, 69
Uekama, K., **2**, 241
Uemura, S., **4**, 70, 353; **11**, 108; **12**, 64, 79
Uenishi, J., **7**, 133
Ueno, A., **14**, 802
Ueno, K., **4**, 421; **13**, 157
Ueno, Y., **4**, 215
Ueoka, R., **2**, 220, 224, 226
Ugen, D., **10**, 249
Ugi, I., **7**, 135
Uhlar, L., **12**, 101
Uhm, T. S., **2**, 165
Ujjainwalla, M., **12**, 58
Ulrici, B., **3**, 1009
Uma, K. V., **4**, 237
Umaeva, V. S., **13**, 98
Umano, K., **13**, 58; **14**, 824
Umeda, M., **3**, 624
Umehara, Y., **3**, 578
Umeyama, H., **2**, 269
Umezu, M., **14**, 17
Umminger, I., **3**, 958
Ung-Chhun, S. N., **1**, 26
Uno, T., **1**, 178
Uosaki, Y., **13**, 114
Urabe, H., **4**, 207
Urabe, K., **12**, 91
Uray, G., **1**, 16; **6**, 110; **9**, 273
Urbanski, J., **9**, 68
Uribe, G., **12**, 128
Uryadova, L. F., **13**, 94
Usha, R., **14**, 490, 491
Usmani, A. A., **13**, 178
Ustavshchikov, B. F., **4**, 488
Ustynyuk, N. A., **10**, 50, 51; **14**, 599
Ustynyuk, Y. A., **10**, 50, 51; **14**, 599
Utamapanya, S., **11**, 94
Utkin, O. V., **9**, 90
Utley, J. H. P., **3**, 748, 754, 760
Uyehara, T., **3**, 571; **11**, 58
Uzhik, O. N., **9**, 244
Uzienko, A. B., **3**, 731

V'yunov, K. A., **9**, 53
Vaccher, C., **6**, 52
Vachkov, K. V., **12**, 52
Vaden, A. K., **8**, 141; **9**, 9
Vaidya, S. R., **11**, 56
Vairamani, M., **14**, 403
Valko, J. T., **14**, 453
Vallari, R. C., **4**, 757
Vallés, A., **14**, 778
Valnot, J. Y., **6**, 106
Valt, M.-H., **6**, 10
Van Bekkum, H., **4**, 687
Van Cantfort, C. K., **14**, 435
Van Den Bril, M., **12**, 257, 258
Van Der Weerdt, A. J. A., **14**, 857
Van Derveer, D., **1**, 85; **10**, 181
Van Doorn, R., **8**, 129; **13**, 42
van Eikeren, P., **1**, 3
Van Engen, D., **11**, 61
Van Etten, R. L., **2**, 16, 401, 452
Van Heerden, C., **1**, 68
Van Horn, D. E., **12**, 109
Van Koten, G., **3**, 60
Van Rantwijk, F., **4**, 687
Van Rijn, P. E., **10**, 47
Van Swaaj, W. P. M., **7**, 97
Van Tamelen, E. E., **4**, 694
Van Tilborg, M. W. E. M., **8**, 129; **13**, 42
Van Vliet, N. P., **14**, 524
Van de Griendt, F., **7**, 47
Van der Kerk, G. J., **4**, 699
Van der Ven, J., **10**, 47
Vancheesan, S., **4**, 685
Vandenbulck-Coyette, B., **13**, 149

Vandendungen, G., **2**, 455
Vander Wegen, P. G., **9**, 31
Vanderese, R., **4**, 596
Vandewalle, M., **14**, 287
Varah, J., **7**, 105
Varelas, M. A., **14**, 687
Varescon, F., **4**, 120
Vargaftik, M. N., **4**, 164, 484
Varma, M., **11**, 18
Varma, R. S., **10**, 317
Vartanian, T., **3**, 1046
Vartanyan, M. M., **14**, 749
Varughese, K. I., **14**, 491
Varveri, F. S., **9**, 135
Vasaros, L., **6**, 64
Vasconi, G., **4**, 419
Vasil'ev, G. N., **3**, 461
Vasil'ev, V. P., **4**, 10
Vasil'eva, V. N., **4**, 10
Vassilatos, S., **10**, 258
Vasvari, G., **3**, 156
Vathke-Ernst, H., **13**, 133, 134, 172
Vaughan, J., **7**, 75
Vaughan, K., **3**, 210
Vaultier, M., **14**, 452
Vavilova, A. N., **12**, 90
Vazeux, M., **14**, 669
Večeřa, M., **2**, 34, 122, 123, 125; **11**, 22
Vedejs, E., **11**, 109
Vedeneev, V. I., **3**, 347
Veenstra, S. J., **14**, 265
Veeranagaiah, V., **1**, 51
Veiner, L. M., **3**, 18
Velibekova, D. S., **3**, 710; **4**, 196; **14**, 625
Velichko, L. I., **7**, 98
Veltman, J., **4**, 396
Velyutin, L. P., **3**, 116; **4**, 476
Venkatarao, K., **2**, 254
Venkatasubban, K. S., **9**, 186
Venkatasubramaniam, K. G., **10**, 301
Venkatasubramanian, N., **4**, 235; **7**, 24
Venkatesan, K., **14**, 342, 490, 491
Venkateshwar, S., **11**, 151
Venugopal, V. K., **14**, 571
Venzo, A., **10**, 49
Verbicky, J. W., **2**, 189, 190
Verhe, R., **14**, 378
Verhoeven, J. W., **3**, 555, 774
Verkruijsse, H. D., **10**, 124; **14**, 348
Verma, R. D., **12**, 16
Verma, R. G., **4**, 91
Verma, S. M., **12**, 28
Vermeer, P., **9**, 18; **10**, 121; **12**, 226, 243
Vernon, C. A., **2**, 371
Vernon, J. M., **6**, 171; **14**, 327
Veronese, A. C., **10**, 141
Veschambre, H., **4**, 730
Veselova, M. E., **4**, 686
Veslov, V. V., **9**, 67
Vesnovskaya, G. I., **3**, 94, 94; **10**, 92
Vessiere, R., **14**, 192
Vest, R. D., **6**, 128
Vetter, W., **14**, 821
Vevert, J.-P., **14**, 142
Vial, M. V., **9**, 243
Viallefont, P., **14**, 509
Viatle, M. A., **14**, 129
Vidal, M., **1**, 105; **3**, 293
Vidari, G., **14**, 503
Viehe, H. G., **11**, 59; **13**, 149
Vieira, R. A., **1**, 113
Vieta, R. S., **14**, 133
Vietti, D. E., **1**, 37
Vigalok, A. A., **9**, 99
Vigalok, I. V., **2**, 450; **9**, 99
Viger, A., **14**, 511
Vigolok, A. A., **2**, 450
Vijayalakshmi, **4**, 244, 245
Viktorova, E. A., **14**, 162, 163
Viladoms, P., **3**, 806
Vilarrasa, J., **5**, 10
Villa, V., **9**, 121
Villain, G., **2**, 110
Villareal, J. A., **3**, 502
Villaveces, J.-L., **1**, 70
Villenave, J. J., **3**, 99, 100, 125, 146
Vincens, M., **1**, 105; **3**, 293
Vincent, C. A., **3**, 772
Vincent, M. A., **3**, 109; **11**, 156; **14**, 770
Vinnick, F. S., **4**, 418
Vinnik, M. I., **2**, 260
Vinogradova, L. E., **8**, 50
Vinokurov, A. I., **2**, 384; **12**, 53
Viola, A., **14**, 184, 305, 306
Viout, P., **2**, 217, 218
Virtanen, P. O. I., **8**, 80
Vishwakarma, L. C., **12**, 125
Vismara, E., **3**, 226
Visser, A. J. W. G., **3**, 962
Visser, R. G., **10**, 120
Vitale, A. A., **10**, 107
Vitale, M. A., **13**, 123
Viteva, L., **12**, 167, 244
Viteva, L. Z., **12**, 245
Vittimberga, B. M., **3**, 911
Vittinghoff, K., **14**, 420
Vitullo, V. P., **8**, 131; **9**, 136
Vivekananda, S., **2**, 254
Vivona, N., **14**, 95–97
Vizgert, R. V., **2**, 421, 422, 430, 437
Vladuchick, S. A., **6**, 128; **14**, 92
Vlasov, V. M., **10**, 306
Vlasova, N. N., **12**, 186
Voet, J. G., **4**, 753
Voevodskaya, M. V., **3**, 735
Vogel, E., **1**, 91; **10**, 188
Vogel, P., **9**, 32; **13**, 111; **14**, 295, 412
Vogt, H.-H., **1**, 107; **10**, 19, 162
Vol'eva, V. B., **3**, 867
Volkov, A. N., **12**, 90
Volkov, M. N., **4**, 489
Volkov, N. D., **3**, 375
Volkov, R. N., **3**, 158
Vollano, J. F., **2**, 454
Vollhardt, K. P. C., **6**, 84; **13**, 209–211
Volodarskii, L. B., **3**, 600
Volodkovich, S. D., **2**, 160
Volosov, A. P., **10**, 23
Volovik, S. V., **3**, 357, 370, 436, 533, 535
Volz, H., **8**, 78
Von Szentpaly, L., **7**, 2
von Wallis, H., **3**, 113
Vorbruggen, H., **1**, 28
Vorkunova, E. I., **3**, 1072
Vorob'ev, N. K., **2**, 164
Vorob'eva, N. S., **4**, 299
Voronkov, M. G., **2**, 343, 344; **3**, 430; **9**, 212; **12**, 186
Vorotyntsev, V. M., **4**, 527
Voskanyan, V. S., **14**, 352
Voss, J., **3**, 1018
Vrba, Z., **6**, 85
Vrieze, K., **3**, 60
Vyas, D. N., **4**, 16
Vyazankin, N. S., **3**, 1023
Vylyudnova, S. M., **4**, 116
Vysotskii, Yu. B., **9**, 237, 238
Vysotsky, Yu., **7**, 8
Vyunov, K. A., **12**, 9, 13–15

Willhalm, B., **4**, 627
Waali, E. E., **5**, 12, 22
Wachi, K., **14**, 65
Wada, F., **6**, 15, 16
Wada, M., **6**, 108
Waddington, D. J., **3**, 322
Wade, T. N., **14**, 421
Wadsworth, W. S., **2**, 358
Waegell, B., **3**, 454
Waeschke, H., **2**, 336
Wagner, G., **14**, 78, 102

Wagner, H. G., **3**, 242, 302, 311
Wagner, P. J., **3**, 488, 582, 583; **11**, 154; **14**, 887
Wagner, R. D., **3**, 181
Wagner, S., **6**, 144
Wagner-Jauregg, T., **13**, 159
Waigh, R. D., **14**, 114
Wakamatsu, S., **14**, 297
Wakatsuki, Y., **14**, 830
Wake, S., **5**, 127
Wakselman, C., **3**, 374
Walborsky, H. M., **3**, 8, 861a
Waldner, A., **8**, 3, 12; **9**, 19, 20, 22, 25, 215
Waley, S. G., **2**, 303
Walker, A., **10**, 104
Walker, D. G., **10**, 91
Walker, G., **8**, 127
Walker, G. E., **8**, 62
Walker, R. W., **3**, 160
Walker, T., **3**, 1019, 1021
Wallenfels, K., **3**, 335
Wallerberg, G., **2**, 351
Walling, C., **3**, 707; **4**, 132
Wallis, D. I., **3**, 445; **14**, 164
Wallis, J. D., **8**, 108
Walsh, E., **10**, 312
Walsh, P. A., **1**, 172
Walsh, R., **5**, 16; **14**, 698
Walsh, T. D., **3**, 736; **4**, 633
Walter, **14**, 725
Walters, C. P., **4**, 582
Walton, D. J., **3**, 772
Walton, D. R. M., **10**, 154, 291
Walton, J. C., **3**, 25, 48, 238, 284, 372, 373, 479, 491
Walz, P., **1**, 104
Wamhoff, H., **14**, 82
Wan, P., **4**, 504
Wan, J. K. S., **3**, 44, 78, 79, 1176, 1177
Wan, P., **2**, 88, 91
Wang, A., **9**, 158
Wang, C.-H., **4**, 343
Wang, C.-L. A., **2**, 277
Wang, H.-C., **3**, 36
Wang, J. K., **1**, 44
Wang, J. T., **3**, 955, 956, 983, 1025
Wang, K. K., **12**, 126
Wang, T., **13**, 200
Wang, T.-F., **14**, 201
Wang, X., **14**, 582
Wang, X.-M., **14**, 619
Wang, Y.-P., **2**, 331
Wann, S. R., **1**, 10; **4**, 570
Wanner, I., **13**, 176
Ward, A. D., **14**, 181
Ward, C. E., **8**, 26
Ward, D. R., **3**, 582; **11**, 154
Ward, K. B., **2**, 317
Ward, M. D., **4**, 691
Ward, R. S., **4**, 273; **14**, 116, 510
Warden, J. T., **3**, 924
Warin, R., **5**, 112
Warin, V., **6**, 52
Waring, A. J., **14**, 112, 113
Warkentin, J., **3**, 208, 209, 494; **5**, 28; **11**, 62, 65
Warner, L. G., **2**, 244
Warner, P., **5**, 81; **11**, 85
Warnhoff, E. W., **1**, 121; **10**, 40; **12**, 39; **14**, 336, 568
Warren, C. T., **2**, 41
Warren, S., **1**, 97; **14**, 547
Warren, S. E., **3**, 425, 426
Warrener, R. N., **4**, 313; **13**, 107, 187
Warrick, P., **10**, 273
Warshel, A., **2**, 268
Wartski, L., **12**, 219, 224
Washburne, S. S., **4**, 267; **11**, 105
Washida, N., **4**, 382
Washio, M., **3**, 885a
Wasiowich, C. A., **14**, 621
Wasserman, H. H., **4**, 390, 418; **13**, 3; **14**, 729, 746
Wastell, A., **2**, 278
Watabe, T., **10**, 64
Watanabe, K., **10**, 157; **14**, 689
Watanabe, K. A., **6**, 124; **14**, 77
Watanabe, M., **2**, 367; **4**, 310; **9**, 81; **14**, 550
Watanabe, S., **13**, 157
Watanabe, T., **14**, 81
Watanabe, Y., **3**, 714; **14**, 354
Watanage, H., **4**, 595
Watanuki, T., **3**, 297
Watarai, H., **1**, 161
Waterman, P., **14**, 243
Waters, B. W., **4**, 741
Waters, D. N., **7**, 126
Watkin, D. J., **8**, 108
Watson, C. R., **3**, 1034
Watson, P. L., **13**, 187
Watson, R. T., **3**, 151
Watson, T. W., **1**, 137
Watson, W. H., **14**, 441
Watt, I., **14**, 340, 859, 861
Wattanasin, S., **3**, 755; **4**, 638; **7**, 116; **14**, 440
Watts, H. P., **11**, 67
Watts, R. D., **1**, 168
Watts, W. E., **2**, 56; **6**, 83; **8**, 46, 47; **11**, 139
Wautier, H., **1**, 170; **8**, 144
Wayda, A. L., **10**, 211
Weavers, R. T., **3**, 862
Webb, T. R., **4**, 682
Webber, A., **3**, 748
Webber, B. D., **4**, 694
Weber, G., **14**, 87
Weber, J. D., **14**, 685
Weber, R., **2**, 397
Weber, R. V., **6**, 8
Weber, W. P., **3**, 749; **4**, 635; **5**, 148, 153; **14**, 717
Webster, O. W., **3**, 1017; **6**, 128; **14**, 92
Wedemeyer, K., **4**, 502
Weed, G. C., **3**, 1160
Weedon, A. C., **14**, 712
Weeks, D. P., **2**, 42
Wege, D., **5**, 91, 114; **6**, 174; **14**, 700
Wegmann, H., **12**, 250
Wegner, G., **3**, 948
Wei, W., **7**, 3
Weichsel, C., **14**, 90
Weidig, C., **1**, 72
Weidmann, K., **13**, 212
Weierhasuer, G., **14**, 292
Weigand, J., **14**, 153
Weimgartner, M., **13**, 60
Weiner, M., **14**, 110
Weinstock, J., **10**, 133
Weir, D., **3**, 78
Weisleder, D., **3**, 171
Weisman, G. R., **4**, 544
Weiss, R., **4**, 749
Weissenfels, M., **3**, 1009
Weitz, E., **14**, 805
Welch, S. C., **4**, 646
Wellander, L. L., **11**, 110
Weller, T., **10**, 166
Welsby, S. E., **3**, 702
Welsh, K. M., **2**, 412
Welsh, W. A., **4**, 125
Wender, P. A., **13**, 206; **14**, 194, 195
Wenderoth, B., **10**, 155
Wendisch, D., **3**, 570
Wenkert, E., **12**, 218; **14**, 477
Wenninger, J., **5**, 136
Wentrup, C., **14**, 814
Werbin, H., **4**, 429
Wermeckes, B., **10**, 112
Werstiuk, N. H., **14**, 12
Wertheimer, V., **14**, 474
Wertz, P. W., **1**, 20
Wertz, W., **1**, 20
West, R., **10**, 214, 215; **14**, 885
Westdemiotis, C., **8**, 126

Westdrop, I., **4**, 643
Westermann, J., **10**, 155
Westheimer, F. H., **1**, 157; **2**, 320, 322
Westley, J., **2**, 405
Westman, T. R., **11**, 40
Westmijze, H., **9**, 18; **10**, 121
Wettach, R. H., **12**, 161
Wetter, W. P., **5**, 54
Weyerstahl, P., **9**,74
Wharry, D. L., **3**, 870
Wharton, C. W., **2**, 292
Wheeler, D. M. S., **4**, 72; **14**, 51
Wheeler, M. M., **4**, 72
White, A. H., **14**, 328
Wite, C. T., **1**, 85; **10**, 181, 182
White, E. H., **2**, 283; **14**, 543
White, J. D., **14**, 488
White, K. B., **4**, 225
White, M. R., **5**, 80
White, R. E., **4**, 754
White, S., **10**, 178, 179
White, V. A., **14**, 23
Whiteley, C. G., **10**, 159
Whiteside, R. A., **8**, 154, 156
Whitesides, G. M., **4**, 711
Whitham, G. H., **9**, 61
Whitney, D. B., **2**, 42
Whittaker, D., **14**, 442, 443
Whittaker, G., **3**, 355
Whittall, J., **14**, 326
Whittle, E., **3**, 343, 344
Whybrow, D., **14**, 883
Wiberg, K. B., **1**, 9; **3**, 1115; **9**, 38; **14**, 188, 539
Wickham, G., **10**, 339
Wickremesinghe, L. K. G., **11**, 137
Widdowson, D. A., **10**, 318
Wiegers, K. E., **4**, 547
Wiering, P. G., **13**, 87
Wiggins, P. L., **8**, 24
Wight, C. A., **10**, 56
Wightman, R. H., **6**, 114; **14**, 56
Wilbur, D. S., **13**, 30
Wildbredt, D.-A., **11**, 66; **13**, 168; **14**, 838
Wilde, H., **14**, 87
Wildi, E. A., **11**, 61
Wilemon, G. M., **12**, 105
Wilgis, F. P., **8**, 131
Wilhelm, E., **13**, 173; **14**, 26, 702
Wilkins, C. W., **12**, 201, 203
Willcott, M. R., **12**, 39
Willen, B. H., **5**, 2
Willhalm, B., **3**, 752
Williams, A., **2**, 187, 417–420; **10**, 255; **11**, 23–26
Williams, D. H., **14**, 173, 396
Williams, D. J., **1**, 18; **5**, 138; **10**, 318; **14**, 269
Williams, D. L. H., **7**, 86; **10**, 329; **14**, 3
Williams, D. R., **1**, 108; **10**, 178, 179, 245
Williams, E., **14**, 503
Williams, F., **3**, 955, 956, 983, 1025, 1044
Williams, G. H., **3**, 517
Williams, I. H., **1**, 115; **9**, 139, 234; **11**, 144
Williams, J. C., **14**, 289
Williams, J. L., **14**, 614
Williams, J. R., **14**, 855, 856
Williams, K. P. J., **3**, 428, 961
Williams, L., **2**, 189, 190
Williams, M. R., **4**, 286
Williams, N. E., **2**, 375; **14**, 742
Williams, P. A., **8**, 29
Williams, P. D., **4**, 336
Williams, P. S., **3**, 321, 451, 495, 546, 968
Williams, R. C., **2**, 191
Williams, R. P. J., **3**, 950
Williams, R. V., **14**, 671
Willis, J. P., **4**, 14
Willis, W. W., **10**, 225
Willson, R. L., **3**, 878, 1139
Wilson, A. A., **7**, 16
Wilson, C. E., **3**, 107; **14**, 708
Wilson, D. A., **1**, 73
Wilson, D. R., **14**, 733
Wilson, I., **6**, 65
Wilson, J. C., **9**, 241; **14**, 408
Wilson, J. W., **4**, 559
Wilson, P., **13**, 23
Wilson, P. M., **6**, 150, 162
Wilson, S. R., **10**, 235; **11**, 155; **12**, 187
Wilson, T., **3**, 106
Wilt, J. W., **3**, 442; **9**, 40
Wimalasena, J. H., **3**, 1102
Winans, R. E., **7**, 19
Wine, P. H., **3**, 312, 538
Wink, D. J., **10**, 211
Winter, G., **14**, 165
Winter, J. G., **14**, 23
Winter, W. J., **3**, 1080
Winter-Mihaly, E., **3**, 597
Winterbourn, C. C., **4**, 769
Winterfeldt, E., **12**, 253
Wintermayr, H., **3**, 1033; **14**, 638
Winters, J. N., **3**, 1175
Wirth, D., **8**, 162; **14**, 304
Wirth, D. D., **3**, 977; **13**, 121
Wiseall, B., **3**, 1148
Wistuba, E., **14**, 424, 432
Witczak, Z., **14**, 752
Witholt, B., **4**, 737
Witkiewicz, P. L., **4**, 61
Witko, M., **4**, 441
Wladislaw, B., **14**, 548
Wlodawer, A., **2**, 411
Wolf, H. R., **3**, 1070; **5**, 30; **14**, 316, 653–657, 888
Wolf-Rüdiger, K., **11**, 78
Wolfbeis, O. S., **1**, 16
Wolfe, J. F., **3**, 807; **6**, 20; **10**, 132
Wolfe, N. L., **2**, 338
Wolfe, S., **4**, 30
Wolfert, P., **14**, 676
Wolff, S., **3**, 438; **14**, 540
Wolinsky, J., **4**, 535
Wolsehann, P., **1**, 118
Wolters, F. C., **14**, 699
Wolters, J., **7**, 141
Wolve, S., **4**, 27
Wong, C.-H., **4**, 711
Wong, C. K., **14**, 67
Wong, J. L., **13**, 146
Wong, J. P. K., **13**, 74
Wong, M. Y., **5**, 18
Wong, P. C., **3**, 489, 750, 903; **5**, 5
Wong, S. K., **3**, 245, 1171, 1174
Woo, S. O., **3**, 366
Wood, D. L., **3**, 787; **5**, 150
Wood, I. T., **3**, 1135
Woodgate, P. D., **3**, 865; **9**, 114, 115; **12**, 40, 56; **14**, 171, 172, 683
Woodgate, S. D., **12**, 40
Woodroffe, D., **5**, 82
Woods, G. F., **14**, 386
Woods, S. P., **7**, 69
Woodthorpe, K. L., **4**, 103; **5**, 69, 129
Woolard, F. X., **4**, 265
Woolley, G. T., **9**, 17
Word, J. M., **10**, 299
Worley, S. D., **9**, 253
Worsfold, D. J., **10**, 34
Woynar, H., **3**, 65, 352, 958
Wrackmeyer, B., **8**, 59; **12**, 122, 123
Wright, C. W., **5**, 22
Wright, J. N., **4**, 742
Wright, J. S., **10**, 299
Wright, T. L., **7**, 139
Wrobel, J. E., **4**, 562
Wroblewski, J., **14**, 138

Wroblowsky, H.-J., **14**, 468, 469
Wu, A., **10**, 140
Wu, H.-L., **2**, 278, 279
Wu, S.-L., **11**, 9
Wu, S.-P., **3**, 592
Wubbels, G. G., **3**, 92
Wuest, J. D., **6**, 117
Wunderlich, H., **4**, 63
Wyckoff, H. W., **2**, 408
Wykypiel, W., **14**, 37
Wyman, L., **3**, 1066
Wynberg, H., **1**, 8; **3**, 701; **4**, 137, 449, 737; **12**, 191, 192

Xu, Y.-D., **14**, 619
Yabe, A., **5**, 121
Yablokov, V. A., **14**, 395
Yablokova, N. V., **3**, 93
Yabushita, S., **4**, 401
Yadav, G. D., **9**, 185
Yadav, J., **10**, 131
Yadav, J. R., **7**, 39
Yadav, R. L., **4**, 91
Yadava, K. L., **7**, 39
Yaeger, D. B., **3**, 409; **12**, 21, 36
Yaggi, N. F., **11**, 21
Yagi, H., **4**, 315
Yagoub, A. K., **11**, 134
Yagupol'skii, L. M., **3**, 375; **6**, 145, 146
Yagyu, M., **14**, 575
Yakobi, V. A., **4**, 293
Yakobson, G. G., **3**, 704; **10**, 306
Yakushkina, N. I., **11**, 135
Yamabe, S., **4**, 136
Yamada, A., **10**, 90
Yamada, F., **3**, 650
Yamada, H., **2**, 258
Yamada, K., **1**, 158; **2**, 221; **4**, 561; **14**, 338
Yamada, M., **4**, 623
Yamada, T., **2**, 367; **9**, 180
Yamada, Y., **3**, 578; **4**, 67; **12**, 153
Yamaguchi, K., **3**, 716; **4**, 401, 402
Yamaguchi, M., **6**, 57
Yamaguchi, R., **3**, 296; **10**, 340, 341; **14**, 623
Yamaguchi, S., **9**, 203
Yamaguichi, M., **13**, 55
Yamaichi, A., **10**, 46; **14**, 833
Yamakado, Y., **1**, 100; **11**, 104
Yamakawa, M., **13**, 20
Yamaki, T., **14**, 332
Yamamoto, G., **3**, 294
Yamamoto, H., **3**, 871; **4**, 702, 704; **11**, 104
Yamamoto, J., **14**, 17
Yamamoto, K., **12**, 112
Yamamoto, M., **3**, 918
Yamamoto, T., **11**, 89
Yamamoto, Y., **1**, 96, 97; **4**, 283, 564; **10**, 186, 193
Yamamura, K., **2**, 237
Yamamura, S., **4**, 74
Yamana, T., **2**, 181
Yamanaka, T., **7**, 129; **12**, 99
Yamane, N., **11**, 47, 48
Yamashita, A., **6**, 82
Yamashita, T., **4**, 729
Yamashita, Y., **3**, 914
Yamataka, H., **9**, 137, 241; **11**, 12; **14**, 408, 409
Yamato, A., **14**, 645
Yamato, H., **4**, 710
Yamato, T., **7**, 67
Yamauchi, M., **10**, 37, 38
Yamauchi, Y., **9**, 226
Yamawaki, J., **14**, 409
Yamazaki, A., **14**, 101
Yamazaki, H., **14**, 830
Yamazaki, N., **4**, 557, 558, 560
Yamishita, T., **1**, 144
Yan, J. S., **1**, 83
Yanada, R., **2**, 376
Yanchuk, N. I., **6**, 59
Yáñez, R., **14**, 508
Yang, D.-D. H., **13**, 200
Yang, G. C., **3**, 602, 643
Yang, K., **2**, 82; **9**, 229
Yang, N. C., **13**, 200
Yang, S.-Y., **4**, 343
Yang, Y.-C., **9**, 168
Yannoni, C. S., **8**, 61
Yano, H., **3**, 599
Yano, K., **5**, 15; **9**, 41; **14**, 101
Yano, T., **3**, 71
Yapp, C. J., **2**, 258
Yarham, A. C., **5**, 18
Yarkov, S. P., **3**, 636
Yashima, H., **8**, 18; **14**, 595
Yashima, T., **3**, 900
Yasman, Ya. B., **8**, 99, 101; **12**, 20
Yasmenko, A. I., **3**, 1087
Yasnikov, A, A., **3**, 731; **4**, 142
Yasuda, A., **12**, 194
Yasuda, H., **10**, 37, 38; **12**, 164
Yasuda, M., **3**, 897, 898; **4**, 549, 565, 705; **11**, 49; **13**, 117, 180, 183, 184
Yasui, S., **4**, 713, 715
Yasunami, S., **2**, 202
Yatagagi, H., **1**, 96
Yatagai, H., **10**, 186
Yatchishin, I. I., **4**, 508
Yatchishin, I. K., **4**, 358
Yates, K., **2**, 17, 89–92; **4**, 504; **12**, 31
Yates, P., **2**, 88; **13**, 113; **14**, 471
Yathirajan, H. S., **4**, 238
Yatsimirskii, A. K., **2**, 174; **4**, 162, 163
Yatsimirskii, K. B., **4**, 82, 121
Yavari, F., **13**, 115
Yazawa, T., **10**, 157
Yeh, M. K., **10**, 55
Yeung Lam Ko, Y. Y. C., **13**, 56
Yeung, C. Y. A., **14**, 473
Yeung, D. W. K., **3**, 208
Yianni, P., **13**, 195
Yip, R. W., **3**, 1074
Yoda, N., **4**, 668
Yoder, C. Y., **2**, 83
Yoh, S.-D., **9**, 266
Yokoe, I., **14**, 889
Yokoi, T., **14**, 519
Yokomatsu, T., **14**, 156
Yokomichi, Y., **3**, 738; **8**, 93, 94
Yokota, K., **7**, 22
Yokoyama, H., **14**, 553
Yokoyama, K., **6**, 76; **14**, 27
Yokoyama, M., **1**, 180; **14**, 553
Yokoyama, Y., **14**, 517, 518
Yonebayashi, F., **14**, 527
Yoneda, F., **4**, 167, 710, 719
Yoneda, Y., **12**, 19
Yonowich-Weiss, M., **14**, 57
Yoo, H. S., **11**, 71
Yoon, S. K., **2**, 165
York, J. L., **1**, 24
Yoshida, C., **3**, 519
Yoshida, M., **3**, 624, 746, 1168; **14**, 822
Yoshida, N., **4**, 174
Yoshida, T., **3**, 571; **7**, 53, 77, 78; **11**, 58; **12**, 110
Yoshida, Y., **12**, 209
Yoshida, Z-I., **14**, 158, 159
Yoshida, Z., **1**, 109; **10**, 194, 195; **12**, 225; **14**, 613
Yoshifuji, M., **10**, 130, 266
Yoshihara, M., **3**, 886
Yoshii, E., **2**, 376
Yoshikawa, S., **4**, 677
Yoshimatsu, A., **2**, 227, 230
Yoshimine, M., **3**, 23, 26; **14**, 534, 771
Yoshimizu, N., **8**, 114
Yoshimura, M., **7**, 88

Yoshimura, T., **4**, 217; **9**, 182; **11**, 118
Yoshinaga, H., **9**, 186
Yoshinaga, K., **4**, 678
Yoshino, S., **13**, 106
Yoshioka, Y., **5**, 139
Yoshiura, N., **3**, 746
Yoshiya, H., **3**, 700; **4**, 138
Young, C. M., **5**, 47
Young, D. W., **14**, 91
Young, F., **10**, 315
Young, P. R., **1**, 119
Young, R., **3**, 964
Young, S. D., **1**, 103; **2**, 37
Youssef, A.-H. A., **9**, 11
Yranzo, G. I., **14**, 825
Yu, L. C., **10**, 176
Yu, M., **9**, 113; **14**, 319
Uu, L., **4**, 764
Yudin, L. G., **7**, 15
Yufit, S. S., **12**, 265
Yukawa, Y., **2**, 29; **9**, 241; **14**, 408, 409
Yukhnovski, I., **10**, 29, 62
Yuki, H., **14**, 747
Yuminov, V. S., **2**, 46
Yun, S. S., **4**, 682
Yus, M., **10**, 13; **12**, 141, 143, 144

Zabel, R. W., **14**, 712
Zabel, V., **14**, 441
Zaby, G., **9**, 12
Zagorodnikov, V. P., **4**, 162
Zagorsky, V. V., **3**, 853
Zagulyaeva, O. A., **1**, 162
Zahra, J.-P., **3**, 751; **4**, 644
Zahradník, R., **14**, 376
Zaidi, J. H., **14**, 112, 113
Zaika, T. D., **2**, 120
Zaikov, G. E., **3**, 168; **4**, 299
Zaitseva, G. I., **3**, 462
Zaitseva, V. E., **14**, 576
Zajac, W. W., **12**, 156
Zajdel, W. J., **10**, 26
Zakhariev, A., **4**, 689
Zakharov, V. I., **12**, 72
Zakharova, V. D., **7**, 76
Zaki, A. B., **2**, 21; **3**, 139
Zakrzewski, J., **12**, 49
Zalewski, R. I., **2**, 45; **9**, 211
Zamarlik, H., **12**, 60, 61
Zamashchikov, V. V., **9**, 242, 244, 245, 260
Zamorano, P., **14**, 750
Zamyslov, R. A., **3**, 391
Zanasi, R., **8**, 160
Zander, F., **14**, 266
Zander, M., **13**, 156; **14**, 864
Zander, W., **14**, 1
Zanette, D., **3**, 868; **5**, 159; **11**, 98
Zang, B., **14**, 469
Zanin, A. M., **3**, 405
Zann, D., **3**, 1047
Zanocco, G., **3**, 614
Zapata, A., **10**, 89
Zaraiskii, A. P., **7**, 98
Zard, S. Z., **12**, 256
Zaripov, I. N., **1**, 173
Zaripova, V. G., **1**, 173
Zarkadis, A., **3**, 46
Zarubin, M. Y., **3**, 699; **4**, 105
Zarytovskii, V. M., **3**, 287, 288, 529
Zaslavskaya, T. N., **12**, 114, 117
Zaslavskii, V. G., **2**, 442
Zatorski, A., **9**, 218
Zavada, J., **11**, 1–3, 11
Zavgorodnii, V. S., **12**, 273
Zavorotnyi, V. A., **9**, 50
Zawadzki, S., **3**, 414
Zaworska, A., **14**, 562
Zbaida, D., **5**, 65
Zeelen, F. J., **14**, 524
Zefirov, N. S., **12**, 41, 55; **14**, 841
Zeib, H-J., **10**, 187
Zeigler, J. P., **1**, 39
Zeldis, I. M., **3**, 689, 690
Zelenova, G. P., **12**, 71
Zelent, B., **3**, 934
Zeller, K. P., **14**, 814
Zellmer, V., **14**, 469
Zemskii, B., **9**, 237
Zemskii, B. P., **9**, 238
Zeng, G. Z., **1**, 83
Zepp, C. M., **3**, 575; **13**, 147
Zerkalenkov, A. A., **7**, 123
Zerner, B., **2**, 276
Zhang, C.-J., **1**, 150
Zhang, J.-C., **7**, 3
Zhao, J., **4**, 305
Zhavoronkov, A. P., **4**, 496
Zhdankovich, E. L., **2**, 343, 344; **9**, 212
Zhdanov, A. A., **4**, 606
Zhdanov, Y. A., **14**, 119, 840
Zhen-Dong, H., **14**, 171
Zheng, X.-M., **1**, 150
Zhidomirov, G. M., **7**, 5; **8**, 164; **14**, 5
Zhil'tsov, N. P., **2**, 106
Zhil'tsov, S. F., **3**, 874
Zhiltsov, N. P., **2**, 168
Zhulin, V. M., **3**, 466
Zhuravlev, M. V., **3**, 136, 371
Ziegler, D., **1**, 23
Ziegler, E., **13**, 169
Ziegler, F. E., **9**, 59; **10**, 236; **12**, 229; **14**, 201
Zielinski, A. Z., **4**, 319
Zielinski, M., **4**, 39
Zielinski, W., **14**, 557
Zil'berman, E. N., **2**, 147
Zilkha, A., **2**, 156, 157
Zilliken, F., **14**, 719
Zimanyova, E., **3**, 332
Zimina, G. M., **3**, 241
Zimmer, O., **1**, 79; **14**, 330
Zimmerman, H. E., **14**, 105, 106, 873, 890
Zimmermann, G., **13**, 92
Zinbo, M., **4**, 497
Zinner, K., **3**, 108
Zippel, M., **10**, 199; **11**, 44; **14**, 826
Zlobin, V. A., **2**, 169
Zlotskii, S. S., **3**, 31, 115, 187, 188, 273, 274, 317, 325–327, 334, 345, 455–458, 463, 466, 467, 1129; **4**, 298
Zmuda, H., **14**, 14
Zoch, H.-G., **5**, 32
Zoch, H.-H., **14**, 642
Zoller, L. W., **7**, 115
Zollinger, H., **3**, 213–216; **6**, 4, 8, 71; **8**, 134; **14**, 16
Zoltewicz, J. A., **6**, 110; **9**, 273
Zorin, V. V., **3**, 273, 274, 317, 325, 327, 345, 463, 466, 467
Zubarev, V. E., **3**, 638
Zubareva, T. M., **2**, 443
Zuber, M., **4**, 684
Zubova, T. P., **12**, 273
Zucco, C., **1**, 137
Zueva, T. S., **4**, 84
Zuman, P., **3**, 753
Zupan, M., **7**, 17; **12**, 11; **13**, 205
Zupancic, J. J., **5**, 6
Zvilichovsky, B., **2**, 179
Zwaard, A. W., **10**, 68
Zwanenburg, B., **13**, 53; **14**, 493, 660
Zwart, L., **14**, 498
Zweifel, G., **10**, 39, 122
Zwierzak, A., **3**, 414
Zyulkovskii, Yu. Yu., **4**, 121

Subject Index

Acetals, 1–4
  cyclic conformation, 3
  cyclic, bromination of, 3
  hemithio, 51
  thio, 370
Acetolysis,
  of bromocholestenones, 328
Acetone, enolization of, 14
Acetophenones, iodination of, 14
Acid halides,
  aminolysis, 38
  solvolysis, 34, 38
Acid proteinases, 52
Acidity,
  absolute and relative, 374–376
  carbon acids, 369, 374
  determination, 375
  equilibrium acidity, 374
  gas-phase, 376
Acree relation, 345
Acrylate anion equivalent, 365
Acrylates, dimerization of, 361
Actinidine, 472
Activation of double bonds by Thallium(III), 420
Activation parameters, solvent dependence, 338
Acyl transfer, 29
Acyl-anion-equivalent, 360, 371, 373
Acylation, aromatic, 291
Acylium ion, 27, 36, 313
Acyloin rearrangement, 464
Adamantanes, 507
  proto, 415
Adamantyl cations, 302, 307
Adamantyl derivatives,
  elimination reactions, 403
  solvolysis of, 326, 327
  trioxa, 25
Addition reactions, polar,
  stereoselective, 409
Addition to alkenes, 99–103, 409
  carbenes, 243
  electrophilic, 409, 421
  intramolecular, 103
  nitrenes, 245
  nucleophilic, 373
  silylenes, 256
Addition, 1, 2-, 423
Addition, 1,4-dipolar, 423, 425
Addition, asymmetric, 409
Addition-elimination reactions, 36, 324, 363, 365
Ad_E2 reactions, 410, 413, 416
Adriamycinone, 464
Alcohol carbonates, homoallylic, 422
Alcohols,
  dehydration, 400
  epoxy, 360
  formation, 371
  olefinic, base-catalysed cyclization, 421
  propargylic, 357
  vinylallenic, 520
Aldehydes,
  acetylenic, 426
  addition reactions, 12–14
  bridgehead aldehydes, 375
  cinnamaldehyde, 1,4-addition of enolates, 423
  crotonaldehyde, 1,4-addition of enolates, 423
  deprotonation, 375
  formation of allenic aldehydes, 362
  α-hydroxy, 370
  oxidation, 17
  reaction with:
    carbanions, 371
    isocyanates, 7
    sulphones, vinyl, 416
  sulphenylated, 369
  synthesis, 364, 369, 371
Aldol condensation, 10–12, 365, 366, 425

stereoselectivity, 10
Alkaloids
  synthesis of, 479
Alkenamidines, addition of amines, 425
Alkenes,
  acetoxylation, 415
  bridgehead, 397, 478
  carbalumination, 503
  carbonylation, 416
  chlorination, 410
  epoxidation, 189
  fluorination, 409
  formation, 364
  halogenation, 412, 415
  hydroboration, 418
  hydroformylation, 416
  hydrohalogenation, 414
  hydrosilylation, 417
  mercuration, 418, 419
  oligomerization, 421
  perfluoro, 360
  phospha, 479
  photoreduction, 221
  prochiral, 418
  radical addition, 99–103
  reaction with alkanesulphinyl halides, 412
  reaction with peracids, 412
Alkylation of carbanions, 359
Alkylation, aromatic, 291
  with dimethylchloronium ion, 291
Alkynes,
  addition, 99
    of propylene sulphide, 420
    radical, 99
  carbonylation, 416
  halogen addition, 411
  hydration, 416
  hydroboration, 418
  hydroformylation, 416
  hydrohalogenation, 414
  hydrosilylation, 417
  radical addition, 99
  reaction with organolanes, 417
  reactions in superbasic media, 409
Allenes,
  addition,
    of ArSCl, 413
    of ArSeBr, 413
  cycloaddition, 436, 445, 452
  formation of twisted, 397
  hydroxyacetylation, 416
  tautomerism of, 475
Allyl cations, 450
Allyl compounds, nucleophilic displacement, 325
Allyl halides,
  deprotonation, 361
Allylic carbanions, 356, 358, 375, 376
  rearrangement of, 477
Aluminium alkyls,
  reaction with alkenes, 503
Aluminium enolates, 366
Ambident nucleophiles, 334, 335
  2-aminothiazole, 268
Amides,
  aminolysis, 37
  deoxygenation, 221
  deprotonation, 363
  hydrolysis, 32, 34
    micellar, 44
  proton exchange, 31
Amidoximes, 540
Amine oxides
  pyrolysis, 395
Amines,
  allyl amines, 419
  as antioxidants, 208
  oxidative deamination, 198
Amines, oxidation of, 188
Amino acids, formation, 371
Aminolysis, intramolecular, 39
Aminomercurials, 419
Aminophenalone, 291
Ammonium ion,
  elimination reactions of, 403
Anhydrides, hydrolysis, 30
Anilines
  nitration, 291
  reaction with base, 276
  sulphonation, 291
Anisole, halogenation, 291
Anthracenes, alkylation, 291
Anthracyclinones, 362
Anthraquinones, 471
Aphidicolin, 476, 539
Apicophilicity, kinetic, 60
Aprotic solvents, 36–39
Arbusov reaction, 341
Argentates, organo-, 424
Armilenium cations, 301
Aromaticity, 312
Arsabenzene, 291
Arylmethylsulphonates, 389
Aryloxenium ions, 293
Aspidospermine, 305
Asymmetric cyclizations, 332
Asymmetric induction, 365, 366, 367, 370,

371, 416, 422
by cyclodextrins, 45
in Diels-Alder reactions, 447
in micellar reactions, 44
Asymmetric synthesis, 417
ATP, 15
acid catalysed hydrolysis of, 59
Autoxidation, 208–211
of aldehydes, 208, 210
of alkanes, 87, 208
of alkenes, 210
of aromatic hydrocarbons, 208
of enol ethers, 209
of imines, 149
of ketones, 90, 208
of natural products, 208
of phenols, 89, 210
of styrene, 209
role of organometallic complexes, 208
styrenes, 89
Azaheptafulvenes, 498
Azahomospirostananones, formation, 515
Azatrienes, 480
Azepin-3-ones, 371
Azepines,
from anilides, 460
pyrido, 474
Azetidin-3-ones, synthesis, 360
Azetidiones, 359
Azides,
addition to olefins, 530
cycloaddition, 439, 440
hydrolysis, 34
photolysis of, 251
Aziridines,
from nitrenes, 245
ring-opening, 245, 330, 442
thermolysis of, 242
Aziridinium ions, ring opening, 524
Azirines,
as intermediates, 490
formation, 524
intermediates, 536
Azlactones, aminolysis, 37
Azo-compounds, 90–93
decomposition, 90
Azo-coupling reactions, 291
Azolidines, reactivity and tautomerism, 335
Azomethines, formation, 359
Azoxy compounds,
reaction with benzene, 293
Azulenes,
formation, 494
hydrogen exchange reactions, 291
Baeyer-Villiger reaction, 202, 515, 524, 528
Bamberger rearrangement, 461
Barbaralane dicarbonitrile, 477
Barbarallyl cation, 301
Beckmann rearrangement, 514
catalysis of,
by silylpolyphosphates, 514
photo-, 514
Belousov-Zhabotinskii reaction, 188
Benzazonines, 497
Benzene, protonation of, 291
Benzimidazoles, 374
Benzoate as a Michael acceptor, 462
Benzofurans, 372
Benzofuroxan,
4,6-dinitro, 16
reaction with base, 276
Benzonitriles, 497
Benzophenones, 369
Benzoquinones, 471
Benzoselenophen, 362
Benzothiazoles, 374
Benzothiazolines, 488
Benzothiophens, 277
Benzoxazepines, 531
Benzoxonium ions, 513
Benzo[9]annulenone, 494
Benzyl derivatives, solvolyses, 335
Benzylic cations, 317
Benzynes, 277
cycloaddition, 437
from halobenzamides, 277
from halogenated reactants, 462
reaction with:
benzoselenadiazoles, 494
reactions with heterocyclic compounds, 277
Betweenanenes, 481
Bicyclic alkenes, cycloaddition, 415
Bicyclic[3.2.N] dienones, 483
Bicycloalkanes, 506
Bicyclohexyl derivatives, solvolysis of, 327
Bicyclooctenyl derivatives, 327
Bicyclo[10.7.0]nonadec-1(12)-enes, 481
Bicyclo[2.1.0]pentane, 377, 536
Bicyclo[2.1.1]hexane, 526
Bicyclo[2.2.0]hexenones, 528
Bicyclo[2.2.1]hept-2-ene, 529
Bicyclo[2.2.1]heptane, 473, 506
Bicyclo[2.2.2]octadienones, 540
Bicyclo[3.1.0]hex-2-en-4-yl derivatives, 506
Bicyclo[3.1.0]hex-3-ene, 521
Bicyclo[3.1.0]hexa-3-ols, 503
Bicyclo[3.1.0]hexan-2-ones, 540

Bicyclo[3.1.0]hexen-2-ones, 469
Bicyclo[3.1.0]hexenone, 483
Bicyclo[3.2.0]dienone, 483
Bicyclo[3.2.0]heptane-3,4-diones, 528
Bicyclo[3.2.1]oct-8-yl cation, 301
Bicyclo[3.2.1]octa-3,6-dienes, 358
Bicyclo[3.2.1]octadienone, 483
Bicyclo[3.2.1]octane, 506
Bicyclo[3.2.2]nona-2,6,8-triene, 535
Bicyclo[3.3.0]octanes, 520
Bicyclo[4.1.0]heptadiene, 252
Bicyclo[4.1.0]heptatriene, 241
Bicyclo[4.2.0]octa-2,4,7-triene, 535
Bicyclo[4.3.1]nona-2,4-dienes, 495
Bicyclo[4.3.2]undecatetraen-7-yl cation, 301
Bicyclo[4.4.0]decan-2-ones, 369
Bicyclo[5.1.0]oct-4-en-3-ones, 499
Bicyclo[6.1.0]nonane, 518
Biradicals, 113–116, 153
  trimethylenemethane, 113
Birch reduction, 216, 217
Bischler-Napieralski reaction, 36
Boranes,
  oxidation, 364
  vinyl, 364
Boron enolates, 367
Brendone derivatives, 508
Bridgehead halides, solvolysis, 327
"Broken window" compounds, 513
Bromination,
  electrophilic aromatic, 291
    with potassium bromate, 291
  radical, 96
Bromination of:
  acetophenones, 376
  cyclohexene, 410
  enol ethers, 410
  norbornene derivatives, 410
  organomercury compounds, 377
  styrenes, 410
Bromine, addition to alkenes, 410
Bromophenol blue, 311
N-bromosuccinimide, 85
  addition promoted by, 410
  bromination by, 291
Brønsted coefficient for $B_{AC}2$ mechanism, 389
Brønsted equation for:
  elimination reactions, 403
  formation of cyclic hemiacetals, 1
  hydrolysis of esters, 30, 56
  lactonization, 47
  nucleophilic substitution reactions, 340
  substitution at sulphur, 331
Butadiene, palladation, 359
Buttressing effect, 29
Tert-butyl chloride, solvolysis, 337
Butyrolactones, 481

Cadinane derivatives, 481
Calorimetry, differential scanning, 465
Calythrone, 528
Canesolide, 462
Cannizaro reaction, 13
Carbamates,
  dilithio derivatives, 363
  elimination reactions, 390
  hydrolysis, 36
  N,N-diphenyl, 391
  pyrolysis, 391
  vinyl, 255
Carbanions,
  acyl, 361, 375
  addition to enones, 364
  allenic, 357, 362
  allylic, 355, 357, 358, 361, 363, 364, 372, 375, 376, 487
  α-amino, 364
  anthracenide, 358
  aromaticity, 358, 359, 368
  azo-cinnyl, 359
  benzyl, 356, 357
  bicyclic, 358
  bicyclodecatetraenyl, 358
  boron-stabilized, 357
  conformations, 355
  cycloheptatrienyl, 357,
  cyclohexadienyl, 353, 356
  cyclooctatrienyl, 357
  cyclopentadienyl, 374
  cyclopropenyl, 358
  cyclopropyl, 360
  dianions, 356, 358, 361, 367, 368, 369
    addition to α-enals, 425
    allenic, 360
    from acepleiadylene, 368
    from allylated ketenes, 365
    from furan carboxylic acids, 369
    from hydroxylactones, 365
    from methacrylic acid, 369
    from propargyl selenide, 368
    from pyrrole-2-carboxylic acid, 369
    from tetrahydropyrimidines, 369
    from tosylhydrazones, 361
    from vicinal diesters, 369
    pseudooxocarbons, 368
    vinyl, 369

dipole-stabilized, 355, 363
dithianyl, 370
ESR spectra, 354
fluorenyl, 354, 355, 357, 359, 364, 486
fluorine containing, 373, 375, 376
formation from coumarins, 376
formation from flavones, 376
formyl anion equivalent, 370
from bicyclo[5,2,0]nonatetraenes, 358
from cyanoacetic acid, 355
from cycloheptadienes, 368
from cycloheptatrienes, 358
from norbornadiene, 357
gas-phase, 354, 357, 363, 370, 374
geometry, 356
helical, 495
heterosubstituted, 371–373
indanide, 356
indenyl, 358
infrared spectra, 357
ion-pair effects, 353, 355, 359, 370
isomerization, 357
E–Z, 356
lithium carbanions of sulphones, 423
MO calculations on, 353, 355, 356, 358
NMR spectra, 354, 355, 357, 358, 375
norbornadienyl, 358
nucleophilic substitution by, 357
optical stability, 371
perfluoro, 360
propargyl, 357
propargylic, 357, 362
proton affinities, 353
proton transfer, 373, 374
reactions of, 359–364
rearrangement, 356, 359, 495
resonance energy, 368
ring current effects, 358
silicon stabilized, 361
α-silyl, 363, 370, 371
rearrangement, 360
solvent effects, 355, 356, 357, 361
stability and structure, 353–358
α-sulphonyl, 371, 498
sulphur stabilized, 369–371
tetraanions, 368
thienyl, 374
thio, 369
trianion, 368
vinyl, 353, 356, 361, 376, 497, 511
visible spectra, 368

Carbenes,
2-nitrophenyl, 242
abstraction reactions, 246–248
acyl-, 249
adamantylidene, 244
addition reactions, intermolecular, to alkenes, 243
addition to,
aromatics, 252
bicyclobutanes, 244
alkyl group migration to, 517
aryl, 242, 246
attack on divalent sulphur, 255
benzocyclobuten-1-yl, 248
chloro(phenyl), 244
chloro(trifluoromethyl), 243
conformational barriers, 239
cyclic, 241
cyclobutylidene, 250
cyclodextrin complexation, 252
cycloheptatrienylidene, 240, 241, 246
cyclopentadienylidene, 241, 246
cyclopropenylidene, 240
cyclopropyl(phenyl), 244
cyclopropylidene, 247
dihalo-, 244, 246
dimethoxy-, 243
dimethyl, 242, 243
diphenyl-, 239, 255
diphenylphosphonyl(phenyl), 250
fluorenylidene, 239
generation, 240–242
halo-, 243, 369
imidazoylylidene, 240
imino, 249
insertion reactions, 246–248
into metal-H bonds, 246
intramolecular reactions, 246
insertion, 252
isopropylidene, 246, 255
keto-, 249, 513
metal complexes, 209
methoxy(methyl), 242
methoxycarbonyl, 246
methylene-, 240, 248
migration of fluorine, 248
migration of hydrogen, 248
migration of phenyl group, 248
naphthyl, 239
oxa-, 242, 512
phenyl-, 246
quinolyl, 239
rearrangements of, 248–250
selectivity, 243
silyl, 250
singlet triplet gaps, 239
transition metal complexes, 257

vinyl, 242, 246
iron complexes, 517
vinylcyclopropylidene, 250
vinylidene, 248
Wolff rearrangement, 513
Carbinolamines, dehydration, 7
Carboalumination, of propargyl derivatives, 417
Carbonate esters, 31
Carbonic anhydrase, 48
Carbonium ions,
acetylated, 473
alkoxycarbenium ions, 312
alkyl cations, 1-phenyl, 308
allylic, 314
aryl cations, 311, 314
benzenium cation, 309
benzyl cation, 311
bicyclobutonium, 508
bicyclo[2.1.1]hexenyl, 509
bicyclo[3.3.0]oct-1-yl cation, 503
butyl, 291
sec-butyl, 309
carbinyl cations, *p*-anisyl, 311
α-cyano-stabilized, 317
cycloalkyl cations, 1-phenyl, 308
cycloalkyl, 314
cyclopentenium ions, 503
dications, 311
from benzylic and diphenic acid, 312
of ether salts, 312
dioxo-, 25
dioxolenium cations, 312
ethyl cations, 314
fluoren-9-yl cations, 9-aryl, 312
isocyano stabilized, 317
isomerization, 314
metal-stabilized, 307
metallocenyl-, 307
methyleneporphyrin cation, 313
naphthenium cation, 309
NMR spectra, 312
organometallic,
reaction with nitromethyl anion, 306
reaction with phosphines, 307
reactions with nucleophiles, 305
reduction, 305
oxo-, 3
oxoniomethylene, 533
oxyallyl, 450
phenanthrenium cations, 309
prop-2-enyl cation, 416
pyramidyl cations, 316
pyrenium cations, 309
pyrylium, 506
relative stabilities, 317
selenium stabilized, 315
α-silyl, 315
spiro, 309, 503
spirocyclic, 505
stabilized, α-cyano, 302
stable, 307–314
thiiranium, 524
thiophenium cation, 313
triaryl, 200
triarylcyclohexenyl, 500
triarylmethyl carbenium ions, 311
trimethylsilyl cation, 315
vinyl, 501
Carbonyl group,
nucleophilic attack on, 13
Carbostyrils, 528
Carboxylic acids, 23
addition to bicyclic alkenes, 415
by $S_{RN}1$ mechanism, 266
decarboxylation, 49
α-methylene, 366
reaction with nucleophiles, 36
Catalysis,
acid, in
hydrolysis of ketenes, 16
rearrangement of N-bromo-4-chloroacetanilide, 461
ring opening of Meisenheimer complexes, 276
alumina, in
rearrangement of aryl sulphides, 465
aluminium halides, in
Claisen rearrangements, 472
cyclization of unsaturated carbonyl compounds, 500
Diels-Alder reactions, 446
rearrangement of 1,2-endo-trimethylene norbornane, 506
rearrangement of lactams, 528
rearrangement of polysilanes, 519
antimony compounds in,
rearrangement of butylbenzene, 464
association-prefaced, 42–49
base, in
autoxidation of phenols, 210
nucleophilic aromatic substitution, 267
nucleophilic substitutions, 267
S_NAr reactions, 272
bifunctional, 30
boron trifluoride in,
rearrangement of epoxy cyclopropyl

ketones, 522
rearrangement of epoxy steroids, 512
by dimethyl formamide, 38
by dimethyl sulphoxide, 38
by thiocyanate ion, 537
thiocyanidate-isothiocyanidate isomerization, 537
by WCl_6, 291
carboxypeptidase, in elimination reactions, 398
copper in:
decomposition of diazocompounds, 246
oxidative coupling of phenols, 186
peroxydisulphate oxidations, 195
Sandmeyer reaction, 266
transformation of β-alkoxycyclopropane esters, 516
Wagner-Meerwein rearrangements, 506
DDQ in rearrangement of paulownin, 510
enzymic in,
1,2-bond shifts, 516
carbon skeleton rearrangements, 516
elimination reactions, 398
enzymic, 50–54
enzymic, by:
acetylcholinesterase, 54
acid phosphatase, 61
adenylate cyclase, 61
alcohol dehydrogenase, 7
alkaline phosphatase, 62
carbamoyl phosphate synthetase, 62
carboxypcptidasc, 52
creatine kinase, 61
galactosidase, 4
glyoxalase, 15
hexokinase, 61
phosphofructokinases, 61
phosphoglucomutase, 61
phospholipase, 54
polynucleotide kinase, 61
polynucleotide phosphorylase, 61
polynucleotide polymerase, 61
ribonuclease, 62
general acid-base, 26
general base, in:
addition of alcohols to aldehydes, 1
aminolysis of esters, 30
cyclodimerization of butenenitriles, 361
elimination reactions, 389, 402
homoketonization of cyclopropanols, 368
hydrolysis by papain, 51
nucleophilic aliphatic substitution, 328
rearrangement of aryl hydrazones, 467
S_NAr reactions, 272
general-acid, in:
hydrolysis of formamide, 32
Grignard reagents in asymmetric reductions, 221
halide ion, in:
conversion of acylsilanes to carbonyl compounds, 360
phosphoryl rearrangements, 499
rearrangement of lithium alkyls, 501
rearrangement of N-acyl benzylaziridines, 525
rearrangement of nitrosamines, 461
hydrophobic complexing, 42
hypochlorite in rearrangement of indoles, 467
imidazole in, 397
in radical reactions, 98
intramolecular, 26, 38, 39–41
by phosphonate, 41
general acid, 40
in oxidations by iodine, 196
nucleophilic, 48
Lewis acid, in:
alkyl halide addition to alkenes, 416
decomposition of unsaturated α-diazo ketones, 505
Diels-Alder reactions, 447
manganese in bromate oxidations, 198
manganese(II) in hydrolysis of geranyl pyrophosphate, 343
mercury in:
hydrochlorination of alkynes, 415
rearrangement of cephalosporins, 528
metal halide, in:
reaction of propargyl halides with dienes, 416
Stevens' rearrangement, 495
metal ion, in:
autoxidation of cyclohexane, 210
elimination reactions, 398, 399
epoxidations, 202
epoxide ring opening, 329
oxidation of alcohols, 210
phosphate ester hydrolysis, 59
methanesulphonic acid in isomerization of alkenes, 415
micellar, in:
enolization of acetone, 14
ester hydrolysis, 42

photolytic deprotonation of benzoin, 360
S_NAr reactions, 269
molybdenum(V) in epoxidation of alkenes, 190
nickel(O) complexes,
in addition of HCN to alkenes, 421
in synthesis of trisubstituted alkenes, 370
nickel(I) in *cis*-isomerization, 517
nitrosophenols in nitrosation of cyclic amines, 377
nitrous acid in nitration of phenols, 291
nucleophilic in:
aminolysis of esters, 30
deacylation of esters, 45
hydrolysis of episulphoxides, 329
hydrolysis of nitriles, 39
palladium complexes in alkene reactions, 438, 517
palladium(O) complexes,
in allylic alkylations, 365
in reduction of nitroaromatics, 216
reaction of diazonium salts, 266
palladium(II) salts,
in acetoxylation of alkenes, 415
in Cope rearrangement of 1,5-dienes, 477
in nucleophilic additions, 426
in rearrangement of *S*-allylthioimidates, 475
in [3,3]-sigmatropic rearrangement, 473
phase-transfer, 56, 291, 338, 339, 359, 422
in hydroxyacetylation of allenes, 416
in permanganate oxidation of alkenes, 185
in S_NAr reactions, 269
in superoxide oxidations, 204
of allylbenzene rearrangement, 465
platinum in hydroformylation of alkenes, 416
pyridoxal in β-elimination, 398
rhodium(I) in double bond shifts, 517
rhodium(II), 252
ruthenium
in hydroformylation of alkenes, 416
in aniline oxidation, 188
in bromate oxidations, 198
silver ion, in:
hydrogen-deuterium exchange reactions, 376
isomerization of cyclopropenes, 518
peroxydisulphate oxidations, 195
reaction of alkynes with acetic acid, 420
silylpolyphosphates in the Beckmann rearrangement, 514
sodium in dimerization of acrylates, 361
thiourea in nitrosation of morpholine, 377
tin halides in hydrochlorination of cyclohexene, 414
titanium(IV) complexes in stereoselective reductions, 214
transition metal complexes in,
catalytic hydrogenations, 516
decomposition of diazocompounds, 246
hydrogenation reactions, 220
iridium in oxidation of alkanes, 184
isomerization of 3,3-dimethyl-1-butene, 517
lithiation of alkenes, 361
oxidation of carpanone by oxygen, 209
porphyrinogen to corphin tautomerization, 537
S_NAr reactions, 270
transition metal:
in ring closure of cyclooctatetraenes, 516
in transesterification, 29
ruthenium in isomerization of dioxepines, 517
vanadium in bromate oxidations, 198
Cefadroxil, aminolysis, 40
Cephams, 524
Chalcones,
addition to, 424
condensation with ethyl cyanoacetate, 425
hydrolysis, 416
reaction with pyrrolidene, 425
Charge-relay systems, 50
Charge-transfer complexes,
in addition to alkenes, 411
in Diels-Alder reactions, 434, 445
Chelidonine, 504
Chemiluminescence,
dioxetanes, 209
Chichibabin amination, 273, 277
Chirality, transfer of, 473, 491
Chlorination,
electrophilic aromatic,
with *N*-chlorosaccharin, 291
with *N*-chlorosuccinimide, 291

of ethanol, 376
of sulphoxides, 371
radical, 95, 112
Chlorine, addition to alkenes, 410
Chlorotelluration, 420
Cholesteryl derivatives, formation, 212
Chroman-6-ols, 204
Chromium tricarbonyl complexes, 357, 376
Chromones, 376
Chrysene oxides, 536
Chrysenes, 291, 504
Chymotrypsin, inhibitors of, 51
Chymotrypsin, x-ray structure, 50
Chymotrypsins, acyl, deacylation of, 50
CIDEP, 141, 157
CIDNP, 85, 86, 140, 141, 147, 154, 156, 157, 291
Cine-substitution in heteroaromatics, 273
Cinnamates, formation, 29
Citral, 518
Claisen rearrangement, 367, 472
amido-, 474
carbon-Claisen, 472
ortho ester, 473
photo, 472
tandem, 474
thio-, 475
ynamine claisen, 475
Claisen-Eschenmoser rearrangement, 476
Clavulanic acid, 53
Clemmensen reduction, 217
Coates, cation, aryl-substituted, 307
Cobaloximes, 109
Cobalttetracarbonyl, 217
Cope rearrangement,
heteroaromatic, 477
monoaze, 479
oxy-Cope, 477, 478
phospho, 479
tandem-Cope-Claisen, 477
Coumaran-3-ones, 522
Cram's rule, 366
Cresols, bromination of, 464
Criege intermediates, 200
Criegee rearrangement, 502, 512, 526
Crotepoxide, 528
Crown ethers,
activation of anions, 269
complexes with diazonium ions, 265
effect on:
alkylation of enolate anions, 365
birch reductions, 217
decarboxylation, 392
fluorenyl chromium tricarbony
anions, 357
ion pairs, 355
in carbene insertion reactions, 246
in elimination reactions, 385, 386
in nucleophilic aliphatic substitution, 332, 338
Cryoenzymology, 50
Cryptands, 355, 365
Cryptates, 354
Cumulene, 521
Cuprates,
addition to thioates, 364
alkylation, 363
dialkenyl, from dialkyl, 424
dialkyl, addition to enones, 424
halomagnesium diorgano, 424
reaction with epoxides, 329
trimethylstannyl, 424
Curtin-Hammett equation, 342
Curtius rearrangement, 516
Cyanates, aryl, 35
Cyclization of hydroxy olefins, 332
Cyclization, intramolecular, 43
Cycloadditions,
1 + 2, 450
1 + 7, 450
1,3-anionic, 364
1,3-dipolar, 249
1,4-dipolar, 450
2 + 2, 433–437
2 + 2 + 2,
cobalt catalysed, 452
2 + 3, 314, 437
thermal, 443
2 + 4, 443–449, 461, 484
cationic, 448
3 + 8, 451
4 + 3, 450
4 + 4, 451, 452
6 + 4, 451
8 + 2, 498
metal catalysed, 452
reaction of azaheptafulvenes with sulphenes, 498
Cycloalkyl cations, hydrido bridging, 309
Cyclobutadiene, dications, 317
Cyclobutenes,
formation, 486
from cyclobutane acetate, 391
ring-opening, 371, 402
Cyclodecadienes, 477
Cyclodecadienones, 488
Cyclodextrins,
catalysis by, 45

functionalized, 47
Cycloheptadienylium iron cations, 307
Cycloheptapyrrol-6(2H)-ones, 470
Cycloheptatriene,
formation, 477
rearrangement, 535
Cyclohexadienes,
energy data, 353
rearrangement, 469–472
Cyclohexadienones,
carbene adducts, 469
from azidoformates, 255
Cyclohexenones, 365, 367
Cyclooctatrienes, rearrangement, 535
Cyclopentadienyl cation, 314
pentamethyl, 509
Cyclopentadienyl cobalt
cyclononatetraene, 516
Cyclopentenediols, 361
Cyclopentenones, 361, 370, 505, 538
reaction with carbanions, 369
Cyclopentenophenanthrenes, 504
Cyclopropanation of alkenes, 243
Cyclopropanation-rearrangement, 520
Cyclopropane,
allyl, 482, 483
vinyl, 520
Cyclopropanes,
alkylidene, 481
formation, 360
ring-opening, 333, 360, 371, 520
vinyl, 526
Cyclopropanimines, nucleophilic ring
opening, 329
Cyclopropanols, 368
formation, 359
Cyclopropanones,
rearrangement, 521
Cyclopropenes,
from carbenes, 244
thermolysis of, 246
Cyclopropenium cations, 311, 314, 317
Cyclopropenones,
reaction with carbanions, 370
Cyclopropyl cations, 315
Cyclopropyl derivatives,
nucleophilic substitution of, 329
Cyclopropylacetaldehydes, 485
Cyclopropylcarbinyl cations, 309
β-cyclopropylidenic alcohols, 329

Deamination of phenylethylamines, 255,
501
Decarbonylation, thermal, 392
Decarboxylation, 43, 48–50, 392, 393
in presence of micelles, 392
isotope effects, 49
of cinnamic acids, 392
oxidative, 49, 506
Dediazoniation, 314
Dehydration of alcohols, 400
Dehydrocyanation, 395
Dehydrohalogenation, 386
Denitrosation, 461
Deoxybenzoins, 497
Desulphonation,
of sulphonium salts, 393
Dewar benzenes,
addition of sulphenium ions, 509
Diarylmethanes, 504
Diazenes, thermal fragmentations of, 394
Diazepines,
cycloaddition, 450
Diazo-compounds,
cycloaddition, 440, 450
decomposition, 252
diazodiphenylmethane, 148
reaction with chloranil, 17
rearrangement, 490
Rh(II) catalysed decomposition, 246
ring opening, 394
Diazo-ketones,
decomposition of, 240, 249
photolysis, 248, 513
rearrangement of, 508, 520
Diazomethane, 438
Diazonium salts,
aryl cations from, 265
aryl radicals from, 91, 92, 112
aryl, 314
base addition, 265
crown ether complexes, 265
decomposition, 265
S_NAr reactions, 265
Diazophospholanones, 531
Diazotization by propyl nitrite, 291
Dibenzene osmium cations, 307
Dibenzene ruthenium cations, 307
Dibenzeneiron cations, 307
Diels-Alder reactions,
asymmetric induction, 447
catalysis in, 446
diradical character, 433
endo-rule, 445
intramolecular, 486
Lewis acid catalysed, 447
molecular orbital calculations, 443
of acrolein, 447

of acrylonitrile, 446
of anthracene, 443
of benzofurans, 473
of cyclohexa-1,3-diene, 144, 446, 449, 452
of cyclopentadiene, 444, 446, 449
of cyclopentadienone, 449
of fulvenes, 451
of furan, 426
of heterodienes, 448, 449
of heterodienophiles, 449
of quinones, 444, 448, 539
of tetracyanoethylene, 449
of ylides, 450
photochemical, 449
radical cations in, 446
regiospecificity, 444, 447
solvent effects, 446
stereochemistry, 445, 447, 450
Dienes,
conjugated, formation of, 357
rearrangement, 475
Dienone-phenol rearrangement, 291
Diformamide, hydrolysis, 31
Dihydrofurans, 492
Dihydrooxepins, 492
Dihydroquinolinones, 528
Diiodomethane, 243
Diketopiperazines, 40
Dimroth rearrangement, 465
Diols, vicinal, 366
Dioxenes, hydrolysis of, 17
Dioxetanes, 471
decomposition of, 83
Dioxirane, 524
Dioxolanes, 3
thermolysis and photolysis, 84
Diradical, 206, 484
bicyclo[4.2.0]octane-2,5-diyl, 526
cyclopentadiyl, 536
dihydrobenzene diradicals, 533
from spirocyclopropaneanthrones, 520
merostabilization, 536
Disulphides, nucleophilic cleavage, 373
Disulphides, azetidinone, 530
Disulphones, 390
Dithanes, ring opening, 370
Dithiane anions, 370
Dithia[13]annulene cation, 313
Dithiepins, 370
Dithiocarbonates, 505
1,3-dithiolanes, 4
Dithiole, 505
Doering-Zeiss mechanism, 340

Dyotropic rearrangement, 488, 490

Effective molarities, 39
Electrochemical reactions:
one-electron transfer reactions, 364
reduction, 359, 371
Electrocyclic reactions, 491–495
ring opening, laser-induced, 492
Electrocyclic rearrangements, 491–495
Electron superdelocalizability, 291
Electron transfer, 9, 78, 82, 85, 118, 131–141
from carbanions, 363
Electrophilic additions, 409–421
halogen addition, 409
mechanistic studies, 409
Electrophilic aliphatic substitution, 376, 377
Electrophilic aromatic substitution of:
aromatic hydrocarbons, 131
heteroaromatics,
indole derivatives, 291
thiophenes, 291
silylbenzenes, 291
Electrostatic effects,
effect on elimination, 385
Electrostatic stabilization, 14
Elimination reactions,
1,4-elimination, 362
carbonyl forming, 402
cis elimination, 397
cis/trans ratios, 385
dehydrofluorination, 373
deoxysilylation, 400
desilylation, 400
desulphonation, 371
E1 mechanism, 371
chiral, 370
E1cb mechanism, 31, 43, 63, 65, 388–390
E2 reactions, 385
halide ion-promoted, 400
Hofmann, 386, 387
leaving group effects, 387, 388
of ammonium salts, 388, 403
of arylmethanesulphonates, 389
of β-alkylated 2-cyclohexenones, 425
of carbamates, 390
of DDT-type compounds, 390, 404
of dithanes, 370
of esters, 390
of ethers, 396
of nitrites, 398
of phosphonium ylides, 372
of selenoxides, 396

of sulphoxides, 480
pyrolytic, 390–396
solvent effects, 387, 400, 403
steric effects in, 385–388, 387
strain effects in, 397
syn-elimination, 385, 399, 403
Ellman's reagent, hydrolysis of, 331
Enamines, 8
lithio derivatives, 360
Michael addition, 426
protonation of, 7
Enaminothiones, 484
Endoperoxide intermediates, 209
Endoperoxides, 87
Ene reactions, 206, 505
aziridines with nitrostyrene, 492
of a thioaldehyde, 491
retro-
pericyclic reactions of acetylenes, 491
Ene-allene, 485
Enol esters,
electrophilic addition to, 420
formation, 363, 420
Enol ethers,
bromination, 410
hydrolysis, 14, 16, 17
reaction with carbanions, 364
Enol phosphates, 15
Enolate anions, 365–368
addition to chiral aldehydes, 10
alkylation, 365
as protecting groups, 367
boron, 367
charge distribution, 354
from azo compounds, 367
geometry of, 355, 366, 367
ion-pairing effects, 355, 365
oxidation, 366
substituent effects, 365
theoretical calculations on, 355
Enolates, 426
boron, 11
lithium enolates, 423, 424
lithium, 10
silyl, 365
stereospecific reactions of, 11
triphenyltin, 11
zirconium, 11
Enolization, 14–16, 368, 372
of acetone, 14
Enols,
addition of halogen, 14
Enones, addition to, 371, 424
Entropy of activation for:
nucleophilic aliphatic substitution, 330, 339
Enzymic reactions, 61, 62
Episulphonium ions, 412
Episulphoxides, hydrolyses, 329
Epoxidation,
metal catalyzed, 202
of alkenes, 202
of azaarenes, 204
stereoselective, 202
vanadium(V) catalysed, 189, 328
Epoxides,
deoxygenation, 257
α-halo, 411
hydrogenolysis, 221
in nucleophilic aliphatic substitution, 328, 329
in oxidation of alkoxystyrenes, 209
reaction with:
amines, 328
ammonia, 328
cresols, 328
cuprates, 329
organametallics, 328
ozone, 201
phosphonic acid chloride, 328
ring-opening, 242, 328, 329
Epoxysilanes, ring opening, 329
Esterification,
benzoic acids with
diphenyldiazomethane, 29
Esters,
aminolysis, 30, 36
general acid-base catalysis, 30
micellar, 43
preassociation mechanism, 30
carbanions from, 363
formation, 29
hydrolysis, 23, 26, 30
$A_{AC}2$, 26
$A_{AL}1$, 26
benzhydryl benzoates, 27
catalysed by cyclodextrins, 45
E1cb mechanism, 28
effect of ultrasound, 28
linear free energy relationships, 27
spontaneous, 29
β-hydroxy, 365
in Friedel-Crafts reactions, 291
α-methylene, 366
pyrolysis, 36
reaction with acetylhypofluorite, 409
Estrone, synthesis of, 478
Ethers,

cleavage, 360
elimination reactions of, 396
from dienes, 414
Ethylene oxide,
reaction with hydrogen halides, 328
Ethylenimine, ring opening, 330

Farnesyl pyrophosphate, cyclization of, 304
Favorskii rearrangement, 359, 499
Ferrocene
mercuration of, 292
Ferrocenes,
electrophilic substitution, 292
Ferrocenylcarbonium ions, 307
elimination reactions of, 403
Flavins, 138, 149
Flavonols, 522
Fluorenes,
from diazonium salts, 265
Fluorenones, reaction with carbanions, 370
Fluorination,
aromatic, 291
with acetylpofluorite, 291
with fluoroxysulphate ion, 291
Fluorine addition, 409
Force-field calculations, 343
Formyl anion equivalent, 424
Fortimicins, 539
Friedel-Crafts reaction, 291, 292
of ethylenimines, 464
Fries rearrangement, 464
photo, 464
Fujiwara reaction, 17
Fulvenes,
6-sila, 250
acetylation of, 416
addition of carbenes, 243
cycloaddition, 443
deprotonation, 372
rearrangement of, 492, 535
Furans,
cycloaddition, 452
dihydro, 372
formation, 471
nucleophilic substitution, 273
reaction with acetylenic aldehydes, 426
Furoxans, 533

Galactopyranose, 4
Geranyl pyrophosphate, cyclization of, 304
Germacranes, synthesis of, 478
Gibberellins, 510
Glutathione, 15
Glycosides,
hydrolysis, 2, 4
mutarotation of neuraminic acid derivatives, 4
tautomerism of, 4
Gomberg reaction, 112
Grignard reagents, 13, 15
allyl, 423
reaction with allylic amines, 423
reaction with propargylic amines, 423
radical reactions, 134
reaction with alkenyl sulphides, 370
Grunwald-Winstein equation, 337

Halide exchange,
in chloropyridines, 272
Halogen dance, 464
Halogenation, 95, 96
aromatic,
of benzocyclopropenes, 291
radical, 101
Halonium ions, 410
in gas phase, 336
Hammett ρ-σ relationship for:
addition of diazomethanes to chloranil, 17
aromatic thalliation reactions, 292
cleavage of coumarins by cyclodextrins, 45
cycloaddition reactions, 443
elimination reactions, 387, 388, 389, 390, 394, 398, 400, 402
E2 mechanism, 404
esterification reactions, 29
hydrolysis of aryl esters, 28
hydrolysis of azides, 34
hydrolysis of hemiacetals, 1
iodination of acetonaphthones, 377
methanolysis of aromatic nitrile oxides, 35
micellar reactions, 44
nucleophilic aliphatic substitution, 325, 330, 335, 340, 341
nucleophilic substitution at sulphur, 331
radical reactions, 78, 85, 95, 127, 135, 137
reaction of acid chlorides with azide ion, 38
Hammett ρ-σ relationship, 293
non-linear, 333
solvent dependence, 337
Heat capacity of activation, 345
Hemiacetals, 25
decomposition of, 17

Hemiketals, 12
Hemiorthoesters, 23–25
  generation of, 23
  hydrolysis, 16
Hemithioacetals,
  rearrangement, 15
Hexachloromelamine, 410
Hexahelicene, 291
Hexamethylenetetramine, hydrolysis, 8
Hexaquinane, 523
Hirsutene, 510
Hofman rearrangement, 516
Homoaromaticity, 317
  bicyclo, 301
  in diazo-tetrazines, 274
Homograyanotoxane, 509
Homoisodrin, 506
Homotetrahedryl cation, 317, 500
HSAB principle, 325
Hunsdiecker reaction, 135
Hydration parameter *R*, 132
Hydrazones, 8–10
  coordination to metals, 9
  formation, 9
  photochromic properties, 534
  ring closure reactions, 10
Hydride, transfer, 2, 316
  intramolecular, 12
  to imines, 8
Hydride-ion shift, 315
  1,2-, 317, 327, 523
  1,3-, 506
  1,5-, 486
  1,5-, 343, 500
  in naphthalenonium ions, 460
Hydroalumination, of alkynes, 417
Hydroazulenes, 488
Hydroboration, 418
Hydrogen bonding, intramolecular in
    haloazines, 272
Hydrogen exchange,
  aromatic, 291
    of trimethoxybenzene, 291
  base-catalysed, 368
  gas phase, 376
  of heteroaromatic compounds, 291, 376
Hydrogen migrations,
  1,2-, 513
  1,3-, 478
  1,5-, 107, 485, 494
  1,6-, 107
  1,7-, 107, 489
  1,8-, 107
Hydrogenation, 219–221
  asymmetric, 220
  of amino acid precursors, 220
  of carbon monoxide, 220
  of carboxylic acids, unsaturated, 220
  of Dewar benzenes, 221
  of ketones, 221
  of steroidal alkenes, 220
  rhodium complexes in, 219
Hydromagnesiation, 214
Hydroperoxides, pyrolysis, 84
Hydroxamic acids,
  formation, 30
Hydroximoyl chlorides, 9
Hydroxylamines,
  N-aryl, 293
  rearrangement, 513
Hydroxylation,
  by peroxymonophosphoric acid, 203
  electrophilic, 203, 292
  with H_2O_2 in SbF_5-HF, 293
  with peroxymonophosphoric acid, 293
  of phenol, 203
  of phenylalanine, 203
  of xylenes, 203
  radical,
    by peroxydiphosphate, 203
Hyperconjugation, 291
  in carbanions, 375

Imidates, 32
  esters, 25
Imidazo thiazine, 524
Imidazoles,
  elimination reactions, effect on, 397
Imidazolidines, ring opening, 5
Imidazolium ions, hydrolysis, 34
Imines, 6
  chiral, 12
  hydrogen transfer, 8
  hydrolysis, 8
  unsaturated, 371
Iminium ions, 449, 509
  acyl, 479
Immonium ions, 317
1,3-Indandione, 10
Indenes, 518
  bromination, 410
Indoles,
  by $S_{RN}1$ mechanism, 266
  formation, 9
  hydrolysis, 44
  nitration, 291
  pyrrolo, 369
Indoles, oxidative cyclization, 189

Inorganic pyrophosphates, 62
Intramolecular substitution, 332, 333
Iodination,
  aromatic,
    with hydrated monopositive iodine, 291
    with iodine monochloride, 291
    with iodosuccinimide, *N*-, 291
  of acetonaphthones, 377
  of organotin compounds, 377
  radical, 96
Iodonium ions, 412
Ion pairs, 275, 353, 501, 505
  in elimination reactions, 403
  in nucleophilic aliphatic substitution, 323, 330
  in nucleophilic aromatic substitution, 269
  in polar additions, 410
  in stereoselective reductions, 213
  racemization of, 536
Ion-cyclotron resonance spectroscopy, 355, 357, 363, 374
Ipso-attack of carbon radicals, 267
Ipso-attack of electrophiles, 464
  in alkylation, 291
  in nitration, 291
Ipso-attack, radical, 112, 138
Iron carbonyl complexes, 305
  ligand substitution, 307
  reaction with diazepines, 487
  tricarbonyl(cyclohexadienylium)iron salts, 305
Isobasic plots, 28, 388
Isocyanates,
  aminocyclopropyl, 529
  aminolysis, 38
  as intermediates, 35
  reaction with aldehydes, 7
  solvolysis, 38
Isocyanides, metallation,, 369
Isomerism, translational, 535
Isomerization of alkanes, cyclopropanoid mechan, 518
Isomerization, E-Z, 355, 534
Isomerization, permutational, 537
Isoquinolines, 372
Isotope effects,
  apparent maximum in, 403
  carbon, in enzyme-catalyzed decarboxylations, 49
  chlorine, 388
  cis-trans equilibrium, 533
  deuterium, 28, 61, 293, 355, 388, 395, 425, 471
    in carbanion reactions, 373, 374
    in phosphorolysis of inosine, 5
    in reaction of acetone with alkoxide ion, 15
    in thermal Claisen rearrangements, 386, 472
    primary, 14, 343
    secondary, 8, 29, 30, 334
    temperature dependence, 66
  heavy-atom, 26
    in elimination reactions, 385–390, 395, 399, 400, 403
    in halogenation of anisole, 291
    in nucleophilic aliphatic substitution, 335
    in radical reactions, 94, 111
    in solvolysis reactions, 500
    in thio-Claisen rearrangements, 476
    model calculations, 335
    nitrogen, 388
    radical reactions, 114
    radical, 100
    solvent, 3, 45, 48, 374, 388
  deuterium on singlet oxygen lifetimes, 206
  deuterium, 28, 65
    in hydrolysis of keten selenoacetals, 16
  temperature dependence, 387

Janovsky reaction,
  mechanism of, 276
cis-Jasmone, 424

Ketals, 1–4
  hydrolysis, 3
  spiro, 2
Ketene adduct, 398
Ketenes, 16, 28
  acyl, 16
  addition of imines, 249
  cycloaddition, 16, 436, 437
  cyclopentadiene, 483
  hydration, 16
  hydrolysis, 31
  oxidation, 197
  vinyl, 494
Ketenimines, 437, 487
  dimerization of, 461
  rearrangement of, 499
  synthesis, 364
Keto esters, enolization of, 16
α-Ketols, 12

Ketones,
addition reactions, 12–14
α-amino, 480
cyclopropyl ketones, 372
deoxygenation, 221
deuteration, 375
dianions from, 362
enantioselective alkylation, 366
enolization, 15, 368
micellar catalyzed, 14
from alkynes, 416
halo-, 359
halogenation, 15
hydration, 12
oxidation, 17
synthesis:, 360, 364, 369, 371
of 1,3-diketones, 365
of α-hydroxy ketones, 366
of α-methylene ketone, 363
of allenyl trimethylsilyl ketones, 362
of β,γ-unsaturated ketones, 366
tetra, 369
thio-, 369
vinyl, 425
Ketothiolate, 524
Knorr-Paal cyclo-condensation, 10
Kukhtin-Ramirez reactions, 499

β-Lactamases, 52
Lactams,
azosteroid, 195
bridged bicyclic, 515
γ-, 491
enol lactams, 373
hydrolysis, ring size effects, 32
of β-lactams, 32
Lactol ring opening, 334
Lactone annulation, 474
Lactones,
β-, 536
carboxybutyrolactones, 521
decarboxylation of, 392
deprotonation, 365
dithio, γ-, 362
epimerization, 367
formation, 197, 529
hydrolysis, 27, 29
macrocyclic, 207, 367, 528
methylene, α-, 369
phenylsulphonyl, 495
propella-ε-lactones, 507
ring-opening, 360
seleno-, 414
spiro-, 504, 512
thio-, 527
vinyl, 492
Lactonization of ω-halo fatty acids, 332
Lactonization, 25
Leaving group effects in nucleophilic substitution, 28, 30, 324, 337
Linear free energy relationships, 27
Lithium alkyls, 353
addition reactions, 423
reaction with peroxides, 360
Longifolene, 418

Malondialdehyde, 6
Mannich bases, 8
Markownikoff amidation, 419
Markownikoff regioselectivity, 418
Markownikoff's rule, 410
Meerwein-Ponndorf-Openauer conditions, 216
Meisenheimer complexes, 274–277
from phenoxide ions, 274
in S_NAr reactions, 267
metal containing adducts, 276
NMR studies, 274
spectroscopic studies, 275
spiro-, 275, 461
Meisenheimer rearrangement, 465
Menschutkin reactions, 339, 343
reverse, 342
Mercuration,
of dienes, 419
of norbornenes, 418
Metallation,
aromatic mercuration, 292
aromatic plumbylation, 292
aromatic thalliation, 292
of alkynes, 362
of amides, 363
of anisoles, 363
of diazepinones, 363
of dihydropyrazines, 372
of halostyrene oxides, 522
of isocyanides, 369
of oxazines, 363
of oxazoles, 363
of pyridines, 357, 362
of quinolines, 357
of thiophens, 362
substituent effects, 362
Metallo-proteinases, 52
Metaphosphate, 54, 59
Methoxybenzene cation, synthesis equivalent, 306
Methoxymercuration of styrenes and

stilbenes, 418
Methylthiolation, 494
Micelles,
anionic, 311
cationic, 43, 311, 360
effect on proton transfer from nitroalkanes, 373
in oxidation of benzylamine, 195
effect on phosphate ester hydrolysis, 56
functionalized, 44, 45
in electrophilic aromatic substitution, 291
in hydrolysis, 42
in nucleophilic aliphatic substitution, 331
in reactions of aldehydes and ketones, 12
reverse, 42
Michael reaction, 12, 367, 421
addition of thiophenoxide, 197
addition to enones, 422
hydroxy, hydrolysis of, 27
Migration of:
acyl groups, 189, 506, 538, 539
intramolecular, 39
alkoxy group, 472
alkyl groups, 362, 391, 423, 470, 471, 500, 510, 517, 524
allyl groups, 472, 482
aryl groups, 248, 250, 323, 486, 501, 504
benzyl groups, 461, 467
carbamoyl groups, 462
cyano groups, 495, 518
double bonds, 355, 395
ester groups, 291, 467
halogen, 248, 269, 461, 466, 469, 485, 501
hydride, 248
naphthyl groups, 460
oxygen, 391, 495
phenyl thio groups, 501
phenylseleno groups, 513
silicon, 207, 250, 485, 488, 498
thioether, 488
triple bond, 498
tripticenyl group, 515
vinyl group, 485, 501, 503
Migratory aptitude, 486, 501, 502
in carbene processes, 513
Molecular-orbital calculations, on:
acidity of carbon acids, 375
addition reactions, 416
fluorination of ethene, 421
carbanions, 353, 355, 356, 358, 368, 375
carbene reactions, 243
carbene rearrangements, 245, 248
carbenes, 239, 240
carbonium ions, 299, 316, 500
carbonyl group of asymmetric ketones, 212
cyanide ion exchange reaction, 360
electrocyclic reactions, 491
electrophilic aromatic substitution, 291
elimination reactions, 395
hydrogen-exchange in arenes, 460, 488
lithium alkyls, 353
meisenheimer complexes, 274
nitrile oxides, 439
nucleophilic aliphatic substitution, 325, 328, 375
CNDO/2 calculations, 342
using PMO method, 343
photoisomerization of isoxazoles, 524
photorearrangement of aci-nitrate anion, 495
photorearrangement of amino-cyanoethylene, 524
proton affinities of carbanions, 375
protonation:, 416
of arenes, 291
of ethene, 416
quinones, benzo and naphtho, 426
radical anions, 145, 147, 336
radical cations, 462
radical reactions, 78, 79, 111, 132
radicals, 80
rearrangement of formamide molecular cations, 500
rearrangement,
aminonitrene-diimide, 533
cyanamide-isocyanamide, 533
of benzo[β]bicyclo[3.1.0]hex-2-en-4-yl derivatives, 506
of ethylidene, 533
sigmatropic migrations of hydrogen, 482
silylenes, 256
S_N2 reactions, 330
tautomeric interconversions, 537
transition metal-alkene complexes, 421
[1,2]- and [1,3]-shifts in silylcyclopentadienes, 488

Naphthalene anions, reduction, 367
Naphthalenes,
adducts from, S-, 275
nitration, 291
rearrangement of dihydro, 485
Naphthofurans, 472

Naphthpyridines, amide addition, 277
Neber rearrangement, 524
Neighbouring-group participation by:
  acetoxy groups, 329
  alkene group, 395
  alkoxy groups, 333
  alkyl groups, 324, 330
  amide group, 421
  aryl groups, 323, 333
  carbonyl group, 410
  carboxyl group, 334
  cyclopropyl group, 326
  double bonds, 411
  ether groups, 334
  hydroxyl group,
    in oxidation of hydroxysulphides, 196
  methoxy group, 329
  nitrogen bridge, 509
  nitrogen, 525
  oxime group, 187
  oxygen, 523
  quaternary ammonium centres, 37
  transition metals, 517
Neighbouring-group participation in:
  addition to alkenes, 410
  ester hydrolysis, 500
  gas-phase, 336
  glycoside hydrolysis, 334
  nucleophilic aliphatic substitution, 326
  solvolysis reactions, 333, 334
Neighbouring-group participation, 39–41
  homoallylic, 327
Nitration,
  aromatic, 291
    encounter controlled, 291
    mass diffusion effects, 291
    with ammonium nitrate, 291
    with dinitrogen tetroxide, 291
    with metal nitrates, 291
    with N-nitropyrazole, 291
    with nitric acid, 291
    with nitronium hexafluorophosphate, 291
  of cinnamic acids, 291
  of indoles, 291
  of styrene derivatives, 291
  of sulphones, 291
  of thiophenes, 291
Nitrenes,
  abstraction reactions, 246–248
  addition to alkenes, 245
  alkoxycarbonyl, 245
  alkyl, 251
  aminonitrenes, 242, 246, 251, 255
    from quinazolones, 189
  aryl, 251, 253
  arylsulphenyl, 245
  attack on divalent sulphur, 255
  cyano, 242
  cycloaddition, 245
  from azides, 253
  generation, 242, 243
  insertion reactions, 246–248
  intramolecular reactions, 247, 255
    with sulphur, 255
  phenyl, 242
  phthalimido, 242
  rearrangement, 251
  singlet-triplet equilibrium, 247
  tert-octyl, 247
  tosyl, 255
  vinyl, 254, 525
Nitrenium ions, 317
  from hydroxylamines, 513
Nitrile imides, 439
Nitrile oxides,
  cycloaddition, 439, 440
Nitriles,
  hydrolysis, 34
    metal catalyses in, 48
  δ-ketoalkyl, 423
  α-lithio, 423
Nitrilium ions, 9, 514
  addition of nucleophiles, 426
Nitrite esters, 65
Nitroalkanes,
  deprotonation of, 373, 375
Nitrogen insertion, 515
Nitrolysis of N,N-dialkylcarboximides, 377
Nitrones,
  cycloaddition, 440
  formation, 359
Nitronium ions,
  in electrophilic aliphatic substitution, 377
Nitrosation,
  aromatic, 291
  by nitrosylhalides, 291
  of 1-indanones, 528
  of aminopyridines, 377
  of secondary amines, 377
Nitroso-compounds,
  reaction with,
    carbanions, 359
N-nitrosoamines, 78, 377
Nitrosobenzene, 293
Nonacetic acids, 473
Norborn-2-yl cations, 299

2-aryl, 299
3-oxo, 300
Norbornadienes, 518
addition of carbenes, 244
Norbornadienyl cation, 301
Norbornadienyl derivatives, 358
Norbornenes, 524
additions, 413
electrophilic addition to,
mercuration, 418
Norbornenyl derivatives, 506
Norbornyl cations, aryl, 308
Norbornyl derivatives,
substitution reactions, 325, 326
Normant reagent, 424
Nucleofugalities, 388
Nucleophiles,
reaction with carbenes and nitrenes, 255
Nucleophilic additions, 354, 421–426
anilines to alkenes, 426
to nitrilium ions, 426
Nucleophilic aliphatic substitution,
activation parameters, 337
carbanion formation, 357
gas phase reactions, 336
intramolecular, 332
alkylation of bromoalkoxyphenoxides, 332
effective molarities, 332
isotope effects, 335
one-electron processes, 132
radical anions in, 132, 336
regioselectivity, 339
S_N3, 331
solvent effects, 332, 335, 337, 338
stereoselective, 339
steric effects, 341, 342
ultrasonic effects, 337
Nucleophilic aromatic substitution,
catalysis,
by base, 272
by copper, 271
by palladium, 271
by transition metal complexes, 271
cine substitution, 273
intramolecular, 270
of halobenzenes, 268
of heterocyclic systems, 272, 273
of indanes, 268
of nitrophenyl esters, 268
of pyridines, 273
of pyridium ions, 272
of pyrilium ions, 272, 273
of pyrimidines, 272
of quinolines, 272
solvent effects, 268, 269
substituent effects, 268
Nucleophilicity, 342
Nucleosides,
hydrolysis, 4
synthesis, 4

Organoalanes, 417
Organoaluminium compounds, 366, 472
Organoboranes, 364
in asymmetric synthesis, 417
Organocopper compounds, 364
Organogermanium compounds,
substitution reactions, 331
Organomercury compounds, 377
Organometallics,
in organic synthesis, 364
Organotin compounds, 360, 364, 369
base cleavage, 374
electrophilic cleavage, 377
in S_N2 reactions, 344
substitution reactions, 331
Organotitanium compounds, 364
Orthoesters, 23–25
Oxabicyclohexyl derivatives, solvolysis of, 334
Oxabicyclo[2.2.1]heptane, 474
Oxaheptatrienyl dipoles, 492
Oxametallacyclohexanes, 472
1,3-Oxathiolans, 4
Oxazepines, 426, 465
Oxazines, 465
Oxaziridines,
hydrolysis, 5
reaction with nucleophiles, 330
Oxazoles, 360
Oxazolidones, 367
Oxazolin-5-ones, 425
Oxazolines, 367
Oxazolinones, 363
Oxenium ions, 216
Oxepinones, 486
Oxetanes, formation of, 360, 386
Oxetenes, 491
Oxidation by:
azodicarboxylates, 193
benzene selenic acid, 195
bismuth(V), 189
bromosuccinimide *N*-, 189, 198
cerium(IV), 122, 188, 189
aldehydes, α,β-unsaturated, 189
steroidal alcohols, 189
chloramin-*T*, 198

chlorine dioxide, 196
chlorine, 197
chromium(VI), 183–185
cobalt(III), 123, 191
copper(II), 185
dimethyl sulphoxide, 194
Fenton's reagent, 125
Fremy's radical, 194
gold(III), 186
halogens, 196–199
hexabutyldistannoxane-bromine, 196
hydrogen peroxide, 190, 202, 203
hypohalite, 196
iodine, 196
iodosobenzene diacetate, 197
iodosuccinimide, *N*-, 197
iridium, 193
iron(III), 122, 124, 125, 191, 192
lead tetra-acetate, 189
lead(IV), 188, 189
manganese dioxide, 124, 185
manganese(III), 123, 185
manganese(VII),
  alkenes, 184
  allylic alcohols, 185
  cinnamic acid, 185
  phenols, 185
  uracils, 184
mercury(I), 186
mercury(II), 187
metal ions, 183–193
molybdenum, 190
nickel(III), 191
nitric acid, 193
nitro-compounds, 194
nitrous acid, 193
osmium(VIII),
  alcohols, 193
  carboxylic acids, 193
  ketones, 193
oxygen, 206–208, 210
palladium(II), 188, 194
  alkenes, 193
  arenes, 193
peracids, 202–205, 204, 210
  of terminal acetylenes, 523
  of trimethylsilyl ketene acetals, 524
periodate, 198
peroxides, 202
peroxomonosulphate, 194
peroxydisulphate, 113, 122, 123, 124, 125, 195
  of cyanohydrins, 519
quinones, 200
ruthenium,
  alkanes, 192
  allylic alcohols, 192
  aromatic aldehydes, 192
selenium compounds, 195
selenium dioxide, 195
silver(I), 186
silver(II), 186
silver(III), 186
sulphonic acids, 198
sulphur and its compounds, 194, 195
superoxide, 204, 206
tellurium(IV), 195
telluroxides, 195
thallium(III), 187, 188
thionyl chloride, 194
trifluoromethanesulphonic anhydride, 194
uranium(III), 189
vanadium(V), 189
Oxidation of:
  adenosines, 185
  alcohols, 124, 183, 185, 193, 194, 197
  aldehydes, 17, 187, 193, 210
  aliphatic amines, 189
  alkenes, 184, 185, 187
  alkyl halides, 204
  alkynes, 196, 206, 523
  allylic ethers, 200
  amides, 186, 195, 204
  amines, 126, 196
  amino-acids, 196, 198
  amino-alcohols, 195
  anthracene, 203
  aromatic amines, 183, 188, 198
  aromatic hydrocarbons, 122, 123, 188
  ascorbic acid, 184
  benzoic acids, 189
  benzoin, 189 192
    to benzil, 194
  benzoquinones, 203
  carbanions, 364
  carboxylic acids, 125, 189, 193, 195, 204
  catechols, 185, 186
  cinnamic acid, 185
  cyanohydrins, 195, 519
  diazirines, 204
  diazoquinones, 207
  diols, 186, 192
  disaccharides, 192
  epoxyalcohols, 189
  esters, 185
  flavins, 204
  furans, 203

furfural, 186
furoxans, 204
glycine, 196
hydrazides, heterocyclic, 192
hydrazines, 188, 198, 200
to tetrazenes, 195
hydrazones, 188
hydroquinones to 1,4-benzoquinones, 185
α-hydroxy-acids, 196
imines, 204
indole derivatives, 203
isoindene, 208
isoquinoline alkaloids, 194
isoquinolines, 189, 206
ketenes, 197
ketenimines, 191
ketones, 17, 124, 185, 189, 191, 202
lactic acid, 195
malonic acid, 184
mercapto compounds, 192, 200
naphthalenes, 204
oxalic acid, 193
oxazoles, 207
paraquat, 203
peptides, 186
phenols, 124, 185, 191, 193, 198
phenylalanine, 185
phosphines to phosphine oxides, 195
piperidines, 204, 210
piperidones, 185, 189
pyrazinones, 207
pyridine, 195
pyridines, 204
pyruvate, 193
quinazolones, 189
quinolines, 204
selenides, 204
silanes, 204
silyl enol ethers, 190
steroids, 187, 204
sulphides, 190, 196, 204, 207
sulphilimines, 196
sulphoxides, 183, 198, 204
tartaric acid, 185
thiocarbonyl compounds, 195
thioketones, 208
thiols, 125
to disulphides, 195
thiosulphinates, 204
thiourea, 195
thioureas, 204
toluenes, 191, 197
triazoles, 207
Oxidation, anodic, 123
Oxidation, enzymic, 125
Oxidation, allylic, 195, 209
Oxidation, selective, 192
Oximes, 8–10
conversion to nitriles, 397
fragmentation of, 397
hydrolysis, 9
isomerization, 9
reaction with:
acetates, catalysed by micelles, 44
tetracyclooctan-3-one oxime, 515
tricyclodeca-7-en-4-one oxime, 515
tricyclonona-3-en-6-one oxime, 515
Oxiran, O-corner-protonated, 328
Oxirane formation, 201, 519
effect of substituents on, 523
Oxiranes, vinyl, 422
Oxirenes, 249
Oxygen insertion, 515
Oxymercuration,
of dihydronoriceane, 377
Oxythallation, 503
Ozonation of alkynes, 201
Ozonation of arenes, 201
Ozone,
reaction with alkanes, 201
reaction with alkenes, 201
reaction with cyclohexanol, 201
reaction with epoxides, 201
Ozonides, 200
Ozonolysis, 200–202
carbonyl compounds, α,β-unsaturated, 201
of alkenes, 200
of vinyl sulphide, 201

Palladium complexes, 359
Papain, 51
Paracyclophanes, 45
Passerini reaction, 13
Penicillanic acid, 53
Pentacovalent intermediates, 56, 57
Pentacyclo[5.3.1.0.0.0]undecane, 518
Pentalenes, 509, 516
Pentaprismane, 499
Pepsin, 52
Peracids,
addition to alkenes, 412
decomposition of, 86
Perepoxidation, 420
Perester,
decomposition, 85
Pericyclic reactions, 364, 491

Perkin reaction, 12
Perkow reaction, 15, 499
Peroxidations, 204
Peroxides, 82–90
  decomposition, 82, 112, 113
Peroxycarbonates, decomposition of, 86
Peroxysulphur intermediates, 204
Persulphones, 208
Persulphoxides, 208
Peterson reaction, 360
Phase-transfer catalysis, 122, 131
  in elimination reactions, 400
Phenanthrene, hydrogen-exchange, 291
Phenol red dimethyl ether, 311
Phenols,
  acetylation, 291
  alkylation, 291
  bromination, 291
  nitration, 291
  oxidation, 186
  sulphonation, 291
Phenonium ions, 501
Phenyl cation, x-route to, 304
Phenyl glycidyl ether, reaction with nucleophil, 329
Phospha alkenes, 479
Phospha alkynes, 479
Phosphate transfer, stereochemistry of, 61
Phosphates,
  chiral, 59, 61
  enol phosphates, 372
  enzymic reactions, 61
Phosphazenes, 372
Phosphene, 514
Phosphonates, 415
  anions from, 372
Phosphoramidates, hydrolysis, 59
Phosphoranes,
  acyloxy, 54
  alcoholyes, 55
  hydrolysis, 54
  intermediates, 537
Phosphorocreatine, hydrolysis, 59
Phosphorus,
  nucleophilic displacement at, 330
Photoaddition,
  of carbonyl compounds to alkenes, 152
  radical, 139
Photocycloaddition,
  intramolecular, 451
  of alkenes, 451
Photodecarboxylation, 151
Photolysis of:
  alkenes, 191
  alkyl halides, 141, 149
  amides, 151
  amino acids, 149
  aryl halides, 111, 149
  azibenzil, 249
  azides, 251
  ketones, 116, 126, 151
  lactams, 151
  peroxides, 82, 85
  propellanes, 242
  triazoles, 243
Photooxidation of:
  alkenes, 138
  aromatics, 138
  carboxylic acids, 150
  ketones, 157, 531
  steroids, 531
Photorearrangement of:
  acrylaldehyde, 491
  allylbenzene, 540
  amino acids, 540
  anilino alkanes, 461
  azepino indoles, 509
  azibenzil, 513
  cyclo-octene, 534
  cycloheptatrienes, 489
  cyclohex-2-en-1-ones, 540
  cyclohexadienones, 469
  diepoxides, 494
  enones, 469
  epoxyenones, 521
  epoxytrienes, 521
  flavonols, 540
  heptatrienes, 485
  heterocyclic derivatives, 465, 529
  hexadienes, 482
  hexulofuranose, 513
  indenes, 483
  naphthalenes, 487
  naphthobarrelenes, 540
  norbornadienenes, 540
  oxazin-6-ones, 528
  photosantonin, 487
  pyrazine-2,3-dione, 466
  pyridine, 2-azido, 465
  pyridones, 465
  thiazines, 467
  thienothiazines, 487
  thiophene 2,2-dioxides, 486
  three membered rings, 519
  trienes, 540
Photoreduction, 98, 138, 141, 152
  of azo-compounds, 143
  of benzil, 151

of carbonyl compounds, 142
Photosolvolysis, 336
Photosubstitution, radical, 139, 152
Phthalamic acids, dehydration, 40
Picryl halides, 268
substitution by hydroxide, 274
Pinacol derivatives, 23
Pinacol rearrangement, 501
Polyalkylbenzenes,
nitration, 291
Polyene cyclization, 304
Porphyrins,
molybdenum, 190
Potential energy diagrams, More O'Ferrall type, 340
Prelog-djersassi lactonic acid, 485
Prevost reaction, 524
Prins reaction pathway, 420
Propargyl alcohols, 413
Propargyl anions, 357
Propargyl cations, 314
Propellanes, 206, 495
photolysis of, 242
Prostacyclins, biologically active, 414
Prostaglandins, 370
synthesis of precursor, 485
Protease, 50
Protease, α-lytic, 51
Protiodesilylation, 293
Protoadamantyl cation, 303
Protodesilylation, 399
Proton transfer, 1
in nitroanilines, 276
of TNT derivatives, 276
rate-limiting, 274
Proton tunnelling, 373
Protostephanine, 471
Pseudoguianolides, synthesis of, 478
Pseudoindene, 248
Pseudorotation, 57, 537
Pummerer rearrangement, 513
Pyramidal cation intermediates, 314
Pyran intermediates, 531
Pyrazoles,
rearrangement, 490
Pyrazolines,
pyrolysis, 393
Pyrethroids, 499
Pyridine N-oxides,
rearrangement of, 465
Pyridines,
dihydro, 370
formation, 359
synthesis of 4-substituted pyridines, 272
Pyridinium ions,
nucleophilic substitution, 273
Pyridones, 475
Pyridoxal in dealdolation reactions, 6
Pyridoxal in elimination reactions, 6
Pyridoxal phosphate, 49
Pyridyl cation, electrophilic substitution, 277
Pyrilium ions, nucleophilic substitution of, 273
Pyrimidines,
benzothieno, 291
nucleophilic substitution, 272
rearrangement, 467
ring transformations, 273
Pyrolysis of:
alkanes, 153
alkenes, 153
alkyl halides, 395
amine oxides, 395
azidothiophens, 398
aziridines, 254, 393
azo-compounds, 90
carbamates, 391
cyclohexadienes, 485
dihydropyrans, 395
dioxolanones, 392
enolates, silymethyl, 365
epoxy-trifluoropropane, 523
esters, 391, 533
mercapto compounds, 491
methylenephthalide, 531
neopentyl bromide, 500
oxazolin-5-ones, 529
peresters and peroxides, 83, 85
phospholes, 488
spiroalkanes, 153
thietan 1-oxide, 481
thiophen 2,2-dioxides, 393
Pyrones, deprotonation, 376
Pyrroles,
lithiation, 362
Pyrolline-1-oxides, 488
Pyrrolines, 514

Quainolides, pseudo, 370
Quinolines,
tetrahydro, 362
Quinoneimine *N*-oxides, 533
Quinones,
addition to carbanions, 359
from phenols, 193

Racemization,

pyridoxal in, 6
Radical anions, 130, 134, 145–149, 534
as reaction intermediates, 147, 149, 364
fragmentation, 147
in nucleophilic aliphatic substitution, 336
in nucleophilic aromatic substitution, 131
in reductive coupling of carbonyl compounds, 217
in $S_{RN}1$ reactions, 266
ion pairs, 355, 461
ketyl, 127, 129, 146, 155
semidiones, 146, 148
trimethylene, 520
Radical cations, 142–145, 465, 534
aromatic, 195, 291
combination, 144
disproportionation, 144
in Diels-Alder reactions, 446
in electrophilic aromatic substitution, 292
in oxidation: 193
of aromatic compounds, 123
reaction with nucleophiles, 144, 206, 213
rearrangement, 144
Radical pairs, 480, 481, 492
Radical reactions,
addition, 99–103
intramolecular, 103
aromatic substitution, 111–115, 138, 267
$S_{ON}2$ mechanism, 113, 266
atom abstraction: 87, 94, 96, 97, 154
halogen, 98
bromination, 96
chlorination, 95
combination, 86, 87, 92, 93
cyclization: 78, 87, 92, 103, 136
disproportionation, 87
fragmentation, 83, 105–107
Grignard reagents, 134
intramolecular hydrogen transfer, 107
isotope effects in, 94
magnetic effects in, 78
oxidation: 210
by cerium(IV), 189
of enolate anions, 366
polar effects in, 78, 87
rearrangements, 107–111
reduction: 130
ring opening, 78
solvent effects, 93, 112
study by ESR spectroscopy, 117, 124, 132, 134, 145, 146
Radicals,
acetoxyalkyl, 82
acyl, 112
acyloxy,
acetoxyl, 97
cumyloxy, 97
alkoxy, 83, 97, 105, 186
butoxyl-*tert*, 97
alkoxyalkyl, 105, 112
alkyl, 79, 90, 92, 94, 98, 112
allyl, 80, 512
allylic alkoxy, 519
amino, 540
aminoalkyl, 80, 126
aminyls, 80
arsinyls, 82
aryl, 131
aryloxy, 89
capto-dative stabilized, 80, 127
cubyl, 79
cyanoalkyl, 109
cycloalkyl, 79
cyclobutylmethyl, 109
cyclohexadienyl, 111, 112
cyclopropyl, 99, 215
cyclopropylcarbinyl, 107
diazenyl, 81
diphenylcyclobutanes, 102
1,2,3-dithiazoyl, 81
electron-transfer chain catalysis, 266
ESR spectra, 79, 81
germanium-containing, 82
haloalkyl, 79, 93, 107
hydroxyalkyl, 127
hydroxyl, 84, 96, 97, 112, 122, 125, 155
iminoxy, 9, 82
iminyl, 106
naphthylmethyl, 80
nitroxides, 116–120
norbornenyl, 108, 133
norbornyl, 79
nortricyclyl, 108, 133
α-oxyalkyl, 79
pentadienyl, 80
peroxy, 86, 87, 89, 97
phenoxy, 82, 93
phenyl, 111
phosphinyl, 82, 102
phosphoranyl, 120–122
pyridinyl, 93
silyl, 82, 98, 102, 112, 154
sulphuranyl, 122
thioalkyl, 80
thiyl, 102, 103

tin, 82, 102, 106
triazenyl, 81
trifluoroalkyl, 100
verdazyl, 81, 102
Radiolyses, 154–156
Reactive intermediates, 396
Reactivity-selectivity principle, 243, 316, 342, 343
Rearrangement of:
2,5-dienones, 4-bromo, 291
acetamides, 462
N-alkylcyano, 465
acetimidates, 474
adamantanes, 506, 507
aldosterone, 499
aldoximes, 534
alkynes, 487, 491
nitroacetylene, 533
allenes, 495
allyl amines, 482
allyl ethers, 472
silicon and germanium analogues, 472
allyl sulphides, 476
allyl vinyl ethers, 517
allyl-boranes, 482
allylbenzene, 465
allylic acetates, cyano, 517
allylic esters, 473
allylsilane, 488
amides, 363
amine oxides, 461, 497
aminobenzenethiols, 461
aminonitriles, 470
ammonium salts, 495
anthraquinones, 476
arsenines, 471
arylsulphenates, 477
azabicycloalkanes, 509
azadi-methane, 515
azawurtzitanes, 525
azepinones, tetrahydro, 477
azides, 516
azidochalchones, 504
aziridines, 474, 491, 525, 536
basketanone, 508
benzene derivatives, 460–465
benzene, butyl, 464
benzenonium ions, 460
benzothiepins, 528
benzvalene, 506
bicyclic cyclobutenes, 492
camphene, 506
carbanions, 359
carbenes, 248–250, 513
carbonitriles, 533
carbonium ions, 500
silicon controlled, 315
chromanones, 515
chromenes, 531
chrysanthemic acid, 518
chrysanthemylamines, 503
cobaloximes, 516
cobalt complexes, 537
codeines, 476
coenzymes, 516
coordination complexes, 516
coumarones, 531
cyclobutane, tetrakis(phenylimino), 526
cyclobutanones, 510, 538
cycloheptatrienes, 535
cyclopentanols, 501
cyclopropanes, 490, 519
allylvinyl, 526
cyclopropanic dihydroanthracenes, 521
decyn-1-ols, 497
dehydrolinalool, 518
diallyldixanthogen, 476
diamides, *N*-chlorophosphonic, 525
diazen-isodiazen *N*-oxide, 530
diazene-*N*-oxides, 529
diazenes, 535
diazepine derivatives, 487
dienes, 477
dienones, 291
dihydroazete 1-oxide, 492
dihydrosemibullvalene derivatives, 484
dioxans, 531
dioxetane, 526–528
dioxiranes, 207
dioxolanes, 537
divinylcycloalkanols, 478
enals, 485
enamines, 475
enones, 473
epiperoxide to hydroxy enone, 517
epoxides, 522
propellane triepoxide, 523
epoxy alcohols, 516
epoxyketones, 522
epoxypyrones, 522
epoxysteroids, 512
feniculin, 476
fulvalenes, 494
fulvenes, 535
tricyclic, 518
furanones, 531
furylethylidene-succinic anhydrides, 494
heterocyclic derivatives, 465–469

furazan oxides, 465
indazole, 465
hetisine, 533
hydrazones, 467
hydroperoxides, α,β-unsaturated, 526
imididates, 482
imines, 465
iminoethers, 480
indenes, 486
indoles, 467, 476
indolines, 501, 522
iron complexes, 517, 526
ketones, 495
ketones, α-halogeno, 472
ketones, hydroxy, 464
lactams, 528
morpholine, *N*-allyl, 524
naphthalenonium ions, 460
naphthyl esters, 464
naphthylamines, 474
natural products, 509–513
anhydrolycoctonan, 513
cephalosporins, 528
cholest-5-ene, 511
coumarin benzoates, 464
coumarins, 476
flavones, 476
friedooleananes, 511
galactopyranosides, 513
himachalene, 510
holamine, 510
hydroxyaplysistatin, 510
isoflavones, 476
kaurene, 509
longifolene, 509
lupanes, 511
paulownin, 510
picasenone derivatives, 509
protoilludane, 510
purines, 466
spectinomycins, 531
terpenoids, 509
nitrenes, 251, 513
nitrite esters, 291
nitrites, 533
nitrophenoxides, 461
nitrosamines, aryl, 461
nitrosophenylalanine, 495
norbornenols, 506
oxadiazoles, 467
oxadiazolones, 467
oxathioles, 530
oxazines, 488
oxazines, cyano, 465
oxazolines, 524
oxazolinones, 474
oxime derivatives, 514
oxiranes, 492, 501
oxirenes, 524, 533
penams, 524
penems, 537
penicillin sulphoxide, 524
pentadienethioates, 485, 531
pentenals, ethylthio, 484
perezone, 510
perfluorotrienes, 499
peroxides, 465, 500, 528
phenylalanine, 521
phosphane oxide, 513
phosphate carbanions, 372
phosphates, 462
phosphinates, 477, 480, 499
phospholines, 531
phosphoranes, 461, 537
phosphorin derivatives, 489
phosphorin, 472
phosphorothionates, 505
phthalimido-azimines, 535
piperazinium salts, 497
platinum complexes, 518
polylithium organic compounds, 519
propargyl alcohols, 537
propellanes, 507
pseudoureas, 474
pyrazines, azidodiphenyl, 466
pyrazoles, 467
pyridazines, perfluoro, 466
pyridines, 371
pyridinium salts, 465, 497, 498
pyrimidines, 465
pyrroles, 467, 521, 539
quinolizinium ions, 467
quinones, 515
radical cations, 144
radicals, 107–110, 115
allylic alkoxy, 519
reissert compounds, 497
selenides, 509
selenoxides, 482
semiaminals, 533
silanes, 519
siloxyvinylcyclopropanes, 485
silylenes, 256
spirostanones, 515
steroids, 537
stilbenes, nitro, 534
sulphamides, 495
sulphenamides, 534

sulphides, 370, 465, 481, 494, 524
sulphonamides, 513
sulphonates, 524
sulphonium salts, 462, 480
sulphoxides, 513, 533
sulphoxides, allyl, 505
tetrahydropyrans, 531
tetrazines, 495
thianaphthalenes, 488
thiiranium ion, 509
thiirenes, 524
thioacetamides, 467
thioesters, 539
thioimidates, s-allylthio, 518
thiolane-2,4-diones, 531
thiones, 531
thioxanthen-methanides, 499
thujan-3-ols, 504
triazenes, 501
triazines, 466
triazoles, 467
triazolines, 465
tricyclenes, 477
triphosphazenes, 499
tungsten complexes, 517
uracil derivatives, 462
vinyl halides, 501
vinyl silicon compounds, 531
vinyl sulphides, 475
vinylcyclopropanes, 477
vitamin D3 derivatives, 504
wheland intermediates, *ipso*-, 464
xanthates, 476, 505
ylidenebutenolides, 528
ylides, 480, 481, 488, 497
zirconium complexes, 517

Rearrangement,
α-ketol, 531
acylpyrone-benzene, 471
anionic, 495–500
aspidospermane-eburnane, 533
azulene-naphthalene, 535
backbone, 511
benzidine, 461
cationic, 500–506
cycloheptatriene-norcaradiene, 535
cyclopropenylcarbinyl-allenyl, 520
cyclopropylcarbinyl, 503
diazoamino, 461
dienol-benzene, 471
dienone-phenol, 469
enamine, 465
Fischer-Hepp, 461
McLafferty type, 500
Meisenheimer, 497
metal-catalysed, 516–519
norbornene-norpinene, 506
polytopal, 537
radical pair mechanism, 110
radical, 132
ring-opening ring-closure, 519–526
silanotropic, 538
sulphoxide-sulphenate, 392, 481
thiazoline-thiazine, 530
triazene, 461
vinylcyclopropane-cyclopentene, 469
vinylidene-cyclobutene, 511
vinylidene-norcaradiene, 535
Wallach, 461
zwitterion, 469

Reduction by:
alkali metals, 127, 217, 368
aminoboranes, 212
bis(trifluoroacetoxy)borane, 214
bis(triphenylphosphine)copper(I)-tetrahydroborate, 214
borohydride, 212–216
carbon, 218
chromium(II), 130
copper hydrides, 214
cyanoborohydride, 214
diiodosamarin, 217
formate ion, 216
formic acid, 216
hydriodic acid, 219
hydroquinones, 218
metal complexes, low valent, 217
metal hydrides,
aluminium, 211, 212
lithium aluminium hydride, 211
lithium tetrakis(*N*-dihydropyridyl)aluminate, 211
tert-butoxyaluminium hydride, 212
metals,
magnesium, 217
potassium, 217
sodium, 217
sodium-potassium alloy, 217
NADH, 221
nicotinamides, 221
nitrites, 219
phosphines, 219
phosphites, 219
phosphorus halides, 218
phosphorus, 219
selenols, 218
silanes, 215, 216, 217
sodium dithionite, 218

sodium naphthenide, 217
thiols, 218
tin hydrides, 132, 519
titanium(III), 129
titanium, 217
tributyltin hydride, 214
triphenylphosphine borane, 214
vanadium(II)-catechol complexes, 217
zinc, 129, 217
opening of aryl-substituted cyclopropyl rings, 217
Reduction of:
acetophenones, 221
acid chlorides, 215, 217
acridinium ions, 221
alcohols, 134
aldehydes, 127, 212, 214, 218
alkenes, 211
alkyl halides, 211, 217
alkynes, 214
allyl ether, 217
amides, 132, 213, 214, 217
amino-acids, 219
anhydrides, 212
anthraquinones, 212
aromatic hydrocarbons, 213
azides, 214
azines, 217
benzoin, 217
carbazoles, 130
carbon dioxide, 214
carbonium ions, 305
cyclooctadiene, 368
diketones, 212
α,β-unsaturated, 217
dioxans, 217
enones, 213
esters, 129, 212, 213, 217
fluoronorcaranes, 214
imides, 211
imines, 212
isoquinolinium cations, 221
α-keto-imines, 217
ketoacids, 219
ketoesters, 221
ketones, 127, 130, 211, 212, 218
stereoselective, 214
steroidal, 217
lactams, 217
nitrilium ions, 215
organomercury compounds, 215
oximes, 212, 214, 217
perylene, 368
pyrene, 368
pyruvamides, 220
sila cyclopentadienes, 368
sulphides, 130
vinyl, 216
sulphilimides, 219
sulphonates, 213
sulphones, 221
sulphoximides, 219
sultones, 211
thioesters, 130
Reduction, asymmetric, 212, 221
by chiral boranes, 214
in borohydride reductions, 212
of ketones, 214
Reduction, cathodic, 127, 130
Reduction, electrochemical, 127, 130, 133
Reduction, stereoselective,
of allyl halides, 213
of benzyl halides, 213
of carbonyl compounds, 212
of enones, 212
of imines, 212
of isoxazolines, 212
Reduction, oxido, 216
Retinoic acid, synthesis of, 518
Ring closure,
of iodoisothiocyanates, 333
stereoselective, 413
Ring contraction, 251
Ring expansion, 415
of anilides, 460
of vitamin D3, 328
Ring opening, eliminative nucleophilic, 255
Ring opening, five membered rings, 528–531
Ring opening, four-membered rings, 526–528
Ring opening, six membered rings, 531–533
Ring transformation, pyrimidine to pyridine, 273
Ring-opening, three membered rings, 519
Ritter reaction, 307
Rubradirin antibiotics, 474

Sachs' reaction, 359
Salt effects, 26
S_AN, addition elimination, 331
Schiff bases, 5–8
condensation reactions, 7
enzymes, 7
formation, 7
hydrogen exchange, 6
hydrolysis, 7

intermediates, 6
methanolysis, 8
reaction with carbanions, 7
Schmidt fragmentation, 304
Schmidt reaction, 515
Schmidt rearrangement, 508
S_E1 mechanism, 377
Secoalkylation, 521
Secodine, 476
Selenides,
allyl, 369
hydroxy, 366
methyl vinyl selenides, 315
propargyl, 482
vinyl, 369
Seleniranium ions, 414
Selenocyanates, 494
Selenolactonization, 414
Selenoxides,
elimination reactions of, 396
fragmentation, 204
oxidation by, 195
rearrangement, 482
Semibullvalene, 535
Semibullvalenylcarbinyl cation, 301
Serine proteinases, 50, 51
Sesquifulvalene, 518
Setoglaucin, 311
S_H2 reactions, 78, 86
Shapiro reaction, 360
Sharpless reaction, 442
Sigmatropic rearrangements, 472–488
[1,2], 480
[1,3], 482–485, 536
[1,4], 469
[1,5], 485–488
[1,7], 489
[2,3], 195, 257, 393, 477, 480–482
[3,3], 472–479
activation parameters for, 476
[3,4]-, 489
[3,5], 489
[5,5]-, 478, 489
[5,7], 490
Silanes,
alkoxy, 216
allyl, 417
allylic, 365
chloro, substitution reactions, 330
oxidation, 204
oxidative desilylation, 200
protodesilylation, 400
reaction with arylsulphenyl chlorides, 480
rearrangement, 485
vinyl, 370, 413, 415
Silene, 250
Silicenium ions, 317
Silicon,
nucleophilic displacement at, 330
Siloxybutadienes, 485
Silyl compounds, 364
base cleavage, 373
deprotonation, 375
reaction with phenylsulphenyl chloride, 484
Silyl derivatives,
elimination reactions of, 399
Silyl enol ethers, 426
oxidation, 190
Silyl ethers, cleavage, 361
Silylenes, 255–257
addition to alkenes, 256
dichlorosilylene, 256
difluorosilylene, 256
dimesitylsilylene, 257
insertion of, 256
methyl(phenyl)-silylene, 256
methylsilylene, 255
molecular orbital calculations, 256
rearrangement of, 256
Silylpolyphosphates, 514
Singlet oxygen, 137, 138, 433, 436
addition to alkenes, 206
addition to furans, 207, 528
formation of endoperoxide polymers, 206
in organic syntheses, 531
reaction with benzenesulphonates, 207
reaction with biological systems, 206
reaction with carbanions, 363
reaction with nitrones, 206
reaction with thiophenolates, 207
Smiles rearrangement, 273, 461
of cysteine, 269
S_N1 reactions, 328, 330, 331, 343
S_N2 reactions, 120, 330, 331, 339, 344
at phosphorus, 55
at sulphur, 413
intramolecular, 332
S_N2' reactions, 325
$S_N2(C+)$ mechanism, 323
S_NAr reactions, 267–271, 274, 344
of diazonium ions, 265
of superoxide ions, 267
Solvent donicity parameters, 367
Solvent effects, 27
in base-catalysed hydrogen exchange, 373

in carbanion addition to fulvenes, 372
in carbanion reactions, 369, 371
in enolate alkylations, 365
in nucleophilic aliphatic substitution, 337, 338
in S_NAr reactions, 268, 269
in stability of carbanions, 355
on eliminations, 387
solvatochromic comparison, 337
Solvolysis,
exo-endo rate ratios, 325, 326
of bicyclooctenyl derivatives, 327
of bridgehead chlorides, 327
of norbornyl tosylates, 325
of tricyclo octane derivatives, 326
Sommelet rearrangement, 387
Sommelet-Hauser rearrangement, 497
Spin-labelling, 117
Spin-trapping, 119, 126, 149
Spiran-intermediate, 470
Spiro cycloheptatriene intermediate, 252
Spiro-compounds,
difluorospiropentane, 519
oxazoles, 188
rearrangement, 467, 485, 490, 519
sesquiterpenes, 508
spirocyclic sulphonium ylides, 481
spirocyclopropylpyrrolium ion, 521
spiroketones, 538
spirothioindoxyl, 524
trienes, 526
triepoxide, 494
Squaric acid, 488
S_{RN} reactions, 131
$S_{RN}1$ mechanism, 78, 131, 266
Stereo-differentiation, 10
Stereochemistry of S_N2(P) reactions, 57–59
Stereoelectronic control, 9, 251
antiperiplanar, 2
in hydride transfer, 216
in nucleophilic substitution reactions, 267
in solvolysis reactions, 328
Stereoelectronic effects, 25
Steric acceleration, 37, 255
Steric approach control, 31
Steric effects, 27
in nucleophilic aliphatic substitution, 325, 338, 341, 342
in nucleophilic substitution reactions, 268
in solvolyses of tetracyclododecane derivatives, 326
in solvolysis of adamantyl derivatives, 327
relative size, 3
Steroids,
addition to unsaturated, 421
Beckmann rearrangement, 514
elimination reactions of, 403
ene reactions of, 491
epoxy, ring opening, 329
photooxidation, 531
photo, 538
rearrangement of, 499
steroid hypoiodites, 512
Stevens rearrangement, 80, 110, 495
Stibenes, 465
Stilbenes,
methoxymercuration, 418
Structure-reactivity relationships, 373
Styrenes,
addition of ArSSCN, 413
bromination, 410
formation, 471
methoxymercuration, 418
Substituent constants, ortho E_S^o, 27
Substitution at phosphorus(III), 61
Subtilisin, 50
Sucrose, hydrolysis of, 5
Suicide inhibitors, 51
Sulphenes,
intermediates in elimination reactions, 63, 64, 371, 389, 390
Sulphenium ions, 509
Sulphides,
allyl, 369, 419
cyclic, 369
vinyl, 369
Sulphilimines, 371
Sulphines, penicillin derived, 528
Sulphonamides,
aminolysis, 64
hydrolysis, 64
Sulphonates,
aminolysis, 63
aromatic, 291
hydrolysis, 63, 371
Sulphonation,
aromatic, 291
with chlorosulphonic acids, 291
Sulphones,
allyl, 371
β-aminoalkyl, 371
carbanions from, 423
α-halo,
coupling of, 402
nitration of, 291

seleno, 414
α,β-unsaturated, 424
vinyl, 423
vinyl, addition to, 324
Sulphonimides, 530
Sulphonium ions, cyclic, 412
Sulphonium salts,
acylamino, 514
elimination reactions, 393
Sulphonyl halides,
aminolysis, 65
solvolysis, 64
Sulphoxides,
alkenyl, 371
allyl, 370
allylic, 485
deoxygenation, 255
elimination reactions, 399
α,β-ethylenic, 424
halogenation, 371
lithio, 422
oxidation, 204
rearrangement of allylic, 485
thermolysis of, 395
trapping of persulphoxides, 208
vinyl, 422
Sulphoximides, 371
Sulphur,
nucleophilic displacement at, 331
Sulphuranes, 63, 413
Superoxide, 137, 149
Surfactants, anionic, 419
Swain-Scott equation, 342

Taft equation for:
elimination reactions, 391
hydrolysis, 29
σ* values, 27
nucleophilic aliphatic substitution, 328, 340, 341
pyrolysis of alkyl halides, 395
Tautomerism of
β-diketones, 16
imazoles, 537
oxalacetate, 368
pyrazoles, 537
pyridines, 537
pyridone, 537
pyrimidines, 537
thiolan-3-ones, 538
Tautomerism, 537, 538
annular, 537
ketimine-enamine, 537
keto-enol, 411
ring chain, 538
Telluroxides, oxidation by, 195
Template effects,
on the formation of benzo-18-crown-6-ether, 332
Terpene synthesis, 369
Tetracyclononenyl derivatives, solvolysis of, 327
Tetracyclooctyl systems, 508
Tetracyclo[5.3.1.0.0]undec-4-ene, 518
Tetrahedral intermediates, 23–26, 47, 50
Tetrahydrofuran-2-ones, synthesis, 360
Tetrahydrofurans, formation, 371
Tetralones, 503
Tetrazolospirostans, 515
Thiazepine derivatives, 528
Thiazines, 462
Thiazolidinones, 363
Thiazolium salts, 416
Thietanone, 505
Thiiran, 524
Thiiranium intermediates, 524
Thiiranium ions, 413, 509
Thioacids, reaction with nucleophiles, 36
Thioamides,
hydrolysis, 31
α,β-unsaturated, 367, 424
Thioamides, allyl, 475
Thiocarbamates, 370
elimination reactions of, 389
Thiocyanogen, addition to alkenes, 413
Thioesters, 539
metal catalyzed hydrolysis, 48
Thioketenes,
reaction with ozone, 201
Thiol proteinases, 51, 52
inhibitors of, 51
Thiolesters, hydrolysis, 31
Thiols,
addition to indene, 422
Thiolsubtilisin, 52
Thiones, 530
Thiophenols, 480
Thiophens,
azido, 398
from benothiepins, 528
nitration, 291
nucleophilic substitution, 272
Thiopyrans, 467, 484
Thiosulphonium ions, 314, 494
Thioureas, cyclization, 27
Thioxanthens, 499
Tin enolates, 367
TNT, reaction with base, 276

Tocopherol, 204
Transamination, 6
Triamantane, 507
Triazenes, from diazonium ions, 265
Triazines, 91
  hydrolysis of chloro-derivatives, 272
Triaziridine, 245
Triazoles, photolysis of, 243
Tributyltin alkoxides, 499
Tricyclodecan-7-ones, 488
Tricyclodecanes, 477
Tricyclodecatriene, 477
Tricyclonona-3,9-dienyl cation, 477
Tricyclonona-4-ones, 513
Tricyclooct-3-en-8-one oxime, 515
Tricyclooctane derivatives, solvolysis of, 327
Tricyclooctyl derivatives, solvolysis of, 327
Tricycloundecene diazonium ions, solvolysis of, 327
Tricyclo[3.2.0.0]hept-5-ene, 486
Tricyclo[3.2.0.0]heptenes, 518
Tricyclo[4.1.0.0]hept-1-enes, 521
Tricyclo[4.2.1.0]nona-2,7-dienyl cations, 301
Tricyclo[4.2.1.0]nona-3,7-diene, 477
Tricyclo[4.2.2.0]octane, 526
Tricyclo[5.3.1.0]undecan-8-ones, 494
Trioxatriquinane, 524
Triplet oxygenations, 208
Trithian complexes, 518
Troponones, 370
Tropylium cations, 311, 312
  homo-, 312
Trypsin, 50
Tunnelling, proton, 398

Ullman reaction, 271
Umplong reactivity, 365
Ureas, dehydration of, 255
Ureas, reaction with amines, 35
Urethanes, formation, 38

Van Alphen-Huttel rearrangement, 467
Veratrole, 291
Vilsmeier-Haack reagent, 292
Vinyl alcohol, 14
Vinyl anion equivalent, 372
Vinyl anions, 353, 357, 361
Vinyl cations, 304, 315, 316, 323
  alkynyl stabilized, 315
  silicon stabilized, 315
Vinyl ethers, 360
  cycloaddition, 434
  hydrolysis, 17
Vinylenethioindigo, 524
Vinylic carbon, nucleophilic displacement at, 323, 324
Vinylindene, 486
Volume of activation for:
  adduct formation, S-, 275
  cleavage of diazonium ions, 265
  proton transfer from nitroalkanes, 373

Wagner-Meerwein rearrangement, 415, 500, 509
Wharton reaction, 519
Wheland intermediates, 291
  Ipso, 291
Winstein-Holness equation, 342
Wittig reaction, 11, 372
  intermediates, 400
Wittig rearrangement, 357, 480
Wolff rearrangement, 246, 248, 249
Woodward-Hoffmann rules, 452
Wurster's cation, 313

X-ray structure,
  of 2,6-dimethoxyphenyl-lithium, 354
  of lithium alkyls, 353
  of stable carbonium ions, 312
Xanthones, 369
Xenon difluoride, aromatic fluorination, 291
Xylylene, 526

Ylides,
  ammonium, 495
  azomethine, 525
  carbonyl, 492
  Diels-Alder reactions, 450
  nitrile ylides, 440
  phosphorus, 372, 499
  sulphonium, 480, 497
  thiocarbonyl, 530
Ynamines,
  cycloaddition, 437

Zirconium enolates, 367
Zwitterions, 426